BIOCHEMICAL ENGINEERING AND BIOTECHNOLOGY HANDBOOK

BIOCHEMICAL ENGINEERING AND BIOTECHNOLOGY HANDBOOK

BERNARD ATKINSON
FERDA MAVITUNA

M
The Nature Press

DEDICATION

This Handbook is dedicated to all those unknown and unnamed people whose efforts over the years have created Biochemical Engineering and Biotechnology, and who have opened up the possibilities which lie before us.

Published in the United Kingdom by MACMILLAN PUBLISHERS LTD
(Journals Division), 1983
Distributed by Globe Book Services Ltd
Canada Road, Byfleet, Surrey, KT14 7JL, England

ISBN 0 333 33274 1

Published in the USA and Canada by THE NATURE PRESS, 1983.
15 East 26th Street, New York, N.Y. 10010

Library of Congress Cataloging in Publication Data
Atkinson, Bernard, 1936-
Biochemical engineering and biotechnology handbook.
Includes index.
1. Biochemical engineering. 2. Bioengineering.
I. Mavituna, Ferda, 1951– II. Title.
TP248.3.A853 1983 660'.6 82-14462
ISBN 0-943818-02-8

Printed in Great Britain by
St Edmundsbury Press, Bury St Edmunds, Suffolk.

Filmset by Filmtype Services Limited, Scarborough, North Yorkshire.

CONTENTS

CHAPTER 3
STOICHIOMETRIC ASPECTS OF MICROBIAL METABOLISM

CHAPTER 4
MICROBIAL ACTIVITY

CHAPTER 5

PRODUCT INFORMATION

CHAPTER 6

ENZYME ACTIVITY

CHAPTER 7

REACTORS

CHAPTER 8

FLOW BEHAVIOUR OF FERMENTATION FLUIDS

CHAPTER 9
GAS-LIQUID MASS TRANSFER

CHAPTER 10
SOLID AND LIQUID-PHASE MASS TRANSFER

CHAPTER 11
HEAT TRANSFER

CHAPTER 12
DOWNSTREAM PROCESS ENGINEERING

CHAPTER 13
PRODUCT RECOVERY PROCESSES AND UNIT OPERATIONS

CHAPTER 14
PROCESSES

INDEX 1115

Preface

The current interest in biotechnology and the belief that it will expand very considerably is based upon three factors:

(i) The raw materials can be obtained from renewable resources.

(ii) Biotechnological processes appear likely to be economical against the chemical processing of vegetable materials.

(iii) A wide range of possibly valuable products is being defined by both traditional biological methods and through genetic manipulation.

As with most areas of technology, local and global political considerations distort both the areas of development and the rates of progress, e.g. alcohol fermentations are important independent of the overall economics. In general biotechnology can be divided into three areas defined as large, medium and small; terms which refer to the scale of industrial development rather than the size of the individual production units which might be built:

(i) Large scale, where biotechnology must compete with petroleum and coal, as a primary source of carbon compounds for fuels and high tonnage industrial products.

(ii) Medium scale, where biotechnology must compete with both petroleum-style technology (whether the carbon source is petroleum, coal or vegetable) to produce either the commodity chemicals currently in use or substitutes for them, and with agriculture to produce natural products such as proteins and lipids.

(iii) Small scale where specifically biochemical products are produced for which no other routes can currently be foreseen.

Small scale processes seem certain to continue their present rapid proliferation and growth; their products offer such improvements in medical practice and for industrial processes, that the cost of the products is not a deciding factor in whether the products are developed but in which manufacturers succeed. The medium and large scale processes are probably not currently viable economically except in a few cases, e.g. some organic acids where production is truly economic, alcoholic drinks, and acetic acid for human consumption. However, it is likely that the next twenty years will see the establishment of large scale microbiological processes with vegetable materials as feedstock (Fig. 1). These processes will:

(i) Produce primary materials as fuels and feedstocks for conversion to commodity chemicals, though whether the later conversion is chemical or microbiological will depend upon processing economics.

(ii) Produce a range of commodity chemicals directly from vegetable substances rather than via some primary product such as ethanol.

At the large and medium scale the 'market' value of many potential biological products is well-established, since the objective is either direct substitution or provision of material with properties similar to an available product. The cost of raw materials is also likely to be known. The difference between cost and value defines the allowable process and marketing expenditure, and serves to target research and development, whether biological or process engineering in emphasis, to meet these needs.

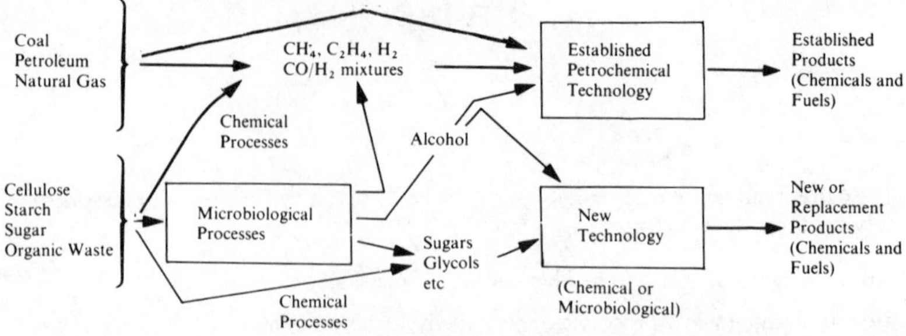

Figure 1. The competitive routes to large tonnage products.

The *Biochemical Engineering and Biotechnology Handbook* is intended to assist the development of both large and small biotechnological processes by establishing for the unit processes and unit operations involved:

1) their scientific and engineering basis;
2) their performance and operating characteristics;
3) the factors which influence their performance;
4) their integration into complete processes.

Such information allows scientists and engineers:

1) to ensure the appropriateness of research and development in meeting overall process objectives;
2) to estimate the extent to which particular processes are sub-economic and the possibilities for research and development to change this situation;
3) to establish the appropriate methodology for particular process development;
4) to identify equipment needs.

The Handbook has been structured into fourteen chapters detailed contents of each chapter are given on p.v, and a brief glossary appears at the beginning of the chapter (as appropriate). Extensive sub-headings have been provided and these have been used to form the basis of an index (p.1115). Within each chapter information has been presented predominantly as tables and figures so that those knowledgeable in a discipline appropriate to biotechnology can readily identify relevant subject matter. The emphasis is placed on the 'what' and 'how' of biological processes, for those who require detailed explanation of the 'why' literature references and bibliography sources have been provided.

Biotechnology will become an important, as opposed to peripheral, part of the industrial scene if it meets the needs of the community at a price the community can afford. Frenzy, fashion and invocation of crisis, whether of food or energy, are not sustainable. Rather facts, hard work, commitment, confidence, professionalism and the pleasure of invention are the substance by which biotechnological processes can make a modest, though significant contribution to material well-being.

B. Atkinson
F. Mavituna
August 1982.

Acknowledgements

In 1976 Mr. A.N. Emery of the Department of Chemical Engineering, University of Birmingham, produced *Biochemical Engineering – a Report Prepared for the Science Research Council and the Institution of Chemical Engineers*. One of the recommendations contained in the report highlighted the need for a handbook of biochemical engineering data. Subsequently the Science and Engineering Research Council, formerly SRC, with the encouragement of the Research Committee of I.Chem. E., provided funding which allowed one of the authors (F.M.) to lay the groundwork for the *Biochemical Engineering and Biotechnology Handbook*.

The authors would like to express their appreciation for the encouragement and direct assistance given by members of the various SERC/I.Chem.E. Committees, and in particular to the Science and Engineering Research Council and Mr. Emery.

The awareness of the authors as to the importance of downstream processing within biochemical engineering was greatly enhanced by the work of one of them (B.A.) in conjunction with Mr. Philip Sainter, on a project entitled 'Technological Forecasting for Downstream Processing in Biotechnology' which formed part of the EEC Forecasting and Assessment in the Field of Science and Technology (Biosociety Sub-Programme). The authors would particularly like to thank Mr. Sainter for his help in conjunction with Chapters 12 and 13.

The form of the Handbook developed over an extensive period but was helped particularly by Dr. Ann Ralph, whose knowledge of biochemistry and previous work on chemical engineering texts was invaluable. The Handbook owes much to Dr. Ralph's patience and attention to detail and to the way in which she undertook the difficult editorial task of converting the manuscript into a harmonious whole.

Appreciation is also due to Rosemary Foster of Macmillan whose charming perseverance contributed greatly to maintaining the momentum of the project.

Thanks are also due to Mrs. Nora Taylor, the Department of Chemical Engineering Librarian at UMIST who was a mine of information on the complexities of library and retrieval systems and with the checking of occasional abstruse references.

Finally, the authors would like to express their gratitude to individuals and organisations too numerous to detail, who by discussion and direct assistance helped them considerably in clarifying their ideas and hopefully achieving a rounded view of the problems inherent in the industrial application of biochemical engineering and biotechnology.

CHAPTER 1

PROPERTIES OF INDUSTRIALLY IMPORTANT MICROORGANISMS

GLOSSARY

Activated sludge Material containing a very large, active microbial population, used in the purification of waste water.
Adaptation Ability to exist in a changed environment.
Aerobe Microorganism that requires molecular oxygen.
Algae Group of simple, mainly aquatic plants capable of **photosynthesis**.
Anaerobe Microorganism that grows in the absence of molecular oxygen.
Bacillus Rod-shaped bacterium.
Budding Form of asexual reproduction.
Capsule Gelatinous layer surrounding the cell wall of many bacteria.
Chlorophyll Green pigment of plants consisting of closely related colouring components, chlorophyll *a* and chlorophyll *b*, etc.
Chloroplast Body in a plant cell containing **chlorophyll**.
Chromosome Thread-shaped structure bearing genes located in the nucleus.
Coccus Spherical bacterium.
Culture Population of **microorganisms**.
Cytoplasm Fluid contained within the cell membrane excluding the **nucleus**.
Diploid Used to describe cells having **chromosomes** in pairs.
Ecology Study of an **organism** in relation to its environment.
Enzyme Protein-based catalyst produced within an organism.
Facultative anaerobe Bacterium that grows under either anaerobic or aerobic conditions.
Fermentation Commonly: any industrial microbiological process. Specifically: anaerobic microbiological processes.
Fermenter Industrial microbiological reactor.
Gram's stain Used for the differentiation of bacteria.
Growth curve Graphical representation of the growth of an **organism** in nutrient medium in a batch reactor.
Haploid Used to describe cells having a single set of **chromosomes**.
Heterotroph Microorganism that uses organic compounds as a carbon source.
Medium Mixture of **nutrient** substances.
Meiosis Form of cell division in which a **diploid** cell divides giving rise to two **haploid** cells.
Metabolism Overall process by which an **organism** uses **nutrients**.
Metabolite Product of biochemical activity.
Microaerophile Bacterium that grows most rapidly in the presence of small amounts of molecular oxygen.
Microbial film Adherent aggregate of **microorganisms** attached to a supporting surface.
Microbial floc Adherent aggregate of **microorganisms** in suspension.
Microorganism Form of life of microscopic dimensions.
Mitochondria Minute granular, rod-like structures contained within the **cytoplasm**.
Mitosis Process by which cell divides into two daughter cells each having an identical complement of genetic information.
Mixed culture Two or more species of **microorganisms** living in the same medium.
Morphology Structure and forms of an

organism.
Mutation Stable change of gene inherited on reproduction.
Nucleus Dense inner mass of the cell enclosed within a membrane.
Nutrient Substance used as food.
Obligate Necessary or required.
Organism Living biological specimen.
Parasite Living **organism** deriving its nutrition from another living organism.
Photosynthesis Production of glucose from carbon dioxide and water in the presence of **chlorophyll** with absorption of light.
Physiology Study of the function of **organisms**.
Plasmid Hereditary unit contained in the **cytoplasm**.
Protein Class of organic compounds associated with living matter, based on a combination of amino acids.
Protozoa Unicellular animals.
Pure culture Culture containing only one species of **microorganism**.
Respiration Any chemical reaction whereby energy is released for use by the **organism**.
Sacrophyte Organism that utilizes organic matter in solution from dead or decaying plant or animal tissue.
Slime layer Gelatinous covering of the cell wall, used synchronously with capsule.
Species One kind of **organism**.
Spore Minute, thick-walled, resistant body that forms within the cell and is considered as the resting stage.
Sterile Free of **viable microorganisms**.
Strain Pure culture of **microorganisms** composed of the descendants of a single **organism**.
Substrate Substance acted upon by an **enzyme**.
Taxonomy Classification of **organisms**.
Tissue Collection of cells forming a structure.
Vacuole Droplet in a cell often containing reserve food material.
Viable Capable of growth.
Virus Parasitic microorganism smaller than a bacterium.
Zoogloeal masses Microbial film or microbial floc, normally associated with biological waste water treatment.

GENERAL CLASSIFICATION OF MICROORGANISMS

A classified arrangement of microorganisms according to their mutual affinities or similarities permits a logical and informative system of naming. It may also be used as a key for the identification of organisms. Depending on the criteria used for grouping, a particular classification may show the evolutionary relationship between organisms (i.e. natural classification) or may be made for a single, defined purpose, e.g., to assist in finding the answers to specific questions (i.e special or artificial classification).

For each of the main groups of microorganisms, the taxonomic hierarchy is a system in which a single, all-inclusive category is divided and subdivided into progressively smaller and less-inclusive categories. Scheme 1 shows the most important categories of the taxonomic hierarchy arranged in the order in which they must be employed.

SCHEME 1

Kingdom→Division (or Phylum)→Class→Order→Family→Genus→Species→Strain

The kingdom Protista (*see* Fig. 1) comprises unicellular organisms capable of self-duplication or of directing their own replication. Procaryotes do not possess a true nucleus or a nuclear membrane, while eucaryotes do have a true nucleus enclosed within a distinct nuclear membrane. Some features distinguishing between procaryotic and eucaryotic cells are summarized in Table 1. The noncellular protists (i.e. viruses) do not undergo self-replication, instead they direct their reproduction within another cell termed the host.

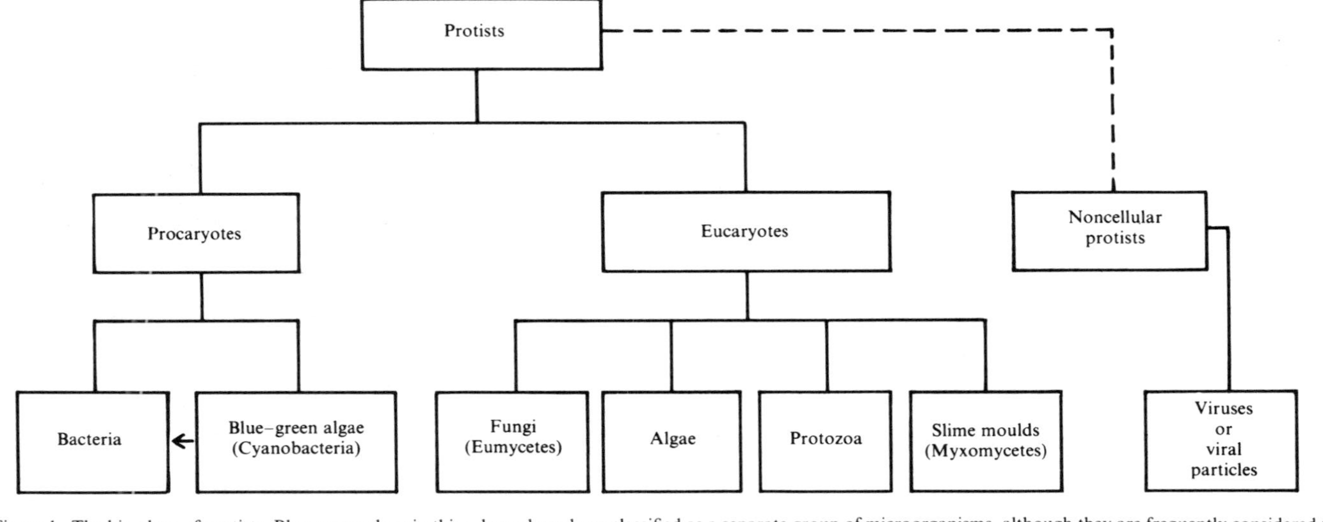

Figure 1. The kingdom of protists. Blue–green algae in this scheme have been classified as a separate group of microorganisms, although they are frequently considered to be cyanobacteria and included with other bacteria. Alternatively blue–green algae are classified as cyanophycophyta (*see* Table 6.).

Table 1. Features distinguishing procaryotic and eucaryotic cells.

Feature	Procaryotic cells	Eucaryotic cells
Size of organism	$< 1\text{–}2 \times 1\text{–}4$ μm	> 5 μm in width or diam
Genetic system		
Location	Nucleoid, chromatin body or nuclear material	Nucleus, mitochondria, chloroplasts
Structure of nucleus	Not bounded by nuclear membrane. One circular chromosome	Bounded by nuclear membrane. One, or more, linear chromosome
	Chromosome does not contain histones. No mitotic division	Chromosomes have histones. Mitotic nuclear division
	Nucleolus absent. Functionally related genes may be clustered	Nucleolus present. Functionally related genes not clustered
Sexuality	Zygote is partially diploid (merozygotic)	Zygote is diploid
Cytoplasmic nature and structures		
Cytoplasmic streaming	Absent	Present
Pinocytosis	Absent	Present
Gas vacuoles	Can be present	Absent
Mesosome	Present	Absent
Ribosomes	70S,[a] distributed in the cytoplasm	80S arrayed on membranes (e.g., endoplasmic reticulum). 70S in mitochondria and chloroplasts
Mitochondria	Absent	Present
Chloroplasts	Absent	May be present
Golgi structures	Absent	Present
Endoplasmic reticulum	Absent	Present
Membrane-bound (true) vacuoles	Absent	Present
Outer cell structures		
Cytoplasmic membranes	Generally do not contain sterols. Contain part of respiratory and, in some, photosynthetic machinery	Sterols present. Do not carry out respiration and photosynthesis
Cell wall	Peptidoglycan (murein or mucopeptide) as component	Absence of peptidoglycan
Locomotor organelles	Simple fibril	Multifibrilled with microtubules
Pseudopodia	Absent	May be present
Metabolic mechanisms	Varied, particularly that of anaerobic energy-yielding reactions. Some fix atmospheric N_2. Some accumulate poly-β-hydroxybutyrate as reserve material	Glycolysis is pathway for anaerobic energy-yielding mechanism
DNA base ratios (G + C%)	28–73	About 40

[a] S refers to the Svedberg unit, the sedimentation coefficient of a particle in the ultracentrifuge.

Table 2. Some criteria used for classifying microorganisms.

Microorganisms	Criteria
Bacteria	Morphology, staining reactions, spore formation, motility, antigenic structure, metabolism
Fungi	Nature of the thallus, form of sexual/asexual reproductive structures
Algae	Types of pigment, number and arrangement of flagella on motile cells, nature of storage carbohydrates
Protozoa	Presence of flagella or cilia, modes of locomotion and/or reproduction, presence and nature of specialized intracellular structures or skeletal structures
Viruses	Type of genome (DNA or RNA), size of the virion, presence of envelope

Detailed Classification of Bacteria

Type 1 Phototrophic bacteria

Order I Rhodospirillales
Suborder Rhodospirillineae
Family I Rhodospirillaceae
Genus I *Rhodospirillum*
Genus II *Rhodopseudomonas*
Genus III *Rhodomicrobium*
Family II Chromatiaceae
Genus I *Chromatium*
Genus II *Thiocystis*
Genus III *Thiosarcina*
Genus IV *Thiospirillum*
Genus V *Thiocapsa*
Genus VI *Lamprocystis*
Genus VII *Thiodictyon*
Genus VIII *Thiopedia*
Genus IX *Amoebobacter*
Genus X *Ectothiorhodospira*
Suborder Chlorobiineae
Family III Chlorobiaceae
Genus I *Chlorobium*
Genus II *Prosthecochloris*
Genus III *Chloropseudomonas*
Genus IV *Pelodictyon*
Genus V *Clathrochloris*
Addenda
Genus *Chlorochromatium*
Genus *Cylindrogloea*
Genus *Chlorobacterium*

Type 2 Gliding bacteria

Order I Myxobacterales
Family I Myxococcaceae
Genus I *Myxococcus*
Family II Archangiaceae
Genus I *Archangium*
Family III Cystobacteraceae
Genus I *Cystobacter*
Genus II *Melittangium*
Genus III *Stigmatella*
Family IV Polyangiaceae
Genus I *Polyangium*
Genus II *Nannocystis*
Genus III *Chondromyces*

Order II Cytophagales
Family I Cytophagaceae
Genus I *Cytophaga*
Genus II *Flexibacter*
Genus III *Herpetosiphon*
Genus IV *Flexithrix*
Genus V *Saprospira*
Genus VI *Sporocytophaga*
Family II Beggiatoaceae
Genus I *Beggiatoa*
Genus II *Vitreoscilla*
Genus III *Thioploca*
Family III Simonsiellaceae
Genus I *Simonsiella*
Genus II *Alysiella*
Family IV Leucotrichaceae
Genus I *Leucothrix*
Genus II *Thiothrix*
Incertae sedis
Genus *Toxothrix*
Familiae incertae sedis
Achromatiaceae
Genus *Achromatium*
Pelonemataceae
Genus I *Pelonema*
Genus II *Achroonema*
Genus III *Peloploca*
Genus IV *Desmanthos*

Type 3 Sheathed bacteria

Genus *Sphaerotilus*
Genus *Leptothrix*
Genus *Streptothrix*
Genus *Lieskeela*
Genus *Phragmidiothrix*
Genus *Crenothrix*
Genus *Clonothrix*

Type 4 Budding and/or appendaged bacteria

Genus *Hyphomicrobium*
Genus *Hyphomonas*
Genus *Pedomicrobium*
Genus *Caulobacter*
Genus *Asticcarcaulis*
Genus *Ancalomicrobium*
Genus *Prosthecomicrobium*
Genus *Thiodendron*
Genus *Pausteria*
Genus *Blastobacter*
Genus *Seliberia*
Genus *Gallionella*
Genus *Nevskia*
Genus *Planctomyces*
Genus *Metailogenium*
Genus *Caulococcus*
Genus *Kusnezonia*

Type 5 Spirochetes

Order I Spirochaetales
Family I Spirochaetaceae
Genus I *Spirochaeta*
Genus II *Cristispira*
Genus III *Treponema*
Genus IV *Borrelia*
Genus V *Leptospira*

Type 6 Spiral and curved bacteria

Family I Spirillaceae
Genus I *Spirillum*
Genus II *Campylobacter*
Incertae sedis
Genus *Bdellovibrio*
Genus *Microcyclus*
Genus *Pelosigma*
Genus *Brachyarcus*

Type 7 Gram-negative, aerobic rods and cocci

Family I Pseudomonadaceae
Genus I *Pseudomonas*
Genus II *Xanthomonas*
Genus III *Zoogloea*
Genus IV *Gluconobacter*
Family II Azotobacteraceae
Genus I *Azotobacter*
Genus II *Azomonas*
Genus III *Beijerinckia*
Genus IV *Dorxia*
Family III Rhizobiaceae
Genus I *Rhizobium*
Genus II *Agrobacterium*
Family IV Methylomonadaceae
Genus I *Methylomonas*
Genus II *Methylococcus*
Family V Halobacteriaceae
Genus I *Halobacterium*
Genus II *Halococcus*
Incertae sedis
Genus *Alcaligenes*
Genus *Acetobacter*
Genus *Brucella*
Genus *Bordetella*
Genus *Francisella*
Genus *Thermus*

Type 8 Gram-negative, facultatively anaerobic rods

Family I Enterobacteriaceae
Genus I *Escherichia*
Genus II *Edwardsiella*
Genus III *Citrobacter*
Genus IV *Salmonella*
Genus V *Shigella*
Genus VI *Klebsiella*
Genus VII *Enterobacter*
Genus VIII *Hafnia*
Genus IX *Serratia*
Genus X *Proteus*
Genus XI *Yersinia*
Genus XII *Erwinia*
Family II Vibrionaceae
Genus I *Vibrio*
Genus II *Aeromonas*
Genus III *Plesiomonas*
Genus IV *Photobacterium*
Genus V *Lucibacterium*
Incertae sedis
Genus *Zymomonas*
Genus *Chromobacterium*
Genus *Flavobacterium*
Genus *Haemophilus*
Genus *Pasteurella*
Genus *Actinobacillus*
Genus *Cardiobacterium*
Genus *Streptobacillus*
Genus *Caiymmatobacterium*

Type 9 Gram-negative, anaerobic bacteria

Family I Bacteriodaceae
Genus I *Bacteriodes*
Genus II *Fusobacterium*
Genus III *Leptotrichia*
Incertae sedis
Genus *Desulfovibrio*
Genus *Butyrivibrio*
Genus *Succinivibrio*
Genus *Succinimonas*
Genus *Lachnospira*
Genus *Selenomonas*

Type 10 Gram-negative cocci and coccobacilli

Family I Neisseriaceae
Genus I *Neisseria*
Genus II *Branhamella*
Genus III *Moraxella*
Genus IV *Acinetobacter*
Incertae sedis
Genus *Paracoccus*
Genus *Lampropedia*

Type 11 Gram-negative, anaerobic cocci

Family I Veillonellaceae
Genus I *Veillonella*
Genus II *Acidaminococcus*
Genus III *Megasphaera*

Type 12 Gram-negative, chemolithotrophic bacteria

Family I Nitrobacteraceae
Genus I *Nitrobacter*
Genus II *Nitrospina*
Genus III *Nitrococcus*
Genus IV *Nitrosomonas*
Genus V *Nitrospira*
Genus VI *Nitrosococcus*
Genus VII *Nitrosolobus*
Organisms metabolizing sulphur
Genus 1 *Thiobacillus*
Genus 2 *Sulpholobus*
Genus 3 *Thiobacterium*
Genus 4 *Macromonas*
Genus 5 *Thiovulum*
Genus 6 *Thiospira*
Family II Siderocapsaceae
Genus I *Siderocapsa*
Genus II *Naumanieila*
Genus III *Ochrobium*
Genus IV *Siderococcus*

Type 13 Methane-producing bacteria

Family I Methanobacteriaceae
Genus I *Methanobacterium*
Genus II *Methanosarcina*
Genus III *Methanococcus*

Type 14 Gram-positive cocci

Family I Micrococcaceae
Genus I *Micrococcus*
Genus II *Staphylococcus*
Genus III *Planococcus*
Family II Streptococcaceae
Genus I *Streptococcus*
Genus II *Leuconostoc*
Genus III *Pediococcus*
Genus IV *Acrococcus*
Genus V *Gemella*
Family III Peptococcaceae
Genus I *Peptococcus*
Genus II *Peptostreptococcus*
Genus III *Ruminococcus*
Genus IV *Sarcina*

Type 15 Endospore-forming rods and cocci

Family I Bacillaceae
Genus I *Bacillus*
Genus II *Sporolactobacillus*
Genus III *Clostridium*
Genus IV *Desulphomaculum*
Genus V *Sporosarcina*
Incertae sedis
Genus *Oscillospira*

Type 16 Gram-negative, asporogenous rod-shaped bacteria

Family I Lactobacillaceae
Genus I *Lactobacillus*
Incertae sedis
Genus *Listeria*
Genus *Eysipelothrix*
Genus *Caryophanon*

Type 17 Actinomycetes and related organisms

Coryneform group of bacteria
Genus I *Corynebacterium*
Genus II *Arthrobacter*
Incertae sedis
Genus I *Brevibacterium*
Genus II *Microbacterium*
Genus III *Cellulomonas*
Genus IV *Kurthia*
Family I Propionibacteriaceae
Genus I *Propionibacterium*
Genus II *Eubacterium*

Order I Actinomycetales
Family I Actinomycetaceae
Genus I *Actinomyces*
Genus II *Arachnia*
Genus III *Bifidobacterium*
Genus IV *Bacterionema*
Genus V *Rothia*
Family II Mycobacteriaceae
Genus I *Mycobacterium*
Family III Frankiaceae
Genus I *Frankia*
Family IV Actinoplanaceae
Genus I *Actinoplanes*
Genus II *Spirillospora*
Genus III *Streptosporangium*
Genus IV *Amorphosporangium*
Genus V *Ampullariella*
Genus VI *Pilimelia*
Genus VII *Planomonospora*
Genus VIII *Planobispora*
Genus IX *Dactylosporangium*
Genus X *Kitastoa*
Family V Dermatophilaceae
Genus I *Dermatophilus*
Genus II *Geodermatophilus*
Family VI Nocardiaceae
Genus I *Nocardia*
Genus II *Pseudonocardia*
Family VII Streptomycetaceae
Genus I *Streptomyces*
Genus II *Streptoverticillium*
Genus III *Sporichthya*
Genus IV *Microellobosporia*
Family VIII Micromonosporaceae
Genus I *Micromonospora*
Genus II *Thermoactinomyces*
Genus III *Actinobifida*
Genus IV *Thermomonospora*
Genus V *Microbispora*
Genus VI *Micropolyspora*

Type 18 Rickettsias

Order I Rickettsiales
Family Rickettsiaceae
Tribe I Rickettsieae
Genus I *Rickettsia*
Genus II *Rochallmaea*
Genus III *Coxiella*
Tribe II Ehrlichieae

Genus IV *Ehrlichia*
Genus V *Cowdria*
Genus VI *Neorickettsia*
Tribe III Wolbachieae
Genus VII *Wolbachia*
Genus VIII *Symbiotes*
Genus IX *Blattabacterium*
Genus X *Ricketsiella*
Family Bartonellaceae
Genus I *Bartonella*
Genus II *Grahamella*
Family Anaplasmataceae
Genus I *Anaplasma*
Genus II *Paranaplasma*
Genus III *Aegyptionella*
Genus IV *Haemobartonella*
Genus V *Eperythrozoon*

Order II Chlamydiales
Family I Chlamydiaceae
Genus I *Chlamydia*

Type 19 Mycoplasma

Class Mollicutes

Order I Mycoplasmatales
Family I Mycoplasmataceae
Genus I *Mycoplasma*
Family II Acholeplasmataceae
Genus I *Acholeplasma*
Incertae sedis
Genus *Thermoplasma*
Incertae sedis
Genus *Spiroplasma*

Table 3. Principal characteristics of different types of bacteria [Hawker and Linton (1979)].

Type[a]	Morphology	Gram stain	Movement	Division	Spores	Metabolism[b]
1	Spiral, spheres, ovoid	Negative	Flagella, if motile	Most binary, some budding	None	Phototrophic
2	Rods may aggregate and differentiate to form fruiting body and myxospores or may form microcysts	Negative	Gliding	Binary	Myxospores or microcysts	Chemoorganotrophic
3	Rods in chains within sheaths in which Fe or Mn may be deposited	Negative	Some have single cells motile by flagella	Transverse and/or longitudinal fission	None	Chemoorganotrophic
4	Cells may bear buds on appendages, be appendaged but not bud, bud or excrete an appendage	Negative	Daughter cells may be motile by flagella	Budding or binary	None	Chemoorganotrophic
5	Helices, protoplasmic cylinder intertwined with axial fibres contained within the outer envelope	Negative	Flexion by axial fibres, rotation	Binary	None	Chemoorganotrophic
6	Rigid, helically curved cells	Negative	Flagella	Binary	None	Chemoorganotrophic
7	Straight or curved rods. A few may form branches, others are cocci	Negative	Flagella	Binary	None, but some form cysts	Chemoorganotrophic
8	Rods	Negative	Flagella	Binary	None	Chemoorganotrophic
9	Rods	Negative	Flagella, if motile	Binary	None	Chemoorganotrophic
10	Cocci or coccobacilli	Negative	Some show twitching motility. No flagella	Binary	None	Chemoorganotrophic
11	Cocci	Negative	Nonmotile	Binary	None	Chemoorganotrophic
12	Rods and cocci	Negative	Flagella	Binary	None	Chemolithotrophic oxidize NH_4 or NO_2. Metabolize S or deposit iron or manganese oxides
13	Rods or cocci	Positive or negative	Motile or nonmotile	Binary	None	Strictly anaerobic producing methane due to CO_2 reduction with hydrogen, formate, acetate etc, as electron donor
14	Cocci	Positive	Flagella, if motile	Binary	None	Chemoorganotrophic
15	Rods or cocci	Mostly positive	Flagella, if motile	Binary	Endospores	Chemoorganotrophic
16	Rods	Positive	Flagella, if motile	Binary	None	Chemoorganotrophic
17	Irregular rods, filaments, branching	Positive	Flagella, if motile	Binary	Conidia	Chemoorganotrophic
18	Rods, cocci often	Negative	Nonmotile	Binary	None	Obligate parasites
19	Highly pleomorphic. Outer layer is triple-layered membrane	Negative	Nonmotile usually	Elementary body formation and release, binary	None	Chemoorganotrophic

[a] *See* Table 1 (p.4).
[b] *See* Table 2, Chapter 2. (p.74)

Classification of Fungi

Table 4

Fungi (Eumycetes)		
Lower fungi (Phycomycetes)	Higher fungi (*see* Classification of Yeasts)	Deuteromycetes[a] (Fungi imperfecti)
Chytridiomycetes	Ascomycetes	Sphaeropsidales
Hyphochytridiomycetes		Melanconiales
Oomycetes	Basidiomycetes	Moniliales
Plasmodiophoromycetes	Heterobasidiomycetes	Mycelia sterilia
Zygomycetes	Homobasidiomycetes	
Trichomycetes		

[a] Commercially important *Aspergillus* and *Penicillium* species belong to this class.

Classification of Yeasts

The following classification of yeasts has been proposed by Phaff *et al.* (1978).

Type 1 Ascosporogenous yeasts

Class Ascomycetes
  Subclass Hemiascomycetes
    Order Saccharomycetales
      Family Saccharomycetaceae
      Subfamily 1 Eremascoideae
        Genus *Eremascus*
      Subfamily 2 Endomycetoideae
        Genus *Endomyces*
      Subfamily 3 Schizosaccharomycetoideae
        Genus *Schizosaccharomyces*
      Subfamily 4 Saccharomycetoideae
        Genus *Endomycopsis*
        Genus *Saccharomyces*
        Genus *Fabospora*
        Genus *Kluyveromyces*
        Genus *Schwanniomyces*
        Genus *Debaryomyces*
        Genus *Citeromyces*
        Genus *Pichia*
        Genus *Hansenula*
        Genus *Pachysolen*
        Genus *Dekkera*
        Genus *Saccharomycodes*
        Genus *Hanseniaspora*
        Genus *Wickerhamia*
        Genus *Saccharomycopsis*
        Genus *Nadsonia*
      Subfamily 5 Lipomycetoideae
        Genus *Lipomyces*
      Subfamily 6 Nematosporoideae
        Genus *Nematospora*
        Genus *Metschnikowia*
        Genus *Coccidiascus*

Type 2 Ballistosporogenous yeasts

  Family Sporobolomycetaceae
    Genus *Sporobolomyces*
    Genus *Bullera*

Type 3 Asporogenous yeasts

  Family Cryptococcaceae
    Genus *Cryptococcus*
    Genus *Rhodotorula*
    Genus *Pityrosporum*
    Genus *Schizoblastosporion*
    Genus *Kloeckera*
    Genus *Trigonopsis*
    Genus *Brettanomyces*
    Genus *Torulopsis*
    Genus *Candida*
    Genus *Trichosporon*

Type 4 Yeast-like organisms

    Genus *Pullularia*
    Genus *Geotrichum*
    Genus *Ashbya*
    Genus *Eremothecium*
    Genus *Taphrina*
    Genus *Prototheca*

Table 5. Principal characteristics of different genera of yeasts [Phaff *et al.* (1978)].

Type[a]	Genus	Morphology	Division	Spores[b]	Metabolism
1	*Eremascus*	True mycelium with cross walls, mould-like		No arthrospores, 8 ovoid ascospores per ascus	Fermentation absent, high sucrose tolerance
	Endomyces	Septate mycelium		1–4 hat-shaped or spherical (with smooth or wrinkled membrane) arthrospores per ascus	Fermentation absent
	Schizosaccharomyces	Elongated cells, primitive mycelium in 1 species	Cross wall formation	4 or 8 spores per ascus	Fermentation present
	Endomycopsis	True mycelium, pseudomycelium	Budding	Arthrospores rare, 1–4 spherical, ellipsoid or hat-, sickle- or Saturn-shaped spores	Fermentation variable, weak if present
	Saccharomyces	Spherical cells	Budding	1–4 ovoid or spherical spores per ascus, ascus wall not lyzed on maturity	Fermentation strong
	Fabospora	Similar to *Saccharomyces*		1–4 kidney-shaped or, sometimes, spherical spores, asci rupture on maturity	Fermentation strong
	Kluyveromyces	Ovoid cells	Multilateral budding	Numerous kidney-shaped or ovoid spores per ascus	Fermentation present
	Schwanniomyces	Spherical cells	Budding, meiosis buds common	1 or 2 walnut-shaped ascospores per ascus	Fermentation present
	Debaryomyces	Spherical cells	Multilateral budding	1 or 2 warty spores per ascus	Fermentation variable, high NaCl tolerance common
	Citeromyces	Ovoid cells	Multilateral budding	1 or 2 warty, spherical spores per ascus	Fermentation present, nitrate assimilation
	Pichia	Ovoid to elongated cells, pseudomycelium variable, true mycelium rare	Multilateral budding	2–4 helmet- or hat-shaped spores per ascus	Fermentation variable
	Hansenula	Polymorphic cells, pseudomycelium, true mycelium sometimes	Budding	2–4 hat- or Saturn-shaped spores per ascus	Fermentation variable, nitrate assimilation

continued overleaf

Table 5 – (continued)

Type[a]	Genus	Morphology	Division	Spores[b]	Metabolism
	Pachysolen	Ovoid to elongated cells	Budding	4 hat-shaped spores per ascus, asci formed at tip of thick-walled special structure	Fermentation weak, nitrate assimilation
	Dekkera	Similar to *Brettanomyces*		1–4 hat-shaped or spherical spores	Fermentation slow, ethanol oxidized to acetic acid, nitrate assimilation
	Saccharomycodes	Lemon-shaped cells	Bipolar budding on broad base	4 spherical spores per ascus	Fermentation in 1 species
	Hanseniaspora	Lemon-shaped or ovoid cells	Bipolar budding	2–4 hat-shaped or 1–2 spherical spores per ascus	Fermentation present
	Wickerhamia	Lemon-shaped or ovoid cells	Polar budding on broad base	1–16 cap-shaped ascospores per ascus	Fermentation in 1 species
	Saccharomycopsis	Large, elongated cells	Budding	1–4 ovoid spores per ascus	Fermentation weak, 1 species requires CO_2, amino acids and 35–40°C for growth
	Nadsonia	Ovoid, lemon-shaped or elongated cells	Bipolar budding on broad base (bud fission)	1 or 2 brown, spherical, spiny spores per ascus	Fermentation present
	Lipomyces	Capsulated, polymorphic cells found in soil	Multilateral budding	4–16 ovoid, amber-coloured spores per ascus	Fermentation absent
	Nematospora	Polymorphic cells, true mycelium sometimes, plant parasites	Multilateral budding	8 needle-shaped spores with whip-like appendages per ascus, large asci	Fermentation present
	Metschnikowia	Ovoid cells	Multilateral budding	1 needle-shaped spore without appendage per ascus, club-shaped asci	Fermentation variable
	Coccidiascus	Ovoid cells observed in *Drosophila*, but not cultivated	Multilateral budding	4 needle-shaped spores without appendages per ascus	

2	*Sporobolomyces*	True mycelium	Budding	Asymmetrical ballistospores forcibly discharged	Fermentation absent, nitrate assimilation variable
	Bullera	Ovoid cells	Multilateral budding	Symmetrical ballistospores forcibly discharged	Fermentation absent
3	*Cryptococcus*	Polymorphic cells, often surrounded by capsule	Multilateral budding	Ascospores absent	Fermentation absent, nitrate assimilation variable, starch-like polysaccharide formed in suitable media
	Rhodotorula	Ovoid to elongated cells usually	Multilateral budding	Ascospores absent	Fermentation absent, nitrate assimilation variable, pink carotenoids present
	Pityrosporum	Ovoid or flask-shaped cells	Budding and fission	Ascospores absent	Fermentation absent, requires lipids for growth
	Schizoblastosporion	Ovoid or flask-shaped cells	Budding and fission	Ascospores absent	Fermentation absent, no lipid requirement
	Kloeckera	Lemon-shaped or ovoid cells, imperfect form of *Hanseniaspora*	Bipolar budding	Ascospores absent	Fermentation present
	Trigonopsis	Triangular or ellipsoidal cells	Budding at corners	Ascospores absent	Fermentation absent in 1 species
	Brettanomyces	Ovoid and ogival cells	Multilateral budding	Ascospores absent	Fermentation slow, ethanol oxidized to acetic acid, nitrate assimilation variable
	Torulopsis	Pseudomycelium absent	Multilateral budding	Ascospores absent	Fermentation variable, nitrate assimilation variable, no starch-like polysaccharides formed
	Candida	Pseudomycelium, true mycelium sometimes	Budding	Arthrospores absent, ascospores absent	Fermentation variable, nitrate assimilation variable
	Trichosporon	Pseudomycelium, true mycelium	Budding	Arthrospores present, ascospores absent	Fermentation absent or weak, nitrate assimilation variable

continued overleaf

Table 5 – (continued)

Type[a]	Genus	Morphology	Division	Spores[b]	Metabolism
4	*Pullularia*	Pseudomycelium, true mycelium, dark brown or black in culture	Budding	Arthrospores present, ascospores absent	Fermentation absent
	Geotrichum	Septate mycelium, mould-like, spreading white, powdery growth		Arthrospores present, ascospores absent	Fermentation absent or very weak
	Ashbya	Sparsely septate, multinucleate mycelium		8–32 needle-shaped spores with appendages in groups in mycelium	Fermentation absent, high riboflavin levels sometimes
	Eremothecium	Septate mycelium		Numerous arcuate ascospores in groups in mycelium	Fermentation absent, high riboflavin levels
	Taphrina	Mycelium, parasitic on plants	Multilateral budding in culture media	8 spores per ascus	Fermentation absent
	Prototheca	Ovoid to spherical cells, colourless algae derived from *Chlorella*	Internal partition, forming 2 to numerous new cells		

[a] See p.10.
[b] See Figs. 26 and 29 for morphology.

Classification of Algae

Table 6. Some properties of major algal taxonomic groups [Pelczar *et al.* (1977)].

Taxonomic group (division)	Chlorophyll	Carotenoids	Biloproteins	Storage products	Flagellation and details of cell structure
Rhodophycophyta (Red algae)	*a*, rarely *d*	β-Carotene, zeaxanthine, ±α-carotene	R-phycocyanin, R-phycoerythrin, C-phycocyanin, C-allophycocyanin, C-phycoerythrin	Floridean starch, oils	Flagella absent

Cyanophycophyta (Blue–green algae)	a	β-Carotene, zeaxanthine, echinenone, myxoxanthrophyll	C-phycocyanin, C-phycoerythrin, allophycocyanin	Glycogen-like cyanophycean starch, proteins	Flagella absent, procaryotic cells
Xanthophycophyta (Yellow–green algae)	a, c, rarely e	β-Carotene, diadinoxanthin, heteroxanthin, vaucheriaxanthin ester		Chrysolaminarin, oils	Flagella (2 unequal, apical)
Chrysophycophyta (Golden algae)	a, c_1, c_2	β-Carotene, fucoxanthin		Chrysolaminarin, oils	Flagella (1 or 2 equal or unequal, apical). Cell surface covered by characteristic scales sometimes
Phaeophycophyta (Brown algae)	a, c_1, c_2	β-Carotene, ±α-carotene, rarely ε-carotene, fucoxanthin		Laminarin, soluble carbohydrates, oils	Flagella (2 lateral)
Bacillariophycophyta (Diatoms)	a, c, c_2	β-Carotene ±α-carotene, rarely ε-carotene, fucoxanthin		Chrysolaminarin, oils	Flagella (1 in male gametes, apical). Cell in two halves, the walls silicified with elaborate markings
Euglenophycophyta (Euglenoids)	a, b	β-Carotene ±γ-carotene, diadinoxanthin		Paramylon, oils	Flagella (1, 2, or 3 equal, slightly apical). Gullet present
Chlorophycophyta (Green algae)	a, b	β-Carotene ±α-carotene, rarely γ-carotene and lycopene, lutein		Starch, oils	Flagella (1, 2, 4 or many, equal apical or subapical)
Cryptophycophyta (Cryptomonads)	a, c_2	α-Carotene ±β-carotene, rarely ε-carotene, alloxanthin	Phycoerythrin, phycocyanin	Starch, oils	Flagella (2 lateral: gullet present in some species)
Pyrrophycophyta (Dinoflagellates, phytodinads)	a, c_2	β-Carotene, peridinin		Starch, oils	Flagella (2 lateral, 1 trailing, 1 girdling). Longitudinal and transverse furrow and angular plates in most

Classification of Protozoa

Table 7. Major groups of protozoa [Pelczar *et al.* (1977)]

Major groups	Characteristics
Sarcomastigophora	Flagella and/or pseudopodia. Single type of nucleus except in developmental stages of certain Foraminiferida. No spore formation. Sexuality, when present, essentially syngamy (union of gametes in fertilization)
Mastigophora (flagellates)	One or more flagella present in trophozoites. Solitary or colonial. Asexual reproduction basically by symmetrogenic binary fission. No sexual reproduction in many groups. Nutrition is phototrophic and/or heterotrophic
Sarcodina (amoebas)	Pseudopodia. Flagella, when present, restricted to developmental stages. Cortical zone of cytoplasm relatively undifferentiated. Body naked or with external or internal tests or skeletons of various types and chemical composition. Asexual reproduction by fission. Sexual reproduction, if present, with flagellate or, more rarely, ameboid gametes. Mostly free-living
Sporozoa	Spores simple, without polar filaments and with one or several sporozoites. Single nucleus. No cilia and flagella, except for flagellated microgametes in some groups. Sexuality, when present, syngamy. Parasitic
Cnidospora	Spores with one or more polar filaments and one or more sporoplasms. Parasitic
Ciliophora (ciliates)	Simple cilia or compound ciliary organelles in at least one stage of life cycle. Subpellicular infraciliature present even when cilia absent. Two types of nucleus, except in a few homocaryotic forms. Binary fission in asexual reproduction. Sexual reproduction by conjugation. Heterotrophic nutrition. Mostly free-living

Classification of Slime Moulds

Table 8. Different types of slime moulds.

Type	Features
Myxomycetes	True plasmodium. Sexual fusion
Acrasiales	Cellular slime moulds. Incomplete plasmodium. No sexual stage. Stalked mass of spores
Labyrinthulales	Net slime moulds. Parasites on algae and aquatic plants
Plasmodiophoromycetes	Endoparasitic slime moulds. Plant parasites, cause of disease in many crops

MORPHOLOGY

Procaryotes, Eucaryotes and Viruses

Biochemical and functional differences between procaryotes and eucaryotes are reflected in their structural features. These major groups are distinguished and classified more by their morphological and ultrastructural differences than by metabolic and biochemical characteristics.

The four major eucaryotic subkingdoms and procaryotes vary in their shapes and dimensions as shown in Table 9.

Table 9. Shape and size of representative microorganisms compared with protein molecules.

Object	Shape	Size
Bacteria		
Bacillus megaterium	Rod-shaped	2.8 × 1.2–1.5 μm
Serratia marcescens	Rod-shaped	0.7–1.0 × 0.7 μm
Staphylococcus albus	Spherical	1.0 μm
Streptomyces scabies	Filamentous	0.5–1.2 μm diam.
Fungi		
Botrytis cinerea, conidiophores	Filamentous	11–23 μm diam.
B. cinerea, conidia	Ellipsoidal	9–15 × 5–10 μm
Rhizopus nigricans, sporangiophores	Filamentous	24–42 μm diam.
R. nigricans, sporangiospores	Almost spherical	11–14 μm
R. nigricans, zygospores	Spherical	150–200 μm
Saccharomyces carlsbergensis, vegetative cells	Ellipsoidal	5–10.5 × 4–8 μm
Algae		
Chlamydomonas kleinii	Ovoid	28–32 × 8–12 μm
Ulothrix subtilis	Filamentous	4–8 μm diam.
Protozoa		
Amoeba proteus	Amorphous	< 600 μm diam.
Euglena viridis	Spindle-shaped	40–65 × 14–20 μm
Paramecium caudatum	Slipper-shaped	180–300 μm length
Slime moulds		
Badhamia utricularis, plasmodia	Amorphous	Indefinitive, extensive
B. utricularis, spores	Spherical	9–12 μm
Protein molecules		
Egg albumin		4.0 nm
Serum albumin		5.6 nm
Serum globulin		6.3 nm
Haemocyanin		22.0 nm

Table 10. Size and composition of various parts of bacteria.

Part	Size	Composition and comments
Slime layer		
Microcapsule	5–10 nm	Protein–polysaccharide–lipid complex responsible for the specific antigens of enteric bacteria and of other species
Capsule	0.5–2.0 nm	Mainly polysaccharides (e.g., *Streptococcus*), sometimes polypeptides (e.g., *Bacillus anthracis*)
Slime	Indefinite	Mainly polysaccharides (e.g., *Leuconostoc*), sometimes polypeptides (e.g., *B. subtilis*)
Cell wall		Confers shape and rigidity to the cell
Gram-positive	10–20 nm	20% dry weight of the cell. Consists mainly of macromolecules of a mixed polymer of *N* -acetylmuramic acid–peptide, teichoic acid and polysaccharides
Gram-negative	10–20 nm	Consists mostly of a protein–polysaccharide–lipid complex with a small amount of the muramic polymer
Cell membrane	5–10 nm	Semipermeable barrier to nutrients. 5–10% dry weight of the cell, consisting of protein (50%), lipid (28%) and carbohydrate (15–20%) in a double-layered membrane
Flagellum	0.01–0.02 × 4–12 μm	Myosin–keratin–fibrinogen-type protein (M. W.40 000). Arises from the cell membrane. Responsible for motility
Inclusions		
Spore	1.0–1.5 × 1.6–2.0 μm	One spore per cell formed intracellularly. Spores highly resistant to heat, dryness and antibacterial agents. Spore walls rich in dipicolinic acid
Storage granules	0.5–2.0 μm	Glycogen-like, sulphur or lipid granules found in some species
Chromatophores	50–100 nm	Organelles in photosynthetic species. *Rhodospirillum rubrum* contains about 6000 per cell
Ribosomes	10–30 nm	Organelles for protein synthesis. About 1000 ribosomes per cell containing RNA (63%) and protein (37%)
Volutin	0.5–1.0 nm	Inorganic polymetaphosphates which stain metachromatically
Nuclear material	About half cell volume	DNA that functions genetically as if the genes were arranged linearly on a single, endless chromosome but which appears by light microscopy as irregular patches with no nuclear membrane or distinguishable chromosomes. Autoradiography confirms the linear arrangement of DNA (M.W. $\geq 10^9$)

Table 11. Size and properties of parts of fungal hyphae.

Part	Size (μm)	Comments
Outer fibrous layer	0.1–0.5	Very electron-dense material
Cell wall	0.1–0.25	*Zygomycetes*, *Ascomycetes*, and *Basidiomycetes* contain chitin (2–26% dry wt). *Oomycetes* contain cellulose not chitin. Yeast cells contain glucan (29%), mannan (31%), protein (13%) and lipid (8.5%)
Cell membrane	0.007–0.01	Much-folded, double-layered membrane. Semipermeable to nutrients
Endoplasmic reticulum	0.007–0.01	Highly invaginated membrane or set of tubules, probably connected with both the cell membrane and the nuclear membrane and concerned in protein synthesis and probably other metabolic functions
Nucleus	0.7–3	Surrounded by a double membrane (10 nm), containing pores (40–70 nm wide). Nucleus is flexible and contains cytologically distinguishable chromosomes. Nucleolus about 3 nm. In Actinomycetes there is no nuclear membrane. The nucleus can migrate
Mitochondria	0.5–1.2 × 0.7–2	Analogous to those in animal and plant cells, containing electron-transport enzymes and bounded by an outer membrane and an inner membrane forming cristae. They probably develop by division of existing mitochondria
Inclusions		Lipid and glycogen-like granules are found in some fungi. Ribosomes in all fungi

Table 12. Size and properties of typical viruses.

Bacteriophage	Dimensions (nm)	Characteristics
Double-stranded DNA		
T-even of *Escherichia coli*	Head: 90×60 Tail: 100×25 Tail fibrils: 130×2.5	Tadpole-shaped phage with DNA (1.3×10^{10} daltons) confined to the head. Tail is protein, some of which is contractile. Long tail fibrils are involved in attachment to the host cell. Particle 2.5×10^{8} daltons
Single-stranded DNA		
ϕX174 of *Escherichia coli*	22	Dodecahedron with 12 subunits. DNA (1.6×10^{6} daltons) 25% dry weight. Particle 6.2×10^{6} daltons
RNA		
f2 of *Escherichia coli*	20	Polyhedron containing RNA (3×10^{-12} μg/virus) and protein. Nucleic acid content is probably similar to that of ϕX174

Microbial Flocs

Many organisms form flocs or pellets over a range of sizes (*see* Table 14). In a fermenter, such as a stirred tank, a floc may exist and this in turn may be influenced by the shear conditions.

Table 13. Characteristic size of flocs [Atkinson and Daoud (1976)].

Microorganism	Floc size range (mm)
Agaricus blazei (mushroom)	2.8–25 diam.
Aspergillus niger	0.2–0.5 diam.
Asp. niger	0.75–2.0 diam.
Brewers' yeast	
Physically limited	< 2 diam.
Fermentation limited	< 13 diam.
Mixed bacterial cultures	0.025–5

Examination of yeast samples from a tower fermenter reveals spherical flocs (*see* Fig. 2).

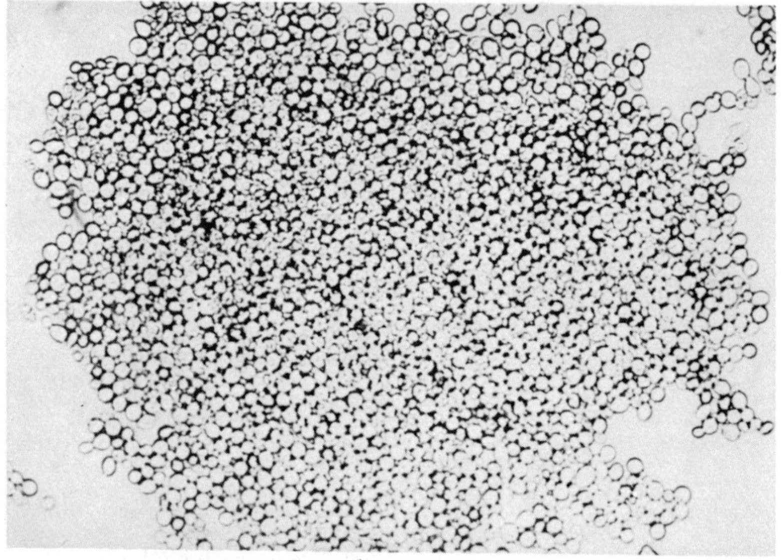

Figure 2. Yeast floc taken from a tower fermenter.

The morphology of *Aspergillus niger* varies with age. Initially, the mould is filamentous and growth is dispersed, but with increasing time individual pellets are formed with a diameter of 0.2–0.5 mm after 16–18 h.

Similar changes in morphology of *Penicillium chrysogenum* occur with pH (6.0–7.4) with a transition from individually distinguishable hyphae to flocculation, with the interlocking of hyphae [Pirt and Callow (1960)]. Clearly visible pellets of *Asp. niger* may be formed in a shake flask after incubation for 4 days (*see* Fig. 3).

The morphology of flocs varies with shear rate. At low shear rates, large flocs are formed that are not apparent at high shear rates. The size distribution of activated sludge floc is shown in Fig. 4.

Figure 3. Flocs of *Aspergillus niger* prepared by batch culture for 4 days of a spore inoculum at pH 5.6 and 25°C, using a 250 ml flask in a shaking incubator.

Table 14. Floc-producing bacteria [McKinney (1956)].

Bacterium	Habitat
Alcaligenes faecalis	Intestinal tract
A. metalcaligenes	Intestinal tract
Bacillus cereus	Soil
B. lentus	Soil
Escherichia coli	Intestinal tract
E. freundii	Soil
E. intermedium	Soil
Flavobacterium sp.	Water
Nocardia actinomorpha	Soil
Paracolobactrum aerogenoides	Soil
Pseudomonas fragi	Soil
Ps. ovalis	Soil
Ps. perlurida	Soil
Ps. segnis	Trickling filter
Ps. solaniolens	Soil
Zoogloea ramigera	Activated sludge

Table 15. Pellet-forming fungi [Whitaker and Long (1973)].

Lower fungi	Higher fungi	Fungi imperfecti
Absidia glauca	*Agaricus bisporus*	*Alternaria solani*
A. spinosa	*Ag. blazei*	*Alt. tenuissima*
Basidiolus ranarum	*Ag. campestris*	*Aspergillus candidus*
Chlamydomucor javanicus	*Ag. rodmanni*	*Asp. clavatus*
Conidiolus sp.	*Cantharellus cibarius*	*Asp. flavicepes*
Cunninghamella bainieri	*Chaetomium caprinum*	*Asp. flavus*
Phycomyces blakesleeanus	*Collybia umbulata*	*Asp. giganteus*
Rhizopus acidus	*C. velutipes*	*Asp. nidulans*
R. chinensis	*Fomes pinicola*	*Asp. niger*
R. oryzae	*Gibberella fujikuroi*	*Asp. stellatus*
	G. saubinetti	*Asp. ochraceus*
	Guignardia bidwelli	*Asp. stellatus*
	Hebeloma sinapizans	*Asp. terreus*

continued overleaf

Table 15 – (continued)

Lower fungi	Higher fungi	Fungi imperfecti
	Hypholoma fasiculare	*Asp. versicolor*
	Lentinus clodes	*Asp. wentii*
	Lenzites trabea	*Cephalosporium ulimi*
	Lepiota naucina	*Dactylium dendroides*
	L. procera	*Fusarium oxysporum*
	Lycoperdon umbrinum	*F. solani*
	Merulius tremellosus	*Gliocladium roseum*
	Morchella conica	*Macrosporium sarcinaeforme*
	M. crassives	*Myrothecium verrucaria*
	M. esculenta	*Penicillium brevicaule*
	M. nortensis	*Pen. camenbertii*
	M. rimosipes	*Pen. chrysogenum*
	M. rotunda	*Pen. clavariforme*
	M. semilibera	*Pen. digitatum*
	M. vulgaris	*Pen italicum*
	Neurospora crassa	*Pen. notatum*
	Pleurotus corticatus	*Pen. roquefortii*
	Pl. ostreatus	*Pen. urticae*
	Polyporus sulphureus	*Trichoderma viridae*
	Poris subacida	
	P. undora	
	Psalliota campestris	
	Psilocybe sp.	
	Schizophyllum commune	
	Sordaria fimicola	
	Trametes heteromorpha	
	T. seriales	

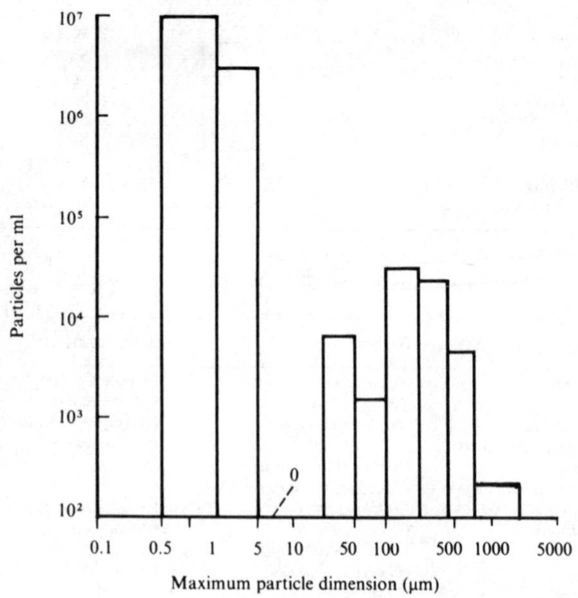

Figure 4. Size distribution of activated sludge floc in a stirred tank (high shear) [Parker *et al.* (1971)].

Microbial Films

In a continuous stirred tank fermenter, it is not unknown for microbes to accumulate around and attach to the impeller. An extreme example of accumulation after operation of such a fermenter for three months using a mixed microbial population is illustrated in Fig. 5.

Fig. 6 shows a stirred fermenter containing glass slides supported by a frame; extreme microbial growth is clearly visible. Such visible film formation is prevented by the introduction of glass beads maintained in suspension (Fig. 7).

The effect of surface area on the rate of nutrient removal can be used to establish the presence of a microbial film that is not normally visible (Fig. 8). Using Fig. 9, it can be seen that nutrient uptake increases with the total surface in the fermenter.

The controlled formation of a film may also be achieved using a grid (*see* Fig. 10).

Commercially, a large surface area for the attachment of a microbial film is achieved using beds of porous material, such as broken brick or slag, or prefabricated sections.

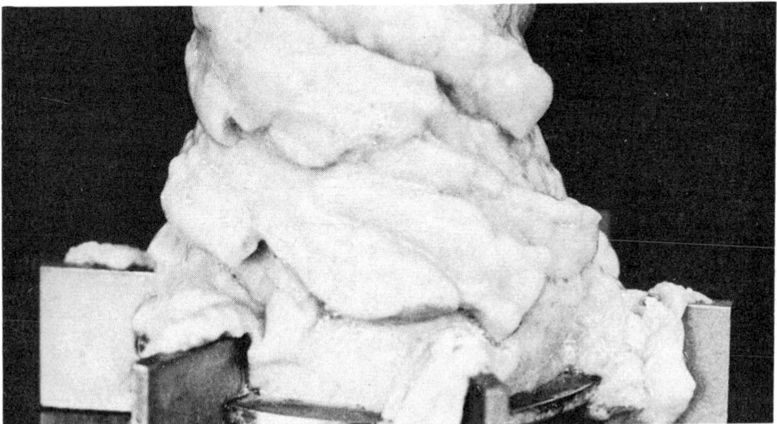

Figure 5. Growth on an impeller (rotation speed 500 rpm)

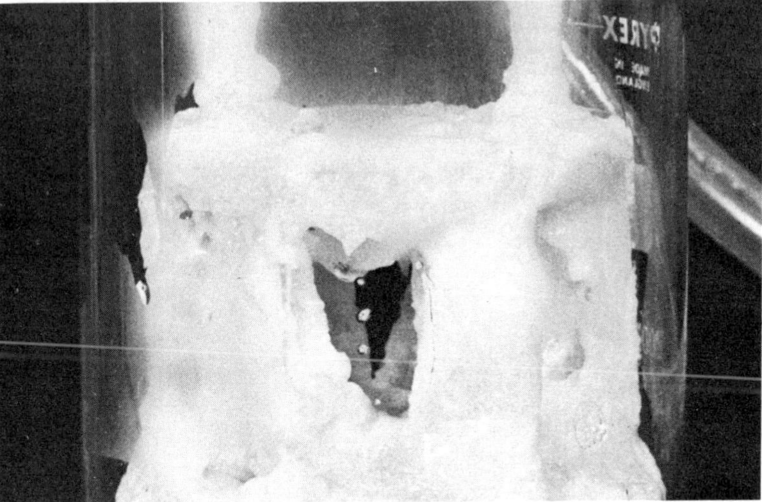

Figure 6. Extensive growth over glass slides and frame at pH 7, without beads

Table 16. Summary of processes using flocs.

Process objectives and culture	Research objectives and factors investigated	Fermentation methods and conditions	Characteristics and control of size of flocs
Biomass production using *Aspergillus niger* [Morris *et al.* (1973), Smith and Greenshields (1974)]	Use of tower fermenters. Aeration, physicochemical conditions	Supplemented molasses medium; 1.0–6.0 cm s^{-1}, air velocity: pH 3.25–6, < 48 h; biomass concn, 0.25–1.125% by wt; tower fermenters (1, 10, 50, 1000 ℓ, aspect ratio 10:1); batch and continuous operation	Filamentous threads through degrees of hairy pellets to smooth rounded colonies. ≤35 mm diam.
Biomass production using *Saccharomyces cerevisiae*, NCYC 1250 and 1251 [Smith and Greenshields, (1974)]	Use of tower fermenters. Yeast strain, media composition, gas and liquid flow rates	Medium: malt extract-based wort (SG 1.03 and 1.014); air flow rates; 2 and 7 ℓ min^{-1}; media flow rates; 12 and 15 ℓ day^{-}; max. yeast concn: 3.1% by wt (NCYC 1250), 17.5% by wt (NCYC 1251); tower fermenters (5, 6 ℓ, aspect ratio 10:1), batch and continuous operation	5-ℓ tower, flocs [0.1–1.0 cm diam. (NCYC 1251)] of various shapes, many disc-like; 6-ℓ tower, small flocs [<0.1 cm diam. (NCYC 1250)] readily washed out. Strain selection for size control
Brewing using *Saccharomyces cerevisiae* [Greenshields and Smith, (1971)]	Use of tower fermenter. Yeast strain, wort density and flow rate	Wort medium (SG 1.040); space velocity, 0.3–0.2 day^{-1}; mean yeast concn. 25% by wt centrifuged wet weight; tower fermenter, continuous operation	a) Physically limited, attains yeast concn of 20–30 by wt, ≤0.2 cm diam b)Fermentation limited, attains yeast concn of 25–40% by wt, <1.3 cm diam. Size control by strain selection, wort SG, superficial liquid velocity
Citric acid production using *Asp. niger* [Steel *et al.* (1955)]	Scale up and factors affecting production in submerged culture. Ferrocyanide treatment and phosphate supplement of molasses, inoculum level, pH, temperature and aeration rate	Inoculum, 1.2–2.8 × 10^4 pellets $ℓ^{-1}$; aeration rate, 500 ml min^{-1} or 6400 ml min^{-1}; initial pH 5.0–7.0; 31°C; 116 h; tower fermenters 2.5 and 36–40 ℓ batch operation	Varied in shape and density, optimum for citric acid production, 1–2 mm diam. Size control by size of spore inoculum, pH and ferrocyanide level
Enhancement of cell recovery in microbial protein production using *Pseudomonas fluorescens*, *Escherichia coli* and *Lactobacillus delbrueckii* [McGregor and Finn (1969)]	Use of chemical flocculants. Bacterial genus, chemical nature and concentration of a cationic polyamine, and small, positively charged alumina fibrils	Neutral pH; flocculation test; 2-ℓ suspension agitated on a shaker for 20 min; flocculation vessel; 2.5 ℓ, batch	Size control by type and concentration of flocculants

Enhancement of cell recovery in microbial protein production using *Candida intermedia* [Gasner and Wang (1970)]	Use and evaluation of 50 chemical flocculants. Chemical nature and concentration of additive. Effect of shear	Air flow, 1.0 min^{-1}; pH 3; 30°C; yeast concn, 4.0 g ℓ^{-1}; 500-ml flocculation vessel; flocculation test, 150-ml broth agitated on a shaker for 5 min; annular flow viscometer for fluid mechanics test; stirred fermenter (14 ℓ), semi continuous operation	Size control by type and concentration of additive; shear
Mushroom mycelium production using *Agaricus blazei* [Block *et al.*, (1953)]	Cultivation on citrus press water in submerged culture. pH, size of fermenter, media	pH 5.5–8.0, 250-ml flask, 1-ℓ and 4-ℓ bottles; batch operation; flask agitated on shaker; bottles stirred and aerated	Shake flasks; 3.2 mm diam. balls; aerated bottles; 2.54 mm diam. clumps, but much less dense than the smaller clumps. Size control by type of agitation, aeration, sugar concentration of medium
Plant tissue culture using carrot [Smith and Street (1974)]	Decline in embryogenic potential in culture. Subculturing every 21 days	pH 5.5, 25°C; biomass concentration, 6.4–10.0 g ℓ^{-1}; continuous light; 100-ml shake flask; batch operation	Tissue grew as embryogenic clumps 100–500 μm diam.
Plant tissue culture using endosperm of corn [Graebe and Novelli, (1966)]	Method for large-scale production. Time of growth, procedure of cultivation	Synthetic medium based on sucrose; pH 6.0; 25–28°C; 71 days; tissue concentration; 24–168 gℓ^{-1}; 6-ℓ flasks; semicontinuous operation; flasks aerated and stirred intermittently	Tissue grew as aggregates few mm diam. when intermittent stirring was used, otherwise $\leq$ 3 cm diam. Size control by intermittent stirring
Tissue culture of embryonic cells. Liver, retina, kidney and limb-bud from chick and mouse embryos [Moscona, (1961)]	Factors affecting aggregation of trypsin-dissociated cells. Type of organism and cell, age of embryo, temperature, rotation speed, presence of inhibitors or promoters	pH 8.0; 15–38°C; 24 h; 25-ml shake flask; batch operation	$\leq$1 mm diam. Size control by age of embryo, temperature, shaker rpm
Waste water treatment using activated sludge [Pavoni *et al.*, (1972)]	Mechanism of flocculation. Physiological state of microorganisms, accumulation of exocellular polymers, pH, agitation	Initial COD; 2 g ℓ^{-1}; pH 1–12; $\leq$ 140 h; biomass concn., 1–17 g^{-1} batch (15 ℓ); aeration and agitation by air diffusers	
Waste water treatment using activated sludge [De Walle and Chian, (1974)]	Formation and effect of humic substances on flocculation	Inoculum, 20.1 g ℓ^{-1}; pH 0.2–8; $\leq$ 53h; suspended solids; 140–260 mg ℓ^{-1}; dissolved oxygen 1.3–8 mg ℓ^{-1}; initial total organic carbon 91 mg ℓ^{-1}; batch (80 ℓ); aeration rate: 28 ℓ min^{-1}	When carbohydrate content of humic substances was low adsorption onto cells decreased with impairment of floc formation and settleability

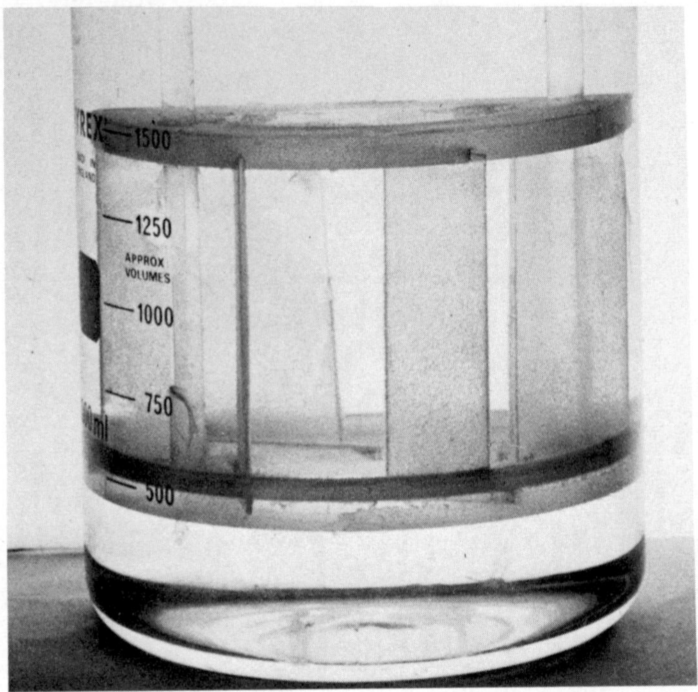

Figure 7. Growth over glass slides and frame at pH 9, with beads in suspension

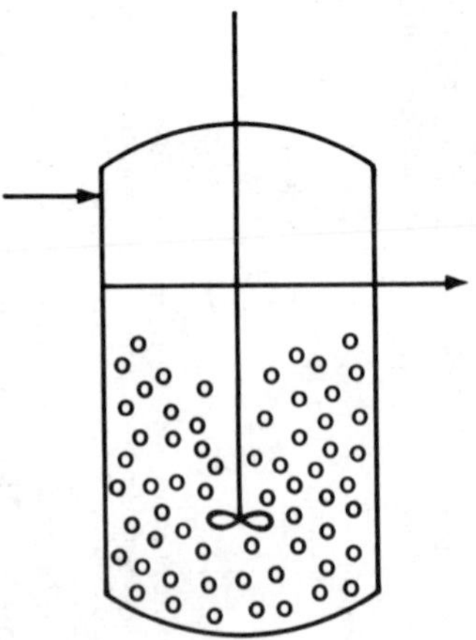

Figure 8. Continuous stirred tank fermenter containing inert particles

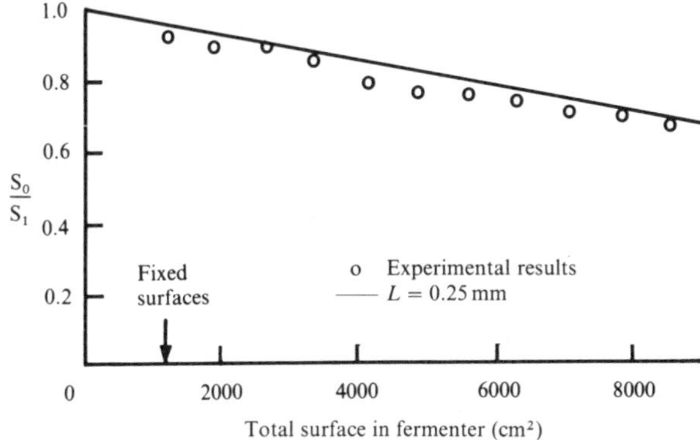

Figure 9. Continuous stirred tank fermenter with particles added – effect of area. (S_0, S_1 are the outlet and inlet substrate concentration respectively).

Figure 10. Supported microbial film for use in a biological film reactor. The microorganisms are predominantly gram-negative rod-shaped bacteria derived from garden soil and are contained by an aluminium grid (thickness 0.2 cm, perforations 1.27 cm × 1.27 cm) fixed to an aluminium sheet (thickness 0.63 cm).

Table 17. Ranges of microbial film thickness.

Description	Film thickness (mm)
Uncontrolled zoogloeal film	0.2 – 4.0
Zoogloeal film, subject to mechanical or hydrodynamic control	0.07 – 0.2
Pure cultures	0.001 – 0.01
Casual deposition	<0.001

Table 18. Summary of studies on microbial films

Research objectives	Conditions	Surface on which film grown	Hold-up values converted to wet film weight (kg m^{-2})	Remarks
Film fermenters Kinetics of fixed-film biological reactors [Kornegay and Andrews (1968)]	Mixed population using minerals and glucose (27–440 mg ℓ^{-1})	Plastic reactor with rotating drum	0.25	Dissolved oxygen and substrate usage constant when wet film thickness > 0.07 mm
Simulation of trickling filter [Maier *et al.* (1967)]	Mixed population using minerals and glucose (25–1000 mg ℓ^{-1})	Plastic plate	0.48–1.4	Film thickness controlled by scraping above fibre-glass meshes of various thicknesses
Diffusion within microbial films [Atkinson *et al.* (1968), Atkinson and Daoud (1970)]	Mixed population using minerals and glucose (< 10 000 mg ℓ^{-1})	1) Roughened plate glass 2) Aluminium plate	1) 0.073 2) 2.0	Film thickness controlled by scraping
Oxidation of sewage by films using a rotating tube reactor [Tomlinson and Snaddon (1966)]	Mixed population using sewage (BOD, 200)	Kieselguhr-coated perspex tube	1.15–3.8	Respiration absent approximately 0.2 mm below film surface
Packed-bed fermenters High-rate biological filtration [Bruce and Merkens (1970), Bruce *et al.* (1970)]	Mixed population using sewage (BOD, 250–300)	Slag, rock, various plastic packings	0.24–1.7	Thickness is 0.24–0.51 mm on plastic packing and 1.3–1.7 mm on slag and rock
Plastic media for trickling filters [Rincke and Wolters (1971)]	Mixed population using minerals and molasses (BOD, < 3000)	Plastic tubes, plastic corrugated sheets	0.118–2.55	Dry film approximately 9.5% by wt of wet film
Oxygen uptake in trickling filter [Monadjemi and Behn (1971)]	Mixed population using minerals and glucose (180–330 mg ℓ^{-1})	4-in diam. tube packed with table tennis balls	0.705	

Continuous stirred tank fermenter				
Attached growth of *Sphaerotilus* and mixed populations [Dias *et al.* (1968)]	Mixed population or pure *Sphaerotilus natans* culture using various media (5–100 mg ℓ$^{-1}$). 150 ml fermenter	Glass continuous culture vessel	$4–12.2 \times 10^{-3}$	
Attachment and growth of bacteria on surface of continuous culture vessel [Larsen and Dimmick (1964)]	Pure culture of 8 different microbes in nutrient broth. 18 ml fermenter. Substrate (5–100 mg ℓ$^{-1}$)	Glass continuous culture vessel	1×10^{-3}	Wall area = 3×10^9 mm^3, 1 organism per μm, *Bacillus cereus* and *B. subtilis* did not adhere
Study of 'takeover' by prototrophs, effect of wall growth	*Escherichia coli* in nutrient broth with substrate (5–100 mg ℓ$^{-1}$) 9 ml fermenter	Grass continuous culture vessel	3.3×10^{-5}	
Tissue culture				
Multilayer perfusion tissue culture [Elsworth (1970)]	Dissociated tissue cells in complex media 9 ml fermenter	Glass bottle	≤ 13 monolayer equivalents	Medium changes controlled film thickness
Glass slides				
Microprobe determination of oxygen profiles in microbial slime [Bungay *et al.* (1969), Whalen *et al.* (1969)]	Mixed population in nutrient broth	Glass slide immersed in continuous culture flask	0.2	Oxygen profile minimum at 50 μm (concentration = 20 ppm) to 150 μm (concentration = 500 ppm)
Adsorption of microorganisms by glass surface [Zuyagintsev (1959)]	Pure culture of 30 different microorganisms in various nutrient media	Various glasses	1×10^{-3}	Some organisms did not adhere
Marine microbiology [ZoBell (1946)]	Mixed population of marine microorganisms in sea water. Total organic matter, 4–5 mg ℓ$^{-1}$	Glass slides	1.2×10^{-4}	
Role of bacteria in underwater fouling [Wood (1950)]	Mixed population of marine microorganisms in sea water	1) Glass slide 2) Ground glass slide	2) 1.23×10^{-4} 2) 5.6×10^{-3}	

Table 19. Industrial fermentations using biological films.

Process	Features studied	General characteristics	Control of film thickness
Waste water treatment			
Trickling filter	Biological oxidation of industrial and domestic effluent [Abson and Todhunter (1967)]	Nonaseptic, microbial growth in packed bed. Waste water distributed intermittently over packing. Aerobic process – packing supported on grid, enhancing aeration by natural convection	Self-regulating. Thick film sloughs off or consumed by insects or worms
Rotating disc	Biological oxidation of industrial and domestic effluent [Borchardt (1971)]	Microbial growth on discs rotating in vertical plane, with discs dipping into trough of water. Microbial growth is alternately in contact with nutrients and air	Excess sloughed from discs
'Quick' vinegar process	Oxidation of ethanol by acetic acid-producing bacteria [Beaman (1967)]	Similar principle to trickling filter, but forced aeration. Wine (or other feed liquor) recirculated over packing (e.g., beechwood chips). Batch process lasting 4–5 days	Low growth rates. High substrate conversion. Microbial film accumulates. Packing discarded after several years
Animal tissue culture	Growth of animal tissue in surface layer for viral cultures [Telling and Radlett (1970)]	Tissue minced. Single cells formed by enzymatic digestion. Cells adhere to surface and grow as film in presence of suitable medium. Subsequently used for virus culture. Strictly aseptic	Use of appropriate media
Bacterial leaching of ores	Recovery of metals from sulphide ores using iron and sulphur-oxidizing bacteria [Trudinger (1971)]	*In situ* bacteria in clumps of low-grade or waste ores. Possibility of tank-leaching methods	Low growth rates. Film accumulates

Table 20. Microbial films in laboratory fermenters.

Fermenter	References	General characteristics	Control of film thickness
Film fermenter			
Annular biological reactor	Kornegay and Andrews (1968)	Cylindrical vessel housing rotating drum. Small annular gap providing large area-to-volume ratio, Impeller and draft tubes, to improve mixing and enhance hydrodynamic shear forces, incorporated into drum	Fluid shear forces encourage sloughing of film
Biological film reactor	Atkinson and Daoud (1970)	Nutrient solution flows over microbial film growing on an inclined plane	Film removed mechanically by scrapers
Rotating-tube reactor	Tomlinson and Snaddon (1966)	Used in experimental studies of waste water treatment. Similar to biological film reactor, but microbial film on inner surface of inclined rotating tube	None specified. Some film disintegration due to action of nematodes
Packed-bed reactor	Atkinson and Williams (1971)	Microbial mass grows in packed bed of wooden spheres. Nutrient flows over film, as in trickling filter. Applicable to other packings	Manual control with water jets
Rotating-disc fermenter (Bio-disc)	—	As industrial version (*see* Table 19)	
Continuous stirred tank fermenter	Anderson (1953) Northrop (1954)	Agitated tank with inlet and outlet flows. May be anaerobic or aerobic, may be aseptic	Presence of film often ignored. Mechanical scrapers have been used
Sponge fermenter	Freeman (1961)	Microorganisms grown on viscous nutrient in cellulose sponge. Sponge enclosed in cylinder and squeezed by piston, thus absorbing medium and aerating	Mechanically removed by compression of sponge

Immobilized Microorganisms

Immobilized microorganisms may be supported on a variety of materials, some which are illustrated in Fig. 11. Calcium carbonate pellets may also be used. Whole microorganisms may be immobilized and used effectively as immobilized enzymes. An example of this is glucose isomerase which is used in the production of fructose from corn syrup.

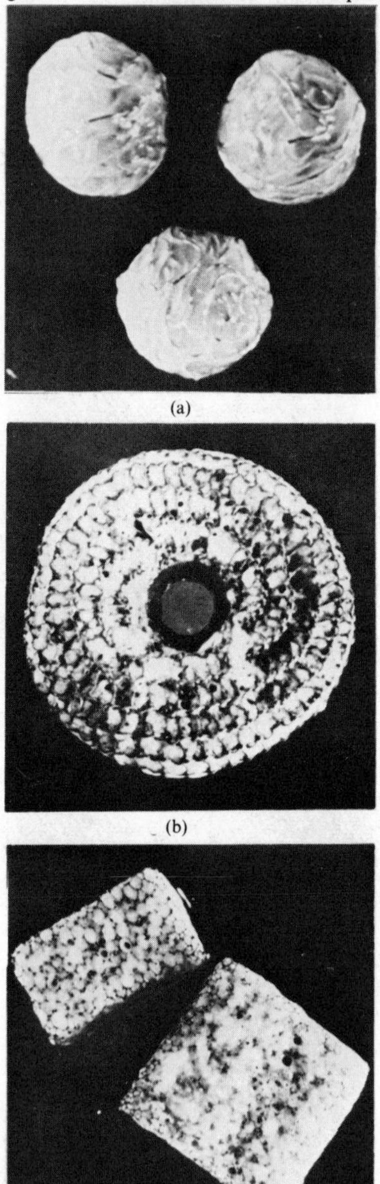

Figure 11. Biomass support particles for mixed microbial culture. (a) Stainless steel wire spheres, 6 mm diameter particles. (b) Polypropylene toroids, 53 mm overall diameter with 20 mm diameter torus. (c) Surface and side view of reticulated polyester foam particles, 25 × 25 × 10 mm.

LIFE CYCLES OF MICROORGANISMS

Bacteria

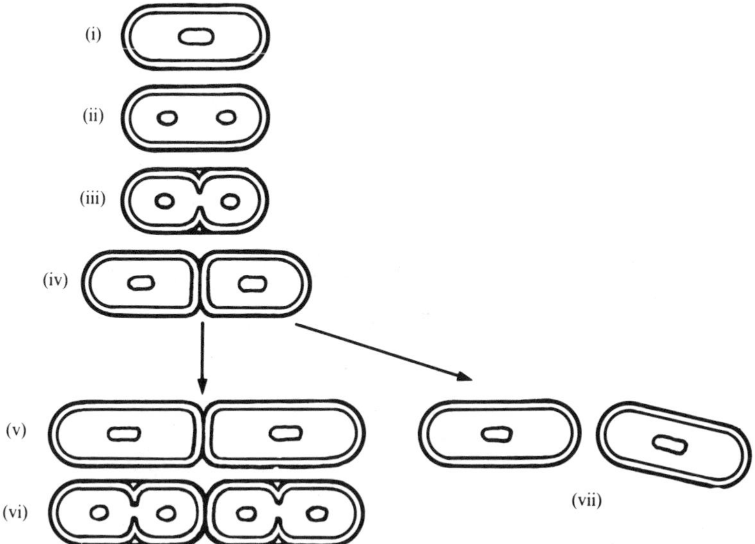

Figure 12. Cell division of bacilli under favourable conditions [Hawker *et al.* (1960)]

The mature cell is ready to divide [*see* stage (i), Fig. 12] with the chromatinic body dividing at an early stage (ii). At a later stage (iii), the cell wall grows inwards at the centre of the bacillus. The cell wall in stage (iv) completely grows across the bacillus and the two daughter cells are formed, although still remaining attached. Daughter cells then grow to full size but remain attached to the chain [stage (v)]. In stage (vi), mature cells begin to divide as in stages (i–iv), but remain attached to form a chain. Such cells when stained by nuclear stains, appear multinucleate. Many bacilli do not remain in chains but separate after division [stage (iv)], grow to full size and then divide [*see* stage (vii)].

The stages of cell division under favourable conditions in a streptococcal chain are similar (*see* Fig. 13), but the cell walls remain attached forming a longer chain.

Cell division in micrococci and staphylococci under favourable conditions is shown in Fig. 14. In mature cocci, in-growths of the cross walls occur in two planes at right angles to each other. A tetrad comprising four cells of equal size and maturity are then formed [stage (iii)]. In staphylococci, a cross wall occurs in a single plane initially [stage (iv)]. Then a further cross wall, at right angles to the first, divides one of the daughter cells only [stage (v)]. Unequal stages of maturity and irregular arrangements result, owing to the nonsynchronized division in the two planes.

Asymmetrical, rather than symmetrical, binary fission occurs in *Caulobacter* (in which there is morphological differentiation). Ternary fission is found in *Pelodictyon*.

Asexual binary fission may be preceded by a mating or conjugation of cells in some species, e.g., *Azotobacter*, *Escherichia*, *Klebsiella*, *Proteus*, *Pseudomonas*, *Proteus*, *Salmonella*, *Serratia*, *Shigella* and *Vibro*. Conjugation may occur between species of different genera, e.g., *Escherichia* × *Salmonella*, *Salmonella* × *Shigella*, *Escherichia* × *Serratia*, *Salmonella* × *Vibro* and *Escherichia* × *Shigella*.

Nitrobacter winogradskyi reproduces by budding (*see* life cycles of yeasts, p.40). *Streptomyces* species produce reproductive spores, each spore giving rise to a new organism. *Nocardia* produce extensive networks of filaments, followed by break-up of these fragments with the formation of bacillary or coccoid cells, each of which gives rise to new growth. A development cycle involving morphological differentiation occurs in *Chlamydia*, while species of Myxobacterales form fruiting bodies.

Under unfavourable environmental conditions, bacteria form spores (*see* Table 21). Sporulation and the vegetative cycles of a spore-forming bacterium are diagrammatically illustrated in Fig. 15.

With the exception of conidia formed by Actinomycetes, one resting cell, in the form of a spore, is produced from each vegetative cell and germinates to give a single cell. The process is therefore not a multiplicative one but only a means of protection and survival under unfavourable conditions.

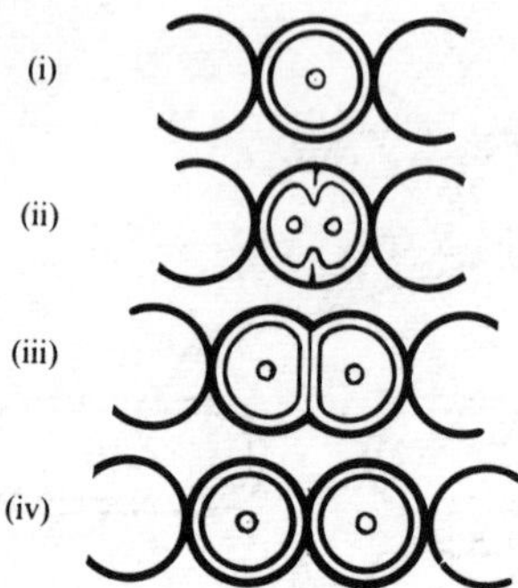

Figure 13. Cell division in streptococci under favourable conditions [Hawker *et al.* (1960)].

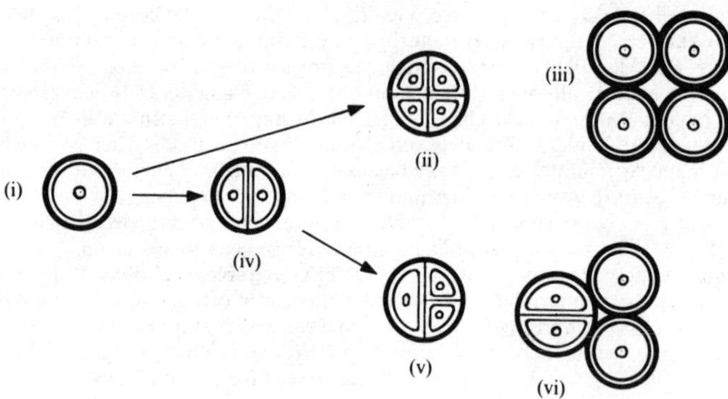

Figure 14. Cell division in micrococci [stages (i)–(iii)] and staphylococci [stages (i) and (iv)–(vi)] under favourable conditions [Hawker *et al.* (1960)].

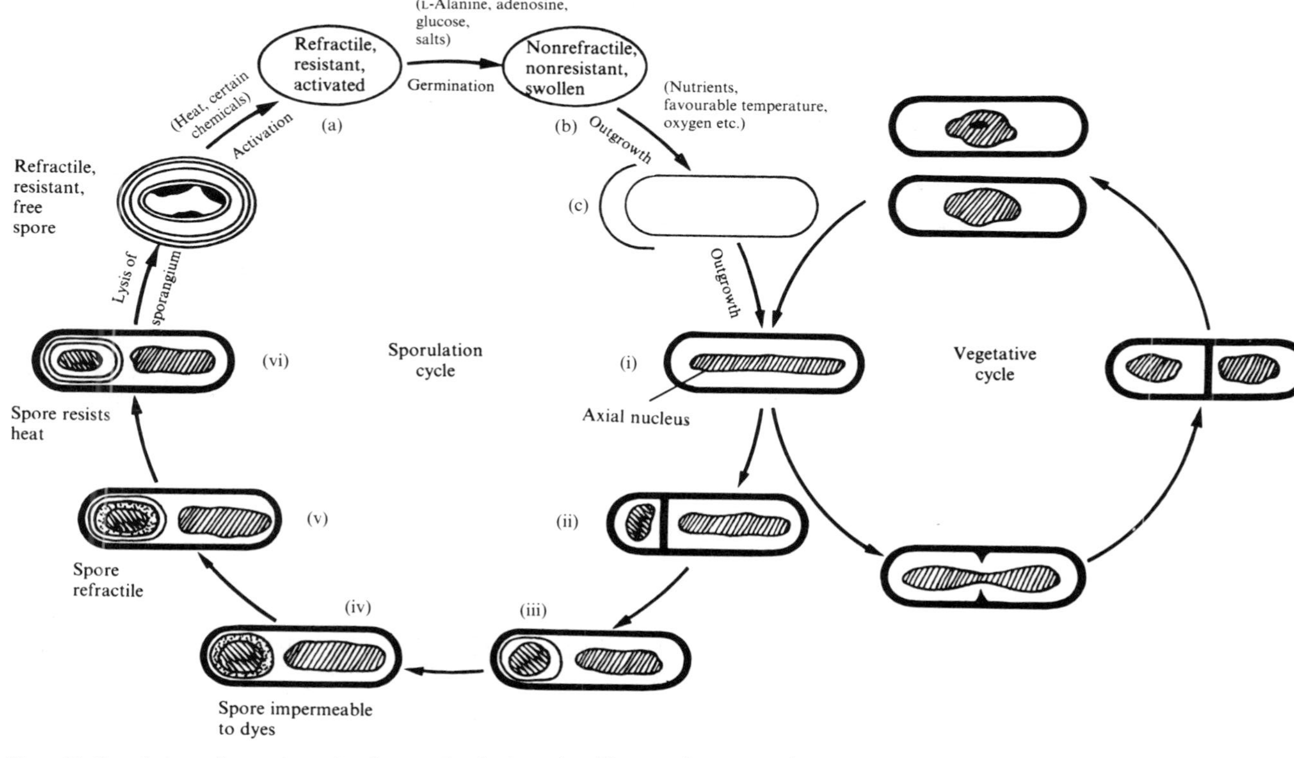

Figure 15. Sporulation and vegetative cycles of a spore-forming bacterium. The stages in sporogenesis are designated (i)–(vi). Stages in return of the dormant free spore to vegetative multiplication are indicated at (a), (b) and (c).

Table 21. Life cycle of bacteria under unfavourable conditions leading to spore formation.

Bacteria	Type of dormancy	Description
Bacillus, Clostridium Sporolactobacillus, Desulphotomaculum, Sporosarcine	Endospores	Nuclear material is arranged as filaments. Cytoplasma surrounded by thin membrane with a cortex. Outside the cortex is a stratified coat. The outermost layer is the exosporium. Mature spores resistant to heat, drying, pH variations and disinfectants
Azotobacter	Cysts	Formed by rounding off of the cell and development of a thick refractive wall, comprising the intine and extine
Myxobacteria	Myxospores	Migrating colony of vegetative cells forms a simple mound or stalk, with fruiting body developing (*see* Fig. 24). Within the fruiting body some cells shorten forming thick-walled myxospores. Other cells autolyze and form a slime enclosing the spores
Actinomycetes	Conidia	Specialized hyphae (sporophores) formed singly in groups or chains. Resting cells have a hydrophobic external sheath surrounding the cell wall

Cyanobacteria

Sexual reproduction has not been detected among the cyanobacteria. Unicellular species reproduce by binary fission. In some species, two or more cells form a group surrounded by a gelatinous envelope. A small number of genera form exospores.

Some filamentous genera develop small segments of these filaments (i.e. hormogonia) that are more motile and propagate vegetatively. Other filaments are surrounded by a thicker wall and are immobile (i.e. hormocysts). Both hormogonia and hormocysts eventually germinate by out-growth of the filament. Some species can form thick-walled spores frequently larger than the vegetative cells (i.e. akinetes) adjacent to the heterocysts. Akinetes are resistant to desiccation and germinate to form a hormogonium.

Fungi

Reproduction is usually by spores. In many fungi, reproduction involves the formation of specialized structures, such as fruiting bodies, which bear sexually or asexually derived spores.

Asexually derived spores are produced in large numbers when conditions are favourable and serve to spread the fungus rapidly. When conditions are less favourable, spores are produced as a result of fusion of sexual cells or branches, or after a period of secondary growth resulting from such a fusion.

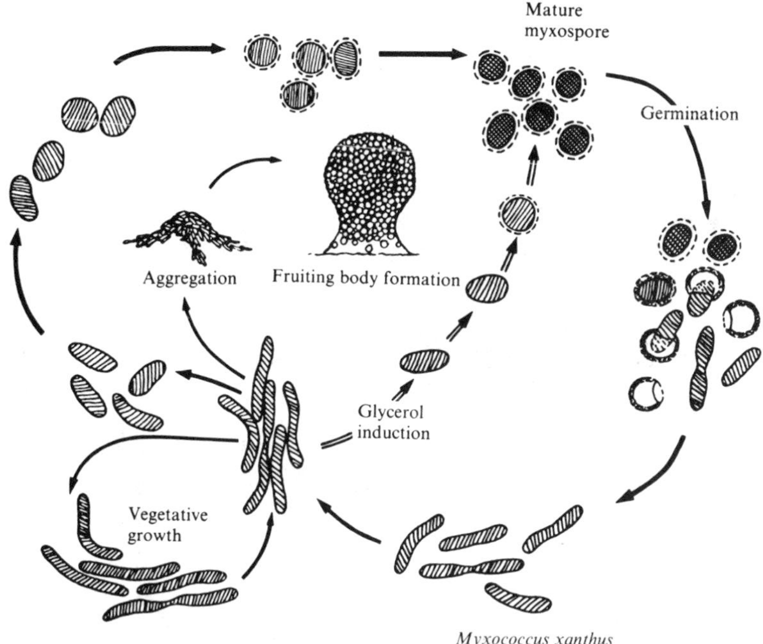

Figure 16. Life cycle of *Myxococcus xanthus*

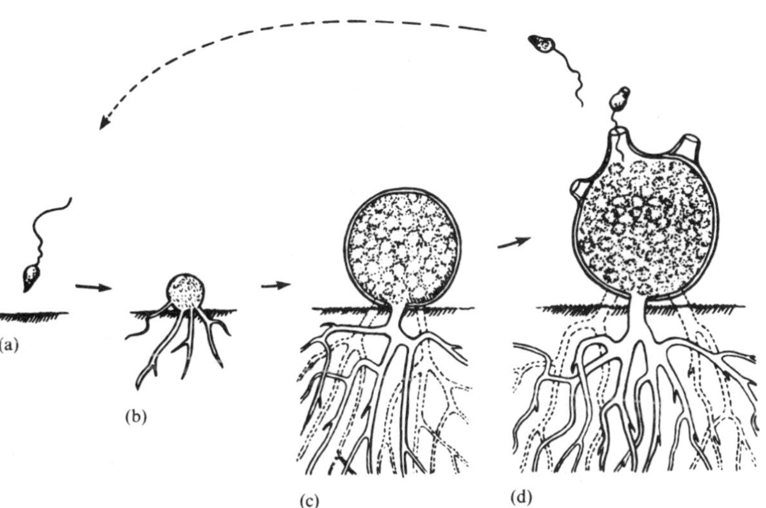

Figure 17. The life cycle of a primitive fungus.

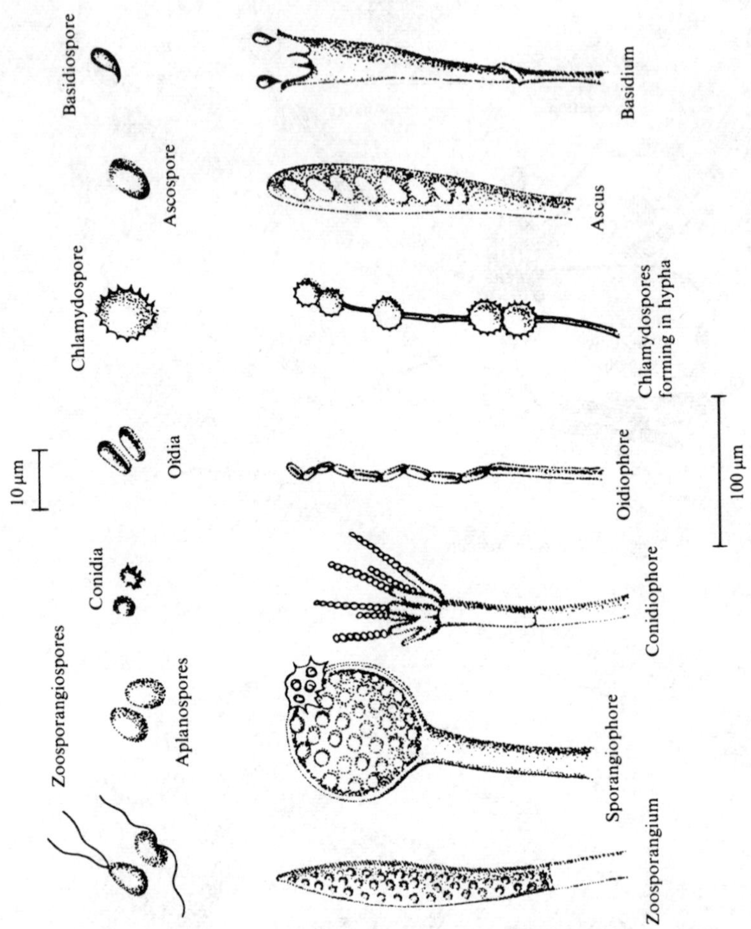

Figure 18. Different types of spores in fungi.

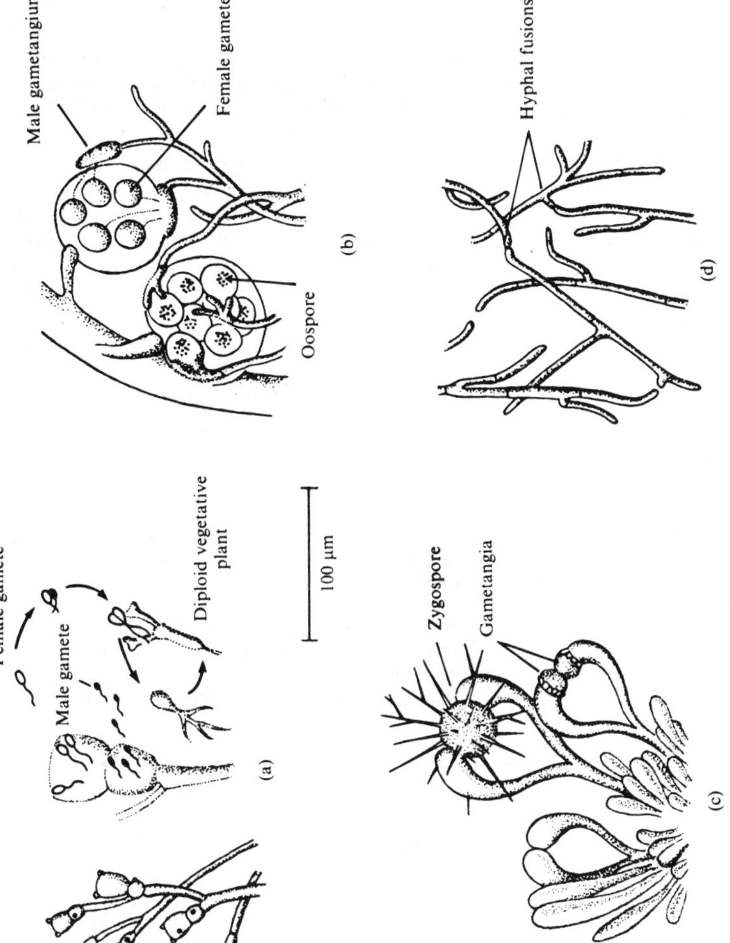

Figure 19. Sexual reproductive mechanisms in fungi.

Primitive aquatic or semi-aquatic fungi produce motile asexual spores (zoospores) (*see* Fig. 17). In Zygomycetes, the spores are not motile and are formed in multispored sporangia. In higher fungi, sporangia are not formed, instead conidia are the typical asexual spores. Fig. 18 illustrates different types of spores found in different genera of fungi.

A sexual spore is formed during sexual reproduction. There are three stages involved in sexual reproduction.

1) Plasmogamy – the bringing together of two compatible nuclei in a single protoplast.
2) Caryogamy – the actual fusion of the two nuclei, giving rise to a diploid zygote nucleus.
3) Meiosis – a reductive division which restores the nucleus to a haploid rather than a diploid state.

Some sexual mechanism in fungi are illustrated in Fig. 19. In gamete copulation [*see* Fig. 19 (a)], pairs of sexual cells fuse forming specialized sporangia-like gametangia. Gamete-gametangial copulation, illustrated in Fig. 19 (b), involves the fusion of a differentiated gamete of one sex with a gametangium of another sex. The direct fusion of gametangia without differentiation of gametes takes place in gametangial copulation [*see* Fig. 19 (c). In somatic copulation [*see* Fig. 19 (d)], sexual fusion of undifferentiated vegetative cells takes place.

Yeasts

Yeasts may reproduce by budding, fission or sporulation, the most common process being budding. In budding, the nucleus, which is initially present in the parent cell, enlarges and then extends into a bud forming a dumb-bell shaped structure that later divides giving rise to one nucleus in the parent cell and one in the daughter cell.

Sporulation in yeasts usually involves the formation of sexual spores (ascospores or basidospores) associated with differentiated cells (i.e. asci or basidia) by meiosis (*see* Fig. 20). Ascospores occur in a variety of forms (*see* Fig. 21). Asexual spores are also formed, such as conidia, arthrospores, blastospores, ballistospores and chlamydospores.

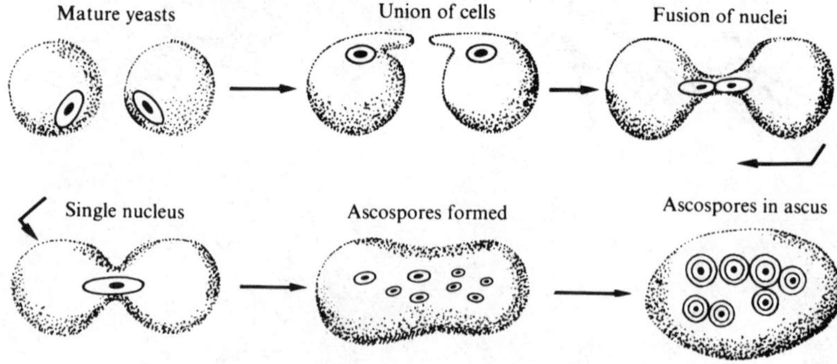

Figure 20. Ascospore formation in *Schizosaccharomyces octosporus*.

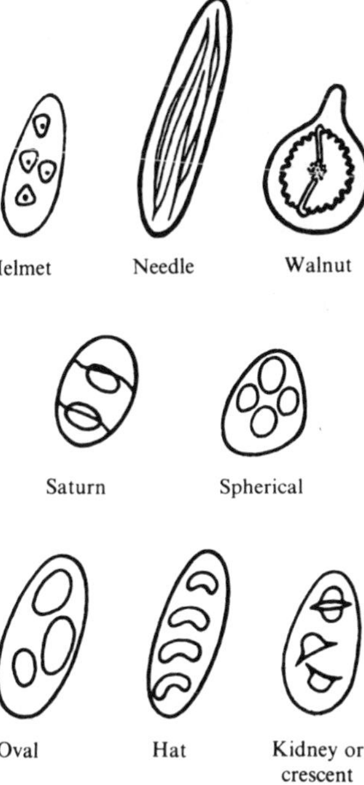

Figure 21. Typical ascospore morphology.

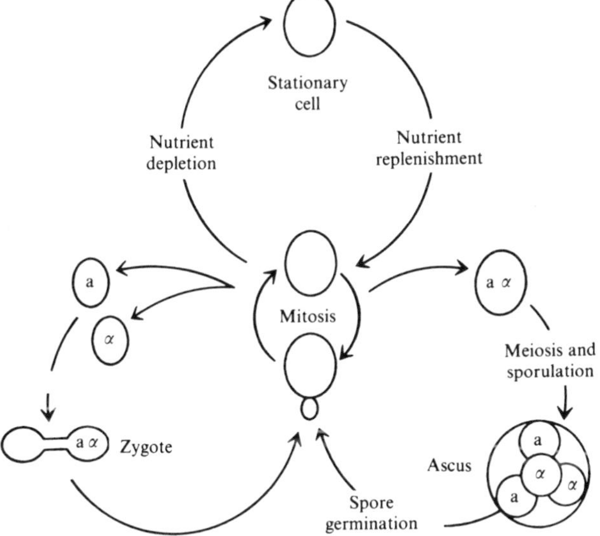

Figure 22. The life cycle of *Saccharomyces cerevisiae*.

The overall life cycle of *Saccharomyces cerevisiae* is summarized in Fig. 22. Mitotic haploid cells of mating type **a** or α may enter stationary phase when nutrients are depleted or may fuse with haploid cells of the opposite mating type to create a diploid zygote, which in turn produces **a**α-diploid mitotic cells. These may enter the stationary phase if there is no nutrient or may undergo meiosis and sporulation. The resulting haplid **a** and α spores germinate under favourable conditions to produce haploid mitotic cells. Haploid stationary cells or spores may fuse to form a zygote without passing through the mitotic cycle, and diploid stationary cells may undergo meiosis and sporulation without the mitotic cycle. These last pathways of development are not shown in Fig. 22.

Algae

Algae may reproduce sexually or asexually. Some species are limited to only one of these processes, but many have complicated life cycles employing both processes.

Asexual reproduction includes purely vegetative cell division. A new algal colony or filament may arise from a fragment of an old multicellular type from which it has broken. Most asexual reproduction in algae is more complex than this and involves unicellular spores, motile zoospores and nonmotile aplanospores.

All forms of sexual reproduction are found in algae. These are illustrated in Fig. 23.

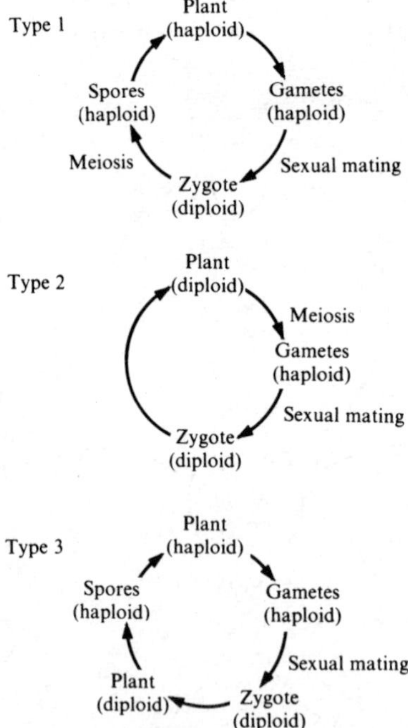

Figure 23. Types of sexual reproduction in algae.

Viruses

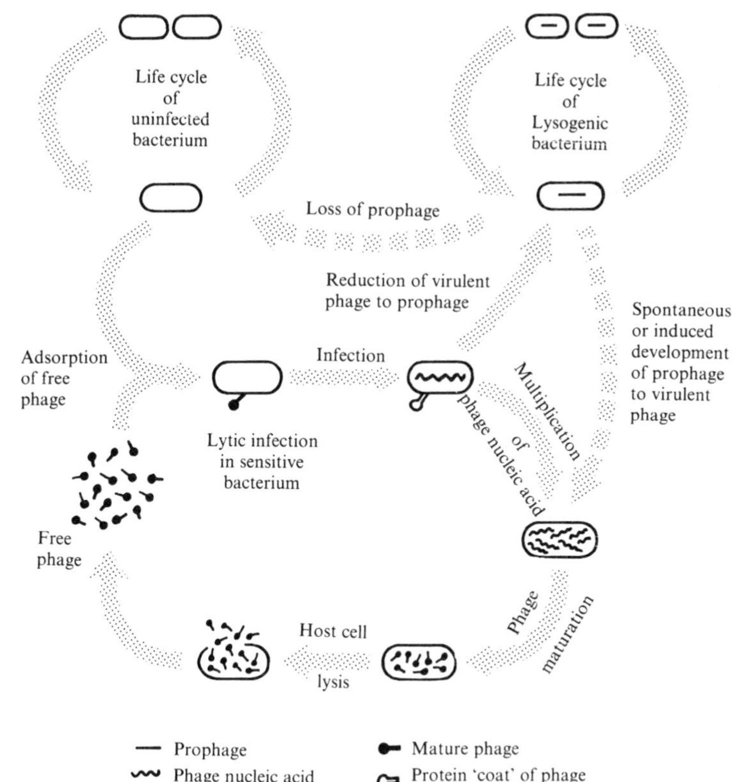

Figure 24. Phage–host life cycle [Hawker and Linton (1979)].

MACROCOMPOSITION OF MICROORGANISMS

Table 22. Mass and RNA content of cells of a number of bacterial species [Herbert (1961)].

Organism	Medium[a]	Cell Mass Resting cells ($\times 10^{-12}$g)	Cell Mass Log phase cells ($\times 10^{-12}$g)	RNA content Resting cells (%)	RNA content Log phase cells (%)
Aerobacter aerogenes	CCY	0.11	0.40	4.4	26.6
Bacillus anthracis	TMB			1.5	24.0
B. cereus	TMB	1.97	3.77	3.9	31.5
Chromobacterium prodigiosum	CCY	0.12	0.35	7.8	32.1
Chrom. violaceum	TMB	0.17	0.56	7.2	30.3
Clostridium welchii	TMB	0.91	2.19	32.2	42.2
Corynebacterium hofmannii	CCY			25.4	51.0
Salmonella typhi	CCY	0.19	0.34	10.5	35.9
Escherichia coli	CCY	0.12	0.41	15.5	37.0
Pasteurella pestis	TMB	0.13	0.15	5.9	20.1
Proteus vulgaris	TMB	0.18	0.36	12.6	35.0
Staphylococcus aureus	CCY	0.19	0.24	5.2	10.0

[a] CCY, casein–yeast extract medium; TMB, tryptic meat digest medium.

Table 23. Chemical analyses, dry weights and the populations of different microorganisms obtained in culture [Aiba *et al.* (1973)].

Organism	Composition Protein (% dry wt)	Composition Nucleic Acid (% dry wt)	Composition Lipid (% dry wt)	Population in culture (numbers ml^{-1})	Dry weight in culture (g 100 ml^{-1})	Comments
Viruses	50–90	5–50	< 1	10^8–10^9	0.0005[a]	Viruses with a lipoprotein sheath may contain 25% lipid
Bacteria	40–50	13–25	10–15	2×10^8–2×10^{11}	0.02–2.9	*Mycobacterium* may contain 30% lipid
Filamentous fungi	10–25	1–3	2–7		3–5	Some *Aspergillus* and *Penicillium* sp. contain 50% lipid
Yeast	40–50	4–10	1–6	1×10^8–4×10^3	1–5	Some *Rhodotorula* and *Candida* sp. contain 50% lipid
Small unicellular algae	10–60 (50)	1–5 (3)	4–80 (10)	4×10^7–8×10^7	0.4–0.9	Figure in parantheses is a commonly found value. Composition varies with the growth conditions

[a] For a virus 200 nm diam.

Table 24. Composition of a rapidly-growing *Escherichia coli* cell [Watson (1970)].

Component	Percentage of total cell weight (%)	Average molecular weight (daltons)	Approximate number per cell	Number of different kinds
H_2O	70	18	4×10^{10}	1
Inorganic ions (Na^+, K^+, Mg^{2+}, Ca^{2+}, Fe^{2+}, Cl^-, PO_4^{4-}, SO_4^{2-}, etc.)	1	40	2.5×10^8	20
Carbohydrates and precursors	3	150	2×10^8	200
Amino acids and precursors	0.4	120	3×10^7	100
Nucleotides and precursors	0.4	300	1.2×10^7	200
Lipids and precursors	2	750	2.5×10^7	50
Other small molecules[a]	0.2	150	1.5×10^7	200
Proteins	15	40 000	10^6	2000–3000
Nucleic acids				
DNA	1	2.5×10^9	4	1
RNA	6			
16S rRNA		500 000	3×10^4	1
23S rRNA		1 000 000	3×10^4	1
tRNA		25 000	4×10^5	40
mRNA		1 000 000	10^3	1000

[a] Heme, quinones, breakdown products of food molecules, etc.

EXAMPLES OF SUBSTANCES PRODUCED BY MICROBIAL ACTIVITY

Antimetabolites

Antibacterial agents

Ampicillin, aureothricin, cephalosporins, chloramphenicol, cycloserine, erythromycin, gentamicin, gramicidins, kanamycins, neomycins, penicillins, streptomycins, tetracyclines, tobramycin.

Antifungal agents

Amphotericin, anisomycin, antifungone, blastomycin, griseofulvins, nystatin.

Anthelminthics

Cinnamic acid, helenin.

Antineoplastic agents

Actinomycin, azaserine, bleomycins, daunomycins, 5-fluorouracil, mithramycins, streptozotocin.

Antiprotozoal agents

Antiamoebin, antiprotozin, monomycin, paromomycin, trichomycin.

Toxins

Aflatoxin, α-toxin (lecithinase), aspergillic acid, cholera toxin, diethylarsine, cytochalasins, ochratoxins, sporidesmin, T-2 toxin, tremorgen toxin.

Pharmaceutically Significant Compounds

Acetylcholine, lysergic acid derivatives, ergot alkaloids, histamine, insulin, interferon, rugulovasines.

Steroids

Acetylated 3-β sterols, anabolic and estrogenic steroids, asposterol, 1,4-diene-3-ketosteroids, 7-cyanosteroids, 2-β and 16-α hydroxylated steroids.

Plant Growth Factors

Auxins, cytokinin, ethylene, gibberellins, helminthosporal, PA-180.

Foods

Animal feedstuffs (single-cell protein), beer, cheese, cider, soy sauce, vinegar, wine, yoghurt.

Amino Acids

Alanine, arginine, aspartic acid, citrulline, glutamic acid, histidine, isoleucine, leucine, lysine, methionine, ornithine, phenylalanine, proline, serine, threonine, tryptophan, tyrosine, valine.

Carbohydrates

Cellulose, dextrans, erythrulose, fructose, glucose, glycogen, maltose, mannitol, ribose, ribulose, sorbose, starch, xylitol, xylose, xylulose, xanthans.

Lipids

Phospholipids, sphingolipids, wax.

Vitamins and Cofactors

Ascorbic acid, β-carotene, cobalamin, coenzymes A and Q, biotin, FAD, NAD, D-pantoic acid, riboflavin, thiamin.

Nucleotides and Precursors

Adenosine, ADP, AMP, cAMP, ATP, cCMP, guanosine, inosine, orotidine, xanthine.

Enzymes

Acetamidase, agarase, alcohol dehydrogenase, alkyl sulphatase, amylases, catalase, collagenase, cytochrome b_4, DNase, fatty acid synthetase, galactosidases, glucose isomerase, glucose oxidase, hyaluronidase, β-lactamase, lipases, maltase, pectinase, peroxidases, phosphatidase, proteases, RNase, superoxide dismutase.

Dyes and Cosmetics

Anthroquinones, benzophenone, caprylic acid, carotenoid pigments, *o*-dihydroxyacetone, gallic acid, indigotin.

Organic Synthesis Intermediates

Acetic acid, acetone, adipic acid, butanol, citric acid, dialkyl resorcinol, ethanol, formic acid, glycerol, methane, oxalic acid, propionic acid, salicylic acid.

CHEMICAL COMPOUNDS SUSCEPTIBLE TO MICROBIAL SPOILAGE OR AMENABLE TO DEGRADATION

Petroleum-Based Compounds

n-Alkanes, aniline, anthracene, asphalt, benzenesulphonates, benzoates, biphenyl, cresols, cutting oils, cyclohexanol, cyclopentanol, detergents, dinitrophenol, emulsion paint thickener, 4-hydroxybenzoate, 4-hydroxypyridine, jet fuel, kerosene, methanol, naphthalene, *p*-naphthol, nitrobenzoates, nitrophenols, petroleum, phenanthrene, phenol, β-phenylpropionic acid, PVC, salicylic acid, vanillate, xylenols.

Animal and Plant Products

Barley husk, camphor, caproic acid, D-catechin, cholesterol, α-conidendrin, ferrichrome, keratin, lignin, malic acid, oxalate, pectin, RNA, rutin and related flavenoids, tannins, thymol, uric acid, wood, xylan.

Pesticides

Alachlor, chlorinated phenol fungicides, diazinon, fluoroacetamide, parathion, trifluoralin, urea-based herbicides.

Drugs

Acetylsalicyclic acid, chloramphenicol, heparin, heparitin, β-lactams, nicotine, phenacetin, puromycin amino nucleoside, steroids, tetracyclines.

Foodstuffs

L-Aspartate, beer, canned foods, cider, eggs, fish jelly, pantothenic acid, riboflavin, thiamin, wine.

Inorganic Compounds

Arsenite, cyanides in waste water, nitriles in waste water, thiocyanate, thiosulphate.

Carbohydrates

Carboxylmethyl cellulose, cellulose, chitin, chondroitin, galactose, glucose, melibiose, *Pneumococcus* polysaccharide (type III).

EXAMPLES OF GENERA CAPABLE OF PRODUCING IMPORTANT INDUSTRIAL FERMENTATION PRODUCTS

Acetic Acid (Vinegar)

Acetobacter, Lactobacillus, Polyporus.

Acetone

Clostridium.

Amylase

Aspergillus, Bacillus, Paecilomyces, Rhizopus, Saccharomycopsis, Streptomyces.

Cellulase

Aspergillus, Cellulomonas, Chaetomium, Fomes.

Dextranase

Aspergillus, Fusarium, Paecilomyces, Penicillium.

Ethanol

Clostridium, Saccharomyces.

Fructose

Bacillus, Gluconobacter, Pseudomonas.

Glucose Isomerase

Arthrobacter, Nocardia, Microbispora, Microellabosporia, Micromonospora, Streptomyces.

L-Glutamic acid

Achromobacter, Arthrobacter, Aspergillus, Bacillus, Brevibacterium, Candida, Corynebacterium, Escherichia, Methylomonas, Microbacterium, Micrococcus, Nocardia, Pichia, Protaminobacter, Pseudomonas, Streptomyces.

Lactic Acid

Bacillus, Lactobacillus, Rhizopus, Streptococcus.

Lipase

Aspergillus, Beauvaria, Candida, Geotrichium, Metarrhizium, Myriococcus, Paecilomyces, Penicillium, Pseudomonas, Saccharomycopsis, Serratia, Verticillium, Xanthomonas.

L-Lysine

Arthrobacter, Bacillus, Brevibacterium, Corynebacterium, Kloeckera, Microbacterium, Nocardia, Protaminobacter.

Polysaccharides

Bacillus, Scherotium, Stromatinia.

Protease

Alternaria, Aspergillus, Bacillus, Beauvaria, Cephalosporium, Conidiobolus, Coprinus, Metarrhizium, Mycelia, Neurospora, Oidiodendron, Paecilomyces, Penicillium, Saccharomyces, Saccharomycopsis, Scytalidium, Serratia, Verticillium.

Acid protease

Actinomucor, Aspergillus, Cladosporium, Penicillium, Polyporus, Rhizopus.

Alkaline protease

Acremonium, Bacillus, Coprinus, Fomitopsis, Fusarium, Gibberella, Irpex, Lenzites, Serratia, Streptomyces.

Protein

Arthrobacter, Aspergillus, Aureobacterium, Candida, Cellulomonas, Corynebacterium,

Kluyveromyces, Morchella, Mycobacterium, Nocardia, Pediococcus, Saccharomycopsis, Staphylococcus.

Steroids (By Conversion)

Actinomucor, Aspergillus, Bacillus, Beauvaria, Botrytis, Colonectria, Chaetomella, Cladosporium, Coryneum, Cunninghamella, Curvularia, Cylindrocapron, Cylindrocephalum, Drechslera, Epicoccum, Gibberella, Cliocladium, Hansenula, Helicostylum, Hormodendrum, Humicola, Kloeckera, Micromonospora, Mortierella, Mucor, Mycobacterium, Nocardia, Pachybasium, Pestalotia, Phoma, Pichia, Pseudomonas, Rhizoctonia, Rhizopus, Rhodoseptoria, Rhodotorula, Saccharomyces, Schizosaccharomyces, Septomyxa, Serratia, Spicaria, Spondylocladium, Stachylidium, Streptomyces, Thamnidium, Thanatephorus, Torulaspora, Verticillium, Zygodesmus.

SOURCES OF INDUSTRIALLY IMPORTANT MICROORGANISMS

ACC Akers Culture Collection of Imperial Chemical Industries, Ltd., Alderley Park, Macclesfield, Cheshire, UK.
AMC Walter Reed Army Medical Center (formerly Army Medical School and Army Medical Department Research and Graduate School), Washington DC, USA
AMIF American Meat Institute Foundation, Chicago, Illinois, USA
AMNH American Museum of Natural History, New York, New York, USA
ARL See ACC
ATCC American Type Culture Collection, Rockville, Maryland USA
ATU See FAT
BBL Baltimore Biological Laboratory, Baltimore, Maryland, USA
BKM All-Union Collection of Microorganisms, USSR
BRL See ACC
BU See BUCSAV
BUCSAV Biologicky Ustav. Ceskoslovenska Akademie Ved, Prague, Czechoslovakia
CASE Case Laboratories, Chicago, Illinois, USA
CBS Centraalbureau voor Schimmelcultures, Baarn, Netherlands
CCAP Cambridge Collection of Algae and Protozoa, Cambridge, UK
CCEB Culture Collection of Entomogenous Bacteria, Prague, Czechoslovakia
CCF Culture Collection of Fungi, Department of Botany, Charles University, Prague, Czechoslovakia
CCM Czechoslovak Collection of Microorganisms, J.E. Purkyne University, Brno, Czechoslovakia (same as MDB)
CCTM Centre de Collection de Type Microbienne, Lille, France
CDA Canadian Department of Agriculture, Ottawa, Canada
CDC Center for Disease Control, Atlanta, Georgia, USA
CI Carnegie Institute, Cold Spring Harbor, Long Island, New York, USA
CIP Collection of the Institute Pasteur, Paris, France
CLMR Central Laboratory, South Manchurian Railway Co., Ltd.
CMI Commonwealth Mycological Institute, Kew, UK [strain numbers formerly prefixed IMI (Imperial Mycological Institute) are included here]
CSIRO Commonwealth Science and Industrial Research Organization, Sydney, Australia
CU See CCAP
DAOM Plant Research Institute, Department of Agriculture, Mycology, Ottawa, Canada
DIFCO Difco Laboratories, Detroit, Michigan, USA
DMUR Department of Mycology, University of Recife, Brazil
DSM Deutsche Sammlung von Mikroorganismen, Gottingen, West Germany

EPA Environmental Protection Agency, Washington DC
ETH Eidgenosische Technische Hochschule, Zurich, Switzerland
FAT Faculty of Agriculture, Tokyo University, Japan
FDA Food and Drug Administration, Washington DC, USA
FERM Fermentation Research Institute, Agency of Industrial Science and Technology, Chiba, Japan
FGSC Fungal Genetic Stock Center, Humboldt State College, Arcata, California
FI Farmitalia SpA, Milan, Italy
FRR Division of Food Research, Food Research Laboratory, CSIRO, Sydney, Australia
GRIF Government Research Institute of Formosa
HACC Hindustan Antibiotics Ltd., Pimpri, Poona, India
HUT Hiroshima University, Faculty of Engineering, Hiroshima, Japan
IAM Institute of Applied Microbiology, University of Tokyo, Japan
IAUR Instituto de Antibioticos da Universidade de Recife, Brazil
IAW Institute of Antibiotics, Warsaw, Poland
ICI Imperial Chemical Industries Ltd., Butterwick Research Laboratories, Welwyn, UK
ICPB International Collection of Phytopathogenic Bacteria, University of California Davis, California, USA
HEM Institute of Epidemiology and Microbiology, Prague, Czechoslovakia
IFM Institute of Food Microbiology, Chiba University, Chiba, Japan
IFO Institute for Fermentation, Osaka, Japan
IHM Instituto de Higiene Experimental, Montevideo, Uruguay
IMCAS Institute of Microbiology, Czechoslovak Academy of Sciences, Prague, Czechoslovakia
IMI See CMI
IMRU Institute of Microbiology, Rutgers–The State University, New Brunswick, New Jersey, USA
IMUR Instituto de Micologia, Universidade de Recife, Brazil
INA Institute for New Antibiotics, Moscow, USSR
INMI Institute for Microbiology, USSR Academy of Sciences, Moscow
IOC Instituto Oswaldo Cruz, Rio de Janeiro, Brazil
IP Institute Pasteur, Paris, France
IPV Istituto de Patologia Vegetale, Milan, Italy
ISC International Salmonella Center
ISL International Subcommittee on Lactobacilli and Closely Related Organisms
ISP International Streptomyces Project
ITCC Indian Type Culture Collection, New Delhi, India
ITCCF Indian Type Culture Collection of Fungi, New Delhi, India
IU See UTEX
KCC Kaken Chemical Company, Ltd., Tokyo, Japan
MDB See CCM
MTHU See MTU
MTU Faculty of Medicine, University of Tokyo, Japan
NADC National Animal Disease Center, Ames, Iowa, USA
NADL See NADC
NBL Naval Biological Laboratory, Oakland, California, USA
NCA National Canners' Association, Washington DC, USA
NCAIA National Center for Antibiotics and Insulin Analysis, FDA, Washington DC, USA
NCDC See CDC
NCDO National Collection of Dairy Organisms, Reading, UK
NCIB National Collection of Industrial Bacteria, Aberdeen, UK
NCIM National Collection of Industrial Microorganisms, National Chemical Laboratory, Poona, India
NCMB National Collection of Marine Bacteria, Aberdeen, UK
NCPPB National Collection of Plant Pathogenic Bacteria, Harpenden, UK
NCTC National Collection of Type Cultures, London, UK
NCYC National Collection of Yeast Cultures, Food Research Institute, Norwich, Norfolk, UK
NDRC National Defense Research Committee, Washington DC, USA
NEA Nobel Explosives Co., Ardeer, UK
NI Nagao Institute, Tokyo, Japan
NIH National Institutes of Health, Bethesda, Maryland, USA.
NIHJ National Institutes of Health, Japan
NIRD National Institute for Research in Dairying, Reading, UK
NRC See NRCC
NRCC National Research Council, Ottawa, Canada

NRRL Northern Utilization Research and Development Division, US Department of Agriculture, Peoria, Illinois, USA
OUT Osaka University, Faculty of Engineering, Osaka, Japan
PD Plantenziektenkundige Dienst, Wageningen, Netherlands
PRL Prairie Regional Laboratory, Saskatoon, Canada
PSA Progetto Sistematica Actinomiceti, Istituto "P. Stazzi," Milan, Italy
QM Quartermaster Research and Development Center, U.S. Army, Natick, Massachusetts, USA
RIA See USSR RIA
SAUG Sammlung von Algenkulturen, University of Gottingen, West Germany
SMG See DSM
TMC Trudeau Mycobacterial Culture Collection, Trudeau Institute, Saranac Lake, New York, USA
TRTC Department of Botany, University of Toronto, Toronto, Canada
UAMII University of Alberta Mold Herbarium and Culture Collection, Alberta, Canada
UC Upjohn Company, Kalamazoo, Michigan, USA
U. Md. University of Maryland, College Park, Maryland, USA
USDA United States Department of Agriculture
USDI United States Department of the Interior
USPHS United States Public Health Service
USSR RIA USSR Research Institute for Antibiotics, Moscow, USSR
UTEX University of Texas, Culture Collection of Algae, Austin, Texas, USA
UTMC University of Texas, Myxomycete Collection, Austin, Texas, USA
VPI Virginia Polytechnic Institute and State University, Blacksburg, Virginia, USA
WB University of Wisconsin, Bacteriology Department, Madison, Wisconsin, USA
WHO World Health Organization
WLRI Warner Lambert Research Institute, Morris Plains, New Jersey, USA
WRAIR Walter Reed Army Institute of Research, Washington DC, USA
WRRL Western Utilization Research and Development Division, U.S. Department of Agriculture, Albany, California, USA

COMPOSITION OF CULTURE MEDIA

Examples of generally applicable laboratory media

Corn meal, yeast, glucose agar (CMYG)

Corn-steep liquor (50% solids)	9.0 g
Glucose	2.0 g
Yeast extract	1.0 g
Distilled water	1.0 ℓ

Corn-steep agar

Corn-steep liquor (50% solids)	9.0 g
Sucrose	0.25 g
$(NH_4)_2HPO_4$	2.0 g
Agar	30.0 g
Distilled water up to	1.0 ℓ

Final pH 6.5.

Czapek's agar

$NaNO_3$	3.0 g
K_2HPO_4	1.0 g
$MgSO_4.7H_2O$	0.5 g
KCl	0.5 g
$FeSO_4.7H_2O$	0.01 g
Sucrose (commercial grade)	30.0 g
Agar	15.0 g
Distilled water	1.0 ℓ

Sucrose is added prior to dispensing.

Czapek's dox agar

Czapek's dox broth (Difco 0338)	35.0 g
Agar	15.0 g
Distilled water	1.0 ℓ

Czapek's dox agar with 3% glucose

Glucose	30.0 g
$NaNO_3$	3.0 g
K_2HPO_4	1.0 g
$MgSO_4.7H_2O$	0.5 g
KCl	0.5 g
$FeSO_4.7H_2O$	0.01 g
Agar	15.0 g
Distilled water	1.0 ℓ

Freezing agar

Glucose	8.0 g
Yeast extract	1.0 g
Activated carbon	0.5 g
Agar	20.0 g

200 g of unpeeled potatoes are cut in slices and autoclaved in 500 ml of tap water for 15 min. at 120°C. The potato infusion is then mashed and filtered through gauze. The volume of the filtrate is made up to 1.0 ℓ and the ingredients listed above are added. Reautoclave for 15 min. at 120°C.

Glucose, yeast, calcium medium (GYC)

Glucose	20.0 g
Yeast extract	10.0 g
$CaCO_3$	20.0 g
Agar	20.0 g
Distilled water	1.0 ℓ

Autoclave 20 minutes at 116°C.

Mannitol agar

Yeast extract	5.0 g
Peptone	3.0 g
Mannitol	25.0 g
Agar	15.0 g
Distilled water	1.0 ℓ

Methanol medium

KH_2PO_4	2.0 g
K_2HPO_4	7.0 g
$(NH_4)_2SO_4$	3.0 g
$MgSO_4.7H_2O$	0.5 g
Yeast extract	0.2 g
Biotin	0.02 g
Thiamin HCl	0.2 g
$FeSO_4.7H_2O$	2.0 µg
$MnSO_4.H_2O$	2.0 µg
Agar	15.0 g
Distilled water	1.0 ℓ

Adjust pH to 7.0. Autoclave and allow to cool to about 50–55°C before aseptically adding methanol (1 ml per 100 ml of the medium).

Peptone, yeast, glucose, maltose agar medium (PYGM)

Peptone	5.0 g
Yeast extract	2.5 g
Glucose	1.25g
Maltose	1.25 g
Cysteine	0.125 g
Salts solution (see below)	10.0 ml
Resazurin (0.05% aqueous solution) (see p.)	1.0 ml
Distilled water	250.0 ml
Salts solution	
K_2HPO_4	0.1 g
KH_2PO_4	0.1 g
$NaHCO_3$	1.0 g
NaCl	0.2 g
$CaCl_2$ (anhydrous)	0.02 g
$MgSO_4$	0.02 g
Na_2MoO_4	Trace
$CoCl_2.6H_2O$	Trace
H_2SO_4(50%)	0.3 ml
Distilled water	100.0 ml

Photosynthetic medium

NH_4Cl	1.0 g
KH_2PO_4	1.0 g
$MgCl_2.6H_2O$	0.5 g
$CaCl_2$	0.02 g
Fe citrate	0.006 g
NaCl	15.0 g
Agar	15.0 g
Distilled water	955.0 ml

15.0 ml of each of the following solutions, prepared and sterilized separately, are aseptically added:

1. 10% solution of $NaHCO_3$
2. 10% solution of Na_2S
3. 10% solution of $Na_2S_2O_3.5H_2O$

Polypeptone (PP) starch

Polypeptone	10.0 g
Soluble starch	10.0 g
K_2HPO_4	3.0 g
$MgSO_4.7H_2O$	1.0 g
Distilled water	1.0 ℓ

Potato–glucose agar

Potato	500.0 g
Glucose	20.0 g
Agar	15.0 g

Slice potatoes thinly and add distilled water immediately to prevent oxidation of juice. Steam for 20 min. or let simmer in water bath at 60°C for 1 h. Filter through cheesecloth. Make up volume to 1.0 ℓ and add other ingredients. Cook one hour. Filter through cotton and tube. Autoclave 20 min. at 115°C.

Sea water yeast extract agar

Seven Seas Mix (Utility Chem. Co.)	37.9 g
Yeast extract	3.0 g
Proteose peptone	10.0 g
Distilled water	1.0 ℓ

Adjust pH to 7.2–7.4.

Soil extract–potato extract medium

Malt extract	10.0 g
Yeast extract	4.0 g
Soil extract	250.0 ml
Potato extract	250.0 ml
Distilled water	500.0 ml

Adjust pH to 7.0.
Soil extract: Autoclave 400 g of soil in 1.0 ℓ of distilled water for 45 min. at 121°C. Filter through gauze.
Potato extract: Boil 400 g of peeled, diced potatoes in 500 ml distilled water for 15 min. Filter through gauze and make up volume to 1.0 ℓ.

Tryptone agar

Tryptone	8.0 g
NaCl	8.0 g
Agar	15.0 g
Distilled water	1.0 ℓ

Examples of Media Formulated for Specific Microorganisms

Acetobacter medium

Autolyzed yeast	10.0 g
$CaCO_3$	10.0 g
Agar	15.0 g
Glucose	3.0 g
Distilled water	1.0 ℓ

Sterilize. Mix $CaCO_3$ thoroughly and cool rapidly. Slant as the agar hardens so as to keep the $CaCO_3$ in suspension.

Acetobacter medium

3–5% alcoholic mash with 2% agar. For preparation of this medium, Löwenbrau beer was used successfully.

Achromobacter methanol medium

Methanol	20.0 ml
NH_4Cl	5.0 g
KH_2PO_4	2.0 g
NaCl	0.5 g
$MgSO_4$	0.2 g
$FeSO_4$	2.0 mg
$MnCl_2$	2.0 mg
Yeast extract	0.2 g
Distilled water	1.0 ℓ

Adjust pH to 7.0

Anaerobes E medium

Rumen fluid (see below)	30.0 ml
Glucose	0.05 g
Maltose	0.05 g
Soluble starch	0.05 g
Peptone	0.05 g
Yeast extract	0.05 g
$(NH_4)_2SO_4$	0.05 g
Resazurin solution (see below)	0.4 ml
Salts solution (see below)	50.0 ml
L-Cysteine.H_2O	0.05 g
Distilled water	20.0 ml

Rumen fluid: Filter rumen contents, obtained from a cow fed on a lucerne–hay concentrate ration, through two layers of cheesecloth to remove large particles. Store under CO_2 in the refrigerator. Much of the particulate matter settles. Use only the supernatant.
Resazurin Solution: Dissolve one resazurin tablet (Allied Chemical no. 506; about 11 mg) in 44 ml of distilled water.

Salts Solution:

$CaCl_2$ (anhydrous)	0.2 g
$MgSO_4$	0.2 g
K_2HPO_4	1.0 g
KH_2PO_4	1.0 g
$NaHCO_3$	10.0 g
NaCl	2.0 g

Mix $CaCl_2$ and $MgSO_4$ in 300 ml of distilled water until dissolved. Add 500 ml water and add the remaining salts while swirling slowly. Add 200 ml of distilled water, mix, and store at 4°C. Mix all of the ingredients in a conical flask. Flask should have a small head space to minimize air volume that must be purged during cooling. Fit a removable chimney to the boiling flask to prevent media from boiling over. Boil (10–20 min.) until medium changes from pink to yellowish. Cool in ice water bath under oxygen-free carbon dioxide. The flow of carbon dioxide should cause gentle bubbling (sufficient to exclude air). Remove from ice bath and add 0.05 g L-cysteine. H_2O. Adjust pH to 7.0 with 8 N NaOH or 5 N HCl. Change to oxygen-free nitrogen gas and dispense media into tubes. Stopper with no. 1 butyl rubber stoppers (or black rubber stoppers, Fisher no. 14–130). The nitrogen prevents a pH change during storage. Place the rack of tubes in a press (so that stoppers do not blow off) and sterilize for 12–15 min. at 121°C.

Aspergillus medium

Yeast extract	1.0 g
Peptone	1.0 g
$NaNO_3$	6.0 g
Casamino acids (Difco 0230)	1.0 g
Agar	15.0 g
Adenine	0.15 g
Vitamin solution	10.0 ml
Distilled water up to	1.0 ℓ

Adjust pH to 6.0. Sterilize at 121°C for 10 min.

Vitamin solution:

Biotin	0.01 g
Pyridoxine.HCl	0.01 g
Thiamin.HCl	0.01 g
Riboflavin	0.01 g
p-Aminobenzoic acid	0.01 g
Nicotinic acid	0.01 g
Distilled water up to	100.0 ml

Sterilize at 121°C for 10 min.

Azotobacter medium

KH_2PO_4	0.2 g
K_2HPO_4	0.8 g
$MgSO_4.7H_2O$	0.2 g
$CaSO_4.2H_2O$	0.1 g
$FeCl_3$	Trace
Na_2MoO_4	Trace
Yeast extract	0.5 g
Sucrose	20.0 g
Agar (if needed)	15.0 g
Distilled water	1.0 ℓ

Adjust pH to 7.2.

Azotomonas medium

Urea	2.0 g
KH_2PO_4	0.3 g
$MgSO_4$	0.2 g
$FeSO_4$	0.01 g
Glucose	20.0 g
Agar (if needed)	15.0 g
Distilled water	1.0 ℓ

Filter–sterilize urea in 10% solution and aseptically add 20 ml/100 of the medium.

Bacillus medium

Peptone	10.0 g
Beef extract	3.0 g
Lactose	5.0 g
NaCl	5.0 g
K_2HPO_4	2.0 g
Distilled water	1.0 ℓ

Adjust pH to 7.2.

Bacterium medium

Beef extract	1.5 g
Yeast extract	3.0 g
Peptone	6.0 g
Dextrose	1.0 g
Agar	20.0 g
Distilled water	1.0 ℓ

Blue–green algae medium BG-11

$NaNO_3$	1.5 g
K_2HPO_4	0.04 g
$MgSO_4.7H_2O$	0.075 g
$CaCl_2.2H_2O$	0.036 g
Citric acid	0.006 g
Ferric ammonium citrate	0.006 g
EDTA (disodium salt)	0.001 g
Na_2CO_3	0.02 g
Trace metal mix A5	1.0 ml
Agar (if needed)	10.0 g
Distilled water	1.0 ℓ

The pH should be 7.1 after sterilization.

Trace metal mix A5:

H_3BO_3	2.86 g
$MnCl_2.4H_2O$	1.81 g
$ZnSO_4$	0.222 g
$Na_2MoO_4.2H_2O$	0.39 g
$CuSO_4.5H_2O$	0.079 g
$Co(NO_3).6H_2O$	49.4 mg
Distilled water	1.0 ℓ

Escherichia medium

K_2HPO_4	7.0 g
KH_2PO_4	3.0 g
Sodium citrate (anhydrous)	0.4 g
$MgSO_4.7H_2O$	0.1 g
$(NH_4)_2SO_4$	0.1 g
Glycerol	2.0 ml
Casein hydrolysate (NBCo)	2.0 g
Agar	15.0 g
Distilled water	1.0 ℓ

Escherichia medium

K_2HPO_4	7.0 g
KH_2PO_4	3.0 g
Sodium citrate (anhydrous)	0.4 g
or Sodium citrate.$2H_2O$	(0.45 g)
$MgSO_4$	0.1 g
$(NH_4)_2SO_4$	0.1 g
Glycerol	2.0 g
Casein hydrolysate (NBCo)	2.0 g
Agar	15.0 g
α, ε-Diaminopimelic acid	0.1 g
Distilled water	1.0 ℓ

Adjust pH to 7.3.

Escherichia medium

Tryptone	10.0 g
Yeast extract	1.0 g
NaCl	8.0 g
Distilled water	1.0 ℓ

Sterilize and add 10.0 ml of 10% glucose, 2.0 ml of 1 M $CaCl_2$, and 1.0 ml of thiamin (10 mg ml^{-1}) If agar is desired, sterilize separately.

Escherichia medium

Casein hydrolysate (NBCo)	6.0 g
K_2HPO_4	0.2 g
$MgSO_4.7H_2O$	0.2 g
$FeSO_4.7H_2O$	Trace
Glycerol	2.0 ml
Asparagine	0.15 g
Agar	15.0 g
Distilled water	1.0 ℓ

Steam the casein hydrolysate, salts and asparagine. Adjust pH to 7.0. Filter. Add glycerol and agar. Steam to dissolve. Prepare a cobalamine concentrated solution containing 40 μg ml^{-1} vitamin B_{12} and add 1 ml of this solution.

Lactic bacteria broth

Trypticase (BBL)	10.0 g
Yeast autolysate	5.0 g
Sodium acetate	12.0 g
Glucose	10.0 g
Solution A (see below)	5.0 ml
Solution B (see below)	5.0 ml

Add distilled water to 1.0 ℓ. Adjust pH to 5.1–5.33.

Solution A:

K_2HPO_4	10.0%
KH_2PO_4	10.0%

Solution B:

$MgSO_4$	4.0%
NaCl	0.2%
$FeSO_4$	0.2%
$MnSO_4$	0.2%

Lactobacillus medium

Peptonized milk	15.0 g
Yeast extract	5.0 g
Glucose	10.0 g
KH_2PO_4	2.0 g
Tomato juice (adjust pH to 7.0)	100.0 ml
Tween 80 (10.0% soln. in ethanol)	10.0 ml
Agar	10.0 g

Add distilled water to 1.0 ℓ.

Lactobacillus medium

Agar	15.0 g
Tryptone	20.0 g
Tryptose	5.0 g
Yeast extract	5.0 g
Tomato juice (filtered, pH 7.0)	200.0 ml
Liver extract concentrate (NBCo)	1.0 g
Tween 80	0.05 g
Glucose	3.0 g
Lactose	2.0 g

Add distilled water to 1.0 ℓ. Adjust pH to 6.5.

Lactobacillus medium

Trypticase (BBL)	10.0 g
Yeast extract (Difco or BBL)	5.0 g
Tryptose (Difco)	3.0 g
KH_2PO_4	3.0 g
K_2HPO_4	3.0 g
Salt solution R (see below)	5.0 ml
Tween 80	1.0 ml
Na acetate	1.0 g
L-Cysteine.HCl	200.0 mg
Agar	20.0 g
H_2O	1.0 ℓ
Glucose	0.5%

Adjust pH to 6.3.

Salt solution R:

$MgSO_4.7H_2O$	11.50 g
$FeSO_4.7H_2O$	0.68 g
$MnSO_4.2H_2O$	2.40 g
H_2O	100.0 ml

Store in the cold.

Lactobacillus milk with 10% glucose

Tomato juice (filtered, pH 7.0)	100.0 ml
Skim milk	100.0 g
Yeast extract	5.0 g
Glucose	100.0 g

Add distilled water to 1.0 ℓ.

Methanomonas medium

KH_2PO_4	1.4 g
Na_2HPO_4	2.1 g
$(NH_4)_2SO_4$	0.5 g
$MgSO_4.7H_2O$	0.2 g
$CaCl_2$	0.01 g
$FeSO_4$	0.005 g
$MnSO_4$	0.003 g
$Na_2MoO_4.2H_2O$	0.25 mg
Distilled water	1.0 ℓ

Aseptically add 1.0 ml of methanol to each 100 ml of the above medium.

Nitrosomonas medium

$(NH_4)_2SO_4$	3.0 g
K_2HPO_4	0.5 g
$MgSO_4$	0.05 g
$CaCl_2$	0.004 g
Chelated iron (Chel-138-HFe, Geigy Chem. Co.)	0.1 mg Fe
Cresol red	0.05 mg
Distilled water	1.0 ℓ

Autoclave calcium and magnesium salts separately to avoid precipitation. Maintain pH during growth with sterile 50% K_2CO_3. pH should be maintained at 8.2–8.4

Pseudomonas medium

NH_4Cl	5.0 g
Yeast extract	0.5 g
$MgSO_4$	0.2 g
K_2HPO_4	1.5 g
KH_2PO_4	0.5 g
L-Tryptophan	1.0 g
Distilled water	1.0 ℓ

Pseudomonas nitrifying medium

$MgSO_4.7H_2O$	0.2 g
K_2HPO_4	1.0 g
$FeSO_4.7H_2O$	0.05 g
$CaCl_2$	0.02 g
$MnCl_2.4H_2O$	0.002 g
$Na_2MoO_4.2H_2O$	0.001 g
Mannitol	4.0 g
Agar	15.0 g
Distilled water	1.0 ℓ

Adjust pH to 7.2.

Sporulation Agar

Yeast extract	1.0 g
Beef extract	1.0 g
Tryptose	2.0 g
$FeSO_4$	Trace
Glucose	10.0 g
Agar	15.0 g
Distilled water	1.0 ℓ

Adjust pH to 7.2. For broth, eliminate agar and reduce concentration to 1/3 of the given quantities.

Streptomyces medium

Glucose	5.0 g
L-Glutamic acid	4.0 g
KH_2PO_4	1.0 g
NaCl	1.0 g
$FeSO_4.7H_2O$	0.003 g
$MgSO_4.7H_2O$	0.7 g
Agar	25.0 g
Distilled water	1.0 ℓ

Adjust pH to 7.0.

Thiobacillus medium

$Na_2HPO_4.7H_2O$	7.9 g
KH_2PO_4	1.5 g
NH_4Cl	0.3 g
$MgSO_4.7H_2O$	0.1 g
Trace metal solution (see below)	5.0 ml
Phenol red	2.0 mg
$Na_2S_2O_3.5H_2O$	10.0 g
Distilled water	1.0 ℓ

Adjust pH to 8.5. Maintain pH at this level by the addition of sterile 10% Na_2CO_3: the indicator should remain pink in colour.

Trace metal solution:

EDTA	50.0 g
$ZnSO_4$	22.0 g
$CaCl_2$	5.54 g
$MnCl_2$	5.06 g
$FeSO_4$	4.99 g
Ammonium molybdate	1.10 g
$CuSO_4$	1.57 g
$CoCl_2$	1.61 g
Distilled water	1.0 ℓ

Adjust pH to 6.0 with KOH.

Thiobacillus medium

$(NH_4)_2SO_4$	0.2 g
$MgSO_4.7H_2O$	0.5 g
$CaCl_2$	0.25 g
KH_2PO_4	3.0 g
$FeSO_4$	0.005 g
Tap water	1.0 ℓ

1.0 g sulphur (precipitated) is placed in each dry flask. The salt solution is prepared, and 100 ml amounts are carefully poured down the side of the flask without wetting the sulphur. Flasks are then sterilized in flowing steam on three consecutive days, ½-h per day. Care must be taken to ensure that the sulphur remains on the surface throughout sterilization.

Thiobacillus medium

Solution 1:	
K_2HPO_4	2.0 g
$CaCl_2$	0.1 g
$MgSO_4$	0.1 g
$FeSO_4$	Trace
$MnSO_4$	Trace
Tap water	900.0 ml
Adjust pH to 7.8.	
Solution 2:	
$Na_2S_2O_3$	10.0 g
Tap water	50.0 ml
Solution 3:	
$(NH_4)_2SO_4$	0.1 g
Tap water	50.0 ml

The three solutions are sterilized separately, aseptically mixed and dispensed in 100.0 ml amounts.

Thiobacillus medium B

NH_4Cl	0.1 g
KH_2PO_4	3.0 g
$MgCl_2$	0.1 g
$CaCl_2$	0.1 g
$Na_2S_2O_3.5H_2O$	5.0 g
Oxoid ionager no. 2 (if needed)*	15.0 g
Distilled water	1.0 ℓ

Adjust pH to 4.2. Sterilize by steaming for 30 min. on three successive days.

*Oxoid ionagars are no longer commercially available. Other types of agar are currently being tested as substitutes.

Examples of Industrially Used Media

Barley

Protein	11.5%
Carbohydrate	68.0%
Fat	1.8%
Fibre	7.0%
Ash	2.5%
Solids	90.0%
Riboflavin	0.7 mg lb^{-1}
Thiamin	2.5 mg lb^{-1}
Pantothenic acid	2.8 mg lb^{-1}
Niacin	24.0 mg lb^{-1}
Pyridoxine	1.6 mg lb^{-1}
Biotin	0.1 mg lb^{-1}
Choline	500.0 mg lb^{-1}
L-Arginine	0.5%
L-Cystine	0.26%
L-Glycine	0.45%
L-Histidine	0.3%
L-Isoleucine	0.5%
L-Leucine	0.8%
L-Lysine	0.42%
L-Methionine	0.22%
L-Phenylalanine	0.6%
L-Threonine	0.4%
L-Tryptophan	0.17%
L-Tyrosine	0.4%
L-Valine	0.6%

Barley malt

Protein	13.0%
Carbohydrate	70.0%
Fat	2.0%
Fibre	3.5%
Ash	2.5%
Solids	96.0%
Riboflavin	1.3 mg lb^{-1}
Thiamin	1.7 mg lb^{-1}
Pantothenic acid	3.6 mg lb^{-1}
Niacin	25.0 mg lb^{-1}
Pyridoxine	2.5 mg lb^{-1}
Choline	400.0 mg lb^{-1}
L-Arginine	0.4%
L-Glycine	0.1%

L-Histidine	0.1%
L-Isoleucine	0.1%
L-Leucine	0.1%
L-Lysine	0.1%
L-Methionine	0.1%
L-Phenylalanine	0.1%
L-Threonine	0.1%
L-Tryptophan	0.1%
L-Tyrosine	0.1%
L-Valine	0.1%

Blood meal

Protein	80.0%
Carbohydrate	2.5%
Fat	<1.0%
Fibre	<1.0%
Ash	3.0%
Solids	93.0%
Riboflavin	0.7 mg lb^{-1}
Thiamin	0.1 mg lb^{-1}
Pantothenic acid	0.5 mg lb^{-1}
Niacin	14.3 mg lb^{-1}
Choline	344.0 mg lb^{-1}
L-Arginine	3.4%
L-Cystine	0.2%
L-Glycine	4.2%
L-Histidine	4.2%
L-Isoleucine	1.2%
L-Leucine	10.0%
L-Lysine	6.5%
L-Methionine	1.1%
L-Phenylalanine	5.4%
L-Threonine	4.1%
L-Tryptophan	1.3%
L-Tyrosine	2.4%
L-Valine	6.7%

Cerelose

Carbohydrate	91.5%
Solids	91.5%

Corn meal

Protein	8.9%
Carbohydrate	68.9%
Fat	3.9%
Fibre	2.0%
Ash	1.3%
Solids	85.0%
Riboflavin	0.6 mg lb^{-1}
Thiamin	2.0 mg lb^{-1}
Pantothenic acid	2.4 mg lb^{-1}
Niacin	9.7 mg lb^{-1}
Biotin	0.03 mg lb^{-1}
Choline	280.0 mg lb^{-1}
L-Arginine	0.4%
L-Cystine	0.2%
L-Glycine	0.4%
L-Histidine	0.2%
L-Isoleucine	0.5%
L-Leucine	1.2%
L-Lysine	0.3%
L-Methionine	0.3%
L-Phenylalanine	0.5%
L-Threonine	0.3%
L-Tryptophan	0.1%
L-Tyrosine	0.5%
L-Valine	0.5%

Corn gluten meal

Protein	62.0%
Carbohydrate	20.0%
Fat	1.8%
Fibre	2.0%
Ash	1.8%
Solids	88.0%
Riboflavin	0.7 mg lb^{-1}
Thiamin	0.1 mg lb^{-1}
Pantothenic acid	1.3 mg lb^{-1}
Niacin	25.0 mg lb^{-1}
Choline	150.0 mg lb^{-1}
L-Arginine	2.2%
L-Cystine	0.9%
L-Glycine	2.5%
L-Lysine	0.8%
L-Methionine	1.6%
L-Tryptophan	0.3%

Corn-steep liquor

Protein	24.0%
Carbohydrate	5.8%
Fat	1.0%
Fibre	1.0%
Ash	8.8%
Solids	50.0%
Riboflavin	2.5 mg lb^{-1}
Thiamin	0.4 mg lb^{-1}
Pantothenic acid	34.0 mg lb^{-1}
Niacin	38.0 mg lb^{-1}
Pyridoxine	8.8 mg lb^{-1}
Biotin	0.4 mg lb^{-1}
Choline	315.0 mg lb^{-1}
L-Arginine	0.4%
L-Cystine	0.5%
L-Glycine	1.1%
L-Histidine	0.3%
L-Isoleucine	0.9%
L-Leucine	0.1%
L-Lysine	0.2%
L-Methionine	0.5%
L-Phenylalanine	0.3%
L-Tyrosine	0.1%
L-Valine	0.5%

CFS concentrate

Protein	27.0%
Carbohydrate	50.0%
Fat	0.2%
Fibre	7.0%
Ash	8.5%
Solids	95.0%
Riboflavin	3.0 mg lb^{-1}
Thiamin	1.2 mg lb^{-1}
Pantothenic acid	12.0 mg lb^{-1}
Niacin	40.0 mg lb^{-1}

Pyridoxine	6.5 mg lb^{-1}
Biotin	0.2 mg lb^{-1}
Choline	1.4 mg lb^{-1}
L-Arginine	1.2%
L-Cystine	0.4%
L-Glycine	1.1%
L-Histidine	0.8%
L-Isoleucine	0.8%
L-Leucine	2.5%
L-Lysine	0.9%
L-Methionine	0.4%
L-Phenylalanine	0.9%
L-Threonine	0.9%
L-Trytophan	0.2%
L-Tyrosine	0.8%
L-Valine	1.4%

Cotton-seed meal

Protein	41.0%
Carbohydrate	28.0%
Fat	1.5%
Fibre	13.0%
Ash	6.5%
Solids	90.0%
Riboflavin	2.0 mg lb^{-1}
Thiamin	6.5 mg lb^{-1}
Pantothenic acid	20.0 mg lb^{-1}
Choline	1.2 g lb^{-1}
L-Arginine	3.3%
L-Cystine	1.0%
L-Glycine	2.4%
L-Histidine	0.9%
L-Isoleucine	1.5%
L-Leucine	2.2%
L-Lysine	1.6%
L-Methionine	0.5%
L-Phenylalanine	1.9%
L-Threonine	1.1%
L-Tryptophan	0.5%
L-Tyrosine	1.0%
L-Valine	1.8%

Dried distillers' soluble

Protein	26.0%
Carbohydrate	45.0%
Fat	9.0%
Fibre	4.0%
Ash	8.0%
Solids	92.0%
Riboflavin	7.0 mg lb^{-1}
Thiamin	2.5 mg lb^{-1}
Pantothenic acid	9.0 mg lb^{-1}
Niacin	50.0 mg lb^{-1}
Biotin	1.3 mg lb^{-1}
Choline	2.0 g lb^{-1}
L-Arginine	1.0%
L-Cystine	0.6%
L-Glycine	1.1%
L-Histidine	0.7%
L-Isoleucine	1.6%
L-Leucine	2.1%
L-Lysine	0.9%
L-Methionine	0.6%
L-Phenylalanine	1.5%
L-Threonine	1.0%
L-Tryptophan	0.2%
L-Tyrosine	0.7%
L-Valine	1.5%

Edamine

Protein	7.5%
Ash	6.0%
L-Arginine	3.1%
L-Cystine	2.2%
L-Glycine	1.7%
L-Histidine	1.9%
L-Isoleucine	5.4%
L-Leucine	10.0%
L-Lysine	10.0%
L-Methionine	1.8%
L-Phenylalanine	3.4%
L-Threonine	3.8%
L-Tryptophan	2.1%
L-Tyrosine	3.3%
L-Valine	4.1%

Enzose

Carbohydrate	70.0%
Ash	3.5%
Solids	75.0%

Fermamin

Protein	50–60%
Ash	8–22%
Solids	92–96%

Fish solubles

Protein	32.0%
Carbohydrate	7.0%
Fat	3.0%
Fibre	1.0%
Ash	10.0%
Solids	50.0%
Riboflavin	4.5 mg lb^{-1}
Thiamin	2.5 mg lb^{-1}
Pantothenic acid	3.6 mg lb^{-1}
Niacin	100.0 mg lb^{-1}
Pyridoxine	5.7 mg lb^{-1}
Choline	1.07 g lb^{-1}
L-Arginine	2.2%
L-Cystine	1.4%
L-Glycine	5.3%
L-Histidine	2.4%
L-Isoleucine	1.5%
L-Leucine	2.2%
L-Lysine	7.6%
L-Methionine	2.4%
L-Phenylalanine	1.3%
L-Threonine	1.1%
L-Tryptophan	0.7%
L-Valine	1.4%

Fish meal

Protein	72.0%

Carbohydrate	5.0%
Fat	1.5%
Fibre	2.0%
Ash	18.1%
Solids	93.6%
Riboflavin	4.9 mg lb^{-1}
Thiamin	0.5 mg lb^{-1}
Pantothenic acid	4.1 mg lb^{-1}
Niacin	14.3 mg lb^{-1}
Pyridoxine	6.7 mg lb^{-1}
Choline	1.66 g lb^{-1}
L-Arginine	4.9%
L-Cystine	0.8%
L-Glycine	3.5%
L-Histidine	2.0%
L-Isoleucine	4.5%
L-Leucine	6.8%
L-Lysine	6.8%
L-Methionine	2.5%
L-Phenylalanine	3.1%
L-Threonine	3.4%
L-Tryptophan	0.8%
L-Tyrosine	2.3%
L-Valine	4.7%

Linseed meal

Protein	36.0%
Carbohydrate	38.0%
Fat	0.5%
Fibre	9.5%
Ash	6.5%
Solids	92.0%
Riboflavin	1.4 mg lb^{-1}
Thiamin	4.0 mg lb^{-1}
Pantothenic acid	6.5 mg lb^{-1}
Niacin	13.5 mg lb^{-1}
Choline	540.0 mg lb^{-1}
L-Arginine	2.5%
L-Cystine	0.6%
L-Histidine	0.5%
L-Isoleucine	1.3%
L-Leucine	2.1%
L-Lysine	1.0%
L-Methionine	0.8%
L-Phenylalanine	1.8%
L-Threonine	1.4%
L-Tryptophan	0.7%
L-Tyrosine	1.7%
L-Valine	1.8%

Meat and bone meal

Protein	50.0%
Fat	8.0%
Fibre	3.0%
Ash	29–33%
Solids	92.0%
Riboflavin	2.0 mg lb^{-1}
Thiamin	0.5 mg lb^{-1}
Pantothenic acid	1.7 mg lb^{-1}
Niacin	21.7 mg lb^{-1}
Choline	993.0 mg lb^{-1}
L-Arginine	4.0%
L-Cystine	1.4%
L-Glycine	6.6%
L-Histidine	0.9%
L-Isoleucine	1.7%
L-Leucine	3.1%
L-Lysine	3.5%
L-Methionine	0.7%
L-Phenylalanine	1.8%
L-Threonine	1.8%
L-Tryptophan	0.2%
L-Valine	2.4%

Black-strap molasses

Protein	3.0%
Carbohydrate	54.0%
Ash	8.0–10.0%
Solids	78.0%
Riboflavin	2.0 mg lb^{-1}
Pantothenic acid	19.5 mg lb^{-1}
Niacin	21.3 mg lb^{-1}
Pyridoxine	20.0 mg lb^{-1}
Choline	300.0 mg lb^{-1}

NZ- Amine B

Protein	80.0%
Ash	5.5%
Solids	96.0%
L-Arginine	3.9%
L-Cystine	0.5%
L-Glycine	1.8%
L-Histidine	2.9%
L-Isoleucine	4.6%
L-Leucine	19.4%
L-Lysine	7.6%
L-Methionine	2.4%
L-Phenylalanine	5.0%
L-Threonine	4.2%
L-Tryptophan	1.6%
L-Tyrosine	3.2%
L-Valine	6.1%

Oats

Protein	12.0%
Carbohydrate	54.0%
Fat	4.5%
Fibre	12.0%
Ash	4.0%
Solids	86.5%
Riboflavin	0.8 mg lb^{-1}
Thiamin	3.0 mg lb^{-1}
Pantothenic acid	6.6 mg lb^{-1}
Niacin	8.2 mg lb^{-1}
Biotin	0.1 mg lb^{-1}
Choline	500 mg lb^{-1}
L-Arginine	0.8%
L-Cystine	0.2%
L-Glycine	0.2%
L-Histidine	0.2%
L-Isoleucine	0.6%
L-Leucine	1.0%
L-Lysine	0.4%
L-Methionine	0.2%
L-Phenylalanine	0.7%

Peanut meal and hulls

Protein	45.0%
Carbohydrate	23.0%
Fat	5.0%
Fibre	12.0%
Ash	5.5%
Solids	90.5%
Riboflavin	2.4 mg lb^{-1}
Thiamin	3.3 mg lb^{-1}
Pantothenic acid	22.0 mg lb^{-1}
Niacin	76.0 mg lb^{-1}
Choline	760 mg lb^{-1}
L-Arginine	4.6%
L-Cystine	0.7%
L-Glycine	3.0%
L-Histidine	1.0%
L-Isoleucine	2.0%
L-Leucine	3.1%
L-Lysine	1.3%
L-Methionine	0.6%
L-Phenylalanine	2.3%
L-Threonine	1.4%
L-Tryptophan	0.5%
L-Valine	2.2%

Pharmaceutical media

Protein	61.0%
Carbohydrate	23.0%
Fat	4.0%
Fibre	2.0%
Ash	7.1%
Solids	97.0%

Rice

Protein	8.0%
Carbohydrate	65.0%
Fat	2.0%
Fibre	10.0%
Ash	4.5%
Solids	89.5%
Riboflavin	0.6 mg lb^{-1}
Thiamin	1.4 mg lb^{-1}
Pantothenic acid	5.0 mg lb^{-1}
Niacin	15.0 mg lb^{-1}
Choline	450.0 mg lb^{-1}
L-Arginine	0.5%
L-Cystine	0.1%
L-Histidine	0.1%
L-Isoleucine	0.4%
L-Leucine	0.6%
L-Lysine	0.3%
L-Methionine	0.3%
L-Phenylalanine	0.4%
L-Threonine	0.3%
L-Trytophan	0.1%
L-Tyrosine	0.7%
L-Valine	0.6%

Rice bran

Protein	13.0%
Carbohydrate	45.0%
Fat	13.0%
Fibre	14.0%
Ash	16.0%
Solids	91.0%
Riboflavin	1.2 mg lb^{-1}
Thiamin	10.0 mg lb^{-1}
Pantothenic acid	10.5 mg lb^{-1}
Niacin	135.0 mg lb^{-1}
Choline	570 mg lb^{-1}
L-Arginine	0.5%
L-Cystine	0.1%
L-Glycine	0.9%
L-Histidine	0.2%
L-Isoleucine	0.4%
L-Leucine	0.6%
L-Lysine	0.5%
L-Methionine	0.2%
L-Phenylalanine	0.4%
L-Threonine	0.4%
L-Tryptophan	0.1%
L-Valine	0.6%

Soyabean meal

Protein	51.0–52.0%
Fat	1.0%
Fibre	3.0%
Ash	5.7%
Solids	92.0%
Riboflavin	1.4 mg lb^{-1}
Thiamin	1.1 mg lb^{-1}
Pantothenic acid	6.5 mg lb^{-1}
Niacin	9.5 mg lb^{-1}
Choline	1.25 g lb^{-1}
L-Arginine	3.2%
L-Cystine	0.6%
L-Glycine	2.4%
L-Histidine	1.1%
L-Isoleucine	2.5%
L-Leucine	3.4%
L-Lysine	2.9%
L-Methionine	0.6%
L-Phenylalanine	2.2%
L-Threonine	1.7%
L-Tryptophan	0.6%
L-Tyrosine	1.4%
L-Valine	2.4%

Wheat

Protein	13.2%
Carbohydrate	69.0%
Fat	1.9%
Fibre	2.6%
Ash	1.8%
Solids	90.0%
Riboflavin	0.5 mg lb^{-1}
Thiamin	2.3 mg lb^{-1}
Pantothenic acid	6.0 mg lb^{-1}
Niacin	28.0 mg lb^{-1}
Choline	400 mg lb^{-1}
L-Arginine	0.8%
L-Cystine	0.2%
L-Histidine	0.3%
L-Isoleucine	0.6%

L-Leucine	1.0%
L-Lysine	0.5%
L-Methionine	0.2%
L-Phenylalanine	0.7%
L-Threonine	0.4%
L-Tryptophan	0.2%
L-Tyrosine	0.5%
L-Valine	0.6%

Whey—dried

Protein	12.0%
Carbohydrate	68.0%
Fat	1.0%
Ash	9.6%
Solids	95.0%
Riboflavin	9.0 mg lb^{-1}
Thiamin	1.8 mg lb^{-1}
Pantothenic acid	22.0 mg lb^{-1}
Niacin	5.0 mg lb^{-1}
Pyridoxine	1.3 mg lb^{-1}
Biotin	0.2 mg lb^{-1}
Choline	1.1 g lb^{-1}
L-Arginine	0.4%
L-Cystine	0.4%
L-Glycine	0.7%
L-Histidine	0.2%
L-Isoleucine	0.7%
L-Leucine	1.2%
L-Lysine	1.0%
L-Methionine	0.4%
L-Phenylalanine	0.5%
L-Threonine	0.6%
L-Tryptophan	0.2%
L-Tyrosine	0.5%
L-Valine	0.6%

Yeast—brewers'

Protein	43.0%
Carbohydrate	39.5%
Fat	1.5%
Fibre	1.5%
Ash	7.0%
Solids	95.0%
Riboflavin	16.0 mg lb^{-1}
Thiamin	34.0 mg lb^{-1}
Pantothenic acid	55.2 mg lb^{-1}
Niacin	226.5 mg lb^{-1}
Pyridoxine	22.6 mg lb^{-1}
Biotin	0.5 mg lb^{-1}
Choline	2.2 g lb^{-1}
L-Arginine	2.2%
L-Cystine	0.6%
L-Glycine	3.4%
L-Histidine	1.3%
L-Isoleucine	2.7%
L-Leucine	3.3%
L-Lysine	3.4%
L-Methionine	1.0%
L-Phenylalanine	1.8%
L-Threonine	2.5%
L-Tryptophan	0.8%
L-Tyrosine	1.9%
L-Valine	2.4%

Yeast hydrolysate

Protein	40.0–65.0%
Fibre	0.0–3.0%
Ash	9.0–11.0%
Solids	93.0–96.0%
L-Arginine	3.3%
L-Cystine	1.4%
L-Histidine	1.6%
L-Isoleucine	5.5%
L-Leucine	6.2%
L-Lysine	6.5%
L-Methionine	2.1%
L-Phenylalanine	3.7%
L-Threonine	3.5%
L-Tryptophan	1.2%
L-Tyrosine	4.6%
L-Valine	4.4%

Yeast—torula

Protein	50.0%
Carbohydrate	32.0%
Fat	4.8%
Fibre	0.8%
Ash	9.0%
Solids	94.0%
Riboflavin	20.0 mg lb^{-1}
Thiamin	2.8 mg lb^{-1}
Pantothenic acid	37.0 mg lb^{-1}
Niacin	225.0 mg lb^{-1}
Pyridoxine	13.0 mg lb^{-1}
Biotin	0.2 mg lb^{-1}
Choline	132.0 mg lb^{-1}
L-Arginine	2.6%
L-Cystine	0.6%
L-Glycine	2.7%
L-Histidine	1.4%
L-Isoleucine	2.9%
L-Leucine	3.5%
L-Lysine	3.8%
L-Methionine	0.8%
L-Phenylalanine	3.0%
L-Threonine	2.6%
L-Tryptophan	0.5%
L-Tyrosine	2.1%
L-Valine	2.9%

BASIC TECHNIQUES IN GENE MANIPULATION

Repression

Repression is the inhibition of a metabolic pathway by products of that, or a related, pathway. This results in reduced synthesis of the enzymes catalyzing that pathway. The inhibition may be due to end-products, intermediates or, occasionally, substrates.

Derepression by environmental manipulation

Derepression may be achieved by one or both of the following processes.
1) Repression by products of the pathway is reduced by limiting accumulation of the product.
2) Catabolite repression is the reduced synthesis of, for example, amylases or proteases by products formed when rapidly utilizable nutrients, e.g., glucose or ammonia, are available. It is limited by adding less readily used nutrients, such as starch or soyabean meal. Catabolite repression is important in the commercial production of xanthans by *Xanthomonas campestris* and dextrans by *Leuconostoc mesenteroides* and related lactic acid bacteria. It is a general effect and not pathway specific.

Derepression by genetic selection

This involves the selection of mutants that are not subject to product repression. Examples of depression by this technique are listed in Table 25.

Table 25.

Microorganism(s)	Product	Yield
Ashbya gossypii	Riboflavin	20 000-fold overproduction
Propionibacterium shermanii + *Pseudomonas denitrificans*	Cobalamin	50 000-fold overproduction
Corynebacterium glutamicum + *Brevibacterium flavum*	Lysine	$75\ g\ \ell^{-1}$

In the case of *Coryn. glutamicum* and *Brev. flavum*, yield is increased by the following 2 methods.

1) Lysine and threonine act together in exerting feedback inhibition on a common enzyme – aspartate kinase – in their biosynthetic pathways. Mutants deficient in the enzyme homoserine dehydrogenase cannot produce threonine. Lysine alone does not inhibit aspartate kinase which retains full activity in the absence of threonine, leading to overproduction of lysine.
2) Mutation of the aspartate kinase gene results in a lack of sensitivity to lysine with its function being unaffected by lysine accumulation.

Mutagenesis

Induction of mutations by sequential or repeated X-irradiation, UV-irradiation and nitrogen mustard exposure have increased penicillin yields from *Penicillium chrysogenum* cultures 10 000-fold after optimization of the medium composition.

Mutations may result in derepression, gene amplification or biosynthesis of modified products.

The inherent randomness of mutations is reduced by extracting purified genes, exposing them to mutagens *in vitro* and then using recombinant DNA techniques to replace the altered gene into the bacterial genome.

Gene Amplification

Genes are transferred to plasmids by recombinant DNA techniques. These plasmids are introduced into bacterial cells and stimulated to reproduce up to 100-fold. Protein production increases with the number of corresponding genes being transcribed.

Bacteriophages reproduce up to 1000-fold and can be employed instead of plasmids for the transmission of selected genes and their subsequent amplification.

Hybridoma Formation

This technique involves the fusion of eucaryotic cells, which are normally difficult to culture *in vitro*, with easily maintained tumour cells, e.g., myelomas. The hybrid cells thus formed, which are known as hybridomas, are easily cultured and secrete the product of the specialized cell.

Monoclonal antibodies have been produced by fusing B lymphocytes with myeloma cells from mice (*see* Fig. 26).

Recombinant DNA Methods

Genes from donor cells are isolated by separation from the donor genome using restriction endonucleases or by synthesis from the relevant mRNA using reverse transcriptases. These genes are then incorporated into plasmids (previously broken by restriction endonucleases) using DNA ligases. Bacteria (typically *Escherichia coli*, because of its well-established genetics and the presence of high plasmid levels) are transformed with recombined plasmids. Incorporation of an antibiotic-resistant gene into the plasmids with the required gene and subsequent exposure to the antibiotic can be used to assist the selection of recombinant bacteria. Products and possible applications are listed in Table 26.

Table 26.

Product	Use
Human insulin	Treatment of diabetes melitus
Human interferon	Treatment of viral diseases and, possibly, cancer
Monoclonal antibodies	Diagnostic testing
Encephalin and endorphin-related compounds	Analgesics
Vaccines	Preparations not contaminated with inflammatory constituents, e.g., an improved whooping cough vaccine
Enzymes	Industrial fermentation
Somatostatin	Biomedical research

Protoplast Fusion

In this technique, cell walls are removed, resulting protoplasts are mixed and the fusion product is allowed to regenerate its cell wall as shown in Fig. 25. Protoplast fusion increases the frequency of recombination from 1:10^6, after crossing different strains of *Streptomyces*, to 1:4, after protoplast fusion. Applications of this technique are listed below.

1) Combination of slow-growing, high-producing mutants with wild-type fast-growing, low-producing strains to give fast-growing, high-producing strains.
2) Combination of several high-producing mutants to achieve additive or synergistic amplification of production.
3) Fusion of related strains, which produce different antibiotics, to create new, hybrid antibiotics.

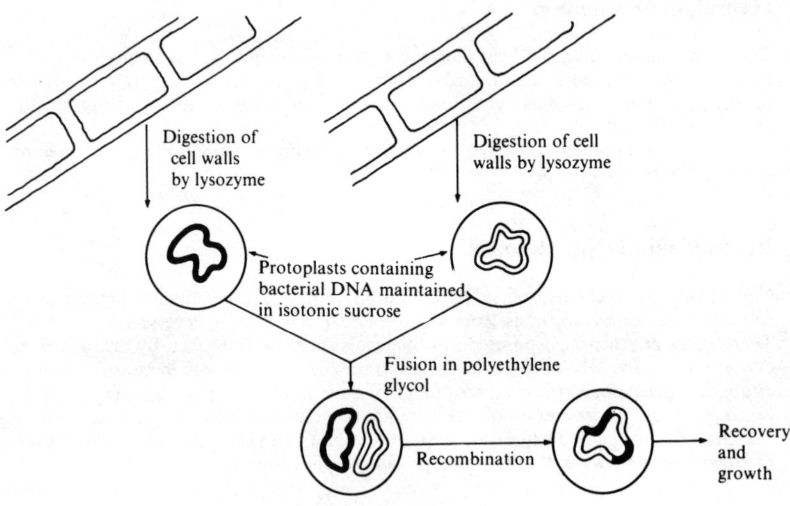

Figure 25. Generalized method for protoplast fusion.

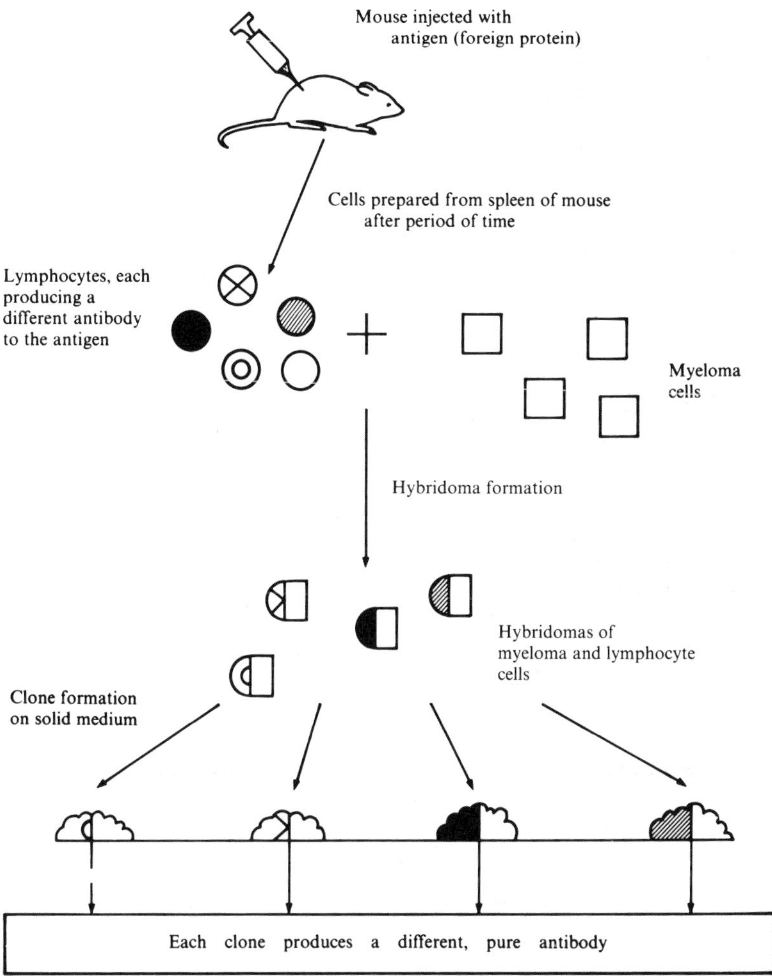

Figure 26. Formation of monoclonal antibodies.

There are, however, some problems associated with recombinant DNA technology.

1) As stated above *E. coli* is almost exclusively used. However, this bacterium is not the most suitable organism for growth in large industrial fermenters.
2) Most eucaryotic genes are not expressed in bacteria because they contain 'introns' sequences of bases which are removed from mRNA copies of the gene before translation into amino acid sequences (*see* Fig. 27). Bacteria lack the enzymes to excise these introns and would translate the entire base sequence.
3) Genes must be inserted together with suitable promoter and initiator sequences before they will be used for protein synthesis.
4) Bacteria digest foreign DNA. To prevent this destruction, mutants deficient in the relevant enzymes are used.
5) Chemical synthesis of long DNA sequences which code for specific proteins is difficult to achieve.

Hybridization by Sexual Reproduction

Any method that brings genes from different organisms into the same cell will result in genetic recombination, i.e. the breakage of DNA strands and the incorporation of foreign genes in the form of DNA. Sexual reproduction evolved as a means of bringing genes from different organisms into a common cell and consequently increasing the natural recombination frequency.

Closely related species of many types can be induced to form hybrids. This technique has been employed to produce yeast hybrids capable of fermenting all the carbohydrates found in malt wort. The carbohydrate composition of malt wort is listed in Table 26.

The process of hybridization in yeasts is outlined in Fig. 28. Initially, yeast cells from two strains are induced to form spores, which are then removed by microdissection using a micromanipulator. Each spore is implanted into culture medium and allowed to multiply with the production of separate spore cultures. The presence of desirable characteristics is assessed and two spore cultures of different sexes are then selected. Selected spores are brought together, when they fuse in pairs to produce new, stable hybrids with inherited qualities. The hybrid thus produced is allowed to grow by budding.

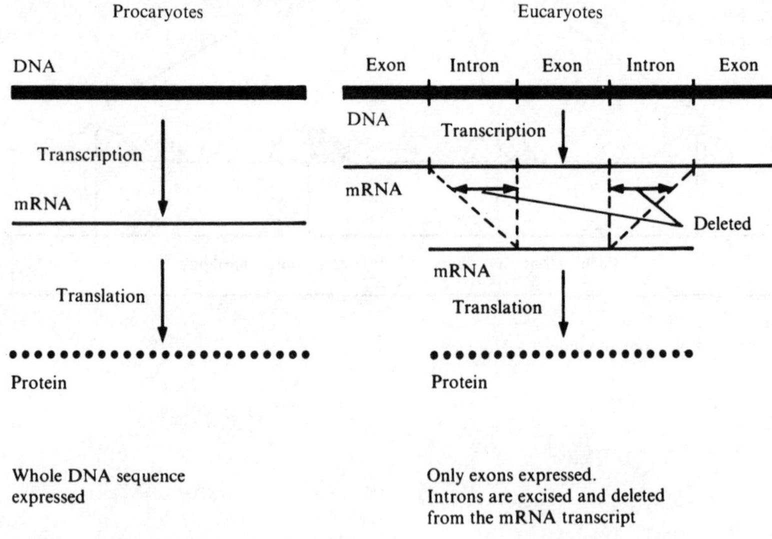

Figure 27. Comparison of protein synthesis in procaryotes and eucaryotes.

Table 27. Carbohydrate content of malt wort

Carbohydrate	Amount present (%)
Maltose	53
Glucose	12
Maltotriose	13
Dextrins	22

Saccharomyces cerevisiae is capable of fermenting all the carbohydrates listed in Table 27 except dextrin. In contrast, *Sacch. diastaticus* can ferment dextrin but produces 4-vinylguaiacol, which gives the product a phenolic taste. Further techniques are required to remove this characteristic from the hybrid (*see* Fig. 29).

Introduction of New Genes on Plasmids and Viruses

Plasmids and temperate bacteriophages, which do not kill the host cell but become incorporated into its DNA, can carry genes into cells where they are expressed and produce protein products. The method is exemplified in Fig. 30 by the introduction of hydrocarbon-metabolizing genes into *Pseudomonas putida* on plasmids.

Energy-conserving genes for glutamate synthesis have been transferred from *E. coli* into *Methylophilus methylotrophus* on a plasmid. Resultant transformants use less energy for nitrogen incorporation from ammonia and produce protein-rich animal feedstuffs supplements in higher yields.

Further Applications of Gene Manipulation

As yet largely hypothetical applications include:

1) Establishment of symbiotic relationships between cereal crops and ammonia-excreting strains of *Azotobacter vinelandii*
2) Detoxification of specific pollutants, e.g., polychlorinated biphenyls, by pure cultures of micoorganisms with enhanced activities.
3) Consumption of petroleum waste and oil spills by gene-amplified strains of *Ps. putida*.
4) Increased resistance to heat, acid and heavy metal toxicity in microorganisms (mainly by *Thiobacillus* species) used for leaching of low-grade ores.
5) Synthesis of new interferons and antibiotics by hybridizing genes from different producing cells before introduction into bacteria by recombinant techniques so that products are formed by enzymes from mixed sources.

Strain A
Sporulation
Selected spore culture
Mating
New hybrid strain
Strain B
Sporulation
Selected spore culture

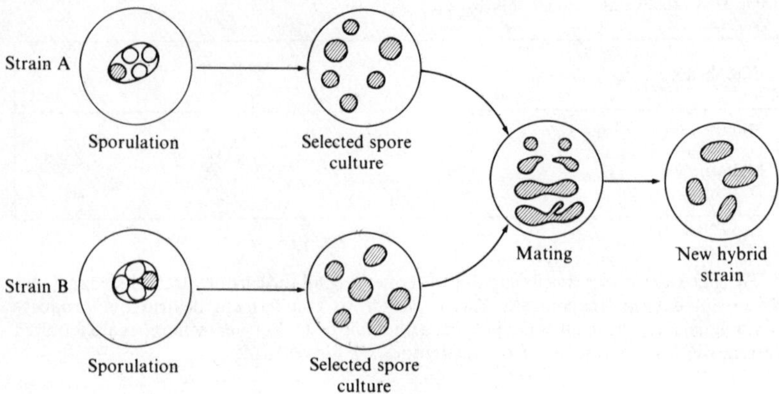

Figure 25. Hybridization in yeasts.

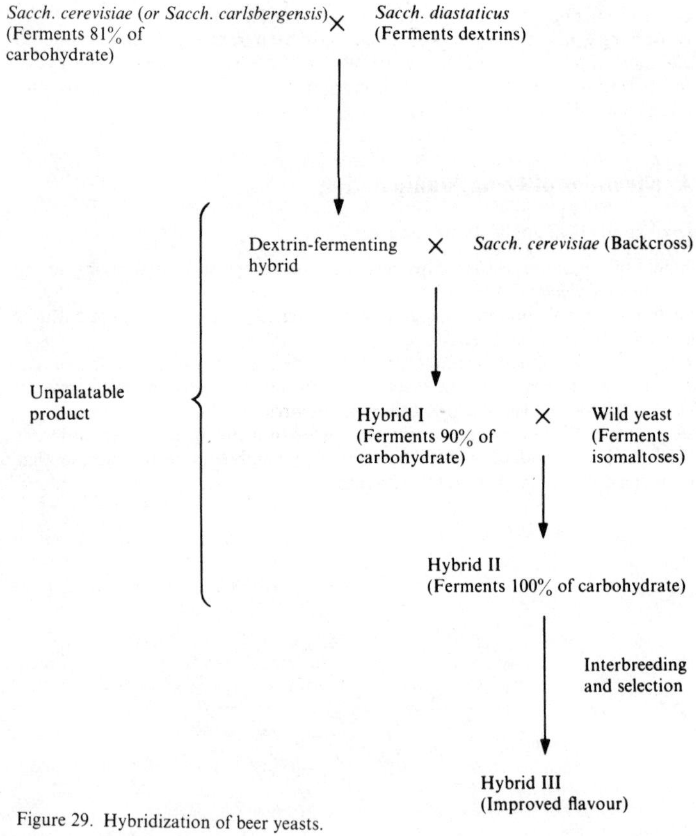

Sacch. cerevisiae (or Sacch. carlsbergensis) (Ferments 81% of carbohydrate)
×
Sacch. diastaticus (Ferments dextrins)
Dextrin-fermenting hybrid
×
Sacch. cerevisiae (Backcross)
Unpalatable product
Hybrid I (Ferments 90% of carbohydrate)
×
Wild yeast (Ferments isomaltoses)
Hybrid II (Ferments 100% of carbohydrate)
Interbreeding and selection
Hybrid III (Improved flavour)

Figure 29. Hybridization of beer yeasts.

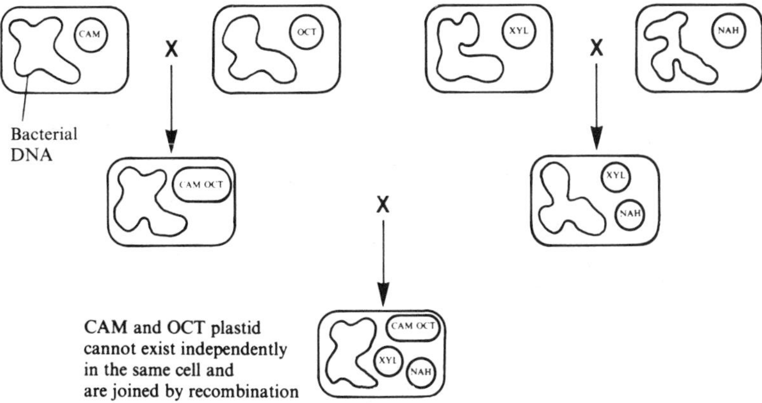

Figure 30. Derivation of hydrocarbon-metabolizing mutants of *Ps. putida*. CAM, plastid contains genes for camphor metabolism; OCT, plastid contains genes for octane, hexane and decane metabolism; XYL, plastid contains genes for xylene and toluene metabolism; NAH, plastid contains genes for naphthalene metabolism.

REFERENCES

Abson, J.W. and Todhunter, K.N. (1967) Effluent Disposal, in *Biochemical and Biological Engineering Science*, vol. 1, ed. by N. Blakebrough (Academic Press, New York).

Aiba, S., Humphrey, A.E. and Millis, N.F. (1973) *Biochemical Engineering*, 2nd edn. (Academic Press, New York).

Anderson, P.A. (1953) Automatic Recording of the Growth Rates of Continuously Cultured Microorganisms. *J. Gen Physiol.*, **36**, 733.

Atkinson, B. and Daoud, I.S. (1970) Diffusion Effects within Microbial Films. *Trans. Instn Chem. Engrs*, **48**, 245.

Atkinson, B., Daoud, I.S. and Williams, D.A. (1968) A Theory for the Biological Film Reactor. *Trans. Instn Chem. Engrs*, **46**, 245.

Atkinson, B. and Daoud, S. (1976) Microbial Flocs and Flocculation. *Advances in Biochemical Engineering*, vol. 4, ed. by T.K. Ghose, A. Fiechter and N. Blakebrough (Springer–Verlag, Berlin).

Atkinson, B. and Williams, D.A. (1971) Performance of a Trickling Filter with Hold-up of Microbial Mass Controlled by Periodic Washing. *Trans Instn. Chem. Engrs*, **49**, 215.

Beaman, R.G. (1967) Vinegar Fermentation, in *Microbial Technology*, ed. by H.J. Peppler (Reinhold, New York).

Block, S.S., Stearns, T.W., Stephens, R.L. and McCaudless, R.F.J. (1953) Experiments with Submerged Culture of Mushroom Mycelium. *J. Agr. Fd Chem.*, **1**, 890.

Borchardt, J.A. (1971) Biological Waste Treatment using Rotating Discs. *Biotechnol. Bioengng Symp.*, **2**, 131.

Bruce, A.M. and Merkens, J.C. (1970) Recent Studies of High-Rate Biological Filtration. *J. Inst. Water Pollut. Control*, **2**, 3.

Bruce, A.M., Merkens, J.C. and Macmillan, S.M. (1970) Research Developments in High-Rate Biological Filtration. *Inst. Publ. Hlth Engrs J.*, **69**, 178.

Bungay, H.R., Whalen, W.J. and Sanders, W.M. (1969) Microprobe Techniques for determining Diffusivities and Respiration Rates in Microbial Slime Systems. *Biotechnol. Bioengng*, **11**, 765.

DeWalle, F.B. and Chian, E.S.K. (1974) Kinetics of Formation of Humic Substances in Activated Sludge Systems and Their Flocculation. *Biotechnol. Bioengng*, **16**, 739.

Dias, F.F., Dondero, N.C. and Finstein, M.S. (1968) Attached Growth of *Sphaerotilis* and Mixed Populations in a Continuous Flow Apparatus. *Appl. Microbiol.*, **16**, 1191.

Elsworth, R. (1970) Multilayer Perfusion Tissue Culture. *Process Biochem.*, **5**, 21.

Freeman, R.R. (1961) Production of Concentrated Cell Suspensions: The Sponge Fermentor. *J. Biochem. Microbiol. Technol. Engng*, **3**, 339.

Gasner, L.L. and Wang, D.I.C. (1970) Microbial Cell Recovery Enhancement through Flocculation. *Biotechnol. Bioengng*, **12**, 873.

Graebe, J.E. and Novelli, G.D. (1966) Amino Acid Incorporation in a Cell-Free System from Submerged Tissue Cultures of *Zea mays*. *Exptl Cell Res.*, **41**, 521.

Greenshields, R.N. and Smith, E.L. (1971) Tower Fermentation Systems and their Applications. *Chem. Engr*, No. 249, 182.

Hawker, L.E. and Linton, A.H. (eds) (1979) Microorganisms – Function, Form and Environment, 2nd edn. (Edward Arnold, London).

Hawker, L.E., Linton, A.H. Folkes and Carlisle (1960) An Introduction to Biology of Microorganisms (Edward Arnold, London).

Herbert, D. (1961) Chemical Composition of Microorganisms as a Function of their Environment. *Symp. Soc. Gen. Microbiol.*, p.391.

Kornegay, B.H. and Andrews, J.F. (1968) Kinetics of Fixed-Film Biological Reactors, *J. Water Pollut. Control Fed*, **40**, R460.

Larsen, D.M. and Dimmick, R.L. (1964) Attachment and Growth of Bacteria on Surfaces of Continuous-Culture Vessels. *J. Bacteriol.*, **88**, 1380.

McGregor, W.C. and Finn, R.K. (1969) Factors affecting the Flocculation of Bacteria by Chemical Additives. *Biotechnol. Bioengng*, **12**, 873.

McKinney, R.E. (1956) Biological Flocculation, in *Biological Treatment of Sewage and Industrial Wastes*, ed. by B.J. McGrabe and W.W. Eckenfelder, jun., p.88 (Reinhold, New York).

Maier, W.J., Behn, V.C. and Gates, C.D. (1967) Simulation of the Trickling Filter Process. *Proc. Am. Soc. Civil Engng, J. Sanit. Engng Divn*, 91.

Monadjemi, P. and Behn, V.C. (1971) Oxygen Uptake and Mechanisms of Substrate Purification in a Model Trickling Filter, in *Advances in Water Pollution Research*, vol. 1, II-12, p.116 (Pergamon Press, Oxford).

Morris, G.G., Greenshields, R.N. and Smith, E.L. (1973) Aeration in Tower Fermenters containing Microorganisms. *Biotechnol. Bioengng Symp.*, No. 4, 535.

Moscona, A.A. (1961) Inhibition by Trypsin of Dissociation of Embryonic Tissue by Trypsin, *Nature (London)*, **199**, 379.

Northrop, J.H. (1954) Apparatus for maintaining Bacterial Cultures in the Steady State. *J. Gen. Physiol.*, **38**, 105.

Parker, D.S., Kaufman, W.J. and Jenkins, D. (1971) Physical Conditioning of Activated Sludge Floc. *J. Water Pollut. Control Fed.*, **43**, 1817.

Pavoni, J.L., Tenney, M.W. and Echelberger, W.F. (1972) Bacterial Exocellular Polymers and Biological Flocculation. *J. Water Pollut. Control Fed.*, **44**, 414.

Pelczar, M.J., Reid, R.D. and Chan, E.C.S. (1977) *Microbiology*, 4th edn. (McGraw-Hill Book Co., London).

Phaff, H.J., Miller, M.W. and Mrak, E.M. (1978) *The Life of Yeasts*, 2nd edn. (Harvard University Press, Cambridge).

Pirt, S.J. and Callow, D.S. (1960) Studies of the Growth of *Penicillium chrysogenum* in Continuous Flow Culture With Reference to Penicillin Production. *J. Appl. Bacteriol.*, **23**, 87.

Rincke, G. and Wolters, N. (1971) Technology of Plastic Medium Trickling Filter, in *Advances in Water Pollution Research*, vol. 1, II-15, p.19 (Pergamon Press, Oxford).

Smith, E.L. and Greenshields, R.N. (1974) Tower Fermentation Systems and their Applications to Aerobic Processes. *Chem. Engr*, No. 249, 28.

Smith, S.M. and Street, H.E. (1974) The Decline of Embryogenic Potential as Callus and Suspension Cultures of Carrot are Serially Sub-cultured. *Ann. Bot.*, **38**, 223.

Steel, R., Leutz, C.P. and Martin, S.M. (1955) Submerged Citric Acid Fermentation of Sugar-Beet Molasses. Increase in Scale. *Can. J. Microbiol.*, **1**, 299.

Telling, R.C. and Radlet, P.J. (1970) Large-Scale Cultivation of Mammalian Cells, in *Advances in Applied Microbiology*, vol. 13, ed. by D. Perlman, (Academic Press, New York).

Tomlinson, T.G. and Snaddon, D.H.M. (1966) Biological Oxidation of Sewage by Films of Microorganisms. *Air Water Pollut.*, **10**, 865.

Trudinger, P.A. (1971) Microbes, Metals and Minerals. *Miner. Sci. Engng*, **3**, 13.

Watson, J.D. (1970) *Molecular Biology of Genes* (W.A. Benjamin, New York).

Whalen, W., Bungay, H.R. and Sanders, W.M. (1969) Microelectrode Determination of Oxygen Profiles in Microbial Slime Systems. *Environ. Sci. Technol.*, **3**, 1297.

Whitaker, and Long (1973). Fungal Pelleting. *Biochem.*, **8**, 27.

Wood, E.J.F. (1950) Investigation on Underwater Fouling. I. The Role of Bacteria in the Early Stages of Fouling. *Austral. J. Marine Freshwater Res.*, **1**, 85.

Zobell, C.E. (1946). *Marine Microbiology* (Chronica Botanica, Waltham).

Zuyaginstev, D.G. (1959) Absorption of Microorganisms by Glass Surface. *Microbiology*, **28**, 104.

CHAPTER 2

THERMODYNAMIC ASPECTS OF MICROBIAL METABOLISM

GLOSSARY

Anabolism Building up of complex organic compounds from simpler material with absorption and storage of energy.
Catabolism Breakdown of complex living matter into simpler compounds with the release of energy.
Citric acid cycle *See TCA cycle.*
Coenzyme Organic compound playing an essential part in a reaction catabolized by an enzyme or system of enzymes, without being consumed in the process.
Endergonic Applied to a chemical reaction in which heat is absorbed.
Entropy The molecular chaos of a system.
Enzyme Protein-based catalyst produced within an organism.
Exergonic Applied to a chemical reaction in which energy in the form of heat is released.
Glycolysis Metabolic pathway involving the conversion of glucose to lactic acid.
Hexose Sugar with six carbon atoms.
Krebs cycle *See TCA cycle.*
Nucleotide Compound comprising a pentose, phosphoric acid and a nitrogen-containing base.
Pentose Sugar with five carbon atoms.
TCA cycle (Tricarboxylic acid cycle) Metabolic pathway involving the breakdown of pyruvic acid to carbon dioxide in the presence of oxygen.

FLOW OF ENERGY IN THE BIOLOGICAL WORLD

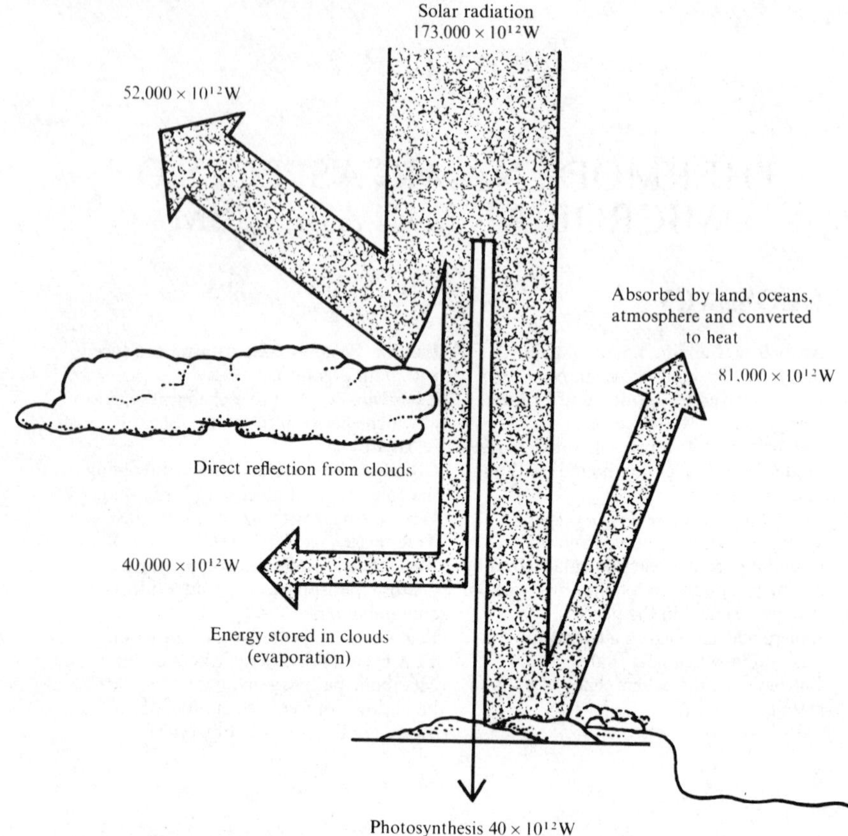

Figure 1. Destination of solar radiation. The thickness of the arrows represents the amount of energy absorbed, reflected or stored.

As can be seen from Fig. 1 only a small amount of the total solar energy reaching the earth is fixed by photosynthesis.

Table 1. Annual fixation of carbon by photosynthesis [Darlington (1964)].

Region	Area (km^2)	Carbon Fixed (tons km^{-2}. yr^{-1})	Total Carbon Fixed (tons yr^{-1})
Forest	44×10^6	250	11×10^9
Cultivated land	27×10^6	160	4.3×10^9
Grassland	31×10^6	36	1.0×10^9
Desert	47×10^6	7	0.3×10^9
Total land	149×10^6	453	16.6×10^9
Total ocean	361×10^6	340	122.6×10^9

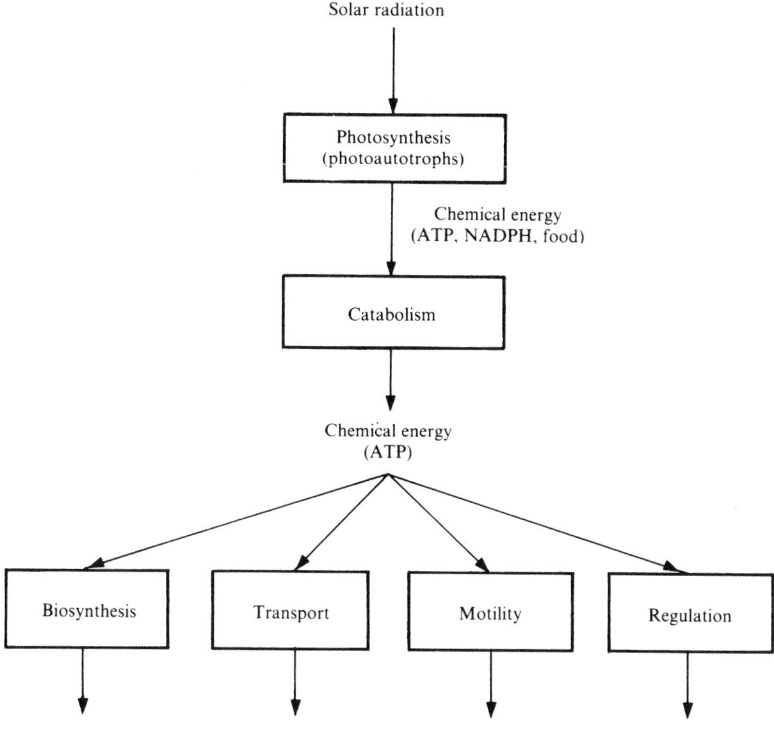

Figure 2. Destination of photosynthetic energy. Solar energy is the origin of all cellular energy. Energy is not recycled in the biological world, it only flows in one direction.

CLASSIFICATION OF MICROORGANISMS BY THEIR CARBON AND ENERGY SOURCES

In addition to the general classification systems for microorganisms based upon structure and function (*see* General Classification of Organisms, Chapter 1), it is helpful (*see* p. 19) to add further descriptors (*see* Table 2) which refer to the source of energy (i.e. light or chemical) and the carbon source (i.e. carbon dioxide or fixed carbon in organic compounds).

Table 2

Carbon Source			Energy source: Chemical	Energy source: Light	
Organic	Heterotrophs		Chemoorganotroph	Photoorganotroph	
			All higher animals, most microorganisms, nonphotosynthetic plant cells, also photosynthetic cells in the dark	Nonsulphur purple bacteria	Species
			Organic compounds (e.g., glucose)	Organic compounds	Electron donors
Carbon dioxide	Autotrophs		Chemolithotroph	Photolithotroph	
			Hydrogen, sulphur, iron, denitrifying bacteria	Green cells of higher plants (in the light), blue–green algae, photosynthetic bacteria	Species
			Inorganic compounds (H_2, S, H_2S, Fe(II), NH_3 etc.)	Inorganic compounds (H_2O, H_2S, S, etc.)	Electron donors

CYCLING OF MATTER IN THE BIOLOGICAL WORLD

Carbon and Oxygen Cycle

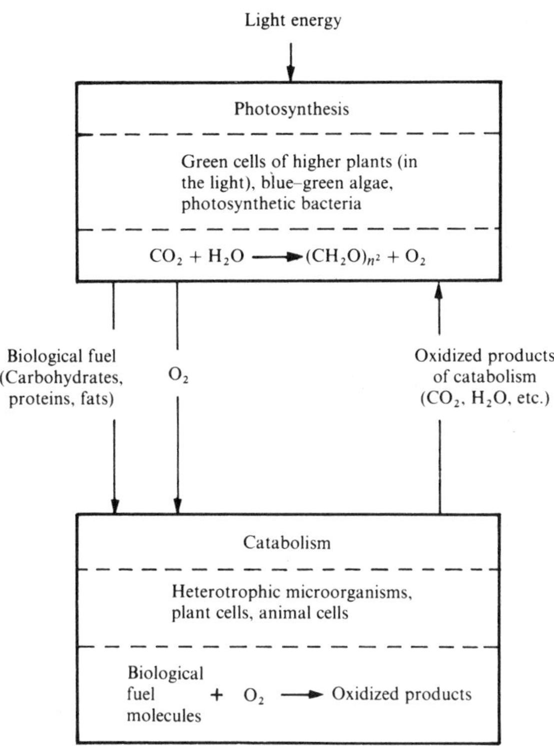

Figure 3

Nitrogen Cycle

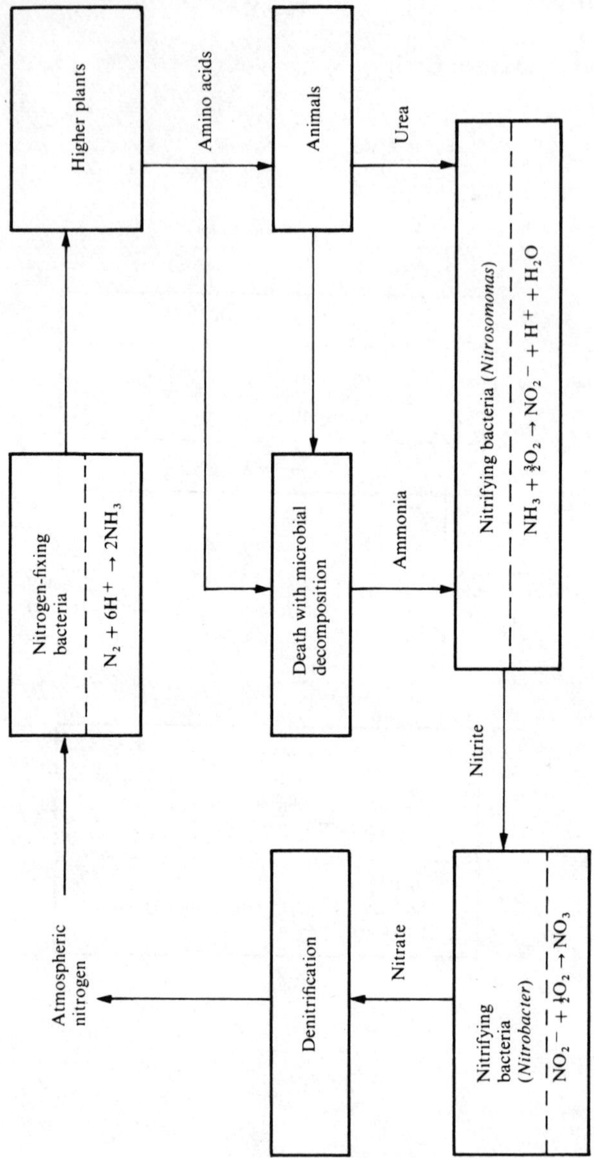

Figure 4

Sulphur Cycle

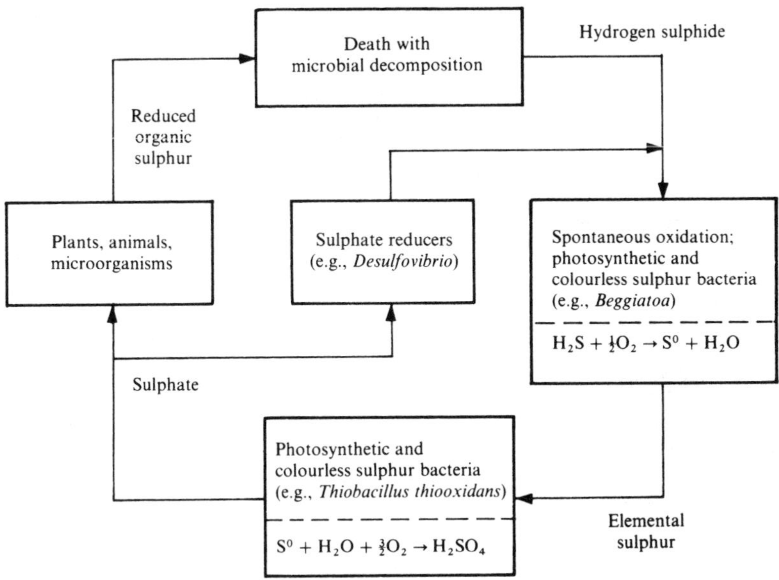

Figure 5

THERMODYNAMIC CONCEPTS IN THE ANALYSIS OF BIOLOGICAL SYSTEMS

Conventions

In thermodynamic analysis of biochemical systems, the following conventions are accepted.

1) The distinction between molal (m) and molar (M) quantities is not strictly observed.
2) In dilute aqueous systems, the concentration or thermodynamic activity of water is taken as 1.0 M, even though its actual value is about 55.5 M.
3) pH 7.0 is used as the standard condition rather than pH 0.0, which corresponds to a hydrogen ion concentration of 1.0 M. In the case of free energy change, ΔG^0 refers to pH 0.0 and $\Delta G^{0\prime}$ to pH 7.0.
4) The standard state of each reactant and product capable of ionization is that mixture of its un-ionized and ionized forms that exist at pH 7.0. Consequently, values of free energy change, for instance, based on pH 7.0 may not necessarily be used at other pH values, because the extent of ionization of one or more components may change with pH. The variation in free energy change with pH for some biochemical reactions may be quite large and difficult to calculate.
5) The principles of classical or equilibrium thermodynamics apply only to closed systems. However, living cells are open systems since they exchange matter with their surroundings.

True thermodynamic equilibrium does not exist in living cells. Although the concentrations of chemical components may appear to be constant, the metabolic processes are in a dynamic state, in which the rate of formation of a given component is exactly counterbalanced by the rate of its removal or breakdown. The application of irreversible or nonequilibrium thermodynamics to biological processes should result in a more profound understanding of these phenomena.

Chemical Work and Energy

At constant temperature and pressure, the only useful work that can be done is electrochemical work. In particular, no heat can be used to do work under these conditions. For this reason, living cells use chemical energy to do work.

The main requirements for driving a chemical reaction in a direction which it would not go spontaneously (i.e. ΔG^0 the free energy change of the reaction is positive) are:

1) there must be another reaction that proceeds with a larger decrease in free energy, and
2) the two reactions must be coupled through a common intermediate.

These stem from the additive nature of the free energy change of reactions.

Nearly all the chemical processes in the cell proceed by sequential reactions. The common intermediate for the transfer of energy can be a phosphate group P_i.

Free Energy of Formation of Some Biochemical Compounds

Table 3. Standard free energies of formation for 1 M aqueous solutions at pH 7.0 and 25°C [Van der Kroon (1969)].

	ΔG^0	
Substance	(kcal mol^{-1})	(kJ mol^{-1})
Acetate$^-$	−68.99	−372.3
cis-Aconitate^{3-}	−220.51	−922.61
L-Alanine	−88.75	−371.3
Ammonium ion	−19.00	−79.50
L-Aspartate$^-$	−166.99	−698.69
Bicarbonate ion	−140.33	−587.14
Carbon dioxide (gas)	−94.45	−395.2
Ethanol	−43.39	−181.6
Fumarate^{2-}	−144.41	−604.21
α-D-Glucose	−219.22	−917.21
Glycerol	−116.76	−488.64
Hydrogen ion	−9.55	−39.96
Hydroxide ion	−37.60	−157.3
α-Ketoglutarate^{2-}	−190.62	−797.56
Lactate$^-$	−123.76	−517.81
L-Malate^{2-}	−201.98	−845.08
Oxaloacetate^{2-}	−190.53	−797.18
Pyruvate$^-$	−113.44	−474.63
Succinate^{2-}	−164.97	−690.23
Water (liquid)	−56.69	−237.2

Free Energy Change of Some Biochemical Reactions

Table 4. Standard free energy changes of some biochemical reactions in dilute aqueous systems at pH 7.0 and 25°C [Van der Kroon (1969)].

Reaction	$\Delta G^{0\prime}$ (kcal mol^{-1})	$\Delta G^{0\prime}$ (kJ mol^{-1})
Hydrolysis		
Acid anhydrides		
Acetic anhydride + H_2O → 2 acetate	−21.8	−91.2
Pyrophosphate + H_2O → $2H_3PO_4$	−8.0	−33.4
Esters		
Ethyl acetate + H_2O → ethanol + acetate	−4.7	−19.7
Glucose 6-phosphate + H_2O → glucose + H_3PO_4	−3.3	−13.8
Amides		
Glutamine + H_2O → glutamate + NH_4^+	−3.4	−14.2
Glycylglycine + H_2O → 2 glycine	−2.2	−9.2
Glycosides		
Sucrose + H_2O → glucose + fructose	−7.0	−29.3
Maltose + H_2O → 2 glucose	−4.0	−16.7
Esterification		
Glucose + phosphate → glucose 6-phosphate + H_2O	+3.3	+13.8
Rearrangement		
Glucose 1-phosphate → glucose 6-phosphate	−1.7	−7.11
Fructose 6-phosphate → glucose 6-phosphate	−0.4	−1.67
Elimination		
Malate → fumarate + H_2O	+0.75	+3.14
Oxidation		
Glucose + $6O_2$ → $6CO_2$ + $6H_2O$	−686	−2870
Palmitic acid + $23O_2$ → $16CO_2$ + $16H_2O$	−2338	−9782
Lactic acid + $3O_2$ → $3CO_2$ + $3H_2O$	−320	−1339

Table 5. Heats of reaction of some representative chemical reactions, under conditions of constant temperature and pressure [Darlington (1964)].

Reaction	Heat change (kcal mol^{-1})
Oxidation	
Glucose + $6O_2$ → $6CO_2$ + $6H_2O$	−673
Hydrolysis	
Sucrose + H_2O → glucose + fructose	−4.8
Glucose 6-phosphate + H_2O → glucose + H_3PO_4	−3.0
Neutralization	
NaOH + HCl → NaCl + H_2O	−13.8

Heats of Combustion of Major Biological Fuel Molecules

Table 6

Fuel	Molecular weight	Heat of combustion (kcal mol^{-1})	(kcal g^{-1})
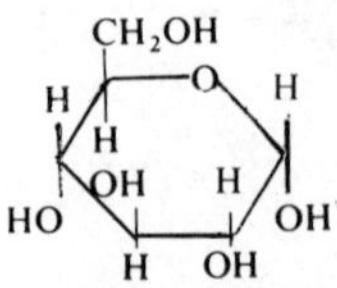 D-Glucose, a carbohydrate	180	673	3.74
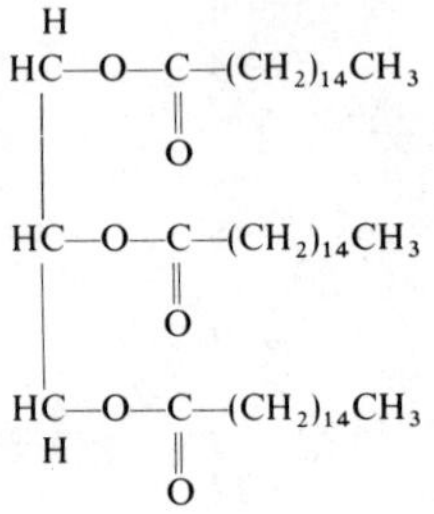 Tripalmitin, a fat	809	7510	9.3
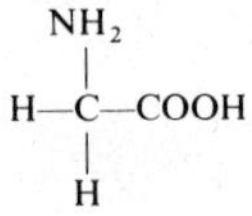 Glycine, an amino acid	75	234	3.12

CHEMICAL ENERGY: PRODUCTION, CONSERVATION AND UTILIZATION IN THE CELL

Biological fuel molecules (carbohydrates, lipids and proteins) contain high levels of chemical energy because of their high degree of structural order; they have relatively little randomness, or entropy. During catabolism, they are degraded, i.e. oxidized, into smaller molecules, like carbon dioxide, water, alcohols, etc. As a result of this transformation, which means an increase in the randomness or entropy of their constituent atoms, the fuel molecules undergo a loss of free energy, i.e. that form of energy capable of doing work at constant temperature and pressure.

The energy thus released during the exergonic reactions of catabolism cannot be utilized in the form of heat since metabolic reactions in living organisms are essentially isothermal. Instead, the free energy of biological fuels is conserved as the chemical energy inherent in the covalent bonding structure of certain compounds. This conservation of energy is achieved by the reactions chemically coupled to the exergonic reactions of catabolism.

Biosynthesis (i.e. anabolism), active transport through the cell membrane and cell mobility involve endergonic biochemical reactions. The energy needed for these is supplied again in the form of chemical energy by high-energy compounds through chemically coupled reactions.

Energy Coupling Through ATP Systems

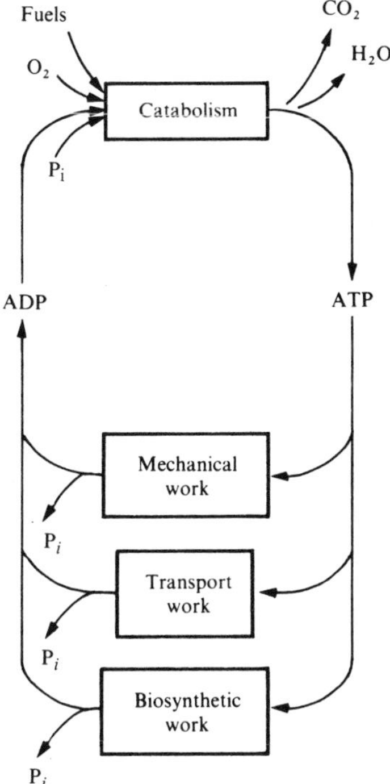

Figure 6. The ATP–ADP cycle. The high-energy phosphate bonds of ATP are used in coupled reactions for carrying out energy-requiring functions; ultimately, inorganic phosphate is released. ADP is rephosphorylated to ATP during energy-yielding reactions of catabolism.

The free energy of cellular fuels is conserved as the chemical energy inherent in the covalent bonding structure of the terminal phosphate groups in adenosine triphosphate ATP (see Fig. 7). ATP is enzymatically generated from adenosine diphosphate ADP and inorganic phosphate P_i in enzymatic phosphate group transfer reactions which are chemically coupled to specific oxidative steps during catabolism.

The ATP–ADP system plays an important role in the transfer of energy in cells for the following reasons.

1) ATP occupies an intermediate position in the thermodynamic scale of phosphate compounds (see Table 7). Because of this position, ATP–ADP systems provide an intermediate link between phosphate compounds with high phosphate group transfer potential and other compounds with low phosphate group transfer potential. ADP acts as the specific enzymatic acceptor of phosphate groups and ATP thus formed acts as a donor.

2) ATP and ADP are obligatory reactants in nearly all the enzymatic phosphate transfer reactions in the cell. One set of phosphate-transferring enzymes catalyses the transfer of phosphate groups from compounds of very high potential to ADP. A second set of enzymes catalyzes the reactions in which the phosphate group is transferred from ATP to low energy phosphate acceptors.

There are no cellular enzymes that can transfer phosphate groups directly from high-energy donors to low-energy acceptors.

These considerations imply that the terminal phosphate group of ATP must undergo very rapid turnover in the cell. The half-life for the turnover of ATP in a rapidly respiring bacterial cell, such as *Escherichia coli*, has been shown, using ^{32}P tracer, to be a matter of only seconds and that of the larger, more slowly respiring liver cells, about 1–2 min.

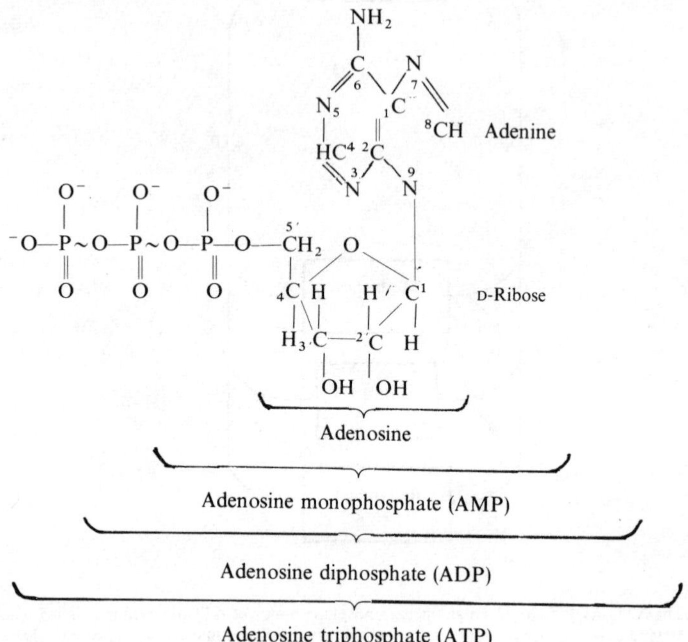

Figure 7. The structure of ATP, showing the ionized form at pH 7.0. The symbol ~ designates the 'high energy' bonds.

Energy Coupling Through NADP and Other Coenzyme Systems

Electrons, or hydrogen atoms, provide another vehicle for transfer of chemical energy from the energy-yielding reactions of catabolism to the energy-requiring reactions in the cell. Electrons are transported enzymatically as carbon–carbon or carbon–oxygen double bonds by means of electron-carrying coenzymes. The most important of such coenzymes is nicotinamide adenine dinucleotide phosphate $NADP^+$.

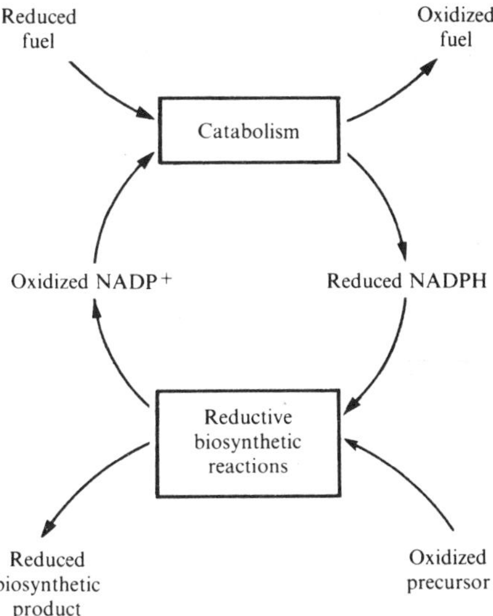

Figure 8. Transfer of reducing power via the nicotinamide adenine dinucleotide phosphate cycle. Other electron-carrying coenzymes, such as flavin nucleotides, also participate in reductive biosynthesis.

ATP SYSTEMS

Properties of ATP, ADP and AMP

1) ATP, ADP and AMP are found in all living cells of microbial, plant and animal origin.
2) These nucleotides are located in the soluble cytoplasm and also in such organelles as mitochondria and nuclei.
3) The total concentration of ATP, ADP and AMP in the soluble cytoplasm of living cells remains relatively constant at between 2 to 10 mM, depending on the species.
4) At pH 7.0, both ATP and ADP are highly charged anions. Three of the four protons in ATP are fully ionized at pH 7.0. The p*K*s of the last ionization steps of ATP and ADP at pH 7.0 are as follows

$$HATP^{3-} \rightleftharpoons ATP^{4-} + H^+ \quad pK' = 6.95$$
$$HADP^{2-} \rightleftharpoons ADP^{3-} + H^+ \quad pK' = 6.88$$

5) ATP and ADP form stable complexes with certain divalent cations found in the cell, such as Mg^{2+}, through reversible reactions.

$$Mg^{2+} + ATP^{4-} \rightleftharpoons Mg.ATP^{2-}$$
$$Mg^{2+} + ADP^{3-} \rightleftharpoons Mg.ADP^{1-}$$

The affinity of ATP for Mg^{2+} is about 10 times that of ADP.

Standard Free Energy of Hydrolysis of Phosphate Compounds

There are several phosphorylated compounds in cells that participate in the transfer of chemical energy. They achieve this by donating their phosphate groups P_i to a low-energy compound and hence raising the energy of the acceptor.

In order to predict in which direction such a phosphate group transfer will take place and at what point the reaction will reach equilibrium, a 'thermodynamic yardstick' is used. It is the tendency of the phosphate group in different phosphorylated compounds to be transferred to a standard, arbitrarily chosen acceptor molecule, namely, water.

$$R - O - PO_3^{2-} + HOH \rightleftharpoons R - OH + HO - PO_3^{2-}$$

This reaction actually represents the hydrolysis of the phosphate ester $R-O-PO_3^{2-}$. The relative transfer tendency of phosphate groups in biological compounds is determined by the standard free energy of hydrolysis of such compounds. However, biological phosphate compounds do not normally undergo simple hydrolysis reactions in intact cells and the choice of water as the standard phosphate acceptor is for convenience and is wholly arbitrary.

Table 7. Standard free energy of hydrolysis of some phosphorylated compounds [Van der Kroon (1969)].

	$\Delta G^{\circ\prime}$	
Compound	(kcal mol^{-1})	(kJ mol^{-1})
Phosphoenolpyruvate	−14.80	−61.9
3-Phosphoglycerol phosphate	−11.80	−49.3
Phosphocreatine	−10.30	−43.1
Acetyl phosphate	−10.10	−42.3
Phosphoarginine	−7.70	−32.2
ATP (→ ADP + P_i)	−7.30	−30.5
Glucose 1-phosphate	−5.00	−20.9
Fructose 6-phosphate	−3.80	−15.9
Glucose 6-phosphate	−3.30	−13.8
Glycerol 1-phosphate	−2.20	−9.2

Fig. 9 and Table 7 give the standard free energy of hydrolysis of some phosphorylated compounds.

Since this scale is arranged in order of the free energy of hydrolysis, those compounds high in the scale undergo more complete hydrolysis at equilibrium than those lower in the scale. In other words, those high in the scale tend to lose their phosphate groups, and those lower in the scale tend to hold on to their phosphate compounds.

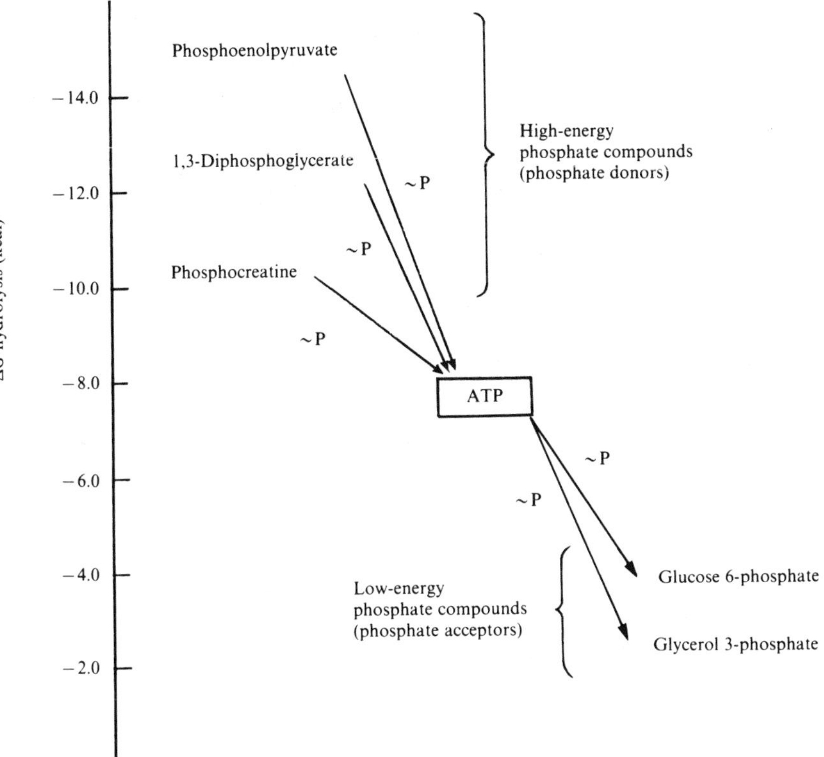

Figure 9. Flow of phosphate groups from high-energy phosphate donors to low-energy acceptors via ATP–ADP system. The direction of flow is toward compounds having a low phosphate group potential, assuming standard conditions, with all reactants and products at 1 M.

Conditions Affecting Free Energy of Hydrolysis of ATP, ADP and AMP

The published values for the standard free energy change of the hydrolysis of ATP vary somewhat. One reason is the analytical difficulties in obtaining precise values for the equilibrium constants of the reactions employed. Another is the fact that measurements have not always been made under precisely the same conditions. However, the values given in Table 8 can be used for the standard free energy change of hydrolysis at pH 7.0.

Table 8.

Reaction	$\Delta G^{0'}$ (kcal mol^{-1})
$ATP + H_2O \rightleftharpoons ADP + P_i$	−7.3
$ADP + H_2O \rightleftharpoons AMP + P_i$	−7.3
$AMP + H_2O \rightleftharpoons Adenosine + P_i$	−3.4

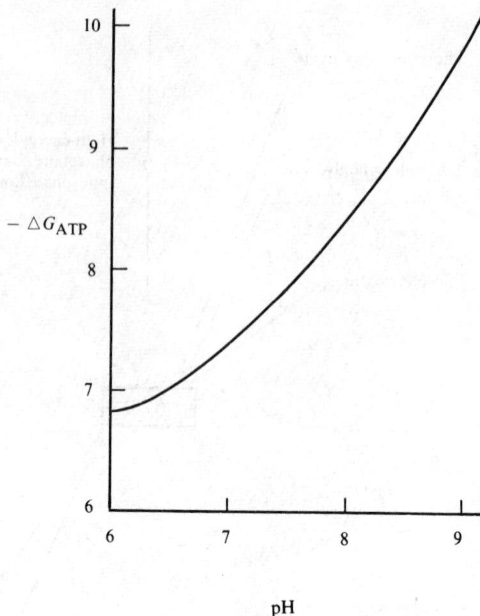

Figure 10. Effect of pH on the free energy of hydrolysis of ATP ΔG_{ATP}.

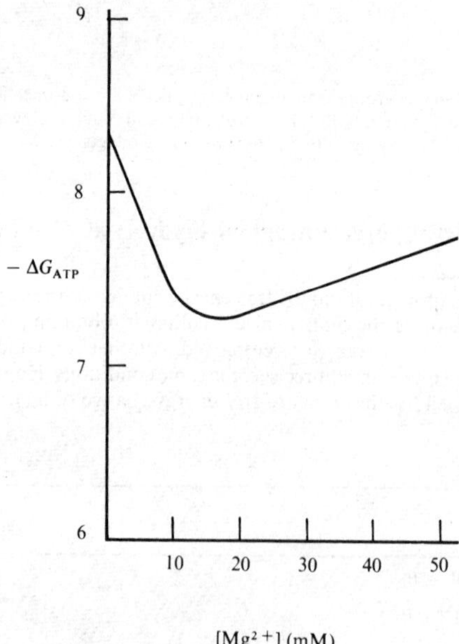

Figure 11. Effect of Mg^{2+} concentration on the free energy of hydrolysis of ATP at pH 7.0.

Temperature, pH and Mg^{2+} concentration can markedly influence the free energy change of hydrolysis of ATP. The influences of pH and Mg^{2+} are shown in Figs. 10 and 11 respectively.

It must be remembered that the concentrations of ATP, ADP and phosphate in intact cells are far from the standard of 1.0 M, on which $\Delta G^{0\prime}$ values are based. After appropriate appropriate corrections for the actual steady-state concentrations of ATP, ADP and phosphate, Mg^{2+} concentration and pH in the intracellular aqueous phase, the free energy of hydrolysis of ATP in intact cells is about $-$ 12.5 kcal mol^{-1}, although even this value can vary from time to time or even with location within the cell.

Examples of Chemical Energy Transfer

Simple transfer of energy

$\Delta G^{0\prime}$ (kcal mol^{-1})

$$\text{D-Glucose} + P_i \longrightarrow \text{D-glucose 6-phosphate} + H_2O \quad +3.3$$

$$ATP + H_2O \longrightarrow ADP + P_i \quad -7.3$$

$$\text{D-Glucose} + ATP \xrightarrow{\text{hexokinase}} \text{D-glucose 6-phosphate} + ADP \quad -4.0$$

Conservation of the energy of oxidation as ATP energy

$$\text{Glyceraldehyde 3-phosphate} + NAD^+ + H_2O \longrightarrow \text{3-phosphoglycerate} + NADH^+ + H^+ \quad -10.3$$

$$ADP + P_i \longrightarrow ATP + H_2O \quad +7.3$$

$$\text{Glyceraldehyde 3-phosphate} + H_2O + NAD^+ + ADP + P_i \longrightarrow \text{3-Phosphoglycerate} + NADH^+ + H^+ + ATP \quad -3.0$$

The actual reaction, involving the common intermediate is

$$\text{3-Phosphoglycerate} + P_i \longrightarrow \text{1,3-diphosphoglyceric acid}$$

$$\text{1,3-Diphosphoglyceric acid} + ADP \longrightarrow \text{3-phosphoglyceric acid} + ATP$$

$$\text{3-Phosphoglycerate} + ADP + P_i \longrightarrow \text{3-phosphoglyceric acid} + ATP$$

Utilization of ATP energy to do chemical work

	$\Delta G^{0\prime}$ (kcal mol^{-1})
Glutamic acid + NH_3 $\rightleftharpoons$ glutamine + H_2O	+ 3.4
+	
ATP + H_2O $\rightleftharpoons$ ADP + P_i	− 7.3
Glutamic acid + NH_3 + ATP $\rightleftharpoons$ glutamine + ADP + P_i	− 3.9

The actual reaction involving the common intermediate is:

Glutamic acid + ATP $\rightleftharpoons$ glutamyl phosphate + ADP

\+

Glutamyl phosphate + NH_3 $\rightleftharpoons$ glutamine + P_i

---

Glutamic acid + NH_3 + ATP $\rightleftharpoons$ glutamine + ADP + P_i

Efficiency of Chemical Work

The free energy changes of a biochemical reaction can be used to give an indication of the efficiency of chemical work done.

In the foregoing example of conservation of the energy of oxidation as ATP energy,

$$\text{Efficiency} = \frac{7.3}{10.3} \times 100 \simeq 71\%$$

For the utilization of ATP energy to form the amide bond of glutamine,

$$\text{Efficiency} = \frac{3.4}{7.3} \times 100 \simeq 47\%$$

Lactic acid fermentation

Theoretically,

$$\underset{\text{Glucose}}{C_6H_{12}O_6} \longrightarrow \underset{\text{lactic acid}}{2C_3H_6O_3} \qquad \Delta G^{0\prime} = -47\ \text{kcal mol}^{-1}$$

However, the following net reaction takes place in living cells.

Glucose + 2ADP + 2P_i $\longrightarrow$ 2 lactate + 2ATP + 2H_2O

$$\Delta G^{0\prime} = -32.4\ \text{kcal mol}^{-1}$$

In the cell, 14.6 kcal of energy is conserved by forming 2ATP per mole of glucose consumed (formation of each ATP requires 7.3 kcal mol^{-1}). Thus

$$\text{Efficiency} = \frac{2 \times 7.3}{47} \times 100 \simeq 31\%$$

Ethanolic fermentation

Theoretically,

$$\underset{\text{Glucose}}{C_6H_{12}O_6} \longrightarrow \underset{\text{ethanol}}{2C_2H_5OH} + 2CO_2 \qquad \Delta G^{0\prime} = -\,56.5\,\text{kcal mol}^{-1}$$

The net reaction in the cell is

$$\text{Glucose} + 2\text{ADP} + 2\text{P}_i \longrightarrow 2\text{ ethanol} + 2CO_2 + 2\text{ATP} \qquad \Delta G^{0\prime} = -\,41.9\,\text{kcal mol}^{-1}$$

$$\text{Efficiency} = \frac{2 \times 7.3}{56.5} \times 100 \simeq 26\%$$

Respiration (oxidative phosphorylation)

Theoretically,

$$\underset{\text{Glucose}}{C_6H_{12}O_6} + 6O_2 \longrightarrow 6CO_2 + 6H_2O \qquad \Delta G^{0\prime} = -\,686\,\text{kcal mol}^{-1}$$

In the cell, the net reaction is

$$\text{Glucose} + 6O_2 + 36\text{ADP} + 36\text{P}_i \longrightarrow 6CO_2 + 42H_2O + 36\text{ATP} \qquad \Delta G^{0\prime} = -\,413\,\text{kcal mol}^{-1}$$

$$\text{Efficiency} = \frac{36 \times 7.3}{686} \times 100 \simeq 38\%$$

It is very likely that all these efficiency figures are minimum values and that given the appropriate corrections for the concentrations of ATP, ADP and phosphate occurring in the cell, the actual efficiency of energy recovery, for example during glucose oxidation, is probably over 60 per cent.

PRODUCTION OF ATP

Substrate Level Phosphorylation (Catabolism)

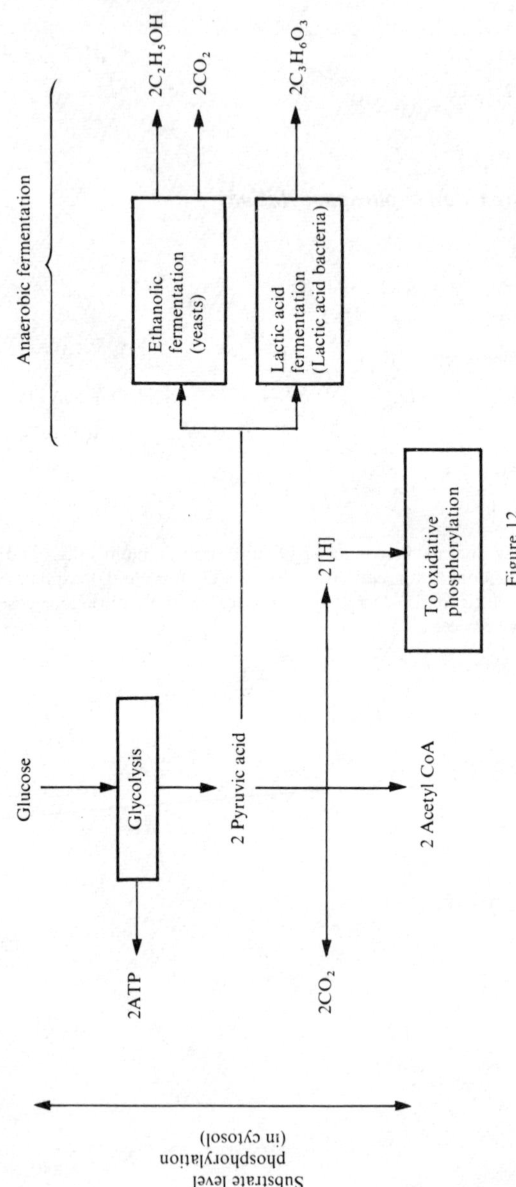

Figure 12

The substrate is partially oxidized during substrate-level phosphorylation with the concomitant production of ATP. The glycolytic pathway is used, ultimately resulting in the formation of organic acids and solvents.

Oxidative Phosphorylation (Respiratory-Chain Phosphorylation)

Under aerobic conditions, the intermediate products of glycolysis are decarboxylated in the tricarboxylic acid cycle and undergo oxidative phosphorylation with the formation of ATP.

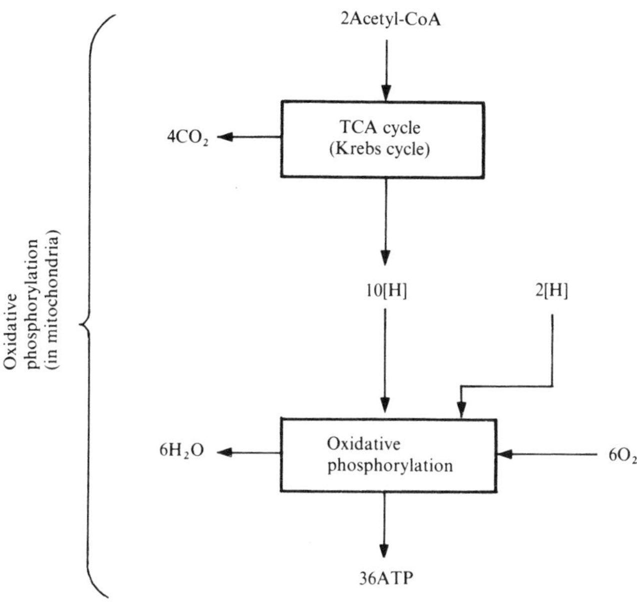

Figure 13

Fig. 14 provides a simple hierarchy for the sequence of hydrolytic, conversion and oxidative reactions that take place during cell metabolism and lead to the formation of ATP.

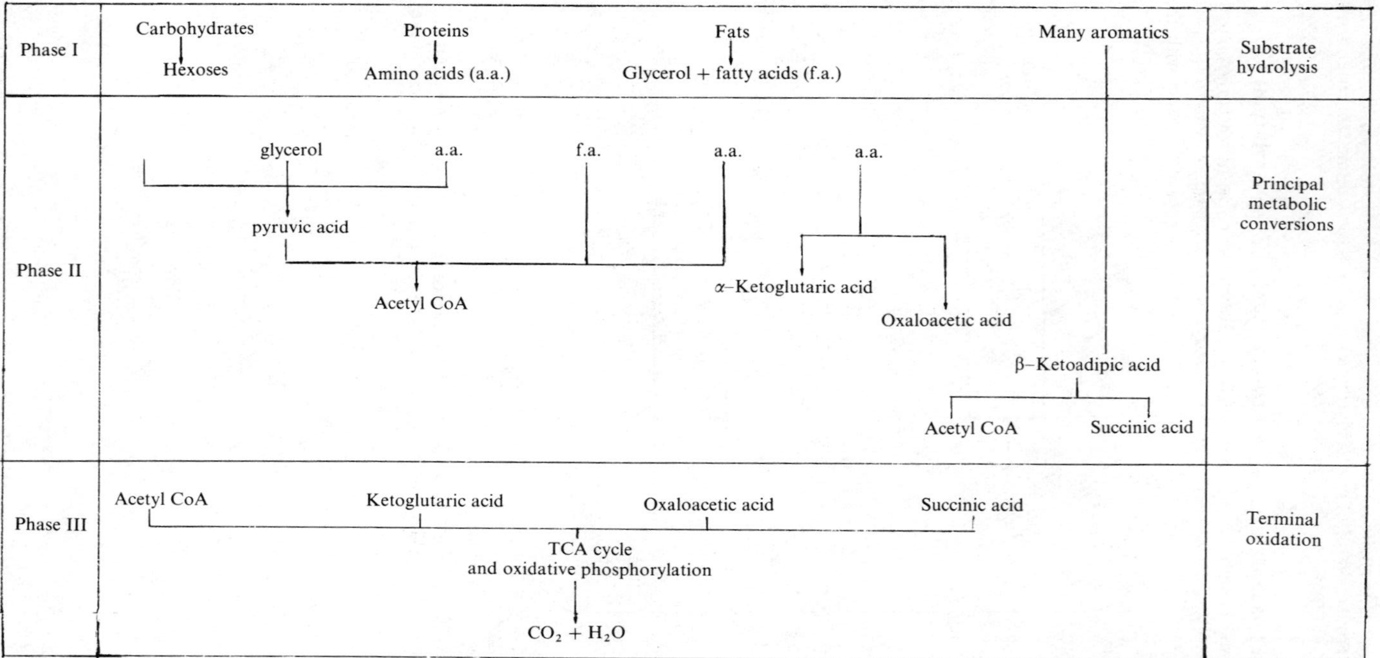

Figure 14. Major metabolic phases after Krebs and Ubreit.

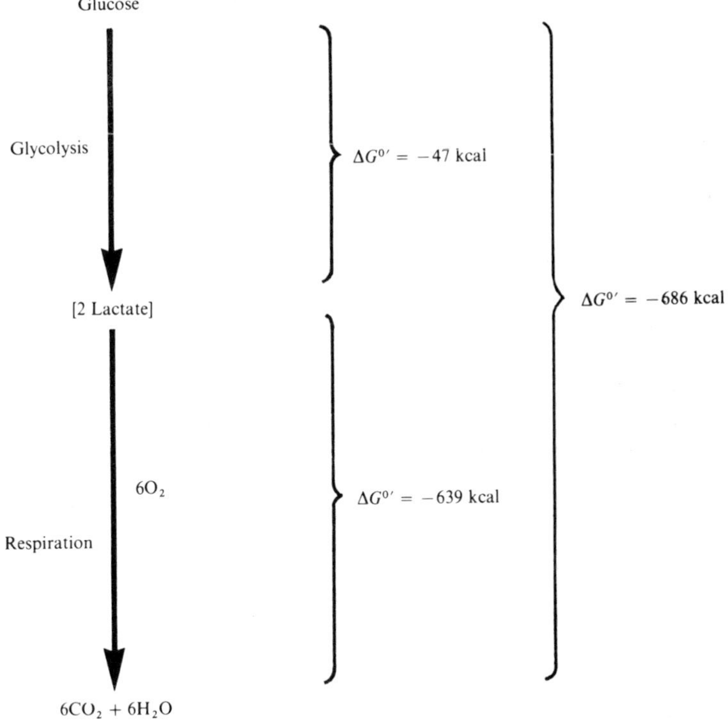

Figure 15. Comparison of the energy yield of glucose breakdown in glycolysis vs respiration.

Table 9. Comparison of the ATP available from the catabolism of various substrates.

Substrate	Conditions	ATP [mole ATP (mole substrate)$^{-1}$]
Glucose	Aerobic (P/O = 3)	38
Glucose	Anaerobic (products, formate + acetate)	3[a]
Glucose	Anaerobic (products, ethanol + CO_2)	2[a]
Glucose	Anaerobic (products, ethanol + CO_2)	1[b]
Pyruvate	Aerobic (P/O = 3)	30
Pyruvate	Anaerobic (products, formate + acetate)	1
Pyruvate	Anaerobic (products, ethanol + CO_2)	0
Palmitate (C_{16})	Aerobic	130

[a] Via the Embden – Meyerhof pathway.
[b] Via the Entner – Doudoroff pathway.

Table 10. Free energy changes of glucose oxidation pathways and corresponding ATP formation

Pathway	ΔG (kcal mol^{-1})	ATP Net gain per mole	ATP ΔG (kcal)	Free energy diverted into ATP (%)
Homolactic fermentation (glucose → 2 lactate$^-$ + 2H^+)				
Via Embden–Meyerhof	−49.7	2	16	32.2
Via Entner–Doudoroff	−49.7	1	8	16.1
Heterolactic fermentation (glucose → ethanol + CO_2 + lactate$^-$ + H^+)				
Via pentose monophosphates	−55.9	1	8	14.3
Alcoholic fermentation (glucose → 2 ethanol + 2CO_2)	−62.2	2	16	25.7
Complete aerobic oxidation of glucose (liver and muscle) (glucose + 6O_2 → 6CO_2 + 6H_2O)	−688.5	38	304	44.2
Oxidative phosphorylation (liver and muscle) (NADH + H^+ + $\frac{1}{2}O_2$ → NAD^+ + H_2O)	−51.9	3	24	46.2

Oxidation/Reduction Reactions

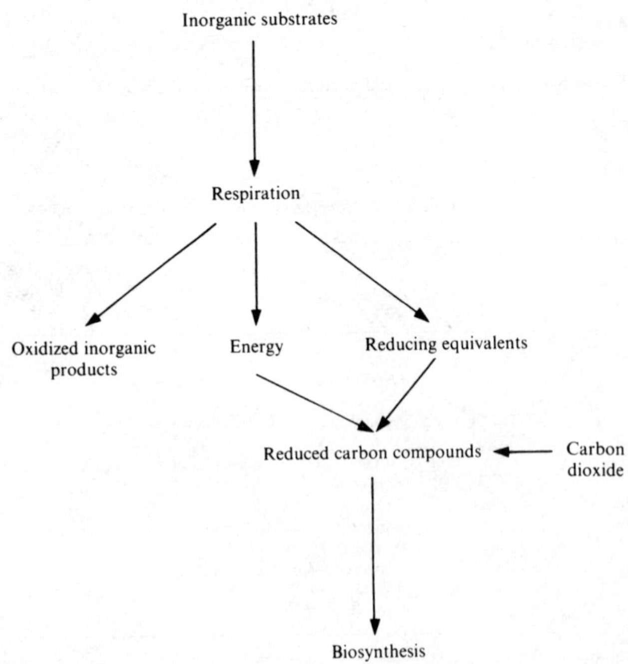

Figure 16. Outline of metabolism in chemolithotrophic bacteria (*see* Table 2).

Chemolithotrophic bacteria are those microorganisms that obtain all energy required for growth from the oxidation of inorganic compounds. Chemolithotrophic bacteria meet all their energy requirements in the form of NADH and ATP formed from NAD^+ and ADP by the oxidation of substances such as hydrogen, inorganic sulphur-containing compounds, nitrite, ammonia or iron in the absence of light (*see* Fig. 16).

Photolithotrophs utilize light energy and oxidize inorganic sulphur-containing compounds or hydrogen to provide electrons (*see* Fig. 17).

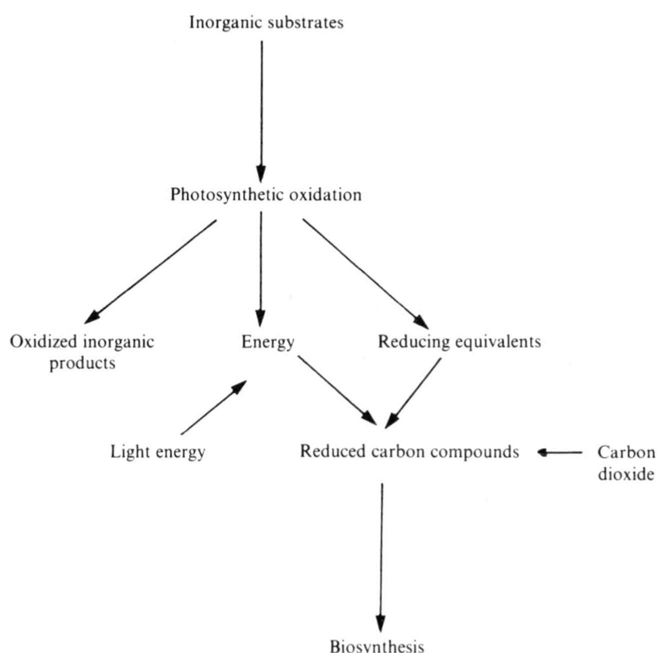

Figure 17. Outline of metabolism in photolithotrophic bacteria (*see* Table 2).

Ammonia and hydroxylamine oxidizers

Nitrosomonas, Nitrosospira, Nitroscoccus, Nitrosolobus and *Nitrosovibrio* are capable of oxidizing ammonia and hydroxylamine with the formation of nitrite.

$$NH_4^+ + \tfrac{1}{2}O_2 \rightarrow NO_2^- + H_2O + 2H^+$$

$$NH_2OH + O_2 \rightarrow NO_2^- + H_2O + H^+$$

All the above-mentioned genera have similar guanine + cytosine contents in their DNA, with values of 50–55 per cent, although their ultrastructural morphology varies widely.

Nitrite oxidation

Nitrite is oxidized to nitrate by *Nitrobacter, Nitrospina* and *Nitrococcus.*

$$NO_2^- + \tfrac{1}{2}O_2 \rightarrow NO_3^-$$

These bacteria possess guanine + cytosine contents of 57.7–61.2 per cent of the total DNA, but again their ultrastructural morphology varies considerably.

Sulphur oxidizers

All species of *Thiobacillus* are capable of oxidizing thiosulphate and sulphide using molecular oxygen with the formation of sulphate.

$$Na_2S_2O_3 + 2O_2 + H_2O \rightarrow Na_2SO_4 + H_2SO_4$$

$$H_2S + 2O_2 \rightarrow H_2SO_4$$

One species, *Thiobacillus denitrificans*, is able to utilize sulphur or thiosulphate under anaerobic conditions with the reduction of nitrate to molecular nitrogen [*see* Taylor and Hoare (1971)].

$$5Na_2S_2O_3 + 8NaNO_3 + H_2O \rightarrow 9Na_2SO_4 + H_2SO_4 + 4N_2$$

Other sulphur-containing inorganic compounds are also oxidized by the following reactions:

$$S_8 + 12O_2 + 8H_2O \rightarrow 8H_2SO_4$$

$$Na_2S_2O_6 + \tfrac{7}{2}O_2 + 3H_2O \rightarrow Na_2SO_4 + 3H_2SO_4$$

$$Na_2S_3O_6 + 2O_2 + 2H_2O \rightarrow Na_2SO_4 + 2H_2SO_4$$

$$NaSCN + 4O_2 + 4H_2O \rightarrow Na_2SO_4 + H_2SO_4 + CO_2 + 2NH_3$$

Thiobacillus species vary in their physiological features [*see* Trudinger (1967)] and their percentage guanine + cytosine content in DNA (*see* Table 12).

Table 11

Species	Culture medium pH	G + C content (%)
T. oxidans	1–5	51
T. neopolitans	3–8	57
T. ferrooxidans	2.5	57
T. thioparus	5–7	70
Thiobacillus A2	7–10	70
T. novellus		70

Beggiatoa, Sulpholobus, Thiobacterium and *Thiomicrospira* are also sulphur and sulphide oxidizers.

Iron oxidizers

Iron-oxidizing bacteria oxidize ferrous to ferric iron in the following way:

$$4FeSO_4 + O_2 + 2H_2SO_4 \rightarrow 2Fe_2(SO_4)_3 + H_2O$$

$$2Fe^{2+} + \tfrac{1}{2}O_2 + 2H^+ \rightarrow 2Fe^{3+} + H_2O$$

T. ferrooxidans uses energy from the oxidation of iron as well as sulphur using minerals such as pyrites and chalcopyrite $CuFeS_2$. Sulphides of other metals, including cadmium, copper, cobalt, lead, nickel and zinc are also oxidized by this bacterium. Other iron-oxidizing bacteria include *Sulpholobus* and *Leptospirillum* species.

Hydrogen oxidation

Hydrogen oxidation to water is performed by facultatively heterotrops under aerobic and occasionally under anaerobic conditions. The anaerobic reaction is as follows:

$$H_2 + \tfrac{1}{2}O_2 \rightarrow H_2O$$

The energy from the hydrogen oxidation is coupled to the fixation of carbon dioxide as the sole carbon source.

Methanobacterium oxidizes hydrogen to methane with the reduction of carbon dioxide.

$$4H_2 + CO_2 \rightarrow CH_4 + 2H_2O$$

Other inorganic oxidizers

Some microorganisms are capable of oxidizing copper, antimony, selenium, manganese and uranium eg.

$$UO_2 + \tfrac{1}{2}O_2 \rightarrow UO_3$$

Energetic balances

Table 12. Free energy efficiency of chemolithotrophic bacteria.

Organism	Reaction	$-\Delta F2$ (kcal)	Free energy efficiency (%)
Hydrogenomonas sp	$H_2 + \tfrac{1}{2}O_2 \rightarrow H_2O$	57.4	30
Thiobacillus thiooxidans	$S^0 + \tfrac{3}{2}O_2 + H_2O \rightarrow H_2SO_4$	118	max 50
T. denitrificans	$5Na_2S_2O_3 + 8KNO_3 + 2NaHCO_3 \rightarrow 6Na_2SO_4 + 2K_2SO_4 + 4N_2 + 2CO_2 + H_2O$	803	max 25
T. thiocyanooxidans	$NH_4CNS + 2O_2 + 2H_2O \rightarrow (NH_4)_2SO_4 + CO_2$	40	25
T. ferrooxidans	$Fe^{++} \rightarrow Fe^{+++} + e$	11.3	3
Nitrosomonas sp	$NH_4^+ + \tfrac{3}{2}O_2 \rightarrow NO_2^- + H_2O + 2H^+$	66.5	max 20

Table 13. Energetic and metabolic balances (*Desulfovibrio desulfuricans*).

Reaction	$\Delta G^{0\prime}$ (kcal mol^{-1})
Fermentation of pyruvate	
Pyruvate $+ \tfrac{1}{4} SO_4^{2-} \xrightarrow{(pH\ 7)}$ acetate $+ CO_2 + \tfrac{1}{4}(S^2 + HS^- + H_2S)$	-22.77
Fermentation of lactate	
Lactate $+ \tfrac{1}{2} SO_4^{2-} \xrightarrow{(pH\ 7)}$ acetate $+ CO_2 + \tfrac{1}{2}(S^{2-} + HS^- + H_2S)$	-24.08
Reaction steps	
Lactate $\rightarrow$ pyruvate $+ 2H^+ + 2e$	-8.78
Pyruvate $+ H_2O \rightarrow$ acetate $+ CO_2 + 2H^+ + 2e$	-32.24
$2H^+ + 2e + \tfrac{1}{4} SO_4^{2-} \rightarrow \tfrac{1}{4} S^{2-} + H_2O$	$+12.37$
$\tfrac{1}{4} S^{2-} \xrightarrow{(pH\ 7)} \tfrac{1}{4}(S^2 + HS^- + H_2S)$	-2.9

The mechanism of reduction of sulphate to sulphite by *Desulfovibrio desulfuricans* under an atmosphere of hydrogen involves the following stages.

1) $H_2 + 2\text{ cytochrome-}Fe^{3+} \xrightarrow{\text{hydrogenase}} 2\text{ cytochrome-}Fe^{2+} + 2H^+$
2) $SO_4^{2-} + ATP^- \; ATP\text{-} \xrightarrow{\text{sulphurglase}}$ adenosine 5′-phosphosulphate + pyrophosphate
3) Adenosine 5′-phosphosulphate + 2 cytochrome-Fe^{2+} + $\xrightarrow{\text{sulphate reductase}}$ AMP + $H/SO_3^- + H_2O + 2\text{ cytochrome-}Fe^{3+}$

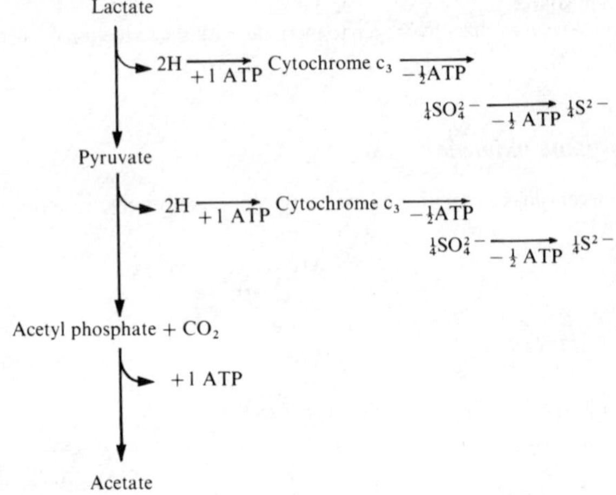

Figure 18. Proposed scheme for ATP production and utilization in *D. desulfuricans*.

Photosynthesis

Mass and energy balances

Cyclic Photophosphorylation (*see* Fig. 19) is

$$\text{Photons} + P_i + ADP \longrightarrow ATP + H_2O$$

while noncyclic Photophosphorylation is

$$\text{Photons} + 2P_i + 2ADP + H_2O + NADP_{ox} \longrightarrow NADP_{red} + \tfrac{1}{2}O_2 + 2H_2O + 2ATP$$

Formation of glucose in the dark phase of photosynthesis is

$$6NADP_{red} + 12H_2O + 12ATP + 6CO_2 \longrightarrow C_6H_{12}O_6 + 6NADP_{ox} + 12ADP + 12P_i$$

Efficiency of photosynthesis

It has been suggested that eight quanta of visible light are required to reduce each module of carbon dioxide, or a total of 48 quanta to form one molecule of glucose. One quantum of light in the red end of the visible spectrum (~ 700 nm) represents about 40 kcal (*see* Fig. 21).

$$6CO_2 + 6H_2O \longrightarrow C_6H_{12}O_6 + 6O_2 \qquad \Delta G^{0'} = +686\text{ kcal/mol}$$

$$\text{Efficiency} = \frac{686}{48 \times 40} \times 100 \simeq 36\%$$

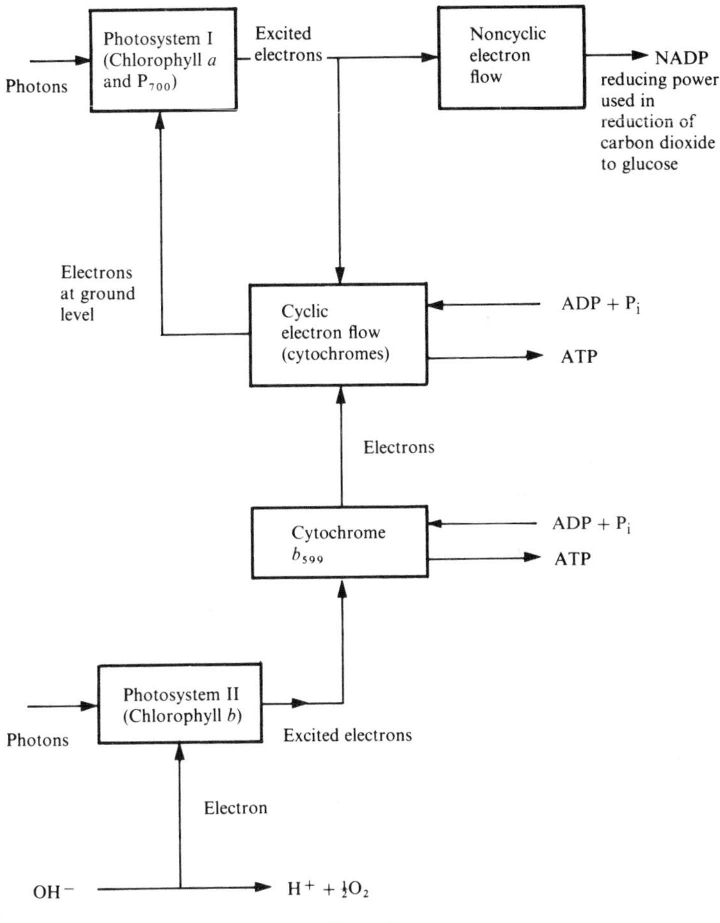

Figure 19

Absorption spectrum

Fig. 20 shows the electromagnetic radiation spectrum and Fig. 21 illustrates the energy content of photons over the wavelengths of visible light. Fig. 22 contains both the absorption spectrum of chlorophyll *a* and the photochemical efficiency over the range of wavelengths. Clearly, the 'red' end of the spectrum (about 680 mm) has both a high absorption and photochemical efficiency. Fig. 23 shows how supplementary light can be used to enhance the absorption of monochromatic light.

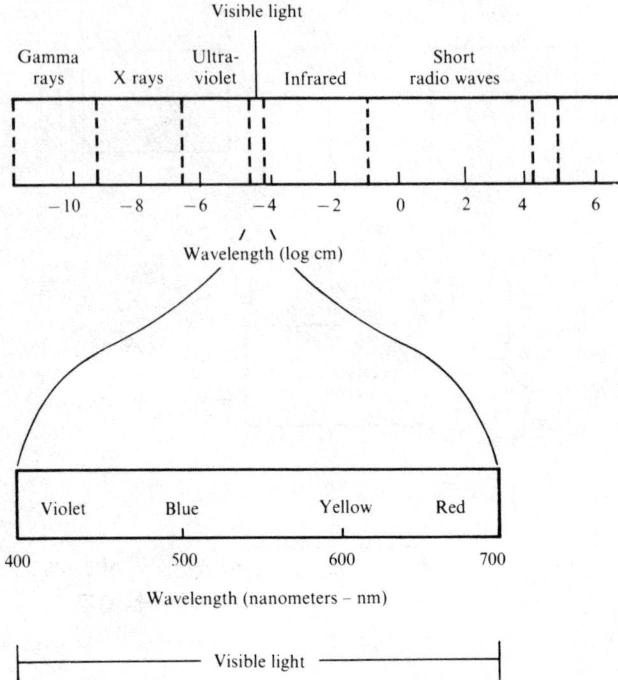

Figure 20. The spectrum of electromagnetic radiation.

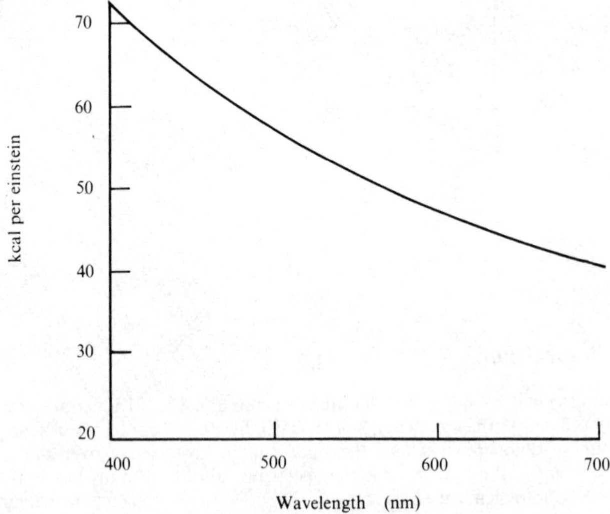

Figure 21. The energy content of photons.

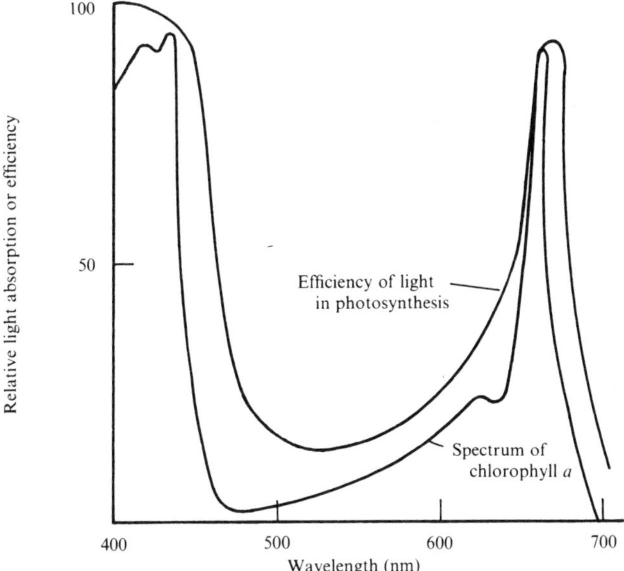

Figure 22. Absorption spectrum of chlorophyll *a*.

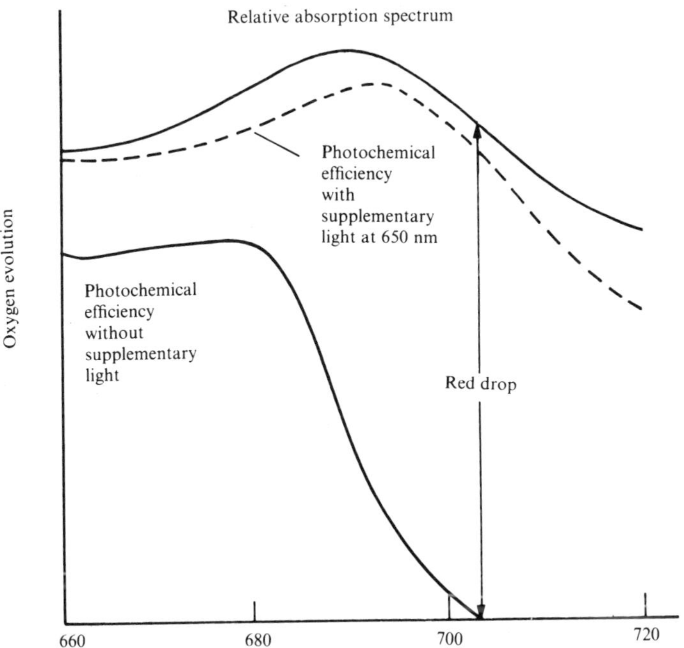

Figure 23. Red drop in *Chlorella*, a green alga. Above about 680 nm, the red region of the spectrum, the efficiency of monochromatic light in supporting oxygen evolution decreases greatly in relation to the absorption spectrum, which is largely due to chlorophyll. This deficit is called red drop. However, if supplementary light at 650 nm is added, the efficiency of the longer wavelengths is restored. Therefore two light-absorbing systems, one absorbing in the region 680 to 720 nm and the other at shorter wavelengths must cooperate to yield maximal rates of photosynthesis.

UTILIZATION OF ATP

Biosynthesis (Anabolism)

Table 14. Energy-rich compounds.

Characteristic linkage	General formula	General designation	Biochemical example	$\Delta G^{0'}$ (kcal mol^{-1})
$-\overset{\|}{C}-\underset{H}{N}\sim$	$R\overset{NH}{\overset{\|}{C}}-\underset{H}{N}\sim$Ⓟ	Guanidinium phosphate	Creatine phosphate	– 10.5
			Arginine phosphate	– 9.0
$-\overset{\|}{C}-O\sim$	$R\overset{CH_2}{\overset{\|}{C}}-O\sim$Ⓟ	Enolphosphate	Phosphoenol pyruvate	– 12.8
$-\overset{\|}{C}-O\sim$	$R\overset{O}{\overset{\|}{C}}-O\sim$Ⓟ	Acyl phosphate	Acetyl phosphate	– 10.1
$-\overset{\|}{\underset{HO}{P}}-O\sim$	$RO\overset{O}{\overset{\|}{\underset{HO}{P}}}-O\sim$Ⓟ	Pyrophosphates	Adenosine diphosphate	– 8.0 to – 12.0
$-\overset{\|}{C}\sim$	$R\overset{O}{\overset{\|}{C}}\sim SR'$	Acyl thioester	Acetyl-CoA	– 10.5

Table 15. Low-energy compounds.

General designation	Structure	Biochemical example	$\Delta G^{0'}$ (kcal mol^{-1})
Peptide bond (within protein)	$-\overset{O}{\overset{\|}{C}}-NH-$	—Tyrosylglycine—	– 0.5
Amides	$-\overset{O}{\overset{\|}{C}}-NH_2$	Glutamine	– 3.4
	O		
Esters, normal	$-\overset{O}{\overset{\|}{C}}-OR$	Ethyl acetate	– 1.8
Sugar phosphate	$>C-O-$Ⓟ	Glucose 6-phosphate	– 2.9
		Glycerol 1-phosphate	– 2.3

Tables 14 and 15 provide examples of the high- and low-energy compounds used in the transfer of energy during biosynthesis. Table 16 illustrates the energy requirements for the formation of cell macromolecules, while Tables 17, 18 and 19 show the rates and amounts of ATP required for cell growth.

Table 16. Chemical work of biosynthesis.

Macromolecule	Building block	Type of bond	$+\Delta G^{0\prime}$ per bond (kcal mol^{-1})
Protein	Amino acid 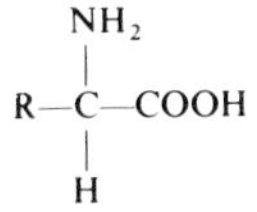 NH_2 R—C—COOH H	Peptide —C—NH— O 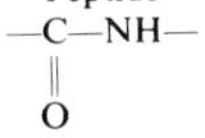	~ 5.0
Nucleic acid	Mononucleotide 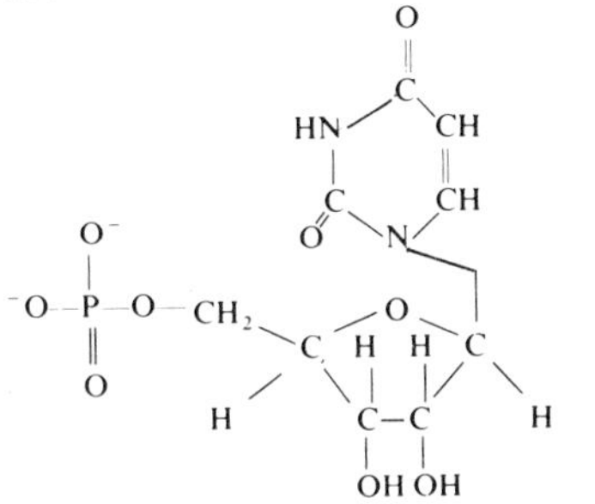 O C HN CH C CH O N O^- ^-O—P—O—CH_2 O C H H C O H C—C H OH OH	Phosphodiester —O—P—O— O O 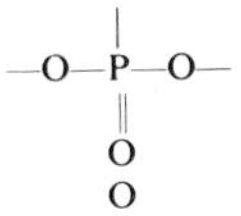	~ 5.0
Polysaccharide	Monosaccharide 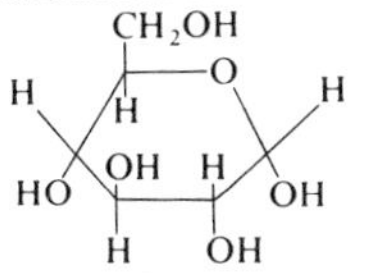 CH_2OH O H H H OH H HO OH H OH	Glycosidic C—O—C 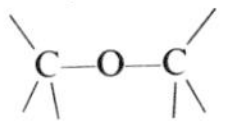	~ 3.0

Table 17. Biosynthetic capacities of a bacterial cell. A cell of *Escherichia coli* is about $1 \times 1 \times 3$ μm in size, has a volume of 2.25 μm^3, a total weight of 10×10^{-13} g and a dry weight of 2.5×10^{-13} g. The rates of biosynthesis given were averaged over a 20-min cell-division cycle.

	Chemical component				
	DNA	RNA	Protein	Lipids	Polysaccharides
% of dry weight	5	10	70	10	5
Approximate molecular weight	2×10^9	1×10^6	6×10^4	1×10^3	2×10^5
Number of molecules per cell	1	1.7×10^3	1.7×10^6	1.5×10^7	3.9×10^4
Number of molecules synthesized per second	0.00083	12.5	1400	12,500	32.5
Number of molecules of ATP required per second	6×10^4	7.5×10^4	2.12×10^6	8.75×10^4	6.5×10^4
% of total biosynthetic energy required	2.5	3.1	88.0	3.7	2.7

Table 18. Synthesis of microbial cells from preformed monomers. [From data of Mandelstam and McQuillen (1968)].

Polymer formed	Amount (g 100 g cells^{-1})	ATP required (mol/g synthesized^{-1})	Total ATP required (mol 100 g cells^{-1})
Mucopeptide	15	0.014	0.21
Protein	60	0.045[a]	2.7
Lipid	6	0.061[b]	0.37
RNA	15	0.017[c]	0.26
DNA	4	0.021[d]	0.08
Total			3.62

[a] Average molecular weight of amino acid taken as 110.
[b] Lipids assumed to contain C_{20} fatty acids.
[c] Three ATP required to convert nucleic acid base to mononucleotide with an average molecular weight of 300.
[d] Four ATP required to convert nucleic acid base to deoxymononucleotide with an average molecular weight of 280.

Table 19. Energy required for synthesis of cell monomers from hexose.

Monomer synthesized	Amount (mol × 10^{-4} g^{-1})	ATP change in synthesis (mols × 10^{-4} g cells produced^{-1})
Alanine	4.54	+4.54
Arginine	2.52	+12.60
Aspartate	2.01	+2.01
Asparagine	1.01	−1.01
Cysteine[a]	1.01	−4.04
Glutamate[b]	3.53	+17.65
Glutamine[b]	2.01	+8.04
Glycine	4.03	−4.03
Histidine	0.50	+2.5
Isoleucine	2.52	0
Leucine	4.03	+12.09
Lysine	4.03	0
Methionine	2.01	0
Phenylalanine	1.51	−3.02
Proline	2.52	+10.08
Serine	3.02	−3.02
Threonine	2.52	−2.52
Tryptophan	0.50	−2.0
Tyrosine	1.01	−2.02
Valine	3.02	+6.04
Hexose	10.26	0
Ribose	4.47	−4.47
Deoxyribose	0.96	−0.96
Thymine	0.24	−0.72
AMP	1.40	−16.8
GMP	1.40	−18.2
CMP	1.40	−7.0
UMP	1.15	−2.3
Glycerol	1.40	−1.4
Fatty acid[c]	2.80	+5.6
Total		+7.64

[a] Formed from glucose and sulphate.
[b] Assumed to be formed from glucose via the tricarboxylic acid cycle.
[c] As acetate.

Transport

Na^+, K^+ – ATPase pumps Na^+ out of the cell to create an inwardly directed Na^+ gradient at the expense of ATP. The inwardly directed Na^+ gradient pulls glucose into the cell by the action of a passive carrier that binds both Na^+ and glucose on the outside and transfers them into the cell.

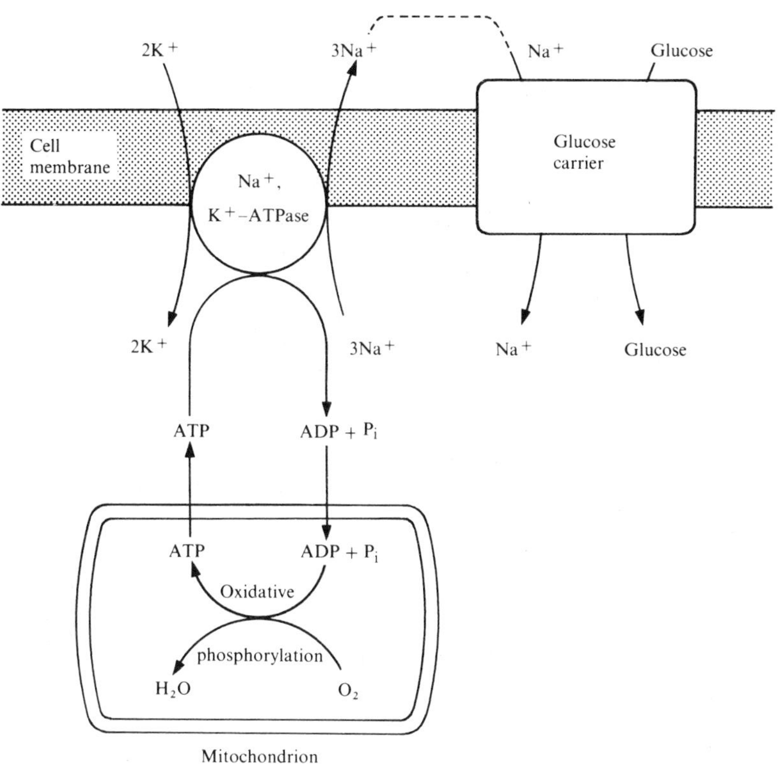

Figure 24

Mobility of ATP within Cells

ATP and ADP are small and highly soluble molecules which diffuse freely in solution and require no specialized mechanisms for their transport within the cytoplasma.

In eucaryotic cells, ATP is generated within mitochondria and used outside them. The mitochondrial membrane represents a barrier to ATP diffusion which is circumvented by a membrane transport function. Fluxes of ATP out of, and ADP into, mitochondria are linked so that uptake of ADP can only proceed if ATP is available and vice versa as shown in Fig. 25.

This mechanism, besides promoting transport across mitochondrial membranes, ensures that efflux of ATP is linked to its cellular utilization and consequent formation of ADP.

The carrier protein has a high affinity for ADP on the cytoplasmic side and a higher affinity for ATP on the mitochondrial matrix side, thus producing a 10-fold greater value for phosphorylation potential in the cytoplasm than in the mitochondrion.

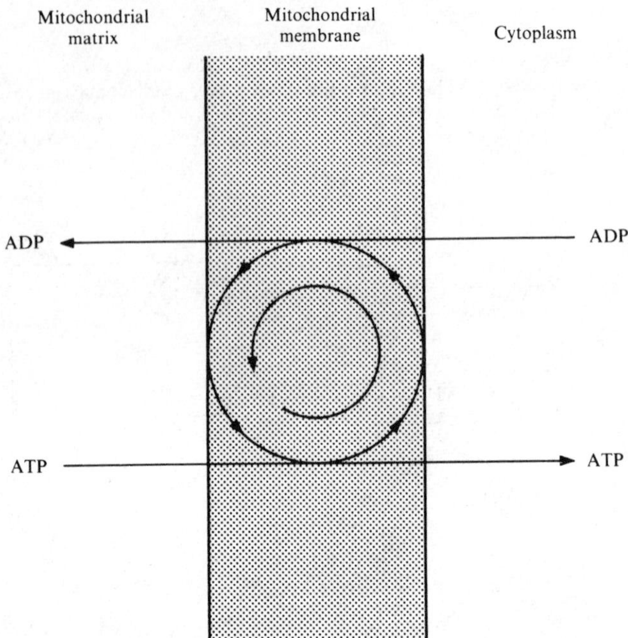

Figure 25. Linked countertransport of ADP and ATP in mitochondrial membranes.

Regulation of ATP Production

ADP and orthophosphate are essential precursors for ATP synthesis. The total number of ATP plus ADP molecules in a cell is constant so that a high ATP concentration means a low ADP concentration and vice versa. If energy is being used at a fast rate, ATP reserves fall and ADP accumulates. As ADP is normally the rate-limiting substrate in oxidative phosphorylation, ATP synthesis is consequently accelerated when ADP concentrations rise. Control of respiratory rate by ADP availability is illustrated in Fig. 26. This immediate and simple feedback control of ATP synthesis ensures that the turnover of ATP increases proportionately with energy demands, while the ATP/ADP ratio remains almost constant.

Accumulation of ATP also inhibits early reactions in the respiratory pathway, providing for a slowing of the whole reaction sequence and preventing the accumulation of reaction intermediates as would occur if only the latter stages (i.e. phosphorylation of ADP) were inhibited by a high cellular ATP/ADP ratio.

ATP availability in a cell is expressed as energy charge, defined as

$$\text{Energy charge} = \frac{[\text{ADP}] + \frac{1}{2}[\text{ADP}]}{[\text{ATP}] + [\text{ADP}] + [\text{AMP}]}$$

This is a measure of pyrophosphate bonds available within the cell and can vary from zero (i.e. all adenylates present as AMP) to one (all adenylates present as ATP). As mass action effects would inhibit ATP-generating pathways and stimulate ATP-utilizing pathways at high energy charges as shown in Fig. 27, the energy charge is stabilized by these feedback effects on ATP generation and utilization at a value of about 0.9. In practice, the energy charge of most cells is in the range 0.80 to 0.95.

An alternative index of ATP synthesis is the phosphorylation potential, which is defined as

$$\text{Phosphorylation potential} = \frac{[\text{ATP}]}{[\text{ADP}][\text{P}_i]}$$

This is a more direct measure of energy available from ATP in the cell.

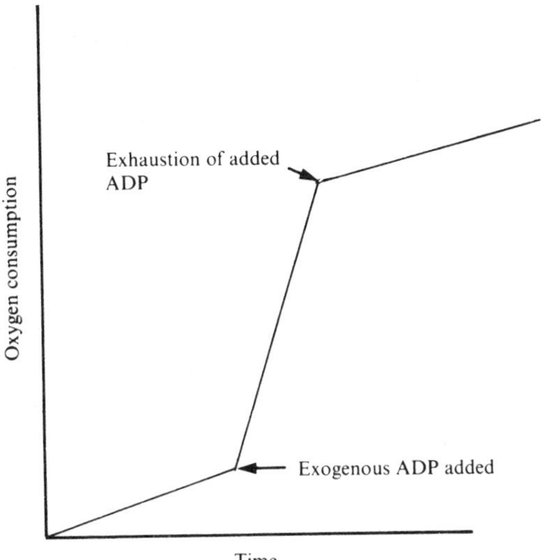

Figure 26. Effect of ADP addition on the respiratory rate of a cell homogenate.

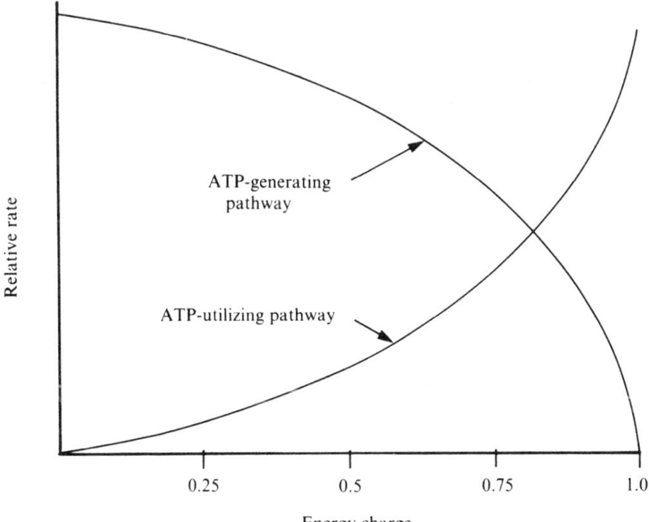

Figure 27. Effect of energy charge on reaction involving ATP.

REFERENCES

Darlington, W.A. (1964) Aerobic Hydrocarbon Fermentation. A Practical Evaluation. *Biotechnol. Bioengng*, **6,** 241.

Kelly, D.P. (1971) Concepts of Lithotrophic Bacteria and their Organic Metabolism. *Ann. Rev. Microbiol.*, **25**, 177.

Taylor, B.F. and Hoare, D.S. (1971) *Thiobacillus denitrificans* as an Obligate Chemolithotroph. II. Cell Suspensions and Enzyme Studies. *Arch. Mikrobiol.*, **80,** 262.

Trudinger, P.S. (1967) The Metabolism of Inorganic Sulphur Compounds. *Rev. Pure Appl. Chem.*, **17,** 1.

Van der Kroon, G.T.M. (1969) The Influence of Suspended Solids on the Rate of Oxygen Transfer in Aqueous Solutions, in *Advances in Water Pollution Research*, p. 219 (Pergamon Press, Oxford).

APPENDIX: THE LAWS OF THERMODYNAMICS

The First Law of Thermodynamics

The first law of thermodynamics is concerned with the conservation of energy. It states that in any chemical or physical process the total energy of the system plus surroundings, that is the total energy of the universe, remains constant. The first law also implies that there is a quantitative correspondence between different kinds of energy.

It can be formulated as

$$\Delta E = Q - W$$

where ΔE = internal energy change between two states
Q = heat added to the system
W = work done by the system.

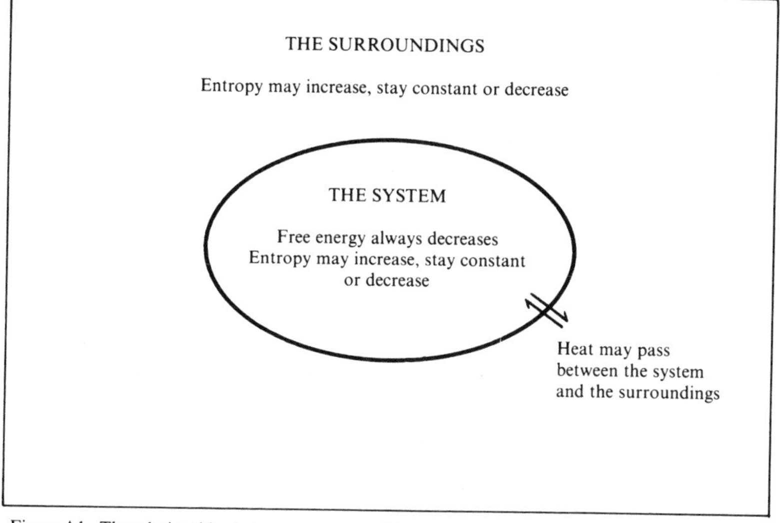

Figure A1. The relationships between entropy and free energy changes in the system and surroundings in processes occurring at constant temperature and pressure. The universe = system + surroundings.

Entropy and the Second Law of Thermodynamics

Spontaneous physical or chemical changes have a direction that cannot be explained by the first law. It also does not provide information as to how efficiently energy transforms into work. These are given by the second law through the recognition of a state or condition of matter and energy called *entropy*, which may be simply defined as randomness or disorder.

The second law of thermodynamics states that all physical and chemical processes proceed in such a direction that the randomness or entropy of the universe increases to the maximum possible; at this point there is equilibrium.

The change in entropy of a reversible process at constant temperature is given by

$$\Delta S_{rev} = \frac{Q_{rev}}{T}$$

where ΔS_{rev} = entropy change

Q_{rev} = heat gained or lost

T = constant temperature of the process

The Third Law of Thermodynamics

According to the third law of thermodynamics the entropy of a perfect crystal of any element or compound at absolute zero temperature is zero.

Free Energy

Entropy or entropy changes are not always easily measured or calculated. Consequently, it is convenient to find other functions of state that can act as indicators of equilibrium and irreversibility and which are related to real measurable quantities. There are a large number of functions of state which can be generated from the internal energy by adding and subtracting variables. The choice between them is dictated by the types of constraints imposed on the system.

For biological transformations, temperature and pressure can be assumed constant. Thus a convenient function of state is the free energy, which can be expressed as

$$\Delta G = \Delta E - T\Delta S$$

at constant temperature, pressure and volume. Free energy has the following properties.

1) In a spontaneous isolated process at constant temperature and pressure, free energy decreases.
2) At equilibrium, there is no change in free energy and it has reached its minimum value.
3) The decrease in the free energy of a system equals the maximum amount of work the system can theoretically perform on the environment. This is reversible work. In actual processes, which are irreversible, the work is much less than this amount.
4) Change in free energy indicates the direction a process will follow, but it does not give any information as to how long it would take to reach the equilibrium state.

Free Energy Change and Work

Since, for a biological system

$$\Delta E = \Delta G + T\Delta S,$$

the free energy change ΔG may be defined as that fraction of the total energy change of a system that is available to do work as the system proceeds to the conditions of equilibrium.

In order to express the free energy change explicitly, it is necessary to define several types of work.

1) Pressure–volume work

$$W = P\Delta V$$

2) Force–distance work

$$W_{\text{force}} = F\Delta\ell$$

3) Electrical work

$$W_{\text{elec}} = -\Psi\Delta g$$

4) Mass transport work

$$W_{\text{mass}} = -\mu\Delta n$$

where P = pressure

ΔV = change in volume

F = force

$\Delta\ell$ = distance

Ψ = electrical potential

Δg = change in electrical charge

μ = chemical potential

Δn = change in mass

At constant temperature and pressure, for a reversible process

$$\Delta G = \mu\Delta n + \Psi\Delta g + F\Delta\ell$$

Functions of State

Thermodynamics distinguishes between two types of properties.
1) Those which do not depend on the past history or path followed in reaching a given state are called the functions of state. Enthalpy, internal energy, entropy and free energy are examples of functions of state.
2) Work and heat on the other hand are not state properties; they are dependent on the path followed.

A state function represents an instantaneous property of a system and always has a value. Work and heat appear only after passage of time during which changes are brought about in a system by a process. Thus, when integrated, the differentials of state functions give a finite change and those of heat and work yield a finite quantity.

Standard state

The absolute values of functions of states are not known. Therefore, some arbitrary base must be selected before the numerical magnitudes can be calculated. The conditions at this datum point are called the standard state. For free energy change of reactions in aqueous solution, the standard state is by convention defined as concentrations of 1.0 m, a temperature of 25°C (or 298 K) and a pressure of 1.0 atm.

Difference between ΔG^0 and ΔG

The standard free energy change ΔG^0 is a constant for any given chemical reaction at a given temperature. Whereas, free energy change ΔG of a chemical reaction varies with the concentrations of reactants and products.

It is the free energy change ΔG that determines whether a chemical reaction will proceed in the direction 'written'. Even if the standard free energy change ΔG^0 of a reaction is positive, the reaction can still go forward as written, provided the concentrations of the reactants and products are such that free energy change ΔG is negative.

Calculation of Free Energy Change of Chemical Reactions

Free energy change and equilibrium constant

For the generalized reaction

$$aA + bB \rightleftharpoons cC + dD$$

the free energy change ΔG at constant temperature and pressure is given by

$$\Delta G = \Delta G^0 + RT \ln \frac{[C]^c [D]^d}{[A]^a [B]^b}$$

where brackets denote molal concentrations. R is the universal gas constant, T is the absolute temperature and ΔG^0 is the standard free energy change of the reaction. ΔG^0 can be calculated from the equilibrium constant, K_{eq}, at standard conditions as

$$\Delta G^0 = - -RT \ln K_{eq}$$

Values are presented in Table A.1. The equilibrium constant K_{eq} changes with temperature according to the following equation

$$\frac{d(\ln K)}{dT} = \frac{\Delta H^0}{RT^2}$$

where ΔH^0 is the standard heat of reaction. If ΔH^0 is assumed to be constant with respect to temperature, integration yields

$$\ln \frac{K_2}{K_1} = -\frac{\Delta H^0}{R}\left(\frac{1}{T_2} - \frac{1}{T_1}\right)$$

Table A1. The numerical relationship between the equilibrium constant and $\Delta G^{0\prime}$ at 25°C and pH 7.0.

K'_{eq}	$\Delta G^{0\prime}$ (kcal mol^{-1})
0.001	+4.09
0.01	+2.73
0.1	+1.36
1.0	0
10.0	−1.36
100.0	−2.73
1000.	−4.09

Calculation of standard free energy change from standard free energies of formation

The standard free energy of formation ΔG_f^0 is defined as the decrease in free energy as 1 mole of the compound of interest is formed from its elements, each in its standard state and in the proper stoichiometric ratio. The standard free energy change of a reaction is given by

$$\Delta G^0 = \sum(\Delta G_f^0)_{\text{products}} - \sum(\Delta G_f^0)_{\text{reactants}}$$

taking into account the actual stoichiometry of the reaction.

Difference between free energy change and heat of reaction

At constant temperature and pressure

$$\Delta G = \Delta H - T\Delta S$$

ΔH denotes the enthalpy change or heat of reaction. Under physiological conditions, there is little difference between the free energy change and heat of reaction.

CHAPTER 3

STOICHIOMETRIC ASPECTS OF MICROBIAL METABOLISM

GLOSSARY

Assimilation Absorption and formation of simple nutrients into the complex constituents of the organism.

Available electrons (av e^-) Those electrons in a substrate that are not involved in orbitals with oxygen and that can be considered as 'available' for (1) transfer of oxygen or (2) synthetic reductive reactions. The number of available electrons per mole of substrate is 4 x TOD, where the number of electrons required to reduce 1 molecule of oxygen is 4.

Biological oxygen demand (BOD) Oxygen consumed by microorganisms during incubation of a sample for 5 days at 20°C.

Biomass Microorganisms produced during fermentation.

Chemostat Fermenter maintaining constant chemical conditions.

Chemical oxygen demand (COD) Oxygen required for the oxidation of a substance using a strong oxidizer at elevated temperature.

Oxidative phosphorylation Linked process of substrate oxidation and synthesis of ATP.

Phosphate/oxygen ratio (P/O ratio) The ratio of inorganic phosphate esterified to the amount of oxygen consumed [i.e. g mol P (g atom O)$^{-1}$].

Proton/oxygen ratio (H^+/O ratio) The ratio of protons released to the amount of oxygen consumed [i.e. g ev H^+ (g atom O)$^{-1}$].

Respiratory quotient (RQ) The ratio of carbon dioxide produced to the amount of oxygen consumed [i.e. g mol CO_2 (g mol O_2)$^{-1}$].

Total carbon The total amount of carbon in a chemical compound.

Total organic carbon (TOC) The total amount of organic carbon in a mixture.

Total oxygen demand (TOD) Oxygen consumed in the combustion of all oxygen-demanding matter by oxygen gas at about 900°C.

IMPORTANCE OF YIELD AND YIELD COEFFICIENTS

Substrates taken up from the medium by microorganisms are incorporated into their cellular material and/or used to provide the energy necessary for biosynthesis and the maintenance of the microorganism.

When a substrate is only incorporated into the biomass (i.e. assimilated), the amount of biomass formed can be estimated from the stoichiometry. If, however, a substrate is used to provide energy for metabolism, the efficiency with which the microorganisms utilize this energy becomes the factor determining the amount of growth.

The yields of biomass and/or other microbial products have significant implications on several aspects of industrial microbiology. Figure 1 illustrates some of the many ways in which both operating and capital costs can be affected.

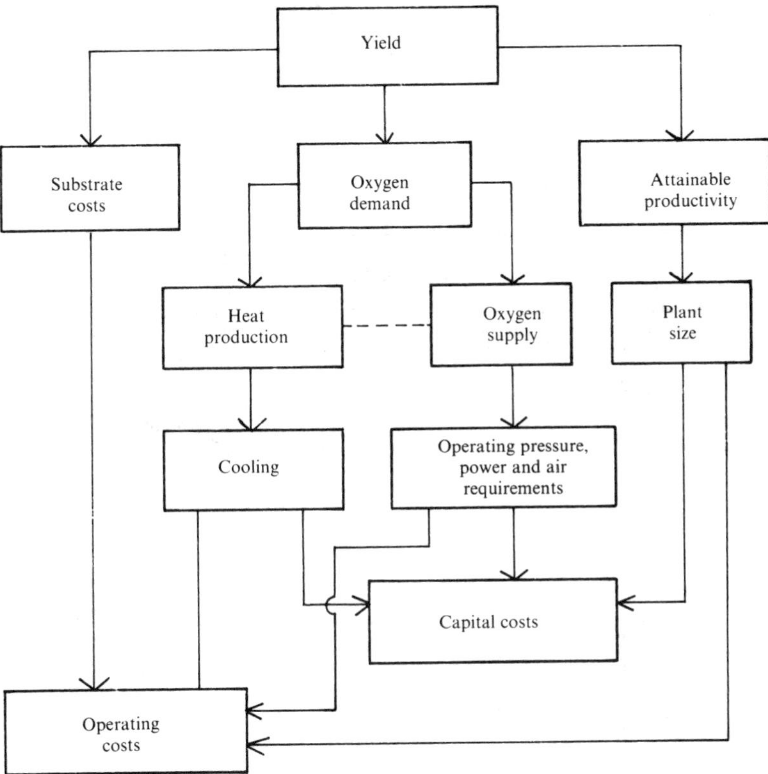

Figure 1. Schematic representation of the influence of yield on costs.

RELATIONSHIP BETWEEN AMOUNT OF GROWTH AND SUBSTRATE UTILIZATION

As demonstrated by Monod (1941, 1949), there is, as a general approximation, a linear relationship between the level of growth and the amount of substrate utilized. This relationship is expressed quantitatively in terms of various yield coefficients for growth (see Definition and Evolution of Various Yield Coefficients for Growth and Product Formation, pp.134–141). Fig. 2–9 are examples of this relationship.

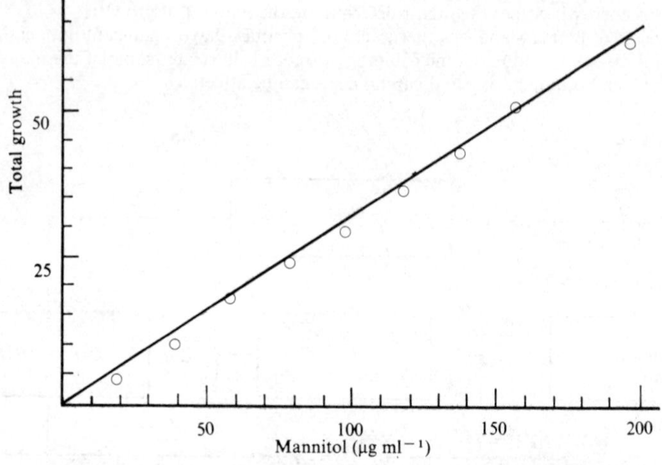

Figure 2. Total growth of *Escherichia coli* in synthetic medium with the organic source (mannitol) as the limiting factor. Ordinates: arbitrary units, where one unit is equivalent to 0.8 μg dry weight ml^{-1} [Monod (1949)].

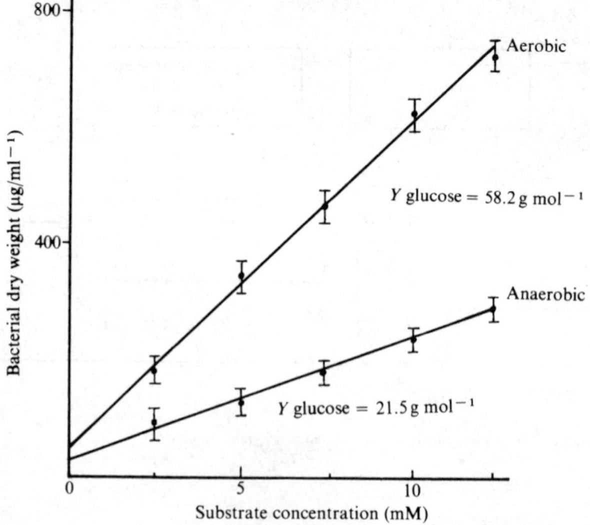

Figure 3. Aerobic and anaerobic growth yields of *Streptococcus faecalis* with glucose as substrate [Smalley *et al.* (1968)]. Molar growth yields were determined according to the method of Bauchop and Elsden (1960). Vertical bars represent the standard deviation of six determinations.

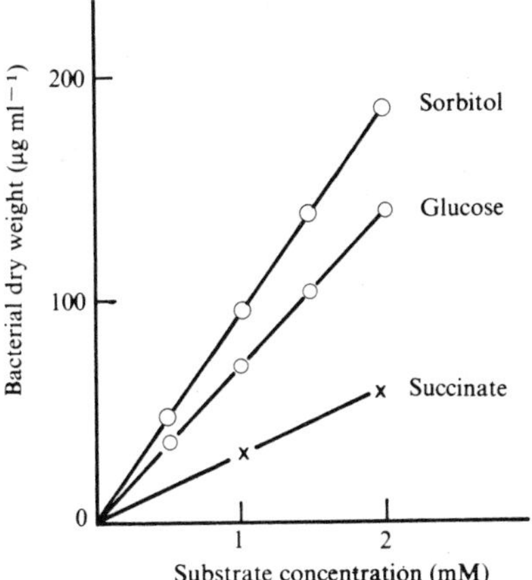

Figure 4. Growth yields with *Aerobacter aerogenes* growing aerobically with the indicated substrates as the sole carbon source in a minimal medium [Hadjipetrou *et al.* (1964)].

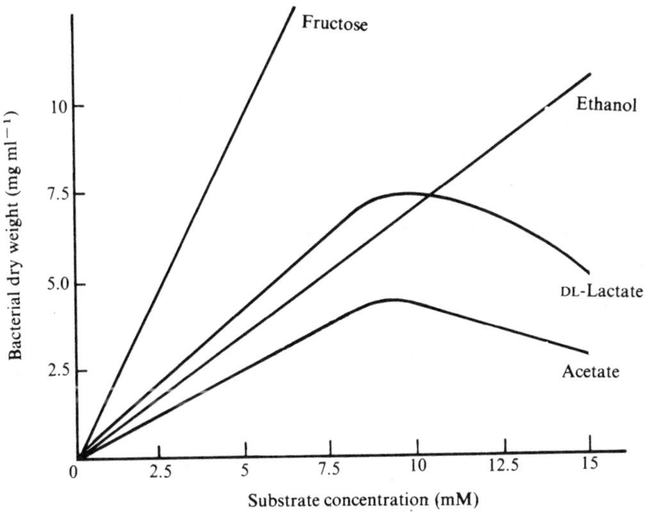

Figure 5. Aerobic growth yield of *Gluconobacter liquefaciens* for growth in synthetic medium with varying amounts of several carbon sources [Stouthamer (1962)].

In the case of *Sacch. cerevisiae* (*see* Fig. 8), it was found that the organism could grow in the absence of oxygen only when the medium was reinforced with ergosterol. The results for *Strep. faecalis* suggest that there was always some growth, even in the absence of added glucose, and indeed experiments at low concentration did lead to growth but in quantities difficult to measure.

Figs. 3, 8 and 9 suggest that the linear approximation of Monod (1949) may break down at low substrate concentrations.

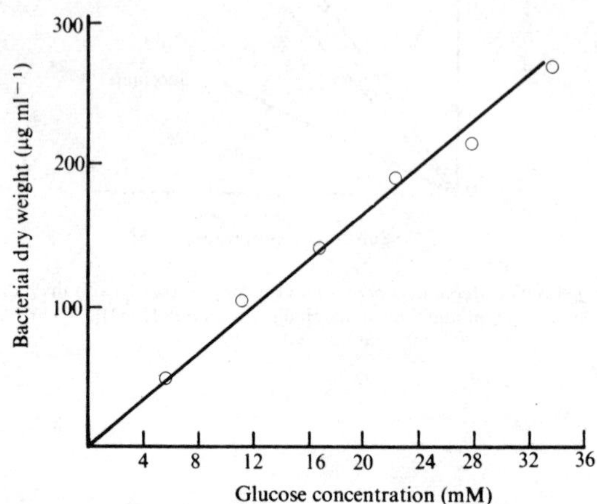

Figure 6. Anaerobic growth of *Pseudomonas lindneri* on glucose, incubated at 30°C. Growth was measured turbidimetrically and the dry weight read from a standard curve relating optical density to dry weight [Bauchop and Elsden (1960)].

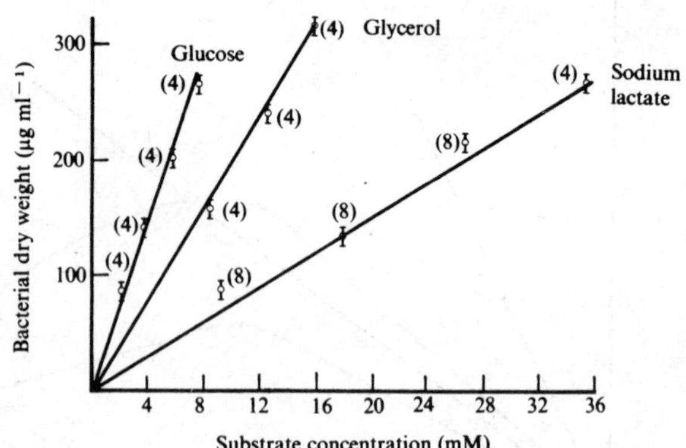

Figure 7. Anaerobic growth of *Propionibacterium pentosaceum* on glucose, glycerol and lactate, incubated at 30°C. At the end of growth, the microorganism was harvested, washed and dried at 100°C to constant weight. Each point is the mean of the number of estimations shown in parentheses; the standard deviations from the mean are represented by the heights of the vertical bars [Bauchop and Elsden (1960)].

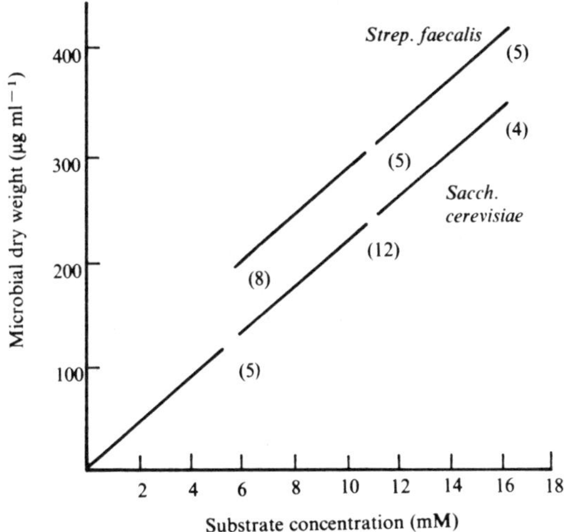

Figure 8. Anaerobic growth of *Streptococcus faecalis* and *Saccharomyces cerevisiae* on glucose. *Strep. faecalis* was grown on partially defined medium and incubated at 37°C; *Sacch. cerevisiae* grown on a partially defined medium and incubated at 30°C. At the end of growth, cells were harvested, washed and dried at 100°C to constant weight. Each point is the mean of the number of estimations shown in parentheses; the standard deviations from the mean are represented by the heights of the vertical bars [Bauchop and Elsden (1960)].

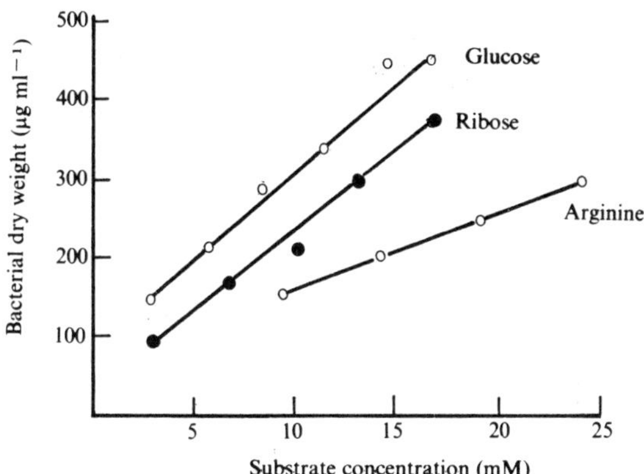

Figure 9. Anaerobic growth of *Strep. faecalis* incubated at 37°C. The bacterium was grown on complex medium with glucose, ribose or arginine as the energy source. Growth was measured turbidimetrically and dry weight read from a standard curve relating optical density to dry weight [Bauchop and Elsden (1960)].

COMPOSITION OF VARIOUS MICROORGANISMS

A knowledge of the composition of microorganisms (expressed elementally or as chemical compounds) is necessary for the stoichiometric calculation of yields. It is also needed for the theoretical estimation of the amount of ATP required for the biosynthesis of a given amount of biomass. The composition of cells varies with growth conditions and substrates used.

Table 1. Elemental composition of microorganisms.

Microorganism	Limiting nutrient	D^a (h^{-1})	Composition (% by wt) C	H	N	O	P	S	Ash	Empirical chemical formula	Formula 'molecular' weight	Reference
Bacteria			53.0	7.3	12.0	19.0			8	$CH_{1.666}N_{0.20}O_{0.27}$	20.7	Abbott and Clamen (1973)
Bacteria			4.7	4.9	13.7	31.3				$CH_2N_{0.25}O_{0.5}$	25.5	van Dijken and Harder (1975)
Aerobacter aerogenes			48.7	7.3	13.9	21.1			8.9	$CH_{1.78}N_{0.24}O_{0.33}$	22.5	Elsworth *et al.* (1968)
Klebsiella aerogenes	Glycerol	0.1	50.6	7.3	13.0	29.0				$CH_{1.74}N_{0.22}O_{0.43}$	23.7	Herbert (1976)
K. aerogenes	Glycerol	0.85	50.1	7.3	14.0	28.7				$CH_{1.73}N_{0.24}O_{0.43}$	24.0	Herbert (1976)
Yeast			47.0	6.5	7.5	31.0			8	$CH_{1.66}N_{0.13}O_{0.49}$	23.5	Darlington (1964), Humphrey (1968)
Yeast			50.3	7.4	8.8	33.5				$CH_{1.75}N_{0.15}O_{0.5}$	23.9	van Dijken and Harder (1975)
Yeast			44.7	6.2	8.5	31.2	1.08	0.6		$CH_{1.64}N_{0.16}O_{0.52}P_{0.01}S_{0.005}$	26.9	Herbert (1976)
Candida utilis	Glucose	0.08	50.0	7.6	11.1	31.3				$CH_{1.82}N_{0.19}O_{0.47}$	24.0	Herbert (1976)
C. utilis	Glucose	0.45	46.9	7.2	10.9	35.0				$CH_{1.84}N_{0.2}O_{0.56}$	25.6	Herbert (1976)
C. utilis	Ethanol	0.06	50.3	7.7	11.0	30.8				$CH_{1.82}N_{0.19}O_{0.46}$	23.9	Herbert (1976)
C. utilis	Ethanol	0.43	47.2	7.3	11.0	34.6				$CH_{1.84}N_{0.2}O_{0.55}$	25.5	Herbert (1976)

[a] Dilution rate in continuous fermentation, generally equal to the specific growth rate μ, and given by the ratio of volumetric throughput to liquid volume in the fermenter.

Table 2. Empirical chemical formula for activated sludge[a] [Servizi and Bogan (1963)].

Empirical formula	Carbon content of formula (%)	Ash content of dry sludge (%)	Carbon content of dry sludge (%)
$C_5H_7NO_2$	53.1	8.6	48.4
$C_7H_{10}NO_3$	53.9	1.3	53.2
$C_5H_8NO_2$	52.6		
$C_9H_{16}NO_5$	49.6		

[a] Biomass in aerated, agitated biological water treatment process.

Table 3. Composition of microbial cells [from Morowitz (1968)].

Compound	Amount (μmol g^{-1})	Compound	Amount (μmol g^{-1})
Alanine	454	Serine	302
Arginine	252	Threonine	252
Aspartate	201	Tryptophan	50
Asparagine	101	Tyrosine	101
Cysteine	101	Valine	302
Glutamate	353	Hexose	1026
Glutamine	201	Ribose	447
Glycine	403	Deoxyribose	96
Histidine	50	Thymine	24
Isoleucine	252	AMP	140
Leucine	403	GMP	140
Lysine	403	CMP	140
Methionine	201	UMP	115
Phenylalanine	151	Glycerol	140
Proline	252	Fatty acid	280

The principal molecules in *Escherichia coli* have been summarized by Watson (1970) (*see* Table 24, Chapter 1).

Table 5. Cell composition expressed as a percentage of dry cell substance as a function of growth rate in batch culture.

Microorganism	Component	Growth rate, μ^a (h^{-1}) 0.139	0.415	0.83	1.66
Aerobacter aerogenes	DNA	4.5	3.8	2.7	
Bacillus megaterium	DNA	3.3	3.0	2.5	
Salmonella typhimurium	DNA	4.0	3.7	3.5	3.0
Torula utilis	DNA	1.3	1.0		
A. aerogenes	Total RNA	7.8	10.4	16.2	
A. aerogenes NTCC 418	Total RNA	9.5	14.1	18.3	
B. megaterium	Total RNA	7.5	9.5	14.0	
S. typhimurium	Total RNA	12.0	18.0	22.0	31.0
T. utilis	Total RNA	8.8	10.9		
S. typhimurium	Ribosomal RNA	3.5	9.0	13.5	25.0
S. typhimurium	tRNA	8.5	9.0	8.5	6.0
A. aerogenes	Protein	62	60	56	
S. typhimurium	Protein	83	78	74	67
T. utilis	Protein	55	61		

[a] Specific growth rate during exponential growth phase, i.e. mass of organisms produced per unit mass of organisms present per unit time.

Table 4. Molecular composition of a typical bacterium.

Component	Weight ($\times 10^{13}$ g)	Percentage of total weight	Molecular weight	Molecules per cell[a]
Entire cell	15	100		
Water	12	80	18	4×10^{10}
Dry weight	3	20		
Protein				
Ribosomal	0.22	1.5	4×10^4	3.3×10^5
Nonribosomal	1.5	10	5×10^4	1.8×10^6
RNA				
Ribosomal, 16 S	0.15	1	6×10^5	1.5×10^4
Ribosomal, 23 S	0.30	2	1.2×10^6	1.5×10^4
t-RNA	0.15	1	2.5×10^4	3.5×10^5
m-RNA	0.15	1	10^6	9×10^5
DNA	0.15	1	4.5×10^9	2
Polysaccharides	0.15	1	1.8×10^2	5×10^7
Lipids	0.15	1	10^3	9×10^6
Small molecules	0.08	0.5	4×10^2	1.2×10^7

[a] Calculated by dividing the weight of a component in grams by its molecular weight to give the number of gram molecular weights (moles) of the component. Since there are approximately 6×10^{23} molecules in one mole of any compound, the number of moles multiplied by 6×10^{23} gives the number of molecules. For example, the number of moles of water in 12×10^{-13} gram of water is $12 \times 10^{-13}/18$ or about 7×10^{-14}; the number of molecules of water per cell is then $7 \times 10^{-4} \times 6 \times 10^{23}$ or about 4×10^{10}.

Table 6. Cell composition of *Klebsiella aerogenes* and *Candida utilis* grown in continuous culture[a].

Microorganism	Limiting nutrient	Dilution rate, D (h^{-1})	Composition of dry ash-free cells (% by wt): Protein	Carbohydrate	RNA	DNA
K. aerogenes NCTC418	Glycerol	0.1	74.1	1.5	11.2	4.4
K. aerogenes NCTC418	Glycerol	0.85	72.0	1.9	21.6	2.5
K. aerogenes NCTC418	Ammonia	0.1	65.6	18.6	8.2	3.1
K. aerogenes NCTC418	Ammonia	0.85	73.4	1.8	19.9	3.2
C. utilis NCYC321	Glucose	0.05	58.8	20.9		
C. utilis NCYC321	Glucose	0.45	55.8	19.5		
C. utilis NCYC321	Ammonia	0.05	43.2	36.2		
C. utilis NCYC321	Ammonia	0.45	56.7	19.0		

[a] Microorganisms were grown on a glycerol–ammonia–salts or a glucose–ammonia–salts medium. *K. aerogenes* was incubated at 37°C, pH 7.0. *C. utilis* was incubated at 30°C, pH 5.5. Total carbohydrate (anthrone) was expressed as equivalent weight of glycogen. Ash content of *K. aerogenes* was 3.6%. Ash content of *C. utilis* was 7.0%.

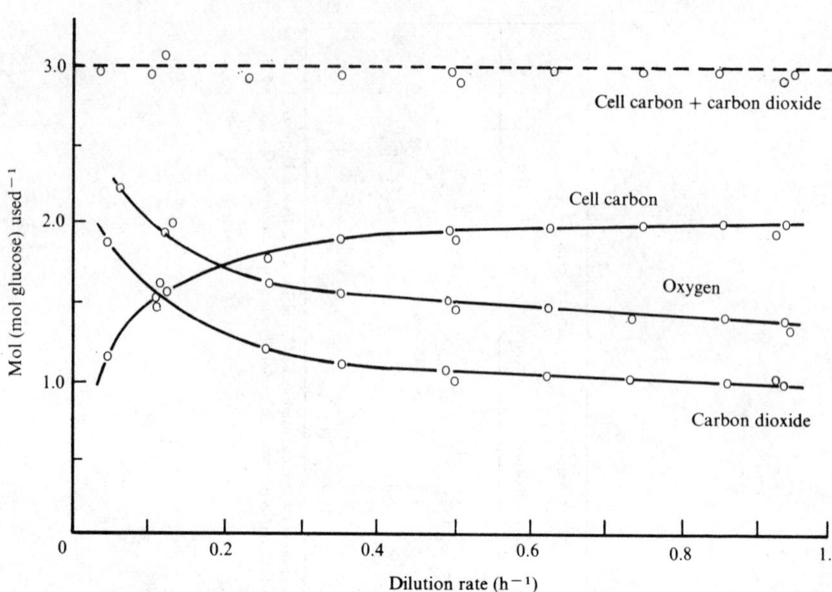

Figure 10. Growth of *K. aerogenes* in continuous culture with glycerol as limiting nutrient. Cell carbon and carbon dioxide formed and oxygen used are expressed as mol per mol glycerol used. Temperature = 37°C, pH = 7.0 [Herbert (1976)].

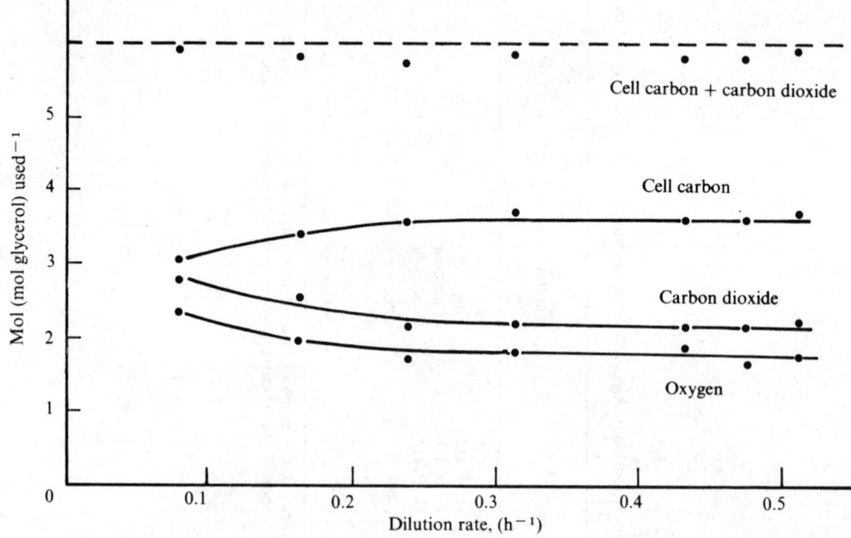

Figure 11. Growth of *C. utilis* in continuous culture with glucose as limiting nutrient. Cell carbon and carbon dioxide formed and oxygen are expressed as mol per mol glucose used. Temperature = 30°C, pH = 5.5 [Herbert (1976)].

Table 7. Overall composition of single-cell proteins and other vegetable proteins.

Type of protein	Substrate	Protein (%)	Fat (%)	Carbohydrate (%)	Ash (%)	Cube fibre (%)
Bacterial	Carbohydrate	40–80	1–30	10–30	1–4	
Bacterial	Methane	59.2	6			
Bacterial	*n*-Alkanes	62–73	10–15	10	6–12	
Bacterial	*n*-Alkanes	65	8.1		6	
Bacterial	Gas oil	70.5	0.45		7.9	
Baker's yeast	Carbohydrate	40–50	1–2	32–40	6–10	< 10
Torula yeast	Sulphite liquor	47	4.8	32	9.1	0.82
Yeast	*n*-Alkanes	54	10	26	10	
Yeast	*n*-Alkanes	43.6	18.5	21.9	4.43	
Yeast	Gas oil	40–50	1–2	12–20	6–12	
Soybean meal		37–53	15–23	3–12	4–6	4–7
Cottonseed meal		39–50	3.5–9.9	23.4–34.5	5.1–8.8	7.3–15.5

Table 8. Amino acid distribution in single-cell and other vegetable proteins[a].

Source of protein	Substrate	Cystine	Isoleucine	Leucine	Lysine	Methionine	Phenylalanine	Threonine	Tryptophan	Valine
Bacterial	Methane			9.1	5.3	3.4	6.2	4.5		8.5
Bacterial	*n*-Alkanes	0.6	3.6	5.6	6.5	2.0	2.9	4.0	0.9	3.5
Bacterial	*n*-Alkanes	1.1	4.5	7.0	7.0	1.8	7.4	4.9	1.4	5.4
Bacterial	Gas oil	0.9	5.3	7.8	7.8	1.6	4.8	5.4	1.3	5.8
Yeast	*n*-Alkanes		3.6	5.9	7.0	1.2	3.7	3.9	0.5	4.0
Yeast	Gas oil		5.1	7.0	6.8	1.3	5.1	5.6	2.3	5.0
Soybean meal		1.2	5.8	7.6	6.6	1.1	4.8	3.9	1.2	5.2
Cottonseed meal			4.0	6.2	4.2	1.5	5.2	3.5	1.6	5.0

[a] Amino acids are expressed as percentage of crude protein

Table 9. Composition of cells grown on various *n*-alkanes

Substrate	Nitrogen added as NH_4OH per 100 g cells (g)	Cell nitrogen (%)	Crude protein yield on substrate (%)	Cell lipid (%)
$C_{12}H_{26}$	9.52	7.10	26.6	5.0
$C_{14}H_{30}$	9.20	7.27	37.7	7.5
$C_{16}H_{34}$	8.60	7.21	36.8	7.2
$C_{17}H_{36}$	8.24	7.26	33.2	10.3
$C_{18}H_{38}$	8.32	6.85	35.5	8.3

Table 10. Essential amino acid contents of microorganisms, and plant and animal samples. The amino acids are those listed in Table 11.

Sample	Content	
	Essential amino acids as dry weight (%)	Nitrogen (%)
Bacteria		
Bacillus subtilis	23.8	10.07
Escherichia coli	33.1	13.19
Staphylococcus aureus	21.6	10.75
Yeasts		
Saccharomyces cerevisiae	17.1–23.8	5.9–8.2
Sacch. cerevisiae	23.1	8.94
Torula yeast	29.5	8.35
Torula yeast	24.4	7.47
Moulds		
Aspergillus niger	9.2	5.21
Penicillium notatum	12.8	6.13
Rhizopus nigricans	9.6	5.80
Mushroom		
Tricholoma nudum	20.8	8.64
Nonmicrobial		
Animal muscle[a]	48.1	15.4
Fish meal	32.1	9.8
Alfalfa meal	6.9	2.72

[a] Average values.

Table 11. Average essential amino acid distribution of the microbial samples of Table 10.

Amino acid	Percentage of total essential amino acids	
Histidine	7.22	±1.05
Arginine	11.18	±0.36
Lysine	15.31	±1.01
Leucine	16.57	±0.53
Isoleucine	11.34	±0.57
Valine	11.85	±0.37
Methionine	3.81	±0.31
Threonine	10.23	±0.33
Phenylalanine	9.43	±0.67

MATERIAL BALANCES FOR GROWTH AND PRODUCT FORMATION (STOICHIOMETRY)

Microbial production of ethanol from glucose can be represented as

$$C_6H_{12}O_6 \xrightarrow{\text{Microorganisms}} 2C_2H_5OH + 2CO_2$$

According to this stoichiometrically balanced equation, on a weight basis, 51.1 per cent of the glucose is converted to ethanol and 48.9 per cent to carbon dioxide. Thus

$$Y'_{\text{EtOHs}} = 0.511 \text{ and } Y'_{\text{CO}_2\text{s}} = 0.489$$

These are the maximal yields theoretically attainable. In practice, yields of only 90–95 per cent of the theoretical value can be achieved because of losses in byproducts formed, entrainment of ethanol within the cells, microbial growth and maintenance.

As is the case for ethanol production, if the product formation route is well defined, a simple stoichiometric equation can be used to estimate the theoretically possible maximum yields on substrate.

However, when microbial growth is the product being considered, an attempt to estimate the theoretical maximum biomass yield on substrate involves considerable oversimplification of the process since it requires the assumption that all the energy contained in the substrate is conserved in the synthesized biomass.

Material balances in terms of stoichiometrically balanced equations can be written for simultaneous microbial growth and product formation if some of the stoichiometric coefficients of components can be determined experimentally or if values can be assumed for them. Examples of this approach are presented below.

Example I
Aerobic growth on carbohydrates [Darlington (1964)]
Microbial composition = $C_{3.92}H_{6.5}O_{1.94}$
Biomass yield $Y's = 0.50$

Thus

$$6.67\,CH_2O + 2.1\,O_2 \longrightarrow C_{3.92}H_{6.5}O_{1.94} + 2.75\,CO_2 + 3.42\,H_2O$$

In this case, nitrogen has been neglected in the material balance.

Example II
Aerobic growth on hydrocarbons [Darlington (1964)]
Microbial composition = $C_{3.92}H_{6.5}O_{1.94}$
Biomass yield $Y's = 1.00$
Thus

$$7.14\,CH_2 + 6.135\,O_2 \longrightarrow C_{3.92}H_{6.5}O_{1.94} + 3.22\,CO_2 + 3.89\,H_2O$$

In this case, nitrogen has been neglected in the material balance.

Example III
Aerobic growth on glucose [Abbott and Clamen (1973)]
Microbial composition = $C_{4.4}H_{7.3}N_{0.86}O_{1.2}$
Biomass yield $Y'_s = 0.66$
Thus

$$C_6H_{12}O_6 + 1.5\,O_2 + 0.77\,NH_3 \longrightarrow 0.90\,C_{4.4}H_{7.3}N_{0.86}O_{1.2} + 2.1\,CO_2 + 3.87\,H_2O$$

Example IV
Aerobic growth on hexadecane [Abbott and Cleman (1973)]
Microbial composition = $C_{4.4}H_{7.3}N_{0.86}O_{1.2}$
Biomass yield $Y'_s = 0.66$
Thus

$$C_{16}H_{34} + 12.5\,O_2 + 2.06\,NH_3 \longrightarrow 2.4\,C_{4.4}H_{7.3}N_{0.86}O_{1.2} + 5.4\,CO_2 + 11.34\,H_2O$$

Example V
Aerobic growth of *Klebsiella aerogenes* in continuous culture [Herbert (1976)].
1) Glycerol-limited growth, $D = 0.1\,h^{-1}$

$$0.68\,C_3H_8O_3 + 1.27\,O_2 + 0.22\,NH_3 \longrightarrow CH_{1.74}N_{0.22}O_{0.43} + 0.97\,CO_2 + 2.20\,H_2O$$

Table 12.

Component	In starting materials (g atom)	In products (g atom)
Carbon	2.04	1.98
Hydrogen	6.10	6.14
Oxygen	4.59	4.57

$Y'_s = 0.393 \qquad Y^x_y = 0.490 \qquad Y^x_y = 0.475 \qquad Y^x_y = 0.401$

2) Glycerol-limited growth, $D = 0.85\,h^{-1}$

$$0.51\,C_3H_8O_3 + 0.76\,O_2 + 0.24\,NH_3 \longrightarrow CH_{1.73}N_{0.24}O_{0.43} + 0.50\,CO_2 + 1.59\,H_2O$$

Table 13.

Component	In starting materials (g atom)	In products (g atom)
Carbon	1.53	1.50
Hydrogen	4.80	4.91
Oxygen	3.05	3.02

$Y'_s = 0.529 \qquad Y^x_y = 0.654 \qquad Y^x_y = 0.327 \qquad Y^x_y = 0.651$

Example VI
Aerobic growth of *Candida utilis* in continuous culture with glucose and ethanol as substrates [Herbert (1976)].

1) Glucose-limited growth, $D = 0.08\,h^{-1}$

$$0.314\,C_6H_{12}O_6 + 0.75O_2 + 0.19\,NH_3 \longrightarrow CH_{1.82}N_{0.19}O_{0.47} + 0.90\,CO_2 + 1.18\,H_2O$$

$$Y'_s = 0.457 \quad Y^C_C = 0.531 \quad Y^{CO_2}_C = 0.478 \quad Y^C_O = 0.667$$

2) Glucose-limited growth, $D = 0.45\,h^{-1}$

$$0.283\,C_6H_{12}O_6 + 0.57\,O_2 + 0.20\,NH_3 \longrightarrow CH_{1.84}N_{0.2}O_{0.56} + 0.66\,CO_2 + 1.04\,H_2O$$

$$Y'_s = 0.540 \quad Y^C_C = 0.589 \quad Y^{CO_2}_C = 0.390 \quad Y^C_O = 0.871$$

3) Ethanol-limited growth, $D = 0.06\,h^{-1}$

$$0.865\,C_2H_6O + 1.12\,O_2 + 0.19\,NH_3 \longrightarrow CH_{1.82}N_{0.19}O_{0.46} + 0.70CO_2 + 1.60H_2O$$

$$Y'_s = 0.664 \quad Y^C_C = 0.578 \quad Y^{CO_2}_C = 0.405 \quad Y^C_O = 0.446$$

4) Ethanol-limited growth, $D = 0.43\,h^{-1}$

$$0.74\,C_2H_6O + 1.00\,O_2 + 0.20\,NH_3 \longrightarrow CH_{1.84}N_{0.2}O_{0.55} + 0.48CO_2 + 1.44H_2O$$

$$Y'_s = 0.803 \quad Y^C_C = 0.676 \quad Y^{CO_2}_C = 0.311 \quad Y^C_O = 0.50$$

Example VII (Table 14)
Influence of aeration on the amounts of glucose and oxygen consumed and carbon dioxide, ethanol, glycerol and acetaldehyde produced during formation of 100 g dry wt. yeast [Dura (1973)]

Table 14.

Oxygen in the gas supply (%)	$C_6H_{12}O_6$ +	NH_3 +	O_2 ⟶	($CH_{1.62}O_{0.53}N_{0.15}$ + 1.29 ash) +	CO_2 +	C_2H_6O +	$C_3H_8O_3$ +	C_2H_4O +	H_2O	C[a]	O[a]
0.0	6.23	0.59	0.00	3.92	9.50	10.95	0.45	0.03	0.38	98.26	90.40
0.2	5.73	0.59	0.23	3.92	9.82	9.55	0.29	—	2.28	98.05	98.79
0.5	5.03	0.59	0.75	3.92	9.51	8.02	0.11	—	3.39	98.74	103.66
1.0	4.65	0.59	0.98	3.92	7.41	7.28	0.05	—	3.57	93.33	93.44
2.1	3.90	0.59	1.32	3.92	6.95	5.83	—	—	3.62	96.28	97.66
5.0	2.83	0.59	1.91	3.92	5.60	3.80	—	—	3.29	100.82	97.93
7.0	2.31	0.59	2.25	3.92	4.78	2.55	—	—	3.92	99.57	93.64
9.6	1.88	0.59	2.33	3.92	4.15	1.60	—	—	4.19	99.91	101.44
14.5	1.62	0.59	2.57	3.92	3.80	0.93	—	—	4.49	99.59	101.95
21.0	1.20	0.59	2.59	3.92	2.95	0.06	—	—	4.73	97.08	103.15
26.4	1.14	0.59	2.56	3.92	2.56	0.02	—	—	4.49	95.32	97.91
28.0	1.20	0.59	2.63	3.92	3.15	—	—	—	4.91	98.18	106.66
30.6	1.12	0.59	2.60	3.92	2.82	0.03	—	—	4.34	101.19	101.43
33.8	1.07	0.59	2.83	3.92	2.33	0.02	—	—	4.07	97.98	89.65
36.8	1.22	0.59	3.96	3.92	4.14	—	—	—	5.03	110.11	100.98
53.2	1.34	0.59	5.66	3.92	3.78	0.023	—	—	5.69	96.27	79.29
63.4	1.22	0.59	4.97	3.92	3.00	0.04	—	—	4.91	95.63	75.49
100.0	3.73	0.59	6.19	3.92	8.22	2.97	—	0.04	10.82	82.57	93.99

[a] Quantities found on the right-hand side of equations expressed as percentages of amounts on the left-hand side.

Example VIII
Anaerobic growth of *Saccharomyces cerevisiae* on glucose [Battley (1979)]

1) Non-growth reaction

$$C_6H_{12}O_6\,(aq) \longrightarrow 2\,CO_2\,(aq) + 2\,C_2H_5OH\,(aq) \quad \Delta G^{0'} = -\,52.02\,\text{kcal}$$

2) Growth reaction
a) Work (catabolic) reaction

$$0.770\,C_6H_{12}O_6\,(aq) \longrightarrow 1.540\,CO_2\,(aq) + 1.540\,C_2H_5OH \quad \Delta G^{0'} = -\,40.05\,\text{kcal}$$

b) Conversion (anabolic) reaction

$$0.230\,C_6H_{12}O_6\,(aq) + 0.240\,C_2H_5OH\,(aq) + 0.118\,NH_3\,(aq)$$
$$\longrightarrow 0.590\,CH_{1.737}N_{0.200}O_{0.451}\,(\text{cells}) + 0.432\,C_3H_8O_3\,(aq) + 0.036\,H_2O(l)$$
$$\Delta G^0 - 0.00\,\text{kcal}$$

c) Overall reaction [i.e. reaction (a) + reaction (b)]

$$C_6H_{12}O_6\,(aq) + 0.118\,NH_3\,(aq) \longrightarrow 0.590\,CH_{1.737}N_{0.200}O_{0.451} + 0.432\,C_3H_8O_3\,(aq)$$
$$+\,1.540\,CO_2\,(aq) + 1.300\,C_2H_5OH\,(aq) + 0.036\,H_2O(l)\,\Delta G^0 = -\,40.05\,\text{kcal}$$

Example IX
Aerobic growth of *Sacch. cerevisiae* [Battley (1979)]

1) Anaerobic growth on glucose (from Example VIII)

$$C_6H_{12}O_6\,(aq) + 0.118\,NH_3\,(aq) \longrightarrow 0.590\,CH_{1.737}N_{0.200}O_{0.451}\,(\text{cells}) + 1.540\,CO_2$$
$$(aq) + 1.300\,C_2H_5OH\,(aq) + 0.432\,C_3H_8O_3\,(aq) + 0.036\,H_2O(l) \quad \Delta G' = -40.05\,\text{kcal}$$

2) Non-growth oxidation of the glycerol produced during reaction (1)

$$0.432\,C_3H_8O_3\,(aq) + 1.512\,O_2\,(aq) \longrightarrow 1.296\,CO_2\,(aq) + 1.728\,H_2O(l)$$
$$\Delta G^{0'} = -\,172.98\,\text{kcal}$$

3) Aerobic growth on the ethanol produced during reaction (1) (from Example VIII)

$$1.300\,C_2H_5OH\,(aq) + 2.406O_2\,(aq) + 0.198\,NH_3\,(aq) \longrightarrow 1.339\,CH_{1.704}N_{0.419}O_{0.408}\,(\text{cells})$$
$$+\,1.262\,CO_2\,(aq) + 3.049\,H_2O(l) \quad \Delta G^{0'} = -258.95\,\text{kcal}$$

4) Overall aerobic reaction [i.e. reaction (1) + reaction (2) + reaction (3)]

$$C_6H_{12}O_6\,(aq) + 3.918\,O_2\,(aq) + 0.316\,NH_3\,(aq) \longrightarrow 1.929\,CH_{1.703}N_{0.171}O_{0.459}\,(\text{cells})$$
$$+\,4.098\,CO_2\,(aq) + 4.813\,H_2O(l) \quad \Delta G^{0'} = -471.98\,\text{kcal}$$

In this example account is taken of the fact that initially an 'anaerobic' reaction occurs with glucose as the substrate, even in the presence of oxygen. This is followed by aerobic growth on the ethanol produced during the anaerobic growth reaction. The glycerol, which is also produced during the anaerobic glucose reaction, is oxidised to carbon dioxide and water.

Example X
Aerobic growth of *Sacch. cerevisiae* on ethanol [Battley (1979)]

1) Non-growth reaction

$$C_2H_5OH\,(aq) + 3\,O_2\,(aq) \longrightarrow 2\,CO_2\,(aq) + 3\,H_2O(l) \quad \Delta G^{0'} = -\,322.84\,\text{kcal}$$

2) Conservative reaction

a) Work (catabolic) reaction

$$0.617\,C_2H_5OH\,(aq) + 1.851\,O_2\,(aq) \longrightarrow 1.235\,CO_2\,(aq) + 1.851\,H_2O(l) \qquad \Delta G^{0'} = -199.19\text{ kcal}$$

b) Conversion (anabolic) reaction

$$0.383\,C_2H_5OH + 0.264\,CO_2\,(aq) + 0.153\,NH_3 \longrightarrow 1.030\,CH_{1.704}N_{0.149}O_{0.408}\text{ (cells)} + 0.495\,H_2O(l) \qquad \Delta G^{0'} = 0.00\text{ kcal}$$

c) Overall reaction [i.e. reaction (a) + reaction (b)]

$$C_2H_5OH\,(aq) + 1.851\,O_2\,(aq) + 0.153\,NH_3\,(aq) \longrightarrow 1.030\,CH_{1.704}N_{0.149}O_{0.408}\text{ (cells)} + 0.970\,CO_2\,(aq) + 2.346\,H_2O\,(l) \qquad \Delta G^{0'} = -199.19\text{ kcal}$$

Table 15. Assimilated and dissimilated carbon and energy substrate [Cheremisinoff and Young (1975)].

Microorganism	Conditions	Glucose carbon assimilated (%)	Glucose dissimilated to provide energy (%)
Aerobacter cloacae	Minimal medium, aerobic	55	45
Saccharomyces cerevisiae[a]	Rich medium, anaerobic	2	98
Sacch. cerevisiae[a]	Rich medium, aerobic	10	90
Streptococcus faecalis[b]	Rich medium, anaerobic	2	98

[a] Kormancikova *et al.* (1969)
[b] Bauchop and Elsden (1960)

Table 16. Comparative carbon balances in the oxidative assimilation of several substrates during the growth of *Escherichia coli*[a] [Siegel and Clifton (1950)].

Substrate (mg carbon)	Succinate	Fumarate	Lactate	Pyruvate	Glycerol	Acetate[b]
Initial substrate carbon	5.51	3.14	3.31	4.45	5.40	
Cell-carbon after assimilation	0.55	0.44	0.37	0.51	0.72	
Cell-carbon before assimilation	0.35	0.21	0.29	0.20	0.51	
Carbon stored	0.20	0.23	0.08	0.31	0.21	
Supernatant-carbon at end of experiment	5.04	2.57	3.19	3.76	5.15	
Carbon dioxide carbon	0.29	0.30	0.12	0.39	0.10	
Total recovered	5.53	3.10	3.40	4.46	5.46	
Total recovered %	100.30	98.70	102.70	100.20	101.10	
Carbon dioxide produced (μl)	536	534	214	724	182	
Oxygen consumed (μl)	380	274	222	394	266	
R.Q.[c] observed	1.41	1.95	0.97	1.84	0.70	

[a] Duration of experiment 4.5 h.
[b] Not utilized by microorganism during the course of the experiment.
[c] Respiratory quotient (see Glossary).

Table 17. Manometric observations on the oxidation of several substrates by washed suspensions of *E. coli* at 30°C in O.67 M phosphate buffer at pH 7.2 [Siegel and Clifton (1950)].

Substrate	Succinate	Fumarate	Lactate	Pyruvate	Glycerol	Acetate
Carbon utilized (mg)	0.883	0.680	0.672	0.720	0.803	0.702
Oxygen consumed (μl)	1030	676	935	755	1465	1014
Carbon dioxide produced (μl)	1152	974	960	950	1175	977
Percentage oxidized	71.5	76.0	69.5	70.5	78.0	74.9
R.Q. observed	1.12	1.44	1.02	1.26	0.80	0.96
R.Q. for complete combustion	1.14	1.33	1.00	1.20	0.86	1.00
R.Q. for oxidative assimilation	1.20	1.50	1.00	1.33	0.80	
Carbon assimilated/Carbon dioxide carbon [a]	0.43	0.35	0.31	0.43	0.28	0.34

[a] Calculated by difference between carbon utilized and carbon dioxide carbon produced, both expressed in milligrams.

Table 18. Calculation of molar growth yields[a] from carbon conversion data [Servizi and Bogan (1963)].

Substrate	Substrate carbon (mg)	Utilized substrate (mmol)	Substrate carbon converted to cell carbon (mg)	Cells synthesized[b] (mg)	Y_s [g cells (mol substrate)$^{-1}$]	$\Delta F°_{ox}$ (kcal mol^{-1})
Succinate	0.47	9.80	0.20	0.40	40.8	363
Fumarate	0.55	11.50	0.23	0.46	40.0	335
Lactate	0.16	4.00	0.08	0.16	40.0	340
Pyruvate	0.70	19.30	0.31	0.62	32.1	282
Glycerol	0.31	8.62	0.21	0.42	48.6	407
Arabinose	0.53	8.84	0.34	0.68	77.0	573
Glucose	0.58	8.06	0.34	0.68	84.3	686
Lactose	0.22	1.53	0.12	0.24	157	1381

[a] For definition of molar growth yields Y_S, see Table 21.
[b] Calculated on the basis of 50 per cent carbon content for dried cell tissue.

Table 19. Comparison of the observed metabolic activities and the extent of assimilation in suspensions and cultures of *E. coli* [Siegel and Clifton (1950)].

Substrate	Metabolic activity per mg carbon (cultures/suspensions)		Assimilation (% carbon assimilated)	
	Oxygen	Carbon dioxide	Washed cells	Cultures
Succinate	0.70	0.87	30	41
Fumarate	0.50	0.65	25	43
Lactate	1.30	1.33	30	40
Pyruvate	0.55	0.80	30	44
Glycerol	0.58	0.49	20	67

Table 20. Efficiencies of synthesis of *E. coli* by several substrates [Siegel and Clifton (1950)].

Substrate	Increase in cell carbon (mg)	Carbon dioxide carbon (mg)	Ratio of stored cell carbon to carbon dioxide carbon	Ratio of stored cell carbon to substrate carbon consumed
Succinate	0.20	0.29	0.69	0.41
Fumarate	0.23	0.30	0.77	0.43
Lactate	0.08	0.12	0.67	0.40
Pyruvate	0.31	0.39	0.79	0.44
Glycerol	0.21	0.10	2.10	0.70

Stoichiometric equations describing the consumption of substrate and the formation of biomass and products can be used in various ways.

1) To obtain relationships between different yield coefficients based on elemental and energy balances describing microbial activity.
2) To obtain the theoretical limits of the maximum yields of biomass and products formed from a given substrate.
3) To estimate yield coefficients from experimentally determined quantities such as substrate consumption, biomass production, oxygen consumption, carbon dioxide production, heat evolution and nitrogen consumption.
4) To explore reasons for discrepancies between measured yield values and those predicted from the stoichiometry of a given system due to systematic errors in measurement and an incorrect system description, such as the presence of an unknown product or substrate.
5) To construct figures that allow quick estimates of yield coefficients as functions of experimentally measureable variables.

Examples I–X illustrate the application of an approximate stoichiometry. A more detailed treatment of stoichiometry is given in the Appendix to this chapter.

DEFINITION AND EVOLUTION OF VARIOUS YIELD COEFFICIENTS FOR GROWTH AND PRODUCT FORMATION

Growth Yield Coefficients

Table 21.

Symbol	Definition	Comments
Y'_s	Gram dry weight of biomass produced per gram of substrate consumed [g (g substrate)$^{-1}$]	Suffers from the fact that the energy and carbon content of different substrates will be different from the same amount of substrate
Y'_s	Molar growth rate. Gram dry weight of biomass produced per gram mole of substrate consumed [g (g mol substrate)$^{-1}$]	An improvement over Y'_s since substrates are brought to a more comparable basis at the molar level

continued

Table 21 – (continued)

Symbol	Definition	Comments
Y_{ATP}	Gram dry weight of biomass produced per gram mole of ATP formed during catabolism of substrate [g (g mol ATP)$^{-1}$]	In principle, superior to Y_s since the amount of growth is directly proportional to the amount of energy, in the form of ATP, obtained from the degradation of the energy source. Accounts for the utilization of different substrates by different microorganisms through different metabolic pathways. Disadvantage is that for certain pathways and especially for aerobic processes the extent of ATP production is not yet known
Y'_{kcal}	Gram dry weight of biomass produced per kilocalorie of heat evolved during fermentation. (Heat of combustion of substrate – heat of combustion of cells, all in stoichiometric amounts) (g kcal^{-1})	Used mainly for mixed substrates and cultures as in waste water treatment. Introduced instead of Y_{ATP} in an attempt to base growth yields on energy concepts that can be calculated by enthalpy balances if all the products and cell contents are known
Y_{kcal}	Gram dry weight of biomass produced per kilocalorie of total energy consumed from the medium. (Energy used for assimilation + energy expended by catabolism) (g kcal^{-1})	As Y'_{kcal}
Y_{O_2}	Gram dry weight of biomass produced per gram mole of oxygen consumed [g (g mol O_2)$^{-1}$]	Commonly used for aerobic microorganisms with the assumption that oxygen uptake is a measure of that part of the substrate that is completely oxidized. The implicit assumption that ATP produced by catabolism corresponds to oxygen is not always valid
Y'_{O_2}	Gram dry weight of biomass produced per gram of oxygen	As Y'_{O_2}
Y_O	Gram dry weight of biomass produced per gram atom of oxygen consumed [g (g atom O)$^{-1}$]	As Y'_{O_2}

continued

Table 21 – (continued)

Symbol	Definition	Comments
$Y_{av\ e-}$	Gram dry weight of biomass produced per equivalent of available electrons [g (av e$^-$)$^{-1}$]	May be more satisfactory than Y_O since it is calculated from the amount of oxygen stoichiometrically required for complete oxidation of the substrate. Does not necessarily correspond to ATP produced in catabolism.
$Y_{NO_3^-}$	Gram dry weight of biomass produced per gram mole of nitrate reduced [g (g mol NO_3^-)$^{-1}$]	Used with nitrate that is a hydrogen (electron) acceptor
$Y_{\text{other hydrogen acceptor}}$	Gram dry weight of biomass produced per gram mole of hydrogen acceptor [g (g mol hydrogen acceptor)$^{-1}$]	Used with substrates that are hydrogen (electron) acceptors
Y_N	Gram dry weight biomass produced per gram or gram atom of nitrogen	
Y_I	Gram dry weight of biomass produced per gram mole of substrate I consumed [g (g I)$^{-1}$ or g (g atom I)$^{-1}$]	I is another substrate, such as sulphur or phosphorous, that is an important constituent of the cell

Product Yields

Table 22.

Symbol	Definition	Comment
$Y_s^{CO_2}$	Gram mole of carbon dioxide produced per gram mole of substrate consumed [g mol CO_2 (g mol substrate)$^{-1}$]	
$Y_{O_2}^{CO_2}$	Gram mole of carbon dioxide produced per gram mole of oxygen consumed [g mol CO_2 (g mol O_2)$^{-1}$]	
Y^{ATP}	Gram mole of ATP produced per gram mole of substrate consumed [g mol ATP (g mol substrate)$^{-1}$]	Known quite accurately for the common anaerobic metabolic pathways. There is uncertainty in the actual amount of ATP produced in oxidative phosphorylation. Used to calculate Y_{ATP} when composition of cells is known

Process Yields

Process yields in the production of industrially important, pharmaceutical and other fine biochemical compounds are expressed in a variety of ways. The most commonly used are:

1) percentage conversion of carbon source.
2) concentration, e.g., g ℓ^{-1}.
3) concentration of assay units (especially for enzymes and vitamins).
4) tonnage, e.g., tonnes yr^{-1}.

In industrial microbiological processes, the percentage conversion of the carbon source to the desired product is often low. This is particularly true for high-value products and has led to the extensive use of product concentration as the basis for monitoring process yield. Product concentration is especially important because of the consequent effect on product recovery (*see* Chapters 12 and 13).

OXYGEN REQUIRED FOR CELL PRODUCTION

The following assumptions have been made by Mateles (1971).

1) The only products of metabolism are cells, carbon dioxide and water.
2) The nitrogen source is ammonia.

Eq. (3) results from a material balance involving the above assumptions and the following relationship between components

$$CH_2O + NH_3 + O \longrightarrow \text{cells } (C', H', O', N') + CO_2 + H_2O + N \quad (1)$$

$$\frac{32}{Y_{O_2}} = \frac{g_{\text{oxygen}}}{g_{\text{cell}}} = 16\left[\frac{2C + (H/2) - O}{Y'_s M} + \left(\frac{N'}{1400} \times \frac{3}{2}\right) - \left(\frac{C'}{1200} \times 2\right) - \frac{H'}{100} \times \frac{1}{2} + \frac{O'}{1600}\right] \quad (2)$$

or

$$\frac{32}{Y_{O^2}} = \frac{32C + 8H - 16O}{Y'_s M} + 0.01\,O' - 0.0267\,C' + 0.01714\,N' - 0.08\,H' \quad (3)$$

where C, H and O represent the number of atoms of carbon, hydrogen and oxygen, respectively, in each molecule of the carbon source. C', H', O' and N' represent the percentages of carbon, hydrogen, oxygen and nitrogen, respectively in the cells. Y'_s is the growth yield and M is the molecular weight of the carbon source (i.e. $Y_s = Y'_s M$).

RELATIONSHIPS BETWEEN YIELD COEFFICIENTS

For defintion of terms consult the Glossary at the beginning of this chapter and Tables 21 and 22.

Y_S and COD

where k_1 is a proportionality constant.

$$Y_s = k_1 Y^{ATP} \quad (4)$$

The yield of ATP can be assumed to be proportional to the 'free energy of oxidation ΔF_{ox} of the substrate, i.e.

$$Y^{ATP} = -k_2 \Delta F_{ox} \quad (5)$$

where k_2 is a proportionality constant. Hence

$$Y_s = -k_1 k_2 \Delta F^0_{ox} \tag{6}$$

where ΔF^0_{ox} is the free energy of oxidation under physiological conditions. The value of ΔF^0_{ox} for a substrate is related to the number of moles of oxygen consumed in oxidizing one molecule of substrate, i.e. the COD. Thus

$$\Delta F^0_{ox} = -k_3 \text{COD} \tag{7}$$

or

$$Y_s = k_1 k_2 k_3 \text{COD} \tag{8}$$

Servizi and Bogan (1963) have shown experimentally that for activated sludge, and for carbohydrates and Krebs cycle intermediates, the yield coefficient Y_s is given by

$$Y_s = 0.38 \text{ COD} \tag{9}$$

and for aromatic and aliphatic acids is given by

$$Y_s = 0.34 \text{ COD} \tag{10}$$

Y'_{kcal}, Y_s and TOD

It is assumed that the only products of metabolism are cells, carbon dioxide and water.

$$Y'_{kcal} = \frac{Y_s}{\begin{pmatrix}\text{heat of combustion of}\\ \text{substrate, kcal mol}^{-1}\end{pmatrix} - \begin{pmatrix}\text{heat of combustion of}\\ \text{cells, kcal g}^{-1}\end{pmatrix}} \tag{11}$$

Heat of combustion of cells $=$ TOD_{cells} $+$ 26.05 kcal (g cells)$^{-1}$

Heat of combustion of substrate $=$ TOD_s $\times$ 26.05 kcal (mol substrate)$^{-1}$

where TOD_{cells} is expressed as moles of oxygen per gram dry weight of cells and TOD_s is expressed as moles of oxygen per mole of substrate. The heat of combustion for the transfer of electrons to oxygen from a 'methane-type' bond is 26.05 kcal (av e$^-$)$^{-1}$.

Y'_{kcal} and Y_{ave-}

$$Y'_{kcal} = \frac{Y_{ave-}}{26.05} \tag{12}$$

Y_{kcal}, Y_s and Y_{O_2}

The total energy consumed E_c from the medium can be considered as the sum of that incorporated biosynthetically into cellular material by assimilation E_a and that required for catabolism E_d.

$$Y_{kcal} = \frac{Y_s}{E_a + E_d} \tag{13a}$$

where

$$E_a = 4 \times \text{TOD}_{cells} \times 26.05 \text{ kcal (g mol substrate)}^{-1} \tag{13b}$$

TOD_{cells} can be calculated from a knowledge of the empirical 'molecular' formula for the cells under consideration. However, E_a may be calculated from the formula

$$E_a = 5{,}3\, Y_s \text{ kcal (g mol substrate)}^{-1} \tag{14}$$

since the average heat of combustion for biomass is 5.3 kcal (g dry wt)$^{-1}$, and

$$E_d = 4 \times \frac{Y_s}{Y_{O_2}} \times 26.05 \text{ kcal (g mol substrate)}^{-1} \tag{15}$$

Therefore

$$Y_{kcal} = \frac{Y_{O_2}}{5.3\, Y_{O_2} + 104.2} \text{ g kcal}^{-1} \tag{16}$$

Y_{ATP} and Y_s

$$Y_{ATP} = \frac{Y_s}{Y^{ATP}} \tag{17}$$

where Y^{ATP} is the amount of ATP formed measured in moles per mole of substrate consumed.

Y_{ATP} and Y_{O_2}

$$Y_{ATP} = \frac{2\, Y_{O_2}}{P/O} \tag{18}$$

where

$$P/O = 2\, Y^{ATP} \tag{19}$$

Y_{O_2}, Y_{ATP} and P/O Ratios in Aerobic Processes

Under anaerobic conditions, if fermentation pathways are known, the amount of ATP produced from the degradation of the substrate Y^{ATP} can be calculated. Hence the yield on ATP produced Y_{ATP} can be determined from Y_s and Y^{ATP} [*see* Eq. (17)].

However, the interpretation of aerobic growth is much more difficult since ATP is produced during substrate-level phosphorylation and during oxidative phosphorylation. Biomass yield on oxygen consumed Y_{O_2} is a measure of that part of the substrate that is completely oxidized. The P/O ratio depends on the nujber of phosphorylation sites in the respiratory chain, i.e. the P/O ratio reflects the efficiency of oxidative phosphorylation.

Several approaches have been used to calculate P/O ratios from aerobic yield studies. The difficulty arises from the fact that either Y_{ATP} or Y_{ATP}^{max} must be known to determine the P/O ratio or, conversely, the P/O ratio must be known to determine Y_{ATP}^{max}.

In continuous culture studies, $Y_{O_2}^{max}$ is determined by Eq. (30), i.e.

$$\frac{1}{Y_{O_2}} = \frac{m_o}{\mu} + \frac{1}{Y_{O_2}^{max}}$$

Under aerobic conditions, the total ATP production is given by Stouthamer (1977)

$$q_{ATP} = \alpha(1 - \beta)\, q_{sub} + q_{O_2}\, 2(\text{P/O}) \tag{20}$$

where α is the number of ATP molecules formed by substrate phosphorylation during the complete oxidation of the substrate, β is the part of the substrate that is assimilated, and

q_{sub} and q_{O_2} are the specific rates of substrate and oxygen consumption, respectively. There is also the relationship [*see* Eq (29)]

$$q_{ATP} = \frac{\mu}{Y_{ATP}} = \frac{\mu}{Y_{ATP}^{max}} + m_e$$

Unfortunately insufficient information is obtained from aerobic chemostat experiments to allow the determination of Y_{ATP}^{max}, m_e and the P/O ratio using the above equations.

DISTINCTION BETWEEN OBSERVED YIELDS AND TRUE GROWTH YIELDS — THE CONCEPT OF MAINTENANCE ENERGY

In the following equations, Y is equivalent to Y'_s, i.e. the cellular material produced per gram of substrate consumed.

Constant Maintenance Energy

According to Pirt (1965), the observed growth yield is

$$Y_{obs} = \frac{\Delta x}{(\Delta s)_G + (\Delta s)_M} \tag{21}$$

where Δx represents the cells produced. $(\Delta s)_G$ and $(\Delta s)_M$ are the substrate consumed for growth and in cell maintenance, respectively. The true growth yield is

$$Y_G = \frac{\Delta x}{(\Delta s)_G} \tag{22}$$

In batch culture

$$\frac{ds}{dt} = \left(\frac{ds}{dt}\right)_M + \left(\frac{ds}{dt}\right)_G \tag{23}$$

where

$$\frac{dx}{dt} = \mu x \quad \text{and} \quad \frac{ds}{dt} = \frac{\mu x}{Y_{obs}} \tag{24}$$

However, defining

$$\left(\frac{ds}{dt}\right)_G = -\frac{\mu x}{Y_G} \tag{25}$$

and

$$\left(\frac{ds}{dt}\right)_M = -m_s x \tag{26}$$

where m_s is the substrate required for cell maintenance. Then

$$\frac{1}{X}\frac{ds}{dt} = m_s + \frac{\mu}{Y_G} \tag{27}$$

and

$$\frac{1}{Y_{obs}} = \frac{m_s}{\mu} + \frac{1}{Y_G} \tag{28}$$

True and Observed Y_{ATP}

According to Stouthamer and Betterhaussen (1976),

$$\left(\frac{1}{Y_{ATP}}\right)_{obs} = \frac{m_e}{\mu} + \frac{1^{-1}}{Y_{O_2}^{max}} \tag{29}$$

where m_e is the ATP required for cell maintenance.

True and Observed Yield Coefficients with a Variable Maintenance Coeffficient that is a Function of Growth Rate

By analogy with the above equations, Neijssel and Tempest (1976) showed that

$$\frac{1}{Y_{O_2}} = \frac{m_o}{\mu} + \frac{1}{Y_{O_2}^{max}} \tag{30}$$

where m_O is the oxygen required for cell maintenance. Introducing an additional growth rate-dependent term gives

$$\frac{1}{Y_{O_2}} = \frac{m_O(1 + c\mu)}{\mu} + \frac{1}{Y_{O_2}^{max}} \tag{31}$$

where c is an empirical constant.

EXPERIMENTAL YIELD COEFFICIENTS

Pure Cultures

Table 23. Biomass yield coefficients for aerobic conditions [from Heijnen and Roels (1981)].

Microorganism	Culture	Temperature (°C)	Nitrogen source	μ (h^{-1})	Y_s[a]	Y_{O_2}[b]	m_s[c]	m_O[d]
Methane-metabolizing								
Methane bacterium	Batch		Ammonia	0.15	0.53	0.42		
Mixed bacterial culture	Continuous	31	Ammonia	0.08	0.59	0.34		
Methylcoccus capsulatus	Continuous	37	Ammonia	0.05	0.65	0.37		
Mixed bacterial culture	Continuous	45	Nitrate	0.09–0.22	0.46	0.33	0.03	0.08
Methylcoccus sp.	Continuous	42	Ammonia	0.04–0.20	0.56	0.39	0.03	0.08
Methylcoccus sp.	Continuous	45	Ammonia	0.34	0.54	0.35		
Methylcoccus sp.	Continuous	45	Nitrate		0.43	0.31		
Methylcoccus sp.	Continuous	45	Nitrogen		0.35	0.22		
Methylosinus trichosporium	Continuous	30	Ammonia	0.07	0.41			
Bacterium 6	Batch	30	Ammonia	0.06	0.36	0.24		
Bacterium HR	Batch	30	Ammonia	0.14	0.43	0.30		
Bacterium HR	Batch	30	Nitrate	0.04	0.35			
n-Alkane-metabolizing								
Candida lipolytica	Batch	30	Ammonia	0.3	0.55	0.65		
C. lipolytica	Continuous	30	Ammonia	0.04–0.18	0.43	0.57	0.02	0.02
C. lipolytica	Batch	18	Ammonia	0.07	0.55	0.71	0.02	0.03
C. lipolytica	Batch	21	Ammonia	0.10	0.60	0.71	0.05	0.06
C. lipolytica	Batch	27	Ammonia	0.15	0.53	0.70	0.08	0.11
C. lipolytica	Batch	30	Ammonia	0.20	0.52	0.69	0.10	0.14
C. lipolytica	Batch	30	Ammonia	0.31	0.56	0.47		
C. tropicalis	Continuous	38	Ammonia	0.18	0.51	0.57		
Yeast	Continuous	30	Ammonia	0.18–0.30	0.65	0.51		
n-Pentane bacterium	Batch	30	Nitrate	0.07	0.49	0.58		
Mixed bacterial culture	Batch	55	Ammonia	0.23	0.58		0.38	

continued overleaf

Microorganism	Culture	Temperature (°C)	Nitrogen source	μ (h^{-1})	Y_s[a]	Y_{O_2}[b]	m_s[c]	m_O[d]
n-Alkane-metabolizing *continued*								
Mixed bacterial culture	Batch	65	Ammonia	0.14	0.58		0.69	
Methanol-metabolizing								
C. boidinii	Batch	32	Ammonia	0.11	0.54	0.50		
C. boidinii	Batch	33	Ammonia		0.52		0.03	
Candida N17	Continuous		Ammonia	0.04–0.08	0.46	0.50		
Hansenula polymorpha	Continuous	37	Ammonia	0.10	0.48	0.49		
H. polymorpha	Continuous	37	Ammonia	0.12–0.15	0.50	0.55		
H. polymorpha	Continuous	35	Ammonia	0.05–0.20	0.52		0.03	
H. polymorpha	Continuous	37	Ammonia	0.05–0.20	0.52		0.05	
H. polymorpha	Continuous	40	Ammonia	0.05–0.20	0.52		0.04	
H. polymorpha	Continuous	43	Ammonia	0.05–0.20	0.52		0.10	
Torulopsis glabrata	Continuous	30	Ammonia	0.10–0.15	0.59			
Torul. glabrata	Continuous		Ammonia	0.11	0.56			
Methylmonas methanolica	Continuous	30	Ammonia	0.10–0.50	0.63	0.69	0.15	0.15
Mixed culture bacterium	Continuous	34	Ammonia	0.26–0.48	0.57	0.79	0.04	0.08
Mixed culture bacterium	Continuous	38	Ammonia	0.26–0.48	0.57	0.79	0.06	
Mixed culture bacterium	Continuous	42	Ammonia	0.26–0.48	0.59	0.90	0.12	0.17
Pseudomonas methyltropha	Continuous	37	Ammonia	0.23	0.57	0.74		
Pseudomonas AM1	Continuous	30	Ammonia	0.10	0.50	0.50		
Bacterium HR	Batch	30	Ammonia	0.35	0.43	0.40		
Pseudomonas C	Continuous	35	Ammonia	0.03–0.40	0.80		0.06	
Ps. methyltropha	Continuous	35	Ammonia	0.03–0.40	0.77		0.10	
Pseudomonas 1	Continuous	35	Ammonia	0.03–0.40	0.60		0.04	
Pseudomonas 135	Continuous	35	Ammonia	0.03–0.40	0.51		0.06	
Ethanol-metabolizing								
C. utilis	Batch	35	Ammonia	0.20	0.64	0.80		
C. utilis	Batch	32	Ammonia	0.50	0.52	0.70		
C. utilis	Continuous	30	Ammonia	0.06–0.43	0.78	1.04	0.02	0.01
Torula utilis	Continous	30	Ammonia	0.05–0.40		1.32		0.02
C. boidinii	Batch	32	Ammonia	0.20	0.64	0.87		
Saccharomyces cerevisiae	Batch	30	Ammonia	0.09	0.40	0.51		
Ps. fluorescens	Batch	35	Ammonia	0.33	0.46	0.55		

Ethanol-metabolizing *continued*								
Bacterium HR	Batch	30	Ammonia	0.13	0.42	0.40		
Bacterium	Batch		Ammonia	0.14	0.51	0.54		
Glycerol-metabolizing								
Escherichia coli		20	Ammonia			1.92		0.003
E. coli		30	Ammonia					
Bacillus megaterium M	Continuous	30	Ammonia	0.05–0.15	0.74	1.92		0.03
B. megaterium D440	Continuous	30	Ammonia	0.10–0.50	0.74	2.17		0.02
Aeromonas punctata	Continuous	30	Ammonia	0.05–0.35	0.73	3.70	0.04	0.02
Aerobacter aerogenes	Continuous		Ammonia		0.69	1.52	0.07	0.06
Ae. aerogenes	Continuous		Ammonia		0.69	1.25	0.06	0.08
Klebsiella aerogenes	Continuous	37	Ammonia		0.69	1.41	0.06	0.06
K. aerogenes	Continuous	35	Ammonia	0.17	0.58	1.47		
Mannitol-metabolizing								
Micrococcus denitrificans	Continuous	37	Ammonia	0.05–0.35	0.56	1.61	0.003	0.02
E. coli	Continuous	37	Ammonia	0.06–0.57		1.64		0.08
K. aerogenes	Continuous	35	Ammonia	0.10–0.50	0.68	1.64		0.02
K. aerogenes	Continuous	35	Ammonia	0.17	0.62	1.30		
Enterobacter aerogenes	Continuous		Ammonia		0.52	2.04	0.28	0.06
Azotobacter chroococcum	Continuous	30	Ammonia	0.05–0.30	0.50		0.03	
Az. chroococcum	Continuous	30	Nitrogen	0.05–0.30	0.33		0.17	
Az. chroococcum	Continuous	30	Nitrogen	0.10–0.22	0.33	0.47		0.11
Acetate-metabolizing								
Chlorella regularis	Continuous	36	Ammonia	0.03–0.25	0.50	0.77	0.02	0.03
C. utilis	Batch	35	Ammonia	0.30	0.44	0.92		
E. coli	Continuous	37	Ammonia	0.14–0.37		1.20		0.31
Ps. fluorescens	Batch	35	Ammonia	0.78	0.34	0.53		
Lactate-metabolizing								
Enterobacter aerogenes	Continuous		Ammonia		0.33	0.83	0.02	0.10
E. coli	Continuous	30	Ammonia		0.48	1.41		
Ae. lwoffi	Continuous	30	Ammonia		0.49	1.15		

continued overleaf

Microorganism	Culture	Temperature (°C)	Nitrogen source	μ (h^{-1})	Y_s[a]	Y_{O_2}[b]	m_s[c]	m_O[d]
Lactate-metabolizing *continued*								
Arthrobacter globiformis	Continuous	30	Ammonia		0.69	1.67		
Paecilomyces denitrificans	Continuous	30	Ammonia		0.71	1.67		
Ps. fluorescens	Batch	35	Ammonia		0.39	1.15		
K. aerogenes	Continuous	35	Ammonia		0.50	1.02		
Glucose-metabolizing								
C. utilis	Batch	35	Ammonia	0.30	0.62	1.72		
C. utilis	Continuous	30	Ammonia	0.08–0.54	0.68	2.04	0.00	0.02
Tor. utilis	Continuous	30	Ammonia	0.05–0.40	0.66	1.64	0.03	0.01
Sacch. cerevisiae	Continuous	30	Ammonia	0.00–0.15	0.67	1.14	0.01	0.02
Penicillium chrysogenum	Continuous	25	Ammonia	0.01–0.11	0.59	1.64	0.02	0.02
P. chrysogenum	Continuous	25	Ammonia	0.02–0.09	0.55	2.00	0.02	0.02
Aspergillus nidulans	Continuous		Ammonia		0.75	2.70	0.02	0.01
Asp. awamori	Continuous		Ammonia		0.65	1.64	0.01	0.02
Trichoderma viride	Continuous	30	Ammonia	0.03–0.14	0.80	1.35	0.04	0.01
Trich. viride	Continuous	30	Ammonia	0.02–0.12	0.64	3.13	0.01	0.02
Pc. varioti	Continuous	38	Ammonia	0.30	0.70	1.85		0.08
Beneckea natriegens	Continuous	31.5	Ammonia	0.10–0.65	0.66	3.33	0.05	0.09
Ar. globiformis	Continuous	25	Ammonia	0.10–0.20	0.67		0.01	
E. coli	Continuous	30	Ammonia	0.05–0.50	0.64	1.69	0.05	0.01
E. coli	Continuous	37	Ammonia	0.10–0.60		1.00		0.01
E. coli	Continuous	17.5	Ammonia	0.09–0.37		2.33		0.02
E. coli	Continuous	23.5	Ammonia	0.13–0.37		2.33		0.04
E. coli	Continuous	30	Ammonia	0.13–0.30		2.27		0.05
E. coli	Batch	15	Ammonia		0.60		0.01	
E. coli	Batch	30	Ammonia		0.60		0.05	
E. coli	Batch	19.34	Ammonia	0.18–0.70	0.54	1.52		
Ps. fluorescens	Batch	35	Ammonia	0.44	0.47	1.10		
Rhodopseudomonas sheperoides	Continuous	30	Ammonia	0.02–0.17	0.60	2.27	0.01	0.01
Az. vinilandii	Continuous	30	Nitrogen	0.07–0.35	0.32	0.53	0.12	0.13
B. coagulans	Continuous	50	Ammonia	0.05–0.40	0.59		0.14	
Bacterium HR	Batch	30	Ammonia	0.28	0.46	0.93		

Glucose-metabolizing *continued*								
K. aerogenes	Continuous	30	Ammonia	0.05–0.60	0.54	1.79	0.05	0.03
K. aerogenes	Continuous	30	Ammonia	0.18	0.71	3.57		
K. aerogenes	Continuous	35	Ammonia	0.10–0.50	0.65	2.04		0.02
K. aerogenes	Continuous	35	Nitrate	0.10–0.50	0.51	1.64		0.02
Ent. aerogenes	Continuous		Nitrate		0.47	1.79	0.04	0.04
Formaldehyde-metabolizing								
Pseudomonas 1	Continuous	35	Ammonia	0.03–0.40	0.49		0.04	
Pseudomonas 135	Continuous	35	Ammonia	0.03–0.40	0.49		0.07	
Gluconate-metabolizing								
Micro. denitrificans	Continuous	37	Ammonia	0.05–0.35	0.53	1.75	0.02	0.02
Succinate-metabolizing								
Micro. denitrificans	Continuous	37	Ammonia	0.05–0.35	0.41	1.41	0.06	0.04
E. coli	Continuous	37	Ammonia	0.08–0.38		0.91		0.15
Citrate-metabolizing								
Ent. aerogenes	Continuous		Ammonia		0.37	1.32	0.05	0.04
Ps. fluorescens	Batch	35	Ammonia	0.35		1.10		
Malate-metabolizing								
Micro. denitrificans	Continuous	37	Ammonia	0.05–0.35	0.42	1.49	0.07	0.05
Ps. fluorescens	Batch	35	Ammonia	1.00	0.36	1.20		
Formate-metabolizing								
Ps. oxalaticus	Continuous	28	Ammonia	0.02–0.20	0.17	0.52	0.08	
Pseudomonas 1	Continuous		Ammonia	0.03–0.40	0.29		0.03	
Pseudomonas 135	Continuous		Ammonia	0.03–0.40	0.32		0.11	
Ps. oxalaticus	Continuous	28	Ammonia	0.02–0.15	0.09	0.55	0.06	

[a] Yield of biomass on substrate, measured as carbon equivalent per carbon equivalent.
[b] Yield of biomass on oxygen, measured as carbon equivalent per mole of oxygen.
[c] Maintenance on substrate, measured as carbon equivalent per carbon equivalent per hour. (*see* pp.140–141)
[d] Maintenance on oxygen, measured as moles of oxygen per carbon equivalent per hour. (*see* p.141)

Table 24. Biomass yield coefficients for anaerobic conditions.

Microorganism	Y_s [g(g mol glucose)$^{-1}$]
Clostridium thermoaceticum	50
Klebsiella aerogenes[a]	23.4 (after 1 day)
K. aerogenes	14.4 (after 3 days)
Lactobacillus plantarum	20.0
L. casei	42.0
Saccharomyces cerevisiae[a]	20.0
Sacch. rosei	23
Streptococcus faecalis[a]	20.0

[a] Bauchop and Elsden (1960). Other references are to be found in reviews by Harrison (1976), Stouthamer (1976), and Forrest and Walker (1971).

Molar growth yields of chemoautotrophic microorganisms

In chemoautotrophic microorganisms, the energy of oxidation of an inorganic substrate is utilized for synthesis of all cellular components by carbon dioxide fixation. The growth yields can be determined by measuring the incorporation of [^{14}C] carbon dioxide into cellular material. Usually the ratio of oxygen consumed to carbon dioxide assimilated is used to express the efficiency of growth. Table 25 gives some examples of the O_2/CO_2 ratio for different microorganisms with several substrates.

Table 25. Relationship between oxygen consumed and carbon dioxide assimilated [Stouthamer (1969)].

Microorganism	Substrate	O_2/CO_2[a]
Thiobacillus thiooxydans	S	18
T. thioparus	$S_2O_3^{2-}$	19
Nitrobacter winogradskyi	NO_2^-	34
N. winogradskyi	NO_2^-	36
Thiobacillus strain C	$S_2O_3^{2-}$	11
T. ferrooxydans	Fe^{2+}	27
T. ferrooxydans	S	7.5
T. neapolitans	$S_2O_3^{2-}$	5.0[b]
T. neapolitans	$S_2O_3^{2-}$	2.4[c]
Hydrogenomonas strain H.16	H_2	2.0[d]

[a] In several cases, these O_2/CO_2 ratios have been calculated from growth yields in g mol^{-1} inorganic substrate oxidized, assuming a carbon content of the bacteria of 50 per cent.
[b] This result is the maximal yield that may be obtained in continuous culture.
[c] This yield is corrected for maintenance requirement.
[d] Value for non-growing cells.

The requirement for maintenance can be especially important in slow-growing chemoautotrophs, such as *N. winogradskyi*.

Mixed Cultures (as in Biological Waste Water Treatment)

Relationships between BOD, COD and TOD

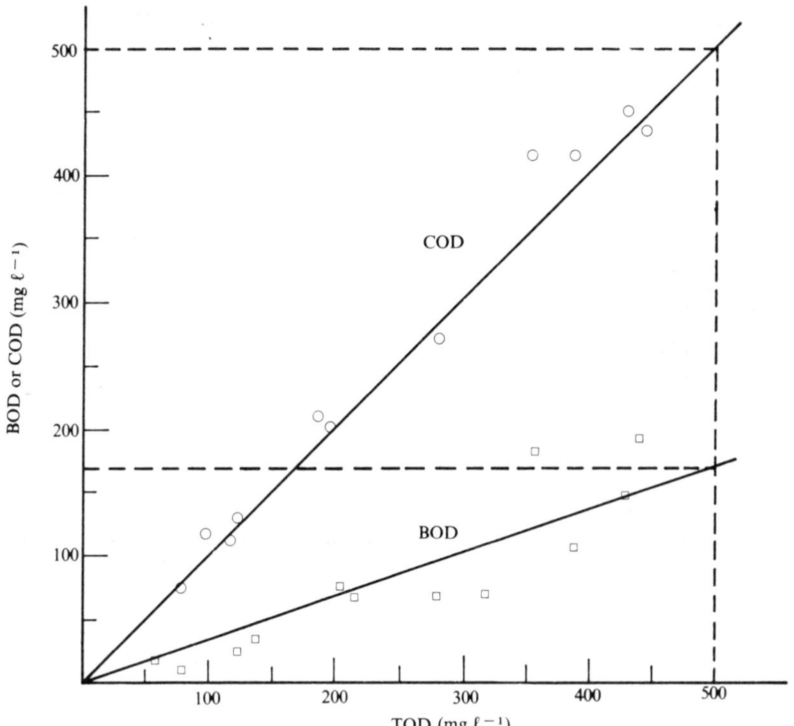

Figure 12. Correlations between TOD and COD or BOD of an effluent in the low range [Cheremisinoff and Young (1975)].

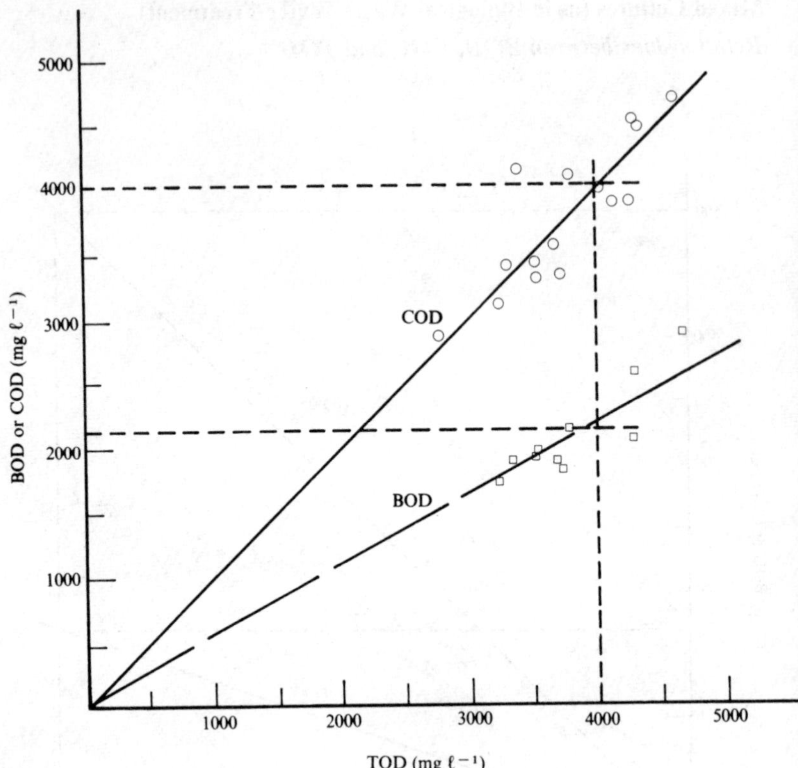

Figure 13. Correlations between TOD and COD or BOD of an effluent in the high range [Cheremisinoff and Young (1975)].

The overall concentration of the complex mixture of components that occurs in a typical effluent is measured in terms of the oxygen requirement under various oxidative conditions. The three analytical methods in common usage determine either the total oxygen demand (TOD), chemical oxygen demand (COD) or biological oxygen demand (BOD) (*see* Glossary, at beginning of this chapter). Experimentally such analyses are either time-consuming or require special equipment. However, since the methods are intended to provide a measure of the same property of the system, it is to be expected that relatively simple correlations should exist between BOD, COD and TOD. Figs. 12 and 13 suggest that

$$\text{COD} = \text{Total organic carbon (TOC)} \tag{32}$$

and

$$\text{BOD} = \text{k TOC} \tag{33}$$

although the proportionality constant *k* appears to differ according to the range of TOC.

In general, Figs. 12 and 13 suggest that TOC, COD and BOD are simply related and that correlations can be achieved between methods. However, such correlations are likely to be dependent on both the nature and the strength of the waste.

Brewers yield coefficients

Table 26. Industrial wastes [Servizi and Bogan (1964)].

Waste	Substrate yield coefficient[a]
Sulphite waste liquor	1.0
Cotton kiering liquor	0.5[b]
Reg rope waste	0.5[b]
Brewery waste	0.5–0.93[b]
Domestic sewage	0.5
Jute cook liquor	0.5
Yeast waste	0.5
Gum waste	0.5
Boardmill white water	0.5
Domestic sewage and glucose	0.35
Spent sulphite liquor	0.38
Penicillin and fine chemical wastes	0.76
Phenolic waste water	0.70
Waste paper repulping and chemical processing	0.77

[a] Kilograms volatile suspended solids per kilogram of BOD removed.
[b] Expressed as kilograms of suspended solids per kilogram of BOD removed.

Table 27. Specific waste components [Stouthamer (1969)].

Substrate	Y_s (g mol^{-1})	(av e$^-$)mol^{-1}	Y_{kcal} (g kcal^{-1})	$Y_{av\,e-}$ [g (ave^{-1})$^{-1}$]
Glucose	81.8	24	0.118	3.41
Glucose	74	24	0.110	3.08
Glycerol	43	14	0.108	3.07
Acetate	22	8	0.105	2.75
Alanine	30	15	0.079	2.00
Benzoate	80	30	0.104	2.68
Anthranilate	74	31	0.092	2.39
Sucrose	142	48	0.105	2.96
Glutamate	52	21	0.096	2.48
Citrate	56	18	0.118	3.12
Hydrocinnamate	111	42	0.102	2.64
Succinate[a]	40.8	14	0.115	2.93
Fumarate[a]	40.0	12	0.125	3.34
Lactate[a]	40.0	12	0.123	3.34
Pyruvate[a]	32.1	10	0.121	3.21
Glycerol[a]	48.6	14	0.122	3.48
Arabinose[a]	77.0	20	0.137	3.85
Glucose[a]	84.3	24	0.125	3.51
Lactose[a]	157.0	48	0.116	3.27
Glycine	14	9	0.087	1.56
Alanine	31	15	0.100	2.06
Alanine	32	15	0.103	2.13
Fumarate	43	12	0.130	3.58
Glycerol	42	14	0.103	3.00
Citrate	64	18	0.125	3.56
Sorbate	76	28	0.103	2.39
Benzoate	82	30	0.106	2.74
Benzoate	77	30	0.100	2.57
Hydrocinnamate	119	42	0.111	2.83
Benzoate, alanine, citrate, glycerol and sucrose[b]	64		0.120	

Table 27 – (continued)

Benzoate, alanine and sucrose	66		0.113	
Benzoate, alanine, sorbate and sucrose[b]	71		0.114	
Glucose	79.6	24		3.32
Glutamate	57.6	21		2.74
Acetate	19.2	8		2.40
Glucose	107.1	24		4.46
Mean	64.72	21.79	0.110	2.94

[a] Based on measurements of cell carbon and the assumption that cells comprise 50% carbon.
[b] Quantities of each not specified.

Hydrocarbons

Table 28. Biomass yields for aerobic growth on hydrocarbons [Stouthamer (1969)].

Microorganism	Hydrocarbon[a]	Y_s (g mol^{-1})	(av e$^-$) mol^{-1}	$Y_{av\,e^-}$ [g (av e$^-$)$^{-1}$]
Mixed bacterial culture	$C_{17}H_{36}$	240.5	104	2.31
Pristane bacterial culture	$C_{18}H_{38}$	245.2	110	2.23
Pristane bacterial culture	$C_{19}H_{40}$	281.1	116	2.42
Candida intermedia	$C_{16}H_{34}$	185.7	98	1.89
C. intermedia	$C_{17}H_{36}$	174.8	104	1.68
C. intermedia	$C_{18}H_{38}$	211.5	110	1.92
C. intermedia	$C_{22}H_{46}$	275.8	134	2.06
C. intermedia	$C_{28}H_{58}$	326.7	170	1.92
C. intermedia + *C. lipolytica*[b]	$C_{16}H_{34}$	177.3	98	1.81
C. intermedia + *C. lipolytica*[b]	$C_{17}H_{36}$	178	104	1.71
C. intermedia + *C. lipolytica*[b]	$C_{18}H_{38}$	215	110	1.95
C. intermedia + *C. lipolytica*[b]	$C_{22}H_{46}$	278	134	2.07
C. intermedia + *C. lipolytica*[b]	$C_{28}H_{58}$	348	170	2.05
C. lipolytica	'$C_{15}H_{32}$'	190.7	92	2.07
C. lipolytica	$C_{18}H_{38}$	217.9	110	1.98
C. tropicalis	$C_{16}H_{34}$	181.8	98	1.86
Micrococcus cerificans	$C_{16}H_{34}$	252.3	98	2.57

Mycobacterium phlei	'$C_{15}H_{32}$'	223	92	2.43
Nocardia NBZ 23	'$C_{15}H_{32}$'	197.7–216.1	92	2.25–2.57
N. opoca	C_{10}–C_{20}	200.8	89	2.26
N. opoca	'$C_{15}H_{32}$'	127	92	1.38
Pseudomonas aeruginosa	$C_{16}H_{34}$	242	110	2.20
Torula sp.	$C_{16}H_{34}$	186.4	98	1.90
Mean		223.2	108.7	2.05

[a] '$C_{15}H_{32}$' represents a C_{14}–C_{17} mixture with a composition such that the average molecular weight is that of a pentadecane (i.e. 212). The average molecular weight of the C_{10}–C_{20} mixture was 205 or 97% of pentadecane and 89 av e^- mol^{-1} was assumed.
[b] Stable mixture.

Table 29. Yields for growth on *n*-alkanes and gas oil [Wang (1968)].

Microorganism	Substrate	Temperature (°C)	Doubling time (h)	Cell yield on substrate, Y_S [g (g substrate)$^{-1}$]	Cell yield on oxygen Y_{O_2} [g (g mol O_2)$^{-1}$]
Torulopsis sp.	*n*-Alkanes (C_{14}–C_{18})	30	7	0.72	0.29
Candida intermedia	*n*-Alkanes (C_{16}–C_{20})	30	4.5–6.5	0.81	0.35
C. intermedia	*n*-Alkanes	32–38	4.3	0.87	0.36–0.39
C. lipolytica	*n*-Alkanes	32–38	3.85	1.02	0.36–0.39
Yeasts	*n*-Alkanes (C_{14}–C_{25})	32–38		0.94–1.08	0.36–0.39
Pseudomonas sp.	*n*-Alkanes (C_{15})			1.03	0.5
Micrococcus cerificans	C_{18}	30	1.4	1.30	1.12
M. cerificans	C_{16}	30	2.8	1.13	0.78
Yeast	Gas oil (32% *n*-alkanes)	30		0.8	
Candida sp.	Gas oil (12.7–18.1% *n*-alkanes)	30	4.0–9.3	0.73–0.96	

Table 30. Growth of *Candida* on petroleum fractions [Johnson (1967)].

Substrate	*n*–Alkane Content (% wt)	*n*–Alkane Used (% of total)	Generation time (hr)	Yield on alkane used (%)	Cell N (%)	Cell Lipid (%)
Virgin gas oil, furnace-boiling range	18.1	96	4.0	73	9.02	1.9
Virgin gas oil, intermediate-boiling range	13.2	90	4.3	90	9.09	1.5
Gas oil from catalytic cracking unit	12.7	97	5.5	76	8.77	2.2
Waxy distillate	10.7	94	6.0	79	9.28	2.3
Viscous high-boiling fraction	6.5	72	9.3	96	8.83	7.2

Table 31. Growth of *Candida* on *n*-alkanes[a] [Johnson (1967)].

Substrate	Generation time (h)	Cell yield on substrate (%)	Carbon dioxide produced [mol^{CO_2}(mol substrate)$^{-1}$]	Carbon recovered as cells and carbon dioxide (%)
$C_{12}H_{26}$	7.0	59.8	4.6	72.7
$C_{14}H_{30}$	6.5	83.0	5.2	84.2
$C_{16}H_{34}$	6.5	81.7	7.5	93.8
$C_{17}H_{36}$	5.0	72.9	4.3	67.6
$C_{18}H_{38}$	4.5	82.5	5.4	76.6

[a] Temperature was 30°C, pH was controlled at 5.5. Cell concentration at the end of the growth period was 10–15 g ℓ^{-1}.

Table 32. Growth of *Candida* on solid alkanes[a] [Johnson (1967)].

Substrate	Generation time (h)	Crude protein yield on substrate (%)
$C_{20}H_{42}$	3.0	45.2
$C_{22}H_{46}$	3.0	44.8

[a] Alkanes were dissolved in 2, 6, 10, 14 – tetramethylpentadecane.

$Y_{av\,e^-}$, Y^{ATP} and P/O Ratios

The various yield coefficients are related by the expressions given earlier in this chapter, (*see* pp.137–140) in particular Y_{ATP}, Y^{ATP} and P/O are related by Eqs. (17) to (19).

Table 33. Growth yields from anaerobic growth of heterotrophs [Hernandez and Johnson (1967)].

Microorganism	Substrate	Y_S [g (mol substrate^{-1})]	y^{ATP} [mol ATP (mol substrate)$^{-1}$]	Y_{ATP} [g (mol ATP)$^{-1}$]
Actinomyces israeli	Glucose	24.7	2.0	12.3
Aerobacter aerogenes	Glucose	26.1	3.0	10.2
A. aerogenes	Fructose	26.7	3.0	10.7
A. aerogenes	Mannitol	21.8	2.5	10.8
A. aerogenes	Gluconate	21.4	2.5	11.0
A. cloacae	Glucose	17.7–27.1	1.5–2.5	11.9
Bifidobacterium bifidum	Glucose	37.4	2.9	13.1
B. bifidum	Lactose	52.8	5.1	10.4
B. bifidum	Galactose	27.8	2.8	9.9
B. bifidum	Mannitol	27.8	2.4	11.8
Clostridium aminobutyricum	γ-Aminobutyrate	7.6	0.5	15.2
C. aminobutyricum	γ-Aminobutyrate	8.9	0.5	17.8
C. glycolicum	Ethylene glycol	7.7	0.5	15.4
C. kluyveri	Crotonate	4.8	0.5	9.6
C. kluyveri	Ethanol + acetate			9.2
C. pasteurianum	Sucrose	73.1	6.6	11.0
C. tetranomorphum	Glutamate	6.8	0.6	10.9
C. thermoaceticum	Glucose	50.0		16.6
Escherichia coli	Glucose	25.8	3.0	11.2
Lactobacillus casei L3	Glucose	42.9	2.1	20.9
L. casei L3	Mannitol	40.5	2.2	18.2
L. casei L3	Citrate	18.2	1.0	19.0
L. plantarum	Glucose	20.4	2.0	9.8
L. plantarum	Galactose	32.5	3.0	10.2
Ruminococcus flavefaciens	Glucose	29.1	2.8	10.6
Saccharomyces cerevisiae	Glucose	18.8–22.3	2.0	10.2
Sacch. rosei	Glucose	22.0–24.6	2.0	11.6
Sarcina ventriculi	Glucose	30.5	2.6	11.7
Streptococcus agalactiae	Glucose	20.8	2.3	9.3
Strep. agalactiae	Pyruvate	7.5	0.7	10.4
Strep. faecalis	Glucose	20.0–37.5	2.0–3.0	10.9
Strep. faecalis	Gluconate	17.6–20.0	1.8	10.4
Strep. faecalis	2-Ketogluconate	19.5	2.3	8.5
Strep. faecalis	Ribose	21.0	1.7	12.6
Strep. faecalis	Arginine	10.2	1.0	10.2
Strep. faecalis	Pyruvate	10.4	1.0	10.4
Strep. lactis	Glucose	19.5	2.0	9.8
Strep. pyogenes	Glucose	25.5	2.6	9.8
Zymomonas anaerobia	Glucose	5.9	1.0	5.9
Z. mobilis	Glucose	8.0–9.3	1.0	8.5

Table 34. Oxygen consumption and accompanying phosphate esterification by a crude cell-free extract of *Gluconobacter liquefaciens* (galactose-grown) [Stouthamer (1962)].

Substrate	Oxygen consumption (μ mol)	Inorganic phosphate esterified (μmol)	P/O ratio
Glucose	5.14	1.70	0.17
Glucose 6-phosphate + $NADP^+$	6.10	4.42	0.36
Citrate + $NADP^+$	5.92	2.55	0.22
Succinate	4.61	1.28	0.14
DL-Lactate	3.08	0.45	0.07
Glycerol	3.43	0.02	0.07
Ethanol			0.10
NAD^+			0.09
$NADP^+$			0.22
α-Ketoglutarate			0.22

Table 35. Influence of pH on the P/O ratios for cell-free extracts of *Gluconobacter liquefaciens* (galactose-grown) [Stouthamer (1962)].

Substrate	P/O ratio pH 6.0	pH 7.2
Glucose	0.12	0.12
Glucose 6-phosphate + $NADP^+$	0.32	0.23
Citrate + $NADP^+$	0.27	0.42

Table 36. Yields and distribution of substrate electrons incorporated into cells or transferred to oxygen [Mayberry *et al.* (1968)].

Microorganism	Substrate	Y_S	O_2/sub	Y_{ave^-}	Equiv e^-/ mole sub (to cells)	Equiv e^-/ mole sub (to O_2)	Total av e^- (calculated)	Av e^-/ mole sub (actual)
Pseudomonas $C_{12}B$	Benzoate	86.8	3.46	2.89	17.3	13.8	31.1	30
Pseudomonas $C_{12}B$	Phenylglyoxylate	102.2	4.06	3.20	20.4	16.2	36.6	32
Pseudomonas $C_{12}B$	Phenylacetate	111.0	3.73	3.08	22.1	14.9	37.0	36
Pseudomonas $C_{12}B$	Succinate + acetate	72.2	2.54	3.28	14.4	10.2	24.6	22
Pseudomonas $C_{12}B$	Acetate	23.5	1.11	2.94	4.7	4.4	9.1	8
Pseudomonas $C_{12}B$	Succinate	42.3	1.40	3.02	8.4	5.6	14.0	14
Pseudomonas $C_{12}B$	1-Dodecanol	217.0	7.41	3.01	43.2	29.6	72.8	72
Bacterium TEG-5	Diethylene glycol	58.0	3.10	2.90	11.6	12.4	24.0	20
Bacterium TEG-5	Triethylene glycol	103.1	4.40	3.44	20.6	17.6	38.2	30
Bacterium TEG-5	Tetraethylene glycol	129.9	6.01	3.25	25.8	24.0	49.8	40

Values for Y_{ave^-} in Table 36 are deduced in the following way. In the case of the oxidation of acetic acid, i.e.

$$CH_3COOH + 2O_2 \longrightarrow 2H_2O + 2CO_2$$

Available electrons (av e^-) may be determined as 4×2 av e^- (mol acetate)$^{-1}$, where 4 is the number of electrons required to reduce 1 mole of oxygen.

Thus

$$Y_{ave^-} = \frac{Y_s}{\text{av e}^- \text{ (mol acetate)}^{-1}} = \frac{23.5}{4 \times 2} = 2.94$$

FACTORS INFLUENCING YIELD—EXPERIMENTAL DATA

A large number of factors influence the biomass yield, oxygen requirement and product yield of a given microorganism. These include the chosen substrate and source of nitrogen, the composition of the medium, the pH and temperature, as well as the growth rate. Further factors include the use of anaerobic or aerobic conditions (*see* Tables 37 and 38), the nature of the hydrogen acceptor (*see* Table 38) and the presence of inhibitors (*see* Tables 39 and 40).

Table 37. Growth of *Streptococcus* sp. under aerobic/anaerobic conditions [Stouthamer (1976)].

Microorganism	Substrate	Conditions	Y_S (g mol^{-1})	Acetate production, $Y_{acetate}$ [mol (mol substrate)$^{-1}$]
Strep. agalactiae	Glucose	Anaerobic	20.8	0.25
Strep. agalactiae	Glucose	Aerobic	53.4	0.85
Strep. agalactiae	Pyruvate	Anaerobic	7.5	0.72
Strep. agalactiae	Pyruvate	Aerobic	11.9	0.64
Strep. faecalis	Glucose	Anaerobic	21.5	
Strep. faecalis	Glucose	Aerobic	58.2	1.70
Strep. faecalis	Pyruvate	Anaerobic	6.0	
Strep. faecalis	Pyruvate	Aerobic	15.5	
Strep. faecalis	Mannitol	Anaerobic	No growth	
Strep. faecalis	Mannitol	Aerobic	64.6	1.83
Strep. faecalis	Glycerol	Anaerobic	No growth	
Strep. faecalis	Glycerol	Aerobic	24.7	
Strep. faecalis	Glucose	Anaerobic	16.0	
Strep. faecalis	Glucose	Aerobic	26.8	
Strep. faecalis	Glucose	Aerobic	55.2	

Table 38. Growth of *Aerobacter aerogenes* under aerobic/anaerobic conditions.

Substrate	Conditions	Y_S [g (mol substrate)$^{-1}$]	Acetate production, $Y^{acetate}$ [mol (mol substrate)$^{-1}$]
Glucose (1.5 mM)	Aerobic + nitrate (35 mM)	52.0	
Glucose (1.5 mM)	Anaerobic + nitrate (35 mM)	43.0	
Glucose (3.0 mM)	Anaerobic	26.1	0.84
Glucose (3.0 mM)	Anaerobic + nitrate (71.33 mM)	45.5	1.63
Glucose (3.0 mM)	Aerobic	72.7	
Glucose (3.0 mM)	Aerobic + nitrite (5 mM)	39	
Glucose (3.0 mM)	Anaerobic + nitrite (5 mM)	16	
Mannitol (3.0 mM)	Anaerobic	21.8	0.42
Mannitol (3.0 mM)	Anaerobic + nitrate (6 mM)	50.6	1.07

Table 38 shows that (1) biomass yield is greater under aerobic conditions, (2) nitrate can substitute for oxygen as the hydrogen acceptor and (3) nitrite is an inhibitor.

Table 39. Growth of *Aerobacter aerogenes* under anaerobic conditions with nitrite as the hydrogen acceptor and glucose (3 mM) as substrate.

Nitrite concentration (mM)	Y_s [g (mol glucose)$^{-1}$]	Acetate production, $Y^{acetate}$ [mol (mol glucose)$^{-1}$]
0.0	26.1	0.84
0.5	27.0	1.20
1.0	24.5	1.58
1.5	22.4	1.60
2.6	20.6	Not determined
5.0	16.0	Not determined
10.0	No growth	

Data in Table 39 show nitrate to be inhibitor for growth beyond 0.5 mM, while the acetate yield improves.

Table 40. The presence of growth inhibitory compounds on molar growth yields of a number of microorganisms growing anaerobically with glucose.

Microorganism	Inhibitor	Inhibitor concentration (mM)	Y_{ATP} [g (g mol ATP)$^{-1}$]
Aerobacter aerogenes	None		10.2
A. aerogenes	Nitrite	0.5	9.3
A. aerogenes	Nitrite	1.0	7.4
A. aerogenes	Nitrite	1.5	6.7
Escherichia coli	None		12.7
E. coli	Ferricyanide	10	10.3
E. coli	Ferricyanide	40	5.9
Proteus mirabilis	None		5.5
P. mirabilis	Thiosulphate	7.5	3.7
Saccharomyces cerevisiae	None		11.1
Sacch. cerevisiae	Dinitrophenol	0.05	8.3
Sacch. cerevisiae	Dinitrophenol	0.1	6.0

Nature of Substrate

Table 41. Growth yields from aerobic growth of heterotrophs in a minimal medium [Stouthamer (1969)].

Microorganism	Substrate	Y_s (g mol^{-1})	av e$^-$ mol^{-1}	Y_{kcal} (g kcal^{-1})	$Y_{av\,e^-}$ [g av e^{-1}]
Aerobacter aerogenes	Maltose	149.2	48	0.14	3.11
A. aerogenes	Sucrose	172.7	48	0.14	3.60
A. aerogenes	Mannitol	95.5	26		3.67
A. aerogenes	Sorbitol	97.4	26	0.14	3.75
A. aerogenes	Galactose	73.5	24		3.06
A. aerogenes	Glucose	72.7	24	0.14	3.03
A. aerogenes	Mannose	69.4	24		2.89
A. aerogenes	Fructose	76.1	24	0.14	3.17
A. aerogenes	Gluconate	62.2	22	0.14	2.83
A. aerogenes	Galactonate	66.3	22	0.14	3.01
A. aerogenes	Ribitol	61.8	22		2.81
A. aerogenes	L-Arabinose	65.3	20	0.13	3.27

(continued)

Table 41 – continued)

Microorganism	Substrate	Y_s ($g\ mol^{-1}$)	av e^- mol^{-1}	Y_{kcal} ($g\ kcal^{-1}$)	$Y_{av\ e^-}$ [$g\ av\ e^{-1}$]
A. aerogenes	Ribose	53.2	20		2.66
A. aerogenes	Xylose	52.2	20		2.61
A. aerogenes	Glucuronate	55.3	20	0.13	2.77
A. aerogenes	Galacturonate	55.6	20	0.14	2.78
A. aerogenes	Citrate	61.8	18		3.43
A. aerogenes	Rhamnose	49.0	26	0.13	
A. aerogenes	Glycerol	41.8	14		2.99
A. aerogenes	Dihydroxyacetone	31.9	12		2.66
A. aerogenes	*myo*-Inositol	52.2	24	0.13	
A. aerogenes	Glucose	72.7	24	0.14	3.03
A. aerogenes	Glycerol	45.0	14		3.21
A. aerogenes	Glycerol	50.0	14		3.57
A. aerogenes	Citrate	19.6	18		1.09
Aerobacter cloacae	Glucose	79.2	24		3.30
A. cloacae	Glycerol	48.2	14		3.44
Arthrobacter globiformis	Glucose	94.0	24		3.92
Azobacter chroococcum	Mannitol	63.5	26		2.44
Bacillus subtilis[a]	Glucose	81.3[c]	24		3.39
Bacterium HR	Glucose	72.0	24	0.11	3.00
Candida utilis	Glucose	91.9	24	0.13	3.83
C. utilis	Acetate	21.6	8	0.10	2.70
C. utilis	Ethanol	31.6	12	0.09	2.63
C. utilis	Glucose	90.7[c]	24		3.78
C. utilis	Glucose	92.7[c]	24		3.86
Escherichia coli	Glucose	67.8	24		2.83
E. coli	Glucose	94.0	24		3.92
E. coli	Glucose	90.0	24		3.75
Hydrogenomonas eutropa	Acetate	20.4	8		2.55
H. eutropa	Lactate	27.8	12		2.32
Neurospora crassa	Sucrose	115.0	48		2.40
Pseudomonas $C_{12}B$	Benzoate	86.8	30	0.11	2.89
Pseudomonas $C_{12}B$	Phenylglyoxylate	102.2	32	0.11	3.20
Pseudomonas $C_{12}B$	Phenylacetate	111.0	36	0.11	3.08
Pseudomonas $C_{12}B$	Succinate	42.3	14	0.11	3.02
Pseudomonas $C_{12}B$	Acetate	23.5	8	0.10	2.94
Pseudomonas $C_{12}B$	Succinate + acetate	72.2	22	0.11	3.28
Pseudomonas $C_{12}B$	Dodecanol	217.0	72	0.11	3.01
Ps. aeruginosa	Glucose (30°C)	78.1	24	0.12	3.25
Ps. aeruginosa	Glucose (37°C)	76.2	24	0.11	3.18
Ps. aeruginosa	Gluconate (30°C)	75.5	22	0.12	3.43
Ps. aeruginosa	Gluconate (37°C)	73.1	22	0.12	3.32
Ps. aeruginosa	2-Oxogluconate (30°C)	64.6	20	0.12	3.23
Ps. aeruginosa	2-Oxogluconate (37°C)	62.0	20	0.12	3.10
Ps. aeruginosa	Glucose	70.4	24		2.93
Ps. aeruginosa	Gluconate	62.0	22		2.82
Ps. aeruginosa	Citrate	57.5	18		3.19
Ps. aeruginosa	α-Oxoglutarate	52.0	16		3.25
Ps. aeruginosa	Succinate	41.6	14		2.97
Ps. aeruginosa	Fumarate	38.0	12		3.17
Ps. aeruginosa	Malate	38.5	12		3.21
Ps. aeruginosa	Pyruvate	30.0	10		3.00
Ps. aeruginosa	Acetate	17.0	8		2.15
Ps. aeruginosa	Malate	44.1	12		3.68
Ps. aeruginosa	Glucose	48.0	24		2.00
Ps. fluroescens	Glucose	69.5	24	0.11	2.90
Ps. fluorescens	Ethanol	22.1	12	0.08	
Ps. fluorescens	Acetate	16.8	8	0.08	

continued

Table 41 – (continued)

Microorganism	Substrate	Y_s (g mol^{-1})	av e$^-$ mol^{-1}	Y_{kcal} (g kcal^{-1})	$Y_{av\ e^-}$ [g av e^{-1}]
Ps. fluorescens	Lactate	28.8	12	0.11	2.40
Ps. fluorescens	Malate	34.8	12	0.11	2.90
Ps. fluorescens	Glucose	119.1	24		4.96
Rhizobium meliloti	Sucrose	144.0	48		3.00
Saccharomyces cerevisiae[a]	Glucose	71.3	24		2.97
Sacch. cerevisae[a]	Glucose	78.4	24		3.27
Salmonella typhimurium	Citrate	21.6	18		1.20
Serratia marcescens	Glucose + citrate	6.9	1.50+ 0.71		3.12
Ser. marcescens	Glucose + citrate	12.6	2.70+ 1.12		3.30
Ser. marcescens	Glucose + citrate	21.4	5.19+ 1.62		3.14
Ser. marcescens	Glucose + citrate	29.0	7.56+ 1.93		3.06
Streptococcus faecalis[b]	Glucose	38.2	24	0.11	
Strep. faecalis	Mannitol	64.6	26	0.10	
Bacterium TEG-5	Diethylene glycol	58.0	20	0.09	2.90
Bacterium TEG-5	Triethylene glycol	103.1	30	0.10	3.40
Bacterium TEG-5	Tetraethylene glycol	129.9	40	0.10	3.25
Mean		65.1	21.51	0.12	3.07

[a] Grown on a supplemented medium.
[b] Grown on a complex medium.
[c] Average value.

Table 42. Growth yield of the autotroph *Desulfovibrio desulfuricans* Canet 41[a] [Payne (1970)].

Substrate	Hydrogen acceptor	Y_s [g(mol^{-1} substrate)$^{-1}$]
Pyruvate	Sulphate	9.4 ± 1.6
Pyruvate	Sulphite	9.5 ± 1.9
Lactate	Sulphate	9.9 ± 1.8

[a] Growth was on synthetic media at 32°C.

An example of growth yields resulting from combinations of microorganisms and substrates that produce a particular product is given in Table 42.

Table 43. Growth yields for microorganisms forming propionate [Stouthamer (1976)].

Microorganism	Substrate	Y_s [g (mol substrate)$^{-1}$]	ATP from substrate level phosphorylation, Y^{ATP}	Y_{ATP} [g (g mol ATP)$^{-1}$]
Anaerovibrio lipolytica	Glycerol	22.0[a]		
A. lipolytica	Fructose	60.0		
Propionibacterium freudenreichii	Glucose	82.0[b]	2.7	16.7
P. freudenreichii	Glycerol	27.0[b]	1.0	13.5
P. freudenreichii	Lactate	10.2[b]	0.3	15.5
P. pentosaceum	Glucose	37.5	2.7	
P. pentosaceum	Glucose	76.3[b]	2.7	16.3
P. pentosaceum	Glycerol	20.0	1.0	
P. pentosaceum	Glycerol	26.3[b]	1.0	13.5
P. pentosaceum	Lactate	7.6	0.3	
P. pentosaceum	Lactate	12.9[b]	0.3	19.5
Selenomonas ruminantium	Glucose	17.0[c]		
S. ruminantium	Glucose	62.0[a]		
S. ruminantium	Pyruvate	21.0[d]		

[a] In chemostat at $D = 0.1$.
[b] In complex medium.
[c] Batch culture.
[d] In chemostat cultures with glucose plus pyruvate.

Table 44. Products and carbon balance for *Streptococcus faecalis* 10C1 grown aerobically [Smalley *et al.* (1968)].

	Substrate			
	Glucose (mM)		Mannitol (mM)	
Products	9.97	59.82[d]	9.75	58.50[e]
Lactate[a]	1.83	5.49	1.62	4.86
Acetoin	0.25	0.99	0.16	0.64
Acetate	16.98	33.96	17.87	35.73
Formate	0.02	0.02	0.04	0.04
Carbon dioxide[b]	0.50	0.50	0.32	0.32
Carbon dioxide[c]	16.98	16.98	17.83	17.83

[a] Results expressed as μ moles of carbon per millilitre.
[b] Estimated from acetoin.
[c] Estimated from acetate.
[d] Total of all products was 57.94 (μmol C)ml^{-1} Carbon recovery = 57.94/59.82 × 100% = 96.9%.
[e] Total of all products was 59.43 (μmol C)ml^{-1} Carbon recovery = 59.43/58.50 × 100% = 101.6%.

The data in Table 44 provide an example of the range of products for a particular microorganism and also show a choice of substrate.

Medium Composition

Table 45. Values of Y_S, Y_{O_2} and $Y_{av\,e^-}$ for various microorganisms grown in minimal media [Mateles (1971)].

Microorganism	Substrate		Y_S		Y_{O_2}	$Y_{av\,e^-}$
		$(g\,g^{-1})$	$(g\,mol^{-1})$	$[g\,(gC)^{-1}]$	$(g\,g^{-1})$	$[g\,(av\,e^-)^{-1}]$
Aerobacter aerogenes	Maltose	0.46	149.2	1.03	1.50	3.11
A. aerogenes	Mannitol	0.52	95.5	1.32	1.18	3.67
A. aerogenes	Fructose	0.42	76.1	1.05	1.46	3.17
A. aerogenes	Glucose	0.40	72.7	1.01	1.11	3.00
A. aerogenes	Ribose	0.35	53.2	0.88	0.98	2.66
A. aerogenes	Succinate	0.25	29.7	0.62	0.62	2.12
A. aerogenes	Glycerol	0.45	41.8	1.16	0.97	2.99
A. aerogenes	Lactate	0.18	16.6	0.46	0.37	1.38
A. aerogenes	Pyruvate	0.20	17.9	0.49	0.48	1.78
A. aerogenes	Acetate	0.18	10.5	0.43	0.31	1.31
Candida utilis	Glucose	0.51	91.8	1.28	1.32	3.82
C. utilis	Acetate	0.36	21.0	0.90	0.70	2.62
C. utilis	Ethanol	0.68	31.2	1.30	0.61	2.60
Klebsiella sp.	Methanol	0.38	12.2	1.01	0.56	2.03
Methylcoccus sp.	Methane	1.01	16.2	1.34	0.29	2.02
Methylomonas sp.	Methanol	0.48	15.4	1.28	0.53	2.56
Penicillium chrysogenum	Glucose	0.43	77.4	1.08	1.35	3.22
Pseudomonas sp.	Methanol	0.41	13.1	1.09	0.44	2.18
Pseudomonas sp.	Methane	0.80	12.8	1.06	0.20	1.60
Pseudomonas sp.	Methane	0.60	9.6	0.80	0.19	1.20
Ps. fluorescens	Glucose	0.38	68.4	0.95	0.85	2.85
Ps. fluorescens	Acetate	0.28	16.8	0.70	0.46	2.10
Ps. fluorescens	Ethanol	0.49	22.5	0.93	0.42	1.87
Ps. methanica	Methane	0.56	9.0	0.75	0.17	1.12
Rhodopseudomonas spheroides	Glucose	0.45	81.0	1.12	1.46	3.37
Saccharomyces cerevisiae	Glucose	0.50	90.0	1.25	0.97	3.75

[a] Media comprised substrate, inorganic nitrogen and phosphorus, plus mineral salts.

Table 46. Molar growth yields for a number of microorganisms growing on various nitrogen sources [Stouthamer (1976)].

Microorganism	Substrate	Growth rate (h^{-1})	Molar growth yield ($g\ mol^{-1}$)		
			Ammonia	Nitrate	Nitrogen
Aerobic conditions					
Aerobacter aerogenes	Glucose	0.67	48.2		
A. aerogenes	Glucose	0.39		24.5	
A. aerogenes	Glucose	Not determined	72.7	52	
Azobacter chroococcum	Mannitol	0.05	58.3		8.5
Az. chroococcum	Mannitol	0.14	60.0		38.2
Az. chroococcum	Mannitol	0.25	55.8		45.3
Anaerobic conditions					
Clostridium pasteurianum	Sucrose	0.22	70.2		39.4
C. pasteurianum	Sucrose	0.39	69.5		43.3
Desulfovibro desulfuricans	Lactate	0.05	9.2		3.9
D. desulfuricans	Lactate	0.10	9.2		5.6
Klebsiella pneumoniae	Glucose	0.13	29.3		11.7
K. pneumoniae	Glucose	0.20	32.7		12.1

Table 47. Influence of hydrogen acceptor on molar growth yield [Hernandez and Johnson (1967)].

Microorganism	Energy source	Hydrogen acceptor	Y_s (g mol^{-1})	Growth rate (h^{-1})
Aerobacter aerogenes	Glucose	None	26.1	
A. aerogenes	Glucose	Nitrate	45.5	
A. aerogenes	Glucose	Oxygen	72.7	
A. aerogenes	Mannitol	None	21.8	
A. aerogenes	Mannitol	Nitrate	50.6	
A. aerogenes	Mannitol	Oxygen	95.5	
Citrobacter freundii	Complex medium	None	45.0	0.51
C. freundii	Complex medium	Nitrate	65.8	0.75
C. freundii	Complex medium	Oxygen	96.1	0.99
C. freundii	Complex medium	Tetrathionate	69.7	0.82
Proteus mirabilis	Minimal medium	None	14.0	0.17
P. mirabilis	Minimal medium	Nitrate	30.1	0.35
P. mirabilis	Minimal medium	Oxygen	58.1	0.69
P. mirabilis	Minimal medium	Tetrathionate	34.8	0.37

The data in Table 47 show that the molar growth yield is strongly affected by the hydrogen acceptor.

Table 48. Influence of nitrate on molar growth yields and products for growth on lactate [Stouthamer (1976)].

Microorganism	Presence of nitrate	Y_S (g mol^{-1})	Fermentation products: Acetate [mol (mol lactate)$^{-1}$]	Carbon dioxide [mol (mol lactate)$^{-1}$]	Hydrogen [mol (mol lactate)$^{-1}$]	Propionate [mol (mol lactate)$^{-1}$]
Propionibacterium pentosaceum	−	7.6	0.33	0.33		0.67
P. pentosaceum	+	20.0	1.0	1.0		
Vibrio alkalescens	−	7.4	0.5	0.5	0.5	0.5
V. alkalescens	+	17.8	1.0	1.0		

Data in Table 48 not only confirm the influence of nitrate on Y_s (*see* Table 47) but also show the concomitant effects on the fermentation products.

Table 49. Influence of amino acids on Y_O values of *Candida utilis* [Hernandez and Johnson (1967)].

Medium components	Growth rate (h^{-1})	Y_O [g (g atom O)$^{-1}$]
Glucose	21.1	0.30
Glucose + amino acids	21.6	0.69
Acetate	11.2	0.30
Acetate + amino acids	13.0	0.43
Ethanol	9.8	0.20
Ethanol + amino acids	14.4	0.38

The presence of amino acids can significantly increase the values of Y_O (*see* Table 49).

Table 50. Values of Y_{O_2} for *Pseudomonas fluorescens* grown on glucose[a] in the presence of amino acids, nucleic acid precursors and yeast extract, with oxygen as the growth-limiting factor [Hernandez and Johnson (1967)].

Additions	Generation time (h)	Y'_{O_2} (g g^{-1})
None	1.6	0.85
Vitamins	0.9	0.81
Nucleic acid precursors	1.3	0.98
Amino acids	0.7	1.06
Vitamins + nucleic acid precursors	1.5	0.77
Amino acids + nucleic acid precursors	0.4	0.92
Amino acids + vitamins	0.6	0.98
Amino acids + vitamins + nucleic acid precursors	0.4	1.09
Amino acids + vitamins + yeast extract	0.5	0.95

[a] Glucose concentration was 0.055 M.

In contrast to Table 49, data in Table 50 suggest that, under oxygen-limited conditions, Y_{O_2} is insensitive to the addition of amino acids, nucleic acid precursors and vitamins.

Table 51. Influence of medium composition on Y_{ATP} [Stouthamer (1976)].

Microorganism	Medium	Specific growth rate (h^{-1})	Y_{ATP} [g (g mol ATP)$^{-1}$]
Escherichia coli	Minimal	Not determined	6.4
E. coli	Minimal + citrate	0.30	4.1
E. coli	Complex[a]	0.81	9.4
Proteus mirabilis	Minimal	0.17	5.5
P. mirabilis	Complex	0.69	12.6
Propionibacterium freudenreichii	Synthetic[b]	Not determined	14.0
P. freudenreichii	Complex	Not determined	16.7
P. pentosaceum	Synthetic	Not determined	11.5
P. pentosaceum	Complex	Not determined	16.3
Saccharomyces carlsbergensis	Minimal + biotin	0.01–0.02	4.7
S. carlsbergensis	Biotin + pyridoxine	0.01–0.03	5.9
S. carlsbergensis	Biotin pyridoxine + thiamine	0.03–0.05	6.5
S. carlsbergensis	9 vitamins	0.11	7.7
S. carlsbergensis	Complex	0.20	10.7
Zymomonas mobilis	Complex without buffer	0.40	8.3
Z. mobilis	Complex without Tris buffer	0.35	6.5
Z. mobilis	Synthetic + pantothenate (5 mg l^{-1})	0.39	6.5
Z. mobilis	Synthetic + pantothenate (4 $\mu g\,l^{-1}$)	0.35	5.2
Z. mobilis	Synthetic + pantothenate (2 $\mu g\,l^{-1}$)	0.28	4.5
Z. mobilis	Synthetic + pantothenate (1 $\mu g\,l^{-1}$)	0.26	4.2
Z. mobilis	Synthetic + pantothenate (0.1 $\mu g\,l^{-1}$)	0.20	3.0
Z. mobilis	Synthetic + pantothenate (50 $\mu g\,l^{-1}$)	0.16	2.5
Z. mobilis	Minimal + pantothenate (5 mg l^{-1})	0.30	4.7
Z. mobilis	Minimal + pantothenate (50 ng l^{-1})	0.28	4.2
Z. mobilis	Minimal + pantothenate (4 $\mu g\,l^{-1}$)	0.18	3.9
Z. mobilis	Minimal + pantothenate (1 $\mu g\,l^{-1}$)	0.15	2.9
Z. mobilis	Minimal + pantothenate (0.5 $\mu g\,l^{-1}$)	0.15	2.5
Z. mobilis	Minimal + pantothenate (50 ng l^{-1})		1.7

[a] Includes preparations obtained from 'natural' sources, e.g. yeast extract.
[b] Comprises substrate, inorganic nitrogen and phosphorus, plus mineral salts.

Data in Table 51 demonstrate that Y_{ATP} is generally increased with the complexity of the medium used, i.e. media from 'natural' sources provide improved conditions for growth.

Specific Growth Rate

Figs. 14 and 15 illustrate the characteristic dependency of the yield coefficients Y_s, Y_{ATP}, etc. on the specific growth rate μ as suggested by Eqs. (21) to (31), i.e.

$$Y_{obs} = \frac{Y^{max}\mu}{m + \mu} \tag{32}$$

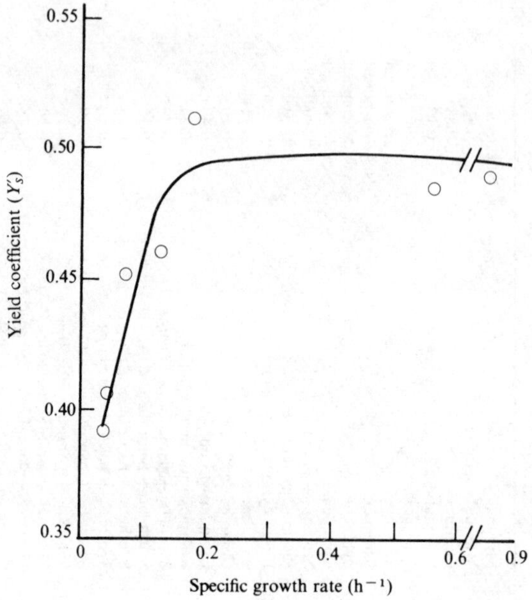

Figure 14. Yield of *Escherichia coli* ML 30 as a function of specific growth rate [Ng (1969)].

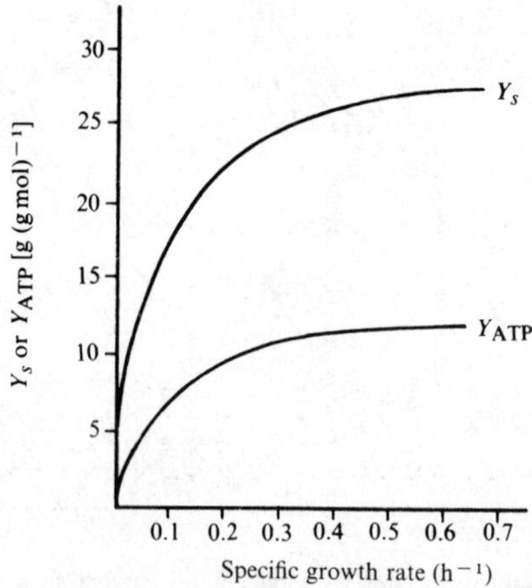

Figure 15. Y_S and Y_{ATP} of *Aerobacter aerogenes* as a function of specific growth rate – glucose-limited anaerobic growth [Stouthamer (1976)].

pH and Temperature

Table 52. Effect of pH and temperature on Y'_S for *Saccharomyces cerevisiae* [Eroshin *et al.* (1976)].

pH	Temperature (°C)	Y'_S [g (g substrate)$^{-1}$]
2.5	29.0	0.58
3.1	22.6	0.58
3.1	35.0	0.57
4.5	20.0	0.54
4.5	29.0	0.54
4.5	29.0	0.58
4.5	29.0	0.62
4.5	29.0	0.64
4.5	29.0	0.67
4.5	38.0	0.48
5.9	22.6	0.55
5.9	35.5	0.55
6.5	29.0	0.55

Data in Table 52 suggest that Y'_S is insensitive to pH and has a weak maximum in the region of 29°C. From Tables 53 and 54, it may be deduced that Y'_S is largely independent of temperature up to a characteristic value beyond which the yield falls rapidly.

Table 53. Effect of temperature on growth of *Aerobacter aerogenes*[a] [Payne (1970)].

Temperature (°C)	Specific growth rate (h^{-1})	Y'_S [g (g substrate)$^{-1}$]
23.0	0.55	0.35
27.0	0.75	0.36
32.0	0.97	0.34
37.0	1.31	0.32
38.8	0.83	0.22
39.7	0.60	0.17
40.8	0.40	0.10
42.0	0.00	0.00

[a] Aerobic cultures in glucose medium.

Table 54. Effect of temperature on growth of *Desulfovibrio desulfuricans* Berre S[a] [Payne (1970)].

	Nitrogen source			
	Nitrogen		Ammonia	
Temperature	Specific growth rate (h^{-1})	Y'_S [g (g substrate)$^{-}$]	Specific growth rate (h^{-})	Y'_S [g (g substrate)$^{-1}$]
24.8	0.06	0.03	0.14	0.07
26.2	0.07	0.03	0.15	0.06
32.0	0.08		0.18	
37.0	0.10	0.03	0.22	0.06
39.2	0.08	0.03		
40.0	0.08	0.02	0.14	0.03
41.7	0.06	0.01	0.13	0.03
42.2			0.07	0.03
43.0	0.00	0.00	0.00	0.00

[a] Lactate (40 mM) limiting medium.

THERMODYNAMICS OF GROWTH

Y_{kcal}

Values for the yield of biomass related to the total energy consumed from the medium (*see* p.135) have been calculated for a range of microorganisms and substrates (*see* Table 55).

Table 55. Y_{kcal} values for heterotrophs growing aerobically in minimal media.

Microorganism	Substrate	Y_{kcal} (g kcal^{-1})
Aerobacter aerogenes	Maltose	0.104
A. aerogenes	Glucose	0.101
Candida utilis	Glucose	0.126
Penicillium chrysogenum	Glucose	0.107
Pseudomonas fluorescens	Glucose	0.096
Rhodopseudomonas spheroides	Glucose	0.112
Saccharomyces cerevisiae	Glucose	0.123
		average 0.116
A. aerogenes	Ribose	0.089
A. aerogenes	Succinate	0.073
A. aerogenes	Glycerol	0.108
A. aerogenes	Lactate	0.050
A. aerogenes	Pyruvate	0.059
		average 0.085
A. aerogenes	Acetate	0.048
C. utilis	Acetate	0.092
Ps. fluorescens	Acetate	0.075
		average 0.077
C. utilis	Ethanol	0.112
Ps. fluorescens	Ethanol	0.077
		average 0.090
Klebsiella sp.	Methanol	0.081
Methylomonas methanolica	Methanol	0.107
Pseudomonas sp.	Methanol	0.088
		average 0.088
Methylococcus sp.	Methane	0.104
Pseudomonas sp.	Methane	0.077
Pseudomonas sp.	Methane	0.054
Ps. methanica	Methane	0.050
		average 0.066

Free Energy of Oxidation

On p.138, linear relationships are developed between Y_s, ΔF^0_{ox} and COD. Some experimental evidence supporting these relationships is given in Figs. 16–19. Values of MF^0_{ox} and COD for various common organic compounds are given in Table 56.

Table 56. ΔF^0_{ox} and COD for common organic compounds [Servizi and Bogan (1963)].

Compound	ΔF^0_{ox} (kcal mol^{-1})	COD [(mol O_2) mol^{-1}]
Acetaldehyde	−270	2.5
Acetic acid	−208	2.0
Alanine	−311	3.0
Aniline	−717	7.0
Arabinose	−573	5.0
Benzene	−765	7.5
Benzoic acid	−773	7.5
n-Butanol	−621	6.0
Butyric acid	−514	5.0
Formaldehyde	−120	1.0
Formic acid	−66	0.5
Fumaric acid	−333	3.0
Glucose	−687	6.0
Glutamic acid	−475	4.5
Glycerol	−407	35.0
Glycine	−161	1.5
Heptanoic acid	−973	9.5
Lactic acid	−340	3.0
Malic acid	−336	3.0
Mannitol	−740	12.0
Methylformate	−231	2.0
Naphthalene	−1216	12.0
Nitrobenzene	−733	7.5
m-Nitrobenzoic acid	−738	7.5
Octanic acid	−1124	11.0
Phenol	−729	7.0
Propionic acid	−360	3.5
Pyrocatechol	−684	6.5
Pyruvic acid	−282	2.5
Resorcinol	−682	6.5
Succinic acid	−369	3.5
Sucrose	−1382	12.0
Toluene	−914	9.0
m-Xylene	−1065	10.5

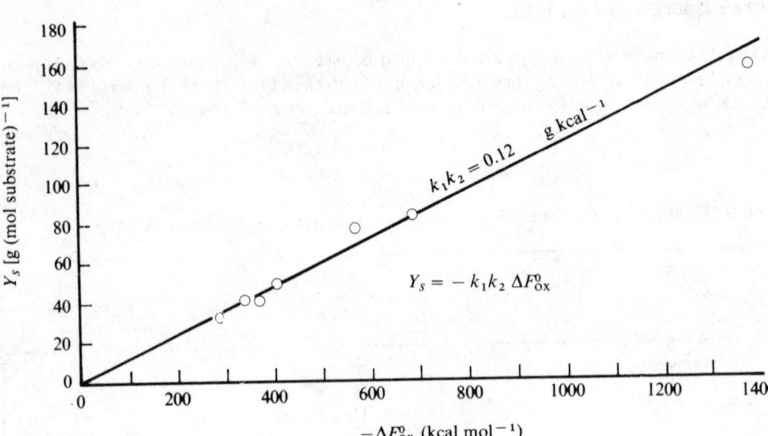

Figure 16. Molar growth yield of *Escherichia coli* as a function of the free energy of substrate oxidation [Servizi and Bogan (1964)].

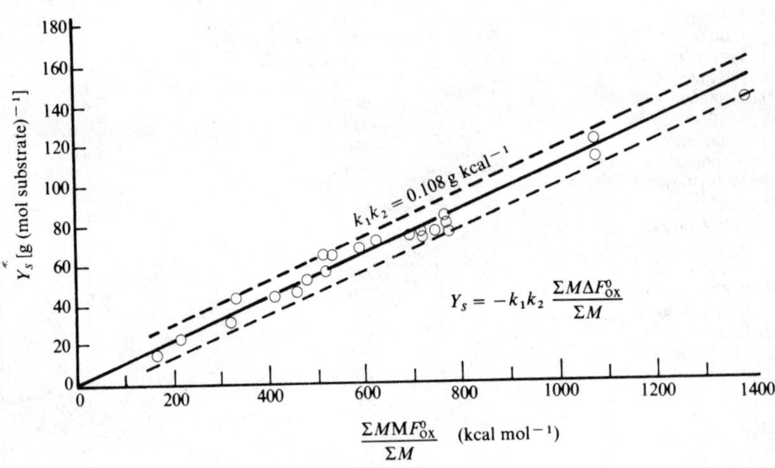

Figure 17. Molar growth yield of activated sludge as a function of the free energy of oxidation of mixed substrates [Servizi and Bogan (1964)].

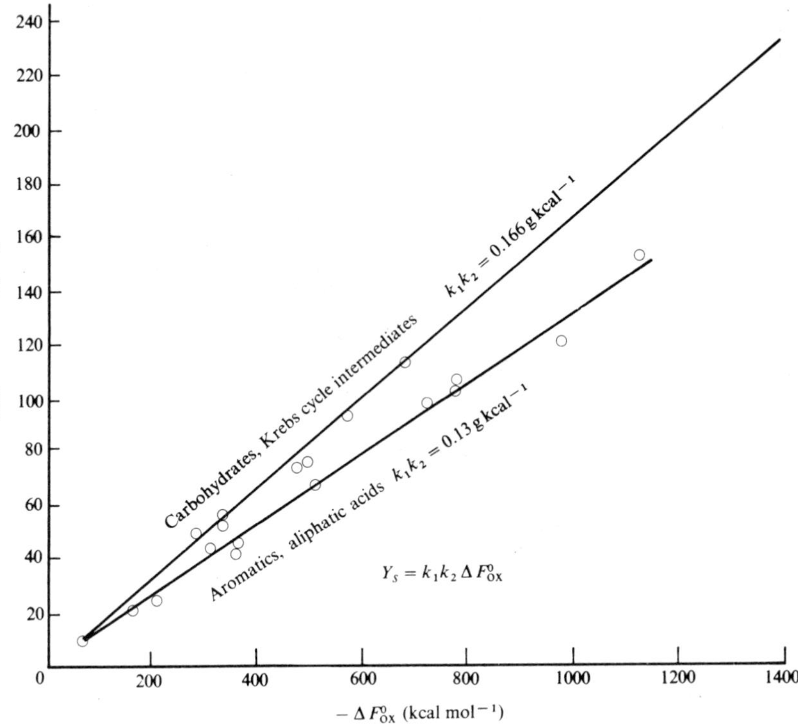

Figure 18. Yield of activated sludge as a function of substrate ΔF^0_{ox} [Servizi and Bogan (1963)].

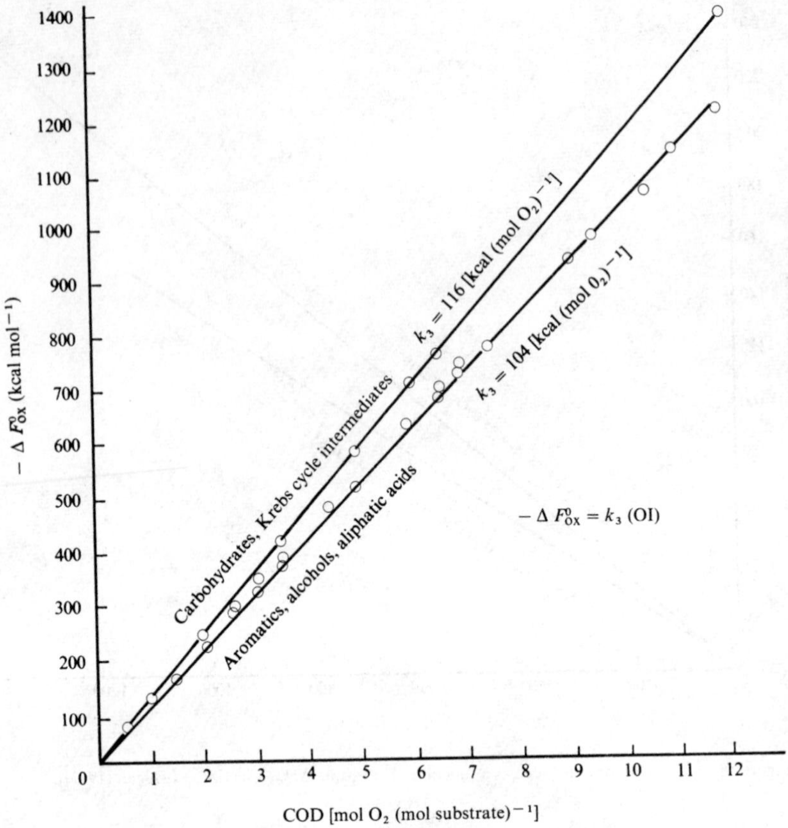

Figure 19. $-\Delta M F^0_{ox}$ as function of theoretical COD [Servizi and Bogan (1963)].

Heat Evolution

Figs. 20 and 21 suggest that the total heat evolved during fermentation is proportional to the oxygen consumed and the carbon dioxide evolved. Fig. 22 suggests similar relationships for the rates of evolution of both heat and carbon dioxide.

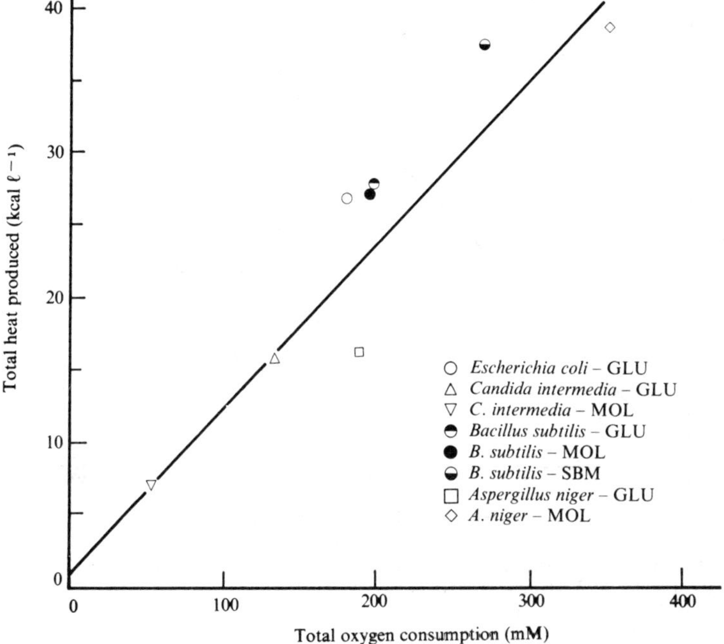

Figure 20. Total heat produced versus the total oxygen consumed. GLU, glucose medium; MOL, molasses medium; SBM, soybean meal medium [Cooney *et al.* (1969)].

Experimentally determined values of various maintenance coefficients are included in Table 23. Tables 57–59 provide additional data.

Table 57. Maximum growth yields and maintenance coefficients of *Paracoccus denitrificans* grown in chemostat cultures [Meijer *et al.* (1977)].

Carbon source	Limiting factor	Y_s^{max} (g mol $^{-1}$)	m_s (m mol g^{-1} h^{-1})	$Y_{O_2}^{max}$ [g (g mol O_2)$^{-1}$]	m_{O_2} (m mol g^{-1} h^{-1})
Succinate	Succinate	40.2	0.48	34.2	1.52
Succinate	Sulphate	37.9	1.04	21.0	1.64
Gluconate	Gluconate	77.8	0.11	43.4	0.67
Gluconate	Sulphate	65.7	0.35	31.3	1.11

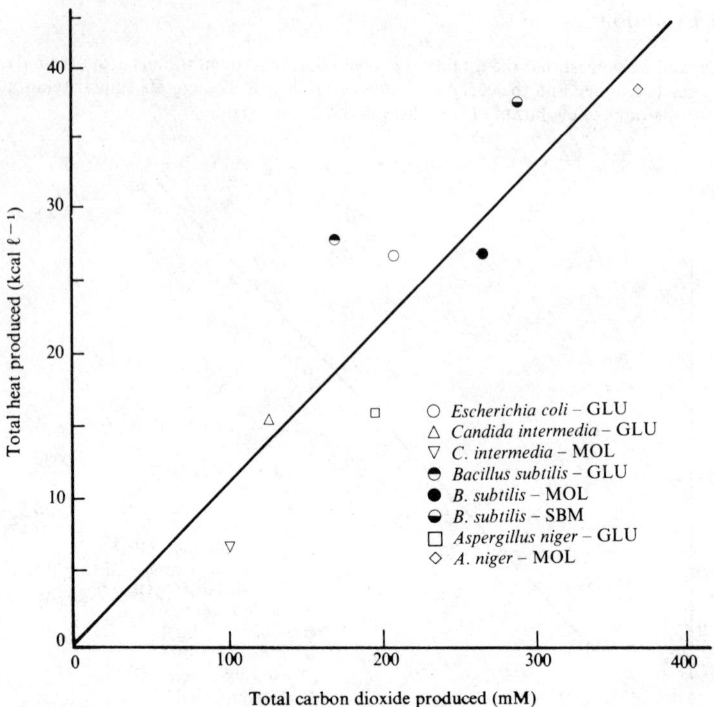

Figure 21. Total heat produced versus the total carbon dioxide evolved. GLU, glucose medium; MOL, molasses medium; SBM, soybean meal medium [Cooney *et al.* (1969)].

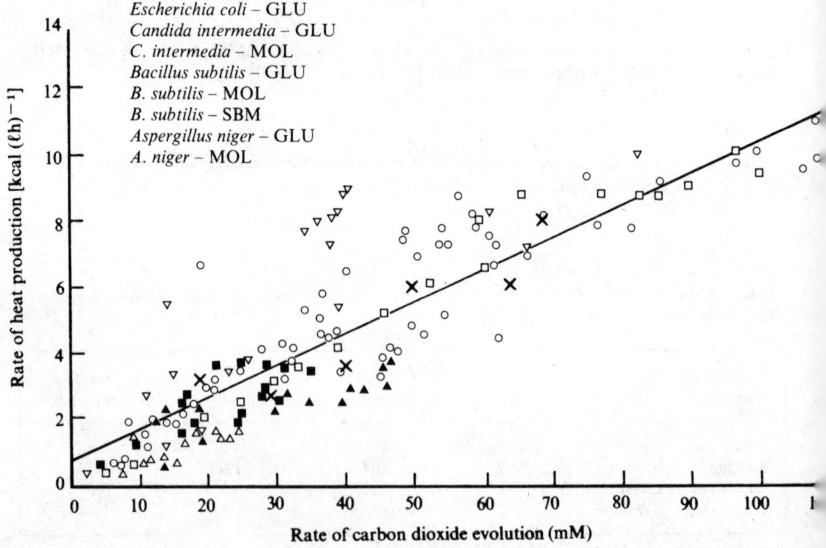

Figure 22. Rate of heat production versus the rate of carbon dioxide evolution. GLU, glucose medium; MOL, molasses medium; SBM, soybean meal medium [Cooney *et al.* (1969)].

MAINTENANCE COEFFICIENTS

The observed and true yield coefficients are shown (*see* p.141) to be related by the formula

$$\frac{1}{Y_{obs}} = \frac{m}{\mu} + \frac{1}{Y_{true}} \tag{33}$$

where Y_{obs} is the 'empirical' yield coefficient, μ is the specific growth rate, Y_{true} is the growth rate-independent system parameter and is the corresponding 'maintenance' coefficient.

Figure 23 illustrates the application of Eq. (33) for Y_{ATP} expressed as

$$Q_{ATP} = m_{ATP} + \frac{1}{Y^{max}_{ATP}}\mu \tag{34}$$

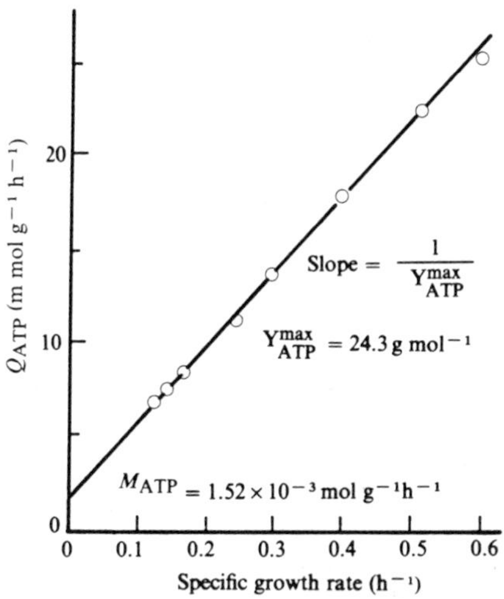

Figure 23. Assessment of maintenance coefficient m_{ATP} and maximum growth yield based on ATP generation Y^{max}_{ATP} of *Lactobacillus casei* anaerobically growing in complex medium when glucose was used as the energy source.

Table 58. Maintenance coefficient m_s for various microorganisms [Aiba and Huang (1969)].

Microorganism	Limiting factor	Growth conditions	m_s (g substrate g^{-1} h^{-1})
Azobacter vinelandii	Nitrogen	Nitrogen fixing, high oxygen	1.50
Az. vinelandii	Nitrogen	Nitrogen fixing, low oxygen	0.15
Klebsiella aerogenes	Glucose	Aerobic	0.04
K. aerogenes	Glucose	Anaerobic	0.50
K. aerogenes	Tryptophan	Anaerobic	3.69
Saccharomyces cerevisiae	Glucose	Anaerobic	0.04

Table 59. Maintenance requirements for various microorganisms with glucose as the energy source [Cheremisnoff and Young (1975)].

Microorganism	Special growth conditions	Maintenance energy m_S (g substrate g^{-1} h^{-1})	m_{ATP} (mmol ATP g^{-1} h^{-1})
Aerobacter cloacae	Aerobic, glucose-limited	0.094	14
Azotobacter vinelandii	Nitrogen fixing, dissolved oxygen tension (0.2 atm)	1.5	220
Az. vinelandii	Nitrogen fixing, dissolved oxygen tension (0.02 atm)	0.15	22
Klebsiella aerogenes	Anaerobic, tryptophan-limited $NH_4Cl(2\ g\ell^{-1})$	2.88	39
K. aerogenes	Anaerobic, tryptophan-limited $NH_4Cl(4\ g\ell^{-1})$	3.69	50
Lactobillus casei		0.135	1.5
Penicillium chrysogenum	Aerobic	0.022	3.2
Saccharomyces cerevisiae	Anaerobic	0.036	0.52
Sacch. cerevisae	Anaerobic NaCl (1.0M)	0.360	2.2

OXYGEN REQUIRED FOR CELL PRODUCTION

The following assumptions have been made by Mateles (1971):

1) The only products of metabolism are cells, carbon dioxide and water.
2) The nitrogen source is ammonia.

Eq. (16) results from a material balance involving the above assumptions and the following relationship between the components:

$$CH_2O + NH_3 + O \longrightarrow \text{cells } (C', H', O', N') + CO_2 + H_2O + N \quad (14)$$

$$\frac{32}{Y_{O_2}} = \frac{g_{oxygen}}{g_{cell}} = 16\left[\frac{2C + H/2 - O}{Y'_s M} + \left(\frac{N'}{1400} \times \frac{3}{2}\right) - \left(\frac{C'}{1200} \times 2\right) \frac{H'}{100.2} + \frac{O'}{1600}\right] \quad (15)$$

or

$$\frac{32}{Y_{O_2}} = \frac{(32C + 8H - 16O)}{Y'_s M} + 0.01\,O' - 0.0267\,C' + 0.01714\,N' - 0.08\,H' \quad (16)$$

$0.01714\,N' - 0.08\,H'$

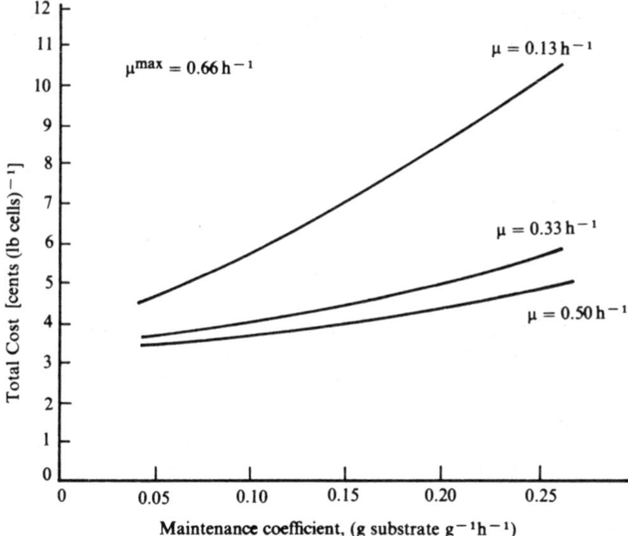

Figure 24. Effect of maintenance coefficient on biomass production costs at different growth rates. Operating costs are for substrate, oxygen transfer and heat removal [Abbott and Clamen (1973)].

Fig. 24 shows that the closer a process is operated to μ_{max} the more attractive are the economics, but the sensitivity of product costs to the growth rate μ are much reduced for small values m_s.

EXPERIMENTAL DETERMINATION OF YIELDS

Y_S' Determinations

An outline of the stages for both batch and continuous cultures is presented [Stouthamer (1969)].

Batch culture

The features of batch culture are:

1) All the nutrients in the medium must be in excess except for the energy source.

2) Several samples of media containing different concentrations of the energy source are inoculated.

3) The variation of the optical densities of the cultures with time is monitored. The maximum optical densities correspond to the complete utilization of the energy source.

4) The dry weights of the microorganisms are determined. These weights correspond to the maximum optical densities for each culture.

5) Mean values of Y_S' and Y_S are obtained from the data expressed as biomass formed and energy source (i.e. substrate) utilized.

Occasionally rapid lysis occurs following complete utilization of the energy source, e.g., for *Bacillus subtilis* or *Selenomonas ruminantium*. In such instances, plots of the increased optical density versus the amount of the energy source utilized [stage (3)] exhibit a degree of variability as the energy source is depleted. When this occurs, the optical density-time data will need evaluation as dry weights of microorganism by calibration or preferably by direct measurement. Plots of biomass formed versus energy source utilized over the course of fermentation lead to Y_s' and Y_s.

Continuous culture

This method offers advantages over the batch culture technique, since the yield coefficient can be determined at any specific growth rate up to and including μ_{max} (*see* Chapter 6) and yield coefficients associated with nutrients other than the limiting substrate are more easily evaluated because of the concentration levels involved.

Since the conditions in a chemostat are uniform, all the microorganisms are exposed to the same environment and the yield coefficients are calculated from the changes in the concentrations of nutrients, limiting substrates and microorganisms between the inlet and outlet.

Use of the equations given on pp.140–41 allows the maintenance coefficient m_s, and the true growth yield Y_G' or Y_G to be evaluated following a series of experiments carried out at different specific growth rates. This is achieved by changing the medium flow rate (*see* Chapter 7).

General Observations

The specific growth rate in a chemostat is below μ_{max} whereas that in batch culture will, for the most part, be in the region of μ_{max}. Therefore the yield coefficients obtained using the two procedures may differ to some extent (due partially, but not totally, to the maintenance coefficient).

Since the conditions in a chemostat are steady, the yield coefficient obtained reflects specific conditions. In contrast, the yield coefficient obtained by batch culture reflects a range of conditions.

The batch culture method cannot be easily adapted to the determination of the maintenance coefficient m_s.

Y_{O_2} Determinations

Batch culture

The oxygen uptake associated with a batch culture can be determined readily by using an apparatus based upon the methodology of that illustrated in Fig. 25 with the growth evaluated as for Y_s'.

Continuous culture

1) At sufficiently low aeration rates, there is no significant dissolved oxygen in the medium and the oxygen supply to the microorganisms is determined by the oxygen transport rate (*see* Chapter 9 for determination).

2) A gas-phase oxygen balance can be made if the oxygen concentration in the inlet and outlet gas streams are determined and the dissolved oxygen level is shown to be steady. Alternatively, a liquid-phase oxygen balance can be made if the oxygen concentration in the feed is enhanced by sparging with an enhanced oxygen/nitrogen mixture.

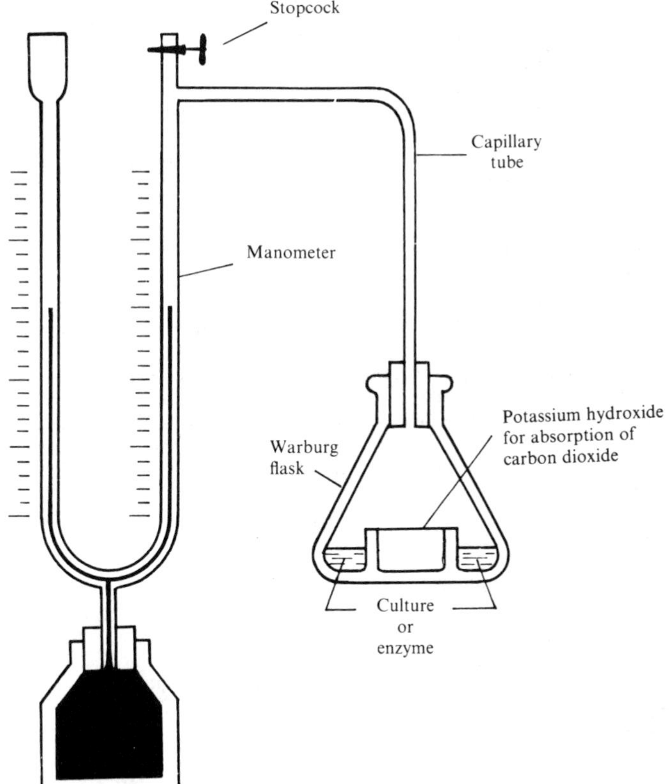

Figure 25. Basic Warburg respirometer.

Factors Associated with the Experimental Determination of Yield Coefficients

1) If the microorganisms are growing anaerobically in complex media, very little of the carbon from the energy source is incorporated into cellular material. It is nearly completely used as an energy source.

2) During aerobic growth, oxygen is taken up and ATP is produced not only by phosphorylation at the substrate level but mostly by oxidative phosphorylation, the efficiency of which is not known.

3) Molar growth yields for anaerobic growth are sometimes higher in the presence of hydrogen acceptors (e.g., nitrate, fumarate).

4) The choice of medium is important. In a complex medium, almost all the carbohydrate is used as an energy source. The amount that is assimilated can be determined using [^{14}C] carbohydrates. In a simple medium, the carbohydrate serves both as the carbon and as the energy source. In this case, the proportion that is assimilated can be determined from the amount of dry weight and the amount of carbon in the cells.

5) In some media, growth may be limited not by the availability of the substrate but by the rate at which some essential nutrients can be synthesized. Under these conditions, uncoupling between growth and energy production may occur because catabolism may proceed at a rate unrelated to the growth rate.

6) It is necessary to determine the molar growth yield at several different substrate concentrations. If a plot of the growth against substrate uptake is linear, the energy source is growth-limiting. If this plot yields a curve that intersects the ordinate at a positive value, a small amount of unknown energy source is present in the medium. However, if the curve yields a curve with an inflection point, it may be that the metabolism is different at lower and higher concentrations of the substrate.

7) In continuous culture, the size of the cells may change with the dilution rate. This is important if the dry weight is determined turbidimetrically.

8) In the determination of Y_{O_2} using a Warburg respirometer, the presence of potassium hydroxide in the centre well (*see* Fig. 25) leads to a depletion of carbon dioxide, which may affect growth. Aeration conditions may also need attention.

9) The time when maximal growth occurs does not necessarily coincide with that of maximum oxygen uptake. Thus it is necessary to follow both growth and oxygen uptake. Usually, Y_{O_2} is determined by using the values that correspond to maximal growth.

10) In most cases, the amount of ATP produced can be determined only for known fermentation pathways. Calculation of ATP is most accurate when the substrate is used only as an energy source and not incorporated into the biomass.

11) There are differences in fermentation balances for growing and resting cells.

12) Changes in temperature may cause the uncoupling of energy production and growth.

13) The following factors determine the molar growth yield for the growth of a given microorganism.

a) Pathway of substrate breakdown
b) Nature of the carbon source
c) Nature of the nitrogen source
d) Complexity of the medium
e) Specific growth rate
f) Maintenance energy requirements
g) Cell composition
h) Possibility of anaerobic electron transport during fermentation
i) Availability and use of external hydrogen acceptors
j) Efficiency of oxidative phosphorylation

PROCESS YIELDS

Information on yields is often closely guarded for commercial reasons and the precise conditions leading to particular levels of production are even more closely guarded. Figs. 26 and 27 illustrate the way in which fermenter productivities can be increased. Tables 60–62 also give some indication of what has been achieved in various areas of commercial fermentation activity.

Table 60. Commercial production of vitamins by microorganisms [Rainbow and Rose (1963a)].

Microorganism	Vitamin	Yield (mg ℓ$^{-1}$)
Eremothecium ashbyii	Riboflavin	2.2
Streptomyces sp.	Vitamin B_{12}	5.7

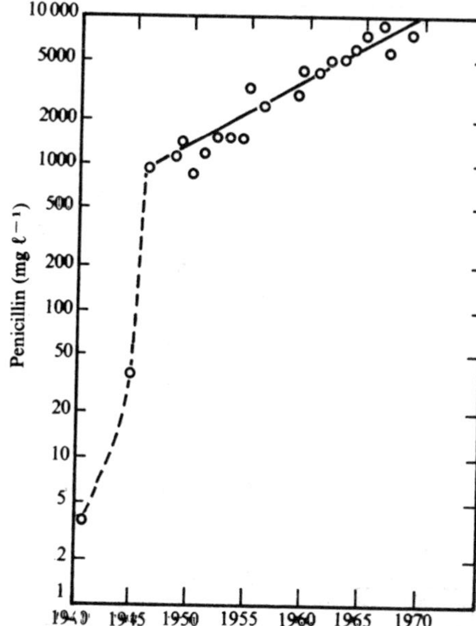

Figure 26. Changes in penicillin broth potencies with time [Demain (1971)].

Table 61. Commercial production of antibiotics.

Microorganism	Antibiotic	Yield (mg ℓ$^{-1}$)
Penicillium chrysogenum	Penicillin	60.1
Pen. patulum	Griseofulvin	1675.0
Streptomyces erythreus	Erythromycin	938.0
Strep. griseus	Streptomycin	1000.0
Strep. niveus	Novobiocin	700.0
Strep. rimosus	Chlorotetracycline	2000.0
Strep. venezuelae	Chloramphenicol	100.0

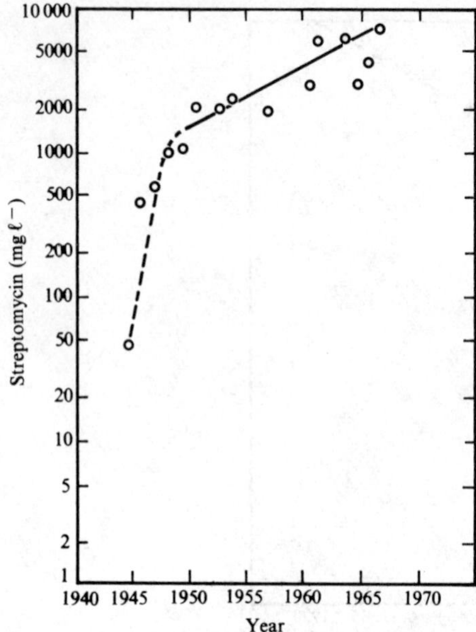

Figure 27. Changes in streptomycin broth potencies with time [Demain (1971)].

It can be seen from Figs. 27 and 28 that antibiotic yields in broths have increased significantly since their commercial introduction. In the case of both penicillin and streptomycin, there was an initial, rapid increase in yield, for penicillin this was spurred on by mounting demand because of the Second World War. Improved yields have been achieved by the selection and isolation of mutants capable of greater antibiotic production.

Table 62. Commercial production of enzymes [(Rainbow and Rose (1963b)].

Microorganism	Enzyme	Yield ($U\ ml^{-1}$)
Aspergillus niger	α-Amylase	12.5
Asp. wentii	Glucoamylase	65.6
Bacillus macerans	Amylase	30.0
Saccharomyces kluyveri	Invertase	800.0

Table 63. Effect of strain differences on enzyme yields [(Rainbow and Rose (1963c)].

Microorganism	Enzyme	Yield (U ml^{-1})
Aspergillus niger 377	α-Amylase	12.5
Asp. niger 605	α-Amylase	0.1
Asp. niger 377	Glucoamylase	21.9
Asp. niger 605	Glucoamylase	61.7
Asp. oryzae 694	α-Amylase	8.3
Asp. oryzae 464	α-Amylase	3.0
Asp. oryzae 694	Glucoamylase	12.2
Asp. oryzae 464	Glucoamylase	64.6
Asp. wentii 377	α-Amylase	0.9
Asp. wentii 378	α-Amylase	2.4
Asp. wentii 377	Glucoamylase	25.4
Asp. wentii 378	Glucoamylase	65.6
Bacillus macerans a	Amylase	0.5
B. macerans b	Amylase	30.0
B. polymyxa a	Glucoamylase	0.7
B. polymyxa b	Glucoamylase	10.0
Saccharomyces kluyveri NRRL Y–4288 bisexual	Invertase	13–26
Sacch. kluyveri NRRL Y–4288 unisexual	Invertase	800.0

Table 64. Yields of organic acids produced by microorganisms.

Microorganism	Substrate	Product	Yield (%)
Acetobacter aceti	Ethanol	Acetic acid	95
Aspergillus niger	Sucrose	Citric acid	85
Asp. niger	Glucose	Gluconic acid	95
Asp. oryzae	Glucose	Kojic acid	50
Asp. terreus	Glucose	Itaconic acid	60
Bacterium succinium	Malic acid	Succinic acid	57
Candida brumptii	Glucose	L(+) isocitric acid	28
C. hydrocarbofumarica	*n*-Alkane	α-Ketoglutaric acid	84
C. lipolytica	*n*-Alkane	Citric acid	140
Gluconobacter suboxydans	Glucose	Tartaric acid	27
G. suboxydans	Glucose	5-Ketogluconic acid	90
Lactobacillus brevis	Glucose	Malic acid	100
L. delbrückii	Glucose	Lactic acid	90
Penicillium notatum	Glucose	D-Araboascorbic acid	45
Pen. purpurogenum	Glucose	L(+)-Alloisocitric acid	40
Propionibacterium shermanii	Glucose	Propionic acid	60
Pseudomonas aeruginosa	Glucose	Pyruvic acid	50
Ps. fluorescens	Glucose	2–Ketogluconic acid	90
Rhizopus delemar	Glucose	Fumaric acid	58

Table 65. Yields of amino acids from various carbon sources [Yamada (1977)].

Microorganism	Substrate	Yield (g ℓ$^{-1}$)	Yield (%)
DL-Alanine			
Brevibacterium flavum	Glucose	14	
Corynebacterium gelatinosum	Glucose	40	40
Coryn. hydrocarboclastus	*n*-Alkane	4	
L-Arginine			
Brev. flavum	Acetic acid	26	10
Brev. flavum	*n*-Alkane	9	8
Brev. flavum	Glucose	29	29
L-Citrulline			
Arthrobacter paraffineus	*n*-Alkane	9	9
Brev. flavum	Acetic acid	15	14
Brev. flavum	Glucose	30	27
L-Glutamic acid			
Arth. paraffineus	*n*-Alkane	62	
Bacillus megaterium	Propylene glycol	27	27
Brevibacterium sp.	Benzoic acid	80	80
Brevibacterium sp.	Ethanol	59	66
Brev. flavum	Acetic acid	98	48
Brev. flavum	Glucose	50	50
Brev. pentosaminoacidum	Xylose	5	10
Brev. thiogentalis	Acetic acid	51	51
Coryn. alkanolyticum	*n*-Alkane	72	
Coryn. glutamicum	Glucose	38	38
Coryn. hydrocarboclastus	*n*-Alkane	84	
Microbacterium ammoniaphilum	Glucose	29	29
L-Glutamine			
Brev. flavum	Acetic acid	25	15
Brev. flavum	Ethanol	35	38
Brev. flavum	Glucose	39	39
L-Histidine			
Brev. flavum	Acetic acid	5	6
Brev. flavum	Ethanol	6	5
Brev. flavum	Glucose	10	10
Coryn. glutamicum	Glucose	8	8
L-Homoserine			
Corynebacterium sp.	*n*-Alkane	12	15
Coryn. glutamicum	Glucose	15	15
Escherichia coli	Glycerol	4	8
L-Isoleucine			
Brev. flavum	Acetic acid	15	11
Brev. flavum	Glucose	15	15
Microb. paraffinolyticus	*n*-Alkane	1.6	
L-Leucine			
Brev. flavum	Ethanol	5	5
Brev. lactofermentum	Glucose	28	21
Coryn. hydrocarboclastus	*n*-Alkane	0.7	
L-Lysine			
Brev. flavum	Acetic acid	61	
Brev. flavum	Glucose	32	32
Brev. lactofermentum	Ethanol	66	28
Coryn. glutamicum	Glucose	39	39
Nocardia sp. No. 258	*n*-Alkane	34	34
L-Methionine			
Coryn. glutamicum	Glucose	2	2
L-Ornithine			
Brev. flavum	Acetic acid	30	25
Coryn. glutamicum	Glucose	30	27
Coryn. hydrocarboclastus	*n*-Alkane	9	32
L-Phenylalanine			
Arth. paraffineus	*n*-Alkane	15	15
Bac. subtilis	Glucose	6	7

continued

Table 66 – (continued)

Microorganism	Substrate	Yield (g ℓ$^{-1}$)	Yield (%)
L-Phenylalanine *continued*			
Brev. flavum	Ethanol	15	12
Coryn. glutamicum	Glucose	9	9
L-Proline			
Brev. flavum	Acetic acid	18	12
Brev. flavum	Glucose	29	29
Coryn. acetoacidophilum	Ethanol	22	14
Coryn. acetoacidophilum	*n*-Alkane	2.2	
L-Serine			
Arth. paraffineus	*n*-Alkane	3	
L-Threonine			
Arth. paraffineus	*n*-Alkane	15	25
Brev. flavum	Acetic acid	27	14
Brev. flavum	Ethanol	33	25
Brev. flavum	Glucose	18	18
L-Tryptophan			
Bac. subtilis	Glucose	6	7
Brev. flavum	Acetic acid	3	2
Coryn. glutamicum	Glucose	12	12
L-Tyrosine			
Arth. paraffineus	*n*-Alkane	18	18
Brev. flavum	Acetic acid	5	5
Coryn. glutamicum	Glucose	17	17
L-Valine			
Brev. flavum	Acetic acid	20	15
Brev. lactofermentum	Glucose	23	29
Coryn. acetoacidophilum	Ethanol	22	14
Coryn. hydrocarboclastus	*n*-Alkane	2	

Table 66. Amino acid production from corresponding intermediates [Yamada (1977)].

Microorganism	Intermediate	Amino acid	Yield (g ℓ$^{-1}$)
Claviceps purpureus	Indole	L-Tryptophan	3
Corynebacterium amagaski	D-Threonine	L-Isoleucine	15
Coryn. glycinophilum	Glycine	L-Serine	16
Hansenula anomala	Anthranilic acid	L-Trytophan	3
Pseudomonas denitrificans	2-Hydroxy-4-methylthiobutyric acid	L-Methionine	11
Serratia marcescens	α-Aminobutyric acid	L-Isoleucine	8

Table 67. Enzymatic production of amino acids [Yamada (1977)].

Microorganism	Substrate(s)	Amino acid	Yield (g ℓ$^{-1}$)
Achromobacter obae + *Citrobacter leurentii*	α-Aminocaprolactam	L-Lysine	140
Erwinia herbicola	Fumaric acid	L-Aspartic acid	168
Er. herbicola	Phenol + pyruvic acid	L-Tyrosine	62
Er. herbicola	Catechol + pyruvic acid	L-DOPA	55
Proteus rettgerii	Indole + pyruvic acid	L-Tryptophan	91
Prot. rettgerii	5-Hydroxyindole + pyruvic acid	5-Hydroxy-L-tryptophan	28
Pseudomonas dacunhae	L-Aspartic acid	L-Alanine	260

Table 68. Production of inosine [Yamada (1977)].

Microorganism	Substrate	Yield	
		(g ℓ$^{-1}$)	(%)
Arthrobacter paraffineus	*n*-Alkane	2.3	5
Bacillus pumilus	Glucose	16	13
B. subtilis	Acetic acid	2.6	5
B. subtilis	Glucose	16	20
B. subtilis	Soluble starch	5	6
Brevibacterium ketoglutamicum	*n*-Alkane	1.5	3
Candida petrophilum	*n*-Alkane	1.5	3

Table 69. Production of 5′-inosine monophosphate [Yamada (1977)].

Microorganism	Substrate	Yield	
		(g ℓ$^{-1}$)	(%)
Bacillus subtilis	Glucose	0.3	3.5
B. subtilis	Glucose	3.5	
Brevibacterium ammoniagenes	Glucose	10.2	10.2
Brev. ammoniagenes	Glucose	19	19
Brev. thiogenialis	Glucose	3.6	3.6
Candida tropicalis	*n*-Alkane	1.5	
Pseudomonas aeruginosa	*n*-Alkane	2.1	

Table 70. Production of guanosine [Yamada (1977)].

Microorganism	Substrate	Yield	
		(g ℓ$^{-1}$)	(%)
Bacillus sp.	Maltose	5.4	5.4
B. subtilis	Glucose	4.3	6
B. subtilis	Glucose	5	6
B. subtilis	Glucose	11	13
B. subtilis	Glucose	10	10

Table 71. Production of nucleic acid-related compounds [Yamada (1977)].

Microorganism	Product	Yield (g ℓ$^{-1}$)
Arthrobacter paraffineus	Orotic acid	20
Bacillus sp.	Adenosine	16
Brevibacterium ammoniagenes	ATP	1.5
Brev. ammoniagenes	NAD^+	1.9
Saccharomyces carlsbergensis	CDP–choline	35[a]
Sarcina lutea	FAD	10

[a] Yield expressed as mM.

REFERENCES

Abbott, B.J., and Clamen, A. (1973) The Relationship of Substrate, Growth Rate, and Maintenance Coefficient to Single Cell Protein Production. *Biotechnol. Bioengng*, **15**, 117.

Aiba, S. and Huang, S.Y. (1969) Oxygen Diffusivity and Permeability in Polymer Membranes Immersed in Water. *Chem. Engng Sci.*, **24**, 1149.

Battley, (1979)

Bauchop, T., and Elsden, S.R. (1960) Growth of Microorganisms in Relation to their Energy Supply. *J. Gen Microbiol.*, **23**, 457.

Cheremisinoff, P.N. and Young, R.A. (eds.) (1975) *Pollution Engineering Handbook*, p. 581 (Ann Arbor Science Publishers Inc., Ann Arbor, Mich.).

Cooney, C.L., Wang, D.I.C. and Mateles, R.I. (1969) Measurement of Heat Evolution and Correlation with Oxygen Consumption during Microbial Growth. *Biotechnol. Bioengng*, **11**, 269.

Darlington, W.A. (1964) Aerobic Hydrocarbon Fermentation. A Practical Evaluation. *Biotechnol. Bioengng*, **6**, 241.

Demain, A.L. (1971) Overproduction of Microbial Metabolites and Enzymes due to Alteration of Regulation, in *Advances in Biochemical Engineering*, vol. 1, ed. by T.K. Ghose *et al*, p. 113 (Spring-Verlag, Berlin).

Elsworth, R., Miller, G.A., Whitaker, A.R., Kitching, D. and Sayer P.D. (1968) Production of *Escherichia coli* as a Source of Nucleic Acids. *J. Appl. Chem.*, **18**, 157.

Erickson, L.E., Minkevich, I.G. and Eroshin, V.K. (1978) Application of Mass and Energy Balance Regularities in Fermentation. *Biotechnol. Bioengng*, **20**, 1595.

Eroshin, V.K., Utkin, I.S., Ladynichev, S.V., Samoylov, V.V., Kurschinnikov, V.D. and Skryabin, G.K. (1976) Influence of pH and Temperature on the Substrate Yield Coefficient of Yeast Growth in a Chemostat, *Biotech. and Bioengng.*, **18**, 289.

Forrest, W.W. and Walker, D.J. (1971) Generation and Utilization of Energy during Growth. *Adv. Microb. Physiol.*, **5**, 213.

Hadjipetrou, L.P., Gerrits, J.P., Tenlings, F.A.G. and Stouthamer, A.H. Relation between Energy Production and Growth of *Aerobacter aerogenes*. *J. Gen. Microbiol.*, **36**, 139.

Heijnen, J.J. and Roels, J.A. (1981) A Macroscopic Model describing Yield and Maintenance Relationships in Aerobic Fermentation. *Biotechnol. Bioengng*, **23**, 739.

Herbert, D. (1976) Stoichiometric Aspects of Microbial Growth in Continuous Culture, in *Applications and New Yields*, ed. by A.R.C. Dean *et al.* (Ellis Horwood, Chichester).

Hernandez, E. and Johnson, M.J. (1967) Energy Supply and Cell Yield in Aerobically Grown Microorganisms. *J. Bacteriol.*, **94**, 996.

Johnson, M.J. (1967) Growth of Microbial Cells on Hydrocarbons. *Science*, **155**, 1515.

Kormancikova, V., Kovac, L. and Vidova, M. (1969) Oxidative Phosphorylation in Yeasts. V. Phosphorylation Efficiencies in Growing Cells determined from Molar Growth Yields. *Biochim. Biophys. Acta*, **180**, 9.

Mateles, R.I. (1971) Calculation of the Oxygen Required for Cell Production. *Biotechnol. Bioengng*, **13**, 581.

Mayberry, W.R., Prochazka, G.J. and Payne, W.J. (1968) Factors derived from Studies of Aerobic Growth in Minimal Media. *J. Bacteriol.*, **96**, 1424.

Meijer, E.M., Van Versevald, H.W., Van der Beek, E.G. and Stouthamer, A.H. (1977) Energy Conservation during Aerobic Growth in *Paracoccus denitrificans*. *Arch. Microbiol.*, **112**, 25.

Monod, J. (1941) Increases in Bacterial Concentration as a Function of the Concentration of the Carbohydrate of the Medium. *Compt. rend.*, **212**, 771.

Monod, J. (1949) The Growth of Bacterial Cultures. *Ann. Rev. Microbiol.*, **3**, 371.

Morowitz, H.J. (1968) *Energy Flow in Biology: Biological Organization as a Problem in Thermal Physics* (Academic Press, New York).

Neijssel, O.M. and Tempest, D.W. (1976) The Role of Energy-Spilling Reactions in the Growth of *Klebsiella aerogenes* NCTC 418 in Aerobic Chemostat Culture. *Arch. Microbiol.*, **110**, 305.

Ng, H. (1969) Effect of Decreasing Growth Temperature on Cell Yield of *Escherichia coli*. *J. Bacteriol.*, **98**, 232.

Oura, E. (1973) Energetics of Yeast Growth under Different Intensity of Aeration. *Biotech. and Bioengng*. Symposium No. 4 *Advances in Microbial Engineering* ed. by B. Sikyta, A. Prokop and M. Novak, p.117 (Wiley, New York).

Payne, W.J. (1970) Energy Yields and Growth of Heterotrophs. *Ann. Rev. Microbiol.*, **24**, 17.

Pirt, S.J. (1965) Maintenance Energy of Bacteria in Growing Cultures. *Proc. Roy. Soc.* (*London*), *Ser. B*, **163**, 224.

Rainbow, C. and Rose, A.H. (1963a) Biochemistry of Industrial Microorganisms, p. 228 (Academic Press, New York).

Rainbow, C. and Rose, A.H. (1963b) Biochemistry of Industrial Microorganisms, p. 103 (Academic Press, New York).

Rainbow, C. and Rose, A.H. (1963c) Biochemistry of Industrial Microorganisms, p. 173 (Academic Press, New York).

Servizi, J.A. and Bogan, R.H. (1963) Free Energy as a Parameter in Biological Treatment. *Proc. Am. Soc. Civil. Engng, J. Soc. Sanit. Engng. Divn.*, **89**, 17.

Servizi, J.A. and Bogan, R.H. (1964) Thermodynamic Aspects of Biological Oxidation and Synthesis. *J. Water Poll. Contr. Fed.*, **36**, 607.

Siegel, B.V., and Clifton, C.E. (1950) Energetics and Assimilation in the Combustion of Carbon Compounds by *Escherichia coli*. *J. Bacteriol.*, **60**, 585.
Smalley, A.J., Jahrling, P. and Van Demark, P.J. (1968) Molar Growth Yields as Evidence for Oxidative Phosphorylation in *Streptococcus faecalis* Strain 10C1. *J. Bacteriol.*, **96**, 1595.
Stouthamer, A.H. (1962) Energy Production in *Gluconobacter liquefaciens*. *Biochim. Biophys. Acta*, **56**, 19.
Stouthamer, A.H. (1969) Determination and Significance of Molar Growth Yields, in *Methods in Microbiology*, vol. 1, ed. by J.R. Norris and D.W. Ribbons, p. 629 (Academic Press, New York).
Stouthamer, A.H. (1976) Biochemistry and Genetics of Nitrate Reductase. *Adv. Microb. Physiol.*, **14**, 315.
Stouthamer, A.H. and Battenhaussen, C.W. (1976) Energetic Aspects of Anaerobic Growth of *Aerobacter aerogenes* in Complex Medium. *Arch. Microbiol.*, **111**, 21.
Van Dijken, J.P., and Harder, W. (1975) Growth Yields of Microorganisms on Methanol and Methane. Theoretical Study. *Biotechnol. Bioengng*, **17**, 15.
Wang, D.I.C. (1968) Proteins from Petroleum. *Chem Engng*, **75**, 99.
Yamada, K. (1977) *Japan's Most Industrial Fermentation Technology and Industry* (The International Technical Information Institute, Tokyo).

APPENDIX

DETAILED STOICHIOMETRY

Degree of Reduction

The degree of reduction is the number of equivalents of oxygen required for the complete oxidation of that quantity of organic compound that contains one gram atom of carbon.

Sample Calculation of the Generalized Degree of Reduction of a Compound

For the reaction

$$CH_aN_bO_cP_dS_e + x\,O_2 \longrightarrow CO_2 + y\,H_2O + b\,NH_3 + d\,H_3PO_4 + e\,H_2SO_4$$

$$y = \tfrac{1}{2}(a - 3b - 3d - 2e)$$

$$x = \tfrac{1}{4}(4 + a - 3b - 2c + 5d + 6e)$$

where x is the number of moles of oxygen required to achieve complete combustion of one mole of the compound. The generalized degree of reduction γ may be determined from x.

$$\gamma = 4x$$

Therefore, for $CH_aN_bO_cP_dS_e$, the generalized degree of reduction is

$$\gamma = 4 + a - 3b - 2c + 6e$$

The coefficients relate to the values of the degree of reduction of C, H, N, O, P and S, i.e. C = 4, H = 1, N = − 3, O = − 2, P = 5 and S = 6

Generalized Degrees of Reduction for Biomass, Product and Substrate

Table A1.

Compound	Generalized degree of reduction	Equation[a]
Biomass	γ_b	$\gamma_b = 4 + bH - \alpha bN - 2bO + 5bP + 6bS$
Product	γ_p	$\gamma_p = 4 + pH - \alpha pN - 2pO + 5pP + 6pS$
Substrate	γ_s	$\gamma_s = 4 + sH - \alpha sN - 2sO + 5sP + 6sS$

[a] *b*H refers to biomass hydrogen, *p*N product nitrogen, *s*O substrate oxygen, etc.
The value of α depends on the nitrogen source, e.g. $\alpha = 3$ for ammonia, $\alpha = 0$ for nitrogen and $\alpha = -5$ for nitrate.

Table A2. Degree of reduction and weight of one carbon equivalent or mole of some substrates and biomass.

Compound	Molecular formula	Degree of reduction γ	Weight, *m*
Biomass	$CH_{1.64}N_{0.16}O_{0.52}$ $P_{0.0054}S_{0.005}$[a]	4.17 (NH_3) 4.65 (N_2) 5.45 (HNO_3)	24.5
Methane	CH_4	8	16.0
n-Alkane	$C_{15}H_{32}$	6.13	14.1
Methanol	CH_4O	6.0	32.0
Ethanol	C_2H_6O	6.0	23.0
Glycerol	$C_3H_8O_3$	4.67	30.7
Mannitol	$C_6H_{14}O_6$	4.33	30.3
Acetic acid	$C_2H_4O_2$	4.0	30.0
Lactic acid	$C_3H_6O_3$	4.0	30.0
Glucose	$C_6H_{12}O_6$	4.0	30.0
Formaldehyde	CH_2O	4.0	30.0
Gluconic acid	$C_6H_{12}O_7$	3.67	32.7
Succinic acid	$C_4H_6O_4$	3.50	29.5
Citric acid	$C_6H_8O_7$	3.0	32.0
Malic acid	$C_4H_6O_5$	3.0	33.5
Formic acid	CH_2O_2	2.0	46.0
Oxalic acid	$C_2H_2O_4$	1.0	45.0

[a] Harrison, (1977)

Table A3.

Compound	Chemical formula (carbon equivalents)[a]	Net conversion[b] mole
Biomass	$CH_{bH}N_{bN}O_{bO}S_{bS}P_{bP}$	Δb
Substrate	$CH_{sH}N_{sN}O_{sO}S_{sS}P_{sP}$	Δs
Product	$CH_{pH}N_{pN}O_{pO}S_{pS}P_{pP}$	Δp
Water	H_2O	Δw
Oxygen	O_2	Δo
Carbon dioxide	CO_2	Δc
Nitrogen source	N_2 NH_3 HNO_3	Δn
Sulphur source	H_2SO_4	Δsu
Phosphorus source	H_3PO_4	Δpho

[a] Molecular formula of the carbon-containing compounds is defined as the carbon equivalent, i.e. the amount that contains one mole of carbon.
[b] Difference between the final and initial amounts.

Elemental Balances on C, H, N, O, S and P for a Heterotrophic Aerobic Fermentation

$$CH_{sH}N_{sN}O_{sO}NP_{sp}S_{sS} + a\,H_2SO_4 + bH_3PO_4 + c\,(\text{nitrogen source}) + d\,O_2$$
$$\longrightarrow e\,CH_{bH}N_{bN}O_{bO}P_{bp}S_{bs} + f\,CH_{pH}{}^{N}{}_{pN}O_{pO}P_{pP}S_{ps} + g\,CO_2 + h\,H_2O$$

$$g = 1 - e - f$$

$$d = \tfrac{1}{4}(\gamma_s - \gamma_b e - \gamma_p f)$$

$$h = \tfrac{1}{4}(\eta_s - \eta_b e - \eta_p f)$$

$$c = -\,sN + bN\,e + pN\,f$$

$$a = -\,sS + bS\,e + pS\,f$$

$$b = -\,sP + bP\,e + bP\,f$$

where γ_s, γ_b and γ_p are generalized degrees of reduction for substrate, bromass and product, respectively.

$$\eta_s = 2sH - \beta\,sN - 4sS - 6sP$$

$$\eta_b = 2bH - \beta\,bN - 4bS - 6bP$$

$$\eta_p = 2pH - \beta\,pN - 4pS - 6pP$$

β depends on the nitrogen source, e.g. $\beta = 6$ for ammonia, $\beta = 0$ for nitrogen and $\beta = 2$ for nitrate. Similarly, if instead of one mole (carbon equivalent) of substrate ΔS mole (carbon equivalent) substrate is used, the net molar conversion become.

$$\Delta c = \Delta s - \Delta b - \Delta p$$

$$\Delta o = \tfrac{1}{4}(\gamma_s \Delta s - \gamma_b \Delta b - \gamma_p \Delta p)$$

$$\Delta w = \tfrac{1}{4}(\eta_s \Delta s - \eta_b \Delta b - \eta_p \Delta p)$$

$$\Delta n = (sN)\Delta s + bN\,\Delta b + pN\,\Delta p$$

$$\Delta su = -(sS)\,\Delta s + (bS)\,\Delta b + (pS)\,\Delta p$$

$$\Delta pho = -\,(sP)\Delta s + (bP)\Delta b + (pP)\,\Delta p$$

where γ_s, γ_b, γ_p, η_s, η_b and η_p are defined as before

In the above treatment

$$e = \frac{\Delta b}{\Delta s} = \text{fractional conversion of substrate carbon to biomass}$$

$$f = \frac{\Delta p}{\Delta s} = \text{fractional conversion of substrate carbon to product.}$$

Theoretical Oxygen Requirement

From the stoichiometry of fermentation, Δo is the moles of oxygen required for biomass growth and product formation. Hence, the theoretical oxygen requirement can be calculated if Δs, Δb, Δp and the elemental composition of compounds involved are known.

Estimation of Stoichiometric Maximum Yields

If the elemental composition of the compounds involved in the heterotrophic aerobic fermentation is known, γ_s, γ_b, γ_p, η_s, η_b and η_p can be calculated from the equations given previously. Δs is usually known or is assigned a desired value. (If one assumes complete exhaustion of substrate supplied in the medium, Δs is the initial amount of substrate.)

The fractional allocation of available electrons in the organic substrate can be written as

$$\frac{4d}{\gamma_s} + \frac{e\gamma_b}{\gamma_s} + \frac{f\gamma_p}{\gamma_s} + \frac{f\gamma_p}{\gamma_s} = 1$$

$\theta_o \equiv \frac{4d}{\gamma_s} =$ fraction of available electrons transferred to oxygen (fraction of energy of substrate evolved as heat)

$\theta_b \equiv \frac{e\gamma_b}{\gamma_s} =$ fraction of available electrons transferred to biomass

$\theta_p \equiv \frac{f\gamma_p}{\gamma_s} =$ fraction of available electrons transferred to product

If σ_s is the weight fraction of carbon in substrate and σ_b is the weight fraction of carbon in biomass

then

$$Y'_s = e\frac{\sigma_s}{\sigma_b}\left(\text{i.e. } \frac{\text{gram of biomass produced}}{\text{gram of substrate consumed}}\right)$$

and

$$Y'_O = \frac{3\theta_b}{2\sigma_b\gamma_b(1-\theta_b-\theta_p)}\left(\frac{\text{i.e. gram of biomass produced}}{\text{gram of oxygen consumed}}\right)$$

The stoichiometric range of θ_b is $0 \le \theta_b \le 1$ whereas in practice $\theta_b < 0.7$. The range of the mass yield coefficient Y'_s is $0 \le Y'_s \le \gamma_s\sigma_s/\gamma_b\sigma_b$. Thus Y'_s depends on the energy content of the organic substrate. If the composition if the cells is unknown γ_b can be taken as 4.2 [Erickson *et al.*, (1978)].

Table A3 shows values of the theoretical maximum mass yield Y'_s and the maximum carbon yield e for a variety of organic substrates for $\theta_b = 1$, $\gamma_b = 4.2$ and $\sigma_b = 0.48$.

Table A3. Mass yields of biomass on organic substrate corresponding to the thermodynamic maximum energetic yield [Erickson *et al.* (1978)].

Substrate	Equivalent available electrons in organic substrate containing 1 g atom carbon, γ_s	Thermodynamic maximum yield corresponding to $\theta_b\ \eta = 1.0,\ \gamma_b = 4.2,\ \tau_b = 0.48$ Carbon yield ε	Weight yield Y'_s
Alkanes			
Methane	8.0	1.9	3.0
Hexane	6.3	1.5	2.7
Hexadecane	6.1	1.5	2.6
Alcohols			
Methanol	6.0	1.4	1.1
Ethanol	6.0	1.4	1.6
Hexadecanol	6.0	1.4	2.4
Ethylene glycol	5.0	1.2	1.0
Glycerol	4.7	1.1	0.9
Carbohydrates			
Formaldehyde	4.0	1.0	0.8
Glucose	4.0	1.0	0.8
Sucrose	4.0	1.0	0.9
Starch, cellulose	4.0	1.0	0.9
Organic acids			
Formic acid	2.0	0.5	0.3
Acetic acid	4.0	1.0	0.8
Propionic acid	4.7	1.1	1.2
Lactic acid	4.0	1.0	0.8
Fumaric acid	3.0	0.7	0.6
Oxalic acid	1.0	0.3	0.1

Table A4 shows that substrates with a high energy content (indicated by a high γ_s value) have high theoretical yields.

The theoretical limits on substrate and oxygen yield factors, which are dictated by the elemental and energy balances, have been explored by Heijnen and Roels (1981). Using their terminology, these limits are

$$Y_{sb} \leq 1$$

$$Y_{ob} \leq \frac{4}{\gamma_s - \gamma_b}$$

where Y_{sb} is yield of biomass expressed as carbon equivalent per carbon, equivalent of substrate and Y_{ob} is the yield of biomass expressed as carbon equivalent per mole of oxygen. Furthermore, if $Y_{ob} > 0$, i.e. there is always consumption of oxygen, then

$$Y_{sb} \leq \frac{\gamma_s}{\gamma_b}$$

Figs. A1 and A2 show these theoretical limits to biomass yields on substrate and oxygen for various degrees of substrate reduction.

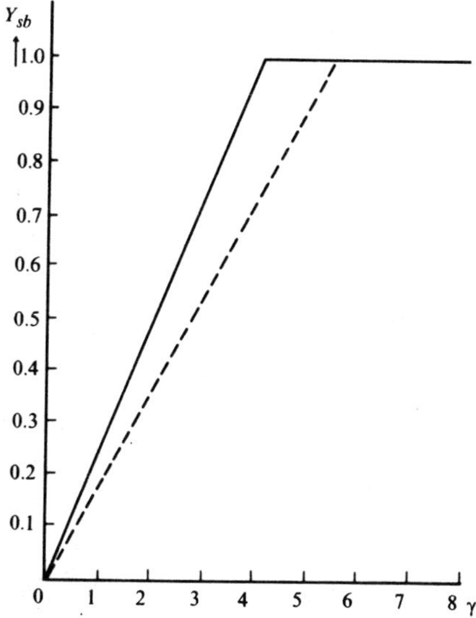

Figure A1. Limits on the biomass yield on substrate for substrates with different degree of reduction. —, nitrogen source ammonia; - - -, nitrogen source nitrate [Heijnen and Roels (1981)].

In Fig. A1, the degree of reduction for biomass γ_b is taken as 4.17 for ammonia and as 5.45 for nitrate as nitrogen sources for degrees of reduction of various substrates are given in Table A2.

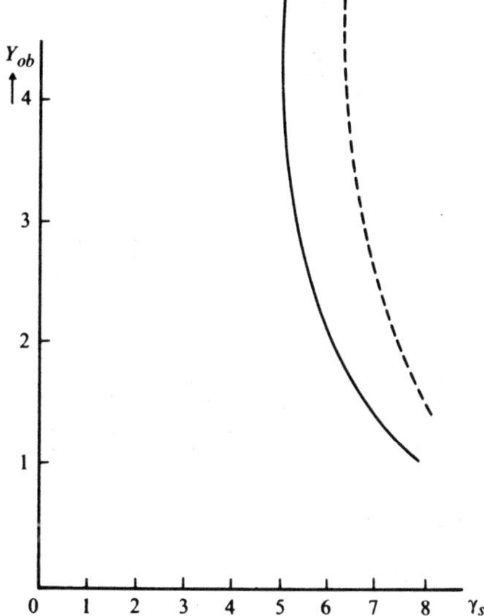

Figure A2. Limit on the biomass yield on oxygen for substrates with different degree of reduction. —, nitrogen source ammonia; - - -, nitrogen source nitrate [Heijnen and Roels (1981)].

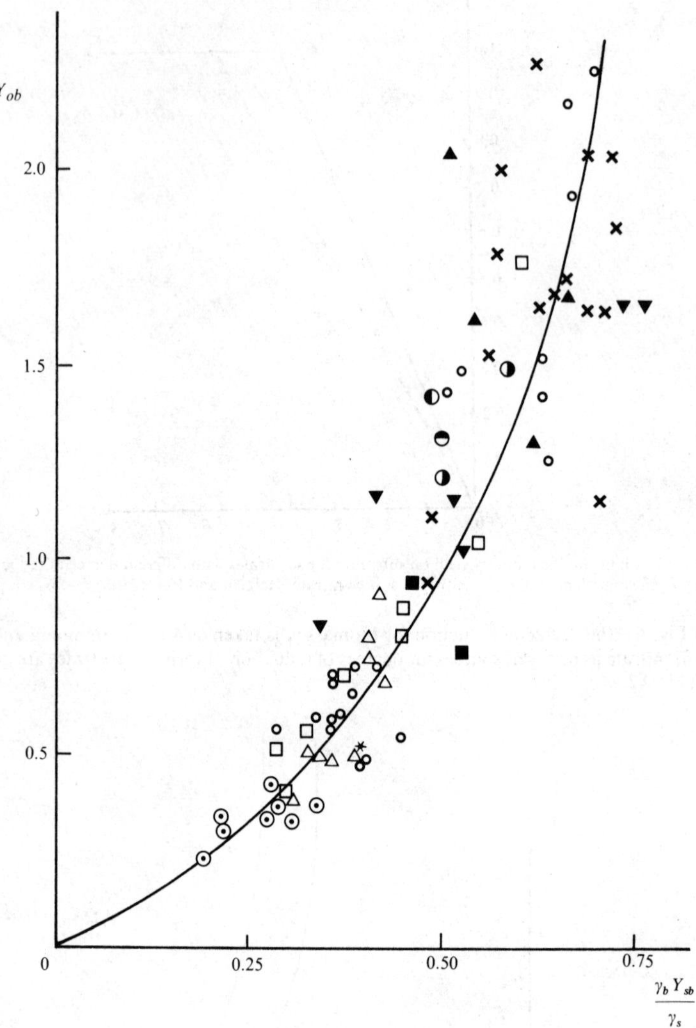

Figure A3. Theoretical relationship between Y_{ob} and Y_{sb} and experimental data for aerobic growth without product formation and ammonia as a nitrogen source with the substrates: ○, glycerol; ▲, mannitol; ■, acetic acid; ▼, lactic acid; ×, glucose; ‘, succinic acid; , citric acid; ○, malic acid; , formic acid; *, oxalic acid; ’, gluconic acid; ⊕, methane; :, *n*-alkanes; △, methanol; □, ethanol [Heijnen and Roels (1981)].

From the elemental balances

$$\frac{1}{Y_{\rho\beta}} = \frac{1}{4}\left(\frac{\gamma_s}{Y_{sb}} - \gamma_b\right)$$

This equation is plotted in Fig. A3 and the experimental data, which can be found in Table 23 of Chapter 3, seem to support the theory.

Estimation of Yield Coefficients

The elemental balances for anaerobic heterotrophic fermentation are given in this appendix dealing with elemental balances for heterotrophic aerobic fermentation six equations with eight unknowns. Therefore some of these unknown parameters have to be assigned a value in order to calculate the yield coefficients.

Fig. A4–A6 have been constructed by Erickson *et al.* (1978) to illustrate how various sets of measured values may be used to estimate the yield coefficients.

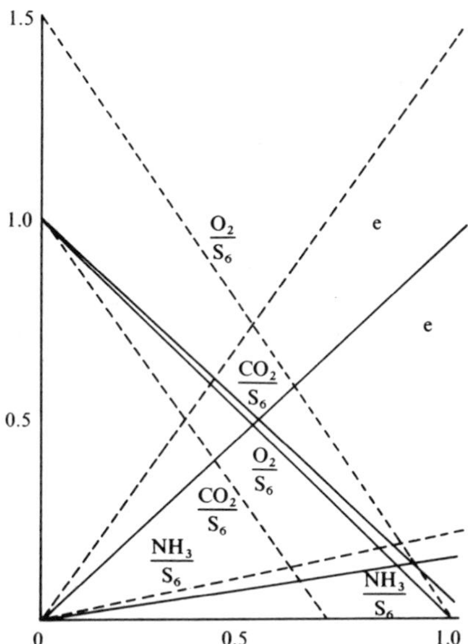

Figure A4. Ratios of $a = NH_3/S_c$, $b = O_2/S_c$, $d = CO_2/S_C$, and Ye_c as a function of the energetic yield, Ob for $\gamma_s = 4.0$, —; $\gamma_s = 6.0$, - - -. S_C is the number of gram atoms carbon in the consumed organic substrate. Product formation is assumed to be zero. $\gamma_b = 4.2$ and the ratio of nitrogen atoms to carbon atoms in the biomass is taken to be 0.16. [Erickson *et al.* (1978)].

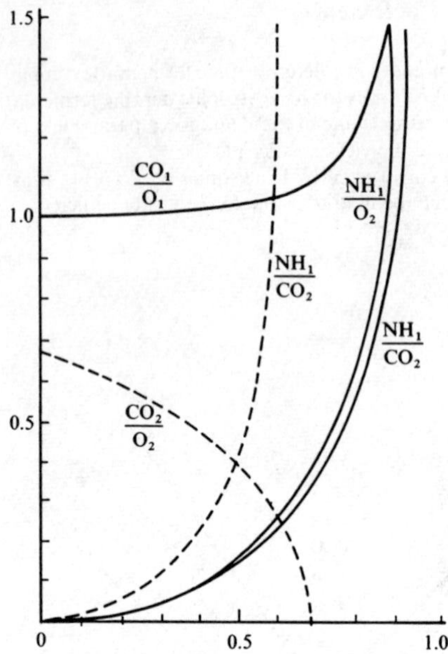

Figure A5. Mole ratios of ammonia/oxygen, carbon dioxide/oxygen and ammonia/carbon dioxide as a function of energetic yield, Ob, for $\gamma_S = 4.0$, —; $\gamma_S = 6.0$, - - -; except for ammonia/oxygen which is identical to that for $\gamma_s = 4.0$. Product formation is assumed to be zero, $\gamma_b = 4.2$ and the ratio of nitrogen atoms to carbon atoms in the biomass is taken to be 0.16 [Erickson *et al.* (1978)].

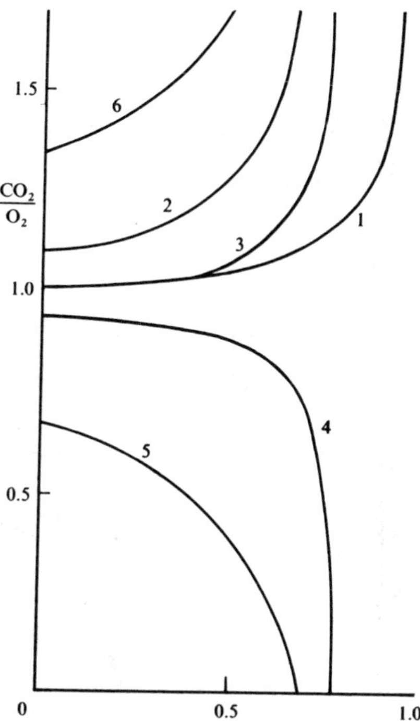

Figure A6. Respiratory quotient versus energetic efficiency, Ob for $\gamma_s = 4.0$ under conditions of (1) no product formation (2) ethanol formation ($\gamma_p = 6.0$, Op $= 0.2$; (3) acetic acid or lactic acid formation ($\gamma_p = 4.0$, $Op = 0.2$) and (4) citric acid formation ($\gamma_p = 3.0$, $Op = 0.2$). If the product is utilized as organic substrate for biomass production, the respiratory quotient is (5) ethanol ($\gamma_s = 6.0$); I) acetic or lactic acid ($\gamma_s = 4.0$), and (6) citric acid ($\gamma_s = 3.0$) [Erickson *et al.* (1978)].

CHAPTER 4
MICROBIAL ACTIVITY

MONOD EQUATION

Fig. 1 shows the time course of a batch fermentation. The similarity of the curves for mycelial dry weight, ethanol production and sugar utilization suggests that the yield coefficients for biomass formation Y'_s and product formation Y^p_s are essentially constant over the course of the fermentation.

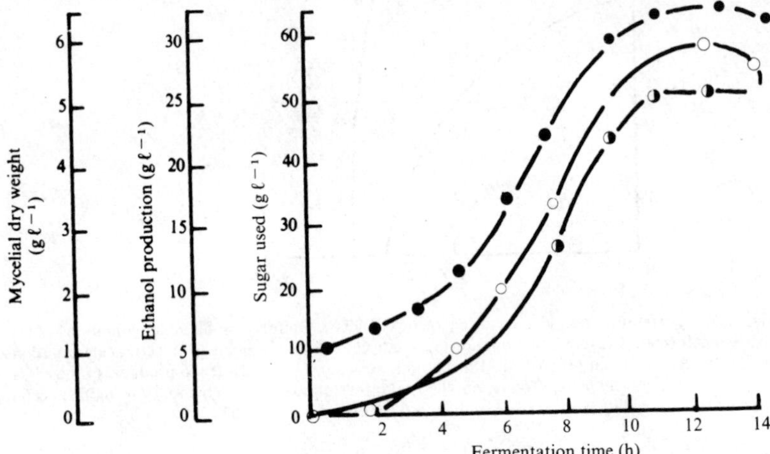

Figure 1. Time–concentration curves for alcohol production. ●, mycelial dry weight; ○, ethanol production; ◑, sugar utilization [Luedeking (1967)].

Fig. 2 contains derived data, i.e. data obtained from the experimental results contained in Fig. 1, and represents the time course of the volumetric rate of substrate uptake R_{vs}, the volumetric rate of biomass formation R_{vb} and the volumetric rate of product formation R_{vp}. Thus

$$R_{vs} = -\frac{ds}{dt} \tag{1}$$

$$R_{vb} = \frac{dx}{dt} \tag{2}$$

$$R_{vp} = \frac{dp}{dt} \tag{3}$$

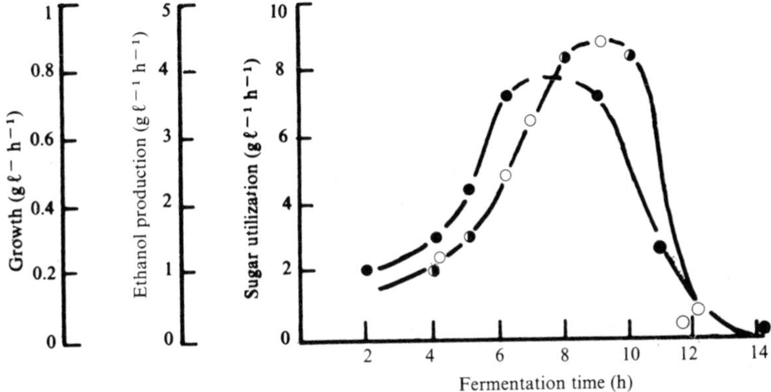

Figure 2. Volumetric rates of alcohol fermentation. ●, mycelial growth; ○, ethanol production. ◑, sugar utilization [Luedeking (1967)].

The volumetric rate of change of a component I can be obtained from the experimental data by a number of techniques of varying sophistication and accuracy. The dimensions of R_{vI} are $M L^{-3} T^{-1}$, namely those of the productivity of unit volume of the fermenter. The volumetric rates of reaction are not intrinsic properties of the reaction, since they can be changed by adding or removing biomass from the system.

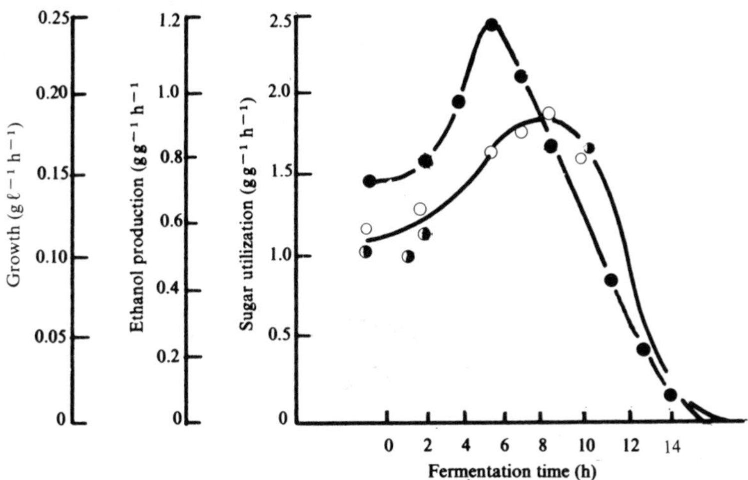

Figure 3. Specific rates of alcohol fermentation. ●, mycelial growth; ○, ethanol production; ◑, sugar utilization [Luedeking (1967)].

Fig. 3 represents the specific rate of substrate uptake R_s, the specific rate of biomass formation μ and the specific rate of product formation R_p, where

$$R_s = -\frac{1}{x}\frac{ds}{dt} = \frac{1}{x} R_{vs} \tag{4}$$

$$\mu = \frac{1}{x}\frac{dx}{dt} = \frac{1}{x} R_{vb} \tag{5}$$

$$R_p = \frac{1}{x}\frac{dp}{dt} = \frac{1}{x} R_{vp} \tag{6}$$

The dimensions of R_s, μ and R_p are $M_s M_o^{-1} T^{-1}$ and $M_p M_o^{-1} T^{-1}$, respectively, i.e. the specific rates relate to the unit mass of microorganism M_o, and represent intrinsic properties of the microbe–substrate system.

However, specific rates are not constant and, in practice, are found to vary between different phases of microbial growth (defined in Table 2). In the industrially important growth/decline phases, which occur at times beyond the maxima in Fig. 3, the specific rates depend upon the concentration of the components in the medium, e.g.

$$\mu = g(s_C, s_N, s_O, s_P, \ldots) \tag{7}$$

Generally, there exists a single limiting component, which is usually the carbon or nitrogen source or the hydrogen acceptor, e.g. molecular oxygen or nitrate.

The relationship between the specific growth rate μ and the limiting component I may be expressed by the Monod equation, i.e.

$$\mu = \mu_{max} \frac{s_I}{K_{m,I} + s_I} \tag{8}$$

In Eq. (8), μ_{max} is the maximum specific growth rate and $K_{m,I}$ is the Monod coefficient corresponding to the component I; usually more simply referred to as K_m. Both are intrinsic parameters of the microbe–substrate system.

The Monod equation was originally considered to be purely empirical, but it does have a partially theoretical grounding in enzyme kinetics and carrier-associated transport across cell membranes.

In general, $K_{m,I}$ depends on the environmental conditions (i.e. pH, temperature, ionic strength, etc.). Graphically, Eq. (8) may be represented in the form shown in Fig. 4.

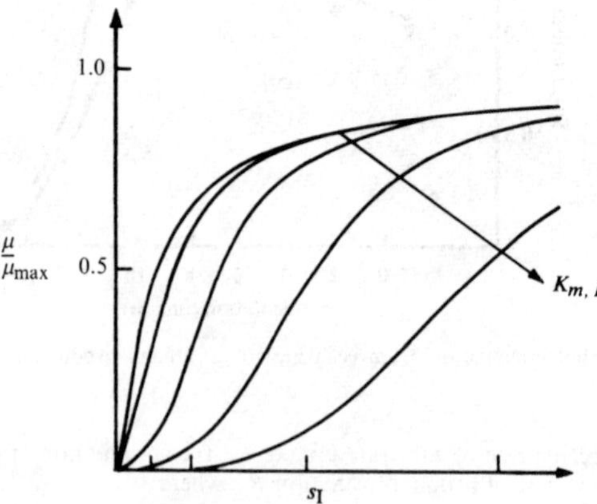

Figure 4. Graphical representation of the Monod equation [Eq. (8)] for various values of $K_{m,I}$.

A typical Monod plot for a mixed microbial population is shown in Fig. 5. From this plot, a value for μ_{max} of 0.46 h^{-1} and a value for $K_{m,s}$ of 55 mg ℓ$^{-1}$ were obtained.

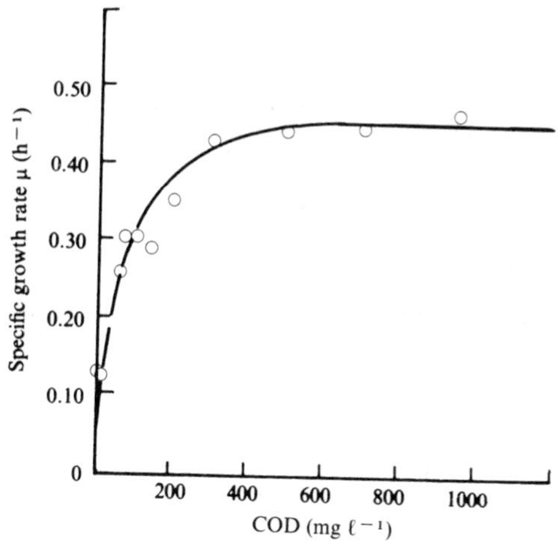

Figure 5. Monod plot for mixed microbial population of sewage origin growing on the soluble fraction of municipal sewage [Gaudy and Gaudy (1972)].

For a given set of experimental results, a graph of μ versus substrate concentration s can be plotted to provide a curve such as shown in Fig. 6.

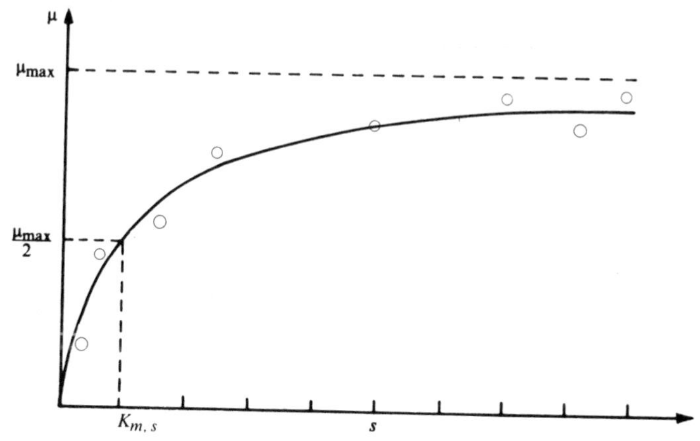

Figure 6. Variation of specific growth rate versus substrate concentration.

From this graph (*see* Fig. 6), approximate values of μ_{max} and consequently, $K_{m,s}$ may be obtained by visually choosing μ_{max} and then establishing the value of s that corresponds to $\mu_{max}/2$, i.e. s', and then setting $K_{m,s} = s'$. The latter relationship follows from the Monod equation, i.e.

$$\frac{\mu_{max}}{2} = \mu_{max} \frac{s'}{K_{m,s} + s'} \tag{9}$$

However, as may be seen from Fig. 6, it is difficult to establish the exact value of μ_{max} because the curve is a rectangular hyperbola. A more accurate graphical method for determining μ_{max} and $K_{m,s}$ involves rearranging the Monod equation in the form

$$\frac{1}{\mu} = \frac{1}{\mu_{max}} + \frac{K_{m,s}}{\mu_{max}} \cdot \frac{1}{s} \tag{10}$$

and either plotting the data in the form shown in Fig. 7 or by using linear regression.

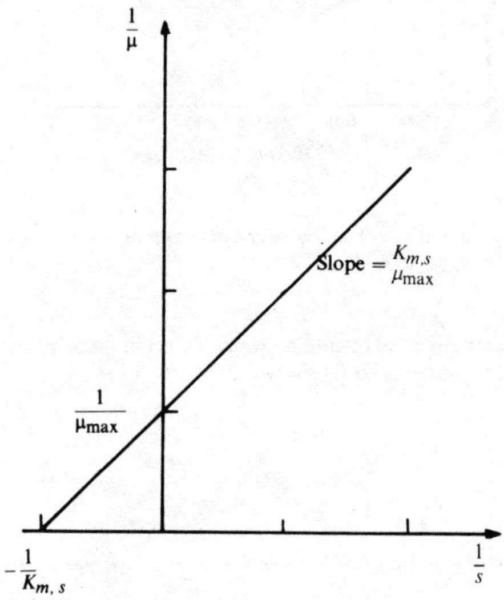

Figure 7. Lineweaver–Burk plot.

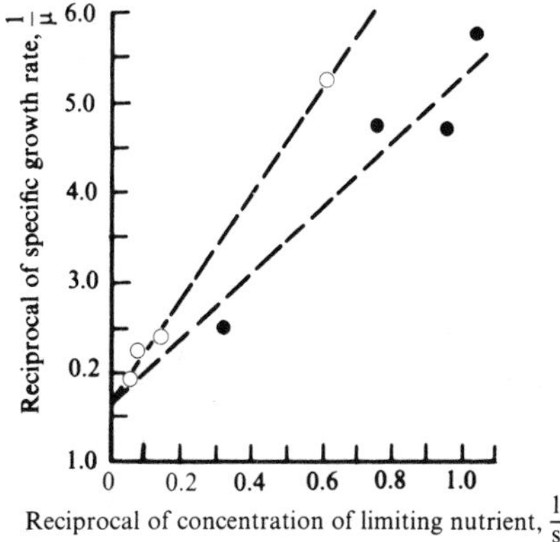

Figure 8. Lineweaver–Burk plot for continuous cultures of *Aerobacter aerogenes* [Contois (1959)].

An example of the double reciprocal plot is shown in Fig. 8. Using this approach, values of $K_{m,s}$ obtained using different substrates are listed in Table 1.

Table 1. Some values for $K_{m,s}$ in the Monod model for growth [Wang *et al.* (1979)].

Microorganism	Substrate	$K_{m,s}$ (mg ℓ^{-1})
Enterobacter aerogenes	Glucose	1.0
Ent. aerogenes	Ammonia	0.1
Ent. aerogenes	Magnesium	0.6
Ent. aerogenes	Sulphate	3.0
Escherichia coli	Glucose	2.0–4.0
Hansenula polymorpha	Ribose	3.0
H. polymorpha	Methanol	120.0
Saccharomyces cerevisiae	Glucose	25.0

A more elegant and satisfactory procedure for the determination of μ_{max}, K_m, etc. has been devised recently by Knights (1982). This method is based upon the empirical equations of Edwards and Wilkie (1968) and simultaneously satisifies the time-course data for the substrate and biomass concentrations such as given in Fig. 1 (*see* Appendix to this chapter).

An easily remembered, but approximate, property of the microbe–substrate system is $s_{CRIT, I}$, i.e. the concentration of the limiting component I required to achieve values of $\mu = 0.99\ \mu_{max}$. From Eq. (8),

$$\mu = 0.99\ \mu_{max} \qquad s_{CRIT,I} = 99\ K_{m,s} \simeq 100\ K_{m,s}$$

$$\mu = 0.999\ \mu_{max} \qquad s_{CRIT,I} = 999\ K_{m,s} \simeq 1000\ K_{m,s}$$

or generally

$$\mu = n\mu_{max} \qquad s_{CRIT.I} = \frac{n}{1-n} K_{m.s} \tag{11}$$

At relatively high substrate concentrations, the specific growth rate is often reduced below its maximum value. This condition is referred to as substrate inhibition and the Monod equation can be extended to accommodate this effect by introducing an inhibition coefficient K_i. Thus

$$\frac{\mu}{\mu_{max}} = \frac{s}{K_{m.s} + s + (s^2/K_i)} \tag{12}$$

Eq. (12) is plotted in Fig. 9 in terms of

$$\frac{\mu}{\mu_{max}} = g\left(\frac{s}{K_{m.s}}, \sqrt{\frac{K_{m.s}}{K_i}}\right) \tag{13}$$

Typical growth inhibition curves are shown in Figs. 10–14 and respiration inhibition curves in Figs. 15–17.

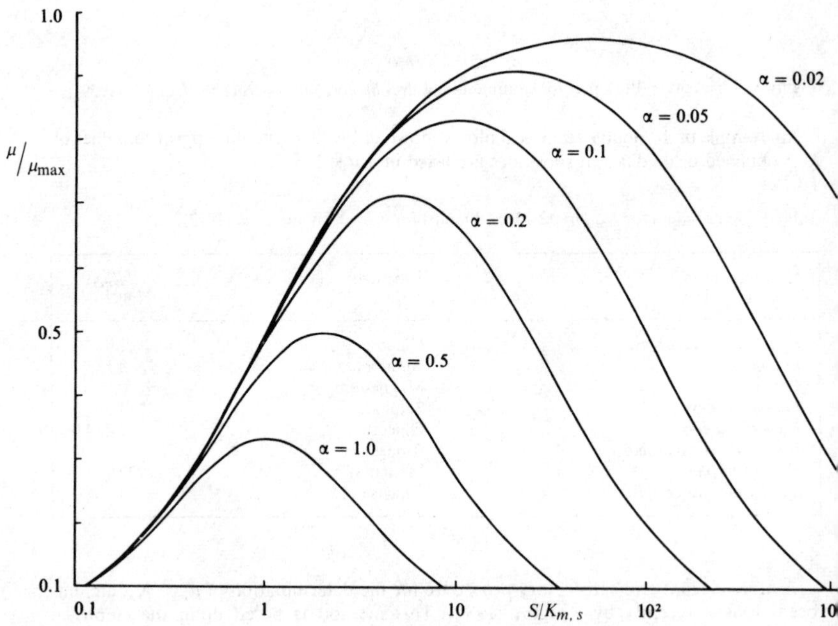

Figure 9. Substrate-inhibited Monod kinetics [Eq. (12)] where $\alpha = (K_{m,s}/K_i)^{1/2}$.

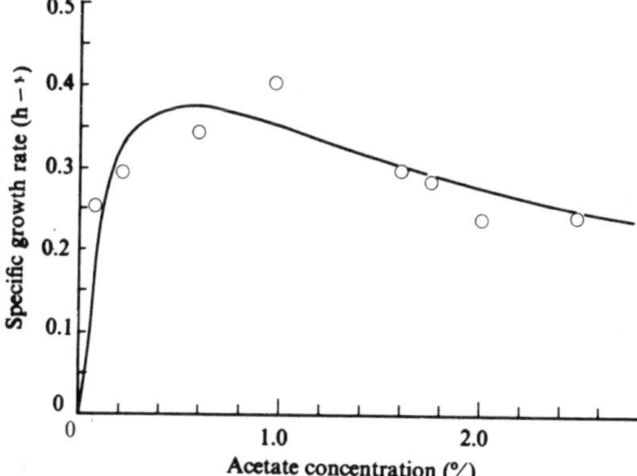

Figure 10. Effect of acetate on the growth of *Candida utilis* [Edwards (1970)].

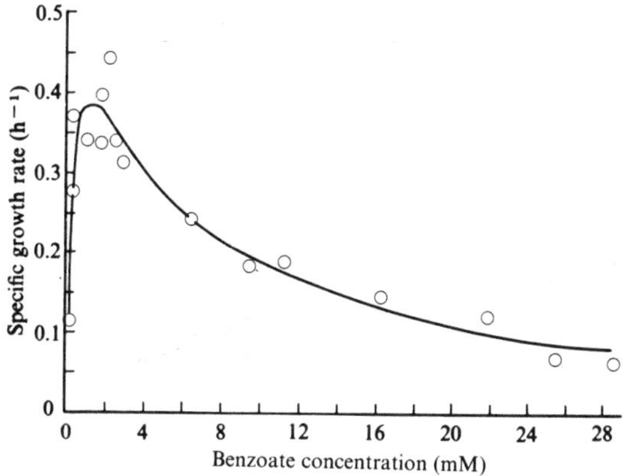

Figure 11. Inhibition of specific growth rate for *Klebsiella aerogences* by sodium benzoate [Edwards (1970)].

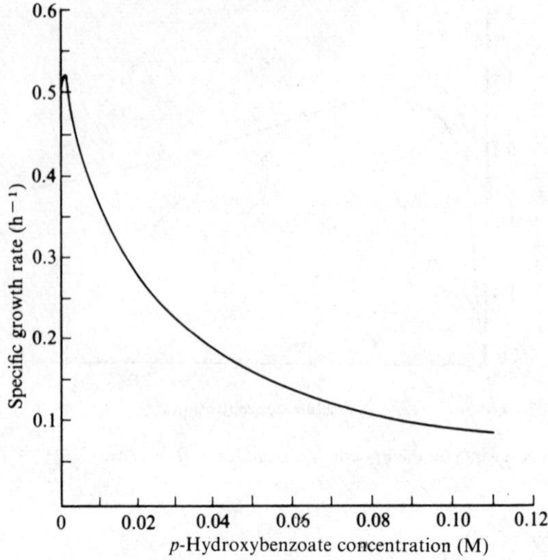

Figure 12. Inhibition of specific growth rate for *Klebsiella aerogenes* by sodium *p*-hydroxybenzoate [Edwards (1970)].

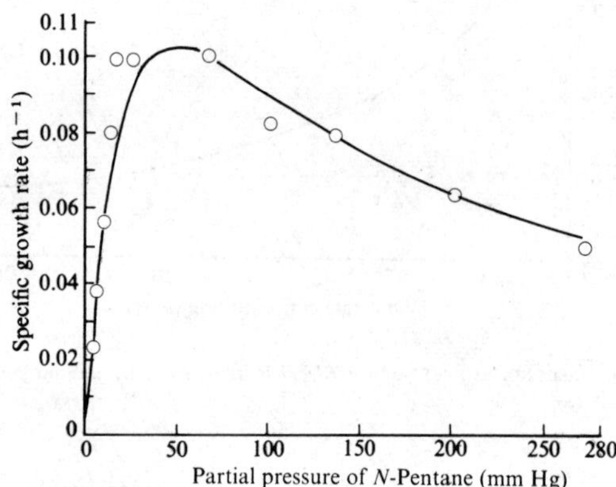

Figure 13. Inhibition by *n*-pentane of *Klebsiella aerogenes* [Edwards (1970)].

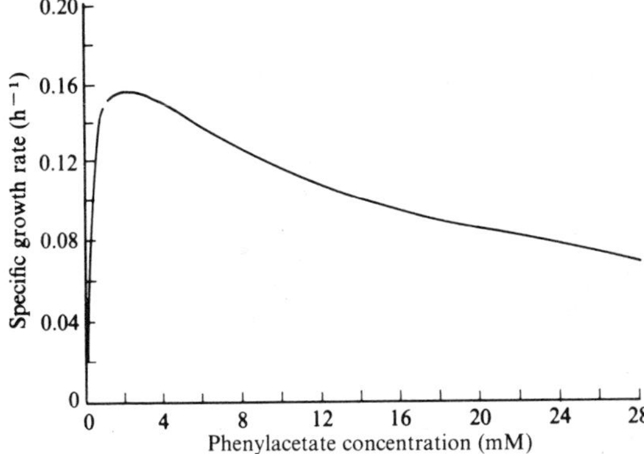

Figure 14. Inhibition of specific growth rate of *Klebsiella aerogenes* by sodium phenylacetate [Edwards (1970)].

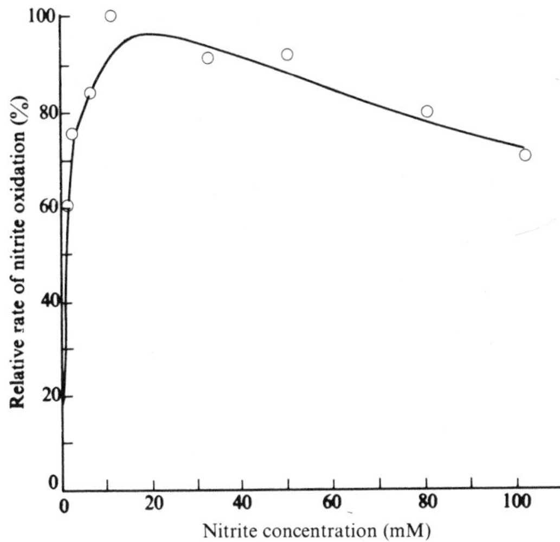

Figure 15. Substrate inhibition of nitrite oxidation [Edwards (1970)].

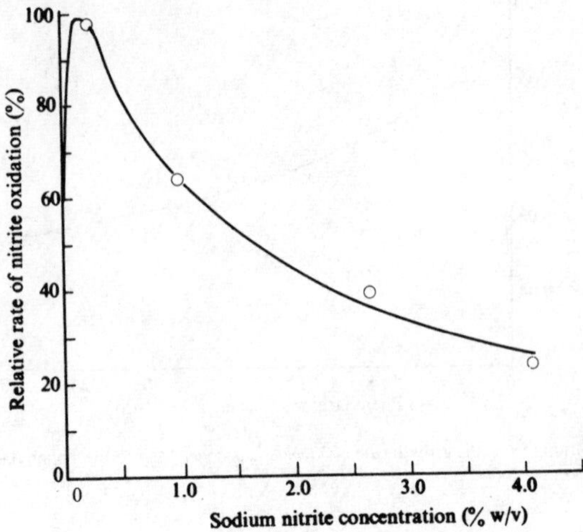

Figure 16. Inhibition of respiration in *Nitrobacter* by sodium nitrite [Edwards (1970)].

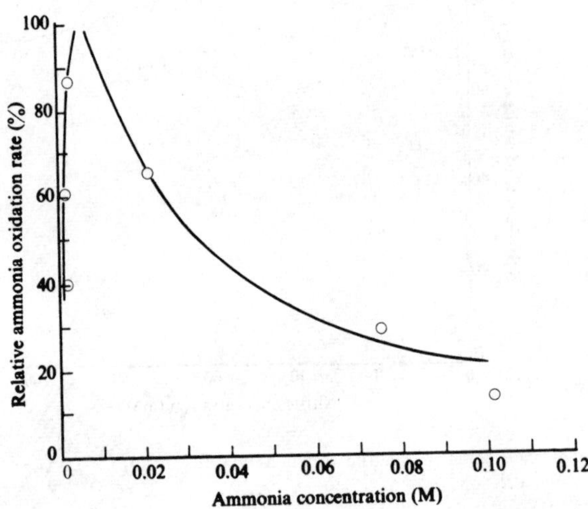

Figure 17. Inhibition of respiration in *Nitrosomonas* by ammonia [Edwards (1970)].

Fig. 18 shows Lineweaver–Burk plots for specific growth rates versus glucose concentrations for various concentrations of the product ethanol. Comparison with Fig. 7 suggests that while $K_{m,s}$ is unaffected by the ethanol concentration μ_{max} is reduced. Fig. 19 shows similar effects for the specific rate of alcohol formation R_p. This phenomenon is referred to as product inhibition.

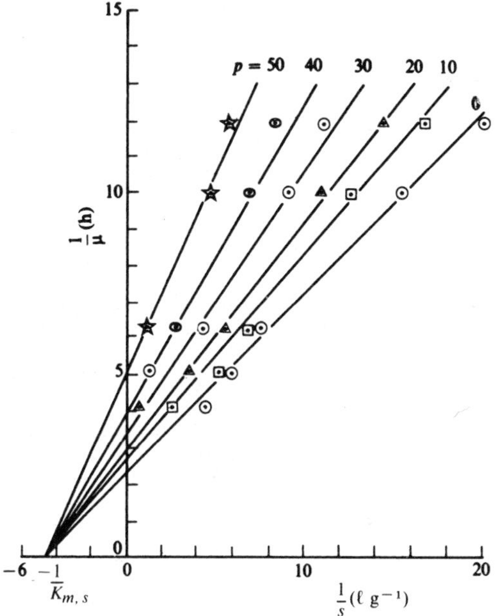

Figure 18. Lineweaver–Burk plot for ethanol production. Ethanol concentration $p = g\,\ell_-{}^{1}$ [Aiba *et al.* (1968)].

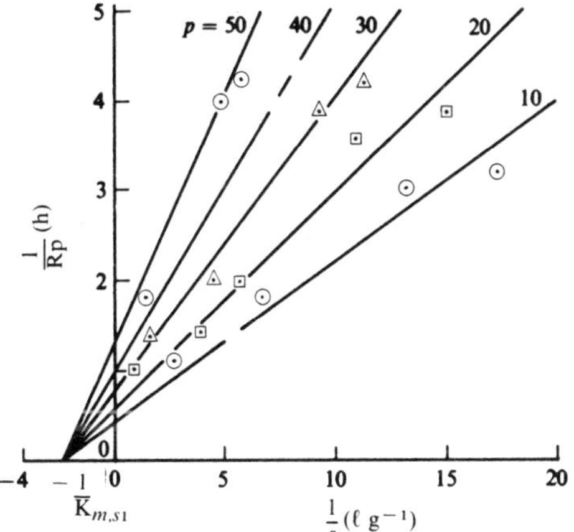

Figure 19. Lineweaver–Burk plot for ethanol production. Ethanol concentration $p = g\,\ell^{-1}$ [Aiba *et et al.* (1968)].

TIME COURSE OF FERMENTATION

Fig. 1 illustrates sigmoid relationships between time and the biomass concentration, sugar utilization and product concentration. The various sections of this typical growth curve have been named. A summary of the terms used is presented in Table 2.

Table 2. Phases of microbial growth.

Phase	Type of growth
Incubation/acclimatization/lag	Physicochemical equilibration between microbe and the environment following inoculation with very little growth (i.e. $\mu \simeq 0$)
Acceleration	Growth starts to occur (i.e. $\mu < \mu_{max}$)
Log (growth)	Growth (i.e. $\mu \simeq \mu_{max}$)
Decline	Growth is limited by the nutrient availability (i.e. $\mu < \mu_{max}$)
Resting/stationary	No net growth occurs as nutrients depleted (i.e. $\mu \leqslant 0$)
Death	Loss of viability and destruction by lysis of biomass (i.e. $\mu < 0$)

For a microbe– substrate system that obeys the Monod equation, and provided that the yield coefficients may be taken as constant, the variation in substrate concentration with time during the log and decline phases can be expressed in the following form

$$\frac{Y'_s + \Delta/K_{m.s}}{Y'_s} \ln \frac{\Delta - Y'_s s}{x_0} - \ln \frac{s}{s_0} = R_{s.\max} \frac{t\Delta}{K_{m.s}} \tag{14}$$

where x = biomass concentration at time t

R_s = specific substrate uptake rate

s = substrate concentration at time t

p = biochemical product concentration at time t

Y_s^p = yield of product per unit of substrate consumed.

Also

$$\Delta = Y'_s s_0 + x_0 \tag{15}$$

where Δ is the biomass equivalent of the initial substrate plus inoculum. Under the conditions defined above, the following equations also apply

$$x - x_0 = Y'_s(s_0 - s) \tag{16}$$

$$p - p_0 = Y_s^p(s_0 - s) \tag{17}$$

From Eq. (16)

$$s = \frac{\Delta - x}{Y'_s} \tag{18}$$

Substitution of Eq. (18) into Eq. (14) gives

$$\frac{Y'_s + \Delta/K_{m.s}}{Y'_s} \ln \frac{x}{x_0} - \ln \frac{\Delta - x}{\Delta - x_0} = \frac{R_{s.\max} t\Delta}{K_{m.s}} \tag{19}$$

which enables one to express biomass formation as a function of time. Similarly, biochemical product formation p can be expressed as a function of time. Diagrammatic plots of these relationships are shown in Fig. 20.

When the specific growth rate is the maximum specific growth rate (i.e. $\mu = \mu_{max}$), which occurs during the log phase of microbial growth, the relationship between the variation in biomass concentration and time may be deduced as follows.

$$\frac{dx}{dt} = \mu x \tag{20}$$

$$\mu = \mu_{max} \tag{21}$$

$$\frac{dx}{x} = \mu_{max} t \tag{22}$$

$$ln \frac{x}{x_0} = \mu_{max} t \tag{23}$$

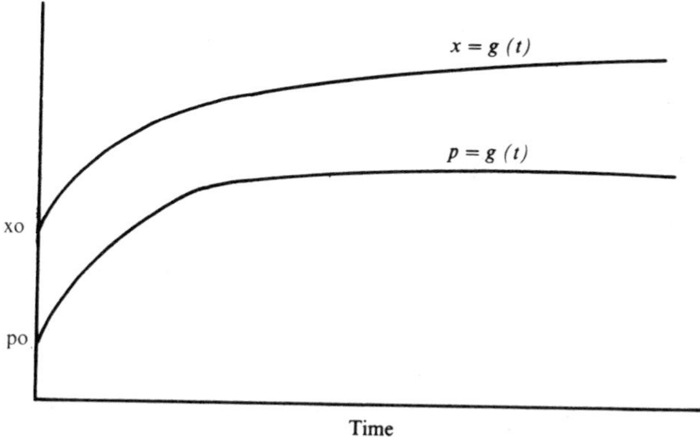

Figure 20. Diagrammatic plots resulting from Eq. (19) for variation in biomass concentration and product formation with time.

For Eqs. (19) and (23) to be compatible, Δ must be very much larger than $K_{m,s}$ [this is equivalent to $s_o \gg s_{CRIT}$, Eq. (11)].

In principle, Eq. (23) may be used to obtain μ_{max} from experimental data on biomass concentration, and Eq. (19) may be used to determine both μ_{max} and $K_{m,s}$ providing the microbe–substrate system follows Monod, kinetics. In practice, because of the degree of scatter associated with typical experimental data and the complexity of Eq. (19), such procedures are not without difficulty or error. A method of calculation is given in the Appendix to this chapter. This method is in effect based upon the simultaneous use of Eqs. (14) and (19) and uses both the experimental substrate and biomass concentration time courses.

Cell doubling time t_D is related to specific growth rate μ by the expression [from Eq. (20)]

$$t_D = \frac{\ln 2}{\mu}$$

Similarly

$$t_{D,\min} = \frac{\ln 2}{\mu_{\max}}$$

Changes in composition of the medium with time are tabulated in Tables 3 and 4. Variations in cell numbers and chemical composition with time are illustrated in Fig. 21. When resting bacterial cells are inoculated into a culture medium there is a rapid increase in cell weight and RNA content per cell. At time a in Fig. 21, when the RNA unit weight and cell weight become constant, the cell concentration increases logarithmically in a similar manner to the dry weight concentration. The DNA unit weight declines initially as the cell numbers are not increasing but cell weight is increasing.

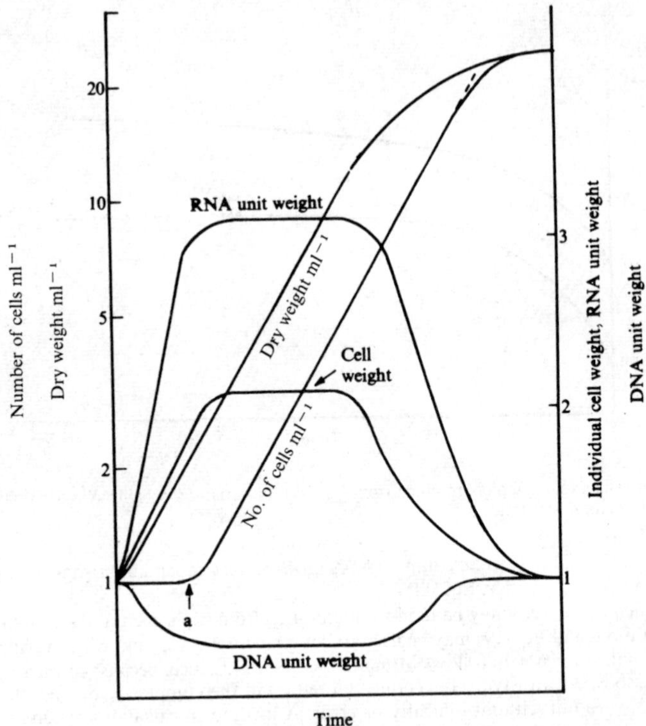

Figure 21. Schematic representation of changes in cell size and chemical composition during growth of bacteria in batch culture [Herbert (1961)].

Table 3. Changes occurring during fermentation of a glucose–meat extract–peptone medium [Hockenhull (1960)].

	Duration of fermentation (h)								
	0	24	48	72	96	120	144	168	192
Mycelium ($g\,\ell^{-1}$)		0.4	5.1	5.8	5.7	4.8	4.6	4.2	3.8
Streptomycin ($mg\,\ell^{-1}$)		0	37	194	198	231	270	186	267
Glucose ($g\,\ell^{-1}$)	9.0	8.8	8.0	2.4	1.2	0.6			
Soluble carbon ($g\,\ell^{-1}$)	10.2	8.6	7.0	5.1	5.0	44	4.6	4.5	4.6
Lactic acid ($mg\,\ell^{-1}$)	292	328	114	13	10	16	12	6	15
Oxygen demand ($Q_{O_2}\,ml^{-1}$)		19	81	82	53	25	5		
Soluble nitrogen ($g\,\ell^{-1}$)	1.48	1.30	1.10	0.67	0.70	0.73	0.90	0.88	1.14
Mycelial nitrogen ($g\,\ell^{-})^{1}$		0.04	0.44	0.62	0.57	0.49	0.40	0.38	0.29
Inorganic phosphorus ($g\,\ell^{-1}$)	118	108	34	1	5	2	19	24	34
Ammonia nitrogen ($mg\,\ell^{-1}$)	66	70	75	63	103	115	170	232	265
pH	7.35	7.30	7.55	7.50	7.75	8.25	8.55	8.65	8.90

Table 4. Changes during fermentations of a glucose–soybean meal–distillers' solubles medium [Hockenhull (1960)].

	Duration of fermentation (h)								
	0	24	48	72	96	120	144	168	192
Reducing sugar ($g\,\ell^{-1}$)	26	25	15	10	7.5	2.5	1.5	1.3	1.3
Streptomycin ($mg\,\ell^{-1}$)			Trace	300	600	1100	1400	1500	1450
Total sugar ($g\,\ell^{-1}$)	30	29	22	14	12.5	10.0	9.0	9.3	8.8
Non-reducing sugar ($g\,\ell^{-1}$)	3.6	4.5	5.0	4.0	5.2	7.5	7.7	8.0	7.5
Soluble nitrogen ($g\,\ell^{-1}$)	0.70		0.80	1.15	1.50	1.80	2.00	2.00	2.00
Ammonia nitrogen ($g\,\ell^{-1}$)	0.70	0.10	0.15	0.20	0.25	0.35	0.45	0.60	0.70
Formol nitrogen ($mg\,\ell^{-1}$)	50	45	15	55	75	60	45	70	87
pH	6.8	7.4	7.9	6.8	6.7	7.7	8.3		

Data corresponding to Fig. 21 are given in Fig. 22 for steady-state conditions showing how the cell mass and complements of protein, DNA and RNA vary with growth rate.

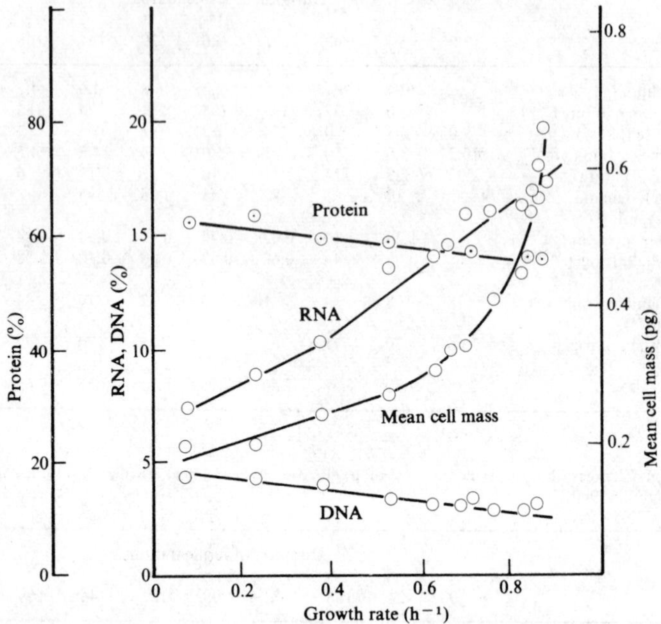

Figure 22. Composition of *Enterobacter aerogenes* grown in a nitrogen-limited chemostat with glycerol as the carbon source [Wang *et al.* (1979)].

SUBSTRATE UTILIZATION AND NUTRIENT REQUIREMENTS

Chemical elements and inorganic ions

Table 5. Nutritional requirements of chemical elements for bacteria and fungi [Spector (1956)].

Species	Chemical element[a]	Importance
Heterotrophic bacteria		
Aerobacter aerogenes	Fe	Growth
	Mg, Mn	Associated with acetoin formation from pyruvate
	Cr or Mn (Al, Cu, Fe, Zn)	Fermentation
Aero. indologenes	Fe	Associated with hydrogenic, formic dehydrogenase and cytochrome
Azotobacter sp.	Fe	Growth
	Ca(Sr), Mo(V)	Nitrogen fixation
	Co, Mg, Mn, Zn	Associated with oxaloacetate decarboxylase
Bacillus anthracis	Ca, Fe, K, Mg, Mn	Growth
B. cereus	K	Spore formation
B. subtilis	Fe, K, Mg, Mn, Zn	Growth, subtilin production
Brucella abortus	Mg or Mn	Growth of non-smooth variants
Bruc. suis	Fe, Mg, Mn	Growth
Cellulomonas sp.	Mg	Growth
Clostridium acetobutylicum	Fe	Fermentation
	Mn, Zn	Associated with phosphatase
	Mo(V)	Nitrogen fixation
Clost. bolutinium	Fe	Associated with polypeptidase
Clost. butyricum	Mo(V)	Nitrogen fixation
Clost. histolyticus	Fe	Associated with polypeptidase
Clost. perfringens	Fe	Fermentation
	Mg	Cell division

continued overleaf

Table 5 – (continued)

Species	Chemical element[a]	Importance
Heterotrophic bacteria *continued*		
Corynebacterium diphtheriae	Fe	Growth, toxin and porphyrin formation
Escherichia coli	Fe	Growth
	Mg, Mn	Associated with pyruvate dehydrogenase and enolase
Haemophilus influenzae	Fe	Growth
Klebsiella pneumoniae	Fe	Growth
Lactobacillis arabinosus	K, Mn	Growth
L. casei	K, Mn	Growth
L. lactis	Co	In cobalamin
L. leichmanii	Co	In cobalamin
Leuconostoc mesenteroides	K, Mn, P	Growth
Propionibacterium jensenii	Mg, Zn	Associated with phosphatase
Pseudomonas sp.	B, Ca, Co, Cu, Fe, Mn, Mo, Zn	Growth
Ps. aeruginosa	Fe, S	Pyocyanine production
	Mg, P, S	Fluorescent pigment production
	Fe	Associated with cytochrome, cytochrome oxidase, catalase and peroxidase
	Mg	Growth
Serratia marcescens	Fe, Mg	Pigment formation
Sporocytophaga myxococcoides	Fe, Mg	Growth and decomposition of cellulose
Streptococcus faecalis	K, Mn, P	Growth
Streptomyces fradiae	Fe, Zn	Neomycin production
Strep. griseus	Co	Cobalamin production
	Fe, Cr, Zn, Co, Cu, Mn, Ni, Se	Streptomycin production
	Fe, Zn	Grisein production
Strep. lavendulae	Fe, Zn	Streptothricin production

continued overleaf

Table 5 – (continued)

Photosynthetic bacteria		
All species	Mg	Bacteriochlorophyll formation
Non-sulphur purple bacteria	H_2	Energy source
Sulphur purple bacteria	S (as S^{2-}, SO_3^{2-}, $S_2O_3^{2-}$)	Reducing agent in photosynthesis
Sulphur green bacteria	S (as S^{2-}, $S_2O_3^{2-}$)	Reducing agent in photosynthesis
Chemoautotrophic bacteria		
Iron bacteria	Fe, Mn	Energy source
Nitrifying bacteria	Cu	Oxidation of ammonia and nitrite
Nitrosomonas sp.	N (as NH_3)	Energy source
Nitrobacter sp.	N (as NO_2^-)	Energy source
Thiobacillus sp.	S (as S^{2-}, $S_2O_3^{2-}$)	Energy source
Thiobacillus denitrificans	S (as S^{2-}, $S_2O_3^{2-}$) N (as NO_3^-)	Energy source Oxidizing agent
Sporovibrio desulfuricans	S (as SO_4^{2-})	Oxidizing agent
Clostridium aceticum	H_2 CO_2	Energy source Oxidizing agent
Hydrogenomonas sp.	H_2	Energy source
Methanobacterium omelianski	H_2 CO_2	Energy source Oxidizing agent
Fungi		
Aspergillus sp.	Fe, Zn	Citrinin production
Asp. fumigatus	Fe, Zn	Gliotoxin and helvolic acid production
Asp. niger	Ag, Al, Co, Cr, Cu, Fe, Mn, Mo, Pb, Sb, Se, Te, U, W, Zn	Citric acid production

continued overleaf

Table 5 – (continued)

Species	Chemical element[a]	Importance
Fungi *continued*		
Asp. niger	Fe	Gluconic acid production
Candida flareri	Fe	Riboflavin production
C. guillermondia	Fe	Riboflavin production
Penicillium citrinum	Fe, Mn	Citrinin production
P. notatum	Fe, Mn	Citrinin production
	Al, Cr, Cu, Fe, Mn, Sn, Zn	Penicillin production
P. patulum	Cu, Fe, Mn	Patulin and gentisic acid production
Rhizopus selemar	Al	Amylase production
R. nigricans	Cu, Fe, Mn, Zn	Ethanol, fumaric acid and lactic acid production

[a] Elements in parentheses may replace the preceding element

Table 6. Inorganic ions commonly required by bacteria [Gunsalus and Stainer (1962)].

Microorganism	Inorganic ion
Aerobacter aerogenes	Mg^{2+}
Azotobacter vinelandii	Ca^{2+}, Mo
Corynebacterium A11	Na
Flavobacterium B9	Cl^-, SO_4^{2-}
Lactic acid bacteria	K^+, PO_4^{3-}
Lactobacillus arabinosus	Mn^{2+}
Marine bacteria	Cl^-, Na^+, SO_4^{2-}
Pseudomonas aeruginosa	Fe^{2+}, Fe^{3+}
Streptomyces griseus	Zn^{2+}

Table 7. Inorganic element requirements by eucaryotes [from Spector (1956)].

Eucaryotes	Element
Algae	B[a], Ca, Co[a], Cu[a], Fe, K, Mg, Mn[a], P, V[a], Zn[a]
Green phytoflagellates	Ba[a], Ca[a], Co[a], Fe, K, Mg, Mn[a], P, K, Zn[a]
Protozoa	Ca, Co[a], Cu[a], Fe, K, Mg, Mn, Na, P, Rb, Zn[a]
Yeasts	Ca[a], Co, Cu[a], Fe, K[a], Mg, Mn[a], Mo[a], P, Zn[a]

[a] Required in some, but not necessarily all, species.

Sulphur Sources

Bacterial and fungal requirements for inorganic sulphur sources are summarized in Table 5.

Algae generally do not utilize or require inorganic sulphur compounds. Exceptions to this are *Synechococcus*, *Oscillaria* and *Scendesmus* which are capable of growing on media containing sulphides. Hydrogen sulphide is toxic to *Chlorella*.

Some species of yeast can utilize elemental sulphur and sulphate.

A summary of the ability of fungi and bacteria to utilize miscellaneous sulphur-containing organic compounds is presented in Table 8.

Table 8. The requirement for and utilization of sulphur-containing organic compounds in some fungi and bacteria [from Spector (1956)].

Microorganisms	Compounds utilized	Compounds required
Yeasts	Alkylsulphides, alkylsufinates, alkylsulphonates, dithionate, ethereal sulphates, glutathione, sulphamate, sulphoxides, taurine, thioacetamide, thioacetate, thiocarbonate, thioglycolate, thiols, thiooxalate, thiourea	Sulphonic acid amides, thioacetate, thiocarbonate, thioglycolate, glutathione
Bacteria	Sulphoxides, thioacetamide, thioacetate, thioglycolate, thiourea	Glutathione, thioctic acid

Requirements for sulphur-containing amino acids are given in Table 12 for bacteria. Sulphur-containing organic compounds are not required by algae, green phytoflagellates and protozoa. Generally, yeasts do not require or utilize sulphur-containing organic compounds, with the exception of *Torula monosa* and *T. dattila*.

Nitrogen sources

Tables 9 and 10 give some indication of the utilization of and requirements for nitrogen sources by microorganisms. It should be stressed that not all species require or utilize the compounds listed but rather that some species have been identified that are able to utilize these compounds. For example, nitrogen gas is only fixed, and therefore utilized, by nitrogen-fixing bacteria.

Table 9. The utilization of and requirement for inorganic nitrogen sources in microorganisms [from Spector (1956)].

Microorganisms	Compounds utilized	Compounds required
Algae	Ammonia, nitrate, nitrite, nitrogen gas	
Bacteria	Ammonia, cyanamide, hyponitrate, nitrate, nitrite, nitrogen gas, thiocyanate	
Fungi	Ammonia, cyanamide, cyanide, nitrate, nitrite, nitrohydroxamate, nitrogen gas	Ammonia, nitrate, nitrite
Green phytoflagellates	Nitrate, nitrogen gas	
Yeasts	Ammonia, nitrate, nitrite, nitrogen gas, thiocyanate	

Table 10. The utilization of and requirement for organic nitrogen sources in microorganisms [from Spector (1956)].

Microorganisms	Compounds utilized	Compounds required
Algae	Acid amides, amines, indoles, urea	
Bacteria	Acid amides, acid imides, amines, imidazoles, oximino compounds, pyridines	Pyridines
Fungi	Acid amides, amines, imidazoles, indoles oximino compounds, pyridines, urea	Indoles
Green flagellates	Urea	
Yeasts	Acid amides, amines, imidazoles, oximino compounds, urea	Pyridines

The bacterial requirements for amino acids, and purines and pyrimidines are considered in Table 12, 13 and 18.

Organic Acids, Alcohols and Fatty Acids

Table 11. Organic acids, alcohols and fatty acids utilized by lower algae and related colourless organisms [from Spector (1956)].

Microorganism	Compound(s)
Photosynthetic forms	
Chlamydomonas algoeformis	Acetic acid, *n*-butyric acid
Chlorogonium elongatum	Acetic acid, *n*-butyric acid
Chlor. euchlorum	Pyruvic acid, succinic acid, acetic acid
Chlorella vulgaris	*cis*-Aconitic acid, citric acid, fumaric acid, lactic acid, malic acid, pyruvic acid, succinic acid, acetic acid, propionic acid
Euglena deses	Ethanol
E. gracilis typica	Lactic acid, malic acid, pyruvic acid, succinic acid, *n*-butanol, ethanol, *n*-hexanol, *n*-propanol, acetic acid, *n*-butyric acid, *i*-caproic acid, *n*-caproic acid, *n*-decylic acid, *n*-nonylic acid, *n*-octylic acid, propionic acid, *n*-valeric acid
E. gracilis bacillaris	Fumaric acid[a], malic acid[a], phosphoglyceric acid[a], succinic acid[a], acetic acid, *n*-butyric acid
E. gracilis urophora	Fumaric acid, citric acid, lactic acid, malic acid[a], pyruvic acid, *i*-butanol, *n*-butanol, ethanol, *n*-hexanol, *n*-proponal, acetic acid, *i*-butyric acid, *n*-butyric acid, *i*-caproic acid, *n*-caproic acid, *n*-decylic acid, *n*-heptylic acid, *n*-nonylic acid, *n*-octylic acid, propionic acid, *n*-valeric acid
Haematococcus pluvialis	Acetic acid, *n*-butyric acid
Colourless forms	
Astasia longa	Lactic acid, malic acid, succinic acid, *i*-butanol, *n*-butanol, ethanol, *n*-hexanol, *n*-propanol, acetic acid, *i*-butyric acid, *n*-butyric acid, *i*-caproic acid, *n*-caproic acid, *n*-decylic acid, *n*-heptylic acid, *n*-nonylic acid, *n*-octylic acid, propionic acid, *n*-valeric acid
Ast. quartana	Lactic acid, pyruvic acid, succinic acid, acetic acid, *i*-butyric acid, *n*-butyric acid, *i*-caproic acid, *n*-caproic acid, *n*-heptylic acid, propionic acid, *i*-valeric acid, *n*-valeric acid
Chilomonas paramecium	Lactic acid, pyruvic acid, succinic acid, *n*-butanol, ethanol, *n*-hexanol, *n*-pentanol, *n*-propanol, acetic acid, *n*-butyric acid, *i*-caproic acid, *n*-caproic acid, *n*-heptylic acid, *n*-nonylic acid, *n*-octylic acid, *i*-valeric acid, *n*-valeric acid
Hyalogonium klebsii	Succinic acid, acetic acid
Polytoma candatum	Pyruvic acid, acetic acid, *n*-butyric acid
P. obtusum	Pyruvic acid, acetic acid, *n*-butyric acid
P. oceliatum	Lactic acid, pyruvic acid, *i*-butanol, *n*-butanol, ethanol, *n*-hexanol, *n*-pentanol, *n*-propanol, acetic acid, *i*-butyric acid, *n*-butyric acid, *i*-caproic acid, *n*-caproic acid, *n*-decylic acid, *n*-heptylic acid, *n*-nonylic acid, *n*-octylic acid, propionic acid, *i*-valeric acid, *n*-valeric acid
P. uvella	Lactic acid, pyruvic acid, succinic acid, acetic acid, *n*-butyric acid, *i*-caproic acid, *n*-caproic acid, *n*-octylic acid, *n*-valeric acid
Polytomella caeca	Pyruvic acid, succinic acid, *i*-butanol, *n*-butanol, ethanol, *n*-hexanol, *n*-propanol, acetic acid, *n*-butyric acid, *i*-caproic acid, *n*-caproic acid, *n*-decylic acid, *n*-heptylic acid, *n*-nonylic acid, *n*-octylic acid, propionic acid, *n*-valeric acid
Prototheca zopfii	Lactic acid[a], pyruvic acid[a], *i*-butanol, *n*-butanol, ethanol, glycerol, *i*-pentanol, *n*-pentanol, *n*-propanol, acetic acid, *i*-butyric acid, *n*-butyric acid, *i*-caproic acid, *n*-caproic acid, *n*-decylic acid, *n*-heptylic acid, *n*-nonylic acid, *n*-octylic acid, propionic acid, *i*-valeric acid, *n*-valeric acid

[a] Utilized only at pH 3.0–5.5.

Table 12. Amino acid requirements of some bacteria[a] [Spector (1965)].

Amino acid	*Aerobacter aerogenes*	*Bacillus anthracis*	*B. licheniformis*	*B. megatherium*	*B. subtilis*	*Brucella suis*	*Erwinia amylovora*	*Escherichia coli*	*Lactobacillus arabinosus*	*Leuconostoc mesenteroides*	*Pasteurella tularensis*	*Salmonella typhosa*	*Serratia marcescens*	*Shigella sonnei*
Alanine	–		–	–	–	–	–	–	–	+			–	–
Arginine	–		–	–	–	–	–	–	+	+			–	–
Aspartic acid									–	+				
Citrulline	–		–	–	–	–	–	–	–	–			–	–
Cystine	–		–	–	–	–	–	–	+	+	+		–	–
Glutamic acid	–		–	–	–	–	–	–	+	+			–	–
Glycine	–		–	–	–	–	–	–	–	+			–	–
Histidine	–		–	–	–	–	–	–	–	+			–	–
Hydroxyproline	–		–	–	–	–	–	–	–	–			–	–
Isoleucine	–		–	–	–	–	–	–	+	+			–	–
Leucine	–		–	–	–	–	–	–	–	+			–	–
Lysine	–		–	–	–	–	–	–	+	+			–	–
Methionine	–		–	–	–	–	–	–	+	+			–	–
Norleucine	–		–	–	–	–	–	–	–				–	–
Phenylalanine														
Proline	–		–	–	–	–	–	–	–	+			–	–
Serine	–		–	–	–	–	–	–	–	+			–	–
Threonine	–		–	–	–	–	–	–	–	+			–	–
Tryptophan	–		–	–	–	–	–	–	+	+		+	–	–
Tyrosine	–		–	–	–	–	–	–	+	+			–	–
Valine	–		–	–	–	–	–	–	+	+			–	–
Other amino acids		+								+	+	+		+
Other nitrogen source	+		+	+	+	+	+	+					+	+

[a] –, amino acid not required; +, amino acid required.

Amino Acids

Amino acids are not generally required by algae, although several algal species are capable of utilizing them. Species of other microorganisms are capable of utilizing all amino acids, except for yeasts where there is no evidence of citrilline being used.

Amino acid requirements for some bacterial species are summarized in Table 12. It is usually the L-form of the acids that are biologically active but, unlike higher animals, some bacteria can also utilize D-amino acids.

Table 13. Growth-limiting concentrations of amino acids for some bacteria [Gunsalus and Stainer (1962)].

Amino acid	Dependent bacterium	Growth limiting concentration[a] (mg ℓ$^{-1}$)
Alanine	*Pediococcus cerevisiae*	0–10
Arginine	*Lactobacillus casei*	0–10
Aspartic acid	*Leuconostoc mesenteroides*	0–10
Cysteine	*Leu. mesenteroides*	0–40
Cystine	*Leu. mesenteroides*	0–40
Glutamate	*L. arabinosus*	0–50
Glycine	*Leu. mesenteroides*	0–10
Histidine	*Streptococcus faecalis*	0–6
Isoleucine	*L. arabinosus*	0–10
Leucine	*L. arabinosus*	0–20
Lysine	*Leu. mesenteroides*	0–20
Methionine	*S. faecalis*	0–10
Phenylalanine	*Leu. mesenteroides*	0–15
Proline	*Leu. mesenteroides*	0–6
Serine	*L. delbrueckii*	0–10
Threonine		0–20
Tryptophan	*S. faecalis*	0–3
Tyrosine	*L. delbrueckii*	0–6
Valine	*L. arabinosus*	0–10

[a] The higher concentration is that at which growth approaches maximum.

Vitamins and Growth Factors

Vitamins, with the exception of choline, pyridoxamine and thiazole, are not generally required by algae or green phytoflagellates. There is considerable species variation in the requirements for vitamins and related factors by other microorganisms. Generally, vitamins A, C, D and K are not necessary for growth. Bacterial requirements are summarized in Table 14. Growth-limiting concentrations required by some microorganisms are presented in Table 15.

Table 14. Summary of vitamin requirements for bacteria [Spector (1956)].

	Vitamin[a]															
	Vitamin A	Biotin	Vitamin C	Choline	Cobalamin	Vitamin D	Vitamin E	Folic acid	Inositol	Vitamin K	Niacin	Pantothenic acid	p-Aminobenzoic acid	Pyridoxals	Riboflavin	Thiamin
Aerobacter aerogenes	−	−	−	−	−	−	−	−	−	−	−	−	−	−	−	−
Bacillus alvei		−									−					+
B. anthracis		−									−					+
B. brevis		−									−					+
B. cereus		−									−					−
B. cereus mycoides		−									−					−
B. circulans		+									−					+
B. coagulans		+									+					+
B. lichenformis		−									−					−
B. macerans		+									−					+
B. magaterium		−									−					−
B. pasteurii		+									+					+
B. polymyxa		+									−					−
B. pumilis		+									−					−
B. sphaericus		+									−					+
B. subtilis		−									−					−
B. subtilis niger		−									−					−
Brucella abortus	−	+	−	−	−	−	−	−	−	−	+	+	−	−	−	+
Bruc. melitensis	−	+	−	−	−	−	−	−	−	−	+	+	−	−	−	+
Bruc. suis	−	+	−	−	−	−	−	−	−	−	+	+	−	−	−	+
Erwinia amylovora	−	+	−	−	−	−	−	−	−	−	+	−	+	−	−	−
Er. tracheiphila	−	−	−	−	−	−	−	−	−	+	+	−	−	−	−	−
Escherichia coli	−	−	+	−	−	−	−	−	−	−	+	−	−	−	−	−
Haemophilus influenzae	−	−	−	−	−	−	−	−	−	−	−	−	−	−	−	−
Lactobacillus																
Heterofermentative		+			+			+			+	+	+	+	−	+
Homofermentative		+			+			+			+	+	+	+	+	−
Leuconostoc citrovorum		+						+			+	+		+	+	+
Leu. mesenteroides		+						+			+	+		+	+	+
Pasteurella multicida		+									+	+				
Past. pseudotuberculosis	−	−	−	−	−	−	−	−	−	−	−	−	−	−	−	−
Past. tularense																+
Proteus morgani											+	+				
Prot. vulgaris											+					
Salmonella cholerae-suis	−	−	−	−	−	−	−	−	−	−	−	−	−	−	−	−
Sal. enteritidis	−	−	−	−	−	−	−	−	−	−	−	−	−	−	−	−
Sal. gallinarum																+
Sal. pullorum	−	−	−	−	−	−	−	−	−	−	+	−	−	−	−	−
Sal. scholtmuelleri	−	−	−	−	−	−	−	−	−	−	−	−	−	−	−	−
Serratia marcescens	−	−	−	−	−	−	−	−	−	−	+	−	−	−	−	−
Shigella alkarescens	−	−	−	−	−	−	−	−	−	−	−	−	−	−	−	−
Shig. paradysenteriae											+	+				
Shig. sonnei											+	+				
Staphylococcus albus			+								+	+		+		+
Staph. aureus			+								+	+		+		+

[a] −, not required; +, required.

Table 15. Vitamins required by various microorganisms [Gunsalus and Stainer (1962)].

Vitamin	Examples of dependent microorganisms	Growth limiting concentration[a] ($\mu g\ \ell^{-1}$)
p-Aminobenzoic acid	*Acetobacter suboxydans*	0–1.0
	Clostridium acetobutylicum	0–0.2
Biotin	*Lactobacillus arabinosus*	0–0.2
	Rhizobium trifolii	0–0.1
	Streptococcus faecalis	0–0.2
Choline	*Pneumococcus* type III	0–6000
	Neurospora crassa cholinelessi	0–2000
Cobalamin	*L. lactis*	0–0.026
	L. leichmannii	0–0.025
Coprogen	*Pilobolus kleinii*	0–5
Ferrichrome	*Arthrobacter* JG-9	0–10
Folic acid	*L. casei*	0–0.15
	Strep. faecalis	0–0.8
Heme	*Haemophilus influenzae*	0–200
Inositol	*Saccharomyces carlsbergensis*	0–100
	Schizosaccharomyces pombe	0–1000
Vitamin K	*Mycobacterium paratuberculosis*	0–1000
	Fusiformis nigrescens	0–1000
Nicotinic acid	*L. arabinosus*	0–40
	Proteus vulgaris	0–20
	Shigella paradysenteriae	0–25
Pantothenic acid	*L. casei*	0–20
	Proteus morganii	0–0.5
Pyridoxal	*L. casei*	0–0.7
Pyridoxal or pyridoxamine	*S. faecalis*	0–0.4
Riboflavin	*L. casei*	0–25
	Cl. tetani	0–100
Terregenes factor	*Arthrobacter terregens*	0–20
Thiamin	*L. fermenti*	0–5
	Staphylococcus aureus	0–0.5

[a] The higher concentration is that at which growth approaches maximum.

Table 16. Summary of miscellaneous growth factor requirements for microorganisms [Spector (1956)].

Nutrient[a]	Fungi	Yeasts	Bacteria	Algae	Green Phytoflagellates
Adenylthiome thylpentose	−	−	−	−	−
Anthranilic acid	+	−	+	−	−
Antibiotics	−	−	+	−	−
Asparagine	−	−	+	−	−
'Bifidus' factor	−	−	+	−	−
Carbon dioxide	+		+	+	+
Carnitine	−	−	−	−	−
Coprogen	+	−	−	−	−
N-D-Glucosylglycine ester	−	−	+	−	−
Glutamine	−	−	+	−	−
Glutathione	−	−	+	−	−
Guanidine	−		−	−	−
Indole-3-acetic acid	+	−	−	−	−
Hematin	+	−	+	−	−
Krebs cycle intermediates	+	−	+	−	+
Mucin	−	−	+	−	−
Mycobactin	−	−	+	−	−
Parahydroxybenzoic acid	+	−	+	−	−
Putrescine	+	−	+	−	−
Quinic acid	+	−	+	−	−
Shikimic acid	+	−	+	−	−
Spermidine	−	−	+	−	−
Strepogenin	−	−	+	−	−
Thioctic acid	−	−	+	−	−
Unidentified factors	+		+		

[a] +, required; −, not required.

Carbohydrates

Carbohydrates are capable of being used by all microorganisms although in no case is there an absolute requirement for this group of organic compounds. The D-forms only are metabolized, except for L-arabinose. Glucose is the most readily metabolized sugar. Most fungi can utilize disaccharides. Sorbose is not metabolized by fungi. Yeasts cannot metabolize lactose, dulcitol or inositol generally [see Spector (1956)].

Lipids

Microbial requirements for steroids, long-chain fatty acids and phospholipids are summarized in Table 17.

Table 17. Summary of lipid requirements [Spector (1956)].

Nutrient[a]	Fungi	Yeasts	Bacteria	Algae	Green phytoflagellates
Steroids					
Cholesterol	+	+	+	–	–
7-Dehydrocholesterol	–	–	–	–	–
Ergostanol acetate	–	–	–	–	–
Ergosterol	–	+	–	–	–
Stigmasterol	–	–	–	–	–
Long-chain fatty acids and their derivatives					
Arachnidonic acid		–		–	–
Linoleic acid	+	–	+	–	–
Linolenic acid		–		–	–
Oleic acid	+	–	+	–	–
Detergents	–	–	–	–	–
Phospholipids					
Lecithin	–	–	–	–	–

[a] +, required; –, not required.

Generally, steroids, other than cholesterol, are not required or utilized by microorganisms. In some bacterial species, it has been reported that some long-chain fatty chains and Tween 80 can stimulate growth.

Purines and Pyrimidines

It is generally only in bacteria that cases of purine and pyrimidine metabolism have been reported. Algae do not utilize these compounds at all, while guanine and hypoxanthine have been reported to stimulate growth in some green phytoflagellates. The requirements for purines and pyrimidines is summarized in Table 18.

Table 18. Requirements for purines, pyrimidines and their derivatives by various microorganisms [Gunsalus and Stainer (1962)].

Compound	Example of dependent microorganism	Growth-limiting concentration (mg ℓ −)[a]
Pyrimidines		
Uracil	*Staphylococcus aureus*	0–6
	Lactobacillus helviticus 335	0–1
Orotic acid	*L. bulgaricus*	0–10
Purines		
Adenine	*Shigella boydii* (9329)	0–10
Guanine	*Leuconostoc mesenteroides*	0–10
Hypoxanthine	*Neisseria gonorrhoeae*	0–2
Xanthine	*Leu. mesenteroides*	0–10
Deoxyribosides		
Thymidine	*L. delbrueckii* 730	0–5
	Thermobacterium acidophilus	0–0.5
Cytosine	*T. acidophilus* R26	0–0.5
deoxyriboside	*T. acidophilus* 200	0–5
Hypoxanthine deoxyriboside	*T. acidophilus* R26	0–0.5
Ribosides		
Adenosine	*Galikya homari*	0–200
Guanosine	*G. homari*	0–200
Inosine	*G. homari*	0–200
Xanthosine	*G. homari*	0–200
Cytidine	*Tetrahymena geleii*	50
	Neurospora sp. 1298	0–100
Uridine	*Neurospora* sp. 1298	0–100
Ribotides and deoxyribotides		
Cytidylic acid	*Tet. geleii*	50
	Neurospora sp. 1298	0–100
Uridine-5′-phosphate	*L. bulgaricus*	0–1000
Adenosine-5′-phosphate	*G. homari*	0–250
Guanylic acid	*G. homari*	0–250
Inosine-5′-phosphate	*G. homari*	0–250
Oligonucleotides		
RNA + thymidine or DNA	*Mycoplasma laidlawii* Strain A	5–50

[a] The higher concentration is that at which growth approaches maximum.

EFFECT OF OXYGEN ON METABOLISM AND STOICHIOMETRY

The availability of oxygen can affect the nature and the rate of metabolic product formation. Also the stoichiometry of product formation at any point during the time course of fermentation varies according to the availability of oxygen. Factors contributing to the availability include the scale of operation as well as the degree of aeration and agitation.

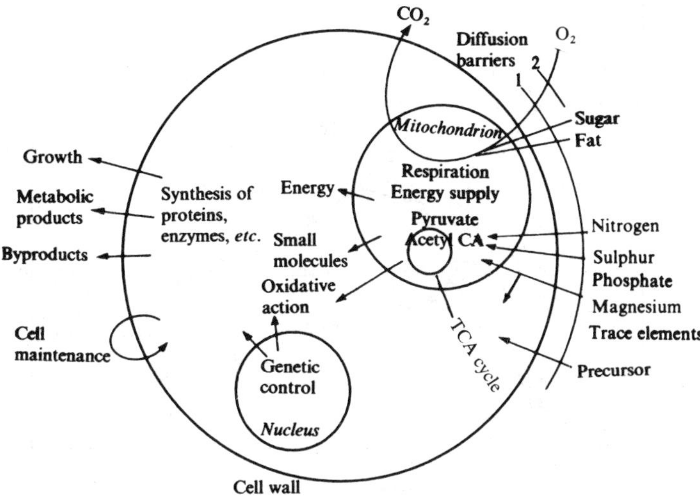

Figure 23. Mechanisms in eucaryotic cell metabolism.

Fig. 23 diagrammatically illustrates eucaryotic cell function, with particular emphasis on respiration in the mitochondrion providing energy to the oxidation of fats and sugars, with the concomitant formation of carbon dioxide.

From Fig. 24, it can be seen that the rates of citric acid and carbon dioxide production and of oxygen consumption vary with the duration of fermentation. Similarly, the respiratory quotient (RQ), i.e. the ratio of carbon dioxide produced to oxygen consumed is not constant. It can therefore be deduced that Y_{O_2} is not constant.

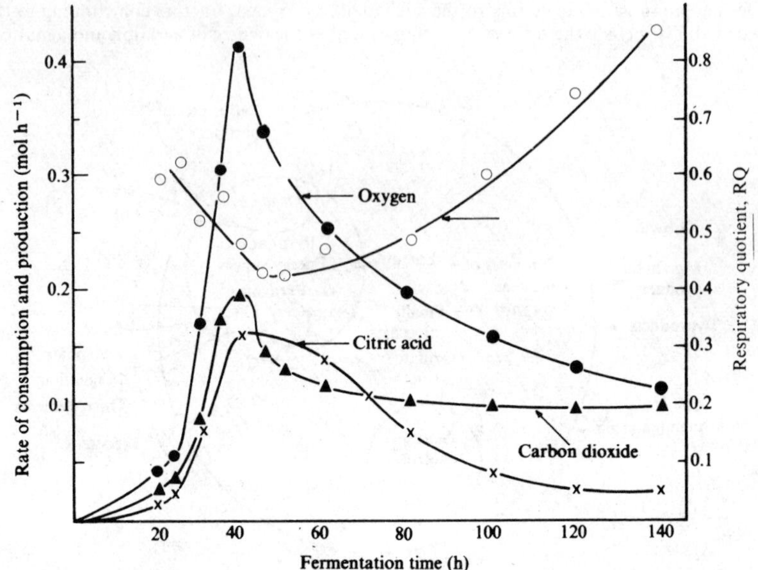

Figure 24. Comparison of rate of consumption of oxygen with rates of production of citric acid and carbon dioxide [Clark and Lentz (1961)].

Steady-state data for a chemostat operating at various oxygen partial pressures in the gas phase and with constant flow rate and inlet glucose concentration are given in Fig. 25.

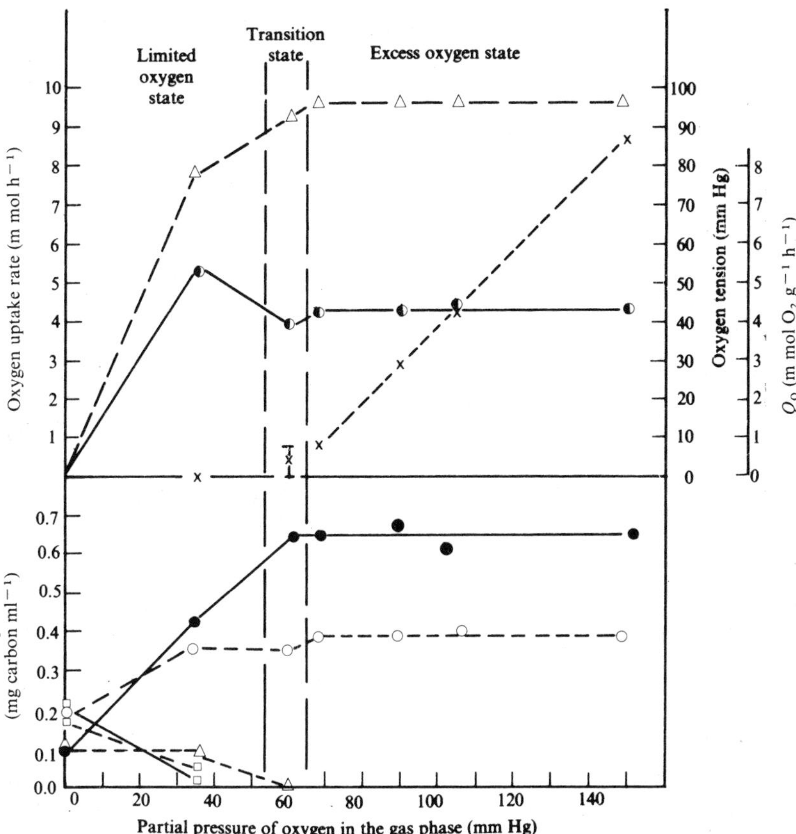

Figure 25. The fate of glucose at different oxygen tensions on glucose-limited growth of *Klebsiella aerogenes* NCIB 8017 at pH 6.0, dilution rate 0.16 h^{-1}. Glucose concentration in entering medium, 2.6 g ℓ^{-1}. ×, dissolved oxygen tension; △, oxygen uptake rate; ○, respiration rate (Q_{O_2}); ●, micro-organism dry weight; ○, carbon dioxide; ■, 2,3-butanediol; □, ethanol; ▲, volatile acid. Glucose utilization was over 97 per cent of that supplied under all conditions [Harrison and Pirt (1967)].

From Fig. 25 it can be seen that dissolved oxygen is critically very low but then increases linearly with the oxygen-partial pressure. The respiratory and oxygen uptake rates both increase and then remain constant above the 'critical' dissolved oxygen level, i.e. the initial oxygen-limited conditions are followed by an excess oxygen state. Similar changes in cell and carbon dioxide concentrations occur. Partially oxidized product concentrations, e.g, butanediol, ethanol and volatile acid, fall to zero values at or before the critical oxygen level.

Using these observations the following deductions can be made.

1) A critical dissolved oxygen level exists above which the respiratory rate, oxygen uptake rate, and cell and carbon dioxide concentrations are independent of the oxygen partial pressure and no partially oxidized products are formed.

2) Below the critical dissolved oxygen level, both activity and the products formed are affected by dissolved oxygen.

The effects of oxygen tension and/or oxygen supply rate on yeast NCYC 239 and 1085 are presented in Tables 19–24.

Table 19. Effect of oxygen tension on growth and uptake of oxygen by yeast NCYC 239 [Cowland (1967)].

Oxygen tension (mm Hg)	Oxygen supply rate (ml ℓ$^{-1}$ h^{-1})	Yeast concentration in culture vessel (mg dry yeast ml^{-1})	Specific rate of oxygen uptake [μl (mg dry yeast)$^{-1}$ h^{-1}]	Specific gravity of cell-free beer
0	0	1.53	0	1.027
0	0.08	2.32	0.03	1.016
0	0.16	2.44	0.06	1.012
0.05	16.80	3.46	2.10	1.011
0.10	210.00	4.56	16.00	1.009

Table 20. Effect of oxygen tension on the utilization of nitrogen by yeast NCYC 239 [Cowland (1967)].

Oxygen tension	Oxygen supply rate	Nitrogen concentration in beer (μg nitrogen ml^{-1})		Gross rate of nitrogen uptake (μg nitrogen ml^{-1} h^{-1})		Specific rate of nitrogen uptake [μg nitrogen (mg dry yeast)$^{-1}$ h^{-1}]		Rate of uptake of nitrogen per unit of yeast nitrogen [μg nitrogen (mg yeast nitrogen)$^{-1}$ h^{-1}]	
(mm Hg)	(ml ℓ$^{-1}$ h^{-1})	α-Amino	Total	α-Amino	Total	α-Amino	Total	α-Amino	Total
0	0	68.4	598	4.32	4.62	2.82	3.02	37.5	40.2
0	0.08	40.0		5.23		2.25			
0	0.16	29.0		5.21		2.13		21.3	
0.05	16.80	23.9	367	5.88	12.79	1.73	3.70	21.8	46.5
0.10	210.00	22.7	330	6.22	14.51	1.36	3.18	19.9	46.5

Table 21. Effect of oxygen tension on the composition of yeast NCYC 239 [Cowland (1967)].

Oxygen tension (mm Hg)	Oxygen supply rate (ml ℓ$^{-1}$ h^{-1})	Total nitrogen content of whole yeast [μg nitrogen (mg dry yeast)$^{-1}$]	Soluble α-amino nitrogen content [μg nitrogen (mg dry yeast)$^{-1}$]	Insoluble carbohydrate of whole yeast [μg glucose (mg dry yeast)$^{-1}$]	Soluble carbohydrate content [μg glucose (mg dry yeast)$^{-1}$]
0	0	75.2	3.56	212.4	32.68
0	0.16			290.9	8.20
0.05	16.80	79.5	3.03	346.7	7.51
0.10	210.00	68.4	1.34	401.1	3.29

Table 22. Effect of oxygen tension on the uptake of glucose and production of ethanol by yeasts NCYC 239 and NCYC 1085 [Cowland (1967)].

Oxygen tension	Oxygen supply rate	Uptake of glucose		Production of ethanol		Sugar converted to ethanol and associated carbon dioxide
		Gross rate	Specific rate	Gross rate	Specific rate	
(mm Hg)	(ml ℓ$^{-1}$ h^{-1})	(μg ml^{-1} h^{-1})	[μg (mg dry yeast)$^{-1}$ h^{-1}]	(μg ℓ$^{-1}$ h^{-1})	[μg (mg dry yeast)$^{-1}$ h^{-1}]	(%)
Yeast NCYC 239						
0	0	1408	921	633	414	87.6
0	0.08	2255	971	1009	434	87.7
0	0.16	2754	1128	1250	512	88.6
0.05	16.80	2667	771	1142	330	83.6
0.10	210.00	2834	621	1071	235	74.0
Yeast NCYC 1085						
9.10	210.00	2854	703	1263	311	86.6
760.0 approx.	1000.00	1750	646	633	234	78.8

Table 23. Effect of oxygen tension on production of acetaldehyde and acetoin by yeast NCYC 239 and NCYC 1085 [Cowland (1967)].

	Acetaldehyde		Acetoin	
Oxygen tension (mm Hg)	Specific rate of production [μg (mg dry yeast)$^{-1}$ h^{-1})	Acetaldehyde production as a proportion of glucose consumption (%)	Specific rate of production [μg (mg dry yeast)$^{-1}$ h^{-1})	Acetoin production as a proportion of glucose consumption (%)
Yeast NCYC 239				
0.0	0.0	0.0	0.0	0.0
0.05	0.0	0.0	0.0	0.0
0.10	2.39	0.36	16.34	2.63
Yeast NCYC 1085				
9.1	6.11	0.87	27.25	3.77
ca. 760	9.75	1.51	15.85	2.45

Table 24. Effect of oxygen tension on production of esters by yeast NCYC 239 [Cowland (1967)].

		Concentration in beer mg ℓ$^{-1}$			Specific rate of production [ng (mg dry yeast)$^{-1}$ h^{-1}]		
Oxygen tension (mm Hg)	Oxygen supply rate (ml ℓ$^{-1}$ h^{-1})	Ethyl acetate	Isoamyl acetate	Ethyl caproate	Ethyl acetate	Isoamyl acetate	Ethyl caproate
0	0	10.9	0.57	0.11	296.8	15.4	3.0
0	0.08	19.0	0.58	0.14	324.2	9.9	6.5
0	0.16	3.2	0.078	0.04	39.4	0.9	0.5
0.05	16.80	2.4	0.027	0.04	21.4	0.3	0.4

Using the data in Tables 19–24, it can be deduced that the yeast concentration, specific oxygen uptake, insoluble carbohydrate content of yeast and specific acetaldehyde and acetoin production increase with respect to dissolved oxygen. The specific gravity of beer, the nitrogen content of beer, specific nitrogen uptake rate, the nitrogen content of yeasts, α-amino acid content of yeast and the soluble carbohydrate content of yeast decrease with respect to dissolved oxygen. Specific glucose uptake rate, specific ethanol production, the percentage of sugar converted to ethanol and carbon dioxide, ester concentrations and specific rates of ester production all exhibit an optimum with respect to dissolved oxygen.

Influence of Scale of Operation, Aeration and Agitation

Tables 25–27 provide an indication of how the biomass yield Y'_S and the protein and vitamin contents of *Saccharomyces cerevisiae, Torula utilis* and *Candida arborea* can be affected by aeration, i.e. agitation and gas flow.

Table 25. Effect of gas flow and agitation on yield, protein and vitamin content of *Saccharomyces cerevisiae*[a] [Singh *et al.* (1948)].

Gas flow ($\ell\ \ell^{-1}\ min^{-1}$)	Agitation (rpm)	Yield on sugar fermented (%)	Protein in dry matter (%)	Vitamins, [μg (g dry weight)$^{-1}$] Thiamine	Riboflavin	Niacin	Folic acid
2.4	0	31.0	57.3	41.3	66.6	323.2	34.1
3.6	0	36.5	56.2	37.2	62.4	376.8	25.1
0	500	20.8	57.3	52.7	84.1	239.5	23.7
0	750	19.6	55.4	52.8	91.3	209.8	26.4
0.1	250	34.7	56.4	38.6	63.6	346.9	24.8
0.1	450	39.7	54.9	35.9	60.0	407.6	25.6
0.4	240	39.3	56.7	40.9	61.7	407.0	27.1
1.2	350	42.3	54.2	34.7	56.3	439.3	21.7

[a] Mason City beet molasses medium.

Table 26. Effect of gas flow and agitation on yield, protein and vitamin content of *Torula utilis*[a] [Singh *et al.* (1948)].

Gas flow ($\ell\,\ell^{-1}\,min^{-1}$)	Agitation (rpm)	Yield on sugar fermented (%)	Protein in dry matter (%)	Vitamins, [μg (g dry weight)$^{-1}$] Thiamine	Riboflavin	Niacin	Folic acid
0	0	3.7	52.1	53.0	108.4	212.0	15.7
0.6	0	7.7	52.4	59.5	104.7	230.4	18.7
1.8	0	22.0	52.0	51.2	97.1	288.2	14.1
3.6	0	33.5	49.5	55.5	90.0	296.5	18.1
0	333	13.1	49.3	41.4	100.1	276.6	18.5
0	500	24.2	54.1	29.6	91.5	290.4	21.3
0	750	62.2	51.3	23.1	66.5	409.9	7.1
0.15	240	40.7	47.8	31.4	95.7	301.6	16.5
0.15	480	59.5	42.0	24.7	64.3	399.0	13.0
0.60	270	60.4	38.8	23.7	64.2	394.2	10.1
0.60	345	63.4	49.8	30.8	64.0	399.8	8.2
0.60	480	57.9	39.4	21.2	67.5	321.1	10.7
0.60	800	64.0	47.3	20.3	63.4	410.4	9.4
1.80	375	58.1	49.6	21.9	65.1	361.7	6.8
1.80	750	66.2	48.4	17.4	62.5	475.0	7.8

[a] Mason City beet molasses medium.

Table 27. Effect of gas flow and agitation on yield, protein and vitamin content of *Candida arborea*[a] [Singh *et al.* (1948)].

Gas flow ($\ell\ \ell^{-1}\ min^{-1}$)	Agitation (rpm)	Yield on sugar fermented (%)	Protein in dry matter (%)	Vitamins, [μg (g dry weight)$^{-1}$] Thiamine	Riboflavin	Niacin	Folic acid
0	0	7.6	42.4	42.9	95.2	316.2	40.6
1.0	0	19.6	42.7	38.6	81.6	316.6	24.3
2.4	0	37.8	40.6	29.2	70.0	327.1	30.6
3.6	0	50.0	40.2	20.2	60.6	350.7	26.4
0	500	19.8	43.1	29.7	82.5	316.6	21.9
0	750	20.0	40.7	31.0	81.1	317.5	21.4
0.15	250	32.0	40.9	21.4	70.0	325.4	16.7
0.15	500	69.2	41.3	14.4	58.4	372.9	16.2
0.60	250	74.9	40.4	13.2	55.8	375.9	16.7
0.60	500	71.9	39.5	12.8	61.6	369.9	18.3

[a] Mason City beet molasses medium.

Fig. 26 shows that the rate of sugar utilization is lowest and reaches a relatively low maximum in the absence of agitation and gas flow. The fermentation is increased when there is gas flow through the medium (*see* curve II) and, over the time course considered, maximum sugar utilization is not attained. Increased fermentation is also achieved as agitation is increased (*see* curve III), the rate being similar to that during gas flow. Curve IV shows that combined gas flow and agitation, both at lower levels, increases the rate of sugar utilization. Biomass production in relation to sugar utilized was greatest when the medium was agitated without aeration and lowest in the absence of gas flow and agitation. Overall, Fig. 26 shows marginal, though useful, increases in sugar utilization due to gas flow/agitation, but very significant changes in the biomass yield coefficient.

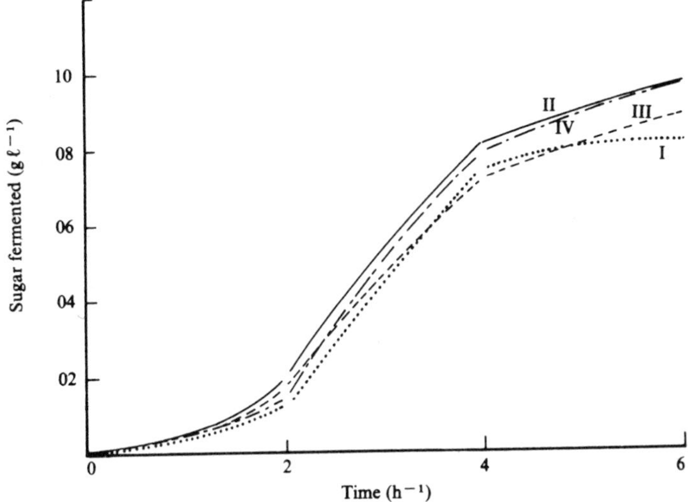

Figure 26. Rate of sugar utilization by *Torula utilis* grown on Mason City beet molasses. Gas flow, vvm; agitation, rpm; yield as percentage of sugar fermented in order. I: 0, 0, 3.7; II: 3.6, 0, 33.5; III: 0, 700, 62.2; IV: 0.6, 480, 57.6 [Singh *et al.* (1948)].

The variation in respiratory rates and penicillin production are summarized in Fig. 27 and Table 28. The respiratory rate in stirred fermenters was greater than that in shake flasks, with increased penicillin production and mycelial dry weight.

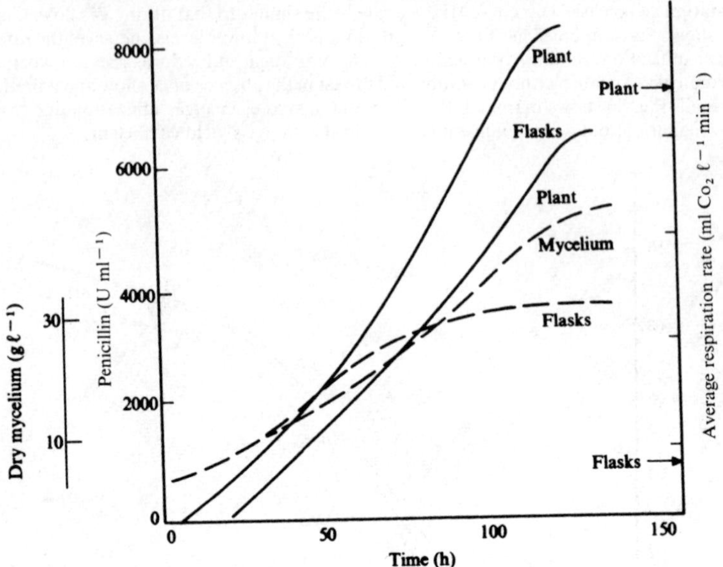

Figure 27. Mycelial weight and penicillin production in shaken flasks and stirred fermenters. —, penicillin production; — — — —, mycelial weight [Nixon and Calam (1968)].

Table 28. Variation in respiration rates and penicillin production according to apparatus used [Nixon and Calam (1968)].

Apparatus	Respiration rate ($mlCO_2$ ℓ^{-1} min^{-1})	Relative penicillin production
5-1 fermenters	8–10	1.2
Pilot plant	15–20	1.2
Plant fermenter	11–16	1.0
Shake flasks	1.6	0.75

Figs. 28 and 29 show that citric acid production, which increases with increasing oxygen flow rate, is also greater at higher oxygen pressures.

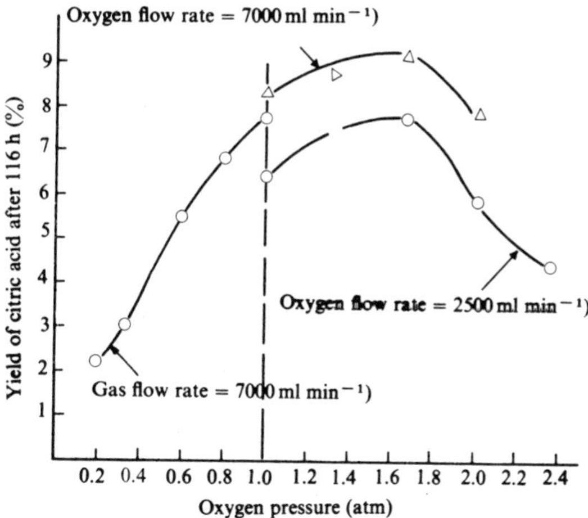

Figure 28. Effect of oxygen partial pressure on yield of citric acid in *Aspergillus niger*. (Total pressure was 1 atm where oxygen partial pressure was less than 1 atm, otherwise the oxygen pressure indicated was the total pressure.) [Clark and Lentz (1961)].

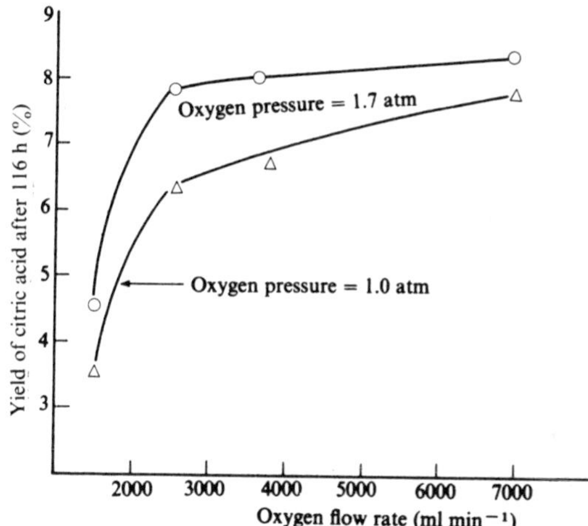

Figure 29. Effect of oxygen flow rate on production of citric acid by *Aspergillus niger* at two oxygen pressures [Clark and Lentz (1961)].

Table 29. Effects of agitation, air flow and power input on the rate of gluconic acid production by *Pseudomonas ovalis* NRRL–B–8 [Tsao and Kempe (1960)].

Agitator speed, N (rpm)	Air flow rate, G (ml ml^{-1} min^{-1})	Rate of acid production (mol ℓ^{-1} h^{-1} $\times 10^{-3}$)	Power input, P (hp ℓ^{-1} h^{-1} $\times 10^{-3}$)	Liquid volume expansion, E (% by vol.)
470	1.16	3.75	10.7	9.3
370	1.16	3.75	7.4	4.9
470	0.31	4.00	12.3	5.8
470	1.81	4.00	10.5	9.5
470	0.73	4.75	11.8	4.9
100	1.16	2.00	2.2	1.5
270	1.16	3.50	4.4	2.6
370	1.16	4.24	7.4	3.5
270	1.81	3.75	4.3	3.0
270	0.31	3.00	4.6	2.5
100	1.81	1.75	2.2	1.5
100	0.73	1.62	2.2	1.3
270	1.81	3.00	4.4	4.8
270	0.73	3.12	4.5	2.4

Dissolved carbon dioxide, which at increasing levels suppresses the respiratory rate (*see* Fig. 30), also inhibits antibiotic production (*see* Fig. 31).

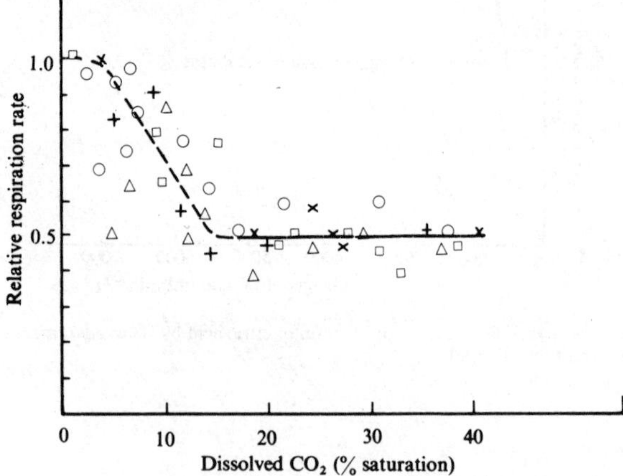

Figure 30. Dependence of respiration intensity of tetracycline- and oleandomycin-producing micro-organisms upon the concentration of dissolved carbon dioxide. □, +, ○, tetracycline; ×, △ oleandomycin [Bylinkina and Birukov (1972)].

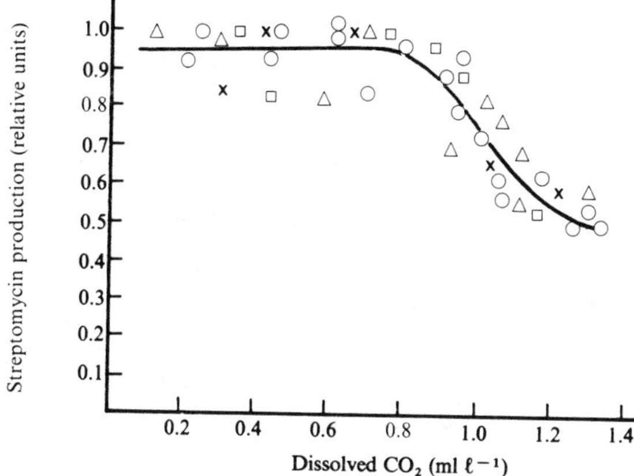

Figure 31. Dependence of streptomycin production upon the concentration of dissolved carbon dioxide within the first 50 h of fermentation. Summary of seven experiments [Bylinkina and Birukov (1972)].

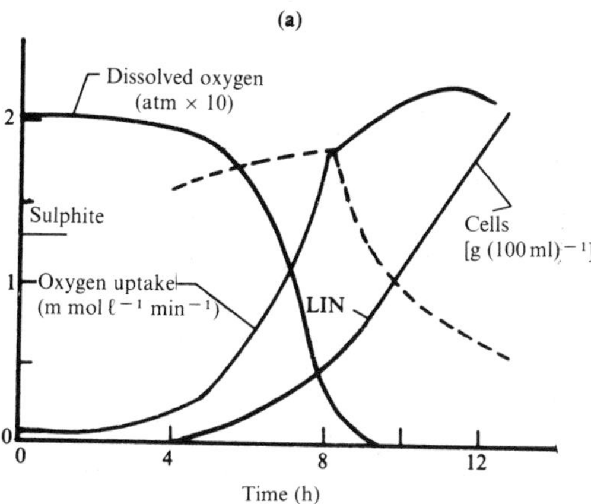

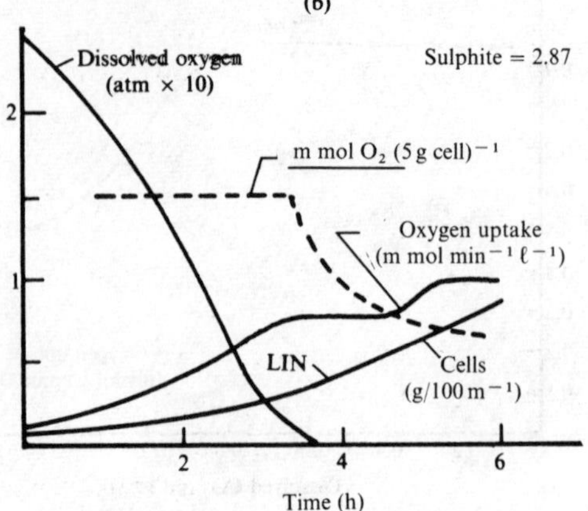

Figure 32. Growth of *Aerobacter aerogenes* in (a) a 3-ℓ fermenter and (b) a 1900-ℓ fermenter. Cell weights are on the dry basis. The dotted line gives the oxygen uptake rate per min per 5 g cells. The point marked 'LIN' indicates the time at which cell mass began to increase at a linear rather than at an exponential rate. The line marked 'Sulphite' indicates the oxygen uptake rate obtained with a sulphite solution in the fermenter, as m mol O_2 ℓ^{-1} min^{-1} [Phillips and Johnson (1961)].

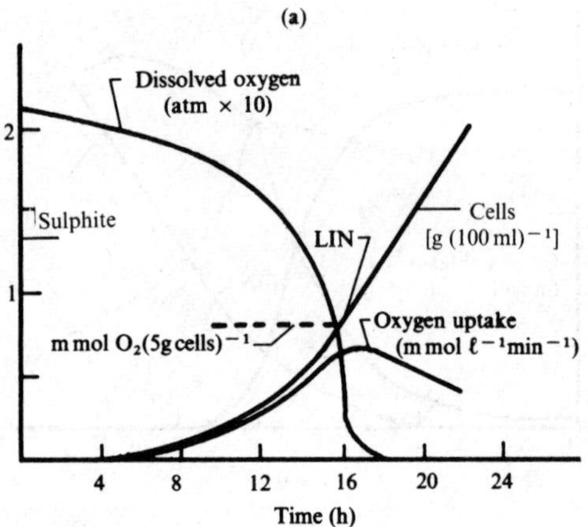

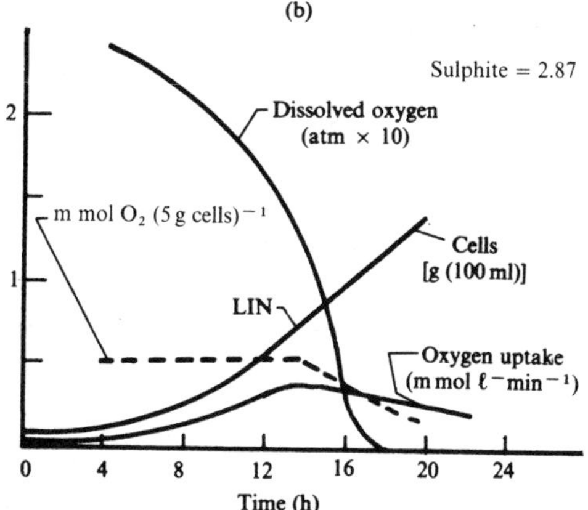

Figure 33. Growth of *Aspergillus niger* in (a) a 3-ℓ fermenter and (b) a 1900-ℓ fermenter. Abbreviations and conditions are as in Fig. 32 [Phillips and Johnson (1961)].

From Figs. 32 and 33, it can be seen that the dissolved oxygen level falls rapidly in large-scale compared with laboratory-scale fermenters. In the case of a large-scale fermenter, using *A. aerogenes* [*see* Fig. 32(b)], the oxygen uptake rate is constant for about 3 h but is further increased on the addition of an antifoam agent with a new, constant rate of oxygen uptake occurring after about 30 min.

Following oxygen depletion, the oxygen uptake rate becomes constant – unless restricted by increasing viscosity – leading to a linear increase in cell concentration with time. These are the conditions where the fermentation is controlled by the supply of oxygen, i.e. aeration or liquid-phase mass transfer controlled.

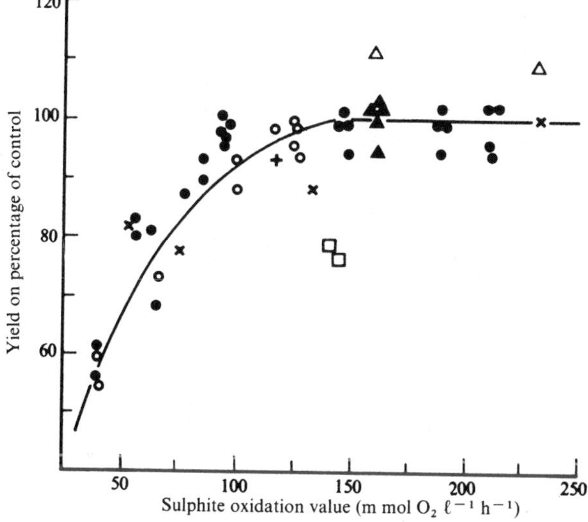

Figure 34. Changes in yield of yeast with degree of aeration, as measured by sulphite oxidation in the apparatus. ×, 190-ml shake flask; □, 19-ℓ fermenter with impeller (600 rpm); ●, 19-ℓ fermenter with agitation (800 rpm); △, 265-ℓ fermenter with agitation (550 rpm); ○, 265-ℓ non-agitated fermenter with small-hole sparger; ▲, 265-ℓ non-agitated fermenter with large-hole sparger; +, 114-m³ non-agitated fermenter [Finn (1967)].

Fig. 34 shows that the yield as a percentage of the control increases with the sulphite oxidation value (*see* Chapter 9) for different fermenters.

OXYGEN REQUIREMENT

Time Course of Oxygen Uptake

The volumetric oxygen uptake rate Q_{O_2} is a measure of the actual rate of oxygen uptake by a unit volume of fermentation broth.

The time courses of *Az. vinelandii* fermentation in a nitrogen-free medium and an ammonia-containing medium are represented in Figs. 35 and 36, respectively, for a 3.5-ℓ fermenter. Variations in agitation and air flow rate during the 28-h fermentation in Fig. 35 are shown in Table 30. Time courses for fermentations using a variety of microorganisms are shown in Figs. 37–40.

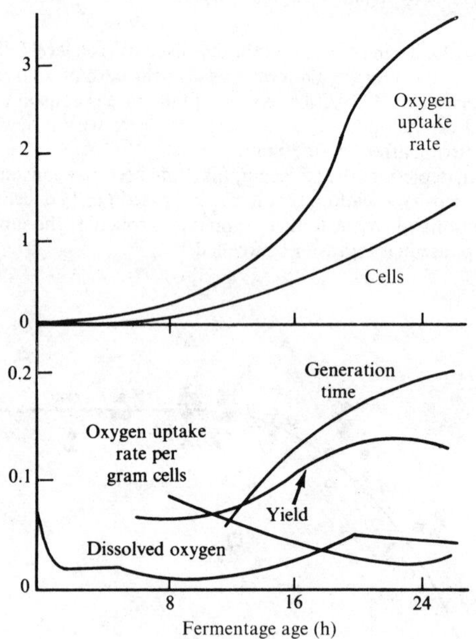

Figure 35. *Azotobacter vinelandii* O fermentation in a 3.5-ℓ fermenter. Cells were grown on a nitrogen-free medium and the dissolved oxygen level was manually controlled. Units: cells, g(500 ml)$^{-1}$; dissolved oxygen, atm; generation time, h × 50; oxygen demand and oxygen uptake rates, m mol O_2 ℓ$^{-1}$ min^{-1}; oxygen uptake rate per g of cells, m mol O_2 (0.1 g cells)$^{-1}$ min^{-1}; yield, g cells produced per g sugar used [Phillips and Johnson (1961)].

Table 30. Agitation and air flow rates used for *Azotobacter vinelandii* fermentation shown in Fig. 35 [Phillips and Johnson (1961)].

Time (h)	Agitation (rpm)	Air flow (ℓ min⁻)
0–1	200	0.16
1–2	280	0.16
2–3	330	0.16
3–6	400	0.16
6–8	480	0.16
8–9	520	0.16
9–11	520	0.78
11–12	580	0.78
12–13	590	1.38
13–14	640	1.38
14–15	760	1.99
15–16	840	1.99
16–18	890	1.99
18–19	890	2.72
19–20	1200	3.29
20–21	1290	3.29
21–22	1560	3.29
22–28	1750	3.29

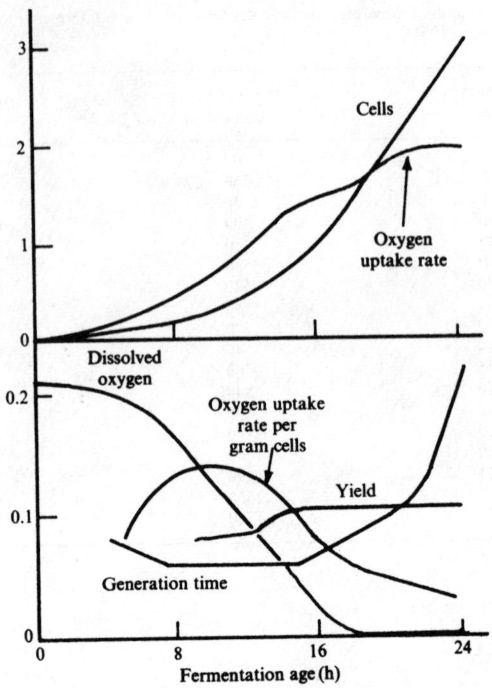

Figure 36. *Azotobacter vinelandii* O fermentation in a 3.5-ℓ fermenter. The cells were grown in the presence of an ammonia nitrogen source. Units: cells, g(500 ml)$^{-1}$; dissolved oxygen, atm; generation time, h × 50; oxygen demand and oxygen uptake rates, each m mol O_2 ℓ$^{-1}$ min^{-1}; oxygen uptake rate per gram cells, m mol O_2 (0.1 g cells)$^{-1}$ min^{-1}; yield, g cells produced per g sugar used [Phillips and Johnson (1961)].

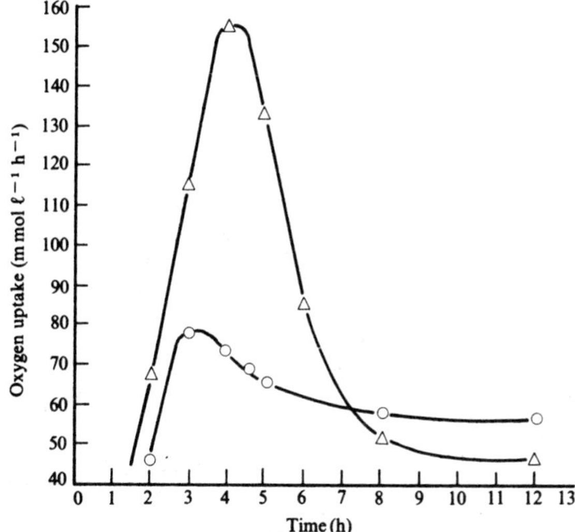

Figure 37. Oxygen uptake of a culture of a *Pseudomonas* sp. (○) and of *Bacilius subtilis* (△) [Sato (1961)].

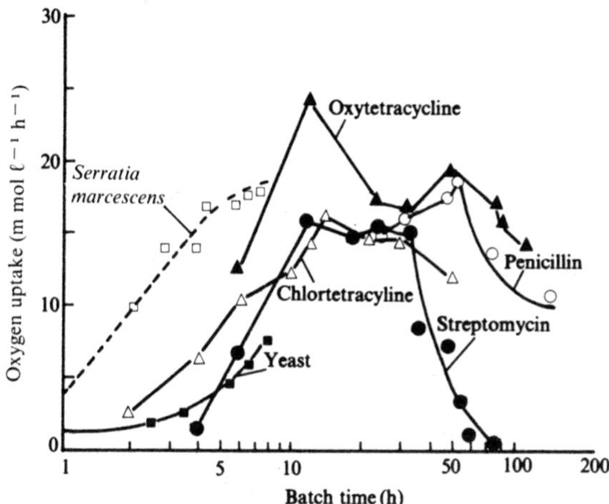

Figure 38. Volumetric oxygen uptake patterns [Gaden (1961)].

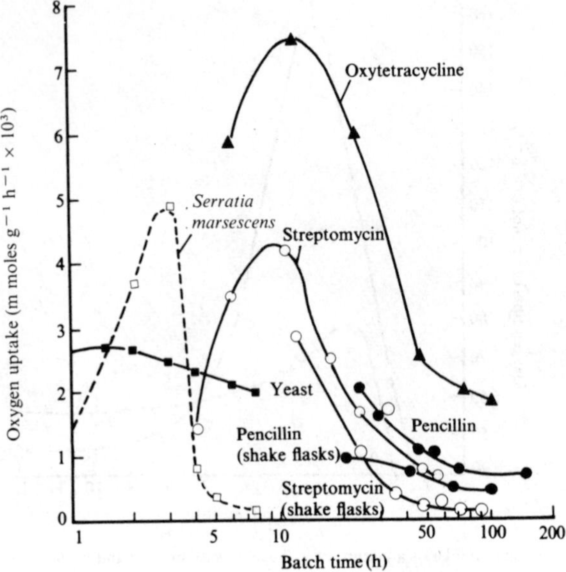

Figure 39. Specific oxygen uptake patterns [Gaden (1961)].

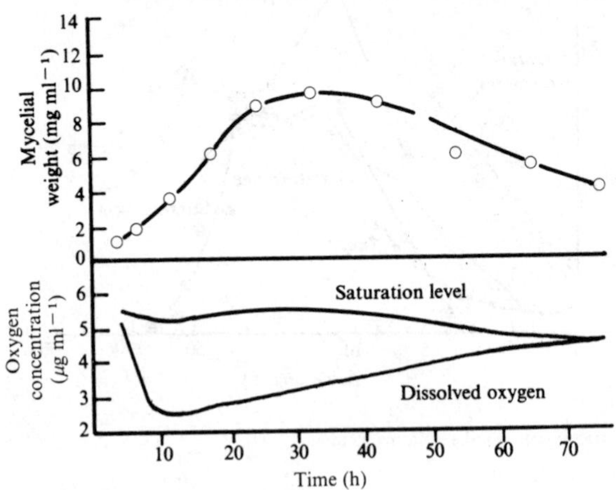

Figure 40. Dissolved oxygen concentrations during *Streptomyces griseus* fermentation in a 5-ℓ fermenter [Bartholomew (1950)].

During the growth phase, Q_{O_2} increases, while at the same time the dissolved oxygen level falls because of a failure to meet the oxygen requirements of the system.

Q_{O_2} can be increased by increased agitation and air flow (*see* Table 30) due to increased k_La (*see* Chapter 9). Typical relationships between specific uptake rates and mycelial concentrations at different agitation speeds are shown in Fig. 41.

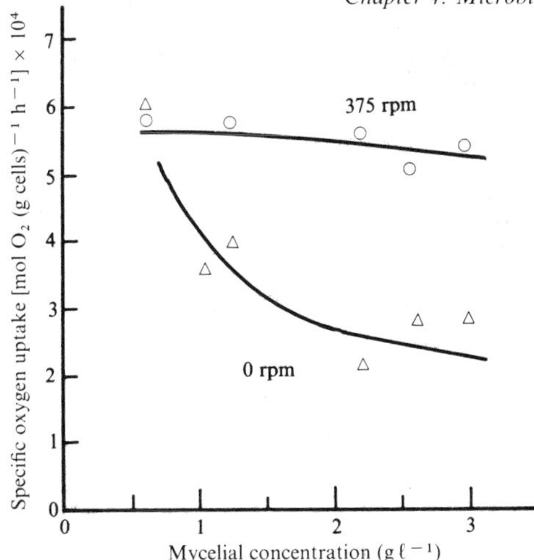

Figure 41. Specific oxygen uptake rate versus mycelial concentration for *Stretomyces griseus* cultured in a 5-ℓ fermenter at 24°C [Bartholomew (1950)].

A decline in Q_{O_2} coincides with increasing viscosity due to accumulation of cells and fermentation products in the broth. This effect is clearly demonstrated in Fig. 41.

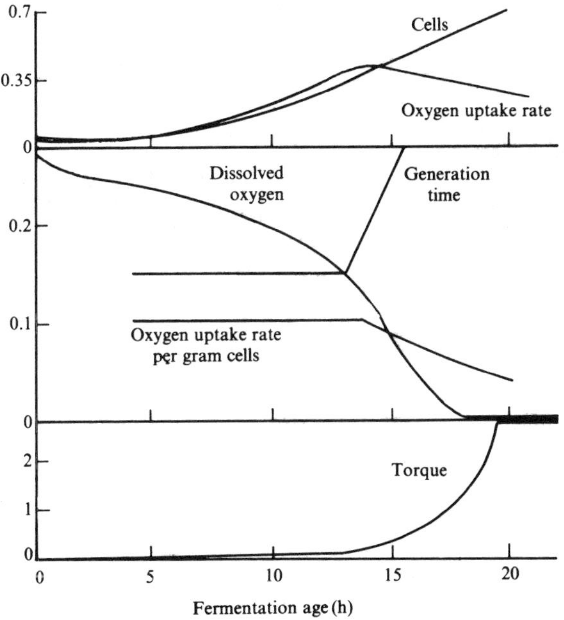

Figure 42. *Aspergillus niger* H72-4 fermentation in a 1,900-ℓ fermenter. Units: cells, $g(50\,ml)^{-1}$; dissolved oxygen, atm; generation time, h × 20; oxygen uptake rate, m mol O_2 ℓ^{-1} min^{-1}; oxygen uptake rate per gram cells, m mol $O_2(2\,g\ cells)^{-1}$ min^{-1}; torque, arbitrary units. The torque is that required to initiate rotation of a small turbine impeller immersed in a sample of culture liquid removed from the fermenter [Phillips and Johnson (1961)].

During the resting or stationary phase, the Q_{O_2} falls (*see* Fig. 42) due to increased oxygen levels in the broth (Fig. 40), which is explained by a reduction in the oxygen demand of cells.

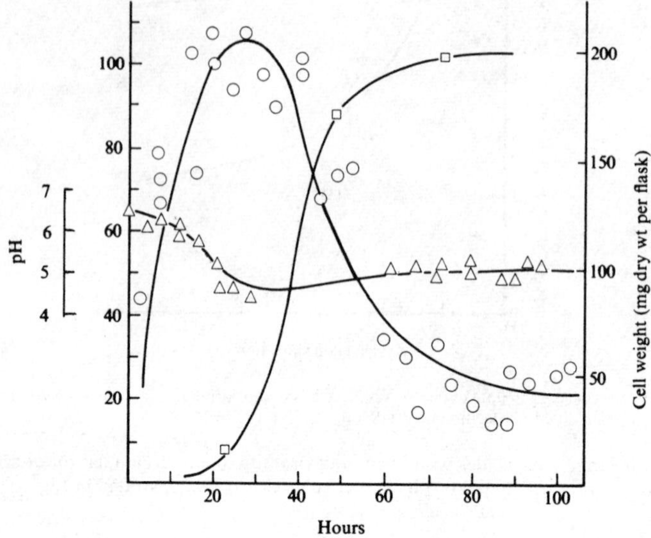

Figure 43. Changes in Q_{O_2} with cell dry weight for *Myrothecium verrucaria*. ○, Q_{O_2}; □, dry weight; △, pH of culture medium [Finn (1967)].

The potential volumetric oxygen demand during the growth phase is based upon $Y_{O_2}\,\mu\,x$ and is often greater than Q_{O_2} because of the inability of the aeration system (i.e. agitation/gas flow) to provide sufficient oxygen transfer. However, in the resting phase, the volumetric oxygen demand is based on the maintenance requirements and is therefore considerably less than $Y_{O_2}\,\mu\,x$. An example of the relationship between the actual oxygen uptake and the oxygen demand is shown in Table 31.

Table 31. Relation between actual oxygen uptake and maximal oxygen demand at different stages of penicillin fermentations[a] [Chain *et al.* (1966)].

Oxygen uptake/oxygen demand			
Between 30th and 70th hour	At the end of the fermentation, 80th hour	Dry weight ($g\,\ell^{-1}$)	Morphology of mycelium
0.80	0.66	0.23	Filamentous
0.85	0.46	0.30	Filamentous
1.00	0.66	0.31	Filamentous
1.00	0.80	0.28	Filamentous
0.80	0.52	0.29	Filamentous
0.98	0.75	0.31	Filamentous
0.80	0.45	0.38	Filamentous
0.77	0.48	0.36	Filamentous
0.90	0.60	0.32	Filamentous
0.75	0.66	0.32	Filamentous
0.81	0.57	0.40	Filamentous
0.80	0.77	0.30	Pellets
	0.89	0.35	Pellets
0.75	0.73	0.35	Pellets
	0.90	0.48	Pellets

[a] Ratio of liquid depth to fermenter diameter = 1.8:1.

Monod Equation

The relationship between the specific oxygen uptake rate (Q_{O_2}) and the oxygen levels in the fermentation medium is shown in Fig. 44.

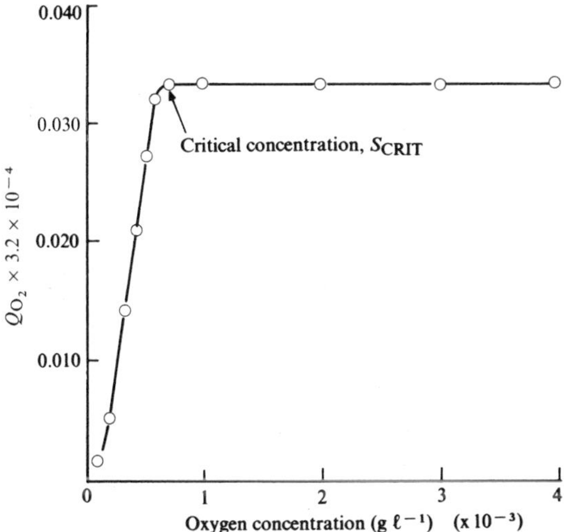

Figure 44. Changes in respiration rate of yeast with oxygen concentration, showing s_{CRIT} [Finn (1967)].

Monod equation plots for a number of microorganisms are shown in Figs. 45–47.

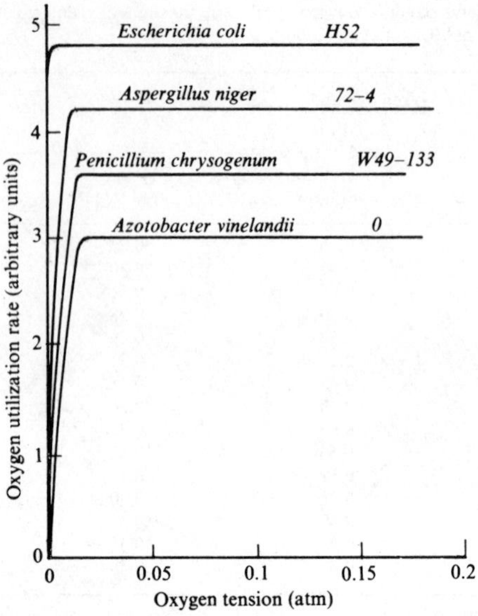

Figure 45. Effect of dissolved oxygen on oxygen utilization rates of several cultures, diluted with fresh, air-saturated medium [Phillips and Johnson (1961)].

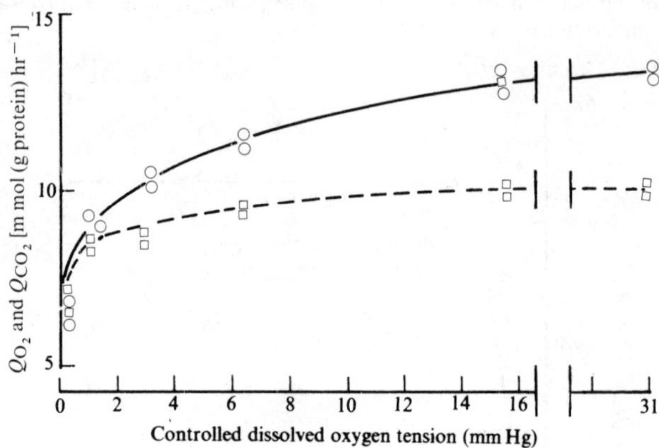

Figure 46. Influence of dissolved oxygen tension on the respiratory rate (Q_{O_2}) and carbon dioxide production (Q_{CO_2}) of *Pseudomonas* DX2 when growing in a chemostat on glucose–ammonia medium with a growth rate of 0.1 h^{-1}. Growth was nitrogen-limited. ○, Q_{O_2}; □, Q_{CO_2} [Harrison *et al.* (1968)].

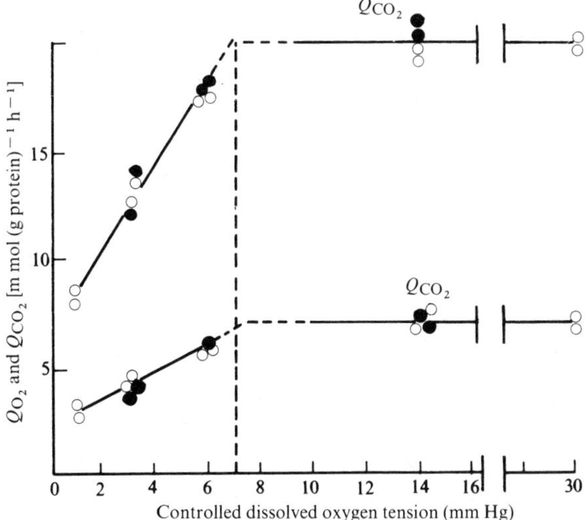

Figure 47. Influence of dissolved oxygen tension on the respiratory rate (Q_{O_2} and Q_{CO_2} of *Pseudomonas* DX2 when growing in a chemostat on decane–ammonia with growth rate of 0.1 h^{-1}. Growth was nitrogen-limited. ○, values obtained when decreasing dissolved oxygen tension; ●, values obtained when increasing dissolved oxygen tension [Harrison *et al.* (1968)].

Lineweaver–Burk plots of the data in Figs. 43–47 may be used to establish the Monod constants, i.e. reciprocals of oxygen concentration and Q_{O_2} or specific growth rate are plotted (*see* Figs. 48–50).

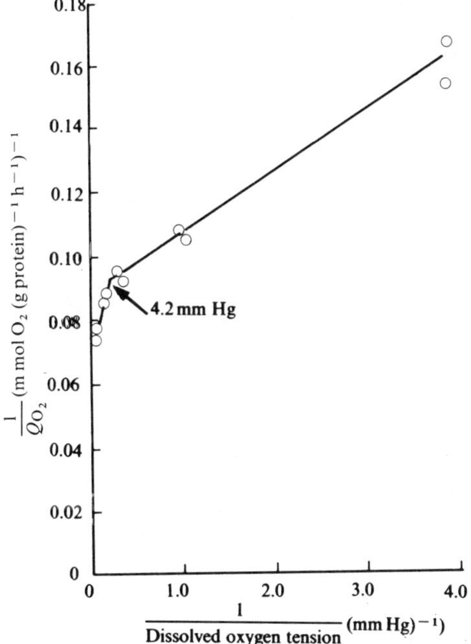

Figure 48. Lineweaver–Burk plot of Q_{O_2} versus dissolved oxygen tension for *Pseudomonas* DX2 growing on glucose–ammonia medium using data from Fig. 46 [Harrison *et al.* (1968)].

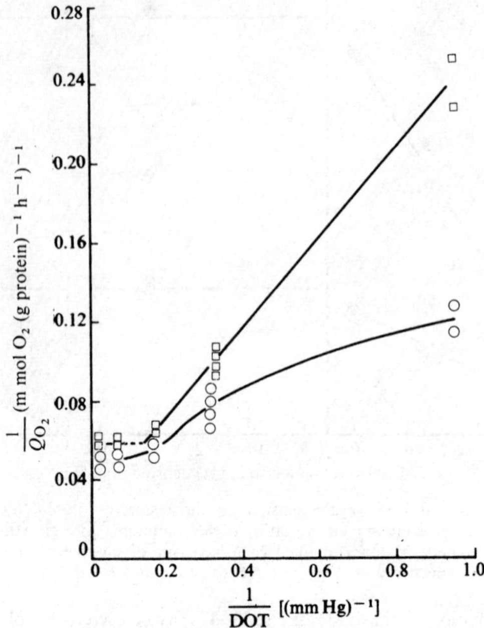

Figure 49. Lineweaver–Burk plot of Q_{O_2} versus dissolved oxygen tension for *Pseudomonas* DX2 growing on decane–ammonia (from data of Fig 47). ○, Q_{O_2}; ◇, Q_{O_2}(respiration) calculated from formula Q_{O_2} (respiration) = $Q_{O_2} - Q_{decane}$ [Harrison *et al.* (1968)].

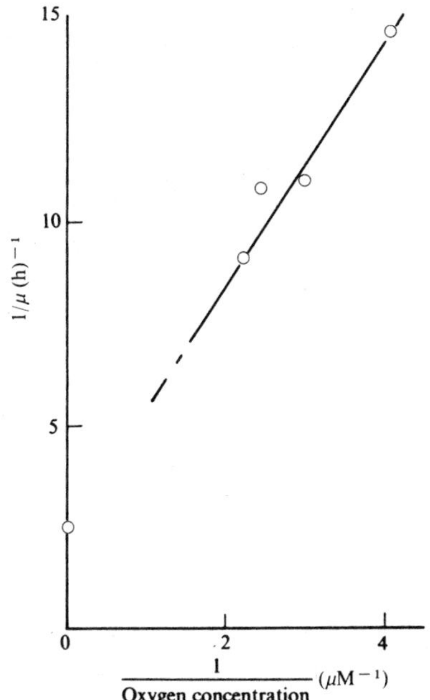

Figure 50. Reciprocal plot of steady-state growth rate as a function of oxygen concentration. The line corresponds to a Monod constant of 1.34 μM and a maximal growth rate of 0.44 h^{-1} [Johnson (1967)].

Figs. 48 and 49 suggest that the Monod equation is rather less satisfactory for the systems considered than indicated in Figs. 45 and 46. Fig. 50 provides supporting evidence for the Monod equation.

Q_{O_2}, max

Maximum oxygen demands (i.e. $Q_{O_2\ max}$) may be determined from the Lineweaver–Burk plots of reciprocal values of Q_{O_2} in relation to dissolved oxygen tensions by analogy with the Monod equation [*see* Eq. (8) of this chapter]. Typical values for $Q_{O_2, max}$ are given in Table 32.

Table 32. Maximum oxygen demands of active cultures [Finn (1954)].

Type of culture	Q_{O_2}, max mM O_2 ℓ h^{-1}
Activated sludge	1–2
Acetobacter sp.	90
Aerobacter aerogenes	5–8
Aspergillus niger	28
Asp. niger	56
Azotobacter vinelandii	260
Escherichia coli	5–8
Penicillium chrysogenum	20–30
Streptomyces griseus	15
Ustilago zeae	16
Yeast	10–15

Oxygen demands of bacteria are presented in Tables 33–35.

Table 33. Hourly oxygen demand per cell and per generation [Rahn and Richardson (1941)].

Microorganism	0.5% peptone			0.5% peptone + 0.5% glucose		
	Generation time (min)	Oxygen per cell per hour (× 10^{-13} g)	Oxygen per generation (× 10^{-13} g)	Generation time (min)	Oxygen per cell per hour (× 10^{-13} g)	Oxygen per generation (× 10^{-13} g)
Aerobacter aerogenes	37.9–38.9	2.9–3.2	1.9–2.0	35.9–37.3	2.2–4.9	1.4–2.9
A. cloacae	33.0–39.8	3.1–3.3	1.7–2.1	40.7–42.3	1.8–3.5	1.3–2.4
A. levans	43.4–46.0	2.9–5.7	2.2–4.1	43.7	6.6–7.1	4.8–5.0
Bacillus cereus	32.4–47.9	18.3–43.1	9.8–31.8	42.9	38.2	27.3
B. megatherium	56.2	28.0	26.1	77.4	102.0	131.5
B. mesentericus	34.2–90.0	6.9–11.0	3.9–12.0	43.2–50.9	5.4–22.0	4.2–18.6
B. peptogenes	57.0	77.3	73.4	82.6	87.5	120.0
B. subtilis	45.5–68.7	16.1–47.8	18.0–46.5	42.0–52.8	9.6–20.7	6.7–18.1
Escherichia coli	38.6–47.6	3.0–4.7	2.2–3.6	39.5–40.7	3.3–3.8	2.2–2.6
E. communior	60.7	3.6	3.6	51.0–53.5	2.2–3.1	1.9–2.3
Proteus vulgaris	41.3–90.0	2.8–9.9	3.0–6.8	44.0–50.5	3.0–7.3	2.5–5.4
Pseudomonas aeruginosa				58.5–64.5	6.5–11.0	6.3–10.9
Ps. fluorescens	59.2–70.2	8.0–8.7	7.9–10.2			

Table 34. *Streptococci* in milk at 30°C [Rahn and Richardson (1941)].

Microorganism	Generation time (min)	Hourly oxygen consumption per cell (× 10^{-3} g)	per ml milk (× 10^{-7} g)
Streptococcus durans	52.3–63.0	1.07	7.02
S. faecalis	49.2–62.0	2.32	8.28
S. lactis 125	42.3–57.9	0.46	8.65
S. lactis L21	38.7–59.1	2.10	7.40
S. liquefaciens	56.7–85.0	7.22	7.83
S. zymogenes	49.8–68.2	2.33	8.28

Table 35. Oxygen and energy requirement for multiplication of bacteria [Rahn and Richardson (1941)].

Microorganism	Cell weight (× 10^{-13} g)	Oxygen per cell per hour (× 10^{-13} g)	Oxygen per 10^{-13} g cells per hour (× 10^{-13} g)	Oxygen required to produce 10^{-13} g cells (× 10^{-13} g)	Living cells per g oxygen (g)	Living cells per calorie of respiration (g)
Acetobacter aerogenes	7.5	3.3	0.44	0.27	3.66	1.14
A. cloacae	7.5	4.0	0.53	0.25	3.95	1.23
A. levans	7.5	5.6	0.75	0.53	1.85	0.58
Bacillus cereus	15.8	32.4	2.05	1.37	0.73	0.23
B. megatherium	32.6	65.0	2.00	2.16	0.46	0.14
B. mesentericus	7.5	9.5	1.27	1.05	0.95	0.30
B. peptogenes	4.5	82.4	18.30	21.4	0.05	0.01
B. subtilis	7.5	20.6	2.74	2.34	0.43	0.13
Escherichia coli	7.5	3.9	0.52	0.36	2.78	0.87
E. communior	7.5	3.0	0.40	0.35	2.88	0.90
Proteus vulgaris	7.5	5.3	0.71	0.60	1.67	0.53
Pseudomonas aeruginosa	7.5	20.6	2.74	2.34	0.43	0.13
Ps. fluorescens	7.5	8.4	1.11	1.13	0.88	0.27

s_{CRIT, O_2}

Table 36. Critical oxygen concentrations and maximum oxygen consumptions for various microorganisms.

Microorganism	Temperature (°C)	s_{CRIT, O_2} (mM)	$Q_{O_2, max}$ (m mol ℓ$^{-1}$ h^{-1})
Acetobacter	30		90
Aspergillus oryzae	30	~ 0.020	
Azotobacter vinelandii	30	0.018–0.049	260
Escherichia coli	15	0.003	
	37	0.008	5–8
Penicillium chrysogenum	24	0.022	20–30
	30	0.009	
Pseudomonas denitrificans	30	0.009	
Ps. oralis		0.034	
Serratia marcescens	31	0.015	
Saccharomyces cerevisiae	30	0.004	10–15
Streptomyces griseus	30		15
Torulopsis utilis		0.063	

Table 37. Consumption of oxygen by yeast cultured in medium containing mineral salts and 1 per cent glucose [Rivière (1977)].

Substrate	Concentration in solution (ppm)	Critical concentration (ppm)	Rate of consumption ($Q_{O_2, max}$) [m mol (g cells)$^{-1}$ h^{-1}]
Glucose	10,000	100	2.8
Oxygen	7	0.8	7.7

K_{m}, O_2

Table 38. K_{m, O_2} for microorganisms

Microorganism	Cell diameter (μm)	Temperature (°C)	K_{m, O_2} (μM)
Acetobacter suboxydans	2.7	19.2	0.002
Aerobacter aerogenes	0.6	19.0	0.003
Azotobacter indicum	1.6	19.6	0.030
Bacillus megatherium	2.0	19.2	0.060
		21.2	0.090
		32.2	0.277
B. megatherium[a]	4.0	20.2	1.21
		34.0	3.57
B. megatherium[b]	2.4	20.0	0.071
Candida utilis			1.34
Escherichia coli	0.6	19.2	0.002
Haemophilus parainfluenzae			2.2–11.1
Micrococcus candicans	0.5	20.2	0.001
Saccharomyces cerevisiae		19.0	0.065
Serratia marcescens	0.7	18.8	0.004

[a] Glycine-containing medium
[b] Lithium chloride-containing medium

TEMPERATURE EFFECTS

Growth Rate, Mutation Rate and Lag Phase

Microorganisms can be classified according to the lowest temperatures at which significant growth occurs (*see* Table 39)

Table 39.

Classification	Lowest growth temperatures
Psychrophiles	< 20°C
Mesophiles	20–45°C
Thermophiles	45–60°C

Table 40. Maximum growth temperature of psychrophilic isolates [Ingraham (1958)].

Maximum growth Temperature (°C)	Number of Isolates
> 40	2
37	6
35.5	8
30	20
< 21	0

The incidence of psychrophiles isolated from spoiled samples varies with the isolation temperature and the source of the samples (*see* Table 41).

Table 41. Percentages of psychrophiles isolated from spoiled chicken, faeces, and soil at various temperatures [Ingraham (1958)].

Temperature of original isolation (°C)	Isolates found to be Psychrophilic (%)		
	Soil	Faeces	Chicken
1			100
3	100		100
12	0	0	83
21	0	0	100
30	0	0	100
40	20	0	0

Generation times and specific growth rates at different temperatures are shown in Figs. 51–53 and Tables 42 and 43 for psychrophilic, mesophilic and thermophilic bacteria. Although psychrophilic strains are capable of growing at low temperatures, specific growth rates can be higher in the temperature range 20–30°C.

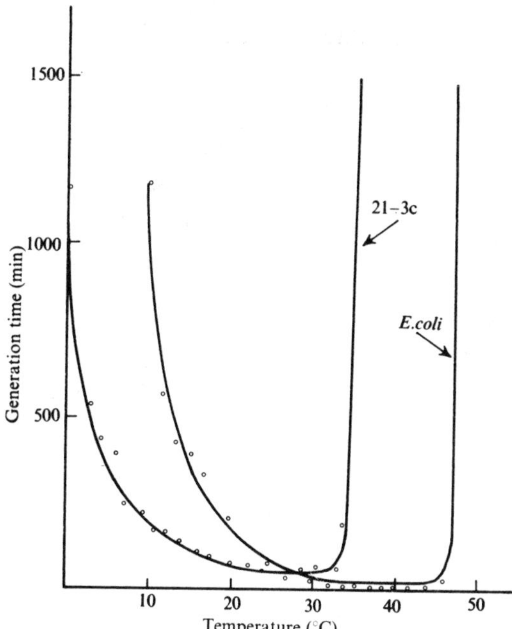

Figure 51. Effect of temperature on the generation time of a typical mesophile (*Escherichia coli*) and a psychrophile (21-3c) [Ingraham (1958)].

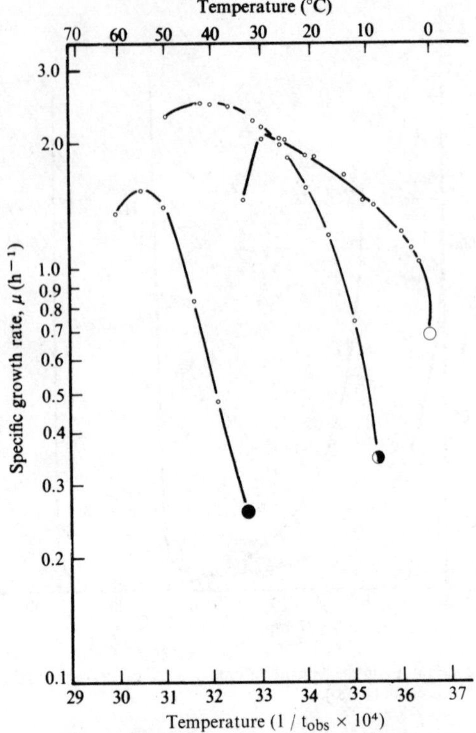

Figure 52. Arrhenius plots of the specific growth rates of a psychrophilic pseudomonad (○), a mesophilic strain of *Escherichia coli* (◑), and a thermophilic strain of *Bacillus circulans* (●) [Farrel and Rose (1967)].

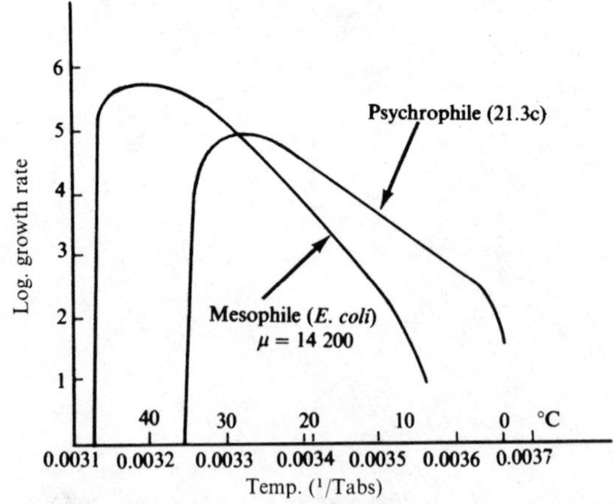

Figure 53. Arrhenius plot of the relationship between growth rate and temperature for psychrophile (21-3c) and a mesophile (*Escherichia coli*) [Ingraham (1958)].

Table 42. Generation times of mesophilic and psychrophilic bacteria [Ingraham (1958)].

Temperature (°C)	Generation time (min)			
	Psychrophilic pseudomonads		Mesophiles	
	1–3b	P-200	*Escherichia coli*	*Pseudomonas aeruginosa*
48			*	
46			32	*
45				46
44			22	
42			21	27
40			21	22
38			22	
36		*	21	
35				30
34	*	60	28	
32	53	43	30	
30	56	51	33	36
28	53		46	
26	59	54	56	
25				56
24	60	68	87	
22	81		96	
20	95	92	170	120
18	100	98	260	
16		110	350	
15				160
14	140	130	400	
12	150	160	580	
10	160	240	1200	470
8	210	330	2500	1400
6	310	450		
4	420	600		
2	480	820		
0	620	2100		

* No growth occurred in 16 h at this temperature.

Table 43. Specific growth rates of coliform and pseudomonad types isolated from waters [Baig and Hopton (1969)].

Microorganism	Growth on nutrient agar[a]	Specific growth rate (h^{-1}) 5°C	10°C	15°C	20°C	25°C	30°C	37°C
C 1	+	0.07	0.13	0.22	0.33	0.44	0.58	
C 2		*	0.06	0.14	0.28	0.36	0.77	0.92
C 4	+	0.06	0.12	0.19	0.25	0.36	0.53	
C 7		*	0.07	0.17	0.32	0.50	0.58	
C 10		*	0.05	0.15	0.23	0.39	0.73	0.95
ML 30		*	0.04	0.10	0.23	0.38	0.59	0.77
EBT	+	0.06	0.13	0.22	0.36	0.44	0.50	
P 11	+	0.05	0.11	0.18	0.23	0.37	0.39	
P 14	+	0.09	0.14	0.21	0.35	0.48	0.73	
P 15		0.02	0.04	0.11	0.23	0.43	0.82	
P 22	+	0.06	0.08	0.20	0.28	0.41	0.69	
P 26	+	0.06	0.11	0.21	0.32	0.43	0.49	
P 27	+	0.03	0.09	0.20	0.32	0.51	0.69	
8602		*	0.03	0.09	0.21	0.37	0.50	0.58

[a] +, visible growth after 7 days at 5°C.
*, mean generation time not determined as growth was too slow.

Mutation produces changes in the sensitivity of bacteria to temperature, reflected in variations in specific growth rates. Such changes are shown in Figs. 54 and 55 and in Table 44.

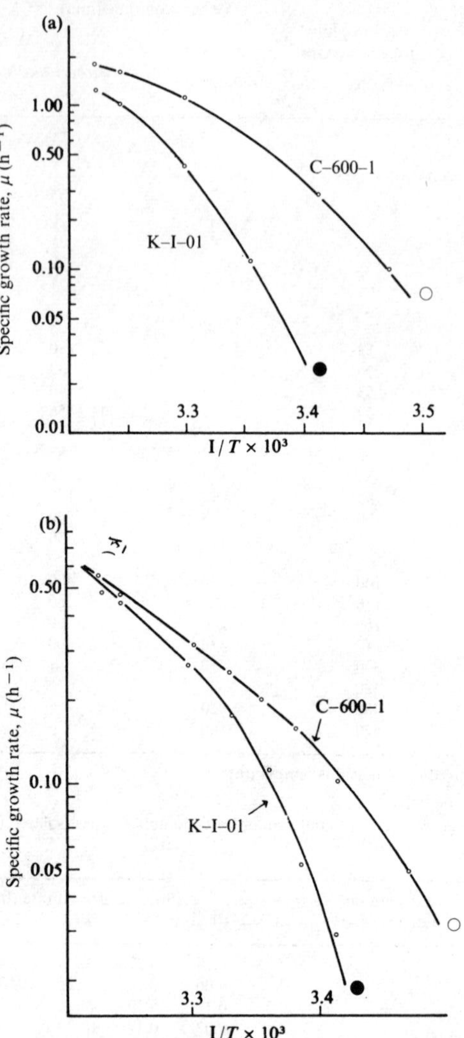

Figure 54. Arrhenius plot of the specific growth rate of temperature mutant K-I-01 (●) and its parent, *Escherichia coli* C-600-1 (○). (a) Growth in complex medium, i.e. 0.5 per cent glucose–yeast extract–0.25 per cent Casamino acids. (b) Growth in glucose minimal medium [O'Donovan *et al.* (1965)].

Different temperatures affect the mutation rate (*see* Fig. 55). The temperature at which the growth rate is maximal need not necessarily be that at which the mutation rate is highest.

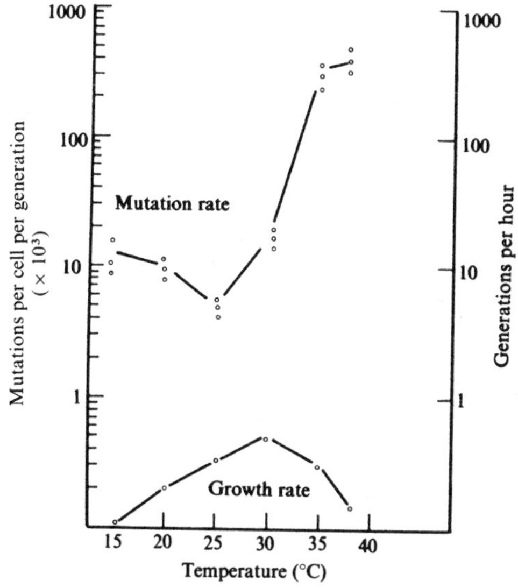

Figure 55. Effect of temperature on the rate of growth and the rate of mutation of *Saccharomyces cerevisiae* strain 14940 [Farrel and Rose (1967].

The duration of the lag phase is inversely related to the temperature of incubation (*see* Fig. 56).

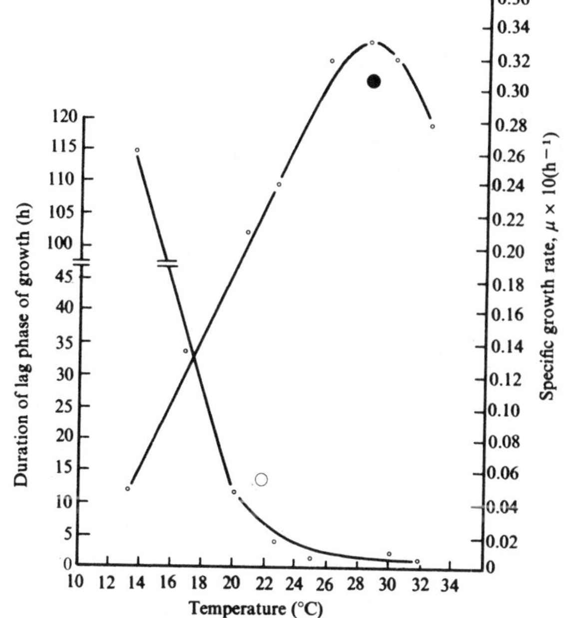

Figure 56. Effect of temperature of incubation on the duration of the lag phase of growth (○) and on the specific growth rate (●) of *Euglena gracilis*. Inoculum organisms were grown at 25°C [Farrel and Rose (1967)].

Temperature Tolerance

Bacteria

The susceptibility of bacteria to high and low temperatures is summarized in Table 44.

Table 44. Effect of temperature on growth and survival of bacteria [Spector (1956)].

Species	Growth temperature (°C) Minimum	Optimum	Maximum	Thermal Temperature (°C)	death Time (min)
Acetobacter roscum	10	30–35	41	50	5
Achromobacter ichthyodermis	−2	25	30		
Actinobacillus lignieresi	20	37		60	15
Actinomyces bovis		37		60	10
Actinom. thermophilus	28	50	65		
Aerobacter aerogenes		30	42–44	60	30
Agrobacterium tumefaciens	0	25–28	37		
Alkaligenes faecalis		37			
Azobacter chroococcum		25–28			
Azotomonas indicum		30			
Bacillus anthracis		35	43	100[a]	10[a]
B. subtilis	15	30–37	55	100[a]	14
B. thermodiastaticus	50	65	75		
Bacteriodes fragilis		37			
Bacterium erthyrogenes		28–35			
Bact. phosphoreum	5	10	25		
Bartonella bacilliformis		28			
Brucella sp.	20	37	40	60	10
Cellfacicula viridis		20			
Cellulomonas biazotea		20			
Cellvibrio ochraceus		20			
Chromobacterium violaceum		25–30			

Clostridium botulinum	18	20–35	55	120[a]	5[a]
Clost. perfringens		38		100[a]	20[a]
Clost. tetani	14	37–38	50		
Coliforms				60–63	30
Corynebacterium diphtheriae	15	34–36	40	54	10
Coxiella burnetii		37		62	30
Cytophaga hutchinsonii		28–30			
Desulfovibrio desulfuricans		25–30	40		
Diplococcus pneumoniae	18	37	42	56	5–7
Erwinia carotovora	4	25–30	39		
Escherichia coli	10	30–37	45	60	10
Flavobacterium aquatile		25			
Haemophilus influenzae	26	37	43	50–55	30
H. pertussis		37		56	60
H. suis		37		60	20
Hydrogenomonas pantotropha		28–30			
Klebsiella pneumoniae	12	37	43	55	30
Lactobacillus casei	10	−30	40		
L. thermophilus	30	50–63	65	71	30
Leptospira icterohaemorrhagiae	25	25–30	37	56	20
Listeria monocytogenes		37		59	10
Methanobacterium omelianskii		37–40	48		
Methanococcus mazei		30–37			
Micrococcus luteus		25			
Micro. pyogenes var. *albus*		37		62	10
Micro. pyogenes var. *aureus*	15	37	40	60	20
Micromonospora chalcea		30–35		70	5
Mycobacterium avium	30	40	44		
Myc. phlei	20	37	58	60	> 60
Myc. tuberculosis	30	37	42	65	15
Neisseria gonorrhoeae	25	37	40	55	< 5
N. meningitidis	25	37	42	50	< 5

continued overleaf

Table 44 – (continued)

Species	Growth temperature (°C) Minimum	Optimum	Maximum	Thermal Temperature (°C)	death Time (min)
Nitrobacter winogradskyi		25–28		50	5
Nitrocococcus nitrosus		20–25			
Nitrosomonas monocella		28			
Nocardia asteroides		37		60	60
Pasteurella multocida		37		60	10
Past. pestis	0	25–30	45	55	5
Past. tularensis	24	37	39	56	10
Pediococcus cerevisiae		25			
Propionibacterium freudenreichii		30			
Proteus vulgaris	10	37	43	55	60
Protaminobacter alboflavum		30			
Pseudomonas aeruginosa	0	37	42	62	10
Rhizobium leguminosarum		25			
Rhodopseudomonas palustris		37			
Rhodospirillum rubrum		30–37			
Saccharobacterium ovale	20	34–35	37	54	10
Salmonella typhimurium	4	37	46	55	24
Sal. typhosa	4	37	46	60	2
Sarcina ventriculi	10	30	45	65	10
Serratia marcescens		25–30	<37	55	60
Shigella dysenteriae		37		60	10
Shig. equuli		37		60	15
Spirochaeta daxensis		42–52		60	30
Sp. plicatilis		20–25			
Sporocytophaga mycococcoides		28–30			
Streptococcus pyogenes	15	37	40	60	15
Streptomyces thermophilus	20	40–45	<53	72–74	10
Streptom. griseus		37			
Thiobacillus thiooxidans		28–30			
Treponema pallidum		37		40	60–180
Vibrio comma	14	37	42	55–60	2
Xanthomonas campestris	5	28–30	38		

[a] Spores.

Resistance of bacteria to high temperatures is greatest in the lag phase and declines rapidly at the start of the log phase of growth (*see* Fig. 57).

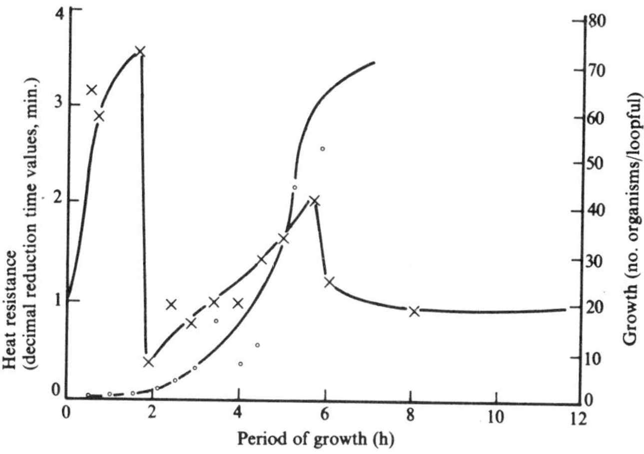

Figure 57. Effect of the age of the culture on the resistance of *Streptococcus faecalis* strain L 5 to heating at 60°C in saline. The bacteria were grown at 37°. Viability was measured by plating dilutions on nutrient agar and incubating at 37°. ×, heat resistance; ○, growth as determined by microscopic counts on a loopful of culture [Farrel and Rose (1967)].

The viability of bacteria declines during storage, although the effect is reduced by exposure to low temperatures, the effect being time-dependent (*see* Fig. 58 and Table 45).

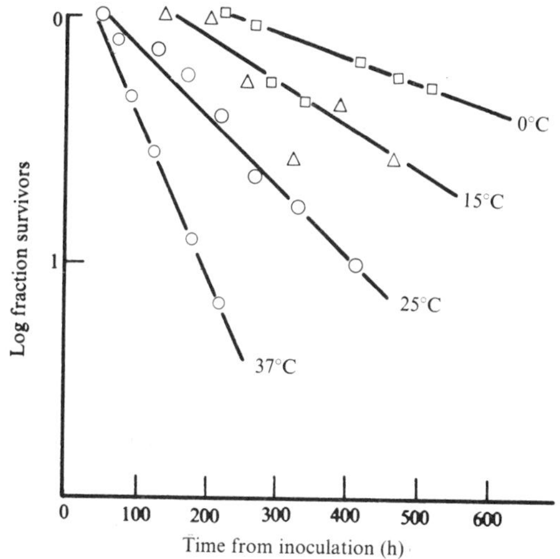

Figure 58. Decrease in viability in stationary-phase culture of *Escherichia coli* grown at different temperatures. Cultures maintained at 37, 25 or 15°C in the stationary phase had been grown at the same temperature. Cultures maintained at 0° were grown at 37°C [Farrel and Rose (1967)].

Table 45. Cold injury of *Pseudomonas fluorescens* as a function of temperature [Ingraham (1962)].

Temperature (°C)	Days in storage	Percentage of initial population		
		Injured	Killed	Unharmed
−7	1	25	51	24
−7	7	17	82	1
−7	15	1.9	98	0.1
−18	1	41	51	8
−18	7	34	63	3
−18	15	12.8	87	0.2
−29	1	35	48	17
−29	7	39	54	7
−29	15	36	59	5

Repeated freezing and thawing of a bacterial suspension reduces the number of viable cells (*see* Fig. 59), this effect being most marked in growing microorganisms.

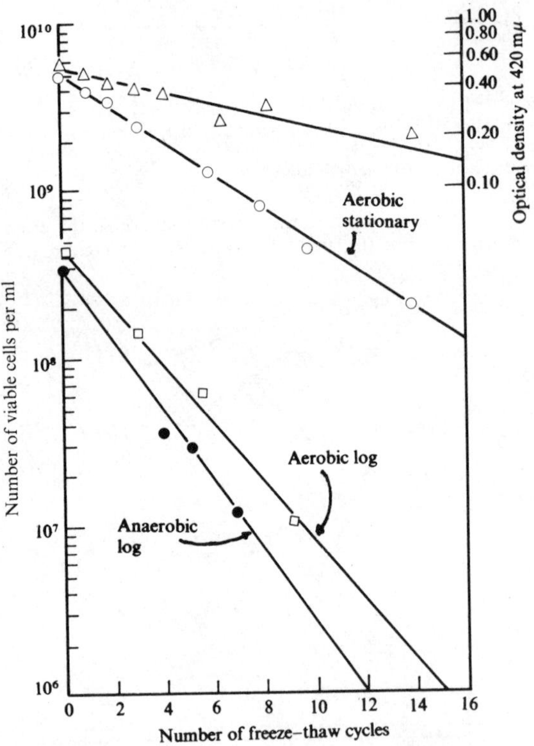

Figure 59. Effect of the phase of growth, and of aeration of the culture, on the turbidity of, and the viability of the organisms in, a suspension of *Escherichia coli* ML 30 after repeated freezing and thawing. △, turbidity; ○, viability of a suspension of aerobically grown, stationary-phase bacteria in 10 per cent spent growth medium. □, viability of aerobically grown, exponential-phase bacteria in 1 per cent spent growth medium; ●, viability of anaerobically grown exponential-phase bacteria in 1 per cent spent growth medium [Farrel and Rose (1967)].

Yeasts

Yeast growth is optimal in the region of 20–30°C for mesophilic and facultative species (*see* Table 46). Maximum temperatures at which growth occurs for these species are in the range 30–45°C. Obligative psychrophilic yeasts are unable to grow above a temperature of 20–30°C.

Table 46. Growth temperatures of yeasts [Stokes (1971)].

	Growth temperatures (°C)		
Species	Minimum	Optimum	Maximum
Mesophilic and facultative psychrophilic yeasts			
Candida macedonensis	5		45
C. parapsilosis	0	20–25	30
C. slooffii	27		45
C. utilis	5–10		
Debaryomyces hansenii	0		35
Hansenula suaveolens	3	30	30–35
H. valbyensis	5		32–33
Kloeckera apiculata	3	30	35
Nadsonia elongata	0		25
Pichia membranaefaciens	3		30
Rhodotorula glutinis	0	23	< 30–39
R. gracilis	5	27	37–42
Saccharomyces carlsbergensis	0	25	33.5
Sacch. cerevisiae	0–5	28–36	40–42
Sacch. fragilis	5		45
Sacch. intermedius	0.5		40
Sacch. ludwigii	1–3		37–38
Sacch. marxianus	0.5		46–47
Sacch. mellis	23		35
Sacch. octosporus	17		33
Sacch. pastorianus	0.5		24
Sacch. turbidans	0.5		40
Sacch. validus	0.5		39–40
Sacchiromycopsis guttulata	35		40
Torulopsis candida	5	22	32
Torulopsis molischiana	5	22	> 42
Obligately psychrophilic yeasts			
C. frigida, strain P8	−7– −5	15	20
C. gelida, strain P16	−7– −5	15	20
C. nivalis, strain P7	−7– −5	15	20
C. scoitii	0	4–15	15–20
Candida sp., strain 3E-2	−7	5–20	20
Candida sp., strain 25	−4.5	15	20
R. infirmo-miniata	−2	14–18	26–28

Comparison of Table 47 with Table 46 suggests that the optimum temperature for sporulation is similar to that for growth.

Table 47. Effect of temperature on sporulation of *Saccharomyces* species [Stokes (1971)].

Species	Sporulation temperatures (°C) Minimum	Optimum	Maximum
Saccharomyces cerevisiae	9–11	24–30	35–37
Sacch. ellipsoideus	4.7–8	25	29.5–33
Sacch. intermedius	0.5–4	25	27–29
Sacch. pastorianus	0.5–4	27.5	29–31.5
Sacch. turbidans	4–8	29	33–35
Sacch. validus	4.8–5	25	27–29

The susceptibility of yeasts to high temperatures is presented in Table 48.

Table 48. Thermal death time of cells and spores of yeasts [Stokes (1971)].

Species	Suspending medium	Lethal temperature (°C)	Death time (min) Cells	Spores
Debaryomyces globosus	Glucose broth	55	25	
		60	5	
Monilia candida	Glucose broth	55	25	
		60	5	
Saccharomyces cerevisiae	Water	54	5	
		62		5
Sacch. ellipsoideus	Beer	54	20	
		56		15
Sacch. ellipsoideus	Grape juice	54	120	
		57.5	10	
Sacch. ellipsoideus	Glucose broth	60	15	
Sacch. odessa	Beer	62–64	20	
		62–64		15
Sacch. turbidans	Beer	52	20	
		58		20
Torula monosa	Glucose broth	60	35	
		50	20	
Willia anomola	Glucose broth	55	10	
		60	5	

Fungi

Table 49. Effects of temperature on growth and survival for fungi [Spector (1956)].

Species	Growth temperature (°C)			Thermal death	
	Minimum	Optimum	Maximum	Temperature (°C)	Time (min)
Absidia corymbifera	20	37	46		
Achorion castellani		20			
Albugo candida	1	10	20		
Allescheria boydii	15	30	45	56	60
Alternia solani	1.5	26–28	45		
Armillaria mellea	15	25	30		
Aspergillus bronchialis		34			
Asp. fumigatus	12	40	> 50		
Asp. nidulans		36–38			
Asp. niger		37	< 60		
Blastomyces brasiliensis		25–130		60	60
B. dermatitidis		35		56	60
Botrytis cinerea	0	20–25	30		
Candida albicans	< 20	30–137	> 40	60	10
C. guilliermondi		30–37			
C. krusei		30–37			
C. tropicalis		30–37		60	10
Castellania hashimotoi		22–25			
Cephalosporium glanulomatis		20–37		53	5
Ceph. recifei		25–30			
Ceratostomelia ulmi	5	22–27	40		
Cerospora beticola	6	25–30	34		
Chaetomium thermophile	30		60		
Cladosporium carpophilum	2	19–28	33		
Clad. gougerotii		25–37			
Clad. mansoni		30–32			
Clad. sphaerospermum	< 18				
Clad. trichoides		30			
Clad. trionumi	6	22–24	30		
Clad. lindemuthiamum	4	20–23	35		
Claviceps purpurea		22–26			
Cronartium ribicola	5	12–18	28		
Coccidiodes immitis	25	30–37	< 42	60	4
Corynebacterium acnes	30	37	< 45	60	10
Coryn. acnes		37			
Cryptococcus neoformans	< 17	25–30	40	50	10
Debaryomyces kloeckeri	3		37		
Deb. laedegaardi	5		37		
Dibotryon morbosum		37			
Diplodia zeae	10	30	35		
Diplocarpon rosae		16–21			
Endomyces pulmonalis	> 5	33–37	< 41		
Endothia parasitica	4	25–30	40		
Epidermophyton floccosum	< 18	27	30	50	10
Erysiphe graminis	5	15–20	29		
Fomes igniarius		30–32	42		
Fusarium cubense var. *oxysporium*	5	20–30	37.5		
F. lycopersici	5	24–30	38		
F. vasifectum	10	28–30	38		
Geotrichum candidum		25–37		56	60
Geo. issavi		22–25			
Gibberella zeae	3	20–30	37		
Glomerella gossypii					
Gymnosporangium juniperi-virginianae	6	20–25	32		
Hansenula anomala	0.5		38		
Helminthosporium gramineum		15–20			
Histoplasma capsulatum	10	25–30	40	55	15
H. farciminosum	15	25–37	40		
Hormodendrum compactum		25–37	37	100	15
Horm. dermatitidis		20–30	< 43	100	15
Horm. pedrosoi		25–37	37	100	15
Humicola grisea var. *thermoidea*	24		56		
H. insolens	23		55		

continued overleaf

Table 49 – (continued)

Species	Growth temperature (°C) Minimum	Optimum	Maximum	Thermal death Temperature (°C)	Time (min)
H. lanuginosa	30		60		
H. stellata	22		50		
Madurella grisea	30	30	60		
Malbranchea purchella var. *thermoidea*	24		56		
Microsporum audouini		25–30	38	60	80
Micro. canis		30–32	40	70	10
Micro. gypseum		25–30	38	60	80
Monilinia fructicola	0	24	32		
Mucor miehei	25		57		
M. pusillus	20		55		
Myciococeum albamyces	26		57		
Nocardia brasiliensis		20–37		60	60
N. caprae		33–37			
N. farcinica	>24	37		70	10
N. intracellularis		37.5		60	10
N. madarae	<20	37	40	60	5
N. paraguayensis		20–37		60	60
Peronospora effusa	3	8–10	30		
Per. parasitica		8–12	29		
Per. tabacina	1	15–23	29		
Phiaiephora jeanselmei		20–37			
Phi. verrecosa		37		100	15
Phomopsi citri	9	24–28	34		
Physalospora malorum	10	20	30		
Phyllostieta solitaria	5	25–30	35		
Phymatotrichum omnivorum	18	29	38		
Physoderma zeae-maydis	23	28–29	30		
Phythium debaryanum	5	27–30	35		
Phytophthera infestans	7	20	30		
Pityrosporium ovale	<22	37		60	30
Plasmodiophora brassicae	10	25–27	30		
Plasmopara vitacola	8	25–28	35		
Podosphaera leucotricha	10	19–20	28		
Pseudoperonospora cubensis		20			
Puccinia antirrhini	5	10	30		
Puc. coronata	2	17–22	35		
Puc. graminis	2	12–20	35		
Puc. triticina	2	16–20	31		
Rhizoctonia solani	8	31	40		
Rhizopus arrhizus	6	32.5–36.5	43		
Rhizopus equinus	>5	37–39		100	20
Rhizopus nigricans	7	25–30	35		
Rhizopus oryzae	7.5	31–34	45.5		
Schizophyllum commune	<16	30	>40		
Scopulariopsis brevicaulis	<6	20–25	37		
Septeria apii-graveolenus	14	22–24	25		
Solerotium cepivorum	5	20–24	29		
Sphaerotheoa humuli		37			
Spongospora subterranea	12	20–23	35		
Sporotrichum schenckii		30–37		59	5
Talaromyces duponti	27		59		
Taphrina deformans	10	20	30		
Thermoascus aurantiacus	22		55		
Thielaviopsis basicola	17		33		
Tilletia caries	4	14	29		
Torula thermophila	23		58		
Trichophyton concentricum		37			
Tric. ferrugineum		25		60	20
Tric. megnini		25		55	10
Tric. mentagrophytes	8	30	40	60–70	10
Tric. schoenleini	<15	33		60	10
Tric. tonsurans		30			
Tric. violaceum		25–30			
Trichosporon beigelii		30–37	<43		
Uromyces cariophyllinus	4	14	29		
Ustilago avenae	4–6	20–28	34–35		
U. tritiei	5–8	20–25	34–40		
U. zeae	8	25–30	40		
Venturia inaequalis	4	20	32		
Verticillium albo-atmim	8	21–26	31		

Algae

Table 50. Temperature tolerance extremes for algae [Spector (1956)].

Species	Temperature (°C) Minimum tolerated	Maximum tolerated	Maximum for habitat
Blue–green algae			
Anabaena sp.			40
Ana. varabilis			35
Anacystis nidulans			41
Anac. thermalis			42
Chroococcus sp.			57
C. minutis fiscus			46
C. yellowstonensis			41
Cylindrospermura stagnate			44.1
Gleocapsa stegophalia			38
Lyngbya sp.			65
L. cutealis	< −17		
Mastigocladus laminosus	−19		65
Microcystis elabens	< −17		
Nostoc kihlmani	< −17		
N. muscorum			32.5
N. sphaericum			30
Oscillatoria amphibia			50
O. filiformis	69	85.2	85.2
O. formosa			50
O. geminata			45
O. okem			44
O. proboscidea			47
O. tennis tergestina			44
Phormidium bijaheresis	38	85.2	60–62
P. geysericola	58	85	75
P. lominosum	58	85	75
P. tenue			47.2
Rivularia globiceps			26.4
Scytonema mirabile	< −17		
Synechococcus eximius	70	84	79
Synechoco. vulcanus	46	85	70
Synechoco. vulcanus bacillariodes	57	70	64
Synechocystis thermalis			62.2
Green algae			
Ankistrodesmus falcatus	< −17		
Chaetomorpha linum	−7 to −2	35	
Chlamydomonas nivalis	−36	10	
Chlamydothris thermalis		72–74	
Cladospora hamosa	< −7	36	
Clad. laetevirens	< −7	36	
Clad. prolifera	−7 to −2		30–35
Clad. rupestris	−20 to −16.8		
Cosmarium consperum	< −17		
Desmidium quadratum	< −17		
Enteromorpha intestinalis	−20 to −18		
Hastrum sublobatum	< −17		
Protococcus botyroides			80
Ulothrix sp.		17	
Golden-brown algae		40.2	
Yellow-green algae		32.5	

continued overleaf

Table 50 – (continued)

Species	Minimum tolerated	Temperature (°C) Maximum tolerated	Maximum for habitat
Brown algae			
Fucus vesiculosus	−20 to −18	30	
Red algae			
Bangia fuseopurpurea	−20 to −18		
Porphyra hiemalis	−20 to −18		

Temperature Changes

A shift in the incubation temperature from the optimum to a lower temperature results in a temperature-dependent reduction in the growth rate (*see* Fig. 60). The growth-reducing effect of a low temperature is rapidly reversible (*see* Fig. 61).

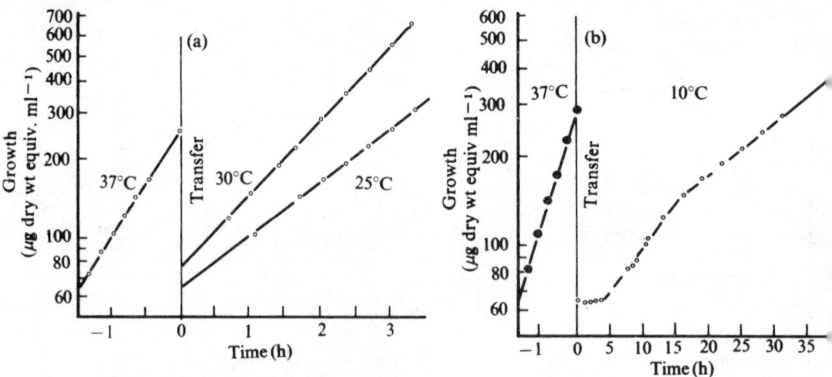

Figure 60. Effect on the growth of *Escherichia coli* ML 30 of a shift in incubation temperature from 37° to (a) 30° or 25°C, and (b) 10°C [Farrel and Rose (1967)].

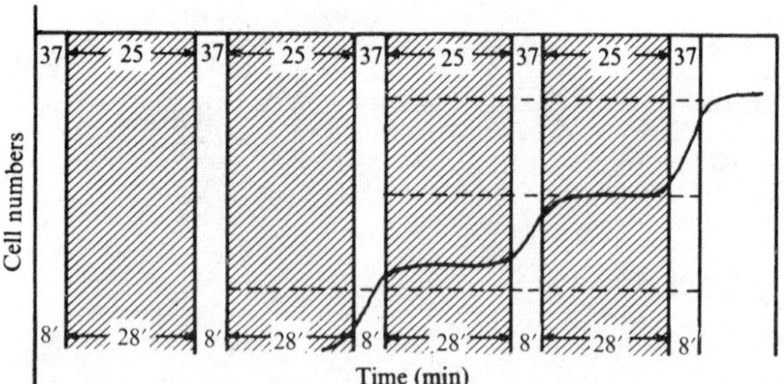

Figure 61. Growth of *Salmonella typhimurium* subjecting the culture to repeated shifts in incubation temperature from 37 (8 min) to 25°C (28 min) [Farrel and Rose (1967)].

Exposure to a lower temperature induces a lag phase (*see* Fig. 62). Fig. 63 shows the duration of this lag phase is directly related to the magnitude of the temperature shift.

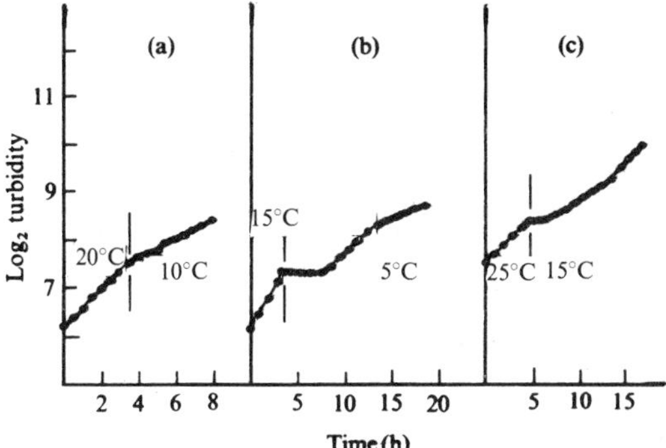

Figure 62. Effect of decreases in temperature on the growth of *Trichosporon pullulans* Y9. (a) 20 to 10°C, (b) 15 to 5°C and (c) 25 to 15°C [Shaw (1967)].

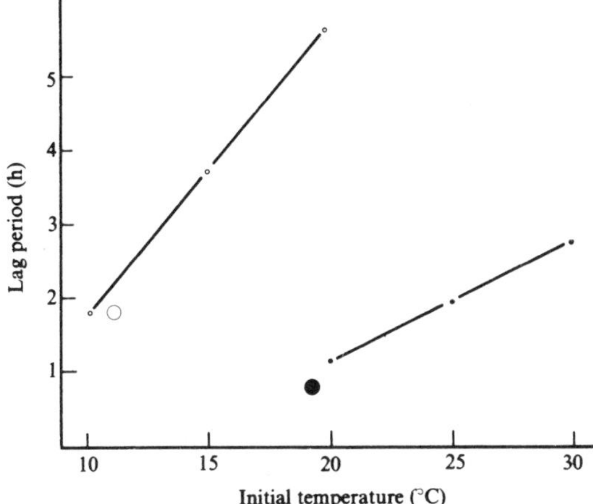

Figure 63. Effect of the magnitude of the temperature shift on the duration of the lag period in step-down experiments, ○ lag periods obtained with *Trichosporon pullulans* (Y9) on transferring to 5°C; ●, lag periods obtained with *Candida tropicalis* on transferring to 10°C [Shaw (1967)].

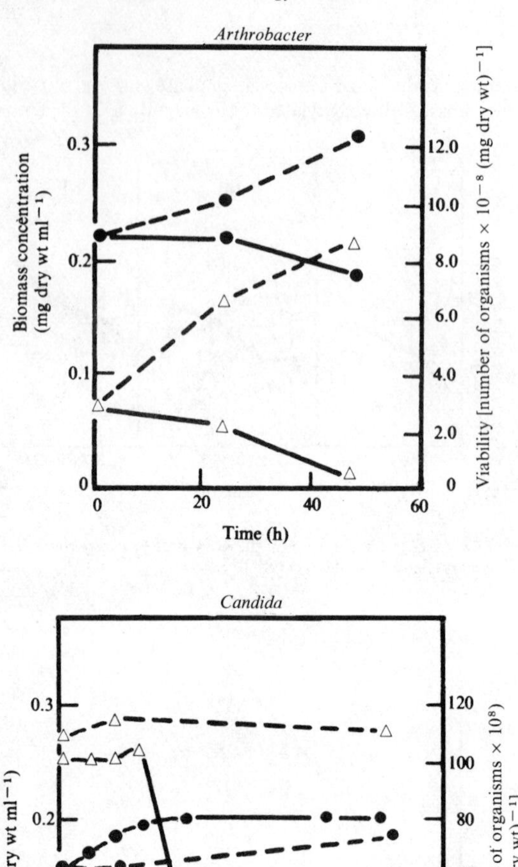
Arthrobacter
Biomass concentration (mg dry wt ml⁻¹)
Viability [number of organisms × 10⁻⁸ (mg dry wt)⁻¹]
0.3
0.2
0.1
0
12.0
10.0
8.0
6.0
4.0
2.0
0
0
20
40
60
Time (h)

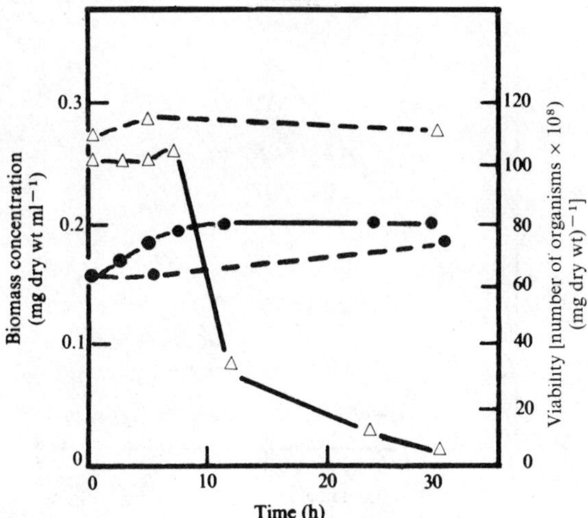
Candida
Biomass concentration (mg dry wt ml⁻¹)
Viability [number of organisms × 10⁸) (mg dry wt)⁻¹]
0.3
0.2
0.1
0
120
100
80
60
40
20
0
0
10
20
30
Time (h)

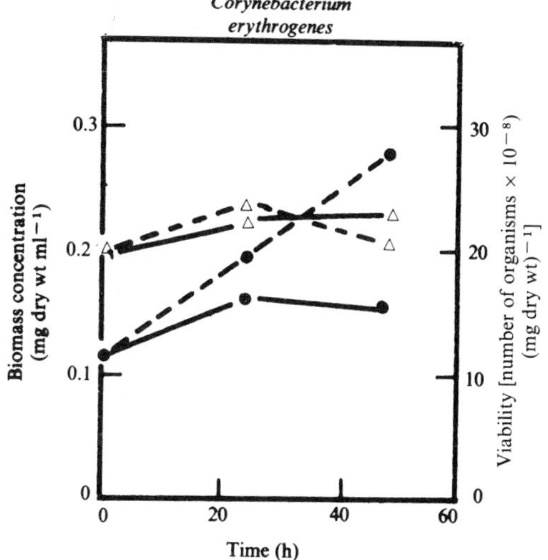

Figure 64. Effect of increase in incubation temperature on biomass concentration (○) and viability (△) of strains of *Arthrobacter, Candida* and *Corynebacterium erythrogenes*. Cultures of the *Arthrobacter* were transferred from 20 to 37°C after 72 h incubation; cultures of *Candida* from 10 to 25°C after 120 h; and *C. erythrogenes* cultures from 15 to 30°C after 120 h. The activities of organisms at the higher temperatures; ----, the activities at the lower temperatures [Farrel and Rose (1967)].

Fig. 64 shows that an increase in the incubation temperature can cause a reduction in both the biomass concentration and viability expressed as active cells per unit of dry biomass.

Nutrition

Tables 51 and 52 provide evidence for the loss of metabolic activity at extremes of temperature.

Table 51. Change in nutrient requirements at high temperature [Langridge (1963)].

Microorganism	Strain	Temperature[a] (°C)	Requirement, stimulant or blocked reaction
Bacteria			
Achromobacter fischeri		>29	Serine, methionine, arginine or glutamic acid
Bacillus coagulans	32	45/36	Desthiobiotin → biotin
B. coagulans	43P	45/36	Pimelic acid → desthiobiotin
B. coagulans	32	45/36	Histidine, folic acid
B. coagulans	1039	55/45	Histidine, methionine, tryptophan
B. globigii	1	45/36	Thiamin, nicotinic acid
B. megatherium	Lysogenic	55/37	Yeast extract, beef extract, protease peptone
B. sphaericus	RA-91	55/37	Yeast extract, beef extract, protease peptone
B. stearothermophilus	5274	55/45	Pimelic acid → desthiobiotin
B. stearothermophilis	1373B	55/45	Pyridoxal

continued overleaf

Table 51 – (continued)

Microorganism	Strain	Temperature[a] (°C)	Requirement, stimulant or blocked reaction
B. stearothermophilus	3084	55/45	Thiamin
B. stearothermophilus	4259	45/36	Methionine, histidine, nicotinic acid
B. stearothermophilus	39	75/55	Histidine, phenylalanine, glutamic acid
B. stearothermophilus	194	80/75	Methionine, leucine, phenylalanine
B. stearothermophilus	1503	65/55	Valine, isoleucine, leucine, glycine
B. subtilis	2 strains	55/37	Yeast extract, beef extract, protease peptone
Escherichia coli		44/37	Glutamic acid, nicotinamide
E. coli		37/24	Methionine or α-aminobenzoic acid
Lactobacillus arabinosus		37/26	Tyrosine
Pasteurella pestis	Virulent	37/27	Calcium
P. pestis	6 strains	36/28	Serine, threonine, biotin, pantothenic acid
P. pestis	6 strains	38/36	Glutamic acid, thiamin, alanine
P. pestis	14 strains	38/36	Glutamic acid, thiamin
Proteus vulgaris		40/25	Nicotinamide, thiamin or 1 of 12 amino acids
P. vulgaris		41/40	Thiamin or 1 of 10 amino acids
P. vulgaris		42/41	Thiamin or 1 of 3 amino acids
P. vulgaris		43/42	Glutamic acid or methionine
Pseudomonas sp.		37/20	Tryptic digest
Rhizobium trifolii	205	30/25	Pantothenic acid
Salmonella typhi		37/25	An organic nitrogen source
S. typhi		40/37	1 of 4 amino acids
S. typhi		43/40	Glutamic acid, arginine, thiamine or lysine
'Flat sour' organism		65/55	Biotin (usually)
Yeasts			
Cryptococcus sp.	Psychrophile	30/16	α-Ketoglutarate
Histoplasma capsulatum		37/25	Cysteine
H. capsulatum	8 strains	37/25	Thiamin
H. capsulatum	1 strain	37/25	Biotin
H. capsulatum	1 strain	37/25	Thioctic acid
Saccharomyces cerevisiae	2 strains	38/30	Pantothenic acid
Sacch. cerevisiae	ATCC2338	38/30	Yeast extract
Sacch. cerevisiae		40/30	Yeast extract
Saccharomycopsis guttulata		41/37	Magnesium
'Distillery type' yeast		38/24	Lipids
Filamentous Fungi			
Coprinus fimetorius		44/40	Methionine
Sclerotinia camelliae		22/10	Inositol
Flagellates			
Crithidia fasciculato		28/26	Glutamic, succinic, lactic acid, or Ca, Mg, Fe and Cu
Ochromonas malhamensis		>26	Vitamins, amino acids
Angiosperms			
Arabidopsis thaliana	P1 & BLA	30/26	Biotin
A. thaliana	LS	30/27	Cytidine
Lycopersicum esculentum		Non optimal	Nicotinic acid
Pisum sativum		High temperature	Sucrose and thiamin or ribosides

[a] The first temperature is that at which the requirement is manifest. The second is that at which the

Table 52. Change in nutrient requirements of bacteria at low temperatures [Ingraham (1962)].

Microorganism	Strain	Temperature[a] (°C)	Requirement, stimulant or blocked reaction
Bacillus coagulans	12	36/45	Valine
B. stearothermophilus	6 strains	37/55	Tryptose and basamin
B. stearothermophilus	2184–2	36/45	Basamin
B. stearothermophilus	3690	36/45	Leucine, nicotinic acid
B. stearothermophilus	5149–5	36/45	Methionine, thiamin
B. stearothermophilus	3084	45/55	Pyridoxamine

[a] The first temperature is that at which the requirement is manifest. The second is that at which the requirement is not manifest.

Table 53. Carbohydrate and protein content in *Escherichia coli* grown at 10 and at 37°C [Ng (1969)].

Growth temperature (°C)	Protein (%)	Carbohydrate (%)	Carbohydrate: protein ratio
10	59.5	21.9	0.37
37	65.5	8.4	0.13

A reduction in the incubation temperature from the optimum reduces slightly the protein content of bacteria but significantly increases the carbohydrate content (*see* Table 53).

Although the cell yield in relation to substrate utilized (i.e. Y_s) is not significantly affected by a reduction in incubation temperature from the optimum (*see* Table 54), Fig. 65 shows that the growth rate and the rates of respiration of endogenous reserves and of exogenous glucose are reduced at the lower temperatures.

Table 54. Effect of growth temperature and dissolved oxygen tension on yields of cells grown under glucose or ammonia limitation[a] [Brown *et al.* (1969)].

Growth temperature (°C)	Dissolved oxygen tension (mm Hg)	$Y_{glucose}$[b]	NH_4^+ $Y_{ammonia}$[b]
30	75	0.54	25
	20	0.52	25
	5	0.45	20
	1	0.35	
	<1	0.19	
	75	0.55	26
	20	0.50	25
20	5	0.47	19
	1	0.32	
	<1	0.17	
	75	0.58	20
	20	0.47	18
15	5	0.48	18
	1	0.31	
	<1	0.17	

[a] Cells were grown at a rate of 0.1 hr^{-1}.
[b] Yields are expressed as grams (dry weight) of cells formed per gram of substrate (glucose or ammonium+) utilized. Values quoted are averages of duplicate measurements on at least two different cultures.

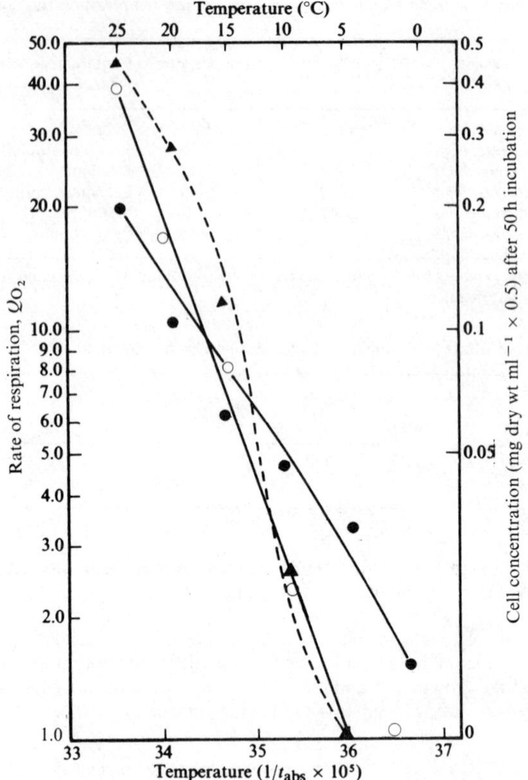

Figure 65. Effect of temperature on the rate of respiration of endogenous reserves (●) and of exogenous glucose (○) and cell concentration (▲) of *Candida utilis* NCYC 321 [Farrel and Rose (1967)].

Cell Composition and Yield

Table 55 shows that a reduction in the incubation temperature produces a change in the fatty acid composition of cells. Comparison of Tables 55 and 56 shows that a variation in medium composition alters the fatty acid composition and its responses to temperature changes.

Table 55. Effect of growth temperature and dissolved oxygen tension on the fatty acid composition of *Candida utilis* grown under glucose limitation[a] [Brown *et al.* (1969)].

Growth temp (°C)	Dissolved oxygen tension (mm Hg)	Fatty acid composition (%)								Total C_{16} acids (%)	Total C_{17} acids (%)
		16:0	16:1	16:2	17:1	18:0	18:1	18:2	18:3		
30	75	12.3	2.3	<0.1	1.2	1.0	33.4	47.6	2.7	14.6	84.2
	30	12.6	2.1	<0.1	0.5	0.5	33.6	50.7	<0.1	14.7	84.8
	5	13.5	2.3	<0.1	0.5	0.5	35.5	47.6	<0.1	15.8	83.7
	1	13.8	4.4	<0.1	2.1	1.1	36.5	42.1	0.0	18.2	79.7
	<1	16.5	8.7	<0.1	2.5	2.5	31.8	38.1	0.0	25.2	72.3
20	75	11.1	4.5	<0.1	2.6	<0.1	27.0	42.9	11.9	15.6	81.8
	20	12.4	5.1	<0.1	2.0	<0.1	26.9	44.6	9.1	17.5	80.5
	5	12.0	5.3	<0.1	2.3	<0.1	28.0	45.6	6.7	17.3	80.4
	1	11.5	5.1	<0.1	1.3	1.3	36.4	44.6	0.0	16.6	82.1
	<1	14.6	9.4	<0.1	1.3	2.3	32.3	40.1	0.0	24.0	74.7
15	75	16.0	8.2	<0.1	<0.1	<0.1	25.5	32.8	17.3	24.2	75.8
	20	14.6	7.9	<0.1	<0.1	< 0.1	22.6	32.3	18.0	22.5	77.5
	5	15.7	8.1	<0.1	2.5	<0.1	22.6	32.3	18.0	23.8	73.7
	1	19.8	8.4	<0.1	1.6	5.6	30.2	34.2	<0.1	28.2	70.2
	<1	19.0	11.3	<0.1	1.6	4.3	37.2	26.6	<0.1	30.3	68.1

[a] Cells were grown at a rate of $0.1\ h^{-1}$

Table 56. Effect of growth temperature and dissolved oxygen tension on the fatty acid composition of *Candida utilis* grown under ammonia limitation[a] [Brown *et al.* (1969)].

Growth temp (°C)	Dissolved oxygen tension (mm Hg)	Fatty acid composition (%)								Total C_{16} acids (%)	Total C_{17} acids (%)
		16:0	16:1	16:2	17:1	18:0	18:1	18:2	18:3		
30	75	17.5	5.5	0.0	1.4	<0.1	41.8	34.4	<0.1	23.0	75.6
	20	14.7	3.4	0.0	<0.1	<0.1	33.8	48.2	<0.1	18.1	81.9
	5	15.1	3.4	0.0	2.4	2.9	29.0	45.0	<0.1	18.5	79.1
20	75	16.9	7.0	0.0	3.2	<0.1	27.4	36.8	8.6	23.9	72.9
	20	17.2	6.8	0.0	3.0	<0.1	27.6	37.9	7.6	24.0	73.0
	5	19.1	6.6	0.0	2.7	<0.1	27.7	37.0	6.7	25.7	71.6
15	75	13.9	7.1	0.0	1.0	<0.1	23.9	32.6	19.2	21.0	78.0
	20	14.8	6.9	0.0	1.1	<0.1	24.2	31.6	21.3	21.7	77.2
	5	15.6	7.5	0.0	<0.1	<0.1	23.3	34.0	20.1	23.1	76.9

[a] Cells were grown at a rate of 0.1 hr^{-1}

When yeast cells are maintained at an above-optimum temperature, in the case of *Candida* sp., the rate of respiration decreases with time of exposure to heat. *Sacch. cerevisiae*, however, is a mesophilic yeast and its rate of respiration is not reduced after exposure to 35°C for up to three hours (*see* Table 57).

Table 57. Effect of heating of cell suspensions of *Candida* sp. P16 and *Saccharomyces cerevisiae* at 35°C on the fermentation and oxidation of glucose [Stokes (1971)].

Organism	Heated at 35° (min)	Q_{CO_2} [a]	Q_{O_2} [a]
Candida sp. P16	0	73	67
	20	3	48
	40	0	37
	60	0	11
Sacch. cerevisiae	0	101	100
	60	107	
	180	113	107

[a] Values for endogenous respiration have been subtracted.

An increase in the incubation temperature induced the changes in the properties of the enzyme content of thermophilic bacterial cells shown in Table 58.

Table 58. Effect of temperature on amylase composition of cells [Ingraham (1962)].

	Bacillus stearothermophilus		*Bacillus coagulans*	
	35°C	55°C	35°C	55°C
Nitrogen (%)	12.0	12.9	13.7	14.0
Optimum temperature	45–55	60–70	45–55	60–70
Stability at 90°C[a]	90	10	92	6
pH stability range	5.0–9.5	5.0–9.5	5.0–9.5	5.0–9.5
Optimum pH	6.5–8.0	6.5–8.0	6.5–8.0	6.5–9.0
Activation by Cl^-	+	+	+	+

[a] Percentage loss in 1 h.

Enzyme Activity

An increase in temperature above the optimum induces a temperature- and exposure time-dependent decrease in enzyme activity (*see* Table 59). In thermophilic bacteria, some enzymes have retained their activity at high temperatures. A summary of enzymes with high thermal stability is presented in Table 60.

Table 59. Death of cells and loss of enzyme activity of *Bacillus cereus* with heat [Ingraham (1962)].

Temperature (°C)	Time of exposure (min)	Percentage of cells or enzymic activity remaining		
		Plate count/ml.	Catalase activity	Dehydrogenase activity
40	10	1.1	80	86
40	20	0.04	65	86
40	40	0.002	48	67
40	80	0.00001	38	46
50	5	2.4		89
50	10	0.006	56	59
50	20	0.00002	56	9
50	40	0.000002	48	

Table 60. Enzymes with unusually high thermal stability from thermophilic bacteria [Langridge (1963)].

Microorganism	Enzyme
Bacillus sp.	α-Amylase, apyrase, ATPase, cellobiase, cytochrome oxidase, malic dehydrogenase, pyrophosphatase
B. coagulans	Amylase
B. stearothermophilus	Aldolase, amylase, maltase
B. subtilis	Malic dehydrogenase
B. terminalis (spores)	Alanine racemase

pH EFFECTS

Many microorganisms display an optimum pH for growth at around 7, with the majority favouring the pH range 5 to 8. However, there are exceptions including acetic acid bacteria oxidizing ethanol to acetic acid, thiobacilli which oxidize sulphur to sulphuric acid and, at the other extreme, urea-decomposing bacteria, many of which cannot grow below pH 8. In addition, numerous algae live in natural waters above pH 10. These alkali-resistant algae have cytoplasmic membranes impermeable to H^+ and OH^-.

Fig. 66 shows a typical growth–pH relationship, the sensitivity to pH being similar at different temperatures. However, pH can have a marked effect on the enzyme content/activity of a microorganism (*see* Fig. 67).

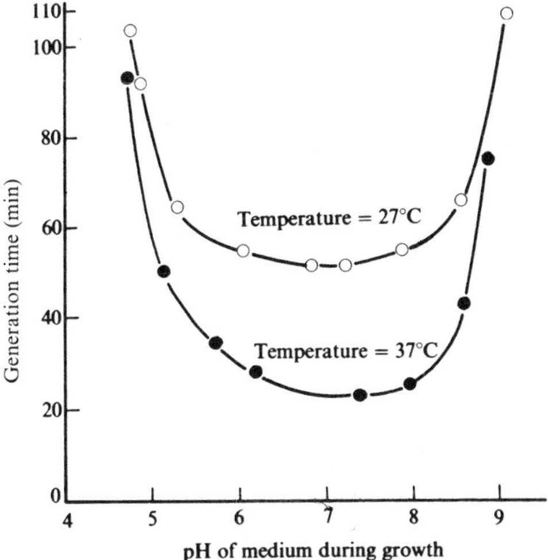

Figure 66. Variation in generation time with pH in *Escherichia coli* [Norris and Ribbons (1970)].

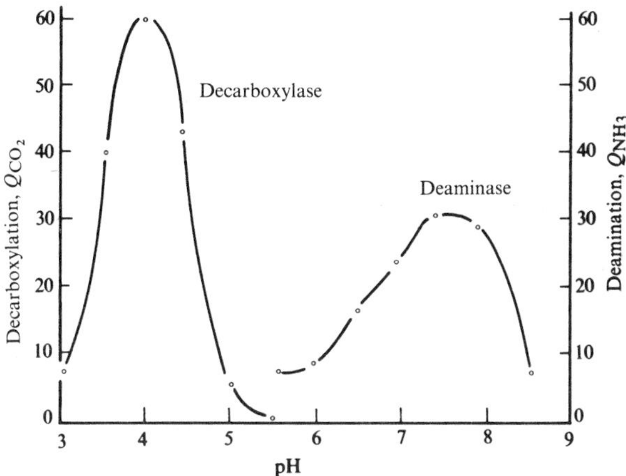

Figure 67. Variation with pH of the activities of the enzymes of *Escherichia coli* which metabolize L-glutamic acid [Norris and Ribbons (1970)].

Studies using automatic pH control demonstrate another distinct effect of pH on the enzyme composition of *Escherichia coli* (*see* Table 61). At pH 6.2, carbon dioxide and hydrogen are produced from glucose fermentation, whereas at pH 7.8 the production of these gases is inhibited due to the inactivation of formic hydrogen lyase and an equivalent amount of formic acid is formed instead. The formation of at least a part of the hydrogen lyase system is suppressed at high pH.

Table 61. Mixed acid fermentation of *Escherichia coli* [Norris and Ribbons (1970)].

Products	Amount formed [m mol (mol glucose fermented)$^{-1}$] pH 6.2	pH 7.8
Acetic acid	3.65	3.87
Acetoin	0.006	0.019
2, 3-Butanediol	0.030	0.026
Carbon dioxide	8.80	0.175
Ethanol	4.98	5.05
Formic acid	0.243	8.60
Glycerol	0.142	0.032
Hydrogen	7.50	0.026
Lactic acid	7.95	7.00
Succinic acid	1.07	1.48
Percentage carbon recovered	91.2	94.7

The products of ethanol fermentation from glucose shows how different pH values affect the metabolic activities of yeast (*see* Table 62).

Table 62. Ethanol fermentation using yeast [Norris and Ribbons (1970)].

Product	Amount formed [m mol (mol glucose fermented)$^{-1}$]				
	pH 3.0	pH 6.0	pH 6.0[a]	pH 7.6	pH 7.6[a]
Acetic acid	0.052	0.403	0.427	0.151	0.916
Acetoin	0.000	0.000	0.000	0.019	0.001
2,3-Butanediol	0.075	0.063	0.039	0.068	0.033
Butyric acid	0.013	0.036	0.039	0.021	0.035
Carbon dioxide	18.08	17.70	17.80	14.85	16.78
Ethanol	17.15	16.05	16.59	12.99	14.80
Formic acid	0.036	0.082	0.046	0.049	0.043
Glycerol	0.616	1.62	1.04	3.23	2.51
Lactic acid	0.082	0.163	0.173	0.137	0.087
Succinic acid	0.053	0.049	0.114	0.068	0.043
Glucose carbon assimilated	1.24	1.24			
Fermentation time, h	29.0	15.5	16.0	25.0	32.0
Percentage glucose fermented	98.5	98.0	98.5	60.3	98.1
Percentage glucose recovered	93.8	96.4	94.0	91.3	94.1

[a] Automatic pH control.

A further example of an ethanol fermentation is given in Table 63. Because of pH, there is a reduction in the ATP produced per mole of glucose fermented and this accounts for the lower cell yield.

Table 63. *Aerobacter cloacae* grown anaerobically on glucose [Norris and Ribbons (1970)].

pH	Generation time (h)	Total cell yield (g)	Products formed [m mol (mol glucose fermented)$^{-1}$]						ATP (mol)	Y_{ATP} [g (mol ATP)$^{-1}$]
			Acetic acid	Acetoin	2,3-Butanediol	Ethanol	Lactic acid	Succinic acid		
5.0 ± 0.1	4	1.77	0.50	0.00	3.88	6.18	0.00	0.33	0.153	11.6
7.2 ± 0.1	2	2.58	6.95	0.15	0.23	6.75	0.36	0.64	0.224	11.5

Table 64 provides an example of how various fermentation conditions can affect the formation of a particular desired product (in this case, 2,5-diketo-D-gluconic acid) and how the pH at harvest provides a measure of the conversion efficiency.

Table 64. Effect of fermentation conditions on the conversion of D-glucose to 2,5-diketo-D-gluconic acid by *Aerobacter melanogenus* ATCC 9937[a] [Stroshane and Perlman (1977)].

Volume of medium in 2 ℓ flask (ml)	Incubation temperature (°C)	Shaking rate (rpm)	Oxygen uptake rate ($mMO_2\ \ell^{-1}\ h^{-1}$)	pH at harvest	Percentage conversion of glucose (%)	Other products
200	25	300	33.4	2.6	93	5-Keto-D-gluconic acid, 2-keto-D-gluconic acid
500	25	300	16.8	2.8	57	D-Gluconic acid
500	30	300	9.9	3.0	24	D-Gluconic acid
500	30	120	4.2	5.0	23	D-Gluconic acid

[a] Incubation for three days with 10 per cent inoculum in cotton-plugged shake flask.

REFERENCES

Aiba, S., Shoda, M. and Nagatani, M. (1968) Kinetics of Product Induction in Alcohol Fermentation. *Biotechnol. Bioengng*, **10**, 845.

Baig, I.A. and Hopton, J.W. (1969) Psychrophilic Properties and the Temperature Characteristic of Growth of Bacteria, *J. Bacteriol*, **100**, 552.

Bartholomew, W.H., Karow, E.O., Sfat, M.R. and Wilhelm, R.H. (1950) Oxygen Transfer and Agitation in Submerged Fermentations, *Ind. Engng Chem.*, **42**, 1801.

Brown, C.M. and Rose, A.H. (1969) Fatty Acid Composition of *Candida utilis* as *Affected* by Growth Temperature and Dissolved Oxygen Tension *J. Bacteriol.* **99**(2), 317.

Bylinkina, E.S. and Birukov, V.V. (1972) The Problem of Scale-Up in Antibiotic Biosynthesis. *Proceedings 4th International Fermentation Symposium*, p.105.

Chain, E.B., Gualandi, G. and Morisi, G. (1966) Aeration Studies. 4. Aeration Conditions in 3000-Litre Submerged Fermentations with Various Microorganisms. *Biotechnol. Bioengng*, **8**, 595.

Clark, D.S. and Lentz, C.P. (1961) Submerged Citric Acid Fermentation of Sugar Beet Molasses. Effect of Pressure and Recirculation of Oxygen. *Can. J. Microbiol.*, **7**, 477.

Contois, D.E. (1959) Kinetics of bacterial growth. Relationship between Population Density and Specific Growth Rate of Continuous Cultures. *J. Gen. Microbiol.*, **21**, 40.

Cowland, T.W. (1967) Some Effects of Aeration on the Continuous Fermentation of Hopped Wort. *J. Inst. Brew.*, **75**, 542.

Edwards, V.H. (1970) The Influence of High Substrate Concentration on Microbial Kinetics. *Biotechnol. Bioengng*, **12**, 679.

Farrel, J. and Rose, A.H. (1967) Temperature Effects on Microorganisms, in *Thermobiology*, ed. by A.H. Rose, p.147 (Academic Press, New York).

Finn, R.K. (1954) Agitation–Aeration in the Laboratory and in Industry, *Bacteriol. Rev.*, **18**, 254.

Finn, R.K. (1967), in *Biochemical and Biological Engineering Science*, vol. 1, ed. by N. Blakebrough, p.69 (Academic Press, New York).

Gaden, E.L. jun. (1961) Aeration and Agitation in Fermentation, *Sci. Rpt. Super Sanita*, **1**, 161.

Gaudy, A.F. and Gaudy, E.T. (1972). Mixed Microbial Populations, in *Advances in Biochemical Engineering*, vol. 2, ed. T.K. Ghose *et al.*, p.97 (Springer-Verlag, Berlin).

Gunsalus, I.C. and Stainer, R.Y. (eds.) (1962). *The Bacteria*, vol. 4, The Physiology of Growth (Academic Press, New York).

Harrison, D.E.F., MacLennan, D.G. and Pirt, S.J. (1968) Responses of Bacteria to Dissolved Oxygen Tension, *Proceedings 3rd International Fermentation Symposium*, p.117.

Harrison, D.E.F. and Pirt, S.J. (1967) The Influence of Dissolved Oxygen Concentration on the Respiration and Glucose Metabolism of *Klebsiella aerogenes* during Growth. *Appl. Microbiol.*, **16**, 193.

Herbert, D. (1961) *11th Symposium of the Society for General Microbiology*, p.395.

Hockenhull, D.J.D. (1960) The Biochemistry of Streptomycin Production, in *Progress in Industrial Microbiology*, vol. 2, (Heywood, London).

Ingraham, J.L. (1958) Growth of Psychrophilic Bacteria, *J. Bacteriol.*, **76**, 75.

Ingraham, J.L. (1962) Temperature Relationships, in The Bacteria, vol. 4, The Physiology of Growth, ed. by I.C. Gunsalus and R.Y. Stainer, p.265 (Academic Press, New York).

Johnson, M.J. (1967) Aerobic Microbial Growth at Low Oxygen Concentrations, *J. Bacteriol.*, **94**, 101.

Langridge, J. (1963) Biochemical Aspects of Temperature Response, *Ann. Rev. Plant Physiol.*, **14**, 441.

Luedeking, R. (1967) Fermentation Process Kinetics, in Biochemical and Biological Engineering Science, vol. 1, ed. by N. Blakebrough (Academic Press, New York.).

Ng, N. (1969) Effect of Decreasing Growth Temperature on Cell Yield of *Escherichia coli*. *J. Bacteriol.*, **98**, 232.

Nixon, I.S. and Calam, C.T. (1968) Problems of applying Theory to Industrial Fermentation Development. *Chemy Ind.*, May, 604.

Norris, J.R. and Ribbons, D.W. (eds.) (1970) Methods in Microbiology, vol. 2, (Academic Press, New York).

O'Donovan, G.A., Kearney, C.L. and Ingraham, J.L. (1965) Mutants of *Escherichia coli* with High Minimal Temperatures of Growth, *J. Bacteriol.*, **90**, 611.

Phillips, D.H. and Johnson, M.J. (1961) Oxygen Transfer in Fermentation, *Sci. Rpt. Super Sanita*, **1**, 140.

Rahn, O. and Richardson, G.L. (1941) Oxygen Demand and Oxygen Supply, *J. Bacteriol*, **41**, 225.

Rivière, J. (1977) Industrial Applications of Microbiology, (Sussex University Press, London).

Sato, K. (1961) Rheological Studies of Some Fermentation Broths, in Kanamycin and Streptomycin Fermentation. *J. Ferment. Technol.*, **39**, 347.

Shaw, M.K. (1967) Effect of Abrupt Temperature Shift on the Growth of Mesophilic and Psychrophilic Yeasts, *J. Bacteriol.*, **93**, 1332.

Singh, K., Agarush, P.N. and Peterson, W.H. (1948) Influence of Aeration and Agitation in the Yield, Protein and Vitamin Content of Food Yeasts. *Arch. Biochem.*, **18**, 181.

Spector, W.S. (ed.) (1956) *Handbook of Biological Data* (W.B. Saunders, Philadelphia).

Stroshane, R.M. and Perlman, D. (1977) Fermentation of Glucose by *Acetobacter melanogenus*. *Biotechnol. Bioengng*, **19**, 459.

Stokes, J.L. (1971) Influence of Temperature on the Growth and Metabolism of Yeasts, in *The Yeast*, vol. 2, ed. A.H. Rose and D.E.F. Harrison, p.119 (Academic Press, New York).

Tsao, G.T. and Kempe, L.K. (1960) Oxygen Transfer in Fermentation System, in Use of Gluconic Acid Fermentation for Determination of Instantaneous Oxygen Transfer Rates. *J. Biochem. Microbiol. Technol. Engng*, **2**, 129.

Wang, D.I.C., Cooney, C.L., Demain, A.L., Dunnill, P., Humphrey, A.E. and Lilly, M.D. (1979) *Fermentation and Enzyme Technology* (John Wiley, New York).

APPENDIX: DETERMINATION OF μ_{max}, R_{max} AND K_m FROM A SINGLE SET OF BATCH CULTURE DATA

The equations describing the time course of batch fermentations when Monod kinetics apply are given on p.205 as

$$-\frac{1}{x}\frac{ds}{dt} = R = R_{max}\frac{s}{K_{m+s}} \qquad \text{(A1)}$$

$$\frac{1}{x}\frac{dx}{dt} = \mu = \mu_{max}\frac{s}{K_{m+s}} \qquad \text{(A2)}$$

At high and low substrate concentrations (i.e. when $K_m/s \ll 1$ and $K_m/s \gg 1$, respectively), Eqs. (A1) and (A2) are greatly simplified and asymptotic solutions result. These asymptotes can be used to determine μ_{max}, R_{max} and K_m, although values of low accuracy are to be expected. Of the asymptotes that involve Eq. (A2), when $K_m/s \ll 1$, the most commonly used leads to

$$x = x_0 \exp(\mu_{max} t) \qquad \text{(A3)}$$

which describes the observed exponential growth of a batch culture (*see* p.216).

A problem associated with the use of Eq. (A3) results from the fact that there is no means of checking whether the value obtained for μ_{max} is the true value. In general, values of μ_{max} obtained using Eq. (A3) in conjunction with experimental data are found to be low. Additionally, Eq. (A3) provides no information on K_m, R_{max}, etc. Only if the yield coefficient Y_s is constant over the course of the fermentation can Eqs. (A1) and (A2) be solved simultaneously.

In view of the above consideration, any preferred method of calculation should use simultaneously the whole time courses of the substrate and biomass concentrations and make no assumptions regarding the constancy of the yield coefficient.

Eqs. (A1) and (A2) are applicable from the beginning of the exponential growth phase, i.e. $t = 0, x = x_0, s = s_0$, until the beginning of the stationary phase, i.e. $dx/dt = 0, s \longrightarrow 0$ or $ds/dt = 0$. Integration of Eq. (A1) within this interval leads to

$$\frac{K_m}{s_o} \ln \frac{1}{1-F} + F = \frac{R_{max}}{s_o} \int_o^t x \, dt \qquad \text{(A4)}$$

where

$$F = \frac{s_0 - s}{s}$$

To proceed further with Eq. (A4), an analytical expression is required for $x(t)$ for the specified portion of the biomass-concentration curve. Knights (1981) devised such an expression [*see* Eq. (A5)], based upon the earlier work of Edwards and Wilkie (1968), Konak (1975) and La Motta (1976), i.e. the so-called logistic equation.

$$X = \frac{e^\tau}{1 - X_m + X_m e^\tau} \qquad \text{(A5)}$$

where $X = x/x_0$, $\tau = \mu_{max}\, t$, $X_m = x/x'_{max}$, x'_{max} is a specified maximum biomass concentration. The logistic equation [Eq. (A5)] reduces to Eq. (A3) when $X_m \longrightarrow 0$. Fig. A1, on which Eq. (A5) has been plotted, shows that the curves simulate the characteristic sigmoidal shape of batch culture curves and describe the complete range of conditions from experimental to zero growth.

Combining Eqs. (A4) and (A5) followed by integration leads to

$$\frac{K_m}{s_0} \ln \frac{1}{1-F} + F = \frac{R_{max} x_0}{\mu_{max} s_0 X_m} \ln \left[X_m \exp(\mu_{max}\, t) - X_m + 1\right]$$

or

$$F_f = \frac{R_{max}\, x_0}{\mu_{max}\, s_0\, X_m} T \qquad \text{(A6)}$$

Eqs. (A5) and (A6) provide a description of simultaneous microbial mass and substrate concentration data and are the basis for the determination of μ_{max}, R_{max} and K_m without any requirement with regards to the constancy of Y_s.

Calculation Procedure

The methodology developed by Knights (1981) can be used provided that μ_{max}, R_{max} and K_m are constant and the logistic equation describes the observed time course of microbial mass concentration. Under these circumstances, a straight-line relationship passing through the origin results when F_f is plotted against T [Eq. (A6)]. This linearity provides a basis for determining the optimum values of the parameters, such that F_f versus T is linear or that the deviation from linearity is minimal.

The algebraic detail of Eq. (A6) suggests that the optimization parameters should be μ_{max} and K_m/s_0. These parameters fulfill the requirements of biological significance and invariance.

The initial test range for the parameters can be based upon the use of Eq. (A3) to determine an approximate value of μmax corresponding to K_m/s_0 as 0.05.
In general

$$\mu_{max.n+1} = \mu_{max.n}(1 + E_g)$$
$$(K_m/s_0)_{j+1} = (K_m/s_0)_j + E_s \qquad \text{(A7)}$$

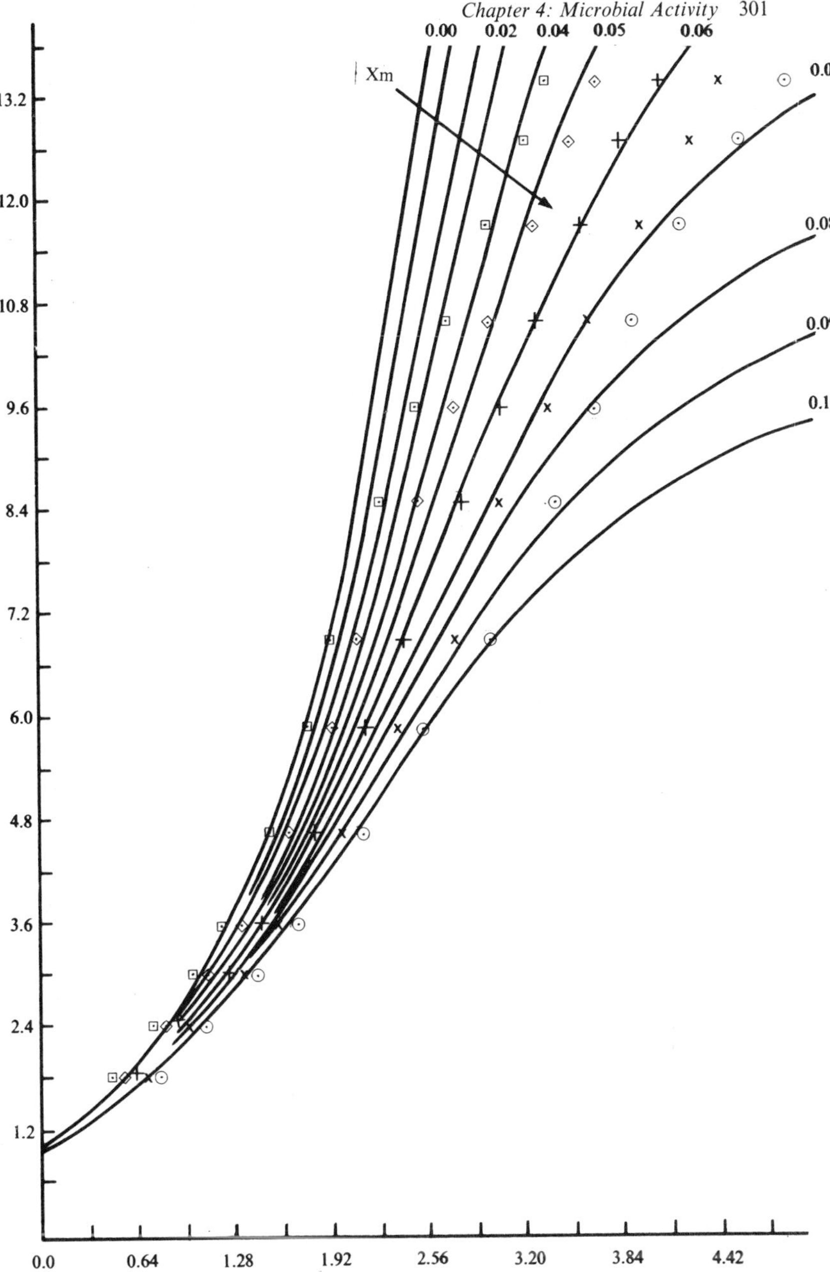

FIGURE A1. Determination of X_m by comparison of the experimental microbial mass data for each $\mu_{max,n}$ with the logistic equation. ⊡ $\mu_{max} = 0.243\ h^{-1}$, $X_m = 0.020$; ◈, $\mu_{max} = 0.268\ h^{-1}$, $X_m = 0.045$; +, $\mu_{max} = 0.296$, $X_m = 0.060$; ×, $\mu_{max} = 0.327\ h^{-1}$, $X_m = 0.075$; ⊙, $\mu_{max} = 0.352$, $X_m = 0.090$.

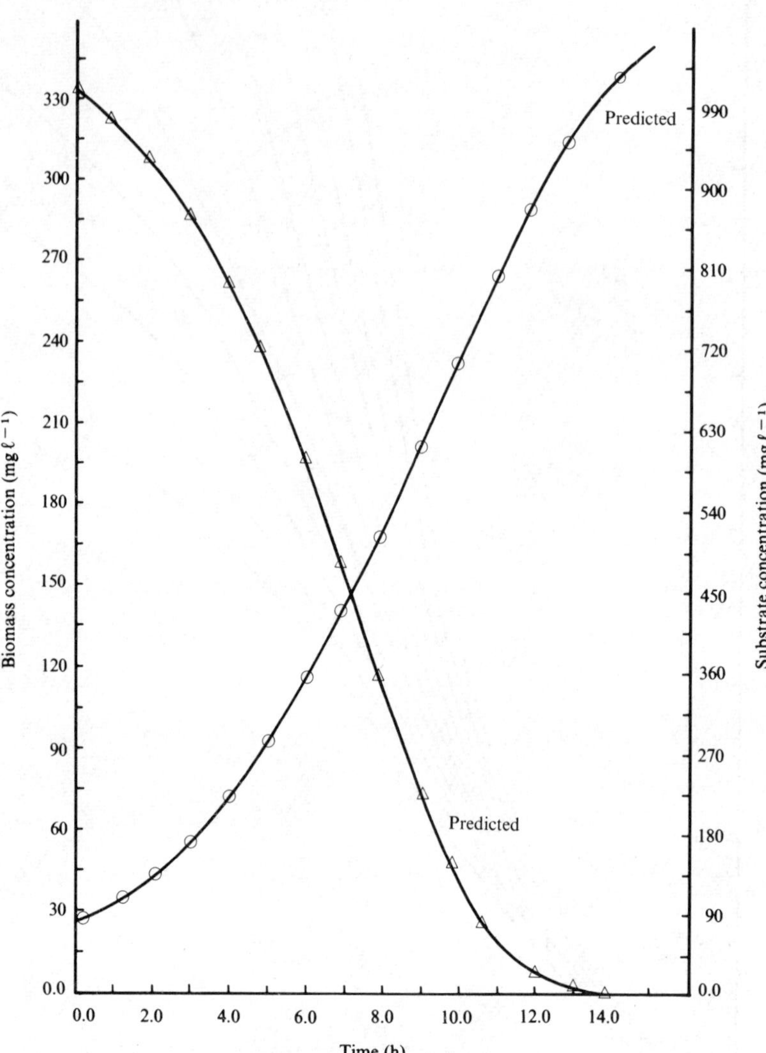

FIGURE A2. Comparison of time-course concentration data predicted using the values given in Eqs. (A9) with the simulated experimental data. ⊙, biomass concentration; data △, substrate concentration data. $x_0 = 25.0\,\text{mg}\,\ell^{-1}$. $s_0 = 1000\,\text{mg}\,\ell^{-1}$.

where E_g and E_s are arbitrary increments.

A simulated set of batch data is given in Fig. A2. This data, in conjunction with Eq. (A3), gives $\mu_{max} = 0.243\,\text{h}^{-1}$, and using this value together with Eq. (A7) suggests the ranges

$$\mu_{max,n} = 0.243, 0.268, 0.296, 0.327, 0.352\,\text{h}^{-1}$$

$$(K_m/s_0)_j = 0.05, 0.10, 0.20, 0.30, 0.40, 0.50, 0.60, 0.70$$

In practice, the number of values tested depends upon the capacity to carry out linear regressions.

Table A1. Dimensionless limiting substrate time-course data [calculated using Eqs. (A6) etc.].

$\mu_{max}(h^{-1})$	0.243	0.268	0.296	0.327	0.352	K_m/s_0	0.05	0.10	0.20	0.30	0.40	0.50	0.60	0.70
X_m	0.020	0.045	0.060	0.075	0.090									
t (h)			T			F				F_f				
0.0	0.0000	0.0000	0.0000	0.0000	0.0000	0.0000	0.0000	0.0000	0.0000	0.0000	0.0000	0.0000	0.0000	0.0000
1.	0.0055	0.0137	0.0205	0.0286	0.0373	0.0350	0.0368	0.0386	0.0421	0.0457	0.0493	0.0528	0.0564	0.0599
2.	0.0124	0.0314	0.0473	0.0669	0.0880	0.0800	0.0842	0.0883	0.0967	0.1050	0.1134	0.1217	0.1300	0.1384
3.	0.0212	0.0541	0.0823	0.1178	0.1559	0.1380	0.1454	0.1529	0.1677	0.1826	0.1974	0.2123	0.2271	0.2420
4.	0.0323	0.0829	0.1276	0.1843	0.2452	0.2110	0.2228	0.2347	0.2584	0.2821	0.3058	0.3295	0.3532	0.3769
5.	0.0463	0.1194	0.1853	0.2698	0.3599	0.3000	0.3178	0.3357	0.3713	0.4070	0.4427	0.4783	0.5140	0.5497
6.	0.0639	0.1652	0.2500	0.3774	0.5031	0.4070	0.4331	0.4593	0.5115	0.5638	0.6160	0.6683	0.7205	0.7728
7.	0.0858	0.2221	0.3481	0.5098	0.6768	0.5280	0.5655	0.6031	0.6782	0.7532	0.8283	0.9034	0.9785	1.0535
8.	0.1131	0.2919	0.4578	0.6684	0.8812	0.6560	0.7094	0.7627	0.8694	0.9761	1.0828	1.1896	1.2963	1.4030
9.	0.1468	0.3764	0.5885	0.8535	1.1146	0.7780	0.8533	0.9285	1.0790	1.2295	1.3800	1.5305	1.6810	1.8316
10.	0.1883	0.4771	0.7409	1.0641	1.3739	0.8770	0.9818	1.0866	1.2961	1.5057	1.7152	1.9248	2.1343	2.3439
11.	0.2388	0.5950	0.9150	1.2979	1.6551	0.9435	1.0872	1.2309	1.5182	1.8056	2.0929	2.3803	2.6676	2.9550
12.	0.2996	0.7308	1.1095	1.5518	1.9541	0.9780	1.1688	1.3597	1.7413	2.1320	2.5047	2.8864	3.2680	3.6497
13.	0.3722	0.8844	1.3228	1.8226	2.2671	0.9925	1.2371	1.4818	1.9711	2.4604	2.9496	3.4389	3.9282	4.4175
14.	0.4577	1.0551	1.5525	2.1071	2.5907	0.9977	1.3014	1.6052	2.2127	2.8208	3.4276	4.0351	4.6426	5.2501

Fig. A3. Optimization rectangle.

		K_m/s_0												
		0.05	0.10	0.20	0.30	0.40	0.50	0.60	0.70	(2.87)	(6.32)			
	0.243	6073	3845	1752	889	480	269	154	90	13552	1882	ELIM at EFR 4		
	0.268	4659	2730	1022	392	139	39	7	7	8992	584	577	ELIM at EFR 2	
μ_{max} (h^{-1})	0.296	3898	2152	679	192	33	1	18	57	7030	301*	244	34*	Less than EFR 2
	0.327	3126	1588	380	54	2	44	118	201	5513	419	218*	46	
	0.352	2478	1136	179	3	50	164	291	414	4715*	922	508	ELIM at EFR2	
	(39.14)	20234	11451	4012	1530	704	517*	588	769	Initial Totals	Total on EFC5	Total on EFC3	Total on EFC2	
					641	224*	248	434	679	Total on EFR4				
					246	35*	45	136	Total on EFR2					
		ELIM at EFC5			ELIM at EFC2	Less than EFC2		ELIM at EFC2	ELIM at EFC2					

* Indicates a minimum in row or column.

The time course of the logistic equation depends upon μ_{max} and X_m. Accordingly, there is for a given set of experimental data a value of X_m corresponding to each assumed value of μ_{max}. In practice, the values of X_m are determined by comparison of the experimental data expressed as x/x_0 and $\mu_{max,n}\,t$ with the logistic equation (*see* Fig. A1).

The simulated experimental data of Fig. A2 combined with $\mu_{max,n}$, $X_{m,n}$ and $(K_m/s_0)_j$ using the definitions of F_f and T [Eq. (A6)] are given in Table A1. The optimization of μ_{max} and K_m/s_0 is achieved by linear regression of the F_f versus T data. The optimum solution occurs when the divergence of the correlation coefficient r from unity is a minimum. The results of the linear regressions expressed as $|(1.0 - r)| \times 10^3$ is given in Fig. A3 (termed the optimization rectangle).

In Fig. A3, the initial total of the rows and columns have been calculated and the ratios of the maximum to the minimum values of the totals given, i.e. 2.87 for rows and 39.14 for columns. An exclusion factor for the columns (EFC) of five is adopted and the three columns $K_m/s_0 = 0.05, 0.10$ and 0.20 eliminated since the initial total/minimum ratio was greater than five for these columns. The row totals were recalculated and gave a maximum/minimum ratio of 6.32. An exclusion factor for rows (EFR) of four was adopted and the row corresponding to $\mu_{max} = 0.243\ \text{h}^{-1}$ eliminated. This procedure was repeated, as shown in Fig. A3, until a reduced matrix was obtained consisting of

$$\begin{aligned} &0.296 < \mu_{max} < 0.327\ \text{h}^{-1} \\ &0.40 < K_m/s_0 < 0.50 \end{aligned} \tag{A8}$$

A second optimization rectangle with values of $\mu_{max,n}$ and $(K_m/s_0)_j$ in the regions defined by Eq. (A8) resulted in a narrower band of values

$$0.296 < \mu_{max} < 0.303\ \text{h}^{-1}$$

$$0.475 < K_m/s_0 < 0.525$$

The corresponding values for R_{max} and Y_s are given in Table A2, with the probable values of all the parameters

$$\begin{aligned} \mu_{max} &= 0.300\ \text{h}^{-1}\ (\pm 1.2\%) \\ R_{max} &= 1.823\ \text{h}^{-1}\ (\pm 5.1\%) \\ K_m &= 500\ \text{mg l}^{-1}\ (\pm 5.0\%) \\ Y_s &= 0.165\ (\pm 6.1\%) \end{aligned} \tag{A9}$$

The values of the parameters given in Eqs. (A9) have been inserted into Eqs. (A5) and (A6) to provide predictions of $x(t)$ and $s(t)$, the resulting data are superimposed on Fig. A2 showing the efficiency of the methodology.

Table A2. Range of values of R_{max} and Y_s[a]

μ_{max}	K_m/s_0	0.475	0.525
0.296	R_{max} (h^{-1})	1.780	1.915
0.296	Y_s	0.166	0.155
0.303	R_{max} (h^{-1})	1.731	1.862
0.303	Y_s	0.175	0.163

[a] $Y_s = \mu_{max}/R_{max}$.

References

Edwards, V. H. and Wilkie, C. R. (1968) Mathematical Representation of Batch Culture Data. *Biotechnol. Bioengng,* **10,** 205.

Knights, A. J. (1981) *Determination of the Biological Kinetic Parameters of Fermenter Design from Batch Culture Data.* Ph.D. Thesis, University of Wales.

Konak, A. R. (1975) An Equation for Batch Bacterial Growth. *Biotechnol. Bioengng,* **17,** 271.

La Motta, E. J. (1976) Kinetics of Continuous Growth Culture using the Logistic Growth Curve. *Biotechnol. Bioengng,* **18,** 1029.

CHAPTER 5

PRODUCT FORMATION

INTRODUCTION

The reaction or fermentation stages of biological processes are considered in this chapter. Table 1 lists the topics that receive attention for any product and the relevant chapters. In contrast to this chapter, Chapter 14 covers the overall process of which the fermentation stage is a crucial part, while Chapter 12 contains information on product recovery. Wherever possible information is given on concentration levels, conversion efficiencies and material balances, together with appropriate flow sheets. General relationships between this chapter and Chapters 12 and 14 are included in Table 1.

Table 1. Interrelationships between Chapters 5, 12 and 14.

Topic	Chapter
Raw materials and their preparation	14
Fermentation	
Medium	5, 14
Inoculum	5, 14
Metabolic pathway	5
Time course of fermentation	5
Broth conditions at time of harvest	14
Product recovery	12, 14
Product preparation for point of sale	14

ACETONE/BUTANOL

Inoculum

The microorganisms used to produce acetone/butanol mixtures are *Clostridium* species, in particular *Cl. acetobutylicum*, *Cl. toanum*, *Cl. saccharobutylacetonicum-liquefaciens* (code C-12), *Cl. celerifactor* and *Cl. madisonii*. These microorganisms produce true spores viable for several years and, consequently, culture collections are maintained in this form.

Media

Spores

The spores of *Clostridium* species are maintained on a mixture of dried soil, sand and volcanic ash or a carbonate salt. The mixture is screened and 250 grams is added to a flask (500 ml), then moistened with water and sterilized for 3-hour periods at 15 psig steam pressure four times at 24-hour intervals. A young culture is added aseptically to the dried sterile soil mixture until it is moderately wet, with the flask then being placed in a desiccator. The resulting soil/spore mixture is used to prepare the inoculum for laboratory and plant fermentations.

Clostridium saccharobutylacetonicum-liquefaciens

A peeled potato medium used for laboratory fermentation or for the inoculum of large-scale processes is prepared by blending 250 grams of potatoes, 5 grams of glucose and 2 grams of calcium carbonate with sufficient water to make 1 litre of medium. After sterilization, the medium is inoculated with spores and incubated at 31°C. Cultures grown on such a medium are used as inoculum for molasses or aided corn fermentations.

Clostridium acetobutylicum

For unaided corn fermentation, when using *Cl. acetobutylicum*, the medium is made with 5 per cent dry corn equivalent in water. The medium is then inoculated with spores and incubated at 37°C. The inoculum medium for aided molasses fermentation contains 5.6 per cent sucrose, 0.3 per cent ammonium sulphate, 0.3 per cent calcium carbonate and 0.01 per cent phosphorus pentoxide (as superphosphate). The fermentation is anaerobic, the gases produced being sufficient to sparge air from the fermenter. All of the microorganisms have a maximum tolerance of approximately 1.2 per cent (by vol.) for butanol. The initial pH of the medium is about 5.4–6.0.

Metabolic Pathway

The conversion of glucose to acetone and butanol via acetoacetyl-CoA is outlined in Fig. 1.

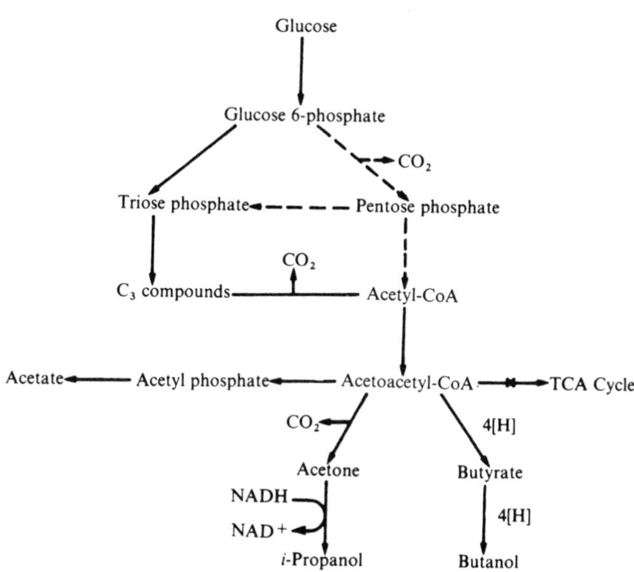

Figure 1. Metabolic pathway for the formation of acetone and butanol.——→, major metabolic routes; – – –→ secondary metabolic pathways.

Time Course

Fig. 2 shows that butyric acid is initially formed from glucose. The acid is then reduced with the formation of butanol, the alcohol formation being related directly to the decline in acid levels. Similarly, the accumulation of acetone is associated with a decline in acetic acid levels.

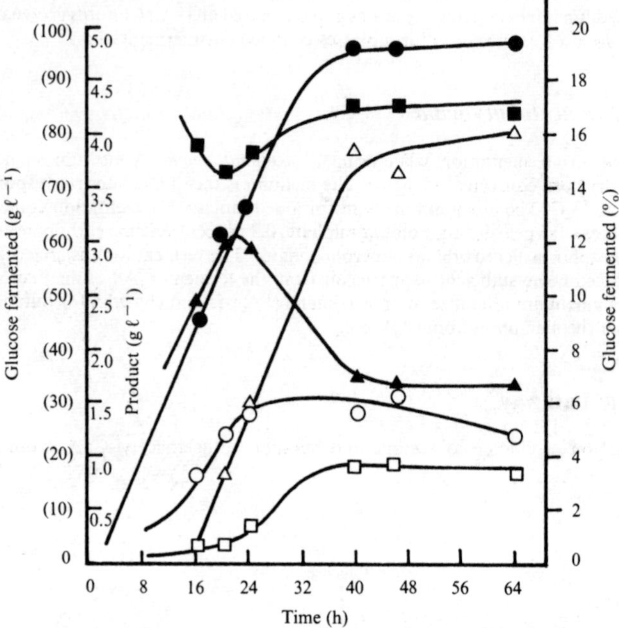

Figure 2. Fermentation of glucose by *Clostridium acetobutylicum* and time course of reaction. ●, glucose; ■, pH; △, butanol; ▲, butyric acid; ○, acetic acid; □, acetone [Davies and Stephenson (1941)].

Table 2. Glucose fermentation balances of three typical *Clostridium* species [Rainbow and Rose (1963)].

Products	Product formation [mM (mM glucose fermented)$^{-1}$]		
	Cl. butyricum	*Cl. acetobutylicum*	*Cl. butylicum*
Butyric acid	76	4	17
Acetic acid	42	14	17
Ethanol		7	
Butanol		56	59
Acetone		22	
i-Propanol			12
Acetoin		6	
Carbon dioxide	188	221	204
Hydrogen	235	135	78
Total carbon (%)	96	100	96

The total yield of acetone and butanol increases with incubation time (*see* Table 3). Table 3 also shows that relatively higher butanol yields are achieved at 32° rather than 26°C; the converse is true for acetone.

Table 3. Effect of the duration of incubation and the incubation temperature on the formation of acetone and butanol [Steel (1958)].

Fermentation time (h)	Temperature (°C)	Total solvents yield (mg ℓ$^{-1}$)	Acetone yield (mg ℓ$^{-1}$)	Butanol yield (mg ℓ$^{-1}$)
17	32	183	38	136
17	26	184	35	140
18	32	186	41	138
18	26	189	36	144

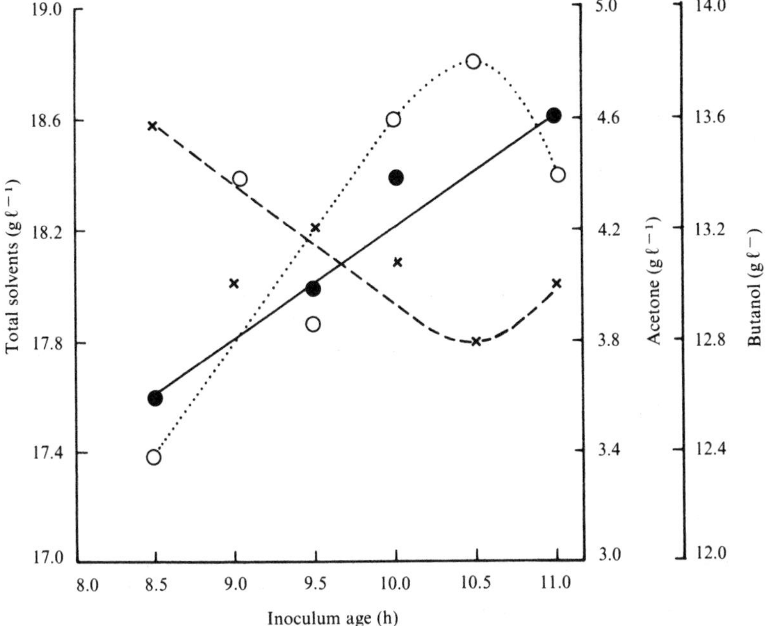

Figure 3. Change of product with age of inoculum cream; initial sucrose concentration 6.0 per cent. ●, total solvents; ○, butanol; ×, acetone. [Steel (1958)].

From Fig. 3, it can be seen that the ratio of acetone to butanol formation from sucrose varies with the inoculum age. Acetone production is maximal and butanol production minimal when the inoculum is 8.5 hours old, while butanol production is maximal and acetone minimal with a 10.5 hour inoculum.

Under continuous culture conditions, acetone production is higher and occurs at a faster rate when the rate of fermentation is slow (*see* Fig. 4). Also in slow fermentations, the inhibitory effect of added butanol is more marked (*see* Fig. 5).

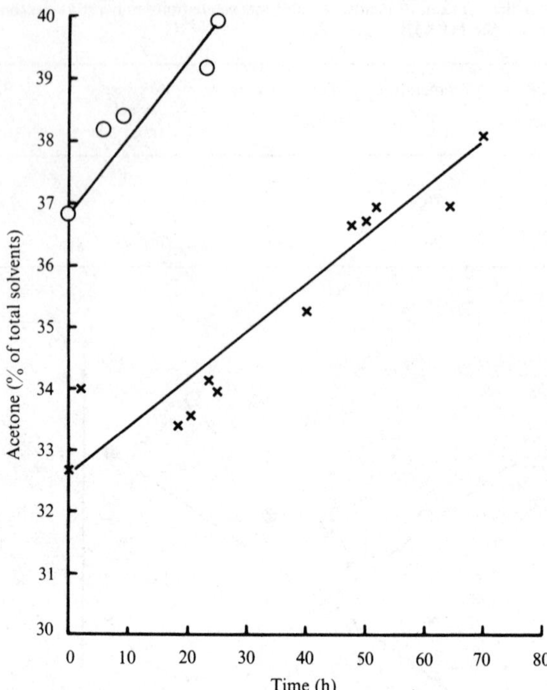

Figure 4. Change of acetone production in continuous fermentations. ○, slow fermentation; ×, fast fermentation [Steel (1958)].

Table 4 and Figs. 6 and 7, respectively, show that relative acetone yields are increased by addition of ammonium sulphate, fluoride and arsenite. Arsenite, which is an inhibitor of the tricarboxylic acid cycle, is more effective when added 16 hours after incubation has commenced.

Table 4. Effect of nitrogen addition on acetone/butanol fermentations [Steel (1958)].

Ammonium sulphate added [g (100 g sugar)$^{-1}$]	Final concentration ($g\ \ell^{-1}$) Ammonia	Total solvents	Acetone	Butanol
4.0	0.00	18.7	5.9	11.8
5.0	0.00	18.8	6.2	11.7
6.0	0.10	18.3	6.6	10.7

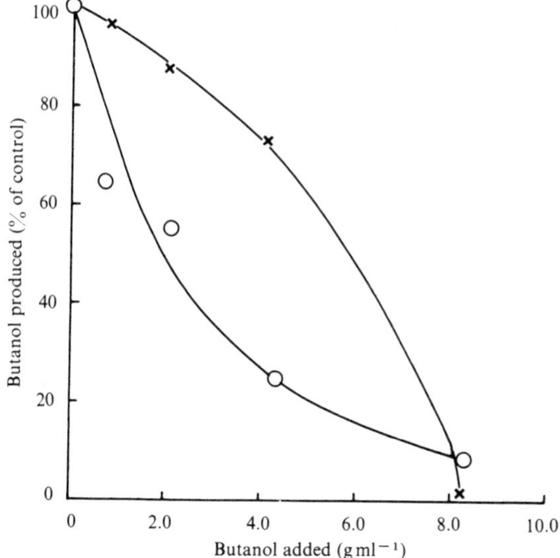

Figure 5. Inhibitory effect of added butanol on fermentations. ○, slow fermentation; ×, fast fermentation [Steel (1958)].

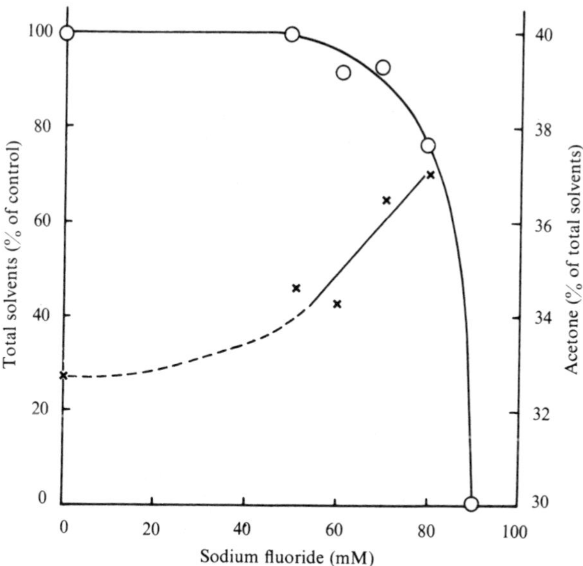

Figure 6. Effect of fluoride on fermentation. ○, total solvents; ×, acetone [(Steel 1958)].

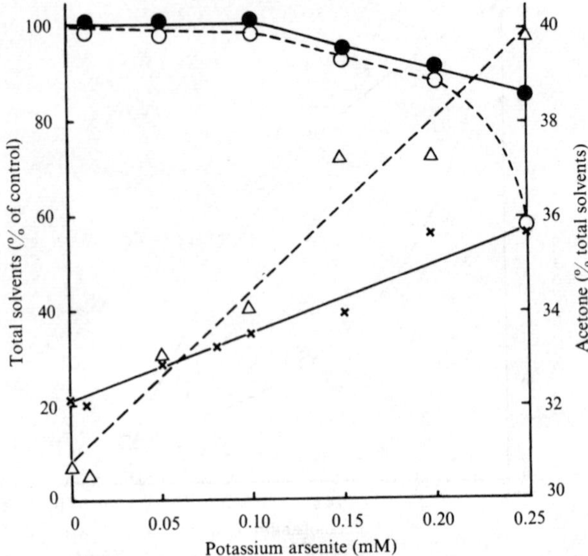

Figure 7. Effect of potassium arsenite on fermentation. ●, Total solvents, arsenite added at seeding; ○, total solvents, arsenite added after 15 hours; ×, acetone, arsenite added at seeding; △, acetone, arsenite added after 15 hours [Steel (1958)].

AMINO ACIDS

General Aspects

Microorganisms

Normally, amino acid synthesis just satisfies the metabolic demand. The cellular accumulation of an amino acid is an abnormal phenomenon occurring in only some wild

Table 5. Amino acids produced by auxotrophic mutants [Hirose and Shibai (1980)].

Amino acids	Microorganism	Requirement
L-Citrulline	*Bacillus subtilis*	Arginine
	Corynebacterium glutamicum	Arginine
L-Glutamic acid	*Brevibacterium thiogentalis*	Oleic acid
L-Leucine	*Coryn. glutamicum*	Histidine, phenylalanine
L-Lysine	*Coryn. glutamicum*	Homoserine
	Brev. flavum	Threonine
L-Ornithine	*Arthrobacter citreus*	Citrulline
L-Phenylalanine	*Coryn. glutamicum*	Tyrosine
L-Proline	*Brevibacterium* sp.	Histidine
	Brev. flavum	Isoleucine
	Kurthia catenoforma	Serine
L-Threonine	*Escherichia coli*	Diaminopimelate, methionine, valine
L-Tyrosine	*Coryn. glutamicum*	Phenylalanine, purine
L-Valine	*Coryn. glutamicum*	Isoleucine, leucine

strains, auxotrophic mutants and regulatory mutants. The amino acids synthesized by wild strains are only those occurring in amphibolic metabolic pathways (i.e. those pathways involved in both biosynthesis and energy production), e.g., L-alanine, L-glutamic acid and L-valine. Some wild strains capable of accumulating L-glutamic acid are given in Table 16. Table 5 lists some auxotrophic mutants producing various amino acids along with their growth requirements, while some of the regulatory mutant microorganisms that accumulate amino acids are presented in Table 6. Certain amino acids can be produced from intermediates of the corresponding amino acid, examples of the microorganisms capable of such conversions being given in Table 7.

Table 6. Amino acids produced by regulatory mutants [Hirose and Shibai (1980)].

Amino acids	Microorganism	Genetic marker (analogue resistance)
L-Arginine	*Brevibacterium flavum*	2-Thiazolealanine
	Corynebacterium glutamicum	Arginine hydroxamate, D-serine, D-arginine
L-Histidine	*Brev. flavum*	2-Thiazolealanine, sulphaguanidine, 2-aminobenzothiazole, ethionine, threonine
	Coryn. glutamicum	1, 2, 4-Triazolealanine, 6-meroaptoguanidine, 8-azaguanidine, 4-thiouracit, 6-mercaptopurine, 5-methyltryptophan
L-Isoleucine	*Brev. flavum*	α-Amino-β-hydroxyvalerate, *O*-methylthreonine
L-Leucine	*Brev. flavum*	2-Thiazolealanine, Met$^-$, Ile$^-$
L-Lysine	*Brev. flavum*	2-Aminoethyl-L-cysteine, α-Chlorocaprolactam
L-Phenylalanine	*Brev. flavum*	*m*-Fluorophenylalanine, *p*-fluorophenylalanine, *p*-aminophenylalanine, Tyr$^-$
L-Serine	*Coryn. glutamicum*	*O*-Methylserine, α-methylserine, isoserine
L-Threonine	*Brev. flavum*	α-Amino-β-hydroxyvalerate, Met$^-$
L-Tryptophan	*Brev. flavum*	5-Fluorotryptophan, 3-fluorotryptophan, Phe$^-$
	Coryn. glutamicum	5-Methyltryptophan, 6-fluorotryptophan, 4-methyltryptophan, *p*-ethylphenylalanine, *p*-aminophenylalanine, tyrosine hydroxamate, phenylalanine hydroxamate, tryptophan hydroxamate
L-Tyrosine	*Brev. flavum*	*m*-Fluorophenylalanine, 3-aminotyrosine, *p*-aminophenylalanine, *p*-ethylphenylalanine, tyrosine hydroxamate
L-Valine	*Brev. flavum*	2-Thiazolealanine

Table 7. Microbial production of amino acids from intermediates of the corresponding amino acids [Yamada (1977)].

Amino acids	Microorganism	Intermediate	Yield ($g\,\ell^{-1}$)
L-Isoleucine	*Bifidobacterium ruminale*	α-Bromobutyrate	
	Bif. ruminale	2-Hydroxybutyrate	
	Corynebacterium amagasaki	D-Threonine	15
	Serratia marcescens	α-Aminobutyric acid	8
L-Methionine	*Pseudomonas denitrificans*	2-Hydroxy-4-methylthiobutyric acid	11
L-Phenylalanine	*Ps. denitrificans*	2-Hydroxy-3-phenylpropionic acid	
L-Serine	*Coryn. glycinophilum*	Glycine	16
L-Tryptophan	*Claviceps purpurens*	Indole	13
	Hansenula anomala	Anthranilic acid	3

Table 8. Amino acids produced by enzymic methods [Hirose and Okada (1979)].

Amino acid	Enzyme	Enzyme source	Reaction	Product concentration ($g\ \ell^{-1}$)	Yield (%)
L-Alanine	Aspartate decarboxylase	*Pseudomonas dacunhae*	L-Aspartate → L-alanine + CO_2	268	100
L-Aspartate	Aspartate	*Escherichia coli*	Fumarate + NH_4^+ → L-asparate	560	99
L-Cysteine	DL2-Aminothiazoline-4-carboxylate racemase, L-2-Aminothiazoline-4-carboxylate hydrolase, *s*-carbamoyl-L-cysteine hydrolase	*Ps. thiazolinophilum*	DL-2-Aminothiazoline-4-carboxylate + $2H_2O$ → L-cysteine + CO_2 + NH_3	30	95
	Cysteine desulphahydrase	*Aerobacter aerogenes*	β-Chloro-L-alanine + Na_2S → L-cysteine + NaCl + NaOH	49	89
L-3-(3,4-Dihydroxyphenyl)-L-alanine (L-DOPA)	Tyrosine phenol lyase	*Erwinia herbicola*	Pyrocatechol + pyruvate + NH_4^+ → L-DOPA + H_2O	59	95
5-Hydroxy-L-tryptophan	Tryptophanase	*Proteus rettgeri*	5-Hydroxyindole + pyruvate + $NH^+{}_4$ → 5-hydroxy-L-tryptophan + H_2O	28	60
L-Lysine	D-Caprolactam racemase, L-Caprolactam hydrolase	*Achromobacter obae* *Cryptococcus laurentii*	DL-Aminocaprolactam + H_2O → L-lysine	100	100
L-Phenylalanine	D-Phenylalanine hydantoin racemase, L-phenylalanine hydantoin hydrolase, *N*-carbamoyl-L-phenylalanine hydrolase	*Flavobacterium aminogenes*	DL-Phenylalanine hydantoin + $2H_2O$ → L-phenylalanine + CO_2 + NH_3	50	100
L-Tryptophan	Tryptophanase	*P. rettgeri*	Indole + pyruvate + NH_4^+ → L-tryptophan + H_2O	63	98
	D-Tryptophan hydantoin racemase, L-tryptophan hydrolase, *N*-carbamoyl-L-tryptophan hydrolase	*Flav. aminogenes*	DL-Tryptophan hydantoin + $2H_2O$ → L-tryptophan + CO_2 + NH_3	50	100
L-Tyrosine	Tyrosine phenol lyase	*Er. herbicola*	Phenol + pyruvate + NH_4^+ → L-tyrosine + H_2O	61	94

Another approach to amino acid production is by microbial transformation, using intact cells or microbial enzymes extracted from these cells. A summary of such enzymic methods is given in Table 8.

Table 9. Microbial production of amino acids from various carbon sources [Yamada (1977)].

Amino acid	Microorganism	Genotype[a]	Carbon source	Product concentration ($g\,\ell^{-1}$)	Yield (%)
L-Alanine	*Brevibacterium flavum*	Met^s	Glucose	14	
	Corynebacterium gelatinosum		Glucose	40	40
	Coryn. hydrocarboclastus		*n*-Alkane	4	
L-Arginine	*Brev. flavum*	Gu^-, TA^r	Acetic acid	26	10
			n-Alkane	9	8
			Glucose	29	29
L-Citrulline	*Arthrobacter paraffineus*		*n*-Alkanes	9	9
	Brev. flavum	Arg^-, SG^r	Acetic acid	15	14
			Glucose	30	27
L-Glutamic acid	*Arthro. paraffineus*		*n*-Alkane	62	
	Bacillus megaterium		Propylene glycol	27	27
	Brevibacterium sp.		Benzoic acid	80	80
			Ethanol	59	66
	Brev. flavum		Acetic acid	98	48
			Glucose	50	50
	Brev. pentosaaminoacidicum		Xylose	5	10
	Brev. thiogentalis		Acetic acid	51	51
	Coryn. alkanolyticum		*n*-Alkane	72	
	Coryn. glutamicum		Glucose	38	38
	Coryn. hydrocarboclastus		*n*-Alkane	84	
	Microbacterium ammoniaphilum		Glucose	29	29
L-Glutamine	*Brev. flavum*	SG^r	Acetic acid	25	15
		SG^r	Ethanol	35	38
		SG^r	Glucose	39	39
L-Histidine	*Brev. flavum*	TA^r, SM^r, Eth^r, ABT^r	Acetic acid	5	6
		TA^r, SM^r, Eth^r, ABT^r	Ethanol	6	5
			Glucose	10	10
	Coryn. glutamicum	1,2,4-Triazole alanine[r]	Glucose	8	8
L-Homoserine	*Corynebacterium* sp.	Thr^-	*n*-Alkane	12	15
	Coryn. glutamicum	Thr^-	Glucose	15	15
	Escherichia coli	Thr^-	Glycerol	4	8
L-Isoleucine	*Brev. flavum*	AHV^r	Acetic acid	15	11
		AHV^r, OMT^r	Glucose	15	15
	Microb. paraffinolyticus	Val^-, Leu^-	*n*-Alkane	1.6	
L-Leucine	*Brev. flavum*	TA^r	Ethanol	5	5
	Brev. lactofermentum	Ile^-, Met^-, TA^r	Glucose	28	21
	Coryn. hydrocarboclastus	Ile^-	*n*-Alkane	0.7	
L-Lysine	*Brev. flavum*	Ala^-, AEC^r	Acetic acid	61	
		AEC^r	Glucose	32	32
	Brev. lactofermentum	AEC^r	Ethanol	66	28
	Coryn. glutamicum	Hom^-, Leu^-,	Glucose	39	39
	Nocardia sp. No. 258	AEC^r	*n*-Alkane	34	34
L-Methionine	*Coryn. glutamicum*	Thr^-, Eth^r, $MetHx^r$	Glucose	2	2

continued overleaf

Table 9. (*continued*)

Amino acid	Microorganism	Genotype[a]	Carbon source	Product concentration ($g\,\ell^{-1}$)	Yield (%)
L-Ornithine	*Brev. flavum*	Arg$^-$	Acetic acid	30	25
	Coryn. glutamicum	Arg$^-$	Glucose	26	26
	Coryn. hydrocarboclastus	Arg$^-$	*n*-Alkane	9	32
L-Phenylalanine	*Arthro. paraffineus*	Tyr$^-$	*n*-Alkane	15	15
	B. subtilis	PFT^r	Glucose	6	7
	Brev. flavum	PFP^r	Ethanol	15	12
	Coryn. glutamicum	Tyr$^-$, PFP^r, PAP^r	Glucose	9	9
L-Proline	*Brev. flarum*	Ile$^-$, SG^r	Acetic acid	18	12
		Ile$^-$, SG^r	Glucose	29	29
	Coryn. acetoacidophilum	Ile$^-$	*n*-Alkane	2.2	
		TA^r	Ethanol	22	14
L-Serine	*Arthro paraffineus*		*n*-Alkane	3	
L-Threonine	*Arthro paraffineus*	Ile^{leaky}	*n*-Alkane	15	25
	Brev. flavum	Met$^-$, AHV^r	Acetic acid	27	14
		AHV^r	Ethanol	33	25
		Met$^-$, AHV^r	Glucose	18	18
L-Tryptophan	*B. subtilis*	FT^r, Leu$^-$	Glucose	6	7
	Brev. flavum	FT^r	Acetic acid	3	2
	Coryn. glutamicum	Phe$^-$, Tyr$^-$, MT^r, FT^r, PAP^r, PFP^r, TyrHx^r, PheHx^r	Glucose	12	12
L-Tyrosine	*Arthro. paraffineus*	Phe$^-$	*n*-Alkane	18	18
	Brev. flavum	PFT^r	Acetic acid	5	5
	Coryn. glutamicum	Phe$^-$, PFP^r, PAP^r, PAT^r, TyrHx^r	Glucose	17	17
L-Valine	*Brev. flavum*	TA^r	Acetic acid	20	15
	Brev. lactofermentum	TA^r	Glucose	23	29
	Coryn. acetoacidophilum	TA^r	Ethanol	22	14
	Coryn. hydrocarboclastus	Ile$^-$	*n*-Alkane	2	

[a]SG^r, sulphaguanidine resistant; TA^r, thiazolealanine resistant; Eth^r, ethionine resistant; MetHx^r, methionine hydroxamine resistant; AEC^r, S-(2-aminoethyl)-L-cysteine resistant; SM^r, sulphamethomidine resistant; ABT^r, 2-aminobenzthiazole resistant; AHV^r, α-amino-β-hydroxyvaleric acid resistant; Met^s, methionine sensitive; OMT^r , O-methylthreonine resistant; MT^r, 5-methyltryptophan resistant; FT^r, 5-fluorotryptophan resistant; PAP^r, *p*-aminophenylalanine resistant; TyrHx^r, tyrosinehydroxamine resistant; PheHx, phenylalaninehydroxamine resistant; PFP^r, *p*-fluorophenylalanine resistant; PAT^r, *p*-aminotyrosine resistant; PFT^r, *p*-fluorotyrosine resistant.

Medium

Various carbon sources and growth requirements for the microbial production of amino acids are listed in Tables 9 and 10. Biotin is essential for growth of wild strains and the effect of the biotin concentration on the levels of amino acid accumulated in *Coryn. glutamicum* is given in Table 11. Since amino acids contain nitrogen, an adequate supply of a suitable nitrogen source is important. Usually gaseous ammonia is used both for this purpose and to control the pH.

Because of high rates of sugar consumption, the oxygen demand is high during fermentation. Fig. 8 demonstrates the effect of the extent of oxygen satisfaction (i.e. the

Table 10. Relationships between amino acid accumulation and growth requirements [Yamada (1977)].

Amino acid accumulated	Microorganism	Substance required for growth
L-Citrulline	*Bacillus subtilis*	Arginine
	Brevibacterium flavum	Arginine
	Corynebacterium glutamicum	Arginine
L-Glutamic acid	*Brev. thiogentalis*	Oleic-acid
	Coryn. alkanolyticum	Glycerol
L-Homoserine	*Coryn. glutamicum*	Threonine
	Escherichia coli	Threonine
	Serratia marcescens	Threonine
L-Leucine	*Coryn. hydrocarboclastus*	Isoleucine
	Serratia marcescens	α-Aminobutyrate
L-Lysine	*Brev. flavum*	Threonine
	Coryn. glutamicum	Homoserine
	Coryn. glutamicum	Methionine + threonine
L-Methionine	*Ustilago maydie*	Leucine
L-Ornithine	*Arthrobacter citreus*	Arginine
	Coryn. glutamicum	Arginine
	E. coli	Arginine
L-Phenylalanine	*Coryn. glutamicum*	Tyrosine
	E. coli	Tyrosine
L-Proline	*Brevibacterium* sp.	Histidine
	Brev. flavum	Isoleucine
		Methionine
	Kurthia catenoforma	Serine
L-Threonine	*E. coli*	Diaminopimelate
		Methionine + valine
L-Valine	*Brevibacterium* sp.	Homoserine
	Coryn. glutamicum	Isoleucine
	Coryn. glutamicum	Leucine
	E. coli	Threonine

Table 11. Effect of biotin on amino acid levels in *Corynebacterium glutamicum* and in culture medium [Moss and Smith (1977)].

Amino acid	Concentration [μmol (g dry wt)$^{-1}$]			
	Biotin-rich medium		Biotin-poor medium	
	Medium	Cells	Medium	Cells
L-Alanine	254	32	1 019	2
L-Aspartic acid	21	0	73	0
L-Glutamic acid	85	49	15 420	3
L-Leucine	7	2	91	0
L-Valine	63	5	9	0

ratio of oxygen uptake rate to microbial oxygen demand) on the relative productivity of leucine, lysine and glutamic acid.

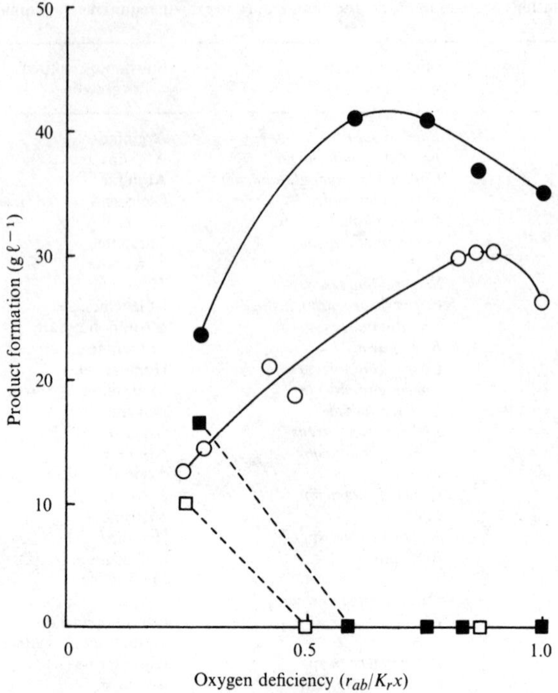

Figure 8. Relationship between microbial products and degree of oxygen deficiency ($r_{ab}/K_r x$) in valine and leucine fermentation, where r_{ab} is the rate of oxygen uptake [mol O_2 (g cell)$^{-1}$ ml^{-}], K_r [mol O_2 (g cell)$^{-1}$ min^{-1}] and x is the cell concentration (g ℓ$^{-1}$). ●, Valine; ○, leucine; ■, lactate in valine fermentation; □, lactate in leucine fermentation [Akashi *et al.* (1978)].

Properties and uses of amino acids

A summary of the chemical structure and physical properties of amino acids is presented in Table 12. An outline of the occurrence and uses of these amino acids is given in Table 13.

Table 12. Formulae and physical properties of amino acids.

Amino acid	Synonym	Formula	Properties
Alanine	α-Aminopropionic acid	$MeCH(NH_2)COOH$	Mol. wt, 89.1
L-Enantiomer			Decomp., 297°C; sp. gr., 1.401; $[\alpha]_D^{20}$ + 13° (5 N HCl); sol. H_2O; slightly sol. EtOH; insol. Et_2O
Arginine	1-Amino-4-guanidino-valeric acid	$HN = C(NH_2)(CH_2)_3CH(NH_2)COOH$	Mol. wt, 174.2
L-Enantiomer			Mp, 207°C; decomp., 244°C; monoclinic plates; $[\alpha]_D^{20}$ + 12.5° (water); sol. H_2O; slightly sol. EtOH; insol. Et_2O
Aspartic acid	α-Aminosuccinic acid	$HOOCCH_2CH(NH_2)COOH$	Mol. wt, 133.1
L-Enantiomer			Mp, 270–1°C; sp.gr., 1.661; orthorhombic plates; $[\alpha]_D^{20}$ + 25° (6 N HCl); sol. acid, alkali and salt soln.; slightly sol. H_2O; insol. EtOH, Et_2O
Citrulline	α-Amino-δ-ureidovaleric acid	$H_2NCONH(CH_2)_3CH(NH_2)COOH$	Mol. wt, 175.2
L-Enantiomer			Mp, 222°C; prisms; $[\alpha]_D^{20}$ + 3.7°; sol. H_2O; insol. MeOH, EtOH
Cysteine	2-Amino-3-mercaptopropionic acid	$HSCH_2CH(NH_2)COOH$	Mol. wt. 121.2
L-Enantiomer			$[\alpha]_D^{25}$ + 6.5° (5N HCl); sol. H_2O, EtOH, AcOH, Me_2CO, NH_4OH; insol. Et_2O, CCl_4, CS_2, EtAc
Glutamic acid	α-Aminoglutaric acid	$HOOC(CH_2)_2CH(NH_2)COOH$	Mol. wt, 147.1
L-Enantiomer			Decomp., 247–9°C; sublimes, 200°C; sp.gr., 1.538; orthorhombic crystals; $[\alpha]_D^{22.4}$ + 31.4° (6 N HCl); sol. H_2O; insol. MeOH, EtOH, Et_2O
D-Enantiomer			Shiny leaflets, $[\alpha]_D^{20}$ −30.5 (6 N HCl)
DL-Enantiomer			Decomp., 225–7°; sp.gr., 1.4601; orthorhombic crystals; sol. H_2O; insol. EtOH, Et_2O
Glutamine	2-Aminoglutamic acid	$HOOCCH(NH_2)(CH_2)_2CONH_2$	Mol. wt, 146.2
L-Enantiomer			Decomp., 185–6°C; needles; $[\alpha]_D^{23}$ + 6.1; sol. H_2O; insol. MeOH, EtOH
DL-Enantiomer			Mp, 185–6°C (or 173–4.5°C); prisms

continued overleaf

Table 12 (*continued*)

Amino acid	Synonym	Formula	Properties
Glutathione	γ-L-Glutamyl-L-cysteinyl glycine	$H_2NCH(COOH)(CH_2)_2$ CONH–$CH(CH_2SH)CONHCH_2COOH$	Mol. wt, 307; mp, 195°C; $[\alpha]^{27}_{D}$ −21°; sol. H_2O, acid, alkali; insol. EtOH, Et_2O, $CHCl_3$
Glycine	Aminoacetic acid	NH_2CH_2COOH	Mol. wt, 75.1; decomp., 233°C; sp.gr., 1.1607; monoclinic prisms; sol. H_2O; slightly sol. EtOH; insol. Et_2O
Histidine	α-Amino-4-imidazole-propionic acid	(imidazol-4-yl)$CH_2CH(NH_2)COOH$	Mol. wt, 307
L-Enantiomer			Mp, 175–85°C; decomp., 287°C; needles or plates; $[\alpha]^{20}_{D}$ −39.74; sol. H_2O; slightly sol. EtOH, insol. Et_2O
Homoserine	2-Amino-4-hydroxybutyric acid	$CH_2OHCH_2CH(NH_2)COOH$	Mol. wt, 119
L-Enantiomer			Decomp., 203°C; flat prisms; $[\alpha]^{26}_{D}$ −8.8° (H_2O)
Isoleucine	2-Amino-3-methylvaleric acid	$EtCH(Me)CH(NH_2)COOH$	Mol. wt, 131.2
L-Enantiomer			Sublimes, 168–70°C; decomp., 284°C; waxy, shiny, rhombic leaflets; $[\alpha]^{20}_{D}$ + 11.29 (H_2O); sol. H_2O; slightly sol. hot EtOH; insol. Et_2O
DL-Enantiomer			Decomp., 282°C; shiny rhombic or monoclinic plates; sol. H_2O
Leucine	2-Amino-4-methylvaleric acid	$Me_2CHCH_2CH(NH_2)COOH$	Mol. wt, 131.2
L-Enantiomer			Sublimes, 145–8°C; decomp., 293–5°C; shiny hexagonal plates; $[\alpha]^{25}_{D}$ − 10.8°; sol. H_2O, EtOH; insol. H_2O
DL-Enantiomer			Decomp., 332°C; leaflets
Lysine	2,6-Diaminohexanoic acid	$H_2N(CH_2)_4CH(NH_2)COOH$	Mol. wt, 146.2
L-Enantiomer			Decomp., 224.5°C; needles; $[\alpha]^{20}_{D}$ + 14.6(H_2O); sol. H_2O, EtOH; insol. Et_2O
Methionine	2-Amino-4-methylthiobutyric acid	$MeS(CH_2)_2CH(NH_2)COOH$	Mol. wt, 149.2
L-Enantiomer			Mp, 280–2°C; hexagonal plates; $[\alpha]^{25}_{D}$ − 8.2(H_2O); sol. H_2O, dil. EtOH; insol. abs. EtOH

DL-Enantiomer			Mp, 281°C; sp.gr., 1.340; plates; sol. H_2O, dil. acid, dil. alkali; slightly sol. abs. EtOH; insol. Et_2O
Ornithine	α-δ-Diaminovaleric acid	$H_2N(CH_2)_3CH(NH_2)COOH$	Mol. wt, 132.2
L-Enantiomer			Mp, 140°C; microcrystals; $[\alpha]_D^{25}$ + 11.5°(H_2O); sol. H_2O, EtOH; slightly sol. Et_2O
DL-Enantiomer			Slightly sol. EtOH
Phenylalanine	α-Amino-β-phenylpropionic acid	$PhCH_2CH(NH_2)COOH$	Mol. wt, 165.2
L-Enantiomer			Mp, 283°C; (decomp.); monoclinic flakes; $[\alpha]_D^{20}$ − 35.1°; slightly sol. H_2O, Et_2O, MeOH, EtOH
Proline	2-Pyrrolidinecarboxylic acid	(pyrrolidine ring, N) COOH	Mol. wt, 115.1
L-Enantiomer			Decomp. 220–2°C; plates; $[\alpha]_D^{20}$ − 52.6 (0.5MHCl); sol. H_2O, EtOH; insol. Et_2O, BuOH
Serine	2-Amino-3-hydroxypropionic acid	$HOCH_2CH(NH_2)COOH$	Mol. wt, 105.1
L-Enantiomer			Decomp., 228°C; hexagonal plates or prisms; $[\alpha]_D^{20}$ − 6.83(H_2O); sol. H_2O; insol. abs. EtOH, Et_2O
DL-Enantiomer			Decomp., 246°; sp.gr., 1.537; monoclinic leaflets; sol. H_2O; abs. EtOH, Et_2O
Threonine	2-Amino-3-hydroxybutyric acid	$MeCH(OH)CH(NH_2)COOH$	Mol. wt, 119.1
L-Enantiomer			Decomp., 255–7°C; $[\alpha]_D^{26}$ − 28.3°(H_2O); sol. H_2O; insol. abs. EtOH, Et_2O, $CHCl_3$
Tryptophan	α-Aminoindole-3-propionic acid	(indole ring, N) $CH_2CH(NH_2)COOH$	Mol. wt, 204.2
L-Enantiomer			Decomp. 289°C; leaflets or plates; $[\alpha]_D^{20}$ + 2.4° (0.5 N HCl); slightly sol. H_2O; insol. $CHCl_3$
Tyrosine	β-(p-Hydroxyphenyl)alanine	HO(benzene ring)$CH_2CH(NH_2)COOH$	Mol. wt, 181.2
L-Enantiomer			Decomp., 342–4°; sp.gr., 1.456; silky needles; $[\alpha]_D^{22}$ − 10.6(1 N HCl); sol. alkali; slightly sol. abs. EtOH, Et_2O
D-Enantiomer			Decomp., 310–4°C; $[\alpha]_D^{25}$ + 10.3° (1 N HCl); slightly sol. H_2O
DL-Enantiomer			Decomp., 316°C; stout needles; slightly sol. H_2O
Valine	2-Amino-3-methylbutyric acid	$Me_2CHCH(NH_2)COOH$	Mol. wt, 117.2
L-Enantiomer			Mp., 315°C; leaflets; $[\alpha]_D^{20}$ − 39.74°(H_2O); sol. H_2O; moderately sol. EtOH, Et_2O, Me_2CO

Table 13. Summary of occurrence and uses of amino acids.

Amino acid	Occurrence	Uses
L-Alanine	Present in silk fibroin	Pharmaceutical agent, additive for brewing
L-Arginine	Protamine	Pharmaceutical agent, food additive
L-Aspartic acid	Unripe sugar canes, sugar beet molasses, germinating beans	Pharmaceutical agent, food additive, additive to biochemical culture media, intermediate in organic synthesis
L-Citrulline	Juice of water melon	Biochemical research
L-Cysteine	Common in most proteins	Biochemical research
L-Glutamic acid	Common in gluten, casein, waste solutions from beet sugar production	Seasoning (sodium salt), antiepileptic
L-Glutamine	Common in many plant, animal and microbial proteins	Pharmaceutical agent, food additive
Glutathione	Common in animals and yeasts	Biochemical research (oxidation/reduction reactions)
Glycine	Gelatin and silk fibroin	Nutrient
L-Histidine	Haemoglobin	Pharmaceutical agent, food additive, biochemical research
L-Homoserine	*Pisum sativum* (as free amino acid)	
L-Isoleucine	Yeast	Pharmaceutical agent, biochemical research
L-Leucine	Gluten, casein, keratin	Food additive, biochemical research
L-Lysine	Common in most proteins, especially casein, fibrin, albumin and muscle protein	Food additive (essential amino acid)
L-Methionine	Casein, ovalbumin	Pharmaceutical agent (hepatic function accelerator)
L-Ornithine	Isolated from proteins after acid hydrolysis	Pharmaceutical agent (anticholesteraemic)
L-Phenylalanine	Fibroin, casein, haemoglobin	Pharmaceutical agent, biochemical research
L-Proline	Gelatin, wheat gliadin	Biochemical research, component of culture media
L-Serine	Silk fibroin, sericine, casein	Pharmaceutical agent, biochemical research, food additive, cosmetic additive
L-Threonine	Eggs, milk, casein, gelatin	Pharmaceutical agent, biochemical research
L-Tryptophan	Casein, serum protein, lactoalbumin	Pharmaceutical agent, biochemical research
L-Tyrosine	Casein, silk fibroin	Pharmaceutical agent, biochemical research
L-Valine	Fibrous proteins, fish proteins	Food additive (essential amino acid)

Arginine

Microorganisms

Regulatory mutants of *Bacillus subtilis*, *Clostridium glutamicum* and *Brevibacterium flavum* are capable of producing arginine from carbohydrates.

Media

Tables 14 and 15 list components of media used for the culture of *B. subtilis* and *Brev. flavum*, respectively.

Table 14. Composition of medium for production of L-arginine using *Bacillus subtilis*.

Component	Amount [% (by wt)]
Glucose	5–15
Glutamic acid and/or aspartic acid	1–5
Calcium carbonate	~2[a]
Ammonium chloride	0.5–4
Peptone	0.02–2
Potassium dihydrogen phosphate	0.5
Magnesium sulphate	0.05

[a] Sufficient to maintain at pH 6–9.

Table 15. Typical medium for culture of *Brevibacterium flavum* for L-arginine production.

Component	Amount
Glucose	10% (by wt)
Ammonium sulphate	6% (by wt)
Calcium carbonate	5% (by wt)
Potassium dihydrogen phosphate	0.1% (by wt)
Magnesium sulphate heptahydrate	0.04% (by wt)
Ferrous ions	2 ppm
Manganous ions	2 ppm
Biotin	50 $\mu g\ \ell^{-1}$
Thiamin.HCl	20 $\mu g\ \ell^{-1}$
Guanine	0.015% (by wt)
Soya protein hydrolyzate	1% (by wt)

Metabolic pathway

Glutamate is converted to arginine via ornithine and citrulline (*see* Fig. 9).

Time course

Production of L-arginine by *Brev. flavum* from glucose is shown in Fig. 10.

Glutamic Acid

Microorganisms

Some wild strains capable of accumulating L-glutamic acid are given in Table 16.

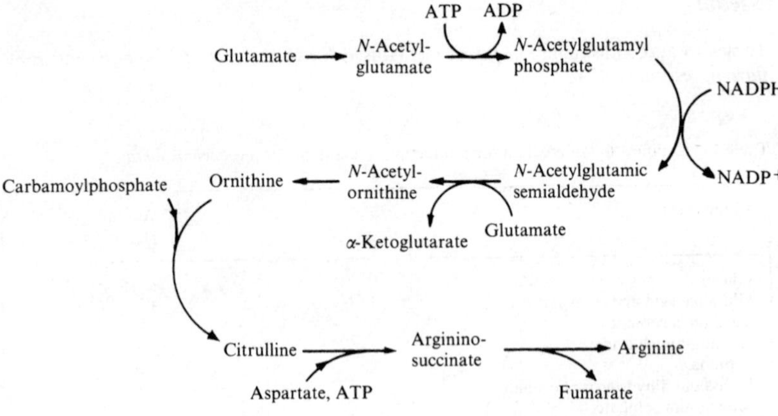

Figure 9. Biosynthesis of L-arginine [Kinoshita and Nakayama (1978)].

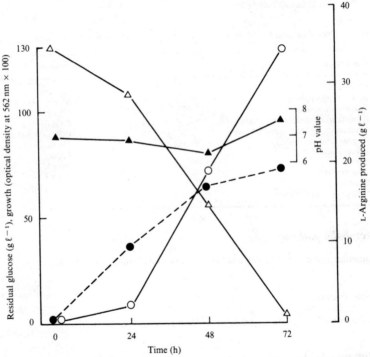

Figure 10. Time course of L-arginine production by *Brevibacterium flavum* No. 352 in a medium containing 13 per cent glucose (w/v). ○, L-arginine production; ●, growth; △, residual glucose; ▲, pH value of the culture [Kubota *et al.* (1971)].

Media

Various carbohydrates can be used as carbon sources, although glucose and sucrose are preferred. Other carbon sources include acetic acid and ethanol.

L-Glutamic acid-producing microorganisms require biotin. However, when sufficient

Table 16. L-Glutamic acid-producing microorganisms (wild strains) [Hirose and Shibai (1980)].

Genus	Species
Arthrobacter	*aminofaciens*
	globiformis
Brevibacterium	*alanicum*
	aminogenes
	ammoniagenes
	divaricatum
	flavum
	immariophilum
	lactofermentum
	roseum
	saccharolyticum
	thiogentalis
Corynebacterium	*callunae*
	glutamicum
	herculis
	lilium
Microbacterium	*ammoniaphilum*
	flavum var. *glutamicum*
	salicinovolum

biotin is supplied for optimal growth, lactate is produced and glutamate is only excreted under conditions of sub-optimal growth, i.e. when the biotin concentration is below 5 μg l^{-1} (*see* Table 18). In the presence of excess biotin, the addition of penicillin during the growth phase results in glutamic acid production. Addition of C_{16}–C_{18} saturated fatty acids also results in the accumulation of glutamic acid, and oleic acid can replace biotin. By including penicillin and saturated fatty acids in the culture medium as antibiotin agents, the industrial use of biotin-rich raw materials, e.g., cane or beet molasses, is possible. Lack of biotin or the presence of antibiotin agents renders the microbial cell walls permeable to glutamic acid with the consequent excretion of the acid and its accumulation in the medium.

Ammonium sulphate, ammonium chloride, ammonium phosphate, ammonia gas and urea can be used as nitrogen sources. A high concentration of ammonium ions is inhibitory to both growth and the production of glutamic acid. Therefore, ammonia, in its various forms, is added as the fermentation progresses.
The concentration of inorganic ions used in culture media for glutamic acid fermentation is presented in Table 17.

The pH for optimum growth and glutamic acid production is 7–8, and the optimum temperature is usually 30–35°C. The optimum overall mass transfer coefficient is about $3 - 5 \times 10^{-6}$ mol O_2 ml^{-1} min^{-1} atm^{-1}.

Table 17. Inorganic salts present in culture media for L-glutamic acid-producing bacteria.

Salt	Amount [% (w/v)]
Calcium carbonate	0.5–4.0
Potassium dihydrogen phosphate	0.05–0.2
Potassium phosphate	0.05–0.2
Magnesium sulphate heptahydrate	0.025–0.1
Ferrous sulphate heptahydrate	0.0005–0.01
Manganese sulphate tetrahydrate	0.0005–0.005

Table 18. Effect of biotin concentration on accumulation of L-glutamate, α-ketoglutarate and lactate by *Micrococcus glutamicusa* [Tanaka *et al.* (1960)].

Biotin concentration ($\mu g\ \ell^{-1}$)	Residual sugar (%)	pH	Maximum product yields ($g\ \ell^{-1}$) L-Glutamate	α-Ketoglutarate	Lactate
0.0	8.5	8.9	1.3	trace	trace
0.5	2.5	8.6	17.0	3.0	7.6
1.0	0.5	8.4	25.0	4.6	7.4
2.5	0.4	8.2	30.8	10.1	6.9
5.0	0.1	8.2	10.8	7.0	13.7
10.0	0.2	8.4	6.7	8.0	20.5
25.0	0.1	8.8	7.5	10.1	23.1
50.0	0.1	8.4	5.7	6.2	30.0

[a] Medium composition: 10 per cent glucose, 0.05 per cent dipotassium hydrogen phosphate, 0.05 per cent potassium dihydrogen phosphate, 0.025 per cent magnesium sulphate heptahydrate, 0.001 per cent ferrous sulphate heptahydrate, 0.001 per cent manganese sulphate heptahydrate and 0.5 per cent urea. Cultures were shaken at 30°C for 72 hours.

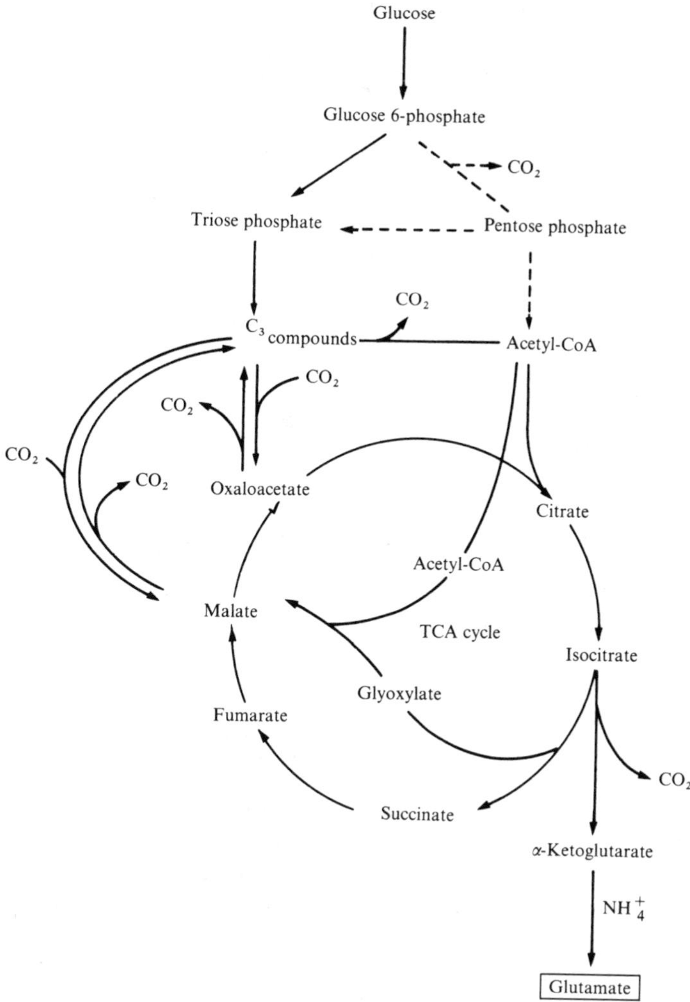

Figure 11. Biosynthesis of glutamic acid from glucose.

Metabolic pathway

The pathways for glutamic acid biosynthesis are shown in Fig. 11. In this figure, the solid lines indicate the Embden–Meyerhof–Parnas (EMP) pathway and the broken lines the hexose monophosphate pathway. Usually the Embden–Meyerhof–Parnas pathway is the principal one under fermentation conditions.

Time course

A typical time course for L-glutamic acid fermentation with intermittent urea feeding is shown in Fig. 12.

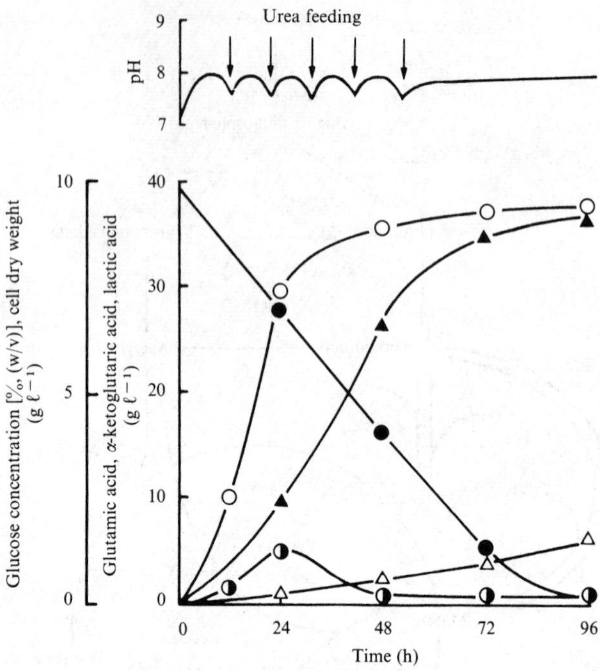

Figure 12. Time course of L-glutamic acid fermentation by *Corynebacterium glutamicum* No. 541. ○, cell dry weight; ●, glucose; ▲, glutamic acid; △, α-ketoglutarate; ◑, lactic acid. The top line on the figure indicates changes in the pH value of the culture which was maintained at 7.5–8.0 by adding urea solutions. Also indicated are the times of urea feeding [Tanaka *et al.* (1960)].

The overall reaction for the production of glutamic acid from glucose is

$$C_6H_{12}O_6 + NH_3 + \tfrac{3}{2}O_2 \rightarrow C_5H_9NO_4 + CO_2 + 3H_2O$$

giving a maximum theoretical yield of 81.7 per cent (by wt) for glutamic acid (i.e. equivalent to 100 per cent molar conversion). Actual yields are usually in the range 50–75 per cent (molar).

Table 19. Effect of various vitamins on accumulation of glutamate; α-ketoglutarate and lactate by *Micrococcus glutamicus* No. 541 [Tanaka *et al.* (1960)].

Vitamin added	Product (g ℓ⁻¹)		
	Glutamate	α-Ketoglutarate	Lactate
None	26.1	2.6	0.5
p-Aminobenzoic acid	26.7	3.3	0.4
Folic acid	27.1	2.5	0.4
Nicotinic acid	25.8	2.5	0.5
DL-Pantothenic acid	22.8	2.8	0.2
Pyridoxal	30.7	3.3	0.4
Pyridoxamine	25.4	3.3	0.3
Pyridoxine	23.6	2.7	0.5
Riboflavin	21.2	2.3	0.5
Thiamin	26.1	2.6	0.5

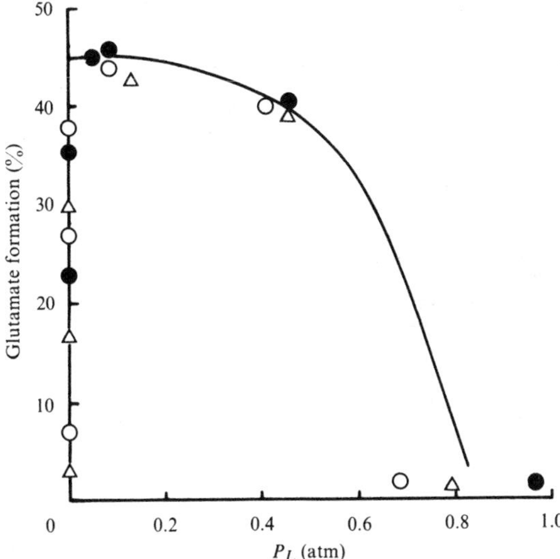

Figure 13. Effect of dissolved oxygen level P_L on the production of L-glutamic acid. P_L was controlled by changing gas-phase oxygen tension. ●, 20-ml working volume in 500-ml flask; ○, 40-ml working volume in 500-ml flask; △, 80-ml working volume in 500-ml flask [Hirose and Okada (1979)].

Low biotin levels stimulate glutamate production, the optimum concentration being 2.5 $\mu g\ \ell^{-1}$. Higher biotin levels suppress glutamic acid production and increase lactate levels (*see* Table 18).

Other vitamins, with the exception of pyridoxal, which is slightly stimulatory, do not affect glutamic acid production (*see* Table 19).

The effect of dissolved oxygen concentration on glutamic acid production is shown in Fig. 13. The product yield decreases with increasing dissolved oxygen concentration.

Glutamic acid can be produced by a two-stage process involving the production of α-ketoglutaric acid by fermentation and subsequent conversion by a second microbial strain or by an enzymic process. Microorganisms capable of producing α-ketoglutaric acid are listed in Table 22. Glucose is the most widely used carbon source. Typical media compositions are given in Tables 20 and 21 for *Serratia marcescens* and *Kluyvera citrophila* var. α, respectively.

Microorganisms capable of converting α-ketoglutaric acid to glutamic acid (*see* Table 22) include species of the genera *Aerobacter*, *Aspergillus*, *Bacillus*, *Debarymomyces*, *Erwinia*, *Escherichia*, *Hansenula*, *Micrococcus*, *Mycotorula*, *Penicillium*, *Pseudomonas*, *Rhizopus*, *Saccharomyces*, *Serratia* and *Xanthomonas*. *Ps. ovalis* gives high conversion yields, approaching 60 per cent of the α-ketoglutaric acid consumed. On the basis of yields

Table 20. Composition of medium used for the production of glutamic acid by *Serratia marcescens*.

Component	Amount
Glucose	4.57%
Calcium carbonate	1.25%
Urea	0.07%
Potassium dihydrogen phosphate	0.05%
Magnesium sulphate heptahydrate	0.01%
Ferric sulphate	8 ppm

Table 21. Composition of medium used for the production of glutamic acid by *Kluyvera citrophila* var. α.

Component	Amount
Fructose	4.93%
Calcium carbonate	4%
Ammonium dihydrogen phosphate	0.12%
Ammonium sulphate	0.12%
Potassium dihydrogen phosphate	0.10%
Magnesium sulphate heptahydrate	0.04%
Sodium chloride	0.30%
Ferrous sulphate heptahydrate	1 ppm

Table 22. Microbial production of α-ketoglutaric acid [Prescott and Dunn (1954)].

Microorganism	Strain number	Approximate maximum reported yield (%)
Aerobacter cloacae	8	40–60
Bacillus megatherium	1	20.6
B. natto (var. *B. subtilis*)		14.6
Bacterium α-ketoglutaricum		50–60
Bact. succinium		13.6
Escherichia coli	3	40–60
	7	40–60
E. freundii	5	40–60
	9	40–60
	12	40–60
Gluconobacter cerinum		~36
Kluyvera citrophila var. α	84C	40–60
K. citrophila	11	40–60
K. citrophila var. β	6	40–60
K. noncitrophila	4	40–60
	10	40–60
Pseudomonas cerinum		16.8[a]
Ps. fluorescens	33F	16.4
Ps. fluorescens (reptilivora)	NRRL B-6	47.9
Serratia marcescens	8UK NRRL B-1418	33.3[b]
	18	49.3

[a] Based on glucose available.
[b] Based on initial glucose.

of 50–60 per cent of α-ketoglutaric acid from glucose, this is equivalent to a 30 per cent yield of glutamic acid from glucose. The fermentation is carried out under anaerobic conditions to prevent the growth of *Ps. ovalis*. The temperature is maintained at about 30°C and the pH is neutral or slightly alkaline.

Leucine

Microorganisms

L-Leucine can be produced by an isoleucine/methionine double auxotroph of *Brevibacterium lactofermentum* 2256 and an auxotrophic mutant of *Corynebacterium glutamicum*.

Media

The composition of media used for *Brev. lactofermentum* and *Coryn. glutamicum* are presented in Tables 23 and 24, respectively.

Table 23. Composition of medium for *Brevibacterium lactofermentum* production of L-leucine.

Component	Amount
Glucose	8%
Calcium carbonate	5%
Ammonium sulphate	4%
Potassium dihydrogen phosphate	0.1%
Magnesium sulphate heptahydrate	0.04%
Biotin	50 $\mu g\ \ell^{-1}$
Thiamin hydrochloride	300 $\mu \gamma g\ \ell^{-1}$
Ferrous ions	2 ppm
Manganous ions	2 ppm
DL-Methionine	40 mg ℓ^{-1}
L-Isoleucine	20 mg ℓ^{-1}

Table 24. Composition of medium for *Corynebacterium glutamicum* production of L-leucine.

Component	Amount
Glucose	13.2%
Calcium carbonate	3%
Ammonium sulphate	1%
Ammonium acetate	2%
Peptone	1%
Meat extract	0.7%
Potassium dihydrogen phosphate	0.15%
Potassium hydrogen phosphate	0.05%
Magnesium sulphate heptahydrate	0.05%
Ferrous sulphate heptahydrate	0.002%
Manganese sulphate tetrahydrate	0.002%
Ammonium heptamolybdate	0.001%
Biotin	50 $\mu g\ \ell^{-1}$
Thiamin hydrochloride	1 mg ℓ^{-1}

Metabolic pathway

The biosynthetic pathway for the production of the amino acids isoleucine, valine and leucine from glucose via pyruvate is outlined in Fig. 14.

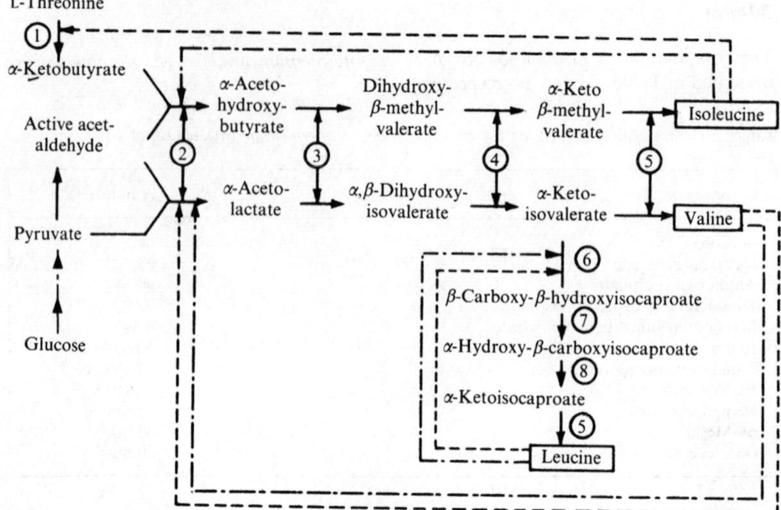

Figure 14. Regulation of the biosynthesis for isoleucine, valine and leucine. – – –, operation of feedback inhibition; –·–, repression of enzyme synthesis. 1, threonine dehydratase; 2, acetohydroxyacid synthetase; 3, dihydroxyacid reductoisomerase; 4, dihydroxyacid dehydratase; 5, branched chain amino acid transaminase; 6, α-isopropylmalate dehydrogenase [Kinoshita and Nakayama (1978)].

Time course

Fig. 15 shows the time course for L-leucine formation from glucose by *Brev. lactofermentum* 2256.

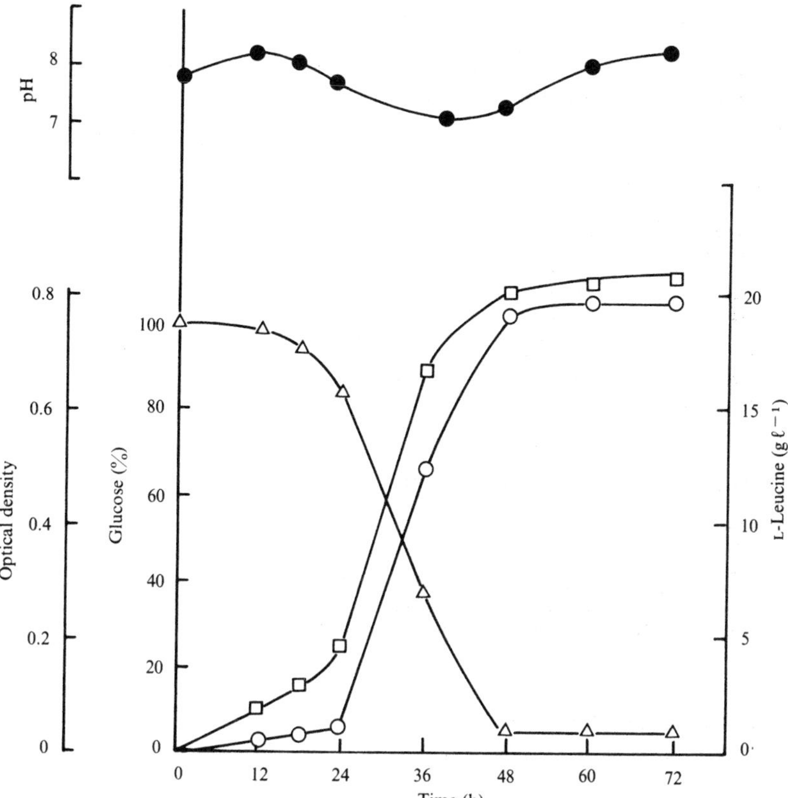

Figure 15. Time course of L-Leucine production by *Brevibacterium lactofermentum*. ○, L-Leucine; □, growth; △, residual glucose; ●, pH, optical density at 562 nm [Tsuchida *et al.* (1974)].

Lysine

Microorganisms

Lysine is produced from carbohydrates by various auxotrophs of *Corynebacterium glutamicum* (synonym *Micrococcus glutamicus*) and homoserine auxotrophs of *Brevibacterium flavum*. Table 25 compares lysine production by various auxotrophic strains of *Micrococcus glutamicus*.

Media

Cane molasses, glucose, acetic acid or ethanol can be used as the carbon source. Ammonium salts are used as the nitrogen source. Urea can also be employed as a nitrogen source if the microorganism possesses urease activity. The composition of the inoculum medium is shown in Table 26.

Table 25. L-Lysine production by various auxotrophs of *Micrococcus glutamicus*[a] [Nakayama *et al.* (1960)].

Strain number	Nutritional requirement	Casein hydrolyzate added (%)	Lysine produced (glysine HCl ℓ^{-1})
534–1	Phenylalanine + tyrosine	1.5	1.4
534–6	Adenine/xanthine	1.0	1.2
534–13	Threonine	1.5	1.8
534–16	Proline	0.5	1.1
534–28	Uracil	1.5	1.7
534–62	Threonine	0.5	2.6
534–75	Leucine	0.2	3.4
534–78	Tryptophan	1.5	1.7
534–Co 127	Arginine/citrulline/ornithine	1.0	0.9
534–Co 155	Homoserine (methionine + threonine)	0.5	7.4
588–219	Phenylalanine	1.5	1.7
615–313	Isoleucine	0.5	2.3
615–338–435	Homoserine (methionine + threonine)	0.2	4.1
615–338–439	Leucine + isoleucine	0.0	1.5
615–396	Isoleucine	0.5	2.3

[a] Culture medium composition: 7.5 per cent glucose, 1.5 per cent ammonium sulphate, 1.0 per cent calcium carbonate, 0.05 per cent potassium hydrogen phosphate, 0.05 per cent potassium dihydrogen phosphate, 0.025 per cent magnesium sulphate heptahydrate, 3 $\mu g\,\ell^{-1}$ biotin.

Table 26. Medium for lysine fermentation.

Component	Amount (%)
Glucose	2.0
Peptone	1.0
Meat extract	0.5
Sodium chloride	0.25

Table 27. Medium for L-lysine fermentation using acetic acid as the carbon source.

Component	Amount
Soya bean meal hydrolyzate	3.5%
Glucose	3%
Acetic acid	0.7%
Potassium dihydrogen phosphate	0.2%
Magnesium sulphate heptahydrate	0.04%
Ferrous sulphate heptahydrate	0.001%
Manganese sulphate monohydrate	0.001%
Biotin	50 $\mu g\,\ell^{-1}$
Thiamin hydrochloride	40 $\mu g\,\ell^{-1}$

The fermentation medium comprises reducing sugars (as invert) (20 per cent) and soya bean meal hydrolyzate (1.8 per cent). Table 27 lists the composition of a medium containing acetic acid as the carbon source. In this fermentation, acetic acid is also supplied in a continuous feed solution to maintain the pH at 7.4.
The feed solution used in conjunction with the medium listed in Table 27 comprises a mixture of acetic acid and ammonium acetate with a molar ratio of 100:25. The temperature of the fermentation is the range 28–33°C.

Yields of L-lysine, as the monohydrochloride, are approximately 30–40 per cent of the initial sugar concentration.

Metabolic pathway

The conversion of aspartic acid to the amino acids lysine, methionine, homoserine, threonine and isoleucine is outlined in Fig. 16.

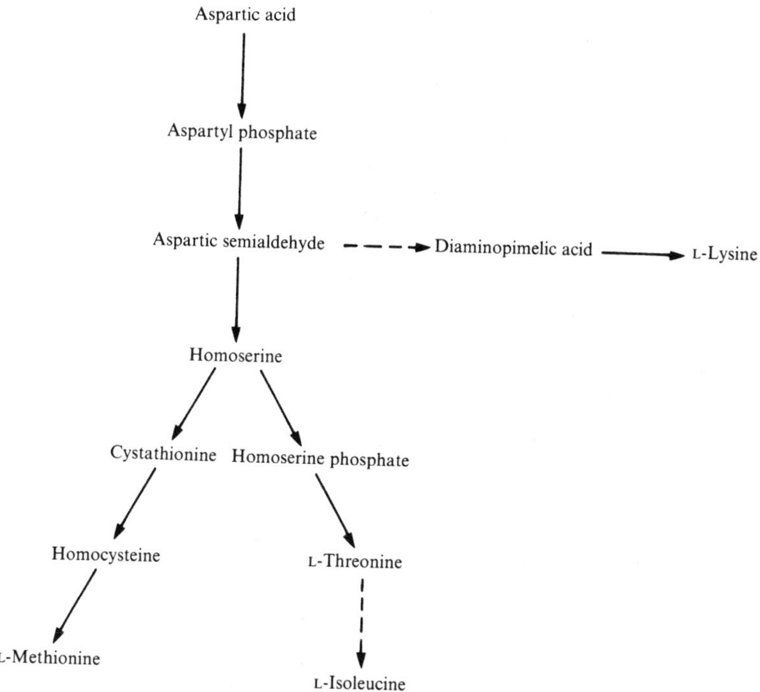

Figure 16. Metabolic pathway of the aspartic acid family of amino acids [Daoust (1976)].

Time courses

Fig. 17 shows the time course of L-lysine fermentation using *Coryn. glutamicum.*

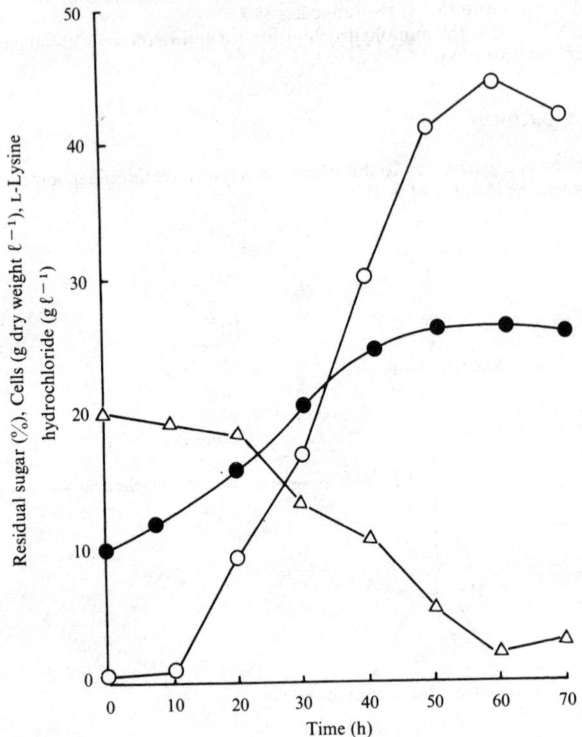

Figure 17. Time course of lysine fermentation using *Corynebacterium glutamicum.* ○, L-Lysine; △, residual sugar; ●, dry cell weight [Nakayama (1972)].

The effects of variations in nutrient composition on lysine accumulation in the presence of *Corynebacterium glutamicus* are presented in Tables 28–30.

More recently, Fukumura (1976) has reported L-lysine production from L-α-aminocaprolactam using the enzyme aminocaprolactam hydrolase from acetone-dried cells of *Cryptococcus laurentii.* The L-enantiomer is converted from D-α-aminocaprolactam by a racemase present in *Achromobacter obae.*

Table 28. Effect of nutrient concentration on accumulation of lysine, glutamate and lactate by *Micrococcus glutamicus*[c] [Nakayama *et al.* (1960)].

Nutrient added (mg ℓ$^{-1}$)					Product accumulated[b] (g ℓ$^{-1}$)		
DL-Homoserine	DL-Threonine	L-Methionine	DL-Isoleucine	Biotin × 10^{-3}	Lysine[a] HCl	Glutamate	Lactate
	200	200	200	20	12.5	Trace	±
	400	200	200	20	7.9	0.27	±
	1000	200	200	20	1.7	Trace	+ +
	400	400	200	20	10.6		
	400	1000	200	20	9.6		+
	400	200	400	20	5.2		+
	200	400	1000	20	6.3		
	1000	1000	1000	20	0.3	Trace	+ +
	1000	1000	1000	5	Trace	Trace	+ +
	1000	1000	1000	2.5	Trace	Trace	39.8
	1000	1000	1000	1.25	Trace	0.8	35.5
	1000	1000	1000	0.6	Trace	7.3	2.3
	1000	1000	1000	0.3	Trace	6.7	0.8
400			200	20	11.1	0.4	0.9
600			200	20	8.9	0.3	3.3
1000			200	20	6.3	Trace	26.7
1000			1000	20	2.9	Trace	32.5

[a] Average value of two flasks on day 4.
[b] Maximum yields during four days.
[c] Basal medium composition: 7.5 per cent glucose, 1.5 per cent ammonium sulphate, 1.0 per cent calcium carbonate, 0.05 per cent dipotassium hydrogen phosphate, 0.05 per cent potassium dihydrogen phosphate, 0.025 per cent magnesium sulphate heptahydrate.

Table 29. Effect of biotin concentration, L-leucine concentration and size of inoculum on accumulation of various amino acids and lactate by *Micrococcus glutamicus* (534–112 Leu)[a] [Nakayama *et al.* (1961)].

Biotin added ($\mu g\ \ell^{-1}$)	Inoculum size (%)	L-Leucine added ($mg\ \ell^{-1}$)	Cell crop (g dry wt)	Product accumulated[b] ($g\ \ell^{-1}$)				
				Valine	Glutamate	Lactate	Alanine	Lysine
10	10	500	7.2	Trace	Trace	5.9	0.61	Trace
0.75	10	500	10.0	Trace	1.38	6.7	2.62	0.36
0.27	10	500	7.7	Trace	3.80	3.74	1.42	Trace
10	1	500	10.7	Trace	Trace	19.1	1.45	Trace
1	1	4	8.9	4.80	0.12	2.2	1.45	0.59
0	1	4	6.1	5.50	1.30	0	1.80	0.53

[a] Basal medium composition: as Table 17.
[b] Measurement made on three-day culture media.

Table 30. Effect of DL-methionine concentration on accumulation of L-homoserine and L-lysine by *Micrococcus glutamicus*[a] [Samejima *et al.* (1960)].

DL-methionine (mg ℓ$^{-1}$)	Amino acid accumulated (g ℓ$^{-1}$) 4-day culture L-Homoserine	4-day culture L-Lysine	5-day culture L-Homoserine	5-day culture L-Lysine
0	13.08	3.89	14.38	9.58
50	12.38	9.48	13.25	9.53
100	10.58	9.38	11.33	15.25
200	4.35	13.10	5.75	16.15
300	3.23	13.70	3.40	17.55
400	4.70	11.65	3.90	14.22
600	4.90	14.65	3.50	14.90
800	5.18	15.15	4.80	14.75
1200	6.75	14.75	5.70	14.45

[a] Culture medium composition: 10 per cent glucose, 2 per cent ammonium sulphate, 2 per cent calcium carbonate, 400 mg ℓ$^{-1}$ L-threonine, 7.5 μg ℓ$^{-1}$ biotin, 0.1 per cent potassium hydrogen phosphate, 0.03 per cent magnesium sulphate heptahydrate. Cultures were incubated at 28°C with shaking.

Threonine

Microorganisms

Threonine can be synthesized from carbohydrates by auxotrophs of *Escherichia coli*, α-amino-β-hydroxyvaleric acid-resistant mutants of *Brev. flavum*, similar mutants of *Proteus rettgeri* and *Coryn. acetoacidophilum* and an auxotroph of *Arthro. paraffineus*.

Table 31. Medium composition for the production of threonine by *Escherichia coli* KY 8280.

Component	Amount (%)
Fructose	7.5
Calcium carbonate	2.0
Ammonium sulphate	1.4
Potassium dihydrogen phosphate	0.3
Magnesium sulphate heptahydrate	0.03

Table 32. Medium composition for the production of threonine by *Brevibacterium flavum* BB82.

Component	Amount
Glucose	10%
Calcium carbonate	5%
Ammonium sulphate	3%
Potassium dihydrogen phosphate	1.5%
Magnesium sulphate heptahydrate	0.04%
Ferrous ions	2 ppm
Manganous ions	2 ppm
Biotin	200 μg ℓ$^{-1}$
Thiamin hydrochloride	300 μg ℓ$^{-1}$
Soya bean meal hydrolyzate	4 ml ℓ$^{-1}$

Media

Typical media compositions for the production of threonine by mutants of *E. coli* and *Brev. flavum* are given in Tables 31 and 32, respectively.

Metabolic pathway

The synthesis of threonine from aspartic acid is summarized in Fig. 16.

Time course

Production of threonine from fructose by *E. coli* KY 8280 over 120 hours is shown in Fig. 18.

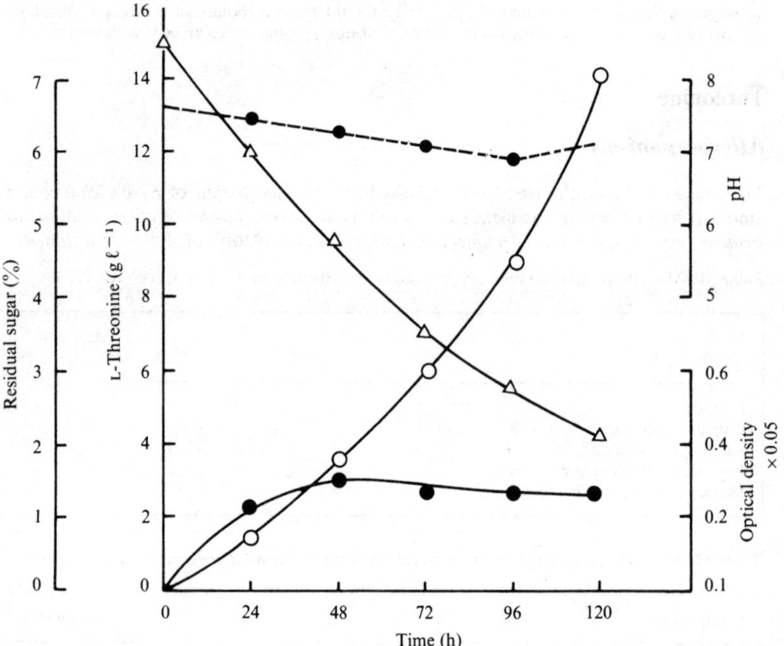

Figure 18. Time course of L-threonine fermentation in *Escherichia coli*. ○, L-Threonine; △, residual sugar; – – –, pH; ——, Optical density at 660 nm [Kase *et al.* (1971)].

Other Amino Acids

With the exception of alanine, aspartate, homoserine, isoleucine, proline, serine and valine, which are considered below, the formation of other amino acids is generally by chemical synthesis or by hydrolysis of protein sources rich in the required amino acid, e.g., alanine from silk fibroin.

Alanine

Alanine is produced by bacteria (e.g., *Brev. monoflagellum* and *B. coagulans*) from glucose. Lower yields of the amino acid are obtained when using *n*-alkanes as substrates.

Aspartate

Aspartate is usually formed from ammonium fumarate in the presence of *E. coli*, *E. freundii* or *Ps. fluorescens*, which all have high aspartase levels. More recently, an immobilized-enzyme technique has been developed for the production of this amino acid [*see* Chibata *et al.* (1974)].

Homoserine

Table 30 shows that, in addition to lysine, homoserine is formed by *M. glutamicus*. Homoserine formation is highest in the absence of DL-methionine in the culture medium.

Isoleucine

No wild-type strains of bacteria are able to produce isoleucine, but mutants of *Serratia marcescens* have been identified by Kisumi *et al.* (1971) that are capable of producing the amino acid from carbohydrates.

Proline

Proline has been produced from cane molasses by many microorganisms including *Brev. flavum* No. 14–5 and *M. glutamicus*. A high ammonium concentration in the nutrient broth is essential.

Serine

Production of serine microbiologically has not been achieved successfully when using a relatively inexpensive starting material.

Valine

Relatively high levels of valine accumulate after inoculation of *M. glutamicus* 534–112 Leu. when biotin is not present in the incubation medium (*see* Table 29).

ANTIBIOTICS

General Aspects

Table 33 contains a list of some antibiotics produced on a commercial scale, the microorganisms capable of producing the antibiotics, the chemical nature of the antibiotics and their therapeutic use.

Table 33. Some antibiotics produced on a commercial scale [Perlman (1979)].

Antibiotic	Microbial source	Antibiotic spectrum[a] G+	G−	My	AF	AT	Other	Chemical type	Therapeutic or other use
Amphomycin	*Streptomyces canus*	+						Peptide	Topical
Amphotericin B	*Strept. nodosus*				+			Polyene	Oral, parenteral
Avoparcin	*Strept. candidus*	+	+					Glycopeptide	Animal growth promotant
Azalomycin F	*Strept. hygroscopicus*	+	+		+				Topical (AF)
Bacitracin	*Bacillus subtilis*	+						Peptide	Topical, animal growth promotant
Bambermycins	*Strept. bambergensis*	+						Phosphoglycolipid	Animal growth promotant
Bicyclomycin	*Strept. sapporonensis*	+	+						Topical
Blasticidin S	*Strept. griseochromogenes*				+			Nucleoside	Agricultural (AF)
Bleomycins	*Strept. verticillus*	+	+			+		Peptide	Parenteral (AT)
Cactinomycin	*Strept. chrysomallus*	+				+		Peptide	Parenteral (AT)
Candicidin B	*Strept. griseus*				+			Polyene	Topical
Candidin	*Strept. viridoflavus*				+			Polyene	Topical
Capreomycin	*Strept. capreolus*			+				Peptide	Parenteral
Cephalosporins	Cephalosporin C is produced by *Cephalosporium acremonium* and converted to 7-aminocephalosporanic acid which is used for preparation of semisynthetic cephalosporins	+	+					Peptide	Oral, parenteral
Chloramphenicol	*Strept. venezuelae*, commercial manufacture is by chemical synthesis	+	+				Rickettsia		Oral, parenteral
Chromomycin A_3	*Strept. griseus*	+				+			Parenteral (AT)
Colistin	*B. colistinus*		+					Peptide	Parenteral, agricultural (AF)
Cycloheximide	*Strept. griseus*				+				
Cycloserine	*Strept. orchidaceus*			+				Amino acid	Parenteral (TB)
Dactinomycin	*Strept. antibioticus*					+		Peptide	Parenteral (AT)
Daunorubicin	*Strept. peucetius*					+			Parenteral (AT)

Doxorubicin	*Strept. peucetius*					+			Parenteral (AT)
Enduracidin	*Strept. fungicidus*	+		+				Peptide	Animal growth promotant
Erythromycin	*Strept. erythreus*	+						Macrolide	Oral, parenteral; animal growth promotant
Fortimicins	*Micromonospora olivoasterospora*	+	+					Aminoglycoside	Parenteral
Fungimycin	*Strept. coelicolor* var. *aminophilus*				+			Polyene	Topical
Fusidic Acid	*Fusidium coelcineum*	+						Steroid	Parenteral
Gentamicins	*Micro. purpurea*	+	+					Aminoglycoside	Parenteral
Gramicidin A	*B. brevis*	+						Peptide	Topical
Gramicidin J (S)	*B. brevis*	+						Peptide	Topical
Griseofulvin	*Penicillium griseofulvum*				+			Spirolactone	Oral
Hygromycin B	*Strept. hygroscopicus*	+	+				Helminths	Aminoglycoside	Animal feed supplement
Josamycin	*Strept. narbonensis*	+						Macrolide	Oral, parenteral
Kanamycins	*Strept. kanamyceticus*	+	+	+				Aminoglycosides	Parenteral
Kasugamycin	*Strept. kasugaensis*	+	+					Aminoglycoside	Agricultural antibacterial
Kitasatamycin	*Strept. kitasatoensis*	+						Macrolide	Oral, parenteral
Lasalocid	*Strept. hazelensis*	+					Coccidia	Polyether	Agricultural use as coccidiostat growth promotant
Lincomycin	*Strept. lincolnensis*	+							Oral, parenteral
Lividomycin		+	+						
Macarbomycins	*Strept. phaechromogenes*	–						Phosphoglycolipid	Animal growth promotant
Mepartricin					+			Polyene	Topical
Midecamycin		–						Macrolide	Oral, topical
Mikamycins	*Strept. mitakaensis*	+						Peptide	Animal growth promotant
Mithramycin	*Streptomyces* species	+		+		+			Parenteral (AT)
Mitomycin C	*Strept. caespitosus*	+	+		+				Parenteral
Mocimycin		+							Animal growth promotant
Monensin	*Strept. cinnamonensis*	+					Coccidia	Polyether	Animal growth promotant
Myxin	*Chromobacterium iodinum* plus chemical modification				+			Phenazine	Topical in veterinary use
Neocarzinostatin	*Strept. carzinostaticus*	+				+		Peptide	Parenteral (AT)
Neomycins	*Strept. fradiae*	+	+					Aminoglycoside	Oral, topical
Nosiheptide	*Strept. actinosus*	+						Peptide	Animal growth promotant
Nisin	*Streptococcus cremoris*	+						Peptide	Food preservative
Novobiocin	*Strept. niveus*	+							Oral, topical
Nystatin	*Strept. noursei*				+			Polyene	Oral, topical
Oleandomycin	*Strept. antibioticus*	+						Macrolide	Oral, parental

continued overleaf

Table 33 (*continued*)

Antibiotic	Microbial source	Antibiotic spectrum[a] G+	G−	My	AF	AT	Other	Chemical type	Therapeutic or other use
Paromomycin	*Strept. rimosus*	+	+				Protozoa	Aminoglycoside	Oral
Penicillin G	*Pen. chrysogenum*	+						Peptide	Oral, parenteral, animal growth promotant
Penicillin V	*Pen. chrysogenum*	+						Peptide	Oral
Pimaricin	*Strept. natalensis*					+		Polyene	Topical, food preservation
Polymyxin B	*B. polymyxa*				+			Peptide	Parenteral, agriculture (AF)
Polyoxins	*Strept. cacaoi* var. *asoensis*								
Pristinamycins	*Strept. pristinaspiralis*	+						Peptide	Parenteral
Quebemycin	*Strept. viridans*	+						Phosphoglycolipid	Animal growth promotant
Ribostamycin	*Strept. ribosidificus*	+	+					Aminoglycoside	Parenteral
Rifamycin SV	*Nocardia mediterranei*			+				Anasamycin	Parenteral
Ristocetin	*N. lurida*	+						Glycopeptide	Parenteral
Sagamycin	*Micro. sagamiensis*	+	+					Aminoglycoside	Parenteral
Salinomycin	*Strept. albus*	+					Coccidia	Polyether	Veterinary use
Siccanin	*Helminthosporium siccans*				+				Veterinary use
Siomycin	*Strept. sioyaensis*	+						Peptide	Animal growth promotant
Sisomicin	*Micro. inyoensis*	+	+					Aminoglycoside	Parenteral
Spectinomycin	*Strept. spectabilis*	+						Aminocyclitol	Parenteral
Spiramycin	*Strept. ambofaciens*	+						Macrolide	Parenteral, oral
Streptomycin	*Strept. griseus*	+	+	+				Aminoglycoside	Parenteral, use in agriculture to control bacteria
Dihydrostreptomycin	*Strept. humidus* (also chemical reduction of streptomycin)	+	+	+				Aminoglycoside	Parenteral
Tetracyclines									
Chlortetracycline	*Strept. aureofaciens*	+	+				Rickettsia	Tetracycline	Parenteral, oral, animal growth promotant
6-Demethyl-	*Strept. aureofaciens*	+	+				Rickettsia	Tetracycline	Parenteral, oral

7-chlortetracycline									
5-Hydroxytetracycline	*Strept. rimosus*	+	+				Rickettsia	Tetracycline	Parenteral, oral animal growth promotant
Tetranactin	*Strept. flaveolus*	+					Insects	Macrotetralide	Insecticide
Thiopeptin	*Strept. tateyamensis*	+						Peptide	Animal growth promotant
Thiostrepton	*Strept. azureus*	+						Peptide	Animal growth promotant
Tobramycin	*Strept. tenebrarius*	+	+					Aminoglycoside	Parenteral
Trichomycin	*Strept. hachijoensis*				+		Trichomonas	Polyene	Topical
Tylosin	*Strept. fradiae*	+					PPLO	Macrolide	Veterinary, animal growth promotant
Tyrothricin	*B. brevis*	+	+					Peptide	Topical
Tyrocidine	*B. brevis*	+	+					Peptide	Topical
Validamycin	*Strept. hygroscopicus* var. *limonesis*	+	+					Aminoglycoside	Parenteral
Vancomycin	*Strept. orientalis*	+						Glycopeptide	Parenteral
Variotin	*Paecilomyces varioti*				+				Topical
Viomycin	*Strept. floridae*	+	+	+				Peptide	Parenteral
Virginiamycin	*Strept. virginiae*	+						Peptide	Animal growth promotant

[a] G+, gram-positive bacteria; G−, gram-negative bacteria; My, mycobacteria; AF, antifungal; AT, antitumour.

Inorganic phosphate affects the biosynthesis of many antibiotics; its inhibitory effects are summarized in Table 34.

Table 34. Antibiotic processes that are inhibited by inorganic phosphate [Martin (1977)].

Antibiotic	Microorganism	Range of inorganic phosphate permitting antibiotic production (mM)
A-9145	*Streptomyces griseolus*	0.28–2.24
Actinomycin	*Strept. antibioticus*	1.4–17
Amphotericin B	*Strept. nodosus*	1.5–2.2
Ayfactin	*Strept. aureofaciens*	1–17
Bacitracin	*Bacillus licheniformis*	0.1–1
Butirosin	*B. circulans*	<5.6
Candicidin	*Strept. griseus*	0.5–5
Candidin	*Strept. viridoflavus*	0.5–5
Cephamycin	*Strept. clavuligerus*	25
Chlortetracyline	*Strept. aureofaciens*	1–5
Coryneicins	*Corynebacterium* sp.	2.24[a]
Cycloheximide	*Strept. griseus*	0.05–0.5
Gramicidin S	*B. brevis*	10–60
Kanamycin	*Strept. kanamyceticus*	2.2–5.7
Levorin	*Strept. levoris*	0.3–4
Monamycin	*Strept. jamaicensis*	0.2–0.4
Mycoheptin	*Streptoverticillium mycoheptinicum*	3.5[a]
Neomycin	*Stept. fradiae*	
Novobiocin	*Strept. niveus*	9–40
Nystatin	*Strept. noursei*	1.6–2.2
Oleandomycin	*Strept. antibioticus*	0.5[a]
Oxytetracycline	*Strept. rimosus*	2–10
Polymyxin	*B. polymyxa*	
Prodigiosin	*Serratia marcescens*	0.05–0.2
Pyocyanine	*Pseudomonas aeruginosa*	0.1[a]
Ristomycin	*Proactynomyces fructiferi*	0.2–5
	P. fructiferi var. *ristomycin*	<2.8
Streptomycin	*Strept. griseus*	1.5–15
Tetracycline	*Strept. aureofaciens*	0.14–0.2
Vancomycin	*Strept. orientalis*	1–7
Viomycin	*Streptomyces* sp.	1–8

[a] Optimal concentration of phosphate for antibiotic production.

Properties and Uses

Table 35 lists the structures and physical properties of some of the principal antibiotics. The therapeutic and other uses of a wide range of antibiotics have been given in Table 33.

Table 35. Structure and physical properties of antibiotics.

Antibiotic	Empirical formula	Structure	Physical properties
Bacitracin	$C_{66}H_{103}N_{17}O_{16}S$	Commercial product is a mixture of at least nine bacitracins, the main one being bacitracin A with the structure 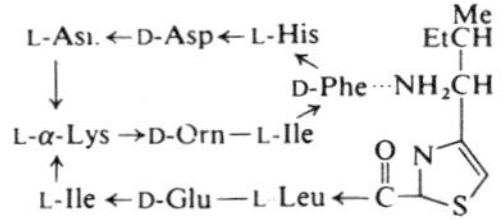	Mol wt, 1411; grey powder; sol. H_2O, EtOH; insol. Et_2O, $CHCl_3$, Me_2CO; aqueous solution stable
Cephalosporin C	$C_6H_{21}N_3O_8S$	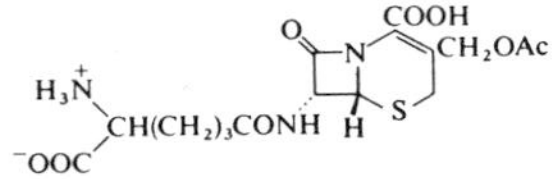 usually as sodium salt	Mol. wt, 415; sodium salt, dihydrate, monoclinic crystals; $[\alpha]_D^{20}$ + 103° (H_2O); sol. H_2O; insol. EtOH, Et_2O
Chlortetracycline (aureomycin)	$C_{22}H_{23}ClN_2O_3$	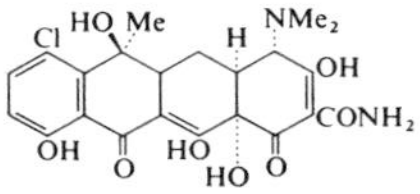	Mol. wt, 478.9; mp, 168–9°C., golden-yellow crystals; $[\alpha]_D^{23}$ − 275° (MeOH); sol. H_2O; slightly sol. MeOH, EtOH, BuOH, Me_2CO, EtOAc, Bz; insol. Et_2O
Gramicidin S	$C_{60}H_{92}N_{12}O_{10}$	L-Pro → L-Val → L-Orn → L-Leu → D-Phe ↑ ↓ D-Phe — L-Leu ← L-Orn ← L-Val ← L-Pro usually as dihydrochloride	Mol. wt, 114.5; dihydrochloride, decomp., 277–8°C; prisms; $[\alpha]_D^{24}$ − 289° (EtOH); sol. EtOH; slightly sol. Me_2CO; insol. H_2O, acid, alkali
Griseofulvin	$C_{17}H_{17}ClO_6$	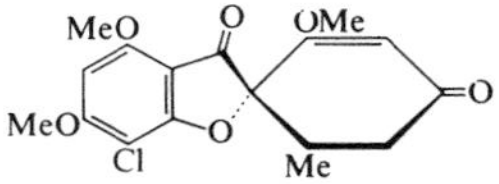	Mol. wt, 352.8; mp, 220°C; octahedral or rhomboid crystals; $[\alpha]_D^{17}$ + 370° ($CHCl_3$); slightly sol. Me_2CO, EtOH, $CHCl_3$; insol. H_2O

continued overleaf

Table 35 (*continued*)

Antibiotic	Empirical formula	Structure	Physical properties
Kanamycins		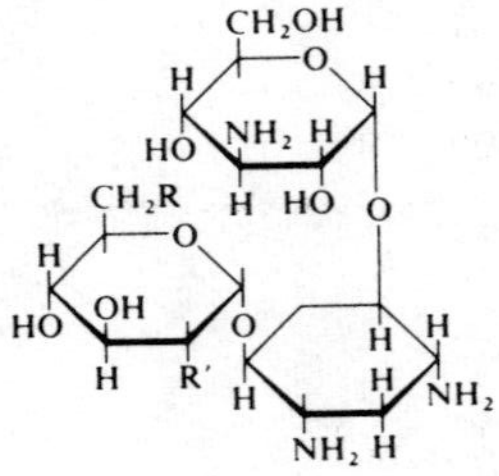	
Kanamycin A	$C_{18}H_{36}N_4O_{11}$	$R = NH_2, R' = OH$	Mol. wt, 484; $[\alpha]_D^{24} + 146°$ (1N H_2SO_4)
Kanamycin B	$C_{18}H_{37}N_5O_{10}$	$R = R^1 = NH_2$	Mol. wt, 483; mp, 178–82° (decomp.); $[\alpha]_D^{21}$ + 114°; sol. H_2O; slightly sol. $CHCl_3$, *i*-PrOH; insol. MeOH, EtOH
Kanamycin C	$C_{18}H_{36}N_4O_{11}$	$R = OH, R' = NH_2$	Mol. wt, 484; decomp. >270°C; $[\alpha]_D^{20} + 126°$ (H_2O); sol. H_2O; insol. MeOH, EtOH
Novobiocin	$C_{31}H_{36}N_2O_{11}$		Mol. wt, 612.7; decomp., 152–6°C; sp.gr. 1.3448; pale yellow rhombic crystals; sol. alkaline soln, Me_2CO, EtOAc, MeOH, EtOH
Oleandomycin	$C_{35}H_{61}NO_{12}$	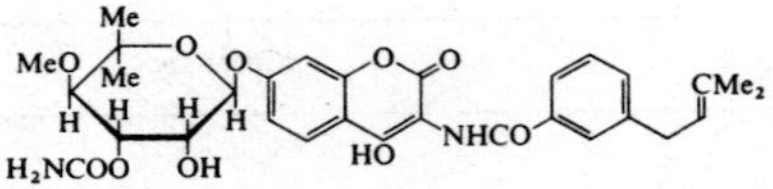	Mol. wt, 687.9; white powder; sol. MeOH, EtOH, BuOH, Me_2CO; insol. CCl_4, Bu_2O

Oxytetracycline	$C_{22}H_{24}N_2O_9$	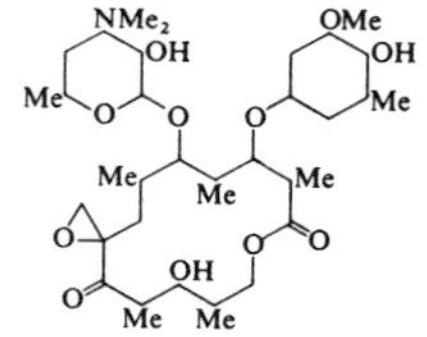	Mol. wt, 496.5; dihydrate, decomp., 181–2°C; needles; $[\alpha]_D^{25}$ − 196.6° (1 N HCl) slightly sol. H_2O, EtOH, MeOH
Penicillin G (benzyl-penicillinic acid)	$C_{16}H_{18}N_2O_4S$	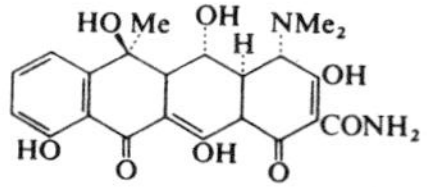	Mol. wt, 334.4; white powder; $[\alpha]_D^{20}$ + 269° (MeOH); sol. MeOH, EtOH, Et_2O, EtOAc, B_3, $CHCl_3$, Me_2CO Potassium salt: decomp., 214–7°C; hygroscopic crystals; sol. H_2O, EtOH, MeOH Sodium salt: sp.gr. 1.41; $[\alpha]_D^{24.8}$ + 301°; sol. H_2O, EtOH, MeOH; insol. Me_2CO, Et_2O, $CHCl_3$, EtOAc
Streptomycin	$C_{21}H_{39}N_7O_{12}$	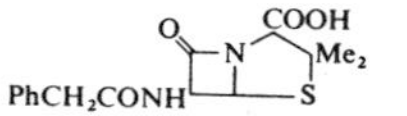 usually as trihydrochloride	Mol. wt, 581.6; Trihydrochloride: pale yellow powder; $[\alpha]_D^{25}$ − 84°; sol. H_2O, MeOH; slightly sol. EtOH
Tetracycline	$C_{22}H_{24}N_2O_3$	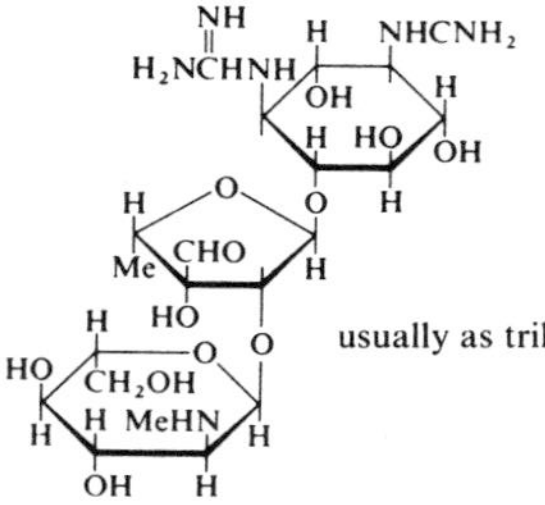	Mol. wt, 444.4; greyish white to yellow powder; sol. EtOH, slightly sol. H_2O; insol. $CHCl_3$, Et_2O

continued overleaf

Table 35 (*continued*)

Antibiotic	Empirical formula	Structure	Physical properties
Viomycin	$C_{25}H_{43}N_{13}O_{10}$	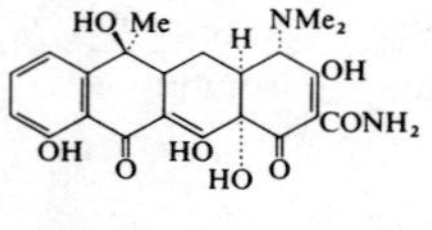	Mol. wt, 685.7; mp, 280°C; purple crystals; sol. H_2O; slightly sol. EtOH
		$C_{25}H_{43}N_{13}O_{10}$ 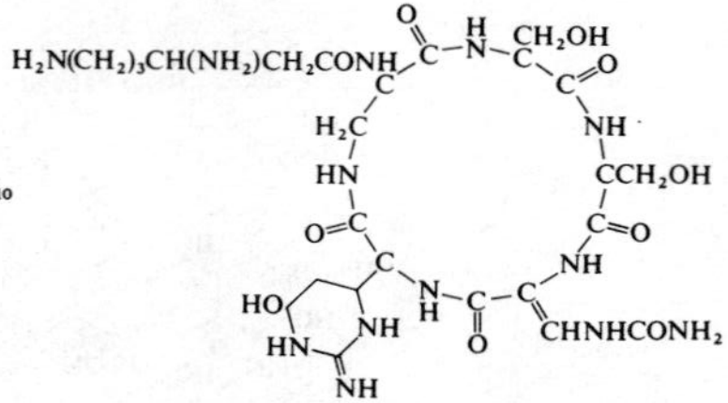	

Penicillin

Microorganisms

The mould used in the early years of development of penicillin was *Penicillium notatum*. Currently, for the submerged culture processes, highly mutated strains of *Pen.*

Table 36. Progress in increasing yields of penicillin by a programme of selection and mutation of strains of *Penicillium* [Moss and Smith (1977)].

Date	Strain identification and origin	Yield (units ml^{-1})	Comments
1929	*Pen. notatum* (Fleming)	2–20	Wild type isolate
1941	*Pen. notatum* (NRRL 832)[a]	40–80	Wild type isolate
1943	*Pen. chrysogenum* (NRRL 1951)[a]	80–100	Wild type isolate from melon
	↓Selection		
1944	NRRL 1951 B25[a]	100–200	
	↓X-ray treatment		
1944	X 1612	300–500	
	↓UV irradiation		
1945	Q 176 (Wisconsin)	800–1000	
	↓UV irradiation		
1947	BL 3D 10 (Wisconsin)	800–1000	Lacks the yellow pigment – chrysogenin
	↓Nitrogen mustard		
1949	49–133 (Wisconsin)	1500–2000	Poor growth
	↓Nitrogen mustard		
1951	51–20 (Wisconsin)	2400	Very poor growth
	↓Selection		
1953	53–399 (Wisconsin)	2700	Selected for improved growth
	↓Selection		
1960	Commercial strains	~5000	Continuous selection for improved behaviour under commercial conditions
	? ↓		
1970	Commercial strains	~10 000	

[a] NRRL, Northern Region Research Laboratories, Peoria.

Table 37. Typical fermentation media for production of penicillin G or V in 1945 and 1967 [Queener and Swartz (1979)].

Component	Amount (%)
1945	
Lactose	3–4
Corn-steep liquor solids	3.5
Calcium carbonate	1.0
Potassium dihydrogen phosphate	0.4
Lard oil antifoam	0.25
1967	
Corn or molasses (by continuous feed)	10
Corn-steep liquor solids	4–5
Phenylacetic acid (by continuous feed)	0.5–0.8
Lard oil (or vegetable oil) antifoam (by continuous addition)	0.5

chrysogenum are employed. Table 36 demonstrates the progress in increasing yields of penicillin through mutation and strain selection.

Media

Simple, chemically defined media, such as that of Czapek and Dox (*see* Chapter 1), were used initially for the surface culture of *Pen. notatum*. Since then several formulations have

Table 38. Composition of a chemically defined medium of penicillin production (Hockenhull (1959)].

Component	Amount ($g\,\ell^{-1}$)
Lactose	30.0
Glucose	10.0
Starch	15.0
Citric acid	10.0
Acetic acid	2.5
Phenylacetic acid	0.5
Ethylamine	3.0
Ammonium sulphate	5.0
Potassium dihydrogen phosphate	1.0
Magnesium sulphate heptahydrate	0.5
Ferrous sulphate heptahydrate	0.05
Zinc sulphate heptahydrate	0.01
Copper sulphate pentahydrate	0.01
Manganese sulphate tetrahydrate	0.01
Cobalt sulphate heptahydrate	0.005
Sodium chloride	0.001

Table 39. Typical medium formulation at the time of changeover from surface to submerged culture in penicillin production [Hockenhull (1963)].

Component	Amount (%)
Corn-steep liquor solids	2–5
Lactose	2–4
Glucose	0–0.5
Sodium nitrate	0–0.5
Calcium carbonate	0–1.0
Phenylacetic acid derivative	0.09–0.10

Table 40. Fermentation medium used for the development of early *Penicillin chrysogenum* production strains [Queener and Swartz (1979)].

Medium	Amount [% (w/v)]
Distilled water	
Corn-steep liquor solids	2.0
Lactose	4.0
Sodium nitrate	0.3
Potassium hydrogen phosphate	0.05
Magnesium sulphate	0.025
Calcium carbonate	0.4
β-Phenylethylamine	0.25
Antifoam	3 drops $(100\,ml)^{-1}$

been tried. Table 37 compares two media used in 1945 and 1967. Compositions of some other media are contained in Tables 38 and 39, while a medium used for the screening of possible penicillin production strains is given in Table 40.

Metabolic pathways

Penicillin, a secondary metabolite, is produced in the idiophase which follows the rapid growth phase (i.e. the trophophase). Two enzymes have been observed to increase markedly at the end of the growth phase: one enzyme activates phenylacetic acid, which forms the side-chain of benzylpenicillin, with the formation of a coenzyme A ester while the other enzyme – penicillin acyltransferase – attaches the activated phenylacetate to the 6-aminopenicillamic acid nucleus of penicillin.

Catabolite regulation is observed in penicillin production. Slow utilization or limited supply of carbon source maintains the catabolites at a low level thus increasing penicillin production. Feedback control also appears to play a role. The inhibition of penicillin production by lysine due to feedback regulation by the amino acid of a branched pathway leading to both lysine and penicillin is illustrated in Fig. 19.

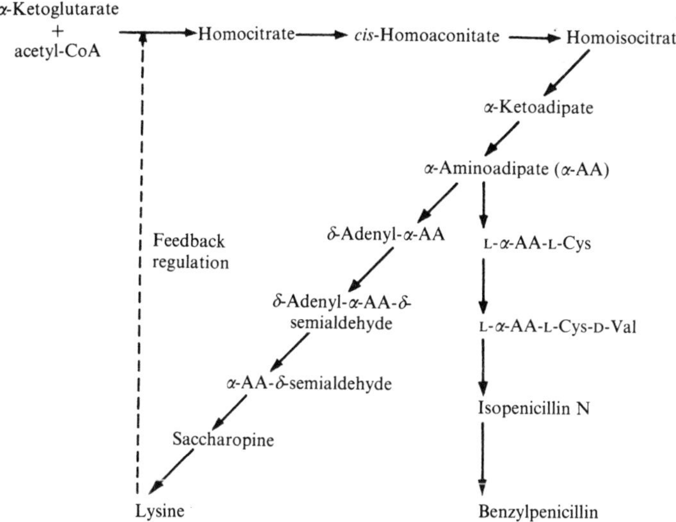

Figure 19. Production of penicillin from α-ketoglutaric acid by Penicillium chrysogenum.

As can be seen from Table 41, the naturally occurring penicillins, whose trivial names are given in Table 42, differ from each other by a single group, all having the β-lactam structure.

COOH
O N Me_2
S
R·CO·NH

Table 41. Structures of naturally occurring penicillins [Perlman (1979)].

R	Chemical name	Empirical formula
EtCH=CHCH$_2$–	6-(2-Hexenamido)penicillin	$C_{14}H_{20}N_2O_4S$
Me (CH$_2$)$_4$–	6-(Hexanamido)penicillin	$C_{14}H_{22}N_2O_4S$
Me (CH$_2$)$_6$–	6-(Heptamido)penicillin	$C_{16}H_{26}N_2O_4S$
Ph CH$_2$–	6-(Phenylacetamido)penicillin	$C_{16}H_{18}N_2O_4S$
p-OH Ph CH$_2$–	6-(*p*-Hydroxyphenylacetamido)-penicillin	$C_{16}H_{18}N_2O_5S$

Table 42. Trivial names of naturally occurring penicillins.

Trivial name	Chemical name
Dihydropenicillin F	6-(Hexanamido) penicillin
Penicillin F	6-(2-Hexenamido) penicillin
Penicillin G	6-(Phenylacetamido) penicillin
Penicillin K	6-(Heptamido) penicillin
Penicillin X	6-(*p*-Hydroxyphenylacetamido) penicillin

Time course

A typical time course of penicillin fermentation is described in Table 43 and shown graphically in Fig. 20.

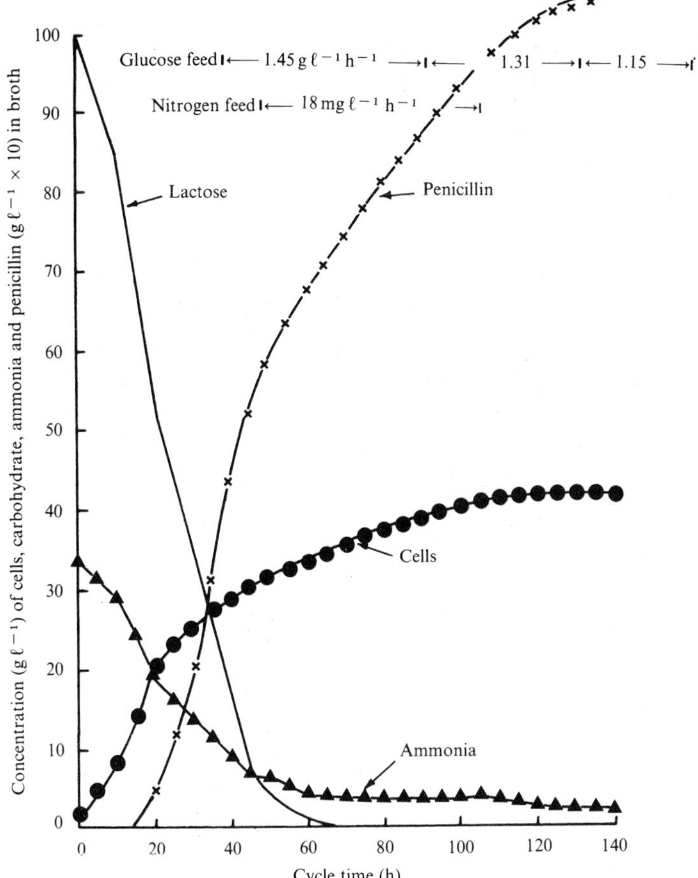

Figure 20. Time course of changes in carbohydrate, nitrogen, penicillin and biomass concentrations during a penicillin fermentation. The biomass curve was drawn arbitrarily but is consistent with data in Table 42. Penicillin production was assumed to begin at 15 hours at specific rates consistent with those reported by Ryu and Hospodka (1980). Values for q_{pen} between 15 and 60 hours were arbitrarily increased by factors of 1.2, 1.4, 1.6, 1.8, 2, 1.8, 1.6, 1.4 and 1.2 for five-hour periods, respectively. Carbohydrate use was calculated based upon three factors. A maintenance-free cell mass yield of 0.45 g (g glucose)$^{-1}$ and a maintenance value of 27 mg (g cell)$^{-1}$ h^{-1} were used together with a theoretical yield of 0.88 g (g glucose)$^{-1}$, which is the average of two cases with and without recycle of α-aminoadipate. Designation of a portion of the carbohydrate as lactose was arbitrary. Nitrogen utilization was based upon an assumed 8 per cent (by wt) content in cells and penicillin. Optimization of the cell mass against time profile to improve penicillin potency or yield of penicillin from carbohydrate might be beneficial [Queener and Swartz (1979)].

Table 43. Changes during fed-batch or semicontinuous penicillin fermentation [Queener and Swartz (1979)].

Criterion	0–40 h	Time 40 h +	Before harvest
Penicillin production	Slight	High but the rate peaks then declines over most of the period: i.e.	Some production occurs but at a low and declining rate: i.e.
Specific growth rate	8–20 h, 0.07 h^{-1}; 20–40 h, 0.01–0.02 h^{-1}	40–75 h, ~0.009 h^{-1}; 90–186 h, ~0.0005 h^{-1}	115–209 h, ~0 h^{-1}
Carbon and energy source used	Organic nitrogen (e.g., corn-steep liquor) and/or carbohydrates, oils	Carbohydrates, triglyceride oils, ethanol, fed or hydrolyzed slowly and generally thought to regulate growth	Various energy sources as available, sometimes insufficient for maintenance. Growth limitation may shift to other nutrients
Nitrogen source used	Mainly organic nitrogen	Organic nitrogen and/or ammonium and/or nitrate salts depending on the process. Used for cell growth and product formation	Nitrogen source may become growth-rate limiting. Cell lysis may occur
Sulphur source used	In crude organic nitrogen and inorganic salts, e.g., ammonium sulphate. Used rapidly for cell growth	Often fed as an inorganic salt. Used for penicillin synthesis and slow cell growth	
Phosphorus source used	In crude organic nitrogen and as inorganic salts	Very little if any is fed as quantitative needs are very low. Growth-rate limitation may be harmful	
Other nutrients	Required for cell growth	Some, like iron, may be regulatory	

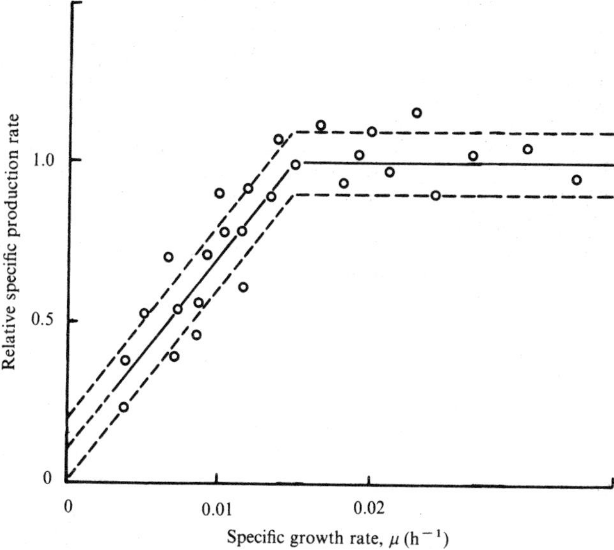

Figure 21. Effect of specific growth rate on the specific production rate of penicillin in glucose – limited growth [Ryu and Hospodka (1980)].

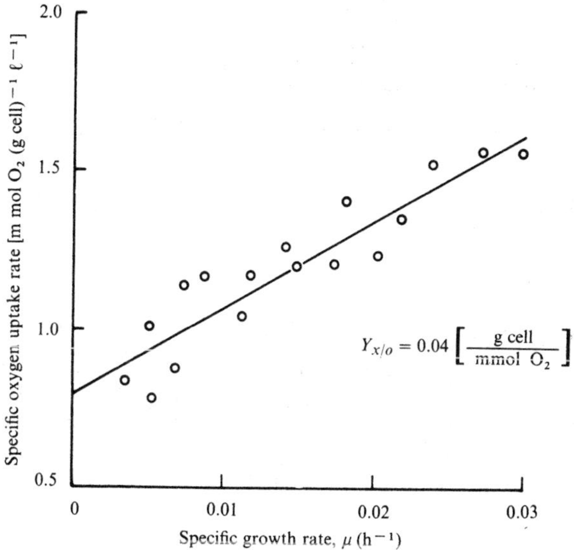

Figure 22. Effect of specific growth rate on the specific oxygen uptake rate in oxygen – limited growth [Ryu and Hospodka (1980)].

Once a large cell mass with a high specific rate of penicillin production is growing at a rate that can be maintained for 120–160 hours, the penicillin fermentation is said to be in the 'production phase'. The culture then functions very much as a continuous one and can be approximated by studies using continuous culture. The actual difference between continuous culture and the productive phase in the fed-batch or repeated fed-batch culture is that a constant cell concentration is maintained in the first case and the cell mass increases slowly in the last two. The quantitative measurements of the specific rates and their interrelationships can be obtained from continuous experiments. An example of such data is given in Table 44. The relevant specific rates from which Table 44 was constructed are given in Figs. 21 and 22.

For these experiments, the optimum specific growth rate for production was $0.015\,h^{-1}$. However, the typical value of the specific growth rate for most industrial fed-batch fermentation is less than $0.01\,h^{-1}$ in the production phase.

Table 44. Specific uptake rates of a mutant strain of *Pen. chrysogenum* KAIS-12690 [Ryu and Hospodka (1980)].

Nutrients and precursor	Specific uptake rates (optimal rate at $\mu = 0.015\,h^{-1}$)	Feed rates required for fed-batch culture (optimal rates at $\mu = 0.015\,h^{-1}$ and $x = 40\,g\,\ell^{-1}$
Carbon (hexose basis)	0.33 [mmol (g cell)$^{-1}$ h^{-1}]	13.2 (mmol ℓ^{-1} h^{-1})
Oxygen	1.6 [optimal, mmol (g cell)$^{-1}$ h^{-1}] 1.2 (at $\mu = 0.015\,h^{-1}$, [mmol (g cell)$^{-1}$ h^{-1}]	64 (mmol ℓ^{-1} h^{-1})
Nitrogen (ammonia basis)	2.0 [mg (g cell)$^{-1}$ h^{-1}]	80 (mg ℓ^{-1} h^{-1})
Phosphorus (phosphate basis)	0.6 [mg (g cell)$^{-1}$ h^{-1}]	24 (mg ℓ^{-1} h^{-1})
Sulphur (sulphate basis)	2.8 (mg (g cell)$^{-1}$ h^{-1}]	112 (mg ℓ^{-1} h^{-1})
Precursor (phenylacetic acid basis)	1.8 (mg (g cell)$^{-1}$ h^{-1}]	72 (mg ℓ^{-1} h^{-1})

Streptomycin

Microorganisms

So far as is known, streptomycin A is produced only by *Streptomyces griseus*, and only a few strains of this species have the ability to produce reasonable amounts. Other less-active forms of streptomycin (*see* Table 45) are also found in nature through fermentations by other streptomycete species. As discussed in Chapter 3 (*see* p.188), an increase in streptomycin yields has been achieved by strain selection and mutation, and by an increased understanding of its biosynthesis.

Media

Both chemically defined (*see* Table 46) and complex media (*see* Table 47) have been used for the study of streptomycin fermentations.

The optimum fermentation temperature is in the range 25–30°C with a pH of 7–8.

Table 45. Natural streptomycins [Claridge (1979)].

Antibiotic	Microorganism	Activity
Streptomycin (streptomycin A)	*Streptomyces griseus*, *Strept. bikiniensis*	
Mannosidostreptomycin (streptomycin B)	*Strept. griseus*	Less active than streptomycin
Hydroxystreptomycin	*Strept. griseocarneus*, *Strept. subrutilus*, *Streptomyces* sp 86	Activity similar to streptomycin and dihydrostreptomycin
Mannosidohydroxystreptomycin	*Streptomyces* sp 86	Less active than streptomycin and hydroxystreptomycin
N-Demethylstreptomycin	*Strept. griseus* (with ethionine)	Only 10% as active as streptomycin
Dihydrostreptomycin	*Strept. humidus*, *Streptomyces* sp.	Activity similar to streptomycin
Bluensomycin	*Strept. bluensis*	

Table 46. Chemically defined medium for streptomycin fermentation using *Streptomyces griseus*.

Component	Amount (%)
Glucose	1.0
Soya bean meal	1.0
Sodium chloride	0.5

Table 47. Complex medium for streptomycin fermentation using *Streptomyces griseus*.

Component	Amount (%)
Glucose	2.5
Extracted soya bean meal	4.0
Distillers' dried solubles	0.5
Sodium chloride	0.25

Metabolic pathway

The structure of streptomycin is provided in Table 35. The antibiotic is synthesized from glucose with the initial formation of *myo*-inositol and subsequent formation of streptidine phosphate. Streptomycin phosphate is then formed which is followed by a phosphate-cleavage step to produce the antibiotic.

Time course

Fig. 23 shows the time course of streptomycin formation in relation to mycelial production and oxygen consumption. Table 48 summarizes the changes that occur during the three phases of fermentation. Streptomycin production is maximal when the respiration rate has peaked.

Table 48. Changes characterizing the three phases of streptomycin production [Hockenhull (1960)].

	Phase 1 Growth	Phase 2 Maturation	Phase 3 Senescence
Streptomycin	Slight production	Maximum rate of production	Streptomycin level ceases to rise or fall
pH	Steady rise	Very slow fall	Rise
Mycelium	Rapid growth	Mycelial weight fairly constant	Mycelial disintegration
Glucose	Used slowly	Used steadily throughout	Usually absent
Ammonia	Released into medium	Utilized	Released
Inorganic phospate	Released	Utilized	Released
Q_{O_2}	High	Moderate	Low
Total oxygen demand	High	High	Low

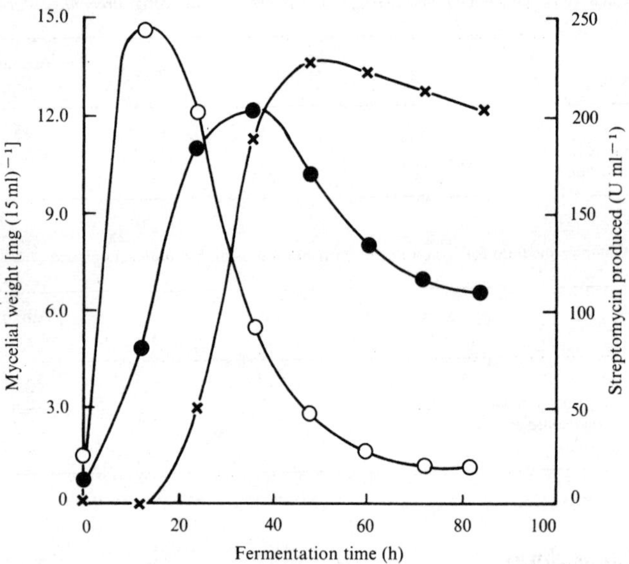

Figure 23. Growth, respiration and production of streptomycin of *Streptomyces griseus* A2. ×, streptomycin; ○, oxygen; ●, mycelium [Steel (1958)].

The concentration of phosphate in the medium is critical for the production of streptomycin. Although phosphate can stimulate growth and sugar utilization, an excess inhibits antibiotic synthesis (*see* Table 49). In defined medium containing glucose, optimum phosphate concentrations are in the range 1–10 mM. In complex media, phosphate is not added generally and inhibition of antibiotic production occurs at about 3 mM. Streptomycin production is stimulated concentration dependently by ammonium phosphate (*see* Table 50) compared with the inhibitory effect of potassium phosphate.

Arsenite also depresses streptomycin yield (*see* Table 49), the effect being most apparent when added earlier in the fermentation.

Table 49. Effect of phosphate or arsenite on streptomycin yields[a] [Hockenhull (1960)].

Addition	Time of addition (day)	Streptomycin yield (mg ℓ^{-1}) Day 6	Day 7	pH Day 6	Day 7
Water			1280		7.39
15 mM Potassium phosphate	0		503		8.07
Water		1250	1040	7.91	8.27
3 mM Sodium arsenite	4	733	627	7.69	7.44
3 mM Sodium arsenite	3	310	310	7.68	8.27
3 mM Sodium arsenite	2	13	12	5.79	5.87

[a] Medium was soya bean meal.

Table 50. Streptomycin production on a proline medium with the amount of diammonium hydrogen phosphate as a variable [Hockenhull (1960)].

Diammonium hydrogen phosphate concentration (g ℓ^{-1})	Streptomycin on day 8 (mg ℓ^{-1})
5.7	225
2.0	170
0.60	570
0.20	675
0.06	530
0.02	390
0.006	200
0.002	130
0.000	25

Tetracyclines

Microorganisms

Streptomycetes species capable of producing various tetracyclines are presented in Table 51.

Media

Strept. aureofaciens is capable of growing and producing chlortetracycline on simple media containing glycerol, for instance, as the only carbon source and ammonium ions as the nitrogen source. The medium whose composition is shown in Table 52 can be used for tetracycline fermentation producing 70 mg ℓ^{-1}.

Table 51. Streptomycetes species producing tetracyclines [Perlman (1979)].

Name	Products
Strept. alboflavus	7-chlortetracycline, tetracycline, oxytetracycline
Strept. antibioticus	Oxytetracycline, tetracycline
Strept. aureofaciens	7-chlortetracycline, tetracycline
Strept. aureus	Tetracycline
Strept. californicus	Actinomycin, tetracycline
Strept. cellulosae	Actinomycin, oxytetracycline
Strept. feofaciens	Tetracycline
Strept. flaveolus	Tetracycline
Strept. flavus	Actinomycin, chlortetracycline, oxytetracycline, tetracycline
Strept. fuscofaciens	Oxytetracycline, quinocycline
Strept. lustanus	Chlortetracycline, tetracycline
Strept. parvus	Actinomycin, oxytetracycline, tetracycline
Strept. platensis	Oxytetracycline
Strept. rimosus	Oxytetracycline, rimocidin, tetracycline
Strept. sayamaensis	Chlortetracycline, tetracycline
Strept. vendargensis	Oxytetracycline, vengacide

Table 52. Composition of medium for tetracycline fermentation.

Component	Amount (%)
Sucrose	1.0
Corn-steep liquor	1.0
Diammonium hydrogen phosphate	0.2
Potassium dihydrogen phosphate	0.2
Calcium carbonate	0.1
Magnesium sulphate heptahydrate	0.025
Zinc sulphate heptahydrate	0.005
Copper sulphate pentahydrate	0.00033
Manganese chloride tetrahydrate	0.00033

Alternatively, the fermentation medium may contain sucrose, peanut oil meal, corn-steep liquor, molasses, ammonium sulphate, calcium carbonate and sodium chloride. It is essential that the medium is free of chloride ions (*see* Figs. 24 and 25), alternatively a compound that inhibits the inclusion of chloride into the antibiotic must be added to the medium. Among such compounds are bromide, fluoride, 2-mercapto-2-thiazoline, 2-benzoxazolethiol, etc.

Yields obtained in tetracycline fermentation range from 1 g ℓ^{-1} for oxytetracycline to 2–9 g ℓ^{-1} for tetracycline.

The aeration requirement for *Strept. rimosus* is around 14–25 mg O_2 ℓ^{-1} min^{-1}. *Strept. aureofaciens* requires more intensive aeration. The production of tetracyclines is affected by the concentration of carbon dioxide in the fermentation medium, e.g., for *Strept. aureofaciens*, the optimum carbon dioxide concentration is 20–80 ml ℓ^{-1}.

The optimum fermentation temperature is about 28°C. For *Strept. aureofaciens*, the pH range is 4.2–8.0 (optimum 6.0–6.8) for growth and 5.5–6.6 (optimum 5.8–6.0) for antibiotic production.

The amount of inoculum is in the range 2–10 per cent of the fermentation medium.

Production of chlortetracycline is stimulated by benzylthiocyanate at a concentration of 0.5–3.0 mg ℓ^{-1}. Arsenate or arsenite partially inhibits growth.

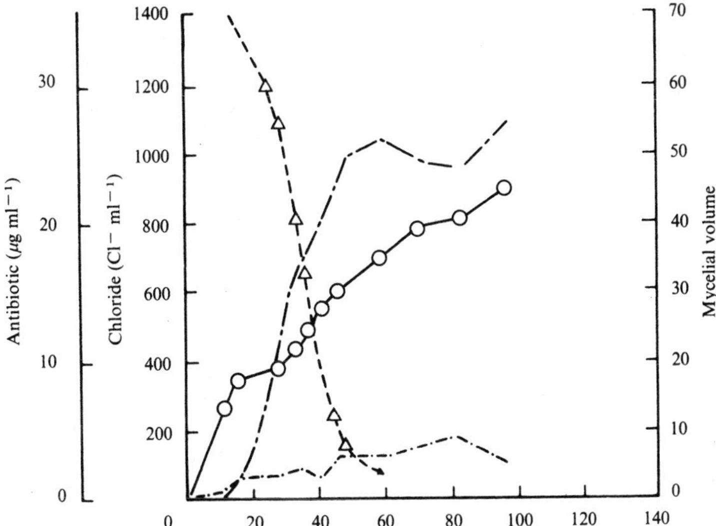

Figure 24. Fermentation of *Streptomyces* strain RO-911 in medium with 100 μg ml^{-1} Cl^{-}. Δ, chloride; –.–.–, tetracycline; ——————, chlortetracycline; ○, mycelial volume [Rolland and Sensi (1955)].

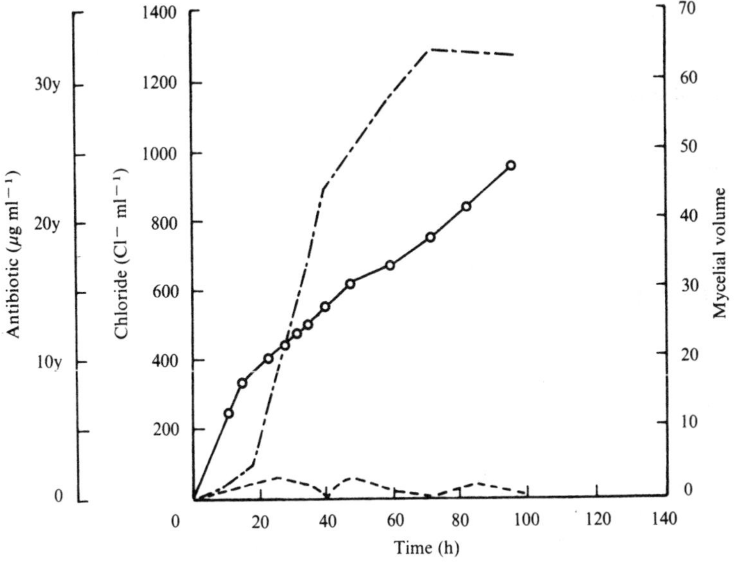

Figure 25. Fermentation of *Streptomyces* strain RO-911 in medium with 2 μg/ml^{-1} Cl^{-} (medium was as follows: 1 per cent cereal extract (Cl^{-} 0.01 per cent), 0.8 per cent ammonium sulphate (Cl^{-} 0.00005 per cent), 0.9 per cent calcium carbonate (Cl^{-} 0.00007 per cent), 5 per cent dextrose (Cl^{-} 0.00005 per cent) and distilled water). – – –, tetracycline; —·—·, chlortetracycline; ○, mycelial volume [Rolland and Sensi (1955)].

Metabolic pathways

The glycolytic pathway provides the building blocks for tetracycline synthesis and the presence of the initial reaction steps, yielding malonyl-CoA, is a prerequisite for high yields. The mechanism of tetracycline biosynthesis is very complex. There is a hypothetical scheme which embodies a total of 72 metabolites, out of which 27 have been identified and 45 are hypothetical, including a nonaketide.

Table 53 lists structures of clinically important tetracyclines with the basic structure

Table 53. Structures of clinically important tetracyclines.

Name	Empirical formula	R_1	R_2	R_3	R_4	Production process
7-Chlortetracycline	$C_{22}H_{23}ClN_2O_8$	H	OH	Me	Cl	Fermentation
6-Demethyl-7-chlortetracycline (demeclocycline)	$C_{21}H_{21}ClN_2O_8$	H	OH	H	Cl	Fermentation
6-Deoxy-5-oxytetracycline (doxycycline)	$C_{22}H_{24}N_2O_8$	OH	H	Me	H	Semisynthetic
7-Dimethylamino-6-deoxy-6-demethyltetracycline (minocycline)	$C_{23}H_{27}N_3O_7$	H	H	H	NMe_2	Semisynthetic
5-Oxytetracycline	$C_{22}H_{24}N_2O_9$	OH	OH	Me	H	Fermentation
Tetracycline	$C_{22}H_{24}N_2O_8$	H	OH	Me	H	Fermentation

Time courses

Figs. 24 and 25 show the time course of tetracycline production and demonstrate the effect of chloride ions. Figs. 26 and 27 give the time courses of chlortetracycline production and Fig. 28 that of oxytetracycline formation.

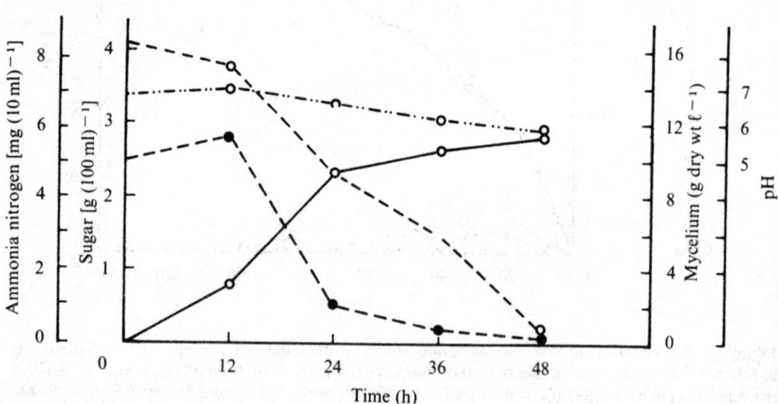

Figure 26. Chemical changes during fermentation of *Streptomyces aureofaciens* in standard medium. – – –, sugar; ——, mycelial dry weight, .●., ammonium nitrogen; –..–.., pH [Biffi *et al.* (1954)].

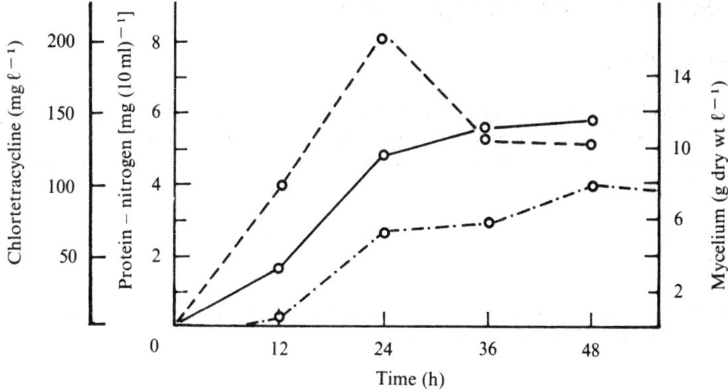

Figure 27. Growth synthesis of proteins and chlortetracycline in standard medium with *Streptomyces aureofaciens*. _ _ _, protein nitrogen; ____, mycelial dry weight; _ . _ ., chlortetracycline [Hockenhull (1959)].

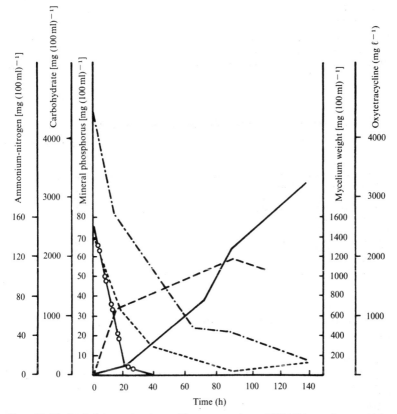

Figure 28. Biochemical changes in cultures of *Streptomyces rimosus* LS-T-118 in a medium containing 5 per cent starch. ____, oxytetracycline; _ _ _ _, mycelial dry weight; _ . _ . _ . _, carbohydrate; ...., ammonium-nitrogen; – ○ ○ – ○ ○ –, inorganic phosphorus [Orlova and Zaytseva (1961)].

2,3-BUTANEDIOL/ACETOIN

Microorganisms

Microorganisms capable of producing 2,3-butanediol are bacteria belonging to the genera *Aerobacter*, *Aerobacillus*, *Aeromonas*, *Serratia* and *Bacillus*. *Aerobacter aerogenes*, *Aeromonas hydrophilia*, *Bacillus polymyxa*, *B. subtilis* and *Serratia marcenscens* are most commonly used.

Media

The composition of the medium depends on the microorganism employed. Some examples of such media are given in Tables 54–58.

Table 54. Medium[a] for 2,3-butanediol/acetoin fermentation using *Aerobacter aerogenes*.

Component	Amount
Acid-hydrolyzed starch	10%
Calcium carbonate	5–10 g ℓ^{-1}
Urea	2 g ℓ^{-1}
Magnesium sulphate heptahydrate	0.25 g ℓ^{-1}
Potassium dihydrogen phosphate	0.60 g ℓ^{-1}

[a] The medium is at pH 5.5–6.0.

Table 55. Inorganic medium for 2,3-butanediol/acetoin fermentation using *Bacillus polymyxa*.

Component	Amount (g ℓ^{-1})
Potassium hydrogen phosphate	1.0
Potassium dihydrogen phosphate	1.0
Magnesium sulphate heptahydrate	0.2
Sodium chloride	0.1
Calcium chloride	0.1
Ferrous sulphate heptahydrate	0.01
Manganese sulphate tetrahydrate	0.01
Zinc sulphate	0.01

Table 56. Medium for 2,3-butanediol/acetoin fermentation using *Bacillus polymyxa*.

Component	Amount (%)
Glucose	2.0
Calcium carbonate	1.0
Peptone	0.5
Potassium hydrogen phosphate	0.2

Table 57. Basal medium for 2,3-butanediol/acetoin fermentation using *Aeromonas hydrophila*.

Component	Amount
Yeast extract	1.25 g
Calcium carbonate	1.0 g
Magnesium sulphate	0.025 g
1 M phosphate buffer (pH 7.4)	0.65 ml
Tap water	250 ml

Aerob. aerogenes can also utilize wood hydrolyzates and glucose (10–25 per cent).

B. polymyxa needs biotin for growth but is capable of fermenting a variety of grain mashes, glucose, xylose, mannitol and other carbohydrates. Mashes containing more than 15 per cent wheat by weight are fermented inefficiently and are usually very viscous. The composition of the inorganic medium used for culture of *B. polymyxa* is given in Table 55, with the composition of an alternative medium being given in Table 56.

The composition of the basal medium for the culture of *Aerom. hydrophila* is given in Table 57.

Carbohydrate at concentration of 1–2 per cent is added to the medium described in Table 57. The carbohydrate used may be glucose, xylose or pyruvic acid.

Table 58. Medium[a] for the culture of *Bacillus subtilis*.

Component	Amount (%)
Glucose	3.0
Yeast extract	1.0
Calcium carbonate	1.0

[a] pH 6.0–6.8.

Metabolic pathway

The biosynthetic pathway for the production of acetoin and 2,3-butanediol is outlined in Fig. 29.

Time course

Fig. 30 contains a time course for fed-batch production of 2,3-butanediol from xylose using *Klebsiella pneumoniae*.

Tables 59–61 indicate the major products of 2,3-butanediol fermentation using a variety of substrates and a range of microorganisms.

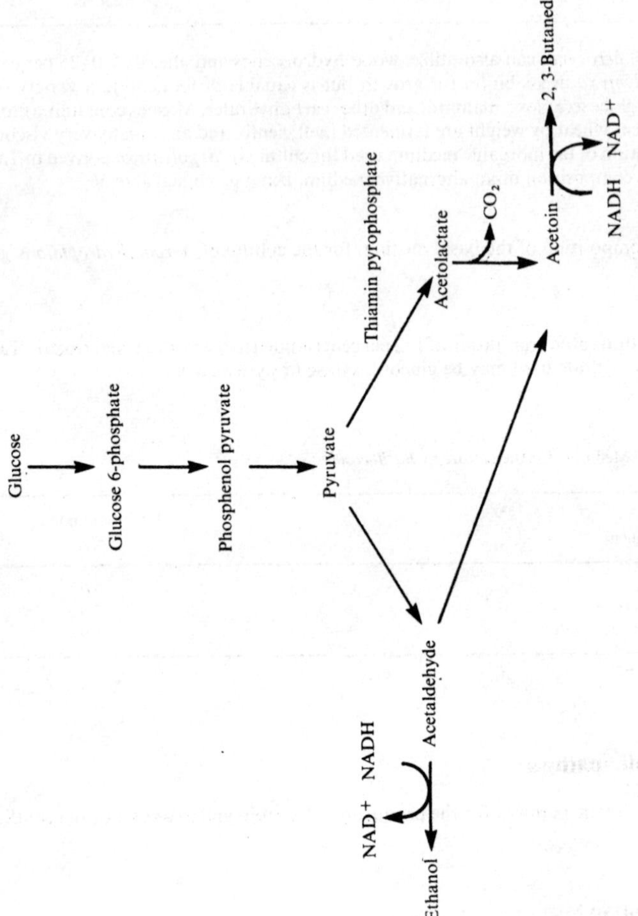

Figure 29. Metabolic pathway for the formation of acetoin and 2,3-butanediol from glucose.

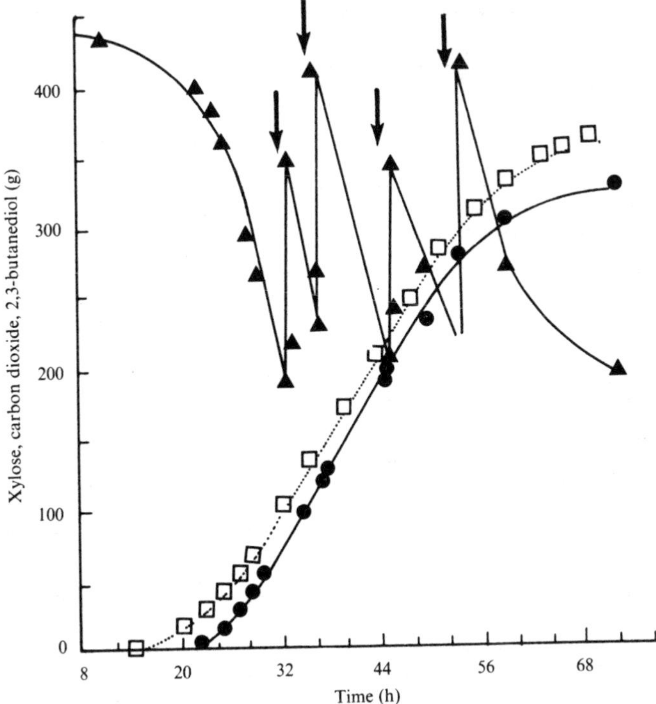

Figure 30. Fed-batch production of 2,3-butanediol from xylose by *Klebsiella pneumoniae* B 199. ▲, Xylose; □, carbon dioxide; ●, 2,3-butanediol [Flickinger (1980)].

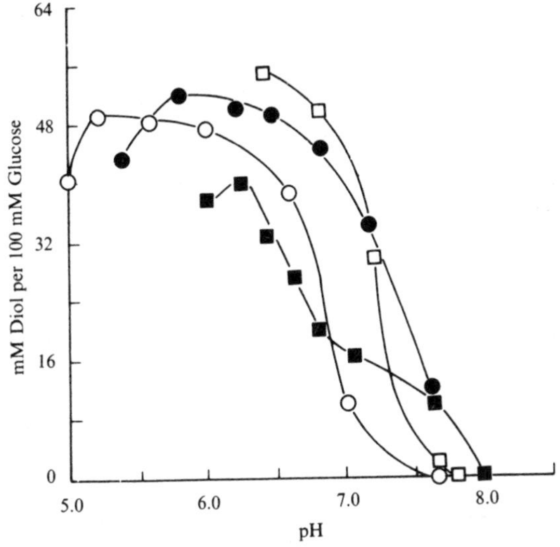

Figure 31. Effect of pH on production of 2,3-butanediol. ○, *Aerobacter aerogenes*; ●, *Bacillus polymyxa*; □, *Serratia marcescens*; ■, *B. subtilis* [Underkofler and Hickey (1954)].

Table 59. Fermentation of glucose, xylose, pyruvic acid, and mannitol by *Bacillus polymyxa*, NRC 25 [Adams and Stanier (1945)].

Raw material	Weights (g)			
	Glucose	Xylose	Pyruvic acid	Mannitol
Amount raw material fermented	5.21	4.20	1.28	2.04
Amounts of end products formed				
2,3-Butanediol	1.70	0.96	None	0.14
Acetoin	0.07	0.06	0.26	0.01
Ethanol	0.88	0.81	Trace	0.52
Acetic acid	0.05	0.13	0.46	0.13
Lactic acid			Trace	0.58
Succinic acid				0.08
Carbon dioxide	2.54	1.99	0.61	0.73
Hydrogen	0.04	0.05	0.02	0.04
Carbon recovery (%)	101.6	92.9	93.0	106.8
Carbon dioxide calculated/observed ratio	1.02	0.94	0.97	0.97
Hydrogen calculated/observed ratio	1.08	0.71	0.95	0.93
O/R index	0.99	0.96	0.97	0.97

Table 60. Fermentation of glucose, xylose and pyruvic acid by *Aeromonas hydrophila* [Stanier and Adams (1944)].

	Glucose		Xylose		Pyruvic acid	
	Weight (g)	Yield (%)	Weight (g)	Yield (%)	Weight (g)	Yield (%)
Amount of raw material fermented	5.49		4.01		3.53	
Amount of end product formed						
2,3-Butanediol	1.50	54.7	0.94	39.0	0.13	3.5
Acetoin	0.05	1.7	0.06	2.6	0.02	0.5
Ethanol	0.73	52.0	0.60	48.9	0.04	2.1
Acetic acid	0.09	4.6	0.15	9.3	1.80	74.8
Lactic acid	0.64	23.3	0.49	20.4	0.78	21.5
Succinic acid	0.13	3.6	0.04	1.1		
Carbon dioxide	2.23	166.2	1.58	134.7	1.48	83.8
Hydrogen	0.04	57.5	0.03	53.9	0.04	43.6
Carbon recovery (%)	98.2		96.6		106.3	
Carbon dioxide calculated/observed ratio	1.00		1.04		1.01	
Hydrogen calculated/observed ratio	1.11		1.13		1.10	
O/R index	1.02		0.99		1.05	
2,3-Butanediol/ethanol (approx.)	1/1		0.8/1		1.6/1	

Table 61. Anaerobic dissimilation of glucose by *Serratia marcescens* [Neish *et al.* (1947)].

Products	mM of products/100 mM. of glucose dissimilated			
	Strain	Strain	Strain	Strain
2,3-Butanediol	57.90	55.20	51.45	42.45
Acetoin	0.25	0.50	0.81	1.14
Glycerol	6.14	4.18	4.54	5.63
Ethanol	40.85	41.30	42.24	25.90
Lactic acid	15.70	26.50	33.09	54.15
Formic acid	48.50	44.00	39.80	27.60
Acetic acid	0.0	0.0	0.0	0.0
Succinic acid	2.98	3.34	3.41	18.80
Carbon dioxide	103.8	102.5	106.1	78.2
Hydrogen	0.0	0.0	0.52	0.27
Fermentation time (days)	17	12	9	7
Glucose dissimilated, (%)	93.2	88.5	99.5	99.9
Carbon accounted for (%)	91.4	93.0	94.6	97.5
O/R index	0.99	1.03	1.04	1.01

Table 62. Effect of pH on the dissimilation of glucose by *Bacillus subtilis*[a, b] [Neish *et al.* (1945)].

Product	Yield (%)	
	Grown at pH 6.2 to 5.8	Grown at pH 7.6 to 6.8
2,3-Butanediol	56.15	36.16
Acetoin	Trace	Trace
Glycerol	26.28	16.39
Ethanol	18.24	28.70
Lactic acid	39.13	53.08
Succinic acid	Trace	5.05
Formic acid	9.97	30.14
Acetic acid	Doubtful	3.98
n-Butyric acid	Doubtful	2.76
Carbon dioxide (calculated)	130.56	101.01
Carbon accounted for	100.0%	97.0%
Glucose dissimilated (4 days)	73.7%	68.2%

[a] 3 per cent glucose, 0.5 per cent potassium dihydrogen phosphate, 0.6 per cent dipotassium hydrogen phosphate, 0.02 per cent magnesium sulphate heptahydrate, 0.1 per cent casein hydrolyzate.
[b] Conditions: grown at 30°C without aeration; the pH was measured with a glass electrode every eight — ten hours and adjusted with *N* sodium hydroxide.

In Fig. 31, the effect of pH on the 2,3-butanediol yield coefficient for a range of microorganisms is shown. In all microbial species studied, the yield declines above pH 6.0. Fig. 32 shows the effect of pH on the fermentation time. Data which illustrate the effect of pH on the relative amounts of various products of a 2,3-butanediol fermentation, using *B. subtilis* with glucose as substrate, are given in Table 62.

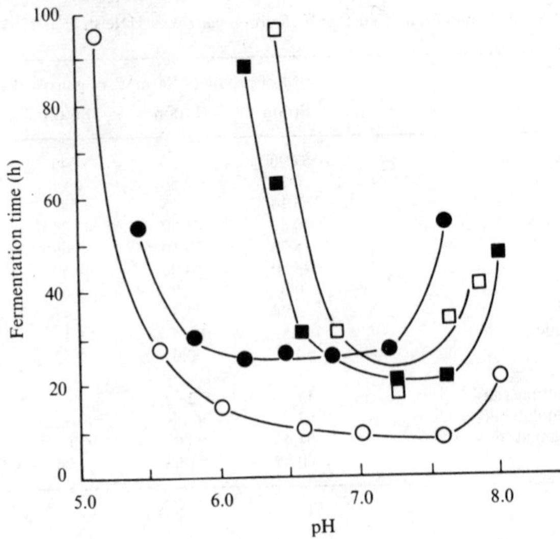

Figure 32. Effect of pH on rate of dissimilation of 5 per cent glucose during production of 2,3-butanediol. ○, *Aerobacter aerogenes*; ●, *Bacillus polymyxa*; □, *Serratia marcescens*; ■, *B. subtilis* [Underkofler and Hickey (1954)].

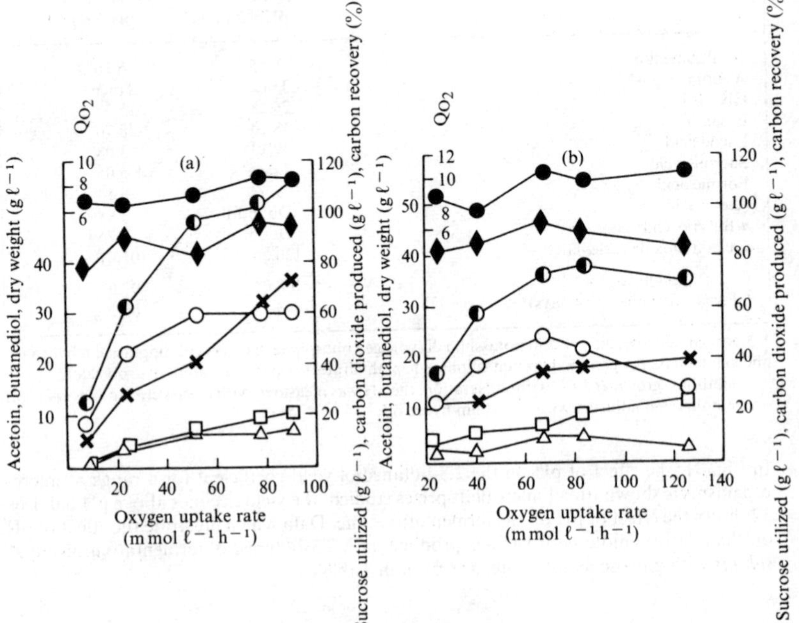

Figure 33. Metabolic product formation versus oxygen transfer rate. (a) Dilution rate, 0.11 h^{-1}. (b) Dilution rate, 0.2 h^{-1}. Temperature, 35°C; pH 5.5. ●, Respiration rate; ◐, sucrose utilized; ◆, percentage of carbon recovered as diol, acetoin, carbon dioxide and bacteria; ○, 2,3-butanediol; □, bacterial dry weight; △, acetoin; ×, carbon dioxide [Pirt and Callow (1958)].

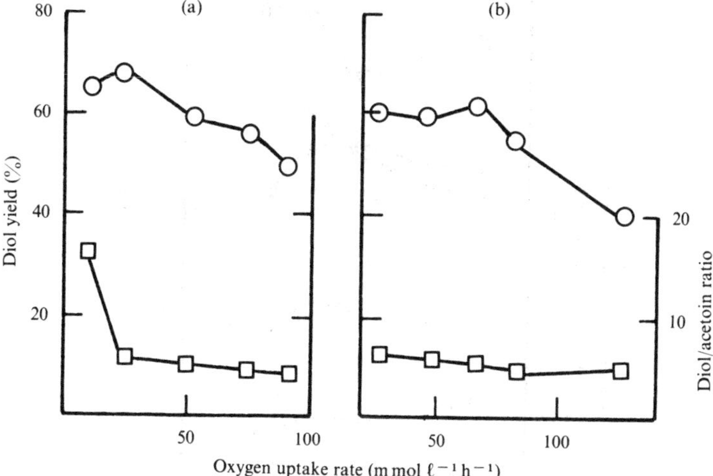

Figure 34. 2,3-Butanediol yield (percentage of theoretical maximum) and the diol/acetoin ratio as a function of oxygen uptake rate. Temperature, 35°C; pH, 5.5. (a) Dilution rate, $0.1\,h^{-1}$. (b) Dilution rate $0.2\,h^{-1}$. ○, Percentage diol yield; □, diol/acetoin ratio [Pirt and Callow (1958)].

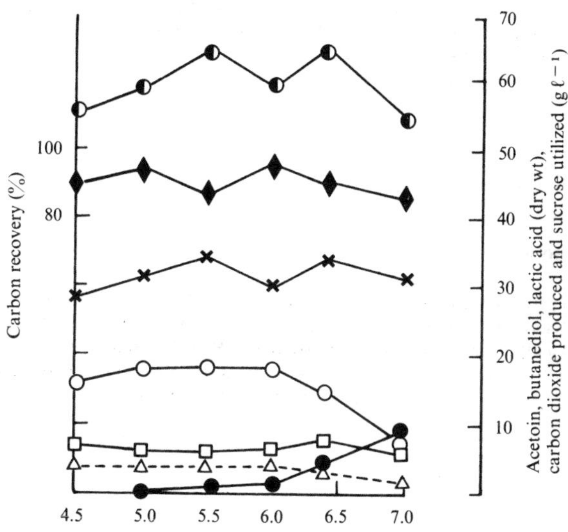

Figure 35. Formation of acetoin and 2,3-butanediol as a function of pH. Dilution rate, $0.2\,h^{-1}$; oxygen uptake rate, $55\,m\,mol\,\ell^{-1}\,h^{-1}$; temperature, 35°C. ◐, Sucrose utilized; ◆, carbon recovery; ×, carbon dioxide; ○, diol; □, bacterial dry weight; △, acetoin; ●, lactic acid [Pirt and Callow (1958)].

Figs. 33–36 contain data obtained in continuous culture, Figs. 33 and 34 being concerned with the effect of the oxygen uptake rate on yield and Figs. 35 and 36 showing the effect of pH and temperature, respectively, on yield.

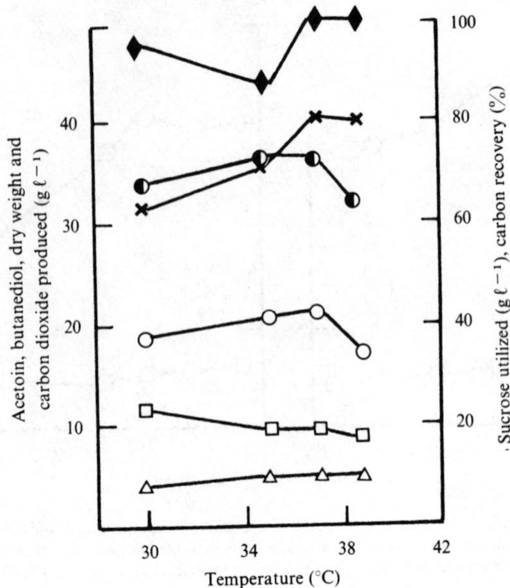

Figure 36. Acetoin and 2,3-butanediol formation as a function of temperature. Dilution rate, $0.2\,h^{-1}$; oxygen uptake rate, 91 m mol $\ell^{-1}h^{-1}$; pH, 5.0. ◆, Percentage carbon recovered as diol/acetoin, bacteria and carbon dioxide; ×, carbon dioxide; ◐, sucrose utilized; ○, diol; □, bacterial dry weight; △, acetoin [Pirt and Callow (1958)].

CAROTENE

Microorganisms

β-Carotene is produced by species of the genera *Choanephora* and *Blakeslea* of the family Choanephoraceae (class Phycomycetes), e.g., *Phycomyces blakesleeanus*, *C. cucurbitarum* and *Bl. trispora* and other fungi, as well as by yeasts, such as *Rhodotorula flava*, *R. gracilis* and *R. sannieli* and by some species of bacteria, algae and lichens. The greatest yields are obtainable using a mixture of + and − strains of *Bl. trispora*.

Media

The composition of a basal medium suitable for the β-carotene-producing microorganisms is listed in Table 63.

Table 63. Basal medium for β-carotene-producing microorganisms.

Component	Amount[a] (g ℓ$^{-1}$)
Distillers' solubles	60
Starch	60
Soya bean oil	35
Cotton seed oil	35
Kenesene	20
Yeast extract	1
Potassium phosphate	0.5
Manganese sulphate	0.2
Thiamin hydrochloride	0.01
Ethoxyquin	0.5

[a] Made up in tap water.

Metabolic Pathway

The formation of carotenes from acetyl-CoA is outlined in Fig. 37.

Yields

Carotene production by some algal species are summarized in Table 64. High carotene yields are achieved in interspecifically mated cultures (*see* Table 65 and Fig. 38).

Tables 66–75 provide data on the effects of various media constituents and additives on carotene production.

Table 64. Carotene contents of some algae [Goodwin and Treble (1958)].

Algae	Carotene [μg (g dry wt)$^{-1}$]	Percentage of total carotenoids (%)
Ceramium rubrum	299	17
Chaetomorpha linum	476	12
Chlorella vulgaris	105	10
Chl. pyrenoidosa	420	16
Euglena gracilis var. *bacillaris*	800	11
Gigartina stellata	86	
Hemiselmis virescens	230	
Microcystis aeruginosa	158	36
Ochromonas dancia	725	17

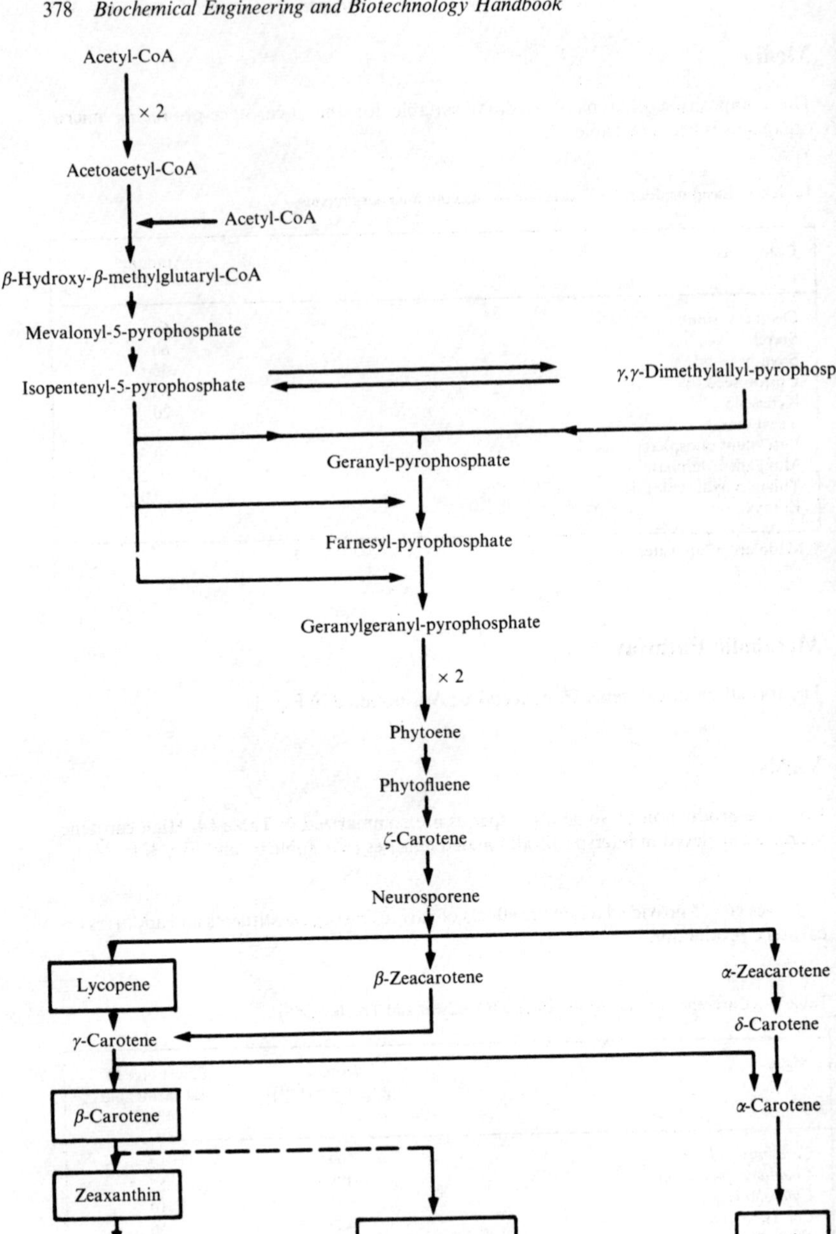

Figure 37. Pathway of carotenoid biosynthesis [Ninet and Renaut (1979)].

Table 65. Carotene synthesis by intraspecifically and interspecifically mated cultures [Anderson *et al.* (1958)].

Culture	Dry weight (g ℓ$^{-1}$)	Carotene production [μg (g dry wt)$^{-1}$]	(mg ℓ$^{-1}$)
Intraspecially mated			
Choanephora circinans			
NRRL A-6680 (+)	27.9	103	2.88
NRRL A-6777 (−)	25.2	90	2.27
A-6680 × A-6777	21.0	240	5.04
C. conjuncta			
NRRL 2560 (+)	14.6	67	0.98
NRRL 2561 (−)	25.5	25	0.64
2560 × 2561	25.0	185	4.83
Interspecifically mated			
Blakeslea trispora			
NRRL 2457 (−)	38.9	138	5.37
C. circinans			
NRRL 2547 (+)	18.4	32	0.59
2547 × 2547	33.6	640	21.50
B. trispora			
NRRL 2456 (+)	37.0	150	5.55
C. conjuncta			
NRRL 2562 (−)	24.8	107	2.66
2456 × 2562	38.4	368	14.12

Table 66. Effect of various hydrocarbons on the production of carotenoids by *Mycobacterium smegmatis*[a] [Tanaka *et al.* (1968)].

Carbon source	Cell yield (g dry wt ℓ$^{-1}$)	Carotenoid produced (μg ℓ$^{-1}$)	[μg (g cells)$^{-1}$]
n-Hexane	No growth		
n-Heptane	No growth		
n-Octane	No growth		
n-Nonane	No growth		
n-Decane	No growth		
n-Undecane	0.99	118	119.1
n-Dodecane	2.00	80	40.0
n-Tridecane	3.21	200	62.4
n-Tetradecane	3.21	305	95.0
n-Pentadecane	2.60	160	61.5
n-Hexadecane	3.74	340	91.7
n-Heptadecane	3.11	167	53.8
n-Octadecane	3.20	180	56.3
n-Nonadecane	3.65	121	33.2
n-Eicosane	No growth		
'Mixture 1' (C_{11-14})	2.48	288	95.9
'Mixture 2' (C_{10-13})	2.95	288	97.6
'Mixture 3' (C_{10-14})	2.95	284	96.3

[a] Cultivation was carried out on a rotary shaker (220 rpm) at 30°C for six days in basal medium except for carbon source.

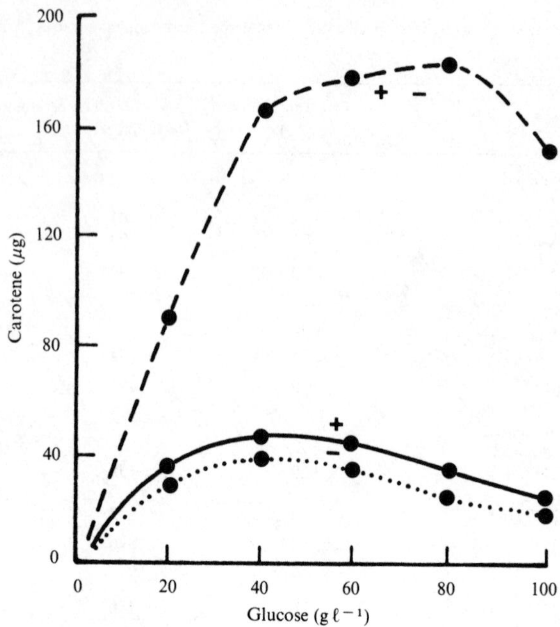

Figure 38. Maximum yields of carotene by +, − and mixed + and − cultures of *Choanephora cucurbitarum* when the glucose concentration in the glutamic acid medium varied between 20 and 100 g ℓ$^{-1}$ [Chu and Lilly (1960)].

Table 67. Effect of nitrogen sources on carotenoid production by *Mycobacterium smegmatis*[a] [Tanaka *et al.* (1968)].

Nitrogen source[b]	Cell yield (g cells ℓ$^{-1}$)	Carotenoid produced (μg ℓ$^{-1}$)
Ammonium carbonate	2.96	360
Ammonium chloride	1.25	112
Ammonium nitrate	1.31	89
Ammonium phosphate	1.34	430
Ammonium sulphate	0.82	250
Casamino acid	1.22	55
Urea	1.91	72
Ammonium carbonate + ammonium phosphate[c]	2.62	361
Ammonium sulphate + ammonium carbonate[c]	1.86	276
Ammonium sulphate + ammonium phosphate[c]	1.71	329
Ammonium sulphate + urea[d]	3.04	245

[a] Cultivation was carried out for five days using 'hydrocarbon mixture 2' as the carbon source.
[b] Unless otherwise stated nitrogen sources were added at a concentration of 0.2 per cent.
[c] Nitrogen sources added at concentrations of 0.1 per cent.
[d] Ammonium sulphate added at 0.15 per cent and urea at 0.005 per cent.

Table 68. Effect of natural nutrients on carotenoid production by *Mycobacterium smegmatis*[a] [Tanaka *et al.* (1968)].

Addition	Amount added (mg ℓ$^{-1}$)	Cell yield (g dry wt ℓ$^{-1}$)	Carotenoid produced (μg ℓ$^{-1}$)	[μg (g cells)$^{-1}$]
None		3.02	336	111
Casamino acid	100	3.09	294	95
Casamino acid	500	3.55	320	90
Corn-steep liquor	100	3.13	252	77
Corn-steep liquor	500	3.55	324	91
Fish meat extract	100	3.30	290	88
Fish meat extract	500	3.31	459	138
Malt extract	100	3.41	264	77
Malt extract	500	3.17	262	83
Peptone	100	3.17	304	96
Peptone	500	3.69	348	94
Yeast extract	100	3.29	348	106
Yeast extract	500	3.61	370	102

[a] Cultivation was carried out on a rotary shaker (220 rpm) for six days using 'hydrogen mixture 2' as a carbon source.

Table 69. Effects of various amino acids on the production of carotenoids by *Mycobacterium smegmatis*[a] [Tanaka *et al.* (1968)].

Amino acid	Amount added (mM)	Cell yield (g dry wt ℓ$^{-1}$)	Carotenoid produced (μg ℓ$^{-1}$)	[μg (g cells)$^{-1}$]
None		2.78	180	65
L-Arginine	1	2.54	220	87
L-Asparagine	1	2.26	140	62
L-Aspartic acid	1	1.79	140	78
DL-Cysteine	2	2.85	190	67
L-Cystine	1	2.80	170	61
L-Glutamic acid	1	1.90	350	184
L-Glutamine	1	1.99	190	96
L-Glycine	1	2.16	270	125
L-Histidine	1	2.38	320	135
L-Homocysteine	1	2.12	200	94
L-Isoleucine	1	2.40	240	100
L-Leucine	1	2.31	300	130
L-Lysine	1	2.35	150	64
L-Methionine	1	0.65	180	277
L-Phenylalanine	1	1.87	140	75
L-Proline	1	2.01	260	129
DL-Serine	2	1.86	360	194
L-Threonine	1	3.06	170	56
DL-Tryptophan	2	2.41	180	75
L-Tyrosine	1	2.07	320	155

[a] Cultivation was over six days using a 'hydrocarbon mixture 2', which is rich in *n*-undecane, as the carbon source.

Table 70. Effect of histidine concentration on the carotenoid production by *Mycobacterium smegmatis*[a] [Tanaka *et al.* (1968)].

Histidine (mM)	Cell yield (g dry wt ℓ$^{-1}$)	Carotenoid produced (μg ℓ$^{-1}$)	[μg (g cells)$^{-1}$]
0	3.18	323	102
0.3	3.16	617	205
3.0	3.28	739	225

[a] Cultivation was carried out on a rotary shaker (220 rpm) for six days using *n*-hexadecane as a carbon source.

Table 71. Effect of water-soluble vitamins on carotenoid production by *Mycobacterium smegmatis*[a] [Tanaka *et al.* (1968)].

Vitamin	Amount added (mg ℓ$^{-1}$)	Cell yield (g dry wt ℓ$^{-1}$)	Carotenoid production (g ℓ$^{-1}$)	[μg (g cells)$^{-1}$]
None		3.04	246	81
Biotin	0.01	3.15	508	129
Folic acid	0.01	3.07	300	98
Riboflavin	1.00	2.96	292	99
Thiamin	1.00	2.89	475	161

[a] Cultivation was carried out for five days using 'hydrocarbon mixture 2' as a carbon source.

Table 72. Effect of β-ionine concentration on production of β-carotene by *Blakeslea trispora* 2456 × 2457[a] [Hanson (1967)].

β-Ionine (mg ℓ$^{-1}$)	Mycelial weight (g ℓ$^{-1}$)	Carotene production [mg (g dry wt)$^{-1}$]	[mg ℓ$^{-1}$]
0.000	61.4	1.65	100.00
0.094	64.9	1.55	100.50
0.940	63.1	1.75	110.30
9.400	58.4	3.36	197.50
94.000	54.5	5.40	294.00
940.000	52.0	7.04	368.20
1880.000	47.8	8.00	382.40

[a] Incubated in 23 g acid hydrolyzed corn, 47 g acid hydrolyzed soya bean oilmeal, 0.5 g potassium dihydrogen phosphate, 1 mg thiamin hydrochloride, 40 ml lipid, 1.2 ml nonionic detergent, water to 1 litre; pH 6.2; 4 per cent cotton-seed oil added and 0.1 ml β-ionine added to each flask after two days' incubation.

Table 73. Effect of nonionic detergents on carotenoid production by *Mycobacterium smegmatis*[a] [Tanaka *et al.* (1968)].

Detergent	Amount added ($\mu g\ \ell^{-1}$)	Cell yield (g dry wt ℓ^{-1})	Carotenoid produced ($\mu g\ \ell^{-1}$)	Carotenoid produced [μg (g cells)$^{-1}$]
None		3.13	328	104.8
Span 20	50	2.77	519	187.4
Span 40	50	2.97	324	109.1
Span 60	5	3.10	270	87.1
Span 60	50	2.66	276	103.8
Span 80	5	3.18	332	104.4
Span 80	50	3.17	325	102.5
Tween 20	5	2.94	620	210.9
Tween 20	50	3.15	354	112.3
Tween 40	5	3.03	514	169.6
Tween 40	50	2.75	580	210.9
Tween 60	5	3.04	530	174.3
Tween 60	50	2.93	309	105.5
Tween 80	50	2.95	644	218.3
Tween 85	5	2.93	290	99.0
Tween 85	50	2.76	252	91.3

[a] Cultivation was carried out on a rotary shaker (220 rpm) at 30°C for five days using 'hydrocarbon mixture 3' as a carbon source.

Table 74. Improvements in β-Carotene production with new activators [Ninet and Renaut (1979)].

	β-Carotene ($mg\ \ell^{-1}$)		
Added substances	None	Ionone ($1\ g\ \ell^{-1}$)	TMACH[a] ($1\ g\ \ell^{-1}$)
None	850	1550	1350
Succinimide	1350	2200	1850
Isoniazid	2200	3050	2950
Iproniazid	2200	3200	3200

[a] TMACH, 2,6,6-trimethyl-1-acetylcyclohexene. Strains: *Blakeslea trispora* NRRL 2456 (+) and NRRL 2457 (−). Basal culture medium: 60 g distillers' solubles; 60 g 'Fox head' starch; 35 g soya bean oil; 35 g cotton seed oil; 1 g yeast extract; 0.5 g potassium phosphate; 0.2 g manganese sulphate; 0.01 g thiamin hydrochloride; 20 g kerosene; 0.5 g ethoxyquin; with tap water to 1 ℓ.

Table 75. Effect of medium adjuncts on carotene production by *Blakeslea trispora*[a] [Hanson (1967)].

Adjuncts[c]	Carotene production[a] (mg ℓ$^{-1}$)		
	NRRL 2456(+)	NRRL 2457(−)	NRRL 2456 × 2457
Control[b]	4.20	4.70	19.80
Oil[c]	5.60	4.00	40.00
β-ionone	3.00	3.20	17.60
Detergent	0.40	3.90	6.40
Oil + β-ionone	3.90	4.20	110.00
Oil + detergent	15.50	10.80	65.00
β-ionone + detergent	0.24	0.75	2.80
Oil + detergent + β-ionone	8.20	14.90	129.60

[a] Production in shake flask for six days.
[b] 75 g ℓ$^{-1}$ acid hydrolyzed corn; 75 g ℓ$^{-1}$ acid-hydrolyzed casein; 5 g ℓ$^{-1}$ corn-steep liquor; 0.5 g ℓ$^{-1}$ potassium dihydrogen phosphate; 1.0 mg thiamin hydrochloride; pH 6.2.
[c] Added aseptically after two days' incubation.

ENZYMES

General Aspects

Examples of increases in enzyme production by mutation are given in Table 76. Table 77 lists a number of systems in which the products of catabolism repress enzyme synthesis requiring a change of substrate. Table 78 provides examples of product induction of enzyme synthesis. Surfactants also have marked effects (*see* Table 79).

Fig. 39 illustrates typical time course data for an extracellular enzyme — α-amylase.

Table 76. Increase in enzyme production by mutation [Demain (1972)].

Enzyme	Microorganism	Ratio of enzyme yield [mutant/control]
α-Amylase	*Aspergillus oryzae*	10
Aspartate transcarbamylase	*Escherichia coli*	500
Cellulase	*Trichoderma viride*	2
Dihydrofolic reductase	*Diplococcus pneumoniae*	200
	Streptoccus faecalis	10–100
β-Galactosidase	*E. coli*	4
Glucoamylase	*Asp. foetidus*	1.6
Homoserine dehydrogenase	*E. coli*	3
Protease	*Bacillus cereus*	10

Table 77. Catabolite repression of certain enzymes [Demain (1972)].

Enzyme	Microorganism	Repressive carbon source
α-Amylase	*Bacillus stearothermophilus*	Fructose
Cellulase	*Trichoderma viride*	Glucose, glycerol, starch, cellobiose
Cellulase	*Pseudomonas fluorescens* var. *cellulosa*	Galactose, glucose, cellobiose
Glucoamylase	*Endomycopsis bispora*	Starch, maltose, glucose, glycerol
Polygalacturonic acid transeliminase	*Aeromonas liquefaciens*	Glucose, polygalacturonic acid
Protease	*Bacillus megaterium*	Glucose

Table 78. Product inducers of enzymes [Wang *et al.* (1979)].

Enzyme	Microorganism	Substrate	Product inducer
Amylase	*Bacillus stearothermophilus*	Starch	Maltodextrins
Dextranase	*Penicillium* sp.	Dextran	Isomaltose
Endopolygalacturonase	*Acrocylindrium* sp.	Polygalacturonic acids	Galacturonic acid
Exopolygalacturonase	*Acrocylindrium* sp.	Polygalacturonic acids	Galacturonic acid
Glucamylase	*Aspergillus niger*	Starch	Maltose, isomaltose
Histidase	*Klebsiella aerogenes*	Histidine	Urocanic acid
Lipase	*Geotrichum candidum*	Lipids	Fatty acids
Pectin esterase	*Acrocylindrium* sp.	Polygalacturonic acids	Galacturonic acid
Pullulanase	*K. aerogenes*	Pullulan	Maltose
Tryptophan oxygenase	*Pseudomonas* sp.	Tryptophan	Kynurenine
Urea carboxylase	*Saccharomyces cerevisiae*	Urea	Allophanic acid

Table 79. Effect of 0.1 per cent Tween 80 on the yield of various enzymes [Moss and Smith (1977)].

Enzyme	Microorganism	Ratio of yield in medium + Tween 80/ medium − Tween 80
Amylase		5
Cellulase	*Trichoderma viride*	20
Dextranase	*Penicillium funiculosum*	2
β-1,3-Glucanase		8
β-Glucosidase		4
Invertase	*Aureobasidium pullulans*	16
Nucleosidase		6
Pullulanase	*Aerobacter aerogenes*	1.5
Xylanase	Numerous fungi	4

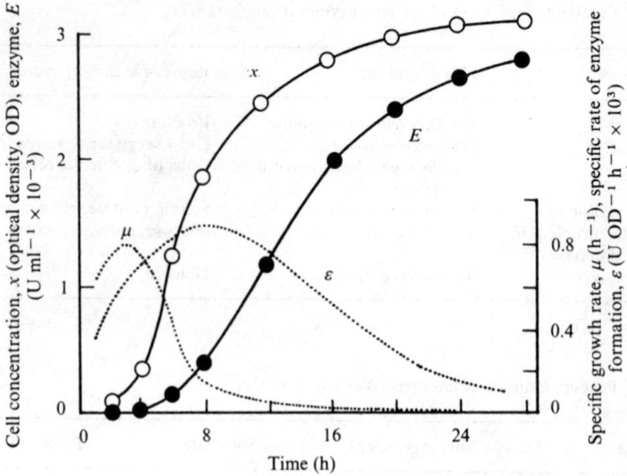

Figure 39. Time course of α-amylase formation by *Bacillus amyloliquefaciens*.

Cellulase

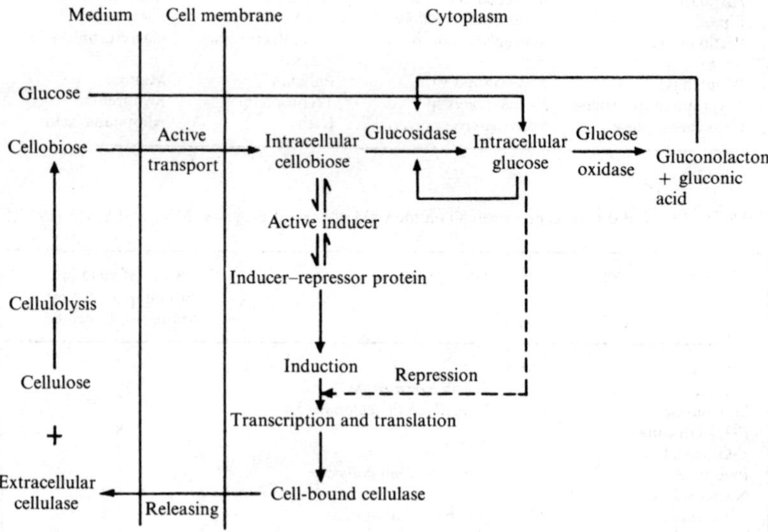

Figure 40. Regulatory model for cellulase biosynthesis [Flickinger (1980)].

A regulatory model for cellulase biosynthesis is illustrated in Fig. 40. The time course of cellulase production by *Trichoderma viride* QM 9414 is given in Fig. 41. Demain (1972) showed that the cellulase yield declined when the concentration of insoluble substrate, in this case cellulose, increases. In contrast, α-galactosidase yield is unaffected by high insoluble substrate levels. Tables 80 and 81 illustrate conditions favourable to cellulase formation, and Table 82 lists enzyme productivities achieved in various fermenter systems. A scheme has been developed for the fractionation of four different cellulolytic enzymes from *T. viride* which is based on ion exchange and molecular sieve chromatography and isoelectric focussing (*see* Fig. 42).

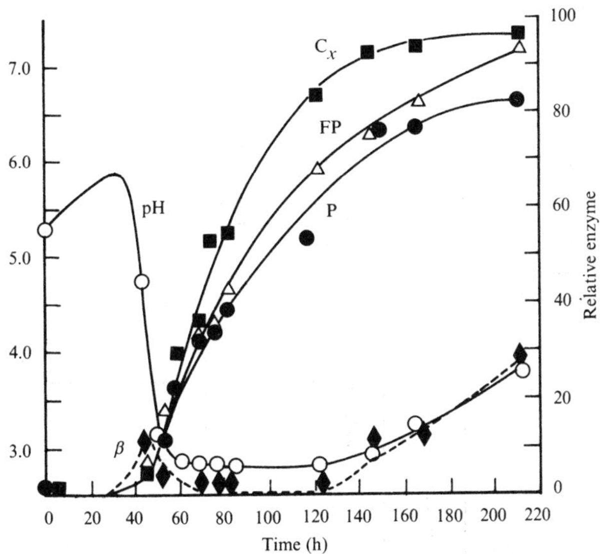

Figure 41. Growth and enzyme production of *Trichoderma viride* QM 9414 on 0.75 per cent (by wt) 200 (ball-milled pulp) + 0.075 per cent proteose peptone, 0.2 per cent Tween 80, spore inoculum. ○, pH; △ saccharifying cellulase (FP) 100 = 1 U ml^{-1}; ■, endo-cellulase (C_x) 100 = 55 U ml^{-1}; ◆, β-glucosidase (β) 100 = 0.10 U ml^{-1}; ●, extracellular protein (P) 100 = 2 mg ml^{-1} [Ghose (1977)].

Table 80. Effect of surfactant on cellulase production by *Trichoderma viride* of growing on 1 per cent cellobiose [Demain (1972)].

Surfactant[a]	Cellulase (μg ml^{-1}) *T. viride* 6a	*T. viride* 9123
None	1	17
Tween 80 (0.2%)	17	34
Sodium oleate (0.1%)	122	85
Sodium oleate (0.2%)	118	94
Sucrose monopalmitate (0.1%)	29	94
Sodium palmitate (0.1%)	2	3

[a] Surfactant added to culture after 24 hours of incubation.

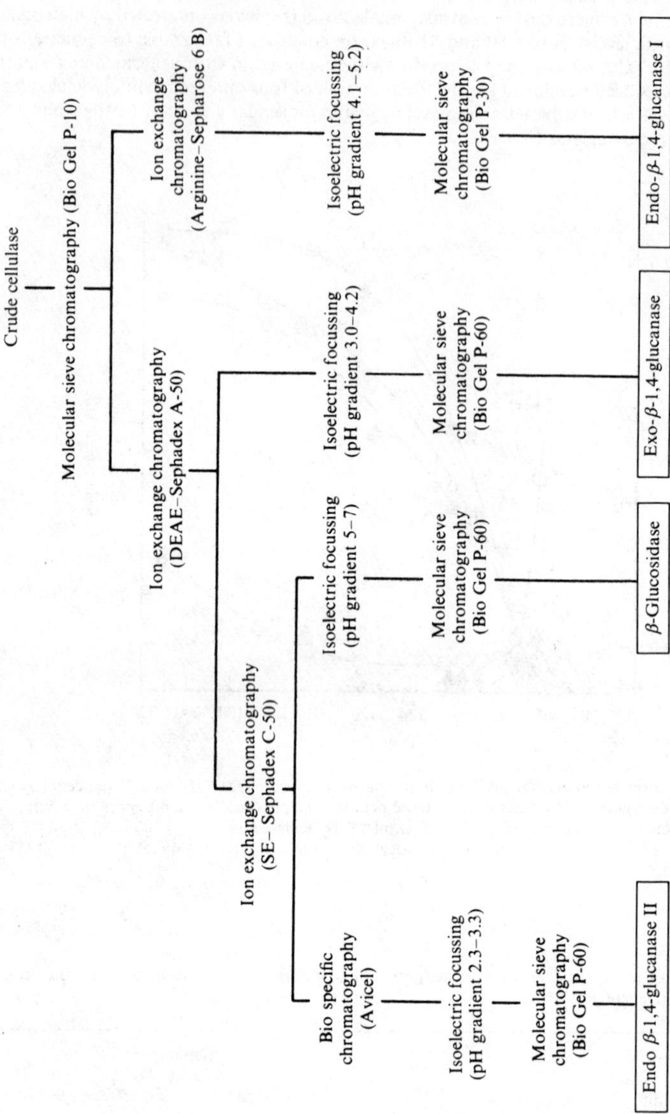

Figure 42. Fractionation scheme for the purification of four different cellulolytic enzymes from *Trichoderma viride* [Ghose (1977)].

Table 81. Conditions favouring induction of cellulase by cellobiose [Demain (1972)].

Microorganism	Maximum cellulase (μg ml^{-1}) Cellulose	Cellobiose	Modification	Yield (μg ml^{-1})
Basidiomycete	113	4.0	Cellobiose, calcium, trace metals	22
Penicillium helicum	32	0.1	Cellobiose octaacetate	28
Pestalotiopsis westerdykii	80	0.3	Cellobiose octaacetate	68
Trichoderma viride	100	16.0	Cellobiose, lower temperature (24°C), unshaken	94

Table 82. Cellulase productivities of different systems [Mukhopadhyay and Malik (1980)].

Culture	Cultivation mode	Cellulose level (%) dilution rate (h^{-1})	Maximum productivity per litre (IU ℓ$^{-1}$ h^{-1})
Trichoderma MCG-77 (hyperproducing)	Two-stage continuous	0.028	53 (90 in second stage only)
Trichoderma	Batch (with pH cycling and temperature profiling)	3	44
T. reesei QM 9414	Batch	5	30
T. reesei QM 9414	Continuous (cell recycle)	0.025	30

Glucose Isomerase

In Table 83, a list of the microorganisms and culture conditions used for the production of glucose isomerase is provided. Specific glucose isomerase activity of *Bacillus coagulans* falls with increasing specific growth rate for both glucose- and glucose/oxygen-limited conditions [*see* Diers (1976)]. Table 84 summarizes the properties of glucose isomerase produced using a variety of substrate–microbe systems, while Table 85 compares three commercial whole-cell glucose isomerase preparations.

Table 83. Cultural conditions for the production of glucose isomerase [Chen 1980a)].

Microorganism	Carbon source	Nitrogen source(s)	Mineral(s)	pH	Temperature (°C)	Time (h)	Yield ($U\ \ell^{-1}$)[a]
Actimoplanes missouriensis NRRL B-3342	Beet molasses	Soy flour or corn-steep liquor + $NaNO_3$	$MgSO_4.7H_2O$ + KCl + $FeSO_4.7H_2O$	7.0	30	72	2500–35 200
Aerobacter aerogenes	Xylose or mannose or lactate or manniol	$(NH_4)_2HPO_4$	$MgSO_4.7H_2O$		28	48	
A. cloacae	Xylose	$(NH_4)_2HPO_4$	$MgSO_4.7H_2O$ + KH_2PO_4	6.8–7.0	30	24	
A. levanicum NRRL B-1678	Xylose or birch wood sulphite liquor	Yeast extract	$CaSO_4.7H_2O$ + KCl	7.5	23	24	300
Arthrobacter NRRL B-3726, 3727, 3728	Glucose	Meat protein + yeast extract + $(NH_4)_2SO_4$	KH_2PO_4 + $MgSO_4.7H_2O$				2220–4720
Arthrobacter NRRL B-3728	Xylose + glucose		Mg^{2+} + Cu^{2+} + Mn^{2+} + Zn^{2+} Ca^{2+} + PO_4^{-3}+ + MnO_4^{-2}	6.9	30		
Bacillus coagulans HN-68	Xylose	Yeast extract + NH_4Cl	$MgSO_4.7H_2O$ + $MnSO_4.4H_2O$ + $CaCO_3$		40	14–16	
B. coagulans NRRL-5650 (mutant)	Glucose	Corn steep liquor + $(NH_4)_2SO_4$	$MgSO_4.7H_2O$ + $MnSO_4.4H_2O$ + K_2HPO_4	6.8	50	13	8800
B. coagulans NRRL 5649-66 (mutant)	Xylose	$(NH_4)_2SO_4$			60		
B. stearothermophilus	Xylose	Peptone + corn-steep liquor + yeast extract + meat extract	$NaCl$ + $MgSO_4.7H_2O$ + $CoCl_2.6H_2O$		55		3890–12 900
Bacillus sp. NRRL B-5350, 5351	Xylose + glucose	Peptone + casein			38	17	
Corynebacterium candidus ATCC-31261	Xylose	Peptone	$MgSO_4.7H_2O$ + K_2HPO_4 + $Na_2HPO_4.10H_2O$		30	48	

Curtobacterium helvolum NCLB-10352	Xylose	Meat extract + yeast extract + peptone	NaCl + $MgSO_4.7H_2O$ + $CoCl_2.6H_2O$		30	24	
Escherichia intermedia	Xylose	NH_4Cl	$MgSO_4.7H_2O$ + K_2HPO_4		28	20	
Flavobacterium arborescens	Lactose	Protein hydrolyzate + yeast extract	K_2HPO_4 + KH_2PO_4		30	72	
Lactobacillus brevis	Xylose	Peptone + yeast extract	Na–acetate + $MgSO_4.7H_2O$ + NaCl + $FeSO_4.7H_2O$		37	16–20	
Microbispora rosea	Xylose + dextrin	Corn-steep liquor + yeast extract + $(NH_4)_2HPO_4$	$MgSO_4.7H_2O$ $CoCl_2.6H_2O$		30	65	160
Norcardia corallia	Xylose	Meat extract + peptone + NH_4NO_2	$MgSO_4.7H_2O$ + KH_2PO_4 + KCl + $CoCl_2.6H_2O$ + $MnSO_4.4H_2O$	7.0	30	72	
N. dassovillei	Xylose + glucose + sorbitol	Corn-steep liquor	$CoCl_2.6H_2O$		30		400
Paracolobactrum aerogenoides	Glucose	NH_4Cl	$MgSO_4.7H_2O$ + KH_2PO_4 + $Na_2HPO_4.12H_2O$ + NaCl		30		
Streptomyces sp. S41-10	Acid hydrolyzate of bagasse pith	Tryptone + yeast extract	$MgSO_4.7H_2O$ + $CoCl_2.4H_2O$		30	24–30	
Streptomyces sp.	Xylose	Tryptone + yeast extract	$MgSO_4.7H_2O$	7.0–7.2	30	96	1100
Strept. albus YT-4	Wheat bran	Corn-steep liquor	$CoCl_2.6H_2O$		30	30	2570
Strept. albus YT-5	Wheat bran or xylose or xylan	Corn-steep liquor	$MgSO_4.7H_2O$ + $CoCl_2.6H_2O$		30	20–24	2800
Strept. albus YT-6	Wheat bran	Corn-steep liquor	$CoCl_2.6H_2O$		45	20	1940
Strept. bikiniensis	Xylose or xylan	Yeast extract + peptone + beef extract	$MgSO_4.7H_2O$ + NaCl			40	2000

continued overleaf

Table 83 (*continued*)

Microorganism	Carbon source	Nitrogen source(s)	Mineral(s)	pH	Temperature (°C)	Time (h)	Yield (U ℓ^{-1})[a]
Strept. bodiliae	Xylose or sorbitol or glycerol	Polypeptone	$MgSO_4.7H_2O$ + K_2HPO_4		30	20	
Strept. flavogriseus	Straw hemicellulose or H_2SO_4 hydrolyzate of straw	Corn-steep liquor	$MgSO_4.7H_2O$	7.0	30	36	3500
Strept. flavovirens	Xylan	Polypeptone	$MgSO_4.7H_2O$ + K_2HPO_4		30		
Strept. galbus	Sorbitol + glucose + xylose	Corn-steep liquor	$CoCl_2.6H_2O$	7.0	28	40	3700
Strept. glaucescens	Xylose + glucose + sorbitol	Yeast extract	$K_2HPO_4.3H_2O$ + $MgSO_4.7H_2O$ + $CoCl_2.6H_2O$		30	46	2220
Strept. olivaceus NRRL B-3916	Xylose + corn starch	Peptone + beef extract + yeast extract	$MgSO_4.7H_2O$ + NaCl	7.0			2960
Strept. olivochromogenes ATCC-21114	Xylose + soluble potato starch	Peptone + meat extract + yeast extract	$MgSO_4.7H_2O$ $CoCl_2.6H_2O$ + NaCl		28	48	200
Strept. phaeochromogenes SK	Xylose	Peptone + meat extract + yeast extract	$MgSO_4.7H_2O$ + $CoCl_2.6H_2O$ + NaCl		28–30	24	5560
Strept. venezuellae ATCC-21113	Xylose + soluble potato starch	Peptone + meat extract + yeast extract	$MgSO_4.7H_2O$ + $CoCl_2.6H_2O$ + NaCl		28	48	2200–6800
Strept. wedmorensis ATCC-21175	Acid hydrolyzate of corncob	Corn-steep liquor	$CoCl_2.6H_2O$	7.0	30	40	3700
Streptosporangium album or *S. oulgare*	Xylose + dextrin	Corn-steep liquor + yeast extract + $(NH_4)_2HPO_4$	$MgSO_4.7H_2O$ + $CoCl_2.6H_2O$		30	65	2780

[a] One unit of glucose isomerase activity was defined as the amount of the enzyme that produced 1 μmole D-fructose per minute.

Table 84. Summary of properties of glucose isomerase [Chen (1980b)].

Microorganism	Substrate Specificity	Substrate K_m (mM)	Temp. optimum (°C)	pH optimum	Metal requirement	Inhibitor	Heat stability (%)	pH stability	Partial specific volume (ml g^{-1})	Molecular weight
Actinoplanes missouriensis	Glucose, Xylose, Ribose			7.0	Mg^{2+}, Co^{2+}					
Aerobacter cloacae	Glucose, Xylose		50	7.6	As^{3+}, Mg^{2+}					
Bacillus coagulans[a] HN-68	Glucose, Xylose, Ribose	90 70	75	7.0	Mg^{2+}, Co^{2+}, Mn^{2+}	Cu^{2+}, Zn^{2+}, Ni^{2+} Ca^{2+}	100 (70°C, 10 min)		0.705	175 000
B. stearothermophilus	Glucose, Xylose, Ribose, Arabinose		80	7.5–8.0	Mg^{2+}, Co^{2+}				0.736	130 000
Escherichia intermedia[a]	Glucose, Xylose, 2-Deoxy-glucose		50	7.0	As^{3+}		0 (60°C, 10 min)	7–9		
Lactobacillus brevis[a]	Glucose, Xylose, Ribose	920 5 670		6.0–7.0	Mn^{2+}, Co^{2+}	Xylitol, Arabitol, Lyxose	10 (60°C, 30 min)			191 000
Streptomyces sp. S41-10	Glucose, Xylose		75	8.5	Mg^{2+}, Co^{2+}					
Streptomyces sp.	Glucose, Xylose	400	80	7.0–8.0	Mg^{2+}		50 (70°C, 120 h)			
Strept. albus YT-5[a]	Glucose, Xylose	160 32	80	8.0–8.5	Mg^{2+} or Co^{2+}	Heavy metals (Ag^{+},Cu^{2+}, Hg^{2+})	90 (70°C, 10 min)	4–11	0.69	157 000

continued overleaf

Table 84 (*continued*)

Microorganism	Substrate Specificity	K_m (mM)	Temp. optimum (°C)	pH optimum	Metal requirement	Inhibitor	Heat stability (%)	pH stability	Partial specific volume (ml g^{-1})	Molecular weight
Strept. albus NRRL B-5778	Glucose, Xylose, Ribose, Arabinose, Rhamnose, Allose	86 93 350	70–80		Mg^{2+}, Co^{2+}	Sorbitol				
Strept. bikiniensis[a]	Glucose, Xylose, Ribose, Rhamnose		80	8.0–9.0	Mg^{2+}, Co^{2+}					52 000
Strept. flavogriseus[a]	Glucose, Xylose	249 78	70	7.5	Mg^{2+}, Co^{2+}	Ag^{+}, Cu^{2+}, Hg^{2+}	100 (70°C, 10 min)	5.0–9.0		171 000
Strept. flavovirens IFO 3197	Glucose, Xylose	500	85	8.5	Ng^{2+}, Co^{2+}	Tris				
Strept. olivochromogenes[a]	Glucose, Xylose, Ribose, Arabinose		80	8.0–9.0	Mg^{2+}, Co^{2+}				0.725	120 000
Strept. phaeochromogenes SK	Glucose, Xylose	300	90	9.3–9.5	Mg^{2+}, Co^{2+}	Tris	96 (80°C, 10 min)			
Strept. phaeochromogenes NRRL B-3559	Glucose, Xylose	250	80	8.0	Mg^{2+}		40 (70°C, 24 h)			

[a] The enzyme is homogeneous.

Table 85. Comparison of three commercial whole cell glucose isomerase preparations [Vaheri and Kauppinen (1977)].

Conditions of activity determination	*Bacillus coagulans* (Novo Industri A/S)	*Actinoplanes missouriensis* (Gist-Brocades NV)	*Streptomyces albus* (Miles Kalichemie KG)
pH	6.5	7.0	7.0
buffer	0.25 M NaOH maleate	0.02 M phosphate	0.1 M phosphate
Glucose concentration	5%	1 M	0.1 M
Mg^{2+} (mM)	100	3	5
Co^{2+} (mM)	1	0.3	
Temperature (°C)	65	70	70
Incubation time (min)	20	10	60
Activity (U g^{-1})	500	1295	278

Proteases

Suitable media and microorganisms for protease production are listed in Tables 86 and 87, respectively. Protease yields achieved in shake flasks are included in Table 87.

Table 86. Composition of media for protease production in shake flasks [Keay *et al.* (1972)].

Medium	pH[a]	Component	Amount (g ℓ^{-1})
Grain (GR)	6.5	Rice bran	10.0
		Wheat bran	30.0
		Corn meal	20.0
Fish meal–Enzose Cerelose (FC)	7.0	Fish meal	15.0
		Enzose	35.0
		Cerelose	26.0
		Corn-steep liquor	43.0
		Ammonium nitrate	0.8
		Calcium carbonate	5.0
		Calcium acetate	1.0
Casein (SAS)	7.0	Casein hydrolyzate	20.0
Soya fluff–starch (SFS)	7.0	Soya fluff	15.0
		Corn starch	20.0
		Yeast extract	2.0
		Potassium dihydrogen phosphate	1.0
		Dipotassium hydrogen phosphate	3.0
		Magnesium sulphate	0.1
		Calcium chloride	0.2
Lactosoya (LS)	7.0	Lactosoya	20.0
Nutrient gelatin (NG)	7.0	Nutrient gelatin	64.0

[a] pH adjustments performed before autoclaving.

Table 87. Protease production in shake flasks[a] [Keay *et al.* (1972)].

Microorganism	Medium[b]	Temperature (°C)	Protease[c] (U ml^{-1})	Time (h)	Yield (g)	Protease[c] (U g^{-1})	Recovery (%)
Bacillus megaterium ATCC 14581	FC	30	4150	50	0.337	505 000	84
		33	3510	50	0.306	438 000	80
B. cereus ATCC 14579	FC	30	1280	48	0.922	73 000	88
B. cereus 1F0 3215	SFS	26	447	69			
B. cereus NCTC 945	LS	30	3140	26	0.159	1 328 000	43
B. polymyxa ATCC 842	FC	30	3580	32	0.712	714 000	66
	SFS	30	4530	26	1.255	485 000	77
	LS	30	2920	30	0.926	300 000	66
Pseudomonas aeruginosa 1F0-3080	NG	37	4450	48	4.119	129 000	91
Ps. aeruginosa 1F0-3454	NG	37	1150	48	3.622	31 000	72
Ps. aeruginosa ATCC 7700	NG	37	1320	22	6.376	24 400	81
B. subtilis SP491	GR	37	1110	24	0.497	131 000	85
	SFS	37	730	24	0.616	42 000	61
Serratia marcescens NCIB 10351	NG	26	1570	27	3.250	59 000	90
	SFS	26	1480	27	2.139	56 000	63
Aeromonas proteolytica ATCC 15338	SFS + 3% NaCl	26	19 970	12	0.842	970 000	58
	CAS + 3% NaCl	26	8730	28	0.985	960 000	58
	LS + 3% NaCl	26	1420	25	0.459	424 000	107
A. oryzae ATCC 11493	SG	26	1190	48	0.164	423 000	77
A. flavus ATCC 11498	SFS	26	1540	72	0.77	683 000	91

[a] 100 ml medium in 500 ml baffled conical flask on rotary shaker at 240 rpm.
[b] Media compositions are given in Table 86.
[c] Casein assay at pH 7.0 and 37°.

ETHANOL

Microorganisms

Yeasts are the most commonly used microorganisms for ethanol production. Selected high-productivity strains come from the species *Saccharomyces cerevisiae* (*see* Table 88), *Sacch. uvarum* (formerly *Sacch. carlsbergensis*) and *Candida utilis*. Other yeasts, such as *Sacch. anamensis* and *Schizosaccharomyces pombe*, are also used. For the fermentation of whey, the most suitable species are *Kluyveromyces fragilis* and *Kl. lactis*. Yeasts generally produce ethanol with high selectivity with only traces of byproducts.

Although other microorganisms produce ethanol, they also yield significant quantities of undesirable byproducts. Some of the bacteria considered for ethanol production are thermophilic, such as *Clostridium thermosaccharolyticum*. For the rapid production of ethanol, the bacterium *Zymomonas mobilis* seems promising. *Sarcina ventriculi* produces ethanol and small quantities of acetate, while *Erwinia amylovora* produces ethanol and small quantities of lactate. These bacteria utilize glucose via the Entner–Doudoroff pathway.

Leuconostoc species and *Lactobacillus* species produce some ethanol via the heterolactic pathway while *Lact. casei* uses the homolactic pathway to produce lactate, acetate, formate and ethanol in an energy-limited continuous culture. The heterolactic bacterium *Thermoanaeronium brockii* produces ethanol as the major product in low yeast extract medium and produces lactate as the major product in high yeast extract medium. Other bacteria that produce ethanol are *Aeromonas*, *Bacillus*, *Klebsiella*, *Clostridium acetobutylicum* and *B. macerans*.

Table 88. Products of the alcoholic fermentation of glucose by *Saccharomyces cerevisiae* at different pH values[a] [Rainbow and Rose (1963)].

Product	Amount produced [mM (100 mM glucose fermented)$^{-1}$] pH 3.0	pH 4.0	pH 5.0	pH 6.0	pH 7.0
Acetic acid	0.52	0.69	0.84	4.03	8.68
Acetoin					0.07
2,3-Butanediol	0.75	0.48	0.46	0.53	0.45
Butyric acid	0.13	0.32	0.25	0.36	0.25
Carbon dioxide	180.8	189.8	187.6	177.0	161.0
Ethanol	171.5	177.0	172.6	160.5	149.5
Formic acid	0.36	0.42	0.63	0.82	0.35
Glycerol	6.16	6.60	7.82	16.2	22.2
Lactic acid	0.82	0.38	0.47	1.63	1.93
Succinic acid	0.53	0.26	0.32	0.49	0.23
Glucose carbon assimilated	12.4	16.1	14.0	12.4	
Fermentation time (h)	29.0	14.5	17.5	15.5	35.0
Percentage glucose fermented	98.5	97.0	96.5	98.0	98.3
Percentage carbon recovered	93.8	98.0	96.3	96.4	92.5

[a] Fermentations were performed in medium containing 5 per cent glucose. pH was maintained by automatic addition of ammonium hydroxide.

Media

One or more of the sugars listed in Table 89 is used as substrate for ethanol production. Table 89 also shows the fermentative ability of *Saccharomyces* and *Kluyveromyces* species.

Table 89. Ability of *Saccharomyces* and *Kluyveromyces* species to ferment sugars [Jones *et al.* (1981)].

Carbon number of basic carbohydrate unit	Type of basic subunit	Sugar	Basic unit	*Sacch. cerevisiae*	*Sacch. uvarum*	*Kl. fragilis*
6		Glucose	Glucose	+	+	+
		Maltose	Glucose	+	+	−
		Maltotriose	Glucose	+	+	−
		Cellobiose	Glucose	−	−	−
	Aldose	Trehalose	Glucose	+/−	+/−	−
	sugars	Galactose	Galactose	+	+	+
		Mannose	Mannose	+	+	+
		Lactose	Glucose, galactose	−	−	+
		Melibiose	Glucose, galactose	−	+	
	Ketose	Fructose	Fructose	+	+	+
	sugars	Sorbose	Sorbose	−	−	−
	Aldoses	Sucrose	Glucose, fructose	+	+	+
	and	Raffinose	Glucose, fructose	+/−	+	+/−
	ketoses		Galactose			
	Deoxy-	Rhamnose	6-deoxymannose	−	−	−
	sugars	Deoxyribose	2-deoxyribose	+/−	+/−	+/−
5	Aldose	Arabinose	Arabinose	−	−	−
	sugars	Xylose	Xylose	−	−	−

Table 90. Carbohydrate raw materials suitable for fermentation to ethanol.

Carbohydrate	Source	Hydrolysis products
Sucrose	Beet sugar	Glucose
	Cane sugar	Fructose
	Cannery wastes	
	Molasses	
Starch	Cassava	Glucose
	Corn	Maltose
	Jerusalem artichokes	Maltotriose
	Other grains	Higher molecular weight dextrins[a]
	Potatoes	
	Sago	
	Sweet sorghum	
	Taro, etc.	
Cellulose	Agricultural residues	Glucose
	(Bagasse, straw, etc)	Mannose
	Municipal wastes	Galactose
	Wood	Arabinose[a]
	Wood wastes	Xylose[a]
Lactose	Whey	Glucose
		Galactose

[a] Not fermentable by most *Saccharomyces* sp.

A list of raw materials that are used as sources of various sugars is contained in Table 90.

At sugar concentrations above 14 per cent (w/v) plasmolysis of the yeast cells occurs due to osmotic effects and the initial rate of fermentation declines. The sugar concentration that can be tolerated depends on the strain; some strains can tolerate up to 22–35 per cent (w/v). However, ethanol inhibition limits the concentration of fermentable substrates in the feed much more significantly than does substrate inhibition.

Since the nitrogen content of yeast is about 10 per cent of the dry weight, nitrogen is an important constituent of any growth medium. It can be supplied as ammonium salts, particularly as ammonium sulphate. Complex nitrogenous sources, e.g., a mixture of amino acids, peptides, nucleic acids and simple bases, as well as fatty acids and lipids enhance growth and fermentation. The uptake of amino acid nitrogen in an amino acid/ammonium salt mixture is five to ten times the uptake of ammonium nitrogen. Amino acids also buffer yeast against ionic inhibition effects such as those of heavy metals, etc. Ammonium sulphate is added to a blackstrap molasses medium at a concentration of about 70–400 mg ℓ^{-1}.

Phosphorus is provided as phosphate and is the most important ionic factor determining the rate of fermentation. It is required at concentrations of about 0.6 mM per gram of cells for optimum fermentation rates.

Sulphur constitutes about 0.4 per cent of yeast dry weight and the preferred source is the amino acid methionine. Inorganic sulphate can also be utilized, but it has to be reduced to methionine within the cell, which is demanding both energetically – two moles of ATP are required per mole of sulphate reduced – and biosynthetically – one mole of aspartate required per mole of sulphate. However, since methionine is relatively expensive, ammonium sulphate is preferred.

Table 91. Major inorganic nutrients [(Jones *et al.* (1981)].

Ion	Role	Concentration (μM)[a]
K^+	Enhances tolerance to toxic ions. Involved in control of intracellular pH. Excretion is used to counterbalance uptake of essential ions, e.g., Zn^{2+}, Co^{2+}. Stabilizes the optimum pH for fermentation.	2000
Mg^{2+}	Levels regulated by divalent cation transport system. Seems to buffer the cell against adverse environmental effects and is involved in activating sugar uptake.	5000
Ca^{2+}	Actively taken up by cells during growth and is incorporated into cell-wall proteins. Buffers the cell against an adverse environment. Counteracts Mg^{2+} inhibition and stimulates the effect of suboptimal concentrations of Mg^{2+}.	1500
Zn^{2+}	Essential for glycolysis and for the synthesis of some vitamins. Uptake is reduced below pH 5, and two K^+ ions are excreted for each Zn^{2+} taken up.	50
Mn^{2+}	Implicated in regulating the effects of Zn^{2+}. Stimulates the synthesis of proteins.	15
Fe^{2+}, Fe^{3+}	In the active site of many yeast proteins.	10
Na^+	Passively diffuses into cells. Stimulates the uptake of some sugars.	0.25
Cl^-	Acts as a counterion to the movement of some positive ions.	0.1
Mo^{2+}, Co^{2+}, B^{2+}	Stimulates growth at low concentrations.	0.5

[a] For 25 g ℓ^{-1} cell growth.

The major inorganic nutrients required by yeasts are listed in Table 91, while Table 92 gives the vitamin requirements. These requirements depend on the strain, e.g., biotin and pantothenate are essential for all strains of *Saccharomyces*, whereas *Candida utilis* does not require biotin.

Table 92. The role of vitamins in yeast metabolism [(Jones *et al.* (1981)].

Vitamin	Active form	Metabolic role	Optimum concentration (mg ℓ$^{-1}$)
Biotin	Biotin	All carboxylation and decarboxylation reactions	0.005–0.5
Pantothenate	Coenzyme A	Keto acid oxidation reactions, fatty acid, amino acid, carbohydrate, choline metabolism	0.2–2.0
Thiamin	Thiamin pyrophosphate	Fermentative decarboxylation of pyruvate, keto acid oxidation and decarboxylation	0.1–1.0
Pyridoxine	Pyridoxal phosphate	Amino acid metabolism, deamination, decarboxylation, racemization reactions	0.1–1.0
p-Aminobenzoic acid and folic acid	Tetrahydrofolate	Transamination, ergosterol synthesis, transfer of one-carbon units	0.5–5
Niacin	NAD^+, $NADP^+$	Dehydrogenation reactions	0.1–1.0
Riboflavin	FMN, FAD	Same flavoprotein, dehydrogenation reactions and some amino acid oxidations	0.2–0.25

Silver, arsenic, barium, mercury, lithium, nickel, osmium, palladium, selenium and tellurium inhibit growth and fermentation at concentrations greater than 10–100 μM. However, these inhibitory effects depend on the medium and in complex media, chelating and sequestering agents, e.g., amino acids tend to buffer the available concentrations.

Table 93. Composition of a typical synthetic medium for ethanol production [Maiorella *et al.* (1981)].

Component	Amount (g ℓ$^{-1}$)
Glucose	100.0
Ammonium sulphate	5.19
Potassium dihydrogen phosphate	1.53
Magnesium sulphate heptahydrate	0.55
Calcium chloride	0.13
Boric acid	0.01
Cobalt sulphate heptahydrate	0.001
Copper sulphate pentahydrate	0.004
Zinc sulphate heptahydrate	0.01
Manganese sulphate monohydrate	0.003
Potassium iodide	0.001
Ferrous sulphate heptahydrate	0.002
Aluminium sulphate	0.003
Biotin	0.000125
Pantothenate	0.00625
Inositol	0.125
Thiamin	0.005
Pyridoxine	0.00625
p-Aminobenzoic acid	0.001
Nicotinic acid	0.005

Copper, for instance, at a concentration of 1 mg ℓ^{-1} completely inhibits growth in a defined sucrose medium whereas copper at concentrations as high as 30 mg ℓ^{-1} can be tolerated in a malt wort medium. In the concentration range of 5–50 mg ℓ^{-1}, there is antagonism between different metal ions, e.g., $Zn^{2+}/Mn^{2+}/Cd^{2+}$ and K^{+}/Li^{+}, the presence of one ion inhibiting the uptake of the other(s). Table 93 contains the composition of a typical synthetic medium.

Substances promoting growth and fermentation include amino acids, nucleic acids, fatty acids and sterols. Inositol is an essential growth factor for many yeasts and plays an important role in maintaining the integrity of the cell membrane.

Although ethanolic fermentation is an anaerobic process, oxygen is required at very low levels for the synthesis of ergosterol, which is an essential component of the yeast membrane, and is necessary for the synthesis of unsaturated fatty acids. At oxygen concentrations of less than 1 ppm, sterols, e.g., ergosterol, and unsaturated fatty acids, e.g., oleic acid or Tween 80, should be added. Table 94 provides some values for the oxygen concentrations required to attain maximum volumetric rates of ethanol production.

Carbon dioxide inhibits the growth of yeast during both aerobic and anaerobic fermentation, e.g., in a glucose-limited, continuous aerobic culture of bakers' yeast, dissolved carbon dioxide in excess of 0.016 M inhibits growth.

In an uncontrolled batch fermentation, the optimum initial pH in slightly buffered media is near 5.5, in more highly buffered media is 4.5–4.7, in molasses-containing media is 4–5 and in grain-based media is 4.8–5.0. The absolute limits of pH for the growth of most yeasts are 2.4 and 8.6, the optimum being in the region of 4.5 with most laboratory-scale continuous fermentation studies employing a pH value between 4.0 and 5.0.

The mesophilic strains of *Saccharomyces* have an optimum temperature range for growth of 28–35°C, while for thermophilic yeasts, the optimum temperature is 40°C.

Ethanol tolerance is strain-dependent, with a maximum allowable concentration of about 10 per cent (w/v) for growth and 20 per cent for ethanol production in the case of the most tolerant strains. In yeasts, only *Saccharomyces* and *Schizosaccharomyces* species have high ethanol tolerances.

Zymomonas mobilis can tolerate higher initial sugar concentrations than yeasts and has been found to produce up to 10 per cent (w/v) ethanol from 25 per cent glucose media. The composition of a typical growth medium for *Z. mobilis* is given in Table 95.

Table 94. Concentration of oxygen required for optimum continuous fermentation using strains of *Saccharomyces cerevisiae* [(Jones *et al.* (1981)].

Strain	Dissolved oxygen concentration	Oxygen uptake	Medium	Fermentation Description
ATCC 239	0.2 mm Hg	9.6 mg O_2 (g dry wt)$^{-1}$ h^{-1}	Defined glucose medium	Continuous
ATCC 239	Below limits of detection	1.5 mg O_2 (g dry wt)$^{-1}$ h^{-1}	Proctor pale ale malt	Continuous
ATCC 4126	0.07 mm Hg dissolved oxygen	0.0015 mg dissolved O_2 (g dry wt)$^{-1}$	Glucose, yeast extract, mineral salts medium	Continuous
ATCC 4126	Aeration rate = 0.14 vvm	100 mg O_2 (g dry wt)$^{-1}$ h^{-1}	Glucose, yeast extract, mineral salts medium	Continuous vacuferm 50 mm Hg total pressure
Respiration-deficient mutant		1.7–9.9 mg O_2 (g dry wt)$^{-1}$ h^{-1}	Defined glucose medium	Continuous

Table 95. Medium for ethanol fermentation using *Zymomonas mobilis*.

Component	Amount ($g\ \ell^{-1}$)
Glucose	100
Yeast extract	5
Ammonium sulphate	1
Potassium dihydrogen phosphate	1
Magnesium sulphate heptahydrate	0.5

Zymomonas grows anaerobically and, unlike yeasts, does not require the controlled addition of oxygen to maintain viability at high cell concentration in continuous cell recycle systems.

Metabolic Pathways

An outline of the fermentation of glucose in some bacteria and yeasts is given in Fig. 43. Under anaerobic conditions, glucose is converted to ethanol and carbon dioxide through glycolysis [*see* Fig. 43(a)].

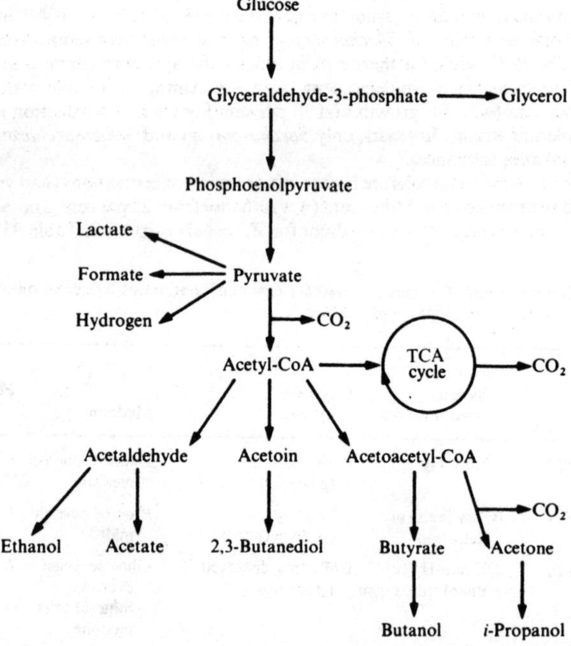

Figure 43(a). Metabolic pathways for ethanol. Outline of anaerobic catabolism of glucose in some bacteria and yeasts.

Fig. 43(b) illustrates both aerobic and anaerobic metabolism in *Sacch. cerevisiae*. The possible effects of carbon dioxide on yeast metabolism are given in Table 96(a); these effects have been verified apart from the first and last reactions.

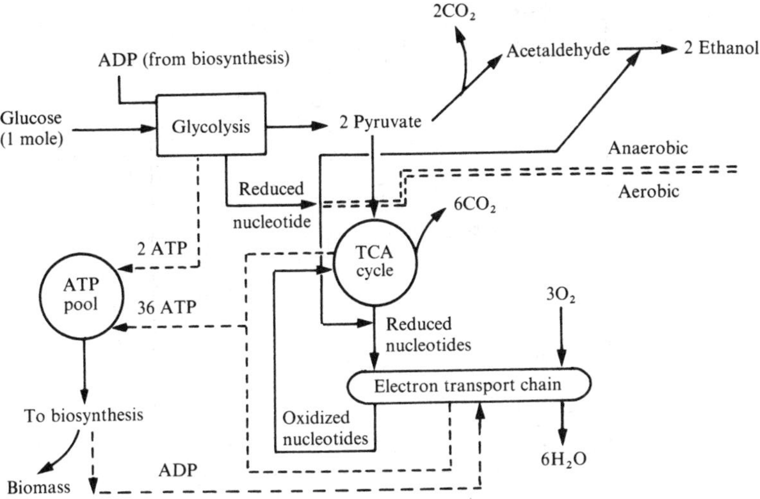

Figure 43(b). Anaerobic and aerobic catabolism of *Saccharomyces cerevisiae* [Maiorella *et al.* (1981)].

Table 96(a). Possible effects of carbon dioxide on key enzymes in *Saccharomyces cerevisiae* [(Jones *et al.* (1981)]

Reaction	Comment
Pyruvate → acetaldehyde → ethanol + CO_2	Reduced production of ethanol
Pyruvate → ATP + O_2 → oxaloacetate + CO_2 → amino acids	Stimulation results in less available pyruvate for ethanol production
Acetyl-CoA + CO_2 → Malonyl-CoA → fatty acids	Stimulation results in less available pyruvate for ethanol production
Pyruvate + CO_2 → malate	Stimulation results in less available pyruvate but malate enzyme level is not high
Phosphoenol pyruvate + ADP → oxaloacetate + ATP + CO_2	Stimulation results in less available pyruvate but enzyme is repressed by glucose
6-Phosphogluconate + NADP → ribulose 5-phosphate + $NADPH_2$ + CO_2	Reduced production of biosynthetic precursors, thus cell yield will decrease. Will reduce rate of production of ethanol

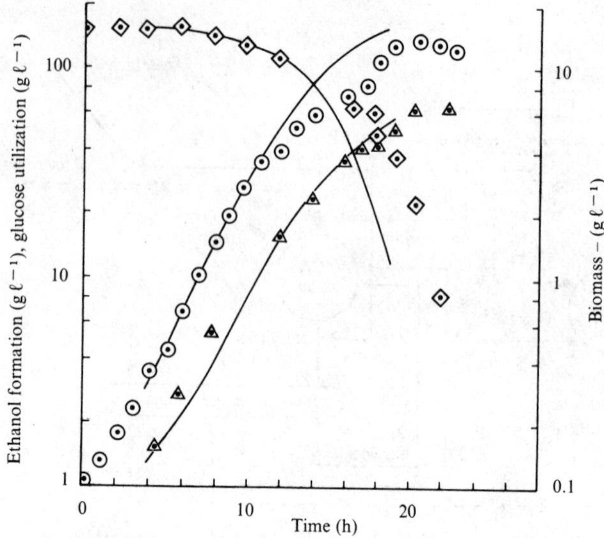

Figure 44. Batch ethanol fermentation by *Saccharomyces cerevisiae*. Observed values; ⊙, biomass; △, ethanol; ◇, glucose [Aiba *et al.* (1968)].

Table 96(b). Steady-state data for continuous ethanol fermentation using *Saccharomyces cerevisiae* [Aiba *et al.* (1968)].

Substrate concentration ($g\,\ell^{-1}$)	Initial substrate concentration ($g\,\ell^{-1}$)	Product concentration ($g\,\ell^{-1}$)	Initial product concentration ($g\,\ell^{-1}$)	Biomass ($g\,\ell^{-1}$)	Specific ethanol production rate (h^{-1})
Dilution rate = 0.084 h^{-1}					
0.054	21.5	8.0	0.0	2.00	0.33
0.096	19.5	13.7	7.9	2.22	0.22
0.122	19.9	21.3	15.1	2.05	0.25
0.127	20.5	28.4	21.9	2.08	0.26
0.118	20.9	42.7	36.9	2.12	0.23
0.212	20.5	57.0	53.3	1.87	0.17
Dilution rate = 0.100 h^{-1}					
0.079	10.9	4.7	0.0	1.20	0.39
0.091	10.0	19.3	15.5	1.37	0.28
0.114	9.6	33.4	30.4	1.40	0.21
0.115	10.7	51.5	48.3	1.22	0.26
Dilution rate = 0.160 h^{-1}					
0.138	21.2	8.6	0.0	2.40	0.57
0.194	21.1	22.7	15.3	2.43	0.49
0.326	21.6	40.0	32.6	2.30	0.52
1.63	20.4	55.4	48.6	1.68	0.65
Dilution rate = 0.198 h^{-1}					
0.186	20.7	8.4	0.0	2.33	0.72
0.723	20.7	24.1	16.3	2.14	0.73
1.82	20.2	37.2	29.8	2.03	0.72
Dilution rate = 0.242 h^{-1}					
0.226	10.8	4.5	0.0	1.25	0.8
1.09	10.4	18.7	15.0	0.92	0.97

Cultures of *Zymomonas* and some other bacteria metabolize glucose via the Entner–Doudoroff pathway, which yields one mole of ATP per mole of glucose utilized, while the glycolytic pathway employed by yeasts yields two moles of ATP per mole of glucose.

Time Course

A time course for a batch ethanol fermentation is given in Fig. 44. The steady-state data for a continuous fermentation are presented in Table 96(b).

Glucose is the primary reactant in the fermentation. At very low glucose concentration (i.e. less than 10 mg ℓ^{-1}) the rate of glucose consumption increases approximately linearly with glucose concentration. At high glucose concentrations (i.e. above 150 g ℓ^{-1}), glucose inhibits the enzymes in both the repressed and oxidative pathways. At intermediate glucose concentration (i.e. 3–100 g ℓ^{-1}) catabolite repression of the oxidative pathways allows the production of ethanol, even if oxygen is present in large amounts. The inhibition of growth and ethanol production by ethanol can be expressed by various equations (*see* Table 97). The inhibitory effects of ethanol are shown in Fig. 45.

Table 97. Ethanol inhibition models for yeast growth and ethanol production.

	Holzberg *et al.* (1967)	Aiba *et al.* (1968)	Bazua and Wilke (1977)
Kinetic equations[a] Growth	$\mu = \mu_0 - k_1 (p - k_2)$	$\mu = \mu_0 e^{-k_1 p} \dfrac{s}{K_s + s}$ $= \dfrac{\mu_0}{1 + P/K_p} \cdot \dfrac{s}{K_s + s}$	$\mu = \mu_0 - \dfrac{k_1 p}{k_2 - p} \cdot \dfrac{s}{K_s + s}$
Ethanol production	$\dfrac{d}{dt} = k_3 \left(\dfrac{\ln x}{\mu}\right) - k_4 p - k_5$	$v = v_0 e^{-k_2 p} \dfrac{s}{K_s' + s}$ $\dfrac{v_0}{1 + p/k_p} \cdot \dfrac{s}{K_s' + s}$	$v = v_0 - \dfrac{k_3 p}{k_4 - p} \cdot \dfrac{s}{K_s' + s}$
Maximum ethanol concentration (g ℓ^{-1})	68.5	76.4	93
Strain	*Saccharomyces cerevisiae ellipsodeus* New York Agricultural Experimental Station No. 223	*Sacch. cerevisiae* Japan Sugar Refinery Ltd. H-1	*Sacch. cerevisiae* ATCC No. 4126
Conditions	21°C, pH 3.6, enriched grape juice	30°C, pH 4.0, glucose	30°C, pH 4.0 glucose
Comments	Equations only apply where fermentation is not nutrient- or substrate-limited	Equations predict continuous cell growth and ethanol productivity	Productivity is affected by ethanol even at low concentrations. Growth and ethanol production cease at finite ethanol concentration

[a] Where s = substrate concentration (g/ℓ),
p = product concentration (g/ℓ),
x = biomass concentration (g/ℓ),
t = time,
μ = specific growth rate (h^{-1}),
v = specific rate of product formation, (g g^{-1} h^{-1}).

Table 98. Systems tested for ethanol production [(Kolot (1980)].

System	Method of immobilization	Sugar concentration (%)	Cell mass concentration ($g\ \ell^{-1}$)	Maximum ethanol production $g\ \ell^{-1}\ h^{-1}$)	Temperature (°C)	Half life t	Percentage of theoretical yield
Batch		10	5.6	2.2	35		
Continuous		10	10–12	7.0	35		
Continuous with recycle		10	50	29.0	35		
Vacuum with recycle		50	124	82.0	35		
Immobilized cells	Entrapment	10	10^{12} cells ℓ^{-1}	100.0	30	3 months	100
Immobilized cells	Co-immobilization with an enzyme	5		20.0	22	20 days	80
Immobilized cells	Adsorption	10			30	1 month	46

Table 99. Comparative ethanol productivities of continuous systems [(Rogers *et al.* (1980)].

Microorganism	System	Optimal Growth Conditions: Input Glucose ($g\ \ell^{-1}$)	Dilution rate (h^{-1})	Cell concentration ($g\ \ell^{-1}$)	Ethanol concentration ($g\ \ell^{-1}$)	Maximum productivity ($g\ \ell^{-1}\ h^{-1}$)
Saccharomyces cerevisiae ATCC 4126	No recycle	100	0.17	12	41	7.0
Sacch. cerevisiae ATCC 4126	Cell recycle	100	0.68	50	43	29
Sacch. cerevisiae NRRL Y-132	Cell recycle	150	0.53	48	60.5	32
Sacch. uvarum ATCC 26602	Cell recycle	200	0.60	50	60	36
Sacch. cerevisiae ATCC 4126	Vacuum (50 mm Hg)	334		50	100–160	40
Sacch. cerevisiae ATCC 4126	Cell recycle, vacuum (50 mm Hg)	334	0.23	124	110–160	82
Zymomonas mobilis ATCC 10988	Cell recycle	100	2.7	38	44.5	120

Table 98 contains a comparison of fermenter systems tested for ethanol production. When maximum ethanol productivity and yields are considered, systems using immobilized cells are superior. A comparison of various continuous systems is given in Table 99. Comparative kinetics of *Z. mobilis* and *Sacch. uvarum* in batch is given in cultures Table 100.

Table 100. Kinetic parameters for *Z. mobilis* and *Saccharomyces uvarum* on glucose (250 g ℓ$^{-1}$) in nonaerated batch culture[a] [(Rogers *et al.* (1980)].

Kinetic parameters	*Z. mobilis*	*Sacch. uvarum*
Specific growth rate (h^{-1})	0.13	0.55[d]
Specific glucose uptake rate (g g^{-1} h^{-1})	5.5	2.1[d]
Specific ethanol production rate (g g^{-1} h^{-1})	2.5	0.87[d]
Cell yield (g g^{-1})[b]	0.019	0.033
Ethanol yield (g g^{-1})[b]	0.47	0.44
Relative ethanol yield (%)[a, b, c]	92.5	86
Maximum ethanol concentration (g ℓ$^{-1}$)	102	108

[a] Temperature, 30°C; pH 5.0.
[b] Based on the difference between initial and residual glucose concentrations.
[c] A molar reaction stoichiometry of 1 glucose → 2 ethanol + $2CO_2$ has been assumed for a theoretical yield.
[d] Kinetic parameters calculated for fermentation run between 16 and 22 hours when the culture was growing fully anaerobically.

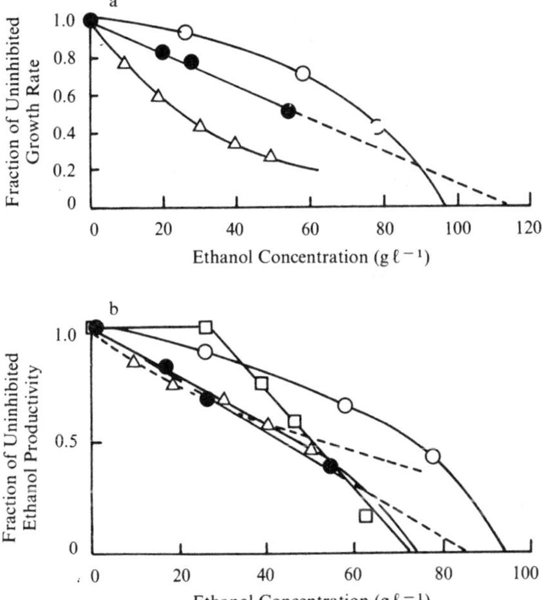

Figure 45(a). Ethanol end product inhibition – comparison of data from different sources. (b) Ethanol inhibition of cell growth – comparison of data from different sources [Maiorella *et al.* (1981)].

MICROBIAL FAT

Microorganisms

All microorganisms contain fatty acids and trigylcerides as cellular constituents. However, to be considered as producers of fats and oils, microorganisms should have a high fat content — about 40 per cent (by wt). Table 101 provides a list of those yeasts and moulds containing significant amounts of fats.

Table 101. Yeasts and moulds as potential fat producers [Ratledge (1978)].

Microorganism	Substrate	Fat content (% dry wt)	Fat coefficient [g fat produced (100 g substrate utilized)$^{-1}$]
Yeasts			
Candida guilliermondii	*n*-Alkanes	30	
C. intermedia	*n*-Alkanes	20	
C. tropicalis	*n*-Alkanes	32	
Candida sp. no. 107	Glucose	42	22.5
	n-Alkanes	15–37	25
Cryptococcus terricolus	Glucose	55–65	21
Hansenula anomala	Glucose	17	
H. ciferrii	Molasses	22	
H. saturnus	Molasses	20	
	Glucose	28	8
Lipomyces lipofer	Glucose	38	
	Peat moss hydrolyzate	48	
Lipomyces sp.	Glucose	67	20
	Xylose	48	17
	Various wastes and Molasses	66	⩽ 24
Lipomyces starkeyi	Lactose	31	10
	Glucose	31–38	9–15
Rhodotorula glutinis		30–35	9–15
R. gracilis	Molasses	40	44 (short time only)
	Glucose	64	
	Sugar cane syrup	67	21
	Glucose	64	15 overall 44 (short time only)
	Ethanol	62	15
	Synthetic ethanol	60	14
	Glucose	66	17
	Alkanes	32	
R. longissima		20	
Moulds			
Aspergillus fischeri	Sucrose	32–53	12–20
Asp. fumigatus	Maltose and other sources	20	
Asp. nidulans	Glucose	27	9
	Glucose	15	7
Asp. ochraceus	Sucrose	48	13
Asp. terreus	Sucrose	51–57	10–13
	Starch	18–24	6
Asp. ustus	Lactose	36	12.7
Chaetomium globosum	Glucose	54	
Cladosporium fulvum	Sucrose	22–14	7
Cladosp. herbarum	Sucrose	20–29	7–11

Table 101 – (continued)

Microorganism	Substrate	Fat content (% dry wt)	Fat coefficient [g fat produced (100 g substrate utilized)$^{-1}$]
Gibberella fujikuroi (*Fusarium moniliforme*)	Glucose	45	7.8
Malbranchea pulchella	Glucose	27	
Mortierella vinacea	Acetate	28	
	Glucose	66	18
	Maltose	34	
Mucor miehei	Glucose	24	
M. pusillus	Glucose	26	
Myrothecium sp.	Not given	30	
Penicillium funiculosum	*n*-Alkanes	22	
Pen. gladioli	Sucrose	32	5.7
Pen. javanicum	Glucose	39	9
Pen. lilacinum	Date extract	23	
	Sucrose	35	25
Pen. soppi	*n*-Alkanes, sucrose	11–25	
	Molasses	19	
Pen. spinulosum	Molasses, sucrose	25–64	6–16
Pythium irregulare	Glucose	30–42	
Py. ultimum	Glucose	48	
Rhizopus sp.	Glucose	27	
Rhizopus arrhizus	Glucose, maltose	20	
Stilbella thermophila	Glucose	38	

Generally, bacteria possess low levels of toxic lipids. When grown in a natural environment, algae also have a low lipid content and a slow growth rate. A list of bacteria and their fat contents is contained in Table 102.

Table 102. Fat and total lipid as a percentage of dry matter for a range of bacteria [Hockenhull (1959)].

Bacterium	Fat (%)	Total lipid (%)
Agrobacterium tumefaciens (Grown on glycerol)	0.9	1.8
(Grown on glucose)	2.1	6.1
Bacillus megaterium		21.0
B. subtilis	2.0	6.1
Bordatella pertussis	22.0	24.0
Brucella abortus	0.9	6.1
Bruc. melitensis	4.8	5.3
Bruc. suis	2.4	5.6
Corynebacterium diphtheriae	6.3	8.8
Escherichia coli	4.0–6.0	4.0–5.0
Lactobacillus acidophilus	4.8	7.0
L. arabinosus		0.5
L. casei		0.6
Malleomyces mallei	5.0–7.0	6.0–8.0
Mycobacterium avium	2.2	15.2
M. phlei	3.0	7.0
M. tuberculosis var. *bovis*	3.3	14.3
M. tuberculosis var. *huminis* Ra.	5.6	16.1
M. tuberculosis var. *huminis* Rv.		21.1
Salmonella paratyphi C	2.8	3.6
S. typhosa	1.3	1.5

The fat content of microorganisms can vary considerably depending on the growth conditions. Apart from the fat content, the quality of the fat, its toxicity, the yields and the rate of formation have to be considered.

Media

Although the biosynthesis of lipids occurs throughout the growth of microorganisms, the accumulation of fat begins only when a nutrient other than the carbon source is exhausted. It follows that fat accumulation occurs usually at the end of growth. Therefore, the medium should be formulated to sustain rapid and extensive growth initially. The carbon source can be glucose, sucrose, or other carbohydrates, ethanol, *n*-alkanes, etc. The nutrient that becomes exhausted from the medium is usually the nitrogen source. Depletion of other nutrients (e.g., phosphate, sulphate and iron) can also induce fat formation in some microbial species. The optimum medium formulations for culture of several microorganisms are given in Tables 103–106.

The glucose in the medium for *R. gracilis* culture (*see* Table 106) can be replaced by ethanol at a concentration not exceeding 2 per cent (by vol) or *n*-alkanes ($20\,g\,\ell^{-1}$).

Table 103. Medium for culture of *Aspergillus fischeri*.

Component	Amount ($g\,\ell^{-1}$)
Commercial glucose	200
Ammonium nitrate	10
Potassium dihydrogen phosphate	6.8
Magnesium sulphate heptahydrate	5.0
Ferric chloride hexahydrate	0.16
Zinc sulphate heptahydrate	0.05

Table 104. Medium for culture of *Aspergillus nidulans*.

Component	Amount ($g\,\ell^{-1}$)
Sucrose	170
Ammonium nitrate	3
Potassium sulphate	0.22
Zinc sulphate heptahydrate	0.05
Sodium dihydrogen phosphate dihydrate	7.3
Magnesium sulphate heptahydrate	5

Table 105. Medium[a] for culture of *Penicillium javanicum*.

Component	Amount ($g\,\ell^{-1}$)
Glucose	200
Ammonium nitrate	2.25–3.4
Potassium dihydrogen phosphate	0.3–1.2
Magnesium sulphate heptahydrate	0.25

[a] pH 3.1–6.8.

Table 106. Medium for culture of *Rhodotorula gracilis*[a].

Component	Amount ($g\,\ell^{-1}$)
Glucose	30–40
Potassium nitrate	1.42
Diammonium hydrogen phosphate	0.33
Magnesium sulphate heptahydrate	1.00
Corn-steep (50% dry weight)	0.05

[a] Air flow rate, 1 vvm; temperature, 28–29°C, pH 5.5–6.0.

Metabolic Pathway

The ultimate precursor for the biosynthesis of saturated fatty acids is acetyl-CoA, which is derived from carbohydrate or amino acid sources. The overall reaction of fatty acid synthesis is catalyzed by a group of seven proteins — the fatty acid synthetase complex — in the cytosol. The usual end product is palmitic acid, which is the precursor of the other long-chain, saturated and unsaturated, fatty acids found in most microorganisms. Acetyl-CoA supplies only one of the eight acetyl units needed for the biosynthesis of palmitic acid [$Me(CH_2)_{14}COOH$]; the other seven are provided in the form of malonyl-CoA. The overall reaction is

$$\text{Acetyl-CoA} + 7\ \text{malonyl-CoA} + 14\text{NADPH} + 14\text{H}^+ \rightarrow \text{palmitic acid} + 7\text{CO}_2 + 8\text{CoA} + 14\text{NADP}^+ + 6\text{H}_2\text{O}$$

The biosynthesis of a triglyceride is outlined in Fig. 46.

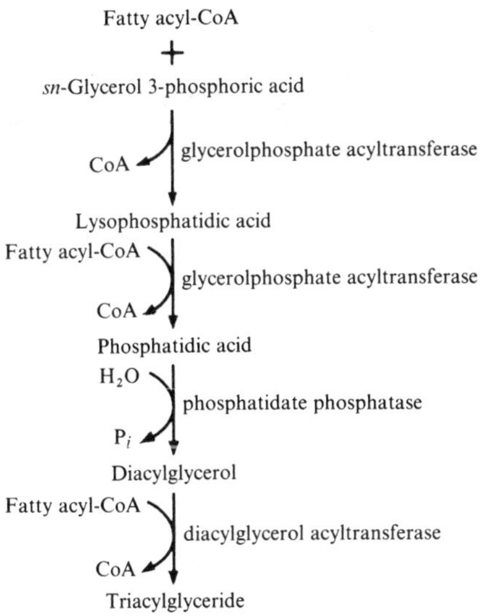

Figure 46. Formation of a triacylglyceride.

Time course

Since most microorganisms start accumulating fat after the initial growth phase, batch culture is usually preferred. Fig. 47 shows the time course of a batch fermentation with *R. gracilis*.

A two-stage continuous fermentation can be used for the production of fat: the first stage for microbial growth and the second for fat accumulation. Using a *Candida* species, fat production is dependent on the growth rate [*see* Ratledge (1978)].

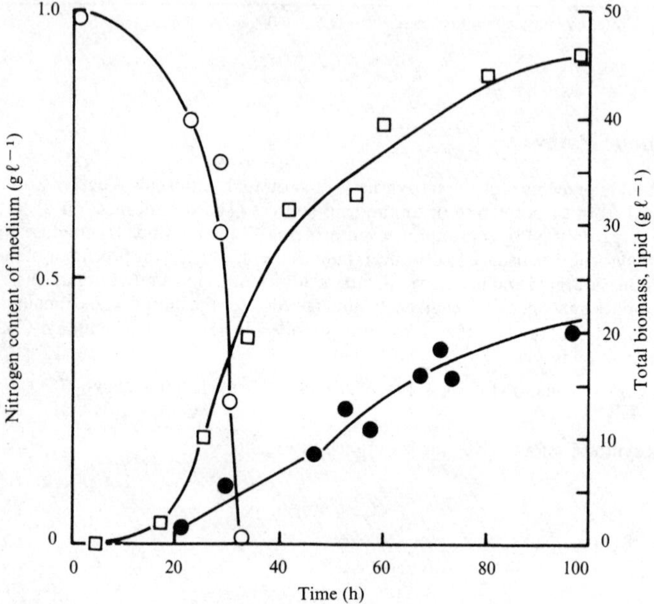

Figure 47. Relationships between nitrogen content of medium (○) and growth (□) and lipid content (●) of *Rhodotorula gracilis*. Cultures were grown in 5-litre baffled fermenters in a medium containing glucose (100 g ℓ^{-1}) and ammonium sulphate (0.6 g ℓ^{-1}) and held at pH 6.0 [Ratledge (1978)].

Tables 107–108 demonstrate the effect of medium components on fat synthesis.

Table 107(a). Fat content of *Rhodotorula gracilis* with different nitrogen sources and under different conditions [Hockenhull (1959)].

Nitrogen source	Aeration	Sugar used (%)	Fat content of medium ($g\,\ell^{-1}$)	Yeast content ($g\,\ell^{-1}$)	Fat content of yeast (%)	Fat coefficient
Ammonium sulphate	–	78.9	3.7	9.0	47	12.3
	+	64.4	3.8	4.0	52	15.3
Asparagine	–	59.5	5.0	7.7	62	14.0
	+	80.4	7.5	11.7	72	18.4
Aspartic acid	–	69.9	5.8	8.9	65	15.8
	+	81.9	9.5	12.7	74	20.9
Urea	–	66.4	6.1	9.7	64	15.9
	+	71.7	6.3	9.6	67	16.2
Uric acid	–	79.2	5.1	7.9	64	11.9
	+	79.1	6.8	10.4	67	16.0

Table 107(b). Effect of inorganic salt supplement on growth, fat content and fat coefficient of moulds grown on whey [Hockenhull (1959)].

Supplement	Mycelium (g)	Fat (%)	Fat coefficient
Aspergillus ustus			
None	0.73	14.0	4.8
0.2 M Potassium sulphate + 0.1 M ammonium nitrate	0.81	20.0	3.8
Penicillium frequentans			
None	0.78	13.1	3.3
0.2 M Potassium sulphate + 0.1 M ammonium nitrate	1.05	24.3	8.2

Table 108. Fatty-acyl compositions of lipids from fat-producing microorganisms or related strains grown on carbohydrates [Ratledge (1978)].

Microorganism	Substrate	Relative proportion of fatty acids [% (by wt)][a]											
		12:0	14:0	14:1	16:0	16:1	17:0, 17:1	18:0	18:1	18:2	18:3	20:0 and 22:0	20+ un-saturated
Yeasts													
Candida sp. no. 107 (Batch culture)	Glucose		1		22	2		8	31	26		3	7
(Continuous culture)	Glucose		1		37	1		14	36	8			4
Hansenula anomala (triglyceride)	Glucose		tr		20	2	1	2	49	26			
H. anomala	Glucose				12	10			28	24	19		
H. anomala	Glucose	tr	3	tr	35	2	2	36	3		3		
Lipomyces lipofer	Glucose		tr		17	4		10	48	16	3		
(triglyceride)	Glucose		tr		12	3		6	77	3	tr	tr	
(triglyceride)	Glucose		2		16	7		3	62	9	1	tr	
L. starkeyi (triglyceride)	Glucose				40	6		5	44	4			
Rhodotorula gracilis	Glucose	tr	1		24	2	1	11	45	12	3	2	
	Glucose		1		20	2	4	1	42	21	8		
	Glucose		1		31			9	53	1	5		
	Ethanol		1		35			11	46	1	6		
R. graminis	Starch, glucose	1	4	1	32	tr		3	37	10	5		3
R. glutinis	Glucose		1		12	2	1	7	50	21	6		
Moulds													
Aspergillus nidulans	Glucose		1		21	1		16	40	17	tr	1	2
Asp. niger	Glucose	1	2		22	3		5	7	46	11	1	1
	Lactose	3	5		50	2	2	10	11	14	2		
Asp. ochraceus	Sucrose		tr		38			tr	15	45	2		
Asp. terreus	Sucrose	tr	2		23	tr		tr	14	40	21		
Chaetomium globosum	Glucose				58	3	4	8	27				
Fusarium moniliforme	Glucose		1	1	14			11	30	42	1		

Malbranchea pulchella	Glucose				11		11	27	51				
Mucor globosus	Glucose	2	8		26	8	7	26	8	16			
M. ramannianus	Glucose		2		19	3		4	28	14	31[b]		
Penicillium chrysogenum	Sucrose	tr	tr		18	1	tr	9	11	53		6	tr
	Glucose				13	1		5	14	61	7		
	Glucose	tr	tr		24	3	tr	9	5	48		5	4
Pen. lilacinum	Glucose		tr		16	3		2	40	13			1
Pen. soppi	Glucose		1		20	2	1	5	9	50	13	tr	
	Lactose	2	2		41	1	tr	8	12	32	3	tr	
Pen. spinulosum	Glucose		tr		18	4		12	43	21	tr	1	tr
Pythium irregulare	Glucose	3	7	2	15	13		3	30	4	2[b]	13	8
Py. ultimum	Glucose		8		23	9		7	22	15	2[b]	5	11
Rhizopus arrhizus	Glucose				21	4		9	42	17	8[b]		
Rhizopus sp.	Glucose		1		21	2		5	30	29	12[b]		
Stilbella thermophila	Glucose		2		43	2	14	25	14				

[a] Tr = trace amounts detected.
[b] γ-Linolenic acid (octadeca-6,9,12-trienoic acid).

NUCLEIC ACIDS AND RELATED COMPOUNDS

Microorganisms

Mutant microorganisms capable of accumulating inosine, guanosine, and 5′-inosine monophosphate (5′-IMP) are given in Tables 109, 110 and 111, respectively. Production of other related compounds is shown in Table 112. Properties and uses of nucleic acid-related compounds are given in Tables 113 and 114, respectively.

Table 109. Fermentative production of inosine [Yamada (1977)].

Microorganism	Carbon source	Yield ($g\ell^{-1}$)	Yield (%)
Arthrobacter paraffineus (Ad⁻)	*n*-Alkane	2.3	5
Brevibacterium ketoglutamicum (Ad⁻)	*n*-Alkane	1.5	3
Corynebacterium petrophilum (Ad⁻)	*n*-Alkane	1.5	3
Bacillus pumilis (Ad⁻)	Glucose	16	13
B. subtilis (Ad⁻, Tyr⁻, His⁻)	Glucose	16	20
B. subtilis (Ad⁻)	Soluble starch	5	6

Table 110. Fermentative production of guanosine[a] [Yamada (1977)].

Microorganism	Carbon source	Yield ($g\ell^{-1}$)	Yield (%)
Bacillus sp. (Ad⁻, red⁻, adenosine^r, 8-AG^r, His⁻)	Maltose	5.4	5.4
B. subtilis (Ad⁻, red⁻, 8-AG^r)	Glucose	4.3	6.0
B. subtilis (Ad⁻, Tyr⁻, red⁻, 8-AX^r)	Glucose	5.0	6.0
B. subtilis (Ad⁻, His⁻, red⁻, MSO^r, SG^r)	Glucose	11	13
B. subtilis (Ad⁻, red⁻, adenase⁻, 8-AG^r, Leu⁻, His⁻)	Glucose	10	10

[a] 8-AG, 8-azaguanine; 8-AX, 8-azaxanthine; MSO, methionine sulphoxide; SG, sulphaguanidine; Enzyme: red, GMP-reductase.

Table 111. Fermentative production of 5′-IMP [Yamada (1977)].

Microorganism	Carbon source	Yield ($g\ \ell^{-1}$)	Yield (%)
Bacillus subtilis (Ad^-, alkaline phosphatase$^-$)	Glucose	0.25	3.5
B. subtilis (Ad^-, nucleotidase$^-$)	Glucose	3.5	
Brevibacterium ammoniagenes (Ad^-)	Glucose	10.2	10.2
Brev. ammoniagenes (Ad^-, Mn^{2+} insensitive)	Glucose	19	19
Brev. thiogentalis	Glucose	3.6	3.6
Corynebacterium tropicalis (Ad^-)	*n*-Alkane	1.5	
Pseudomonas aeruginosa (Ad^-)	*n*-Alkane	2.1	

Table 112. Fermentative production of other nucleic acid-related compounds [Yamada (1977)].

Product	Microorganism	Accumulation
Adenosine	*Bacillus* sp. (xanthine$^-$)	$16\ g\ \ell^{-1}$
ATP	*Brevibacterium ammoniagenes*	$1.5\ g\ \ell^{-1}$
CDP–choline	*Saccharomyces carlsbergensis*	$35\ \mu M$
FAD	*Sarcina lutea*	$1\ g\ \ell^{-1}$
NAD	*Brev. ammoniagenes*	$1.9\ g\ \ell^{-1}$
Orotic acid	*Arthrobacter paraffineus* (uracil$^-$)	$20\ g\ \ell^{-1}$

Table 113. Properties of nucleic acid-related compounds.

Compound	Empirical formula	Physical properties
Adenosine	$C_{10}H_{13}N_5O_4$	Mol. wt, 267.2; mp, 234–5°C; $[\alpha]_D^{11}$ – 61.7°C (H_2O), sol. H_2O; insol. EtOH
5′-GMP	$C_{10}H_{14}N_5O_8P$	Mol. wt, 363.2; decomp., 190–200°C; slightly sol. H_2O
5′-IMP	$C_{10}H_{14}N_4O_8P$	Mol. wt, 348.2; syrup; $[\alpha]_D^{20}$ – 185°(Ba salt in HCl); sol. H_2O; slightly sol. EtOH, Et_2O
Inosine	$C_{10}H_{12}N_4O_5$	Mol. wt, 268.2; mp 90°C (dihydrate); rectangular plates (dihydrate) needles (anhydrous), $[\alpha]_D^{18}$ – 49.2°; sol. H_2O
NAD	$C_{21}H_{27}N_7O_{14}P_2$	Mol. wt, 663.4; hygroscopic powder; sol. H_2O; monobasic salt
Orotic acid	$C_5H_4N_2O_4$	Mol. wt, 156.1; mp, 345–6°C

Table 114. Summary of occurrence and uses of nucleic acid-related compounds.

Compound	Occurrence	Uses
Adenosine	Widely found in nature	Seasoning (sodium salt), biochemical research
5′-GMP	Widely found in nature (formed in hydrolyzates of RNA)	Seasoning (sodium salt)
5′-IMP	Widely found in nature	Seasoning (sodium salt), biochemical research
Inosine	Muscle, yeast and urine	Activates cellular functions, treatment of cardiac conditions
NAD	Fresh bakers' yeast, rabbit muscle	Biochemical research, alcohol and narcotic antagonist
Orotic acid	Milk	Animal feed supplement, uricosuric

Media

Table 115 lists the composition of various media used for the production of inosine by different mutants of *Bacillus* species and *Brevibacterium ammoniagenes.*

Metabolic pathways

The biosynthesis of 5′-inosine monophosphate, 5′-guanosine monophosphate, and the nucleotides, which is complex, is described by Demain (1978).

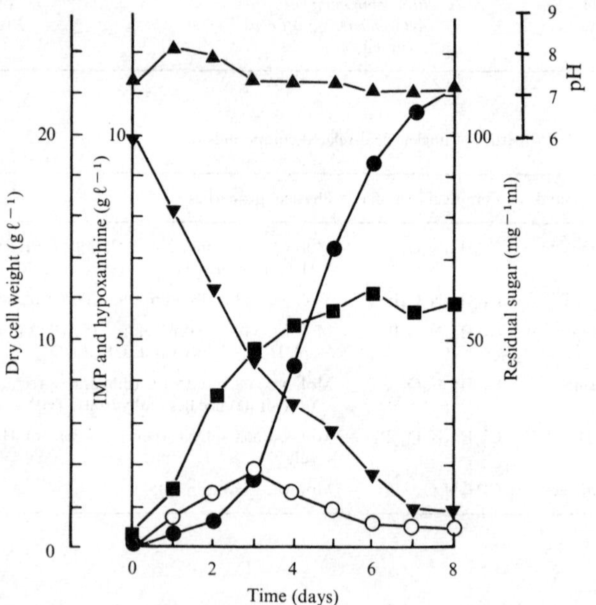

Figure 48. Time course of IMP accumulation by *Brevibacterium ammoniagenes* KY 13102. ●, IMP; ○, hypoxanthine; ▲, pH; ▼, residual sugar; ■, dry cell weight [Nakao (1979)].

Table 115. Composition of media employed for inosine accumulation[a] [Nakao (1979)].

Compounds	*B. subtilis* (Ad$^-$, His$^-$, Tyr$^-$)	*B. subtilis* (Ad$^-$, Try$^-$, Red$^-$, Dea$^-$, 8-AG^r)	*Bacillus* sp. (Ad$^-$, His$^-$, Red$^-$, 8-AG^r)	*Bacillus* sp. (Ad$^-$, His$^-$, Thr$^-$, Dea$^-$)	*Brevibacterium ammoniagenes* (Ad^L, 6-MG^r, Gua$^-$)	*Brev. ammoniagenes* (Ad^L, 6-MG^r, 6-MTP^r, Gua$^-$)
Glucose	60–70 g ℓ^{-1}	70 g ℓ^{-1}		100 g ℓ^{-1}	130 g ℓ^{-1}	
Maltose			100 g ℓ^{-1}			
Inverted molasses						75 g ℓ^{-1} (0 h), 100 g ℓ^{-1} (22 h)
Ammonium chloride	20 g ℓ^{-1}	15 g ℓ^{-1}		20 g ℓ^{-1}		
Ammonium sulphate			20 g ℓ^{-1}			
Urea					4 g ℓ^{-1}	2 g ℓ^{-1}
Potassium dihydrogen phosphate	1 g ℓ^{-1}	1 g ℓ^{-1}		2 g ℓ^{-1}	10 g ℓ^{-1}	1 g ℓ^{-1}
Dipotassium hydrogen phosphate					10 g ℓ^{-1}	1 g ℓ^{-1}
Calcium phosphate (dibasic)			5 g ℓ^{-1}			
Calcium phosphate (tribasic)			5 g ℓ^{-1}			
Magnesium sulphate heptahydrate	0.4 g ℓ^{-1}	0.4 g ℓ^{-1}	2 g ℓ^{-1}	0.5 g ℓ^{-1}	10 g ℓ^{-1}	1 g ℓ^{-1}
Ferrous sulphate heptahydrate	2 ppm (Fe^{2+})	2 ppm (Fe^{2+})			10 mg ℓ^{-1}	
Zinc sulphate heptahydrate					10 mg ℓ^{-1}	
Manganese sulphate (hydrated)	2 ppm (Mn^{2+})	2 ppm (Mn^{2+})			10 mg ℓ^{-1}	
Manganese chloride tetrahydrate				0.01 g ℓ^{-1}		
Cysteine					20 mg ℓ^{-1}	
Tryptophan		300 mg ℓ^{-1}				
Thiamin					5 mg ℓ^{-1}	
Pantothenate					10 mg ℓ^{-1}	
Biotin			0.2 mg ℓ^{-1}		30 μg ℓ^{-1}	
Nicotinic acid					5 mg ℓ^{-1}	
Mieki	4 g ℓ^{-1}	2 g ℓ^{-1}				
Casamino acid				1 g ℓ^{-1}		
Yeast extract				10 g ℓ^{-1}		
Meat extract					10 mg ℓ^{-1}	
Dry yeast	14 g ℓ^{-1}		10 g ℓ^{-1}			
Adenine		100 mg ℓ^{-1}			100 mg ℓ^{-1}	
Guanine					100 mg ℓ^{-1}	

continued overleaf

Table 115 – (continued)

Compounds	*B. subtilis* (Ad^-, His^-, Tyr^-)	*B. subtilis* (Ad^-, Try^-, Red^-, Dea^-, $8\text{-}AG^r$)	*Bacillus* sp. (Ad^-, His^-, Red^-, $8\text{-}AG^r$)	*Bacillus* sp. (Ad^-, His^-, Thr^-, Dea^-)	*Brevibacterium ammoniagenes* (Ad^L, $6\text{-}MG^r$, Gua^-)	*Brev. ammoniagenes* (Ad^L, $6\text{-}MG^r$, $6\text{-}MTP^r$, Gua^-)
Calcium carbonate	2%	2.5%	2%	5%		
pH	6.0	7.0	7.6		8.3	
Max. inosine accumulated	10.5 g ℓ^{-1}	18 g ℓ^{-1}	14.1 g ℓ^{-1}	10.8 g ℓ^{-1}	13.6 g ℓ^{-1}	30 g ℓ^{-1}

[a] Ad^-, adenine-requiring; Gua^-, guanine-requiring; His^-, histidine-requiring; Tyr^-, tyrosine-requiring; Thr^-, threonine-requiring; Try^-, tryptophan-requiring; Ad^L, leaky adenine mutant; Red^-, GMP reductase-deficient; Dea^-, AMP deaminase-deficient; $8\text{-}AG^r$, 8-azaguanine-resistant; $6\text{-}MG^r$, 6-mercaptoguanine-resistant; $6\text{-}MTP^r$, 6-methylthiopurine-resistant.

Time Course

A typical time course for the production of inosine monophosphate by *Brev. ammoniagenes* KY 13102 is shown in Fig. 48.

ORGANIC ACIDS

General Aspects

Table 116 contains a list of the main organic acids produced by microorganisms together with the expected yields. Apart from lactic acid, gibberellic acid and a few others, the majority of organic acids fall into two groups.

1. Compounds belonging to or closely related to the tricarboxylic acid cycle.
2. Compounds obtained directly by oxidation of glucose.

The latter group must be prepared from glucose, while the former can be derived from a variety of starting materials. Most acid-producing microorganisms have similar nutritional requirements.

Table 116. Main organic acids produced by microorganisms [Yamada (1977)].

Acid	Microorganism	Carbon source	Yield (%)
Acetic acid	*Acetobacter aceti*	Ethanol	95
L-Alloisocitric acid	*Penicillium purpurogenum*	Glucose	40
D-Araboascorbic acid	*Pen. notatum*	Glucose	45
Citric acid	*Aspergillus niger*	Sucrose	85
	Candida lipolytica	*n*-Alkane	140
Fumaric acid	*Rhizopus delemar*	Glucose	58
Gluconic acid	*Asp. niger*	Glucose	95
L-Isocitric acid	*C. brumptii*	Glucose	28
Itaconic acid	*Asp. terreus*	Glucose	60
2-Ketogluconic acid	*Pseudomonas fluorescens*	Glucose	90
5-Ketogluconic acid	*Gluconobacter suboxydans*	Glucose	90
α-Ketoglutaric acid	*C. hydrocarbofumarica*	*n*-Alkane	84
Kojic acid	*Asp. oryzae*	Glucose	50
Lactic acid	*Lactobacillus delbrueckii*	Glucose	90
Malic acid	*L. brevis*	Glucose	100
Propionic acid	*Propionibacterium shermanii*	Glucose	60
Pyruvic acid	*Ps. aeruginosa*	Glucose	50
Succinic acid	*Bacterium succinicum*	Malic acid	57
Tartaric acid	*Gluc. suboxydans*	Glucose	27

Properties and Uses

The empirical formulae and physical properties of organic acids are tabulated in Table 117, while occurrence and uses of these acids are given in Table 118.

Table 117. Structure and physical properties of some organic acids.

Compound	Structure	Empirical formula	Physical properties
Citric acid	$HOOCCH_2COH(COOH)CH_2COOH$	$C_6H_8O_7$	Mol. wt, 192.1; mp, 153°C; sp. gr., 1.665; monoclinic holohedra; sol. H_2O
D-Gluconic acid	$HOOC[CH(OH)]_4CH_2OH$	$C_6H_{12}O_7$	Mol. wt, 196.2; mp, 131°C; $[\alpha]_D^{20}$ − 6.1°(H_2O); sol. H_2O; slightly sol. Et_2OH; insol. Et_2O
2-Ketogluconic acid	$HOOCCO[CH(OH)]_3CH_2OH$	$C_6H_{10}O_7$	Mol. wt, 194.2; $[\alpha]_D^0$− 99.6°(HCl)
Itaconic acid	$CH_2C(COOH)CH_2COOH$	$C_5H_6O_4$	Mol. wt, 131.1; mp, 162–4°C (decomp.); sp. gr., 1.63; hygroscopic crystals; sol. H_2O, EtOH; slightly sol. Bz, $CHCl_3$, CS_2 EtO
DL-Lactic acid	MeCH(OH)COOH	$C_3H_6O_3$	Mol. wt, 90.1; mp, 16.8°C; sol. H_2O, EtOH, insol. $CHCl_3$
L-Malic acid	$HOOCCH(OH)CH_2COOH$	$C_4H_6O_5$	Mol. wt, 134.1; mp, 100°C; decomp., 140°C; $[\alpha]_D$ − 2.3°C (H_2O); sol. H_2O, MeOH, EtOH, Et_2O, Me_2CO
L-Tartaric acid	$HOOC[CH(OH)]_2COOH$	$C_4H_6O_6$	Mol. wt, 150.1; mp, 168–70°C; sp. gr., 1.7598; monoclinic prisms; $[\alpha]_D^{20}$ + 12.0°(H_2O); sol. H_2O, MeOH, EtOH, PrOH; slightly sol. Et_2O; insol. $CHCl_3$

Table 118. Occurrence and uses of some organic acids.

Acid	Occurrence	Uses
Citric acid	Widely occurs in animals and plants, exists as free acid in citrus fruits	Flavouring for beverages, confectionery, food, pharmaceutical syrups, production of resins, antifoaming agent, sequestering agent, dye mordant
D-Gluconic acid		Pharmaceutical (antispasmodic), acid detergent for metals
Itaconic acid		Intermediate for organic syntheses
DL-Lactic acid	Rancid milk, molasses, fruits, exercised animal muscle	Seasoning, mordant for dyes, cheese production, confectionery, dehairing and tanning of hides, plasticizer
L-Malic acid	Apples	Food and drinks manufacture
L-Tartaric acid	Fruits	Food additive, photography, tanning, ceramic manufacture

Acetic Acid

Microorganisms

The main acetic acid-producing bacteria are *Acetobacter*, over 100 species, subspecies and varieties of the genus having been classified. The majority of these species are normally acceptable in acetic acid fermenters, with the exception of *Acet. xylinum*, a heavy slime-former, and any *Acetobacter* species that oxidizes acetic acid to carbon dioxide. Some submerged operations start with pure cultures, but aseptic conditions are not usually maintained. *Gluconobacter* species produce acetic acid using glucose as the carbon source. *Acet. schuetzenbachii* or *Acet. curvum* may be used to produce acetic acid from ethanol in the quick vinegar process, while *Acet. orleanense* may be used in either the quick vinegar or the Orleans process. Two other microorganisms capable of producing acetic acid are *Lactobacillus plantarum* and *Polyporus palustris*.

Media

Acetic acid is produced by the oxidation of ethanol at a concentration of up to 15 per cent (w/v). There are specially adapted strains of acetic acid-producing bacteria that can use a pure alcohol/salts medium. *Acet. aceti* cannot grow in an ethanol/mineral salts/ammonium/nitrogen medium unless acetic acid, an acetate or glucose is added. Some other species, such as *Acet. suboxydans* and *Acet. melanogenus*, have the same requirements.

It appears that sugar, or related substances, or acetic acid is required to initiate growth after which ethanol may be used as an additional source of carbon and energy. Typical medium formulations are given in Tables 119 and 120.

Table 119. Medium for acetic acid fermentation.

Component	Amount
Ethanol	30 ml ℓ^{-1}
Ammonium phosphate	1 g ℓ^{-1}
Potassium phosphate	1 g ℓ^{-1}
Magnesium phosphate	1 g ℓ^{-1}
Sodium acetate	1 g ℓ^{-1}

Table 120. Medium[a] composition for acetic acid fermentation.

Component	Amount
Ethanol (95%)	30 ml
Ammonium sulphate	1 g
Potassium dihydrogen phosphate	0.9 g
Dipotassium hydrogen phosphate	0.1 g
Magnesium sulphate heptahydrate	0.25 g
Ferric chloride [hydrate, 1% (w/v) aqueous solution]	0.5 ml

[a] Dissolved in distilled water to make up 1 litre.

Acetobacter species require B complex vitamins in addition to a nitrogen source and basic minerals. The formulation of soluble nutrients that can be added to a pure ethanol/water/acetic acid medium to be used in submerged culture is listed in Table 121.

Table 121. Medium for submerged-culture production of acetic acid.

Component	Amount ($g\,\ell^{-1}$)
Corn sugar	15.8
Magnesium	7.03
Magnesium sulphate	1.76
Potassium citrate	1.76
Calcium *d*-pantothenate	0.013

The optimum temperature is 30–32°C. Full aeration is required; interruption of the air supply for only 30 seconds can kill acetic acid-producing bacteria.

Metabolic pathway

The conversion of ethanol to acetic acid is relatively straightforward compared to other biological processes. During aceticification, the oxidation of ethanol by *Acetobacter* species is thought to occur by a two-stage oxidation. Initially, ethanol is converted to acetaldehyde and concurrently hydrates. Then the hydrated acetaldehyde is dehydrogenated to acetic acid. Thus

$$CH_3CH_2OH + [O] \rightarrow CH_3CHO + H_2O$$

$$CH_3CHO + H_2O \rightarrow CH_3CH(OH)_2$$

$$CH_3CH(OH)_2 + [O] \xrightarrow[\text{dehydrogenase}]{\text{aldehyde}} CH_3COOH + H_2O$$

Citric Acid

Microorganisms

A large number of fungi have the ability to produce citric acid. For example, *Aspergillus niger*, *Asp. clavatus*, *Penicillium luteum*, *Pen. citrinum*, *Paecilomyces divaricatum*, *Mucor piriformis*, *Ustulina vulgaris*, *Trichoderma viride*, *Candida lipolytica*, other *Candida* species and *Saccharomycopsis lipolytica* are capable of producing citric acid. *Asp. niger*, *Asp. fumaricus*, *Asp. japonicus*, *Asp. wentii* and *C. lipolytica* are usually preferred for submerged culture.

Media

Although various carbon sources can be used, sucrose and glucose are preferred. When raw materials are used, they require some purification (*see* Table 122) as some trace elements affect citric acid production. Various media formulations used for citric acid production are given in Tables 123–125.

Table 122. Treatment of raw materials for the citric acid production.

Raw material	Purification process
Pure sucrose[a]	Cation exchanger
Pure starch[b]	Cation exchanger
Pure glucose	Cation exchanger
Raw sugar	Clarification by ferrocyanide
Molasses[a]	Clarification by ferrocyanide
High test molasses	Chemical treatment
Raw starch[b]	Chemical treatment
Maize grits[b]	Chemical treatment

[a] Starch containing materials will be saccharified enzymically.
[b] Raw materials which can also be utilized for the surface process.

Table 123. Media for the culture of *Aspergillus niger* for citric acid production.

Component	Amount ($g\,\ell^{-1}$)	
	Sporulation medium	Fermentation medium
Sucrose	140	140
Bacto agar	20	0.0
Ammonium nitrate	2.5	2.5
Potassium dihydrogen phosphate	1.0	2.5
Magnesium sulphate heptahydrate	0.25	0.25
Copper ions	0.00048	0.00006
Zinc ions	0.0038	0.00025
Ferrous ions	0.0022	0.0013
Manganous ions	0.001	0.001

Table 124. Medium for citric acid production using *Candida lipolytica*.

Component	Amount ($g\,\ell^{-1}$)
Sodium acetate	40
Glucose	20
Magnesium sulphate heptahydrate	1.0
Ammonium chloride	6.0
Potassium dihydrogen phosphate	2.0
Yeast extract	0.5

Using the medium described in Table 124, every 24 to 36 hours during fermentation, sodium lactate is added to a total level of $120\,g\,\ell^{-1}$.

Using the medium described in Table 125, the temperature is maintained at 30°C and the pH at 7.2. At the end of six days, a citric acid concentration of $80\,g\,\ell^{-1}$ is expected.

With most media, the starting pH is 4.0–6.0, but this usually falls to 1.5–2.0 as the citric acid is synthesized and excreted into the medium. The aeration rate is about 0.5–1.5 vvm. Most strains produce citric acid over the temperature range 25–30°C.

Table 125. Medium[a] for citric acid production using *Aspergillus wentii*.

Component	Amount ($g\,\ell^{-1}$)
Sucrose	150
Urea	1.0
Magnesium sulphate heptahydrate	0.5
Potassium chloride	0.15
Manganese sulphate tetrahydrate	0.02
Zinc sulphate heptahydrate	0.01
Potassium dihydrogen phosphate	0.08

[a] pH of medium is 2.0, achieved by adding hydrochloric acid.

Metabolic pathway

In the case of *Asp. niger*, 78 per cent of the sugar consumed passes through the Embden–Meyerhof–Parnas pathway. Both the Embden-Meyerhof–Parnas and hexose monophosphate pathways are active at all times. The greatest activity of the former pathway occurs during the vegetative stages, while that of the latter occurs during conidiation. Pyruvic acid, the product of the Embden–Meyerhof–Parnas pathway, is converted to acetyl-CoA and condenses with oxaloacetate in the presence of the condensing enzyme to form citrate.

Time course

Table 126 lists the general conditions for submerged fermentation, while Table 127 contains a comparison of the surface and submerged-culture processes for the production of citric acid using *Asp. niger*. The effects of substrate purity and of iron and copper on the yields of citric acid are demonstrated in Tables 128–131. The inhibitory effect of manganese ions is overcome by cycloheximide but not by chloramphenicol (*see* Table 131).

Fig. 49 demonstrates a general time course for citric acid production using *Saccharomycopis lipolytica* and Table 132 contains data obtained using *Asp. foetidus*.

Table 126. Conditions for submerged culture citric acid production.

Criterion	Comment
pH value	Normally 1.5–2.8
> 2.8	Oxalic acid formation, danger of infection, decreased citric acid production
< 1.5	Low growth rate of mycelium
Fermentation temperature	Normally 28–33°C
> 33°C	Oxalic acid formation
< 28°C	Slowed down production
Fermentation duration	Dependent on the sugar concentration in the raw material within 6–15 days
Trace elements	Special importance — iron, manganese, zinc. Also important — copper, heavy- and alkaline metals

Table 127. Comparison of surface and submerged citric acid fermentation processes.

Criterion	Surface process	Submerged process
Productivity	0.85–1.0 kg m^{-3} day^{-1}	16.5–18.0 kg m^{-3} day^{-1}
Yield[a]		
Fermentation	65–75%	87–92%
Whole plant	58–70%	81–88%
Energy requirements		
Electricity	1.400–1.600 kWh $tonne^{-1}$	2.500–2.800 kWh $tonne^{-1}$
Saturated steam 9 bar	15–20 $tonne^{-1}$ $tonne^{-1}$	9.7–10.2 tonne $tonne^{-1}$
Space requirements[b]		
Fermentation section	370 m^2 $tonne^{-1}$ day^{-1}	~ 8 m^2 $tonne^{-1}$ day

[a] To fermentable disaccharide available.
[b] For production of citric acid monohydrate.

Table 128. Effect of substrate purity on production of citric acid by *Aspergillus niger* [Moss and Smith (1977)].

Treatment	Yield of citric acid[a] (%)
Sucrose passed through resin	98
Molasses	
Clarified and passed through resin	75
Treated with ferrocyanide	68
Untreated	62

[a] (Weight of acid produced/weight of glucose used) × 100.

Table 129. Effect of iron on the production of citric acid from decationized sucrose by *Aspergillus niger* in submerged culture[a] [Lockwood (1979)].

Iron[b] (mg ℓ^{-1})	Weight yield of citric acid[c] (%)
0.0	67.0
0.05	71.0
0.5	88.0
0.75	79.0
1.00	76.0
2.00	71.0
5.00	57.0
10.0	39.0

[a] Medium composition: sucrose purified by mixed resin bed ion exchange from 3 000 000 Ω resistance at 40 per cent concentration, diluted with deionized water at 18 000 000 Ω resistance to 14.2 per cent sugar content and recrystallized nutrient salts added: 0.014 per cent potassium dihydrogen phosphate; 0.1 per cent magnesium sulphate heptahydrate; 0.2 per cent ammonium carbonate; 0.2 per cent hydrochloric acid to pH 2.6.
[b] Supplied as ferric chloride.
[c] (Grams citric acid formed/equivalent grams hexose moiety used) × 100.

Table 130. Effect of copper and iron ions on citric acid production from glucose by *Aspergillus niger*[a] [Lockwood and Schweiger (1967)]

Fe_3+ (mg ℓ$^{-1}$)	Cu^{2+} (mg ℓ$^{-1}$)	Citric acid yield[b] (%)
10	50	77.8
50	50	69.1
100	50	50.7
150	50	14.2
10	100	77.2
50	100	65.4
100	100	53.9
150	100	29.8
10	500	74.0
50	500	65.4
100	500	60.6
150	500	27.6

[a] Medium composition: approximately 14 per cent glucose (commercial grade); 0.2 per cent ammonium carbonate, 0.014 per cent potassium dihydrogen phosphate, 0.1 per cent magnesium sulphate, Fe and Cu supplied as sulphates.
[b] (Grams citric acid produced/grams glucose supplied) × 100.

Table 131. Influence of cycloheximide and chloramphenicol on manganese-inhibited citric acid accumulation by *Aspergillus niger* [Röhr and Kubicek (1981)].

Manganese (mM)	Cycloheximide (μg ml^{-1})	Chloramphenicol (μg ml^{-1})	Citric acid[a] (mM)
0	0	0	294
0.01	0	0	84
0.01	10	0	128
0.01	20	0	237
0.001	0	0	134
0.001	10	0	258
0.001	20	0	270
0.01	0	50	80
0.01	0	100	84

[a] Measured after 150 hours.

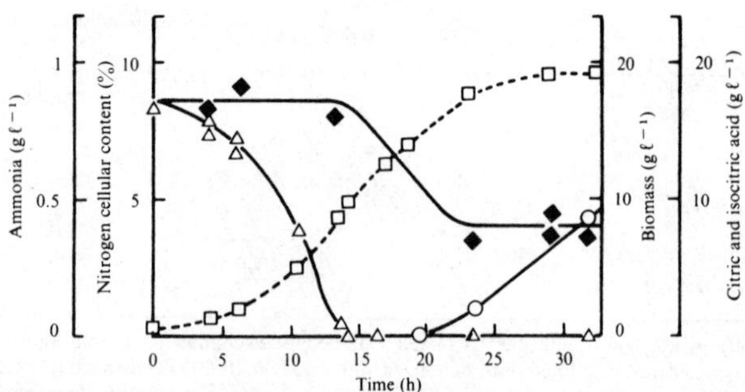

Figure 49. Variation with time of (△) ammonia extracellular concentration, (□) biomass concentration, (○) total citric and isocitric acid concentrations, and (◆) relative nitrogen content of cells (g nitrogen, 100 g biomass) during the growth of *Saccharomycopsis (Candida) lipolytica* on glucose [Briffaud and Engasser (1979)].

Table 132. Citric acid production by *Aspergillus foetidus* [Kristiansen and Sinclair (1978)].

Criterion	Run							
Final dry weight ($kg\,m^3$)	12.5	25.4	34.0	14.5	10.5	16.5	19.0	23.8
Time (h)	130	160	180	115	120	185	240	320
Initial pH	4.0	4.0	5.4	6.1	4.0	6.5	6.5	3.0
Time of initial fall in pH (h)	20	30	20	25	30	40	50	—
pH controlled at	1.80	2.0	2.6	2.2	2.4	2.35	2.5	3.0
Biomass end of fall in pH ($kg\,m^{-3}$)	4.5	13.0	2.0	6.0	3.5	8.0	3.5	—
Nitrogen supply exhausted at (h)	45	40	40	40	40	40	60	35
Biomass at that time ($kg\,m^{-3}$)	9.5	11.0	7.5	10.5	5.5	8.0	5.0	7
Final citric acid ($kg\,m^{-3}$)	8.5	11.0	20.5	9.8	1.1	21.0	17.0	18.4
Production starts (h)	30	35	25	20	25	20	40	25
Biomass at that time ($kg\,m^{-3}$)	7.0	9.5	3.0	3.0	3.0	2.5	2.5	4.0
Inoculum ($kg\,m^{-3}$)	0.10	1.35	1.20	0.15	1.30	0.35	0.07	0.10

Gibberellic Acid

Microorganisms

Gibberellic acid is produced by *Gibberella fujikuroi* (i.e. the conidial stage of *Fusarium moniliforme*). Table 133 contains production data obtained using a number of different strains.

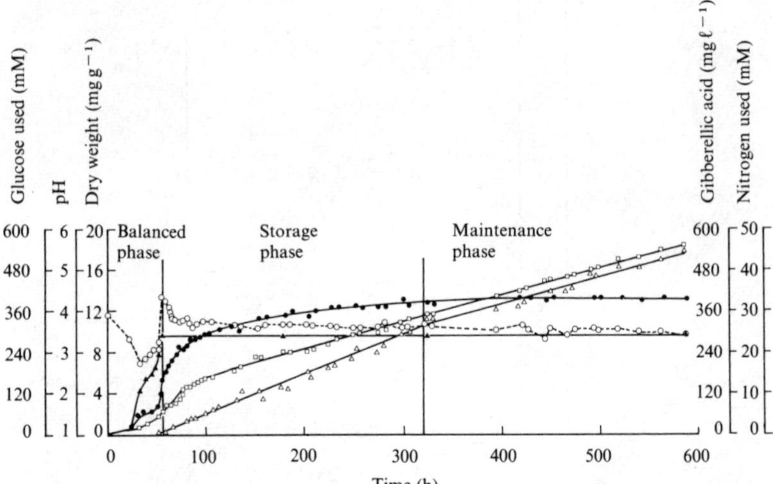

Figure 50. Time course of fermentation by *Gibberella fujikuroi* on ammonium nitrate medium. Initial nitrogen concentration, 24.1 mM. △, Gibberellic acid; ▲, nitrogen used; □, glucose used; ○, pH; ●, dry weight [Borrow *et al.* (1964)].

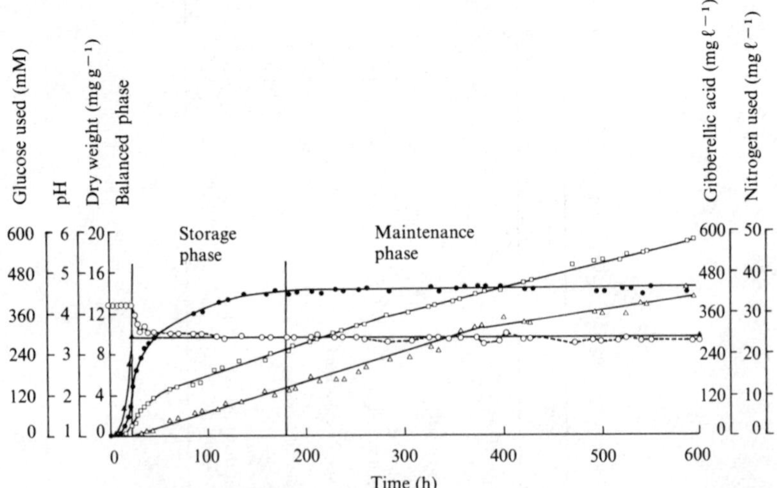

Figure 51. Time course of fermentation by *Gibberella fujikuroi* on glycine medium. Initial nitrogen content of medium, 25.8 mM. △ Gibberellic acid; ▲, nitrogen used; □, glucose used, ○, pH; ●, dry weight [Borrow *et al.* (1964)].

Table 133. Gibberellic acid production by different strains of *Gibberella fujikuroi* [Borrow *et al.* (1964)].

Strain	Nitrogen source	Initial nitrogen concentration (mM)	Carbon source	Initial carbon concentration ($g\ \ell^{-1}$)	Sugar utilized ($g\ \ell^{-1}$)	Rate of gibberellic production		Medium gibberellic acid levels	
						($mg\ \ell^{-1}\ h^{-1}$)	[mg (mmol N) h^{-1}]	($g\ \ell^{-1}$)	[mg (mmol N)$^{-1}$]
ACC 917	Ammonium nitrate	60	Glucose	2.0	0.17	1.25	0.021	544	9.07
ACC 917	Ammonium nitrate	100	Glucose	1.0	0.28	2.34	0.024	1000	10.00
ACC 917	Ammonium tartrate	32	Glucose	0.25	0.09	0.67	0.021	180	5.63
ACC 917	Corn-steep, ammonium sulphate	~100	Glucose	0.20	0.20	4.19	0.042	620	~6.20
ACC 917	Glycine	204	Glucose	1.65	0.55	9.72	0.018		
ACC 1135	Ammonium tartrate	103	Sucrose	0.4	0.28	1.04	0.010	400	3.88
AS 67	Ammonium chloride	56	Glycerol	0.10				8	0.14
NRRL 2284	Ammonium chloride	56	Glucose	0.15	0.15			14	0.25
U	Corn-steep, ammonium sulphate	~40	Sucrose	0.5	0.11	0.44	0.011	137	~3.43
V-V/88	Ammonium acetate	75	Sucrose	0.4	0.21			116	1.55

Medium

A typical medium composition is given in Table 134.

Table 134. Medium for the production of gibberellic acid.

Component	Amount (%)
Glucose	
For inoculum	2.0
For fermentation	1.25–1.5
Ammonium chloride	0.3
Potassium dihydrogen phosphate	0.3
Magnesium sulphate heptahydrate	0.3
Octadecanol	0.025

The temperature range for growth is 31–32°C and for acid production is 29°C. Air flow/agitation should be as vigorous as practicable. Ammonium nitrogen is preferred, with natural plant meals as suitable additional sources. Sucrose is superior to glucose, with magnesium, potassium, phosphate and sulphate all being necessary.

Metabolic pathway

The biosynthesis of gibberellic acid is acetate-based and initially follows the pathways leading to diterpenes. Acetate is accumulated via acetyl-CoA through mevalonate to

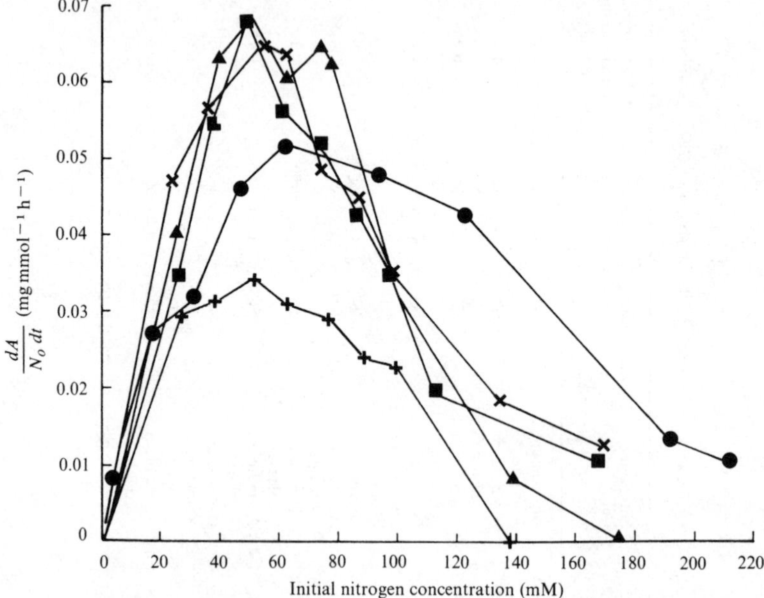

Figure 52. Gibberellic acid productivity. ■, Glycine; ▲, urea; ×, ammonium tartrate; ●, ammonium nitrate; +, ammonium acetate [Borrow *et al.* (1964)].

geranyl–geranyl pyrophosphate. This forms kaurene-related compounds, oxidative conversion proceeding from *ent*-kaur-16-en-oic acid to *ent*-7α-hydroxykaurenoic acid, GA_{12} aldehyde, GA_{14} aldehyde, GA_{14}, GA_4, GA_7 and, finally, to GA_3 (i.e. gibberellic acid).

Time course

Figs. 50 and 51 provide time course data for gibberellic acid formation, while Figs. 52 and 53 demonstrate the effects of the nitrogen source and initial nitrogen and glucose concentrations. Effects of pH and of dissolved carbon dioxide are shown in Fig. 54 and Table 135, respectively.

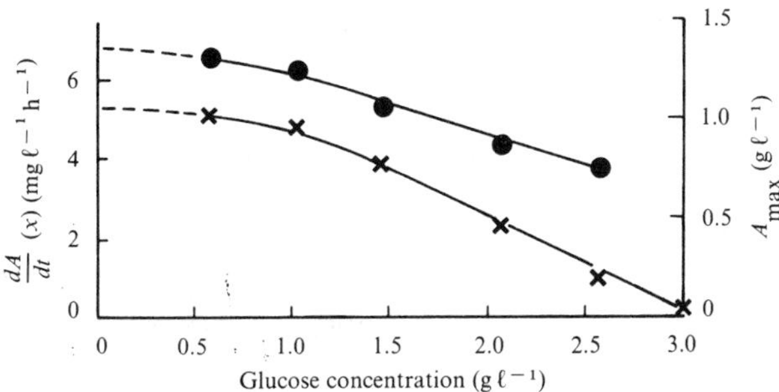

Figure 53. Effect of glucose on production of gibberellic acid on ammonium nitrate media. Initial nitrogen concentration 84 mM [Borrow *et al.* (1964)].

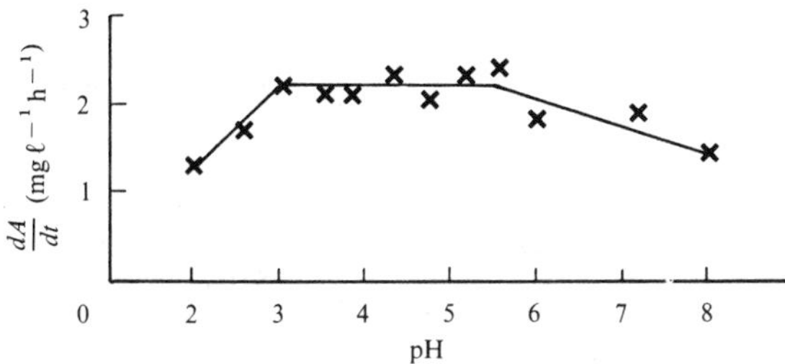

Figure 54. Effect of pH on the rate of production of gibberellic acid by *Gibberella fujikuroi* maintained on ammonium nitrate medium. Initial nitrogen concentration 62 mM; glucose concentration, 1.05 g ℓ$^{-1}$ [Borrow *et al.* (1964)].

Table 135. Gibberellic acid yield with carbon dioxide addition [Hanson (1967)].

Incubation time (days)	Air[a] (mg ℓ$^{-1}$)	Air + carbon dioxide[b] (mg ℓ$^{-1}$)
6.5	nil	8
9.5	70	110
13.5	226	316
20.0	484	516
27.5	498	630
37.5	—	594

[a] Air at rate of 20 ℓ min^{-1}.
[b] Air rate, 19 ℓ min^{-1}; carbon dioxide 1 ℓ min^{-1}.

Gluconic acid

Microorganisms

The most commonly used microorganism is *Aspergillus niger*. However, there are several other microorganisms capable of gluconic acid production, e.g., *Pullularia pullulans*, species of *Penicillium, Gliocladium, Scopulariopsis, Gonatobotrys* and *Endomycopsis*.

Media

Table 136 contains the composition of a typical medium. The pH range is 6.5–10.0, with temperatures in the range 30–34°C.

Table 136. Media used for growing *Aspergillus niger* in the industrial production of gluconic acid [Moyer *et al.* (1937)].

Component	Concentration ($g\ \ell^{-1}$)		
	Growth of spore-bearing mycelium	Spore germination	Gluconate production
Glucose	91.5	120	150
Calcium carbonate			100
Ammonium nitrate	0.45		
Diammonium hydrogen phosphate		0.7	0.388
Potassium dihydrogen phosphate	0.072	0.3	0.188
Magnesium sulphate heptahydrate	0.06	0.25	0.156
Peptone		0.15	
Beer ($ml\ \ell^{-1}$)	60	67	

Metabolic pathway

Gluconic acid and 5-ketogluconic acid production from D-glucose via D-gluco-δ-lactone, which is catalyzed by β-D-glucose: oxygen oxidoreductase. 2-Ketogluconic acid is formed from D-gluconic acid in the presence of *Pseudomonas* species (*see* Fig. 55).

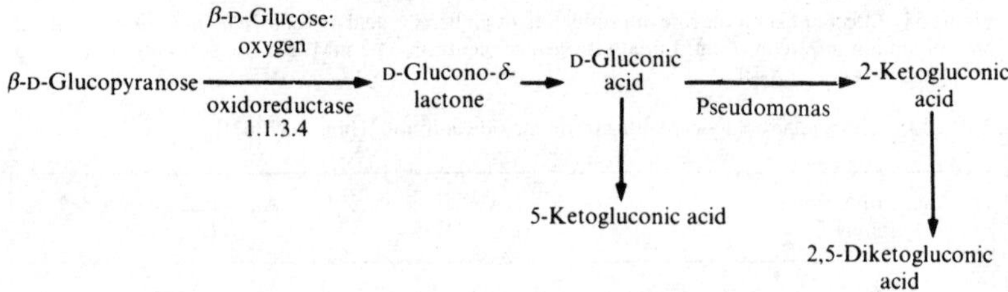

Figure 55. Biosynthesis of gluconic acid from glucose.

Time course

Yields on glucose can be as high as 80–100 per cent (*see* Table 137) at a rate of $15\ g\ \ell^{-1}\ h^{-1}$. Fig. 56 provides time course data of gluconic acid production by *Pseudomonas*.

Table 137. Production of gluconic acid by species of *Pseudomonas* and *Phytomonas*[a] [Lockwood *et al.* (1941)].

Microorganism	Glucose consumed ($g\ \ell^{-1}$)[b]	Gluconic acid formed[c] (%)
Agrobacter tumefaciens A	32	0
Ag. tumifaciens B	41	0
Ag. tumifaciens C	20	0
Corynebacterium michiganense	4	0
Pseudomonas coronafaciens	9	0
Ps. mucidolens 4685	93	78
Ps. mucidolens 4686	84	58
Ps. mucidolens 4687	84	96
Ps. myxogenes 946	9	0
Ps. myxogenes 947	14	0
Ps. oralis	100	92
Ps. striafaciens	94	80
Ps. syringae	72	82
Xanthomonas begoniae	19	52
X. campestris	2	0
X. stewartii	23	40

[a] Medium (200 ml) composition: 5 g calcium carbonate, 1 g corn-steep liquor, 0.4 g urea, 0.12 g potassium dihydrogen phosphate, 0.05 g magnesium sulphate heptahydrate, 1 drop oleic acid. Temperature 30°C; air flow 200 ml min^{-1}, duration, 8 days.
[b] Original concentration 9.3–10.0 per cent.
[c] Based on glucose consumed.

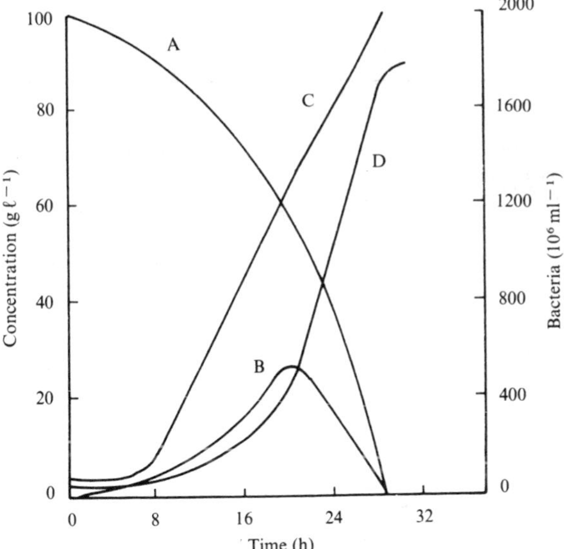

Figure 56. Composition of fermenting culture solution at various ages for 2-ketogluconate production. Curve A, glucose in solution; curve B, glucono-δ-lactone and gluconate in solution; curve C, 2-ketogluconate in solution; curve D, *Pseudomonas* cells [Lockwood (1979)].

Itaconic Acid

Microorganisms

Itaconic acid is produced by *Asp. terreus*, *Asp. itaconicus* and *Ustilago zeae*.

Media

Typical media formulations for surface and submerged cultures are given in Tables 138 and 139, respectively.

Table 138. Medium for itaconic acid production in surface culture.

Component	Amount
Glucose monohydrate	165 g ℓ^{-1}
Ammonium nitrate	2.5 g ℓ^{-1}
Magnesium sulphate heptahydrate	4.4 g ℓ^{-1}
Sodium chloride	0.4 g ℓ^{-1}
Zinc sulphate heptahydrate	4.4 mg ℓ^{-1}
Corn-steep liquor (50% solids)	4.0 ml ℓ^{-1}
Nitric acid	1.6 ml ℓ^{-1}
Iron	5 mg ℓ^{-1}

Table 139. Medium for itaconic acid production in submerged culture.

Component	Amount (g ℓ^{-1})
Glucose monohydrate	66.0
Ammonium sulphate	2.7
Magnesium sulphate	0.8
Corn-steep liquor	1.8

Metabolic pathway

It is probable that the Embden–Meyerhof–Parnas pathway is followed as far as pyruvic acid. Then, as shown in Fig. 57, acetyl-CoA instead of combining with oxaloacetate to form citric acid condenses with pyruvic acid to form citramalic acid, which is converted to itatartaric acid, and this is reduced to itaconic acid.

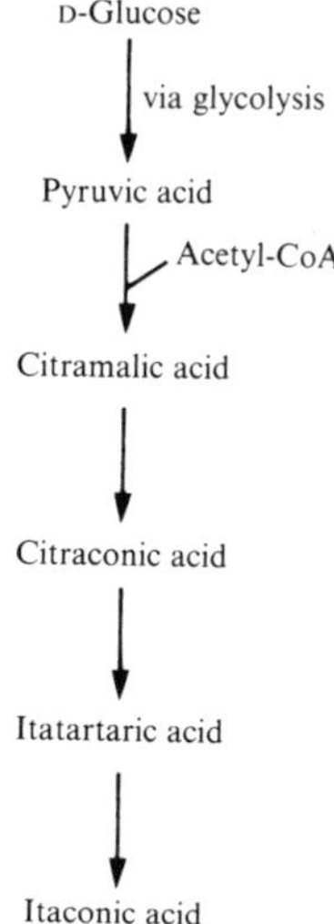

Figure 57. Metabolic sequence in the production of itaconic acid [Lockwood (1979)].

Time course

Figs. 58–60 show the results of extended fermentations achieved by serial transfer of mycelium from batch to batch. The effect of pH is given in Fig. 61. Tables 140, 141 and 142 show the effects of iron, copper and zinc, and alkaline earth metals, respectively, on yields.

Table 140. Effect of iron on the production of itaconic acid by *Aspergillus terreus* [Lockwood and Schweiger (1967)].

Iron ($mg\ \ell^{-1}$)	Weight yield[a] (%)
0	57
1	25
2	17
4	17

[a] (Grams itaconic acid produced/grams sugar supplied) × 100.

Table 141. Effect of ions of copper and zinc on the production of itaconic acid by *Aspergillus terreus* [Lockwood and Schweiger (1967)].

Copper (mg ℓ$^{-1}$)	Zinc (mg ℓ$^{-1}$)	Weight yield[a] (%)
0	0	16
0	0.5	43.3
0	6.0	50.4
0.5	0.5	55.4
1	0.5	51.7
3	0.5	52.6
6	0.5	55.4

[a] (Grams itaconic acid produced/grams sugar supplied) × 100.

Table 142. Effect of additions of alkaline earth metals on itaconic acid production by *Aspergillus terreus* [Lockwood and Schweiger (1967)].

Metallic ion added Type	(mg ℓ$^{-1}$)	Weight yield[a] (%)
None		9
Calcium	337	43
	2700	59
Magnesium	337	20
	2700	48

[a] (Grams itaconic acid produced/grams sugar supplied) × 100.

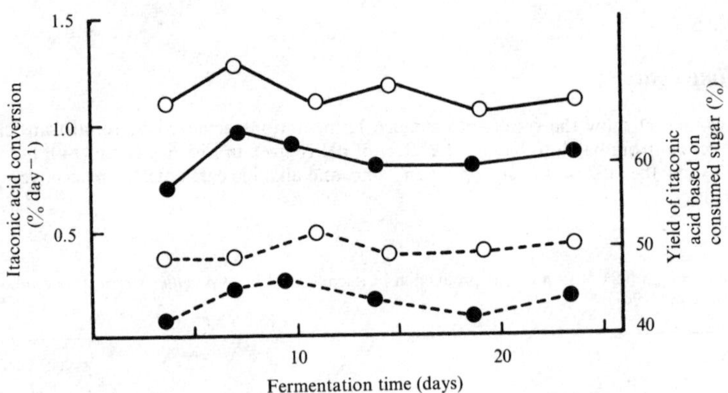

Figure 58. Average fermentation rate (———) and the yield (----) of itaconic acid by *Aspergillus terreus* (○) and *Asp. itaconicus* (●) transferring 30 per cent mycelium from the preceding run [Kobayashi and Nakamura (1961)].

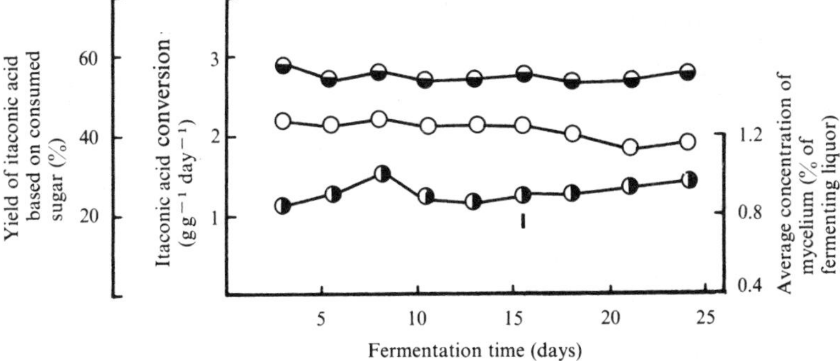

Figure 59. Extended fermentation by using a serial transfer of mycelium of *Aspergillus terreus* u-4–16 and the average fermentation rate, yield, and mycelium concentration. ○, average fermentation rate; ◑, mycelium; ◒, yield of itaconic acid [Kobayashi and Nakamura (1961)].

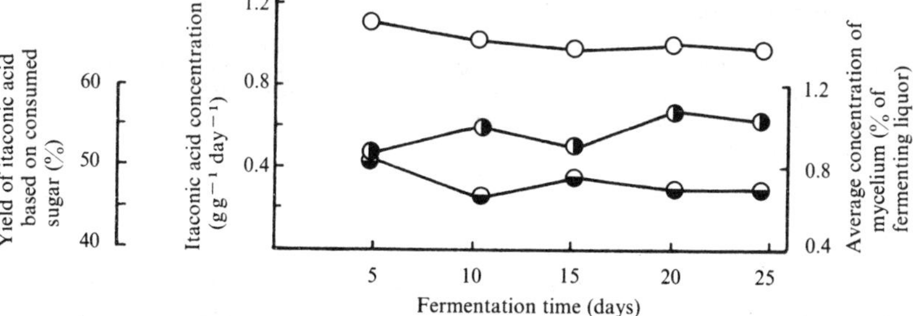

Figure 60. Extended fermentation by using a serial transfer of mycelium of *Aspergillus itaconicus* 34 and the average fermentation rate, yield, and mycelium concentration. ○, average fermentation rate; ◑, mycelium; ◒, yield of itaconic acid [Kobayashi and Nakamura (1961)].

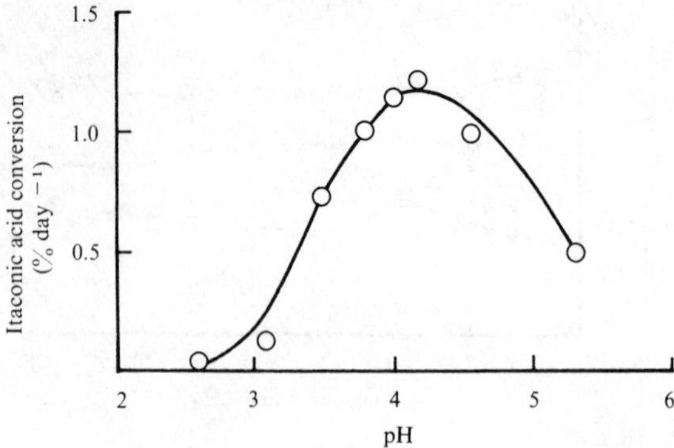

Figure 61. Effect of pH value of fermentation medium on itaconic acid fermentation rate by *Aspergillus itaconicus* 34 [Kobayashi and Nakamura (1961)].

Lactic Acid

Microorganisms

Lactic acid is produced by numerous microorganisms. However, the most commonly used are species of the genus *Lactobacillus*, especially *L. delbrueckii*. Others used include *L. pentosis* and *L. bulgaricus*. The fungus *Rhizopus oryzae* is also capable of producing lactic acid from glucose-containing medium.

Media

The fermentation medium used contains hydrolyzed starch, dextrose syrups, glucose, maltose, lactose or sucrose as the carbon source. The content of a typical medium is given in Table 143.

Table 143. Composition of medium for lactic acid fermentation.

Component	Amount ($g\,\ell^{-1}$)
Glucose	120
Malt sprouts	3.75
Diammonium hydrogen phosphate	2.5

Fermentation is carried out at about 45–50°C and pH is on the acid side of neutrality.

Metabolic pathway

Lactic acid is the end product of glycolysis. Two moles of lactic acid are produced per mole of glucose fermented via glycolysis. In heterolactic fermentation, one mole of lactic acid, ethanol and carbon dioxide constitute the end products.

Fig. 62 shows the time course of lactic acid sythesis using *L. delbrueckii*. Fig. 63 shows that the specific rates of product formation and growth are related by the equation

$$\frac{1}{x}\frac{dp}{dt} = \alpha\mu + \beta$$

Values for α and β are given in Fig. 64 as a function of pH.

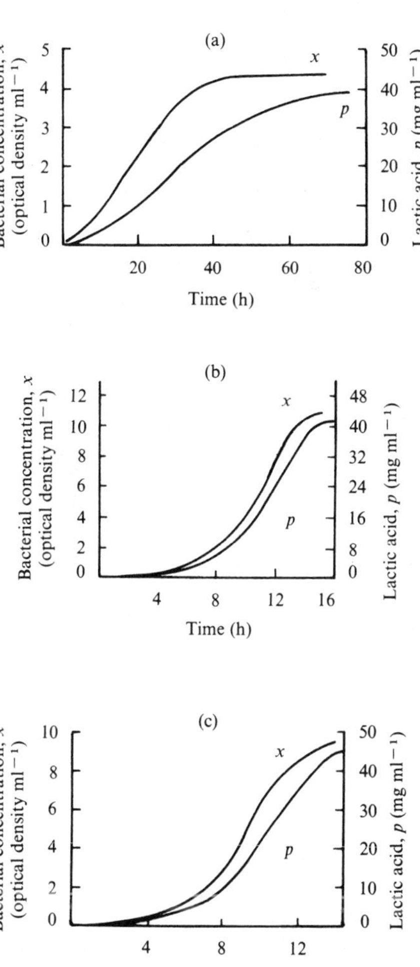

Figure 62. Growth and lactic acid synthesis during batch fermentations using *Lactobacillus delbrueckii* at controlled pH levels. (a) pH 4.5 (b) 5.4. (c) pH 6.0 [Luedeking and Piret (1959)].

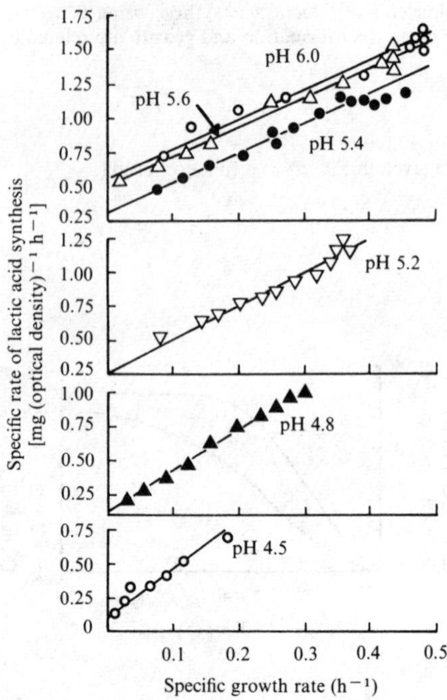

Figure 63. Specific rate of lactic acid synthesis as a function of specific bacterial growth rate during batch fermentation at controlled pH levels [Luedeking and Piret (1959)].

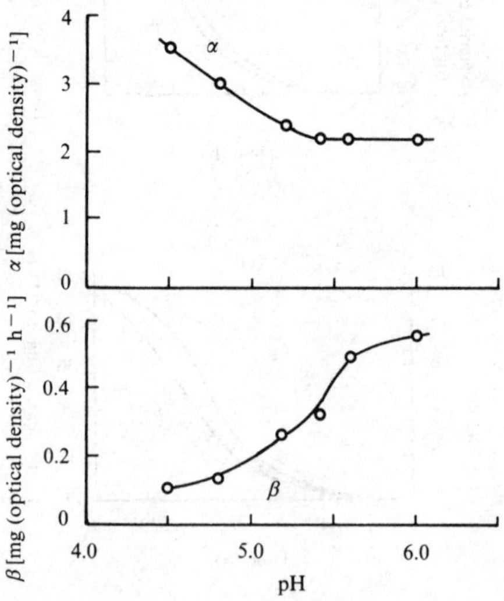

Figure 64. Effect of pH on the coefficients α and β in the equation $1/x \cdot dp/dt = \alpha\mu + \beta$ [Luedeking and Piret (1959)].

Salicyclic Acid

Microorganisms

The microorganisms used for the production of salicyclic acid through the conversion of naphthalene are *Pseudomonas fluorescens* and *Corynebacterium renale.*

Medium

The composition of a typical medium used for *Coryn. renale* is given in Table 144. The characteristics of the emulsion of naphthalene (5 g ℓ^{-1}) prepared by mixing with Span 80 (0.4 per cent), Span 20 (0.07 per cent), lecithin (0.1 per cent) and Tween 80, which is converted to salicyclic acid, is given in Table 145.

Table 144. Salicylate production medium [Tangnu and Ghose (1981)].

Component	Concentration ($g\,\ell^{-1}$)
Ammonium nitrate	2.5
Magnesium sulphate heptahydrate	0.25
Disodium hydrogen phosphate dihydrate	2.0
Potassium dihydrogen phosphate	0.6
Calcium chloride dihydrate	0.04
Ferrous sulphate heptahydrate	0.004
Manganese chloride monohydrate	0.003
Sodium molybdate	0.002

Table 145. Characteristics of emulsion [Tangnu and Ghose (1980)].

Appearance	Viscosity	Feel	Dispersed phase particle size	Stability	Dispersibility	Toxicity
Milky	Thick fluid	Between water and oil, depending upon the dilution	Colloidal particles	No creaming or phase separation even at room temperature for 1 month	Water-soluble as well as oil-soluble. Thus, could be diluted with water or emulsifier	Biologically active and nontoxic to the cells

Metabolic pathway

Naphthalene is converted to salicyclic acid during growth. It can be subsequently converted to catechol.

Time course

Product inhibition is the chief factor in salicylate production. One solution to this problem is the addition of an ion exchange resin to the culture medium to remove salicylate. Another method involves the use of dialysis to remove salicylate during growth.

Figs. 65–68 contain time course data obtained in the presence of resin. Table 146 provides information on the salicylate production in association with dialysis.

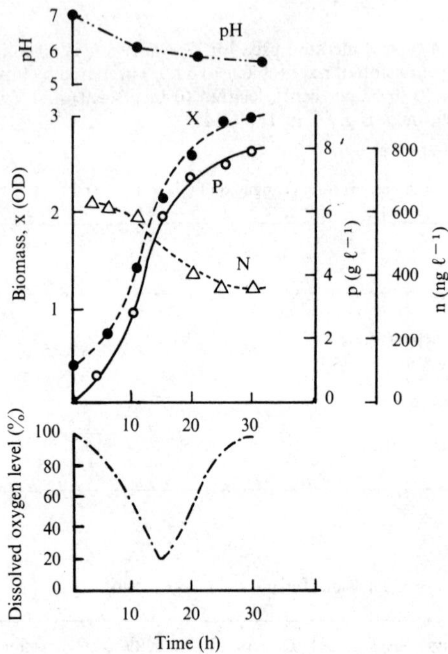

Figure 65. Course of salicylic acid fermentation using *Pseudomonas fluorescens* [Kitai *et al.* (1968)].

Table 146. Effect of environment control manipulation on yield coefficients [Tangnu and Ghose (1980)].

Yield coefficients	Experimental conditions										
	Substrate concentration ($g\ \ell^{-1}$)			pH			Without cell recycle	With cell recycle	Dilution rates, μ (h^{-1})		
	5.0	7.50	10.0	6.0	6.5	7.0			0.02	0.025	0.03
Product/substrate	0.9	0.73	0.55	0.66	0.9	0.78	0.55	0.77	0.95	0.84	0.52
Cell/substrate	0.28	0.24	0.2	0.25	0.28	0.32	0.2	0.33	0.29	0.27	0.20
Product/cell	3.21	3.04	2.75	2.64	3.21	2.43	2.75	2.33	3.28	3.1	2.6

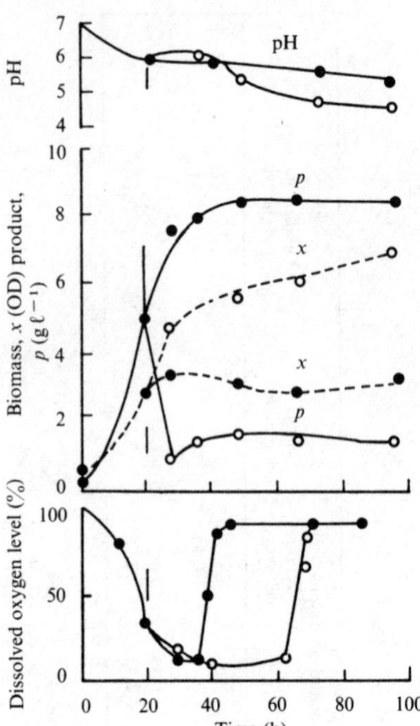

Figure 66. Course of salicylic acid fermentation released from product inhibition in comparison with conventional system using *Pseudomonas fluorescens*. ○, System where salicylic acid was removed; ●, conventional system [Kitai *et al.* (1968)].

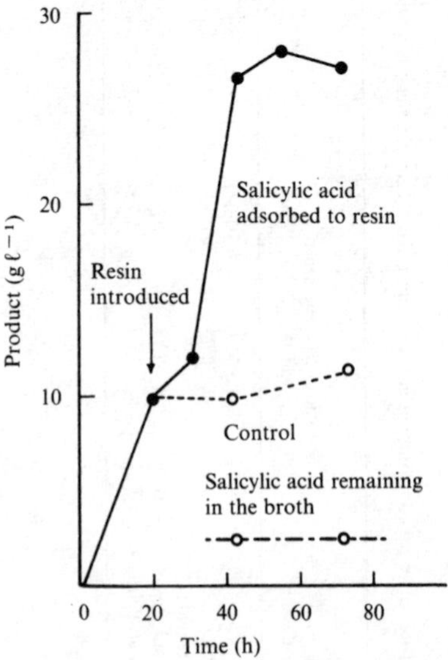

Figure 67. Pattern of salicylic acid formation in the product removing system using *Pseudomonas fluorescens* [Kitai *et al.* (1968)].

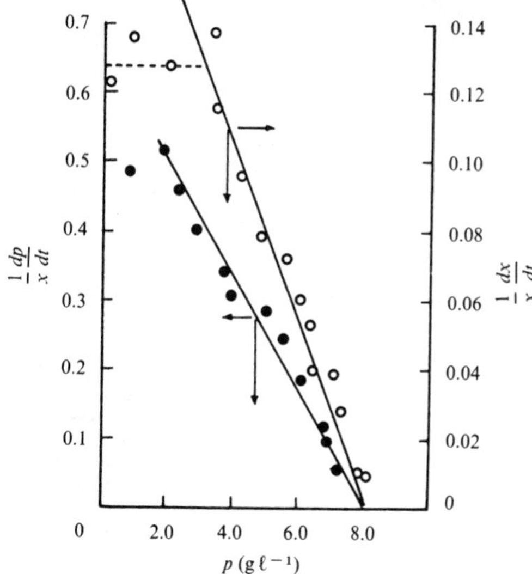

Figure 68. Relation between salicylic acid concentration and specific rates using *Pseudomonas fluorescens* [Kitai *et al.* (1968)].

POLYSACCHARIDES

Microorganisms

A list of some polysaccharide-producing microorganisms is given in Table 147. Some neutral and anionic exopolysaccharides are given in Tables 148 and 149, respectively, along with the producing microorganisms and properties of solutions of the polysaccharides.

Table 147. Some polysaccharide-producing microorganisms [Lawson and Sutherland (1978)].

Group	Microorganism
Bacteria	
Gram-positive	*Bacillus* sp.
	Leuconostoc sp.
	Streptococcus sp.
	Streptococcus mutans
Gram-negative	*Acetobacter xylinum*
	Achromobacter sp.
	Agrobacterium sp.
	Alcaligenes faecalis var. *myxogenes*
	Arthrobacter viscosus
	Azotobacter sp.
	Erwinia sp.
	Escherichia coli
	Klebsiella aerogenes
	Pseudomonas aeruginosa
	Rhizobium sp.
	Sphaerotilus natans
	Xanthomonas campestris
	X. phaseoli
Yeasts	*Aureobasidium pullulans*
	Cryptococcus sp.
	Hansenula capsulata
	H. holstii
	Lipomyces sp.
	Pachysolen tannophilus
	Pichia sp.
	Rhodotorula sp.
	Torulopsis molischiana
	Torulopsis pinus
Other fungi	*Penicillium* sp.
	Tremella mesentaria

Media

For some microorganisms, the carbon source determines both the quantity of the polysaccharide formed and its quality. The carbon source is a carbohydrate (e.g., fructose, galactose, arabinose, etc.). Although a nitrogen source is necessary for growth and the enzyme synthesis associated with the polysaccharide formation, an excess of nitrogen generally reduces the amount of polysaccharide produced. Many polysaccharide-producing microorganisms have a strict requirement for mineral elements, such as potassium, phosphorus, magnesium, manganese and calcium and, possibly, molybdenum, iron, copper and zinc.

Table 148. Some neutral microbial exopolysaccharides [Pace and Righelato (1980)].

Trivial name	Microorganism	Special solution properties; (potential) uses
Dextran	*Acetobacter* sp., *Leuconostoc mesenteroides*	Plastic flow; as source for making dextran derivatives for pharmaceutical uses (transfusion)
Pullulan	*Pullularia pullulans*	Gels on heating, oxygen-impermeable films formed, readily biodegradable
Curdlan	*Agrobacterium*	Gels on heating or acidification; gelled foods, stable salt and acids
Scleroglucan	*Sclerotium glucanicum*	Highly viscous and pseudoplastic; drilling muds, latex points
Cellulose	*Acetobacter* sp.	Water insoluble
Levan	*Bacillus* sp., *Leuconostoc mesenteroides*	
	Serratia marcescens	
	Pseudomonas NCLB 11264	Highly viscous and pseudoplastic

Table 149. Some anionic microbial exopolysaccharides [Pace and Righelato (1980)].

Trivial names	Microorganism	Special solution properties; (potential) uses
Alginate	*Pseudomonas aeruginosa*, *Azotobacter vinelandii*	Range of viscosity types, gels with Ca^{2+}, textile printing, food applications
Xanthan	*Xanthomonas campestris*	Highly viscous and pseudoplastic, gels with galactomannans, resistant to acid, alkali, and biodegradation; oil-well drilling, food industry
Phosphomannan	*Hansenula capsulata*, *H. holstii*	Thixotrophic at high concentration, gels with borax

Typical media for xanthan gum production by *X. campestris* and alginate by *Az. vinelandii* are shown in Tables 150 and 151, respectively.

Table 150. Medium suitable for culture of *Xanthomonas campestris*.

Component	Amount
Glucose	20 g ℓ$^{-1}$
Dipotassium hydrogen phosphate	3 g ℓ$^{-1}$
Diammonium hydrogen phosphate	1.5 g ℓ$^{-1}$
Magnesium sulphate heptahydrate	0.25 g ℓ$^{-1}$
Antifoam agent	2 ml ℓ$^{-1}$
Water	6 ℓ

Table 151. Medium suitable for the culture of *Azotobacter vinelandii*.

Component	Amount (g ℓ$^{-1}$)
Sucrose	8.0
Magnesium sulphate heptahydrate	1.6
Sodium chloride	1.6
Potassium dihydrogen phosphate	0.064
Calcium chloride dihydrate	0.34
Ferrous chloride dihydrate	0.017
Boric acid	0.023
Sodium molybdate	0.008
Zinc sulphate heptahydrate	0.009
Cobalt sulphate heptahydrate	0.009
Manganous chloride tetrahydrate	0.0007
Copper sulphate heptahydrate	0.0008

Using the medium described in Table 151 at pH 6.9–7.5 and 28–30°C the polysaccharide yield is 72 per cent.

Metabolic Pathways

Most exopolysaccharides are presumed to be synthesized via metabolic pathways identical to those involved in cell wall synthesis. However, there are a few, such as dextrans and, probably, levans, which are synthesized extracellularly. A proposed pathway for the biosynthesis of xanthan, using *X. campestris*, is outlined in Fig. 69.

Time Course

Polysaccharide production is reduced in *X. campestris* and *Ps. aeruginosa* under carbohydrate-limited conditions, compared with carbohydrate-excess conditions (*see* Table 152). Table 153 shows the effect of calcium ion concentration on the composition of alginate synthesized. Phosphate (0.3–5.0 mM) suppresses alginate synthesis by *Az.*

vinelandii, while low concentrations increase alginate levels. The alginate/cell concentration ratio is dose dependently suppressed by phosphate (0.0–5.0 mM).

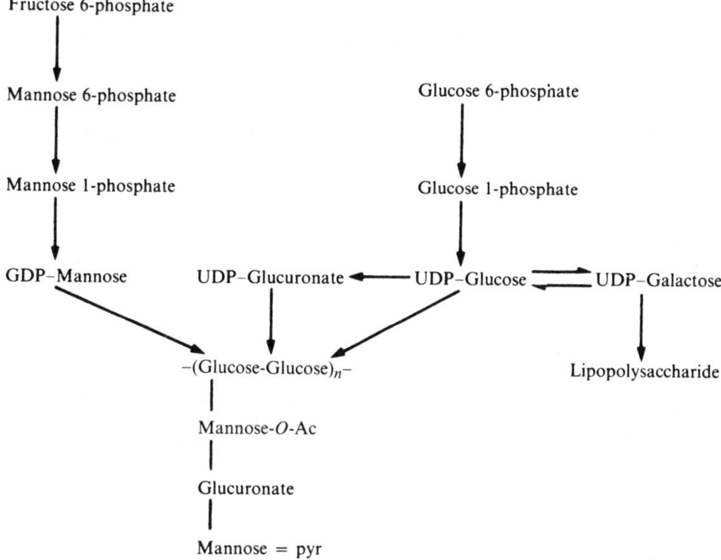

Figure 69. Biosynthesis of xanthan gum by *Xanthomonas campestris* suggested by Pace and Righelato (1980).

Table 152. Exopolysaccharide production in carbohydrate-limited and excess conditions using synthetic media [Pace and Righelato (1980)].

Microorganism	Carbohydrate-limited		Carbohydrate excess (nitrogen-limited)	
	Cell ($g\,\ell^{-1}$)	Polysaccharide ($g\,\ell^{-1}$)	Cell ($g\,\ell^{-1}$)	Polysaccharide ($g\,\ell^{-1}$)
Azotobacter vinelandii[a]	1.3	2.2	1.5	2.0
Pseudomonas aeruginosa[b]	1.5	5.8	1.3	8.9
Xanthomonas campestris[b]	1.1	2.7	1.6	7.0

[a] Dilute rate = $0.15\,h^{-1}$.
[b] Dilution rate = $0.05\,h^{-1}$.

Table 153. Effect of calcium ion concentration on the ratio of mannuronic to guluronic acid residues in alginate from *Azotobacter vinelandii* [Lawson and Sutherland (1978)].

Calcium chloride (mM)	0	0.30	0.50	0.75
Mannuronic acid residues (%)	78	73	55–46	31

VITAMINS

Microorganisms

Yeasts, moulds and bacteria capable of producing various vitamins are listed in Tables 154–157.

Table 154. Production of some vitamins using microorganisms [Moss and Smith (1977)].

Vitamin	Culture	Medium	Fermentation conditions	Extraction	Yield
Carotene (precursor of vitamin A)	*Blakeslea trispora*	Molasses, soya bean oil, β-ionone, thiamin	72 h at 30°C, aerobic	Solvent	$1\ g\ \ell^{-1}$
	Mycobacterium smegmatis				$0.007\ g\ \ell^{-1}$
Riboflavin	*Ashbya gossypii*	Glucose, collagen, soya bean oil, glycine	6 days at 36°C, aerobic	Heated 1 h at 120°C, followed by addition of reagents to precipitate riboflavin	$4.25\ g\ \ell^{-1}$
L-Sorbose (in vitamin C synthesis)	*Gluconobacter oxidans* subsp. *suboxidans*	D-Sorbitol, 30% corn steep	45 h at 30°C, aerobic	Filtration, concentration under vacuum	70% based on substrate used
5-Ketogluconic acid (in vitamin C synthesis)	*Gluconobacter oxidans* subsp. *suboxidans*	Glucose, calcium carbonate, corn steep	33 h at 30°C, aerobic	Filtration, concentration under vacuum	100% based on substrate used

Table 155. Yeasts producing vitamins [Sakai (1980)].

Yeast	Amount (mg ℓ$^{-1}$) Pyridoxine phosphate	Pyridoxal phosphate
Awamori yeast (Sakumoto)	0.3	0.15
Brennereihefe Rasse 12	0.05	0.10
Candida utilis IFO 0619	0.0	0.45
Cryptococcus neoformans IFO 0410	0.0	0.25
Debaryomyces japonicus IFO 0039	0.025	0.10
Endomyces fibuliger IFO 0103	0.275	0.325
E. hordei IFO 0104	0.4	0.35
E. lindneri IFO 0106	0.25	0.25
Hansenula anomala Y-56	0.125	0.20
Nematospora coryli IFO 0658	0.15	0.05
Pichia farinosa var. *japonica* IFO 0459	0.0	0.05
P. polymorpha Kloecker IFO 0195	0.53	0.20
Saccharomyces carlsbergensis Hansen Y-33	0.0	0.075
S. cerevisiae (Oriental)	0.20	0.0
S. logos Y-22	0.10	0.25
S. marxianus IFO 0277	0.05	0.20
S. rouxii IAM 4011	0.0	0.15
S. sake H-31	0.0	0.0
S. sake Hozan	0.025	0.075
S. sake Kyokai 2	0.10	0.10
Saccharomycodes ludwigii IFO 0339	0.2	0.15
Saccharomycopsis capsularis IFO 0672	0.225	0.25
Schizosaccharomyces pombe IFO 0346	0.10	0.20
Sporobolomyces salmonicolor Y.u.	0.0	0.25
Trichosporon beigelii IFO 0598	0.125	0.10
Zygosaccharomyces japonicus IFO 0595	0.05	0.075

Table 156. Moulds producing vitamin B_6 [Sakai (1980)].

Mould	Amount ($mg\ \ell^{-1}$)	
	Pyridoxine phosphate	Pyridoxal phosphate
Aspergillus awamori M-66	0.40	0.20
Asp. candidus M-70	7.30	1.10
Asp. flavus IFO 5839	9.45	1.35
Asp. nidulans IFO 5719	0.70	0.35
Asp. niger 4416	1.30	0.30
Asp. oryzae Cohn IFO 4117	2.20	0.00
Asp. oryzae M-61	3.50	0.80
Asp. oryzae var. *globosus* M-71	7.10	0.90
Asp. tamarii M-68	5.70	0.60
Asp. terreus Thom 6123	0.65	0.35
Asp. usamii IFO 4388	0.20	0.40
Fusarium culmorum IFO 5902	0.125	0.025
F. oxysporum IFO 5942	0.05	0.20
Gibberella fujikuroi IFO 5268	0.00	0.00
Monascus anka IAM 8001	0.125	0.00
M. purpureus IAM 8001	0.10	0.05
Mucor fragilis Bainier IFO 6449	0.375	0.275
Muc. javanicus Wehmer IFO 4570	0.40	0.225
Muc. javanicus Wehmer IFO 4572	0.675	0.275
Neurospora crassa IFO 6068	0.675	0.425
Penicillium chrysogenum Thom IFO 4626	0.45	0.15
Pen. notatum Westling IFO 4640	0.50	0.25
Pen. oxalicum IFO 5750	1.025	0.05
Pullularia pullulans IFO 4464	0.125	0.00
Rhizopus batatas M-24	0.00	0.225
R. javanicus Takeda IFO 5442	0.00	0.20
R. oryzae M-21	0.275	0.225

Table 157. Processes for microbial production of vitamin B_{12}[a] [Perlman (1978)].

Microorganism	Ingredients of medium	Yield of vitamin ($mg\,\ell^{-1}$)	Comments
Bacillus megaterium	Beet molasses, ammonium phosphate, cobalt salt, inorganic salts	0.45	Aerated fermentation (18 h)
Butyribacterium rettgeri	Corn-steep liquor, cobalt salt, glucose (maintained at pH 7 with ammonium hydroxide)	5	4-day anaerobic fermentation
Micromonospora sp.	Soya bean meal, glucose, calcium carbonate, cobalt salt	11.5	7-day aerated fermentation
Propionibacterium freudenreichii	Corn-steep liquor, glucose, cobalt salt (maintained at pH 7 with ammonium hydroxide)	19	3 days anaerobic, 3 days aerobic
P. freudenreichii	Corn-steep liquor (or autolyzed *Penicillium* mycelium), glucose, cobalt salt (maintained at pH 7 with ammonium hydroxide)	8	Continuous 2-stage fermentation, 33-h retention time
P. shermanii	Corn-steep liquor, glucose, cobalt salt (maintained at pH 7 with ammonium hydroxide)	23	3 days anaerobic, 3 days aerobic
Pseudomonas denitrificans	Sucrose, betaine, glutamic acid, cobalt salt, 5,6-dimethylbenzimidazole, salts	15	2-day aerated fermentation
Streptomyces sp.	Soya bean meal, glucose, cobalt salt, dipotassium hydrogen phosphate	5.7	6-day aerated fermentation
Streptomyces olivaceus	Soya bean meal, glucose, distillers' solubles, cobalt salt, inorganic salts	3.3	6-day aerated fermentation

[a] While the cultures are presumed to produce the coenzyme form of vitamin B_{12}(5,6-dimethyl-α-benzimidazolylcobamide-5′-deoxyadenosine), the vitamin is usually isolated in the cyanide form.

Media

Compositions of media used for the production of vitamin B_{12} by various microorganisms are listed in Tables 158–160.

Table 158. Medium for the culture of *Arthrobacter* in vitamin B_{12} production.

Component	Amount
i-Propanol	$10\,ml\,\ell^{-1}$
Calcium carbonate	$6\,g\,\ell^{-1}$
Peptone	$3\,g\,\ell^{-1}$
Yeast extract	$1\,g\,\ell^{-1}$
Corn-steep liquor	$9\,ml\,\ell^{-1}$
Ammonium nitrate	$12\,g\,\ell^{-1}$
Sodium phosphate (dibasic)	$1.5\,g\,\ell^{-1}$
Potassium dihydrogen phosphate	$1.9\,g\,\ell^{-1}$
Magnesium sulphate heptahydrate	$0.5\,g\,\ell^{-1}$
Ferrous sulphate heptahydrate	$10\,mg\,\ell^{-1}$
Zinc sulphate heptahydrate	$10\,mg\,\ell^{-1}$
Cobalt nitrate	$5\,mg\,\ell^{-1}$
Copper sulphate pentahydrate	$50\,\mu g\,\ell^{-1}$
Molybdenum trioxide	$10\,\mu g\,\ell^{-1}$
Antifoam	$0.5\,ml\,\ell^{-1}$

For the medium described in Table 158, *i*-propanol is added during the logarithmic-growth phase at a rate, which ensures that its concentration does not exceed 3 ml ℓ^{-1}. Subsequently, in the vitamin-production phase, the alcohol is added at a rate that maintains the pH of the culture at 7–8.

Table 159. Medium for vitamin B_{12} production using *Propionibacterium* species.

Component	Amount ($g\,\ell^{-1}$)
Glucose or molasses	10–100
Casein hydrolyzate or corn-steep liquor	30–70
Iron, manganese, magnesium, cobalt salts	0.01–0.1

Culturing *Propionibacterium* species in the appropriate medium (*see* Table 159) at pH 6.5–7.0 and 30°C gives a vitamin yield of about 25–40 mg ℓ^{-1}.

Table 160. Medium[a] for vitamin B_{12} production using *Pseudomonas* species.

Component	Amount ($g\,\ell^{-1}$)
Beet molasses	100.0
Yeast extract	2.0
Diammonium hydrogen phosphate	5.0
Magnesium sulphate	3.0
Manganese sulphate	0.2
Cobalt nitrate	0.188
5,6-Dimethylbenzimidazole	0.025
Zinc sulphate	0.020
Sodium molybdate	0.005

[a] pH 7.4.

The composition of a medium used for riboflavin production by *Ashbya gossypii* is given in Table 161.

Table 161. Medium for riboflavin production using *Ashbya gossypii.*

Component	Amount
Corn-steep liquor solids	2.25% (w/v)
Wilson's peptone W-809	3.5% (w/v)
Soya bean oil	4.5% (w/v)
Glycine	1–3 g ℓ^{-1}

Metabolic Pathways

The biosynthetic pathways for cobalamins and riboflavin are outlined in Figs. 70 and 71, respectively.

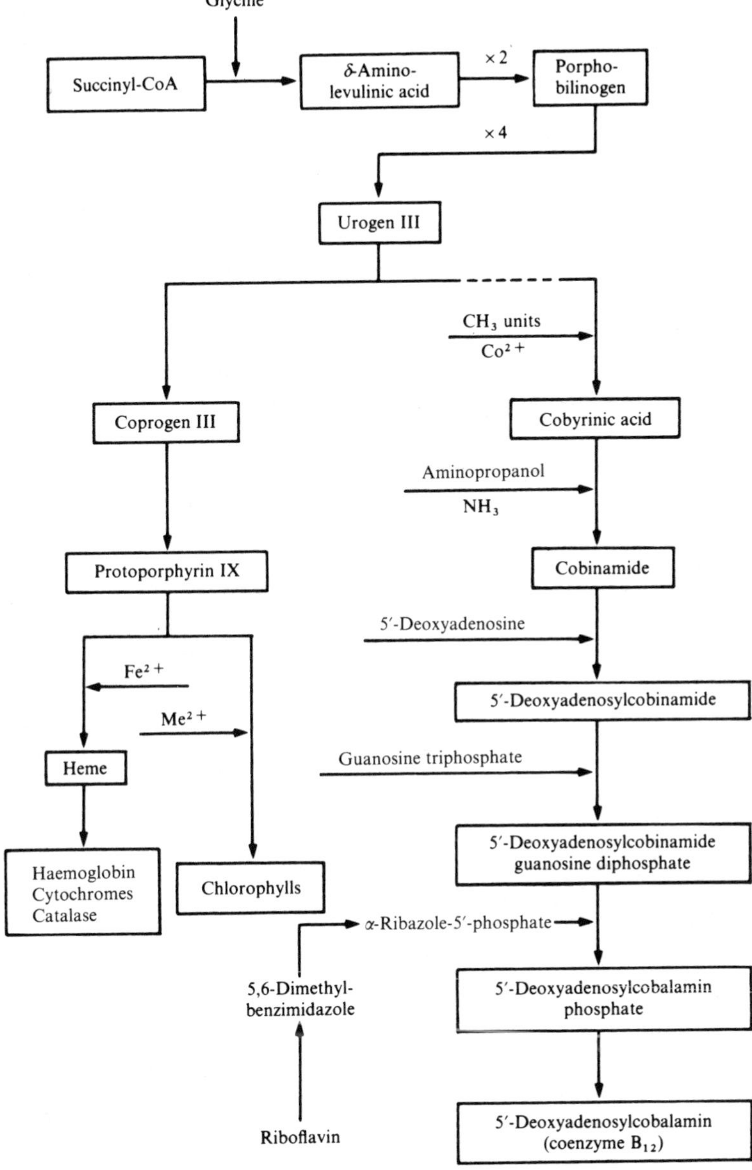

Figure 70. A general pathway for the biosynthesis of cobalamins [Florent and Ninet (1979)].

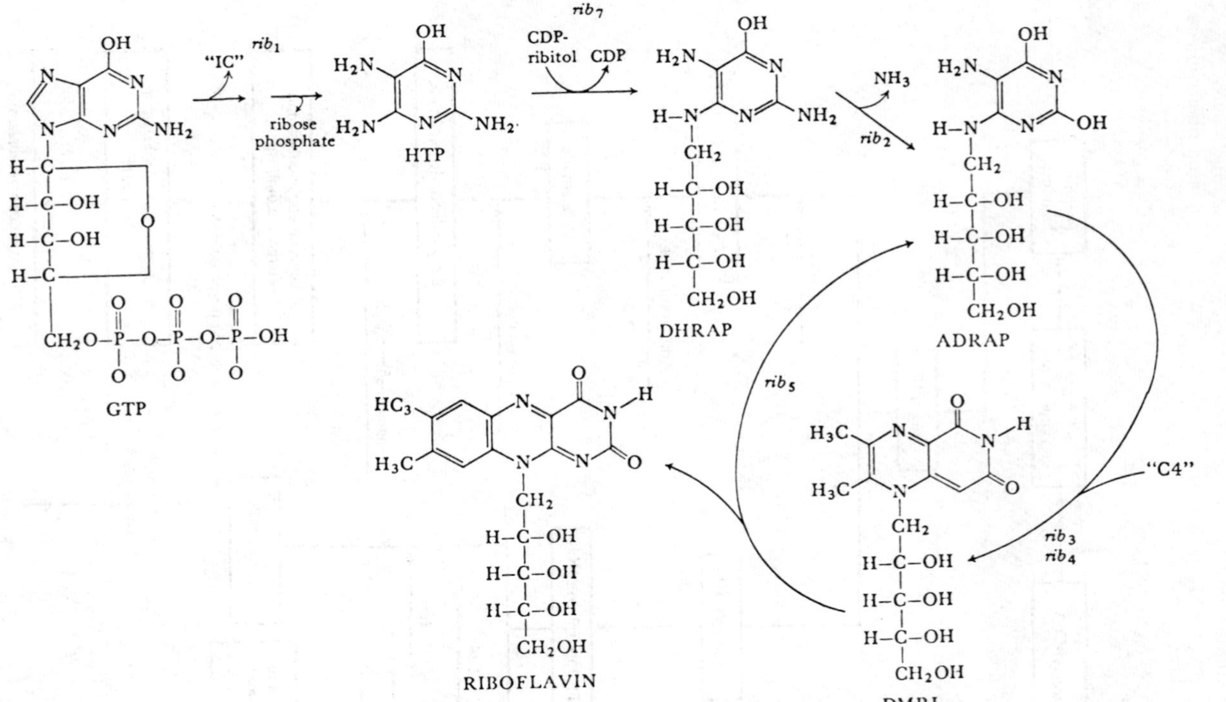

Figure 71. Probable pathway of riboflavin biosynthesis. The genes (rib_1 to rib_5, rib_7) in *Saccharomyces cerevisiae* which correspond to each reaction are shown. HTP, 6-hydroxy-2,4,5-triaminopyrimidine; DHRAP, 2,5-diamino-6-hydroxy-4-ribitylaminopyrimidine; ADRAP, 5-amino-2,6-dihydroxy-4-ribitylaminopyrimidine; DMRL, 6,7-dimethyl-8-ribityllumazine [Perlman (1978)].

Time Course

A time course of vitamin B_{12} fermentation is given in Fig. 72, while time courses of riboflavin fermentation with *Ashbya gossypii* and *Eremothecium ashbyii* are shown in Figs. 73 and 74, respectively. Tables 162–167 provide some information on riboflavin and thiamin yields.

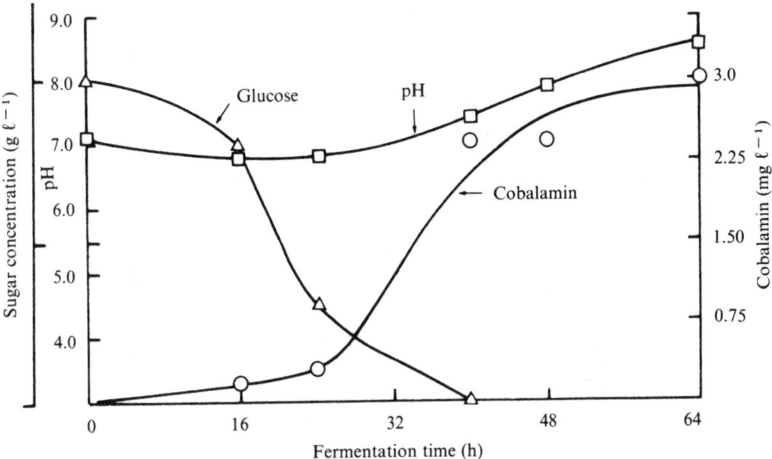

Figure 72. Course of cobalamin fermentation with *Streptomyces olivaceus* [Underkofler and Hickey (1954)].

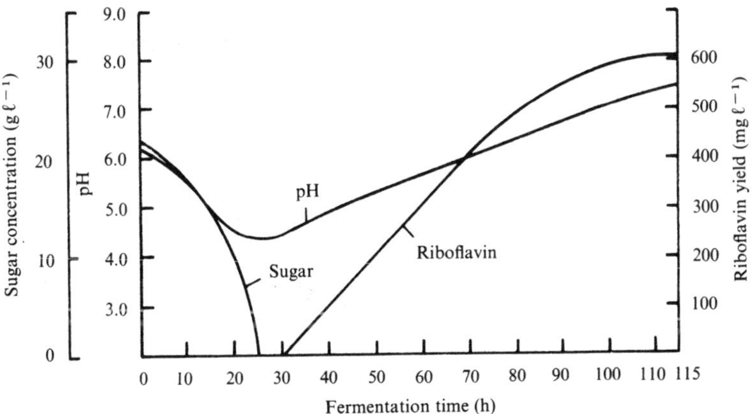

Figure 73. Changes in medium during fermentation of *Ashbya gossypii* in shake culture [Hockenhull (1959)].

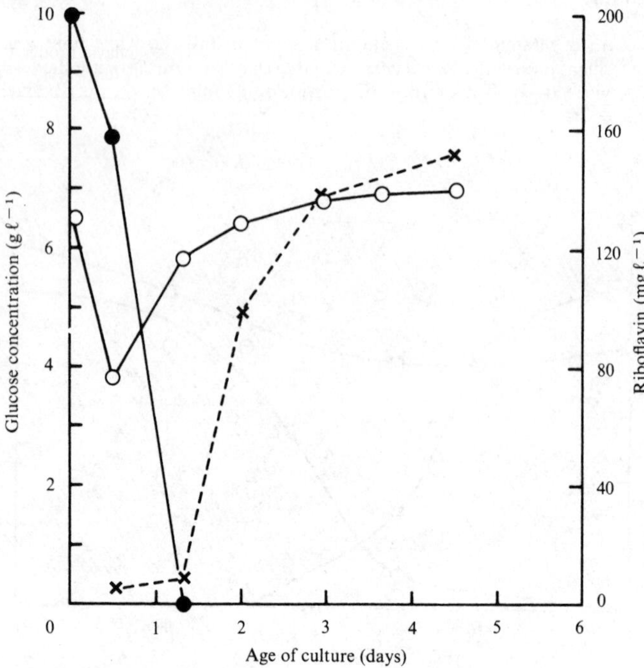

Figure 74. Changes in the medium of shake cultures of *Eremethecium ashbyii* ●———●, glucose concentration; ○———○, pH; ×- - - -×, riboflavin [Hockenhull (1959)].

Table 162. Moderately flavinogenic bacteria [Pridham (1952)].

Microorganism	Riboflavin synthesized ($mg\,\ell^{-1}$)	[μg (g dry wt)$^{-1}$]
Corynebacterium diphtheriae	25	
Escherichia coli	505	106
Pseudomonas fluorescens	22–34	310
Rhizobium trifolii		300
Serratia marcescens		160

Table 163. Flavinogenic microorganisms [Goodwin (1959)].

Microorganism	Riboflavin produced [μg (g dry wt cells)$^{-1}$]
Yeasts	
Hansenula	54
Mycotorula	59
Oidium	55
Saccharomyces	50–100
Torulopsis[a]	50–90
Higher fungi	
Agropyrum	10
Alternaria	10–20
Elymus	57
Fusarium	20–30
Hormodendrum	10–20
Mucor	20
Penicillium	20–40
Phalaris	6
Phragmites	8
Rhizopus	20
Secale	17
Trichoderma	20

[a] Synthesis inhibited by iron.

Table 164. Microorganisms producing considerable amounts of riboflavin and the effects of iron on biosynthesis of the vitamin [Perlman (1978)].

Microorganism	Riboflavin yield (mg ℓ$^{-1}$)	Optimum iron concentration (mg ℓ$^{-1}$)
Ashbya gossypii	6420	Not critical
Candida flareri	567	0.04–0.06
Clostridium acetobutylicum	97	1–3
Eremothecium ashbyii	2480	Not critical
Mycobacterium smegmatis	58	Not critical
Mycocandida riboflavin	200	Not critical

Table 165. Effect of glucose concentration on growth and flavinogenesis in *Eremothecium ashbyii*[a] [Hockenhull (1959)].

Concentration of glucose (% w/v)	Dry weight cells (mg)	Riboflavin (μg)
0.2	9.1	216
0.4	11.8	369
0.6	15.2	545
0.8	18.2	599
1.0	20.4	667
2.0	28.3	574
3.0	39.6	544
5.0	49.9	608
10.0	64.6	587

[a] Amounts produced in 15-ml basal medium; five-day static cultures.

Table 166. Growth and flavinogenesis by *Eremothecium ashbyii* in the presence of various amino acids [Hockenhull (1959)].

Medium	Dry weight (mg)	Riboflavin (μg)
Control A	13.7	230
L-Asparagine	27.2	506
DL-Serine	14.5	342
DL-Threonine	10.9	423
Glycine	13.7	141
Control B	22.6	473

[a] Dry weight and riboflavin production per 15 ml medium in 50-ml conical flasks; temperature 28°C; five-day static cultures; media A and B contain the basal medium 50 mg ℓ^{-1} and 250 mg ℓ^{-1} peptone–nitrogen, respectively; other media contain (5 g ℓ^{-1}) peptone–nitrogen and amino acid–nitrogen (200 mg ℓ^{-1}).

Table 167. Synthesis of thiamin from pyrimidine and thiazole by bakers' yeast [Underkofler and Hickey (1954)].

Intermediate added to the medium			
Pyrimidine (μM)	Thiazole (μM)	Yeast yield (g dry wt ℓ^{-1})	Thiamin content [μg (g dry wt)$^{-1}$]
0.0	0.0	7.27	65
5	5	7.60	261
10	10	7.62	430
15	15	7.55	667
20	20	7.60	802
50	50	7.67	835
100	100	7.77	795
0.0	20	7.80	150
20	0.0	7.27	62

REFERENCES

Adams. G.A. and Stanier, R.Y. (1945) Production and Properties of 2,3-Butanediol. III. Studies on the Biochemistry of Carbohydrate Fermentation by *Aerobacillus polymyxa. Can. J. Res.*, **23B**, 1.

Aiba, S., Shoda, M. and Nagatani, M. (1968) Kinetics of Product Inhibition in Alcohol Fermentation. *Biotechnol. Bioengng*, **10**, 845.

Akashi, K., Ikeda, S., Shibai, H., Kobayash, K. and Hirose, Y. (1978) Determination of Redox Potential Levels Critical for Cell Respiration and Suitable for L-Leucine Production. *Biotechnol. Bioengng*, **20**, 27.

Anderson, R.F., Arnold, M., Nelson, G.E.N. and Ciegler, A. (1958) Microbiological Production of β-Carotene in Shaken Flasks. *Agric. Fd Chem.*, **6**, 543.

Bazua, C.D. and Wilke, C.R. (1977) Ethanol Effects on the Kinetics of a Continuous Fermentation with *Saccharomyces cerevisiae. Biotechnol. Bioengng*, **7**, 105.

Biffi, G.,Boretti,G., De Marco, A. and Pennella, P. (1954) Metabolic Behaviour and Chlortetracycline Production by *Streptomyces aureofaciens* in Liquid Culture. *Appl. Microbiol.*, **2**, 288.

Borrow, A., Brown, S., Jeffereys, E.G., Kessell, R.H.J., Lloyd, P.B., Rothwell, A., Rothwell, B. and Swait, J.C. (1964) The Kinetics of Metabolism of *Gibberella fujikuroi* in Stirred Culture. *Can. J. Microbiol.*, **10**, 407.

Briffaud, J. and Engasser, M. (1979) Citric Acid Production from Glucose. Growth and Excretion Kinetics in a Stirred Fermenter. *Biotechnol. Bioengng*, **21**, 2083.

Chen, W.-P. (1980a) Glucose Isomerase (a Review). *Proc. Biochem.*, (June/July), 30.

Chen, W.-P. (1980b) Glucose Isomerase (a Review). *Proc. Biochem.*, (August/Sept.), 36.

Chibata, I., Tosa, T. and Sato, T. (1974) Immobilized Aspartase-containing Microbial Cells: Preparation and Enzyme Properties. *Appl. Microbiol.*, **27**, 878.

Chu, F.S. and Lilly V.G. (1960) Factors affecting the Production of Carotene by *Choanephora cucurbitarum.Mycologia*, **52**, 80.

Claridge, C.A. (1979) Aminoglycoside Antibiotics, in *Economic Microbiology*, vol. 3, *Secondary Products of Metabolism*, ed. by A.H. Rose, p.151 (Academic Press, London).

Daoust, D.R. (1976) Microbial Synthesis of Amino Acids, in *Industrial Microbiology*, ed. B.M. Miller and W. Litsky, p.106 (McGraw-Hill, New York).

Davies, R. and Stephenson, M. (1941) Studies on the Acetone/Butyl Alcohol Fermentation – Nutritional and Other Factors involved in the Preparation of Active Suspensions of *Clostridium acetobutylicum. Biochem. J.*, **35**, 1320.

Demain, A.L. (1972) Theoretical and Applied Aspects of Enzyme Regulation and Biosynthesis in Microbial Cells. *Biotechnol. Bioengng, Symp.*, **3**, 21.

Demain, A.L. (1978) Production of Nucleotides by Microorganisms, in *Economic Microbiology*, vol. 2, *Primary Products of Metabolism*, ed. by A.H. Rose, p.187 (Academic Press, London).

Diers, I. (1976) Glucose Isomerase in *Bacillus coaguilans*, in *Continuous Culture*, vol. 6, *Applications and New Fields*, ed. by A.C.R. Dean *et al.* (Ellis Horwood, Chichester).

Flickinger, M.C. (1980) Current Biological Research in Convention of Cellulosic Carbohydrates into Liquid Fuels: How far have we come? *Biotechnol. Bioengng*, **22**, 27.

Florent, J. and Ninet, L. (1979) Vitamin B_{12}, in *Microbial Technology*, vol. 1, *Microbial Technology*, ed. by H.J. Peppler and D. Perlman, p.497 (Academic Press, New York).

Fukumura, T. (1976) Hydrolysis of L-α-Amino-ε-caprolactam by Yeasts. *Agric. Biol. Chem.*, **40**, 1695.

Ghose, T.K. (1977) Cellulase Biosynthesis and Hydrolysis of Cellulosic Substances, in *Advances in Biochemical Engineering*, vol. 6, ed. by A. Fiechter *et al.*, p.39 (Springer-Verlag, Berlin).

Goodwin, T.W. (1959) Microbial Fat: Microorganisms as Potential Fat Producers. *Progress in Industrial Microbiology*, vol. 1, ed. by D.J.D. Hockenhull, p.137 (Heywood, London).

Goodwin, T.W. and Treble, D.H. (1958) The Incorporation of [2-14 C] Acetylmethylcarbinol [Acetoin] Ring A of Riboflavin by *Eremothecium ashbyii*, a new route for the biosynthesis of an aromatic ring. *Biochem. J.*, **70**, 14P.

Hanson, A.M. (1967) Microbial Production of Pigments and Vitamins, in *Microbial Technology*, ed. by. H.J. Peppler, p.222 (Reinhold, New York).

Hirose, Y. and Okada, H. (1979) Microbial Production of Amino Acids, in *Microbial Technology*, vol. 1, *Microbial Processes*, ed. by. H.J. Peppler and D. Perlman, p.211 (Academic Press, New York).

Hirose, Y. and Shibai, H. (1980) Amino Acid Fermentation. *Biotechnol. Bioengng*, **22**, 111.

Hockenhull, D.J.D. (ed.) (1959) The Influence of Medium Constituents on the Biosynthesis of Penicillin. *Progress in Industrial Microbiology*, vol. 1 (Heywood, London).

Hockenhull, D.J.D. (ed.) (1960) The Biochemistry of Streptomycin Production, in *Progress in Industrial Microbiology*, vol. 2 (Heywood, London).

Hockenhull, D.J.D. (1963) Antibiotics, in *Biochemistry of Industrial Microorganisms*, ed. by C. Rainbow and A.H. Rose, p.227 (Academic Press, London).

Holzberg, I., Finn, R.K. and Steinkraus, K.H. (1967) A Kinetic Study of the Alcoholic Fermentation of Grape Juice. *Biotechnol. Bioengng*, **9**, 413.

Jones, R.P., Pamment, N. and Greenfield, P.F. (1981) Alcohol Fermentation by Yeasts – The Effect of Environmental and Other Variables. *Proc. Biochem.*, (April/May), 42.

Kase, H., Tanaka, H. and Nakayama, K. (1971) Studies on L-Threonine Fermentation. I. Production of L-Threonine by Auxotrophic Mutants. *Agric. Biol. Chem.*, **35**, 2089.

Keay, L., Moseley, M.H., Anderson, R.G., O'Connor, R.J. and Wildi, B.S. (1972) Production and Isolation of Microbial Proteases. *Biotechnol. Bioengng Symp.*, **3**, 63.

Kinoshita, S. and Nakayama, K. (1978) Amino Acids, in *Economic Microbiology*, vol. 2, *Primary Products of Metabolism*, ed. by A.H. Rose, p.210 (Academic Press, London).

Kisumi, M., Komatsubara, S., Sugiyama, M., and Chibata, I. (1971) Isoleucine Hydroxamate, an Isoleucine Antagonist. *J. Bacteriol.*, **107**, 741.

Kitai, A., Tone, H., Ishikura, T. and Ozaki, A. (1968) Microbial Production of Salicylic Acid from Naphthalene. II. Product Inhibitory Kinetics and Effect of Product Removal on the Fermentation. *J. Ferment. Technol.*, **46**, 442.

Kolot, F.B. (1980) New Trends in Yeast Technology – Immobilised Cells. *Proc. Biochem.*, (Oct/Nov.), 2

Kobayashi, T. and Nakamura, I. (1961) Studies on Itaconic Acid Fermentation. III. Fundamental Studies on Continuous Fermentation, Fermentation Rate and Acid Yield by Shaking Culture using *Aspergillus terreus* and *Aspergillus itaconicus*. *J. Ferment. Technol.*, **39**, 341.

Kristiansen. B. and Sinclair, C.G. (1978) Production of Citric Acid in Batch Culture. *Biotecnol. Bioengng*, **20**, 1711.

Kubota, K., Shiro, T. and Okumura, S. (1971) *J. Gen. Appl. Microbiol.*, **17**, 1.

Lawson, C.J. and Sutherland, I.W. (1978) Polysaccharides, in *Economic Microbiology*, vol. 2, *Primary Products of Metabolism*, ed. by A.H. Rose, p.328 (Academic Press, London).

Lockwood, L.B. (1979) Production of Organic Acids by Fermentation, in *Microbial Technology*, vol. 1, *Microbial Process*, ed. by H.J. Peppler and D. Perlman, p.356 (Academic Press, New York).

Lockwood, L.B. and Schweiger, L.B. (1967) Citric and Itaconic Acid Fermentations, in *Microbial Technology*, ed. by H.J. Peppler, p.183 (Reinhold, New York).

Lockwood, L.B., Tabenkin, B. and Ward, G.E. (1941) The Production of Gluconic Acid and 2-Ketogluconic Acid from Glucose by Species of *Pseudomonas* and *Phytomonas*. *J. Bacteriol.*, **42**, 51.

Luedeking, R. and Piret, E.L. (1959) A Kinetic Study of the Lactic Acid Fermentation. Batch Process at Controlled pH. *J. Biochem. Microbiol. Technol. Engng*, **1**, 359.

Maiorella, B., Wilke, C.R. and Blanch, H.W. (1981) Alcohol Production of Recovery, in *Advances in Biochemical Engineering*, vol. 20, ed. by A. Fiechter *et al.*, p.43 (Springer-Verlag, Berlin).

Martin, J.F. (1977) Control of Antibiotic Synthesis by Phosphate, in *Advances in Biochemical Engineering*, vol. 6, ed. by A. Fiechter *et al.*, p.105 (Springer-Verlag, Berlin).

Moss, M.O. and Smith, J.E. (1977) *Industrial Application of Microbiology* (Surrey University Press, London).

Moyer, A.J., Wells, P.A., Stubbs, J.J., Herrick, H.T. and May, O.E. (1937) Gluconic Acid Production – Development of Inoculum and Composition of Fermentation Solution for Gluconic Acid Production by Submerged Mould Growths under Increased Air Pressure. *Ind. Engng Chem.*, **29**, 777.

Mukhopadhyay, S.N. and Malik, R.K. (1980) Increased Production of Cellulase of *Trichoderma* sp. by pH Cycling and Temperature Profiling. *Biotechnol. Bioengng*, **22**, 2237.

Nakao, Y. (1979) Microbial Production of Nucleosides and Nucleotides, in *Microbial Technology*, vol. 1, *Microbial Processes*, p.312, ed. by H.J. Peppler and D. Perlman (Academic Press, New York).

Nakayama, K. (1972) Lysine and Diaminopimelic Acid, in *The Microbial Production of Amino Acids*, ed. by K. Yamada *et al.*, (Kōdansha, Tokyo).

Nakayama, K., Kitada, S. and Kinoshita, S. (1960) *Amino Acids*, vol. 2, p.105 (Institute of Applied Microbiology, Tokyo).

Nakayama, K., Kitada, S. and Kinoshita, S. (1961) Leucine Production, *J. Gen. Appl. Microbiol.*, **7**, 5.

Neish, A.C., Blackwood, A.C. and Ledingham, G.A. (1945) Dissimilation of Glucose by *Bacillus subtilis* (Ford Strain). *Can. J. Res.*, **23B** 290.

Neish, A.C., Blackwood, A.C., Robertson, F.M. and Ledingham, G.A. (1947) Production and Properties of 2,3-Butanediol. XVIII. Dissimilation of Glucose by *Serratia marcescens*. *Can. J. Res.*, **25B**, 65.

Ninet, L. and Renaut, J. (1979) Carotenoids, in *Microbial Technology*, vol. 1, *Microbial Processes*, ed. H.J. Peppler and D. Perlman, p.529 (Academic Press, New York).

Orlova, N.V. and Zaytseva, Z.M. (1961) *Sci. Rpt. Inst. Super Sanita.* **1**, 370.

Pace, G.W. and Righelato, R.C. (1980) Production of Extracellular Microbial Polysaccharides, in *Advances in Biochemical Engineering*, vol. 15, ed. by A. Fiechter, p.41 (Springer-Verlag, Berlin).

Perlman, D. (1978) Vitamins, in *Economic Microbiology*, vol. 2, *Primary Products of Metabolism*, ed. by A.H. Rose, p.303 (Academic Press, London).

Perlman, D. (1979) Microbial Production of Antibiotics, in *Microbial Technology*, vol. 1, *Microbial Processes*, ed. by. H.J. Peppler and D. Perlman, p.241 (Academic Press, New York).

Pirt, S.J. and Callow, D.S. (1958) Exocellular Product Formation by Microorganisms in Continuous Culture. I. Production of 2:3-Butanediol by *Aerobacter aerogenes* in a Single-Stage Process. *J. Appl. Bacteriol.*, **21**, 188.

Prescott, S.C. and Dunn, C.G. (1954) *Industrial Microbiology*, 3rd edn (McGraw-Hill, New York).

Pridham, T.G. (1952) Microbial Synthesis of Riboflavin. *Econ. Bot.*, **6**, 185.

Queener, S. and Swartz, R. (1979) Penicillins: Biosynthetic and Semisynthetic, in *Economic Microbiology*, vol. 3, *Secondary Products of Metabolism*, ed. by A.H. Rose, p.35 (Academic Press, London).

Rainbow, C. and Rose, A.H. (eds.) (1963) *Biochemistry of Industrial Microorganisms* (Academic Press, London).

Ratledge, C. (1978) Lipids and Fatty Acids, in Economic Microbiology, vol. 2, Primary Products of Metabolism, ed. by A.H. Rose, p.268 (Academic Press, London).

Rolland, G. and Sensi, P. (1955) *Pharmaco (Ed. Sci.)*, 10, 37.

Rogers, P.L., Lee, K.J. and Tribe, D.E. (1980) High Productivity Ethanol Fermentations with *Zymomonas mobilis. Proc. Biochem.*, (Aug./Sept.), 7.

Röhr, M. and Kubicek, C.P. (1981) Regulatory Aspects of Citric Acid Fermentation by *Aspergillus niger. Proc. Biochem.* (June/July), 34.

Ryu, D.D.Y. and Hospodka, J. (1980) Quantitative Physiology of *Penicillium chrysogenum* in Penicillin Fermentation. *Biotecnhnol. Bioengng*, **22**, 289.

Sakai, T. (1980) Microbial Production of Coenzymes. *Biotechnol. Bioengng*, **22**, 143.

Samejima, H., Nara, T., Fujita, C. and Kimoshita, S. (1960) *J. Agric. Chem. Soc., Japan*, **34**, 832.

Stanier, R.Y. and Adams, G.A. (1944) The Nature of the *Aeromonas* Fermentation. *Biochem. J.*, **38**, 168.

Steel, R. (ed.) (1958) *Biochemical Engineering — Unit Processes in Fermentation* (Heywood, London).

Tanaka, A., Nagasaki, T., Inagawa, M. and Fukui, S. (1968) Studies on Formation of Vitamins and Their Functions in Hydrocarbon Fermentation. V. Production of Carotenoids by *Microbacterium smegmatis* in Hydrocarbon Media. *J. Ferment. Technol.*, **46**, 468.

Tanaka, K., Iwasaki, T. and Kinoshita, S. (1960) *J. Agric. Chem. Soc., Japan*, **34**, 593.

Tangnu, S.K. and Ghose, T.K. (1980) Continuous Production of Salicyclic Acid: Effect of Dilution Rate, pH, Substrate Concentration and Cell Recycle. *Proc. Biochem.* (Dec.), 22.

Tangnu, S.K. and Ghose, T.K. (1981) Manipulation of Salicyclic Acid Fermentation. *Proc. Biochem.*, (Aug./Sept.), 25.

Tsuchida, T., Yoshinaga, F., Kubota, K., Momose, H. and Okumura, S. (1974) Production of L-Leucine by a Mutant of *Brevibacterium lactofermentum* 2256. *Agric. Biol. Chem.*, **38,** 1907.

Underkofler, L.A. and Hickey, R.J. (1954) *Industrial Fermentations* (Chemical Publishing Co., New York).

Vaheri, M. and Kauppinen, V. (1977) Improved Microbial Glucose Isomerase Production. *Proc. Biochem.*, (July/Aug.), 5.

Wang, D.I.C., Cooney, C.L., Demain, A.L., Dunnill, P., Humphrey, A.E. and Lilly, M.D. (1979) *Fermentation and Enzyme Technology* (John Wiley, New York).

Yamada, K. (1977) Recent Advances in Industrial Fermentation in Japan. *Biotechnol. Bioengng*, **19**, 1563.

CHAPTER 6
ENZYME ACTIVITY

GLOSSARY

Active site The portion of the enzyme surface involved in the binding of the substrate and where catalysis occurs.
AE Aminoethyl group.
Affinity Degree of binding of a substrate to an active site.
Allosterism Effect of binding of small molecules to an enzyme at a site or sites distant from the active site but which results in changes in the activity of the active site.
AMP Adenosine monophosphate.
ATEE *N*-Acetyl-L-tyrosine ethyl ester.
ATP Adenosine triphosphate.
BAA α-*N*-Benzoyl-L-arginine amide.
BAEE α-*N*-Benzoyl-L-arginine ethyl ester.
CDAPC 1-Cyclohexyl-3-(3'-dimethylaminopropyl) carbodiimide.
CM Carboxymethyl group.
CMC 1-Cyclohexyl-3-(2-morpholinoethyl) carbodiimide metho-*p*-toluene sulphonate.
Coenzyme An accessory substance, not protein, that is necessary for the functioning of the enzyme, frequently involved in transfer of atoms or groups.
CTP Cytidine triphosphate.
DCC *N*, *N'*-Dicyclohexylcarbodiimide.
DEAE Diethyleminoethyl group.
Dialyzable A solute capable of being removed through a semipermeable membrane.
DNA Deoxyribonucleic acid.
L-DOPA L-3,4-Hydroxyphenylalanine.
EDAPC 1-Ethyl-3-(3'-dimethylaminopropyl) carbodiimide.
Enzacryl Trade name for polyacrylamide-based water-insoluble matrix produced by Koch-Light Laboratories Ltd.
Enzyme unit Amount of substrate converted to product per unit time.
FAD Flavin adenine dinucleotide.
FMN Flavin monophosphate or riboflavin 5'-phosphate.
GTP Guanosine triphosphate.
IMP Inosine monophosphate.
Inhibitor An agent that prevents the normal action of an enzyme without destroying it.
ITP Inosine triphosphate.
NAD+ Nicotinamide adenine dinucleotide (oxidized form).
NADH Nicotinamide adenine dinucleotide (reduced form).
NADP+ Nicotinamide adenine dinucleotide phosphate (oxidized form).
NADPH Nicotinamide adenine dinucleotide phosphate (reduced form).
NTP Nucleoside triphosphate, general term for ATP, CTP, GTP, ITP, TTP, UTP.
Poloidal Homopolyer of 4-iodobutyl methacrylate or 2-iodoethyl methacrylate.
RNA Ribonucleic acid.
Sephadex Trade name for spherical dextran gel particles produced by Pharmacia Fine Chemicals Inc.
Sepharose Trade name of spherical agarose gel particles produced by Pharmacia Fine Chemical Inc.
S-MDA Starch–methylenedianiline.
Specific enzyme activity Amount of substrate converted to product per unit mass of protein in the enzyme per unit time.
Specificity The restricted metabolism of certain substrates by a given enzyme.
Substrate Chemical compound on which the enzyme acts with the formation of a product.
TTP Thymidine triphosphate.
Turnover number Number of molecules converted to product by each catalyst molecule per unit time.
UTP Uridine triphosphate.
Wort Unfermented extract of malt or other grains.

NOMENCLATURE AND CLASSIFICATION

Individual enzymes selectively catalyze specific reactions. Absolute specificity, in which an enzyme is only capable of catalyzing a reaction involving a single substrate, does not exist although metabolism of alternative substrates may be at very slow rates (i.e. $\times 10^{-4}$) compared with the natural substrate (*see* Table 7). Enzymes display the following forms of specificity.

1) Reaction specificity
2) Bond specificity
3) Optical specificity

Because of the high degree of specificity, a system has been established by the Commission on Enzymes of the International Union of Biochemistry [*see* Nomenclature Committee of the International Union of Biochemistry]. In this system, enzymes are divided into six major classes (*see* Table 1).

Table 1. Outline of the major enzyme classes.

Major class number	Class	Reactions catalyzed
1	Oxidoreductases	Oxidation–reduction
2	Transferases	Group transfer
3	Hydrolases	Hydrolysis
4	Lyases	Removal of group leaving double bond Addition of group to double bond
5	Isomerases	Isomerization
6	Ligases (synthetases)	Joining of two molecules coupled with cleavage of a pyrophosphate bond of ATP or similar triphosphate

Each of the major classes is further divided into numerical subclasses and sub-subclasses according to the individual reactions involved. Factors taken into consideration when extending this classification include the particular coenzyme or cofactor involved, the nature of the isomerization and the type of bond involved. The fourth number in the classification is the serial number of the enzyme within a subclass. Enzyme classification numbers proposed by the Enzyme Commission (i.e. EC numbers) for some important enzymes are given in Table 2.

Table 2. Classification and nomenclature of some important enzymes.

EC number	Systematic name	Trivial name
Oxidoreductases		
1.1.1.27	L-Lactate: NAD oxidoreductase	L-Lactate dehydrogenase
1.1.1.28	D-Lactate: NAD oxidoreductase	D-Lactate dehydrogenase
1.1.3.4	β-D-Glucose:oxygen oxidoreductase	Glucose oxidase
1.1.3.9	D-Galactose:oxygen oxidoreductase	Galactose oxidase
1.11.1.6	H_2O_2:H_2O_2 oxidoreductase	Catalase
1.11.1.7	Donor:hydrogen peroxide oxidoreductase	Peroxidase
1.13.11.12	Linoleate:oxygen oxidoreductase	Lipoxygenase, lipoxidase
Transferases		
2.4.1.5	α-1, 6-Glucan:D-fructose 2-glucosyltransferase	Sucrose 6-glucosyltransferase, dextran sucrase
2.4.1.19	α-1, 4-Glucan 4-glycosyltransferase	Cyclodextrin glycosyltransferase, *Bacillus macerans* amylase
Hydrolases		
3.1.1.3	Glycerol-ester hydrolase	Lipase
3.1.1.11	Pectin pectylhydrolase	Pectinesterase, pectinmethylesterase
3.2.1.1	α-1,4-Glucan glucanohydrolase	α-Amylase
3.2.1.2	α-1,4-Glucan maltohydrolase	β-Amylase
3.2.1.3	α-1,4-Glucan glucohydrolase	Glucoamylase, amyloglucosidase
3.2.1.4	β-1,4-Glucan glucanohydrolase	Cellulase
3.2.1.15	Poly-α-1,4-galacturonide glycanohydrolase	Polygalacturonase
3.2.1.20	α-D-Glucoside glucohydrolase	α-Glucosidase, maltase
3.2.1.23	β-D-Galactoside galactohydrolase	β-Galactosidase, lactase
3.2.1.26	β-D-Fructofuranoside fructohydrolase	β-Fructofuranosidase, sucrase, invertase
3.4.23.1		Pepsin
3.4.23.4		Rennin
3.4.21.4		Trypsin
3.4.22.2		Papain
3.4.22.3		Ficin
3.4.21.14		Subtilopeptidase A, subtilisin, bacterial protease
3.4.23.6		Aspergillopeptidase A, fungal protease
3.4.22.4		Bromelain
3.5.1.1	L-Asparagine amidohydrolase	Asparaginase
3.5.1.5	Urea amidohydrolase	Urease
Lyases		
4.1.3.6	Citrate oxaloacetate-lyase	Citrate lyase, citrase
4.2.1.3	Citrate (isocitrate) hydrolyase	Aconitate hydratase, aconitase
4.2.2.2	Poly-α-1, 4-D-galacturonide lyase	Pectate lyase, pectate transeliminase
Isomerases		
5.1.2.1	Lactate racemase	Lactate racemase
5.3.1.5	D-Xylose ketol-isomerase	Xylose isomerase, glucose isomerase
Ligases		
6.2.1.1	Acetate:CoA ligase	Acetyl-CoA synthetase
6.2.1.4	Succinate:CoA ligase	Succinyl-CoA synthetase
6.4.1.1	Pyruvate:CO_2 ligase	Pyruvate carboxylase
6.4.1.2	Acetyl-CoA: CO_2 ligase	Acetyl-CoA carboxylase

This classification system is complex and often difficulties are experienced in accurately describing the reaction involved. There is an additional problem that a trivial enzyme nomenclature exists which was well established before the introduction of the international system. A summary of the trivial names for enzymes and the types of reactions they catalyze is given in Table 3.

Table 3. Common categories of enzymes used in trivial enzyme classification.

Name	Reaction catalyzed
Dehydrogenases	Dehydrogenation of substrate with hydrogen transfer to a molecule other than molecular oxygen
Hydroxylases	Hydroxylation of substrate with molecular oxygen as donor
Kinases	Transfer of phosphate from nucleoside triphosphate to substrate
Mutases	Migration of phosphate group from one position to another within the same molecule
Oxidases	Oxidation of substrate with molecular oxygen as donor
Oxygenases	Incorporation of oxygen molecule into substrate with oxidative cleavage of a carbon–carbon bond
Phosphatases	Hydrolysis of phosphate ester
Phosphorylases	Addition of the elements of phosphoric acid across glycosyl or related bond
Synthetases	Condensation of two molecules with cleavage of nucleoside triphosphate
Thiokinases	Formation of thiol ester from carboxylic acid substrate with cleavage of ATP
Transferases	Transfer of group from one substrate to another

A further complication in enzyme classification is that some enzymes, in particular proteolytic enzymes which hydrolyze the peptide bonds of proteins, for example trypsin, pepsin, papain and ficin, possess common names that do not relate to either the substrate or the reaction involved.

Obviously, because of the different systems of nomenclature several names may be applied to one enzyme. An example of this problem is citrate lysase and citrase, both being names for citrate oxaloacetate lyase (EC 4.1.3.6) (*see* Table 2).

Because of the complex nature of carbohydrates, which differ according to the nature of the monomer units and the glycosidic bonds linking these units, a wide range of enzymes exists which catalyze the hydrolysis of these polymers. An outline of the classification of polysaccharide-degrading enzymes is shown in Fig. 1. The degradation mechanisms involved in the hydrolysis of starch are diagrammatically represented in Fig. 2. Principally, the reactions involved include the cleavage of 1,4- or 1,6-glycosidic bonds, enzymes cleaving the latter bonds being described as debranching. Exoenzymes act on the terminal glycosidic bonds of the polysaccharide, in contrast to the endoenzymes which cleave bonds within the polysaccharide chain.

Similarly, the complex nature of proteins and polypeptides demands a large number of enzymes for the hydrolysis of these compounds. Proteolytic enzymes generally display specificity with regard to the amino acids at the carboxy and/or amino side of the amide bond cleaved.

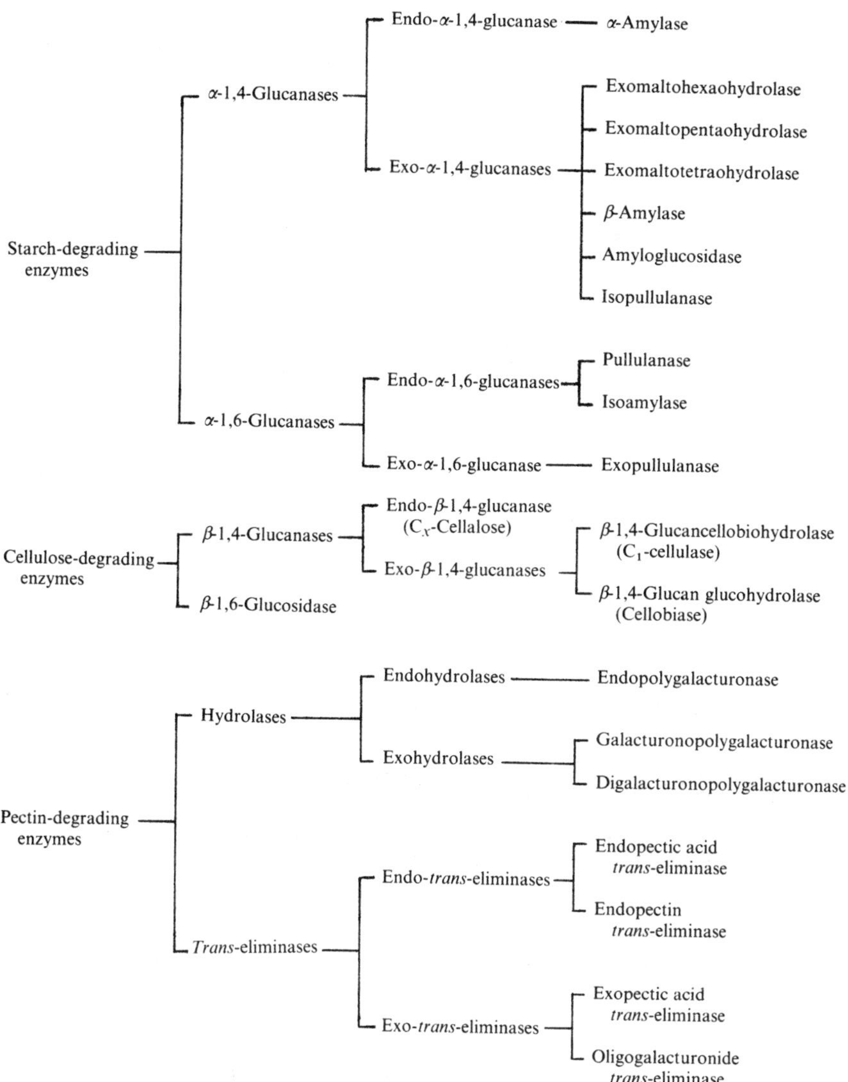

Figure 1. Classification of polysaccharide-degrading enzymes [Böing (1982)].

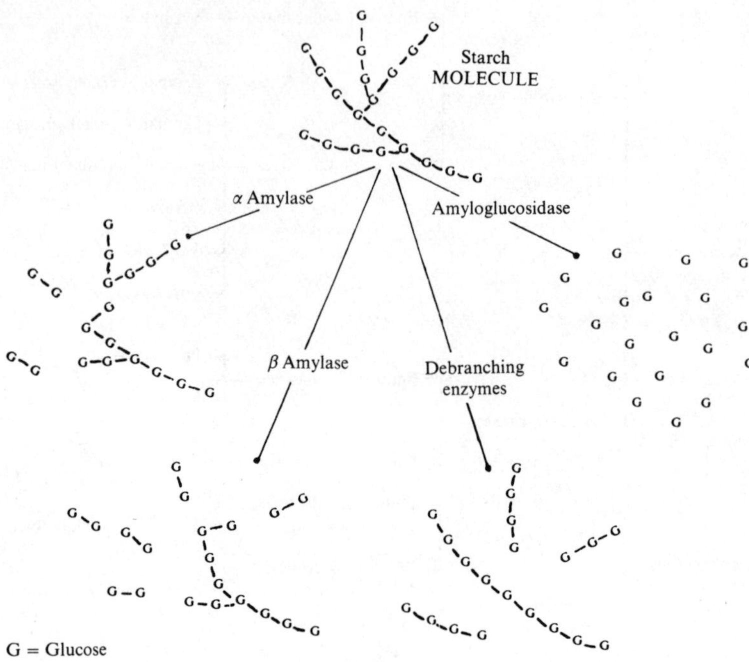

Figure 2. The enzymatic hydrolysis of starch [Stewart *et al.* (1979)].

GENERAL CHARACTERISTICS

All enzymes are proteins produced by living cells, the enzyme activity varying considerably according to the source of the enzyme (*see* Table 4).

Table 4. Amylase activities of various grain sources[a] [Bailey and Ollis (1977)].

	Total relative activity			
	β-Amylase		α-Amylase	
Cereal	Ungerminated	Germinated	Ungerminated	Germinated
Barley	29.8	34.4	0.058	94.0
Wheat	25.1	23.7	0.063	214.7
Rye	17.8	17.6	0.111	119.8
Oats	2.4		0.297	60.3
Maize			0.249	35.6
Sorghum			0.127	75.6
Rice		0.2		2.3

[a] Where no values are given, activity was below the sensitivity of the test.

Table 5 shows that some enzymes contain several similar subunits and these may be significant in allosterism (*see* p.496).

Table 5. Some examples of enzyme molecular weights [Wang *et al.* (1979)].

Enzyme	EC number	Molecular weight	Subunits	
			Number	Molecular weight
Alcohol dehydrogenase (yeast)	1.1.1.1	150 000	4	37 000
Aldolase (yeast)	4.1.2.13	80 000	2	40 000
α-Amylase	3.2.1.1	50 000	2	25 000
Aspartyl transcarbamylase	2.1.3.2	310 000	6	33 000
Catalase	1.11.1.6	232 000	4	57 000
Ribonuclease	3.1.4.22	13 700		
RNA polymerase (*Azotobacter*)	2.7.7.6	782 000	2	391 000
Trypsin	3.4.21.4	23 800		

In four respects, the catalytic activity of enzymes differs from that of other catalysts.

1) Enzymes are highly efficient, with turnover numbers N, i.e. the number of molecules associated with a site per second, frequently higher than those of inorganic catalysts, especially when considering reactions carried out within the physiological temperature range of 25–37°C. Comparison of turnover numbers is presented in Table 6. In addition, enzyme reactions are carried out at pH values close to neutrality (*see* Fig. 3).
2) Enzyme reactions are specific in the nature of the reaction catalyzed and the substrate utilized (*see* pp.468–471). An indication of the extent of the specificity of enzymes may be seen from Table 2, which shows that two lactate dehydrogenases exist that catalyze the conversion of the D- and L-isomers of lactate to pyruvate.
3) Enzymes are extremely versatile catalysts. From Table 2 it can be seen that a very wide range of chemical reactions can be catalyzed by enzymes.
4) Enzymes are subject to cellular control. For example, genetic control can influence the rate of synthesis and final cellular concentration of an enzyme. Using gene manipulation techniques (*see* Chapter 1, pp.62–68), it is possible to produce microbial mutants that are capable of producing large amounts of an enzyme, resulting in high product formation from a readily available substrate. Furthermore, the presence of a certain substrate may induce the conversion of an enzyme from an inactive to an active form, i.e. allosterism (*see* p.496).

Table 6. Some turnover numbers for enzyme- and inorganic-catalyzed reactions [Bailey and Ollis (1977)].

Catalyst	EC number	Reaction	Turnover number	Temperature (°C)
Enzymes				
Bromelain	3.4.22.4	Hydrolysis of peptides	4×10^{-3}–5×10^{-1}	0–37
Carbonic anhydrase	4.2.1.1	Hydration of carbonyl compounds	8×10^{-1}–6×10^{5}	0–37
Fumarase	4.2.1.2	L-Malate ⇌ fumarate + H_2O	10^3 (forward)	0–37
			3×10^3 (backward)	0–37
Papain	3.4.22.2	Hydrolysis of peptides	8×10^{-2}–10	0–37
Ribonuclease	3.1.4.22	Transfer phosphate of polynucleotide	2–2×10^3	0–37
Trypsin	3.4.21.4	Hydrolysis of peptides	3×10^{-3}–10^2	0–37
Inorganic catalysts				
Aluminium trichloride/alumina		n-Hexane isomerization	10^{-2}	25
			1.5×10^{-2}	60
Copper/silver (Cu_3Au)		Formic acid dehydrogenation	2×10^{-7}	25
			3×10^{10}	327
Silica-alumina		Cumene cracking	3×10^{-8}	25
			2×10^4	420
Vanadium trioxide		Cyclohexene dehydrogenation	7×10^{-11}	25
			10^2	350
Zeolite (decationized)		Cumene cracking	$\sim 10^3$	25
			$\sim 10^8$	325

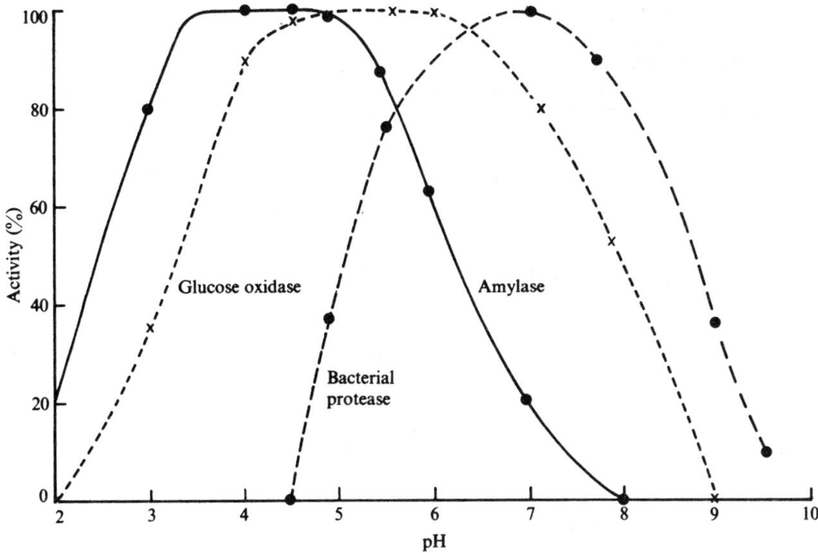

Figure 3. pH–activity curves for glucose oxidase, bacterial protease and amylase.

ENZYME PROPERTIES

Kinetics

At a constant enzyme concentration [E], the rate of an enzyme-catalyzed reaction increases with increasing substrate concentration [S] in the manner shown in Fig. 4.

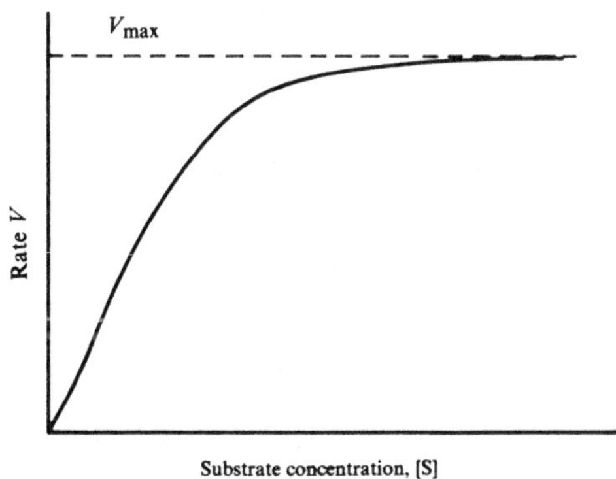

Figure 4. Rate of an enzyme-catalyzed reaction as a function of substrate concentration.

A characteristic of all enzyme-catalyzed reactions is that the rate V (dimensions = M T^{-1}) tends towards a maximum, usually expressed as V_{max}. This is due to the saturation of the active site of the enzyme with the formation of an enzyme–substrate complex ES, which is an essential stage in the formation of the product P. A simplified description of the overall chemical reaction is given by

$$E + S \underset{k_2}{\overset{k_1}{\rightleftharpoons}} ES \overset{k_3}{\rightarrow} E + P$$

In 1913 Michaelis and Menten [Dixon and Webb (1979)] suggested that under steady-state conditions

$$k_1 [E][S] = (k_2 + k_3)[ES] \tag{1}$$

or

$$[ES] = \frac{[E][S]}{K_m} \tag{2}$$

where K_m, the Michaelis constant (dimensions = ML^{-3}) is given by

$$K_m = \frac{k_2 + k_2}{k_1} \tag{3}$$

The total enzyme present in various states $[E_T]$ is

$$[E_T] = [E] + [ES] \tag{4}$$

Substituting Eq. (4) into Eq. (2), followed by rearrangement, gives

$$[ES] = [E_T] \frac{[S]}{[S] + K_m} \tag{5}$$

The rate of the reaction V is dependent on the concentration of the enzyme–substrate complex [ES], i.e.

$$V = k_3 [ES] \tag{6}$$

Thus

$$V = k_3 [E_T] \frac{[S]}{[S] + K_m} \tag{7}$$

The maximal rate of reaction V_{max} is achieved when the enzyme is saturated with substrate due to excess substrate (i.e. $[S] \gg K_m$), so that

$$\frac{[S]}{[S] + K_m} \simeq 1 \tag{8}$$

when

$$V_{max} = k_3 [E_T] \tag{9}$$

Substituting Eq. (9) into Eq. (7) gives the Michaelis–Menten equation

$$V = V_{max} \frac{[S]}{[S] + K_m} \tag{10}$$

Eq. (10) is algebraically identical to the Monod equation for microbial kinetics (*see* Chapter 4, p.205). There are significant implications arising from this situation.

1) Common methods of handling experimental data may be used to obtain the intrinsic parameters V_{max}, K_m, μ_{max}, etc. (*see* pp.206–207),
2) Diffusion limitations, whether liquid phase or 'solid' phase, can be handled by a common methodology (*see* Chapter 10),
3) The development of 'reactor' equations and reactor configurations follows a common course (*see* Chapter 7).

Rearrangement of Eq. (10) gives

$$\frac{1}{V} = \frac{1}{V_{max}} + \frac{K_m}{V_{max}} \cdot \frac{1}{[S]} \tag{11}$$

A plot of $1/V$ versus $1/[S]$ (known as a Lineweaver–Burk plot) (*see* Fig. 5) provides a means of determining the parameters K_m and V_{max} from experimental data. An alternative plot, although less frequently used, is that of Eadie–Hofstee (also shown in Fig. 5) using the rearrangement

$$\frac{V}{[S]} = -\frac{V}{K_m} + \frac{V_{max}}{K_m} \tag{12}$$

Accurate determination of actual values of V_{max} and K_m using either the Lineweaver–Burk or the Eadie–Hofstee equations requires statistical treatment of the data by methods such as that developed by Wilkinson (1961). Such methods are applied to experimentally determined rates less than 90 per cent of the experimental maximum to avoid excessive weighting by values in the saturation range.

Since the rate of reaction V (dimensions = MT^{-1}) is directly proportional to the enzyme concentration $[E_T]$, it follows that the Michaelis–Menten equation, like the Monod equation, can *and should* be expressed in terms of *specific rates of reaction*, i.e. V and V_{max} should be expressed as mass of substrate consumed per unit mass of enzyme present per unit time (dimensions = T^{-1}). This specific 'activity' is independent of enzyme concentration and the benefits that accrue from its use are the same as those identified on pp.204–205 (*see* Chapter 4).

Clearly, care has to be exercised over the units of V and the onus is on the experimenter to identify $[E_T]$. V and V_{max} are variously expressed with dimensions MT^{-1} and T^{-1}, and the symbol k_{max} is sometimes used for the maximum specific rate of an enzyme-catalyzed reaction (k_{max} is the same as k_3 of Eq. (9)).

The values of K_m for some enzyme–substrate systems are given in Table 7. The values of V_{max} vary according to the system considered, the enzyme purity and incubation conditions.

Table 7. Michaelis constants for some enzymes.

Enzyme	EC number	Source	Substrate	K_m (mM)
Acetylcholinesterase	3.1.1.7	Bovine RBC	Acetylcholine	0.27
Alcohol dehydrogenase	1.1.1.1	Yeast	Ethanol	13.0
L-Amino acid oxidase	1.4.3.2	Snake venom	L-Leucine	1.0
		Rat kidney mitochondria	L-Leucine	13.1
α-Amylase	3.2.1.1	*Bacillus stearothermophilus*	Starch	1.0
		Porcine pancreas	Starch	0.4
β-Amylase	3.2.1.2	Sweet potato	Amylose	0.07
Amyloglucosidase	3.2.1.3	*Coniophora cerebella*	Amylopectin	0.0007
	3.2.1.3	*Coniophora cerebella*	Amylose	0.032
	3.2.1.3	*Rhizopus delemar*	Amylopectin	0.0004
			Amylose	0.044
Asparaginase	3.5.1.1	*Pseudomonas* sp.	L-Asparagine	0.1
Aspartase	4.3.1.1	*Bacillus cadaveris*	L-Aspartate	30.0
Bromelain	3.4.22.4	Pineapple	Benzoyl-L-arginine ethyl ester	170
			Benzoyl-L-arginine amide	1.2
Carboxypeptidase	3.4.17.1	Bovine pancreas	Carbobenzoxyglycyl-L-phenylalamine	5.83
Chymotrypsin	3.4.21.1	Bovine pancreas	Acetyl-L-tryptophan ethyl ester	0.09
			Acetyl-L-phenylalanine ethyl ester	1.8
			Acetyl-L-leucine ethyl ester	29.0
			Acetyl-L-alanine methyl ester	129.0
			Acetyl-L-valine methyl ester	112.0
			Acetyl glycine ethyl ester	96.0
		Porcine pancreas	Benzoyl-L-leucine ethyl ester	11.0
			Benzoyl-L-phenylalanine ethyl ester	5.0
Creatine kinase	2.7.3.2	Rabbit muscle	Creatine	16.0
Ficin	3.4.22.3	Fig tree sap	Benzoyl-L-arginine ethyl ester	2.5
Fumarase	4.2.1.2	Porcine heart	L-Fumarate	0.0017
			L-Malate	0.0038
Galactose oxidase	1.1.3.9	*Polyporus circinatus*	D-Galactose	240
Glucose isomerase	5.3.1.5	*Lactobacillus brevis*	D-Glucose	920
			D-Xylose	5.0

Glucose oxidase	1.1.3.4	*Aspergillus niger*	D-Glucose	33.0
		Penicillium notatum	D-Glucose	9.6
Glucose-6-phosphate dehydrogenase	1.1.1.49	Yeast	Glucose-6-phosphate	0.02
Histidinase	4.3.1.3	*Pseudomonas fluorescens*	L-Histidine	8.9
Invertase	3.2.1.26	Yeast	Sucrose	9.1
		Neurospora crassa	Sucrose	6.1
Lactase or β-galactosidase	3.2.1.23	*Escherichia coli*	Lactose	3.85
Lactate dehydrogenase	1.1.1.27	*Bacillus subtilis*	Lactate	30.0
Lipoxygenase	1.13.11.12	Soya bean	Linoleate	1.0
Malate dehydrogenase	1.1.1.37	*Bacillus subtilis*	L-Malate	0.9
Papain	3.4.22.2	Papaya juice	Benzoyl-L-arginine ethyl ester	1.89
			Carbobenzoxylglycylglycine	270
Pepsin	3.4.23.1	Porcine gastric juice	Acetyl-L-phenylalanyldiiodotyrosine	0.075
			Acetyl-L-phenylalanyl-L-phenylalanine	0.43
			Acetyl-L-phenylalanyl-L-tyrosine	2.4
			Carbobenzoxy-L-glutamyl-L-tyrosine	1.89
Penicillinase	3.5.2.6	*Bacillus licheniformis*	Benzylpenicillin	0.049
Pyruvate kinase	2.7.1.40	Rabbit muscle	Pyruvate	10.0
Tryptophanase	4.1.99.1	*Escherichia coli*	L-Tryptophan	0.3
		Bacillus alvei	L-Tryptophan	0.27
Urease	3.5.1.5	Jack bean	Urea	10.5
Uricase	1.7.3.3.	Porcine liver	Urate	0.017
Xanthine oxidase	1.2.3.2	Milk	Xanthine	0.0036

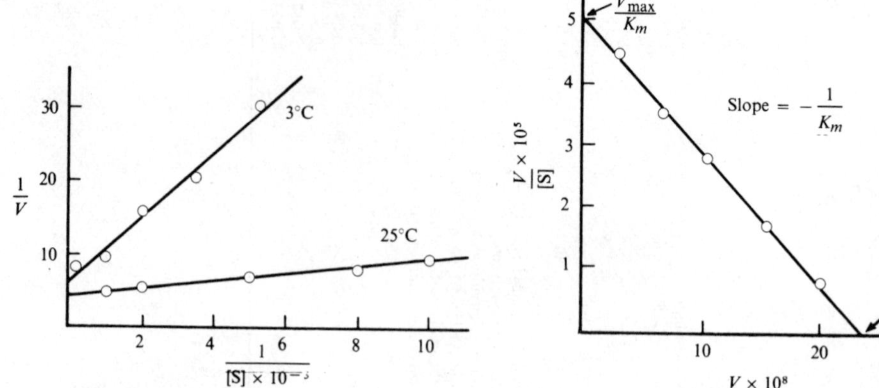

Figure 5. (*a*) Lineweaver–Burk plot of experimental data for pepsin. (*b*) Eadie–Hofstee plot of data for hydrolysis of methyl hydrocinnamate catalyzed by chymotrypsin [Bailey and Ollis (1977)].

Data given in Fig. 5 provides experimental support for the substrate dependency in the Michaelis–Menten equation, while that in Fig. 6 supports the proportionality between the rate of reaction V and enzyme concentration $[E_T]$ at fixed substrate concentration.

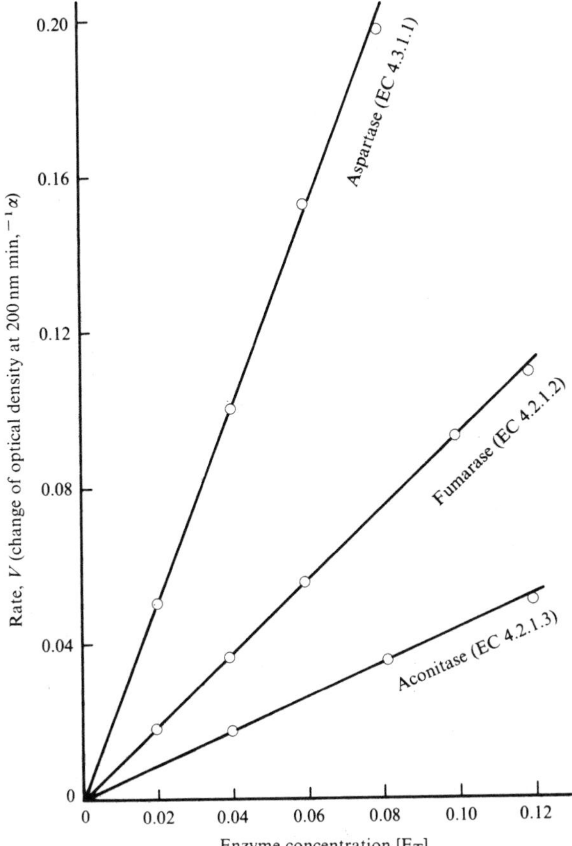

Figure 6. Influence of enzyme concentration on the rate of enzyme-catalyzed reactions [Dixon and Webb (1979)].

Table 8 shows that the kinetic parameters of a particular enzyme–substrate system may vary according to the source of the enzyme and its degree of purification. Obviously, the practical interpretation of $[E_T]$ is based on the mass of an enzymically active preparation that is usually somewhat less than 'pure'. Purification leads to a reduction in material and an increased overall rate of reaction of the preparation without any change in the intrinsic specific rate or actual mass of active enzyme present (assuming that no losses occur on purification). Thus, in most industrial applications, impure enzyme preparations are used since purification is costiy and the reduced amounts of material can actually produce handling problems because of the extra care required to deal with small quantities of highly active material. The purification requirements of enzyme preparations used for research, analytical and medical purposes are largely associated with the effects of impurities on the application of the preparation.

These various factors are illustrated by the wide range of V_{max} compared with the relatively narrow range of K_m in Table 8. The latter largely represents a 'source' effect and the former a 'purity' effect.

Table 8. Michaelis constants and maximum rates for cellulase complexes from various sources and of varying degrees of purity with respect to cellobiose[a] [Klyosov and Rabinowitch (1980)].

Source	Preparation[b]	K_m	V_{max}
Trichoderma lignorum	I	1.3	7
	II	1.3	12
	III	1.2	40
	IV	1.3	550
Geotrichum candidum	I	2.5	5
	II	2.0	45
	III	2.5	180
	IV	2.8	270
	V	2.4	550
T. koningii	I	1.0	6.5
	II	1.8	20
T. reesei	I	2.0	20
	II	1.8	70
Aspergillus niger	I	2.5	35
	II	2.7	1200
Asp. foetidus	I	2.0	75
	II	2.0	10 000
	III	1.8	9000
T. viride		2.0	23
T. longibrachiatum		1.5	10

[a] Measured at pH 4.5 and 40°C.
[b] Preparations of varying degrees of purity and from different manufacturers.

pH

All enzymes are sensitive to pH. Fig. 3 shows typical bell-shaped relationships between V_{max} and pH resulting in an optimum pH (pH_{opt}) and a maximum value of V_{max}. This phenomenon has two important consequences.

1) The need for careful selection of medium pH.
2) The need for pH control, by buffering for example, since most enzymatic reactions result in a change in pH.

A variation in pH can have the following effects on the enzyme.

1) Alteration in the conformation of the protein structure.
2) Ionization of the active site or the substrate.

One, or both, of these factors results in a variation in the formation of the enzyme–substrate complex and can lead to a change in the K_m and/or V_{max}. The pH dependency of K_m is generally more complex than that of V_{max} (see Appendix A2 to this chapter).

Values of the pH optima for various enzymes from different sources are presented in Table 9.

Table 9. pH optima of enzymes.

Enzyme	EC number	Source	pH optimum
Agarase	3.2.1.81	*Bacterium* sp.	7.0
		Vibrio liquefaciens	5.4
Aldose mutarotase	5.1.3.3	*Penicillium notatum*	5.7
Alginase	3.2.1.16	*Cloaca cloacae*	7.0
α-Amylase	3.2.1.1	*Aspergillus niger*	5.0
		Asp. oryzae	5.5–5.9
		Bacillus spp.	5.7–6.0
		Clostridium acetobutylicum	4.8
		Endomycopsis fibuliger	4.8–5.0
		Malt	5.7–6.3
		Streptococcus bovis	6.5
β-Amylase	3.2.1.2	Malt	5.0–5.5
Amyloglucosidase	3.2.1.3	*Asp. niger*	3.8
		Asp. oryzae	4.2–5.2
		Endomycopsis fibuliger	4.8–5.0
		Rhizopus delemar	4.5–5.5
Carrageenanase	3.2.1.83	*Bacterium* sp.	7.0
Cell-wall lyases		*Bacillus cereus*	5.3
		B. polymyxa	8.8
		B. subtilis	6.5
		Streptomyces L_3	7.5–8.5
Cellulase	3.2.1.4	*Myrothecium verrucaria*	5.5
		Pen. pusillum	5.3
		Trichoderma koningii	4.0–6.0
Chitinase	3.2.1.14	*Streptomyces griseus*	6.3
Chitobiase	3.2.1.30	*Strep. griseus*	6.3
Clostridiopeptidase A	3.4.24.3	*Clostridium hystolyticum*	7.6
Clostridiopeptidase B	3.4.22.8	*Clost. hystolyticum*	7.2–7.4
Collagenase	3.4.24.3	*Clost. capitovale*	6.5–7.0
		Clost. perfringens	6.0–7.5
Cyclodextrin glycosyltransferase	2.4.1.19	*B. macerans*	5.2
Dextran sucrase	2.4.1.5	*Leuconostoc mesenteroides*	5.0
		Streptococcus bovis	5.0–6.5
Dextranase	3.2.1.11	*Bacteroides* spp.	5.4
		Cellvibrio fulva	5.2–5.3
		Lactobacillus bifidus	5.4–6.5
		Penicillium sp.	4.0–6.5
		Spicaria violacea	4.0–5.5
Dextrin 6-glucosyltransferase	2.4.1.2	*Acetobacter capsulatum*	4.6
Elastase	3.4.21.11	*B. subtilis*	9.8
		B. mesentericus	9.8
		Flavobacterium elastolyticum	7.5
Endopolygalacturonase	3.2.1.—	*Asp. niger*	3.8–4.6
		Clost. felsineum	5.0–5.6
		Erwinia aroideae	8.9
Exopolygalacturonase	3.2.1.67	*Er. carotovara*	6.5
		Saccharomyces fragilis 3511	6.4
Endopolymethylgalacturonase	3.2.1.—	*Flavorbacterium pectinovorum*	7.4–8.2
Endopolymethylgalacturonase I	3.2.1.—	*Asp. foetidus*	5.3–5.8
		Asp. niger	5.5
		Clost. falsineum	6.0
Endopolymethylgalacturonase II	3.2.1.—	*Er. groideae*	8.5–9.0
Exopolygalacturonase	3.2.1.—	*Asp. foetidus*	3.5
		Asp. niger	3.6–3.8
Exopolymethylgalacturonase	3.2.1.—	*Fusarium moniliforme*	8.0–9.0
		Klebsiella aerogenes	7.5
Galactomannanase	3.2.1.—	*Ruminococcus* sp.	7.0
β-1, 2-Glucan hydrolase	3.2.1.71	*Asp. fumigatus*	3.6–4.4

continued overleaf

Table 9. *continued.*

Enzyme	EC number	Source	pH optimum
β-1, 2-Glucan hydrolase		*Pen. funiculosum*	3.6–4.4
		Pen. javanicum	3.6–4.4
		B. subtilis	6.5–6.6
		Pen. brefeldianum	3.6–4.4
Glucose isomerase	5.3.1.5	*Lactobacillus brevis*	6.0–7.0
Glucose oxidase	1.1.3.4	*Pen. amagasakiense*	5.6
		Asp. oryzae	3.4–3.5
α-Glucosidase	3.2.1.20	*Clost. acetobutylicum*	4.3–4.5
β-Glucosidase	3.2.1.21	*Stachybotrys atra*	4.0–5.0
		Escherichia coli	6.2
α-Glucosyl transferase	2.4.1.31	*Aspergillus* spp.	3.5–4.4
β-Glucosyl transferase	2.4.1.31	*Aspergillus* spp.	5.0
Hyaluronidase	4.2.2.1	*B. subtilis*	4.5–6
Inulinase	3.2.1.7	*Arthrobacter* sp.	5.8
		Azotobacter sp.	5.8
		Bacillus sp.	5.8
		Pseudomonas sp.	5.8
Invertase	3.2.1.26	*B. subtilis*	7.0
		Cunninghamella echinulata	5.0
Keratinase	3.2.1. –	*Strep. fradiae*	8.5–9.0
Laminarinase	3.2.1.6	*Asp. niger*	5.0
		Myrothecium verrucaria	4.8
		Rhizopus arrhizus	4.5
Levan dextrase	2.4.1.10	*B. subtilis*	5.2
Lipase	3.1.1.3	*Asp. malleus*	6.0
		Asp. oryzae	7.5
		Pen. chrysogenum	7.0
		Rhizopus nigricans	4.9
Mannosidostreptomycinase		*Strep. griseus*	6.5–7.4
Neuraminidase	3.2.1.18	*Clost. perfringens*	5.0
		Vibrio cholerae	4.5–5.0
Nuclease	3.1.31.1	*Pen. citrinum*	5.0
		Staphylococcus aureus	8.6–8.8
		Streptococci	7.0
Oligo-1, 6-glucosidase	3.2.1.10	*Asp. awamori*	4.8
Penicillinase	3.5.2.6	*B. cereus*	7.0
		B. subtilis	6.0–7.5
Pectin estarase	3.1.1.11	*Fusarium oxysporum*	5.0–7.0
Pectin transeliminase	4.2.2.2	*Aspergillus* sp.	5.2
		B. polymyxa	8.9–9.4
Peptidase	3.4.11.8	*B. licheniformis*	7.2–7.6
		Clost. acetobutylicum	5.5
		Clost. histolyticum	7.4
		Clost. parabotulinum	7.0
		Clost. perfringens	6.0–7.5
		Ps. fluorescens	8.0
		Ps. putrefasciens	6.5–8.0
Peptidase, λ enzyme	3.4.11.11	*Clost. perfringens*	6.0–7.2
Phospholipase B	3.1.1.5	*Vibrio El Tor*	8.0
Phospholipase C	3.1.4.3	*Clost. perfringens*	7.0–7.6
m-Polyhydroxyphenol oxidase	1.10.3.2	*Pericularia oryzae*	7.0–7.5
		Polyporus versicolor	5.0–5.5
Protease	3.4.24.4	*Asp. ochraceus*	7.4–7.6
		Asp. oryzae	6.0
		B. subtilis	10.0–11.0
		B. cereus	6.0–8.0
		B. stearothermophilus	6.9–7.2
		Clost. histolyticum	7.0
		Malt	5.0–5.5

continued

Table 9. *continued.*

Enzyme	EC number	Source	pH optimum
Protease		*Ps. myxogenes*	7.0–8.5
		Strep. griseus	7.0–8.0
		Strep. proteolyticus	6.8–7.0
		Tetrahymena pyriformis	7.0
Ribonuclease II	3.1.13.1	*B. subtilis*	7.5
Ribonuclease T_1	3.1.27.1	*Asp. oryzae*	7.5
Ribonuclease T_2	3.1.27.1	*Asp. oryzae*	4.5
Sialic acid aldolase	4.1.3.3	*Vibrio cholerae*	6.1
Staphylocoagulase		*Staphylococcus aureus*	7.4
Staphylokinase	3.4.24.4	*Staph. aureus*	7.4
Streptococcus peptidase A	3.4.6.18	*Streptococci* group A	7.5
Thioloxidase	1.8.3.2	*Pericularia oryzae*	6.0–7.0
		Polyporus versicolor	6.0–7.0
β-1, 4-Xylanase	3.2.1.32	*Asp. niger*	4.5
		Bacillus sp.	7.0–7.5
		Fusarium roseum	6.3
		Pericularia oryzae	6.4

Temperature

Fig. 7 shows the effect of temperature on the pH–activity characteristics of amylase from two bacterial sources. In Fig. 7(a), the activity increases with temperature, while Fig. 7(b) shows a temperature optimum.

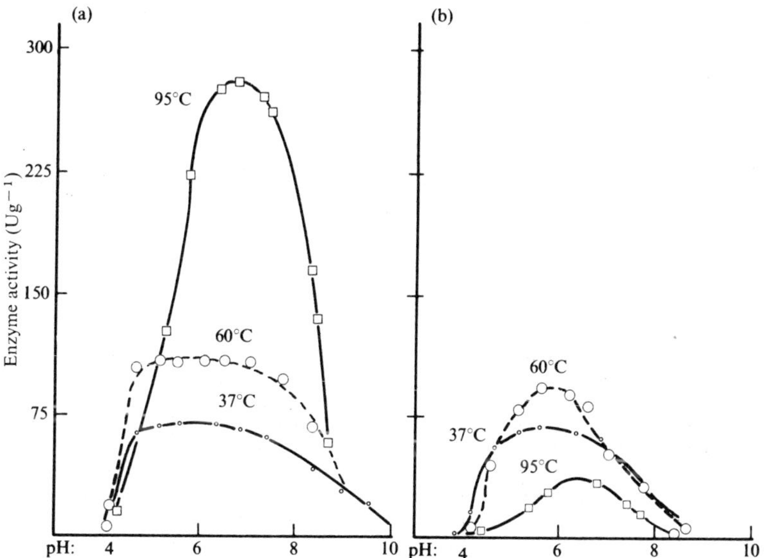

Figure 7. Effect of pH and temperature on amylase activity. (a) *Bacillus licheniformis*. (b) *B. subtilis* [Slott *et al.* (1974)].

The data presented in Fig. 7 represent the net effect of the two opposing phenomena of stimulation or inactivation.

1) An increase in activity with temperature follows laws of the Arrhenius-type, i.e.

$$V_{max} = k e^{-a/T} \tag{13}$$

or qualitative guides such as a ten-degree rise in temperature results in a doubling of the rate of reaction (i.e. Q_{10}).

2) A loss of enzyme activity may be due to enzyme inactivation (denaturation of the protein) according to relationships of the form

$$E = E_0 e^{-bt} \tag{14}$$

Since V_{max} is based upon E_0, it appears that V_{max} decreases with time at constant temperature whereas, strictly, the intrinsic activity is constant but active enzyme is lost.

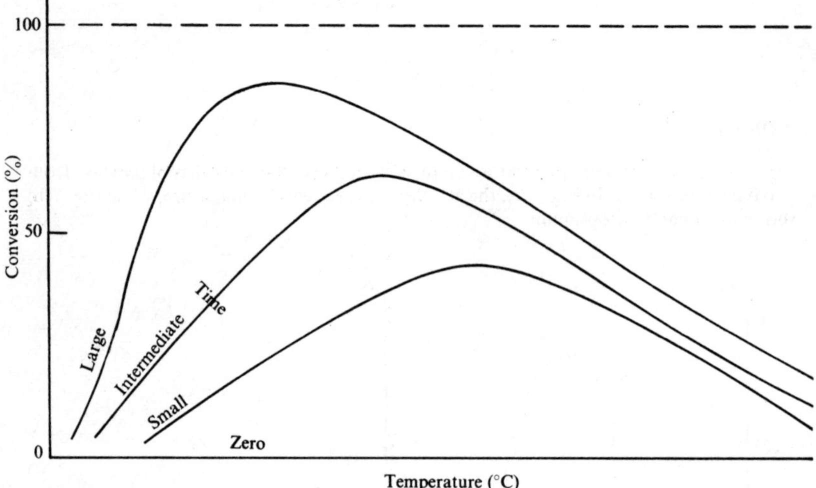

Figure 8. Representation of percentage conversion at various batch reaction times versus temperature.

Fig. 8 shows a sketch of the results of the competing effects of Eqs. (13) and (14) on the percentage conversion achieved during a batch reaction. The curve corresponding to 'large' time suggests that all the enzyme has been deactivated before complete conversion has been achieved. The increase in conversion with temperature on the left-hand side of Fig. 8 shows the effects of Eq. (13), while the decrease in conversion with temperature on the right-hand side shows the effect of rapid enzyme inactivation. The optimum temperatures in Fig. 8 result from dissimilar opposing effects and vary with conversion time.

Fig. 9 shows a set of data similar to that sketched in Fig. 8, but in which each curve has been normalized on the basis of the maximum percentage conversion achieved at the given time. This method of plotting the data emphasizes the changes in the temperature optimum with reaction time but makes the curves themselves difficult to interpret compared with the presentation used in Fig. 8.

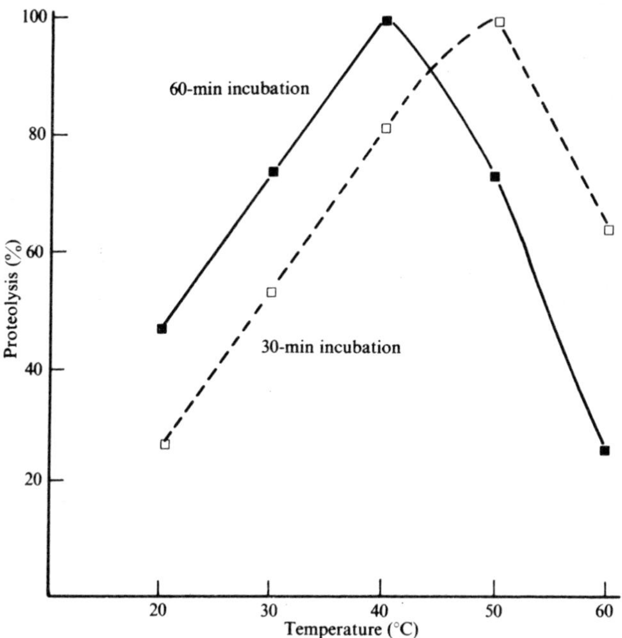

Figure 9. Temperature–activity curves for bacterial protease using gelatin as substrate at pH 7.0.

The optimum temperatures of different enzymes from the same source can vary (*see* Table 10).

Table 10. Optimum temperature of malt enzymes.

Enzyme	EC number	Optimum temperature (°C)
α-Amylase	3.2.1.1	60–70
β-Amylase	3.2.1.2	57–65
Protease	3.4.24.4	38–50

Thermophilic microorgamisms possess enzymes with relatively high optimum reaction temperatures (*see* Table 11).

Table 11. Optimum reaction temperatures of enzymes from thermophilic microorganisms [Doig (1974)].

Enzyme	EC number	Microorganism	Optimum reaction temperature (°C)	Optimum growth temperature (°C)
Aldolase	4.1.2.13	*Thermus aquaticus*	90	70
ω-Amidase	3.5.1.3	*T. aquaticus*	80	70
ATPase	3.6.1.3	*Bacillus stearothermophilus*	65	65
Enolase	4.2.1.11	*Thermus* X − 1	70	70
Malate dehydrogenase	1.1.1.37	*B. stearothermophilus*	62–65	65
Threonine deaminase	4.2.1.16	*Thermus* X − 1	70	70

Table 12. Heat stability of enzymes from mesophiles and thermophiles [Doig (1974)].

Enzyme	EC number	Temperature (°C)	Mesophile Microorganism	Mesophile $t_{\frac{1}{2}}$ (h)	Thermophile Microorganism	Thermophile $t_{\frac{1}{2}}$ (h)
α-Amylase	3.2.1.1	90	*Bacillus subtilis*	0.005	*B. stearothermophilus*	0.4
Asparaginase	3.5.1.1	55	*B. coagulans*	0.3	*B. stearothermophilus*	1.4
Isocitrate lyase	4.1.3.1	55	*Pseudomonas indigofera*	0.05	*B. stearothermophilus*	0.5
6-Phosphonogluconate dehydrogenase	1.1.1.43	45	*Penicillium rotatum*	0.06	*Pen. duponti*	0.1

The stability of enzymes from mesophilic and thermophilic microorganisms to high temperatures is compared in Table 12, where $t_{\frac{1}{2}}$ represents the 'half-life' of the enzyme, i.e. the time required for the loss of 50 per cent of the enzyme [from Eq. (14), $t_{\frac{1}{2}} = -1/b \ln 0.5$].

Coenzymes

Many enzymes require thermostable, dialyzable nonprotein organic compounds known as coenzymes. These compounds are frequently required for group transfer, which may be associated with isomerization reactions and oxidoreductions and for reactions involving the formation of covalent bonds. However, enzymes catalyzing hydrolytic reactions do not generally require coenzymes. Coenzymes may be classified according to their chemical characteristics (*see* Table 13).

A typical reaction involving a coenzyme, in this case NAD^+, with its regeneration from the reduced form, NADH, by association with another oxidoreduction reaction involving a separate substrate is shown in Fig. 10. Similar reactions with coenzyme regeneration occur for the transfer of other groups, such as those listed in Table 13. Because of the coupled system portrayed in Fig. 10, with the consequent coenzyme regeneration, overall coenzyme requirements are generally small.

Table 13. Summary of coenzymes and their functions.

Compound	Function
Compounds containing an aromatic heterocyclic ring	
Purine-containing	
ATP	Phosphate transfer, adenylate transfer, 5-adenosyl transfer, pyrophosphate transfer
GTP	Phosphate transfer
ITP	Phosphate transfer
UDP	Phosphate transfer, glycosyl transfer
NAD^+	Electron transfer
$NADP^+$	Electron transfer
FAD	Electron transfer
S-Adenosyl methionine	Methyl transfer
Coenzyme A	Acyl group transfer
Cobamides	Methyl group transfer
Nonpurine-containing compounds	
FMN	Electron transfer
Thiamin pyrophosphate	Transketolation, decarboxylation, α, β-elimination, β, γ-elimination, dealdolization, racemization, tryptophan synthesis
Folate	C_1 transfer
Compounds containing a nonaromatic heterocyclic ring	
Biotin	Carbon dioxide fixation
Glutathione	Hydrogenation
Lipoic acid	Hydrogenation
Compounds containing no heterocyclic ring	
Sugar phosphates	Glycosyl transfer, phosphate transfer
Coenzyme Q	Electron transfer

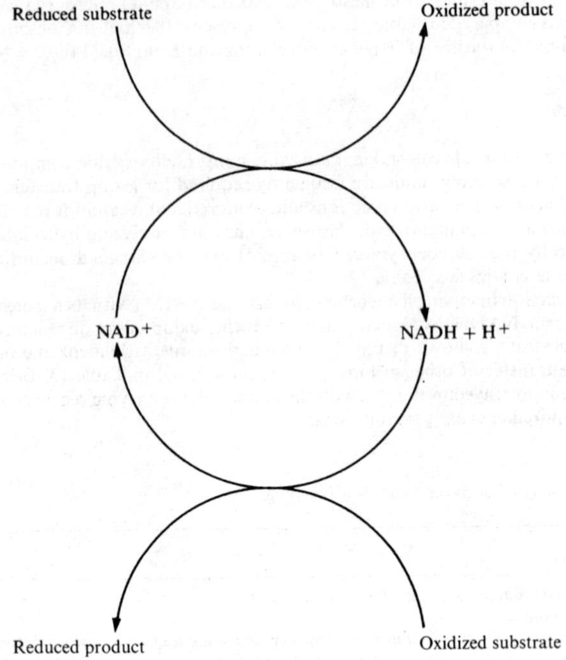

Figure 10. Principle of enzyme function using NAD^+ as an example.

Inorganic Ions

Many enzymes require, in addition to a coenzyme, a specific metal ion to activate the enzyme. A summary of inorganic ion requirements is given in Table 14.

Table 14. Some enzymes requiring metal ions.

Enzyme	EC number	Ionic requirement
Alcohol dehydrogenase	1.1.1.1	Zn^{2+}
Arginase	3.5.3.1	Mn^{2+}
ATPase (plasma membrane)	3.6.1.3	K^+, Mg^{2+}, Na^+
Carbonic anhydrase	4.2.1.1	Zn^{2+}
Carboxypeptidase	3.4.17.1	Zn^{2+}
Catalase	1.11.1.6	Fe^{2+} or Fe^{3+}
Cytochrome *c* oxidase	1.9.3.1	Cu^{2+}
Cytochromes	1.9.—.—	Fe^{2+} or Fe^{3+}
Peroxidase	1.11.1.7	Fe^{2+} or Fe^{3+}
Phosphohydrolases	3.2.1.—	Mg^{2+}
Phosphotransferases	2.7.2.—	Mg^{2+}, Mn^{2+}
Pyruvate phosphokinase	2.7.1.40	K^+, Mg^{2+}

The requirement for metal ions may vary according to the source of the enzyme (*see* Table 15).

Table 15. Inorganic ion requirements of enzymes from different sources.

Enzyme	EC number	Source	Inorganic ion requirement
α-Amylase	3.2.1.1	*Aspergillus oryzae*	Ca (Sr or Mg)
		Bacillus sp.	Ca (+Zn)
		Endomycopsis fibuliger	Ca or Mg
		Streptococci Group A	Ca
Clostridiopeptidase	3.4.24.3	*Clostridium histolyticum*	Ca
Collagenase	3.4.24.3	*Clost. capitovale*	Ca
Cycloheptaglucanase	3.2.1.54	*Asp. oryzae*	Ca or Sr
Cyclohexaglucanase	3.2.1.54	*Asp. oryzae*	Ca or Sr
Dextran branching enzyme	3.2.1.68	*Betacoccus arabinosaceous*	Mg
Dextran sucrase	2.4.1.5	*Leuconostoc mesenteroides*	Ca
Endopolymethyl-galacturonase	3.2.1.90	*Erwinia aroideae*	Ca
Keratinase		*Streptomyces fradiae*	Mg (or Ca)
Lactate racemase	5.1.2.1	*Clost. acetobutylicum*	Fe
Mannosidostrepto-mycinase		*Streptomyces griseus*	Ca + Fe
Neuraminidase	3.2.1.18	*Vibrio cholerae*	Ca or Mn
Neutral protease	3.4.24.4	*Asp. ochraceus*	Ca
		Asp. oryzae	Zn, Co or Mn
Pectin transeliminase	4.2.2.2	*Bacillus polymyxa*	Ca
Peptidase	3.4.11.8	*B. licheniformis*	Co
Peptidase	3.4.11.13	*Clost. histolyticum*	Co
Phospholipase B	3.1.1.5	*Vibrio El Tor*	Ca
Protease	3.4.24.4	*B. cereus*	Ca
		B. stearothermophilus	Ca or Mn
		B. subtilis	Zn, Co, Mn, Mg or Ca
		Clost. botilinum Type B	Fe (II)
		Clost. histolyticum	Fe (II)
		Pseudomonas myxogenes	Ca or Sr
		Strep. griseus	Ca
		Strep. proteolyticus	Zn
β-1, 4-Xylanase	3.2.1.32	*Pericularia versicolor*	Ca (or Mg)

In addition to the requirements for metal ions, the presence of such ions can enhance enzyme stability (*see* Fig. 11), although the concentrations required vary according to the source of the enzyme. The data in Fig. 11 may be related to Eq. (14).

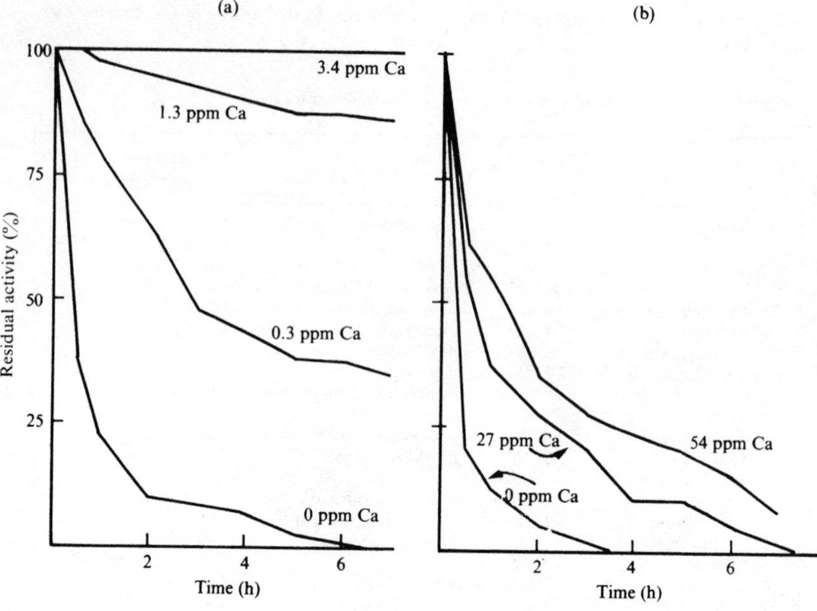

Figure 11. Effect of added calcium on amylase stability at 70°C. (a) *Bacillus licheniformis*. (b) *B. subtilis* [Slott *et al.* (1974)].

It should be pointed out that some metal ions, in particular those of heavy metals, can act as inhibitors (*see* Table 17).

Inhibition

Competitive inhibition

A typical plot of an enzyme-catalyzed reaction in the presence of a competitive inhibitor is shown in Fig. 12. The modification to the Michaelis–Menten equation is shown in Eq. (15). The corresponding Lineweaver–Burk plot is shown in Fig. 13. Inhibitors of this type are similar to the substrate in shape, size and charge distribution and are thus able to bind to the enzyme's active site, preventing the formation of the enzyme–substrate complex. Inhibition of this type is reversible and is overcome in the presence of excess substrate.

$$V = V_{\max} \frac{[S]}{K_m[1 + ([I]/K_i)] + [S]} \tag{15}$$

where [I] is the concentration of the inhibitor and K_i is the inhibition coefficient.

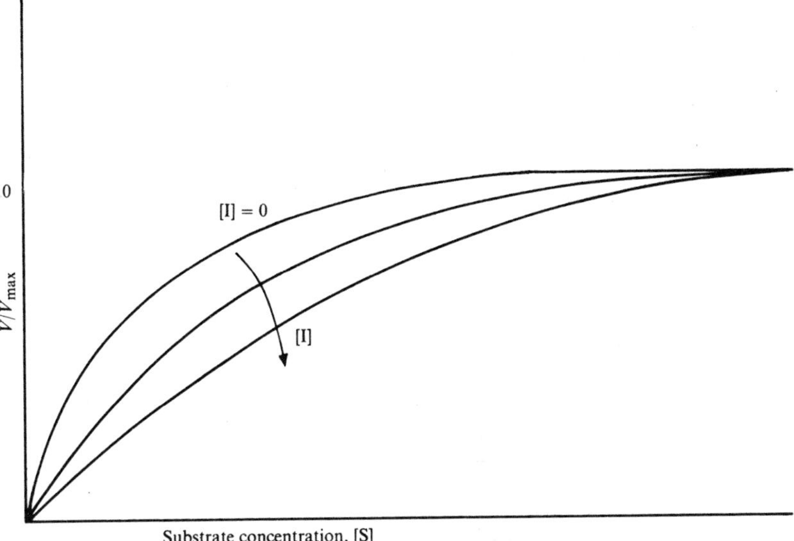

Figure 12. Michaelis–Menten plot for competitive inhibition.

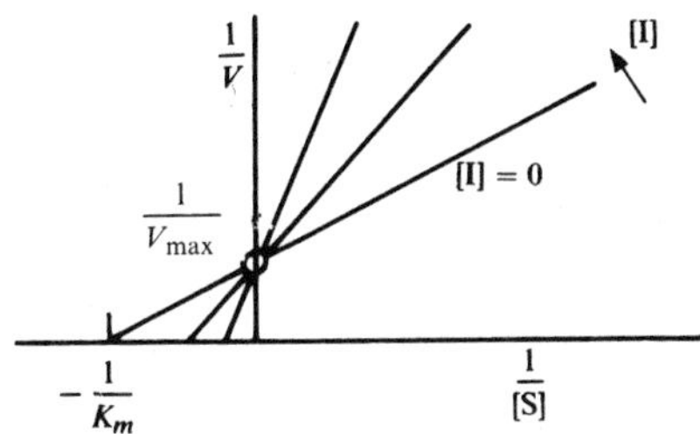

Figure 13. Lineweaver–Burk plot for competitive inhibition. Slope $= 1 + ([I])/K_i$); intercepts: ordinate $= 1/V_{max}$, abscissa $= \dfrac{-[I]/K_m}{1 + ([I]/K_i)}$

Noncompetitive inhibition

Plots for noncompetitive inhibition are shown in Figs. 14 and 15. The modification to the Michaelis–Menten equation is given in Eq. (16). Unlike competitive inhibitors, these inhibitors do not bind to the active site but, instead, at more remote position(s), with the result that the maximum specific rate of reaction is reduced. Noncompetitive inhibition is irreversible.

$$V = \left[\frac{V_{max}}{1 + ([I]/K_i)}\right]\frac{[S]}{K_m + [S]} \tag{16}$$

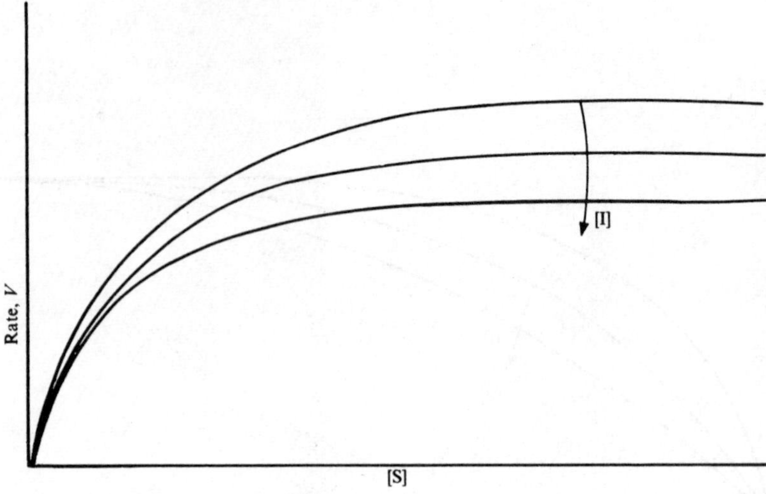

Figure 14. Michaelis–Menten plot for noncompetitive inhibition.

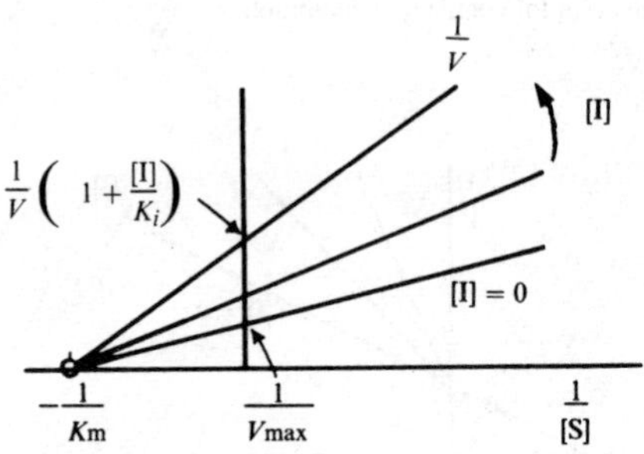

Figure 15. Lineweaver–Burk plot for noncompetitive inhibition. Slope = $[1 + ([I]/Ki)]\ K_m/V$; intercepts: ordinate = $1/V[1 + ([I]/K_i)]$, abscissa = $-1/K_m$.

Uncompetitive inhibition

This form of inhibition is the least common and occurs when the inhibitor binds with the enzyme–substrate complex ES. The modified Michaelis–Menten equation is given by Eq. (17).

$$V = \frac{V_{max}}{1 + [I]/K_i}\ \frac{[S]}{K_m/(1 + [I]/K_i) + [S]} \tag{17}$$

The corresponding graphical relationship is illustrated in Fig. 16 and the Lineweaver–Burk plot is shown in Fig. 17.

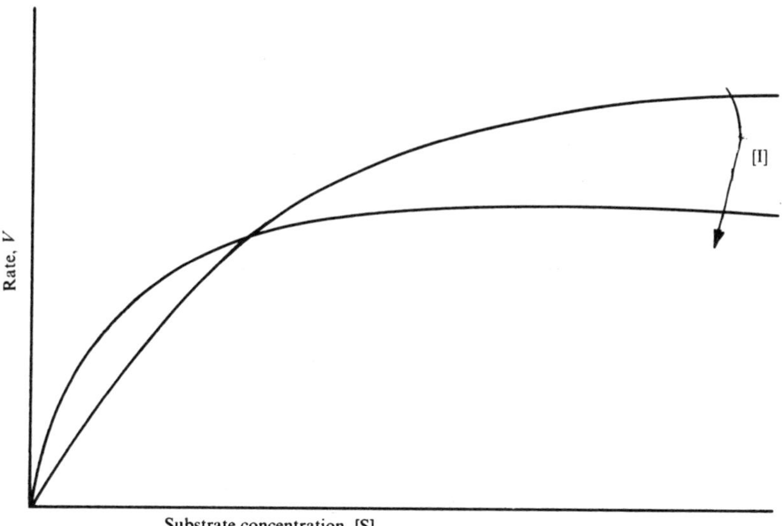

Figure 16. Michaelis–Menten plot for uncompetitive inhibition.

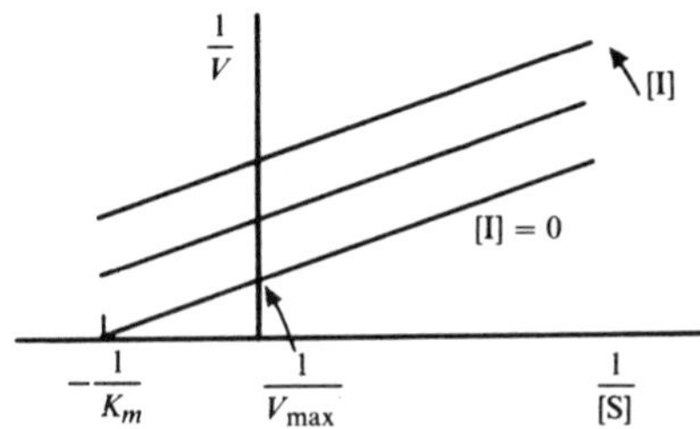

Figure 17. Lineweaver–Burk plot for uncompetitive inhibition.

Substrate inhibition

A high substrate concentration may reduce enzyme activity (*see* Fig. 18). A modification to the Michaelis–Menten equation which accounts for this phenomenon is given in Eq. (18) (*see* also pp.209–213). The corresponding Lineweaver–Burk plot is shown in Fig. 19.

$$V = V_{\max} \frac{[S]}{K_m + [S] + [S]^2/K_i} \tag{18}$$

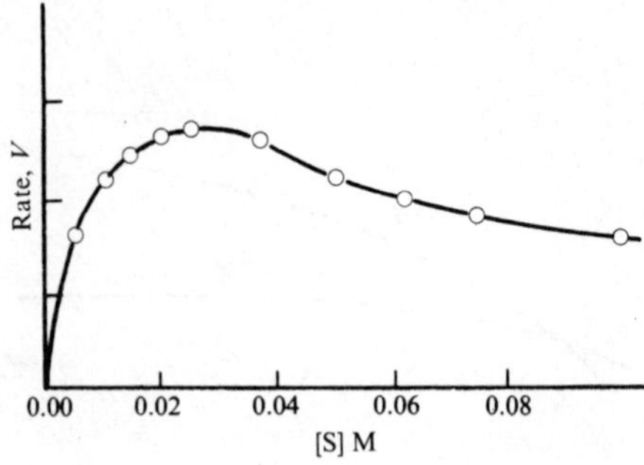

Figure 18. Substrate inhibition of carboxypeptidase [Lumry *et al.* (1951)].

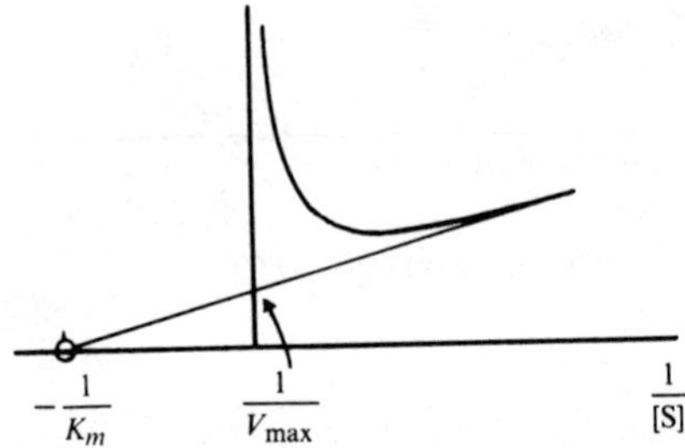

Figure 19. Lineweaver–Burk plot for substrate inhibition. Curve ——, asymptote for large 1/[S]. Slope = K_m/V_{max}; intercepts: ordinate = $1/V_{max}$, abscissa = $-1/K_m$.

End-product inhibition

In a metabolic chain, the final product may inhibit an enzyme earlier in that chain by the end-product, rather than an activator, binding to a second active site within the enzyme molecule. The binding of the activator shifts the enzyme to an active form, while binding of the inhibitor reverses this transition. A typical sigmoid curve for the fraction of active sites that have bound substrate versus substrate concentration is shown in Fig. 20, together with the effect of the activator and inhibitor. The overall system of activation and inhibition is termed allosterism.

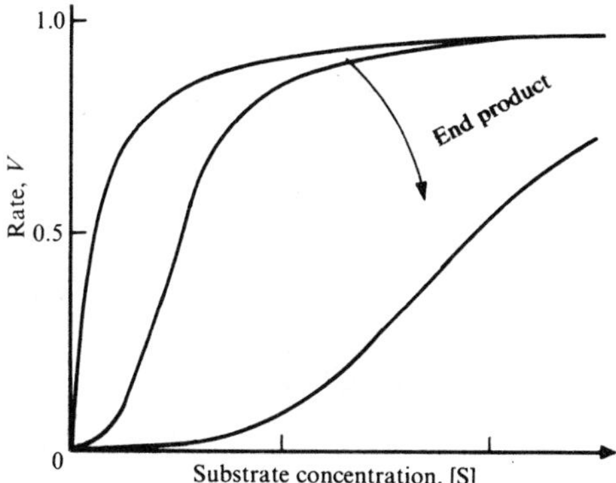

Figure 20. Michaelis–Menten plot showing end-product inhibition.

Inhibitors

Many chemical compounds can act as inhibitors and have been used to inactivate contaminating enzymes that may not easily be removed from a reacting system. A list of some important enzyme inhibitors is presented in Table 16. In addition to synthetic compounds, there are some naturally occurring compounds that act as enzyme inhibitors (*see* Table 17). Table 18 gives K_i values for a variety of industrially important enzymes.

Table 16. Some important metabolic inhibitors [Mahler and Cordes (1966)].

Compound	Reaction or enzyme inhibited	Mode of action
Avidin	All biotin enzymes	Binds to biotin
Azaserine	Amino group transfer reactions in purine synthesis	Competitive inhibitor of glutamine
6-Azauridine-5′-phosphate	Orotidylate decarboxylase i.e. pyrimidine and nucleic acid biosynthesis	Competitive inhibitor of orotidylate
Cadmium ions	Pyruvate dehydrogenase, α-ketoglutarate dehydrogenase, dihydrolipoate dehydrogenase	Formation of metal–disulphide with cysteines and lipoates
Cytosine arabinoside	Reduction of cytidylate to 2′-deoxycytidylate in DNA synthesis	Competitive inhibitor of cytidylate
Deoxypyridoxine phosphate	Reactions involving amino acids	Competitive inhibitor of pyridoxal phosphate
Diguanidines	Diamine oxidase	Competitive inhibitor of diamines
5-Fluorodeoxyuridine-5′-phosphate	Thymidylate synthetase in DNA synthesis	Competitive inhibitor of uridylate
Folic acid analogues	Reduction of folate and dihydrofolate with inhibition of methionine, thymine and purine synthesis	Competitive inhibitors of folate
Fluorocitrate	Aconitase, TCA cycle	Competitive inhibitor of citrate
Heavy metals	Many enzymes, e.g., β-fructofuranosidase	Formation of covalent metal salts with amino acids
Iodoacetate derivatives	Many enzymes, e.g., phosphoglyceraldehyde dehydrogenase	Formation of *S*-alkyl derivatives of cysteine
Long-chain fatty acid amides	Alcohol dehydrogenase	Competitive inhibitor of alcohols
Malonate	Succinate dehydrogenase, TCA cycle	Competitive inhibitor of succinate
Metal-chelating agents		
Diethyldithiocarbamate	Metalloenzymes	Removal of copper from metalloenzymes
α, α-Dipyrridyl	Metalloenzymes	Removal of iron or zinc from metalloenzymes
Ethylenediaminetetraacetate	Metalloenzymes	Formation of complex with metalloenzyme or removal of metal
8-Hydroxyquinoline	Metalloenzymes	Removal of magnesium or zinc from metalloenzymes
O-Phenanthroline	Metalloenzymes	Removal of iron or zinc from metalloenzymes
Metal-complexing agents		
Cyanides, sulphides, azides (Low concentrations)	Cytochrome oxidase, catalase, urate oxidase	Formation of complex with metalloenzymes, especially metalloporphyrins or copper enzymes

Cyanides, sulphides, azides, fluorides (High concentrations)	Metalloenzymes	Formation of complex with metalloenzymes or removal of metal from enzyme
Organic mercurials	Many enzymes	Covalently binds to cysteine
Organophosphorus compounds	Trypsin, cholinesterase, phosphoglucomutase	Covalently binds to hydroxyl group of serine
Oxamate	L-Lactate dehydrogenase	Competitive inhibitor of L-lactate
Puromycin	Protein synthesis	Competitive inhibitor of aminoacyl t-RNAs
Pyridine-3-sulphonate	Nicotinamide coenzyme biosynthesis	Competitive inhibitor of nicotinic acid
Pyrithiamine	Thiamin pyrophosphate formation	Competitive inhibitor of thiamin
Pyrophosphate	Succinate dehydrogenase, TCA cycle	Competitive inhibitor of succinate
Riboflavin monophosphate	FAD synthesis and riboflavin monophosphate-dependent reactions	Competitive inhibitor of riboflavin monophosphate
Sulphonamides	Folic acid formation	Competitive inhibitor of p-aminobenzoate
D-Threose-2, 4-diphosphate	Phosphoglyceraldehyde dehydrogenase	Noncompetitive inhibitor of D-glyceraldehyde-3-phosphate
Trivalent arsenicals	Pyruvate dehydrogenase, α-ketoglutarate dehydrogenase, dihydrolipoate dehydrogenase	Formation of As–mercaptide with cysteines or lipoate

Table 17. Some enzymes inhibited by naturally occurring inhibitors.

Enzyme	EC number	Inhibitor(s)
α-Amylase	3.2.1.1	Many alcohols, ascorbic acid, lactose, maltose, oxalate, phosphates, sucrose
Catalase	1.11.1.6	Hydrogen peroxide, urea
Cellulase	3.2.1.4	Cellobiose, glucose, quinones
Chymotrypsin	3.4.21.1	Protein inhibitor, soya bean trypsin inhibitors
β-Galactosidase	3.2.1.23	Milk protein inhibitor
Glucose isomerase	5.3.1.18	Oxygen
Glucose oxidase	1.1.3.4	D-Arabinose
Lipase	3.1.1.3	Penicillin
Papain	3.4.22.2	Wheat protein inhibitor
Pectinase	3.2.1.23	Ethanol, polyphenols
Protease	3.4.23.3	Phosphate, urea
Subtilisin	3.4.21.14	Barley protein inhibitor, broad bean protein inhibitor, potato protein inhibitor
Thermolysin	3.4.24.4	Phosphate
Trypsin	3.4.21.4	Egg white inhibitor, lima bean protein inhibitor, soya bean protein inhibitor, wheat protein inhibitor

Tabele 18. Inhibitor constants for some enzymes

Enzyme	EC number	Source	Substrate	Inhibitor	K_i (mM)
Acetylcholinesterase	3.1.1.7	*Electrophorus electricus*	Acetylcholine	Carbamate	0.00001
Alcohol dehydrogenase	1.1.1.1	Yeast	Ethanol	Acetaldehyde	0.67
β-Amylase	3.2.1.2	Sweet potato	Amylose	Cyclohexaamylose	0.2
Aspartase	4.3.1.1	*Bacillus cadavaris*	L-Aspartate	Hydroxylamine	30.0
Carboxypeptidase	3.4.2.1	Bovine pancreas	Carbobenzoxyglycyl-L-phenylalanine	Phenylacetate	0.39
			Carbobenzoxyglycyl-L-phenylalanine	D-Histidine	20.0
			Carbobenzoxyglycyl-L-tryptophan	Iodoacetate	0.078
			Carbobenzoxyglycyl-L-tryptophan	Indolepropionate	3.3
Chymotrypsin	3.4.4.5	Bovine pancreas	Benzoyl-L-phenylalanine ethyl ester	Acetyl-D-tryptophan methyl ester	0.09
			Benzoyl-L-phenylalanine ethyl ester	Indole	0.72
			Benzoyl-L-phenylalanine ethyl ester	Acetyl-L-tryptophan	17.5
Creatine kinase	2.7.3.2	Rabbit muscle	Creatine	ATP	0.2
Ficin	3.4.4.12	Figtree sap	Benzoyl-L-arginine ethyl ester	Benzoyl-L-arginine	60
Fumarase	4.2.1.2	Porcine heart	Fumarate	Adipate	100.0
			Fumarate	Malonate	40.0
			Fumarate	Malate	6.3
			Fumarate	*trans*-Aconitate	0.63
Glucose isomerase	5.3.1.5	*Lactobacillus brevis*	D-glucose	Xylitol	4.5
Glucose-6-phosphate dehydrogenase	1.1.1.49	Yeast	Glucose-6-phosphate	Glucosamine-6-phosphate	0.72
β-Glucosidase	3.2.1.21	Limpet	*O*-Nitrophenyl-β-glucoside	Gluconolactone	0.015
			O-Nitrophenyl-β-glucoside	Fucondactone	0.0039
Histidinase	4.3.1.3	*Pseudomonas fluorescens*	L-histidine	L-cysteine	0.8
Lactate dehydrogenase	1.1.1.27	Rabbit liver	Lactate	Pyruvate	0.18
Pepsin	3.4.4.1	Porcine gastric juice	Acetyl-L-phenylalanyldiodotyrosine	Methanol	608.0
Penicillinase	3.5.2.6	*Bacillus licheniformis*	Benzylpenicillin	Methicillin	0.0009
Tryptophanase	4.1.99.1	*Escherichia coli*	Tryptophan	Indole	0.05
			Tryptophan	*S*-methyl-L-cysteine	8.8

USES

Because of their characteristics (*see* pp.472–474), enzymes are, in principal, ideal catalysts for use in process-engineering applications as they can act in aqueous solutions at ambient temperatures and atmospheric pressure. The specificity of enzymes can be both an advantage and a disadvantage. The beneficial effects occur when a specific reaction is required in the presence of species other than the required substrate and when a conversion is required without the formation of byproducts. Specificity is a drawback in that each reaction requires a different enzyme and any given enzyme has little 'development' potential for application to other reactions.

Growth in the use of enzymes is partly reflected in the number of enzymes known to exist. The rapid expansion in the number of known enzymes since 1930 is illustrated in Fig. 21.

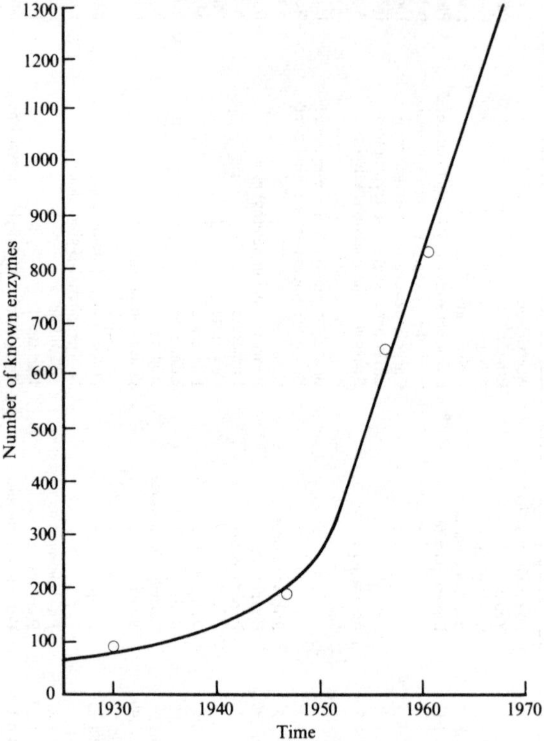

Figure 21. Increasing number of known enzymes [Bailey and Ollis (1977)].

Until recently, industrial enzyme production has been restricted to relatively simple enzymes, such as proteases, pectinases and amylases, for use in food processing either for the speeding up of the process or product improvement. However, with increasing numbers of enzymes being identified, enzymes are now being prepared on an industrial scale for use in the pharmaceutical industry, as well as for analytical purposes and in medical research. Commercial applications of enzymes are listed in Tables 19–24, according to their class (*see* Table 1).

Table 19. Commercial applications of oxidoreductases.

Enzyme	Other names (EC number)	Reaction	Uses
Alcohol dehydrogenase	Alcohol: NAD oxidoreductase (EC 1.1.1.1)	Ethanol + NAD^+ → Acetaldehyde + NADH	Assay of ethanol
L-Amino acid oxidase	L-Amino acid: oxygen oxidoreductase (EC 1.4.3.2)	L-Amino acid + H_2O + O_2 → 2-oxoacid + NH_3 + H_2O_2	Assay of amino acids
Catalase	Hydrogen peroxide: hydrogen peroxide oxidoreductase (EC 1.11.1.6)	$H_2O_2 \rightarrow O_2 + H_2O$	Removal of hydrogen peroxide used in cold sterilization of cheese and milk Removal of hydrogen peroxide used in bleaching of textiles and hair Oxygen production for sterilization, microbial growth, and porous cement, foam rubber and baked goods production
Cholesterol oxidase	(EC 1.1.3.6)		Assay of cholesterol
Cytochrome *c* oxidase	(EC 1.9.3.1)	Complex reaction in terminal oxidation	Medical use
Galactose oxidase	β-D-Galactose: oxygen oxidoreductase (EC 1.1.3.9)	D-Galactose + O_2 + H_2O → D-galactonic acid + H_2O_2	Possible use in dairy industry to remove galactose from milk products Galactose assay
Glucose oxidase	β-D-Glucose: oxygen oxidoreductase (EC 1.1.3.4)	D-Glucose + O_2 + H_2O → D-gluconic acid + H_2O_2	Glucose removal from dried egg to improve colour, flavour and shelf-life Oxygen removal from fruit juices and wines to stabilize terpenes, preventing light-catalyzed oxidation which causes browning Oxygen removal from mayonnaise to prevent rancidity Automatic glucose assay in conjunction with catalase
Glucose-6-phosphate dehydrogenase	D-Glucose-6-phosphate: NADP oxidoreductase (EC 1.1.1.49)	D-Glucose-6-phosphate + $NADP^+$ → glucono-δ-lactone-6-phosphate + NADPH	Generation of $NADP^+$
Laccase	*p*-Diphenol oxidase (EC 1.10.3.2)	2*p*-Diphenol + O_2 → *p*-quinone + $2H_2O$	Drying of lacquers

continued overleaf

Table 19 – (continued)

Enzyme	Other names (EC number)	Reaction	Uses
Lactate dehydrogenase	L-Lactate: NAD oxidoreductase (EC 1.1.1.27)	L-Lactate + NAD^+ → pyruvate + NADH	Assay of pyruvate or lactate Study of isoenzymes Diagnosis of myocardial infarction and leukaemia
Lipoxygenase	Lidoxidase (EC 1.13.11.12)	Carotenoids + linoleic acid + O_2 → peroxidized linoleic acid	Whitens bread Preparation of peroxidized oils used in flavourings
Malate dehydrogenase	L-Malate: NAD oxidoreductase (EC 1.1.1.37)	L-Malate + NAD^+ → oxaloacetate + NADH	Assay of malate Coupling enzyme in aspartate aminotransferase assay
Peroxidase	(EC 1.11.1.7)	Similar reaction to catalase	Similar uses to catalase
Tyrosinase	Polyphenol oxidase (EC 1.14.18.1)	Tyrosine + H_2O → DOPA + O_2	Antihypertensive Darkens beer
Uricase	Uric acid: oxygen oxidoreductase, urate oxidase (EC 1.7.3.3)	Uric acid + H_2O + O_2 → allantoin + H_2O	Assay of uric acid Diagnosis of gout
Xanthine oxidase	Xanthine: oxygen oxidoreductase (EC 1.2.3.2)	Xanthine + H_2O + O_2 → urate + H_2O	Assay of xanthine

Table 20. Commercial applications of transferases.

Enzyme	Other names (EC number)	Reaction catalyzed	Uses
Creatine kinase	ATP: creatine phosphotransferase (EC 2.7.3.2)	Creatine + ATP $\rightleftharpoons$ ADP + phosphocreatine	ATP generation for protein synthesis
DNA polymerase	DNA nucleotidyltransferase (EC 2.7.7.7)	$(DNA)_n$ + dNTP → $(DNA)_{n+1}$ + PP_i	DNA synthesis
Luciferase	(EC 2.8.2.10)		Spectrophotometric ATP assay
Pyruvate kinase	ATP: pyruvate phosphotransferase (EC 2.7.1.40)	ATP + pyruvate $\rightleftharpoons$ ADP + phosphoenolpyruvate	ATP generation for protein synthesis

Table 21. Commercial applications of carbohydrate-hydrolyzing enzymes.

Enzyme	Other names (EC number)	Reaction	Uses
α-Amylase	α-1, 4-Glucanhydrolase (EC 3.2.1.1)	Starch → glucose + maltose + oligosaccharides	Reduction of dough viscosity Improvement of texture and appearance of bread Mashing in alcoholic beverage production Manufacture of syrups in chocolate manufacture Preparation of baby foods Sugar recovery from scrap candy Treatment of pig starter feedstuffs Present in pharmacologically active digestive aids Desizing of fabrics Liquefying of soups and purees Cold water dispersion of laundry starch Wallpaper removal
β-Amylase	(EC 3.2.1.2)	Starch → maltose + dextrin	Maltose production after α-amylase treatment
Amyloglucosidase	Glucoamylase, α-1, 4-Glucanhydrolase (EC 3.2.1.3)	Starch → glucose + oligosaccharides	Production of glucose from corn syrup Production of beer for diabetics
Cellulase	Complex mixture of enzymes including G-cellulase	Cellulose → glucose + polysaccharides + oligosaccharides	Mushroom softening Oil extraction from vegetables Preparation of dehydrated fruits and vegetables Preparation of agar from seaweed Starch recovery from fibrous wastes Dissolution of yeast cell walls Removal of fibrous waste — in drain cleaners Present in pharmacologically active digestive aids
Dextranase	(EC 3.2.1.11)	Dextran → glucose + oligosaccharides	Lowers viscosity of syrup during sugar processing Added to toothpaste to reduce dentral plaque
Diastase	Mixture of α- and β-amylase (EC 3.2.1.1 + 3.2.1.2)	Starch → glucose or dextrin	Production of glucose from corn syrup
β-Glucuronidase	(EC 3.2.1.31)	Cleaves drug conjugates → drug or drug metabolite + glucuronic acid	Medical research — determination of drug metabolism
Hemicellulases	β-Glucanase Pentosanases	Glucans, Arabans, Xylans → Oligosaccharides + monosaccharides	Coffee mucilage liquefaction Preparation of cake frosting spreading agents Shrimp extraction Prevents staling of biscuits

Hyaluronidase	Hyaluronate glycanohydrolase (EC 3.2.1.35)		Preparation of isolated cells for tissue culture techniques Promotion of topical drug absorption Treatment of sprains and bruises
Invertase	Sucrase, β-fructofuranosidase (EC 3.2.1.26)	Sucrose → Glucose + fructose	Prevention of crystallization in jam making Preparation of soft-centred chocolates
Isoamylase	Amylopectin 6-glucanohydrolase (EC 3.2.1.68)	Glycogen → oligosaccharides + polysaccharides	Malting process Reduction of viscosity of beer or wort
Lactase	β-Galactosidase (EC 3.2.1.23)	Lactose → glucose + galactose	Formation of glucose + galactose from milk and whey Prevention of grittiness in ice cream or cheese Sweetens milk bread
Lysozyme	Muramidase, *N*-acetylmuramide glycanhydrolase (EC 3.2.1.17)	Degradation of bacterial cell walls → *N*-acetylglucuronides	Added to baby milk preparations Present in ophthalmic preparations Structural analysis of microbial cell walls
Naringinase	Hesperidinase (EC 3.2.–.–)	Naringin → prunin + rhamnose	Debitter juice and peel of grapefruit
Pectinase	Complex mixture of polygalacturonase	Pectin → galacturonic acid	Coffee bean fermentation Clarification of fruit juices Improvement of juice extraction yields Clarification of wine must Citrus oil recovery Preparation of vegetable hydrolyzates
Phosphomannanase	Complex mixture of enzymes	Dissolves yeast cell walls → mannosides	Structural analysis of yeast cell walls Research uses
Pullulanase	Amylopectin 6-glucanohydrolase (EC 3.2.1.41)	Starch → maltose + maltotriose	Treatment of wort

Table 22. Commercial applications of proteolytic enzymes.

Enzyme	Other names (EC number)	Reaction	Uses
Bacterial proteases	(EC 3.4.24.4)	Hydrolyze protein with broad specificity	Present in 'biological' detergents and presoak products Treatment of barley prior to brewing Removal of gelatin from photographic film prior to silver recovery Preparation of fish solubles Meat tenderizers Removal of hairs from hides Spot removal during dry cleaning
Bromelain	(EC 3.4.22.4)	Hydrolyzes proteins at carboxy lysine, glycine, tyrosine, alanine	Preparation of fish concentrates Liquefying of gelatin desserts
Carboxypeptidase	(EC 3.4.2.1)	Hydrolyzes C-terminal amino-acid unless this is basic or proline	Structural analysis of proteins
Chymotrypsin	(EC 3.4.21.1)	Hydrolyzes proteins at carboxyl tyrosine, phenylalanine and tryptophan	Lens removal for cataract treatment Present in pharmacologically active digestive aids Structural analysis of proteins
Collagenase	Clostridiopeptidase A (EC 3.4.24.3)	Hydrolyzes protein at glycine	Preparation of isolated cells for tissue culture
Elastase	Pancreatopeptidase E (EC 3.4.21.11)	Hydrolyzes proteins at *N*-terminal bonds of aliphatic L-amino acids	Treatment of rheumatoid arthritis Medical research — gerontology
Ficin	(EC 3.4.22.3)	Hydrolyzes proteins with broad specificity	Dissolving scrap film prior to silver recovery Meat tenderizer
Fungal proteases	(EC 3.4.24.3)	Hydrolyzes proteins with broad specificity	Softening bread dough Flavouring in sake Removal of haze in sake Meat tenderizers
Papain	(EC 3.4.22.2)	Hydrolyses proteins with bonds adjacent to basic L-amino acids, leucine or glycine	Meat tenderizer Prevention of chill haze in beer Treatment of wounds – debriding agent Shrink-proofing of wool Anthelmintic
Pepsin	(EC 3.4.23.1)	Hydrolyzes proteins at carboxyl aromatic amino acids, glutamic acid, cysteine and cystine	Digestive aid Structural analysis of proteins Preparation of precooked cereals

Rennin	Rennet (EC 3.4.23.4)	Casein → paracasein	Preparation of baby foods Improvement of cheese flavour Curdling of milk in cheese production Preparation of junket
Streptokinase	Fibrinolysin, plasminokinase (EC 3.4.21.7)	Hydrolyzes proteins at arginine and especially lysine	Anticoagulant Treatment of sprains and bruises Treatment of wounds (internal and external)
Trypsin	(EC 3.4.21.4)	Hydrolyzes proteins at carboxy lysine or arginine	Prevention of chill haze in beer Present in pharmacologically active digestive aids Structural analysis of proteins
Varidase	(EC 3.4.–.–)		Medical use

Table 23. Commercial applications of other hydrolytic enzymes.

Enzyme	Other names (EC number)	Reaction	Uses
Acetylcholine esterase	Choline esterase, acetylcholine acetyl-hydrolase (EC 3.1.1.7)	Acetylcholine + H_2O → choline + acetic acid	Cleaning of drains Neurochemical research Assay of neuroactive peptides
Adenylate deaminase	AMP deaminase (EC 3.5.4.6)	AMP → IMP + NH_3 + H_2O	IMP synthesis
Aminoacylase	(EC 3.5.1.14)	*N*-Acetylamino acids → amino acids	Resolution of L-methionine
Asparaginase	L-Asparagine amidohydrolase, L-asnase (EC 3.5.1.1)		Treatment of acute lymphatic leukaemia
Lipase	(EC 3.1.1.3)	Fats → fatty acids + glycerol	Improvement of flavour of ice cream, margarine, cheese and chocolate Improvement of whipping quality of egg white
Penicillin acylase	(EC 3.5.1.14)	Removal of side chains from penicillins (reversed with change of pH)	Oyster and shrimp extraction Production of semi-synthetic penicillins
Penicillinase	(EC 3.5.2.6)	Destruction of penicillin Penicillin + H_2O → penicin + carboxylic acid amide	Removal of penicillin from contaminated milk Treatment of penicillin allergy
Ribonuclease	(EC 3.1.27.–)	Yeast RNA → 5′-nucleotides (e.g. 5′-UMP)	Flavour enhancer production
Urease	Urea amidohydrolase (EC 3.5.1.5)	Urea + H_2O → CO_2 + $2NH_3$	Assay of urea Removal of urea from blood in renal failure

Table 24. Commercial applications of some other enzymes (classes 4–6).

Enzyme	Other names (EC number)	Reaction	Uses
Aspartase	L-Aspartate ammonia-lyase (EC 4.3.1.1)	Fumarate + $NH_3 \rightleftharpoons$ aspartate	Amino acid production
Fumarase	Fumarate hydratase (EC 4.2.1.2)	Fumarate + $H_2O \rightleftharpoons$ malate	Fumarate production for amino acid synthesis
Glucose isomerase	D-Xyloseketo isomerase (EC 5.3.1.5)	D-Glucose → D-fructose	Improving sweetness of products Fructose production from corn syrup
Histidinase	L-Histidine ammonia lyase (EC 4.3.1.3)	Histidine → urocanoate + NH_3	Production of urocanoic acid (present in sun screen)
Tryptophanase	(EC 4.1.99.1)	Indole + pyruvate + NH_3 → L-tryptophan + H_2O	Amino acid production

SOURCES

Sources of the commercially important enzymes listed in Tables 19–24 are given in Tables 25–30, respectively.

Table 25. Sources of commercially important oxidoreductases.

Enzyme	EC number	Source
Alcohol dehydrogenase	1.1.1.1	Yeast, basidomycetes
L-Amino acid oxidase	1.4.3.2	*Candida* sp., diamond rattle snake, rat kidney
Catalase	1.11.1.6	*Aspergillus niger, Penicillium vitale, Micrococcus lysodeikticus*
Cholesterol oxidase	1.1.3.6	Many microbial sources
Cytochrome *c* oxidase	1.9.3.1	Mitochondria, many microbial sources
Galactose oxidase	1.1.3.9	*Dactylium dendroides*
Glucose oxidase	1.1.3.4	*Asp. niger, Pen. amagasakiense, Pen. chrysogenum, Pen. notatum*
Laccase	1.10.3.2	*Coriolus versicolor*
Lactate dehydrogenase	1.1.1.27	Heart muscle, rat skeletal muscle, Jansen sarcoma, yeast
Lipoxygenase	1.13.11.12	Green peas
Peroxidase	1.11.1.7	Horseradish root, fig tree sap
Tyrosinase	1.14.18.1	Edible mushroom, potato, *Neurospora* sp.
Uricase	1.7.3.3	Bovine kidney, porcine liver
Xanthine oxidase	1.2.3.2	Mammalian liver, milk

Table 26. Sources of commercially important transferases.

Enzyme	EC number	Source
Creatine kinase	2.7.3.2.	Yeast
DNA polymerase	2.7.7.7	*Escherichia coli*
Luciferase	2.8.2.10	Fire fly
Pyruvate kinase	2.7.1.40	Yeast

Table 27. Sources of commercially important carbohydrate hydrolyzing enzymes.

Enzyme	EC number	Source
α-Amylase	3.2.1.1	*Aspergillus oryzae, Bacillus subtilis,* malt
β-Amylase	3.2.1.2	Malt, bacteria, fungi
Amyloglucosidase	3.2.1.3	*Asp. niger, Asp. oryzae, Endomycopsis fibuliger*
Cellulase	3.2.1.4	*Trichoderma viride, T. koningii*
Diastase	3.2.1.1. + 3.2.1.2	Malt
β-Glucosidase	3.2.1.21	Limpet, snail gut
Hemicellulase	3.2.—.—	*B. subtilis*
Hyaluronidase	3.2.1.35	Snake venom
Invertase	3.2.1.26	*Saccharomyces cerevisiae, Sacch. carlsbergensis*
Isoamylase	3.2.1.68	*Cytophaga, Pseudomonas* spp.
Lactase	3.2.1.23	*Sacch. fragilis*
Lysozyme	3.2.1.17	Egg white
Pectinase	3.2.1.15	*Asp. flavus, Asp. niger, Asp. oryzae, Coniothyrium diplodiella, Sclerotina libertina*
Phosphomannanase	3.2.1.—	Snail gut
Pullulanase	3.2.1.41	*Aerobacter aerogenes*

Table 28. Sources of commercially important proteolytic enzyme.

Enzyme	EC number	Source
Bacterial proteases	3.4.24.4	*Bacillus subtilis*
Bromelain	3.4.22.4	Pineapple juice
Chymotrypsin	3.4.21.1	Bovine pancreas
Collagenase	3.4.24.3	*Clostridium histolyticum*
Elastase	3.4.21.11	Bovine pancreas, porcine pancreas
Ficin	3.4.22.3	Fig tree sap
Fungal proteases	3.4.24.3	*Aspergillus oryzae*
Papain	3.4.22.2	Papaya latex
Pepsin	3.4.23.1	Bovine stomach, chicken stomach
Rennin	3.4.23.4	Fourth stomach of unweaned calf, *Mucor* sp.
Streptokinase	3.4.21.7	*Streptococcus* sp.
Varidase	3.4.—.—	*Streptococcus* sp.

Table 29. Sources of commercially important hydrolytic enzymes other than those acting on carbohydrates and proteins.

Enzyme	EC number	Source
Asparaginase	3.5.1.1	*Aspergillus niger, Bacillus coagulans, Escherichia coli, Penicillium camemberti, Pseudomonas fluorescens*
Lipase	3.1.1.3	*Asp. niger, Candida cylindracea, Geotrichum candidum, Humicola lanuginosa, Rhizopus arrhizus*
Penicillin acylase	3.5.1.14	*Escherichia coli*
Penicillinase	3.5.2.6	*B. subtilis*
Phosphodiesterase	3.1.4.1	*B. subtilis, Pen. citrinum, Streptomyces griseus*
Ribonucleases	3.1.27.—	*Pen. citrinum, Strep. griseus*
Urease	3.5.1.5	Jack bean

Table 30. Sources of some other commercially important enzymes (classes 4–6).

Enzyme	EC number	Source
Aspartase	4.3.1.1	*Escherichia coli, E. freundii, Pseudomonas fluorescens*
Fumarase	4.2.1.2	*Brevibacterium ammoniagenes*
Glucose isomerase	5.3.1.5	*Bacillus coagulans, Streptomyces phaeochromogenes*
Histidinase	4.3.1.3	*Achromobacter liquidum*
Tryptophanase	4.1.99.1	*Proteus rettgeri*

IMMOBILIZED ENZYMES

The use of enzymes in free solution is wasteful, although not necessarily uneconomic. To prevent loss, enzymes may be immobilized by association with insoluble materials. The balance of economic factors that have to be taken into account to establish the feasibility of immobilization are listed below.

1. Cost of enzymes
2. Extent of enzyme purification required
3. Cost of immobilization chemicals and the immobilization process
4. Enzyme stability
5. Inhibition and poisoning effects

A summary of the areas of process applicability of free and immobilized enzymes, together with some of the consequences of their use, is presented in Table 31.

Table 31. Areas of process applicability of immobilized enzymes.

Factors	Free enzymes Batch	Immobilized enzymes Batch	Immobilized enzymes Continuous
Enzyme costs			
High	Unsuitable	Suitable	Suitable
Low	Suitable	Unsuitable	Unsuitable
Enzyme reuse	Not possible	Usual	Usual
Enzyme stability			
High	Suitable	Suitable	Suitable
Low	Suitable	Unsuitable	Unsuitable
Enzyme rate			
High	Suitable	Suitable	Suitable
Low	Suitable	Suitable	Suitable
Product			
Yield	Depends on time allowed	Depends on time allowed	Depends on flow rate
Purity	Depends on raw material and enzyme purity	Depends on raw material purity	Depends on raw material purity
Equipment			
Cost	Low	Low	High
Automatic control	Possible	Possible	Possible
Automation	Difficult	Difficult	With ease
Use	Flexible	Flexible	Restricted
Operating costs			
Materials	Low	High	High
Manpower	High	High	Low
Scale of operation	Small	Small	Large

The importance of immobilized enzymes in reducing the cost of industrial production is illustrated in Fig. 22.

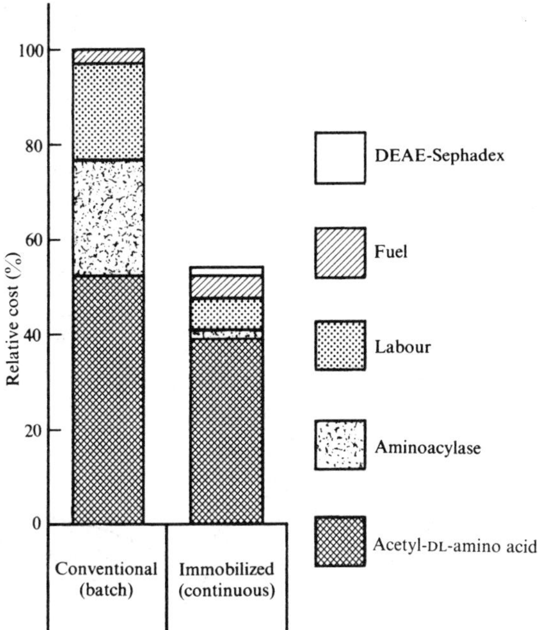

Figure 22. Comparison of relative cost for industrial production of L-amino acids [Chibata and Tosa (1976)].

Techniques used for the immobilization of enzymes may be summarized as follows.

1. Covalent attachment
2. Covalent crosslinking
3. Adsorption
4. Entrapment or encapsulation

Representation of these immobilization techniques is presented in Fig. 23.

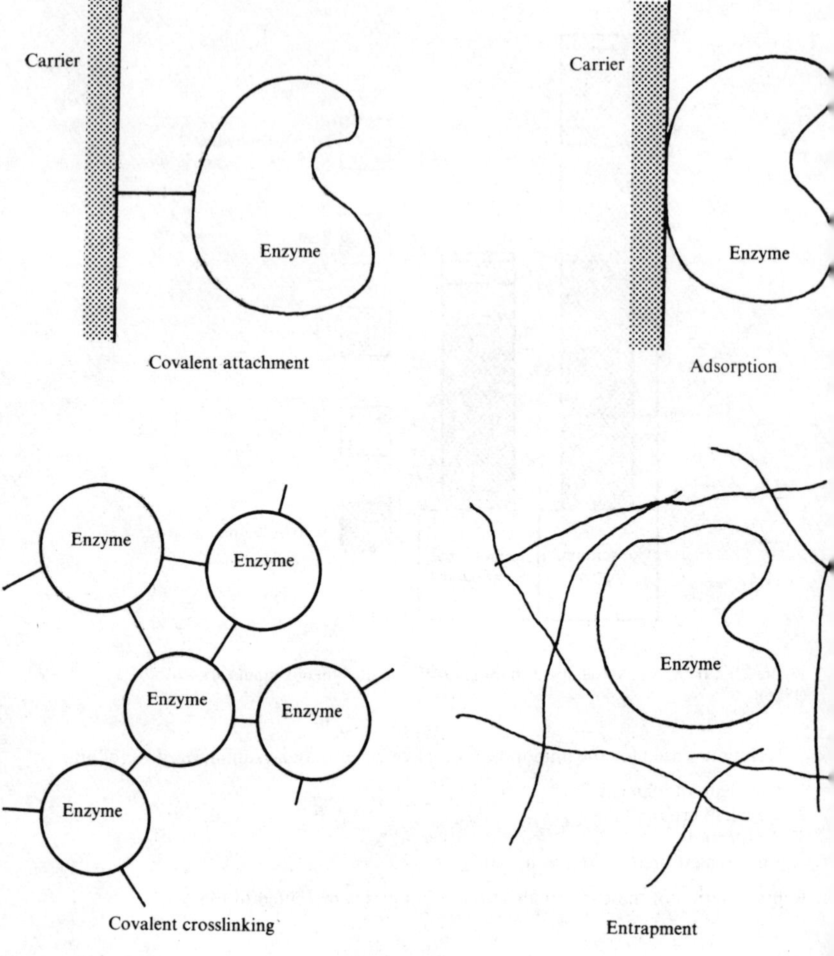

Figure 23. Diagrammatic representation of immobilization techniques.

A summary of the advantages and disadvantages of the four basic methods of immobilization is given in Table 32.

Table 32. Limitations of immobilized enzyme techniques.

Method	Advantages	Disadvantages
Covalent attachment	Not affected by pH, ionic strength of the medium or substrate concentration	Active site may be modified. Costly process
Covalent crosslinking	Enzyme strongly bound, thus unlikely to be lost	Loss of enzyme activity during preparation. Not effective for macromolecular substrates. Regeneration of carrier not possible
Adsorption	Simple with no modification of enzyme. Regeneration of carrier possible. Cheap technique	Changes in ionic strength may cause desorption. Enzyme subject to microbial or proteolytic enzyme attack
Entrapment	No chemical modification of enzyme	Diffusion effects affect transport of substrate to and product from the active site. Preparation difficult and often results in enzyme inactivation. Continuous loss of enzyme due to distribution of pore size. Not effective for macromolecular substrates. Enzyme not subject to microbial or proteolytic action.

Covalent Attachment

The range of carriers that have been used to immobilize enzymes by covalent binding is summarized in Table 33.

Table 33. Supports used for covalent attachment of enzymes.

Synthetic organic supports	Natural organic supports	Inorganic supports
Acrylamide-based polymers	Cellulose	Glass
Maleic anhydride-based polymers	Dextran	Brick
	Sephadex	
Methacrylic acid-based polymers	Starch	

The enzymes are covalently bound to the carrier in the presence of a competitive inhibitor or substrate to prevent the possibility of damage to the enzyme's activity site. Binding is generally by one of the following reactions.

1. Peptide bond formation
2. Alkylation or arylation
3. Diazo linkage
4. Isourea linkage
5. Other reactions

The parent polymer may lack the ability to react directly with the enzyme in which case covalent bonding is facilitated by initial activation of the carrier. Typical reactions are presented below.

1. Peptide bond formation

a. Via azide formation

$$\text{}\{\!-CH_2COOH \xrightarrow[HCl]{MeOH} \}\!-CH_2COOMe \xrightarrow{NH_2NH_2} \}\!-CH_2CONHNH_2 \xrightarrow[HCl]{NaNO_2} \}\!-CH_2CON_3 \xrightarrow{\text{Protein-NH}_2} \}\!-CH_2CONH\text{-Protein}$$

b. Via anhydride formation

$$\text{(cyclic anhydride: } \}\!-C(=O)-O-C(=O)\!-\{) + \text{Protein-NH}_2 \longrightarrow \}\!-C(=O)\text{NH-Protein},\ \}\!-C(=O)-OH$$

c. Using condensing agents (e.g., carbodiimides)

$$\}\!-CH_2COOH + R-N=C^+=N-R' \longrightarrow \}\!-CH_2COOC\begin{matrix}\diagup NHR \\ \diagdown NR'\end{matrix} \xrightarrow[\text{(pH 4–5)}]{\text{Protein-NH}_2}$$

$$\}\!-CH_2CONH\text{-Protein} + O=C\begin{matrix}\diagup NHR \\ \diagdown NHR'\end{matrix}$$

d. Using Woodwards reagent K (*N*-ethyl-5-phenylisoazolium-3-sulphonate)

SO_3^-

—COO⁻ + … N_+Et ⟶ —COOC … CH CONHEt

Protein-NH_2
(pH 4–5)

—CONH-Protein + … SO_3 … CH(OH) CH CONHEt

e. Using *N*-ethoxycarbonyl-2-ethoxy-1, 2-dihydroquinoline

—COO⁻ + … OEt, N, COOEt ⟶ —COOCOOEt + (quinoline) + EtOH

—CONH-Protein + CO_2 + EtOH

2. Alkylation or arylation

a. Using a 3-fluoro-4, 6-dinitrophenyl group

(F, NO_2, O_2N) + Protein-NH_2 ⟶ (NH-Protein, NO_2, O_2N)

b. Using *s*-triazine derivatives

—OH + Cl(triazine: N, Cl, N, N, X) ⟶ —O—(triazine: N, Cl, N, N, X) —Protein-NH_2⟶ —(triazine: N, NH-Protein, N, N)

where X = Cl, NH_2 or chromophore

3. Diazotization

$$\text{—}CH_2\text{—}C_6H_4\text{—}NH_2 \xrightarrow[HCl]{NaNO_2} \text{—}CH_2\text{—}C_6H_4\text{—}N_2^+\,Cl^-$$

$$\downarrow \text{Protein—}C_6H_4\text{—OH}$$

$$\text{—}CH_2\text{—}C_6H_4\text{—}N=N\text{—}C_6H_3(HO)\text{—Protein}$$

Similar reactions occur using the amino group of arginine or lysine, rather than the tyrosine hydroxyl group

4. Isourea linkage

a. Isothiocyanate reaction

$$\text{—}CH_2NH_2 + Cl\text{—}\overset{\overset{S}{\|}}{C}\text{—}Cl \longrightarrow \text{—}CH_2N=C=S \xrightarrow[\text{(pH 8–9)}]{\text{Protein-NH}_2} \text{—}CH_2\text{—NH}\overset{\overset{S}{\|}}{C}\text{—NH-Protein}$$

b. Cyanuric chloride

$$\text{—}CH_2\text{—}C_6H_4\text{—}NH_2 + Cl\text{—}\overset{\overset{O}{\|}}{C}\text{—}Cl \longrightarrow \text{—}CH_2\text{—}C_6H_4\text{—NCO}$$

$$\downarrow \text{Protein-NH}_2\ \text{(pH 8–9)}$$

$$\text{—}CH_2\text{—}C_6H_4\text{—NHCONH-Protein}$$

c. Cyanogen bromide reaction

$$\begin{array}{c}|\\ \text{CHOH}\\ |\\ \text{CHOH}\\ |\end{array} + \text{BrCN} \longrightarrow \begin{array}{c}|\\ \text{CH O—C}\equiv\text{N}\\ |\\ \text{CHOH}\\ |\end{array} \longrightarrow \begin{array}{c}|\\ \text{CH—O}\\ |\\ \text{CH—O}\end{array}\!\!>\text{C}=\text{NH}$$

$$\downarrow \text{Protein-NH}_2\ \text{(pH 8–9)}$$

$$\begin{array}{c}|\\ \text{CH—OCONH-Protein}\\ |\\ \text{CHOH}\\ |\end{array}$$

5. Other reactions

a. Titanium tetrachloride

$$\{\!-CH_2OH + TiCl_4 \longrightarrow \{\!-CH_2OTiCl_3 + HCl \xrightarrow{\text{Protein-NH}_2} \{\!-CH_2OTiCl_2-O-\text{Protein} + HCl$$

b. Ugi reaction

$$\begin{matrix} R \\ R' \end{matrix}\!\!>C=O + NH_2\text{-Protein} \xrightarrow{H^+} \begin{matrix} R \\ R' \end{matrix}\!\!>C=NHR'' + H_2O \xrightarrow[R''''N\equiv C]{R'''COO^{-+}}$$

$$R''''\text{-}N=C\begin{cases} C(R)(R')\text{-NH-Protein} \\ OC(=O)R''' \end{cases}$$

$$\downarrow$$

$$R''''-NHC(=O)-C(R)(R'')-N(\text{Protein})(C(=O)R''')$$

In addition to organic polymeric carriers, some enzymes are attached to inorganic supports

$$\begin{matrix} & | & \\ & O & \\ & | & \\ -O- & Si & -O- \\ & | & \\ & O & \\ & | & \\ -O- & Si & -O- \\ & | & \\ & O & \\ & | & \end{matrix} + H_2N(CH_2)_3Si(OEt)_3 \longrightarrow \begin{matrix} & | & & | & \\ & O & & O & \\ & | & & | & \\ -O- & Si & -O- & Si & -(CH_2)_3NH_2 \\ & | & & | & \\ & O & & O & \\ & | & & | & \\ -O- & Si & -O- & Si & -(CH_2)_3NH_2 \\ & | & & | & \\ & O & & O & \\ & | & & | & \end{matrix}$$

The amino groups generated are then coupled to the carboxyl groups of proteins.

Several amino acid residues possess groups capable of reacting covalently with immobilizing agents (*see* Table 38)

Table 34. Covalently attached enzymes [Zaborsky (1973)].

Enzyme	EC number	Carrier Polymer	Activation process
Acetyl cholinesterase	3.1.1.7	Glass, amino-alkyl derivative	Carbodiimide or nitrous acid
		Sepharose	Cyanogen bromide
Acid phosphatase	3.1.3.2	CM-cellulose	*N, N'*-Dicyclohexylcarbodiimide
		Polyacrylamide	Glutaraldehyde
		Polymethacrylic acid anhydride	None
Alcohol dehydrogenase	1.1.1.1	Methacrylic acid/methacrylic acid *m*-fluoroanilide	Nitric acid
		Polymethacrylic acid anhydride	None
Aldolase	4.1.2.13	Cellulose, aminoethyl ether	Glutaraldehyde
		Maleic anhydride/ethylene	None
		Sepharose	Cyanogen bromide
Alkaline phosphatase	3.1.3.1	Brick	Sulphuryl or thionyl chloride
		CM-cellulose	*N, N'*-Dicyclohexylcarbodiimide
		Glass, aminoaryl derivative	Nitrous acid
		Maleic anhydride/ethylene or maleic anhydride/methyl vinyl ether	None
		Methacrylic acid/methacrylic acid *m*-fluoroanilide	Nitric acid
L-Amino acid oxidase	1.4.3.2	Glass, aminoaryl derivative	Nitrous acid
Aminoacylase	3.5.1.14	Acrylamide/methyl acrylate or poly (methylacrylate)	Hydrazine + nitrous acid
		Cellulose, *p*-aminobenzyl ether or CM-cellulose hydrazide	Nitrous acid
		Cellulose, aminoethyl ether or CM-cellulose	*N, N'*-Dicyclohexylcarbodiimide
		Cellulose, haloacetyl esters	None
		Sephadex	Cyanogen bromide
α-Amylase	3.2.1.1	Cellulose	Transition metal salts
		Cellulose, *p*-aminobenzyl or 3-(*p*-aminophenoxy)-2-hydroxypropyl ether, CM-cellulose hydrazide, Enzacryl AA or AH, polyaminostyrene	Nitrous acid
		Cellulose, 3-(*p*-aminophenoxy)-2-hydroxypropyl ether, or Enzacryl AA	Thiophosgene
		Methacrylic acid/methacrylic acid *m*-fluoroanilide or methacrylic acid/3,5-dinitro-4-fluorostyrene	None
β-Amylase	3.2.1.2	Bentonite	Cyanuric chloride
		Cellulose, 3-(*p*-aminophenoxy)-2-hydroxypropyl ether, or Enzacryl AA	Nitrous acid
		Cellulose, 3(-*p*-aminophenoxy)-2-hydroxypropyl ether, or	Thiophosgene

Amyloglucosidase	3.2.1.3	Cellulose, glass or Nylon	Transition metal salts
		Cellulose, 3-(*p*-aminophenoxy)-2-hydroxypropyl ether, or CM-cellulose	Nitrous acid
		DEAE-Cellulose	2-Amino-4,6-dichloro-*s*-triazine
Apyrase	3.6.1.5	L-Ala/L-Glu copolymer, poly-Asp, CM-cellulose, polygalacturonic acid or poly (methyl methacrylate)	Woodward reagent K
		Cellulose	Cyanuric chloride
		CM-cellulose hydrazide or maleic anhydride/ethylene	None
		CM-Cellulose hydrazide or poly aminostyrene	Nitrous acid
L-Asparaginase	3.5.1.1	Cellulose	Cyanogen bromide
		CM-Cellulose	Carbodiimide
		CM-Cellulose	Woodward reagent K
		Dacron, aminoaryl derivative or CM-dextran hydrazide	Nitrous acid
ATPase	3.6.1.3	Cellulose	Cyanogen bromide
		CM-Cellulose or CM-cellulose hydrazide	None
		CM-Cellulose hydrazide	Nitrous acid
Bromelain	3.4.22.4	CM-Cellulose hydrazide	Nitrous acid
		Polymethacrylic acid anhydride	None
Carboxypeptidase A	3.4.17.1	Polyaminostyrene	Nitrous acid
Carboxypeptidase B	3.4.17.2	*p*-Amino-DL-Phe/L-Leu	Nitrous acid
		Maleic anhydride/ethylene	None
		Sepharose	Cyanogen bromide
Catalase	1.11.1.6	Cellulose	Cyanuric chloride
		Cellulose, *p*-aminobenzyl ether	Nitrous acid
		Cheesecloth	Sodium periodate
		Polyaminostyrene	Phosgene
		Poliodal-4	None
Cellulase	3.2.1.4	Bentonite	Cyanuric chloride
		Methacrylic acid/methacrylic acid *m*-fluoroanilide	Nitric acid
Cholinesterase	3.1.1.8	L-Ala/L-Glu, CM-cellulose, poly-Asp, polygalacturonic acid	Woodward reagent K
		DEAE-Cellulose	Procion brilliant orange
		Sepharose	Cyanogen bromide
Chymotrypsin	3.4.21.1	CM-Agarose, Enzacryl AA, keratin, polyacrylamide, aminoethyl derivative, CM-Sephadex, sepharose, wool	Ugi reaction
		Cellulose, Sephadex, sepharose,	Cyanogen bromide

continued overleaf

Table 34 – (continued)

Enzyme	EC number	Carrier Polymer	Activation process
		Sepharose	Cyanogen iodide
		Cellulose	Cyanuric chloride
		Glass, aminoalkyl derivative, polyacrylamide	Glutaraldehyde
		L-Ala/L-Glu, CM-Cellulose, polyacrylamide, poly-L-Glu	Woodward reagent K
		Bio-Gel P-150-carboxyl, CM-cellulose, CM-Sephadex	EDAPC
		Cellulose, DEAE-cellulose, Sephadex, sepharose	2, Amino-4,6-dichloro-*s*-triazine
		Cellulose	2,4-Dichloro-6-carboxymethylamino-*s*-triazine
		Cellulose, *p*-aminobenzoyl ester, *p*-aminobenzyl ether, 3-aminobenzyloxy methyl ether or 3-amino-4-methoxyphenylsulphonyl ethyl ether, CM-cellulose, CM-cellulose hydrazide	Nitrous acid
		CM-Cellulose	*N,N'*-Dihyclohexylcarbodiimide
		Maleic anhydride/ethylene, polymethacrylic acid anhydride	None
		Poliodal-2, Poliodal-4, Copoliodal-2 or Copoliodal-4	
		Sephadex, 3-*p*-aminophenoxy-2-hydroxypropyl ether	Thiophosgene
δ-Chymotrypsin		Poliodal-4	None
Chymotrypsinogen		Poliodal-4	None
Creatine kinase	2.7.3.2	Cellulose, *p*-aminobenzyl ether, CM-cellulose	Nitrous acid
		DEAE-Cellulose	2,4-Dichloro-6-carboxymethylamino-*s*-triazine
Deoxyribonuclease	3.1.21.1	L-Ala/L-Glu, CM-cellulose, polygalacturonic acid	Woodward reagent K
		CM-Cellulose	*N,N'*-Dicyclohexylcarbodiimide
		CM-Cellulose hydrazide or glass, aminoaryl derivative	Nitrous acid
		Maleic anhydride/ethylene	None
Dextranase	3.2.1.11	Enzacryl Polyacetal	Acid
Ficin	3.4.22.3	CM-Cellulose	*N,N'*-Dicyclohexyldicarbodiimide
		CM-Cellulose hydrazide or glass, aminoaryl derivative	Nitrous acid
Fructose-1,6-diphosphatase	3.1.3.11	Cellulose, amino ethyl ether	Glutaraldehyde
β-Galactosidase	3.2.1.23	Cellulose	Cyanuric chloride
Glucose isomerase	5.3.1.5	Cellulose	Transition metal salts
Glucose oxidase	1.1.3.4	Alumina, glass, aminoaryl derivative, hydroxyapatite, nickel oxide, aminoalkyl derivative	Thiophosgene
		Cellulose	Transition metal salts
		CM-Cellulose	*N,N'*-Dicyclohexylcarbodiimide
		CM-Cellulose	Nitrous acid
		Polyacrylamide	Glutaraldehyde

Glucose-6-phosphate dehydrogenase	1.1.1.49	Acrylamide/acrylic acid	Carbodiimide
β-Glucosidase	3.2.1.21	Cellulose	Ethyl chloroformatetriethylamine
β-Glucuronidase	3.2.1.31	CM-Cellulose	*N,N′*-Dicyclohexylcarbodiimide
Glutamate dehydrogenase	1.4.1.3	CM-Cellulose, CM-Sephadex	EDAPC
		Collagen	Hydrazine + nitrous acid
Glyceraldehyde phosphate dehydrogenase	1.2.1.12	Cellulose, aminoethyl ether	Glutaraldehyde
Hexokinase	2.7.1.1	Cellulose	Cyanuric chloride
		Sephadex	Cyanogen bromide
Hyaluronidase	3.2.1.35	Agarose	2-Amino-4,6-dichloro-*s*-triazine
Invertase	3.2.1.26	Bentonite, brick or glass	Cyanuric chloride, sulphuryl chloride or thionyl chloride
		Cellulose	Transition metal salts
		Methylacrylic acid/methacrylic acid *m*-fluoroanilide	Nitric acid
Isoleucyl t-RNA synthetase	6.1.1.5	Sepharose	Cyanogen bromide
Kallikrein	3.4.21.8	Maleic anhydride/ethylene	None
		Sepharose	Cyanogen bromide
Lactate dehydrogenase	1.1.1.27	Brick	Sulphuryl chloride or thionyl chloride
		DEAE-Cellulose	Cyanuric chloride or Procion brilliant orange
		Polymethacrylic acid anhydride	None
Leucine aminopeptidase	3.4.11.1	Sephadex	Cyanogen bromide
Lipase	3.1.1.3	Polyaminostyrene	Phosgene
Luciferase	2.8.2.10	Polyacrylic acid	Methanol + hydrazine + nitrous acid
Naringinase	3.2—.—	Maleic anhydride/ethylene, maleic anhydride isobutylvinyl ether, or maleic anhydride/styrene	None
Papain	3.4.22.2	*p*-Amino-DL-Phe/L-Leu, cellulose, *p*-aminobenzyl ether, CM-cellulose hydrazide, glass, aminoaryl derivative, or S-MDA	Nitrous acid
		Methacrylic acid/methacrylic acid *m*-fluoroanilide, methacrylic acid/2-,3-, or 4-fluorostyrene	Nitric acid
		m-Aminostyrene/methacrylic acid, polyvinylamine, Lewatit, CA 9119	Thiophosgene
		Polymethacrylic acid anhydride, *m*-isothiocyanatostyrene/methacrylic acid or *m*-isothiocyanatostyrene/acrylic acid	None

continued overleaf

Table 34 – (continued)

Enzyme	EC number	Carrier Polymer	Activation process
		Collagen	Bis diazobenzidine-2,2′- or 3,3′-dicarboxylic acid
		Enzacryl Polyacetal	Acid
		Sepharose	Cyanogen bromide
		Polystyrene	Sulphonation
Penicillin amidase	3.5.1.11	Cellulose	Cyanuric chloride
		CM-Cellulose hydrazide	Nitrous acid
		DEAE-Cellulose	2,4-Dichloro-6-carboxyamino-*s*-triazine
Pepsin	3.4.23.1	Glass, amino alkyl derivative	Carbodiimide
		Methacrylic acid/methacrylic acid *m*-fluoroanilide	Nitric acid
		Polyaminostyrene	Nitrous acid
		Sepharose	Ugi reaction
Peroxidase	1.11.1.7	Cellulose, amino ethyl ether or CM-cellulose	DCC
		CM-Cellulose, benzidine derivative	Nitrous acid
Plasminogen	3.4.21.7	*p*-Amino-DL-Phe/L-Leu	Nitrous acid
Polynucleotide phosphorylase	2.7.7.8	Cellulose	Cyanogen bromide
Prolinase	3.4.13.9	Sepharose	Cyanogen bromide
Pronase	3.4.4.— + 3.4.1.—	Glass, aminoalkyl derivative	Glutaraldehyde
		Cellulose, bromoacetyl ester, or polymethacrylic acid anhydride	None
		p-Amino-DL-Phe/L-Leu or glass, aminoaryl derivative	Nitrous acid
		CM-Sephadex	EDAPC
Prothrombin		*p*-Amino-DL-Phe/L-Leu	Nitrous acid
Pyruvate decarboxylase	4.1.1.1	Polyaminomethylstyrene	CDAPC
Pyruvate kinase	2.7.1.40	Cellulose	Cyanuric chloride
Renin	3.4.99.19	Sepharose	Cyanogen bromide
Rennin	3.4.23.4	Cellulose, aminoethyl ether	Glutaraldehyde
		Sepharose	Cyanogen bromide
Ribonuclease A	3.1.4.22	Cellulose, aminobenzoyl ester or *p*-aminobenzyl ether, CM-cellulose hydrazide, polyaminostyrene	Nitrous acid
		Bentonite	Cyanuric chloride
		CM-Cellulose	*N*,*N*′-dicyclohexylcarbodiimide
		Polyacrylamide	Glutaraldehyde
		Sepharose	Cyanogen bromide
Ribonuclease T	3.1.4.8	Cellulose, *p*-aminobenzyl ether or CM-cellulose	Nitrous acid
		[illegible]	Cyanogen bromide

Staphylococcal nuclease	3.1.31.1	Sepharose	Cyanogen bromide
Steroid esterase	3.1.1.—	Glass, aminoaryl derivative	Nitrous acid
Sterol sulphatase	3.1.6.2	Glass, aminoaryl derivative	Nitrous acid
Streptokinase	3.4.—.—	*p*-Amino-DL-Phe/L–Leu, cellulose, *p*-aminobenzyl ether	Nitrous acid
Subtilopeptidase A	3.4.21.14	Cellulose, *p*-aminobenzyl ether or S-MDA	Nitrous acid
		CM-Cellulose	*N,N′*-dicyclohexylcarbodiimide
Threonine deaminase	4.2.1.16	Brick	Suphuryl or thionyl chloride
Thrombin	3.4.21.5	*p*-Amino-DL-Phe/L/Leu or cellulose, *m*-aminobenzyloxymethyl ether	Nitrous acid
		Cellulose, bromoacetyl ester or maleic anhydride/ethylene	None
Trypsin	3.4.21.4	*p*-Amino-DL-Phe/L-Leu, cellulose, *p*-aminobenzoyl ester, *p*-aminobenzyl ether, *m*-aminobenzyloxymethyl ester, or 3-amino-4-methoxyphenylsulphonylethyl ether, CM-cellulose hydrazide, glass, aminoaryl derivative, Nylon	Nitrous acid
		Polyacrylamide, acrylamide/methyl acrylate	Hydrazine + nitrous acid
		Acrylamide/acrylic acid, cellulose, *p*-aminobenzyl ether or aminoethyl ether	Glutaraldehyde
		Acrylamide/acrylic acid	Carbodiimide
		Acrylamide/hydroxyethyl methacrylate, cellulose, Sephadex, sepharose	Cyanogen bromide
		Bentonite, cellulose	Cyanuric chloride
		L-Ala/L-Glu	Woodward reagent K
		Cellulose	Transition metal salts
		CM-Cellulose	*N,N′*-dicyclohexylcarbodiimide
		Enzacryl Polyacetal	Acid
		Glass, aminoalkyl derivative, Sephadex, 3-*p*-aminophenoxy-2-hydroxypropyl ether	Thiophosgene
		Maleic anhydride/ethylene, polymethacrylic acid anhydride, Poliodal-4, Copoliodal-4	None
Trypsinogen		Maleic anhydride/ethylene	None
		Sepharose	Cyanogen bromide
Tyrosinase	1.10.3.1	DEAE-Cellulose	2-Amino-4,6-dichloro-*s*-triazine
Urease	3.5.1.5	*p*-Amino-DL-Phe/L-Ala, *p*-amino-DL-Phe/Gly, *p*-amino-DL-Phe/L-Leu, glass, aminoaryl derivative	Nitrous acid
		Bentonite	Cyanuric chloride
		Enzacryl Polyacetal	Acid
		Nylon	Glutaraldehyde
		Poliodal	None

Table 35. Capacities of some polymeric supports for enzymes immobilized by covalent attachment [Goldstein and Manacke (1976)].

Polymer	Capacity [mg protein (g conjugate)$^{-1}$]
Acrylamide/acrylic acid	114
Acrylamide/methacrylic acid	200–500
p-Amino-DL-Phe/L-Leu	100–300
Bentonite	3–15
Cellulose	60–300
Cellulose, 3-(*p*-aminophenoxy) 2-hydroxypropyl ether	10–40
CM-Cellulose	40–500
CM-Cellulose hydrazide	50–400
DEAE-Cellulose	130–220
Enzacryl AA	10–30
Enzacryl AA + thiophosgene treatment	20–30
Enzacryl AH	2
Glass, aminoalkyl derivative	12–16
Glass, aminoaryl derivative	10–20
L-Glu/L-Ala	330
Maleic anhydride/acrylamide	0.5–25
Maleic anhydride/ethylene	100–800
Polyacrylamide	30–100
Polyacrylic acid	450
Polygalacturonic acid	600–800
Sephadex	212
Sepharose	70–390

Covalent Crosslinking

The use of low molecular weight multifunctional agents to link enzymes results in a considerable loss of enzyme activity, with the formation of gelatinous products. To avoid loss in activity, the enzyme is frequently adsorbed onto a suitable carrier before the polymerization. Structures of typical crosslinking agents are shown in Fig. 24. In addition to diazobenzidine, derivatives with methoxy, carboxyl or sulphonyl groups at the 2,2′- and 3,3′-positions are used. A summary of enzymes immobilized using this approach is given in Table 36.

N_2^+ N_2^+

Diazobenzidine

$S = C = N$ $N = C = S$

SO_3H SO_3H

4,4′-Diisothiocyanatobiphenyl-2, 2′ disulphonic acid

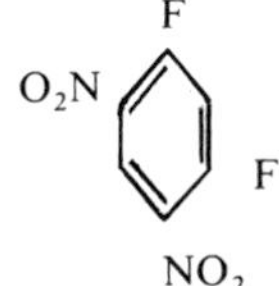

1, 5-Difluoro-2, 4-dinitrobenzene

$$\begin{matrix} O & & O \\ \| & & \| \\ C & (CH_2)_3 & C \\ / & & \backslash \\ H & & H \end{matrix}$$

Glutaraldehyde

$I\ CH_2\ CONH\ (CH_2)_6\ NHCOCH_2I$

N, *N*′-Hexamethylene bisiodoacetamide

$O = C = N(CH_2)_6N = C = O$

Hexamethylenediisocyanate

Figure 24. Multifunctional crosslinking agents.

Table 36. Enzymes immobilized by crosslinking [Zaborsky (1973)].

Enzyme	EC number	Crosslinking agent	Remarks
Alcohol dehydrogenase	1.1.1.1	Glutaraldehyde	Initially adsorbed on cellophane membrane
Aldolase	4.1.2.13	*N,N'*-Hexamethylenebisiodoacetate	
Alkaline phosphatase	3.1.3.1	Glutaraldehyde	Initially adsorbed on collodion membrane
α-Amylase	3.2.1.1	Glutaraldehyde	Initially adsorbed on cellophane membrane
Apyrase	3.6.1.5	Glutaraldehyde	
		Diazobenzidine (or derivative)	Initially adsorbed on filter paper, collodion membrane or Millipore filter
Asparaginase	3.5.1.1	Glutaraldehyde	Initially adsorbed on cellophane membrane or initial microencapsulation
Carbonic anhydrase	4.2.1.1	Diazobenzidine-3,3'-dianisoline	Initially adsorbed on cellophane membrane
		Glutaraldehyde	Initially adsorbed on cellophane membrane
		Glutaraldehyde	Concomitant adsorption on Silastic sheet
Carboxypeptidase A	3.4.17.1	Glutaraldehyde	Initially adsorbed on colloidal silica
Catalase	1.11.1.6	Glutaraldehyde	Initially adsorbed on cellophane membrane, cheesecloth, DEAE-cellulose, or initial microencapsulation
α-Chymotrypsin	3.4.21.1	Glutaraldehyde	Crosslinked with bovine serum albumin or initially adsorbed on cellophane membrane or colloidal silica
		Hexamethylenediisocyanate	
Chymotrypsinogen		Ethyl chloroformate	
		Glutaraldehyde	
β-Galactosidase	3.2.1.23	Glutaraldehyde	Initially adsorbed on cellophane membrane
Glucose oxidase	1.1.3.4	Glutaraldehyde	Initially adsorbed on cellophane membrane
		Glutaraldehyde	Crosslinked with bovine serum albumin or plasma albumin
		Diazobenzidine-3,3'-dianisoline	Initially adsorbed on cellophane membrane
Glucose-6-phosphate dehydrogenase	1.1.1.49	Glutaraldehyde	Initially adsorbed on cellophane membrane
Glucose-6-phosphate isomerase	5.3.1.9	Glutaraldehyde	Initially adsorbed on cellophane membrane
Glutamate–aspartate transaminase	2.6.1.1	Glutaraldehyde	All activity lost even when crosslinked with human albumin or globulin
		Glutaraldehyde	Initially adsorbed on cellophane membrane
		Ethyl chloroformate	All activity lost even when crosslinked with human albumin or globulin

Kallikrein	3.4.21.8	Glutaraldehyde	Initially adsorbed on colloidal silica
Lactate dehydrogenase	1.1.1.27	Glutaraldehyde	Initially adsorbed on cellophane membrane
Lysozyme	3.2.1.17	Glutaraldehyde	Initially adsorbed on colloidal silica
Papain	3.4.22.2	4,4-Diisothiocyanatobiphenyl-2.2'-disulphonic acid	
		Diazobenzidine (or derivatives)	Initially adsorbed on collodion membrane
Peroxidase	1.11.1.7	Glutaraldehyde	Crosslinked with bovine serum albumin
Pepsin	3.4.23.1.	Ethyl chloroformate	Condensed with gelatin or amino acids
Phenylalanine decarboxylase	4.1.1.53	Glutaraldehyde	Initially adsorbed on cellophane membrane
Subtilisin Nova	3.4.21.14	Glutaraldehyde	Crosslinking in the presence of ammonium sulphate or acetone
Triose phosphate isomerase	5.3.1.1	Glutaraldehyde	Initially adsorbed on cellophane membrane
Trypsin	3.4.21.4	Glutaraldehyde	Crosslinking in the presence of ammonium sulphate
		Glutaraldehyde	Initially adsorbed on cellophane membrane or colloidal silica
Urease	3.5.1.5	Glutaraldehyde	Initially adsorbed on cellophane membrane or colloidal silica
		Glutaraldehyde	Initial microencapsulation

Table 37. Capacities of various methods of immobilization with glutaraldehyde crosslinking.

Method of immobilization	Capacity [mg protein (g conjugate)$^{-1}$]
Impregnation of cellophane membranes with enzyme followed by crosslinking	0.1[a]
Enzyme crosslinked in solution, then included in agarose–polyacrylamide gel	7
Enzyme co-crosslinked in solution with inert protein, e.g., albumin, then spread on glass plate to obtain membrane	7–8
Enzyme co-crosslinked in solution with inert protein, e.g., albumin, then frozen at −30°C and warmed slowly to obtain spongelike conjugate	70–80
Enzyme co-crosslinked with inert protein, e.g., gelatin, in the presence of fillers (bentonite, alumina, silica gel or Celite)	50–500
Enzyme co-crosslinked with chitin	30
Enzyme crosslinked with inert protein, e.g., bovine serum albumin	
Enzyme adsorbed on magnetite followed by crosslinking	4–36
Enzyme adsorbed on carbon followed by crosslinking	
Adsorption on colloidal silica particles (210–230 m^2 g^{-1}) followed by crosslinking	300
Activation of amino alkyl derivatives of silanized porous glass beads followed by coupling of protein	12–16
Activation of 1-amino-6-hexamido derivatives of crosslinked ethylene/maleic acid copolymers, followed by coupling of protein	100

[a] Expressed as mg protein (cm^2 membrane)$^{-1}$.

The reactive amino acid residues involved in the formation of covalent attachment or crosslinking are listed in Table 38.

Table 38. Reactive amino acid residues in the common immobilizing methods [Melrose (1971)].

Immobilizing agent	Reactive amino acid residues
Acid azide	Cysteine, Lysine, Serine, Tyrosine
N-Carboxy-α-amino acid anhydride	Lysine
Chloro-*s*-triazinyl derivative	Lysine
Cyclic iminocarbonate	Lysine
Diazo derivative	Arginine, Cysteine, Histidine, Lysine, Tyrosine
Diimide	Aspartate, Cysteine, Glutamate, Lysine, Tyrosine
N-Ethyl-5-phenylisoxazolium-3'-sulphonate	Aspartate, Glutamate, Lysine
Glutaraldehyde	Cysteine, Histamine, Lysine, Tyrosine
Maleic acid/maleic acid anhydride	Lysine

Adsorption

Using this technique, enzymes become associated with the carrier matrix by multiple salt linkages, hydrogen bonding or van der Waals forces. The pH of the medium in which the enzyme and carrier are suspended can influence the formation of the enzyme–carrier complex due to its influence on the ionization states of the individual components. The binding curve for the formation of the collagen–lysozyme complex is presented in Fig. 25.

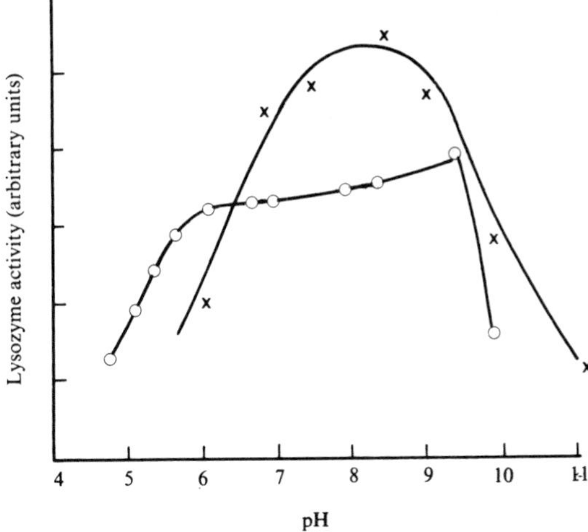

Figure 25. Binding curve for collagen–lysozyme complex formation and lytic activity of lysozyme as a function of pH at ionic strengths of 0.04–0.05. —×—, binding curve for lysozyme–collagen complex formation. —O—, lytic activity of lysozyme [Vieth *et al.* (1972)].

Table 39. Enzymes immobilized by adsorption (Zaborsky (1973)].

Enzyme	E.C. number	Adsorbent
Acid phosphatase	3.1.3.2	Cephalin-coated carbon or silica
		Silica
Aminoacylase	3.5.1.14	Alumina, Amberlite IR-4B, DEAE-cellulose, Dianion SA-11A, CM-Sephadex, DEAE-Sephadex
α-Amylase	3.2.1.1	Alumina, Amberlite CG-50, bentonite, calcium phosphate gel, carbon, clay (acid), collagen, diatomaceous earth, Dowex 1-X4, kaolinite, silica gel
β-Amylase	3.2.1.2	Bentonite, collagen, kaolinite
Amyloglucosidase	3.2.1.3	Amberlite CG-4B type II or IR45 (OH^-), carbon, CM–cellulose, DEAE-cellulose, clay (acid), diatomaceous earth, Dowex-1-X10 (Cl^-), CM-Sephadex
Asparaginase	3.5.1.1	CM-cellulose, DEAE-cellulose
ATPase	3.6.1.3	Millipore filter
ATP deaminase	3.5.4.–	Carbon, DEAE-cellulose, TEAE-cellulose
Catalase	1.11.1.6	Bentonite, Amberlite XE-97, calcium chloride, carbon + lauric acid or cephalin, CM-cellulose, kaolinite, glass or metal plates treated with barium stearate, thorium nitrate or sodium deoxycholate, polyaminostyrene, silica + lauric acid, cephalin or tridodecylamine
Cellulase	3.2.1.4	Cellulose, collagen
α-Chymotrypsin	3.4.21.1	Cellulose nitrate or phosphate, CM-cellulose, kaolinite, glass plates coated with barium stearate, thorium nitrate, sodium silicate or sodium deoxycholate
Deoxyribonuclease I	3.1.21.1	Cellulose
Dextran sucrase	2.4.1.5	DEAE-Sephadex
β-Galactosidase	3.2.1.23	DEAE-cellulose
Glucose oxidase	1.1.3.4	Glass
Glucose-6-phosphate dehydrogenase	1.1.1.49	Colloidion, silica with or without lecithin

Hexokinase	2.7.1.1	Silica gel, silica with lauric acid or cephalin
Invertase	3.2.1.26	Alumina, bentonite, carbon, DEAE-cellulose, collagen
Lactate dehydrogenase	1.1.1.27	Millipore filters
Leucine aminopeptidase	3.4.11.1	Calcium phosphate gel
Lipase	3.1.1.3	Amberlite XE-97, polyaminostyrene
Lysine decarboxylase	4.1.1.18	Alumina
Lysozyme	3.2.1.17	Collagen
Malate dehydrogenase	1.1.1.37	Silica with or without lecithin, cephalin or cholesterol
NAD pyrophosphorylase	2.7.7.1	Hydroxylapatite
D-Oxynitrilase	4.1.2.10	Cellulose, TEAE-epichlorohydrin
Papain	3.4.22.2	Glass
Pepsin	3.4.23.1	DEAE-cellulose, metal or glass plates coated with barium stearate, thorium nitrate, sodium silicate or sodium deoxycholate
Phosphoglucomutase	2.7.5.1	Cephalin-coated carbon or silica, silica
Phosphomonoesterase	3.1.3.–	CM-cellulose
Polynucleotide phosphorylase	2.7.7.8	Millipore filter
Proteases	3.4.–.–	DEAE-cellulose
Ribonuclease A	3.1.4.22	Cationic resin SBS 4 (H), Dowex-2 anion exchange resin, Dowex-50 cation exchange resin, glass
Succinate dehydrogenase	1.3.99.1	Silica gel or carbon with monolayer of cephalin or lecithin
Trypsin	3.4.21.4	Cellulose nitrate or phosphate, CM-cellulose, kaolinite, metal-coated glass plates with barium stearate
Urease	3.5.1.5	Collagen, kaolinite, metal or glass plates treated with barium stearate, thorium nitrate, sodium silicate or sodium deoxycholate

Table 40 shows that enzymes from different sources vary in their ability to be adsorbed on insoluble carriers. Similarly, different resins vary in their ability to adsorb enzymes (*see* Table 41).

Table 40. Endoglucanase adsorption on insoluble cellulose in a column reactor [Klyosov and Rabinowitch (1980)].

Source	Adsorbed endoglucanase (%) Filter paper	Avicel
Geotrichum candidum	30	73
Trichoderma lignorum	64	97
Aspergilus niger	–	13
Rapidase	8	4

Entrapment and Microencapsulation

In lattice-entrapped enzymes, the enzyme is present within the interstitial regions of filamentous structures. The nature of the entrapment matrices is summarized in Table 42.

Table 41. Comparison of different resins for the adsorption of enzymes [Samejima and Kimura (1974)].

Resin	Type of resin		Activity [U (ml resin)$^{-1}$]		
	Matrix[a]	Functional groups	Glucose isomerase	RNase	Aspartase
Duolite A-2	PF	1,2,3-Amines	17.7		9.6
Duolite A-4	PF	2,3-Amines	28.5		9.1
Duolite A-7	PF	1,2,3-Amines	32.6	37.4	9.2
Duolite S-30	PF	Hydroxyl	31.8	12.3	7.6
Duolite ES-104	S	4-Amine			0
Duolite A-57	E	3,4-Amines			0
Diaion WA20	S	1,2-Amines	2.4	2.1	0
Diaion WA21	S	1,2-Amines	1.6	2.3	
Diaion HP20	S			2.7	
Amberlite XAD-7	AE		15.8		3.6
Amberlite IRA-93	P	3-Amine		16.6	
Amberlite IRC-50	A	Carboxyl		8.9	

[a] PF, phenol-formaldehyde; AE, acrylic ester; S, polystyrene; P, polyphenol; A, polyacryl; E, epoxy.

Table 42. Lattice-entrapped enzymes [Zaborsky (1973)].

Enzyme	E.C. number	Entrapment matrix
Acetyl cholinesterase	3.1.1.7	Silastic resin
Alcohol dehydrogenase	1.1.1.1	Polyacrylamide
Aldolase	4.1.2.13	Polyacrylamide, poly(*N*,N^1-methylenebisacrylamide)
Alkaline phosphatase	3.1.3.1	Polyacrylamide
D-Amino acid oxidase	1.4.3.3	Polyacrylamide
L-Amino acid oxidase	1.4.3.2	Polyacrylamide
α-Amylase	3.2.1.1	Poly(*N*,*N*'-methylenebisacrylamide)
β-Amylase	3.2.1.2	Poly(*N*,*N*'-methylenebisacrylamide)
Apyrase	3.6.1.5	Polyacrylamide
Asparaginase	3.5.1.1	Polyacrylamide
Catalase	1.11.1.6	Polyacrylamide, silica gel
α-Chymotrypsin	3.4.21.1	Poly(*N*,*N*'-methylenebisacrylamide), silastic resin
Citrate synthase	4.1.3.7	Polyacrylamide
Cholinesterase	3.1.1.8	Polyacrylamide, silastic resin, starch
Enolase	4.2.1.11	Poly(*N*, *N*'-methylenebisacrylamide)
Glucose isomerase	5.3.1.5	Polyacrylamide
Glucose oxidase	1.1.3.4	Polyacrylamide, starch
Glucose oxidase/peroxidase		Silastic resin
Glucose-6-phosphate dehydrogenase	1.1.1.49	Polyacrylamide
Glucose-6-phosphate isomerase	5.3.1.9	Polyacrylamide
Glutamate dehydrogenase	1.4.1.2	Polyacrylamide
Glutaminase	3.5.1.2	Polyacrylamide
Hexokinase	2.7.1.1	Polyacrylamide
Lactate dehydrogenase	1.1.1.27	Polyacrylamide
Orsellinic acid decarboxylase	4.1.1.58	Polyacrylamide
Papain	3.4.22.2	Poly(*N*,*N*'-methylenebisacrylamide)
Peroxidase	1.11.1.7	Polyacrylamide
Phosphofructokinase	2.7.1.11	Polyacrylamide
Phosphoglycerate mutase	2.7.5.3	Poly(*N*,*N*'-methylenebisacrylamide)
Ribonuclease A	3.1.4.22	Poly(*N*,*N*'-methylenebisacrylamide)
Trypsin	3.4.21.4	Polyacrylamide, poly(*N*,*N*'-methylenebisacrylamide), silastic resin, silica gel
Urease	3.5.1.5	Polyacrylamide, silastic resin, silica gel, starch

In addition to entrapment, some enzymes are enclosed within semipermeable 'solid' or 'liquid' membranes (*see* Fig. 26).

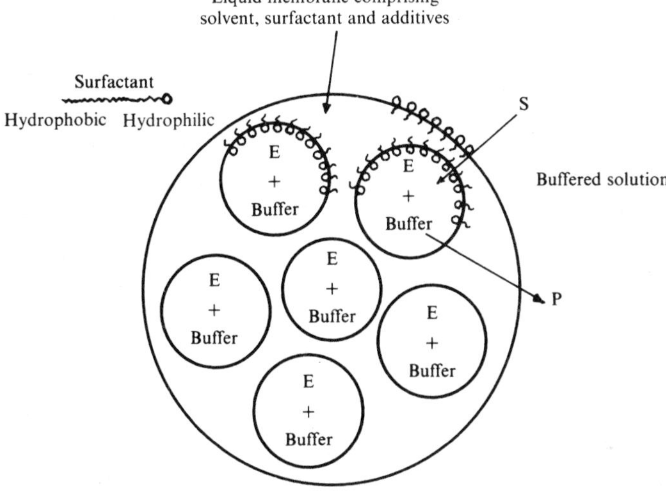

Figure 26. Diagram of liquid membrane with encapsulated enzyme particles [May and Li (1974)].

Examples of microencapsulated enzymes are given in Table 43.

Table 43. Microencapsulated enzymes [Zaborsky (1973)].

Enzyme	E.C. number	Membrane
L-Asparaginase	3.5.1.1	Collodion, Nylon
Carbonic anhydrase	4.2.1.1	Collodion
Catalase	1.11.1.6	Collodion, polystyrene, silicone
Lipase	3.1.1.3	Ethylcellulose, polystyrene, silicone
Trypsin	3.4.21.4	Collodion, Nylon
Urate oxidase	1.7.3.3	Collodion, Nylon
Urease	3.5.1.5	Benzylalkonium–heparin–collodion, ethylcellulose, Nylon, polystyrene, silicone

Table 44. Capacities of supports for the immobilization of enzymes by entrapment [Chibata and Tosa (1976)].

Entrapment matrix	Capacity [mg (g conjugate)$^{-1}$]
Polyacrylamide crosslinked gel	6–100
Polyacrylamide crosslinked beads	2–5
Polyvinyl alcohol, radiation crosslinked gel	5–10

IMMOBILIZED ENZYME PROPERTIES

Introduction

An immobilized enzyme particle prepared by any of the methods illustrated in Fig. 23 is likely to contain enzyme molecules distributed throughout the structure. Most support structures used for covalent attachment or adsorption contain accessible internal structures to which the enzyme molecules are attached/adsorbed [*see* Fig. 27(a)]. Covalent crosslinking, with or without inert material, produces a lattice structure [*see* Fig. 27(b)] not dissimilar to the lattice used in entrapment procedures. Such artefacts can be used in any conventional reactor configuration, e.g., suspended in stirred tanks or fluidized beds, or as stationary particles in fixed beds (*see* Chapter 7). In such reactor configurations, the particles are retained by the usual solid–liquid separation techniques, e.g., sedimentation and centrifugation, which exploit particle size and the difference in density between the fluid and the solid. In order to facilitate the liquid flow patterns and the interrelated mass transfer characteristics, it is clearly advantageous to be able to use discrete particles of any preferred size rather than a given size or restricted size range (*see* Chapter 10).

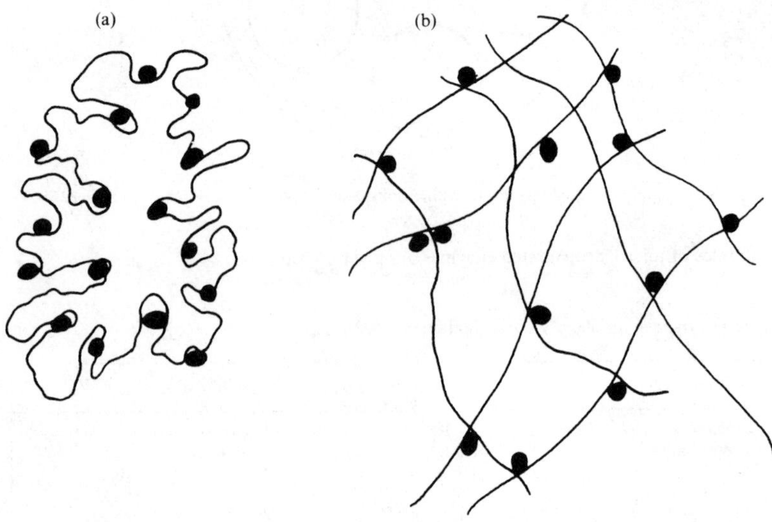

Figure 27. Immobilized enzyme particles (a) prepared by attachment/adsorption or (b) prepared by crosslinking/entrapment.

The reasons for using immobilized enzymes are simply economic and result from consideration of enzyme retention and reuse *see* Table 31). The artefacts illustrated in Fig. 27 lead to high enzyme loadings per unit volume of support and, when the particles are contained in a reactor at high number density (i.e. number of particles per unit reactor volume) they lead to high enzyme loadings per unit reactor volume. This last factor is also of economic benefit.

The particles illustrated in Fig. 27 are usually much larger than a single enzyme molecule and are, in general, subject to diffusion limitations in relation to both the substrates and the products (*see* Chapter 10), i.e. the individual enzymes are exposed to conditions that are different – usually inferior – from those in the bulk fluid. The need to use such large particles mainly stems from the mechanisms used to retain them in reactors and, in the case of fixed-bed reactors, to achieve reasonable throughputs at acceptable pressure drops.

From the standpoint of process engineering, a balance is necessary between increased enzyme loading and the diffusion limitations that result from the use of particles. There are no criteria that can be adopted for optimum particle sizes and acceptable diffusion limitations other than economic ones, although maximization of the volumetric rate of reaction, which simultaneously includes particle size and number density, comes closest to such a criterion. This situation is completely analogous to that found with reactors containing heterogeneous catalytic particles where a balance between diffusional limitations and catalyst inventory has to be struck (*see* Thomas and Thomas, 1967).

In Table 31, high enzyme stability is identified as a requirement of immobilized enzymes. A further requirement of immobilized enzyme particles is a long physical life under the reaction conditions; in particular, under the fluid shear rates encountered in reactors, wear, attrition and breakup should be minimal.

Table 45. Properties of immobilized aminoacylase using acetyl-DL-methionine as substrate [Chibata and Tosa (1976)].

Properties	Native amino-acylase	Immobilized aminoacylase		
		Ionic binding to DEAE-Sephadex	Covalent binding to iodoacetyl-cellulose	Entrapping by polyacrylamide
Optimum pH	7.5–8.0	7.0	7.5–8.0	7.0
Optimum temperature (°C)	60	72	55	65
Activation energy[a] (kcal mol^{-1})	6.9	7.0	3.9	5.3
Optimum Co^{2+} (mM)[a]	0.5	0.5	0.5	0.5
K_m (mM)[a]	5.7	8.7	6.7	5.0
V_{max} (mol h^{-1})[a]	1.52	3.33	4.65	2.33
Heat stability (%)[b] 60°C, 10 min.	62.5	100	77.5	78.5
70°C, 10 min.	12.5	87.5	62.5	34.5
Operation stability (half life[c], days)		65 days, 50°C		48 days, 37°C

[a] All assays done at 37°C and pH 7.0.
[b] Remaining activity.
[c] The time required for 50 per cent of the enzyme activity to be lost.

Table 45 provides a summary of those factors worthy of consideration when evaluating the efficacy of various immobilized enzyme preparations for a given enzyme intended for use under operational rather than laboratory conditions.

Kinetics

The kinetics of enzymes in free solution are often described by the Michaelis–Menten equation and various modifications (*see* pp.492–495). The effects of the immobilization process may result in a change in the intrinsic kinetics, such that the Michaelis–Menten equation is no longer applicable, or the values of the Michaelis constant K_m and V_{max} may be changed, for the reasons given in Fig. 28. In this context, 'immobilization' refers to the interaction between the enzyme molecule and the support at the molecular level, i.e. when the immobilized enzyme particles are *very* small (*see* Chapter 10). Generally, the Michaelis–Menten equation applies although the Michaelis constants are altered.

For particles of finite size and of the type useful in reactors, diffusion limitations are likely to be superimposed on the intrinsic immobilized enzyme kinetics. Such limitations result in an individual enzyme molecule within a particle being exposed to conditions that differ greatly from those of the bulk fluid. Therefore, the particle reacts under the *local* conditions to which it is exposed. The local conditions arise from diffusion limitations on substrates and products as well as the effects of the reaction on such factors as pH (*see* Chapter 10).

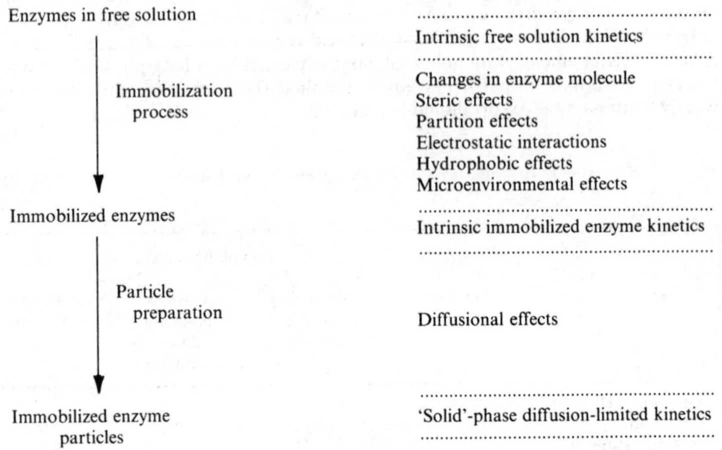

Figure 28. Changes in enzyme kinetics resulting from immobilization and particle preparation.

When immobilized enzyme particles are prepared, the resulting artefacts encompass simultaneously changes in the intrinsic kinetics and kinetic coefficients, as well as diffusion limitations (*see* Fig. 28). The diffusional limitations can be eliminated in research and development studies by reducing the particle size (i.e. below $\alpha k_2 \rho_e V_p / A_p = 0.5$, *see* Chapter 10) in a series of experiments (*see* Fig. 29) and then extrapolating to zero particle size. An advantage of this procedure lies in the fact that no precise measurement of the particle size or shape is required; it is sufficient to achieve conditions in which no further change in the rate takes place as the size is reduced. Failure to achieve such conditions in practice may be due to the fact that enzyme detachment occurs as the particle size is reduced, with the result that the smaller particles may not reflect the enzyme loading of the large particles.

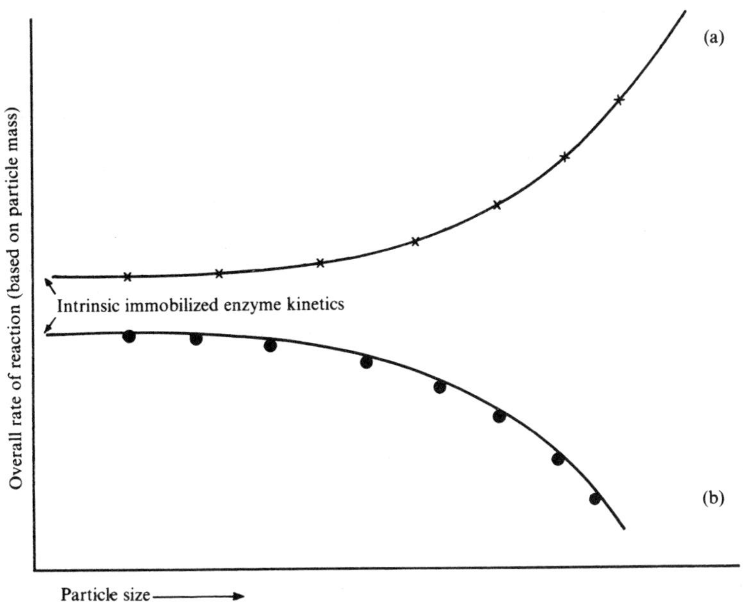

Figure 29. Effect of reduced particle size on the overall rate of reaction of an immobilized enzyme particle. (a) When diffusional limitations are advantageous. (b) When diffusional limitations are disadvantageous.

From the above remarks, it is clear that the intrinsic kinetics of immobilized enzymes are of some importance. When intrinsic kinetics are not available, the free solution kinetics can be used although circumspectly. For diffusion-limited particles, the methodology of Chapter 10 relating the diffusion-limited kinetics to intrinsic kinetics can be adopted.

A further complicating factor is due to enzyme loading of the particle or the amount of enzyme contained per unit volume of particle (i.e. ρ_e). The immobilization procedures described above (*see* pp.515–539) result in less than total uptake of available enzyme because of the conversion efficiencies involved and because of enzyme deactivation due to thermal and steric effects. The actual loading ρ_e is therefore less than the potential loading $\overset{\mathrm{I}}{\rho_e}$, based on the initial quantity of enzyme available, i.e. $\rho_e = \phi \overset{\mathrm{I}}{\rho_e}$, where $\phi < 1$. Consequently, the factor ϕ is of considerable economic importance.

The maximum specific rate of an immobilized enzyme particle $V_{max,\,ie}$ is given by

$$V_{max.ie} = \frac{r_{max}}{\rho_e} = \frac{r_{max}}{\phi \rho_e^{\mathrm{I}}} \tag{19}$$

where r_{max} is the maximum rate of reaction per unit particle volume.

Without an independent, and relevant, method of determination of ϕ (or ρ_e), $V_{max,\,ie}$ cannot be evaluated. Consequently, the Michaelis–Menten equation has to be expressed as

$$\frac{r}{r_{max}} = \frac{V_{ie}}{V_{max.ie}} = \frac{[S]}{K_m + [S]} \tag{20}$$

without specific knowledge of either V_{ie} or $V_{max,\,ie}$.

In any event, the product $\phi V_{max,\,ie}$ is the important kinetic parameter and information regarding the value of $V_{max,\,ie}$ and how it compares with V_{max} for the enzyme in free solution is largely a matter of research interest.

The diffusion parameter devised in Chapter 10 can also be expressed in terms of r_{max}, i.e.

$$\alpha\left(\frac{V_{max,ie}}{D_e K_m}\right)^{1/2} Pe^{1/2}\frac{V_p}{A_p} = \alpha\left(\frac{r_{max}}{D_e K_m}\right)^{1/2}\frac{V_p}{A_p} \tag{21}$$

It follows from Eqs. (19), (20) and (21) that immobilized enzyme particles are most conveniently assessed in terms of the rate of reaction per unit particle volume r_{ie}. An upper limit for $r_{max,\,ie}$ is given by

$$r_{max,\,ie} = V_{max},\ \text{free solution}\ \overset{I}{\rho_e} \tag{22}$$

on the assumption that

$$V_{max,free\,solution} \geq V_{max,ie} \tag{23}$$

Determination of individual values of ϕ and $V_{max,\,ie}$ is largely of research interest. From the standpoint of process engineering, the important parameter is $V_{max,\,ie}\ \phi$ and the preparative conditions (*see* pp.515–539) which lead to high values.

Eq. (20) is sometimes employed to describe diffusion-limited immobilized enzyme kinetics by introducing an 'apparent' value of the Michaelis constant $K_{m,\,app}$.

$$\frac{r}{r_{max}} = \frac{[S]}{K_{m,app} + [S]} \tag{24}$$

Obviously, such a procedure is appropriate on occasions and Eq. (24) may be used to 'curve-fit' empirically experimental data. Reference to Chapter 10 shows that the parameter $K_{m,\,app}$ is size-dependent and that understanding of the various phenomena involved is obscured.

Fig. 30 shows a Michaelis–Menten-type dependency of the rate of reaction on substrate concentration both for free and immobilized glucose oxidase and suggests that

$$V_{max,ie} \simeq V_{max,free\,solution}$$
$$K_{m,ie} > K_{m,free\,solution} \tag{25}$$

The first relationship in Eqs. (25) probably follows from the fact that, in principle, it is possible to load membranes without significant loss of enzyme. The second relationship suggests either a change in K_m due to immobilization or a diffusion limitation, either 'solid' or liquid. Similar phenomena occur in Fig. 31, but with decreased K_m.

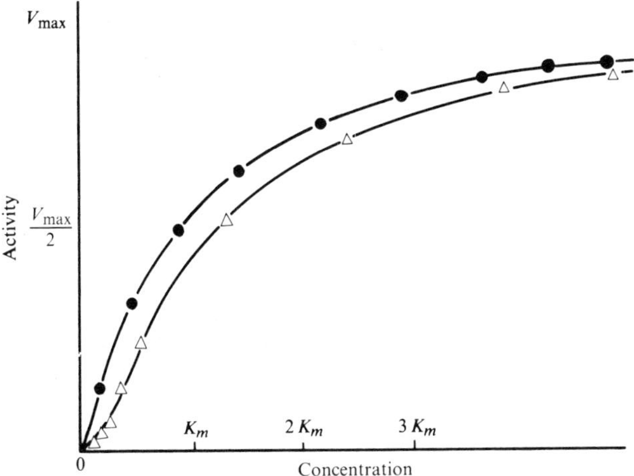

Figure 30. Variation of enzyme activity as a function of substrate concentration expressed as multiples of K_m ●, free glucose oxidase; △, membrane containing glucose oxidase [Thomas *et al.* (1972)].

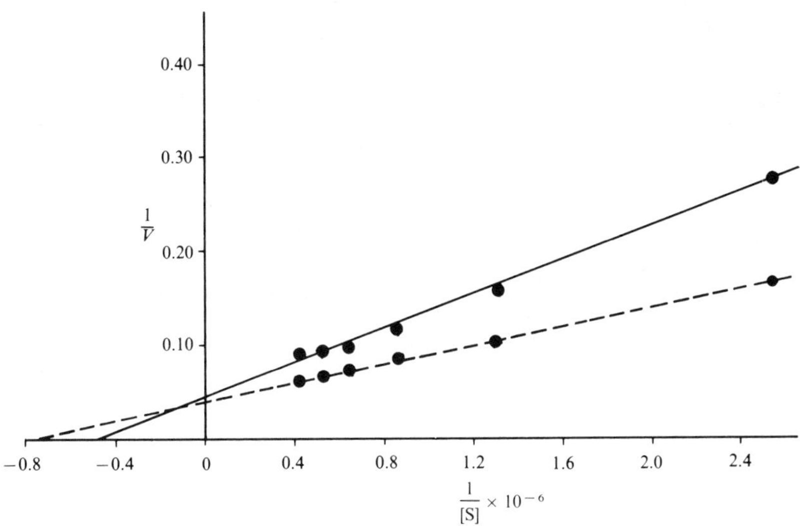

Figure 31. Lineweaver–Burk plot of the effect of concentration of hexa-(*N*-acetylglucosamine). ———, native enzyme; ------, immobilized enzyme [Brown *et al.* (1974)].

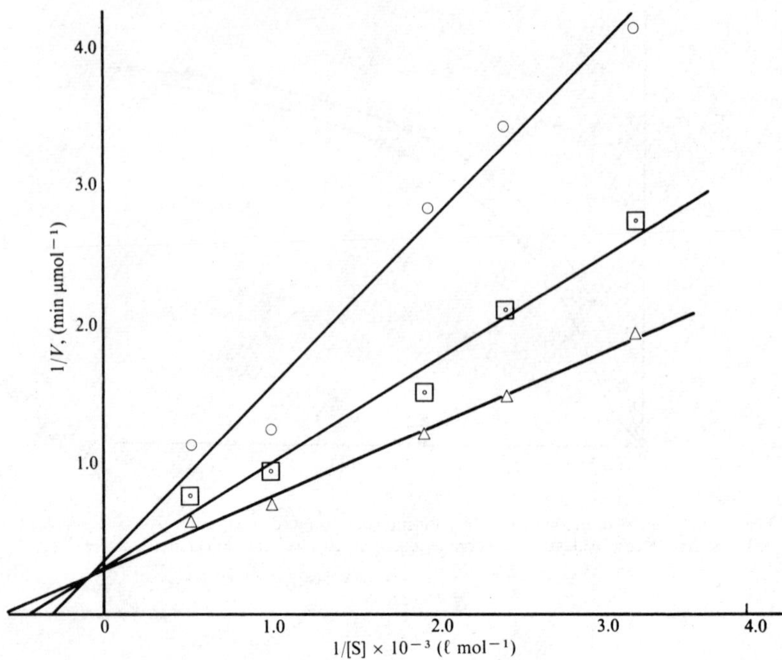

Figure 32. Effect of glucose concentration on activity of a hexokinase column. Flow rates: ○, 0.57 ml min^{-1}; ⊡, 1.20 ml min^{-1}; △, 1.85 ml min^{-1}.

Fig. 32 shows experimental data obtained using a fixed-bed reactor. $K_{m,app}$ decreases with increased flow rate, leading to higher rates of reaction at a given substrate concentration by a reduction in the liquid-phase diffusion limitation. Clearly, care has to be exercised by the experimenter to ensure the absence of liquid-phase diffusion limitations by increasing the flow rate/agitation and of solid-phase diffusion limitations by decreasing particle size (*see* Fig. 29) when determining intrinsic kinetics.

Some typical values of K_m for immobilized enzymes $K_{m,\,ie}$ are presented in Table 46. Overall immobilization appears to produce an increased value of $K_{m,\,ie}$, which is interpreted as a reduction in the affinity of the substrate for the enzyme.

Table 46. Michaelis constants of some free solution and immobilized enzymes.

Enzyme	EC number	Immobilizing agent	Substrate[a]	K_m (M)
Creatine kinase	2.7.3.2	None	ATP	6.5×10^{-4}
		p-Aminobenzylcellulose	ATP	8.0×10^{-4}
		CM-Cellulose-90	ATP	7.0×10^{-4}
α-Chymotrypsin	3.4.21.1	None	ATEE	1.0×10^{-3}
		Heparin	ATEE	1.1×10^{-3}
		Aldehyde Sephadex	ATEE	2.5×10^{-3}
		Soluble aldehyde dextran	ATEE	1.3×10^{-3}
		Soluble acrylic acid copolymer	ATEE	1.2×10^{-3}
Ficin	3.4.22.3	None	BAEE	2×10^{-2}
		CM–Cellulose-70	BAEE	2×10^{-2}
Lactate dehydrogenase	1.1.1.27	None	Pyruvate	1.85×10^{-4}
		Succinylpropyl-glass	Pyruvate	5.1×10^{-6}
		Propyl-glass	Pyruvate	3.6×10^{-6}
		None	NADH	7.8×10^{-6}
		Succinylpropyl-glass	NADH	3.9×10^{-5}
		Propyl-glass	NADH	5.5×10^{-5}
Papain	3.4.22.2	None	BAEE	1.9×10^{-2}
		p-Aminophenylalanine/L-Leu	BAEE	1.9×10^{-2}
Trypsin	3.4.21.4	None	BAA	6.8×10^{-3}
		Maleic acid/ethylene	BAA	2.0×10^{-4}

[a] ATEE, *N*-acetyl-L-tyrosine ethyl ester; BAEE, *N*-benzoylarginine ethyl ester; BAA, *N*-benzoylarginine amide.

Inevitably, most experimentation carried out with immobilized enzymes is not concerned with detailed investigation of either the intrinsic or diffusion-limited kinetics but with the 'activity' of the immobilized enzyme preparation. Such data, while inconclusive in scientific terms from a process-engineering standpoint, provides a matrix of information from which heuristic conclusions can be drawn as regards immobilization procedures and supports. Such typical information is given in Table 47–49.

Table 47. Specific activities of glucose oxidase and D-hydroxynitrile lyase on polyacrylamide matrices with different fixation methods [Jaworek (1974)].

Matrix	Fixation	Specific activity (Ug^{-1})	
		Glucose oxidase	D-Hydroxynitrile lyase
Acrylamide/maleic Acid copolymer	Anhydride	10	1.3
Polyacrylamide	Mechanically entrapped	80	2.3
Polyacrylamide	Copolymerization	500	13

Table 48. Activities of immobilized enzymes.

Enzyme	EC number	Carrier	Substrate	Activity [μmol (mg support)$^{-1}$min^{-1}]
Alcohol dehydrogenase	1.1.1.1	Nylon	NAD	0.013
Amyloglucosidase	3.2.1.3	DEAE-cellulose	Maltose	0.13
Chymotrypsin	3.4.21.1	CM-Cellulose	Acetyl-L-tryptophan ethyl ester	0.4
		Cellulose	Acetyl-L-trytophan ethyl ester	0.26
		DEAE-Cellulose	Acetyl-L-tryptophan ethyl ester	13.5
		Nylon	Acetyl-L-tryptophan ethyl ester	0.049
		Porous glass	Acetyl-L-tryptophan ethyl ester	2.0
Creatine kinase	2.7.3.2	*p*-AB-Cellulose	ATP	0.14
		CM-Cellulose	ATP	0.11
Ficin	3.4.22.3	CM-Cellulose	Benzoyl-L-arginine ethyl ester	0.026
β-Galactoside	3.2.1.23	Nylon	*o*-Nitrophenyl-β-galactoside	0.10
Lactate dehydrogenase	1.1.1.27	Nylon	NAD	0.013
Penicillin amidase	3.5.1.11	DEAE-Cellulose	Penicillin G	0.005

Table 49. Various immobilized aminoacylases and their activities [Chibata and Tosa (1976)].

Immobilization methods and carriers	Aminoacylase used[a] (U[b])	Immobilized aminoacylase[a] Activity (U[b])	Yield of activity (%)
Physical adsorption			
Activated carbon	1210	0	0
Acidic aluminum oxide	1210	13	1.0
Neutral aluminum oxide	1210	10	0.8
Basic aluminum oxide	1210	0	0
Silica gel	1210	0	0
Ionic binding			
PAB-Cellulose	1210	0	0
ECTEOLA-Cellulose	1210	293	24.2
TEAE-Cellulose	1210	623	51.5
DEAE-Cellulose	1210	668	55.2
CM-Sephadex C-50	1210	0	0
SE-Sephadex C-50	1210	0	0
DEAE-Sephadex A-25	1210	713	58.9
DEAE-Sephadex A-50	1210	680	56.2
Amberlite IRC-50	1210	0	0
Amberlite IR-4B	1210	0	0
Amberlite IR-45	1210	0	0
Covalent binding			
Diazotized PAB-cellulose	1210	64	5.3
Diazotized arylaminoglass	1210	525	43.4
Diazotized Enzacryl AA	1210	44	3.6
CM-Cellulose azide	1210	0	0
BrCN-activated cellulose	1210	12	1.0
BrCN-activated Sephadex	1210	15	1.2
Chloroacetyl cellulose	1210	137	11.3
Bromoacetyl cellulose	1210	339	28.0
Iodoacetyl cellulose	1210	472	39.0
Crosslinking using carrier			
AE-Cellulose			
1,4-Butylene dibromide	1440	6	0.4
1,4-Butylene dichloride	1440	6	0.4
Dicyclohexyl carbodiimide	1440	17	1.2
Diiodomethane	1440	5	0.3
Glutaraldehyde	1440	8	0.6
Hexamethylene diisocyanate	1440	23	1.6
Toluene diisocyanate	1440	3	0.2
CM-Cellulose			
Dicyclohexyl carbodiimide	1440	1	0.1
Crosslinking by bifunctional reagent			
Glutaraldehyde	1440	211	14.7
Toluene diisocyanate	1440	18	1.3
Lattice entrapping			
Acrylamide	1000	526	52.6
DEAE-Hydroxypropiomethyl cellulose phthalate	1000	190	19.0
Encapsulation			
Nylon	1000	360	36.0
Polyurea	1000	150	15.0
Ethylcellulose	1000	104	10.4

[a] All immobilized enzyme assays were carried out with the same degree of agitation; the native enzymes were assayed without agitation.
[b] One enzyme unit is defined as that amount of enzyme which hydrolyzes 1 μmol of acetyl-DL-methionine per hour at 37°C.

Figs. 33 and 34 illustrate some effects of the preparatory procedure on enzyme retention. Fig. 33 shows increased loss of enzyme when simple adsorption is used compared with enzyme complexation. Fig. 34 shows that retention of *active* enzyme can reach maximal proportions even though inactive protein continues to be immobilized by exposure to additional free solution enzyme.

Figs. 35 and 36 illustrate the effect of shear on the particle size distribution of AE-cellulase particles. Fig. 35 contains data for a typical stirred-tank operation, while Fig. 36 contains data obtained at a known shear rate. Clearly, the reactor enzyme inventory is likely to be reduced by the effects shown in Figs. 33 and 36, and any immobilization procedure/carrier with similar properties is at a severe disadvantage (*see* Table 48).

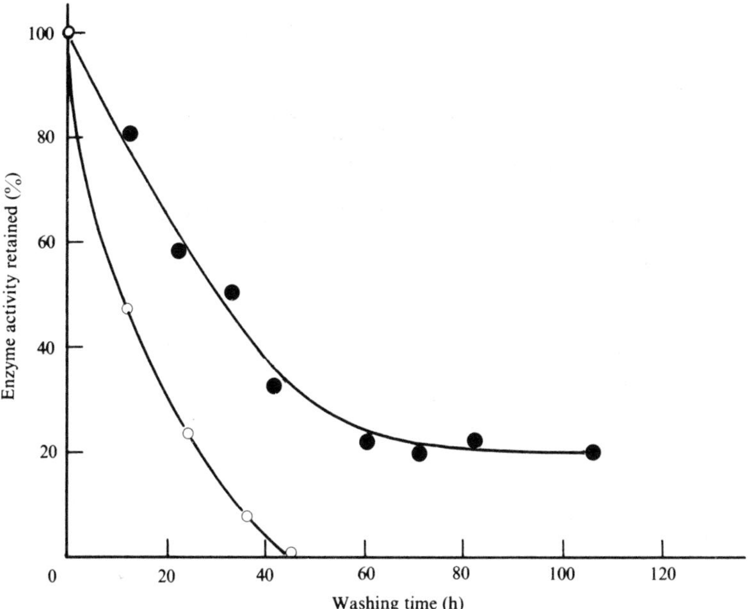

Figure 33. Comparison between enzyme adsorbed (3-h contact) and enzyme complexed (20-h impregnation, 6-h drying) on collagen. Washing solution, 1 M sodium chloride at 25°C. ○, catalase immobilized on collagen by adsorption; ●, catalase immobilized on collagen by complexation [Wang *et al.* (1974)].

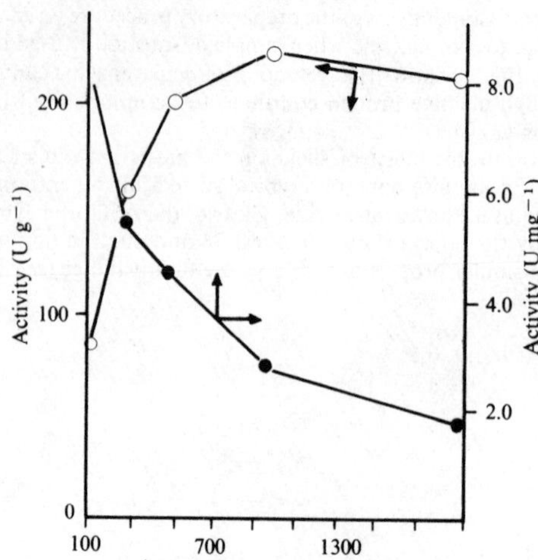

Figure 34. Relationship between the activity of immobilized protein and total units of native amyloglucosidase from *End. bispora*. ●, activity [U (mg immobilized protein)$^{-1}$]; ○, activity [U (g enzyme-carrier)$^{-1}$] [Nakhapetyan and Menyailova (1980)].

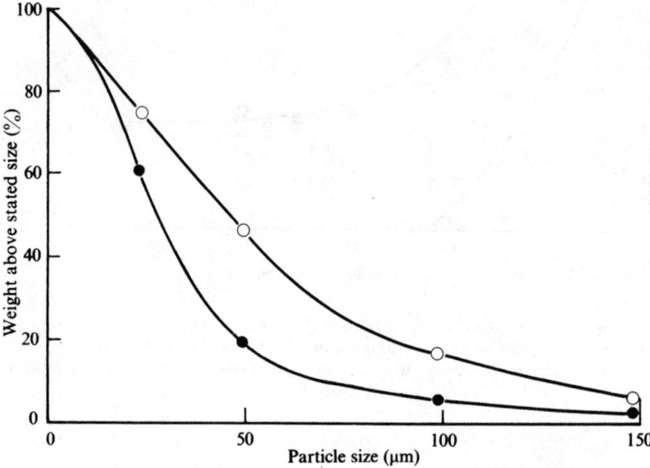

Figure 35. Change in particle size distribution of AE-cellulose–glucose oxidase when stirred in a 100-l aerated reactor. AE-cellulose, 5 g ℓ$^{-1}$; 575 rpm; aeration, 20 ℓ min^{-1}. ○, 0 h; ●, 13 h [Lilly *et al.* (1974)].

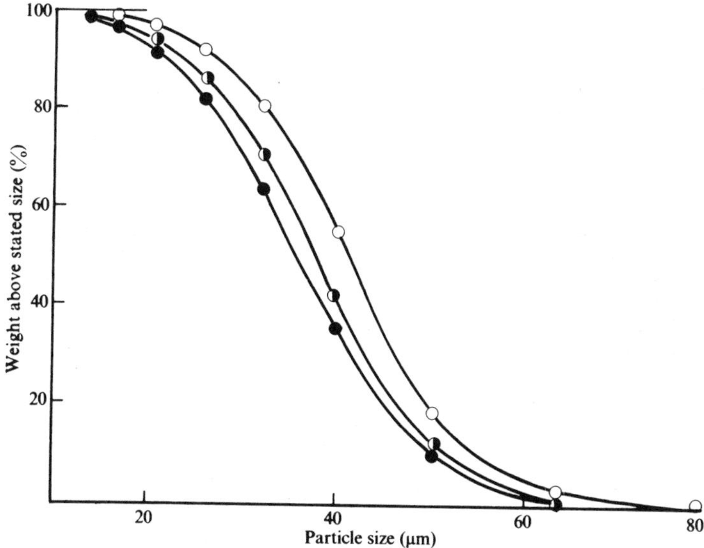

Figure 36. Change in particle size distribution with time of a 3 per cent (by weight) aqueous suspension of AE-cellulose after being subjected to a rate of shear of 702 s^{-1}. ○, 0 h; ◑, 2 h; ●, 6 h. [Lilly *et al.* (1974)].

Fig. 37 compares time courses of starch hydrolysis for batch operation using free and immobilized enzyme. Such a comparison suggests that no significant untoward effects are likely to be encountered when substituting immobilized enzymes for free enzymes in process-engineering applications.

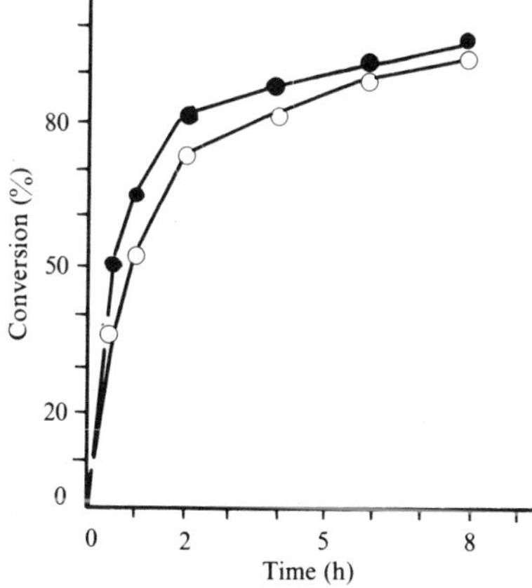

Figure 37. Cumulative hydrolysis of 30 per cent starch solution by amyloglucosidase from *Endomycopsis bispora* immobilized on silochrome C-80 via glutaraldehyde. Hydrolysis was carried out at 50°C and pH 4.7. ●, immobilized enzyme [50 U (g dry starch)$^{-1}$]; ○, soluble native enzyme [50 U (g dry starch)$^{-1}$] [Nakhapetyan and Menyailova (1980)].

pH Effects and Ionic Strength

When bound to negatively charged carriers (e.g., maleic anhydride/ethylene, CM-cellulose, L-Ala/L-Glu, polygalacturonic acid or polyaspartic acid) displacement of the pH–activity profiles is to more alkaline values (*see* Fig. 38 and 39). Conversely, binding to positively charged carriers (e.g., DEAE-cellulose, polyornithyl polymer) shifts the pH–activity profiles to more acidic values (*see* Fig. 38). These displacements are eliminated by increasing the ionic strength (*see* Fig. 39). Since the data in Figs. 38 and 39 are normalized — expressed relative to $V_{max,\,ie}$ evaluated at pH_{opt} for each preparation — they give no information as regards $V_{max,\,ie}$ at pH_{opt} for the various preparations. However, for the reasons given on p.541, it would be difficult to differentiate between the immobilization efficiency and $V_{max,\,ie}$.

Fig. 40 provides data on $K_{m,\,app}$ and suggests an influence of particle size compatible with solid-phase diffusion limitation.

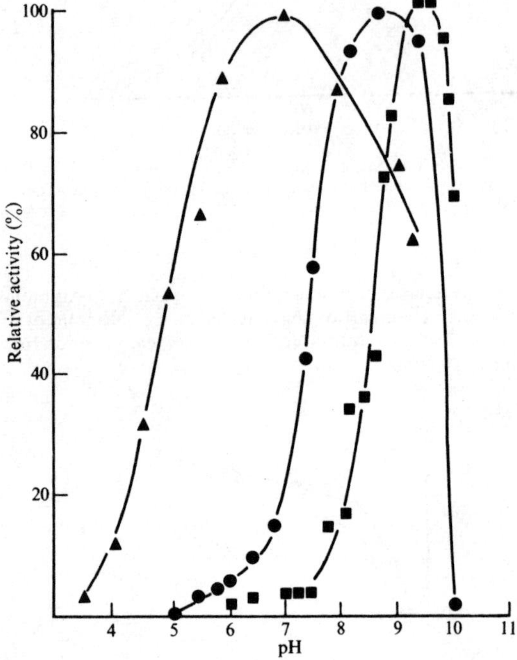

Figure 38. pH–activity curves for chymotrypsin (●), chymotrypsin–maleic anhydride ethylene copolymer (■) and chymotrypsin–polyornithyl (▲) at ionic strength of 0.001 with *N*-acetyl-L- tyrosine ethyl ester as substrate [Zaborsky (1973)].

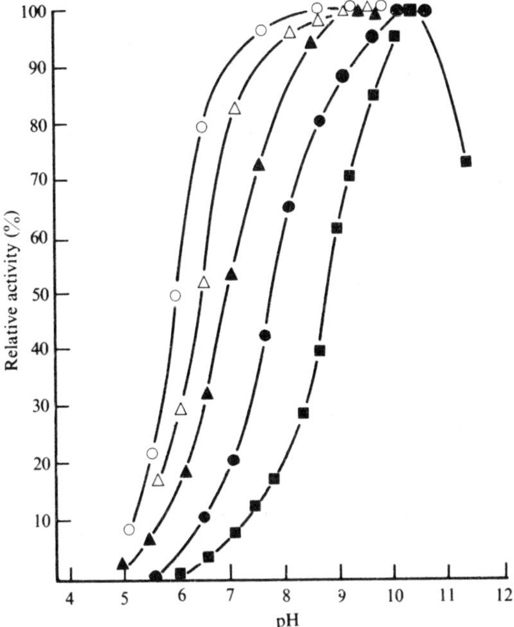

Figure 39. pH–activity curves for trypsin and trypsin–maleic anhydride ethylene copolymer at different ionic strengths with *N*-benzoyl arginine ethyl ester as substrate. Trypsin at ionic strength of 0.035 (○) and 1.0 (△). Trypsin–maleic anhydride ethylene copolymer at ionic strengths of 0.0058 (■), 0.035 (●), and 1.0 (▲) [Zaborsky (1973)].

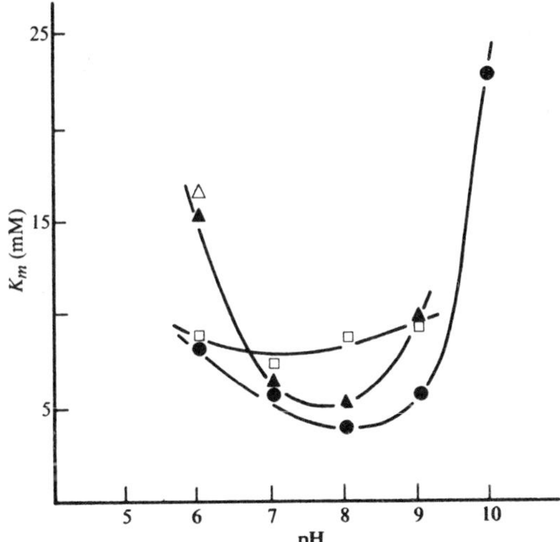

Figure 40. Variation of Michaelis constant (apparent and real) with pH for α-chymotrypsin (●) and α-chymotrypsin–DEAE-cellulose [110 mg (g support)$^{-1}$] before grinding (△) and after grinding (▲). α-Chymotrypsin–DEAE-cellulose [11 mg (g support)$^{-1}$] (□). Activity measured in presence of dioxane, phosphate buffer and NaCl [Zaborsky (1973)].

Figs. 41–43 show a variety of pH effects. Fig. 41 shows no change in pH_{opt} but a reduced overall rate of reaction. Fig. 42 shows both a shift in pH_{opt} and a reduced overall rate of reaction. Fig. 43 shows apparently markedly different pH effects, but since the data are normalized at pH 6 (i.e. the free solution pH_{opt}) no inference can be drawn as regards the relative magnitude of $V_{max,\,ie}/V_{max,\,free\,solution}$. Figs. 41–43 must, in the absence of further information, be taken to represent a combination of the factors listed in Table 32 (*see* Chapter 10 for possible 'solid' diffusion effects on the pH characteristic).

Figs. 44 and 45 show examples of the effect of ionic strength on K_m and activity. An increase in ionic strength results in a reduction in the activity (i.e. increase in K_m) of the immobilized enzyme. Only slight variations in free enzyme activity with ionic strength occur.

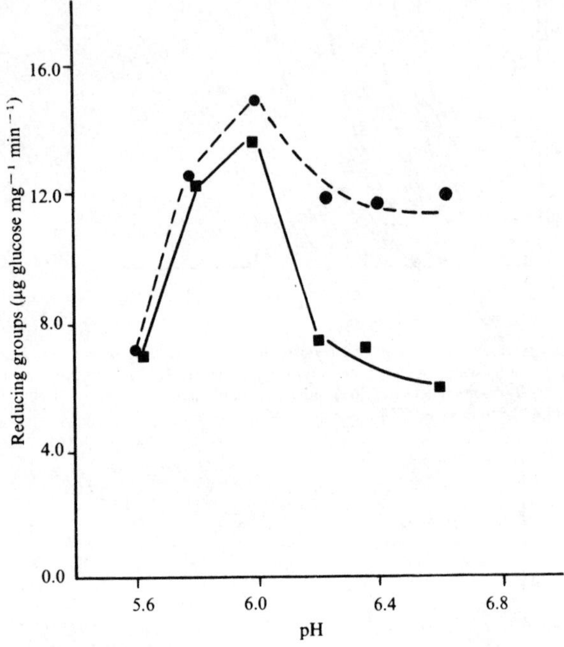

Figure 41. pH Profile of native (●) and crosslinked (condensed) (■) lysozyme in 0.1 M malate buffer [Brown *et al.* (1974)].

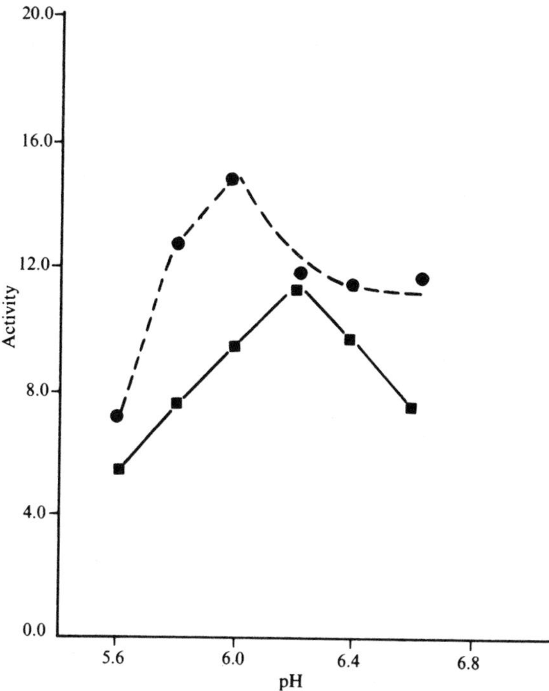

Figure 42. pH Profiles of native (●) and matrix supported lysozyme (■). Activity in μg reducing groups/min/mg protein at 37°C [Brown *et al.* (1974)].

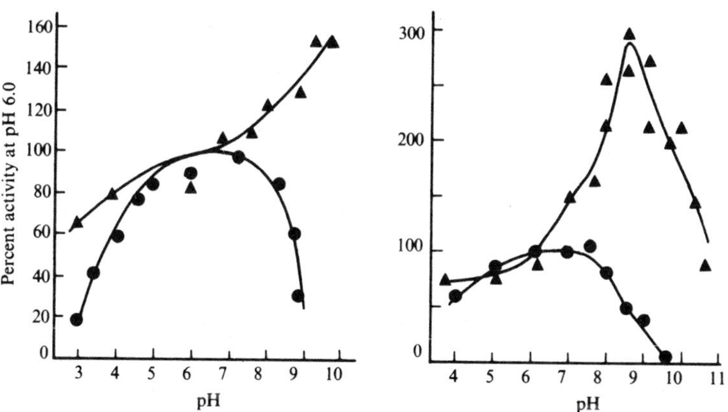

Figure 43. pH–activity profiles obtained with a papain membrane (▲) and with papain in free solution (●). The data on the left-hand graph were obtained by the pH-stat method with a reaction mixture of the following composition: 0.05 M *N*-α-benzoyl-L-arginine ethyl ester, 0.005 M cysteine, and 0.002 M ethylenediaminetetraacetic acid. The data on the right-hand graph were obtained in the same way with the following reaction mixture: 0.015 M *N*-benzoyl-L-glycine ethyl ester, 0.024 M 2.3-dimercapto-propanol, and 0.33 M KCl [Engasser and Horvath (1976)].

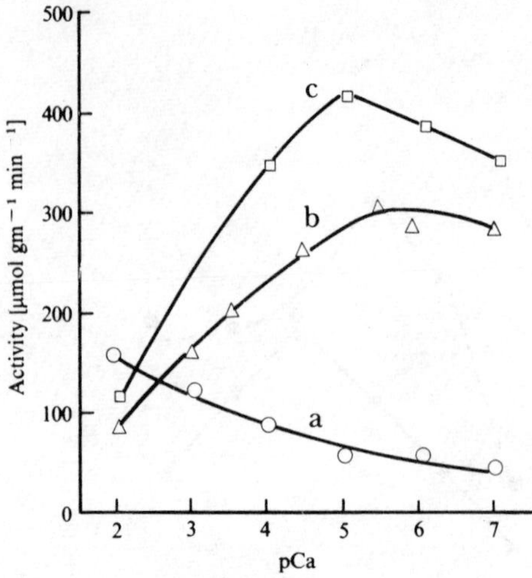

Figure 44. Effect of calcium ion concentration on the activity of trypsin in free solution (a) and in thin and thick polycarboxylic membranes (b and c, respectively). Substrate: 5×10^{-3} M *N*-α-tosyl-L-arginine methyl ester in 10^{-2} M Tris buffer, pH 8.0 [Engasser and Horvath (1976)].

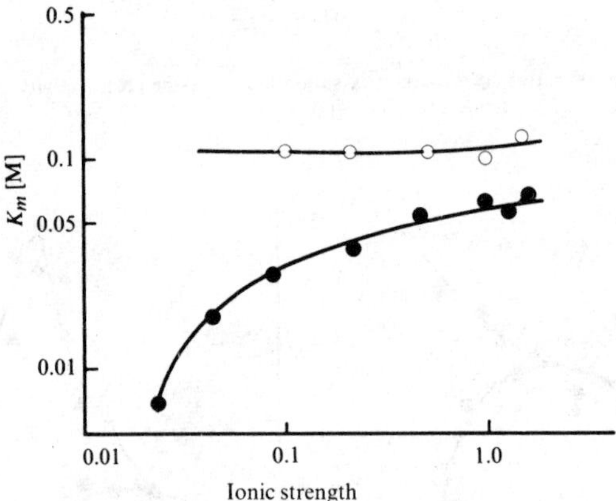

Figure 45. Effect of the ionic strength on the K_m value measured with bromelain in free solution (○) and with bromelain immobilized on carboxymethyl cellulose (●) for the hydrolysis of *N*-α-benzoyl-L-arginine ethyl ester at pH 7.0. The ionic strength was adjusted with potassium chloride [Engasser and Horvath (1976)].

Temperature Effects

Kinetics

The effects of temperature on the activity of soluble and immobilized enzymes are compared for hexokinase (*see* Fig. 46) and glucose oxidase (*see* Fig. 47). Typical Arrhenius plots for free solution and immobilized enzymes are shown in Figs. 48 and 49.

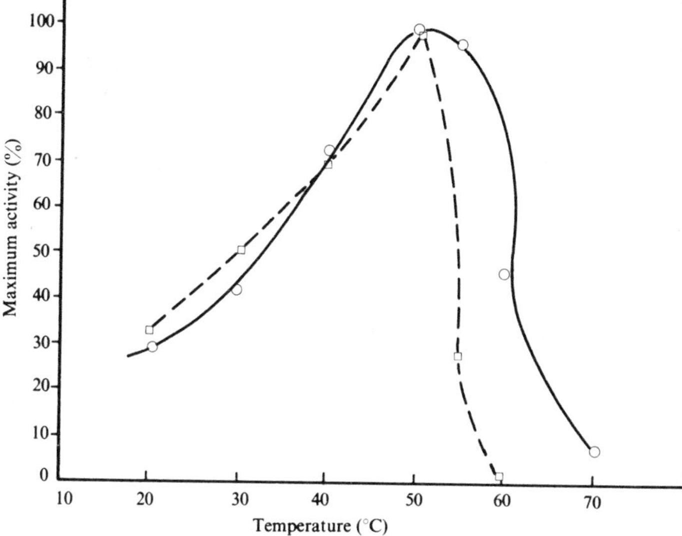

Figure 46. Effect of temperature on activity of hexokinase. □, soluble enzyme; ○, immobilized enzyme [Marshall *et al.* (1972)].

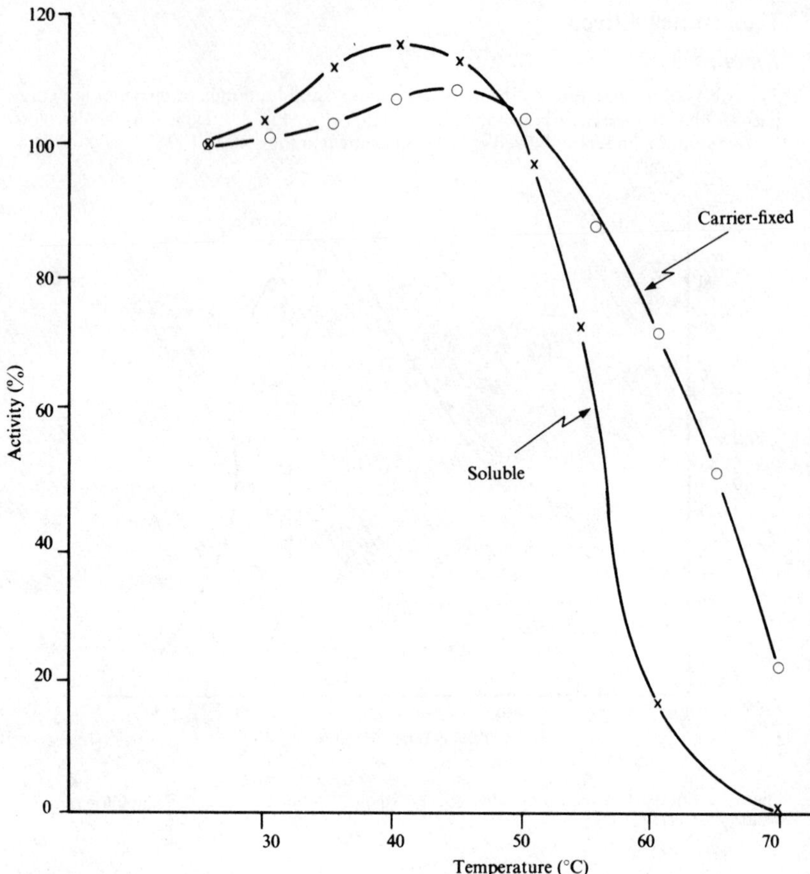

Figure 47. Temperature sensitivity of carrier-fixed and soluble glucose oxidase [Jaworek (1974)].

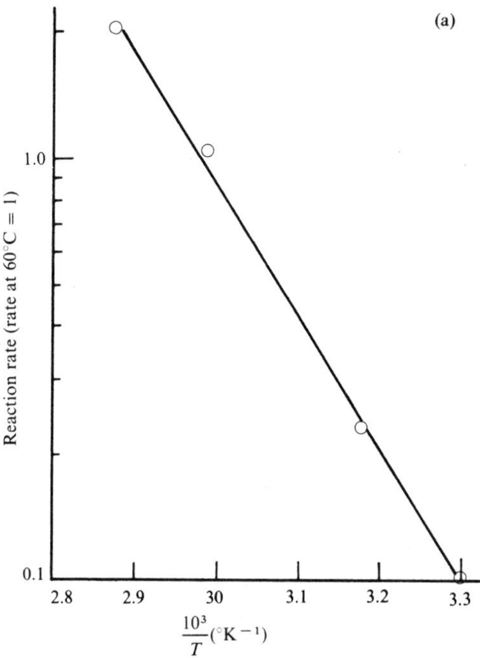

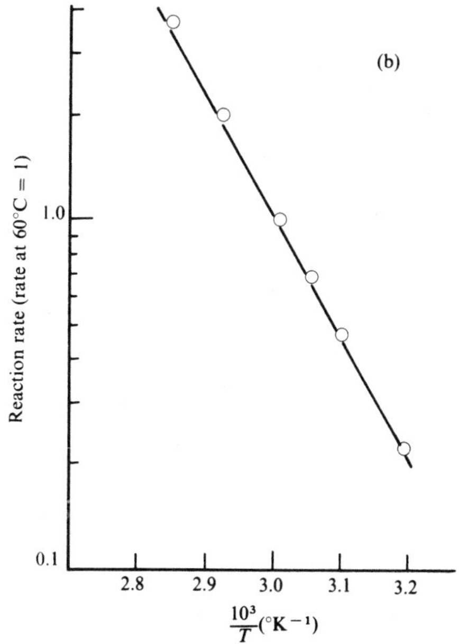

Figure 48. Arrhenius plot for (a) soluble glucose isomerase and (b) immobilized glucose isomerase [Hevewala and Pitcher (1974)].

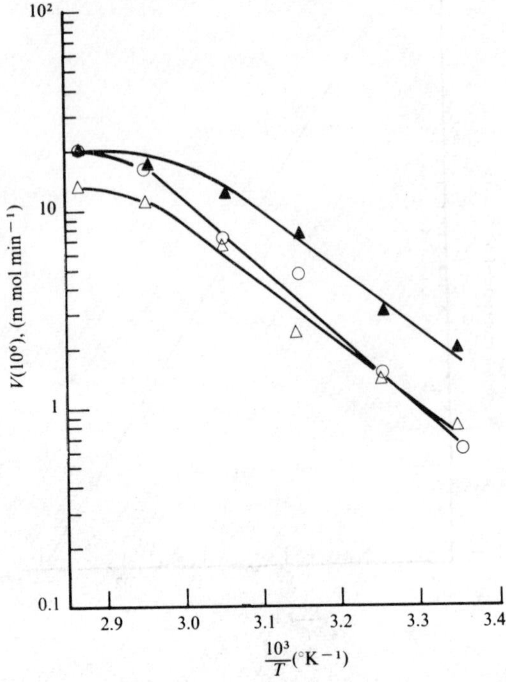

Figure 49. Arrhenius plot for soluble and immobilized enzymes. ○, soluble enzyme; △, crude azo-enzyme; ▲, pure azo-enzyme [Weetall and Havewala (1972)].

Deactivation

The rate of enzyme deactivation can be expressed simply as

$$\frac{d(\text{activity})}{dt} = -k_{\text{in}}t \tag{26}$$

where k_{in} is the deactivation coefficient, which is likely to be temperature-, environment- and immobilization procedure-dependent.

Fig. 50 shows how k_{in} is improved (i.e. reduced) by (1) the presence of substrate, (2) the nature of the substrate and (3) immobilization.

Tables 50–52 provide a range of immobilization techniques and the subsequent influence on k_{in} both in the presence and absence of substrate.

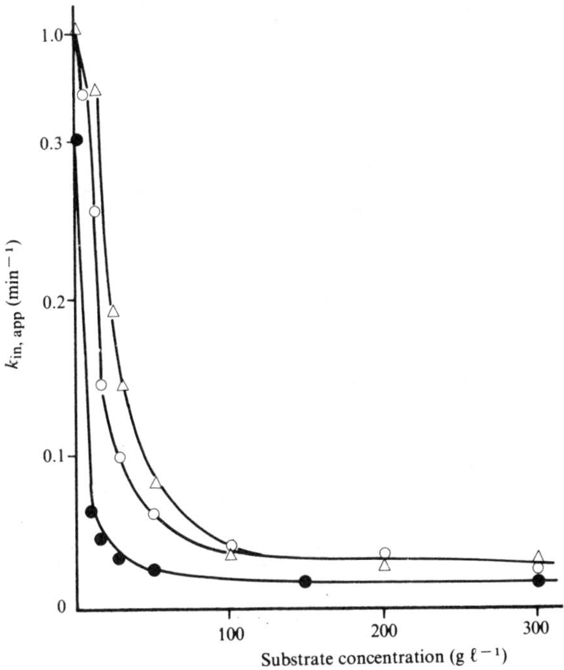

Figure 50. Apparent rate constants for thermal inactivation of glucoamylase at 75°C, 0.1 M acetate buffer (pH 4.5). △, soluble enzyme with maltose substrate; ○, soluble enzyme with maltodextrins as substrate; ●, silica-immobilized enzyme with maltodextrins as substrate [Klyosov *et al.* (1980)].

Table 50. Amyloglucosidase inactivation rate constant in the absence of substrate[a] [Klyosov *et al.* (1980)].

Substrate	k_{in} (min^{-1}) Soluble enzyme	Immobilized enzyme
Maltodextrins	1.0 ± 0.1	0.3 ± 0.05
Maltose	1.0 ± 0.1	

[a] Reaction performed at 75°C, pH 4.5.

Table 51. Thermostability of immobilized amyloglucosidase from *Aspergillus niger*[a] [Klyosov *et al.* (1980)].

Carrier	Immobilization Technique	k_{in} ($min^{-1} \times 10^3$)
	Soluble enzyme	27 ± 3
Silica	Silane glutaraldehyde	9.0 ± 0.5
Silica	Silane glutaraldehyde + acryloylchloride	9.9 ± 0.7
Silica	Silane glutaraldehyde + 30 percent maltodextrins	5 ± 2
Silica	Silane gossipol (dialdehyde)	7.9 ± 0.6
Silica	Silane gossipol (in the presence of 30 percent maltodextrins)	7.3 ± 0.7
20 percent Polyacrylamide gel	Acrolein	13 ± 1
30 percent Polyacrylamide gel	Acrolein	11 ± 1
30 percent Polyacrylamide gel	Acryloylchloride	5.2 ± 0.4
30 percent Polyacrylamide gel	Acryloylchloride (in the presence of 30 percent maltodextrins)	5.2 ± 0.5
Silica + 20 percent polyacrylamide gel	Silane glutaraldehyde + acryloylchloride + covalent coupling to polyacrylamide gel	5.5 ± 0.5

[a] Carried out at 75°C, pH 4.5, in the presence of 30 per cent cornstarch maltodextrins.

Table 52. Inactivation rate constant for amyloglucosidase in acetate buffer at pH 4.7 without substrate[a] [Nakhapetyan and Menyailova (1980)].

Preparation	Rate Constants[a] (h^{-1})			
	40°C	50°C	60°C	70°C
Amyloglucosidase from *Endomycopsis bispora* immobilized on silochrome C-80 via				
Azo coupling	0.03	0.10	0.33	2.9
Glutaraldehyde	0.03	0.08	0.31	2.3
Preoxidized enzyme	0.03	0.10	0.32	2.8
Amyloglucosidase from *Aspergillus niger* immobilized on silochrome C-80 via glutaraldehyde	0.02	0.08	0.27	2.3
Native amyloglucosidase from:				
Asp. niger		0.082		
End. bispora		0.12		

[a] After 4 h at 40–60° or 1 h at 70°C.

IMMOBILIZED WHOLE CELLS

All the processes involved in the immobilization of an enzyme can result in loss of enzyme activity. Fig. 51 compares the enzyme activities of various enzyme preparations and demonstrates that activities can be higher using immobilized cells compared with immobilized enzymes prepared after enzyme isolation. Enzyme activity is also greater using autolysis of whole cells or homogenization of immobilized cells. One can assume that such an increase in enzyme activity is due to the release of diffusion limitations or steric effects.

Possible uses of immobilized cells are listed in Table 53. Immobilization procedures used for *E. coli* are compared in Table 54.

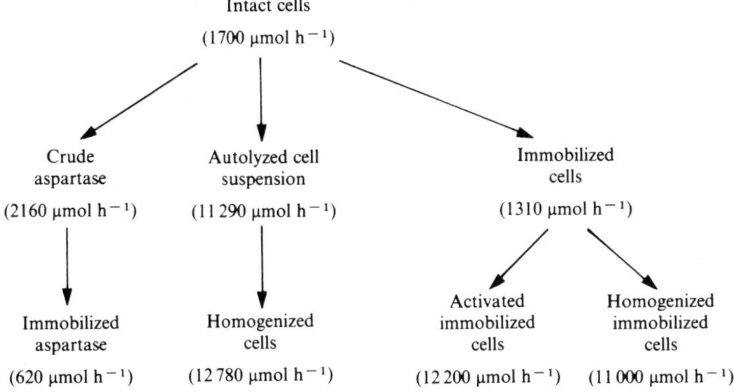

Figure 51. Comparison of aspartase activity of various enzyme preparations per unit of intact cells. One gram (packed wet weight) of inact cells corresponds to 0.2 g of dried cells. Numerical values in parentheses are aspartase activities obtained from 1 g of intact cells. Crude aspartase was immobilized by entrapment in a polyacrylamide gel lattice [Chibata and Tosa (1976)].

Table 53. Use of immobilized cells [Chibata and Tosa (1976)].

Microbial cells	Methods of immobilization	Enzyme system	Products
Achromobacter liquidum	Entrapment by polyacrylamide	Histidine ammonialyase	Urocanic acid
Aspergillus oryzae	Entrapment by cellulose nitrate	Aminoacylase	L-Tryptophan
Brevibacterium ammoniagenes	Entrapment by polyacrylamide	Fumarase	L-Malic acid
		Multiple	Coenzyme A
Corynebacterium glutamicus	Entrapment by polyacrylamide	Multiple	L-Glutamic acid
Curvularia lunata	Entrapment by polyacrylamide	11-β-Hydroxylase	Prednisolone
Escherichia coli	Entrapment by polyacrylamide	Aspartase	L-Aspartic acid
		Penicillin amidase	6-Aminopenicillanic acid
Fungal spore	Adsorption to ECTEOLA-cellulose	Invertase	Glucose and fructose
Lichen	Entrapment by polyacrylamide	Depsidase and orsellinic acid decarboxylase	Orcinol and orcinol monomethyl ether
Pseudomonas putida	Entrapment by polyacrylamide	Arginine deiminase	L-Citrulline
Ps. mephitica	Heat treatment	Lipase	Hydrolyzate of tributyrin and triacetin
Streptomyces sp.	Heat treatment	Glucose isomerase	Fructose

Table 54. Immobilization of *Escherichia coli* [Chibata *et al.* (1974)].

Reagents for immobilization	Aspartase activity (μmol h^{-1})	Activity yield (%)
N,N'-Methylene bisacrylamide	1220	67.0
N,N'-Propylene bisacrylamide	1104	60.7
Diacrylamide dimethylether	1048	57.6
1,2-Diacrylamide ethyleneglycol	1136	62.4
N,N'-Diallyl tartardiamide	1320	72.5
Ethylene urea bisacrylamide	128	7.0
Glutaraldehyde	620	34.2
Toluene diisocyanate	0	0
Toluene diisocyanate + hexamethylene diamine	0	0
1,3,5-Triacryloyl hexahydro-*s*-triazine	128	7.0

Figs. 52–55 provide a variety of data which illustrate the stability of immobilized whole cell preparations.

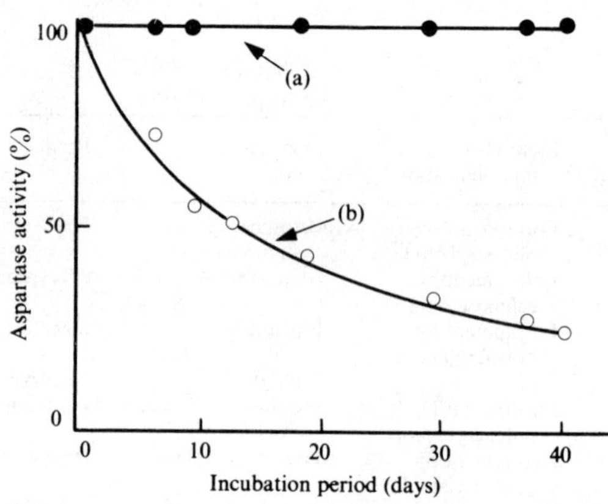

Figure 52. Stability of aspartase activity of immobilized cells and intact cells. Immobilized cells or intact cells were incubated with a solution of 1 M ammonium fumarate (pH 8.5, containing 1 mM Mg^{2+}) at 37°C for 40 days. At appropriate intervals their remaining activities were determined under standard conditions. (a) immobilized cells; (b) intact cells [Chibata and Tosa (1976)].

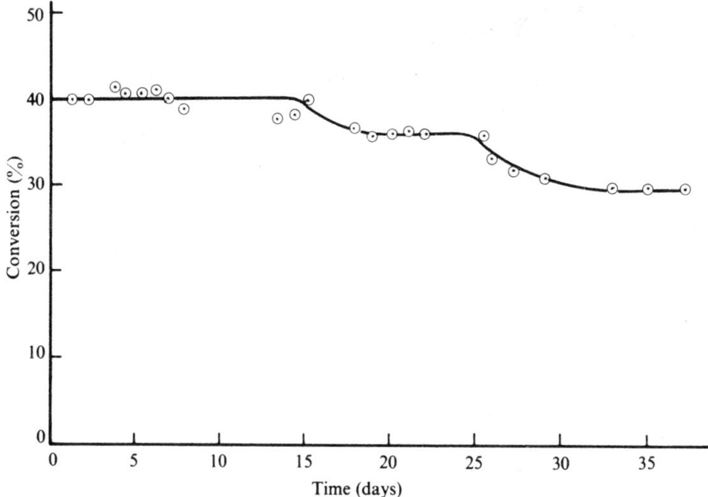

Figure 53. Stability of *Streptomyces phaeochromogenes* immobilized on collagen. Enzymic activity measured was glucose isomerase [Wang *et al.* (1974)].

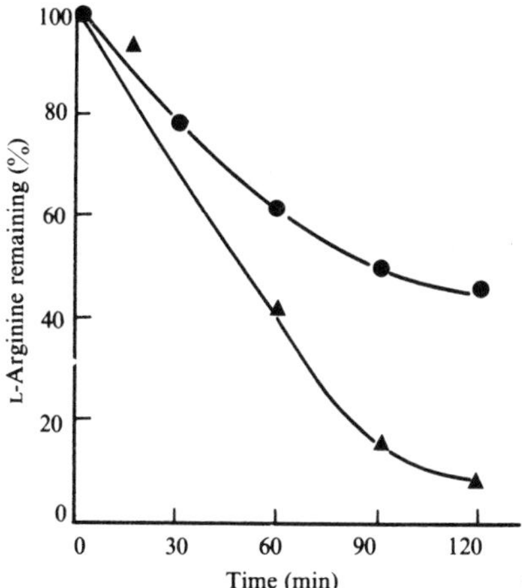

Figure 54. Incubation of arginine with *Streptococcus faecalis* cellular material. Stirring incubation of 400 mg *S. faecalis* ATCC 8043 frozen cell paste (▲) and a 10 per cent (by weight) polyacrylamide gel containing 400 mg of *S. faecalis* ATCC 8043 frozen cell paste (●) with 25 ml of 0.1 M sodium phosphate buffer (pH 7.0) containing L-arginine·HCl 1.2 (5.7 mM) at 30°C.

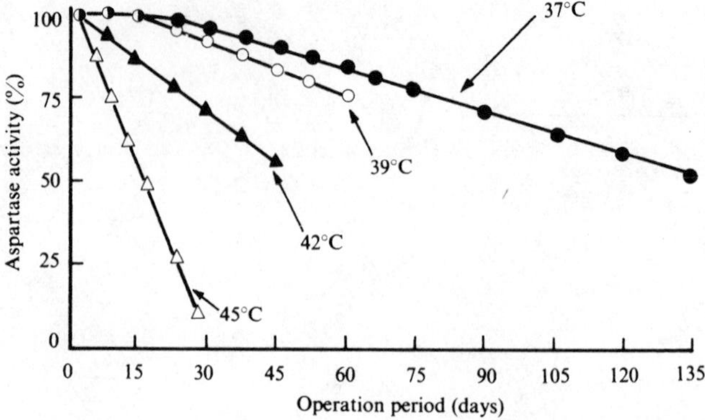

Figure 55. Stability of immobilized *Escherichia coli* cell column at various temperatures. A solution of 1 M ammonium fumarate (pH 8.5, containing 1 mM Mg^{2+}) was applied to the column at a flow rate of S 0.6 h^{-1} at 37°, 39°, 42°, or 45°C for 30–135 days. The activity of the column was determined under standard conditions [Chibata and Tosa (1976)].

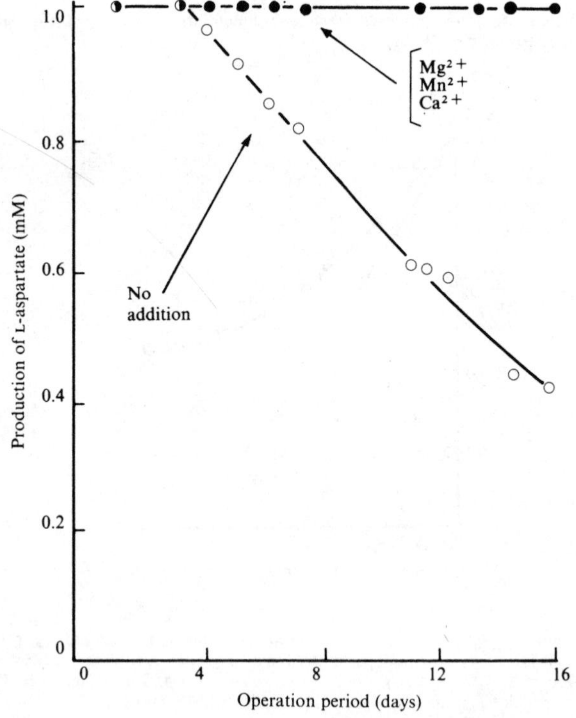

Figure 56. Effect of metal ions on continuous reaction by an immobilized *Escherichia coli* column. Immobilized *E. coli* was packed into a jacketed column (2 × 9.5 cm) and maintained at 37°C. A solution of 1 M ammonium fumarate, pH 8.5, containing 1 mM metal ions was passed through at the flow rate that gave a space velocity of 0.5 [Chibata *et al.* (1974)].

REFERENCES

Bailey, J.E. and Ollis, D.F. (1977) *Biochemical Engineering Fundamentals* (McGraw-Hill, New York).

Böing, J.T.P. (1982) Enzyme Production, in *Prescott and Dunn's Industrial Microbiology*, ed. by G. Reed, p.634 (AVI Publishing, Westport, Macmillan, London).

Brown, H.D., Bartling, G.J. and Chattopadhyay, S.K. (1974) An Organic Millieu in Immobilized Enzyme Synthesis and Catalysis, in *Enzyme Engineering*, vol. 2, ed. by E.K. Pye and L.B. Wingard, jun., p.83 (Plenum Press, New York).

Chibata, I. and Tosa, T. (1976) Industrial Applications of Immobilized Enzymes and Immobilized Cells, in *Applied Biochemistry and Bioengineering*, vol. 1, *Immobilized Enzyme Principles*, ed. by L.B. Wingard, jun. *et al.*, p.329 (Academic Press, New York).

Chibata, I., Tosa, T., Sato, T., Mori, T. and Yamamato, N. (1974) Continuous Enzyme Reactors by Immobilized Microbial Cells, in *Enzyme Engineering*, vol. 2, ed. by E.K. Pye and L.B. Wingard, jun., p.303 (Plenum Press, New York).

Dixon, M. and Webb, E.C. (1979) *Enzymes*, 3rd edn (Longman, London).

Doig, A.R., jun. (1974) Stability of Enzymes from Thermophilic Microorganisms, in *Enzyme Engineering*, vol. 2, ed. E.K. Pye and L.B. Wingard, jun., p.17 (Plenum Press, New York).

Engasser, J.-M. and Horvath, C. (1976) Diffusion and Kinetics with Immobilized Enzymes, in *Applied Biochemistry and Bioengineering*, vol. 1, *Immobilized Enzyme Principles*, p.127 (Academic Press, New York).

Goldstein, L. and Manacke, G. (1976) The Chemistry of Enzyme Immobilization, in *Applied Biochemistry and Bioengineering*, vol. 1, *Immobilized Enzyme Principles*, ed. by L.B. Wingard, jun. *et al.*, p.23 (Academic Press, New York).

Hevewala, N.B. and Pitcher, W.H. (1974) Immobilized Glucose Isomerase for the Production of High Fructose Syrups in Enzyme Engineering, in *Enzyme Engineering*, vol. 2, ed. by E.K. Pye and L.B. Wingard, jun., p.315 (Plenum Press, New York).

Jaworek, D. (1974) New Immobilization Techniques and Supports, in *Enzyme Engineering*, vol. 2, ed. by E.K. Pye and L.B. Wingard, jun., p.105 (Plenum Press, New York).

Klyosov, A.A., Gerasimas, V.B. and Sinitsyn, A.P. (1980) Substrate Stabilization of Soluble and Immobilized Glucoamylase against Heating, in *Enzyme Engineering: Future Directions*, ed. by L.B. Wingard, jun. *et al.*, p.197 (Plenum Press, New York).

Klyosov, A.A. and Rabinowitch, M.L. (1980) Enzymatic Conversion of Cellulose to Glucose: Present State of the Art and Potential, in *Enzyme Engineering: Future Directions*, ed. by L.B. Wingard, jun. *et al.*, p.83 (Plenum Press, New York).

Lilly, M.D., Regan, D.L. and Dunnill, P. (1974) Well-mixed Immobilized Enzyme Reactors, in *Enzyme Engineering*, vol. 2, ed. by E.K. Pye and L.B. Wingard, jun., p.245 (Plenum Press, New York).

Lumry, R., Smith, E.L. and Grantz, R.R. (1951) Kinetics of Carboxypeptidase Action. I. Effects of various extrinsic factors on Kinetic Parameters. *J. Amer. Chem. Soc.*, **73**, 4330.

Marshall, D.L., Walter, J.L., and Fabb, R.D. Preparation and Characterization of Immobilized Hexokinase, in *Enzyme Engineering*, ed. L.B. Wingard, jun., p.195 (Interscience Publishers, New York).

May, S.W. and Li, N.N. (1974) Encapsulation of Enzymes in Liquid Membrane Emulsion, in *Enzyme Engineering*, vol. 2, ed. by E.K. Pye and L.B.Wingard, jun., (Plenum Press, New York).

Mahler, H.R. and Cordes, E.H. (1966) *Biological Chemistry* (Harper & Row, New York).

Melrose, G.J.H. (1971) Insoluble Enzymes. Biochemical Applications of Synthetic Polymers. *Rev. Pure Appl. Chem.*, **21**, 83.

Nakhapetyan, L.A. and Menyailova, I.I. (1980) Immobilized Amyloglucosidase: Preparation, Properties and Application for Starch Hydrolysis, in *Enzyme Engineering: Future Directions*, ed. by L.B. Wingard, jun. *et al.*, p.167 (Plenum Press, New York).

Nomenclature Committee of the International Union of Biochemistry (1979) *Enzyme Nomenclature*.

Samejumi, H. and Kimura, K. (1974) Immobilized Enzymes using Resinous Carriers, in *Enzyme Engineering*, vol. 2, ed. by E.K. Pye and L.B. Wingard, jun., p.131 (Plenum Press, New York).

Slott, S., Madsen, G. and Norman, B.E. (1974) Application of Heat Stable Bacterial Amylase in the Starch Industry, in *Enzyme Engineering*, vol. 2, ed. by E.K. Pye and L.B. Wingard, jun., p.17 (Plenum Press, New York).

Stewart, G.G., Erratt, J., and Garrison, I., Goring, T. and Hancock, (1979) Studies on the Utilization of Wort Carbohydrates by Brewers' Yeast Strains. *MBAA Tech. Quart.*, **16**, 179.

Thomas, J.M. and Thomas, W.J. (1967) *Introduction to the Principles of Heterogeneous Catalysis*. (Academic Press, New York).

Thomas, T., Tran, M.C., Gellf, G., Domurdo, D., Paillot, B., Jacobsen, R. and Broun, G. (19) Films bearing Reticulated Enzymes. Applications to Biological Models and Enzyme Technology, in *Enzyme Engineering*, ed. by L.B. Wingard, jun., p.299 (Interscience Publishers, New York).

Vieth, W.R., Wang, S.S. and Gilbert, S.G. Urea Hydrolysis on Collagen – Urease Complex Membrane, in *Enzyme Engineering*, ed. by L.B. Wingard, jun. p.285 (Interscience Publishers, New York).

Wang, D.I.C., Cooney, C.L., Demain, A.L., Dunnill, P., Humphrey, A.E. and Lilly, M.D. (1979) *Fermentation and Enzyme Technology* (John Wiley, New York).

Wang, S.S., Vieth, W.R. and Constantinides, A. (1974) Complexation of Enzymes or Whole Cells with Collagen, in *Enzyme Engineering*, vol. 2, ed. E.K. Pye and L.B. Wingard, jun., p.123 (Plenum Press, New York).

Weetall, H.H. and Havewala, N.B. Continuous Production of Dextrose from Cornstarch. A Study of Reactor Parameters Necessary for Commercial Application, in *Enzyme Engineering*, ed. L.B. Wingard, jun., p.241 (Interscience Publishers, New York).

Wilkinson, G.N. (1961) Statistical Estimations in Enzyme Kinetics. *Biochem. J.*, **80**, 324.

Zaborsky, O. (1973) *Immobilized enzymes* (CRC Press, Cleveland).

APPENDIX 1: ENZYME MANUFACTURERS

Table A1. Manufacturers of free enzymes.

Enzymes	Manufacturers
Carbohydrate-hydrolyzing enzymes	Aktiebolaget Montoil, Daiwa Kasei K.K., Gist-Brocades N.V., Grindestedvaerket A/S, Hankyu Kyoei Bussan Co Ltd., Laboratories Sanders SA, Meiji Seika Kaisha Ltd., Munton and Fison Ltd., Nagase and Co. Ltd., Röhm GmbH., Royal Netherlands Fermentation Ind. Ltd., Schmitt-Jourdan, G.D. Searle and Co., Societa Prodotti Antibiotici, Yakult Biochemicals Co. Ltd.
Industrial enzymes	Associated British Maltsters Ltd., Hughes and Hughes (Enzymes) Ltd., E. Merck AG, Société Rapidase
Lipase	Nagase and Co. Ltd.
Pharmaceutical enzymes	Armour Pharmaceutical Co. Ltd., CIBA-Geigy Ltd., Diosynth, E. Merck AG, Miles-Seravac (Pty) Ltd., Schering AG, Vifor SA
Proteases	Aktiebolaget Montoil, Akzo Chemie, Daiwa Kasei K.K., Grindestedvaerket A/S, Hankyu Kyoei Bussan Co. Ltd., Meiji Seika Kaisha Ltd., Nagase and Co. Ltd., Novo Industri A/S, Otto Norwald KG, Röhm GmbH, Royal Netherlands Fermentation Ind. Ltd., Schering AG, Schmitt-Jourdan, Yakult Biochemicals Co. Ltd.
Research and analytical enzymes	Boehringer Mannheim GmbH, British Drug Houses, Calbiochem, Gist-Brocades N.V., Hopkins and Williams Ltd., Hughes and Hughes (Enzymes) Ltd., Koch-Light Laboratories Ltd., Laboratories Sanders SA, Miles-Seravac (Pty) Ltd., Otto Norwald KG, PL Biochemicals, G.D. Searle and Co., Sigma Corp., John and E. Sturge Ltd., Whatman Biochemicals Ltd., Worthington Biochemical Corp.

Table A2. Commercially available water-insoluble, polymer-bound enzymes [Zaborsky (1973)].

Immobilized enzyme	EC number	Matrix	Functional group of polymer used for immobilization	Supplier
Acid phosphatase	3.1.3.2	Polyacrylamide	Carbonyl	Worthington Biochemical Corp.
Alcohol dehydrogenase	1.1.1.1	CM-Cellulose	Triazinyl chloride	Boehringer Mannheim Corp.
Alkaline phosphatase	3.1.3.1	Sepharose	Imidocarbonate	Worthington Biochemical Corp.
α-Amylase	3.2.1.1	APHP-Cellulose	Diazonium	Miles Laboratories Inc.
Bromelain	3.4.22.4	CM-Cellulose	Acylazide	EM Laboratories Inc. Gallard-Schlesinger Chemical Mfg. Corp. Miles Laboratories Inc.
Carboxypeptidase A	3.4.17.1	Sepharose	Imidocarbonate	Worthington Biochemical Corp.
α-Chymotrypsin	3.4.21.1	Agarose	Triazinyl chloride	Miles Laboratories Inc.
		CM-Cellulose	Acylazide	EM Laboratories Inc. Gallard-Schlesinger Chemical Mfg. Corp. Miles Laboratories Inc.
		Maleic anhydride/ethylene	Anhydride	Miles Laboratories Inc.
		Polyacrylamide	Unknown	Boehringer Mannheim Corp.
Ficin	3.4.22.3	CM-Cellulose	Acylazide	EM Laboratories Inc. Gallard-Schlesinger Chemical Mfg. Corp. Miles Laboratories Inc.
Glucose oxidase	1.1.3.4	CM-Cellulose	Acylazide	Gallard-Schlesinger Chemical Mfg. Corp. Miles Laboratories Inc.
		DEAE-Cellulose	Triazinyl chloride	Miles Laboratories Inc.
		Sepharose	Imidocarbonate	Worthington Biochemical Corp.
Leucine aminopeptidase	3.4.11.1	DEAE-Cellulose	Triazinyl chloride	Miles Laboratories Inc.
Papain	3.4.22.2	Agarose	Triazinyl chloride	Miles Laboratories Inc.
		CM-Cellulose	Acylazide	EM Laboratories Inc. Gallard-Schlesinger Chemical Mfg. Corp. Miles Laboratories Inc.
		Maleic anhydride/ethylene	Anhydride	Miles Laboratories Inc.
		Maleic anhydride/divinyl ether	Anhydride	EM Laboratories Inc.
		Polyacrylamide	Unknown	Boehringer Mannheim Corp.
		S-DMA	Diazonium	Miles Laboratories Inc.
Peroxidase	1.11.1.7	CM-Cellulose	Acylazide	Miles Laboratories Inc.

continued overleaf

Table A2 – (continued)

Immobilized enzyme	EC number	Matrix	Functional group of polymer used for immobilization	Supplier
Pronase	3.4.4.– + 3.4.1.–	CM-Cellulose	Acylazide	EM Laboratories Inc.
Protease (from *Streptomyces griseus*)	3.4.22.10	Agarose	Triazinyl chloride	Miles Laboratories Inc.
Protease K	3.4.21.14	CM-Cellulose	Acylazide	EM Laboratories Inc.
Ribonuclease A	3.1.4.22	CM-Cellulose	Acylazide	EM Laboratories Inc.
		Maleic anhydride/divinyl ether	Anhydride	EM Laboratories Inc.
		Sepharose	Imidocarbonate	Worthington Biochemical Corp.
Subtilisin (from *Bacillus subtilis*)	3.4.21.14	CM-Cellulose	Acylazide	EM Laboratories Inc.
Subtilopeptidase A	3.4.21.14	Maleic anhydride/ethylene	Anhydride	Miles Laboratories Inc.
Subtilopeptidase B	3.4.21.14	Maleic anhydride/ethylene	Anhydride	Miles Laboratories Inc.
Trypsin	3.4.21.4	Agarose	Triazinyl chloride	Miles Laboratories Inc.
		CM-Cellulose	Acylazide	EM Laboratories Inc.
				Gallard-Schlesinger Chemical Mfg. Corp.
				Miles Laboratories Inc.
		Maleic anhydride/divinyl ether	Anhydride	EM Laboratories Inc.
		Maleic anhydride/ethylene	Anhydride	Miles Laboratories Inc.
		Polyacrylamide	Unknown	Boehringer Mannheim Corp.
		Sepharose	Imidocarbonate	Worthington Biochemical Corp.
Trypsin (polytyrosyl derivative)		S-DMA	Diazonium	Miles Laboratories Inc.
Urease	3.5.1.5	DEAE-Cellulose	Triazinyl chloride	Miles Laboratories Inc.

Table A3. Immobilized glucose isomerase [Marconi (1980)].

Enzyme source	Immobilization method	Company
Streptomyces albus	Temperature fixing inside the cells	Agency of Industrial Science and Technology
Strep. sp.[a]	Adsorption onto DEAE cellulose	Standard brands
Arthrobacter	Aggregation with a flocculating agent	Reynolds Tobacco Co.
Bacillus coagulans	Crosslinking with glutaraldehyde	Novo Industri A/S
Strep. phaeochromogenes[a]	Adsorption on phenol formaldehyde resin	Kyowa Hakko Kogyo
Strep. sp.[a]	Adsorption on special porous alumina	Corning Glass Works
Strep. phaeochromogenes	Adsorption on special ion exchange resin	Mitsubishi Chemical Industries
Actinoplanes missouriensis	Occlusion in gelatin and crosslinking with glutaraldehyde	Gist-Brocades N.V.
Strep. olivaceus	Crosslinking with glutaraldehyde	Miles Laboratories Inc.
Streptomyces sp.[a]	Entrapment in fibres	Snamprogetti S.p.A.

[a] Enzyme from this source; other entries refer to cells.

References

Marconi, W. (1980) Immobilized Enzymes in Nutritional Applications, in *Enzyme Engineering: Future Directions*, ed. by L.B. Wingard, jun. *et al.*, p.465 (Plenum Press, New York).
Zaborsky, O. (1973) *Immobilized Enzymes* (CRC Press, Cleveland).

APPENDIX II: FREE SOLUTION KINETICS OF GLUCOSE OXIDASE [Atkinson and Lester (1974)]

Glucose oxidase catalyzes the formation of δ-gluconolactone from β-D-glucose with the simultaneous production of hydrogen peroxide. The gluconolactone formed is then hydrolyzed to gluconic acid, the overall stoichiometry being expressed as

$$\text{Glucose} + \tfrac{1}{2}O_2 \rightarrow \text{acid} \tag{A1}$$

provided that (1) the glucose available can be assumed to exist entirely in the β form, i.e. the equilibrium between α and β anomers is maintained under all conditions and (2) the hydrogen peroxide is decomposed rapidly and stoichiometrically with the formation of oxygen by adding catalase to the gel.

Thus the rate of consumption of glucose r_g and oxygen r_O, and the rate of formation of acid r_a are related thus

$$r_a = -r_g = -2r_O \tag{A2}$$

To formulate a function which relates the rate of acid formation r_a to the substrate concentrations, etc. requires the most general reaction mechanism as given by Eq. (A3). The mechanisms suggested by Gibson *et al.* (1964), Bright and Gibson (1967) and Duke *et al.* (1969) are special cases of this general scheme.

$$\begin{aligned}
&E_O + G \underset{k_{-1}}{\overset{k_{+1}}{\rightleftharpoons}} E_OG \\
&E_OG \xrightarrow{k_6} E_rL \\
&E_rL \xrightarrow{k_2} E_r + L \\
&E_r + O_2 \xrightarrow{k_3} E_OP \\
&E_OP \xrightarrow{k_4} E_O + P \\
&E_rL + G \xrightarrow{k_5} E_r + L + G \\
&L \xrightarrow{k_H} A
\end{aligned} \tag{A3}$$

where G refers to glucose, P to the product hydrogen peroxide, L to the lactone, A to gluconic acid, E_r to the reduced form of the enzyme–coenzyme complex E–$FADH_2$ and E_O to the oxidized form of the complex E–FAD.

From Eq. (A3) it can be seen that the hydrolysis of the lactone is a chemical step that is relatively slow. However, the preceding step is irreversible and consequently a steady state will be achieved by the lactone concentration rising to a sufficently high level such that the molal rate of consumption of glucose and decomposition of the lactone are equal.

The conventional algebra of enzyme chemistry when applied to Eq. (A3) leads to

$$\frac{1}{r_a} = \frac{k_1 + k_6}{k_1 k_6 s_g} + \frac{1}{k_3 s_O} + \frac{1}{k_2 + k_5 s_g} + \frac{1}{k_4} + \frac{1}{k_6} \tag{A4}$$

where s_g and s_O are the concentrations of glucose and oxygen, and r_a is the specific rate of acid formation.

For the present purposes, it is convenient to use the simplifying assumptions of Gibson *et al* (1964), i.e. $k_5 = 0$ and $k_6 = \infty$, which suggest that the formation of the enzyme–lactone complex is very rapid and that the presence of glucose does not affect its decomposition significantly. Thus Eq. (A4) is reduced to

$$\frac{1}{r_a} = \frac{1}{k_1 s_g} + \frac{1}{k_3 s_O} + \frac{k_2 + k_4}{k_2 k_4} \tag{A5}$$

Eq. (A5) represents a typical 'ping-pong' mechanism and may be written as

$$\frac{r_{a.\max}}{r_a} = \frac{\beta_g}{s_g} + \frac{\beta_O}{s_O} + 1 \tag{A6}$$

where $r_{a.\max}$, β_g and β_O are evaluated at the optimum pH of the enzyme in free solution and

$$r_{a.\max} = \frac{k_2 k_4}{k_2 + k_4}$$

$$\beta_g = \frac{k_2 k_4}{k_1 (k_2 + k_4)}$$

$$\beta_O = \frac{k_2 k_4}{k_3 (k_2 + k_4)}$$

Eq. (A5) has been extended by Bright and Appleby (1969) to include the influence of pH (3–8) on the various constants k_I. The most convenient procedure for generalizing Eq. (A6), using the formulations of Bright and Appleby (1969), is

$$\frac{r_{a.\max}}{r_a} G_I s_H = \frac{\beta_g G_g(s_H)}{s_g} + \frac{\beta_O G_O(s_H)}{s_O} + 1 \tag{A7}$$

where s_H represents the hydrogen ion concentration and $G_I(s_H)$ are dimensionless pH functions given by

$$G_I(s_H) = \frac{1}{1 + s_3 + s_4}$$

$$G_g(s_H) = (1 + s_1) G_1$$

$$G_O(s_H) = (1 + s_2) G_1$$

and (A8a)

$$s_1 = \frac{s_H}{K_1} \qquad S_2 = \frac{K_4}{s_H} \tag{A8b}$$

$$S_3 = \frac{k_2}{k_2 + k_4} \frac{s_H}{K_5} \qquad S_4 = \frac{k_2}{k_2 + k_4} \frac{K_5^1}{s_H}$$

In Eq. (A8), K_1, K_4, K_5 and K_5^1 are the equilibrium constants of the oxidation–reduction reactions

$$\mathrm{H} + \mathrm{E}_O \overset{K_1}{\rightleftharpoons} \mathrm{E}_O + \mathrm{H}^+$$

$$\mathrm{H}^+ + \mathrm{E}_r \overset{K_4}{\rightleftharpoons} \mathrm{E}_r\mathrm{H}^+$$

$$\begin{array}{ccc} \mathrm{E}_O\mathrm{PH}^+ & \overset{K_5}{\rightleftharpoons} & \mathrm{E}_O\mathrm{P} \\ \upharpoonleft\downharpoonright K_5' & & \upharpoonleft\downharpoonright K_5' \\ \mathrm{H}^+\,\mathrm{E}_O\mathrm{PH}^+ & \overset{K_5}{\rightleftharpoons} & \mathrm{H}^+\mathrm{E}_O\mathrm{P} \end{array} \qquad \text{(A9)}$$

pH functions $G_I(s_H)$ are given in Fig. A1 for the values of the coefficients k_I and K_I given in Table A1. Values of k_I were obtained by interpolation from log-linear regressions of each k_I versus temperature, using the data of Gibson *et al.* (1964), Bright and Gibson (1967), Bright and Appleby (1969) and Duke *et al.* (1969). Values of K_I are based on the data of Bright and Appleby (1969).

Table A4. Values of the kinetic coefficients k_I and K_I

Coefficient	Dimensions	Correlation coefficient	Number of points	Value at 25°C
k_1	$\ell\ mol^{-1}\ s^{-1}$	0.904	12	11944
k_2	s^{-1}	0.807	5	5093
k_3	$\ell\ mol^{-1}\ s^{-1}$	0.906	12	2.15×10^6
k_4	s^{-1}	0.979	12	1171
K_1	$mol\ \ell^{-1}$			10^{-4}
K_4	$mol\ \ell^{-1}$			1.26×10^{-7}
K_5	$mol\ \ell^{-1}$			7.94×10^{-5}
K_5'	$mol\ \ell^{-1}$			4.0×10^{-8}

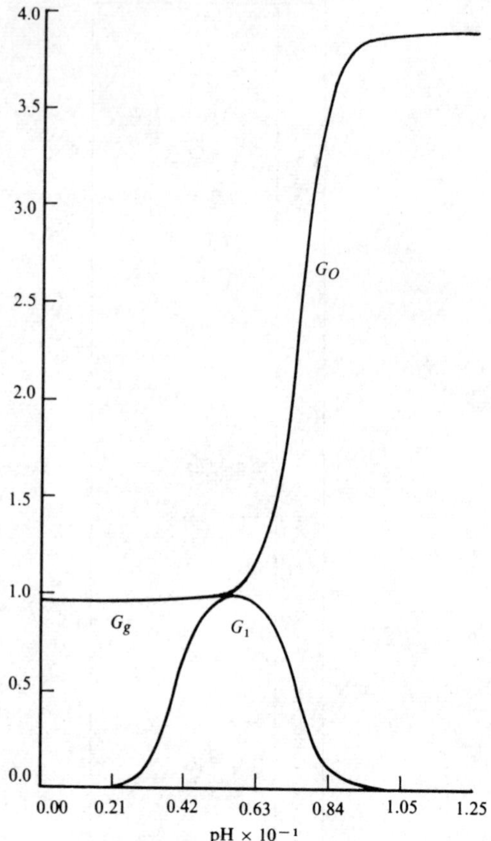

Figure A1. pH functions G_I (S_H)

It can be seen from Fig. A1 that $r_{a,max}$ has a characteristic bell-shaped relationship with pH, while the pH dependencies of the Michaelis coefficients β_g and β_O are quite different.

References

Atkinson, B. and Lester, D.E. (1974) An Enzyme Rate Equation for the Overall Rate of Reaction of Gel-Immobilized Glucose Oxidase Particles under Buffered Conditions. I. Pseudo-One Substrate Conditions. *Biotechnol. Bioengng*, **16**, 1299.

Bright, H.J. and Appleby, M. (1969) pH Dependence of the Individual Steps in the Glucose Oxidase Reaction. *J. Biol. Chem.*, **244**, 3625.

Bright, H.J. and Gibson, Q.H. (1967) The Oxidation of 1-Deuterated Glucose by Glucose Oxidase. *J. Biol. Chem.*, **242**, 994.

Duke, F.R., Weibel, M., Page, D.S., Bulgrin, V.G. and Luthy, J. (1969) Glucose Oxidase Mechanism. Enzyme Activation by Substrate. *J. Amer. Chem. Soc.*, **91**, 3904.

Gibson, Q.H., Swoboda, B.E.P. and Massey, V. (1964) Kinetics of and Mechanism of Action of Glucose Oxidase. *J. Biol. Chem.*, **239**, 3927.

CHAPTER 7

REACTORS

GLOSSARY

Aerobic reactor Reactor operated under aerobic conditions with air being used as the source of dissolved oxygen.

Anaerobic reactor Reactor in which no dissolved oxygen or nitrate is present and microbial activity is due to anaerobic bacteria.

Anoxic reactor Reactor in which no dissolved oxygen is present, but biochemical oxidation takes place by aerobic bacteria (facultative heterotrophs – for definition see Glossary to Chapter 1) using oxygen from the nitrate ion.

Biomass hold-up Biomass contained in a reactor or associated with a support particle.

Expanded bed Bed of small particles (typically, in range 0.2–2.0 mm) in which the biomass grows on the surface of the particles, hence causing expansion of the bed. Bed expansion may also be caused by (1) gas hold-up within the bed or (2) rearrangement of the particles to form a loose packed bed. The particles appear to be stationary but the bed is seen to expand slowly over a period of days.

Fluidized bed Bed of small particles freely suspended in upward flow of liquid or combined liquid and air flow. At the point of minimum fluidization, the pressure drop across the bed equals the total weight of particles in the bed (corrected for their buoyancy in the liquid). Increase in flow beyond the minimum fluidization velocity results in expansion of the bed by increasing the voidage.

Oxygenic reactor Reactor in which commercial-grade oxygen is used to provide the dissolved oxygen necessary for biochemical oxidation using aerobic bacteria.

Particle porosity Space taken up by liquid and microorganism within a biomass support particle.

Voidage Space occupied by gas and liquid in a fixed or fluidized bed.

INTRODUCTION

The total fermentation capacity in the United Kingdom is listed in Table 1. Of these fermentation processes, the majority are operated batch-wise, the only significant exception being biological wastewater treatment. Batch processes provide economic and operational advantages, with single units being used to produce a range of products.

Table 1. Estimated total United Kingdom fermentation capacity [Dunnill (1981)].

Product	Total capacity (m^3)
Waste-water treatment	2 800 000
Beer	128 000
Antibiotics	10 500
Cheese	3000
Bakers' yeast	1900
Bread	700

It has been shown to be technically feasible to establish continuous, aseptic processes, such as in the production of beer described by Ault *et al.* (1969) and by Coutts (1958, 1961) and also of bakers' yeast described by Olsen (1960). However, despite the technical feasibility of these continuous processes, economic factors due to scales of operation have prevented the general changeover to continuous processes.

Outside the food and beverage areas, fermentation processes were initially associated with organic acids and solvents and, subsequently, with the production of low-volume, high-added-value products, such as vitamins, antibiotics and vaccines. More recently, there has been a trend to high-volume, low-added-value products, such as single-cell protein and fuels. The deciding factor as to whether such processes are economically feasible is the capital cost of the single-purpose equipment required.

The following are some of the characteristic features of biochemical process plant.

1. Volumetric rates of reaction are low.
2. Plant is relatively large (*see* Table 2) compared with other process industries of comparable production scale.
3. Power consumption for mixing, aeration, etc. is high (*see* Chapter 9).
4. Separation costs are high as product concentrations are low and cells are present (*see* Chapter 12).
5. The capital costs of product recovery equipment are high (*see* Chapter 13).

Table 2. Size of United Kingdom fermenters [Dunnill (1981)].

Product	Maximum size (m^3)
Wastewater treatment (activated sludge)[a]	27 000
Single-cell protein[a]	1 500
Beer	320
Citric acid	240
Bakers' yeast	200
Antibiotics	200
Cheese	20
Yoghurt	10
Bread	1

[a] Continuous rather than batch-wise process.

The approach, often described as process intensification, for overcoming some of these factors is considered on pp.000–000.

An important requirement is for the design engineer to have available a range of configurations of reactors and separation devices from which to choose in order to meet particular process specifications. The principles for the design of such equipment needs to be established and the methodology of design based upon laboratory and pilot-plant experimentation must be well understood.

Configurations

Fermenters of various types are depicted diagrammatically in Fig. 1. A feature of most reactors is that they must remain aseptic after initial sterilization, thus preventing the introduction of contaminating microorganisms with different physical and chemical properties to the elected strain with its ability to carry out a particular biochemical conversion.

The simplest system available is a batch fermenter, which is presented in Fig. 1(a). The decline in substrate concentration with time for a batch fermenter is shown in Fig. 2(a).

A train of continuous stirred-tank fermenters, as depicted in Fig. 1(b), allows possibilities for

1. High substrate conversion efficiencies.
2. Improved volumetric rates of reaction, because high substrate concentrations can be allowed in the initial reactors, while high biomass concentrations are a feature of the final reactors.
3. Changing the conditions (i.e. pH, temperature, aeration, etc.) from reactor to reactor.

Fig. 2(b) shows a step-wise drop in substrate concentration as the broth passes from fermenter to fermenter and the feed distributed between the fermenters.

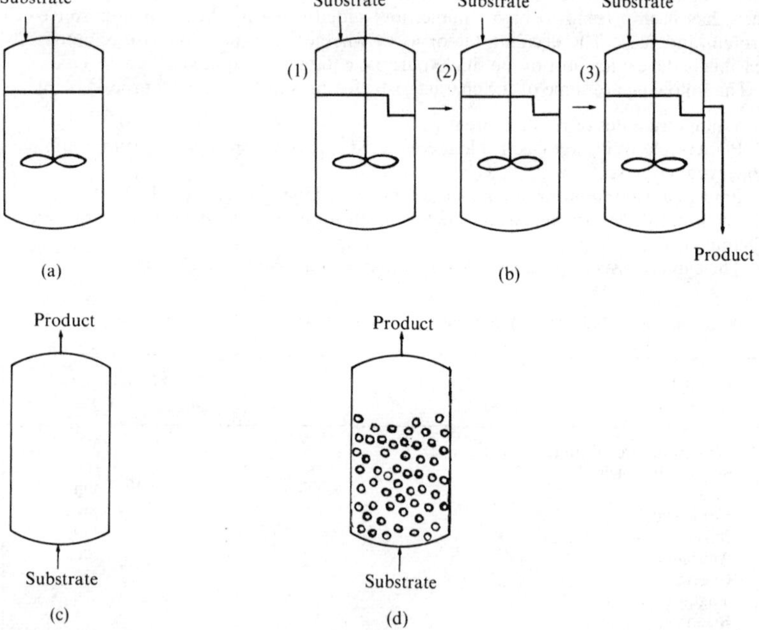

Figure 1. Fermenter configurations. (a) Batch fermenter, (b) continuous stirred-tank fermenters, (c) tubular fermenter, (d) fluidized-bed fermenter [Atkinson (1974)].

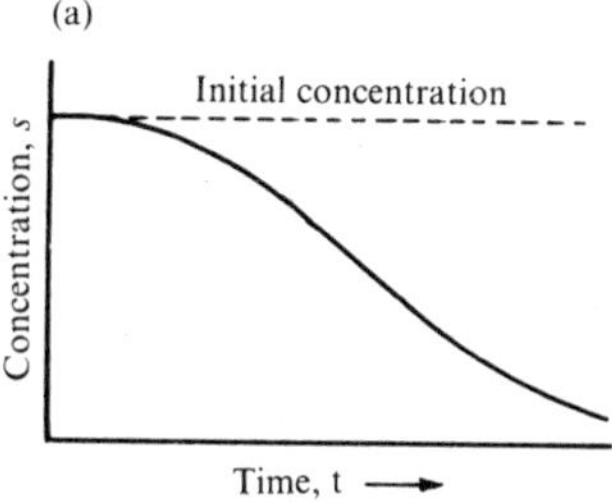

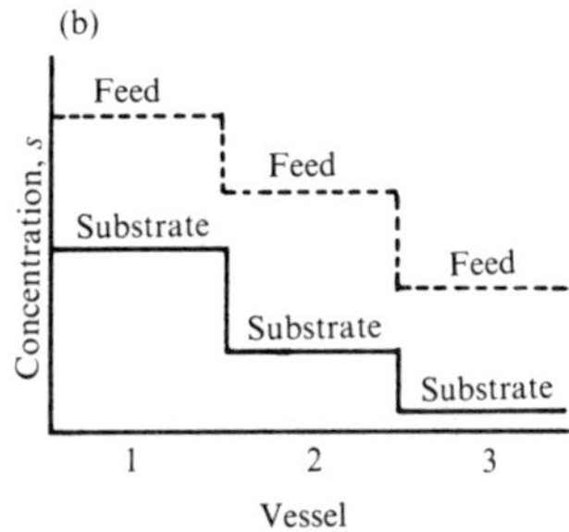

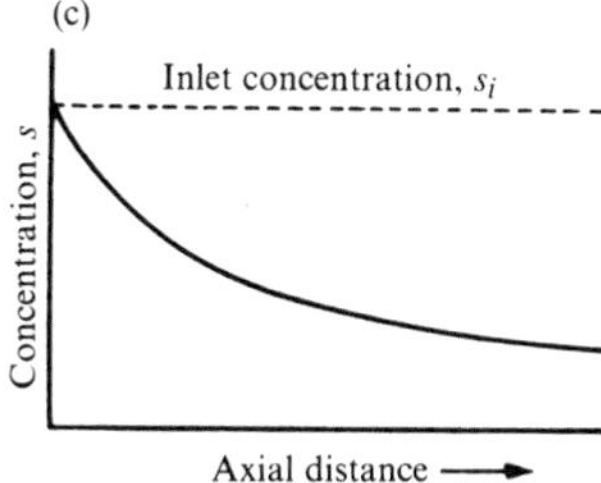

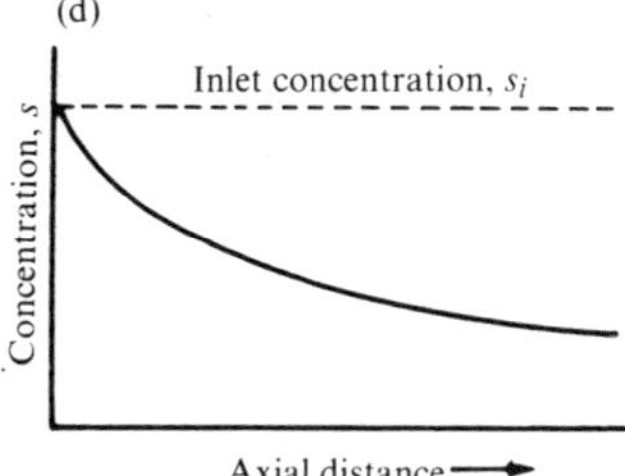

Figure 2. Substrate concentration profiles for different fermenter configurations. (a) Batch fermenter, (b) train of continuous stirred-tank fermenters, (c) tubular fermenter, (d) fluidized-bed fermenter [Atkinson (1974)].

An alternative to the continuous stirred-tank arrangement is a tubular fermenter [*see* Fig. 1(c)]. Tubular fermenters contain either suspended microorganisms or microbial films on support surfaces. The former arrangement requires a constant input of microorganisms because they are elutriated continuously with the product. The change in substrate concentration in the direction of flow through the fermenter is shown in Fig. 2(c).

Microbial flocs, microorganisms associated with particulate supports and immobilized enzyme particles prepared by one of the techniques described in Chapter 6 (*see* pp.540–541) can be arranged as a fluidized bed [*see* Fig. 1(d)]. The substrate concentration profile in such a fermenter is usually similar to that for a tubular fermenter [*see* Fig. 2(d)].

Variations in the substrate, product and biomass concentration with time and position within the fermenter are summarized in Table 3 for the fermenter types represented in Fig. 1. The operational features of these fermenters are considered in Table 4 with their current areas of industrial application.

Table 3. Concentration histories in fermenters [Atkinson (1974)].

Fermenter type	Time dependence of substrate, microbial mass and biochemical product concentrations	Variation of substrate and biochemical product concentrations with position	Variation of microbial mass concentration with position	Environmental history of microbial mass
Batch	Dependent	Completely mixed (ideally)	Completely mixed (ideally)	Varies over course of the fermentation
Continuous stirred-tank	Independent	Completely mixed (ideally)	Completely mixed (ideally)	Constant (all flocs exposed to the same environment at all times)
Tubular				
Containing microbial 'flocs'	Independent	Varies from inlet to outlet	Varies from inlet to outlet	Varies as flocs travel through the fermenter
Containing microbial films	Largely independent but some accumulation of microbial film may occur; this will also affect the substrate and product concentrations	Varies from inlet to outlet	Largely independent of position	Constant (however, the microbial film in different parts of the fermenter is exposed to different environments)
Fluidized bed	Independent	Varies from inlet to outlet	Varies from inlet to outlet	Largely constant but some movement of flocs does take place (flocs in different parts of the fermenter are exposed to different environments)

Table 4. Operational features of fermenters [Atkinson (1974)].

Fermenter type	pH control	Temperature control	Features of industrial importance	Chief industrial applications
Batch	If required	If required	Labour intensive	Most commercial fermentations
Continuous stirred-tank	If required	If required	Flow rate limited by wash-out	Wastewater treatment, microbial protein production
Tubular				
Containing microbial flocs	Difficult to control (except at high recirculation rates)	If required	Requires a constant feed of microorganisms	
Containing microbial films	Difficult to control	If required	Difficult to control microbial hold-up	Wastewater treatment, vinegar production
Fluidized bed	Difficult to control	If required	Flow rate limited by wash-out	Beer, cider and vinegar production, wastewater treatment

The reactor configurations can affect the process yield coefficients (*see* p.137) due to variations in the environment to which the microorganisms are exposed. For example, changes in osmotic pressure, temperature and pH may result in stress to the microorganisms leading to a reduction in the yield coefficient. The relationship between the process yield coefficients and the reactor configuration is shown in Table 5.

Table 5. Interaction between reactor configuration and process yield coefficients [Atkinson and Kossen (1978)].

Environment	Reactor configuration	Process yield coefficient
Steady-state, completely mixed 'chemostat' environment		'Stoichiometric' yield coefficient
Batch, unsteady-state, completely mixed, closed system		Reflects natural time course of the fermentation
Batch, unsteady-state completely mixed, partly open system		Reflects controlled modifications to natural time course of the fermentation
Steady-state, microorganisms pass through a series of zones maintained in different states, e.g., pH, temperature, dissolved oxygen, substrate concentrations		Reflects the 'history' of microorganisms passing through complete system, e.g., carbon conversion followed by extended aeration to destroy biomass in wastewater treatment
As above with recycle of microorganisms		Reflects repeated and rapid exposure of microorganisms to an environment gradient, e.g., dissolved oxygen
Immobilized organisms, exposed to a variety of environments induced by diffusional gradients of reactants/products; microorganism 'age' dependent on growth rates and sloughing/attrition mechanisms		Reflects concentration gradients, organism 'age' and in wastewater treatment, the variety of species

Biomass Hold-Up

The volumetric rates of reaction in fermenters are dependent upon the microbial concentrations. An increase in biomass levels increases the rates of reaction, provided that the consequent increase in viscosity does not lead to reduced aeration (*see* Chapter 9) and other transport problems. Clearly, the biomass concentration must be kept below the limits of fluidity (*see* Chapter 12).

1. In a stirred, batch fermenter [*see* Fig. 1(a)] the biomass concentration at time t is given by

$$x(t) = x_0 + Y'_s\,[s_o - s(t)] \tag{1}$$

The final biomass concentration is

$$x_f \simeq Y_s' s_o \tag{2}$$

i.e. it is proportional to the initial substrate concentration.

2. In a single continuous stirred-tank fermenter or the first one in a series [*see* Fig. 1(b)]

$$x_f = Y_s' (s_i - s_o) \tag{3}$$

where x_f is the biomass concentration in the fermenter. For complete conversion

$$x_f = Y_s' s_i \tag{4}$$

i.e. the biomass concentration is proportional to the inlet substrate concentration s_i.

3. In a train of continuous stirred-tank reactors [*see* Fig. 1(b)] transfer of biomass between tanks enhances the biomass concentration in the downstream tanks. If substrate is added to each tank, this contributes to enhanced biomass concentration by providing additional substrate for conversion to biomass; the effect is mitigated to some extent because the additional flow 'dilutes' the biomass concentration transferred tank to tank. That care has to be exercised is illustrated by the fact that for two tanks in series with the feed split equally

$$\begin{aligned} x_{f,1} &= Y_s' s_i \\ x_{f,2} &= Y_s' s_i \end{aligned} \tag{5}$$

i.e., for complete conversion in each tank, the effect of splitting the flow is to provide a constant biomass concentration tank to tank.

4. In a continuous stirred-tank fermenter with two outlet streams and either a filter (*see* Fig. 3) or a sedimentation zone (*see* Fig. 4)

$$x_f = \frac{Y_s' (s_i - s_o)}{W} \tag{6}$$

where

$$W = 1 - \tau(1 - \gamma) \tag{7}$$

$\tau(< 1)$ represents the flow split between exit streams, and $\gamma(< 1)$ represents the effectiveness of the filter. It follows that $W < 1$ and that the biomass concentration x_f is enhanced by the arrangements in Figs. 3 and 4.

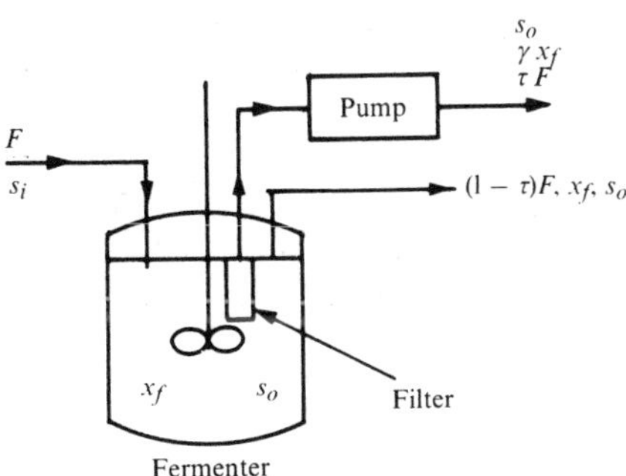

Figure 3. Concentration of biomass achieved by a filter and two effluent streams [Pirt and Kurowski (1970)].

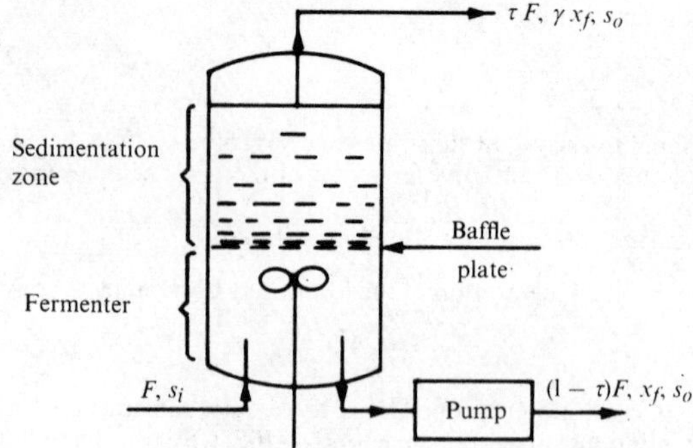

Figure 4. Concentration of biomass achieved by sedimentation zone and two effluent streams [Atkinson (1974)].

$$x_f \simeq \left(\frac{1}{1-\tau}\right) Y_s' (s_i - s_o) \qquad \frac{Y_s' s_i}{1-\tau} \tag{8}$$

5. In a continuous stirred-tank fermenter with recycle (*see* Fig. 5) or a tubular fermenter with recycle (*see* Fig. 6)

$$x_f = \frac{Y_s' (s_i - s_o)}{W} \tag{9}$$

where

$$W = 1 - \tau(\gamma - 1)$$

$\tau(> 0)$ is the recycle ratio and $\gamma(> 1)$ represents the concentration enhancement achieved in the separator, i.e.

$$x_f = \frac{Y_s' s_i}{W} \tag{10}$$

6. In a fermenter containing microbial films, the additional biomass concentration to that given in Eqs. (1)–(10) is

$$x_{f\,\text{film}} = L\, A_{\text{film}} \rho_o \tag{11}$$

where A_{film} is the area of support surface provided per unit volume of fermenter, L is the 'wet' thickness of the microbial films and ρ_o is the microbial density expressed as dry weight per unit wet volume.

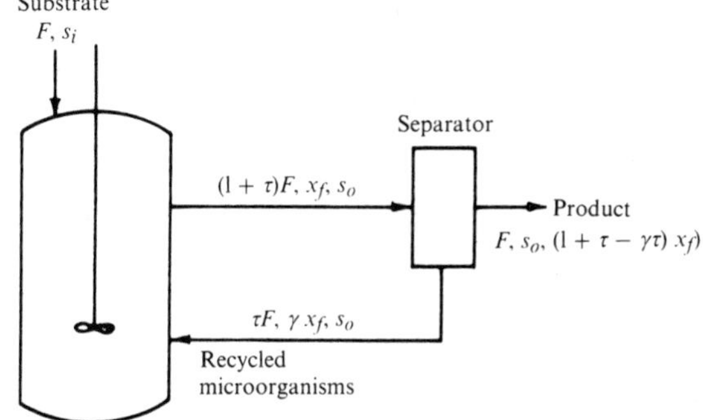

Figure 5. Continuous stirred-tank fermenter with recycle [Atkinson (1974)].

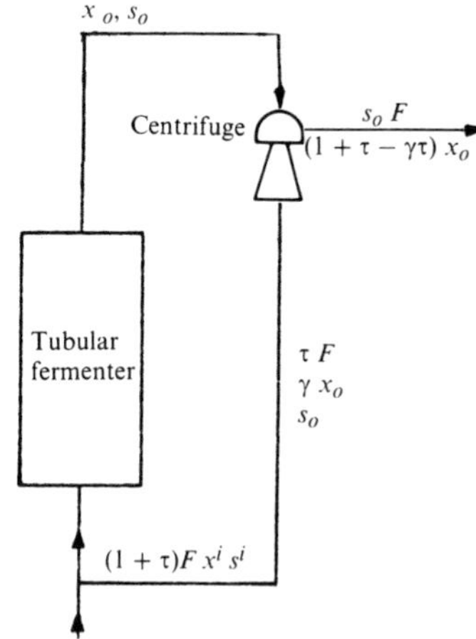

Figure 6. Tubular fermenter with recycle [Atkinson (1974)].

7. In a fermenter containing biomass support particles (*see* p.32) the additional biomass concentration to that given in Eqs. (1) to (10) is

$$x_{f,bsp} = n_p m_p \tag{12}$$

where n_p is the number of particles contained in unit fermenter volume and m_p is the particle biomass hold-up, i.e. the dry weight associated with a particle.
8. In a fluidized-bed fermenter containing microbial flocs [*see* Fig. 1(d)] the additional biomass concentration to that given in Eqs. (1) to (10), is

$$x_{f,fbf} = n_p \rho_o V_p \tag{13}$$

where V_p is the wet volume of a particle and ρ_o is the microbial density.

Aeration

For aerobic fermentation, oxygen is an essential requirement. The relationship between the power input and the volumetric mass transfer k_La is shown in Fig. 4, Chapter 9. Typical aerobic fermenters are shown in Fig. 7. Such fermenters may include a separate stirring system consisting of an impeller and baffles, as in the case of the 'Porton' pot [*see* Fig. 7(a)], or a combined impeller and sparger, in the case of the Vogelbusch fermenter [*see* Fig. 7(f)]. Air-lift fermenters [*see* Figs. 7(b), (c), (d) and (e)] utilize the air circulating within the equipment to bring about the mixing of the broth. A variation on this principle is the tubular loop fermenter [*see* Fig. 5(g)].

Aerators used in the activated sludge process for wastewater treatment are shown in Fig. 8.

An air-lift fermenter of the type represented in Fig. 7(b) is illustrated in Fig. 9, together with the control and measurement equipment. An air-lift fermenter that has reached the commercial stage is the Imperial Chemical Industries pressure cycle fermenter described by Hines (1978) (*see* Figs. 10 and 22). This fermenter contains perforated plates in the upflow tube to counteract coalescence.

All the fermenters depicted in Figs. 5–10 are of the 'mixed' variety [*see* Fig. 2(b)] with mixing being achieved either by internal action (stirrers) or by rapid recycle (i.e. air-lift). The scale of some activated sludge units (*see* Fig. 8) is such as to require multiple aerators located throughout the tank, and the flow pattern from inlet to exit leads to concentration gradients characteristic of tubular fermenters [*see* Fig. 2(c)].

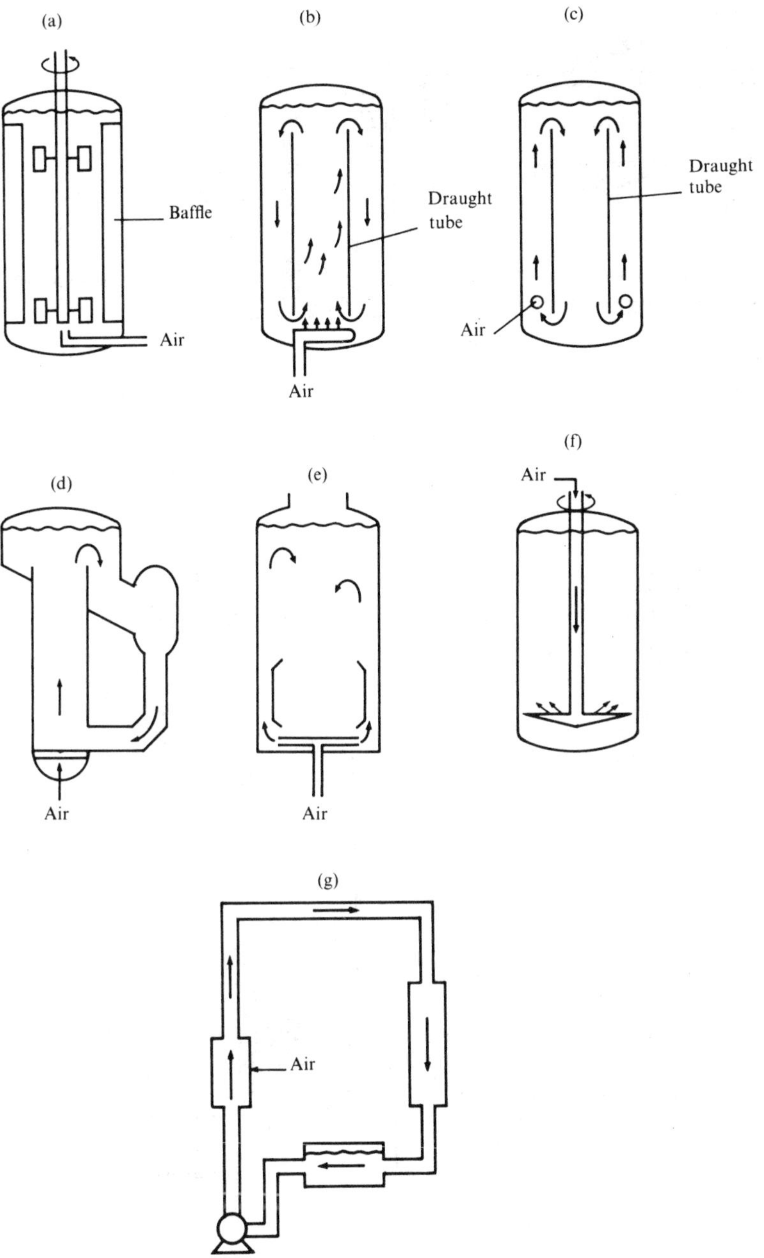

Figure 7. Selection of aerobic fermenter configurations proposed for single-cell protein production (a) stirred, baffled 'Porton' pot, (b) air-lift fermenter, (c) Wasco air-lift, (d) Kanegafuchi air-lift, (e) Lafrancois air-lift, (f) Vogel busch fermenter, (g) tubular loop fermenter [Rose (1979)].

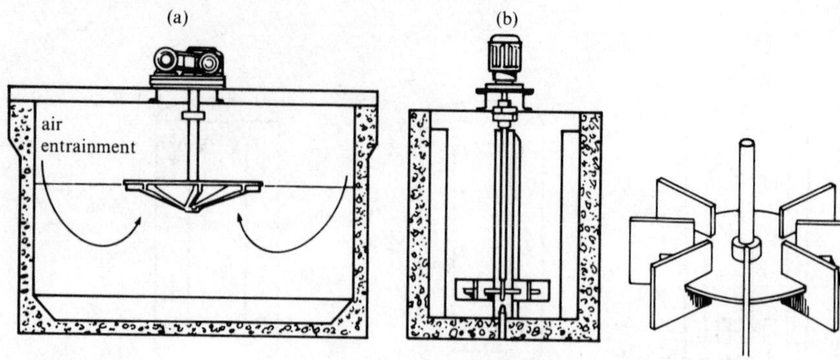

Figure 8. Activated sludge units (a) Simcar aerator, (b) sparger aerator [Abson and Todhunter (1967)]

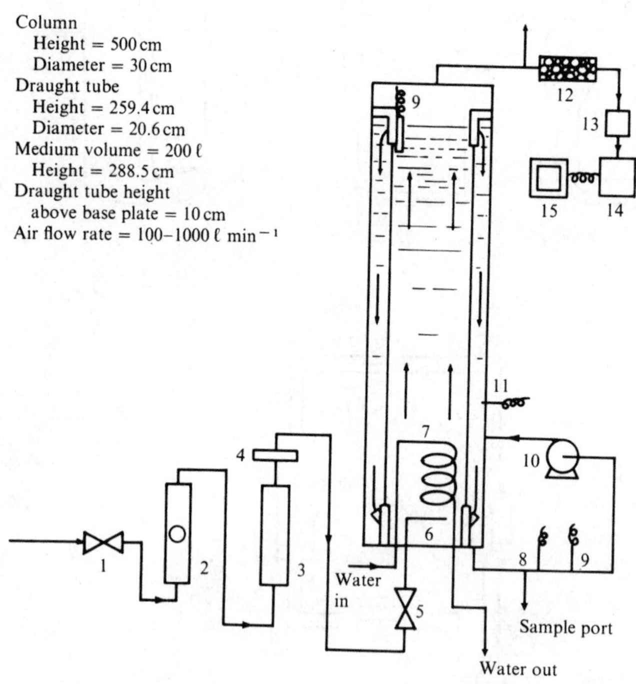

Figure 9. Air-lift fermenter (arrows indicate direction of liquid flow). (1) Air-pressure regulator, (2) rotameter, (3) glass-wool air filter, (4) millipore air filter, (5) check valve, (6) air sparger, (7) heating and cooling coil, (8) pH electrode, (9) oxygen probe, (10) centrifugal pump, (11) thermistor probe, (12) silica-gel bed, (13) Diaphragm pump, (14) paramagnetic oxygen analyzer, (15) recorder. [Wang and Humphrey (1969)].

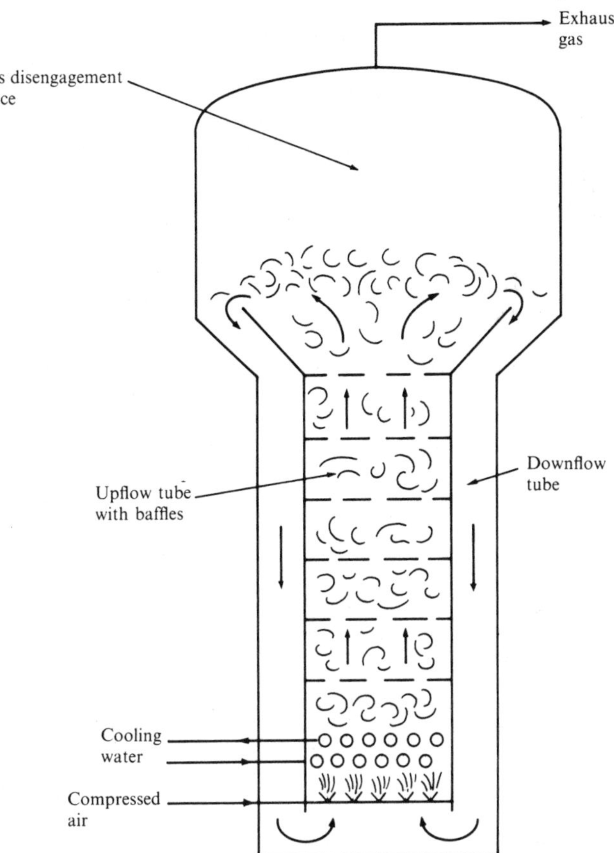

Figure 10. Imperial Chemical Industries' pressure cycle fermenter [Hines (1978)].

DESIGN METHODOLOGY

The design methodology of reactor systems includes the exploration and optimization of the following aspects of process development.

1. Utilization of the extensive literature on process engineering.
2. Selection of appropriate reactor systems after consideration of the process requirements.
3. Effective laboratory and pilot-scale experimentation.
4. Rational extrapolation of experimental data to plant on a commercial scale.
5. Development of design procedures and cost models.
6. Minimization of the delay between initial concept and full scale production.

The cost and time devoted to items 2, 3 and 4 are much reduced by predictive modelling [Atkinson and Kossen (1978)] (i.e. devising *on paper* systems that use, to advantage, the known properties of the microbe–substrate–product reaction). Clearly, many systems can be explored rapidly using those contained in Figs. 1 and 7 as a starting point, and this allows development of the geometric arrangement, the flow patterns and the environment to meet the overall process requirements. In summary, the use of predictive modelling allows the following features.

1. Systems are developed appropriate to the overall process requirements.
2. The advantages/disadvantages of various systems are identified.
3. Past experience and the vast process engineering literature – both biological and chemical – are utilized in an ordered way.
3. Laboratory and pilot-scale equipment is improved.
4. Effective laboratory and pilot-scale experimentation is developed.
5. An orderly system of 'case lore' is evolved for application to future programmes.

Predictive modelling presumes fermenter configurations beyond, but not excluding, stirred tanks and assumes that the designer has the flexibility to choose and develop a system to meet the process requirements. To do all this assumes the availability of enabling technology covering reactors (*see* this chapter) and separation processes (*see* Chapters 12 and 13) and a body of biochemical engineering data and correlations covering fluid mechanics (*see* Chapter 8), heat and mass transfer (*see* Chapters 10 and 11), aeration (*see* Chapter 9), fluidization characteristics, wetted areas, etc. (*see* this chapter).

Intrinsic Kinetics

The intrinsic kinetics of a microbial system that satisfies the Monod equation are μ_{max} and K_m. Values of these parameters are either available from the literature (*see* Chapter 4) or have to be obtained experimentally. Since, presently, there is no convincing evidence to suggest that microorganisms attached to surfaces behave differently, either qualitatively or quantitatively, from freely suspended microorganisms, the experimenter can choose to use either microbial state as the basis of experimentation.

Table 6 summarizes the five apparatuses that can be used to obtain intrinsic kinetic data. The biological film reactor and the fluidized bed containing biomass support particles can also be used to study diffusion-limited conditions (*see* Chapter 10).

The parameters obtained may be integral (i.e. reflecting changes in the microorganisms or environment over the time course of the experiment), initial (i.e. referring to specific controlled conditions) or pseudoinitial (i.e. reflecting concentration and other changes within the reactor).

The methodology for the determination of kinetic parameters follows a number of stages.

1. A mathematical model for the chosen fermenter is devised and solved to provide the analytical performance equation relating conversion efficiency to time or flow rate (as appropriate), s_i, μ_{max}, K_m and, if appropriate, the effective substrate diffusion coefficient in biomass D_e. The mathematical equations corresponding to the reactors in Table 6 are given in the appropriate sections of this chapter.
2. The experimental dependency between s_o and time, flow rate or s_i, as appropriate, is established.
3. Experimental results and theory to calculate the biological parameters μ_{max}, K_m and D_e, if appropriate, are compared.

Table 6. Intrinsic kinetics and experimental methods.

	Biological flocs		Biological films		
	Batch fermenter	Continuous, stirred-tank fermenter	Biological film reactor	Fluidized-bed fermenter – solid particles	Fluidized-bed fermenter – biomass support particles
Conditions	Aseptic	Aseptic	Nonaseptic	Aseptic	Aseptic
Rate coefficients	μ_{max}, K_m	μ_{max}, K_m	μ_{max}, K_m, D_e	μ_{max}/D_e, K_m	μ_{max}, K_m, D_e
Type of rate coefficient	Integral	Initial	Pseudoinitial	Initial	Initial
Mass transfer coefficient	None	None	From theory	By experiment	By experiment
Biomass size requirement	As close to single microbes as possible	As close to single microbes as possible	Know biological film thickness	As close to a monolayer as possible	Support particles of known size and biomass hold-up

Diffusion-Limited Kinetics

The nonaseptic biological film reactor described by Atkinson *et al.* (1967) (*see* Fig. 11) was developed for the determination of the μ_{max}, D_e and the Monod coefficient K_m. The liquid, which is in laminar flow, is in constant contact with a microbial film of thickness L as represented in Fig. 12. The film is supported within a grid that dictates the film thickness (*see* p.27). Using the experimental approach, data are obtained such as those shown in Fig. 13 in the presence or absence of 'solid'-phase diffusional limitations. Data for thin and thick films are given in Figs. 14 and 15, respectively. Flow and other data associated with a thick microbial film in a biological film reactor are given in Table 7. From Chapter 10, the substrate flux N is given by Eqs. (14) – (16).

1. For 'thin' films at low s^*

$$N = \frac{Y'_s \mu_{max} \rho_o}{K_m} \quad Ls^* \tag{14}$$

2. For 'thick' films at low s^*

$$N \longrightarrow \left(\frac{Y'_s \mu_{max} \rho_o D_e}{K_m} \right)^{1/2} s^* \tag{15}$$

3. For all films at sufficiently high s^*

$$N \longrightarrow (Y'_{so} \mu_{max} \rho_o)\ L \tag{16}$$

Eqs. (14) and (16) can be used in conjunction with experimental data for a thin film of unknown thickness to determine K_m. Eq. (16) and a film of known thickness leads to μ_{max}. Eq. (15) and data for a thick film together with values of μ_{max} and K_m lead to D_e [Atkinson (1974)].

The same methodology can be extended to aseptic systems using fluidized beds and biomass support particles.

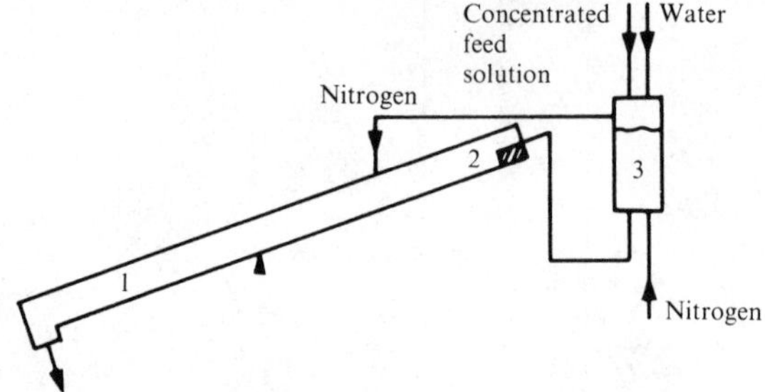

Figure 11. Biological film reactor. (1) Reaction surface, (2) wooden weir, (3) mixer and deaerator [Atkinson (1974)].

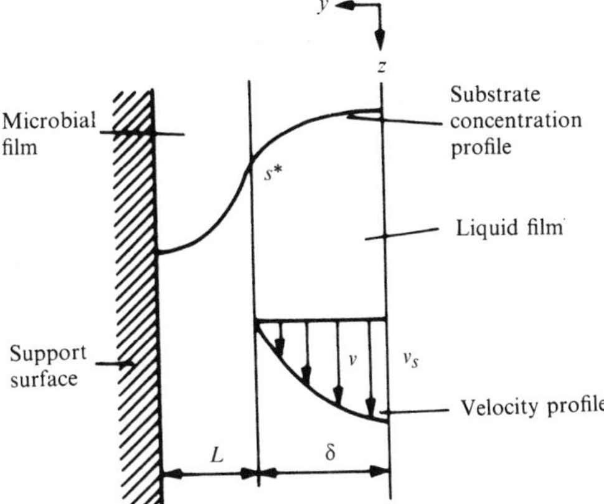

Figure 12. Model of the biological film reactor [Atkinson (1974)].

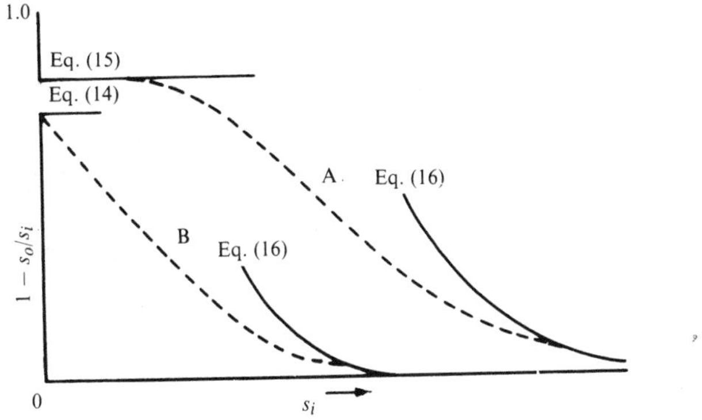

Figure 13. Typical experimental data. Curve A, experimental results ('solid'-phase diffusional limitation); Curve B, experimental results (no 'solid'-phase diffusional limitation) [Atkinson (1974)].

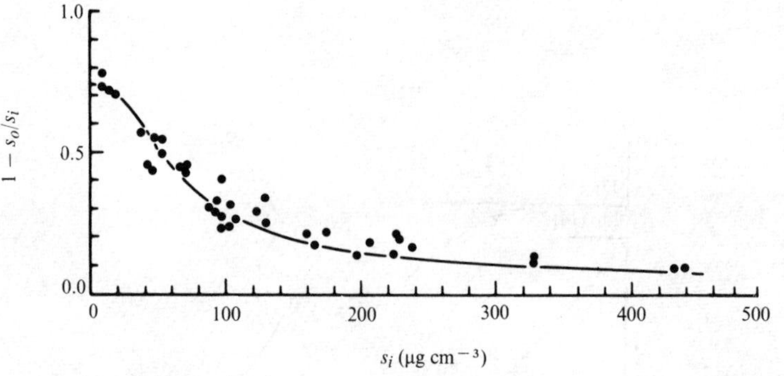

Figure 14. Experimental data obtained using a 'thin' microbial film [Coulson and Richardson (1971)].

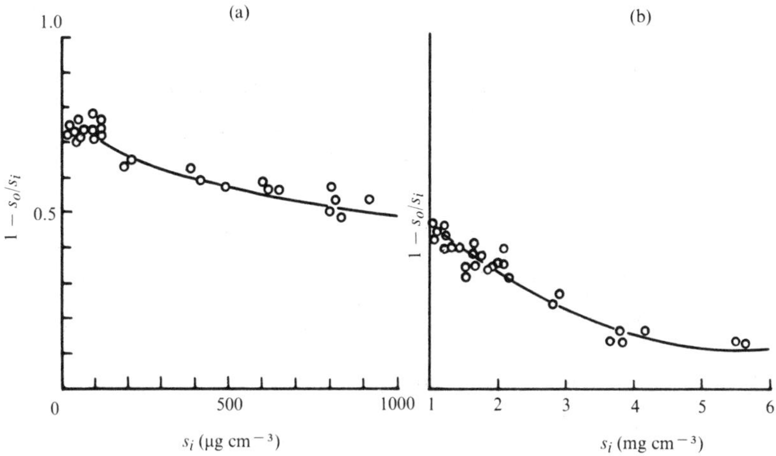

Figure 15. Experimental data obtained using a 'thick' microbial film. (a) Low substrate inlet concentration (0–1000 μg cm^{-3}), (b) High substrate inlet concentration (1–7 mg cm^{-3}) [Coulson and Richardson (1971)].

Table 7. Flow and other data associated with a 'thick' microbial film in a biological film reactor [Coulson and Richardson (1971)].

Parameter	Value
Microbial film thickness	2.032 mm
Liquid film	
Length	2.11 m
Width	182 mm
Microbial surface	
Length	1.83 m
Width	165 mm
Temperature	22 °C
Flow rate	0.82 g s^{-1}
Surface velocity (experimental)	18.6 mm s^{-1}

Scale-Up

Multi-purpose batch fermenters

After the time course of a given fermentation process has been established on a laboratory scale (*see* Chapter 5), development of the corresponding commercial-scale unit involves a number of stages.

1. Provision of aeration, heating and cooling capacity.
2. Control of substrate concentration, biomass, foam, pH, etc.
3. Facilities for aseptic monitoring and operation.

Empirical methodology of this type has permitted industrial development of processes without recourse to detailed studies of complex biochemical reactions involving poorly understood reaction kinetics and variables, such as metabolic control, and substrate- and product-concentration dependence. A consequence of this approach is that many commercial processes are overdesigned in the sense that they are carried out under suboptimal conditions.

In practice, the scale-up of batch fermenters is largely concerned with the provision of aeration capacity in large-scale commercial fermenters that operate at substantially lower power intensity than laboratory units (*see* Chapter 14, Fig. 4).

Predictive modelling in the sense that the time course can be optimized and, in particular, the final product concentration can be maximized by the selection of an optimum strategy for controlling the fermentation is much discussed but is only in its infancy.

Single-purpose continuous equipment

Large-scale fermentation operates under such cost restraints, i.e. allowable processing per unit throughput (*see* Chapter 12), that significant development and design effort are both necessary and tolerable. As an example of this, the Imperial Chemical Industries pressure cycle fermenter is perhaps classic in that not only are the allowable processing costs low, but the development costs were high since, even on the pilot-scale, the major dimension, i.e. the fermenter height, had to be essentially full-scale.

By way of illustration, it is convenient to consider the trickle flow, packed-bed, microbial film fermenter. It can be argued that the outlet substrate concentration s_o from this fermenter depends on the microbial film thickness L, the interfacial area between the microbial and liquid films A_w, the quantity of biomass per unit fermenter volume LA_w, the liquid-phase mass transfer coefficient h', the biological kinetic parameters μ_{max} and K_m, and the effective diffusion coefficient of the substrate D_e within the microbial film, i.e.

$$s_o = g(s_i, Z, F, \mu_{max}, K_m, D_e, L, A_w, h') \tag{17}$$

where: μ_{max}, K_m and D_e are *biological parameters*; L, A_w and k are physical parameters; s_o and s_i are the outlet and inlet substrate concentrations; Z is the bed height; and F the volumetric flow rate.

The methodology of scale-up involves a number of stages.

1. A mathematical model for the fermenter is devised and solved to provide the analytical performance equation corresponding to Eq. (17).
2. The biological parameters are obtained (or estimated) from the literature.
3. The physical parameters are obtained (or estimated) from the general process engineering literature, e.g., gas absorption manuals.
4. A full-scale design is carried out based upon stages 1, 2 and 3.

If the biological parameters are unavailable, these need to be determined by the methods described on pp.596–597. If the physical parameters are unavailable, these can be determined by experimentation using a pilot plant containing the packing of interest and using the microbe–substrate system under consideration (although this is by no means an absolute requirement). Experimental data from the pilot plant, the performance equation and the biological parameters lead to the physical parameters. Sufficient data allow the physical parameters to be related to the flow rate, the packing characteristics, the type of microbial film control, the fluid properties and the factors that influence mass transfer. The information obtained can be used for design purposes (*see* stage 4), above), for the development of appropriate correlations and as a basis for a data notebook.

Development of New Reactor Configurations

New reactor configurations are developed usually to increase the biomass hold-up, increase aeration or reduce power input. In most cases, the need is associated with a particular product, e.g., single-cell protein, power alcohol. In the first instance, the 'new' configuration should be explored algebraically by developing performance equations (i.e. predictive modelling). Consideration can then be given to the configurations with attractive potential performance characteristics. The various configurations can be evaluated at small pilot-scale using nonaseptic versions. Finally, the configuration of choice can be developed as an aseptic pilot-scale unit suited to the microbe–substrate system of interest. Economics of effort, time and cost suggest the development of modular pilot-scale units that can easily be transported to the source of the feedstock for both evaluation and demonstration purposes (*see* Fig. 16).

Figure 16. Alcon's continuous fermentation ethanol pilot plant on test at the British Sugar Corporation plant, Nottingham.

BATCH REACTORS

Configurations

A laboratory-scale (5 ℓ) stirred batch fermenter is shown in Fig. 17. Table 8 lists the routine measurement and control equipment usually associated with such a fermenter. As can be seen from Table 9, the cost of the fermenter is only a small proportion of the total cost of a laboratory-scale fermentation system.

Cross-sections of commercial-scale fermenters with potential capacities of a few hundred to several thousand gallons are shown in Figs. 18–20. Commercial fermenters usually comprise a jacketed pressure vessel that is sealed to maintain asepsis. Because of the

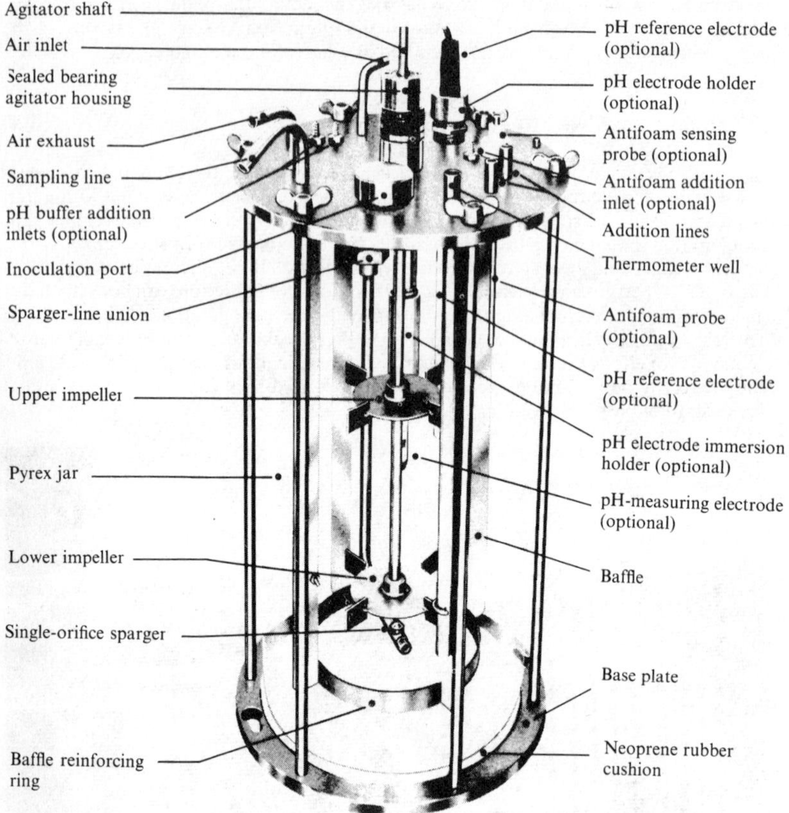

Figure 17. Five-litre capacity fermenter.

intermittent nature of batch operations and the probability of surface fouling by suspended solids, cleaning in place (CIP) facilities are provided. Additional heat transfer surface is often provided by internal coils (*see* Fig. 18). Internal sterilization is achieved by the passage of steam through the jackets and coils and, as appropriate, through the CIP jets.

Pressure can be of importance in large-volume fermenters where the effective pressure comprises the hydrostatic pressure (i.e. the product of the liquid height and the suspension density) and the back pressure on the exit line. Such pressure can lead to excess gas solubility, as in the case of carbon dioxide, and the possible inhibitory effects on microbial activity.

In practice, pH is difficult to measure because of calibration and drift, and there is frequent failure. Instead of measuring gases individually, a variety of gases may be analyzed simultaneously using mass spectrometry. Dissolved oxygen is only occasionally determined. Temperature measurement with an accuracy of $\pm 0.5°C$ is usually adequate.

Because of the problems of mixing in large-scale fermenters, several sensors may be incorporated. Also, because of this problem, heating and cooling jackets may be sectioned and used independently, so as to achieve local control of conditions. Fig. 18 illustrates a fairly standard aerobic fermenter configuration with jackets and coils for heating and cooling, and impellers and spargers for mixing and aeration.

Table 8. Instrumentation associated with a reactor.

Parameter measured	Equipment used
Pressure	Pneumatic or strain gauge sensor
Fermenter contents magnitude	
Volume	Differential pressure cells at top and bottom
Weight	Fermenter mounted on load cell
pH	Glass and reference electrodes
Stirrer speed	Tachometer
Air flow	Variable-area flowmeter, orifice plate
Carbon dioxide production	Infra-red analyzer
Oxygen consumption	Zirconia analyzer
Dissolved oxygen	Polarographic oxygen electrodes
Temperature	Thermocouple, resistance thermometer, thermistor, capillary

Table 9. Cost of 10 ℓ laboratory-scale fermenter and auxiliary equipment, based on 1982 costs.

Equipment	Cost (£)
Fermenter (with recycle loop)	1700
Thermocirculator (-10–$+40$°C)	1300
Instrument console	1750
Service module	1500
pH measurement	375
Temperature measurement	350
pO_2 measurement	550
pO_2, pH, temperature control	1400
Timer	225
pH, pO_2 electrodes	200
Multiplexer	2500
Minicomputer	1000
Disc unit	1000
Printer	450
Pressure control	2000
Foam control	500
Gas monitor	1500
Carbon dioxide probe	400
x–y plotter	800
Total	19 500

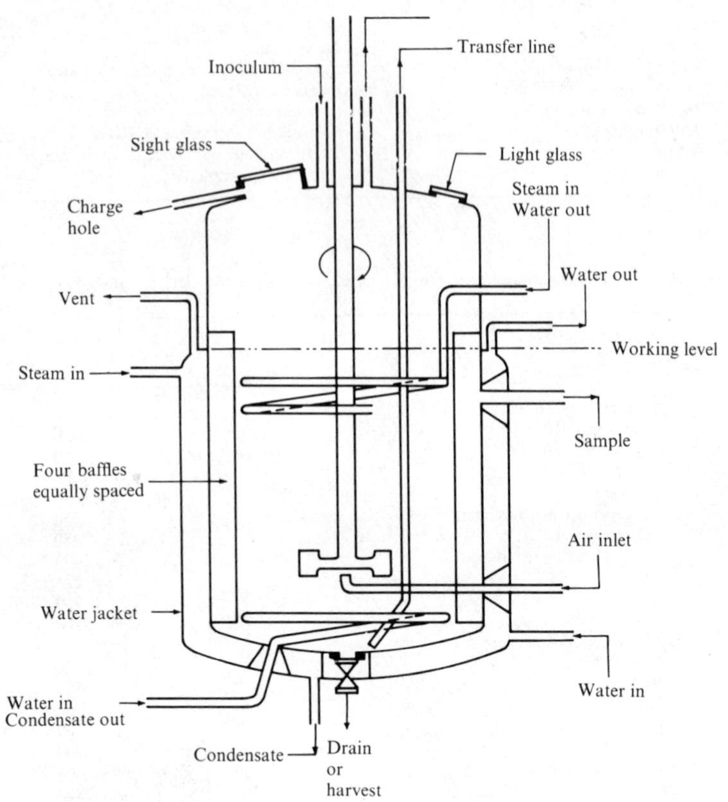

Figure 18. Fermentation vessel [Atkinson (1974)].

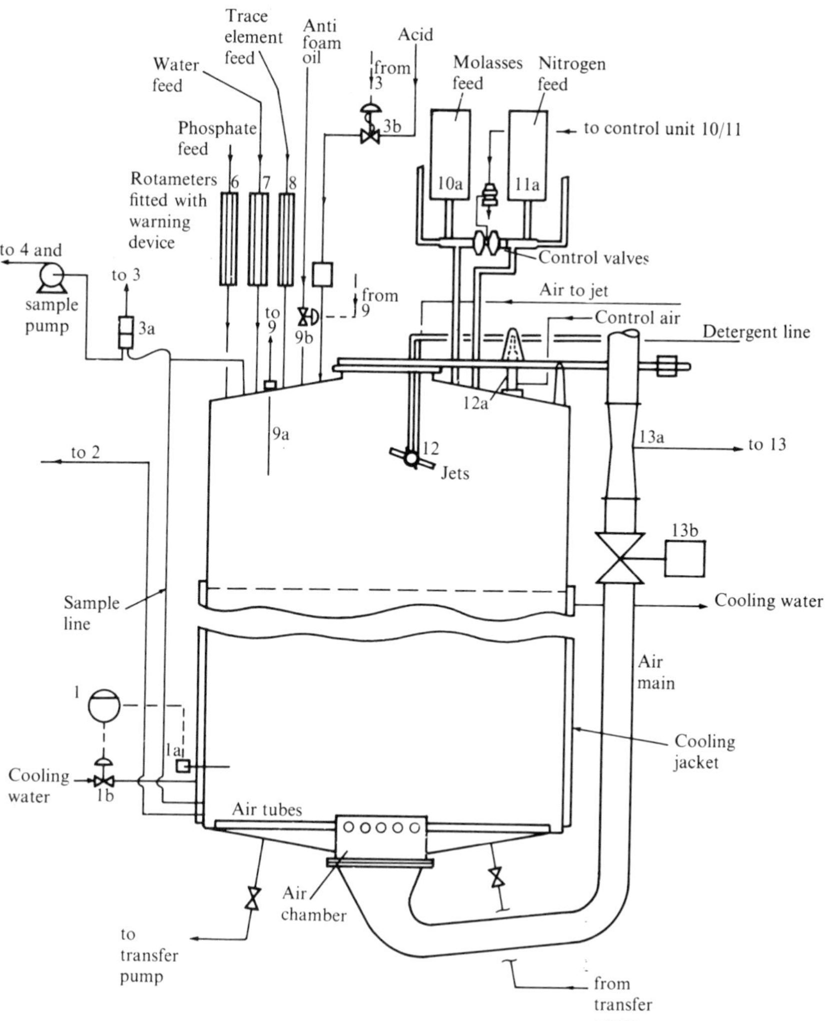

Figure 19. Typical fermenter with auxiliary equipment (1) temperature controller and recorder*, (1a) resistance thermometer, (1b) control valve, (2) fermenter level*, (3) pH recorder and controller*, (3a) pH electrode system, (3b) control valve, (4) yeast concentration recorder*, (5) recorder controller*, (6) phosphate-feed rotameter, (7) water-feed rotameter, (8) trace element-feed rotameter, (9) foam controller*, (9a) foam detector, (9b) control valve, (10/11) dosage control unit, (10a) molasses feed, (11a) nitrogen feed, (12) rotor jet, (12a) power unit, (13) air controller recorder*, (13a) venturi, (13b) power operated air control valve, where * indicates panel-mounted instruments [Olsen (1960)].

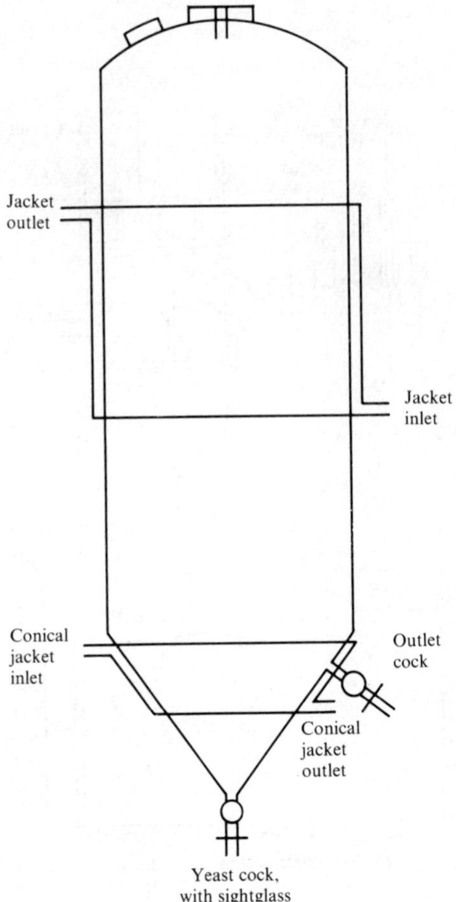

Figure 20. Cylindroconical (Nathan) fermenter [Moss and Smith (1977)].

The quantity of material in a large fermenter is not easy to determine as it consists of gas, liquid and solids in ratios varying from dispersions to foams. Load cells are used; the application of ultrasonic devices is in its infancy and sight glasses are invaluable.

Fermenter measurement and control are largely associated with the maintenance of set conditions or controlled changes, e.g., pH, according to a predetermined schedule. On-line monitoring of fermenter concentrations, rates of reaction, etc. is held back because of the lack of adequate measurement probes that provide a specific determination reliably under harsh conditions. Oxygen consumption and carbon dioxide production can be used to establish an indication of the course of the fermentation provided the stoichiometry has been reasonably well developed (*see* Chapter 3).

Fig. 19 illustrates an aerobic fermenter in which gas dispersion and mixing are achieved by a plenum chamber and a distribution network. An aerobic cylindro-conical fermenter used extensively in brewing is shown in Fig. 20. Mixing is achieved by the carbon dioxide evolved as a fermentation product. Two independent cooling jackets are provided to control the fermentation temperature and to cool the contents following fermentation in order to achieve yeast flocculation and sedimentation. The cone angle is somewhat critical in ensuring the unrestricted flow of sedimented yeast at harvest.

Ideal Time Course Equations

The time courses of an extensive range of fermentations are given in Chapters 4 and 5. Predictive modelling of detailed time courses of these fermentations is in its infancy and awaits the further development of the stoichiometry (*see* Chapter 3) and greater understanding of metabolic activity, particularly with regard to the formation of biochemical products as compared to either substrate utilization or growth.

For a simple growth-associated system, in the absence of diffusional limitations, the time courses of fermentation during the growth and decline phases are given in Chapter 4 by the following equations.

1. For substrate

$$\frac{Y'_s + \Delta/K_m}{Y'_s} \ln \frac{\Delta - Y'_s s}{x_o} - \ln \frac{s}{s_o} = \frac{\mu_{\max}}{Y'_s} \frac{t\Delta}{k_m} \tag{18}$$

2. For growth

$$\frac{Y'_s + \Delta/K_m}{Y'_s} \ln \frac{x}{x_o} - \ln \frac{\Delta - x}{\Delta - x_o} = \frac{\mu_{\max} t\Delta}{Y'_s K_m} \tag{19}$$

3. For product

$$\frac{Y'_s + \Delta/K_m}{Y'_s} \ln \left[1 - \frac{Y'_s (p - p_i)}{Y_p x_i} \right] - \ln \left(1 - \frac{p - p_i}{Y_p s_i} \right) = \frac{\mu_{\max} t\Delta}{Y'_s K_m} \tag{20}$$

where

$$\Delta = Y'_s s_i + x_i$$

A calculation procedure by which the intrinsic kinetics $\mu_{\max}$ and K_m could be obtained from the time course data obtained for both substrate and biomass concentration is given in Appendix I, Chapter 4.

In general, the information necessary to predict the time course of a batch fermentation for a given microbe–substrate system involves a knowledge of the intrinsic kinetics (i.e. $\mu_{\max}$, K_m and D_e), the characteristic size of the microbial particles (V_p/A_p) involved in the process and the liquid-phase mass transfer coefficient h'. These system parameters connect together the hydrodynamic and kinetic aspects of the problem, even when complete mixing is achieved. The reason for this lies in the fact that the internal hydrodynamics of stirred-tank fermenters (i.e. those due to the mixing device) influence both the particle size and the liquid-phase mass transfer coefficient.

$$\frac{1}{x}\frac{dx}{dt} = \mu \left(\frac{V_p}{A_p}, s^* \right) \tag{21}$$

$$-\frac{1}{x}\frac{ds}{dt} = \frac{\mu}{Y'_s} \left(\frac{V_p}{A_p}, s^* \right) \tag{22}$$

$$\frac{1}{x}\frac{dp}{dt} = \frac{\mu}{Y_p} \left(\frac{V_p}{A_p}, s^* \right) \tag{23}$$

$$\mu = \frac{h'}{Y'_s}(s - s^*) \tag{24}$$

The differential equations describing the fermentation under diffusion-limited conditions are given in Eqs. (21), (22) and (23), the liquid-phase diffusional limitation being expressed in Eq. (24). The kinetic function $\mu(V_p/A_p, s^*)$ is given in Chapter 10. Numerical solution of Eqs. (21) to (24) has been discussed by Atkinson (1974).

CONTINUOUS REACTORS CONTAINING FREELY SUSPENDED BIOMASS

Configurations

Continuous fermenters containing freely suspended microorganisms suffer the limitation that the microbes can be 'washed out' from the fermenter. Thus the microbial concentration depends upon flow rate. When the flow rate exceeds a particular value F_w, the microbial concentration falls to zero and the reaction ceases. This flow rate is given by

$$F_w = \mu s_i)\ V \tag{25}$$

where $\mu(s_i)$ represents the actual specific growth rate evaluated at the inlet concentration s_i, thus including any 'solid'- or liquid-phase diffusional limitations that may occur within the fermenter. The maximum value of F_w is given by

$$F_w = \mu_{\max}\ V \tag{26}$$

These continuous fermenters may be operated as chemostats, allowing all conditions to reach a steady state after fixing the inlet flow rate and concentration, in which case, the conditions in the fermenter and outflow are identical. Continuous fermenters may also be operated as turbidostats, which fix the biomass concentration and control the flow rate (*see* Fig. 21). However, such an arrangement brings no significant benefits for the additional complexity and is inappropriate on a large scale.

Figs. 22–27 contain a variety of commercial-scale fermenters and process arrangements.

Fig. 22 shows the installation of a 1500-m^3 pressure cycle fermenter (*see*, in addition, Fig. 10). This is a mixed fermenter with a large throughput due to the large volume involved [*see* Eq. (26)].

In Fig. 23, a train of mixed fermenters used for the production of bakers' yeast is shown [also *see* Figs. 1(b) and 2(b)]. The passage of yeast from fermenter to fermenter enhances biomass concentrations (*see* pp.586–587), but since mixed fermenters usually operate at high conversion efficiencies (*see* p.616), additional feed to each tank is required.

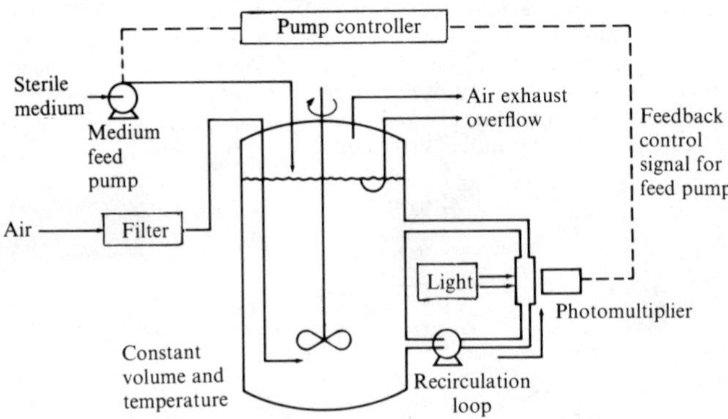

Figure 21. Typical laboratory set-up for a turbidostat [Bailey and Otis (1977)].

Figure 22. Installation of a 1500-m^3 pressure cycle fermenter (courtesy Imperial Chemical Industries Ltd.).

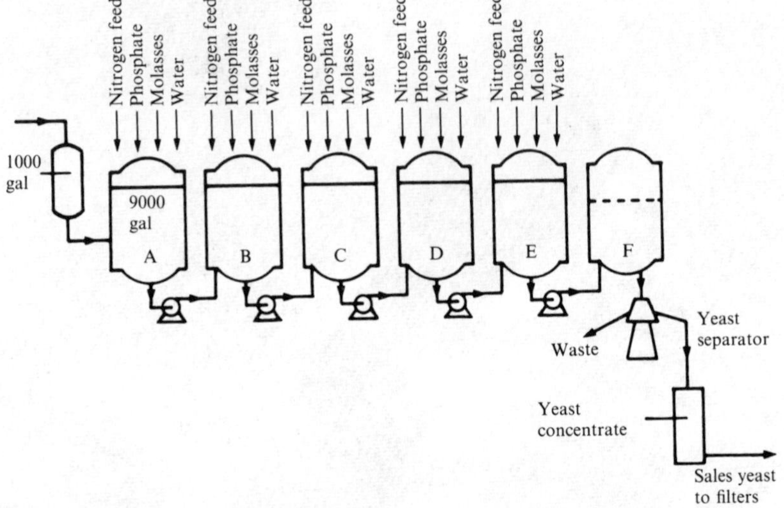

Figure 23. Flow sheet of continuous production of bakers' yeast [Olsen (1960)].

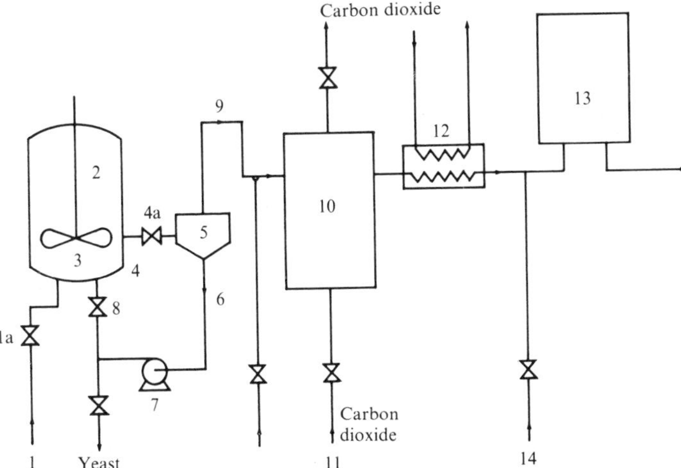

Figure 24. Flow sheet of the continuous, single-stage fermentation for the production of beer. (1) wort inflow, (1a) control valve, (2) fermenter, (3) impeller, (4) beer outflow, (4a) control valve, (5) sedimentation vessel or centrifuge, (6) yeast recirculation, (7) pump, (8) recirculation control valve, (9) clearbeer outflow, (10) washing vessel, (11) carbon dioxide inlet, (12) heat exchanger (cooler), (13) reservoir, (14) finings inlet [Coutts (1958, 1961)].

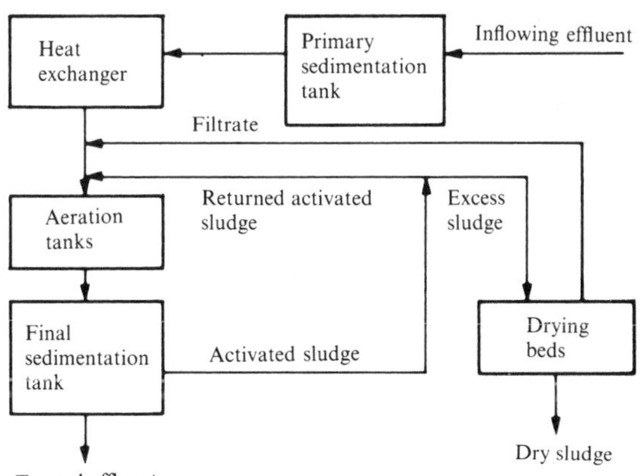

Figure 25. Flow diagram of typical industrial wastewater treatment plant employing activated-sludge tanks [Ainsworth (1970)].

Figs. 24 and 25 illustrate biomass separation and recycle, which have the effect of increasing the biomass concentration in the fermenter (*see* pp.583–584) and increasing the wash-out flow rate beyond that given by Eq. (26).

Two methods of operating an enzyme reactor continuously when the enzyme is in a soluble state are shown in Figs. 26 and 27. The process shown in Fig. 26 retains the enzyme using an ultrafiltration membrane, while that shown in Fig. 27 uses ion exchange recovery methods followed by elution and recycle.

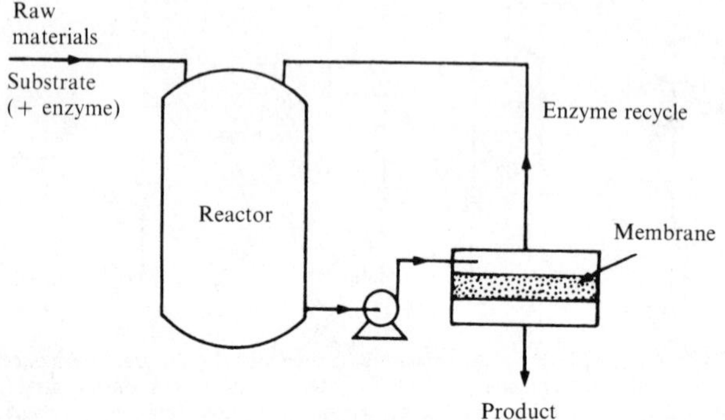

Figure 26. Ultrafiltration reactor [Atkinson (1974)].

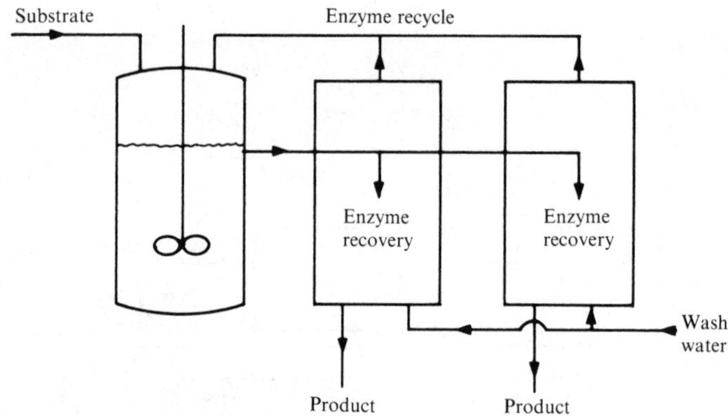

Figure 27. Continuous reactor with enzyme recovery [Butterworth *et al.* (1970)].

Ideal Performance Equations

Completely mixed fermenters

Table 10 describes the combination of conditions that can occur in a completely mixed, continuous fermenter (*see* Fig. 5) and identifies the biological and physical parameters relevant to each combination. The most general case involves both 'solid'- and liquid-

Table 10. Conditions in mixed continuous fermenters.

Conditions	Interfacial limiting substrate concentration, s^*	Biological system parameters required	Physical system parameters required
Liquid-phase diffusion controlled	$s^* = 0$	Y'_s, Y_p, κ	h', s_i, F/V
'Reaction' controlled[a]			
Large flocs	$s^* = s$	Y'_s, Y_p, κ, μ_{max}, K_m, D_e	V_p/A_p, s_i, F/V
Small flocs	$s^* = s$	Y'_s, Y_p, κ, μ_{max}, K_m	s_i, F/V
Liquid-phase diffusion limited			
Large flocs	$0 < s^* < s$	Y'_s, Y_p, κ, μ_{max}, K_m, D_e	h', V_p/A_p, s_i, F/V
Small flocs	$0 < s^* < s$	Y'_s, Y_p, κ, μ_{max}, K_m	h', s_i, F/V

[a] Refers to a case where overall rate of reaction is not limited by liquid-phase diffusion but a 'solid'-phase diffusional limitation is involved.

phase diffusional limitations and, consequently, requires the complete ranges of both sets of parameters.

Reaction control in the absence of 'solid'-phase diffusional limitation and liquid-phase diffusional control leads to analytical solutions because, in the former case, $s^* = s$ and, in the latter, $s^* = 0$. All other cases require numerical iterative solutions to evaluate s^*.

1. For reaction control using small flocs

$$s_o = K_m \frac{W + \kappa V/F}{V/F} \left(\mu_{max} - \frac{W + \kappa V/F}{V/F} \right)^{-1} \tag{27}$$

and

$$F_w = \frac{\mu_{max}}{W} \frac{s_i}{K_m + s_i} V \tag{28}$$

where

$$W = 1 - \tau(\gamma - 1)$$

γ, which is greater than one, is the concentration ratio achieved in the separator (*see* Figs. 24 and 25), τ, which is greater than zero, is the recycle ratio and κ is the endogenous respiration coefficient.

The following points may be deduced from Eqs. (27) and (28)

a. When $\gamma = 1$ (i.e. no microbial concentration stage exists), external recirculation has no effect.
b. The wash-out flow rate F_w is increased by an increase in γ.
c. When $\tau = (\gamma - 1)^{-1}$, the flow rate at wash-out is infinite.
d. In general, for given τ and γ, wash-out remains a feature of the fermenter.
e. τ and γ have to be selected so that $0 < W < 1$, otherwise wash-out will be induced by recirculation.

2. For liquid-phase diffusion control

$$s = \frac{W + V/F}{Y_s' h' V/F} \tag{29}$$

$$F_w = \frac{Y_s' s_i h'}{W} V \tag{30}$$

Comparison of Eqs. (28) and (30) shows that the wash-out flow rate is greatest for reaction-controlled conditions.

Tubular fermenters

In tubular fermenters (*see* Fig. 6) for reaction-controlled conditions with small flocs

$$\frac{K_m}{s_i} \ln \frac{(1 + \tau) s_o/s_i}{1 + \tau s_o/s_i} - \frac{K_m + s_i}{s_i} \ln \frac{1 + \tau}{\tau} = \frac{\mu_{max}}{(1 + \tau)} \frac{V}{F} \tag{31}$$

and

$$F_w = \frac{\mu_{max}}{(1 + \tau) \ln (1 + \tau)/\tau} \frac{s_i}{K_m + s_i} V \tag{32}$$

Eqs. (31) and (32) apply to simple recirculation (i.e. $\gamma = 1$).

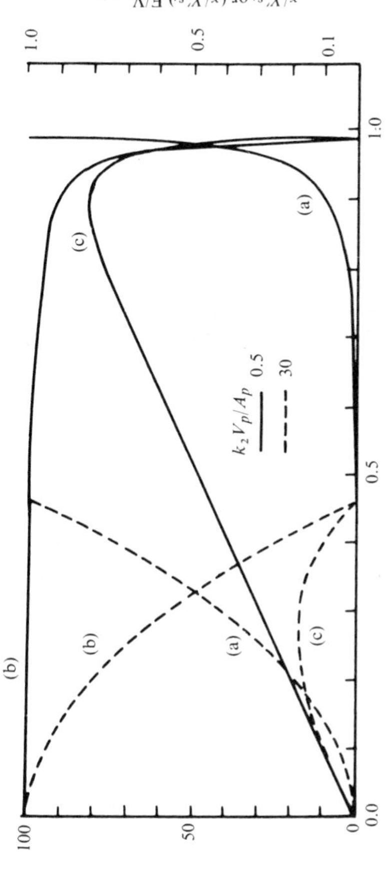

Figure 28. Influence of floc size on the performance characteristics of a continuous stirred-tank fermenter. (a) Substrate concentration, (b) biomass concentration, (c) productivity [Atkinson and Daoud (1976)].

1. When $\tau = 0$, $F_w = 0$.
2. When τ is large, F_w from Eq. (32) is the same as that for a mixed fermenter [*see* Eq. (26)].

It follows from conditions 1 and 2 that recirculation is mandatory and that the range of operation is much the same as that for a completely mixed system.

The equations corresponding to Eqs. (31) and (32), but including a biomass concentration stage, are

$$\frac{K_m(1+\tau)}{s_i[(1+\tau s_o/s_i)+(\gamma\tau/W)(1-s_o/s_i)]}\ln\frac{(1+\tau)s_o/s_i}{1+\tau s_o/s_i} - \left\{\frac{K_m(1+\tau)}{s_i[(1+\tau s_o/s_i)+(\gamma\tau/W)(1-s_o/s_i)]}+1\right\}\ln\frac{1+\tau}{\gamma\tau} = \frac{\mu_{max}}{1+\tau}\frac{V}{F} \tag{33}$$

and

$$F_w = \frac{\mu_{max}}{(1+\tau)\ln[(1+\tau)/\gamma\tau]}\frac{s_i}{K_m+s_i}V \tag{34}$$

Performance Characteristics

Eq. (27) has been plotted in Fig. 28 for $W = 1$ (i.e. no recirculation) and $\kappa = 0$ (i.e. no endogenous respiration). The wash-out flow rate is clearly evident, conforming to Eq. (26). The fact that, over most of the range of flow rates, complete conversion is achieved is also clear. The effect of a 'solid'-phase diffusional limitation (*see* Table 10) is shown in Fig. 28 by an increased particle size reducing the wash-out flow rate, the conversion efficiency and the productivity. Figs. 29–31 contain experimental data on the performance characteristics of continuous fermenters.

Eq. (29) is plotted diagrammatically in Fig. 32 for $\kappa = 0$ to illustrate the linear dependency of conversion efficiency on the flow-rate under liquid-phase diffusion-controlled conditions (*see* Table 10).

Fig. 33 shows the similarity in the performance characteristics of a tubular fermenter and a continuous mixed fermenter for the same values of the recirculation parameters τ and γ. Fig. 34 compares the performance of a tubular fermenter at various values of τ and γ.

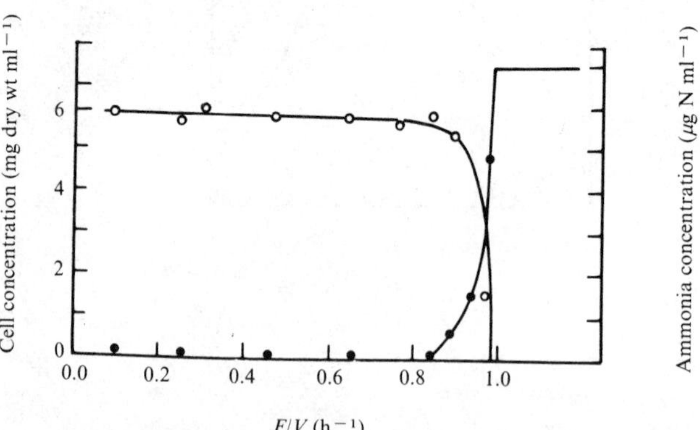

Figure 29. Steady-state growth of *Aerobacter aerogenes* in single-stage continuous culture with ammonia as growth-limiting substrate. ○, Cell concentration; ●, ammonia concentration [Herbert (1958)].

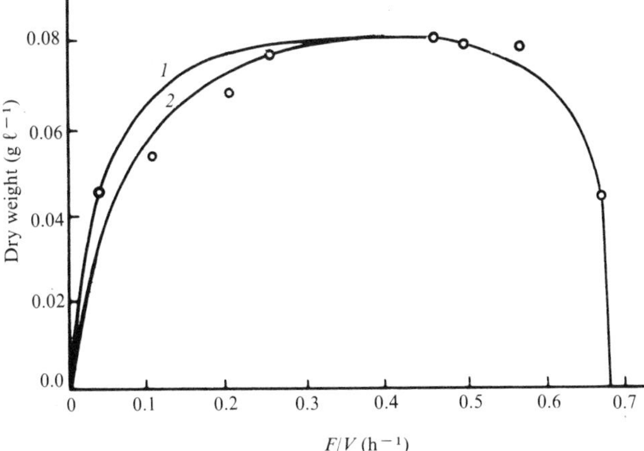

Figure 30. Performance characteristics of a continuous stirred-tank fermenter. Theoretical curve 1, $\kappa = 0.04$, $1/K_m = 111$, $Y_o = 0.396$, $\mu_{\max} = 0.728$; theoretical curve 2, $\kappa = 0.08$, $1/k_m = 111$, $Y_o = 0.435$, $\mu_{\max} = 0.770$; ○, experimental points [Sinclair *et al.* (1971)].

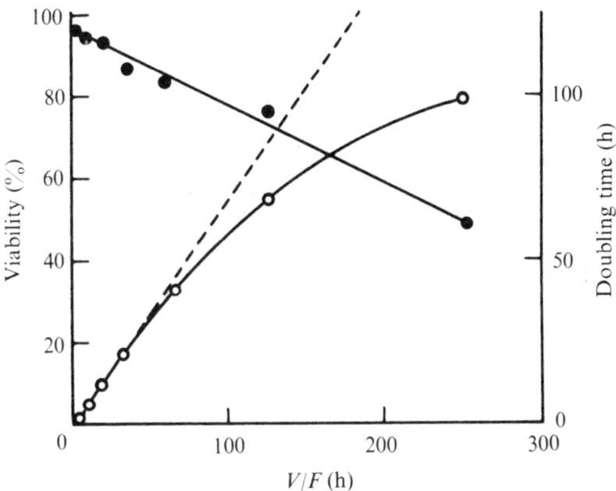

Figure 31. Variation in viability (●) and doubling times (○) of viable organisms with the reciprocal of the dilution rate (i.e. the 'replacement time'). Curve -----, culture doubling time [Tempest *et al.* (1967)].

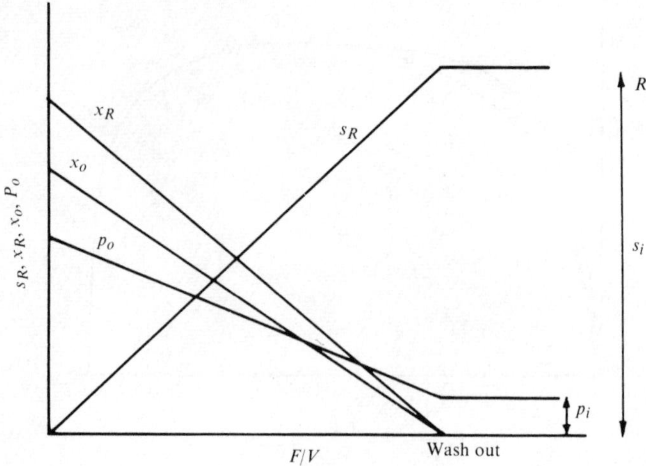

Figure 32. Variation of concentration in a continuous stirred-tank fermenter with liquid-phase diffusion control [Coulson and Richardson (1971)].

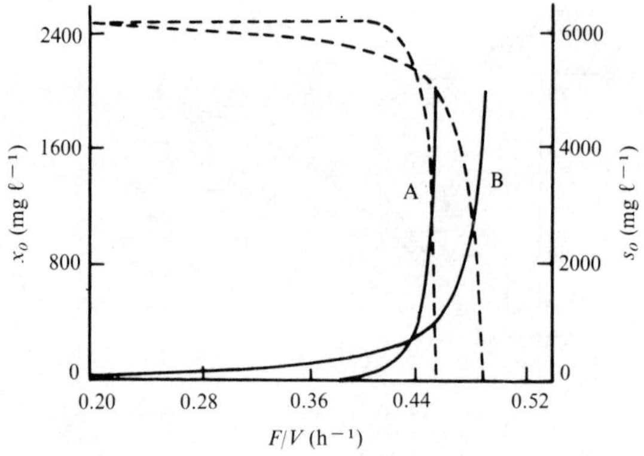

Figure 33. Comparison of performance of fermenters under the same recirculation conditions ($\tau = 0.4$; $\gamma = 3.0$) ———, s_o; – – – – –, x_o. Curve A, tubular fermenter; curve B, continuous, stirred-tank fermenter [Grieves *et al.* (1964)].

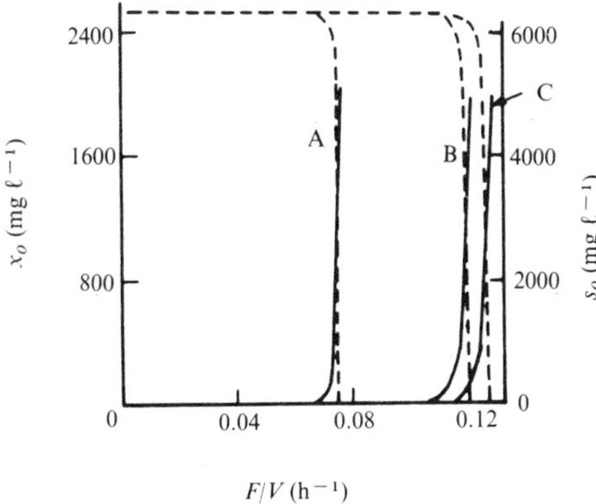

Figure 34. Performance characteristics of a tubular fermenter ———, s; - - - - -, o. Curve A, $\tau = 0.2$, $\gamma = 2.0$; curve B, $\tau = 0.2$, $\gamma = 3.0$; curve C, $\tau = 0.4$, $\gamma = 2.0$ [Grieves *et al.* (1964)].

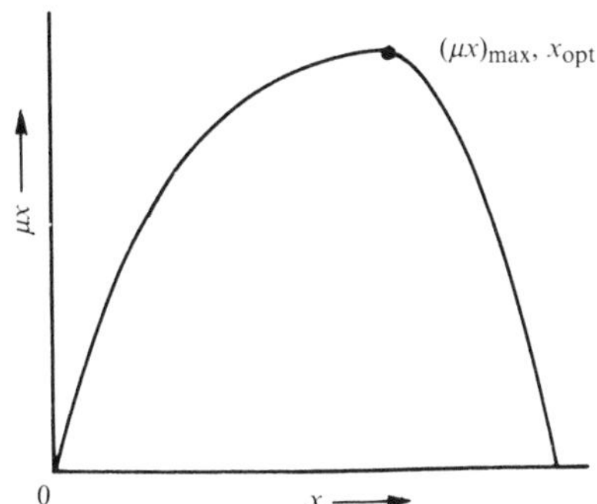

Figure 35. Relation between volumetric rate of reaction and microbial concentration [Eq. (37)].

Graphical Representations

For a simple growth-associated system and small flocs

$$\mu = \mu_{\max} \frac{s}{K_m + s} \tag{35}$$

A balance of the biomass across a fermenter leads to

$$x = Y_s' (s_i - s) \tag{36}$$

The volumetric rate of substrate uptake is given by

$$R_v = \frac{\mu}{Y_s'} x = \frac{\mu_{\max}}{Y_s'} \frac{(Y'_s s_i - x)\, x}{Y_s' (K_m + s_i) - x} \tag{37}$$

Eq. (36) is illustrated diagrammatically in Fig. 35 and shows an optimum biomass concentration and a maximum volumetric rate of substrate uptake.

Completely mixed fermenters

A balance on the biomass in a mixed fermenter leads to

$$F x_o = Y_s' R_v V \tag{38}$$

or

$$\frac{x_o}{Y_s' R_v} = \frac{V}{F} \tag{39}$$

Fig. 36 shows a reciprocal plot of Eq. (37). The rectangle marked *abcd* corresponds to V/F in Eq. (39), i.e. it defines the residence time required to achieve conversion from 0 to x_o. The position marked *b* identifies the conditions in the fermenter. Fig. 36 provides a ready method of identifying fermenter size. The curve in Fig. 35 can be established when Monod kinetics are not applicable by following the algebraic steps in Eqs. (35)–(37). Using the procedure defined in Fig. 36, Figs. 37 and 38 show how the total fermenter volume is minimized by using a single fermenter, when $x_o < x_{\text{opt}}$, and by using a train of fermenters, when $x_o > x_{\text{opt}}$.

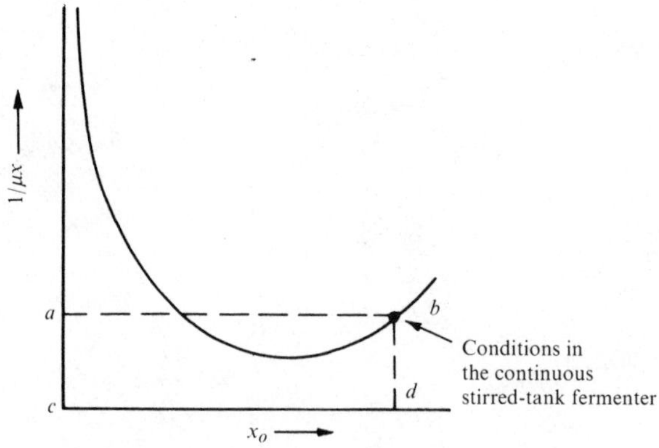

Figure 36. Residence time (V/F) in a x_o continuous stirred-tank fermenter [Atkinson (1974)].

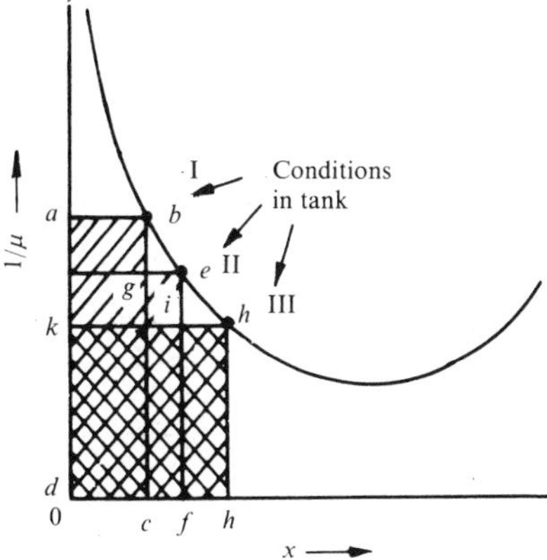

Figure 37. Comparison of fermenter volumes required with n fermenters used in series to achieve a given conversion ($x < x_{opt}$). Area /// indicates volume required for three tanks in series; area \\\ indicates volume for a single tank [Atkinson (1974)].

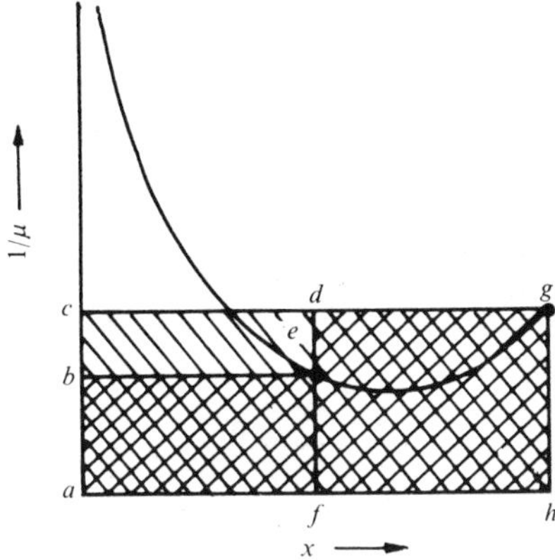

Figure 38. Comparison of fermenter volumes required with n fermenters used in series to achieve a given conversion ($x > x_{opt}$). Area \\\ indicates volume of a single tank; area /// indicates total volume of two tanks [Atkinson (1974)].

Tubular fermenters

The equation corresponding to Eq. (39) for a tubular fermenter is

$$\frac{V}{F} = -\int_{x_i}^{x_0} \frac{dx}{Y'_s R_v} \tag{40}$$

Fig. 39 illustrates the determination of the required fermenter residence time (i.e. the area *efgh* under the curve).

It can be seen from Fig. 40 that, when $x_o < x_{opt}$, a tubular fermenter leads to the largest volume. While when $x_i < x_{opt} < x_o$, an optimization exercise is required to find the minimum volume when using a single mixed or tubular fermenter. However, Fig. 41 shows that, when $x_i < x_{opt} < x_o$, the fermenter volume is minimized by using two fermenters in series – a mixed fermenter operating at x_{opt} followed by a tubular fermenter. Such an arrangement can be achieved in a single vessel by careful arrangement of the mixed and plug flow zones.

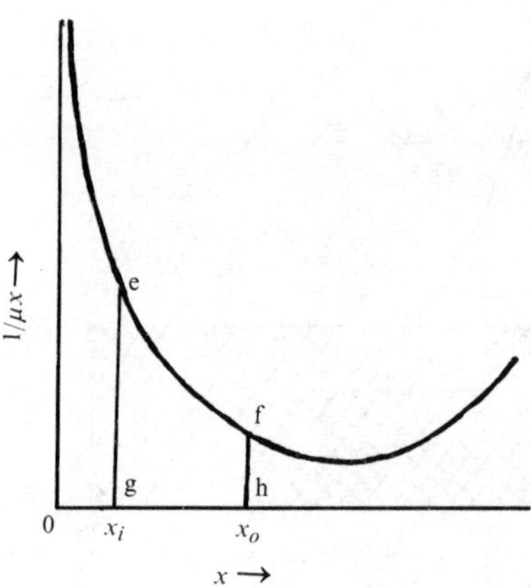

Figure 39. Residence time (V/F) in a tubular fermenter [Atkinson (1974)].

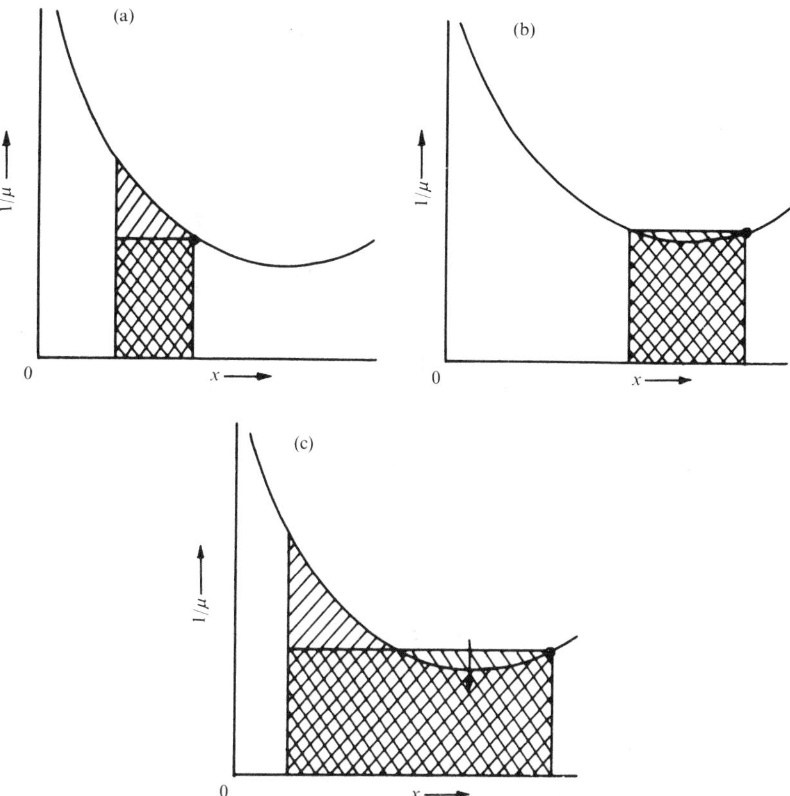

Figure 40. Comparison of the volumes for a continuous stirred-tank fermenter or tubular fermenter. Area /// indicates volume of tubular fermenter; area \\\ indicates volume of a single continuous stirred-tank fermenter. (a) $x_o < x_{opt}$, (b) $x_o > x_{opt}$, (c) $x_o > x_{opt}$ [Atkinson (1974)].

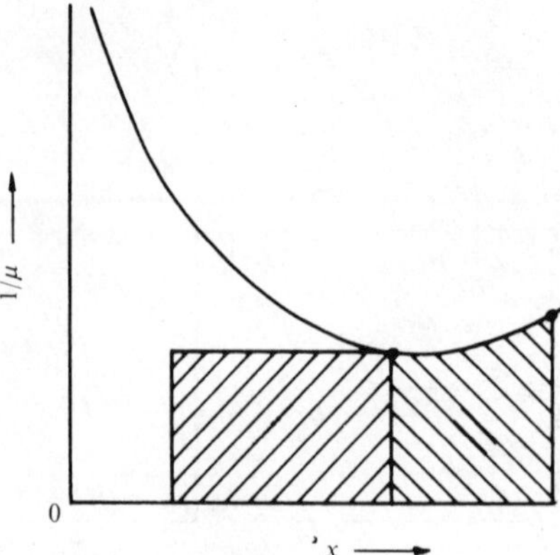

Figure 41. Minimum total fermenter volume for a continuous stirred-tank fermenter and tubular fermenter in series ($x_o > x_{opt}$). [Atkinson (1974)].

CONTINUOUS REACTORS CONTAINING IMMOBILIZED BIOMASS IN SUSPENSION

Configurations

Immobilized biomass may consist of either multilayers of microorganisms adhering to an inert particle (*see* p.26) or biomass contained within the interstices of a biomass support particle (*see* p.32). In both cases, the quantity of biomass associated with an individual particle (*see* Fig. 42) is controlled by three mechanisms.

1. Attrition, e.g., particle–particle and particle–wall contacts, which balance growth.
2. Passing through a zone of high shear, e.g., in the region of an impeller.
3. By removal from the fermenter followed by cleaning and return (*see* p.649).

Mechanisms 1 and 2 lead to essentially constant amounts of biomass associated with an individual particle. Mechanism 3 leads to a range of particle biomass hold-ups, although the fermenter biomass hold-up is steady.

Various fermenter arrangements can be used in conjunction with immobilized biomass particles.

1. Particles can be added to a conventional stirred-tank fermenter and maintained in suspension by the fluid motion from the impeller.
2. Particles can be arranged as a fluidized bed in an external recirculation loop from a conventional fermenter (*see* Fig. 43). This has the advantage that a high number density of particles can be used and that the flow velocity required for fluidization can be achieved independent of fermenter throughput.
3. Particles with a lower density than water can be maintained in suspension by using the air flow of an aerobic fermenter to reduce the bulk fluid density and, therefore, particle buoyancy (*see* Figs. 44 and 45).

There are three important aspects that require assessment when considering the possible use of immobilized biomass technology in conjunction with a given microbe–substrate system.

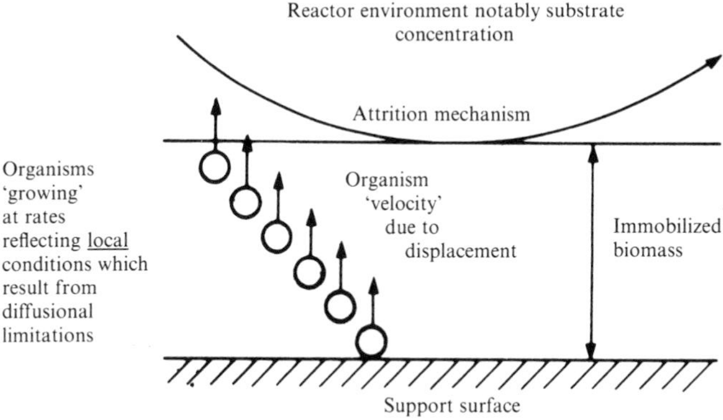

Figure 42. Concept of organism 'age' within a microbial film [Atkinson and Kossen (1978)].

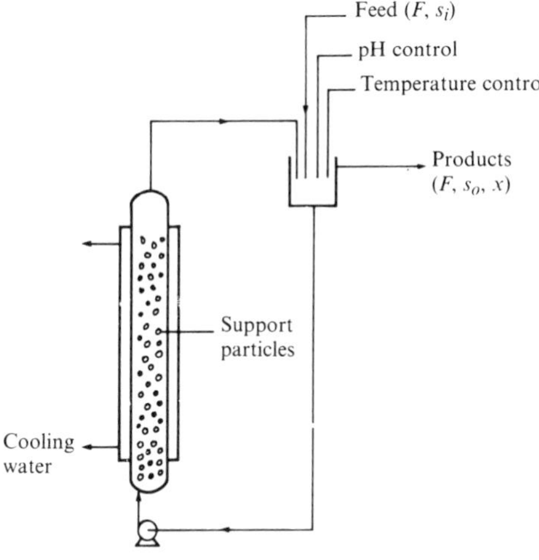

Figure 43. Completely mixed microbial film fermenter [Atkinson (1974)].

1. Does the intended support particle retain the biomass? A convenient experiment is to add particles to a conventional stirred-tank fermenter operating beyond wash-out; improved conversion efficiencies suggest biomass retention (*see* p.27).
2. Does the intended biomass control mechanism lead to steady particle biomass hold-ups? A bed of particles in a recirculation loop connected to a conventional fermenter (*see* Fig. 43) allows identification of the flow velocity required to achieve the necessary level of attrition. This flow velocity may be lower or higher than the incipient fluidizing velocity depending on the morphology of the microorganisms. Fungi present the greatest problems.
3. What number density of particles can be achieved in the fermenter? Fig. 46 shows the number densities and flow regimes associated with biomass support particles (in this case, plastic toroids lighter than water) maintained in suspension by air flow.

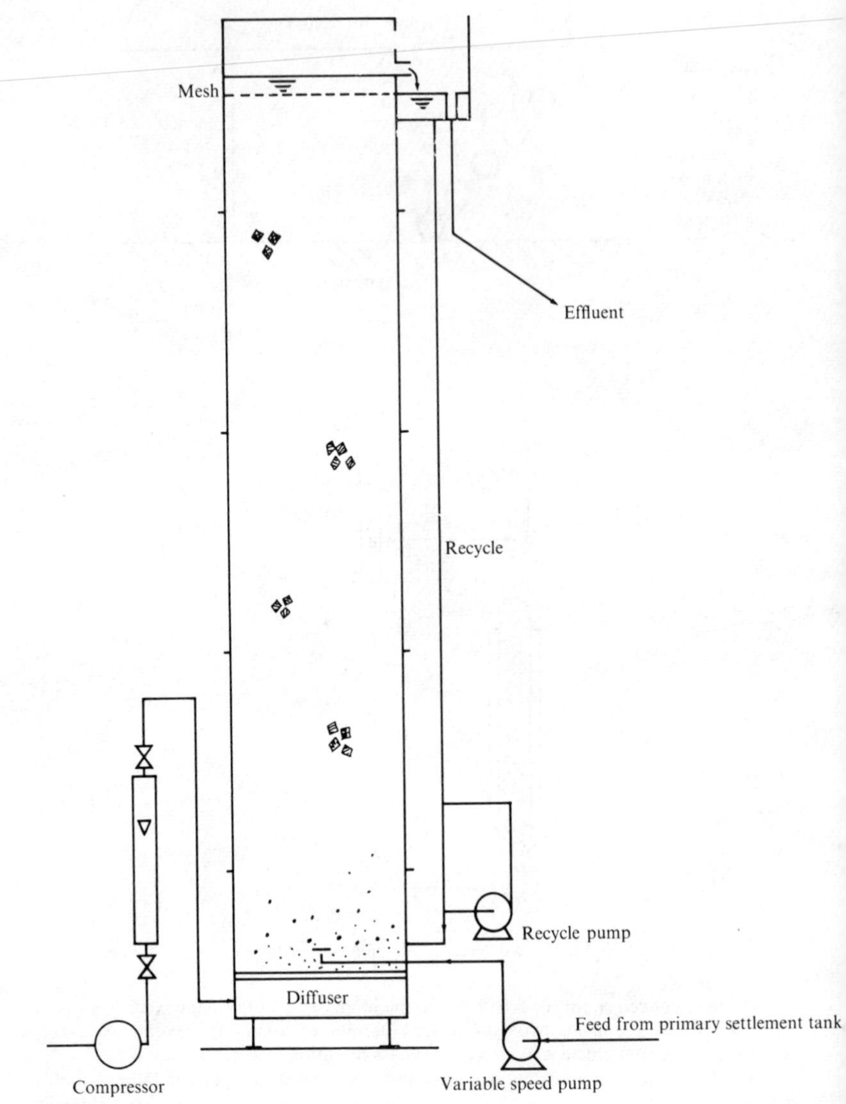

Figure 44. Schematic of 0.6 m³ pilot units. Column diameter, 0.5 m; column height, 3.5 m. [Walker and Austin (1981)].

Figure 45. Pilot plant of Simon Hartley. Stoke on Trent.

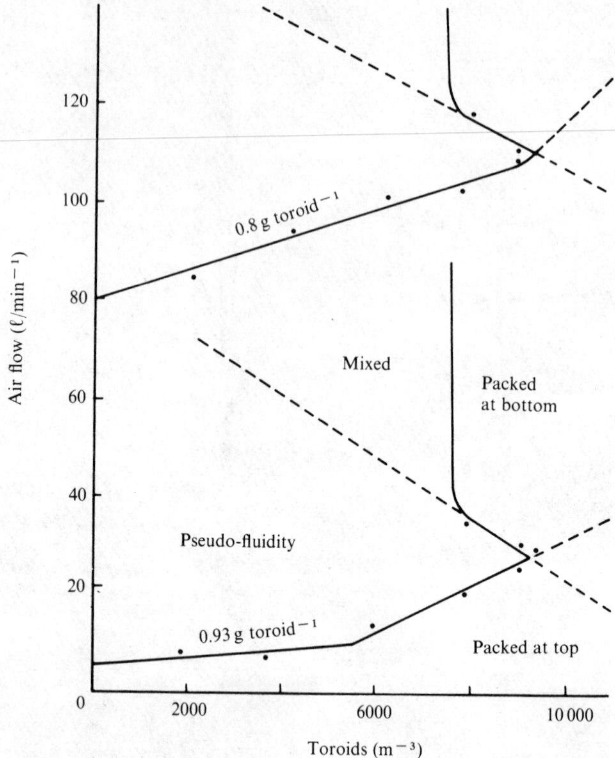

Figure 46. Relationship between air flow and the number of toroids in the reactor at 0.8 g and 0.93 g biomass per toroid [Walker and Austin (1981)].

Ideal Performance Equations

The volumetric rate of substrate uptake for a fermenter containing immobilized biomass in the absence of 'solid'- or liquid-phase diffusional limitations (*see* Chapter 10) is given by

$$R_v = \frac{\mu}{Y'_s}(x + x_I) \tag{41}$$

where x_I is the immobilized biomass concentration. For solid supports

$$x_I = \rho_o L A \tag{42}$$

where A is the area of support surface per unit fermenter volume. For biomass support particles

$$x_I = \rho_o n_p \phi V_p \tag{43}$$

where n_p is the number density of particles in the fermenter and ϕ and V_p are the particle porosity and volume, respectively.

Balances on microorganisms and substrate followed by algebraic combination and rearrangement lead to, according to Atkinson and Davies (1972).

$$\mathscr{A}\left(\frac{s}{s_i}\right)^2 + \mathscr{B}\frac{s}{s_i} - 1 = 0 \tag{44}$$

where

$$0 < \frac{s}{s_i} < 1$$

$$\mathscr{A} = \frac{s_i}{K_{m,}}\left(1 - \frac{\mu_{\max} V}{F}\right)$$

and

$$\mathscr{B} = 1 - \frac{s_i}{K_{m,}} + \frac{\mu_{\max}}{F}\left(\frac{s_i}{K_{m,}} + \frac{x_I}{Y'_s K_{m,}}\right)$$

Eq. (44) provides the performance characteristics of fermenters containing immobilized biomass and shows that

$$\frac{s}{s_i} = g\left(\frac{F}{V\mu_{\max}}, \frac{s_i}{K_{m,}}, \frac{x_I}{Y'_s K_{m,}}\right) \tag{45}$$

i.e. the conversion efficiency depends upon (1) a dimensionless flow rate $F/V\mu_{\max}$, (2) a dimensionless inlet concentration s_i/K_m and (3) a dimensionless biomass hold-up $x_I/Y'_s K_{m,s}$. The algebra of Eqs. (41)–(45) has been extended to include both 'solid'- and liquid-phase diffusional limitations by Fonseca (1978).

Performance Characteristics

Figs. 47–49 contain solutions to Eq. (44). In Fig. 47, the line corresponding to $\beta = 0$ is equivalent to Fig. 28, i.e. the simple chemostat. The features of significance in Fig. 47 are the absence of a wash-out flow rate and increased conversion efficiency β with increased immobilized biomass in the fermenter. Fig. 48 demonstrates that increased inlet concentration α lowers the conversion efficiency and a sufficiently high value can produce performance characteristics very similar to wash-out. The volumetric rate of substrate uptake, i.e. the productivity, for a fermenter containing immobilized biomass is shown in Fig. 49. Productivity is increased by greater β, up to a limiting value characteristic of the particular flow rate. At higher values, the productivity is insensitive to changes in the flow rate.

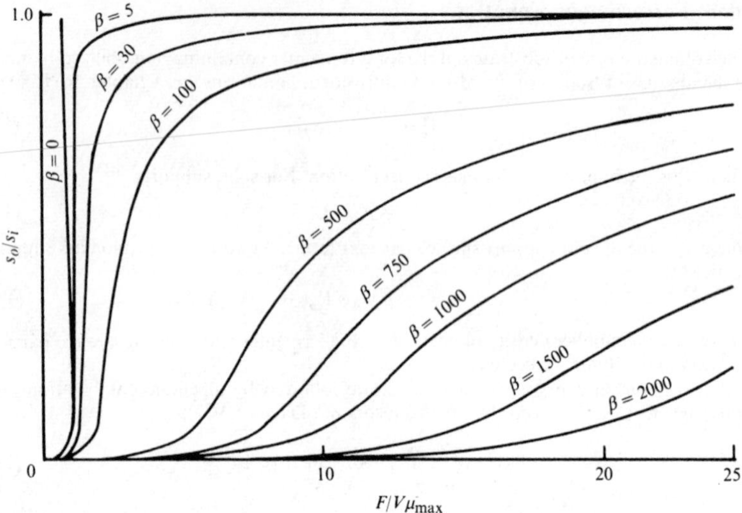

Figure 47. Effect of immobilized biomass on the conversion efficiency of a completely mixed fermenter $s_i/K_m = 100$; $\beta = x_I/Y_s' K_m$.

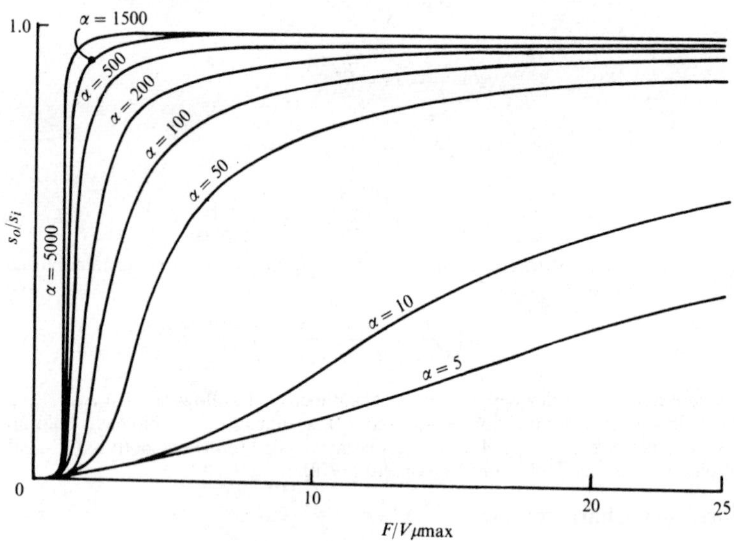

Figure 48. Effect of inlet concentration on the conversion efficiency of a completely mixed fermenter containing immobilized biomass ($x_I/Y_s' k_{m,s} = 100$; $\alpha = s_i/K_m$).

In Fig. 50, lines of constant conversion efficiency obtained from Fig. 47 are superimposed on Fig. 49. It is clear from Fig. 50 that maximum productivities occur at higher flow rates than do the highest conversions (also *see* Fig. 28), and judgement has to be exercised in selecting the appropriate compromise between conversion efficiency and productivity.

The effects of the particle size [i.e. 'solid'-phase diffusional limitations (*see* Chapter 10)] on conversion efficiency and productivity are shown in Figs. 51 and 52. The effects are greatest at low fermenter concentrations. A large number of combinations of particle size and fermenter biomass hold-up can be used to achieve a required conversion efficiency or productivity.

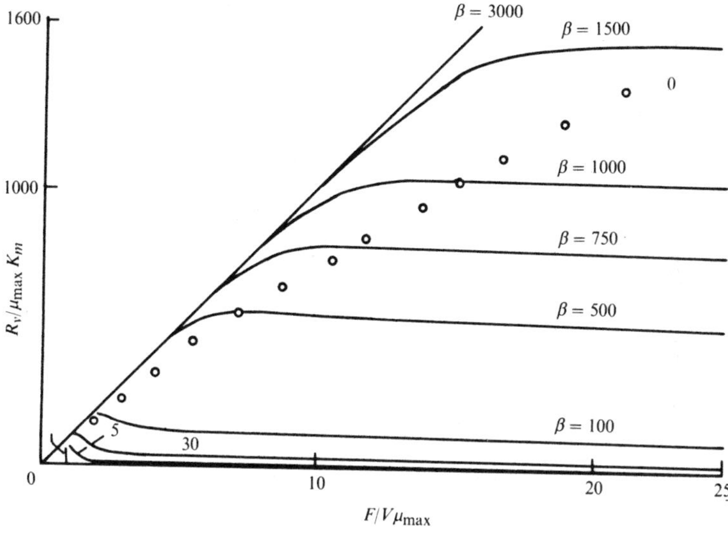

Figure 49. Dimensionless volumetric rate of reaction of a completely mixed fermenter containing immobilized biomass ($s_i/K_m = 100$; $\beta = x_I/Y_s' K_m$). ○, Locus of $R_{v,\,max}/\mu_{max}\,K_m$.

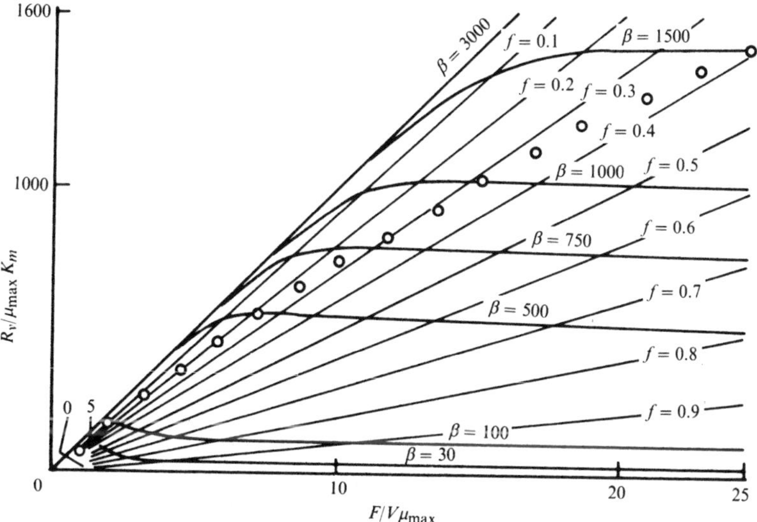

Figure 50. Relationship between volumetric rate of reaction and conversion efficiency of a completely mixed fermenter containing immobilized biomass. $f = s_o/s_i$, $s_i/K_m = 100$, ○. Locus of $R_{v,\,max}\,K_m$.

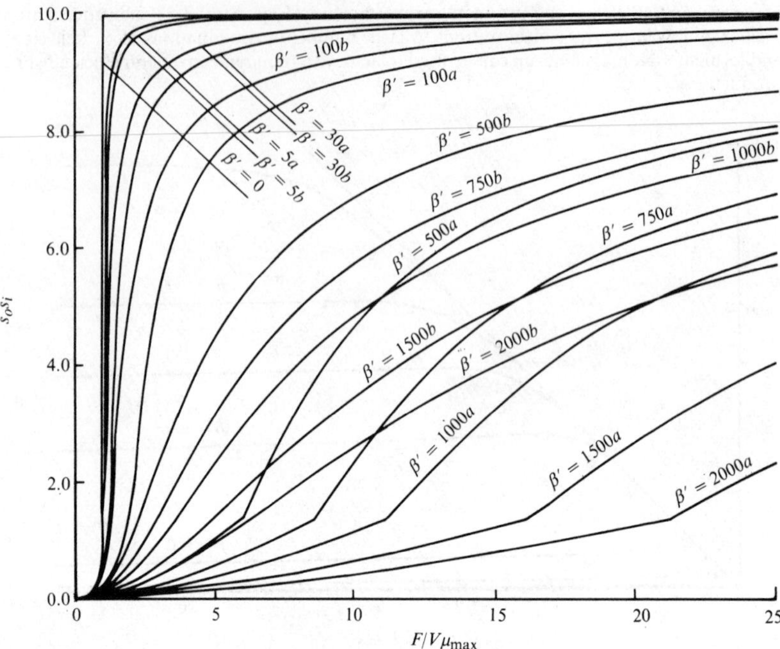

Figure 51. Effect of 'solid'-phase diffusional limitations on the substrate concentration leaving a fermenter containing biomass–support particles. $s_i/K_m = 100$, $a = k_2' V_p/A_p = 5$, $b = k_2' V_p/A_p = 20$. and $\beta' = x_I/Y_s' K_m$.

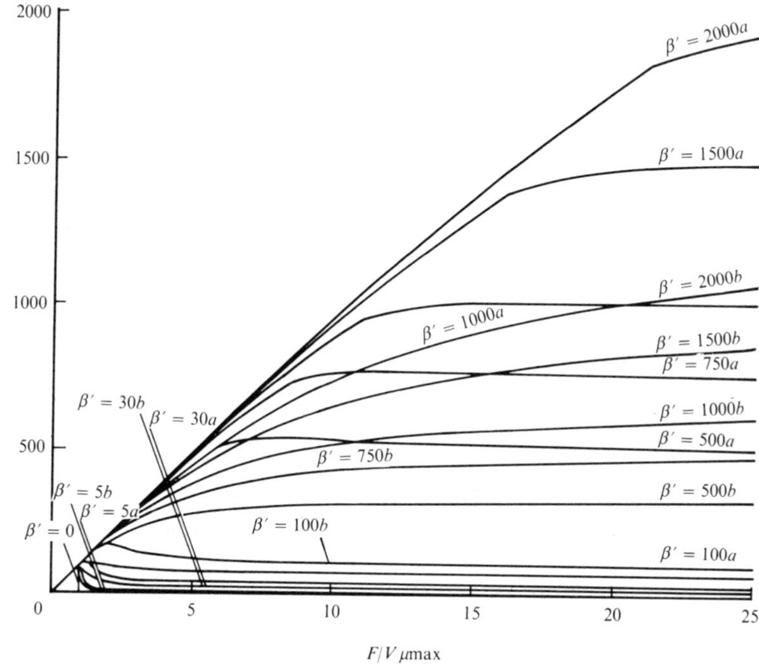

Figure 52. Effect of 'solid'-phase diffusional limitations on the productivity of a fermenter containing biomass–support particles $s_i/k_m = 100$, $a = k_2' \, V_p/A_p = 5$, $b = k_2' \, V_p/A_p = 20$ and $\beta' = x'/Y_s' K_{m,s}$.

FIXED-BED REACTORS

Configurations

The trickling filter, as illustrated in Fig. 53, which has been used extensively for the last 80 years in wastewater treatment, is essentially a tubular fermenter [*see* Fig.1(c)] consisting of a fixed bed of large inert particles, such as broken brick or slag, on which microbial films develop. The trickling filter has the advantage that the throughput is not restricted by wash-out, making it ideal for water treatment, but its applications beyond this field are restricted for a number of reasons.

1. Only microbial strains that can adhere to surfaces can be used.
2. Accumulation of biomass on the support particles must be controlled. In the case of trickling filters used for wastewater treatment, this is achieved by a combination of sloughing, erosion and the action of worms and other small animals. If the biomass is not removed effectively, its accumulation will tend to block the bed.
3. Because the biomass hold-up is variable, the efficiency of conversion can also vary considerably (*see* Figs. 59 and 60).
4. Air circulates through the bed by natural convection, making aseptic operation difficult, if not impossible.
5. It is not possible to recover the biomass from the bed.

The sole commercial application of trickle-flow technology resides in the various forms of vinegar towers (acetifiers) (*see* e.g., the Fring's acetifier, Chapter 14), where low growth rates prevent blockage and low pH prevents contamination.

Atkinson and Williams (1971) developed a trickle-flow fermenter (*see* Fig. 54) that overcomes the problems of accumulating biomass hold-up by periodic washing of the wooden spheres used as support particles. Anoxic operation was achieved using nitrate as a chemical oxidant and by passing nitrogen through the packed bed and feed vessels. This fermenter was developed for reasons of design methodology described on pp.600–601.

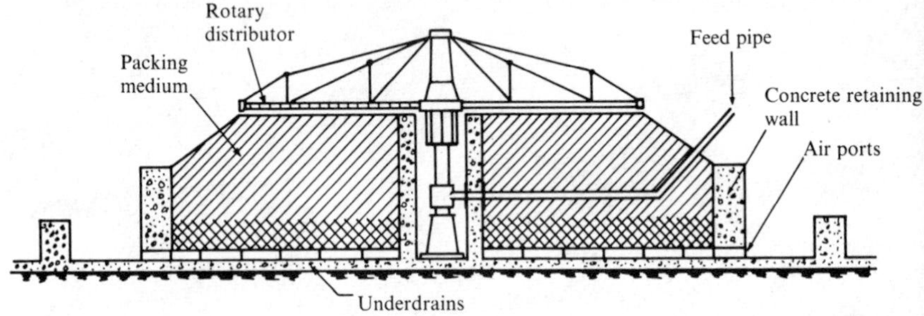

Figure 53. Trickling filter [Atkinson (1974)].

It has been suggested (*see* p.583) that a tubular fermenter containing microbial flocs in suspension requires a flow of microorganisms to the inlet otherwise total elutriation would occur and the biomass would be washed out completely. An alternative to recirculation as a way of producing a supply of microorganisms for the feed is shown in Fig. 55. In this arrangement, the film fermenter generates microorganisms that are then introduced into the tubular fermenter. The two fermenter components are all part of a single vessel. The two regions have different diameters because of the different flow velocities required in what are essentially two reactors. The high velocity in the lower part provides for attrition of the growing biomass and the low velocity in the upper part provides the residence time to achieve the required conversion.

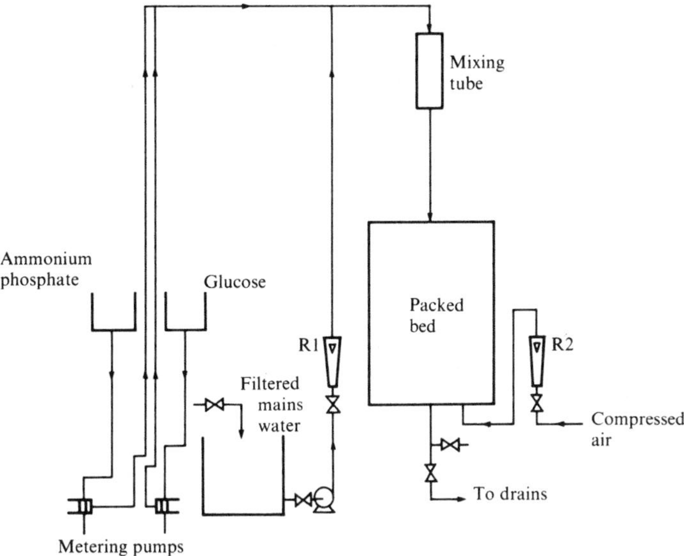

Figure 54. Trickle-flow pilot-scale fermenter [Atkinson and Ali (1976)].

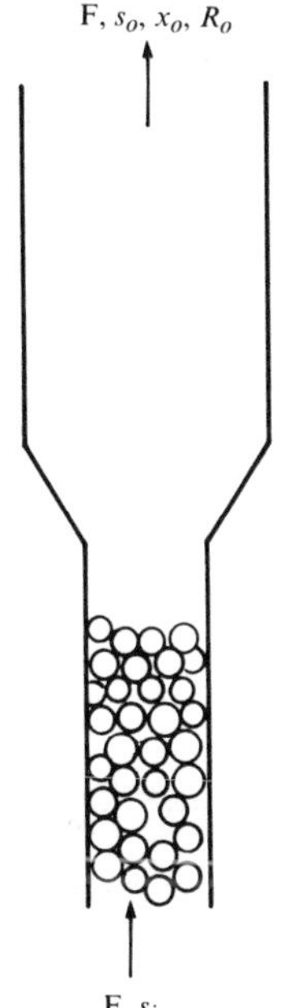

Figure 55. Film fermenter feeding a tubular fermenter [Atkinson (1974)].

Ideal Performance Equations

A range of assumptions covering tubular fermenter containing support packing materials, together with their implications are given in Table 11.

Table 11. Conditions in a packed-bed biological film reactor.

Assumption	Implications
Steady state	Constant input conditions (i.e. composition, concentration, temperature), constant liquid hold-up and wetted area, no net accumulation of microbial mass, constant net metabolic oxidation rate
No reaction in liquid phase	No chemical oxidation, no significant biological oxidation by microorganisms in suspension
Soluble substrate removal only	Contribution of suspended solids to slime thickness as a result of filtration and subsequent biooxidation ignored (includes redistribution of slime by sloughing)
Excess oxygen available	Gas–liquid interfacial area and liquid-phase mass transfer coefficients are sufficiently large so that the dissolved oxygen concentrations at the liquid–slime interface result in biooxidation reactions that are zero order with respect to oxygen
Slime thickness uniform	Slime thickness independent of position in filter and therefore of local conditions
Uniform temperature	Temperature independent of position in filter (i.e. local equilibrium between heat generation by oxidation reactions and heat loss to the atmosphere)
No longitudinal mixing	Flow through bed is uniform everywhere (i.e. plug flow) therefore no longitudinal mixing due to velocity distribution, diffusion or channelling effects

Submerged flow

The microbial film thicknesses in a submerged flow reactor are most likely 'thin' on account of the liquid shear forces and bereft of 'solid'-phase diffusional limitations. The corresponding reactor equation is given by Atkinson (1974) as

$$\frac{1}{\beta + s_i/K_{m.}} \ln \frac{s_o}{s_i} - \frac{1 + \beta + s_i/K_{m.}}{\beta + s_i/K_{m.}} \ln \left[1 + \frac{s_i/K_m}{\beta} = \left(1 + \frac{s_i/K_m}{\beta} \left(1 - \frac{s_o}{s_i} \right) \right) \right] - \frac{\mu_{\max} V}{F} \quad (46)$$

where

$$\beta = \frac{\rho_o L A}{Y'_s K_{m,s}}$$

Eq. (46) accounts for the contribution of freely suspended microorganisms.

Trickle flow

The microbial film thickness in a trickle-flow reactor is most likely to be 'thick' and to exert a strong diffusional limitation. The substrate flux N is, therefore, given (*see* Chapter 10) by

$$N = \left(\frac{\mu_{max} K_m D_e \rho_o}{Y'_s}\right)^{1/2} \sqrt{2}\left(\frac{s}{K_m}\right)^{1/2} \tag{47}$$

and the corresponding reactor equation [Atkinson and Ali (1976)] is given by

$$\sqrt{2}\left(\frac{s_i}{K_m}\right)^{1/2}\left[\left(\frac{s_o}{s_i}\right)^{1/2} - 1\right] = -\left(\frac{\mu_{max}\rho_o D_e}{Y'_s K_m}\right)^{1/2}\frac{A_w V}{F} \tag{48}$$

where A_w is the interfacial area between microbial and liquid films per unit fermenter volume (i.e. $A_w \leq A$). Eq. (48) does not contain a biomass hold-up term because the microbial film thickness does not feature in Eq. (47). Also, Eq. (48) does not account for any contribution to the conversion efficiency by suspended microorganisms as these have only a marginal effect.

Performance Characteristics

It can be seen from Fig. 56 that a consistent performance can be achieved in a packed-bed biological film reactor, in which the biomass hold-up is controlled by periodic hydraulic washing.

Figs. 57(a) and (b) show the characteristic influence of inlet substrate concentration on conversion efficiency (*see*, in addition, Figs. 14 and 15), when the microbial film thickness is held constant. Data on A_w, L and h' obtained from Figs. 57(a) and (b) are contained in Table 12, using a knowlege of the biological parameters following the procedure outlined on pp.600–601. These data suggest the absence of any significant liquid-phase diffusional limitations. The wetted area data are plotted in Fig. 58 and show a monotonic increase with flow rate up to the specific surface area of the particular packing material.

The variation of performance of a trickling filter in which the biomass is 'naturally' controlled is shown in Fig. 59. Equivalent data for the variation of biomass hold-up are given in Fig. 60.

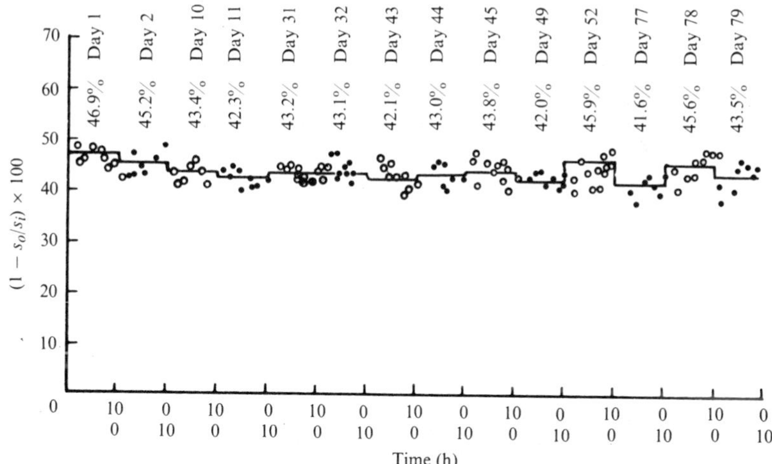

Figure 56. Consistency and reproducibility of data obtained from a trickle-flow fermenter. Percentages represent values for mean efficiency: Oxidant, dissolved nitrate; feed flow, 0.00755 cm s^{-1}; temperature 20°C $s_i \ell_o$, 100 mg ℓ^{-1} [Atkinson and Williams (1971)].

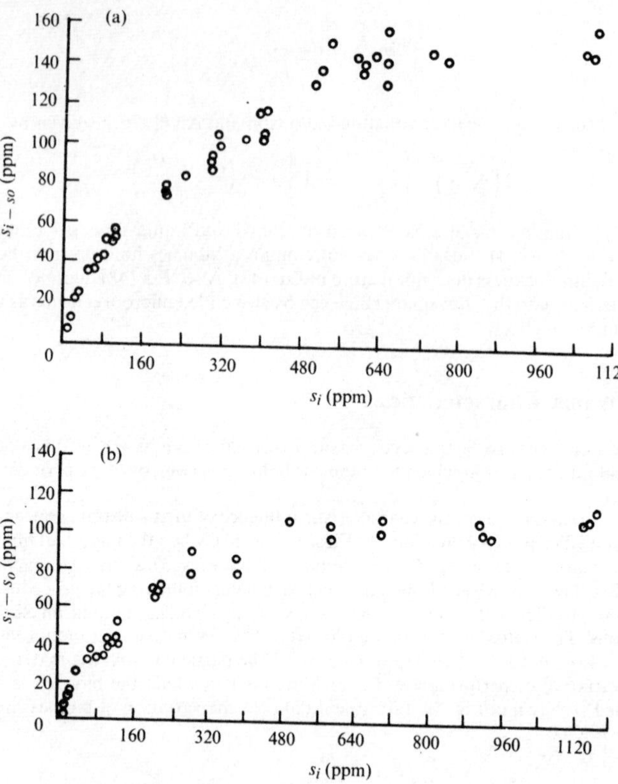

Figure 57. Glucose uptake data for a trickle flow fermenter containing 2-in. Biopac. (a) Flow rate, $1.6 \times 10^{-4}\ m^3\ m^{-2}\ s^{-1}$. (b) Flow rate, $2.0 \times 10^{-4}\ m^3\ m^{-2}\ s^{-1}$ [Atkinson and Ali (1976)].

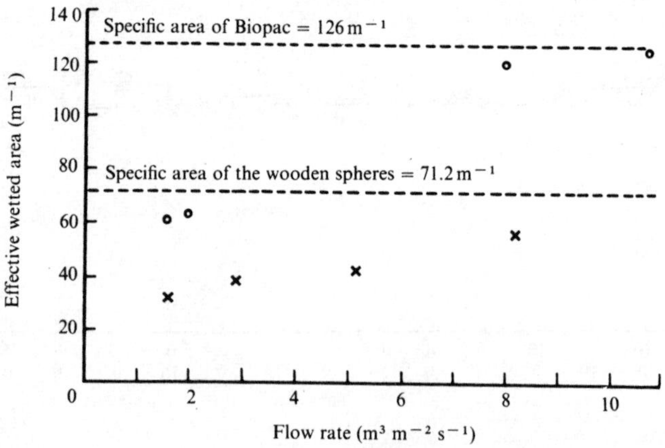

Figure 58. Effect of flow rate on effective wetted area in a trickle-flow fermenter. ○, Biopac; ×, wooden spheres [Atkinson and Ali (1976)].

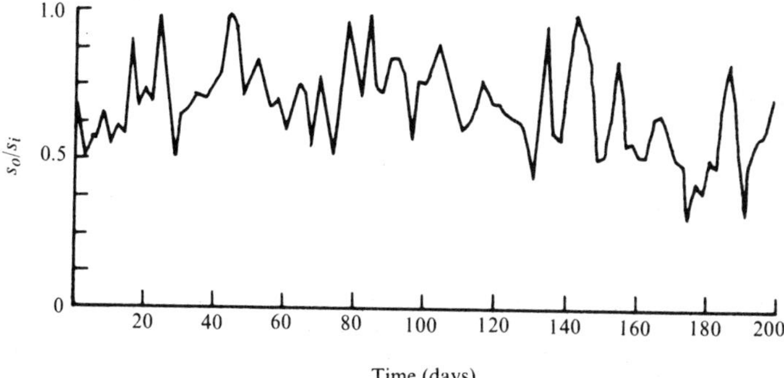

Figure 59. Efficiency data for Stevenage sewage using crinkle-close Surfpac [Bruce and Merkens (1973)].

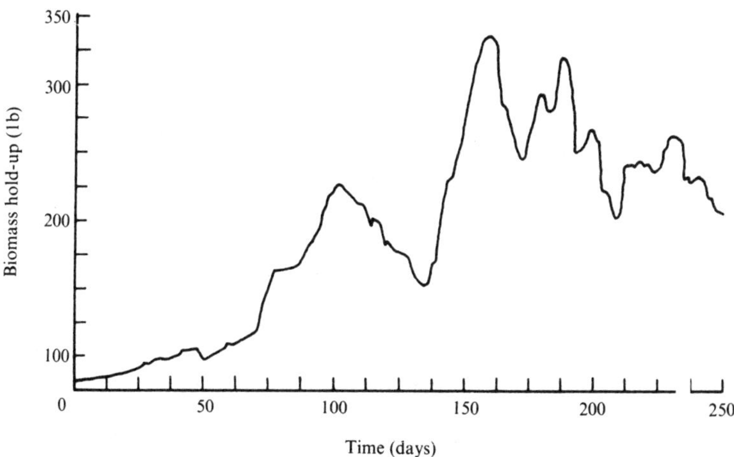

Figure 60. Variation of biomass hold-up in an experimental filter containing Flocar [Reynolds (1974)].

Table 12. Wetted area, biological film thickness and liquid-phase mass transfer coefficients in a trickle-flow fermenter.

Flow rate, F ($m^3/m^{-2}\ s^{-1} \times 10^{-2}$)	Wetted area, A_w (m^{-1})	Biological film thickness, L (mm)	Mass transfer coefficient, h ($m\ s^{-1} \times 10$)
0.0148	30.5	0.60	1215.0
0.0270	36.0	0.61	3674.0
0.0540	43.2	0.63	1525.0
0.0810	54.7	0.57	1209.0
0.0158	61.9	0.94	1323.0
0.0202	62.9	0.835	2197.0
0.0807	123.7	0.73	2597.0
0.1076	124.0	0.955	2380.0

Figs. 61 and 62 present data on the same packed-bed, biological film reactor used to obtain the data given in Figs. 57(a) and (b). These data, which were obtained after biomass had been allowed to accumulate, demonstrate that the interaction between the biomass and the trickle flow is such as to reduce the interfacial area A_w, with a corresponding reduction in conversion efficiency.

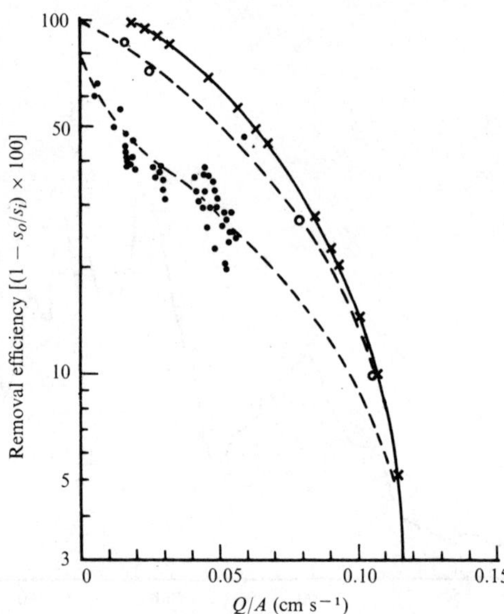

Figure 61. Dependency of removal efficiency on flow rate for 2-in. Biopac – effect of biomass hold-up. +, Experimental data; ×, maximal theoretical efficiency (i.e. $A_W = A$); ○, theoretical efficiency based on results of Atkinson and Ali (1976) [Atkinson and Ali (1978)].

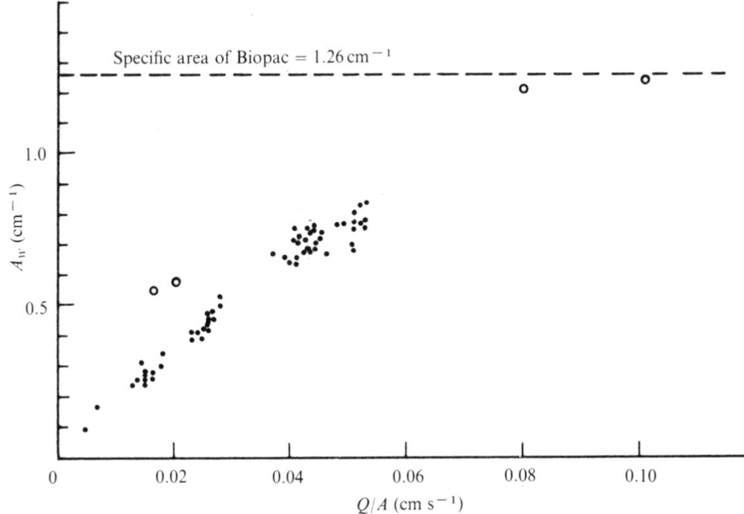

Figure 62. Wetted area for 2-in. Biopac – effect of biomass hold-up: ○, Data of Atkinson and Ali (1976), where $L \simeq 1$ mm; ●, L is very large [Atkinson and Ali (1978)].

Graphical Representation

Eq. (40) may be expressed in terms of the substrate concentration, i.e.

$$\frac{V}{F} = -\int_{s_I}^{s_0} \frac{ds}{R_v} \tag{49}$$

For a submerged, packed-bed, biological film reactor

$$R_v = NA + Rx$$

or

$$R_v = \frac{\mu_{\max} s}{1 + s/K_m}\left(\beta + \frac{s_i}{K_m} - \frac{s}{K_m}\right) \tag{50}$$

Eq. (50) is plotted diagrammatically in Fig. 63 to show the effect of support surface area. Graphical integration of Eq. (49) follows the procedure of Fig. 39.

Fig. 64 demonstrates the fermenter volumes required when using a packed-bed film fermenter in series with a tubular fermenter (*see* Fig. 55).

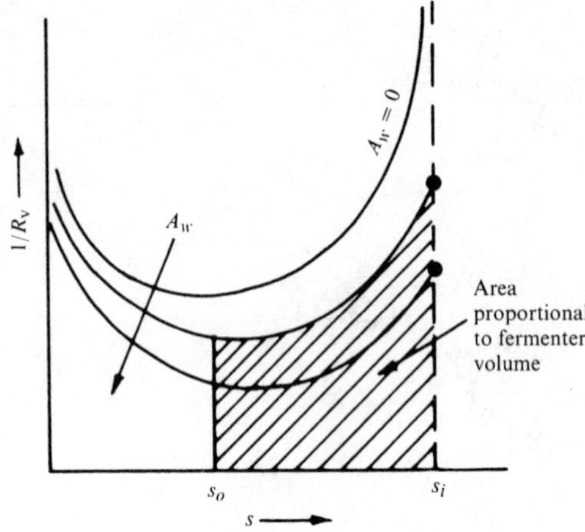

Figure 63. Effect of surface area on the required volume of a tubular fermenter [Atkinson (1974)].

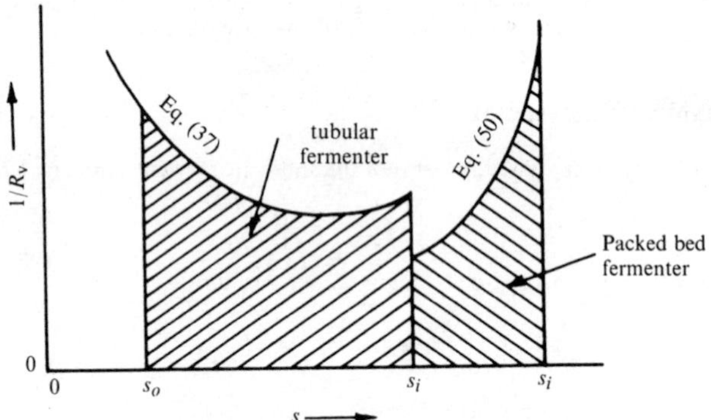

Figure 64. Design of a film fermenter in series with a tubular fermenter (Atkinson (1974)].

FLUIDIZED BEDS

Fig. 65 illustrates the effect of increased flow rate on a bed of discrete particles. Fluidization occurs initially at the incipient fluidization velocity when the pressure drop across the bed is equal to the weight of the bed. Any further increase in velocity leads to bed expansion and ultimately to elutriation when the force exerted on an individual particle exceeds its weight. These conditions are controlled by Stokes's law, which relates the force exerted by the fluid on a particle under fluidization/elutriation conditions to the particle size and density.

A feature of fluidized beds is that particles of mixed size and/or density stratify, i.e. the larger and/or denser particles collect towards the bottom of the bed while the smaller, lighter particles migrate towards the top [*see* Coulson and Richardson (1968)]. A consequence of these features is that very small particles, e.g., single microorganisms, are elutriated from an otherwise stable bed.

There are three types of particle that can be used in a biological fluidized bed.

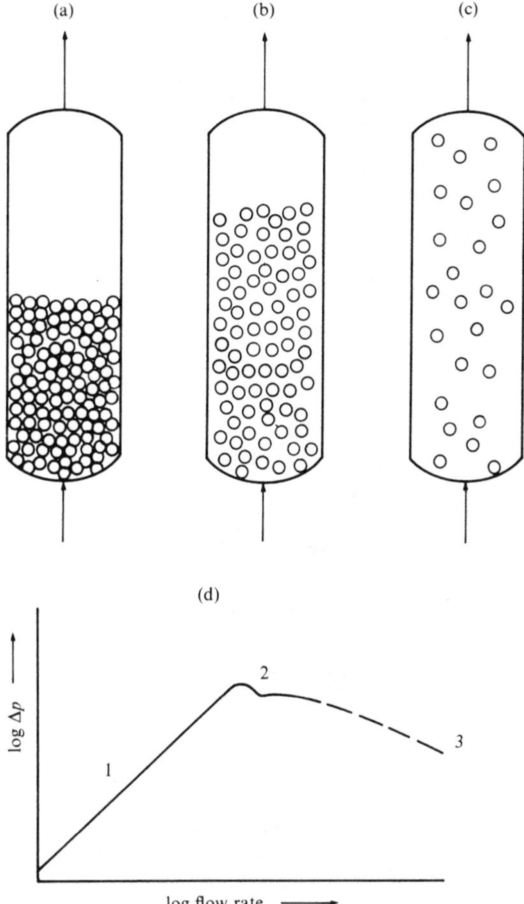

Figure 65. Effect of flow rate on the particles in tubular fermenters. (a) fixed-bed fermenter, (b) fluidized-bed fermenter, (c) elutriation, (d) pressure losses [Atkinson (1974)].

1. Flocculated microorganisms. Such particles have a low incipient fluidizing velocity and are subject to rupture by vigorous fluid action with the resultant small particles being elutriated.
2. Solid supports with attached biomass. Two conditions can be achieved, depending on the size and density of the support particles.

a. Large, relatively dense particles provide a balance between growth and attrition. This results in a steady biomass hold-up where the biomass film thickness is small compared with the size of the support particle.

b. Small, relatively light particles accumulate biomass so that the film thickness becomes larger than the support particles.

The accumulation of biomass on the inert particles can result in unsteady-state fermenter biomass hold-ups and particle size distributions, with the largest, most overgrown particles being at the top and the smallest at the bottom.

3. Biomass support particles (*see* Fig. 66 and p.32).

Figure 66. Biomass-filled mesh particles in a fluidized bed [Atkinson and Lewis (1980)].

Growth can initiate fluidization in a bed of inert particles when the flow is below the incipient fluidizing velocity of the uninoculated bed. The stages involved are listed below.

1. At initial growth following inoculation, attachment of microorganisms produces an expanded stationary bed.
2. Further growth results in fluidization caused by the reduction in the overall particle density.
3. Elutriation occurs if particle–particle–wall contacts produce insufficient attrition to balance growth.

Therefore, it follows that biological fluidized beds can be operated with steady or variable particle biomass hold-up. Table 13 summarizes various features of reactors operated in the two modes.

Table 13. Characteristics of fluidized-bed fermenters containing support particles.

Factor	Steady particle biomass hold-up	Variable particle biomass hold-up
Biomass control	Particle–particle–wall attrition	Continuous/periodic particle removal and cleaning
Biomass recovery	Conventional centrifuge/clarifier	At high concentration external to the reactor (*see* Fig. 72)
Bed biomass hold-up	Steady	Steady (variable for periodic particle removal)
Reactor conversion efficiency	Steady	Steady (variable for periodic particle removal)
Design	Facilitated by the fact that all aspects of performance are steady and both particle and biomass hold-ups can be preselected (as with biomass support particles)	Requires prediction of the variation of particle biomass hold-up with time (*see* Fig. 77)

Configurations

Figs. 67–74 cover a range of fluidized-bed fermenters developed for use in the brewing, vinegar and wastewater treatment industries. The tower fermenter (*see* Figs. 67–79) has been developed to a commercial level but, since it is based upon flocculant particles, it is restricted by the following features.

1. Flocculant microbial strains are essential.
2. Floc size distributions lead to variations in biomass hold-up throughout the tower.
3. Only low throughputs (i.e. superficial velocities) are possible due to floc size and densities.
4. A dynamic balance has to be achieved between growth, attrition and elutriation, leading to a virtually constant biomass concentration.
5. Fluid mechanical restrictions and the residence time necessary to achieve the required conversion are not always compatible.
6. Floc breakdown and flotation can occur due to the gas formed during the biochemical reaction coming out of solution to form large bubbles and intense agitation.
7. The non-bed volume required is particularly large to provide for gas disengagement and sedimentation of particles that have escaped the bed.

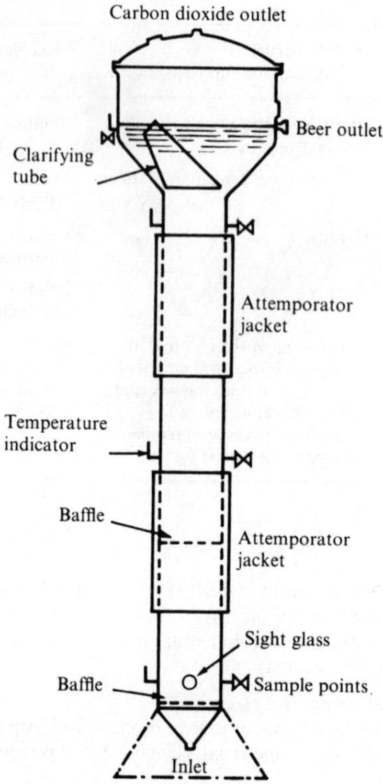

Figure 67. Diagram of the tower fermenter [Shore and Royston (1968)].

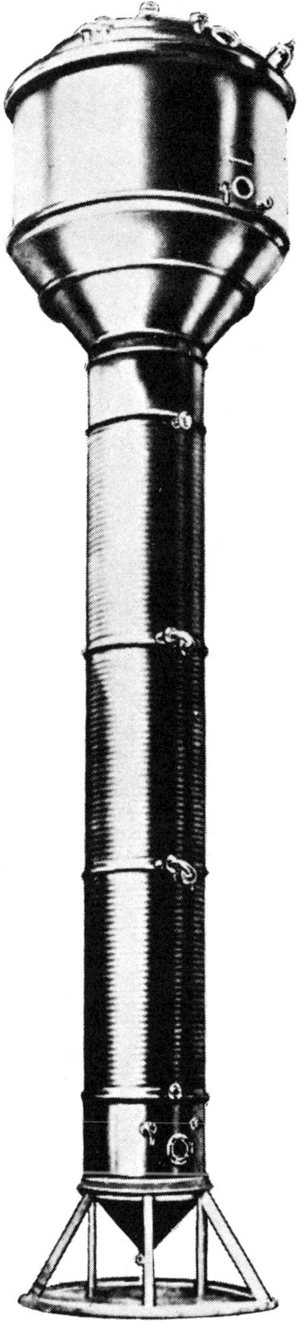

Figure 68. Tower fermenter (courtesy of APV Co. Ltd.).

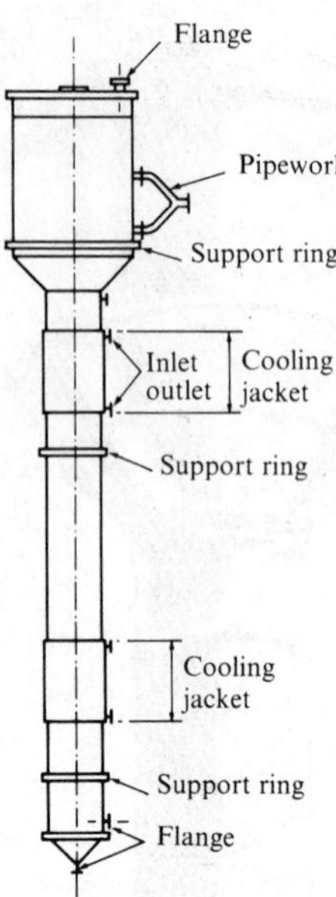

Figure 69. Diagram of a tower acetifier of a type installed in a UK malt vinegar factory for fully continuous acetification [Greenshields (1978)].

A plan view of a large fluidized-bed unit proposed by Jeris and Owens (1981) for use in a municipal sewage works is shown in Fig. 70. Sand, carbon, anthracite or similar media in the size range 0.2–0.3 mm provide very large surface areas per unit bed volume and are excellent supports for the mixed microbial populations associated with biological wastewater treatment. Aeration demands are high and usually are achieved using air and occasionally oxygen in external devices maintained at elevated pressure by a hydrostatic leg or compressor. External aeration has the advantage of not subjecting the particle-bound biomass to erosion by gas bubbles.

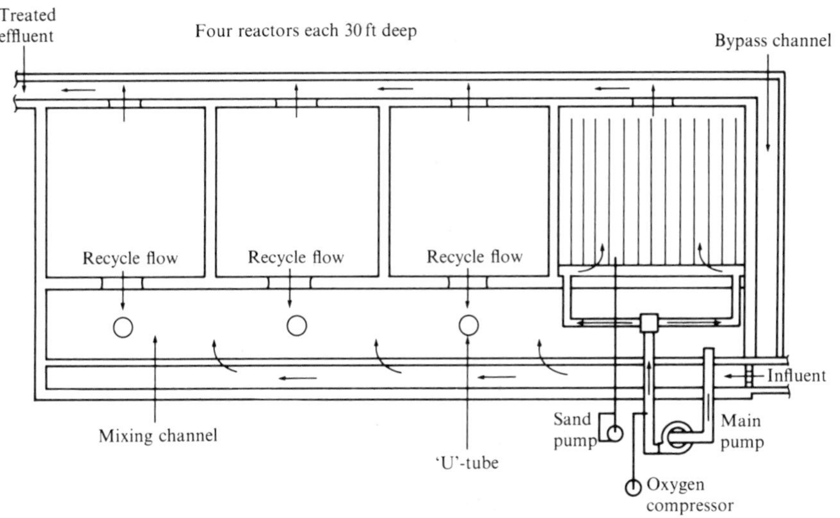

Figure 70. Aerobic fluidized-bed reactor (plan view) [Jeris and Owens (1981)].

Figs. 71 and 72 show two methods of controlling particle biomass hold-up external to the fluidized bed. Sufficient bed agitation can, of course, control the hold-up *in situ*, or a high-speed agitator can be located in the top of the bed to provide the necessary shear. However, with the latter, some particle carry-over and losses are inevitable. In Fig. 71, both liquor and particles are caused to overflow into a common vessel containing a vibrating screen, with the liquor and detached biomass passing through the screen into a conventional sedimentation tank, while the 'clean' particles are returned to the base of the bed. A development of the principles of Fig. 71 is shown in Fig. 72 where the particles that overflow are collected in a sedimentation chamber – effectively a decanter. When sufficient particles have been collected, they are subjected to intense shear, with the concentrated sludge being recovered and the clean particles returned.

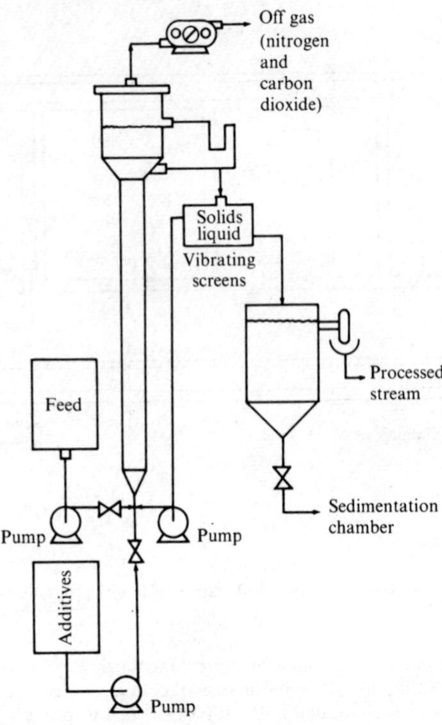

Figure 71. Basic process flow-sheet for biological denitrification in a fluidized-bed reactor [Francis and Hancher (1981)].

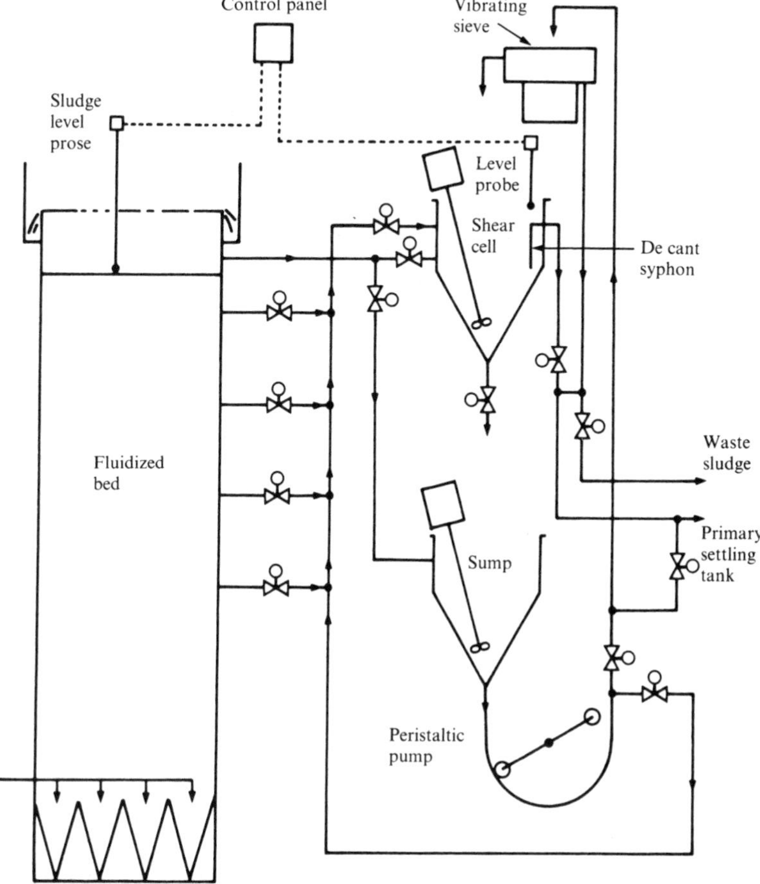

Figure 72. Simplified diagram of sand/biomass separation system for fluidized-bed pilot plant at WRC Coleshill Experimental Plant [Hawkins *et al.* (1981)].

A two reactor fluidized-bed system for carbonaceous oxidation, nitrification and dentrification is shown in Fig. 73. The first reactor oxidizes incoming sewage using nitrate from the second reactor, which aerobically oxidizes carbonaceous compounds and produces a nitrified effluent. The significance of this system lies in the two different microbial populations in the separate reactors, which result from different local chemical environments even though the reactors are connected by flow streams.

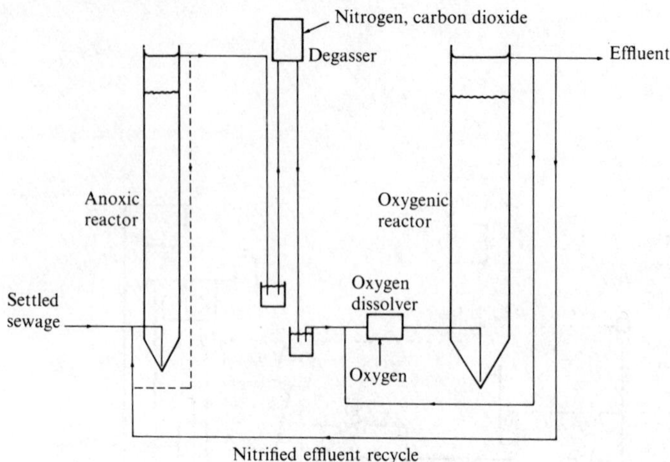

Figure 73. Two-reactor fluidized-bed system for complete sewage treatment including denitrification, using the carbon present in settled sewage [Cooper and Wheeldon (1981)].

Ideal Performance Equations

Particle properties

When considering particle properties both overall particle density and particle growth are important factors.

1. The overall particle densities of solid support particles, biomass support particles and associated biomass are given in Figs. 74 and 75. Fig. 74 shows the sensitivity of the overall density to microbial film thickness, particularly for heavy particles. This is a contibutory factor to stratification in the fluidized beds illustrated in Figs. 71 and 72. The equivalent overall particle density changes for biomass support particles as they become filled with biomass are associated with the fraction of the porosity ϕ taken up by biomass. The largest changes in overall particle density of biomass support particles between empty and full occur with particles of high porosity prepared from material of high specific gravity.

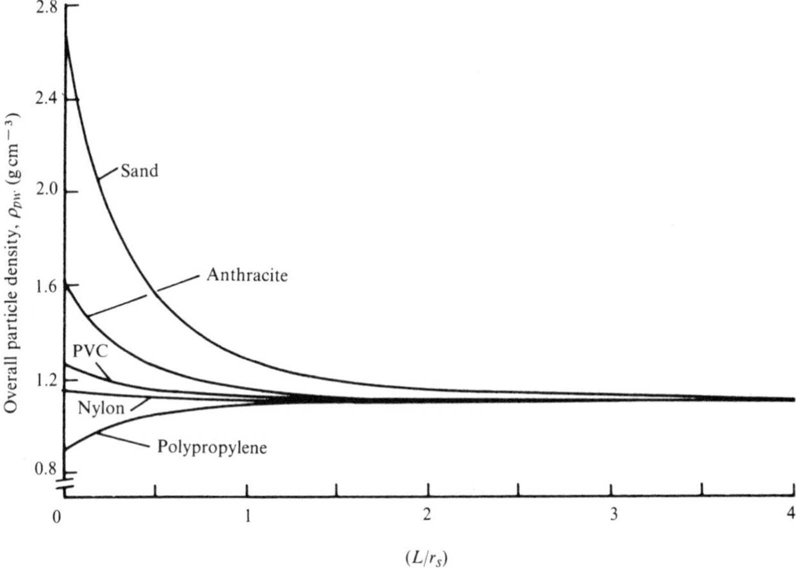

Figure 74. Overall densities of solid support particles with attached biomass of wet density 1.1 g cm^{-3}, L, microbial film thickness; r_s, support particle radius [Atkinson *et al.* (1981)].

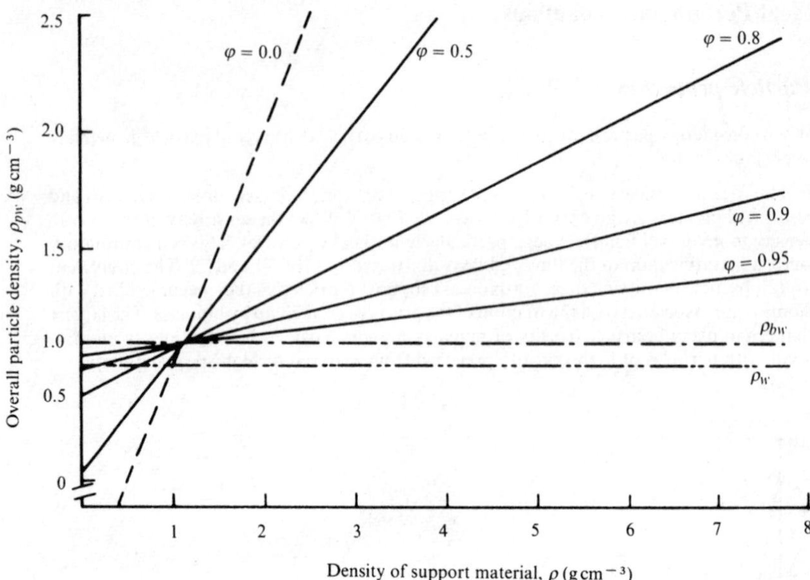

Figure 75. Overall densities of biomass support particles completely filled with biomass, φ, particle density [Atkinson *et al.* (1981)].

2. In the absence of attrition, solid support particles accumulate biomass and the overall particle increases in size. The conditions within the microbial film change from reaction-controlled to strongly 'solid'-phase diffusion-limited (*see* Chapter 10). Fig. 76 shows the time course of the overall increase in the size of a particle exposed to a constant fermenter substrate concentration. In principle, Fig. 76 can be used to estimate the length of the time the solid supports need to reside in a fluidized bed to achieve a required overall size.

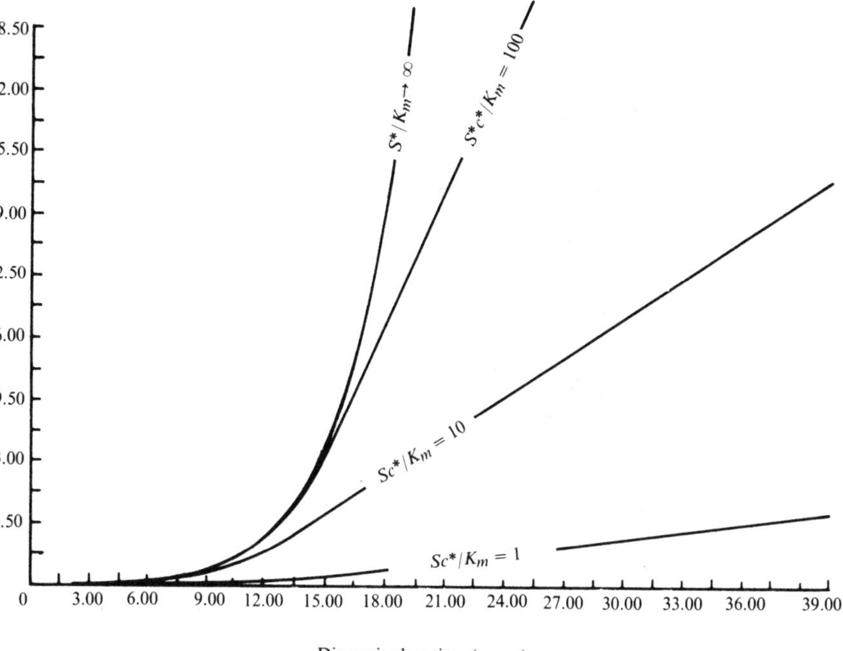

Figure 76. Growth of particles exposed to a constant substrate concentration [Atkinson and Daoud (1976)].

Bed biomass hold-up

The maximum bed biomass hold-up corresponds to incipient fluidization, i.e. the maximum number density of particles with sufficient movement to produce abrasion.

Fig. 77 gives the volume of bed occupied by wet biomass assuming a bed voidage of 40 per cent for a range of support particle sizes and microbial film thickness. In the absence of support particles, the biomass takes up 60 per cent of the volume independent of 'floc' size. As the support particle size increases, the volume occupied by wet biomass decreases rapidly, particularly for 'thin' films.

For spherical biomass supports full of biomass

$$\text{Percentage volume of bed occupied by wet biomass} = \phi \times 60 \qquad (51)$$

using the same basis for bed voidage as in Fig. 77. It follows from Eq. (51) that the biomass hold-up is independent of particle size when using biomass support particles.

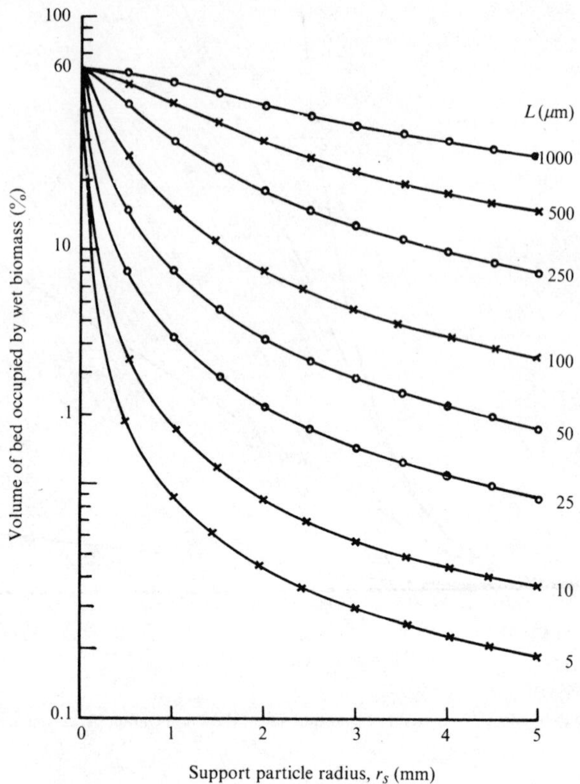

Figure 77. Immobilized biomass hold-ups in reactors containing solid supports ($\varepsilon = 0.4$) wet biomass is expressed as g dry wt (C reactor volume)$^{-1}$ when biomass density is 0.1 g cm^{-3}. (Atkinson *et al.* (1981)].

Volumetric rates of substrate uptake

1. Atkinson *et al.* (1981) have provided the following equations [Eqs. (52)–(54)] when considering particle biomass hold-up, on the assumption that strongly diffusion-limited conditions prevail and that these are described by 'half'-order kinetics (*see* Chapter 10).

a. For spherical supports

$$\frac{R_v}{(1-\varepsilon)R_{\max}\rho_o} = \frac{3}{1.16}\frac{1}{r_s(1+L/r_s)}\left(\frac{2D_e}{R_{\max}\rho_o}\right)^{1/2} s^{*1/2} \tag{52}$$

b. For spherical biomass support particles

$$\frac{R_v}{(1-\varepsilon)R_{\max}\rho_e} = \frac{3}{1.16}\frac{1}{r_p}\left(\frac{2D_e}{R_{\max}\rho_e}\right)^{1/2} s^{*1/2} \tag{53}$$

where ρ_e is the effective biomass concentration within a biomass support particle and r_p is the radius of the biomass support particles

c. For cuboid biomass support particles

$$\frac{R_v}{(1-\varepsilon)R_{\max}\rho_e} = \frac{1}{V_p/A_p}\left(\frac{2D_e}{R_{\max}\rho_e}\right)^{1/2} s^{*1/2} \tag{54}$$

At high substrate concentrations, 'zero'-order conditions prevail throughout the immobilized biomass and results in the following limiting values.

a. For spherical solid supports

$$\frac{R_v}{(1-\varepsilon)R_{\max}\rho_o} = \frac{(1+L/r_s)^3 - 1}{(1+L/r_s)^3} \tag{55}$$

b. For biomass support particles

$$\frac{R_v}{(1-\varepsilon)R_{\max}\rho_e} = \phi \tag{56}$$

Eqs. (52)–(56) are compared in Fig. 78. The sharp change from zero- to half-order kinetics is evident as is the effect of the increasing 'solid'-phase diffusion limitation with increasing particle size.

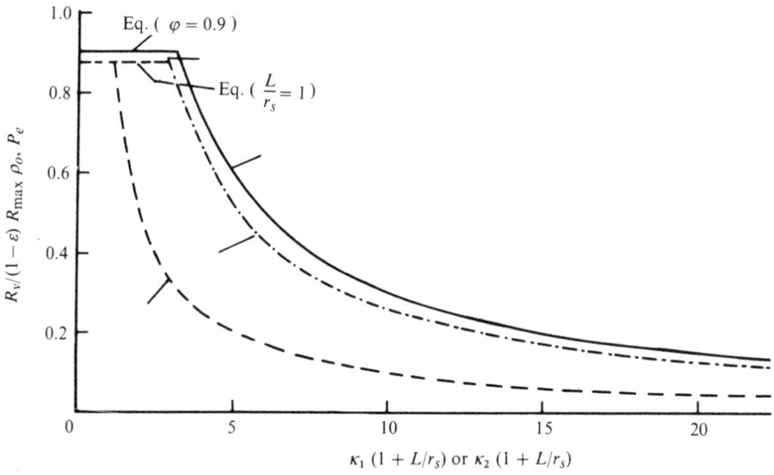

Figure 79. Overall volumetric rates of carbon conversion (steady particle biomass hold-up). Curve _ _ _ _, biomass support particle slabs [Eqs. (54) and (56)]; curve _ . _ . _ ., biomass support particle spheres [Eqs. (53) and (56)]; curve ______, biomass support particle cubes [Eqs. (54) and (56)], where $\kappa_1 = (R_{\max\,\rho o}/2\,D_e s^*)^{\frac{1}{2}}$, $\kappa_2 = (R_{\max \rho e}/2\,D_e s^*)^{\frac{1}{2}}$ [Atkinson *et al.* (1981)].

2. Utilization of the information leading to Fig. 76 together with Eqs. (52)–(54) lead Atkinson *et al.* (1981) to deduce Eqs. (57)–(59), when considering accumulating biomass hold-up.

a. For spherical solid supports

$$\bar{R}_v \simeq \frac{2}{3} R_v \tag{57}$$

b. For spherical biomass support particles

$$\bar{R}_v = \frac{1}{2} R_v \tag{58}$$

c. For cuboid support particles

$$\bar{R}_v = \frac{1}{2} R_v \tag{59}$$

where $\overline{R}_v$ is the time average volumetric rate of substrate uptake by a bed of particles with accumulating biomass, the particles being continuously removed, cleaned and returned.

Eqs. (57)–(59) suggest the extent to which the volumetric rate of reaction is reduced because of the reduced bed biomass hold-up.

Performance Characteristics

At present, it is reasonable to assume that the liquid phase in a fluidized bed is completely mixed due to action of particles and any gas bubbles that exist. With this assumption Eqs. (41)–(45) are applicable to fluidized beds and the performance characteristics are those given in Figs. 47–52.

As fluidized-bed technology develops, the mixing conditions in the liquid phase will require attention under the particular operating regimes that prove to have greatest commercial benefit.

REACTOR INTENSIFICATION

The history of the development of large-scale process engineering suggests the existence of a regular pattern or 'learning curve' [Malpas (1978)]. The beginnings of a given 'learning curve' are the discovery of a reaction system, e.g., acetic acid production by immobilized biomass and the subsequent path is characterized by three broadly identifiable areas of activity.

1. Economies of scale, i.e. an increase in the throughput of a given works by the use of larger process units.
2. Process intensification, i.e. a reduction in the size of individual process units while maintaining performance.
3. The discovery of an alternative reaction system after which the cycle is repeated.

The drive for process intensification is recognized as a powerful factor in the continuing development of large-scale processes, although it is accepted that many forms of such intensification can be devised that do not lead to reduced costs. Nevertheless, overall experience is that economic benefit and process intensification are inextricably linked with organizations that fail to participate successfully not being able to enter or having to give up the particular business area. Since the achievement of process intensification is demanding in terms of research and development time and costs, it follows that extremely careful targeting is required, coupled with well-led, thoughtful and dedicated research effort [Malpas (1977)]. Furthermore, it follows that extensive use must be made of modelling for the evaluation of process concepts in a predictive sense, prior to the construction of laboratory- and pilot-scale devices (*see* p.593).

Process intensification without any increase in the complexity of the hardware requires the exploitation of fluid mechanics phenomena as the basis for elegant solutions to the process engineering problems that result from the need to contact the biomass and liquid phases and the need to retain and control the biomass, as well as biomass recovery. Immobilized biomass and fluidized-bed biological reactors are examples of technical developments which meet some of these requirements, i.e. immobilized biomass provides the basis of a significant contribution to the development of the process engineering associated with biological reactors. The extent to which this need can be satisfied is summarized in Table 14.

The development of the process engineering (i.e. new approaches both to the reactor arrangement and biomass recovery) involves the expenditure of very great effort over a long period. Such effort for biological systems is justified. The potential for exploiting the interaction between the reactor configuration and the reaction, with the forcing of the *overall* rates of reaction and yields in an advantageous direction, is probably greater for biological than for chemical systems.

Table 14 contains a list of those aspects of biological reactor and biomass recovery processes that, if achieved, would contribute significantly to process intensification; the nature of the improvements are also identified. Tables 15 and 16 show the extent to which the objectives of Table 14 have been achieved for tower fermenters (*see* p.645) and trickling filters (*see* p.634). These two particular fermenter configurations have been chosen because they are reasonably well understood, differ in operational principles and exemplify the methodology of process intensification.

Table 15. Process intensification characteristics of tower fermenters.

Technical improvement	Achievement	Comment
Increased biomass concentration (i.e. hold-up)	Yes	Use of flocculated microorganism in a fluidized bed
Predetermined biomass concentration (hold-up)	Partially over restricted range	Use of knowledge of liquid-phase fluidization
Biomass concentration (i.e. hold-up) independent of process throughput	No	Bed expansion and elutriation depend on process throughput
Predetermined biomass size and shape	Partially	By selection of strains which give a restricted range of size and shape, although there will be a maximum size for a given strain
Improved biomass recovery	Yes	Because flocculant strains are used
Improved yield coefficients		No information
Biomass/sludge 'age' independent of process throughput	Partially	Due to retention of biomass by flocculation
Multiple reactions in a single reactor due to local variations in environment and/or species	Yes	Because liquid phase is in plug flow and carbon compounds are attacked sequentially
Retention by fluid phase of dilute, suspension-free, aqueous properties, especially viscosity	Yes	Because liquid and 'solid' phases are segregated

Table 14. Areas leading to reactor intensification.

Technical improvements	Benefits					
	Increased volumetric rates of reaction	Reduced reactor size	Improved separation processes (cell/product/liquid)	Reduced number of vessels in the total process	Enhanced design capability	Enhanced design flexibility
Increased biomass concentration (hold-up)	√	√	√	√		
Predetermined biomass concentration (hold-up)					√	√
Biomass concentration (hold-up) independent of process throughput	√	√		√	√	√
Predetermined biomass size and shape					√	√
Improved biomass recovery			√	√	√	√
Improved yield coefficients	√	√	√			
Biomass/sludge 'age' independent of throughput				√	√	√
Multiple reactions in a single reactor due to local variations in environment and/or species	√	√		√		√
Retention by fluid phase of dilute, suspension-free aqueous properties, particularly viscosity	√	√			√	√

Table 16. Process intensification characteristics of trickling filters.

Technical improvement	Achievement	Comment
Increased biomass concentration (hold-up)	Yes	By use of immobilized biomass, (i.e. adherent microbial films)
Predetermined biomass concentration (hold-up)	Partially	From the specific surface area of the support media
Biomass concentration (hold-up) independent of process throughput	Partially	Although liquid flow provides a scouring action
Predetermined biomass size and shape	Size–none, shape–partially	Biological films are essentially planar due to the relative sizes of the films and the support media; film thickness indeterminate
Improved biomass recovery	Yes	Because the films are largely composed of flocculent microorganisms and because of the sloughing of large agglomerates
Improved yield coefficients		No information
Biomass/sludge 'age' independent of process throughout	Partially	Due to retention of biomass by adhesion
Multiple reactions in a single reactor due to local variations in environment and/or species	Yes	Because liquid phase is in 'plug' flow the vertical distribution of soluble components leads to vertical distribution of species
Retention by the fluid phase of dilute, suspension-free, aqueous properties, especially viscosity	Yes	Because liquid and 'solid' phases are segregated

The complete achievement of the developments identified in Table 14 requires radical improvement in the present technology which, on the basis of past experience, must be expected to be simple both in concept and execution. There are requirements for new process hardware, which are listed below.

1. Robustness, i.e. relatively insensitive to operator error, equipment failure and variations in reactant input or process conditions.
2. Applicability to the greatest range of processes in the medium term.
3. Potential for long-term development.
4. Cost.

Against the above background, the contributions that immobilized biomass can make to reactor intensification are listed in Table 17. Table 17 also gives an indication of the present position of the technology. A summary of the characteristics of reactors containing immobilized biomass is given in Table 18, with fixed, expanded, agitated and fluidized-bed types of reactors being considered.

Table 17. Present position and potential of immobilized biomass [Atkinson (1981)].

Areas of improvement	Advantages	Disadvantages	Present position
Reactor biomass hold-up	Increased	Not all usable due to diffusion limitations	Increased by factors of 10 or more compared with the activated sludge process
Overall volumetric rates of carbon conversion	Increased due to increased reactor biomass hold-up		Achieved
Biomass recovery at high solids content	Reduced solids-handling costs	Reduction in overall volumetric rate of reaction compared with operation at steady particle biomass hold-up	Technology under development
Effect of particle/particle contacts during particle biomass accumulation	Allows particle biomass accumulation or steady particle biomass hold-up as required	Biomass losses from support particles during biomass accumulation	Largely achieved – technology under development
Predetermined reactor biomass hold-up	Accurate design once further knowledge available on number/density of particles in fluidized beds	Requires extensive pilot-scale experimentation	Under investigation
Steady particle biomass size and shape	Accurate design when based upon predetermined biomass size and shape	Restricted to 'large' particles/intense agitation	Steady particle biomass hold-up achieved; attempts to determine hold-up a priori
Exploitation of diffusional limitation	Process intensification due to multiple species/reactions within a single particle	Requires steady particle biomass size and shape	Under investigation

Table 18. Characteristics of reactors containing support particles for biomass [Atkinson *et al.* (1981)].

	Fixed/expanded beds	Fluidized beds/agitated beds	
	Variable particle biomass hold-up	Variable particle biomass hold-up	Steady particle biomass hold-up
Biomass control	Back-washing	Continuous particle removal and cleaning (particles could be removed periodically)	Particle/particle/wall attrition within fluidized beds when using some particles; particle/particle/agitator attrition
Biomass recovery	Washings collected in a humus tank	At high concentration external to reactor	Conventional clarifier
Bed biomass hold-up	Variable	Steady (variable for periodic particle removal)	Steady
Reactor performance efficiency, e.g., carbon removal	Variable	Steady (variable for periodic particle removal)	Steady
Design	Requires prediction of the variation of the bed biomass hold-up with time	Requires prediction of the variation of the bed/particle biomass hold-up with time	Facilitated by fact that all aspects of performance are steady and both particle and bed biomass hold-ups preselected

The performance efficiency of any biological reactor containing support particles depends on a combination of the following four factors.

1. The number of support particles per unit reactor volume – a largely fluid mechanics phenomenon (e.g., *see* Figs. 46 and 65).
2. The average biomass hold-up per particle – a balance between growth and attrition.
3. The average overall specific rates of reaction of the immobilized biomass – a matter of advantageous/disadvantageous diffusional limitations (*see* Chapter 10).
4. The average overall yield coefficient of the immobilized biomass.

However, the potential of support particles includes more subtle features than those given above.

1. Completely filled particles removed from the fermenter can be passed into a second reactor environment in order to produce the required product or to remove or destroy the biomass.
2. Different species can be located at various positions within a fermenter.
3. Any combination of changing environment and variety of microbial species can be envisaged within a single process by using counterflow of the particles and the fluid or by using a series of beds through which the fluid is passed.
4. Different species can be induced to occur at different locations within the immobilized biomass under the 'pressure' of the differences in the local environment brought about by diffusional limitations, leading to an advantageous range of reactions and overall yield coefficients.

REFERENCES

Abson, J.W. and Todhunter, K.H. (1967) Effluent Disposal, in *Biochemical and Biological Engineering*, vol. 1, ed. by N. Blakebrough, p.309 (Academic Press, New York).

Ainsworth, G. (1970) The Activated Sludge Process, in *Water Pollution Control Engineering*, ed. by A.L. Downing, p.31 (HMSO, London).

Atkinson, B. (1974) *Biochemical Reactors* (Pion, London).

Atkinson, B. (1981) Immobilized Biomass – A Basis for Process Development in Wastewater Treatment, in *Biological Fluidised Bed Treatment of Water and Wastewater*, ed. by P.F. Cooper and B. Atkinson, p.22 (Ellis Horwood, Chichester).

Atkinson, B. and Ali, M.E.A.R. (1976) Wetted Area, Slime Thickness and Liquid Phase Mass Transfer in Packed Bed Biological Film Reactors (Trickling Filters). *Trans. Instn Chem. Engrs*, **54**, 239.

Atkinson, B. and Ali, M.E.A.R. (1978). Biomass Hold-Up and Packing Surface in Trickling Filters. *Water Res.*, **12**, 147.

Atkinson, B., Black, G.M. and Pinches, A. (1981) The Characteristics of Solid Supports and Biomass Support Particles when used in Fluidized Beds, in *Biological Fluidised Bed Treatment of Water and Wastewater*, ed. by P.F. Cooper and B. Atkinson, p.75 (Ellis Horwood, Chichester).

Atkinson, B., Busch, A.W., Swilley, E.L. and Williams, D.A. (1967) Kinetics, Mass Transfer and Organism Growth in a Biological Film Reactor. *Trans. Instn Chem. Engrs*, **45**, 257.

Atkinson, B. and Daoud, (1976), in *Advances in Bioengineering*, vol. 4, ed. by T.K. Ghose *et al.*, (Springer-Verlag, Berlin).

Atkinson, B. and Davies, I.J. (1972) The Completely Mixed Microbial Film Fermenter – a Method of Overcoming Wash-Out in Continuous Fermentation. *Trans. Instn Chem. Engrs*, **50**, 208.

Atkinson, B. and Kossen, N.W.F. (1978) Fermenter Design and Modelling of Continuous Processes, in *Biotechnology*, p.37 (Verlag Chemie, Weinheim).

Atkinson, B. and Lewis, P.J.S. (1980) The Development of Immobilized Fungal Particles and their Use in Fluidised Bed Fermenters, in: *Fungal Biotechnology*, ed. by J.E. Smith *et al.* (Academic Press, London).

Atkinson, B. and Williams, D.A. (1971) The Performance Characteristics of a Trickling Filter with Hold-up of Microbial Mass controlled by Periodic Washing. *Trans. Instn Chem. Engrs*, **49**, 215.

Ault, R.G., Hampton, A.N., Newton, R. and Roberts, R.H. (1969) Biological and Biochemical Aspects of Tower Fermentation. *J. Inst. Brewing*, **75**, 260.

Bailey, J.E. and Ollis, D.F. (1977) *Biochemical Engineering Fundamentals* (McGraw-Hill, New York).

Bruce, A.W. and Merkens, J.C. (1973) Further Studies of Partial Treatment of Sewage by High-Rate Biological Filtration. *Water Pollut. Cont.*, **72**, 499.

Butterworth, T.A., Wang, D.I.C. and Sinskey, A.J. (1970) Application of Ultrafiltration for Enzyme Retention during Continuous Enzymatic Reaction. *Biotechnol. Bioengng*, **12**, 615.

Cooper, P.F. and Wheeldon, D.H.V. (1981) Complete Treatment of Sewage in a Two-Fluidised Bed System, in *Biological Fluidised Bed Treatment of Water and Wastewater*, ed. by P.F. Cooper and B. Atkinson, p.121 (Ellis Horwood, Chichester).

Coulson, J.M. and Richardson, J.F. (1968) *Chemical Engineering*, vol. 2, 2nd edn. (Pergamon Press, Oxford).

Coulson, J.M. and Richardson, J.F. (1971) *Chemical Engineering*, vol. 3, 2nd edn. (Pergamon Press, Oxford).

Coutts, M.W. (1958) *Austral. Pat.*, 216 618.

Coutts, M.W. (1961) *Brit. Pat.*, 872391–9, 872 400.

Dunnill, (1981) Biotechnology and Industry. *Chemy Ind.*, No. 7, 204.

Fonseca, M.M.R. (1978) Performance Characteristics of Biological Reactors contaning Biomass Support Particles. M.Sc. thesis, UMIST.

Francis, C.W. and Hancher, C.W. (1981) Biological Denitrification of High-Nitrate generated in the Nuclear Industry, in *Biological Fluidised Bed Treatment of Water and Wastewater*, ed. by P.F. Cooper and B. Atkinson, p.234 (Ellis Horwood, Chichester).

Greenshields, R.N. (1978) Acetic Acid: Vinegar, in *Economic Microbiology*, vol. 2. *Primary Products of Metabolism*, ed. by A.H. Rose, p.121 (Academic Press, London).

Greenshields, R.N. and Smith, E.L. (1971) Tower Fermentation Systems and Their Applications. *Chem. Engr*, No. 249, 182.

Grieves, R.B., and Pipies, W.O., Milbury, W.F. and Wood, R.K. (1964) Piston-Flow Reactor Model for Industrial Fermentations. *J. Appl. Chem.*, **14**, 478.

Hawkins, J.E., Stott, D.A., Stokes, R.L. and Clennett, A. (1981) Denitrification of Sewage Effluent on a Full-Scale using an Expanded Bed, in *Biological Fluidised Bed Treatment of Water and Wastewater*, ed. by P.F. Cooper and B Atkinson, p.357 (Ellis Horwood, Chichester).

Herbert, D. (1958) Some Principles of Continuous Culture, in *Recent Progress in Microbiology*, 7th Int. Congr. Microbiol., Stockholm.

Hines, D.A. (1978) The Large-Scale Pressure Cycle Fermenter Configuration, in *Biotechnology*, p.55 (Verlag Chemie, Weinheim).

Jeris, J.S. and Owens, R.W. (1981) Secondary Treatment of Municipal Wastewater with Fluidised Bed Technology, in *Biological Fluidised Bed Treatment of Water and Wastewater*, ed. by P.F. Cooper and B. Atkinson, p.112 (Ellis Horwood, Chichester).

Malpas, R. (1977) Chemical Technology – scaling Greater Heights in the Next Ten Years. *Chemy Ind.*, 111.

Malpas, R. (1978) Engineering Excellence in Manufacturing Processes, CEI (North West) 11th Annual Lecture

Moss, M.O. and Smith, J.E. (ed.) (1977) *Industrial Applications of Microbiology* (Surrey University Press, London).

Olsen, A.J.C. (1960) Manufacture of Bakers' Yeast by Continuous Culture. Part I. Plant and Process. *Chemy Ind.*, 416.

Pirt, S.J. and Kurowski, W.M. (1970) An Extension of the Theory for the Chemostat with Feed-back of Organisms. Its Experimental Realization with a Yeast Culture. *J. Gen. Microbiol.*, 63, 357.

Reynolds, L.F. (1974) Private communication.

Rose, A.H. (1979) History and Scientific Basis of Large-Scale Production of Microbial Biomass, in *Economic Microbiology*, vol. 4, *Microbial Biomass*, ed. by A.H. Rose, p.1 (Academic Press, London).

Shore, D.T. and Royston, M.G. (1968) Chemical Engineering of the Continuous Brewing Process. *Chem. Engr*, No. 218, 99.

Sinclair, C.G., Topiwala, H.H. and Brown, D.E. (1971) An Experimental Investigation of a Growth Model for *Aerobacter aerogenes* in Continuous Culture. *Chem. Engr*, No. 249, 194.

Tempest, D.W., Herbert, D. and Phipps, D.J. (1967) Studies on the Growth of *Aerobacter aerogenes* and Low Dilution Rates in a Chemostat, in *Microbial Physiology and Continuous Culture*, ed. by E.O. Powell *et al.*, p.97 (HMSO, London).

Walker, I. and Austin, E.P. (1981) The Use of Plastic, Porous Biomass Supports in a Pseudo-plastic Bed for Effluent Treatment, in *Biological Fluidised Bed Treatment of Water and Wastewater*, ed. by P.F. Cooper and B. Atkinson, p.272 (Ellis Horwood, Chichester).

Wang, D.I.C. and Humphrey, A.E. (1969) Biochemical Engineering, *Chem. Engng*, **76**, 108.

NOMENCLATURE

A_{film} = Area of support surface provided per unit volume of fermenter
A_w = Interfacial area
D_e = Effective substrate diffusion coefficient in biomass
F = volumetric flow rate
F_w = Volumetric flow rate at wash-out
h' = Liquid-phase mass transfer coefficient
K_m = Monod coefficient
L = 'Wet' thickness of microbial film
m_p = Particle biomass hold-up
N = Substrate flux
n_p = Number of particles contained in unit fermenter volume
p = Product concentration
R_v = Overall rate of uptake per unit liquid volume in a fermenter
r_p = Radius of biomass support particle
r_s = Radius of the support particle
s = Substrate concentration
s^* = Limiting substrate concentration at the interface between biomass and the adjacent solution
s_i = Inlet substrate concentration
s_o = Outlet substrate concentration
t = Time
V_p = Volume of a single 'wet' biological particle or floc
W = Recycle parameter
x = Concentration of 'dry' microbial mass
x_o = Biomass concentration at zero time
x_f = Final biomass concentration
x_I = Immobilized biomass concentration
x_i = Initial biomass concentration
x_{opt} = Optimum biomass concentration
Y_s' = Microbial yield per unit of substrate consumed

Z = Bed height
γ = Microbial concentration coefficient
δ = Liquid film thickness
κ = Endogenous respiration coefficient
μ = Specific growth rate
μ_{max} = Maximum specific growth rate
ρ_e = Effective biomass concentration
ρ_o = Microbial density
τ = Recycle ratio
φ = Particle porosity

APPENDIX: EVALUATION OF MICROBIAL ADHESION

Adhesion of microbial cells, which is important in biological film fermenters involves a two-stage process, with the initial physicochemical attachment being followed by biological stabilization.

The LH–Fowler cell adhesion measurement module shown in Fig. A1 has been developed to take account of the two stages of attachment, and provides information on the forces involved in both stages.

The essential feature of the apparatus is a radial flow chamber which is used in conjunction with a fermenter system. The radial flow chamber consists of two parallel discs with only a narrow separation h between them. A microbial suspension is pumped at constant volumetric flow rate through the centre of the back disc. The suspension then flows out radially between the discs to a collection manifold at the periphery, with the subsequent return of the suspension to the fermenter or a reservoir. Flow between the front and back discs sets up a velocity profile (*see* Fig. A2) and results in a hydrodynamic shear stress at the surfaces. Since the flow area increases with increasing radius, the fluid velocity (and hence the surface shear stress) decreases. This gives a continuous range of surface shear stress conditions from the centre to the periphery under which the attachment of the microbial cells can be studied. At an area close to the inlet there is a flow disturbance, resulting from the change of flow direction as the fluid enters the chamber. This area should be avoided in studying the microbe/surface interaction, although the inlet is profiled in order to minimize the disturbance.

It can be seen in Fig. A3 that there is a clear central zone around the inlet, while in the outer zone microbial film has accumulated. The junction between the two zones is well-defined. The radius of the junction between the two zones r is the radius at which there is a critical fluid shear stress that is just overcome by the attraction between the cells and the surface. By varying the composition of disc surface one can assess the adhesion of cells quantitatively, since the critical shear stress can be estimated from the velocity profile (*see* Fig. A2).

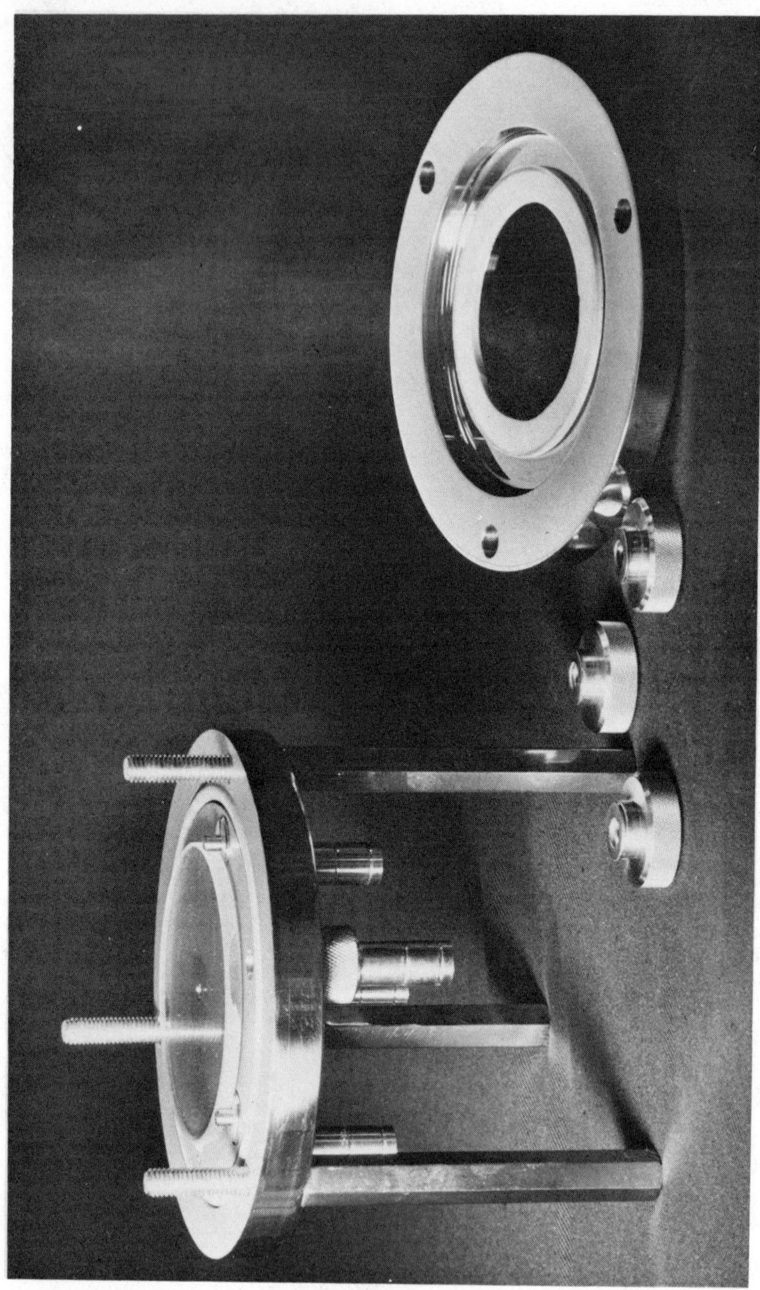

Figure A1. The LM – Fowler cell adhesion measurement module.

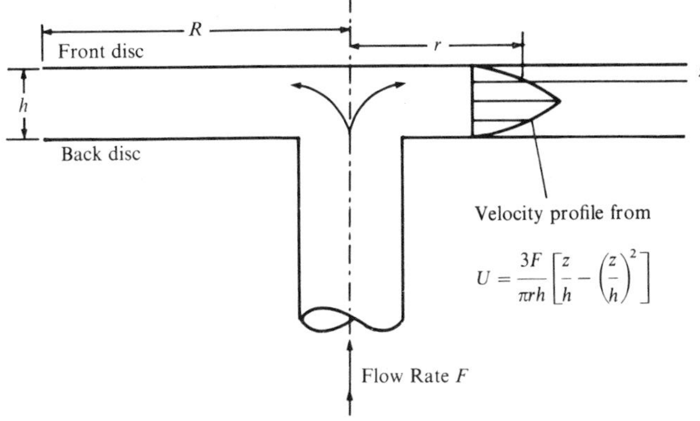

Figure A2. Flow profile between parallel discs.

Figure A3. Glass disc from the radial flow growth chamber, showing clear zone round the inlet.

CHAPTER 8
FLOW BEHAVIOUR OF FERMENTATION FLUIDS

GLOSSARY

Apparent viscosity Ratio of **shear stress** to **shear rate** when this ratio is dependent on the rate of shear, μ_a ($ML^{-1}T^{-1}$).
Coefficient of viscosity *See* **viscosity**.
Consistency General term for the property of a material by which it resists permanent change of shape.
Constitutive equation Equation relating **stress**, **strain**, time and, sometimes, other variables, such as temperature. Also called rheological equation of state.
Dynamic viscosity *See* **viscosity** μ, η ($ML^{-1}T^{-1}$).
Elongation viscosity Ratio of the tensile **stress** to the rate of extension, μ_E, η_E ($ML^{-1}T^{-1}$).
Flow Deformation of which at least part is nonrecoverable.
Kinetic viscosity Ratio of dynamic **viscosity** to the density of the material, both measured at the same temperature, ν (L^2T^{-1}).
Relative viscosity Ratio of the dynamic **viscosity** of a solution or of a dispersion to that of the solvent or continuous phase, each measured at the same temperature, η_r, μ_r.
Rheogram Graph of a rheological relationship, e.g., **shear stress** versus **shear rate**.
Rheology Science of deformation and **flow** of matter.
Rheopectic fluid Fluid that when subjected to a constant **shear stress** exhibits an **apparent viscosity** that decreases with time.
Shear Movement of a layer relative to a parallel adjacent layer.
Shear rate Rate of change of **shear**, i.e. **velocity gradient**, du/dy, γ (T^{-1}).
Shear stress Component of stress parallel or tangential to the area under consideration, τ ($ML^{-1}T^{-2}$).
Specific viscosity Difference between viscosity of a solution or dispersion and that of the solvent or continuous phase, all measured at the same temperature.
Strain Measurement of deformation relative to a reference length, area or volume, γ.
Strain rate Rate of change of **strain**, $\dot{\gamma}$ (T^{-1}).
Stress Force per unit area, τ ($ML^{-1}T^{-2}$).
Stress tensor Tensor indicating the orientation of the surface on which the **stress** acts (first subscript) and the direction of action of this stress (second subscript), τ_{ij} ($ML^{-1}T^{-2}$).
Tension/pressure Force normal to the surface on which it acts, τ_{ii} ($ML^{-1}T^{-2}$).
Thixotropic fluid Fluid that when subjected to a constant **shear stress** exhibits an **apparent viscosity** that increases with time.
Torque In rotary motion of a body around an axis, the torque equals the rate of increase of the moment of momentum, which is the product of the momentum (mass × velocity) and the normal distance from the axis, G_T (ML^2T^{-2}).
Trouton viscosity See **elongation viscosity**.
Turgor pressure Hydrostatic pressure within a cell providing rigidity (ML^{-2}).
Velocity gradient Derivative of velocity of a fluid element with respect to a space coordinate, i.e. **shear rate**, du/dy, γ (T^{-1}).
Velocity profile Velocity distribution normal to the direction of **flow**.
Viscoelasticity Partial elastic recovery upon removal of a deformed **shear stress**.
Viscosity (dynamic) Ability of a material to resist deformation, i.e. resistance to **flow**. A measure of ability to resist deformation, defined as the ratio of **shear stress** to **shear rate** in steady flow, μ, η ($ML^{-1}T^{-1}$).
Yield stress Value of **shear stress** below which there is no flow, τ_0 ($ML^{-1}T^{-2}$).

IMPORTANCE OF VISCOSITY IN BIOLOGICAL PROCESS ENGINEERING

Viscosity is the most significant property affecting the flow behaviour of a fluid. Such behaviour has a marked effect on pumping, mixing, heat and mass transfer, and aeration, and these, in turn, contribute significantly to process design and economics. Biological reactions can strongly affect viscosity due to the presence of microorganisms, substrates, metabolites and air. Morphology, concentration and chemical type are contributing factors. These interactions are illustrated schematically in Fig. 1.

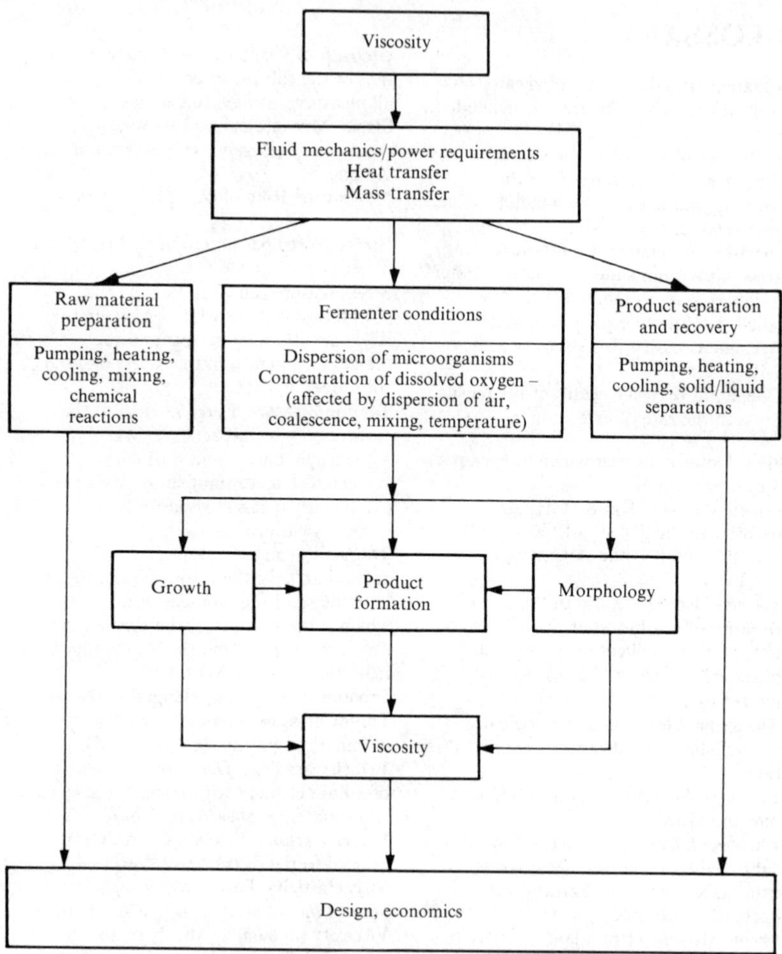

Figure 1. Schematic representation of the effect of viscosity on various processing steps.

VISCOSITIES OF SOME COMMONLY OCCURRING MATERIALS

Table 1.

Material	Viscosity (cP)	Temperature (°C)	Conditions
Air	1.7×10^{-2}	0	
	2.2×10^{-2}	100	
Water	1.0	20	
Milk	1.4	25	
Sucrose solution	2.0	20	20% w/w
in water	6.2	20	40% w/w
	57	20	60% w/w
Kerosene	2.5	20	
Egg albumin	12	20	Shear rate 100 s^{-1}
Linseed oil	33	30	
	45	20	
Olive oil	84	20	
Glycerol	100	60	
	15×10^3	20	
Light machine oil	114	16	
Heavy machine oil	661	16	
Castor oil	986	20	
	2.4×10^3	10	
Glucose	2.5×10^4	100	

SHEAR RATES

Table 2. Some shear rates of engineering interest.

Shearing conditions (materials more viscous than water)	Shear rates (s^{-1})
Molten polymers, under laminar flow conditions	10–5000
Common mixers	100–10 000
Boundary layers and turbulent flow in ducts	1000–500 000
Roll and knife coating devices	$\leq 5 \times 10^6$
Lubricating applications	Probably $\leq 10^7$–10^8

In stirred tanks, which are often used in fermentation, the maximum shear rate occurs at the tips of the impeller or blades and has to be estimated from velocity profiles measured in the vicinity of the impeller (*see* Fig. 2).

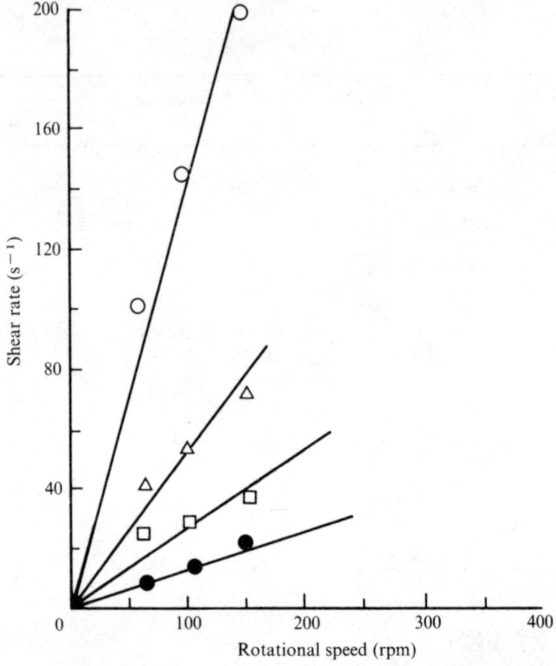

Figure 2. Shear rate versus rotational speed at various distances from the tip of a turbine impeller using carboxymethylcellulose (1.2 per cent by weight), n = 0.53. ○, 0.0 in. from impeller tip; △, 0.1 in.; □, 0.2 in.; ●, 0.5 in. [Metzner and Taylor (1960)].

The average shear rate can be determined from the power required to rotate the impeller at a given speed. The typical dependency of shear rate on impeller speed is shown in Fig. 3.

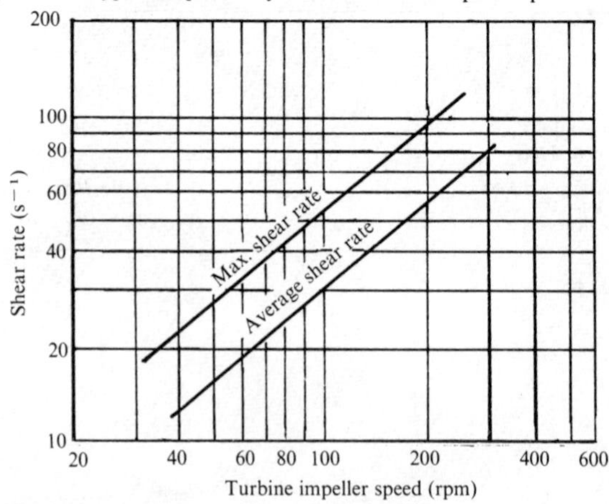

Figure 3. Shear rate characteristics of a 6-in. six-blade turbine impeller rotating in 18-in. tank containing water [Oldshue (1966)].

CLASSIFICATION OF FLUIDS ACCORDING TO THEIR FLOW BEHAVIOUR

Table 3.

Description	Rheograms	Constitutive equation	Apparent viscosity ($\mu_a = \tau\gamma^{-1}$)	Examples
Newtonian	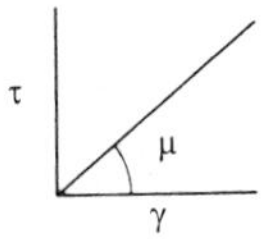 τ; μ; γ	$\tau = \mu\gamma$	Constant $\mu_a = \mu$	All gases, water, dispersions of gases in water, low molecular weight liquids, aqueous solutions of low molecular weight components
Pseudoplastic (power law)	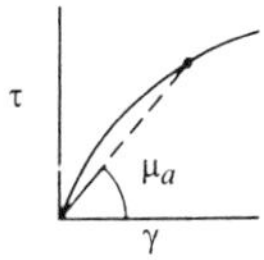 τ; μ_a; γ	$\tau = K\gamma^n$ $n < 1$	Decreases with increasing shear rate $\mu_a = K\gamma^{n-1}$	Rubber solutions, adhesives, polymer solutions, some melts, some greases, starch suspensions, cellulose acetate, mayonnaise, some soap and detergent slurries, some paper pulps, paints, biological fluids
Dilatant (power law)	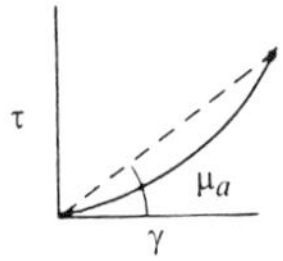 τ; μ_a; γ	$\tau = K\gamma^n$ $n > 1$	Increases with increasing shear rate $\mu_a = K\gamma^{n-1}$	Some cornflour and sugar solutions, starch, quicksand, wet beach sand, iron powder dispersed in low-viscosity liquids
Plastic Bingham plastic	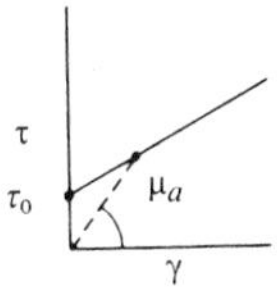 τ; τ_0; μ_a; γ	$\tau = \tau_0 + K_p\gamma$	Decreases with increasing shear rate when yield stress τ_0 is exceeded $\mu_a = \dfrac{\tau_0}{\gamma} + K_p$	Some plastic melts, margarine, cooking fats, some greases, chocolate mixtures, toothpaste, some soap and detergent slurries, some paper pulps
Casson plastic	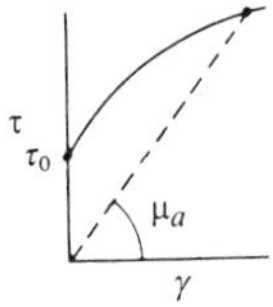 τ; τ_0; μ_a; γ	$\tau^{\frac{1}{2}} = \tau_0^{\frac{1}{2}} + K^P\gamma^{\frac{1}{2}}$	Decreases with increasing shear rate when yield stress τ_0 is exceeded $\mu_a = \left[\left(\dfrac{\tau_0}{\gamma}\right)^{1/2} + K_p\right]^2$	Blood, tomato ketchup, orange juice, melted chocolate, printing ink

GENERAL RELATIONSHIPS APPLICABLE TO STIRRED TANKS

Average Shear Rate versus Impeller Speed

According to Calderbank and Moo-Young (1959), the average shear rate can be expressed as

$$\gamma_{av} = K\,N \tag{1}$$

where K is a constant independent of the rheological characteristics of the fluid but is dependent on the measuring system used. N is the impeller speed (*see* Fig. 3). Eq. (1) applies to the laminar flow regime only, i.e. when $\mathrm{Re} < 10$. The Reynolds number Re may be defined as follows

$$\mathrm{Re} = \frac{\rho\,N\,d_i}{\mu_a} \tag{2}$$

where ρ = density of the fluid
d_i = impeller diameter
μ_a = apparent viscosity
Values of K are listed in Table 4.

Table 4. Experimental values of K for pseudoplastic fluids [Margaritis and Zajic (1978)].

Impeller system	K
Curved-blade paddle	7
Paddle	10–13
Six-blade turbine	10–13
Propeller	10
Anchor	20–25
Helical ribbon	30

Power Input versus Torque

The power input P is expressed as

$$P = 2\pi NM \tag{3}$$

where M is the torque. In the laminar flow regime (i.e. $\mathrm{Re} < 10$)

$$N_p = \frac{64}{\mathrm{Re}} \tag{4}$$

where N_p is the power number.

$$N_p = \frac{P}{\rho\,N^3\,d_i^5} \tag{5}$$

Combining these equations, one obtains

$$M = \frac{64}{2\pi}\,\mu_a\,N\,d_i^3 \tag{6}$$

At high impeller speeds, turbulent flow predominates and the linear relationship between torque and stirrer speed [Eq. (6)] does not apply. The power number is essentially constant under these conditions and combination of Eqs. (3) and (5) indicates that the torque on the impeller is proportional to the square of the impeller speed, i.e.

$$M \propto N^2 \tag{7}$$

The effect of the transition from laminar to turbulent flow is shown in Fig. 4.

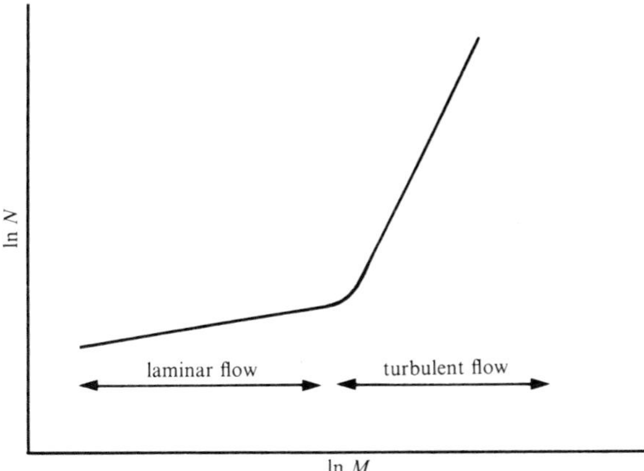

Figure 4. Torque versus impeller speed for the impeller viscometer.

Shear Stress versus Torque

By definition, the shear stress is

$$\tau = \mu_a \gamma \tag{8}$$

For the laminar flow regime, combining the previous equations one obtains

$$\tau = \frac{2\pi K}{64\, d_i^3} M \tag{9}$$

TEST METHODS FOR MEASUREMENT OF VISCOSITY

Table 5. Test methods.

Fluid characteristics	Viscometer
Newtonian	Capillary, coaxial cylinder, cone and plate, variable pressure capillary, vibrating reed, rolling or falling sphere, rotating spindle, orifice
Nonnewtonian	Coaxial cylinder, cone and plate, variable pressure capillary
Viscoelastic	Coaxial cylinder in normal way, or with appropriate modifications to permit study of creep compliance–time response, stress relaxation, dynamic rheological properties; cone and plate in simple or oscillatory shear

The different types of apparatus used for the measurement are described diagrammatically in Figs. 5–8. The principles of their operation are summarized in Table 6.

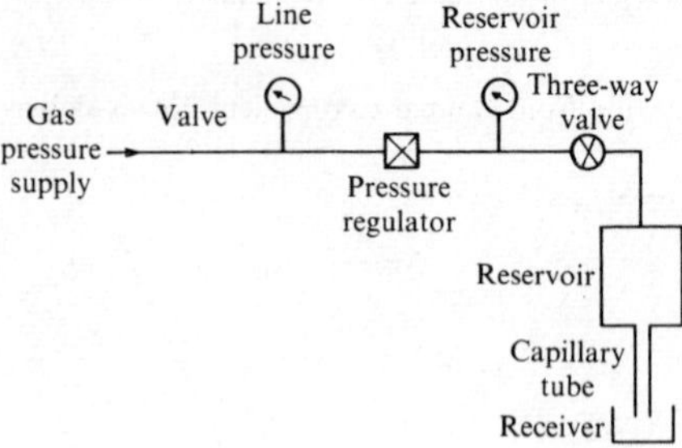

Figure 5. Schematic diagram of a capillary tube viscometer.

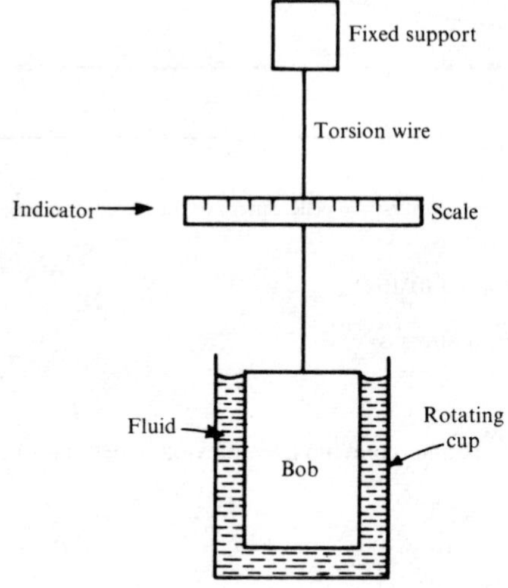

Figure 6. Schematic diagram of a coaxial cylinder rotary viscometer.

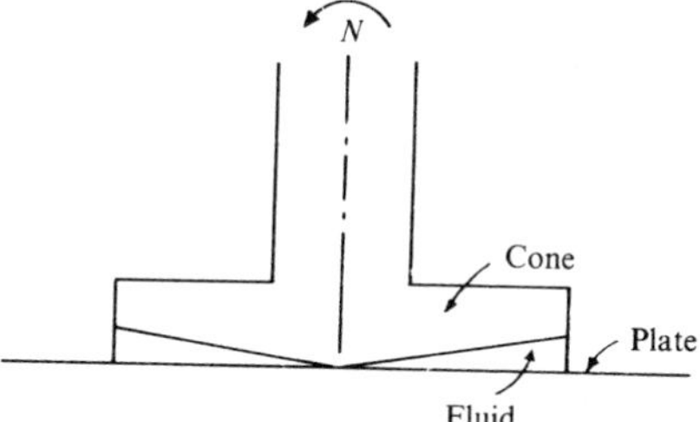

Figure 7. Schematic diagram of a cone and plate viscometer.

Figure 8. Schematic diagram of a falling sphere viscometer.

Table 6. Commercial viscometers — principles of operation [Sherman (1970)].

Type	Principles of operation	Applications and limitations
Capillary viscometers		
Glass capillary		
Ostwald U-tube	Reservoir bulb from which fixed volume of sample flows down through capillary to receiver bulb at lower level in other arm of U-tube	Small driving force since operates by force of gravity. Low shear stress. Suitable for measuring newtonian viscosity. Wide range of viscosities covered by capillaries of different dimensions
Cannon–Fenske	Reservoir and receiving bulbs lie in same vertical axis. Sample of fixed volume flows from reservoir bulb through capillary to receiving bulb. Range of capillary lengths and diameters available	Design reduces errors introduced when viscometer not aligned correctly in vertical plane. Only suited in this form to newtonian viscosity measurement. In conjunction with external source of pressure nonnewtonian viscosities can be measured
Ubbelohde	U-tube viscometer with third arm. Principles of flow as above. Range of capillary lengths and diameters available	When liquid emerges from capillary flows only over walls of lower bulb forming a 'suspended level'. Not necessary to use fixed volume of sample for test. Suited to newtonian viscosity measurements
Variable pressure		
Instron rheometer	Sample forced out of container chamber through capillary by plunger fastened to moving crosshead which can move at various speeds. Series of capillaries with different diameters and lengths. Force required to move plunger at each speed detected by load cell situated on crosshead	Wide range (five decades) of shear rates and shear stress available. Suitable for measuring newtonian and nonnewtonian viscosities. Suitable for high shear rate studies.
Techne printing viscometer	Sample forced by air pressure through horizontal capillary tube. Air displaced by this motion moves drop of water along another horizontal tube; time taken to pass between two graduations noted. Air pressure, or suction, produced by vibrated dead weight gauge. In automated model, photoelectric cells control electric timer and printer	Suitable for newtonian viscosities within range 0.003–2000 P. Only small volumes of sample necessary. Using alternately air pressure and suction, repeated tests can be made on the same sample
Bingham	Sample extruded through capillary by air pressure, or other pressure source, into receiver. Rate of flow measured by air displacement from receiver which is recorded by flowmeter. Flow rates noted over wide range of pressure	Suitable particularly for high shear studies, but not for low shear work

Rotational viscometers		
Coaxial cylinders		
Stormer	Stationary outer cup, inner rotor. Other geometric designs of rotor available. Rotor driven by weights and pulley. Stress varied by applying different weights	End and edge effects. Also distorted streamlines of flow. Instrument calibrated with fluid of known viscosity. Derived instrument constant includes factors for these effects
Haake rotovisco	Fixed outer cup and inner rotor. Several combinations of cups and rotors of different dimensions available. Rotor driven through torque dynamometer. Ten basic speeds from 3.6–582 rpm; using reduction gears can be reduced by a factor of 10 or 100. No provision for end effects other than overflow from annulus into cup-shaped top of rotor. Some rotors ribbed to prevent slippage. Others made of plastic for high-temperature studies	Within restricted viscosity range rates of shear down to $10^{-2}\,s^{-1}$ available. Newtonian and nonnewtonian viscosities can be measured. Structure recovery at low shear rate after subjecting to high shear rate also can be determined. Temperature control unsatisfactory when operating at high shear
Epprecht (Drage) rheomat	Fixed outer cup and inner rotating bob of cylindrical shape, but with conical top and bottom. Several combinations of cups and rotors of different dimensions. Viscous drag of inner rotating bob measured via rotation of motor assembly suspended by wire free to rotate against torque spring. Angular deflection of spring recorded. Fifteen driving speeds available. Conical ends reduce turbulent end effects	End effects present. Width of gap between bob and cup rather large for many combinations so large velocity gradients. Suspension system has a large moment of inertia; difficult to measure time-dependent effects or yield value
Portable Ferranti	Inverted outer rotating cylinder driven by synchronous motor. Viscous drag on inner cylinder measured through calibrated spring with pointer. Nine cylinder combinations of different dimensions available. Five-speed gear box gives speeds from 1–300 rpm. End effect almost eliminated. Drag on inner cylinder minimized since both cylinders end in same plane	End effect almost eliminated. Drag on base of inner cylinder negligible. Disagreement between viscosity data obtained with different cylinder combinations when testing low-viscosity dispersions at same mean rate of shear. Not a very robust instrument
Rotating spindle		
Brookfield synchro-lectric	Measures viscous traction on spindle rotating in sample. Spindle (up to seven models available) driven through a beryllium–copper spring by synchronous motor. Geometry of spindle results in wide range of shear rates in sample	Geometry of spindle makes it extremely difficult to calculate shear rate. Not suitable for absolute measurements of viscosity. End and edge effects. Distorted flow streamlines

continued overleaf

Table 6 – (continued)

Type	Principles of operation	Applications and limitations
Cone and plate		
Ferranti–Shirley	Rotating, small-angled cone and stationary lower flat plate. Three cones available – large, medium, small. Apex of cone just touches plate; sample sheared in gap between. Viscous drag on cone exerts torque on electromechanical torque dynamometer. Less than 0.5 ml sample required	Uniform shear rate throughout sample, provided cone angle small. Suitable for measuring newtonian and nonnewtonian flow at shear rates exceeding few s^{-1}. Particularly suited to high shear rate measurements. Practically no end effect; sample held in gap by surface tension
Weissenberg rheogoniometer	Cone rigidly fixed, flat lower plate rotates. Selection of cone angles and plate diameters. Flat plate rotates and torsion (i.e. tangential stress) imparted to cone measured; also normal force acting on plate. Two synchronous motors drive 60-speed gearbox. Very small volume of sample necessary. In oscillatory tests, platen oscillates about its axis; oscillatory motion transmitted to cone via sample	Uniform shear rate throughout sample. Suitable for measuring newtonian and nonnewtonian viscosities. Range of shear rates from about 10^{-4}–$2 \times 10^{-2}\,s^{-1}$. Oscillatory shear can be used to determine elastic and viscous components
Haake rotovisco	Rotary cone and stationary plate. Three cones in stainless steel available. Torque developed when cone rotates measured with dynamometer	Suitable for measuring newtonian and nonnewtonian viscosities, particularly at very high rates of shear. Under these conditions, danger of frictional heating effects
Orifice viscometers		
Redwood Saybolt Engler	Consists essentially of reservoir, orifice and receiver. Time measured for fixed volume of sample to flow through orifice	Efflux time taken as arbitrary measure of viscosity, although they are not related in a simple way. Cannot be used for absolute measurements
Rolling sphere		
Höppler	Cylindrical glass tube contains sample. Time taken for steel ball bearing to roll through middle, graduated section of tube. This ensures that ball has steady velocity before reaching test section. Set of six ball bearings of different diameters available. If liquid sample is opaque, time can be determined electrically	Difficult to define shear stress and rate of shear. Suited only to newtonian viscosity measurements
Vibrating reed		
Ultraviscoson	Thin alloy steel blade at end of probe vibrates at ultrasonic frequency. Ultrasonic shear waves develop in sample surrounding blade. Computer translates energy required to produce this motion into viscosity. Automatic temperature compensator	Only one rate of shear, so suited only to newtonian viscosities or for detecting deviation from fixed viscosity. Multiprobe computer available for taking several viscosity measurements at turn of switch

Calculation of Parameters for Tube or Capillary Viscometers

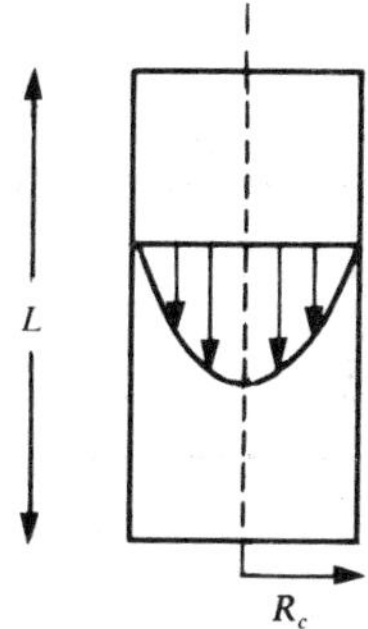

Figure 9. Laminar velocity profile for tube or capillary viscometer.

The following general equation was established by Whorlow (1980) for a tube or capillary viscometer (*see* Fig. 9):

$$\frac{Q}{\pi R_c^3} = \frac{1}{\tau_w^3} \int_0^{\tau_w} \tau^2 f(\tau_0)\, d\tau \qquad \text{and } \gamma = f(\tau) \tag{10}$$

where Q = volumetric flow rate
τ_w = shear stress at wall

Equations and graphical representations applicable to tube or capillary viscometers are presented in Table 7.

Table 7

	Newtonian	Power law	Bingham plastic
Rheological equation	$\tau = \mu\gamma$	$\tau = K\gamma n$	$\tau = \tau_0 + K_p\gamma$
Flow equation	$\frac{Q}{\pi R_c^3} = \frac{\tau_w}{4\mu}$	$\frac{Q}{\pi R_c^3} = \frac{1}{K^{1/n}} \frac{1}{3 + 1/n} \tau_w^{1/n}$	$\frac{Q}{\pi R_c^3} = \frac{\tau_w}{4K_p}\left[1 - \frac{4}{3}\left(\frac{\tau_0}{\tau_w}\right) + \frac{1}{3}\left(\frac{\tau_0}{\tau_w}\right)^4\right]$
Graphical representation	$\frac{Q}{\pi R_0^3}$; slope $= \frac{1}{4\mu}$; τ_w	$\ln \frac{Q}{\pi R_c^3}$; slope $= \frac{1}{n}$; $\ln \tau w$; $\ln\left(\frac{1}{K^{1/n}} \frac{1}{3 + 1/n}\right)$	$\frac{Q}{\pi R_c^3}$; slope $= \frac{1}{4K_p}$; τ_0; τ_w
Wall shear stress[a]	$\tau_w = \frac{R_c \Delta P}{2L}$	$\tau_w = \frac{R_c \Delta P}{2L}$	$\tau_w = \frac{R_c \Delta P}{2L}$

[a] ΔP = pressure over tube length L

Calculation of Parameters for Coaxial Cylinder Viscometers

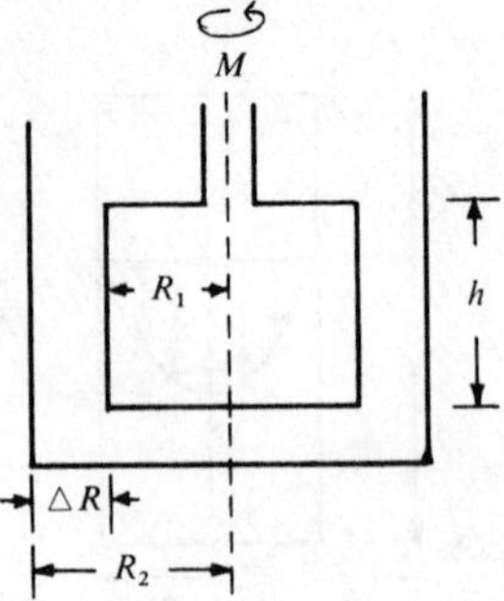

Figure 10. Coaxial cylinder viscometer. M = torque, R_1 = 'bob' radius R_2 = 'cup' radius, $\triangle R$ = gap between bob and cup where $\triangle R \ll R$ and h = bob height.

The following general equations have been established [Whorlow (1980)] for the coaxial cylinder viscometer illustrated in Fig. 10:

$$\tau = \frac{M}{2\pi R^2 L} \tag{11}$$

$$\gamma = R\frac{dw}{dr} \tag{12}$$

A summary of equations applying to coaxial cylinder viscometers are given in Table 8.

Table 8

	Newtonian	Power law	Bingham plastic
Rheological equation	$\tau = \mu\gamma$	$\tau = K\gamma^n$	$\tau = \tau_0 + K_p\gamma$
Flow equation	$w = \frac{M}{4\pi\mu h}\left(\frac{1}{R_1^2} - \frac{1}{R_2^2}\right)$	$w = \frac{n}{2K^{1/n}}\left(\frac{M}{2\pi h}\right)^{1/n}\left(\frac{1}{R_1^{2/n}} - \frac{1}{R_2^{2/n}}\right)$	$w = \frac{M}{4\pi K_p hc}\left(\frac{1}{R_1^2} - \frac{1}{R_2^2}\right)$
Graphical representation	w vs. M; slope $= \frac{1}{4\pi\mu h}\left(\frac{1}{R_1^2} - \frac{1}{R_2^2}\right)$	$\ln w$ vs. $\ln M$; slope $= \frac{1}{n}$; intercept $\ln\left[\frac{n}{2K^{1/n}}\left(\frac{1}{R^{2n}} - \frac{1}{R_2^{2n}}\right)\right] - \frac{1}{n}\ln 2\pi h$	w vs. M; slope $= \frac{1}{4\pi K_p hc}\left(\frac{1}{R_1^2} - \frac{1}{R_2^2}\right)$; intercept $\frac{\tau_0}{K_p}\ln\frac{R_2}{R_1}$
Shear rate	$\gamma = \frac{2R_2^2 w}{R_2^2 - R_1^2}$	$\gamma = \frac{2w}{n[1 - (R_1/R_2)^{2n}]}$	$\gamma = \frac{M}{2\pi K_p h R_1^2} - \frac{\tau_0}{K_p}$

Calculation of Parameters for Cone and Plate Viscometers

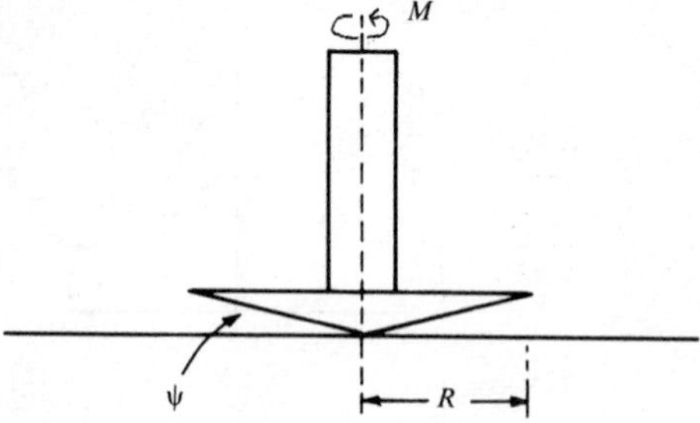

Figure 11. Cone and plate viscometer. M = torque, R = cone radius and Ψ = cone angle.

Sherman (1970) has established the following general equations for the cone and plate viscometer illustrated in Fig. 11:

$$\tau = \frac{3M}{2\pi R^3} \tag{13}$$

$$\gamma = \frac{w}{\psi} \tag{14}$$

where w = angular velocity.

Rheological equations and graphical representations for cone and plate viscometers are given in Table 9.

Table 9

	Newtonian	Power law	Bingham plastic
Rheological equation	$\tau = \mu\gamma$	$\tau = K\gamma^n$	$\tau = \tau_0 + K_p\gamma$
Graphical representation	τ vs. γ; slope = μ	$\ln\tau$ vs. $\ln\gamma$; intercept $\ln K_p$; slope = M	τ vs. γ; intercept τ_0; slope = K_p

Table 10. Tube viscometers [Whorlow (1980)].

Name and manufacturer of instrument	Flow rate control	Tube dimensions (mm)	Flow rate measurement
Extrusion rheometer A120, Surrell	Pressure	1.5–6 × 50	Cut and weigh
Extrusion rheometer A250, Surrell	Pressure	1.5–6 × 50	Cut and weigh
Rheograph 2000, Göttfert	Ram speed or pressure	0.5–2 × 5–30	Ram travel
Rheograph 1000, Göttfert	Ram speed	0.5–2 × 5.30	Ram travel
Extrusiometer 45, Göttfert	Screw speed	Tube or slit dies	
Capillary rheometer 3211, Instron	Ram speed	0.76–1.52 × 25–100	Ram travel
High-shear viscometer, Pressure Products	Pressure		Cut and weigh
Singlaff–McKelvey viscometer, Tinlus Olsen	Ram speed or pressure	0.25–2 × 25.4	Ram movement
Extrusion viscometer, Davenport	Ram speed	10 × 20	Ram movement
High-shear viscometer, Davenport	Pressure	0.5 × 5	Cut and weigh
Automatic rheometer, Mansanto	Pressure	1–2 × 8–16	Ram movement
Atkinson–Nancarrow, Mansanto	Ram speed	1.5–3 × 8–25	Ram movement

Calculation of Parameters for Falling Sphere Viscometers

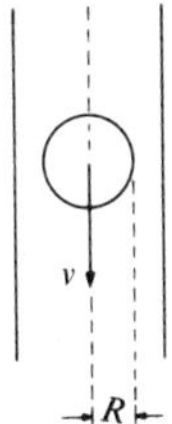

Figure 12. Falling sphere viscometer. v = falling velocity of sphere and R = radius of sphere.

According to Sherman (1970) $$\mu = \frac{2}{9}\frac{\rho_s - \rho}{v} g R^2 \tag{15}$$

$$\gamma_{\text{equator}} = \frac{3v}{2R} \tag{16}$$

$$\gamma_{\text{leading edge}} = 0 \tag{17}$$

where g = acceleration due to gravity, ρ_s = density of sphere and ρ = density of fluid

The falling sphere viscometer illustrated in Fig. 12 cannot be used to determine the parameters associated with nonnewtonian fluids because of a lack of mathematical theory.

CHARACTERISTICS OF COMMERCIAL VISCOMETERS

Pressure range (Pa)	Shear rate (s^{-1})	Shear stress (Pa)	Apparent viscosity (Pa s)	Temperature (°C)
$0–6.9 \times 10^5$	$50–10^3$		$10^{-3}–10^3$	20
$0–3.5 \times 10^6$	$50–5 \times 10^3$		$10^{-3}–10^4$	20–250
$10^6–2 \times 10^8$	$0.7–6 \times 10^5$	$2 \times 10^4–10^7$	$3.5 \times 10^{-2}–1.4 \times 10^7$	20–400
$10^6–10^8$	$0.7–3 \times 10^5$	$2 \times 10^4–5 \times 10^6$	$7 \times 10^3–7 \times 10^6$	20–400
	10	80		20
$3 \times 10^5–2.8 \times 10^8$	$0.4–1.1 \times 10^4$	$5.7 \times 10^2–4.3 \times 10^4$	$5 \times 10^{-2}–10^7$	40–350
$3 \times 10^5–1.5 \times 10^7$			$10–10^6$	100–300
$10^5–10^7$	$0.2–7 \times 10^5$	$5 \times 10^3–10^7$	$7 \times 10^{-3}–5 \times 10^7$	50–426
1.4×10^8	$62–1.2 \times 10^4$			100–300
$0–2 \times 10^7$				100–300
1.1×10^6	$0.1–10^6$	2×10^6		30–425
5×10^7				20–200

Table 11. Steady rotation viscometers [Whorlow (1980)].

Name (type[a]) and manufacturer of instrument	Controlled variable	Normal speed (rpm)	Normal torque (Nm)
Rheotron (coax. cyl or cone/pl.), Arabender	Speed	5×10^{-2}–10^3	10^{-4}–0.5
Synchro-Lectric (coax. cyl or coni. cyl.), Brookfield	Speed	0.5–100	6.7×10^{-5}–5.7×1
Wells-Brookfield (cone/pl.), Brookfield	Speed	0.5–100	6.7×10^{-5}–$5.7 \times$ [illegible]
Rheomat 15 (coax. cyl. or cone/pl.), Contraves	Speed	5.6–352	10^{-4}–$3.9 \times 10^{-}$
Rheomat 30 (coax. cyl. or cone/pl.), Contraves	Speed	5×10^{-2}–350	10^{-4}–$4.9 \times 10^{-}$
Ferranti-Shirley (cone/pl.), Ferranti	Speed	1–10^3	4×10^{-3}–0.12
Portable VL (coax. cyl.), Ferranti	Speed	100–300	4×10^{-4}–2×10
Portable VM (coax. cyl.), Ferranti	Speed	15–160	4×10^{-4}–2×10
Portable VH (coax. cyl.), Ferranti	Speed	1–12	4×10^{-4}–2×10
MacMichael (coax. cyl.), Fisher	Speed	10–38	
Kanawez rheometer (special rotor), Göttfert	Speed	0.02–20	2–200
Rotovisco RV3 (coax. cyl. or cone/pl.), Haake	Speed	0.1–10^3	4.9×10^{-4}–0.4
Rotovisco RV12 (coax. cyl. or cone/pl.), Haake	Speed	10^{-2}–512	4.9×10^{-3}–0.1
Rotovisco RV100/CV100 (coax. cyl.), Haake	Speed	0.20–200	10^{-6}–10^{-3}
Viscotester VT24 (coax. cyl. or cone/pl.), Haake	Speed	5.7–23	2.5×10^{-3}–$2.5 \times$
Viscotest VT181 (coax. cyl. or cone/pl.), Haake	Speed	45–181	10^{-3}–10^{-2}
Rotary rheometer (cone/pl. or par. pl.), Instron	Speed	10^{-3}–2.5×10^3	5×10^{-2}–2
Mechanical spectrometer (cone/pl., coax or par. pl.), Rheometrics	Speed or torque	10^{-2}–2.4×10^3	0.01, 0.1 or 1
Fluid rheometer (cone/pl., coax. cyl. or par. pl.), Rheometrics	Speed	0.1–10^3	10^{-5}–10^{-2}
Weissenberg (cone/pl. or par. pl.) Rheogoniometer R19, Sangamo	Speed	4.7×10^{-3}–375	5×10^{-7}–5
Stormer (coax. cyl.), Thomas	Torque	≥ 600	

[a] coax. cyl., coaxial cylinder; cone/pl., cone and plate; par. pl., parallel plate.

Shear rate (s^{-1})	Shear stress (Pa)	Viscosity (Pa s)	Temperature (°C)
10^{-2}–2×10^4	0.1–2.5×10^5	10^{-3}–10^6	300
$\times 10^{-2}$–100		10^{-3}–6.4×10^4	100
$^{-3}$–1.5×10^3		10^{-3}–7.5×10^3	100
1.3–4×10^3		10^{-3}–1.3×10^4	80
10^{-2}–3.9×10^3		10^{-3}–1.7×10^7	80
18–1.8×10^3	44–1.1×10^5	2.4×10^{-3}–6.2×10^4	200
43–950	0.8×390	8×10^{-4}–9.0	200
4.7–508	0.8×610	1.6×10^{-3}–130	200
0.3–37	0.8–6.1×10^3	2.1×10^{-2}–2×10^4	200
		$\geq 10^3$	
$\times 10^{-3}$–0.4			400
10^{-4}–4×10^4	0.2–10^6	1.5×10^{-3}–10^6	−60 – 30
10^{-4}–2×10^4	2×10^{-1}–10^5	1.5×10^{-2}–10^6	−60 – 30
$\times 10^{-2}$–10^3	10^{-3}–3×10^2	10^{-4}–10^3	100
.5–4.7×10^2	10–10^5	7×10^{-2}–360	−30 – 150
.2–3.7×10^3	2–4.6×10^3	2×10^{-3}–18	−30 – 150
10^{-6}–2.6×10^4	0.9–9.6×10^5	3.5×10^{-5}–1.2×10^{11}	400
10^{-3}–10^4	0.1–10^7	10^{-3}–10^{10}	−150 – 400
0.1–10^4	0.1–10^5	10^{-3}–10^4	−20 – 250
10^{-3}–9×10^3	1.9×10^{-3}–1.2×10^4	10^{-4}–5×10^6	−50 – 400

EXPERIMENTAL PHENOMENA ARISING FROM THE USE OF FERMENTATION BROTHS IN VISCOMETERS

The time-dependent behaviour of a filamentous *Penicillium chrysogenum* suspension is illustrated in Fig. 13. These data were obtained using a 50 mm six-blade Rushton turbine impeller rotating in a 75 mm container. Filtered mycelium was suspended in water to give a concentration of 9 kg m^{-3}. The torque was measured at stirrer speeds of 0.9 s^{-1} and 0.45 s^{-1}.

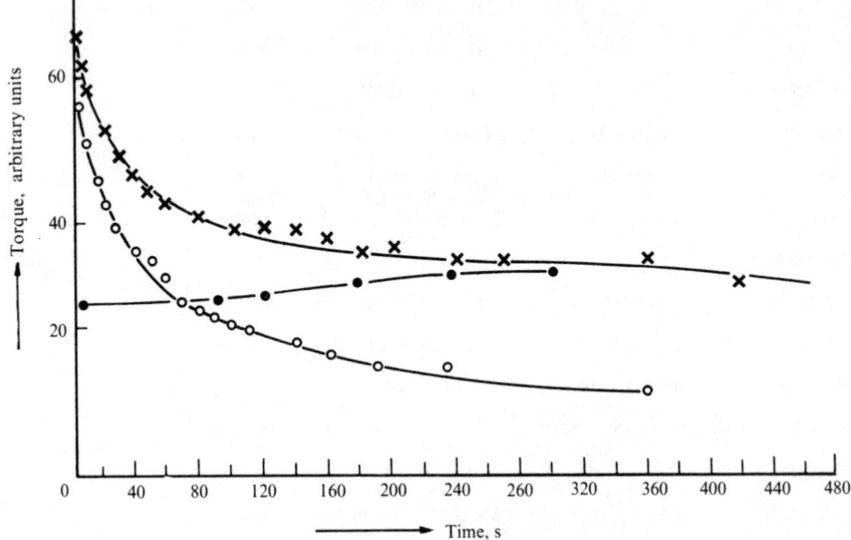

Figure 13. The time dependence of torque in an impeller viscometer containing a filamentous suspension. ×, $N = 0.9\,s^{-1}$ with mycelial concentration of 9 kg m^{-3} suspended in water; ○, $N = 0.45\,s^{-1}$ with mycelial concentration of 9 kg m^{-3} suspended in water; ●, $N = 0.45\,s^{-1}$ with mycelial concentration of 13.8 kg m^{-3} suspended in glycerol-water $\rho = 1060$ kg m^{-3}

It can be seen from Fig. 13 that the time dependence of torque is very pronounced. The reasons for this phenomenon could include

1) Flocculation/deflocculation equilibrium.
2) Orientation of the particles in the laminar flow field.
3) Centrifugation of the suspension so that a region with lower concentration develops around the impeller.
4) Separation of the suspension behind the stirrer blades, resulting in the measurement of a low viscosity.
5) Sedimentation of the suspension.

The possibility of sedimentation was investigated in the following experiment. Filtered mycelium was redispersed in a glycerol–water mixture to give a density of 1060 kg m^{-3} (i.e. a concentration of 19.8 kg m^{-3}), which is about equal to the density of mycelium. In a system with the same tank-to-impeller ratio of 1.5 as before and an impeller speed of 0.45 s^{-1}, no serious effect could be detected.

Data contained in Fig. 14 were obtained when *P. chrysogenum* pellets were suspended in water to give a concentration of 60 per cent pellets by volume. Using a six-blade Rushton turbine impeller, with a tank-to-impeller diameter ratio of 1.5 at impeller speeds of 1.35 and 0.15 s^{-1}, the same time-dependent behaviour of torque was observed with pellets as in Fig. 13, although less pronounced than that with mycelium.

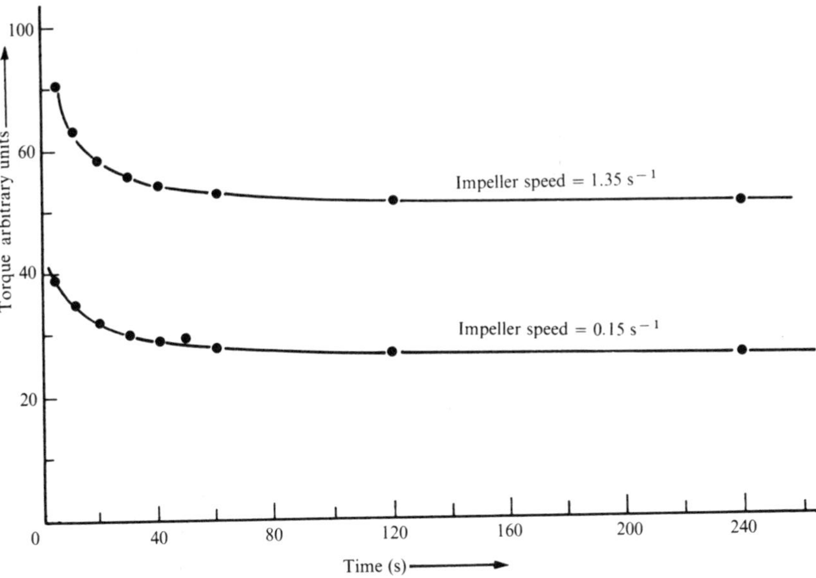

Figure 14. Time dependence of torque in impeller viscometer containing a pellet suspension [Metz *et al.* (1979)].

Phase separation, i.e. the formation of a less-dense region in the immediate vicinity of a rotating bob is illustrated in Fig. 15 by a plot of viscosity as a function of time for a 1 per cent by volume culture of the mycelium *Aspergillus niger*. Similar behaviour is exhibited by yeasts suspensions of about 4 per cent by volume, although to a much smaller extent.

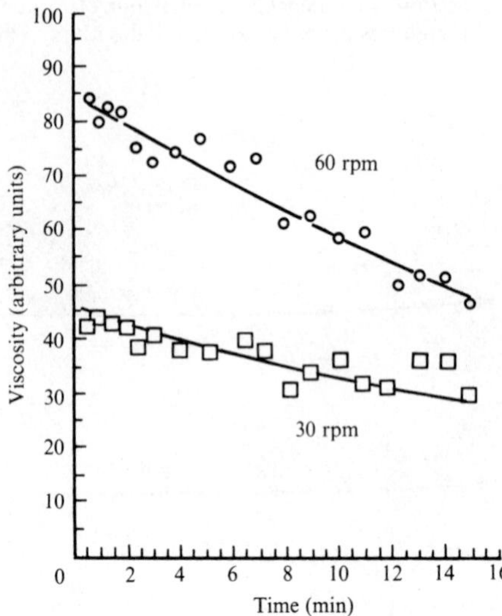

Figure 15. Effect of phase separation on Brookfield viscometer reading [Charles (1978)].

Table 12. Effect of storage of a filamentous mould suspension (at room temperature) upon rheological behaviour [Metz *et al.* (1979)].

N (s^{-1})	Torque (arbitrary units) Directly	After 24 h
0.05	6.6	6.2
0.10	7.6	6.6
0.15	8.2	7.8
0.30	9.2	8.7
0.45	10.1	9.5
0.90	12.5	11.7
1.35	15.0	14.0

Table 13. Effect of storage of a pellet suspension upon rheological behaviour [Metz *et al.* (1979)].

	Torque (arbitrary units)			
	Suspension 1		Suspension 2	
N (s^{-1})	Measured Directly $\phi_s = 40\%$	Measured After 24 h $\phi_s = 26\%$	Measured Directly $\phi_s = 60\%$	Measured After 24 h $\phi_s = 39\%$
0.1	3	—	22	4.5
0.15	3.5	1	25	5
0.3	4	1.5	29	5.5
0.45	4.5	1.5	33	7
0.9	6.5	turb.	42	9.5
1.35	9	turb.	49	11.5

ϕ_s = volume fraction of pellets.

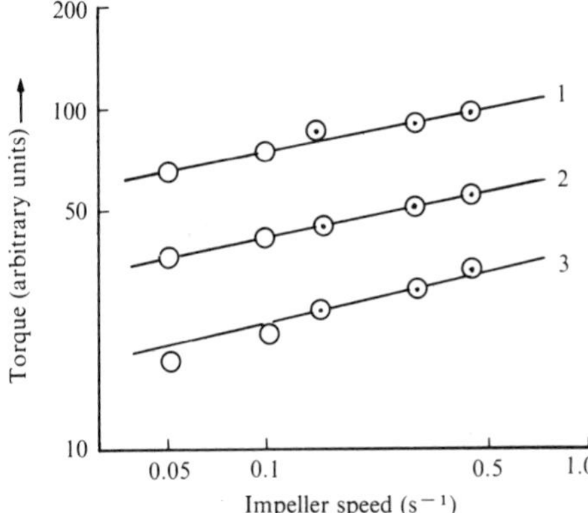

Figure 16. Influence of baffles upon viscosity measurements. Lines 1 and 2, filamentous suspensions, with concentrations of 14.8 and 8.2 kg m^{-3} respectively; line 3, pellet suspension of 60 per cent by volume fraction. ○, with baffles; ⊙, without baffles [Metz *et al.* (1979)].

Possible slip effects at the wall due to the formation of water layers in an impeller viscometer have been investigated by the addition of four baffles with a width of 0.1 times the vessel diameter. As can be seen from Fig. 16, the baffles had no influence upon the measured torque values.

The time that has elapsed since sampling can influence viscosity measurements. To investigate this, the torque for a suspension of mycelium *P. chrysogenum* was measured directly after standing for 24 h at room temperature without aeration. The results are given in Table 12. Although differences are small, the deviation is systematic. In contrast, for pellet suspensions, the interval between sample collection and measurement has a considerable influence, as may be seen from Table 13.

The explanation for this phenomenon is probably that during the 24 hours autolysis occurs, thus leading to collapse of the pellets with a reduction in the volume fraction of the pellets and hence resulting in lower torque values.

FACTORS AFFECTING THE VISCOSITY OF FERMENTATION BROTHS

Yeasts

Fermentation conditions

The apparent viscosity of *Saccharomyces cerevisiae* suspensions in water was measured using a Brookfield rotational viscometer as a function of shear rate, temperature and cell concentration. From Fig. 17 it can be seen that above 10 per cent solids, the suspension displays pseudoplastic behaviour ($K\gamma^n$, where $n < 1$); values of the parameters K and n are given in Table 14.

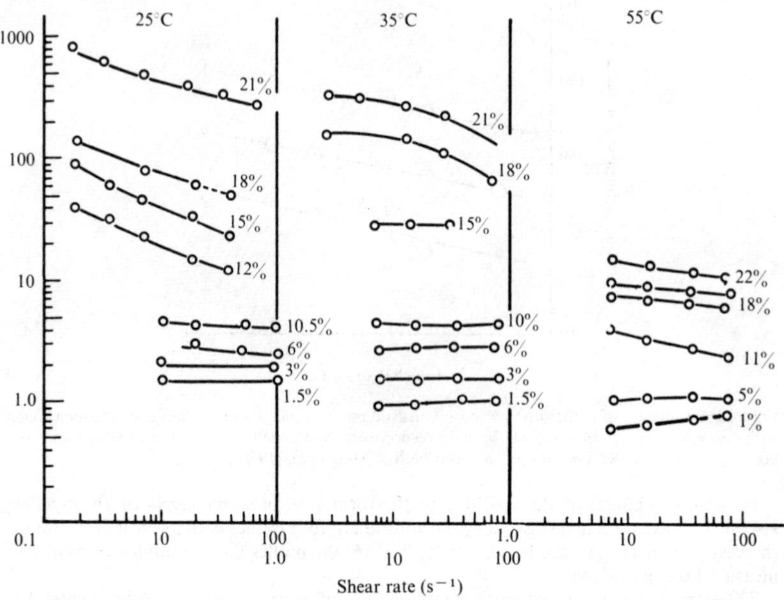

Figure 17. Apparent viscosity of yeast suspensions as a function of temperature and solids concentration [Labuza *et al.* (1970)].

Table 14. Fluid viscosity properties of yeast suspensions.

Temperature (°C)	Solids (% by wt)	*n*	*K*
Temperature, 25°C	12	0.60	8.80
	15	0.51	18.70
	18	0.68	33.00
	21	0.68	42.20
Temperature, 35°C	18	0.68	34.00
	21	0.87	51.00
Temperature, 55°C	11	0.79	7.66
	15	0.88	24.00
	18	0.95	35.30
	22	0.90	42.70

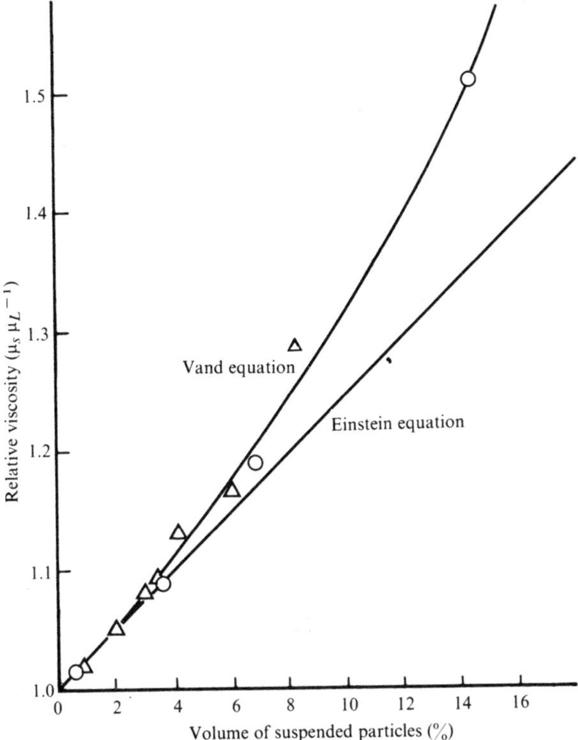

Figure 18. Newtonian behaviour of yeast cells (○) and *Lycoperdon bovista* (△) spores [Blanch and Bhavaraju (1976)].

Einstein's equation

$$\mu_s = \mu_L (1 + 2.5\phi) \tag{18}$$

may be applied to yeast suspensions at low volume fraction, whereas the Vand equation

$$\mu_s = \mu_L (1 + 2.5\phi + 7.25\phi^2) \tag{19}$$

gives a better agreement for yeast and spore suspensions up to a 14 per cent volume fraction (*see* Fig. 14). In Eqs. (18) and (19), μ is the viscosity of suspension, μ_L is the viscosity of the suspending liquid and ϕ is the volume fraction of particles.

Biomass recovery conditions

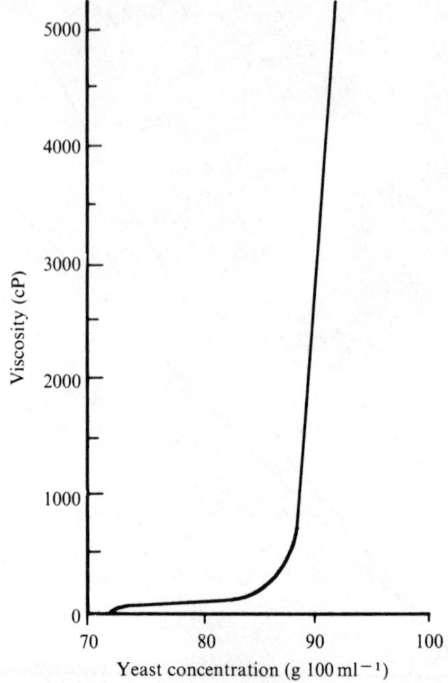

Figure 19. Viscosities of concentrated yeast suspensions [Charles (1978)].

While it is unlikely the concentrations, in the range considered, will ever occur during growth, it is quite likely that they do so during recovery.

Mycelia

Figure 20 shows the effect of temperature on apparent viscosity of kanamycin fermentation broths of various ages. The results suggest that the relationship between apparent viscosity and temperature can be expressed as an Arrhenius-type equation

$$\mu_a = A\, e^{E/RT} \qquad (20)$$

where μ_a = apparent viscosity
T = temperature
E = activation energy
R = gas constant
A = empirical constant

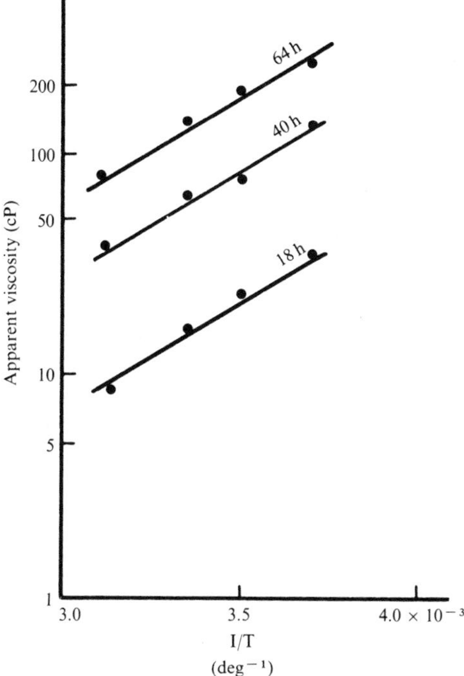

Figure 20. Effect of temperature on apparent viscosity of kanamycin fermentation broths at various ages [Taguchi (1971)].

In Fig. 21, the torque applied by the turbine impeller of 3.5 cm diameter was measured as the angular rotation of a cup of 5 cm diameter. The latter floated in a larger vessel filled with glycerol and was attached to the bottom of this vessel by a torsion wire. The different concentrations of mycelium were obtained by dilution of the original concentration C_0 with physiological saline. The results are described by the Casson equation (*see* 675).

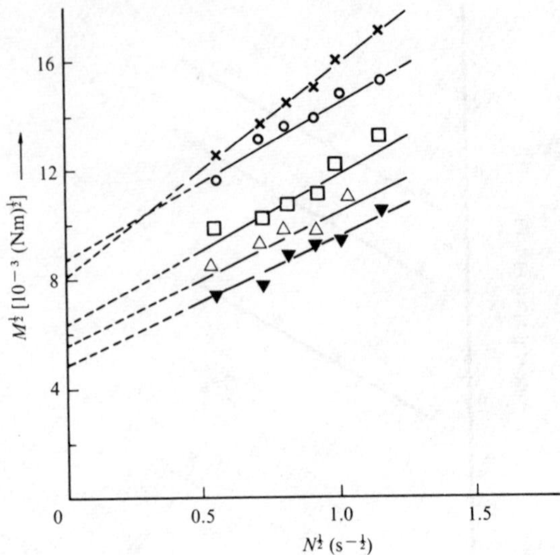

Figure 21. Viscosity measurements of *Penicillium chrysogenum* broths with different mycelial concentrations. Relative mycelial concentrations: ×, 1; ○, 0.88; □, 0.78; ▲, 0.70; ▼, 0.64.

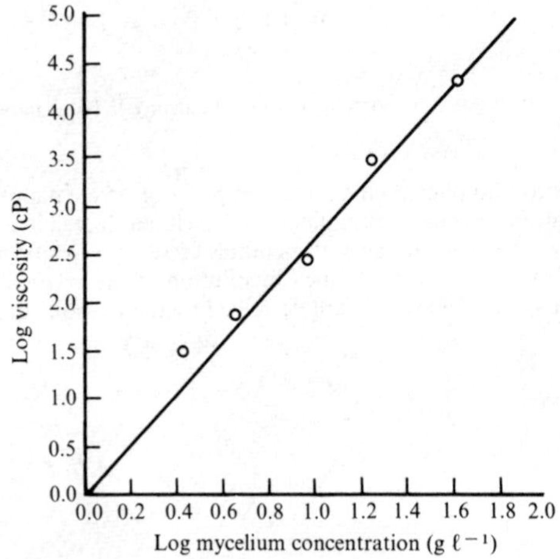

Figure 22. The apparent viscosity of *A. niger* broth as a function of mycelium concentration. The viscosity was measured at a shear rate of 1 s^{-1} [Sato (1961)].

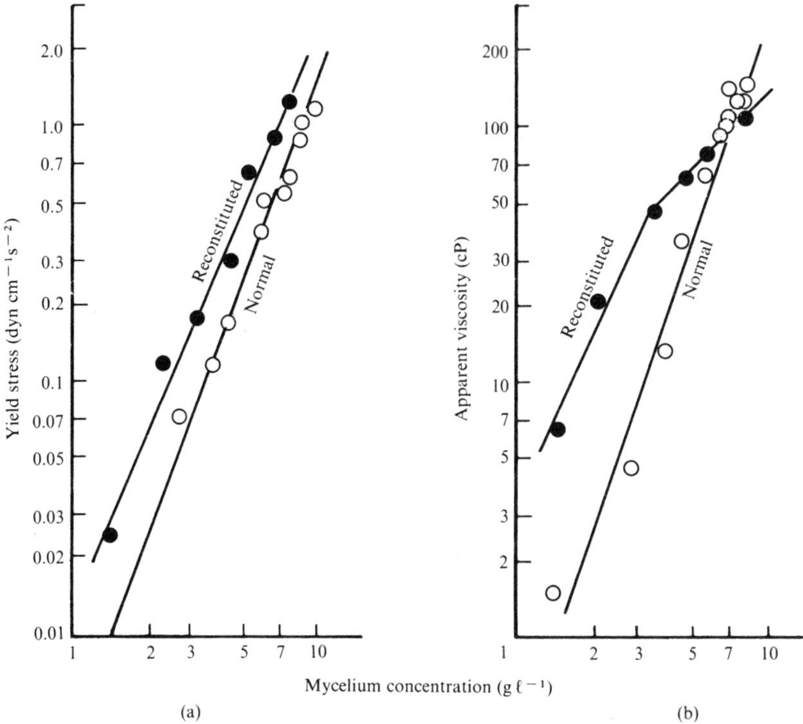

Figure 23. Effect of mycelium concentration on (a) yield stress and (b) apparent viscosity [Deindorfer and Gaden (1955)].

In Fig. 23, the line 'normal' is for *P. chrysogenum* suspensions removed from the fermenter after 40 h. The same mycelium was poisoned with sodium azide to inhibit respiration, and this was used for reconstituting the filtered broth. Both suspensions showed Bingham plastic behaviour. The differences between 'normal' and 'reconstituted' broths are due to disruption of the original mycelial structure and the extent of physical interlacing of hyphal networks during filtering and resuspending.

The parameters K and n display pseudoplastic behaviour and are dependent on the rotational speed of the impeller (*see* Fig. 24).

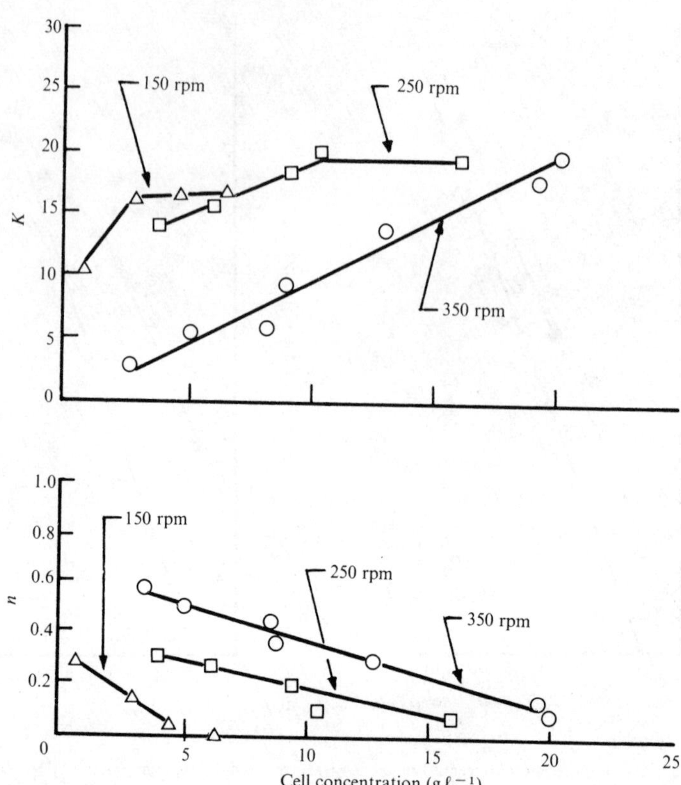

Figure 24. Viscosity characteristics of *Streptomyces niveus* culture broth measured *in situ* [Charles (1978)].

Pellets

During mould fermentations in tower fermenters, the morphology of the microorganism varies considerably as fermentation progresses from filamentous mycelium to smooth, rounded pellets. These changes in morphology can alter apparent viscosity (*see* Fig. 25).

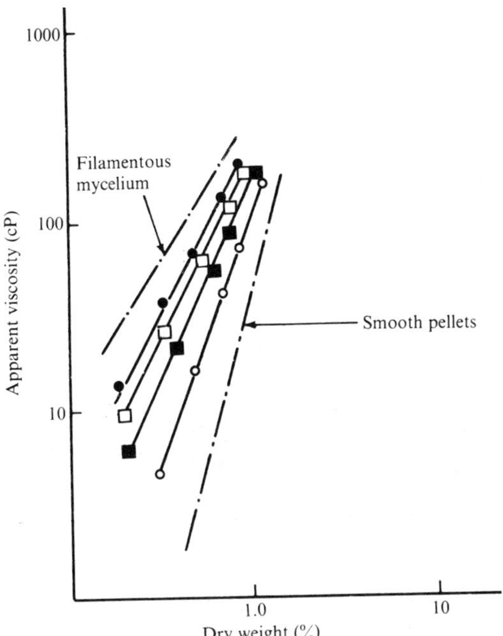

Figure 25. Effect of the morphology of *Aspergillus niger* on apparent viscosity. Mycelial age: ●, 22 h; □, 24 h; ■, 29 h; ○, 28 h [Morris *et al.* (1973)].

Table 15. Influence of pellets upon rheological behaviour of filamentous *Penicillium chrysogenum* suspensions [Metz *et al.* (1979)].

Pellet concentration (% volume)	Yield stress (arbitrary units)	Casson viscosity (arbitrary units)
0	13.24	2.48
1	13.07	2.52
2	13.31	2.45
3	13.08	2.49
4	13.00	2.50
6	12.64	2.50
8	12.83	2.47
10	12.68	2.36
12	11.38	2.43

In Table 15, pellets with a diameter greater than 0.55 mm were added to a filamentous suspension.

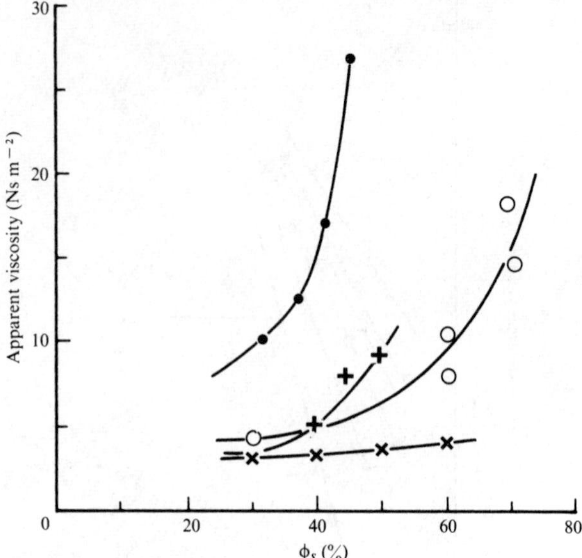

Figure 26. Apparent viscosity as a function of the volume fraction of pellets ϕ_s for a shear rate of 0.91 s^{-1}. (●) Suspension of loose 'hairy' pellets of *Sporotrichum pulverulentum* of diameter 3.60 mm; (+) suspension of smooth dense pellets of the same organism of diameter 2.35 mm; (○) suspension of pellets of *P. chrvsogenum* of diameter 0.81 mm; (×) suspension of polystyrene spheres of diameter 1.35 mm [Metz *et al.* (1979)].

As can be seen from Fig. 26, there is no distinct relationship between the apparent viscosity and the volume fraction of pellets for different suspensions.

Fluid Properties

Physicochemical conditions

In Fig. 27, the osmotic pressure was varied by adding sodium chloride or distilled water to a filamentous *Penicillium chrysogenum* suspension. The influence of osmotic pressure upon the yield stress is through its effect on hyphal flexibility which is brought about by changes in turgor pressure of cells.

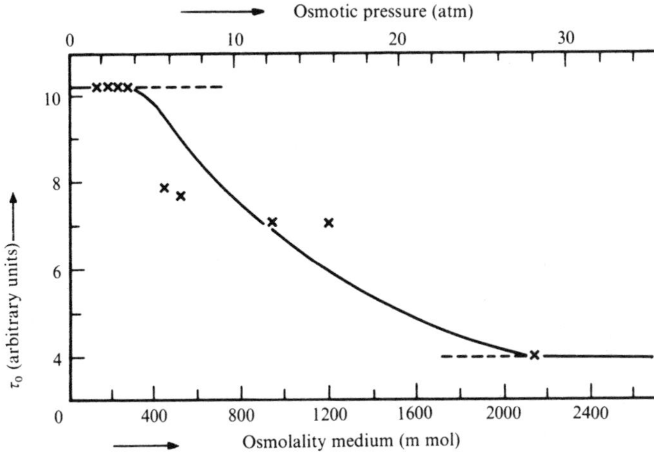

Figure 27. Influence of the osmotic pressure of suspending medium upon the yield stress of a mould suspension [Metz *et al.* (1979)].

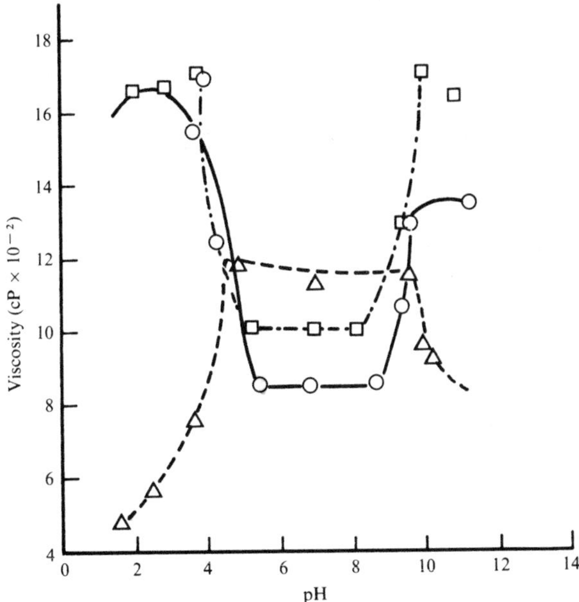

Figure 28. Effect of pH on the viscosity of B-1459 polysaccharide solutions (□) PS–L, (△) PS–S_{in} and (○) PS–P at 0.5 per cent concentration produced by *Xanthomonas campestris* NRRL B–1459. [Margaritis and Zajic (1978)].

The data contained in Fig. 28 show that solutions of xanthan gum and xanthan gum-containing fluids exhibit pronounced pseudoplasticity. They are also thixotropic.

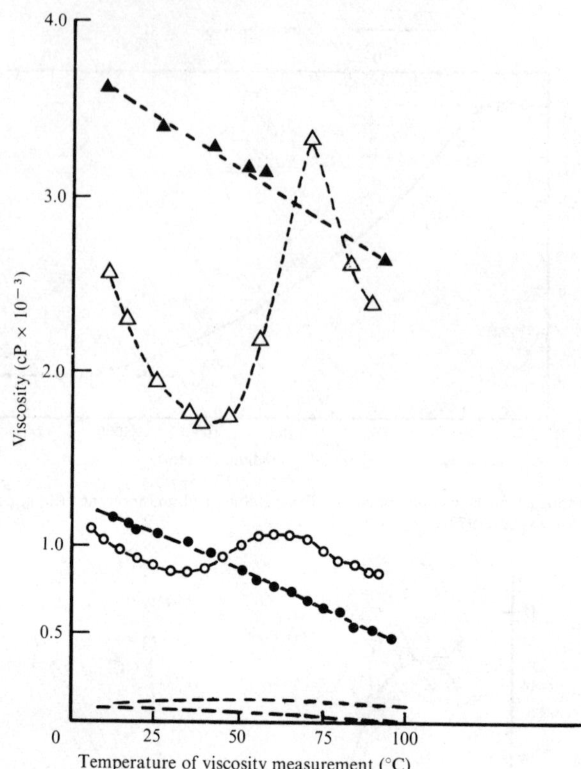

Figure 29. Effects of temperature and potassium chloride on the viscosity of polysaccharide (PS) B–1459 solutions produced by *X. campestris* NRRL B–1459. (▲) PS (1.0%), KCl (0.1–1.0%); (△) PS (1.0%), KCl (0.0%); (□) PS (0.5%), KCl (0.0%); (■) PS (0.5%), KCl (0.1–1.0%); (○) PS (0.1%), KCl (0.0%); (●) PS (0.1%), KCl (0.1%) [Margaritis and Zajic (1978)].

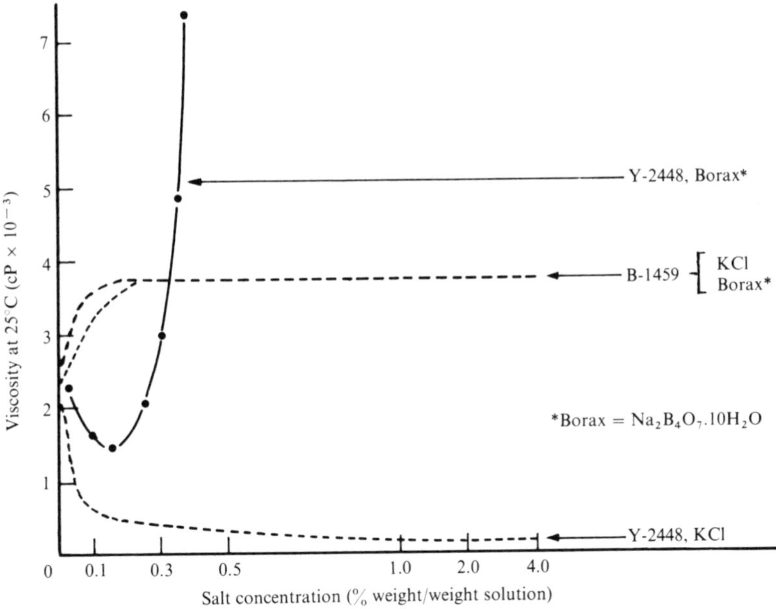

Figure 30. Effect of salt on viscosity of polysaccharides NRRL Y–2448 produced by *X. campestris* NRRL B–1459 and *Hansenula holstii* [Margaritis and Zajic (1978)].

Products

The cells of *Xanthomonas campestris* have very little influence on the viscosity of the broth (*see* Fig. 31).

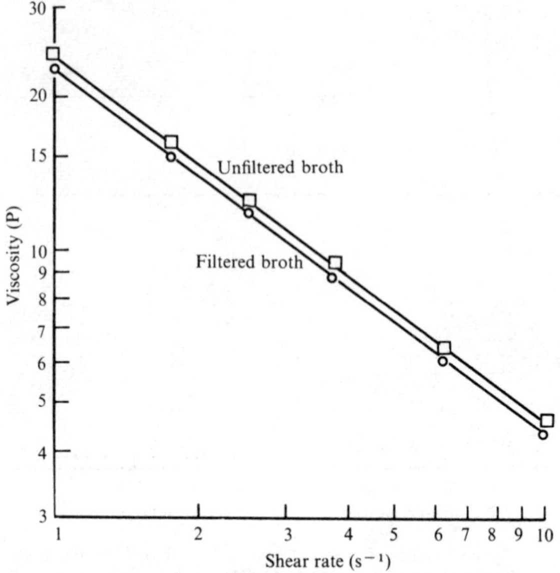

Figure 31. Effect of cell mass on viscosity of culture fluid containing xanthan gum [Charles (1978)].

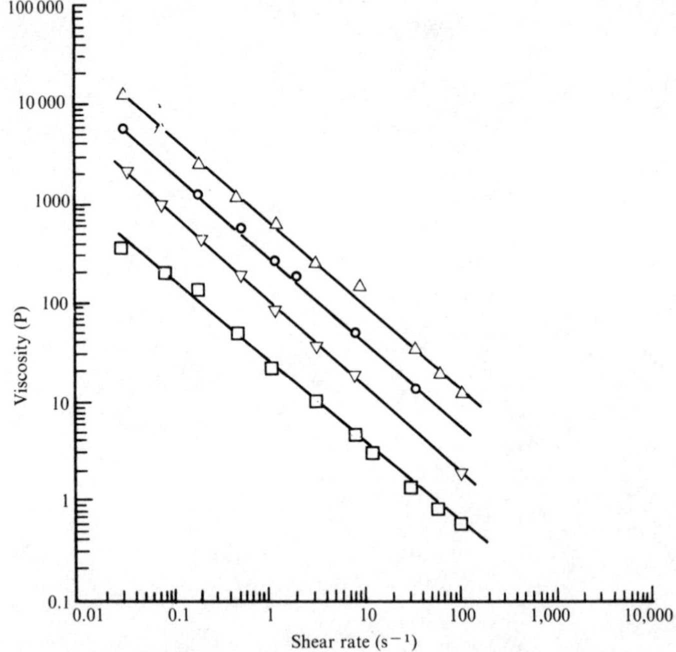

Figure 32. Viscosity characteristics of culture fluids containing various concentrations of xanthan gum. □, 0.5 per cent; ▽, 1.0 per cent; ○, 3.0 per cent; △, 5.0 per cent [Charles (1978)].

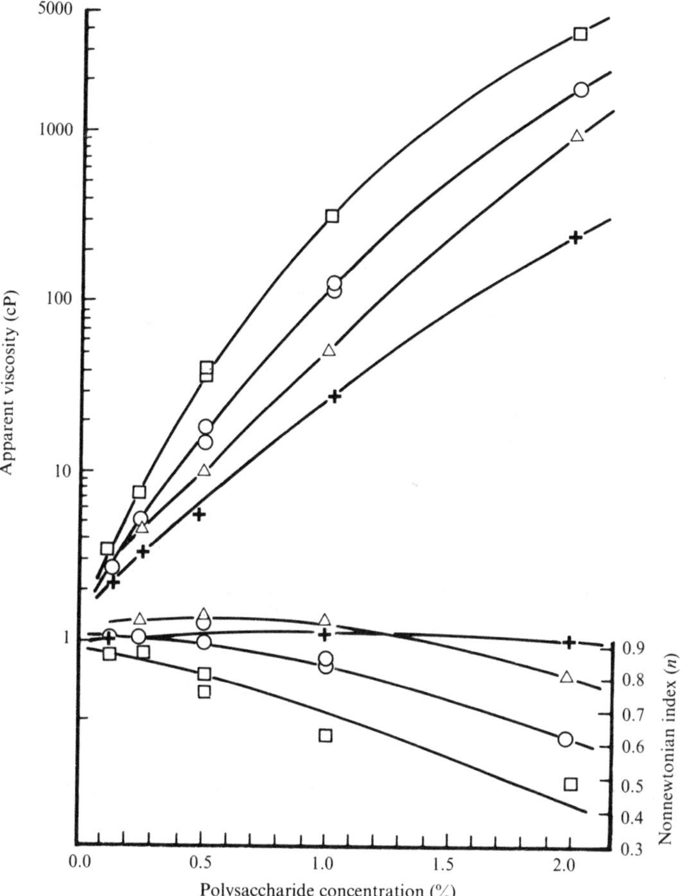

Figure 33. Rheological properties of a polysaccharide isolated at different stages of fermentation. □, 1 day old; ○, 3 days old; △, 5 days old; +, 8 days old [Leduy *et al.* (1974)].

The large deviations of power-law index *n* from 1.0 at high concentrations shown in Fig. 33 indicates pseudoplastic behaviour.

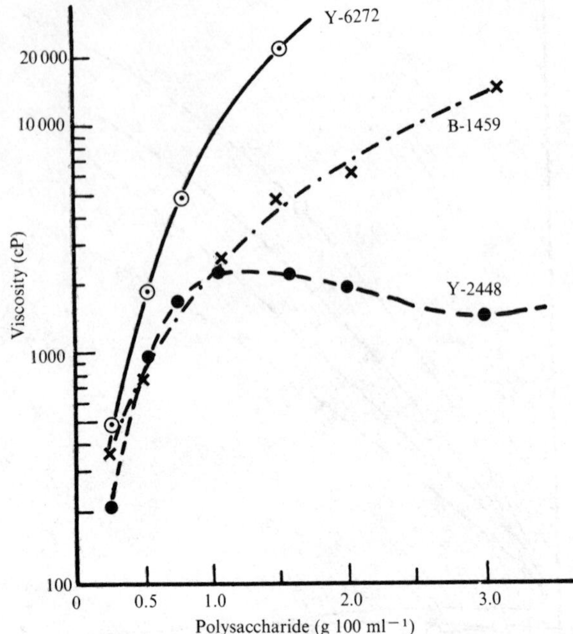

Figure 34. Viscosity–concentration curves of microbial polysaccharide: ⊙ *Rhinocladiella mansonii* NRRL Y–6272; × *X. campestris*, NRRL B–1459 (xanthan gum); ●, *H. holstii* NRRL Y–2448 (phosphomannan).

SUMMARY OF EXPERIMENTAL VISCOMETRIC RESULTS

Rheological data for various cultures without and with extracellular polysaccharides are listed in Tables 16 and 17 respectively.

SUMMARY OF EXPERIMENTAL VISCOMETRIC RESULTS

Table 16. Rheological properties of mycelial culture fluids [Charles (1978)].

Culture	Shear rate[a] (s^{-1})	Viscometer	Flow curves given	Comments
Aspergillus niger (Washed cells)	Not given	Ferranti VM (concentric cylinder)	No	Bingham plastic behaviour mentioned, but not quantified. Gives K values for various cell mass concentration (C) $K = 2.65 \log C$ ($g\,\ell^{-1}$)
Penicillium chrysogenum (Whole broth)	1–10	MacMichael (concentric cylinder)	Yes	Flow curves demonstrate Bingham plastic behaviour. τ_0 and K_p vary directly with cell mass concentration and/or fermentation time
P. chrysogenum (Whole broth)	<0.2[b]	Brookfield (guard removed)	Yes	Flow curve seems to indicate both yield stress and some other pronounced structure effect. Instrument artefact or error in calculation of γ suspected
P. chrysogenum (Whole broth)	0.1–1.5	'Turbine'	Yes	Casson behaviour observed. Yield stress and Casson viscosity increases with reaction time and cell concentration
P. chrysogenum (Whole broth and diluted whole broth)	<15	'Turbine'	Yes	Observations parallel those of preceeding example. Model incorporating morphological concepts developed

continued overleaf

Table 16 – (continued)

Culture	Shear rate[a] (s^{-1})	Viscometer	Flow curves given	Comments
Endomycopses sp. (Whole broth)	Not given	Ferranti coaxial cylinder	No	Power law behaviour. K and n vary appreciably but not monotonically over course of batch
Candida hellebori (Whole broth)	<0.2[b]	Brookfield (guard removed)	Yes	Bingham plastic behaviour early and late in batch. Yield stress (equal to Bingham yield stress) late in batch and pseudoplasticity at intermediate times
Streptomyces griseus (Whole broth)	<0.2[b]	Brookfield (guard removed)	Yes	Bingham plastic behaviour early in fermentation. Newtonian behaviour late in batch: newtonian viscosity increases with time but does not become as large as K_p for plastic behaviour period
S. griseus (Whole broth)	1–10	Brookfield	Yes	Pseudoplasticity exhibited throughout batch. Apparent Bingham plastic behaviour for $\gamma > 5\ s^{-1}$. τ_0 and K_p for apparent plastic behaviour both exhibit maxima
S. aureofaciens (Whole broth)	2–3	Brookfield and Contraves rheomat	Yes	Initial viscosity high due to high starch concentration but decreases due to starch hydrolysis. Newtonian behaviour early in batch
S. noursei	<0.2[b]	Brookfield (guard removed)	Yes	Newtonian behaviour throughout batch, despite mycelial character of organism. Viscosity increases with fermentation time
S. niveus (Whole broth)		*In situ*	No	Power law behaviour throughout batch

[a] Shear rate was assumed equal to 10 N (s^{-1}) when not given explicitly (*see* Eq. 1).

[b] Shear rate assumed as in footnote a, but differs greatly from values cited by authors. Error is suspected in reported values.

Table 17. Rheological properties of culture fluids containing extracellular microbial polysaccharides.

Culture	Shear rate[a] (s^{-1})	Viscometer	Flow curves given	Comments
Aureobasitium pullulans	10.2–1020	Fann V–G (model 35)	Yes	Power law behaviour throughout batch. Both n and K vary with fermentation; K exhibits a maximum, n a minimum. Variation in rheological conditions of batch. Some evidence for depolymerase activity near end of batch
X. campestris	0.0035–100	Weissenberg rheogoniometer (cone and plate)	Yes	Power law behaviour; K continually increases, but n levels off to a constant value after concentration of xanthan becomes greater than 0.5%. Cell mass (0.6% max.) had little or no effect on viscosity measurements
X. campestris (Whole broth and filtered broth)	1.0–10.0	Brookfield (guard removed)	Yes	Whole and filtered broth obeyed power law. Cells had minor effect on K, virtually no effect on n
X. campestris (Whole broth)	5 or 10	Brookfield (guard removed)	No	Viscosity increases monotonically with fermentation time. Nature of rheological behaviour not discussed
Au. pullulans (Whole broth)	0.17	Brookfield	No	Viscosity increases monotonically with fermentation time. Nature of rheological behaviour not given
P. funiculosum	Not given	Brookfield	No	Medium contains dextran. Viscosity measurement provides excellent measure of dextranase production. Nature of rheological behaviour not discussed
H. holstii	5.0	Brookfield	No	Viscosity increases monotonically with fermentation time. Nature of rheological properties not discussed
R. mansonii	5.0; 1 rpm (γ not given)	Brookfield; Brookfield–Wells cone and plate	No	Viscosity shows extreme dependence on concentration; gentle increase for concentration up to 0.3% then drastic rise

[a] Shear assumed equal to 10 N (s^{-1}) when not given explicitly (see Eq. 1).

Table 18. Correlation of experimental results using constitutive equations [Charles (1978)].

Organism	Constitutive equation	Effect of cell concentration	Comments
P. chrysogenum	$\sqrt{\tau} = \sqrt{\tau_0} + K_e\sqrt{\tau}$	$\tau_0 \propto X^2$	Based on fibre-like morphology; Casson equation fits entire shear range
P. chryosgenum	$\tau = \tau_0 + K\gamma$	$\tau_0 \propto X$	$\tau_0 = \alpha X^2$ also agrees with experimental data
Kanamycin broth	$\tau = \tau_0 + K\gamma$		$0.3 \leqslant n \leqslant 0.9$
Endomyces sp.	$\tau = K(\gamma)^n$		$2 \leqslant, K \leqslant 35$ dyn^{-1} cm^{-1}
P. chrysogenum			
P. chrysogenum	$\tau = \tau_0 + K\gamma$		
S. grisens	$\tau = K(\gamma)^n$		τ_0 and plastic viscosity varied sinusoidally with fermentation age
S. aurofaciens	$\tau = \tau_0 + K\gamma$		
Digested sewage sludge	$\tau = K(\gamma)^n$	Included in constitutive equation	First equation gave better agreement with data than second equation
Digested sewage sludge	$\tau = A \cdot X^{a_1} \cdot (\gamma)^{a_2 + a_3 \ln X}$		
P. pullulans	$\tau = A \cdot e^{a_1 x} \cdot (\gamma)^{a_2 + a_3 X}$		n and K functions of broth age and polysaccharide concentration
X. campestris	$\tau = \tau_0 + K(\gamma)$		Only apparent viscosity data reported with fermentation age

VARIATION OF RHEOLOGICAL PROPERTIES DURING FERMENTATION

Time Course Data — Apparent Viscosity

Figures 35–38 show data for the time courses of four highly viscous fermentations. The data in Figs. 37 and 38 suggest a strong relationship between polymer concentration and viscosity, while Fig. 36 indicates some polymer degradation or reorientation towards the end of fermentation. In contrast, Fig. 35 suggests that the changes in viscosity were dominated by changes in the cell concentration and morphology.

Overall, Figs. 35–38 suggest that viscosity is influenced by cell concentration and morphology, but that these effects can be overshadowed by the inherent viscosities of aqueous solutions of the products.

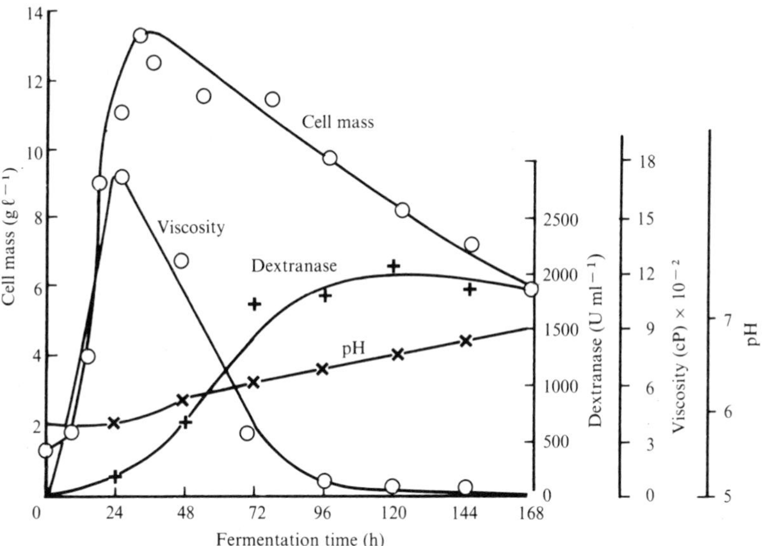

Figure 35. Relationship of cell mass, viscosity, dextranase activities and pH with fermentation time [Kosaric *et al.* (1973)].

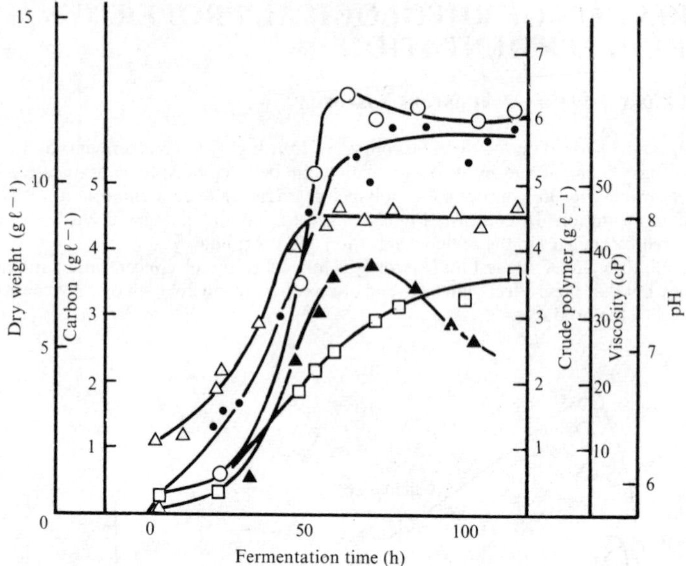

Figure 36. Fermentation of *Corynebacterium hydrocarboclastus* for polymer production. Typical 15-ℓ batch fermentation with $NaNO_3$ (0.5 per cent) and yeast extract (0.3 per cent by wt) as a nitrogen source. ●, dry weight of cell mass; ○, dry weight of crude polymer; △, terminal pH; □, carbohydrates in fermentation broth; ▲, viscosity of fermentation broth [Knetting and Zajic (1972].

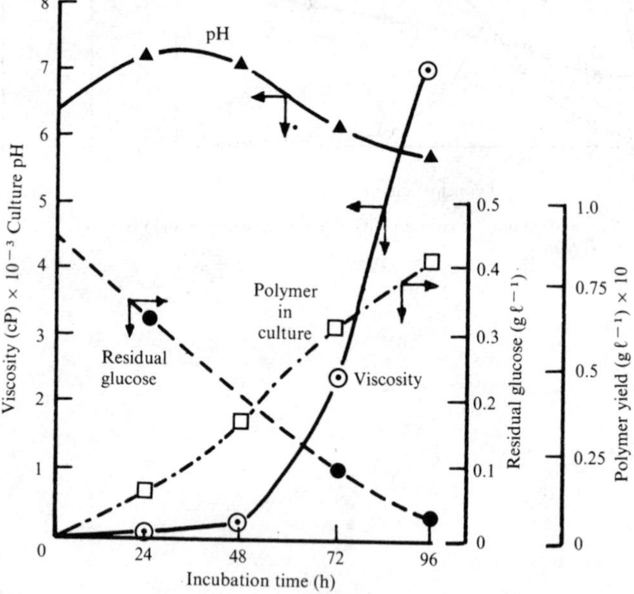

Figure 37. Time course of *Rhinocladiella mansonii* NRRL Y–6272 20-ℓ fermentation, modified urea medium at 25°C [Burton *et al.* (1976)].

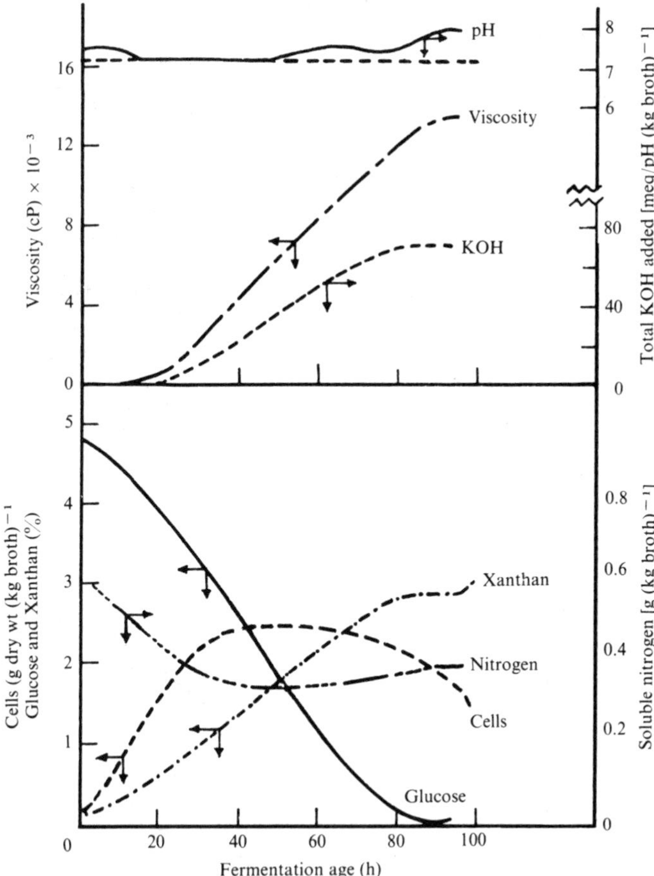

Figure 38. Time course of fermentation for xanthan production, pH control with potassium hydroxide [Moraine and Rogovin (1971)].

Rheological Characteristics of Broths

Figs. 39–45 contain rheograms obtained on various fermenter broths and show the effect of fermentation time on the rheological characteristics. Table 19 summarizes the salient features of Figs. 39–45.

Table 19. Shear stress–shear rate characteristics of broths.

Fig.	Fermentation	Feature
39	*Streptomyces noursei*/nystatin	Newtonian. Viscosity increases with time
40	*Aspergillus griseus*	Bingham plastic
41	*Penicillium chrysogenum*	Suggests Casson plastic behaviour. Yield stress depends on fermentation time
42, 43	Penicillin	Suggests Bingham plastic behaviour. Yield stress and Bingham viscosity depend on fermentation time
44, 45	Streptomycin	Highly variable pseudoplastic behaviour

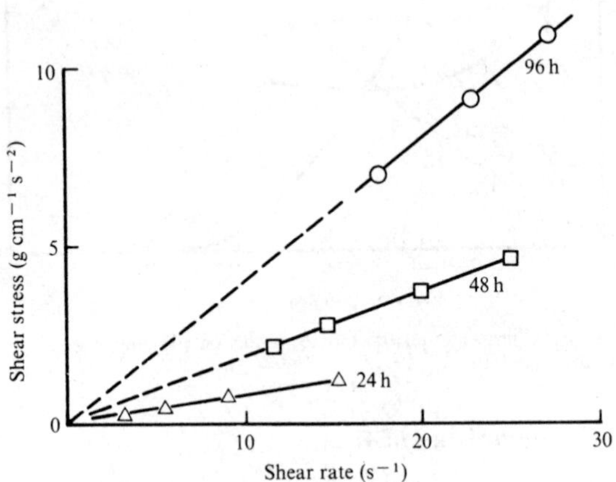

Figure 39. Newtcnian behaviour of nystatin broth at various broth ages.

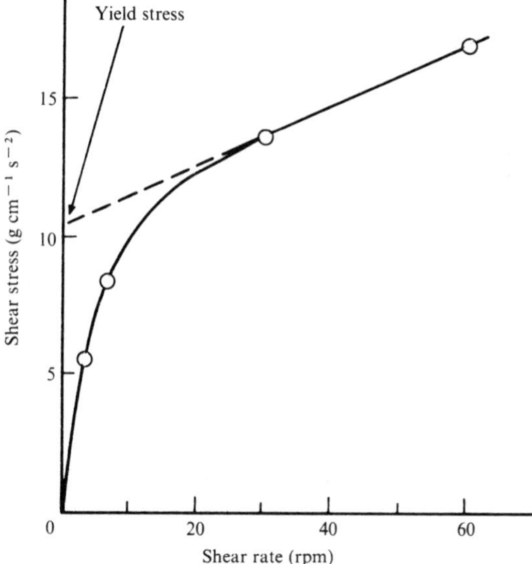

Figure 40. Flow curve and 'yield-stress' determination for *Aspergillus griseus* [Charles (1978)].

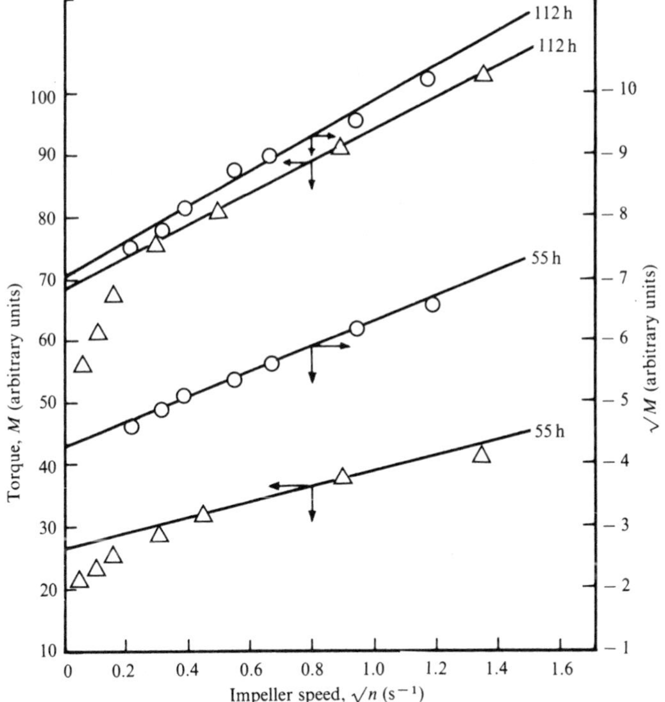

Figure 41. *Penicillium chrysogenum* showing Bingham and Casson model equations.

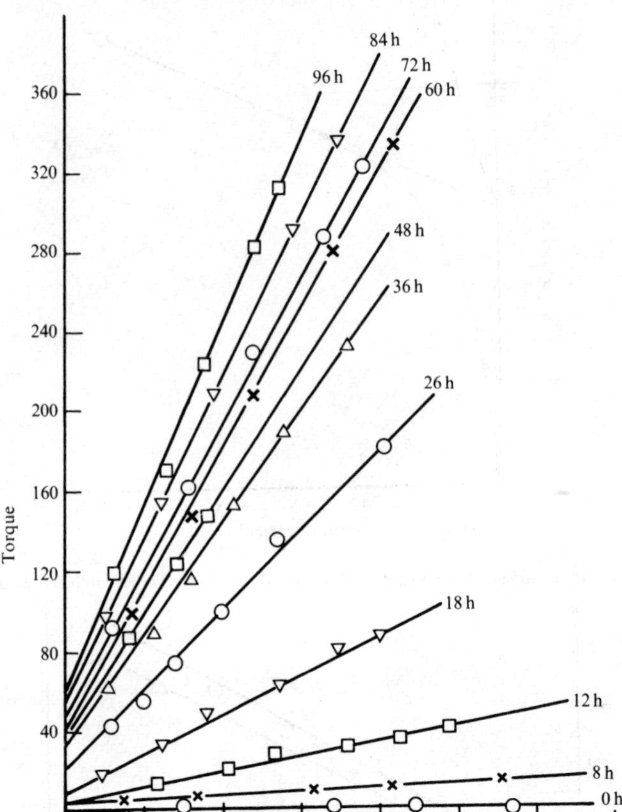

Figure 42. Shear diagrams for penicillin broth during fermentation [Deindorfer and Gaden (1955)].

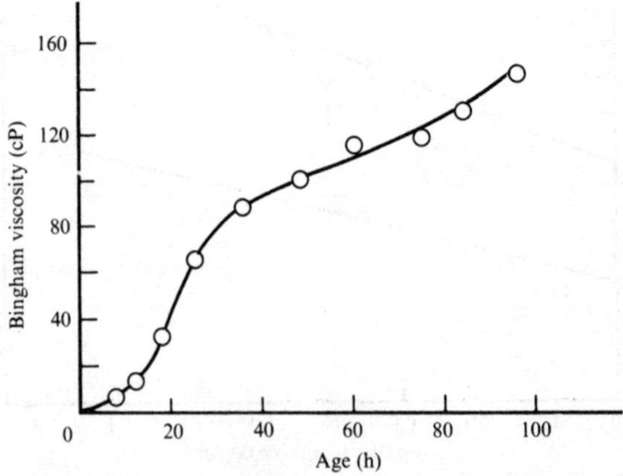

Figure 43. Bingham viscosity of penicillin broth during fermentation [Deindorfer and Gaden (1955)].

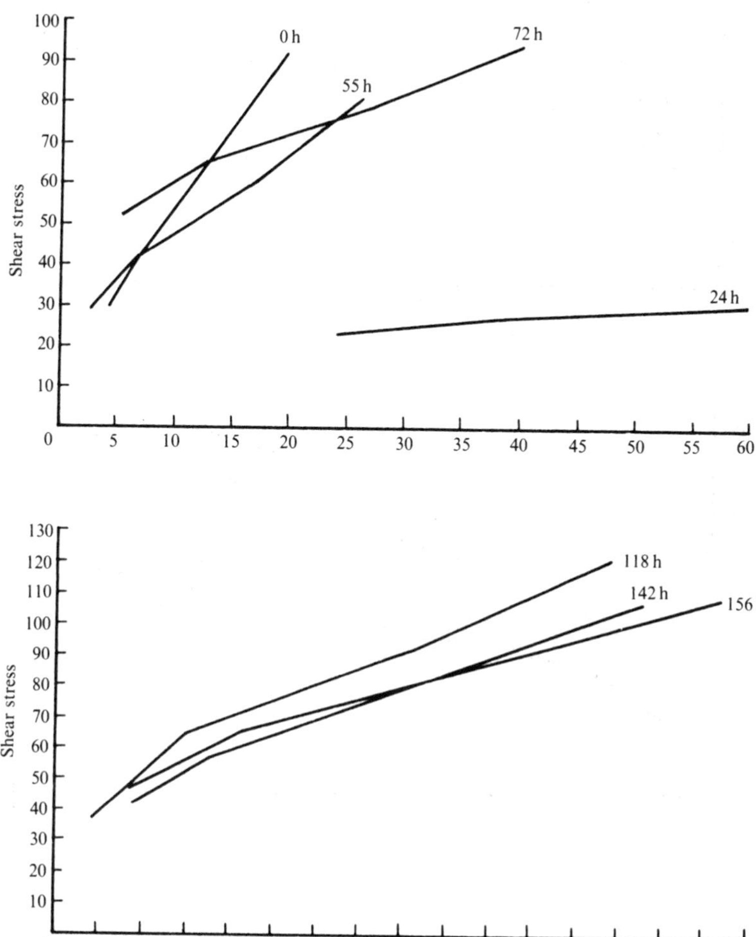

Figure 44. Rheograms for a typically viscous streptomycete fermentation [Tuffile and Pinho (1970)].

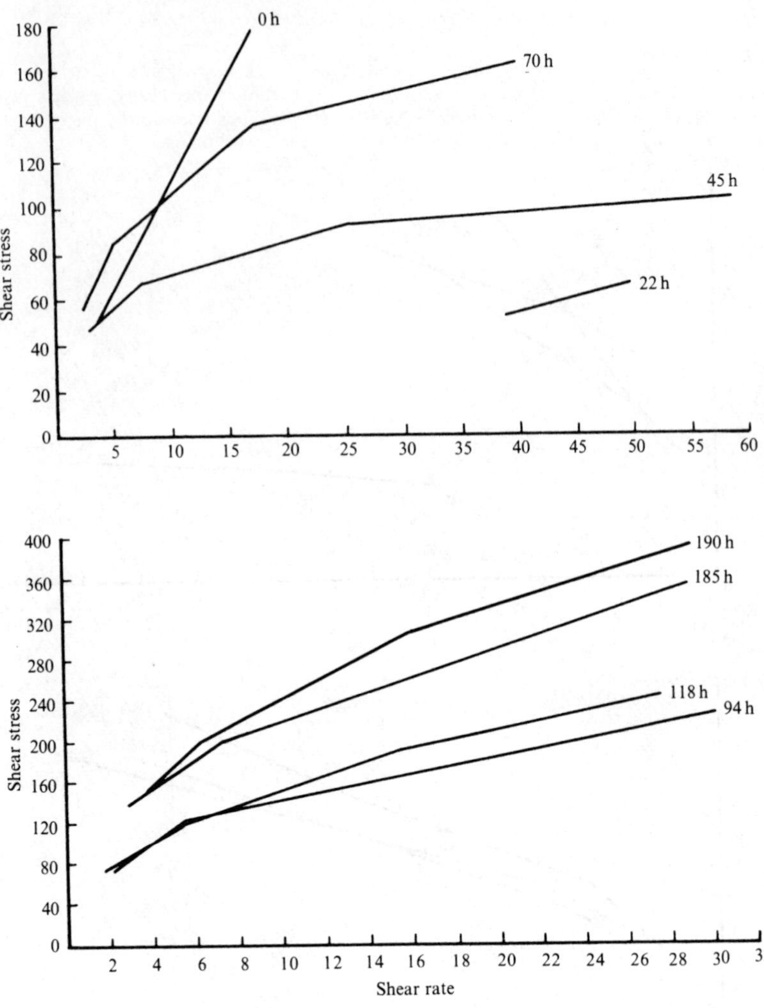

Figure 45. Rheograms for an extremely viscous streptomycete fermentation [Tuffile and Pinho (1970)].

Time Course Data — Rheological Parameters

Figs. 46–52 demonstrate how the power law index n and the consistency K exhibit a characteristic pattern over the time course of fermentations. Specifically, n starts just below unity (i.e. Newtonian conditions), falls rapidly and then rises slowly. In contrast, K, which is a measure of the apparent viscosity, increases with time.

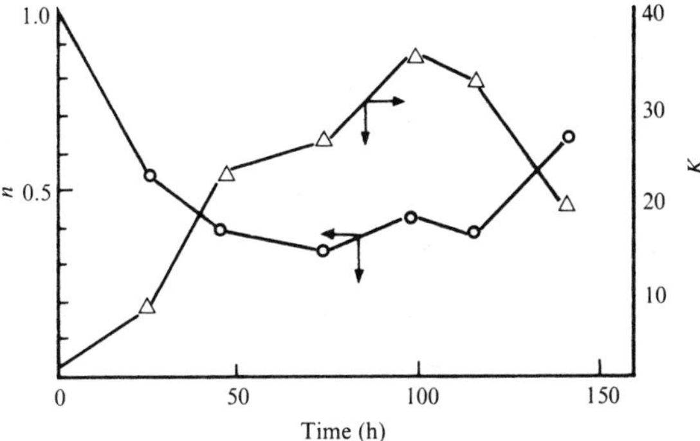

Figure 46. Variation of pseudoplastic characteristics with fermentation time, gluc-amylase broth [Taguchi and Miyamoto (1966)].

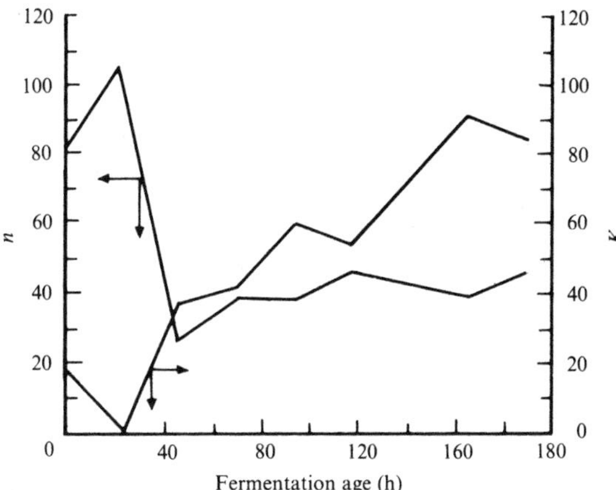

Figure 47. Variation of pseudoplastic characteristics with fermentation time for *Streptomyces aureofaciens*.

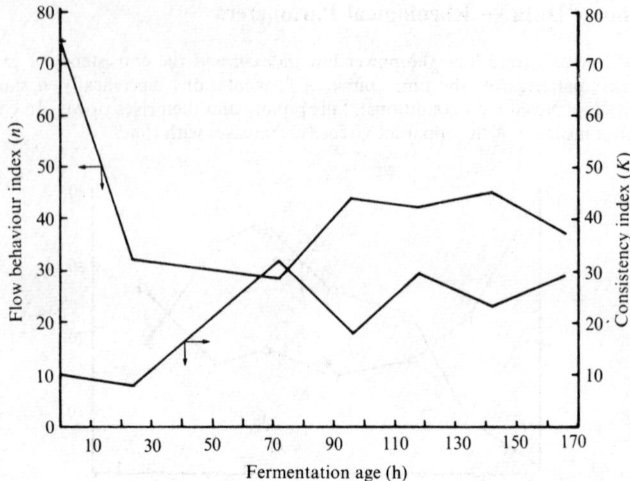

Figure 48. Time course of consistency and flow indices for a typically viscous streptomycete fermentation [Tuffile and Pinho (1970)].

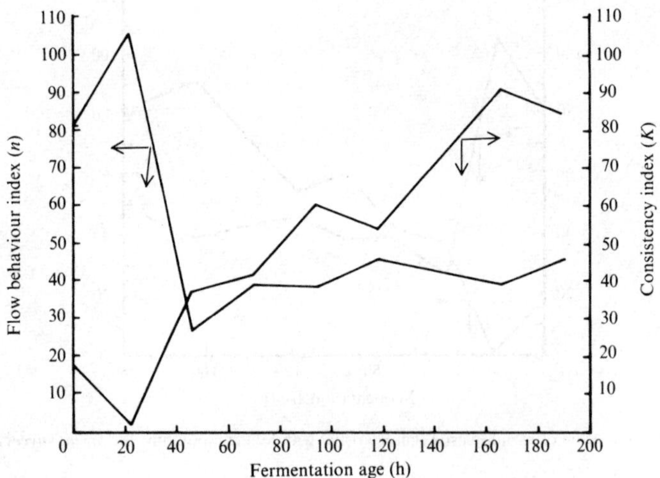

Figure 49. Time course of consistency and flow indices for an extremely viscous streptomycete fermentation [Tuffile and Pinho (1970)].

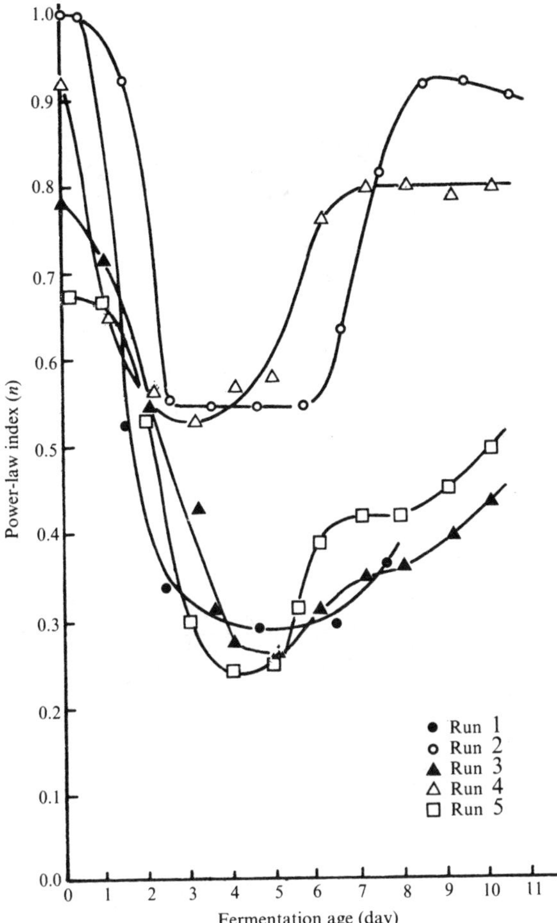

Figure 50. Development of the power-law index of the pullulan broths under various conditions for *Pullularia pullulans*.

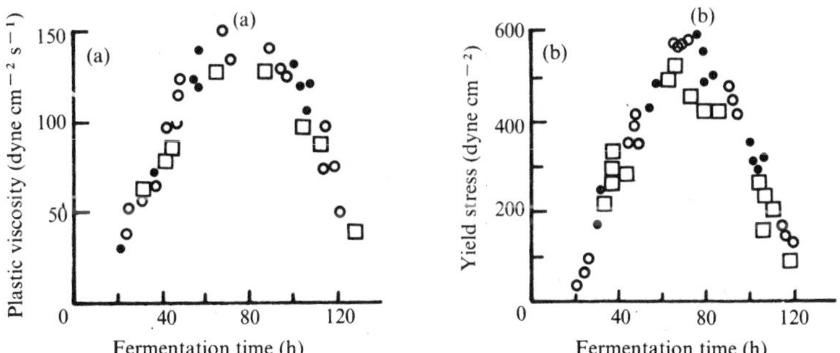

Figure 51. (a) Bingham plastic viscosity and (b) 'yield stress' versus fermentation time for *Streptomyces griseus* [Charles (1978)].

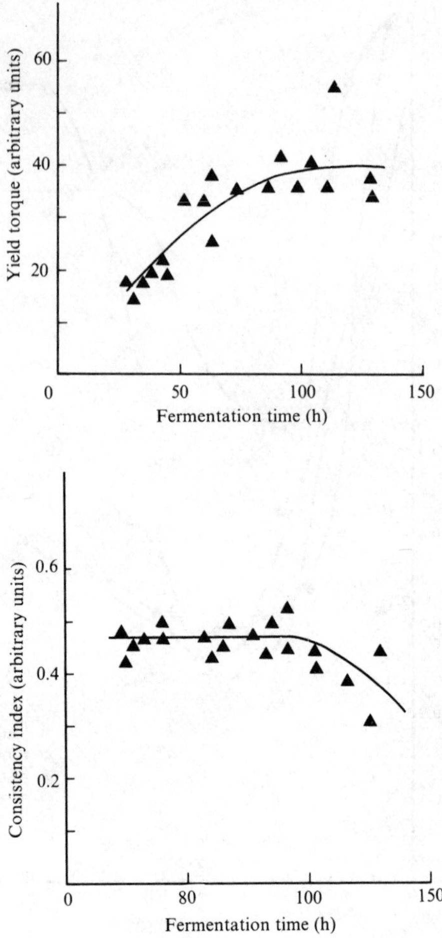

Figure 52. Yield torque and the Casson consistency index for *Penicillium chrysogenum* as functions of fermentation age (*see* Fig. 41) [Metz *et al.* (1979)].

BIBLIOGRAPHY

Metzner, A.B. (1956), Non-Newtonian Technology, in *Advances in Chemical Engineering*, Vol. 1, ed. by T.B. Drew and T.B. Hoopes, Chap. 2 (Academic Press, New York).

Skelland, A.H.P. (1967), *Non-Newtonian Flow and Heat Transfer* (John Wiley and Sons, New York).

Smith, J.M. (1979), Non-Newtonian Technology, in *Chemical Engineering*, 2nd edn., Vol. 3, ed. by J.M. Coulson, J.F. Richardson and D.G. Peacock (Pergamon Press, Oxford).

Wilkinson, W.L. (1960), *Non-Newtonian Fluids* (Pergamon Press, Oxford).

REFERENCES

Blanch, H.W. and Bhavaraju, S.M. (1976) Non-Newtonian Fermentation Broths: Rheology and Mass transfer. *Biotechnol. Bioengng*, **18**, 745.

Burton, K.A., Cadmus, M.C., Lagoda, A.A., Saudford, P.A., and Watson, P.R. (1976) A Unique Biopolymer from *Rhinocladiella mansonii* NRRL Y-6272: Production in 20-Liter Fermentors. *Biotechnol. Bioengng*, **18**, 1669.

Calderbank, P.H. and Moo-Young, M.B. (1959) Prediction of Power Consumption in Agitation of Non-Newtonian Fluids. *Trans. Instn Chem. Engrs*, **37**, 26.

Charles, M. (1978) Technical Aspects of the Rheological Properties of Microbial Cultures. *Adv. Biochem. Engng*, **8**, 1.

Deindorfer, F.H. and Gaden, E.L. (1955) Effects of Liquid Physical Properties on Oxygen Transfer in Penicillin Fermentation. *Appl. Microbiol.*, **3**, 253.

Knetting, E., and Zoyic, J.E. (1972) Flocculant Production from Kerosene. *Biotechnol. Bioengng*, **14**, 379.

Kosaric, N., Yu, K. and Zoyic, J.E. (1973) Dextranose Production from *Penicillium funiculosium*. *Biotechnol. Bioengng*, **15**, 729.

Leduy, A., Marsan, A.A. and Coupal, B. (1974) Rheological Properties of Non-Newtonian Fermentation Broth. *Biotechnol. Bioengng*, **16**, 61.

Labuza, T.P., Santos, D.B. and Roop, R.N. (1970) Engineering Factors in Single-Cell Protein Production. I. Fluid Properties and Concentration of Yeast by Evaporation. *Biotechnol. Bioengng*, **12**, 123.

Margaritis, A. and Zajic, J. (1978) Mixing, Mass Transfer and Scale-Up of Polysaccharide Fermentations. *Biotechnol. Bioengng*, **10**, 939.

Metz, B., Kossen, N.W.F. and van Sunijdam, J.C. (1979) The Rheology of Mould Suspensions. *Adv. Biochem. Engng*, **11**, 103.

Metzner, A.B., and Taylor, J.S. (1960) Flow Patterns in Agitated Vessels. *A.I.Ch.E. J.*, **6**, 109.

Moraine, R.A., and Rogovin, P. (1971) Xanthan Biopolymer Production at Increased Concentration by pH Control. *Biotechnol. Bioengng*, **13**, 381.

Morris, G.G., Greenshield, R.N. and Smith, E.L. (1973) Aeration in Tower Fermentors containing Microorganisms. *Biotechnol. Bioengng. Syrup.*, No. 4, 535.

Oldshue, J.Y. (1966) Fluid Mixing, Heat Transfer and Scale-Up. *Chem. Process Engng*, **47**, 183.

Sato, K. (1961) Rheological Studies of some Fermentation Broths. I. Kanamycin and Streptomycin Fermentation Broths. I. *J. Ferm. Technol.*, 39, 347

Sherman, P. (1970) *Industrial Rheology: With Particular Reference to Food, Pharmaceuticals and Cosmetics* (Academic Press, New York).

Taguchi, H. (1971) The Nature of Fermentation Fluids. *Adv. Biochem. Engng*, **1**, 1.

Taguchi, H. and Miyamoto, S. (1966) Power Requirement in Non-Newtonian Fermentation Broth. *Biotechnol. Bioengng*, **8**, 43.

Tuffile, C.M. and Pinho, F. (1970) Determination of Oxygen Transfer Coefficients in Viscous Streptomycete Fermentations. *Biotechnol. Bioengng*, **12**, 849.

Whorlow, R.W. (1980) *Rheological Techniques* (Ellis Horwood, Chichester).

APPENDIX

Units and factors of conversion for terms describing flow behaviour

Table A1.

Term	SI units	Equivalent in cgs units	Equivalent in fps units	Equivalent in other units
Viscosity (dynamic or absolute)	1 kg m^{-1} s^{-1} or 1 N s m^{-2} or 1 Pa s	10 g cm^{-1} s^{-1} or 10 P or 1000 cP	0.672 lb ft^{-1} s^{-1}	3500 kg m^{-1} h^{-1} or 2419.1 lb ft^{-1} h^{-1}
Kinematic viscosity	1 m^2 s^{-1}	1×10^4 cm^2 s^{-1} or 1×10^4 stokes	10.764 ft^2 s^{-1}	
Stress	1 kg m^{-1} s^{-2} or 1 N m^{-2} or 1 Pa	10 g cm^{-1} s^{-2} or 10 dyne cm^{-2}	0.672 lb ft^{-1} s^{-2}	
Torque	1 kg m^2 s^{-2} or 1 N m or 1 J	1×10^7 g cm^2 s^{-2} or 1×10^7 erg	23.731 lb ft^2 s^{-2}	

CHAPTER 9

GAS–LIQUID MASS TRANSFER

GLOSSARY

Maximum oxygen transfer rate R **Volumetric mass transfer coefficient** times the oxygen solubility. The value, which is often obtained by the sulphite method (*see* p.739), is used for the evaluation of **oxygen transfer rate** capacity of a fermenter.

Oxygen transfer rate This is equivalent to the oxygen uptake rate and is given by Eq. (1).

Oxygen uptake rate The actual volumetric oxygen utilization and is given by $(\mu/Y_{O_2})x$.

Volumetric mass transfer coefficient k_La The proportionality coefficient in Eqs. (1) and (2). k_La reflects both molecular diffusion and 'turbulent' mass transfer (*see* Appendix 2) and area for mass transfer (*see* Fig. 1). Many factors that affect k_La also affect the solubility c_s, e.g., substrate concentration affects both fluid viscosity and oxygen solubility.

Volumetric oxygen demand The maximum potential oxygen requirement by a fermenter broth and is given by $(\mu_{max}/Y_{O_2})x$, which is often measured using a Warburg respirometer.

GENERAL ASPECTS

The interrelationship between the power input to a gas–liquid system and the overall rate of mass transfer from the gas to the liquid phase is illustrated in Fig. 1.

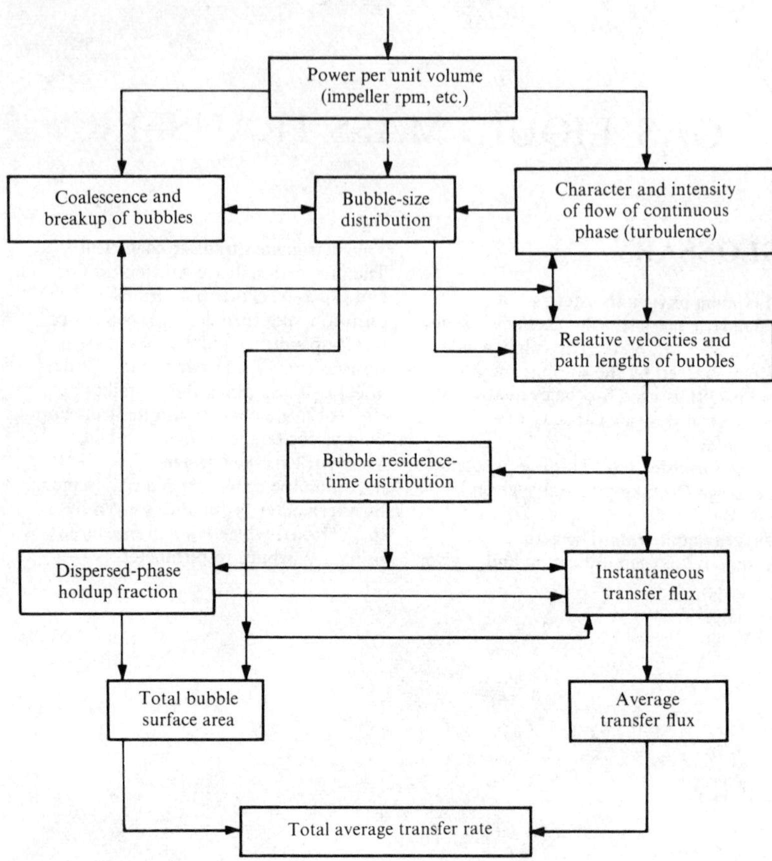

Figure 1. Relationships between agitation intensity and transfer rates at constant gas flow [Resnick and Gal-Or (1968)].

Fig. 1 suggests the relationship

$$R = k_L a\,(c^* - c) \tag{1}$$

where R is the overall volumetric mass transfer rate (dimensions $\mathrm{M\,L^{-3}\,T^{-1}}$), k_L is the mass transfer coefficient, based on the liquid–phase resistance to mass transfer (*see* Appendix 1 to this chapter) (dimensions $\mathrm{L\,T^{-1}}$), a is the bubble surface area per unit volume (dimensions $\mathrm{L^{-1}}$) and c^* and c are the oxygen solubility and dissolved oxygen concentration, respectively.

All the terms in Eq. (1) refer to time average values of a dynamic situation. The volumetric basis for R and a is properly the bubble-free liquid volume, although, in view of the fact that the gas hold-up is low in many devices, the volume of the dispersion is often used in practice.

Eq. (1) is used to calculate $k_L a$ values from R versus c data determined by the methods described on pp.737–739. Where values of a are measured directly at the same time as the data leading to $k_L a$, then the values of k_L can be obtained by calculation. In general, data are reported as $k_L a$ with units of min^{-1}, h^{-1}, lb mol ft^{-3} h^{-1} atm^{-1}, g mol ℓ^{-1} h^{-1} atm^{-1}, etc. Insight into the physical phenomena involved in gas–liquid mass transfer has been provided by Higbie (1935) and Danckwerts (1951) (*see* Appendix 2).

It can be deduced from Fig. 1 that $k_L a$ values will differ for the various contacting devices, and that both k_L and a will be affected by the presence of materials in the liquid phase which influence coalescence, break up and bubble size, i.e. materials such as fermentation nutrients and products. This latter phenomenon is illustrated by the data contained in Table 1, where the purity of the water used has a marked effect on the oxygen transfer rate.

Table 1. Effect of the purity of water on the oxygen transfer rate [Bell and Gallo (1971)].

Water used	pH	Oxygen transfer rate (mmol ℓ^{-1} h^{-1})	Formation of foam
Normal laboratory distilled water	4.0	51	None
	7.5	120	Yes
	8.5–9.0	126	Yes
Double-distilled water before cleaning of reactor	7.5	105	Slight
Double-distilled water after cleaning of reactor	4.0	44	None
	7.5	54	Trace

The gas phase in a gas–liquid contacting device may be dispersed or continuous. The former situation is usually associated with low gas hold-up except in plate columns, where the plates repeatedly break up a coalescing gas stream. Table 2 shows that, while gas hold-up can be increased markedly by a change in fermenter configuration, the corresponding improvements in specific bubble surface depend on whether the gas phase is dispersed or continuous.

Table 2. Comparison of gas–liquid contacting equipment [Phillips (1968)].

Equipment classification	Typical fermenter type	Gas phase	Gas rate ($m\ s^{-1}$)	Specific surface (m^{-1})	Gas hold-up (%, based on vessel volume)
Plate column	ICI pressure cycle fermenter (*see* Fig. 10, (Chapter 7)	Dispersed	0.6	50	50–90
Packed column	Trickling filter fermenter (*see* Fig. 53, Chapter 7)	Continuous	0.9	16	90
Wetted-wall column	Biological film reactor (*see* Fig. 11, Chapter 7)	Continuous	2.1	5	95
Bubble column	Bubble column reactor	Dispersed	0.02	7	8
Stirred tank	Stirred tank reactor	Dispersed	0.06	25	15

Table 3. Effect of aeration device on specific surface areas in various fermenter types [Schügerl *et al.* (1977)].

Fermenter type	Aeration device	Specific surface (m^{-1})	Power input per unit volume ($kW\ m^{-3}$)	Gas rate ($cm\ s^{-1}$)
Stirred tank	Sparger	300	1	
		400	1	
		600	2	
	Perforated plate	1000	4	
	Porous plate	2200	10	
Air lift with draught tube	Ejector nozzle	1300	0.9	3
		1300	0.9	8
		1800	1.8	8
		2000	7.2	3
		2500	7.2	8
Bubble column	Perforated plate	650	0.6	4.5
	Porous plate	2000	0.9	4
	Injector nozzle	6000	1.5	3
	Ejector nozzle	8000	2.2	7

Table 3 shows that quite disparate power inputs are associated with the same specific surface areas created in different combinations of fermenter/aerator.

It follows from Eq. (1) that the magnitude of k_La is of considerable importance. Since oxygen is a sparingly soluble gas, i.e. c^* is small, the maximum driving force is also small and the maximum possible oxygen transfer rate is given by

$$R = k_La\, c^* \tag{2}$$

For reactors in which the air flow is dispersed (*see* Table 2), the fluid mechanics suggest two extreme types according to Andrew (1982). These types are

1. Free-bubble rise reactors, i.e. bubble columns, tower fermenters, air-lift fermenters and virtually all large agitated tanks.
2. High-turbulence reactors, i.e. small laboratory agitated tanks.

Commercial-scale agitated tanks fall into the first category because the bubbles formed by the agitator rise freely through the liquid under the action of buoyant forces. Unfortunately, laboratory fermenters operate under a completely different regime where the bubbles are highly dispersed and frequent coalescence and break-up occurs. Thus, high-turbulence reactors are suitable for the study of microbial kinetics (*see* Chapter 4), but are unsuitable for studies on flow and mass transfer regimes relevant to large-scale fermenters. Until recently, research literature has been virtually exclusively restricted to high-turbulence reactors. The literature on gas–liquid reactors is extensive but not particularly informative, except with regard to the general principles and detailed correlations of data for specific situations. The reasons for the slow progress on the subject against a background of great effort are (1) the wide variety of reactor geometries and internals, and (2) the complexity of the surface chemistry of bubbles in anything but single-component, 'pure' liquids. This complexity is particularly important in cell cultures which contain not only a variety of foam-producing and foam-stabilizing products but also added antifoams.

Sufficient surface-active material is likely to be present to make all but the largest bubbles (i.e. those with diameters of greater than 0.3 cm) act as rigid spheres, in which case, during free rise at low air hold-ups, at 20°C

$$\frac{k_La}{\varepsilon}\mu^{0.5} = 0.4\,\mathrm{s}^{-1} \tag{3}$$

where ε is the gas hold-up and μ the viscosity of the liquid in centipoise.

Fig. 2 shows the effect of bubble diameter on k_L in a variety of systems. Conservative interpretation and design in fermentation systems suggests the use of Eq. (3). The contribution of the medium components is illustrated in Fig. 3. Fig. 4 demonstrates the

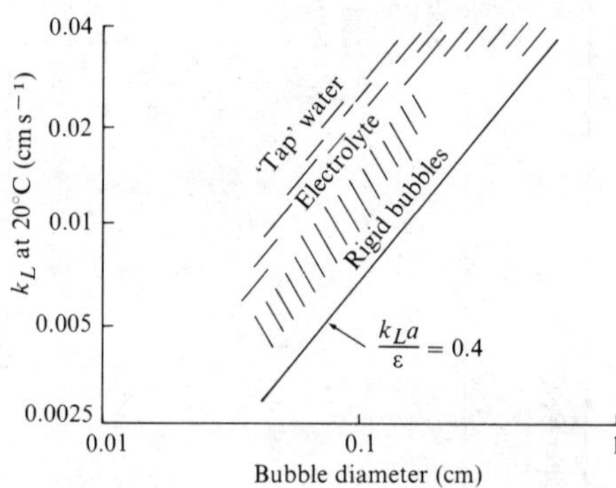

Figure 2. Mass transfer coefficient for oxygen absorption from bubbles in free rise [Andrew (1982)].

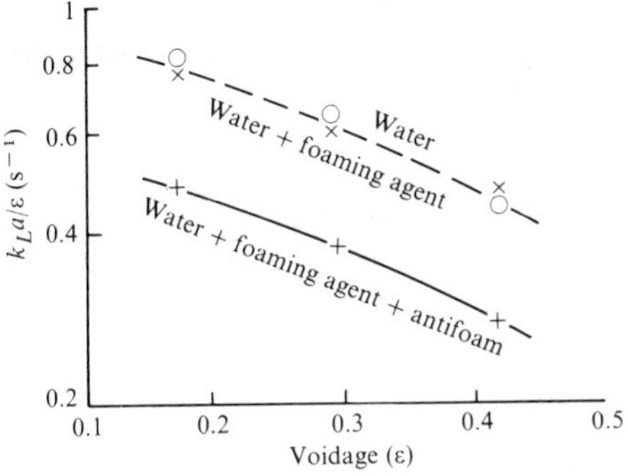

Figure 3. Effect of foaming agent with or without antifoam on carbon dioxide absorption into water at 20°C, 1130 rpm [Andrew (1982)].

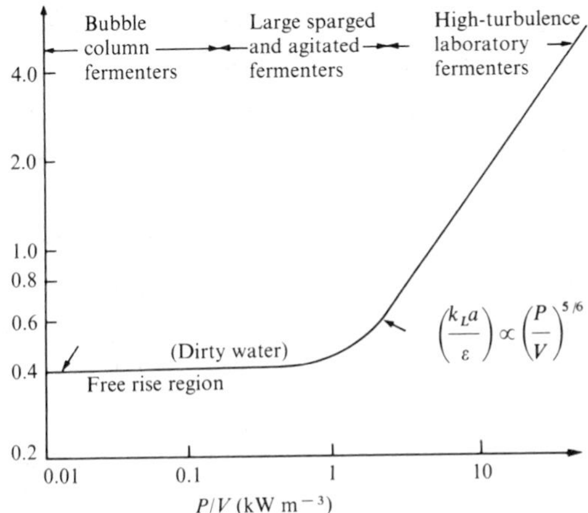

Figure 4. Fermenter characterization [Andrew (1982)].

increasing influence of power input on k_La and emphasizes the intermediate status of large fermenters. This figure also underlines the fact that the use of correlations of mass transfer data available in the literature should be restricted only to the range of parameters investigated and extrapolation should be avoided.

SOLUBILITY DATA

The solubility of oxygen in aqueous solutions in equilibrium with air at fermentation temperatures is in the region of 10 mg ℓ^{-1} (i.e. 10 ppm). The main factors affecting the solubilities of sparingly soluble gases are their partial pressures, the presence of solutes and temperature.

The effect of the partial pressure p_G of a given component in the gas phase on its solubility c_s is expressed by Henry's law

$$C_s = \frac{p_G}{H} \tag{4}$$

where H is the Henry's law constant in appropriate units.

Table 4 provides an example of the use of Eq. (4) and shows the dependency of the solubility of oxygen in pure water on temperature, solubility falling with increasing temperature.

Table 4. Oxygen solubility and Henry's constant for pure water.

Temperature (°C)	Oxygen concentration[a] (mg ℓ$^{-1}$)	Henry's constant (atm l mg^{-1})
25	8.10	0.0258
35	6.99	0.0299

[a] In equilibrium with air at 1 atm (p_{O_2} = 0.209 atm), calculated from Eq. (4).

Oxygen solubility data over a temperature range of 0–40°C is provided in Table 5. The variation in solubility with temperature is also given by Eq. (5).

$$c_s = 14.16 - 0.394\,T + 0.007714\,T^2 - 0.000646\,T^3 \tag{5}$$

where c_s has units of milligrams per litre of liquid and T is expressed in degrees centigrade.

Table 5. Solubility of oxygen at 1 atm at various temperatures in water [Finn (1967)].

Temperature (°C)	Oxygen concentration in water (mmol ℓ$^{-1}$)
0	2.18
10	1.70
15	1.54
20	1.38
25	1.26
30	1.16
35	1.09
40	1.03

The influence of acids and salts on oxygen solubility is indicated by the data presented in Table 6 and that of sugars on oxygen solubility in Table 7. These data suggest that oxygen solubility is decreased by components in the aqueous phase.

Table 6. Solubility of oxygen at 1 atm in solutions of salt and acids at 25°C [Finn (1967)].

Concentration (M)	Oxygen solubility (mM O_2 ℓ$^{-1}$)		
	HCl	$\frac{1}{2}$ H_2SO_4	NaCl
0.0	1.26	1.26	1.26
0.5	1.21	1.21	1.07
1.0	1.16	1.12	0.89
2.0	1.12	1.02	0.71

Table 7. Solubility of oxygen at 1 atm in solutions of sugars [International Critical Tables (1928)].

Sugar	Concentration (mM)	Temperature (°C)	Oxygen solubility (ℓ O_2 ℓ^{-1})
Glucose	0.7	20	2.70
	1.5	20	2.55
	3.0	20	2.45
Sucrose	0.4	15	2.97
	0.9	15	2.43
	1.2	15	2.16

Solubility data for a range of sparingly soluble gases and hydrocarbons of significance in fermentation are given in Tables 8 and 9. The temperature dependency of these solubilities is given in International Critical Tables (1928).

Table 8. Solubilities of gases in water at 20°C and 1 atm [Phillips (1968)].

Gas	Solubility (mmol ℓ^{-1})
Butane	6.7
Carbon dioxide	40.0
Ethane	2.1
Methane	1.5
Nitrogen	0.76
Oxygen	1.4
Propane	2.9

Table 9. Solubility of *n*-alkanes in water at 25°C [Moss and Smith (1977)].

n-Alkane	Solubility (mmol ℓ^{-1})
Hexane	0.11
Octane	0.0058
Decane	0.00031
Dodecane	0.000017
Tetradecane	0.00000098

Data given in Tables 5–9 provide an indication of the order of magnitude of solubilities and demonstrate the importance of direct and careful measurement of values for individual fermentation broths. Also, it should be stressed that, other than for the simplest media, generalized data on the detailed effects of the presence of solutes is unlikely to be available.

In Table 10, solubility data for various organic substances associated with fermentation, either as substrates or products, are listed.

Table 10. Solubilities of substrates and metabolites in water [Windholz (1976)].

Substance	Temperature (°C)	Solubility (g ℓ$^{-1}$)
Alanine	25	147
L-Asparagine	25	24.6
Glucose	25	590
Glycine	25	21.7
Lactose	25	170
L-Leucine	25	22.4
Sucrose	25	909
Tartaric acid	15	769
Urea	25	620
DL-Valine	25	66.8

Fig. 5 shows dissolved oxygen concentration profiles within a layer of microorganisms and the adjacent nutrient solution. The gradients in the liquid and biomass phases suggest liquid- and 'solid'-phase diffusion limitations (*see* Chapter 10). Within the accuracy of the measurement, the dissolved oxygen concentration profiles are continuous at the liquid–'solid' interface. This suggests that there is no strong equilibrium phenomenon at the interface (i.e. the situation is not analogous to Henry's Law) and that the solubilities of gases in aqueous solutions and biomass are very likely similar.

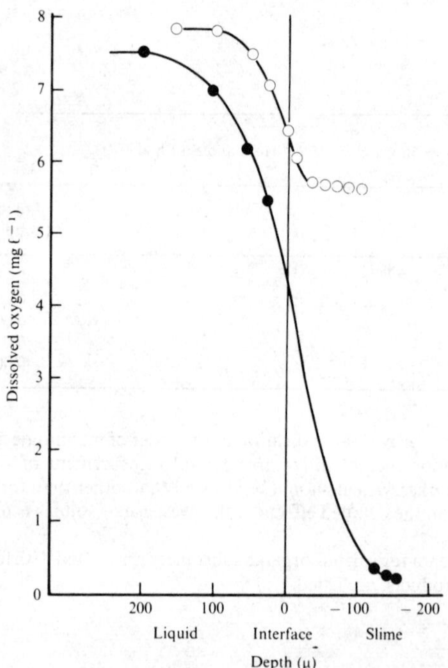

Figure 5. Dissolved oxygen profiles determined with microelectrode for bacterial slimes in a continuous culture system. Nutrient flow was 0.457 m s^{-1} and temperature 27°C. With nutrient broth 20 mg^{-1} (○), the oxygen in the slime stablilized at 5.5 ppm below 75 μm in depth. With 500 mg^{-1} nutrient broth (●), the profile reached 150 μm before stabilizing at 0.25 mg^{-1} [Sanders *et al.* (1971)].

EXPERIMENTAL DETERMINATION OF k_La

Biological Gassing-Out Methods

Static gassing-out methods

In static gassing-out methods, aeration is discontinued at a time during the fermentation at which a k_La value is required and the dissolved oxygen is allowed to fall to zero. The microorganisms are killed or respiration is stopped by the addition of a respiratory inhibitor, such as fluorocitrate or malonate. The dissolved oxygen concentration is followed for a period after the start of aeration.

Eq. (6) describes the time course of dissolved oxygen [i.e. $c(t)$] from the start of aeration.

$$\ln\left(1 - \frac{c}{c^*}\right) = k_La\ t \tag{6}$$

Fig. 6 illustrates a graphical procedure for the determination of k_La from the slope of the $\ln[1 - (c/c^*)]$ versus t graph.

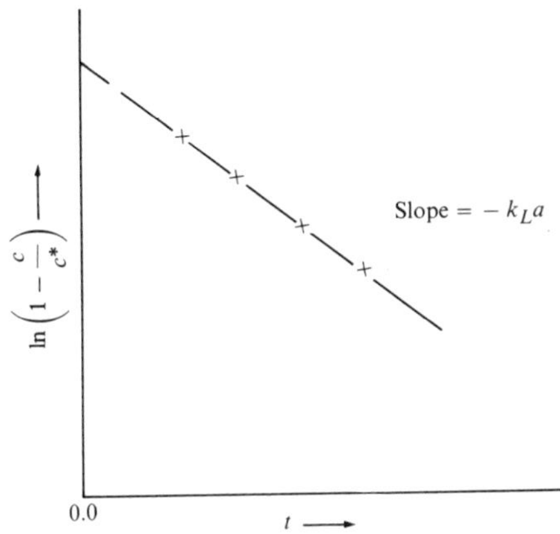

Figure 6. Graphical representation of static gassing-out data.

The limitations of the static gassing-out method lie in the facts that a nonrespiring system is involved, which may differ from the actual fermentation conditions and the fermentations cannot be restarted following the test. However, this approach may be used for simulated fermentation fluids with a nitrogen purge.

Dynamic gassing-out method

In dynamic gassing-out, the aeration of an actively respiring system is stopped and the time course of the consequent decrease in dissolved oxygen due to respiration is monitored. Aeration is resumed before the critical oxygen concentration is reached (*see* p.209), and the increasing dissolved oxygen concentration is followed.

The time course of the dissolved oxygen concentration is described in Eqs. (7) and (8).

Falling disssolved oxygen
$$-\frac{dc}{dt} = Q_{O_2}\,x = \text{constant} \tag{7}$$

Increasing dissolved oxygen
$$c = -\frac{1}{k_La}\left(\frac{dc}{dt} + Q_{O_2}\,x\right) + c^* \tag{8}$$

Fig. 7 represents typical data and the determination of $Q_{O_2}x$ and Fig. 8 illustrates a graphical procedure for the determination of k_La. In Eqs. (7) and (8), $Q_{O_2}x$ is constant because (1) $c > c_{CRIT}$ and (2) $x \simeq$ constant over the period of the experiment.

The dynamic method has the advantage that it may be applied to fermentation systems without extensively disrupting the metabolism. Furthermore, this technique has been widely applied when using yeast. However, application to viscous mycelial fermentations is questionable since rapid disengagement of air bubbles is necessary after discontinuing the air supply. This is not normally the case with highly viscous nonnewtonian broths. If significant surface aeration occurs, a nitrogen blanket is necessary.

Eq. (8) may be integrated to give

$$\ln\left(\frac{c_{INIT} - c}{c_{INIT} - c_1}\right) = -k_La(t - t_1) \tag{9}$$

since

$$Q_{o_2}x = k_La(c^* - c_{INIT}) \tag{10}$$

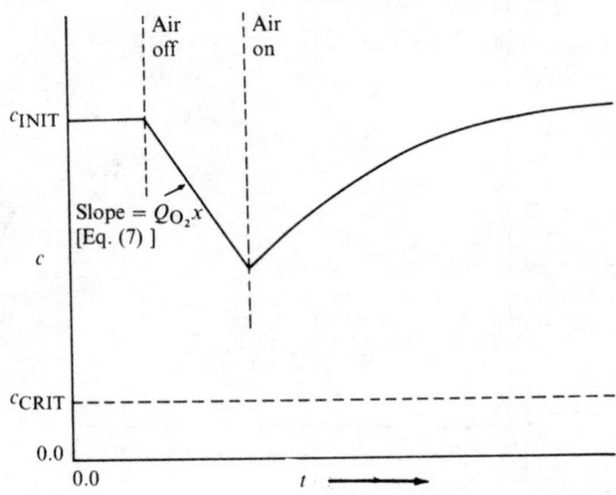

Figure 7. Time course of dissolved oxygen and determination of $Q_{O_2}x$ for the dynamic gassing-out method.

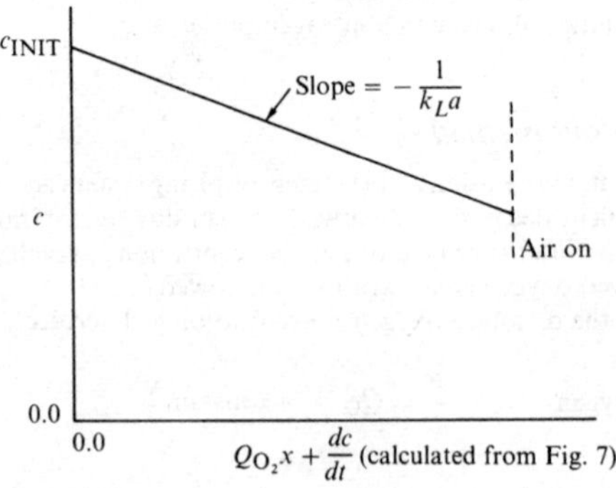

Figure 8. Graphical determination of k_La by the dynamic gassing-out method with increasing dissolved oxygen [*see* Eq. (8)].

It follows from Eq. (9) that the data in Fig. 7 can be used to determine k_La by the method illustrated in Fig. 9.

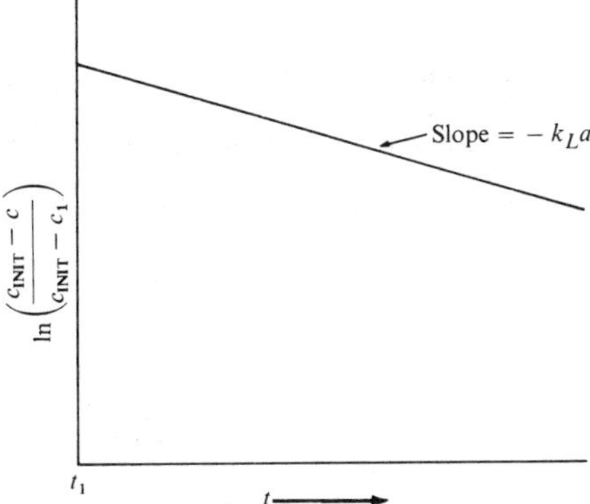

Figure 9. Graphical representation of Eq. (9) for the determination of k_La in the dynamic gassing-out method.

Oxygen Balance Method

Determination of the oxygen concentrations in the inlet and outlet gas streams (and liquid streams where appropriate), together with the corresponding flow rates, allows oxygen balances to be developed leading to k_La using Eq. (1). Using the oxygen balance method, fermentation is neither terminated nor disturbed. There are no inherent shortcomings using this approach, but accurate measurement of low flow rates or small differences in concentrations places demands on both the equipment and the experimenter.

Sulphite Oxidation Method

The method is based upon the oxidation of sodium sulphite to the sulphate in the presence of a catalyst, usually a divalent cation, the reaction involved being

$$Na_2SO_3 + \tfrac{1}{2}O_2 \xrightarrow{Cu^{2+} \text{ or } Co^{2+}} Na_2SO_4 \tag{11}$$

The kinetics of the sulphite oxidation are independent of the reactant concentration (i.e. zero order) and sufficiently fast that the overall rate of oxidation is controlled by the oxygen transfer rate. Experimentally, 1 N sodium sulphite solution containing 1 mM Cu^{2+} reacts with the dissolved oxygen. Samples are removed at intervals and assayed for residual sulphite by reacting with excess iodine followed by back-titration with sodium thiosulphate, using starch as an indicator. k_La is then calculated using Eq. (2). Since the rate of sulphite oxidation is very rapid, the dissolved oxygen concentration in the liquid is zero at all times. Care has to be taken to prevent the samples absorbing oxygen.

Although the sulphite method is used extensively, the results obtained appear to depend on the type of equipment, the nature of the catalyst, the pH of the solution and the purity of the sulphite solution. The method gives higher k_La values than other techniques, while not simulating fermentation broth in many of the factors that can strongly affect k_La. Fig. 22 provides an example of the variation in results obtained by the sulphite method and those determined polarographically.

EXPERIMENTAL DETERMINATION OF DISSOLVED OXYGEN CONCENTRATIONS

Oxygen electrodes

The reduction of dissolved oxygen at a noble metal surface negatively polarized with respect to a reference electrode forms the basis of the design and operation of oxygen electrodes. Although this phenomenon was first observed as early as 1897. it was not until 1956 that the first membrane-covered dissolved oxygen electrode was introduced. Since then many different types of electrode have been developed. An historical review is given by Lee and Tsao (1979).

Polarographic electrodes

When an electrode of a noble metal is made 0.6–0.8 V negative with respect to a suitable reference electrode in a suitable electrolyte solution, the dissolved oxygen is reduced at the electrode's surface. The cathode may be platinum or gold, the reference electrode (i.e. the anode) may be calomel or silver/silver chloride and the electrolyte may be potassium chloride. The construction of a typical polarographic electrode is shown in Fig. 10.

The reduction of the dissolved oxygen produces a current. Fig. 11 shows how this phenomenon can be observed in the form of a current–voltage diagram (i.e. a polarogram) for the electrode.

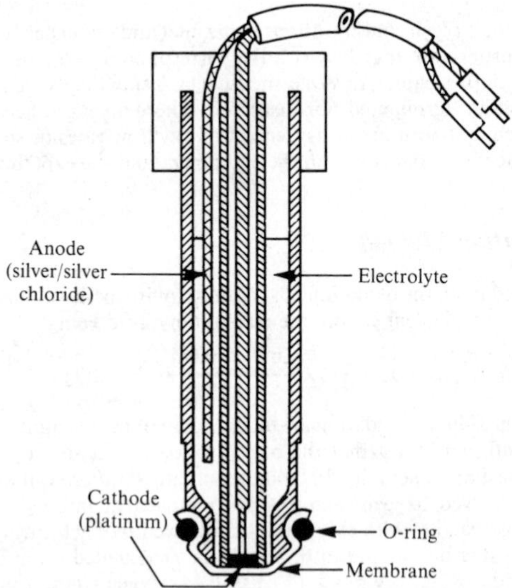

Figure 10. Construction of a polarographic electrode [Lee and Tsao (1979)].

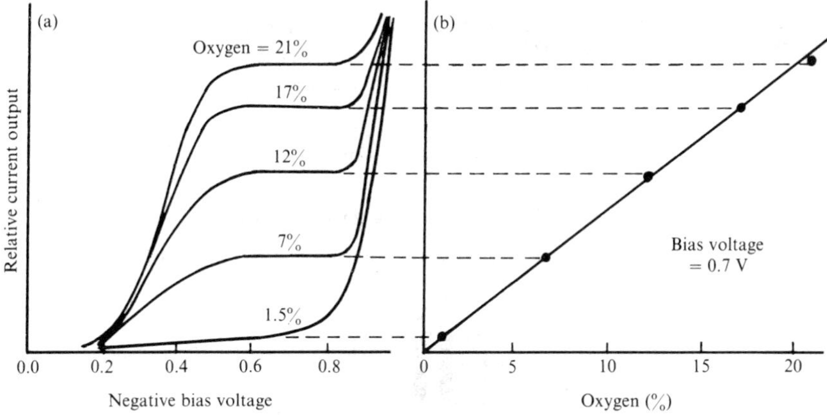

Figure 11. (a) Polarogram and (b) calibration curve of an oxygen electrode. If a fixed voltage in the plateau region of the current–voltage diagram (a) is applied to the cathode, the current output of the electrode can be calibrated against the dissolved oxygen concentration (b) [Cobbold (1974)].

As can be seen from Fig. 11(a), the current increases initially with an increase in the negative bias voltage. Then a plateau is reached where the reaction of oxygen at the cathode is so fast that the rate of reaction is limited by the diffusion of oxygen to the cathode. If the negative bias voltage is further increased, other reactions, mainly the electrolysis of water with hydrogen production, start and these increase the current output of the electrode. In the plateau region, the current is directly proportional to the activity or equivalent partial pressure of the dissolved oxygen (i.e. oxygen tension).

For polarographic electrodes, the reactions at the cathode are

$$O_2 + 2H_2O + 2e^- \rightarrow H_2O_2 + 2OH^-$$
$$H_2O_2 + 2e^- \rightarrow 2OH^-$$

and the reaction at the anode is

$$Ag + Cl^- \rightarrow AgCl + e^-$$

Thus the overall reaction is

$$4Ag + O_2 + 2H_2O + 4Cl^- \rightarrow 4AgCl + 4OH^-$$

The overall reaction tends to produce alkaline conditions together with a small amount of hydrogen peroxide and also consumes chloride ions. Therefore, chloride ions have to be replenished when the electrode becomes depleted. A number of other mechanisms have been suggested for the electrode reactions [*see* Lee and Tsao (1979) and Beechey and Ribbons (1972)]. The precise stoichiometry probably depends on the nature of the electrode, the applied voltage, the electrode surface and the conditions to which the electrode is exposed. This final factor may be the cause of the time-dependent drift of the polarographic probe readings.

Galvanic Electrodes

The main difference between galvanic and polarographic electrodes is that the former do not require an external voltage source of the reduction of oxygen at the electrode. This is due to the fact that when a relatively basic metal, such as zinc, lead or cadmium, is used as the cathode the voltage generated by such an electrode pair is sufficient for the spontaneous reduction of oxygen at the cathode surface. The reaction for the silver–lead galvanic probe at the cathode is

$$O_2 + 2H_2O + 4e^- \rightarrow 4OH^-$$

and at the anode is

$$Pb \rightarrow Pb^{2+} + 2e^-$$

Thus the overall reaction is

$$O_2 + 2Pb + 2H_2O \rightarrow 2H_2O + 2Pb(OH)_2$$

The construction of galvanic electrodes in common use is presented in Fig. 12 (a), (b) and (c).

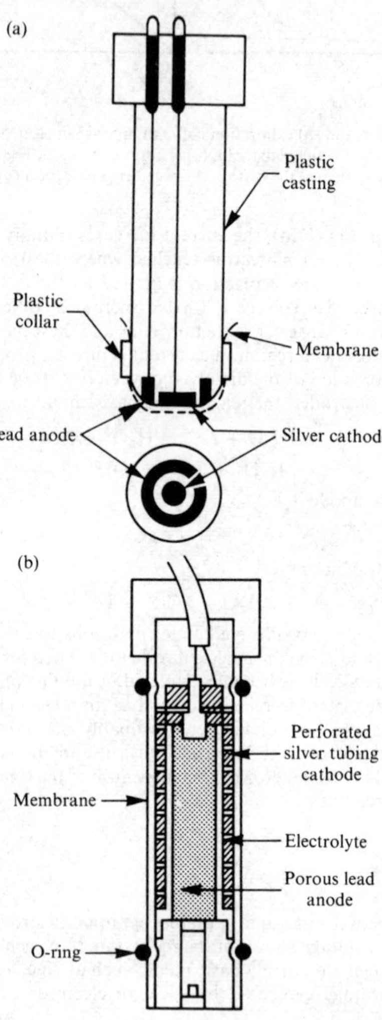

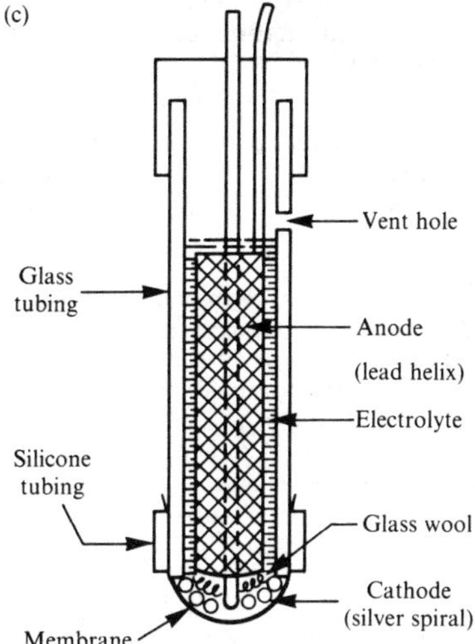

Figure 12. Construction of various galvanic electrodes [Lee and Tsao (1979)].

Theory of operation of membrane-covered dissolved oxygen probes

The cathode, anode and the electrolyte of polarographic or galvanic electrodes are separated from the measuring medium by a membrane permeable to oxygen but impermeable to most ions. Provided that the rate-controlling step in the diffusion of oxygen from the liquid medium to the surface of the cathode is diffusion through the membrane, the current output of the probe is proportional to the activity or the partial pressure of the oxygen in the liquid medium.

A review of the different mathematical models for the response of the dissolved oxygen probe is given by Lee and Tsao (1979).

A simple one-layer model involves the assumptions that

1) The thickness of the electrolyte layer between the membrane and the cathode can be neglected.
2) The partial pressure of oxygen at the membrane surface is the same as that of the bulk liquid (i.e. the liquid around the probe is well agitated).
3) Diffusion is in one direction – perpendicular to the electrode surface.

The current output of the electrode is proportional to the oxygen flux at the cathode surface. Using Fick's second law of diffusion, this model yields the following equations. The current output of the electrode as a function of time is

$$I_t = n F A \frac{P_m}{d_m} P_o \left(1 + 2 \sum_{n=1}^{\infty} (-1)^n \exp \frac{-n^2 \pi^2 D_m t}{d_m^2}\right) \tag{12}$$

The current output of the electrode under steady-state conditions is

$$I_s = n F A \frac{P_m}{d_m} P_o \tag{13}$$

where I_t = transient current output
I_s = steady-state current output
n = number of electrons involved per mole of oxygen reduced
F = Faraday's constant
A = cathode surface area

P_m = oxygen permeability of the membrane
D_m = oxygen diffusivity in the membrane
d_m = membrane thickness
P_o = oxygen partial pressure in the liquid

From Eq. (12) for the transient current output, the time response of the probe can be characterized by a probe constant k as

$$k = \frac{\pi^2 D_m}{d_m^2} \tag{14}$$

A large value for k corresponds to a thin membrane and/or a high oxygen diffusivity in the membrane, leading to a fast probe response. However, such a condition weakens the validity of the assumption that the rate of oxygen transfer is membrane-controlled.

If there is a significant mass transfer resistance in the liquid film around the membrane, the steady-state current output decreases and the probe response time increases. The response time of the probe is taken as the time required to reach a given fraction of the steady-state current following a step change in oxygen concentration. The fractional response t can be determined experimentally using the definition

$$t = 1 - \frac{I_t}{I_s} \tag{15}$$

Response times are shown in Fig. 13 for a platinum electrode.

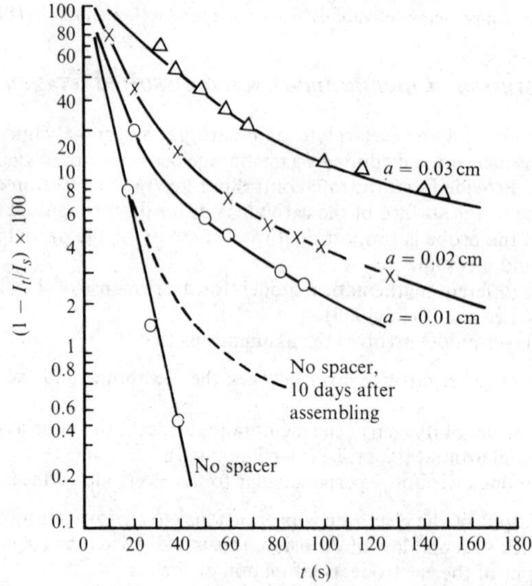

Figure 13. Indicial response of platinum electrode with various thickness of electrolyte layer [Aiba and Huang (1969)].

Preferred characteristics of oxygen electrodes

1) The current output should be large and linear with dissolved oxygen tension.
2) The calibration should be stable over a long period.
3) The response time should be rapid.
4) The effect of hydrodynamic conditions in the liquid on the probe performance should be small.
5) The probe response should be independent of the temperature of the liquid. Fig. 14 demonstrates the variation in the sensitivity of a galvanic electrode with temperature.

6) The probe should withstand high pressure and prolonged autoclaving. The effect of repeated sterilization is shown in Fig. 15.
7) The residual current (i.e. the current output at zero oxygen level) should be small.
8) The polarizing voltage should be stable and correctly chosen (*see* Fig. 11).
9) Oxygen should not back-diffuse from the internal electrolyte.
10) The membrane should be mechanically strong, chemically inert and of low permeability to carbon dioxide.

Use of dissolved oxygen electrodes

The following points should be considered when using dissolved oxygen electrodes.

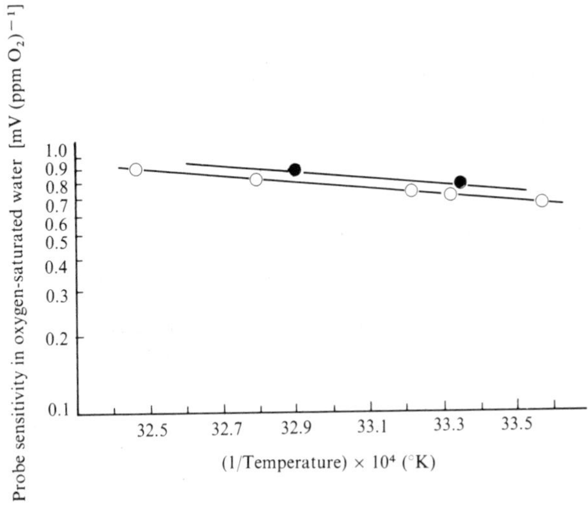

Figure 14. Changes in galvanic–electrode probe sensitivity to oxygen partial pressure with temperature variations [Tuffile and Pinho (1970)].

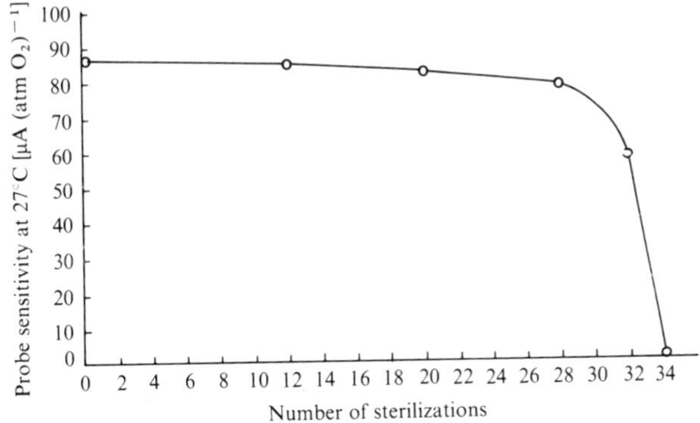

Figure 15. Galvanic electrode dissolved oxygen probe sensitivity following repeated sterilizations [Tuffile and Pinho (1970)].

1) Growth of microorganisms may occur on the outer surface of the membrane.
2) The presence of gases that are reduced at 0.6–1.0 V (e.g., chlorine, bromine, iodine, oxides of nitrogen, etc.) may interfere with the measurements.
3) Hydrogen sulphide, sulphur dioxide and thio-organic compounds have been reported to be poisonous to cathodes.
4) Silver ion deposition on platinum, oxidation of the surface or excessive deposition of silver chloride on the reference electrode may cause the ageing of polarographic probes.
5) Any change in electrolyte concentration due to evaporation or diffusion of water through the membrane may affect probe stability.
6) Response hysteresis may occur (i.e. there may be a marked difference in the response time between the upstep response and the downstep response).
7) Changes in temperature can affect the measurements.
8) Care has to be taken not to include air bubbles inside the electrode chamber when filling the electrolyte solution.
9) Probe sensitivity should be checked frequently.
10) The position of the probe in a fermenter has to be chosen carefully to ensure that a characteristic dissolved oxygen concentration is measured.

Tubing Method

In this method of measuring the dissolved oxygen, a coil of 10 m or more comprising thin-walled Teflon or silicone tubing is immersed in the fermentation broth [*see* Phillips and Johnson (1961)]. A slow, metered stream of pure nitrogen, or other oxygen-free gas, is passed through the tubing. Oxygen diffuses from the broth through the walls of the tubing into the stream of carrier gas. A paramagnetic analyzer is used to monitor the concentration of oxygen in the exit gas, which can be related to the oxygen concentration in the medium. The effects of the length of the tubing and gas velocity on oxygen uptake by the carrier gas are shown in Figs. 16 and 17, respectively.

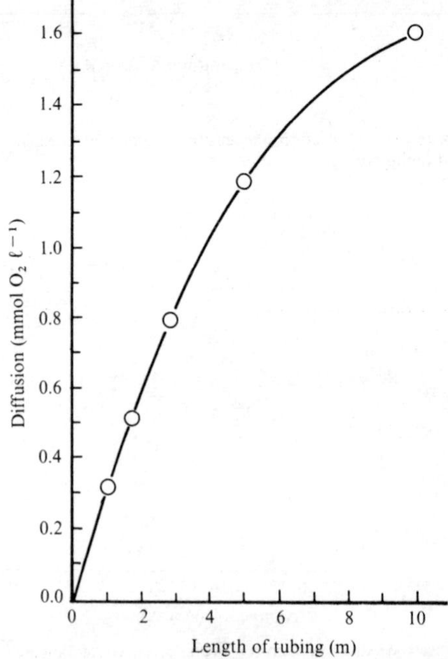

Figure 16. Oxygen diffusion through silicone tubing measured at the end of the tubing by means of an oxygen probe. Gas velocity 20 ml min^{-1} and temperature 25°C [Houkili and Lee (1971)].

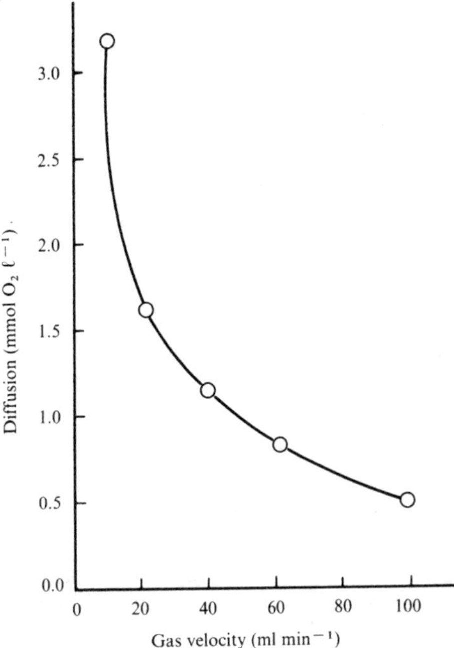

Figure 17. Oxygen uptake by nitrogen in a 10 m silicone tube as a function of gas velocity at 25°C [Houkili and Lee (1971)].

Chemical Methods

Various chemical methods can be used for the calibration of dissolved oxygen electrodes or for the analysis of samples taken from a fermentation medium. These methods give the actual concentration of oxygen as opposed to the oxygen tension as is measured using the dissolved oxygen electrodes.

Winkler method

This method [*see* ASTM Standards (1978)] is used widely for determining the solubility of oxygen in water and waste water. An excess of a standard solution of manganous ions is added to the sample and then the unoxidized manganous ions remaining are back-titrated by an iodometric procedure. It should be noted that fermentation broth usually contains substances that can interfere with this method.

Oxidation of NADH

NADH concentrations are determined spectrophotometrically [*see* Beechey and Ribbons (1972)]. The assay usually involves the use of partially disrupted mitochondria from either liver or heart or NADH oxidase preparations from microorganisms. As an alternative, a nonenzymic method involves the use of phenazonium methosulphate.

Oxidation of phenylhydrazine

This method has been described by Misra and Fridovich (1976).

Other methods

Gas chromatography, mass spectrometry, manometry and volumetric techniques have also been applied to the determination of dissolved oxygen levels [*see* Battino and Clever (1966)].

STIRRED TANKS

Oxygen Transfer Rates

Oxygen transfer rates and efficiencies vary according to the scale of operation of the fermentation (*see* Table 11).

Air flow, impeller speed and tank internals

Oxygen transfer rates OTR increase with air flow (Figs. 18, 19, 21, 27 and 30) and impeller speed (*see* Figs. 20, 21, 22, 23, 26 and 27). Rates are also influenced by the internal arrangements within the tank, e.g., baffles, impeller configuration and size, location of the sparger, etc. (*see* Figs. 23, 26, 27, 28, 29, 30 and 31 and Tables 11 and 12).

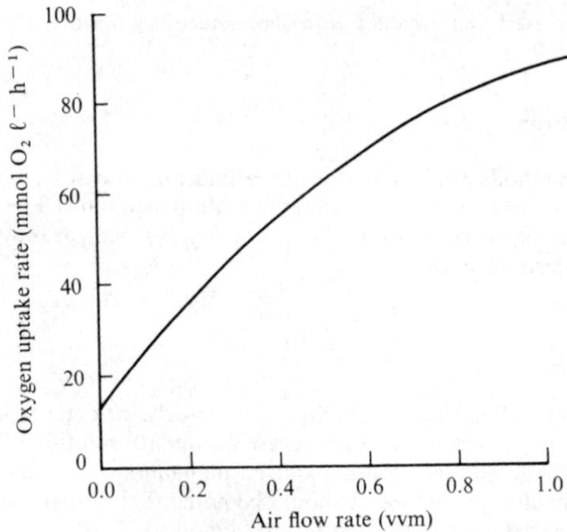

Figure 18. Effect of air sparging on oxygen uptake rate in a baffled stirred tank [Finn (1967)].

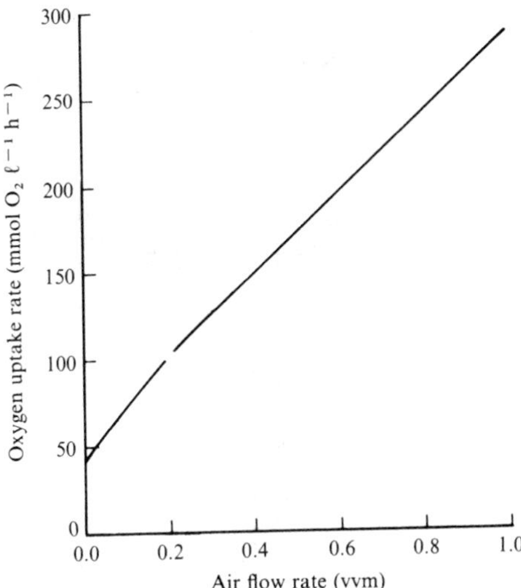

Figure 19. Effect of air sparging on oxygen uptake rate of sulphite solution, using copper catalyst in vortex system, single orifice sparger, with vaned disc impeller (diameter 102 mm) in vessel of 228 mm diameter; liquid volume, 12 ℓ [Finn (1967)].

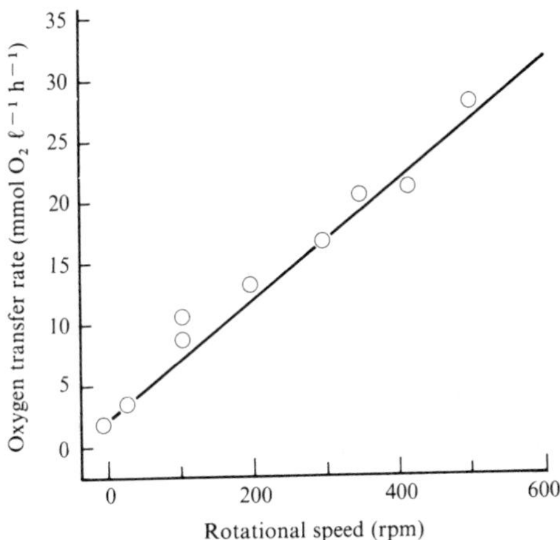

Figure 20. Relation between oxygen transfer rates for *Candida utilis* fermentations and rotational speed in 7.62 cm diameter fermenter [Phillips (1968)].

Air flow is usually expressed as the volumetric rate of air flow per unit volume (i.e. vvm) or as the volumetric rate of air flow per unit tank cross-sectional area (i.e. superficial velocity).

Table 11. Comparison of oxygen transfer rates and efficiencies at various scales of operation.

Liquid volume in fermenter (ℓ)	Type of fermentation system	Oxygen transfer rate ($mMO_2\ \ell^{-1}\ h^{-1}$)	Power per unit volume [hp $(1000\ \ell)^{-1}$]	Superficial gas velocity (cm/min^{-1})	Oxygen transfer efficiency		
					Agitation energy [kWh $(kgO_2)^{-1}$]	Compression energy [kWh $(kgO_2)^{-1}$]	Total energy [kWh $(kgO_2)^{-1}$]
60	Sulphite	60	0.63	32	0.23		0.23
4200	Sulphite	60	2.3	32	0.90	0.03	0.93
42 000	Sulphite	60	2.3	32	0.90	0.05	0.95
20 000	Yeast (newtonian)	22	0.49	60	0.49	0.05	0.54
30 000	Endomyces (nonnewtonian)	22	1.73	60	1.75	0.06	1.81
42 000	Sulphite (newtonian)	22	0.46	60	0.46	0.08	0.55

In addition to illustrating the effects of air flow rate and impeller speed, Fig. 21 shows the increase in the oxygen transfer rate achieved through increased solubility by use of oxygen instead of air. The data suggest no change in k_La but a five-fold improvement in the oxygen transfer rate due to a five-fold increase in the partial pressure and therefore in solubility [*see* Eqs. (4) and (5)]. There is also the suggestion of a limiting impeller speed beyond which no further improvement in k_La occurs.

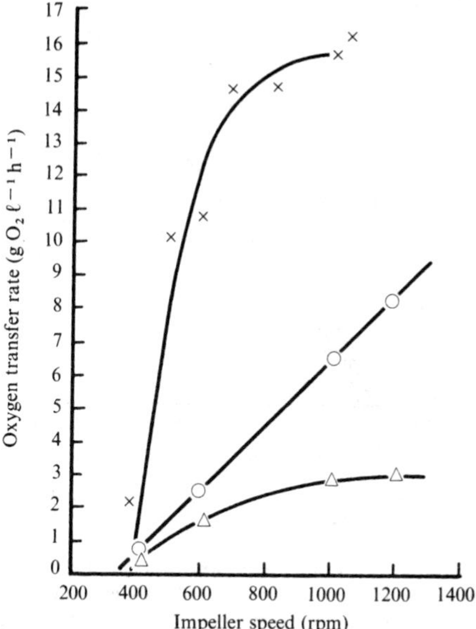

Figure 21. Effect of air flow rate and impeller speed on oxygen transfer rate for a 10.5 cm (1 ℓ) fermenter, single-bore sparger, 5.25 cm impeller, fully baffled. ×, 0.2 ℓ O_2 min^{-1}; ○, 1.0 ℓ air min^{-1}; △, 0.2 ℓ air min^{-1} [Solomons and Perkin (1958)].

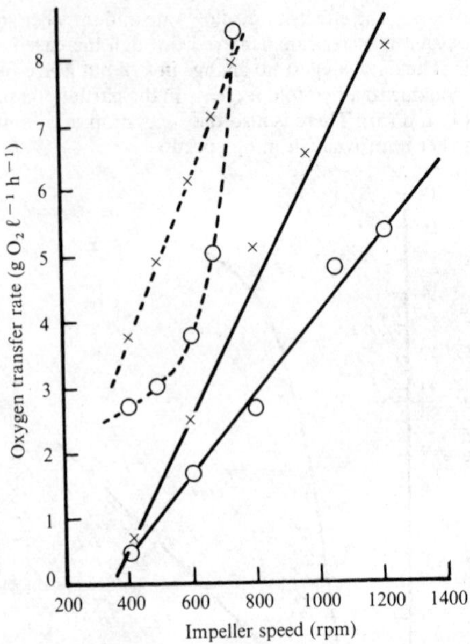

Figure 22. Oxygen transfer rate versus impeller speed. ×, sulphite method; ○, polarographic method; —, 10.5 cm (1 ℓ) fermenter; ...., 20.0 cm (6.3 ℓ) fermenter [Solomons and Perkin (1958)].

Fig. 22 shows that while the sulphite method can detect performance differences between vessel arrangements and can provide a guide for dependencies, such as that between the oxygen transfer rate and impeller speed, care has to be exercised as to the absolute values obtained.

The advantages that can accrue by including baffles in a stirred tank are shown in Fig. 23. The baffles serve to disrupt the vortex pattern that develops around a single-shaft impeller rotating in an unconstrained fluid. The baffles produce a largely planar liquid surface and a more uniform flow pattern, as well as reducing the unproductive gas volume, i.e. baffles increase the liquid hold-up for a given fermenter volume.

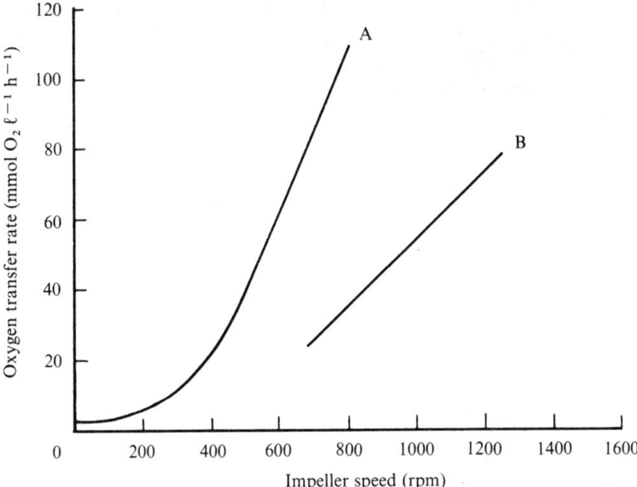

Figure 23. Effect of stirrer rate on oxygen transfer rate in baffled and vortex aeration systems with the same vaned disc impeller in a laboratory fermenter. Curve A, baffled system; curve B, vortex system.

An empirical attempt to correlate the combined effects of air flow rate and impeller speed on oxygen transfer rate for a given system is shown in Fig. 24. The resulting equation is given by

$$\text{Oxygen transfer rate} = 3.43 \times 10^{-6} N^{1.2} G^{0.2} \phi^{-0.5} \tag{16}$$

where the oxygen transfer rate has units of mol O_2 ℓ$^{-1}$ h^{-1}, N is the impeller speed (i.e. rpm), G is the air flow rate (i.e. vvm) and ϕ is the fractional liquid hold-up expressed as a percentage.

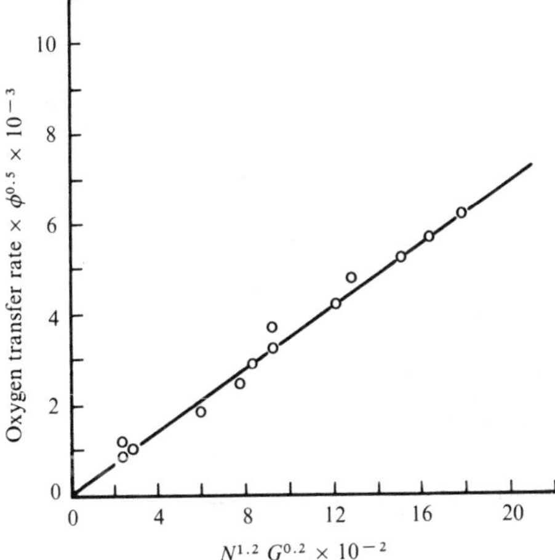

Figure 24. Effects of agitation and air flow on the oxygen transfer rate in a *Pseudomonas ovalis* system [Tsao and Kempe (1960)].

The power required to achieve a given impeller speed in a given system can also be related to the oxygen transfer rate as illustrated in Fig. 25. The empirical equation describing the data in Fig. 25 is expressed by

$$\text{Oxygen transfer rate} = 0.2\, P^{0.8}\, G^{0.2}\, \phi^{-0.5} \tag{17}$$

where, in addition to the variables defined in Eq. (16), P is the power input (i.e. hp ℓ^{-1}).

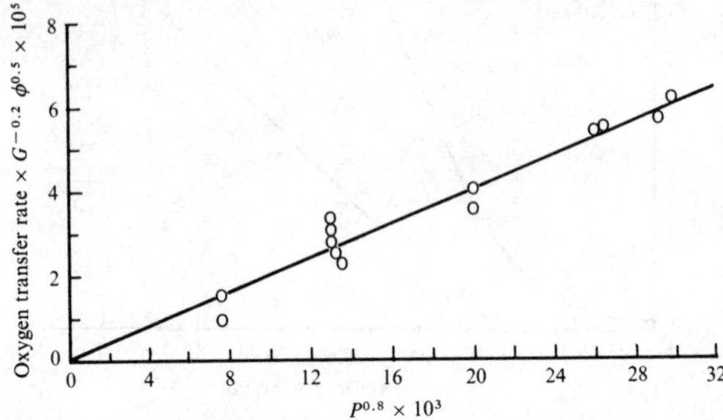

Figure 25. Effects of power input and air flow on the oxygen transfer rate in a *Pseudomonas ovalis* system [Tsao and Kempe (1960)].

Fig. 26 shows that two impellers rotating on the same shaft can increase the oxygen transfer rate by counteracting coalescence and increasing the bubble residence time (*see* Fig. 1) at the expense of increased power consumption.

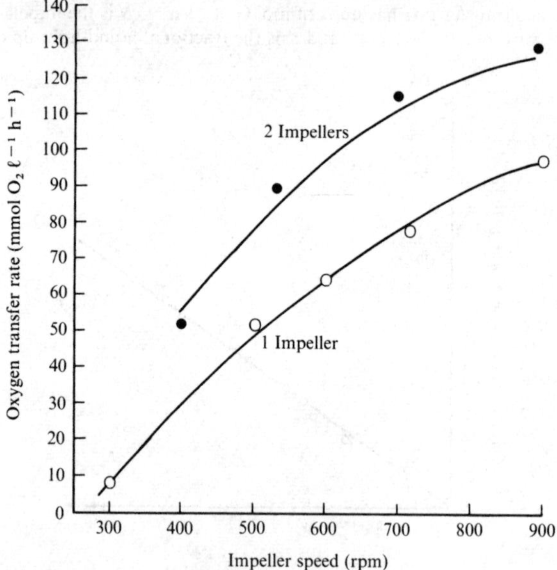

Figure 26. Oxygen transfer rate versus impeller speed for a 15 ℓ fermenter, using *Corynebacterium glutamicum* molasses medium (10 ℓ) [Hunt *et al.* (1971)].

Table 12 and Fig. 27 demonstrate that an increase in the impeller diameter relative to the tank diameter at constant impeller speeds leads to increased oxygen transfer rates — at the penalty of increased power consumption.

Table 12. Effect of turbine size on the oxygen transfer rate, measured by the sulphite method, using a 16 cm diameter vessel [Sato (1961)].

Turbine diameter (cm)	Turbine speed (rpm)	Oxygen transfer rate (mmol O_2 ℓ^{-1} h^{-1})
5.3	850	95
8.0	850	223

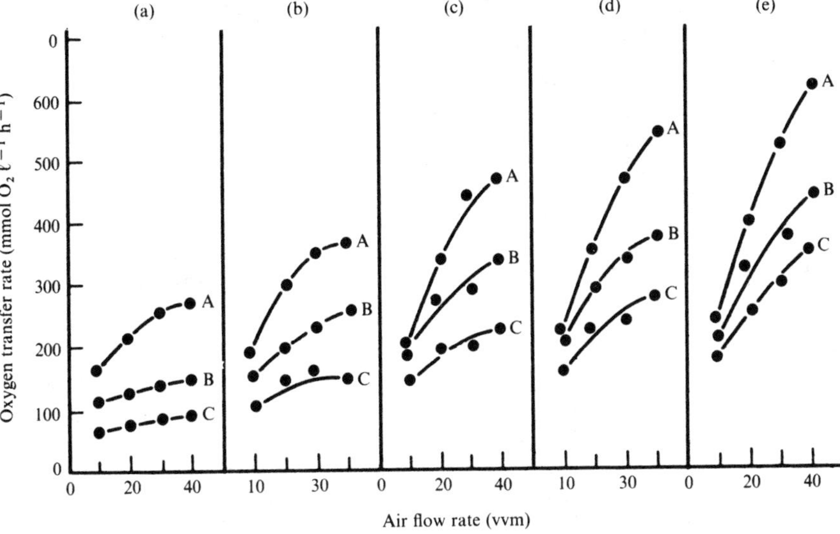

Figure 27. Effects of air flow and impeller speed on the oxygen transfer rate measured by the sulphite method in 20 ℓ baffled fermenters. A, 6 in. impeller; B, 5 in. impeller; C, 4 in. impeller. Impeller speeds: (a) 500 rpm, (b) 620 rpm, (c) 730 rpm, (d) 840 rpm, (e) 955 rpm [Finn (1967)].

It can be seen from Fig. 28 that for a given power input the impeller speed is greater for smaller impellers, leading to larger oxygen transfer rates. Thus Fig. 28 establishes the primacy of impeller speed as regards oxygen transfer rate [*see* Eq. (16)].

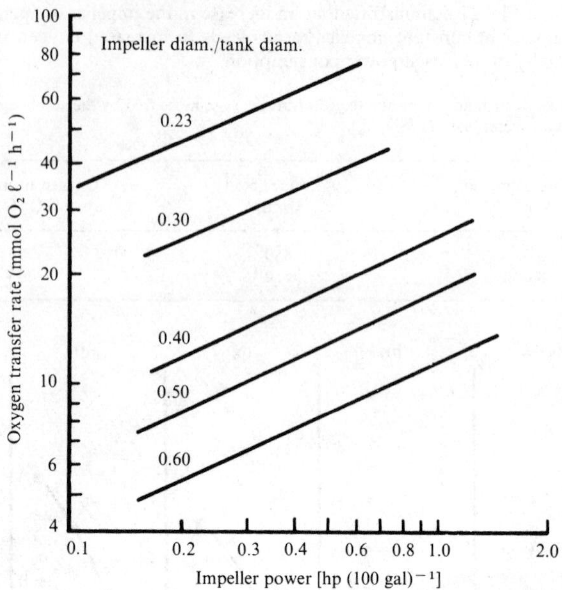

Figure 28. Effect of power on the oxygen transfer rate in novobiocin fermentations [Taguchi (1971)].

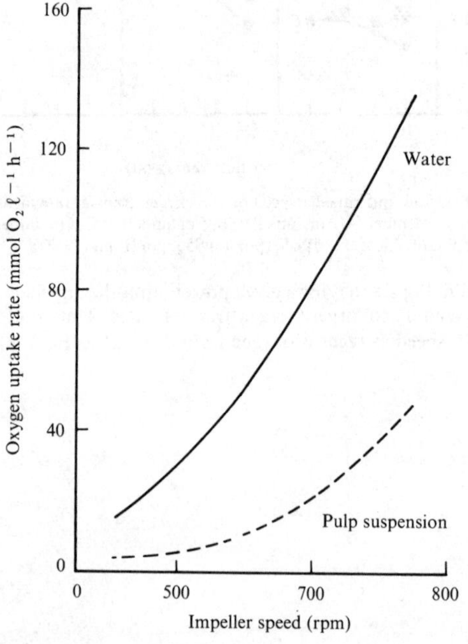

Figure 29. Effect of baffle position on oxygen uptake rate. Air flow rate, 1 vvm. [Blakebrough and Sambamurthy (1964)].

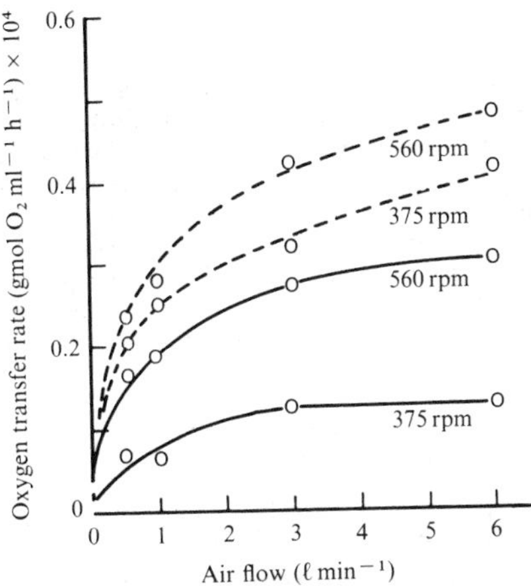

Figure 30. Oxygen transfer rate versus air flow determined in a laboratory fermenter using a galvanic electrode and a streptomycin soya bean medium. ----, sintered sparger; —, constricted pipe [Bartholomew *et al.* (1950)].

Fig. 29 shows that the location of the baffles does not affect the oxygen transfer rate, but that, particularly at lower speeds, the presence and location of the sparger can have an important influence on the oxygen transfer rate achieved (*see* Figs. 30 and 31, respectively).

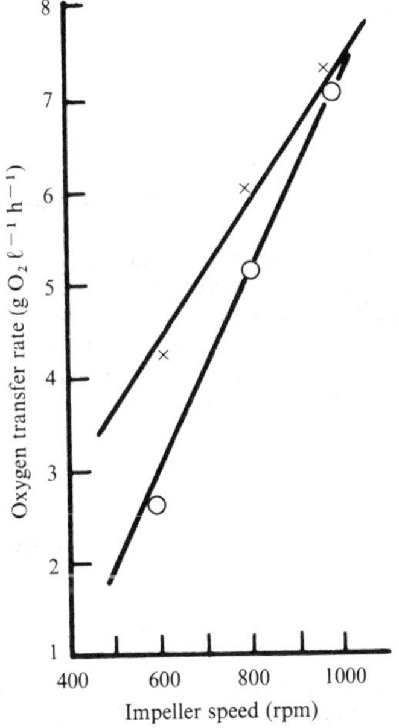

Figure 31. Effect of position of the sparger for a 10.5 cm (1 ℓ) fermenter with single-hole sparger 1 mm diam. (fully baffled). Air flow: 1 ℓ min^{-1}. ×, sparger at the bottom of vessel; ○, sparger directly beneath the impeller [Elsworth *et al.* (1957)].

Microorganisms

The oxygen transfer rate is affected by the microbial species, morphology and concentration. Fig. 32 shows how significantly different oxygen transfer rates can result from similar physical conditions but with different microbial species. Those species with more complex morphology lead to lower oxygen transfer rates. The presence of microorganisms and microbial products affects coalescence and break-up, while morphology influences viscosity. Table 13 demonstrates that although the oxygen demand and oxygen uptake rate are largely independent of *Penicillium* species morphology (i.e. pellets or filaments) the flow behaviour (*see* Chapter 8) is strongly affected and the oxygen transfer rates are greatly reduced at constant power input by the increased viscosity implied.

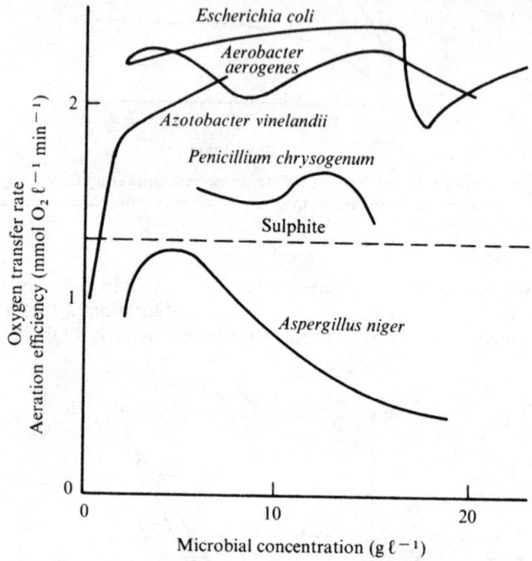

Figure 32. Oxygen transfer rate in a 3.5 ℓ fermenter [Phillips and Johnson (1961)].

Tables 14, 15 and 16 show how the aeration efficiency (i.e. the rate of oxygen transfer per unit power input) is decreased when microorganisms of increasing complexity are used under the same physical conditions. These tables also demonstrate how aeration efficiency is enhanced by the use of larger aspect ratios (i.e. the ratio of liquid height to tank diameter) and by increased numbers of impellers on the shaft. In these tables, an indication is also provided of the relative contributions of power consumption by the impeller and by air compression.

Viscosity

As indicated above, increased viscosity leads to reduced oxygen transfer rates. Studies of the influence of viscosity on oxygen transfer rates have often been carried out using stimulated broth, e.g., using sucrose (*see* Fig. 33) or, more frequently, using paper pulp (*see* Fig. 34). Both Figs. 33 and 34 demonstrate a marked reduction in oxygen transfer rates, such reductions being concentration- and, consequently, viscosity-dependent. Fig. 35 contains data for a pseudo-plastic mould with the viscosity altered by dilution (*see* Chapter 8), showing effects similar to those seen in Figs. 33 and 34.

Table 17 illustrates that slight dilution of a mycelial suspension can result in a considerable increase in both the maximum oxygen transfer rate and the dissolved oxygen concentration. These effects are compatible with a reduction in viscosity and a largely unchanged oxygen demand and uptake rate.

Table 13. Effect of morphological form of mycelium on aeration in penicillin fermentations[a] [Chain *et al.* (1966)].

Mycelial dry weight ($g\,\ell^{-1}$)	Oxygen demand ($ml\,O_2\,\ell^{-1}\,h^{-1}$)	Oxygen uptake ($ml\,O_2\,\ell^{-1}\,h^{-1}$)	Maximum oxygen transfer rate[b] ($ml\,O_2\,\ell^{-1}\,h^{-1}$)	Power consumption ($W\,\ell^{-1}$) Mechanical agitation	Power consumption ($W\,\ell^{-1}$) Mechanical agitation and air compression	Flow behaviour index	Consistency index[d] (cP)
Pellets							
0.34	8.8	7.2	15.0	0.70	3.70	0.4	500
0.33[c]	8.1	7.5	21.3	1.25	4.25		500
Pettels and filamentous hyphae							
0.29	7.4	6.2	9.2	0.65	3.65	0.2	950
0.30	8.5	6.5	11.0	0.70	3.70		700
Filaments							
0.25	7.2	5.4	6.3	0.75	3.75	$\simeq 0$	6400
0.28	7.7	5.0	5.9	0.70	3.70		4100

[a] Ratio of liquid depth to fermenter diameter, 1.2; 2 impellers, 50 cm diameter; impeller speed, 120 rpm.
[b] Calculated from respiration rate (measured by polarographic determination of oxygen levels in outgoing air) and concentration of dissolved oxygen, using Eqs. (1) and (2).
[c] Impeller speed, 140 rpm.
[d] *See* Chapter 8.

Table 14. Aeration and power consumption in penicillin fermentations with *Penicillium chrysogenum*[a] [Chain *et al.* (1966)].

Impeller diameter (cm)	Impeller speed (rpm)	Air flow (vvm)	Maximum oxygen transfer rate[b] (ml O_2, ℓ^{-1} h^{-1})	Power consumption (W ℓ^{-1}) Mechanical agitation	Mechanical agitation + air compression	Aeration efficiency (ℓ O_2 W^{-1} h^{-1})
Fermenter 1[c]						
45	210	1.0	20	1.15	4.15	0.45
45	210	0.5	19	1.25	2.75	0.70
45	210	0.3	13	1.40	2.40	0.50
45	160	1.0	17	0.50	3.50	0.45
45	160	0.5	11	0.60	2.10	0.50
45	160	0.3	10	0.70	1.70	0.60
45	124	1.0	11	0.25	3.25	0.30
45	124	0.5	9	0.30	1.80	0.45
45	124	0.3	6	0.35	1.35	0.45
40	210	1.0	17	0.65	3.65	0.45
40	210	0.5	15	0.70	2.20	0.65
40	210	0.3	12	0.75	1.75	0.65
40	165	1.0	11	0.25	3.25	0.35
40	165	0.5	8	0.30	1.80	0.45
40	165	0.3	6	0.40	1.40	0.45
40	135	1.0	9	0.20	3.20	0.30
40	135	0.5	5	0.25	1.75	0.30
40	135	0.3	4	0.30	1.30	0.35
30	310	1.0	16	0.35	3.35	0.50
30	310	0.5	9	0.40	1.90	0.50
Fermenter 2[d]						
45	140	1.0	62	0.60	3.60	0.15
45	140	0.5	55	0.65	2.17	0.25
45	140	0.3	50	0.70	1.70	0.30
45	170	1.0	9	0.95	3.95	0.25
45	170	0.5	6	1.00	2.50	0.25
45	210	1.0	13	1.80	4.80	0.30
45	210	0.5	11	2.10	3.60	0.30

45	210	0.3	9	2.30	3.30	0.30
50	130–140	1.0	12	1.05	4.05	0.30
50	130–140	0.5	9	1.30	2.80	0.35
50	130–140	0.3	7	1.50	2.50	0.30
30	310	1.0	11	0.65	3.65	0.30
30	310	0.5	8	0.75	2.25	0.35
30	310	0.3	5	0.85	1.85	0.30
45–50	130	1.0	8	0.80	3.80	0.20
Fermenter 3[e]						
45	140	1.0	9	0.50	3.50	0.25
45	140	0.5	9	0.70	2.20	0.40
45	140	0.3	7	0.80	1.80	0.40
45	170	1.0	11	0.85	3.85	0.30
45	170	0.5	8	1.30	2.80	0.30
45	170	0.3	7	1.40	2.40	0.30
45	210	1.0	13	1.45	4.45	0.30
45	210	0.5	8	1.80	3.30	0.25
45	210	0.3	5	1.85	2.85	0.20
35	240	1.0	10	0.65	3.65	0.30
35	240	0.5	8	1.05	2.55	0.35
35	240	0.3	8	1.15	2.15	0.35
50	120	1.0	11	0.75	3.75	0.30
50	120	0.5	7	0.90	2.40	0.30
50	120	0.3	7	0.95	1.95	0.35

[a] Average fermentation time, 30–70 h; average dry weight, 0.25–0.30 g ℓ^{-1}.
[b] Calculated from respiration rate (measured by polarographic determination of oxygen levels in outgoing air) and concentration of dissolved oxygen using Eqs. (1) and (2).
[c] Ratio of liquid depth to fermenter diameter, 2.8; fully baffled; 3 impellers on shaft.
[d] Ratio of liquid depth to fermenter diameter, 1.8; fully baffled; 3 impellers on shaft.
[e] Ratio of liquid depth to fermenter diameter, 1.2; fully baffled; 2 impellers on shaft.

Table 15. Aeration and power consumption in 2-ketogluconic acid fermentations with *Pseudomonas fluorescens*[a] [Chain *et al.* (1966)].

Impeller diameter (cm)	Impeller speed (rpm)	Air flow (vvm)	Maximum oxygen transfer rate[b] (ml O_2, ℓ^{-1} h^{-1})	Power consumption ($W \ell^{-1}$): Mechanical agitation	Power consumption ($W \ell^{-1}$): Mechanical agitation + air compression	Aeration efficiency (ℓ O_2 W^{-1} h^{-1})
Fermenter 1[c]						
40	215	1.0	25	0.60	3.60	0.65
40	215	0.5	20	0.70	2.20	0.90
40	215	0.3	18	0.80	1.80	1.00
40	165	1.0	22	0.25	3.25	0.70
40	165	0.5	18	0.35	1.65	1.00
40	165	0.3	15	0.40	1.40	1.10
40	135	1.0	13	0.25	3.25	0.40
40	135	0.5	10	0.30	1.80	0.55
40	135	0.3	7	0.35	1.35	0.55
30	310	1.0	22	0.35	3.35	0.65
30	310	0.5	19	0.40	1.90	1.00
30	310	0.3	15	0.44	1.45	1.00
45	125	1.0	14	0.30	3.30	0.40
45	125	0.5	11	0.35	1.85	0.60
45	125	0.3	7	0.40	1.40	0.50
Fermenter 2[d]						
45	140	1.0	11	0.55	3.55	0.30
45	140	0.5	8	0.65	2.50	0.35
45	140	0.3	6	0.75	1.75	0.35
45	170	1.0	15	0.95	3.95	0.40
45	170	0.5	10	1.15	2.65	0.35
45	170	0.3	8	1.30	2.30	0.35
45	210	1.0	22	1.70	4.70	0.45
45	210	0.5	15	2.15	3.65	0.40
45	210	0.3	11	2.35	3.35	0.35
45	285	1.0	26	4.70	7.70	0.35
45	285	0.5	21	5.40	6.90	0.30
50	130–140	1.0	15	1.00	4.00	0.35
50	130–140	0.5	11	1.25	2.75	0.40

50	130–140	0.3	8	1.40	2.40	0.35
50	200	1.0	31	3.15	6.15	0.50
50	200	0.5	23	3.75	5.25	0.45
50	200	0.3	15	4.00	5.00	0.30
30	310	1.0	12	0.60	3.60	0.30
30	310	0.5	9	0.70	2.20	0.45
30	310	0.3	8	0.90	1.90	0.45
Fermenter 3[e]						
45	140	1.0	14	0.50	3.50	0.40
45	140	0.5	11	0.70	2.20	0.50
45	140	0.3	10	0.80	1.80	0.55
45	170	1.0	15	0.80	3.80	0.40
45	170	0.5	12	1.20	2.70	0.45
45	170	0.3	11	1.30	2.30	0.45
45	210	1.0	22	1.50	4.50	0.50
45	210	0.5	18	2.25	3.75	0.50
45	210	0.3	17	2.35	3.35	0.50
35	240	1.0	13	0.70	3.70	0.35
35	240	0.5	11	0.90	2.40	0.45
35	240	0.3	10	1.10	2.10	0.50
50	120	1.0	16	0.65	3.65	0.45
50	120	0.5	12	0.85	2.35	0.50
50	120	0.3	10	1.00	2.00	0.50

[a] Average fermentation time, 12 h; average dry weight, 0.25 g ℓ^{-1}.
[b] Calculated from respiration rate (measured by polarographic determination of oxygen levels in outgoing air) and concentration of dissolved oxygen using Eqs. (1) and (2).
[c] Ratio of liquid depth to fermenter diameter, 2.8; fully baffled; 3 impellers on shaft.
[d] Ratio of liquid depth to fermenter diameter, 1.8; fully baffled; 3 impellers on shaft.
[e] Ratio of liquid depth to fermenter diameter, 1.2; fully baffled; 2 impellers on shaft.

Table 16. Aeration and power consumption in yeast fermentations with *Torula utilis*[a] [Chain *et al.* (1966)].

Impeller diameter (cm)	Impeller speed (rpm)	Air flow (vvm)	Maximum oxygen transfer rate[b] (ml O_2, ℓ^{-1} h^{-1})	Power consumption (W ℓ^{-1}) Mechanical agitation	Power consumption (W ℓ^{-1}) Mechanical agitation + air compression	Aeration efficiency (ℓ O_2 W^{-1} h^{-1})
Fermenter 1[c]						
40	215	1.0	39	0.50	3.50	1.05
40	215	0.5	33	0.65	2.15	1.50
40	215	0.3	26	0.75	1.75	1.50
40	165	1.0	36	0.25	3.25	1.10
40	165	0.5	28	0.30	2.80	1.50
40	165	0.3	22	0.40	1.40	1.60
40	135	1.0	22	0.20	3.20	0.70
40	135	0.5	18	0.25	1.75	1.00
40	135	0.3	15	0.30	1.30	1.15
30	310	1.0	37	0.30	3.30	1.10
30	310	0.5	27	0.40	1.90	1.40
30	310	0.3	22	0.45	1.45	1.50
45	125	1.0	25	0.25	3.25	0.80
45	125	0.5	20	0.30	1.80	1.15
45	125	0.3	16	0.35	1.35	1.20
45	210	1.0	54	0.95	3.95	1.40
Fermenter 2[d]						
45	140	1.0	28	0.45	3.45	0.80
45	140	0.5	23	0.60	2.10	1.10
45	140	0.3	13	0.70	1.70	0.75
45	165	1.0	31	0.80	3.80	0.80
45	165	0.5	23	1.00	2.50	0.90
45	165	0.3	10	1.30	2.30	0.45
45	210	1.0	40	1.30	4.30	0.95
45	210	0.5	30	1.50	3.00	1.00
45	210	0.3	17	1.80	2.80	0.60
50	130	1.0	31	0.75	3.75	0.80
50	130	0.5	22	0.95	2.35	0.90

50	130	0.3	16	1.10	2.10	0.75
30	310	1.0	30	0.50	3.50	0.85
30	310	0.5	20	0.65	2.15	0.95
30	310	0.3	17	0.80	1.80	0.90
Fermenter 3[e]						
45	135	1.0	24	0.35	3.35	0.70
45	135	0.5	17	0.40	1.90	0.90
45	135	0.3	14	0.55	1.55	0.85
45	170	1.0	28	0.65	3.65	0.75
45	170	0.5	22	0.75	2.25	0.95
45	170	0.3	18	1.00	2.00	0.85
45	205	1.0	35	1.45	4.45	0.75
45	205	0.5	27	1.85	3.35	0.80
45	205	0.3	22	2.05	3.05	0.70
50	120	1.0	25	0.70	3.70	0.65
50	120	0.5	17	0.80	2.30	0.75
50	120	0.3	15	0.90	1.90	0.80
35	240	1.0	24	0.60	3.60	0.65
35	240	0.5	19	0.75	2.25	0.80
35	240	0.3	17	0.90	1.90	0.90

[a] Average fermentation time, 10–12 h; dry weight, 0.10–0.13 g ℓ^{-1}.
[b] Calculated from respiration rate (measured by polarographic determination of oxygen levels in outgoing air) and concentration of dissolved oxygen using Eqs. (1) and (2).
[c] Ratio of liquid depth to fermenter diameter, 2.8; fully baffled; 3 impellers on shaft.
[d] Ratio of liquid depth to fermenter diameter, 1.8; fully baffled; 3 impellers on shaft.
[e] Ratio of liquid depth to fermenter diameter, 1.2; fully baffled; 2 impellers on shaft.

Table 17. Effect of diluting suspensions of filamentous mycelium on viscosity, oxygen transfer rate and oxygen uptake [Chain *et al.* (1966)].

Dilution (%)	Mycelial dry weight ($g\ \ell^{-1}$)	Concentration of dissolved oxygen (% saturation)	Consistency index (cP)	Oxygen demand ($ml\ O_2\ \ell^{-1}\ h^{-1}$)	Oxygen uptake ($ml\ O_2\ \ell^{-1}\ h^{-1}$)	Maximum oxygen transfer rate ($ml\ O_2\ \ell^{-1}\ h^{-1}$)
0	1.84	0	6500	5.3	2.9	2.9
10	1.74	10	3600	4.9	4.2	4.6
0	2.69	15	4900	6.8	5.5	6.5
15	2.33	47	2000	6.2	5.8	11.0
0	2.72	10	5000	6.8	5.0	5.5
10	2.50	33	2900	6.2	5.0	7.5
0	2.50	10	5000	6.6	4.1	4.5
10		37	3200	6.0	4.5	7.1

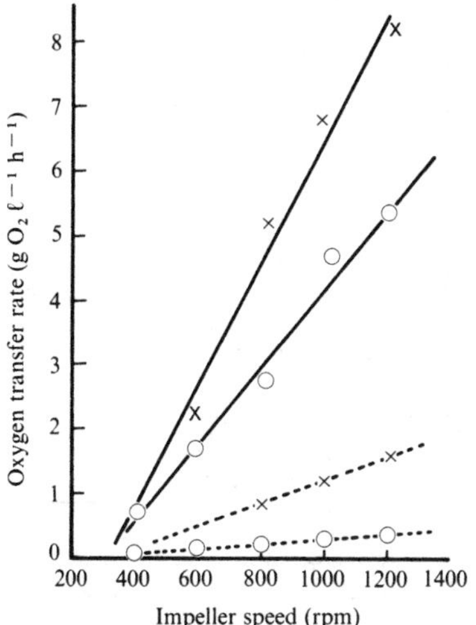

Figure 33. Effect of Newtonian solutions on oxygen transfer rate in a 10.5 cm (1 ℓ) fermenter; single-hole sparger; air flow, 1 ℓ air min^{-1}; 5.25 cm impeller (fully baffled). ×, sulphite method; ○, polarographic method; —, 1 cP solution (water at 20°C); ----, 10 cP solution (45% sucrose at 20°C) [Solomons and Perkin (1958)].

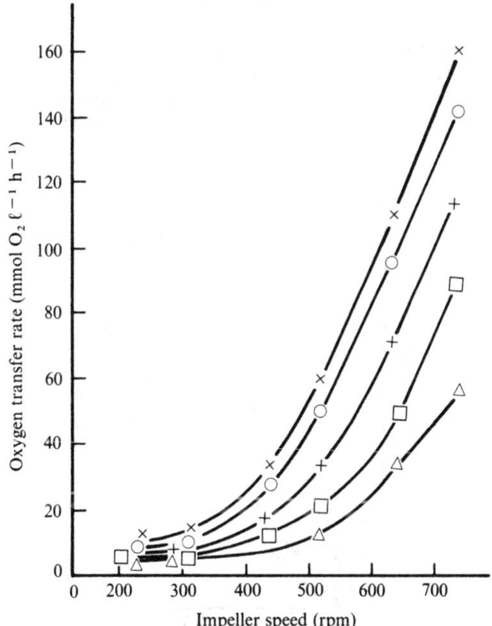

Figure 34. Effect of impeller speed and paper pulp on rate of oxygen transfer. ×, 0.0 per cent; ○, 0.2 per cent; +, 0.4 per cent; □, 0.6 per cent; △, 0.8 per cent [Blakebrough and Hamer (1963)].

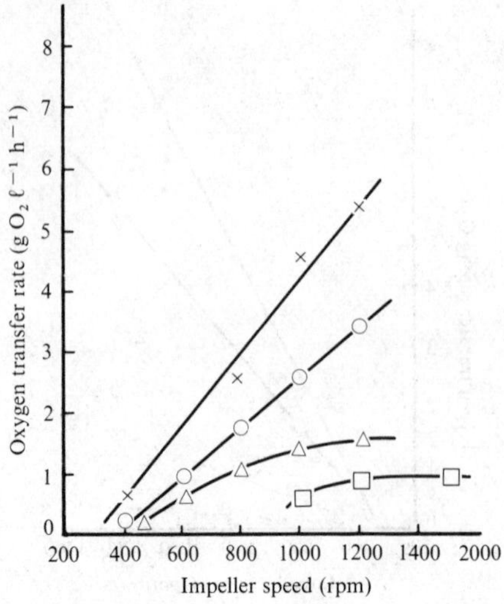

Figure 35. Effect of mould viscosity on oxygen transfer rates in a 10.5 cm (1 ℓ) fermenter; single-hole sparger; air flow, 1 ℓ min^{-1}; 5.25 cm impeller (fully baffled) oxygen measured by the polarographic method. ×, 1 cP (water); ○, mould suspension, viscosity at 1 dyne cm^{-1} s^{-1} 100 cP; △, mould suspension, viscosity at 1 dyne cm^{-1} s^{-1} 1000 cP; □, mould suspension, viscosity at 1 dyne cm^{-1} s^{-1} 10 000 cP [Solomons and Perkin (1958)].

Antifoam agents

Many antifoam agents reduce oxygen transfer rate (*see* Figs. 36 and 37 and Tables 18 and 19). In contrast, detergents usually enhance oxygen transfer rates (*see* Table 19). When antifoam agents and detergents are present simultaneously, they exert compensating effects (*see* Table 19). Oxygen transfer rates depressed due to the presence of antifoam agents result in suppressed dissolved oxygen levels for a given oxygen demand (*see* Table 18).

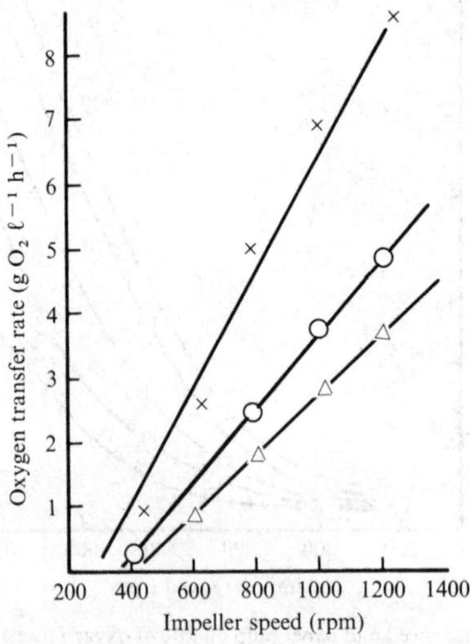

Figure 36. Effect of antifoam agents on oxygen transfer rates for 10.5 cm (1 ℓ) fermenter with single-hole sparger; air flow, 1 ℓ min^{-1}; 5.8 cm impeller (fully baffled) oxygen measured by the sulphite method. ×, control; ○, 2 ml 10% Silicone A in castor oil; △, 2 ml 15% Silicone B in water [Solomons and Perkin (1958)].

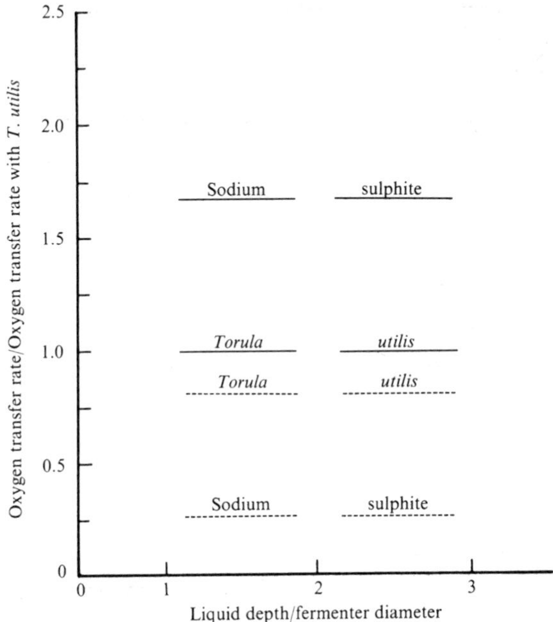

Figure 37. Effect of antifoam agent on aeration rates in suspensions of yeasts and in sodium sulphite solutions. ——, without oil; -----, with oil; [Chain *et al.* (1966)].

Table 18. Effect of antifoam additions on the oxygen uptake rate and dissolved oxygen concentration in penicillin fermentation [Chain *et al.* (1966)].

Mycelial morphology	Oxygen demand (ml O_2 ℓ^{-1} h^{-1})	Oxygen uptake rate (ml O_2 ℓ^{-1} h^{-1})		Concentration of dissolved oxygen (% saturation)	
		Before antifoam addition	After antifoam addition	Before antifoam addition	After antifoam addition
Filamentous	8.2	7.0	4.2	52	0
Filamentous	8.4	6.0	3.6	8	0
Filamentous	9.4	8.2	7.1	5	0
Filamentous	9.3	7.3	6.1	30	0
Filamentous	9.8	8.4	6.8	30	0
Filamentous	8.0	7.3	6.5	10	0
Filamentous	6.4	6.2	5.7	50	5
Filamentous	8.1	7.1	4.4	50	0
Filamentous	9.0	7.2	3.6		0
Filamentous	7.0	7.5	5.8	50	0
Filamentous	8.1	7.6	4.5	50	0
Filamentous	8.4	7.4	4.5	50	0
Filamentous	9.6	7.3	5.9	54	0
Pellets	5.6	5.1	4.6	57	31
Pellets	5.9	5.0	4.8	62	42
Pellets		5.9	4.6	47	21

Table 19. Effect of surface-active and antifoam agents on oxygen transfer rates measured by the sulphite method [Solomons and Perkin (1958)].

Agent	Oxygen transfer rate ($g\ O_2\ \ell^{-1}\ h^{-1}$)	Change (%)
Shake flask, 132 rpm; 2 in. throw		
100 ml in 1 ℓ conical flask		
None	1.70	100
0.02% Sodium lauryl sulphate	3.83	225
0.10% Sodium lauryl sulphate	5.25	309
0.02% Tween 80	1.04	61
0.10% Tween 80	1.75	103
1 ℓ in 5 ℓ bolthead flask		
None	0.43	100
0.02% Sodium lauryl sulphate	1.44	343
1 ℓ fermenter; impeller, 5.8 cm; air flow, 1 vvm		
850 rpm; no baffles		
None	2.11	100
0.02% Sodium lauryl sulphate	4.22	200
0.01% Silicone B	0.89	42
0.02% Sodium lauryl sulphate + 0.01% Silicone B	2.21	104
1000 rpm; baffled		
None	6.16	100
0.03% Silicone B	2.90	47
0.03% Silicone B + 0.01% Sodium lauryl sulphate	3.88	63
0.03% Silicone B + 0.01% Tween 80	2.80	45.5
0.03% Silicone B + 0.01% cationic detergent	2.60	42
0.03% Silicone B + 0.01% calsolene oil	3.04	49.5

Volumetric Mass Transfer Coefficients k_La

Table 20 provides a selection of data on k_La values for a number of vessel configurations at the laboratory and pilot-plant scale. k_La values in this table cover the range 5–2650 h^{-1}, suggesting an equivalent wide range for the maximum oxygen transfer rates based upon Eq. (2). The footnotes to Table 20 illustrate some of the problems associated with the assessment of data contained in the literature.

Table 20. Typical values of k_La for various types of fermenter [Finn (1954)].

Fermentation apparatus	Agitation	Air rate ($m\ h^{-1}$)	k_La[a] (h^{-1})
Shake fermenters			
Respirometer vessel, 25 ml containing 3 ml (liquid depth, 1 mm)	2.4 cm throw, 150 throw min^{-1}		8.3[b]
Conical flask, 1 ℓ containing 300 ml	7 cm throw, 9 throw min^{-1}		24
Conical flask, 250 ml containing 75 ml	3.8 cm eccentric, 220 rpm		26
Conical flask, 500 ml containing 50 ml	3.0 cm eccentric, 210 rpm		90[c]
Conical flask, 500 ml containing 50 ml	2.9 cm eccentric, 253 rpm		~200
Bubble fermenters (unstirred)			
Sewage aerator model (porous plate in tank)	None	18.3[d]	5–9[e]
Glass column, 5 cm diam. (2.5 cm coarse sinter disc)	None	45.7	30
Glass vessel, 2 ℓ 15 cm diam.,	None	18.3	13
single orifice 1.5 mm diam.	None	36.6	22
Glass vessel, 2 ℓ 15 cm diam.,	None	18.3	23
fritted stainless steel 6.3 cm diam. 65 μm openings	None	36.6	30
Single orifice, uniform stream of single bubbles (3 mm diam.)	None	0.0017	0.086

Bubble fermenters (stirred)			
Standard design vessel, 15.2 cm	Single impeller, 7.6 cm diam., 500 rpm [0.5 hp (100 gal^{-1})]	18.3	420
Standard design vessel, 15.2 cm[f]	Single impeller, 10.2 cm		
	500 rpm	18.3	325
	750 rpm	18.3	1000
	1680 rpm	18.3	2650
Standard design vessel, 15.2 cm	Single impeller, 10.2 cm 740 rpm	19.8	420
Standard design vessel, 91.4 cm (120 gallon working capacity)	Single impeller, 25.4 cm 300 rpm	15.2	600
Vessel, 15 000 gallon	0.2 hp gal^{-1}	54.3	370

[a] Values are only approximate. They were often obtained by interpolation of published diagrams. Most experimenters used the sulphite method of evaluation.
[b] Value seems extremely low. The experimental technique may be subject to large errors.
[c] With mycelium present.
[d] Based on the area of the sparger. A superficial velocity cannot be calculated. Air rates three to four times as high are used in activated sludge tanks.
[e] Depends on the permeability rating of the porous plate. The sulphite method used to determine k_La was not well conducted.
[f] Liquid height was only half the vessel diameter. If the height of the liquid had been equal to the diameter of the vessel, k_La would have been about 60 per cent of the value shown.

Table 21 shows the relative benefits of aeration/agitation for a variety of fermenters and conditions compared with the use of the same vessels operated under quiescent conditions.

Table 21. Improvements in k_La values as a result of aeration/agitation compared with quiescent conditions [Bylinkina and Birukov (1972)].

Equipment	Relative k_La value
Apparatus without aeration and agitation	1.0
15 ℓ fermenter; 2-stage impeller (200 rpm)	3.3
15 ℓ fermenter; 2-stage impeller (400 rpm)	5.0
15 ℓ fermenter; 2-stage impeller (600 rpm)	7.8
15 ℓ fermenter; 2-stage impeller (800 rpm)	9.5
100 ℓ fermenter; 2-stage impeller (170 rpm)	2.2
100 ℓ fermenter; air flow, 50 ℓ min^{-1} (170 rpm)	3.6
100 ℓ fermenter, (500 rpm)	4.4
Conical flask; liquid volume, 50 ml (220 rpm)	7.0

Air flow, impeller speed and tank internals

The effects of air flow and impeller speed on k_La values are shown in Figs. 38 and 39, while Fig. 40 demonstrates the effect of the speed of the impeller on the mass transfer coefficient k_L. Increased numbers of impellers can result in greater power requirements without there being any concomitant increase in k_La values (*see* Figs. 41 and 42).

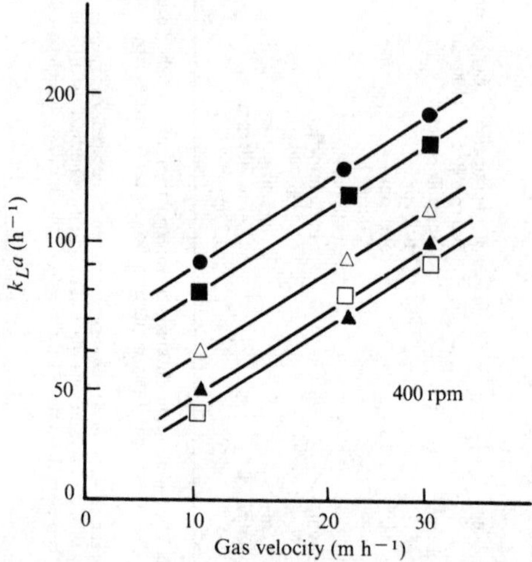

Figure 38. Variation in k_La with gas velocity in a sparged stirred fermenter. ●, water; ■, dead cells (2 g dry wt $ℓ^{-1}$); △, 0.5 ppm Tween 85; ▲, 2.5 ppm Tween 80; □, 140 ppm KM 70 + dead cells (3.1 g dry wt $ℓ^{-1}$) [Yagi and Yoshida (1974)].

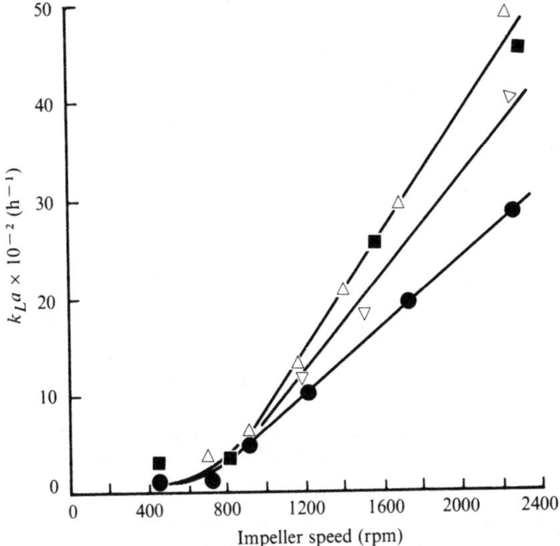

Figure 39. Effect of impeller rotational speed and gas superficial velocity on k_La. Oxygen measured by sulphite oxidation. ●, 0.00375 s^{-1}; ▽, 0.0075 s^{-1}; ■, 0.01125 s^{-1}; △, 0.0150 s^{-1} [Robinson and Wilke (1973)].

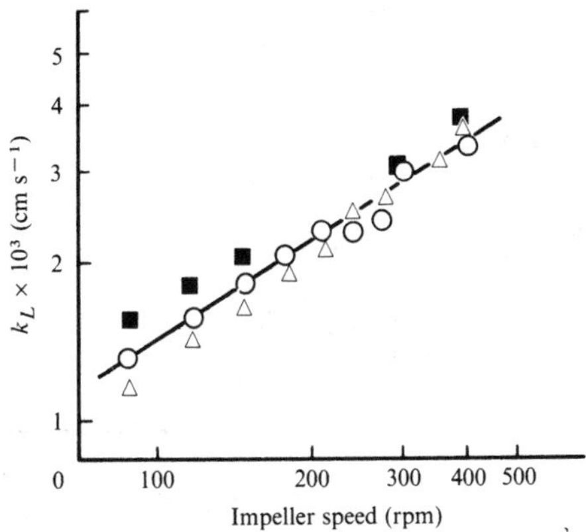

Figure 40. Values of k_L in stirred vessel with known interfacial area plotted against stirrer speed. △, Tween 85 (40 ppm); ■, cells (2.5 g dry wt ℓ^{-1}); ○, pure water [Yagi and Yoshida (1974)].

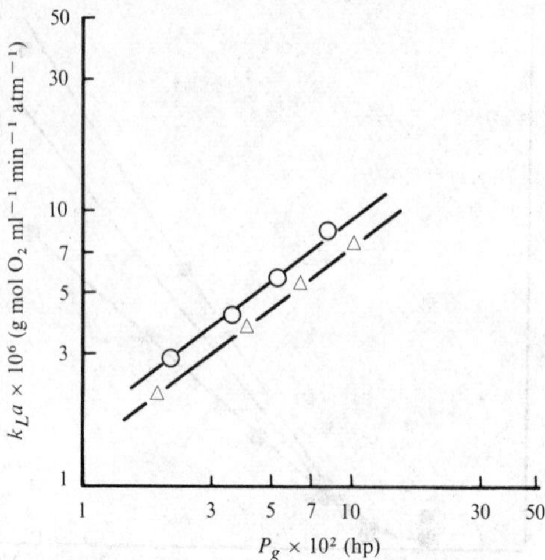

Figure 41. Effect of the number of impeller stages on the volumetric oxygen transfer coefficient k_La in water. Liquid depth/tank diameter = 2.41. ○, 2 impellers, impeller spacing/impeller diameter = 2.66; △, 3 impellers, impeller spacing/impeller diameter = 1.80 [Taguchi (1971)].

(a)

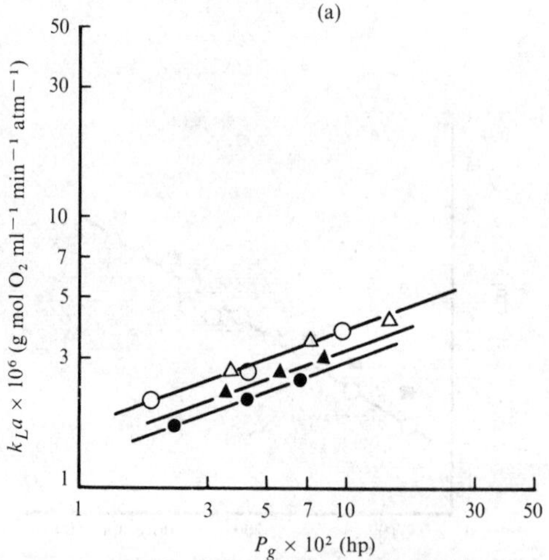

Figure 42. Effect of number of impeller stages on the volumetric oxygen transfer coefficient in viscous mycelial suspensions. Liquid depth/tank diameter = 2.41. △, 3 impellers, impeller spacing/impeller diameter = 1.37, 500 cP; ○, 2 impellers, impeller spacing/impeller diameter = 2.06, 500 cP; ▲, 3 impellers, impeller spacing/impeller diameter = 1.37, 700 cP; ●, 2 impellers, impeller spacing/impeller diameter = 2.06, 700 cP [Taguchi (1971)].

Table 22 provides some information on the relative positions of the impeller and sparger without affect on k_La under mycelia-free and mycelia-containing conditions.

Table 22. Effects of impeller and sparger position on oxygen absorption in the absence or presence of mycelium (0.15 per cent dry weight)[a] [Brierly and Steele (1959)].

Distance of sparger from base (cm)	k_La (h^{-1}) Impeller 5 cm from base		Impeller 6.98 cm from base		Impeller 8.89 cm from base	
	Without mycelium	With mycelium	Without mycelium	With mycelium	Without mycelium	With mycelium
2.22	315	174	320	178	320	175
4.13	310	182	320	178	320	167
6.03					315	175

[a] Impeller speed, 800 rpm; air flow rate, 2.5 ℓ min^{-1}.

Both k_La and a can be correlated by the empirical equations

$$k_La = \alpha_1 V_g^{a_1} N^{b_1} \left(\frac{P}{V}\right)^{c_1} \tag{18}$$

$$a = \alpha_2 V_g^{a_2} N^{b_2} \left(\frac{P}{V}\right)^{c_2} \tag{19}$$

where V_g, N and P/V are the air flow, the impeller speed and the power consumption per unit volume, respectively. Tables 23 and 24 provide values for the coefficients in Eqs. (18) and (19), respectively, for a variety of tanks and impellers. In Table 25, values of the coefficients for given geometries and a variety of additives are listed.

Table 23. Empirical correlations of the volumetric mass transfer coefficient k_La [Resnick and Gal-Or (1968)].

Liquid volume, L (ℓ)	Tank diameter, D_T (m)	$\frac{L}{D_T}$ (ℓ m^{-1})	Impeller Type	Number of blades	Value of exponent of V_g (a_1)	N (b_1)	P/V (c_1)
2.7, 67	0.15, 0.44	0.4	Vaned disc	16	0.67		0.95
11, 11 000	0.24, 2.44	0.25	Paddle	2	0.67		0.57
1.9	0.13	0.7	Turbine	2	0.75	1.67	
0.8	0.10	0.5	Turbine	6	0.21	2.04	
0.8	0.10	0.5	Turbine	6	0.13	2.40	
0.2	0.06	0.47	Vaned disc	4	0.14	2.17	
1.4	0.12	0.30	Paddle	2	0.05	1.79	
1.4	0.12	0.30	Paddle	4	0	1.26	
2.7	0.15	0.20, 0.5	Paddle	4	0	3.00	
1.5	0.13	0.5	Turbine	6	0.68	2.00	
2.7, 41	0.15, 0.38	0.40	Turbine	12	0.67	2.00	
2.7, 41	0.15, 0.38	0.40	Vaned disc	16	0.40–0.84	1.29–2.05	
1.5			Propeller		0.40	1.70	
20	0.33	0.30, 0.46	Turbine	4		2.4	
26,600	3.05	0.20	Turbine	8	0.40		0.53
1.9	0.13	0.7	Turbine	8	1.00	3.00	
3.95	0.16, 0.50	0.24-0.39	Various				0.43–0.95
22.3	0.31	0.33	Turbine	6			0.78
6.3–98	0.2–0.5	0.1–0.63	Turbine		0.33–0.75	< 1.55	

Table 24. Empirical correlations of the specific interfacial area a [Resnick and Gal-Or (1968)].

Liquid volume, L (ℓ)	Tank diameter, D_T (m)	$\frac{L}{D_T}$ (ℓ m^{-1})	Impeller Type	Impeller Number of blades	Value of exponent of V_g (a_2)	Value of exponent of N (b_2)	Value of exponent of P/V (c_2)
12.3, 104	0.25, 0.51	0.6	Paddle	4		1.5	
6.3, 104	0.20, 0.51	0.33	Turbine	6	0.50		0.40
6.3, 104	0.21, 0.51	0.33	Turbine	6	0.50		0.40
12.3, 155	0.25, 0.58	0.40	Turbine	12	0.75	1.1	
12.3, 155	0.25, 0.58	0.40	Vaned disc	16		0.7–0.9	
44	0.38	0.33, 0.40	Paddle	5	0.60		0.35
2.2–570	0.14–0.90	0.2–0.7	Turbine	6	0[a]	1.0	
2.7–170		0.2–0.7	Turbine	4	0[a]	1.0	
5.5	0.19	0.5–0.7	Paddle	2	0[a]	1.0	
5.5	0.19	0.4–0.7	Propeller	3	0[a]	1.0	

[a] At agitation rates above some minimum agitation rate.

Table 25. Table of $k_L a$ correlations for stirred tanks with six-blade turbines using Eq. (18). Impeller diameter/tank diameter, 0.333 [Moo-Young and Blanch (1981)].

Liquid	Method	Tank diameter, D_T (m)	$P/V \times 10^{-2}$ ($W\,m^{-3}$)	$V_g \times 10$ ($m\,s^{-1}$)	a_1	c_1	$\alpha_1 N^{b_1}$
Water	Physical absorption/desorption (Winkler method)	0.15–0.5	2.6–5.3	0.3–1.8	0.5	0.4	0.024
Electrolyte solutions	Physical absorption/desorption (Winkler method)	0.15–0.5	2.6–5.3	0.3–1.8	0.26	0.74	0.018
Water	Physical absorption/desorption (oxygen probe)	0.15	0.3–180	0.1–0.5	0.35	0.4	
Potassium chloride (0.22 N)	Physical absorption/desorption (oxygen probe)	0.15	0.3–180	0.1–0.5	0.36	0.71	
Potassium chloride (0.10 N)	Physical absorption/desorption (oxygen probe)	0.15	0.3–180	0.1–0.5	0.62	0.63	
Water	Physical absorption/desorption (oxygen probe)	0.15	4.4–100	0.37–1.11	0.43	0.42	0.0275
Sodium sulphate + potassium hydroxide	Physical absorption/desorption (oxygen probe)	0.15	4.4–100	0.37–1.11	0.43	0.52	0.017

Microorganisms

The effects of the presence of microorganisms on k_La are likely to result from physical factors, such as viscosity, and the presence of metabolic products rather than the fact that the microorganisms are 'alive' per se [*see* Fig. 43].

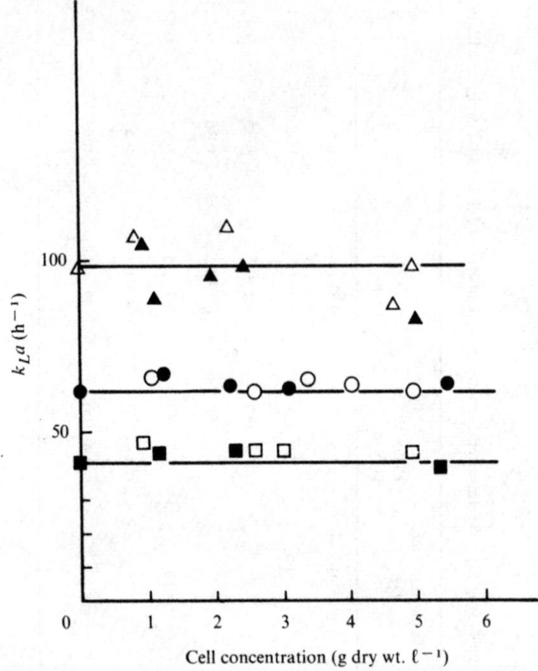

Figure 43. Values of k_La in sparged stirred fermenter, effects of concentrations of living and sterilized microorganisms. Impeller speed, gas velocity, viability of cells: △, 480 rpm, 20.4 m h^{-1}, living cells; ▲, 480 rpm, 20.4 m h^{-1}, dead cells; ○, 400 rpm, 20.4 m h^{-1}, living cells; ●, 400 rpm, 20.4 m h^{-1}, dead cells; □, 400 rpm, 10.2 m h^{-1}, living cells; ■, 400 rpm, 10.2 m h^{-1}, dead cells [Yagi and Yoshida (1975)].

Fig. 44 shows a strong effect on k_La caused by the viscous nonnewtonian characteristics which are a feature of mycelial suspensions (*see* Chapter 8), while Fig. 45 illustrates how paper pulp can be used to simulate the fermentation broth.

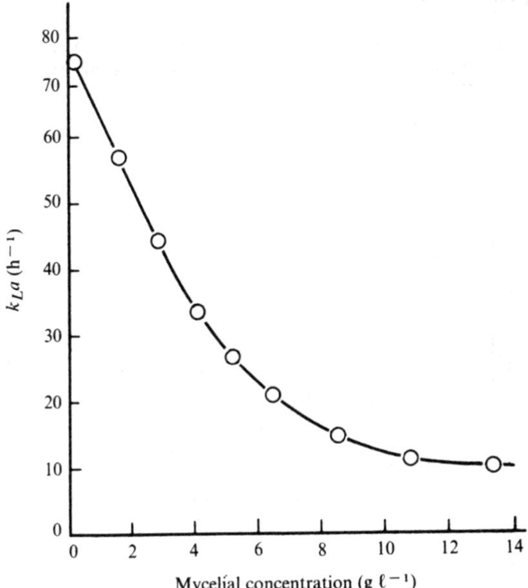

Figure 44. Effect of mycelial concentration of *Penicillium chrysogenum* on the absorption rate of a typical 5 ℓ stirred fermenter [Finn (1954)].

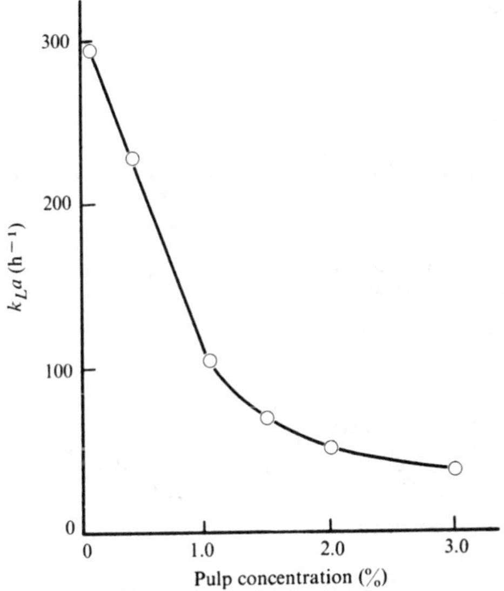

Figure 45. Effect of paper pulp concentration on the rate of oxygen absorption. Impeller speed, 800 rpm; air flow, 2.5 ℓ min^{-1} [Brierly and Steele (1959)].

Antifoam agents

Fig. 46 demonstrates that the surface tension is decreased due to addition of antifoam, resulting in reduced k_La values (*see* Fig. 3). Fig. 47 suggests that these phenomena are likely to occur over the time course of a fermentation but to a variable degree.

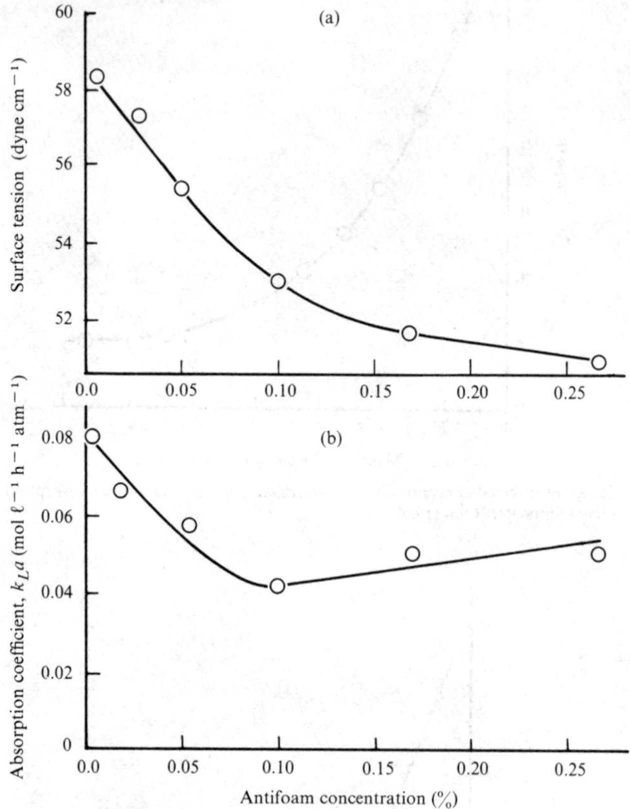

Figure 46. Effect of antifoam agent concentration on (a) surface tension and (b) absorption coefficient of 62 h filtered broth [Deindorfer and Gaden (1955)].

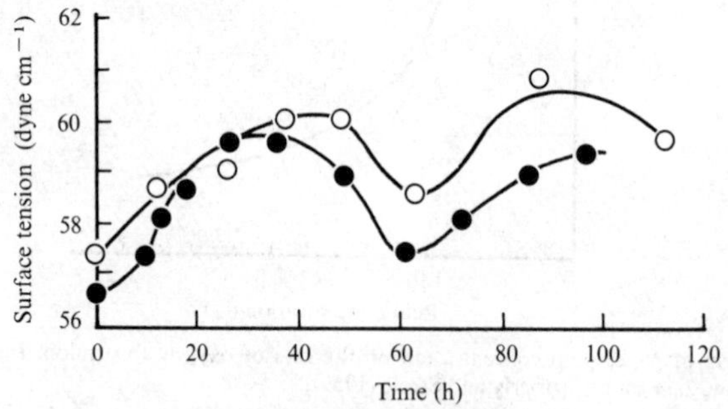

Figure 47. Variation in the surface tension of filtered fermentation broth [Deindorfer and Gaden (1955)].

A linear relationship seems to exist between k_La and the gas hold-up (*see* Fig. 48) similar to that suggested by Eq. (3) with the k_La values depressed by increased surfactant concentrations. Fig. 49 shows that presence of microorganisms, in this case *Candida tropicalis*, did not significantly affect the relative k_L values in a stirred vessel.

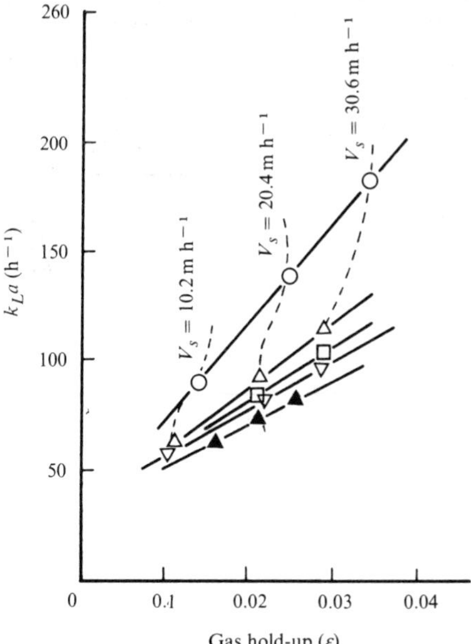

Figure 48. Relationship between k_La and gas hold-up in a sparged, stirred fermenter for various surfactant concentrations. Impeller speed, 400 rpm. Surfactant concentration: ○, 0 ppm; △, 0.5 ppm; □, 1.0 ppm; ▽, 2.5 ppm; ▲, 5.0 ppm [Metz *et al.* (1979)].

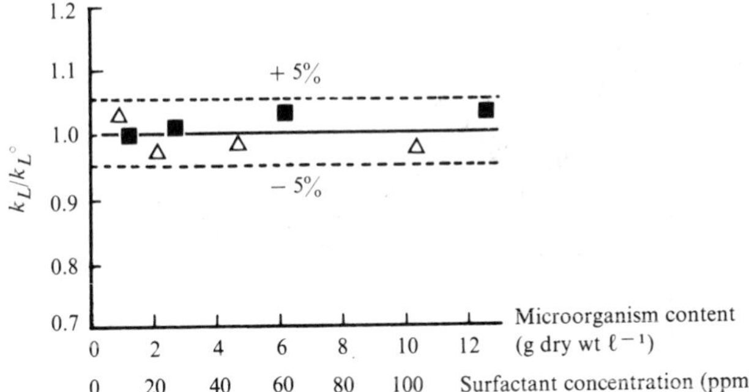

Figure 49. Relative values of k_L in stirred vessel with known interfacial area, effects of surfactant and the presence of microorganisms. Impeller speed, 180 rpm; ■, *Candida tropicalis*; △, Tween 85 [Metz *et al.* (1979)].

Variation during fermentation

Table 26 shows that the k_La of filtered broth, i.e. broth free of viscous effects due to mycelium, can be largely steady over the time course of a fermentation. In contrast, Figs. 50–53 cover a variety of fermentations and demonstrate that the actual k_La values within a fermenter can vary markedly as the fermentation proceeds.

Table 26. Absorption coefficient of filtered broth during fermentation [Deindorfer and Gaden (1955)].

Fermentation time (h)	Absorption coefficient, k_La (g mol O_2 ml^{-1} h^{-1} atm^{-1})
0	0.89
12	0.88
24	0.87
36	0.85
48	0.82
62	0.84
86	0.88
111	0.88

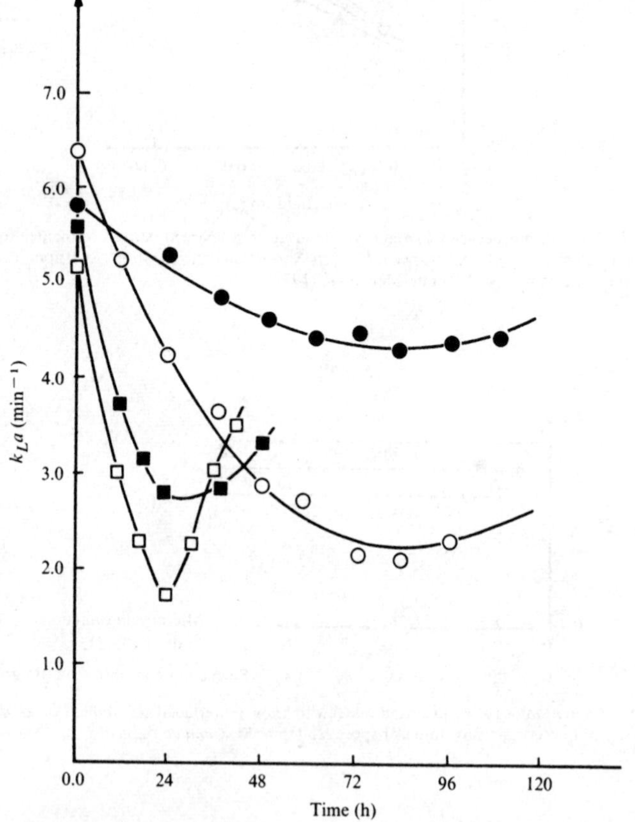

Figure 50. Variation of k_La values during primary metabolite fermentations. ●, cellulose; ○, acid protease; ■, neutral protease; □, alkaline protease [Jarai (1972)].

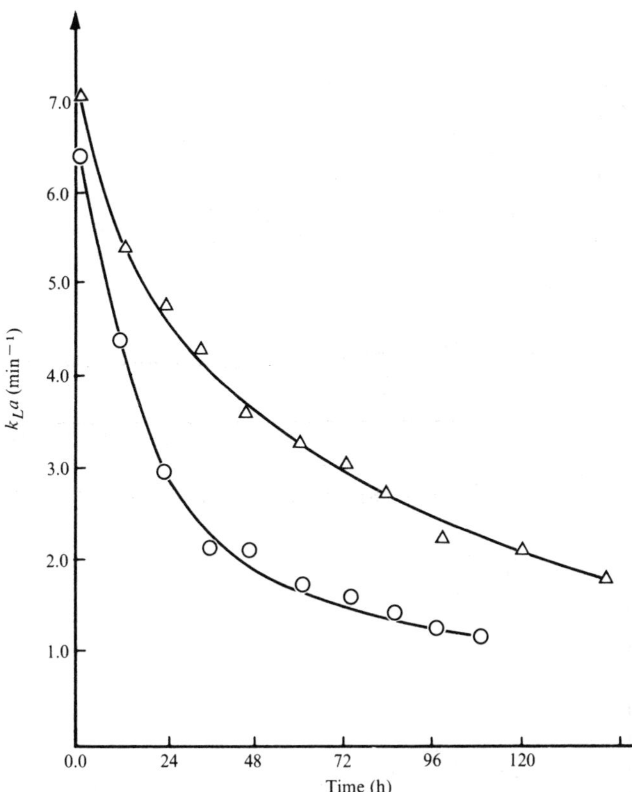

Figure 51. Variation of k_La values during secondary metabolite fermentations. △, fumagillin; ○, nystatin [Jarai (1972)].

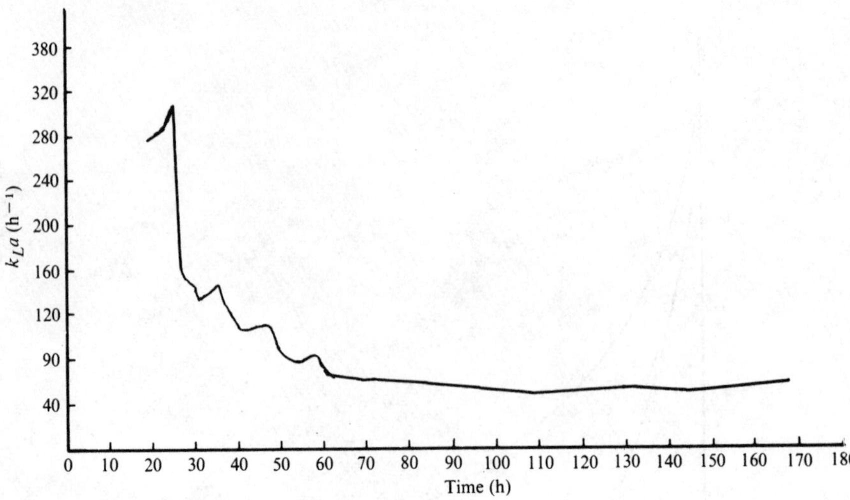

Figure 52. Time course of k_La for a 300 ℓ streptomycete fermentation [Tuffile and Pinho (1970)].

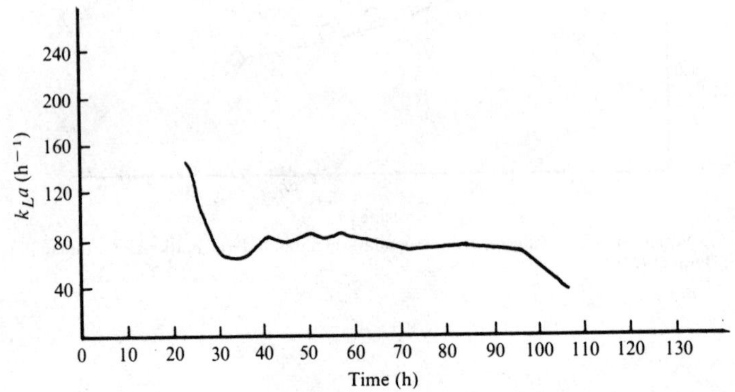

Figure 53. Time course of for a 300 ℓ streptomycete fermentation [Tuffile and Pinho (1970)].

OTHER CONFIGURATIONS

Shakers

Tables 27 and 28 and Figs. 54 and 55 show the characteristic effects of liquid volume, flask size, shaking action, the presence of plugs and cell concentration.

Table 27. Oxygen transfer rates in open-top, conical flasks for a reciprocating shaker measured by the sulphite method [Solomons (1969)].

		Oxygen transfer rate (mM O_2 ℓ^{-1} h^{-1})							
		86 cycles min^{-1} Amplitude (cm)				120 cycles min^{-1} Amplitude (cm)			
Liquid volume	Flask volume	2	4	6	8	2	4	6	8
50	500	11.2	18.0	26.6	42.1	29.3	54.6	83.7	109.8
100	500	4.0	7.2	11.5	23.0	22.3	37.2	57.8	81.8
100	1000	14.2	25.3	34.1	53.7	18.1	50.6	61.9	80.0
200	2000	4.1	11.8	17.2	31.2	22.4	32.9	40.4	59.4

Table 28. Effect of closures on oxygen transfer rate of shaker flasks. Reciprocating shaker; amplitude, 5 cm; 120 cycles min^{-1}, 100 ml sulphite, 1 ℓ plain conical flask [Solomons (1969)].

	Oxygen transfer rate (mmol O_2 ℓ$^{-1}$ h^{-1})	Reduction (%)
Control, open top	53.7	
Plugged with cotton wool	23.8	56
Covered with lint top	29.8	44.5

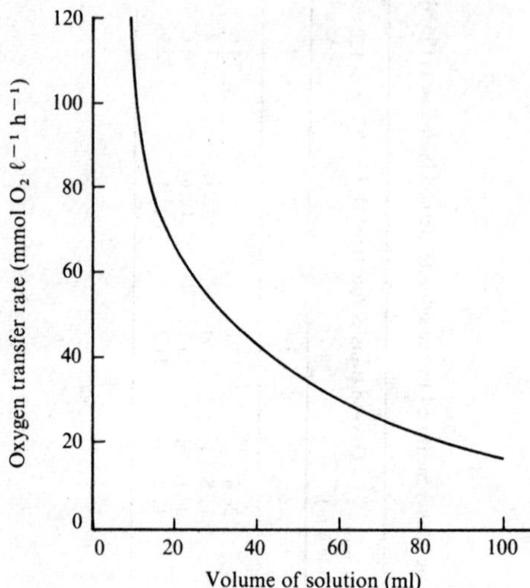

Figure 54. Effect of liquid volume on oxygen transfer rate of sulphite solution (copper catalyst) in a shake flask. Flask capacity, 500 ml; rotary shaking at 250 rpm; throw, 50 mm [Pirt (1966)].

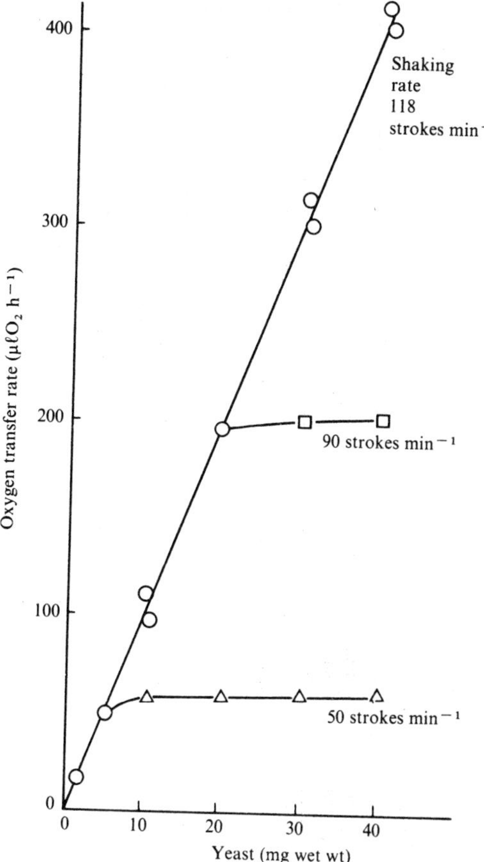

Figure 55. Influence of shaking rate, strokes per minute, on oxygen transfer for 3 ml of yeast suspension in Warburg flasks [Finn (1967)].

Tower (Bubble Column) Fermenters

In Fig. 56, the characteristic logarithmic relationship between $k_L a$ and the gas flow rate in a tower fermenter is shown. Table 29 provides a summary of the coefficients in the relationship

$$k_L a = \alpha V_g^b \tag{20}$$

where V_g is the superficial gas velocity expressed in terms of metres per second. The values of β are in the region of unity, whereas the values of α depend on the nature of the additives.

Table 29. Summary of k_La correlations for bubble columns using Eq. (20) [Moo-Young and Blanch (1981)].

Liquid	Sparger type	Column diameter (cm)	Column length (m)	Gas velocity ($cm\,s^{-1}$)	Flow model	α	β
Sulphite (0.1–0.4 N)	Porous plate	6.3	0.96	0.2–2.8	Well mixed		1.00
		15.2	1.92	0.2–2.8			1.00
Sulphite (0.3–1.0 N)	Multiorifice	10	1.53	3.0–22.0	Well mixed		1.00
		15	1.53	3.0–22.0			1.00
Sulphite (0.3 N)	Single orifice	7.7–60	0.9–3.5	11.7	Well mixed	0.42	0.90
Water	Single orifice	15.2	4.00	3.0–22.0	Well mixed	0.24	0.90
Water	Multiorifice	20	7.23	0.2–9.0	Axial dispersion	0.73	0.96
Sodium sulphate (0.7 N), sodium chloride (0.17 N), sodium sulphate (0.225 N)	Multiorifice	20	7.23	0.2–9.0	Axial dispersion	0.75	0.89

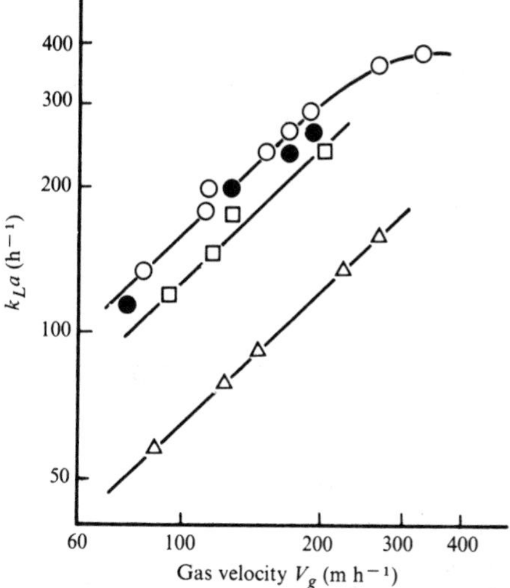

Figure 56. Values of k_La in a tower fermenter plotted against superficial gas velocity. ○, tap water; ●, distilled water; □, *Candida tropicalis* (1.1 g ℓ$^{-1}$); △, Tween 85 (10 ppm) [Yagi and Yoshida (1974)].

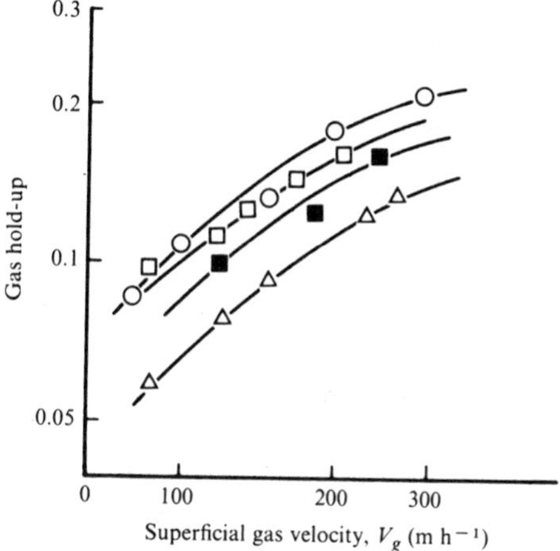

Figure 57. Fraction gas hold-up in a tower fermenter plotted against superficial gas velocity. ○, tap water; □, *Candida tropicalis* (2.8 g ℓ$^{-1}$); ■, *C. tropicalis* (6.2 g ℓ$^{-1}$); △, Tween 85 (10 ppm) [Yagi and Yoshida (1974)].

Since $a = 6\,\varepsilon/d_b$, where d_b is the bubble diameter, and assuming d_b is essentially constant, the dependencies of k_La and gas hold-up on the gas velocity in Figs. 56 and 57 suggest that k_L is largely independent of the gas velocity as shown in Fig. 58.

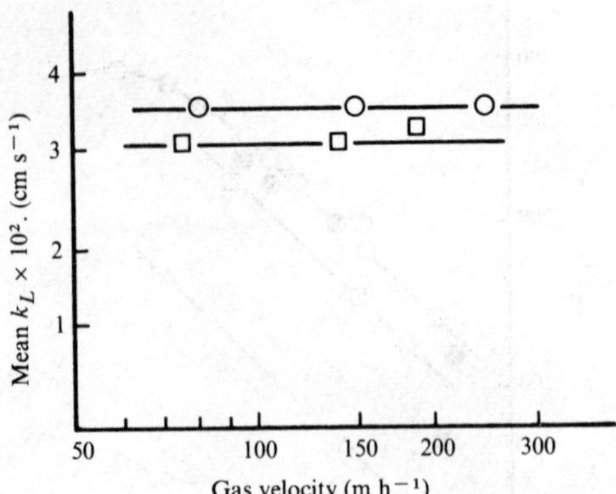

Figure 58. Values of k_L in tower fermenter plotted against superficial gas velocity. ○, tap water; □, cell suspension (0.4 g ℓ$^{-1}$) [Yagi and Yoshida (1974)].

The effects of cell concentrations and of antifoam on k_La and gas hold-up values are shown in Figs. 59 and 60, respectively.

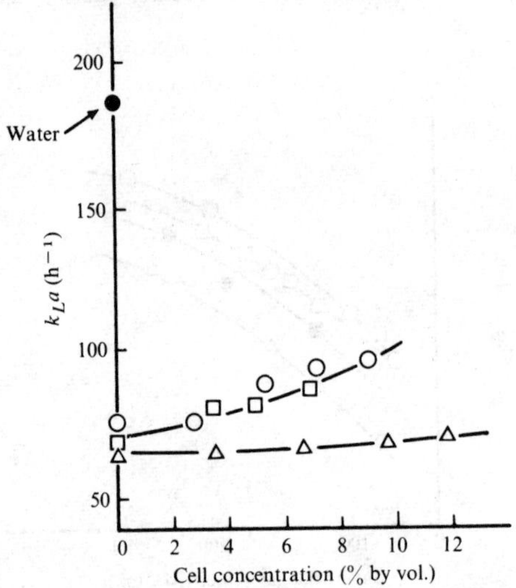

Figure 59. Values of k_La in tower fermenter, effects of concentrations of antifoam agent and sterilized cells. Antifoam agent concentration: □, 10 ppm; ○, 50 ppm; △, 100 ppm [Yagi and Yoshida (1974)].

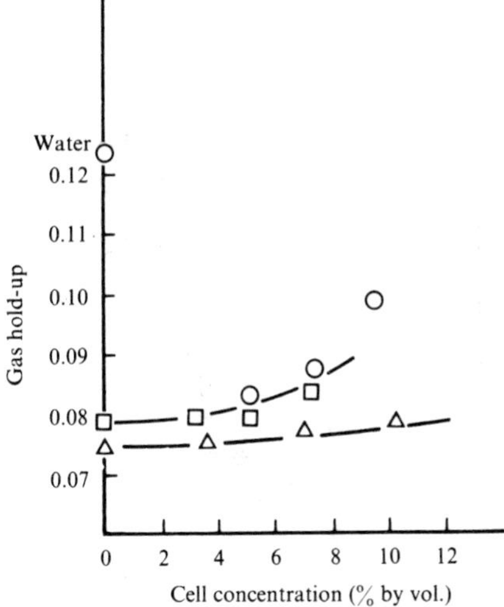

Figure 60. Fractional gas hold-up in tower fermenter, effects of concentrations of antifoam agent and sterilized cells. Antifoam agent concentration: □, 10 ppm; ○, 50 ppm; △, 100 ppm [Yagi and Yoshida (1974)].

Fluidized Beds

Figs. 61 and 62 show the $k_L a$ values for bubble and fluidized beds containing 1 mm and 6 mm particles. Fig. 61 demonstrates that beds containing 1 mm particles are inferior to bubble columns, while beds of 6 mm particles are superior. The reasons for this phenomenon lie with the interactions between particles and the large bubbles produced by coalescence. Local values can vary with axial position in fluidized beds as a result of bubble/particle interactions (*see* Fig. 63).

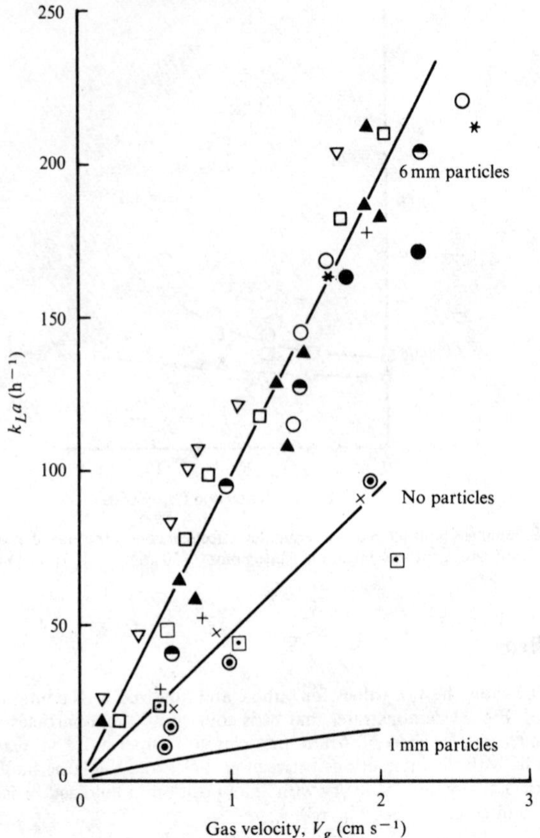

Figure 61. k_La values for bubble columns and gas–liquid fluidized beds of 1 mm and 6 mm ballotini. Percentage carbon dioxide, liquid velocity (cm s^{-1}): ▽, 100 per cent, 6.8; □, 100 per cent, 8.4; ✱, 50 per cent, 11.9; ▲, 80 per cent, 11.9; †, 100 per cent, 11.9; ○, 50 per cent, 17; ◓, 80 per cent, 17; ●, 100 per cent, 17; +, 100 per cent, 6.8; ×, 100 per cent, 8.4; ◉, 100 per cent, 11.9; □, 100 per cent, 17 [Østergaard and Suchozebrski (1968)].

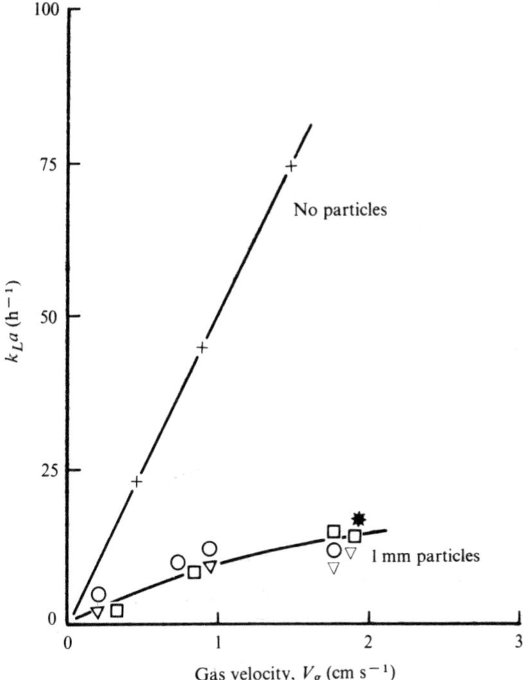

Figure 62. k_La values for bubble columns and for gas–liquid fluidized beds of 1 mm ballotini. Percentage carbon dioxide, liquid velocity (cm s^{-1}): +, 100 per cent, 1.5; ○, 100 per cent, 1.2; □, 100 per cent, 1.5; ▽, 100 per cent, 2.0; * 100 per cent, 3.1 [Østergaard and Suchozebrski (1968)].

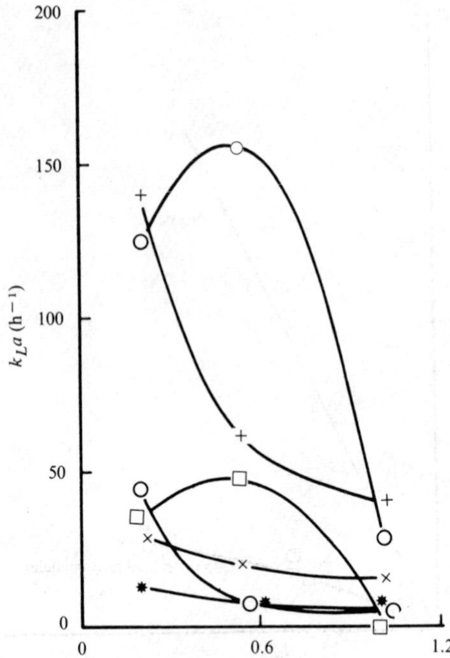

Figure 63. Variation in k_La with distance from the distributor plate. Beds of 6 mm particles, liquid velocity (cm s^{-1}), gas velocity (cms^{-1}): ○, 8.4, 2.4; □, 8.4, 0.82. Bubble columns, liquid velocity (cm s^{-1}), gas velocity (cm s^{-1}): +, 8.4, 2.7; ×, 8.4, 0.78. Beds of 1 mm particles, liquid velocity (cm s^{-1}), gas velocity (cm s^{-1}): ▽, 2.0, 1.6; *, 2.0, 0.83 [Østergaard and Suchozebrski (1968)].

REFERENCES

Aiba, S. and Huang, S.Y. (1969) Oxygen Diffusivity and Permeability in Polymer Membranes immersed in Water. *Chem. Engng Sci.*, **24**, 1149.

Andrew, S.P.S. (1982) Gas–Liquid Mass Transfer in Microbiological Reactors. *Trans. Instn Chem. Engrs*, **60**, 3.

ASTM (1978) *Annual Book of ASTM Standards*, **D1589–60**, p.438.

Bartholomew, W.H., Karow, E.O., Sfat, M.R. and Wilhelm, R.H. (1950) Oxygen Transfer and Agitation in Submerged Fermentations. *Ind. Engng Chem.*, **42**, 1801.

Battino, R. and Clever, H.L. (1966) Solubility of Gases in Liquids. *Chem. Rev.*, **66**, 395.

Beechey, R.B. and Ribbons, D.W. (1972) Oxygen Electrode Measurements, in *Methods in Microbiology*, vol. 6B, ed. by J.R. Norris and D.W. Ribbons, p.25 (Academic Press, New York).

Bell, G.H. and Gallo, M. (1971) Effects of Impurities on Oxygen Transfer. *Process Biochem.*, **6** (4), 33.

Blakebrough, N. and Hamer, G. (1963) Turbine Impellers as Gas–Liquid Contacting Devices. *J. Appl. Chem.*, **13**, 517.

Blakebrough, N. and Sambamurthy, K. (1964) Performance of Turbine Impellers in Sparger Aerated Fermentation Vessels. *J. Appl. Chem.*, **14**, 413.

Brierly, M.R. and Steele, R. (1959) Effects of Solid Disperse Phase on Oxygen Absorption in a Fermenter. *Appl. Microbiol.*, **7**, 57.

Bylinkina, E.S. and Birukov, V.V. (1972) The Problem of Scale-Up in Antibiotic Biosynthesis, in *Proc. 4th International Fermentation Symposium*, p.105.

Chain, E.B., Gualandi, G. and Morisi, G. (1966) Aeration Studies. IV. Aeration Conditions in 3000-Litre Vessel. *Biotechnol. Bioengng*, **8**, 595.

Cobbold, R.S.C. (1974) *Transducers for Biomedical Measurements* (John Wiley, New York).

Danckwerts, P.V. (1951) Significance of Liquid-Film Coefficients in Gas Absorption. *Ind. Engng Chem.*, **43**, 1460.

Deindorfer, F.H. and Gaden, E.L., jun. (1955) Effect of Liquid Physical Properties on Oxygen Transfer in Penicillin Fermentations. *Appl. Microbiol.*, **3**, 253.

Elsworth, R., Williams, V. and Harris-Smith, R. (1957) A Systematic Assessment of Dissolved Oxygen Supply in a Twenty-Litre Vessel. *J. Appl. Chem.*, **7**, 261.

Finn, R.K. (1954) Agitation–Aeration in the Laboratory and in Industry. *Bacteriol. Rev.*, **18**, 254.

Finn, R.K. (1967) Agitation and Aeration, in *Biochemical and Biological Engineering Science*, vol. 1, ed. by N. Blakebrough, p.1 (Academic Press, New York).

Higbie, R. (1935) The Rate of Absorption of a Pure Gas into a Still Liquid during Short Periods of Exposure. *Trans. Amer. Inst. Chem. Engrs*, **35**, 365.

Houkili, A.D. and Lee, S. (1971) Controlled Supply of Trace Amounts of Oxygen in Laboratory-Scale Fermentation. *Biotechnol. Bioengng*, **13**, 619.

Hunt, G., Reisman, H.B. and Lago, J. (1971) Oxygen Mass Transfer in Biological Systems at Zero Dissolved Oxygen Concentration. *Chem. Engng Prog. Symp. Ser.*, **67**, 60.

International Critical Tables (1928) vol. 3, (McGraw-Hill, New York).

Jarai, M. (1972) Oxygen Transfer in the Fermentations of Primary and Secondary Metabolites, in *Proc. 4th International Fermentation Symposium*, p.97.

Lee, Y.H. and Tsao, G.T. (1979) Dissolved Oxygen Electrodes, in *Advances in Biochemical Engineering*, vol. 13, ed. by T.K. Ghose *et al.*, p.35 (Springer Verlag, Berlin).

Metz, B., Kossen, N.W.P. and van Suijdam, J.C. (1979) The Rheology of Mold Suspensions, in *Advances in Biochemical Engineering*, vol. 11, ed. by T.K. Ghose *et al.*, p.103 (Springer Verlag, Berlin).

Misra, H.P. and Fridovich, I. (1976). A Convenient Calibration of the Clark Oxygen Electrode. *Anal. Biochem.*, **70**, 632.

Moo-Young, M. and Blanch, H.W. (1981) Design of Biochemical Reactors. Mass Transfer Criteria for Simple and Complex Systems, in *Advances in Biochemical Engineering*, vol. 19, ed. by A. Fiechter, p.1 (Springer Verlag, Berlin).

Moss, M.O. and Smith, J.E. (eds.) (1977) *Industrial Applications of Microbiology* (Surrey University Press, London).

Østergaard, K. and Suchozebrski, W. (1968) Gas–Liquid Mass Transfer in Gas–Liquid Fluidized Beds, in *Proc. 4th European Symposium on Chemical Reaction Engineering* (Pergamon Press, Oxford).

Phillips, D.H. and Johnson, M.J. (1961) Aeration in Fermentations. *J. Biochem. Microbial. Technol. Engng*, **3**, 277.

Phillips, K.L. (1968) Fermentation Advances, in *Proc. 3rd International Fermentation Symposium*, p.465.

Pirt, S.J. (1966) *Proc. Roy. Soc., London, Ser. B*, **166**, 369.

Resnick, W. and Gal-Or, B. (1968) Gas–Liquid Dispersions. *Adv. Chem. Engng*, **7**, 295.

Robinson, C.W. and Wilke, C.R. (1973) Oxygen Absorption in Stirred Tanks. A Correlation for Ionic Strength Effects. *Biotechnol. Bioengng*, **15**, 755.

Sanders, W.M., Bungay, H.R. and Whalen, W.J. (1971) Oxygen Microprobe Studies of Microbial Slime Films. *Chem. Engng Prog. Symp. Ser.*, **67**, 69.

Sato, K. (1961) Rheological Studies of Some Fermentation Broths. I. Kanamycin and Streptomycin Fermentation Broths. *J. Ferment. Technol.*, **39**, 347.

Schügerl, K., Lucke, J. and Oels, U. (1977) Bubble Column Bioreactors, in *Advances in Biochemical Engineering*, vol. 7, ed. by T.K. Ghose *et al.*, p.1 (Springer Verlag, Berlin).

Solomons, G.L. (1969) *Materials and Methods in Fermentation* (Academic Press, New York).

Solomons, G.L. and Perkin, M.D. (1958) The Measurement and Mechanism of Oxygen Transfer in Submerged Culture. *J. Appl. Chem.*, **8**, 251.

Taguchi, H. (1971) The Nature of Fermentation Fluids, in *Advances in Biochemical Engineering*, vol. 1, ed. by T.K. Ghose *et al.*, p.1 (Springer Verlag, Berlin).

Tsao, G.T. and Kempe, L.K. (1960) Oxygen Transfer in Fermentation Systems. I. Use of Gluconic Acid Fermentation for Determination of Instantaneous Oxygen Transfer Rates. *J. Biochem. Microbiol. Technol. Engng*, **2**, 129.

Tuffile, C.M. and Pinho, F. (1970) Determination of Oxygen Transfer Coefficients in Viscous Streptomycete Fermentations. *Biotechnol. Bioengng*, **12**, 849.

Windholz, M. (ed.) (1976) *The Merck Index* (Merck, Rahway).

Yagi, H. and Yoshida, F. (1974) Oxygen Absorption in Fermenters. Effect of Surfactants, Antifoaming Agents and Sterilized Cells. *J. Ferment. Technol.*, **52**, 905.

Yagi, H. and Yoshida, F. (1975) Enhancement Factor for Oxygen Absorption in Fermentation Broth. *Biotechnol. Bioengng*, **17**, 1083.

NOMENCLATURE

A = cathode surface area
a = bubble surface area per unit volume
c = dissolved oxygen concentration
c^* = oxygen solubility
c_{CRIT} = critical oxygen concentration
c_{INIT} = initial oxygen concentration
c_s = saturation solubility
D_m = oxygen diffusivity in the membrane
d_b = bubble diameter
d_m = membrane thickness
F = Faraday's constant
G = air flow rate
H = Henry's constant
I_O = current output at time zero
I_s = steady-state current output
I_t = transient current output
k = probe constant
k_L = mass transfer coefficient
k_La = volumetric mass transfer coefficient
N = impeller speed
n = number of electrons involved per mole of oxygen reduced
P_m = oxygen permeability of the membrane
P_o = oxygen partial pressure in the liquid
P_g = partial pressure of gas
Q_{O_2} = specific oxygen consumption rate
$\dot{R}$ = overall volumetric mass transfer rate
T = temperature
t = time
V = volume
V_g = superficial gas velocity
x = microorganism concentration
ε = gas hold-up
μ = viscosity

APPENDIX 1

The overall oxygen transfer rate is given by

$$R = k_G a (p_G - p_i) = k_L a (c_i - c_L) \tag{A1}$$

where $k_G a$ and $k_L a$ are the gas film and liquid film volumetric mass transfer coefficients, respectively, p_G is the oxygen partial pressure in the gas phase, c_L is the oxygen concentration, and p_i and c_i are the equilibrium conditions at the gas–liquid interface given by Henry's law [*see* Eq. (4)], i.e. $c_i = p_i/H$.

Eq. (A1) can also be expressed as

$$R = K_G a (p_G - p_e) = K_L a (c_e - c_L) \tag{A2}$$

where $K_G a$ and $K_L a$ are the overall volumetric mass transfer coefficients, p_e is the oxygen partial pressure in equilibrium with c_L (i.e. $p_e = c_L H$) and c_e is the dissolved oxygen concentration in equilibrium with p_G (i.e. $c_e = p_G/H$).

Combination of Eqs. (A1) and (A2) leads to

$$\frac{1}{K_L a} = \frac{1}{k_L a} + \frac{1}{H k_G a} \tag{A3}$$

For a sparingly soluble gas, such as oxygen,

1. the Henry's law is relatively large
2. $k_La \langle\langle k_Ga$, since $k_L \alpha D_L$, $k_G \alpha D_G$ and $D_G \rangle\rangle D_L$, where D_G and D_L are the oxygen diffusivities in the gas and liquid phases, respectively.

In these conditions

$$K_La = k_La \tag{A4a}$$

and from Eqs. (A1) and (A2)

$$c_i = c_e \tag{A4b}$$

Thus

$$R = k_La\,(c^* - c) \tag{A5}$$

where c^* is the solubility of oxygen in equilibrium with p_G, (i.e. $c^* = p_G/H$) and c is the dissolved oxygen concentration in the bulk liquid. The physical condition defined by Eqs. (A4) and (A5) is referred to as liquid-film control.

When the oxygen concentration in the liquid phase is measured as dissolved oxygen with units typically of mol m^{-3}, k_La has units typically of h^{-1} and oxygen transfer rate of mol m^{-3} h^{-1}. When the oxygen concentration is determined by electrode methods (*see* pp.740–746 i.e. as oxygen tension (partial pressure typically in atmosphere) then k_La has units of mol m^{-3} h^{-1} atm^{-1}.

APPENDIX 2

Penetration Theory of Higbie (1935)

If the liquid immediately adjacent to a rising bubble is assumed to rise with the bubble but otherwise to be quiescent, the mass transfer conditions are those of unsteady-state molecular diffusion. Mathematical solution of this problem leads to

$$k_L = 2\sqrt{\frac{D_L}{\pi\theta}} \tag{A6}$$

where k_L is the liquid-film mass transfer coefficient over the life time of the bubble θ.

Values of k_L calculated from Eq. (A6) decrease with increasing life time due to the decreasing oxygen concentration gradients in the liquid resulting from increased oxygen penetration into the liquid.

For given hydrodynamic conditions, θ may be taken as a constant reflecting the time between bubble formation and coalescence. The latter may occur due to contact with another bubble or with the liquid surface.

Surface Renewal Theory of Danckwerts (1951)

If turbulent mixing occurs in the liquid in the proximity of a bubble, it is likely that unsteady molecular diffusion will take place in individual eddies. In these conditions, mathematical analysis leads to

$$k_L \alpha \sqrt{D_L s} \tag{A7}$$

where s is the rate of surface renewal, i.e. the rate at which eddies sweep the boundary layer. Eq. (A7) effectively contains the eddy life time in the surface renewal rate and it is to be expected that s will increase with increased turbulence intensity just as θ in Eq. (A6) will decrease under similar conditions.

REFERENCES

Higbie, R. (1935) The Rate of Absorption of a Pure Gas into a Still Liquid during Short Periods of Exposure. *Trans. Amer. Inst. Chem. Engrs*, **35**, 365.

Danckwerts, P.V. (1951) Significance of Liquid-Film Coefficients in Gas Absorption. *Ind. Engng Chem.*, **43**, 1460.

CHAPTER 10
SOLID- AND LIQUID-PHASE MASS TRANSFER

GLOSSARY

Lineweaver–Burk plot Method of recasting kinetic data (growth rates or enzyme-catalyzed reactions) in linear form using a double reciprocal plot of 1/rate versus 1/substrate concentration.
Michaelis–Menten kinetics Study of rate of conversions in enzyme-catalyzed reactions.
Microbial film Thin layer of microorganisms adsorbed onto a supporting surface.
Microbial floc Adherent mass of microorganisms.
Monod kinetics Study of the rate of growth of microbial systems.
Specific rate of reaction Rate of reaction (i.e. substrate utilization or product formation) per unit mass of microorganism ($M_s\ M_o^{-1}\ T^{-1}$).
Volumetric rate of reaction Rate of substrate utilization or product formation (i.e. productivity) per unit volume ($M_s\ L^{-3}\ T^{-1}$).

INTRODUCTION

Biological particles and artefacts, such as microbial flocs (*see* Fig. 2, Chapter 1), immobilized microorganisms (*see* Fig. 11, Chapter 1), microbial films (*see* Fig. 10, Chapter 1) and immobilized enzymes (*see* Fig. 27, Chapter 6), are often of a characteristic size such that they are visible by the naked eye and are consequently of a scale significantly larger than that of an enzyme molecule or even a single cell. Such relatively large particle sizes facilitate mechanical handling and sedimentation, etc., and aid the development of the industrial configurations of fermenters, e.g., fluidized beds (*see* Fig. 65, Chapter 7). While large particles can lead to high biomass concentrations (*see* Chapter 7), this advantage can be offset by the reduced effectiveness of the biomass because of diffusional limitations on the ingress of the substrate.

Increased size leads to diffusional limitations on substrate ingress and/or product egress. Usually, this leads to *reduced* overall specific rates of reaction. However, there are cases when diffusional limitations can *increase* the overall specific rate of reaction, e.g., when substrate inhibition occurs and when the rate of reaction depends upon a property affected by product formation such as pH.
Fig. 24 and pp.28–32 of this chapter).

When diffusional limitations lead to reduced rates of reaction, the optimum particle size, from the standpoint of the overall specific rate of reaction, is clearly the smallest one. In practice, there is the criterion of 'sufficiently small' [*see* Eq. (19)]. If maximization of the overall volumetric rate of reaction is the objective (i.e. $R_v = R_x$ where x = biomass concentration), the optimum particle size is that which leads to both a high overall specific rate of reaction and a high biomass holdup.

When diffusional limitations lead to increased rates of reaction, optimum particle sizes exist that result in maximum specific rates of reaction (*see* Fig. 20).

Understanding these phenomena is made more complex because of the often irregular shapes and size distributions involved (*see* Table 1) as well as the tendency of the biomass to become deformed under fluid shear. However, theoretical consideration of the problems of diffusion in relation to biological reactions allows general characteristics to be established even when the intrinsic kinetics used are greatly simplified. This, in turn, provides guidelines for both particle and fermenter development and design.

Since both the intrinsic kinetics of both microbial and enzyme systems are described by algebraically identical equations, i.e. Monod and Michaelis–Menten kinetics, it follows that a common set of principles and equations may be used to describe the range of diffusion-limited particles and artefacts described in Table 1.

Table 1. Geometric characteristics of biological particles and artefacts.

	Freely suspended microbial flocs	Immobilized microorganisms	Microbial films	Immobilized enzymes
Size	Effective size unknown (to be determined experimentally)	Predetermined by support	Thickness unknown (to be determined experimentally)	Predetermined by preparatory method
Size distribution	*See* Fig. 10	Supports of uniform size	Depends on fermenter configuration and operation, e.g., tubular fermenters with low fluid shear contain a range of thicknesses	Predetermined by preparatory method
Shape	Irregular	Regular, e.g., spherical or cuboid	Follows shape of support, e.g., planar or near-spherical	Predetermined by preparatory method

NATURE OF DIFFUSIONAL LIMITATIONS

Liquid Phase

For Monod/Michaelis–Menten kinetics

$$R = h(s - s^*) = R_{max} \frac{s^*}{K_m + s^*} \tag{1}$$

where R is the overall specific rate of substrate uptake, h is the liquid-phase mass transfer coefficient with dimensions $L^3 M^{-1} T^{-1}$ and s^* is the substrate concentration adjacent to the particle.

Fig. 1 illustrates the effect of an increased mass transfer coefficient (achieved by increased liquid turbulence) on the overall specific rate of substrate uptake.

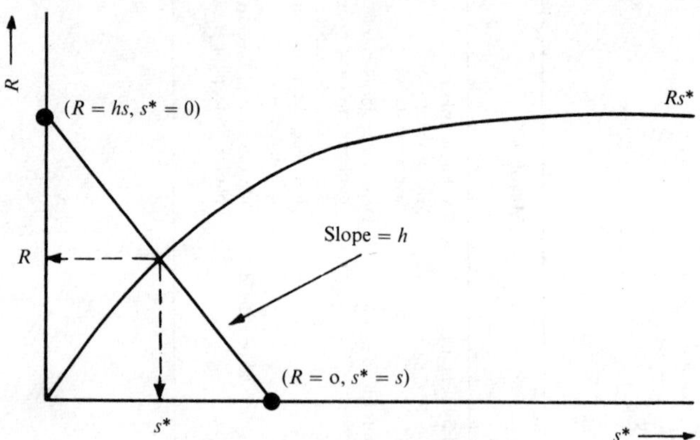

Figure 1. Graphical solution of Eq. (1).

The following three situations may be considered.

1. Reaction control, when h is large, $s^* \longrightarrow s$, thus

$$R \longrightarrow R_{max} \frac{s}{K_m + s} \tag{2}$$

2. 'Liquid-phase diffusion' control, when h is small, thus

$$R = h(s - s^*) \longrightarrow \frac{R_{max}}{K_m} s^*$$

or

$$R = \frac{1}{(K_m/R_{max}) + (1/h)} s$$

and when $h \ll R_{max}/K_m$

$$R \longrightarrow hs \tag{3}$$

3. Minimum mass transfer coefficient h_{min} compatible with zero-order conditions (i.e. $s^* = s^*_{CRIT}$)

$$R = h_{min}(s - s^*_{CRIT}) \longrightarrow R_{max}$$

$$h_{min} = \frac{R_{max}}{(s - s^*_{CRIT})} \tag{4}$$

When $h < h_{min}$, zero-order conditions cannot be achieved even though $s > s^*_{CRIT}$.

'Solid' Phase

Fig. 2 shows that for constant h a 'solid'-phase diffusion limitation results in

1. a reduced overall specific rate of substrate uptake R,
2. an increased substrate concentration adjacent to the particle s^*, and
3. an increase in the substrate concentration s^*_{CRIT} to achieve zero-order conditions.

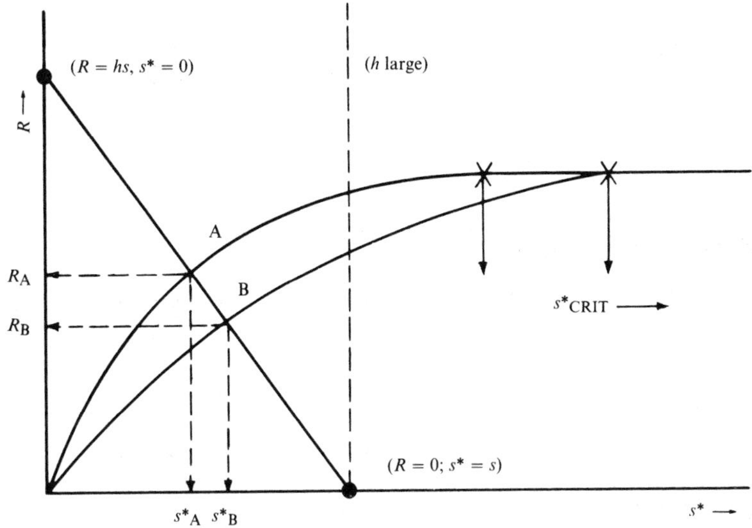

Figure 2. Effect of a 'solid'-phase diffusional limitation. Curve A, Monod/Michaelis–Menten kinetics; Curve B, 'solid'-phase, diffusion-limited kinetics.

BIOLOGICAL RATE EQUATION

Eq. (1) can be generalized to account for the existence of a solid-phase diffusion limitation by the introduction of an effectiveness factor λ, where $0 < \lambda \leqslant 1$. Thus

$$R = h(s - s^*) = \lambda R_{max} \frac{s^*}{K_m + s^*} \tag{5}$$

Atkinson and Daoud (1968) devised an algebraic equation describing λ using the concepts of molecular diffusion and Monod/Michaelis–Menten kinetics. Subsequently, these equations were generalized by Atkinson and Rahman (1979) to cover the regular geometries of a slab, a cylinder and a sphere.

For the predominantly reaction-controlled region, i.e. for relatively small particles defined by $\phi_P \leqslant 1$, where ϕ_P is a dimensionless parameter,

$$\lambda = 1 - \frac{\tan h\,(\alpha k_2 V_p/A_p)}{\alpha k_2 V_p/A_p}\left(\frac{\phi_p}{\tan h\,\phi_p} - 1\right) \tag{6}$$

where V_p is the particle volume and A_p is the external particle area.

For the predominantly diffusion-limited region, i.e. relatively large particles defined by $\phi_P \leqslant 1$

$$\lambda = \frac{1}{\phi_p} - \frac{\tan h(\alpha k_2 V_p/A_p)}{\alpha k_2 V_p/A_p}\left(\frac{1}{\tan h\phi_p} - 1\right) \tag{7}$$

In Eqs. (6) and (7)

$$\phi_P = \frac{(\alpha k_2 V_p/A_p)s^*}{(K_m + s^*)\{2[s^*/K_m - \ln(1 + s^*/K_m)]\}^{1/2}} \simeq \frac{\alpha k_2 V_p/A_p}{(1 + 2s^*/K_m)^{1/2}} \tag{8}$$

Particle Size and Shape

V_p/A_p and α in Eqs. (6) to (8) represent the characteristic size and shape, respectively, and are given in Table 2.

Table 2. Values of V_p/A_p and α.

	V_p/A_p[a]	α
Films	L	1.00
Cylindrical particles	$r_o/2$	1.06
Spherical particles	$r_o/3$	1.16

[a] L is the film thickness and r_o is the particle radius.

Intrinsic Kinetics

Microbial systems

R_{max} = maximum specific substrate uptake rate
K_m = Monod coefficient

$$k_2 = \left(\frac{\rho_o \mu_{max}}{K_m D_e Y_s}\right)^{1/2} \tag{9}$$

where ρ_o is the cell density expressed as dry weight per wet particle volume and D_e is the effective diffusion coefficient of the substrate within the wet biomass.

Enzyme systems

$$R_{max} = \frac{V_{max}\rho_e}{\rho_p} \tag{10}$$

where R_{max} is based on the unit mass of particles, V_{max} is based on the unit mass of the enzyme per unit particle volume and ρ_p and ρ_e are the particle and enzyme densities, respectively.

$$K_m = \text{Michaelis constant}$$

$$k_2 = \left(\frac{\rho_e V_{max}}{K_m D_e}\right)^{1/2} \tag{11}$$

Extent of diffusion limitation

For given kinetics and substrate concentration s^*, the value of the effectiveness factor λ is determined by the dimensionless parameter $\alpha\, k_2\, V_p/A_p$, which is traditionally referred to as the Thiele modulus. This parameter largely determines the value of ϕ_p [Eq. (8)], characterizes particles as large or small [Eqs. (6) and (7)] and identifies that range for which the particles are 'sufficiently small' for solid-phase diffusional limitations to be neglected, i.e. $\lambda \simeq 1$ [*see* Eq. (19)].

Graphical representation of the biological rate equation

Figs. 3, 4 and 5 represent the biological rate equation for films, cylindrical and spherical particles, respectively. The data points are values calculated using Eqs. (5) to (8), while the continuous lines are the accurate numerical solution to the problem of molecular diffusion with biological reaction described by Monod/Michaelis–Menten kinetics.

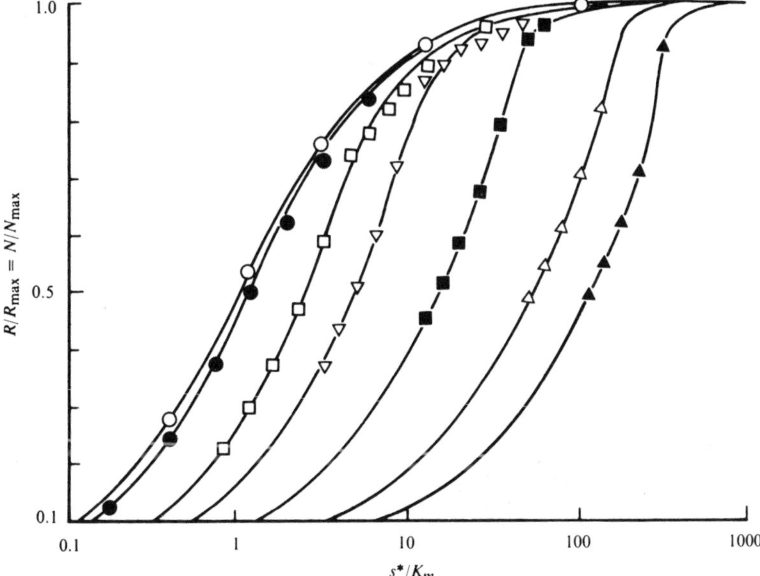

Figure 3. Biological rate equation for films. N represents the flux of substrate based upon the area of film exposed to substrate. k_2L values: ○, 0.5; ●, 1.0; □, 3.0; ▽, 5.0; ■, 10.0; △, 20.0; ▲, 30.0

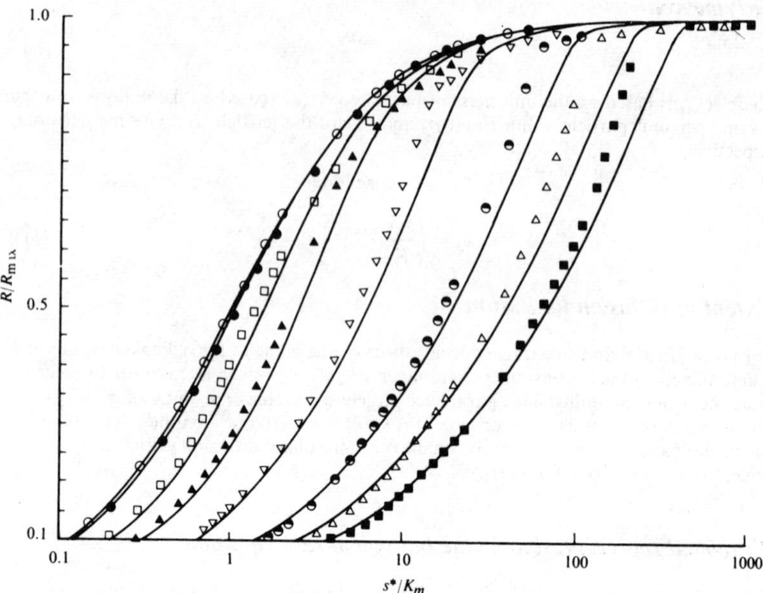

Figure 4. Biological rate equation for cylindrical particles. $\alpha k_2 V_p/A_p$ values: ○, 0.26; ●, 0.5; □, 1.59; ▲, 2.65; ▽, 5.30; ◓, 10.60; △, 15.90; ■, 21.20.

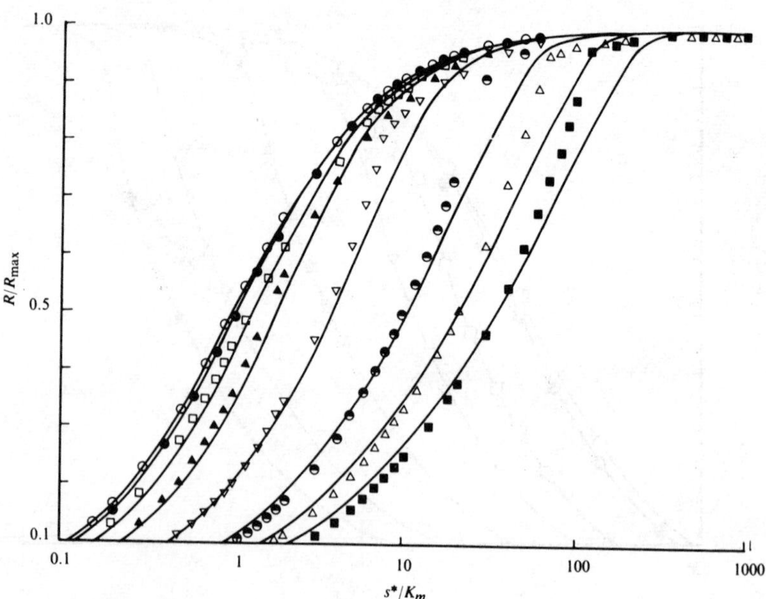

Figure 5. Biological rate equation for spherical particles. $\alpha k_2 V_p/A_p$ values: ○, 0.193; ●, 0.5; □, 1.16; ▲, 1.93; ▽, 3.87; ◓, 7.73; △, 11.60; ■, 15.47.

Utilization of the biological rate equation

For microbial systems, Eqs. (5) to (8) or Figs. 3 to 5 require values of the intrinsic kinetic parameters (R_{max}, K_m and D_e) and the physical parameters (V_p/A_p, s^* and ρ_o). Fig. 6 represents data calculated using the biological rate equation for a microbial film and shows clearly the effect of film thickness on the substrate flux.

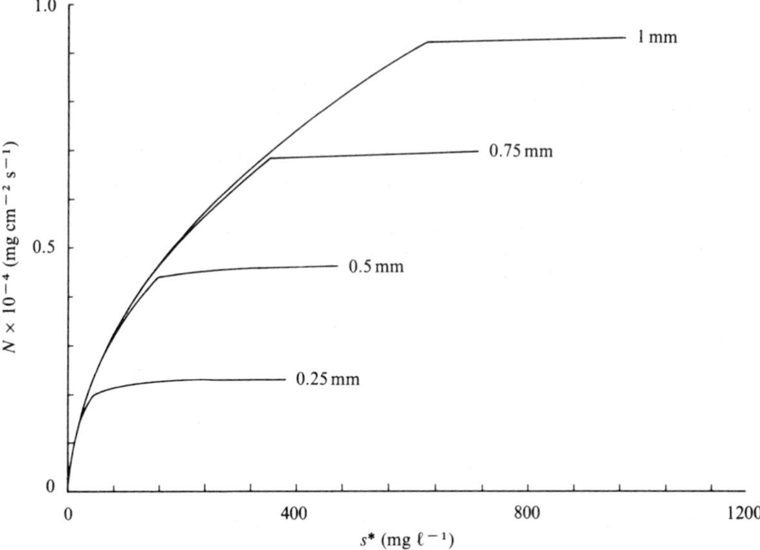

Figure 6. Relationship between flux N and interfacial substrate concentration s^* for a microbial film calculated using the biological rate equation. $\rho_o R_{max} = 1 \times 10^{-6}$ g cm^{-3} s^{-1}, $K_m = 5 = 10^{-6}$ g cm^{-3} and $k_2 = 150$ cm^{-1}.

For enzyme systems, Eqs. (5) to (8) or Figs. 3 to 5 require values of the intrinsic kinetic parameters (V_{max}, K_m and D_e) and the physical parameters (V_p/A_p, s^* and ρ_e). Fig. 7 represents data calculated using the biological rate equation for enzyme particles containing glucose oxidase. The data show that an increase in enzyme concentration (ρ_e) can increase the diffusional limitation through the effect on the Thiele Modulus ($\alpha k_2 V_p/A_p$) [*see* Eq. (11)].

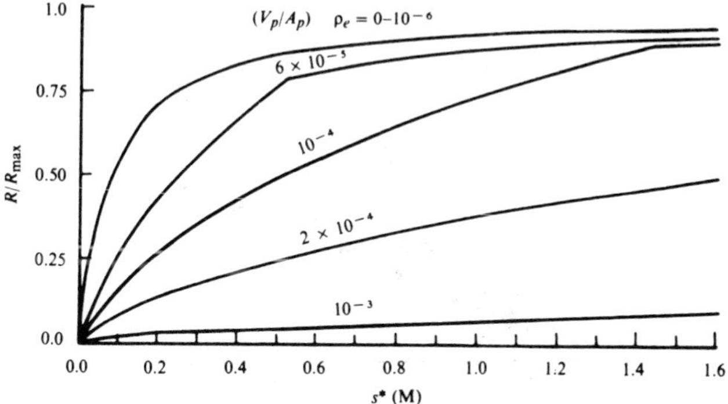

Figure 7. Effect of particle size and enzyme concentration on the overall rate of reaction of gel-immobilized glucose oxidase particles (calculated using the biological rate equation) [Atkinson and Lester (1975)].

LINEWEAVER–BURK PLOT OF THE BIOLOGICAL RATE EQUATION

The Lineweaver–Burk procedure is used in Chapter 4 (*see* p.207) to obtain values of K_m and R_{max}. Rearrangement of Eq. (5) leads to

$$\frac{R_{max}}{R} = \frac{1}{\lambda} + \frac{K_m}{\lambda}\frac{1}{s^*} \tag{12}$$

Since λ is dependent on s^*, Eq. (12) will not lead to a straight line if $1/R$ is plotted against $1/s^*$. Furthermore, Eq. (12) does not yield a single line, even for the same biological system, since λ depends upon $\alpha k_2 V_p/A_p$.

A plot of Eq. (12), using Eqs. (5) to (8), in terms of R_{max}/R versus K_m/s^*, leads to a family of curves with $\alpha k_2 V_p/A_p$ as a parameter (*see* Fig. 8). Since $\lambda \to 1$ as s^* is increased, it follows that all the curves coincide at $K_m/s^* = 0$, $R_{max}/R = 1$. For 'sufficiently small' values of $\alpha k_2 V_p/A_p$, $\lambda \simeq 1$, the corresponding curves are essentially linear for all values of $1/s^*$; the intercept and slope may be used to calculate R_{max} and K_m. For large values of $\alpha k_2 V_p/A_p$ and relatively small values of s^*, the biological rate equation reduces to

$$\frac{R}{R_{max}} \simeq \frac{s^*/K_m}{\alpha k_2 V_p/A_p}$$

or

$$\frac{1}{R} \simeq \left(\frac{\alpha k_2 V_p/A_p}{R_{max}/K_m}\right)\frac{1}{s^*} \tag{13}$$

Eq. (13) is linear and applies for large values of $1/s^*$, as can be seen from Fig. 8.

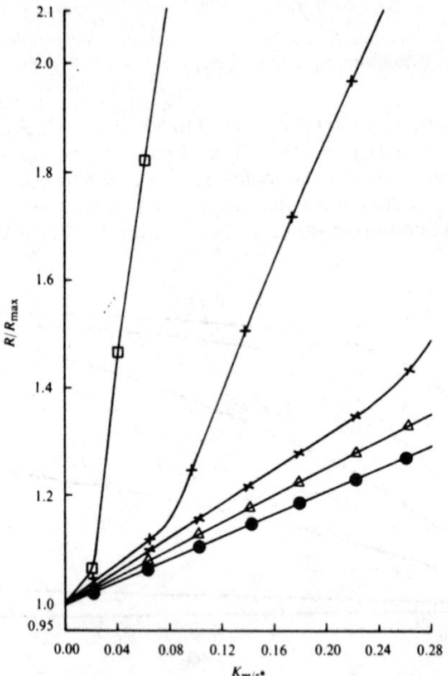

Figure 8. Lineweaver–Burk plot of the biological rate equation. $\alpha k_2 V_p/A_p$ values: ●, 0.5; △, 1.5; ×, 3.0; +, 5.0; □, 10.0

The dangers involved in plotting experimental data over a restricted concentration range are apparent from examination of Fig. 8. Clearly, approximate linearity of experimental data is insufficient to justify the absence of a 'solid'-phase diffusion limitation. The responsibility rests with the experimenter to show that size has no effect or, alternatively, to reduce V_p/A_p to 'sufficiently small'.

It is also clear from Fig. 8 that where the particle size is not fixed a priori (*see* Table 1) values of $\alpha k_2 V_p/A_p$ can only be obtained by parameter-fitting nonlinear experimental data.

ASYMPTOTES OF THE BIOLOGICAL RATE EQUATION

Empirically, the overall rate of substrate uptake is related to s^* by the equations

$$n = f\frac{s^*}{g + s^*} \tag{14a}$$

$$n = a \tag{14b}$$

$$n = bs^* \tag{14c}$$

$$n = d(s^*)^{1/2} \tag{14d}$$

where n is the rate of substrate uptake in appropriate units, and a, b, d, f and g are coefficients with units corresponding to those of n and s^*.

Experience of chemical reaction engineering suggests that it is unlikely that completely different equations would apply if the substrate concentration were changed or if the geometry were altered, e.g., from floc to film, for the same biological system.

Table 3 summarizes various asymptotic forms of the biological rate equation at high and low substrate concentrations for small and large characteristic sizes.

Comparison of Eqs. (14) and Table 3 suggests the following points.

1. Eq. (14a) represents Monod/Michaelis–Menten kinetics and corresponds to small characteristic sizes for all values of s^*.
2. Eq. (14b) represents 'zero-order' kinetics and corresponds to large s^* for all characteristic sizes (in the absence of substrate inhibition, *see* Figs. 3–5).
3. Eq. (14c) represents 'first-order' kinetics and corresponds to large characteristic sizes and low, or very low, substrate concentrations.
4. Eq. (14d) represents 'half-order' kinetics and corresponds to large characteristic sizes and moderate substrate concentrations.

The algebraic definitions of a, b, d, f and g can be obtained from Table 3.

Table 3. Reduced forms of the biological rate equation [Eqs. (5) to (8)].

Conditions	Films	Particles
All s^*, all V_p/A_p	$\frac{N}{N_{max}} = \lambda \frac{s^*}{K_m + s^*}$ $N_{max} = \rho_o L R_{max}$	$\frac{R}{R_{max}} = \lambda \frac{s^*}{K_m + s^*}$
All s^*, small V_p/A_p	$\frac{N}{N_{max}} = \frac{s^*}{K_m + s^*}$ [Eq. (14a)]	$\frac{R}{R_{max}} = \frac{s^*}{K_m + s^*}$ [Eq. (14a)]
Large s^*, all V_p/A_p	$\frac{N}{N_{max}} = 1$ [Eq. (14b)]	$\frac{R}{R_{max}} = 1$ [Eq. (14b)]
Small s^*, large V_p/A_p	$N = \frac{\rho_o R_{max}}{k_2} \frac{s^*}{K_m}$ [Eq. (14c)]	$\frac{R}{R_{max}} = \frac{1}{\alpha k_2 V_p/A_p} \frac{s^*}{K_m}$ [Eq. (14c)]
Small/moderate/large s^*, large V_p/A_p	$N = \frac{\rho_o R_{max}}{k_2} \frac{[1 + (2s^*/K_m)]^{1/2}}{[1 + (s^*/K_m)]} \frac{s^*}{K_m}$	$\frac{R}{R_{max}} = \frac{1}{\alpha k_2 V_p/A_p} \frac{[1 + (2s^*/k_m)]^{1/2}}{[1 + (s^*/K_m)]} \frac{s^*}{K_m}$
Moderate/large s^*, large V_p/A_p	$N = \frac{\rho_o R_{max}}{k_2} \left(\frac{2s^*}{K_m}\right)^{1/2}$ [Eq. (14d)]	$\frac{R}{R_{max}} = \frac{1}{\alpha k_2 V_p/A_p} \left(\frac{2s^*}{K_m}\right)^{1/2}$ [Eq. (14d)]

SIZE CRITERIA FOR EXISTENCE OF DIFFUSIONAL LIMITATIONS

The mass transfer coefficient h of Eq. (1) is related to the conventional mass transfer coefficient based on transfer area h_a by the equation

$$h = \frac{h_a}{\rho_p V_p / A_p} \tag{15}$$

where h_a has the dimensions L T^{-1}.

Data on mass transfer to single spheres immersed in a flowing fluid are correlated by the equation proposed by Frössling (1938), i.e.

$$\text{Sh} = 2.0 + 0.552\,\text{Re}^{1/2}\,\text{Sc}^{1/2} \tag{16}$$

where Sh is the Sherwood number (Sh $= h_a d_p/D$), Re is the Reynolds number (Re $= \rho u d_p/\mu$) and Sc is the Schmidt number (Sc $= \mu/\rho D$). In the definition of these engineering numbers, d_p is the particle diameter, D is the diffusivity of the species transferred, ρ and μ are the density and viscosity, respectively, of the fluid and u is the fluid velocity.

Microbial particles in suspension tend to follow the fluid flow pattern relatively closely, since the density difference between particle and fluid is often small. Under these circumstances, Eq. (16) reduces to

$$\text{Sh} = 2.0 \tag{17}$$

Combining Eqs. (15) and (17) leads to

$$h = \frac{D}{3\rho_p (V_p/A_p)^2} \tag{18}$$

Values of h calculated from Eq. (18) are independent of the mixing device and fluid flow pattern, providing a lower limit for liquid-phase mass transfer.

Eqs. (5), (6), (7) and (18) show that both 'solid'- and liquid-phase mass transfer limitations depend upon characteristic size (i.e. V_p/A_p).

'Solid' Phase

The Thiele modulus (i.e. $\alpha\, k_2\, V_p/A_p$) identifies the extent of the 'solid'-phase diffusion limitation. Inspection of Figs. 3, 4 and 5 shows that when

$$\frac{\alpha k_2 V_p}{A_p} < 0.5 \tag{19}$$

the 'solid'-phase diffusional limitation is negligible, i.e. $\lambda \simeq 1$ for all values of s^*/K_m.

Liquid Phase

A guide to the occurrence of a liquid-phase diffusional limitation is less straightforward to develop than that for the 'solid'-phase [*see* Eq. (19)]. However, Atkinson and Rahman (1979) have used Eq. (15) in conjunction with the biological rate equation to establish the minimal Sherwood number required for a given biological particle exposed to a given substrate concentration (*see* Fig. 9).

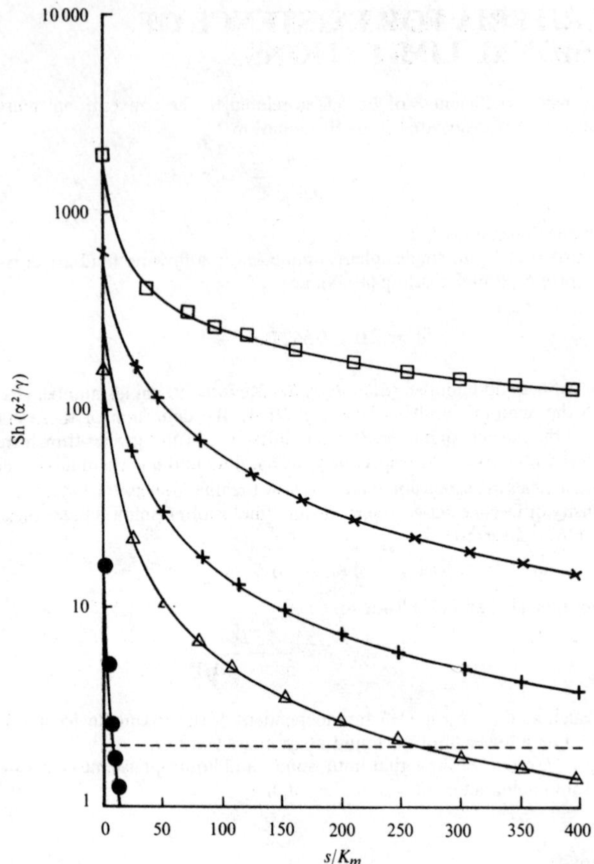

Figure 9. Minimal Sherwood number Sh required for the absence of liquid-phase diffusion limitation. $\gamma D_e/D$. $\alpha k_2 V_p/A_p$ values: ●, 0.5; △, 3.0; +, 5.0; ×, 10.0; □, 30.0. ----, $Sh(\alpha^2/\gamma) = 2.0$.

When Eq. (17) is applicable, $Sh(\alpha^2/\gamma)$ is also close to 2.0, since γ is probably only slightly lower than unity and α is slightly greater than unity. This value is identified in Fig. 9 and is clearly applicable over the whole range of s/K_m only for the smallest particles, i.e. when

$$\frac{\alpha k_2 V_p}{A_p} < 0.2 \tag{20}$$

If Eq. (20) is satisfied, there is no liquid-phase diffusional limitation even under quiescent conditions and, since Eq. (19) is also satisfied, no solid-phase limitation either.

FLOC SIZE DISTRIBUTION AND MEAN SIZE

The average overall specific rate of substrate uptake $\bar{R}$ of a number of particles n of various shapes and sizes exposed to the same concentration s^* is given by

$$\bar{R} = \frac{\sum_{i=1}^{n} \rho_o V_{p.i} \lambda(\alpha_i k_2 V_{p.i}/A_{p.i}, s^*/K_m) R_{max}[s^*/(K_m + s^*)]}{\sum_{i=1}^{n} \rho_o V_{p.i}}$$

$$\text{or } \bar{R} = R_{max} \frac{s^*}{K_m + s^*} \frac{\sum_{i=1}^{n} V_{p.i} \lambda(\alpha_i k_2 V_{p.i}/A_{p.i}, s^*/K_m)}{\sum_{i=1}^{n} V_{p.i}}$$

For the special case of spherical particles (where $V_p/A_p = d_p/6$)

$$\frac{\bar{R}}{R_{max}} = \frac{s^*/K_m}{1 + s^*/K_m} \frac{\sum_{i=1}^{n} (\alpha k_2 d_{p.i}/6)^3 \lambda(\alpha k_2 d_{p.i}/6, s^*/K_m)}{\sum_{i=1}^{n} (\alpha k_2 d_{p.i}/6)^3} \tag{21}$$

Fig. 10 shows an experimental floc size distribution and Fig. 11 contains a number of assumed distributions (defined in Table 4).

Table 4. Assumed floc size distributions.

Curve	Size distribution	Figure number
A	Bimodal	Fig. 10
B	Wide gaussian (large α)	Fig. 11
C	Rectangular	Fig. 11
D	Narrow gaussian (small α)	Fig. 11
E	Positively skewed	Fig. 11
F	Negatively skewed	Fig. 11

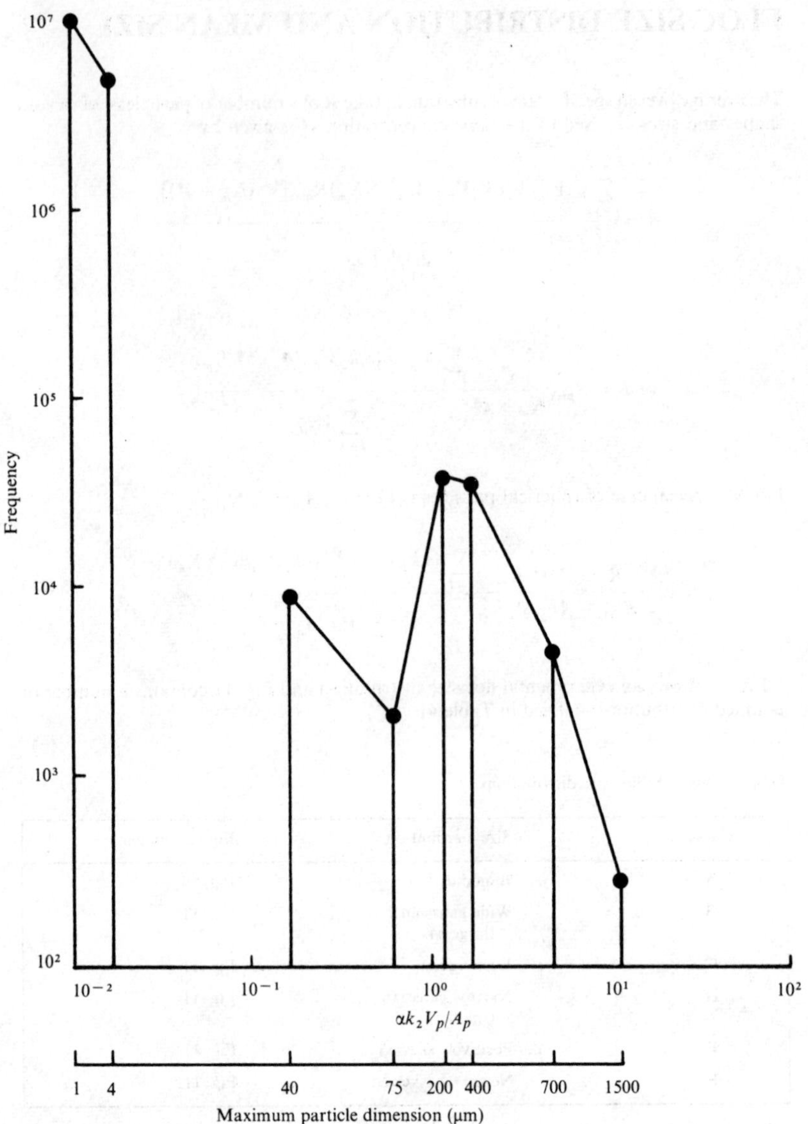

Figure 10. Size distribution for activated sludge floc [From Parker *et al.*) (1971)].

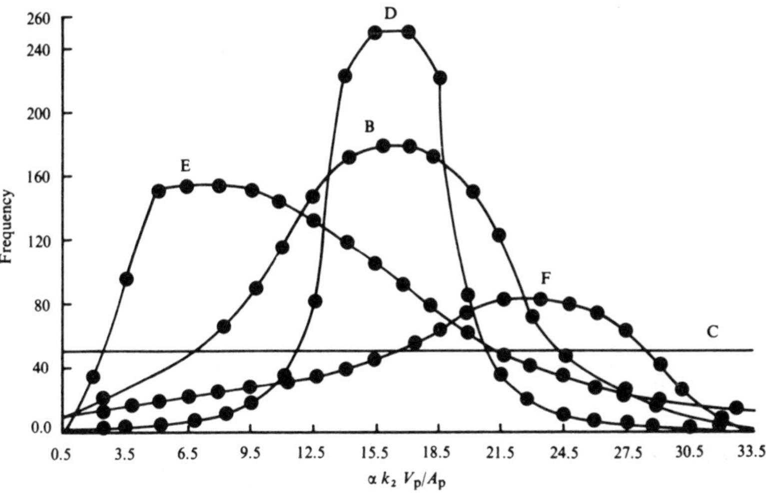

Figure 11. Assumed floc size distributions. Definitions of curves B–F are given in Table 4.

Fig. 12 is a Lineweaver–Burk plot obtained by inserting the assumed particle size distributions of Figs. 10 and 11 into Eq. (21). Superimposed on Fig. 12 are data points for 'best' fit obtained using the biological rate equation.

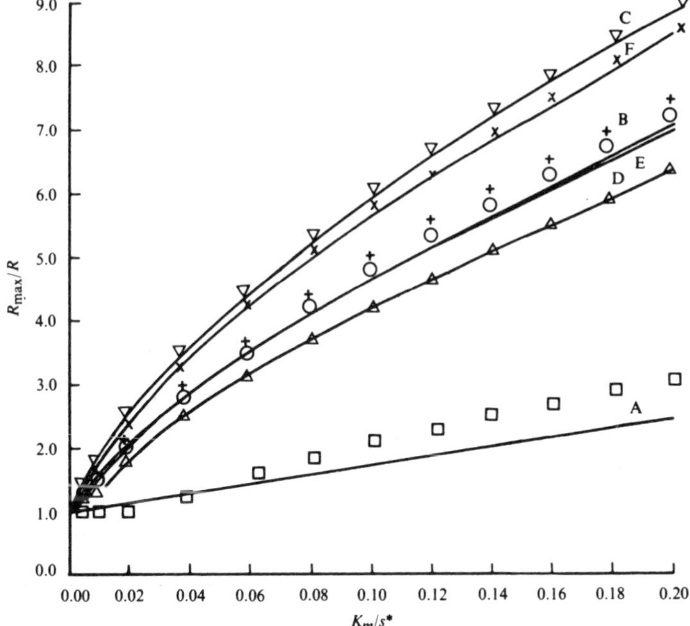

Figure 12. Lineweaver–Burk plot of rate data from Eq. (21). From biological rate equation: □, 8.28; △, 17.64; +, 19.75; ○, 20.15; ×, 23.65; ▽, 24.85. ———, particle size distributions.

Table 5 provides a comparison between the surface means calculated from the particle size distributions and the 'best' fit $\alpha k_2 V_p/A_p$ values.

Table 5. Surface means and best fit $\alpha k_2 V_p/A_p$ values.

Curve	Surface mean $\left(\sum_{i=1}^{n} n_i x_i^3 \Big/ \sum_{i=1}^{n} n_i x_i^2\right)$	Best fit
A	6.33	8.28
B	19.49	19.75
C	24.55	24.85
D	17.56	17.64
E	19.38	20.15
F	23.51	23.65

PARTICLE GROWTH

The increase in the size of an individual spherical microbial particle exposed to a constant substrate concentration s^*, neglecting any loss of mass due to attrition, is given by

$$\int_{m_o}^{m} \frac{dm}{m\lambda(\alpha k_2 V_p/A_p, s^*/K_m)} = \mu_{\max} \frac{s^*/K_m}{1 + s^*/K_m} t \tag{22}$$

where m is the particle mass.

Integration of Eq. (22) gives the results shown in Fig. 13.

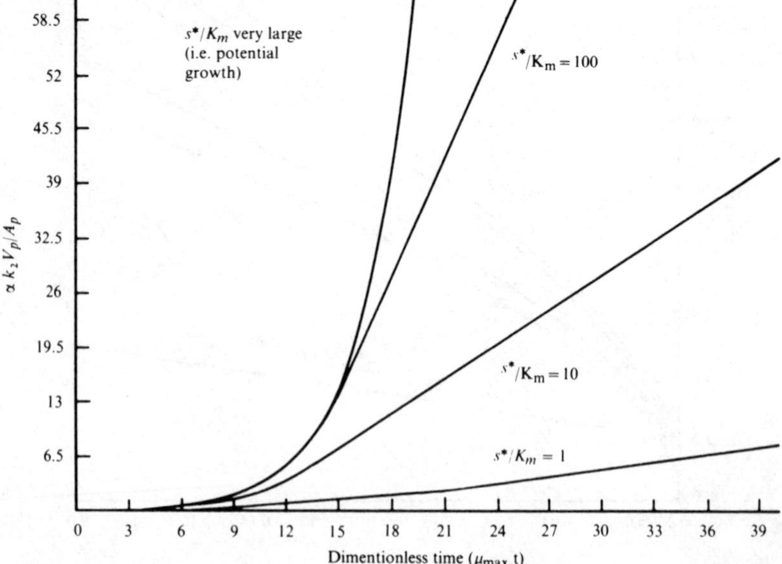

Figure 13. Time course of growth of particles exposed to a constant substrate concentration [Atkinson and Daoud (1976)].

EFFECT OF PARTICLE SIZE ON CRITICAL SUBSTRATE CONCENTRATION

The critical substrate concentration s^*_{CRIT} is generally described as that which leads to maximum specific rate of reaction, e.g., $R = n\ R_{max}$, $\mu = n\,\mu_{max}$, etc., where n is selected arbitrarily ($0.95 < n < 0.99$). An appreciation of mass transfer suggests that for a diffusion-limited system s^*_{CRIT} will depend upon particle size. Thus from the biological rate equation.

$$\frac{R}{R_{max}} = n = \frac{s^*_{CRIT}/K_m}{1 + s^*_{CRIT}/K_m} \tag{23}$$

Eq. (23) can be solved in terms of s^*_{CRIT}/K_m for various values of $k_2\ V_p/A_p$ and n. The corresponding data are given in Fig. 14, which shows the increase in s^*_{CRIT} as the particle size is increased. Chapter 4 gives the limiting values of the data in Fig. 14 for small $k_2\ V_p/A_p$, i.e. $k_2 V_p/A_p > 0.5$

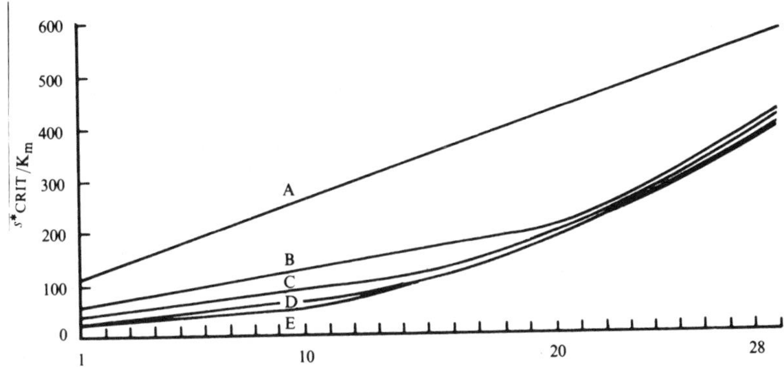

Figure 14. Dependency of critical substrate concentration S^*_{CRIT} on particle size calculated from the biological rate equation. Curve A, $R/R_{max} = 0.99$; curve B, $R/R_{max} = 0.98$; curve C, $R/R_{max} = 0.97$; curve D, $R/R_{max} = 0.96$; curve E, $R/R_{max} = 0.95$

IDEAL FILM THICKNESS AND CRITICAL PARTICLE SIZE

Ideal Film Thickness

As a microbial film increases in size, the substrate concentration at the support decreases and ultimately a stage is reached where the corresponding local rate of reaction is practically zero. These conditions lead to endogenous respiration and lysis.

These phenomena were observed by Sanders (1966), who found that as a microbial film grows the rate of uptake of substrate increases up to a critical film thickness beyond which no further increases in substrate uptake occur. Instead undesirable metabolic products are formed. Tomlinson and Snaddon (1966) suggested that for aerobic reactions the active part of the film was the aerobic region, while Kornegay and Andrews (1968) considered this region as a basic property of a particular microbe–substrate system. Atkinson and Fowler (1974) proposed that the 'ideal' film thickness L_{ideal} for use in film fermenters is equal to the penetration depth of the substrate, since uptake rates (and growth rates) increase until the film thickness exceeds the penetration depth. The ideal film thickness is defined as that thickness which results in an overall rate of substrate uptake (expressed as flux) which is 99 per cent of that given by a very thick film, i.e. $N = 0.99\ N_L \rightarrow \infty$ (*see* Fig. 6).

Atkinson and Knights (1975) presented the following formula as a rough estimate of the ideal film thickness of a microbial film when used in continuous fermentation.

$$\left(\frac{\mu_{\max}}{Y_s D_e K_m}\right)^{1/2} L_{\text{ideal}} = \frac{s^*}{K_m} \frac{(1 + 2s^*/K_m)^{1/2}}{(1 + s^*/K_m)} \tag{24}$$

Fig. 15 provides a more accurate estimate of $k_2 L_{\text{ideal}}$. These data were obtained by Rahman (1982) by numerical solution of the problem of diffusion with Monod reaction. Eq. (24) and Fig. 15 can be applied to immobilized enzymes if the parameters are reinterpreted using Eqs. (10) and (11). In the latter case, the L_{ideal} represents the 'reactive' thickness, i.e. the zone in which the reaction is taking place.

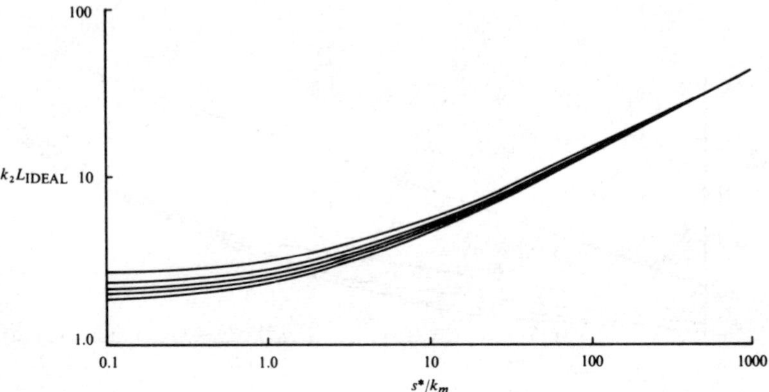

Figure 15. Ideal film thickness for a microbial film ($n = N/N_{L\rightarrow\infty}$).

Critical Particle Size

Endogenous respiration and lysis take place in the centre of a particle that is larger than the critical particle size. The concentration at the centre can therefore be taken as the lowest allowable for a given microbe–substrate system, i.e. that which avoids endogenous respiration and lysis within the biomass.

Figs. 16(b) and 17(b) provide data on the critical particle size for given s^* and centre line concentration s_0 for both spheres and cylinders. Figs. 16(a) and 17(a) provide the corresponding specific rates of substrate uptake. These data were obtained by Rahman (1982) by numerical solution of the problem of diffusion with Monod reaction.

Figs. 16 and 17 can be used for the determination of the 'reactive' radius of enzyme particles if the parameters are reinterpreted using Eqs. (10) and (11). Particle radii determined in this manner ensure that the enzyme activity is being used throughout the particle. In particular, if s_0 is chosen as s^*_{CRIT} of the Michaelis equation, then zero-order conditions will occur throughout.

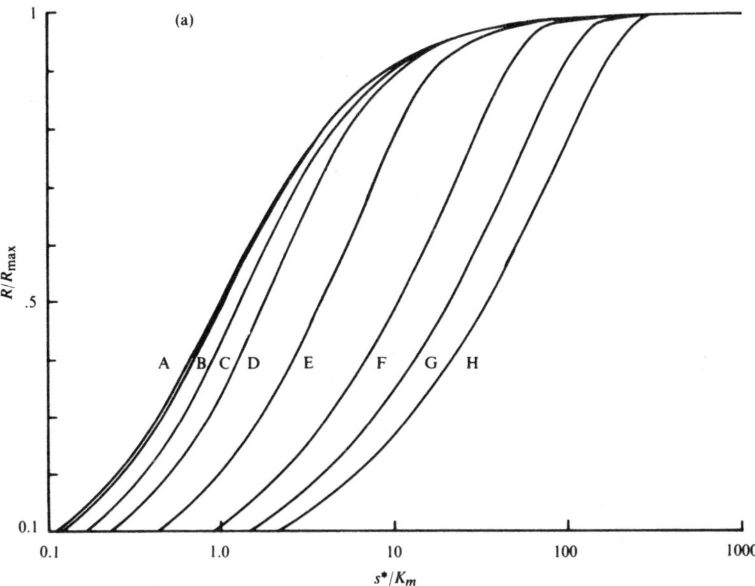

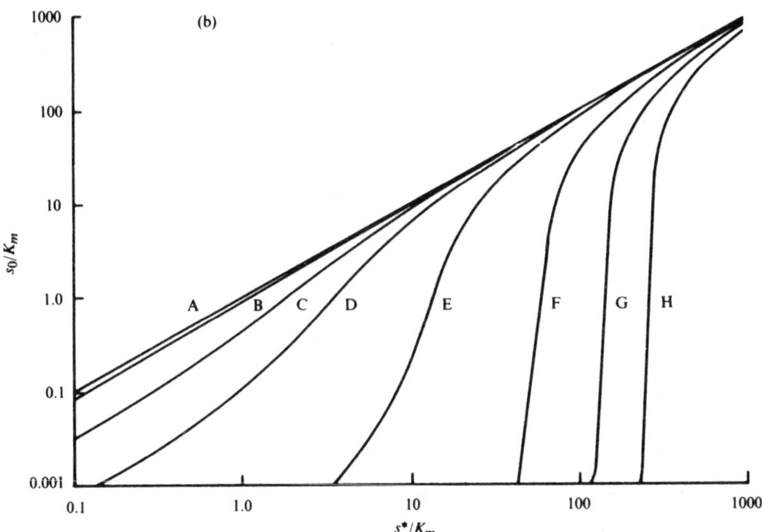

Figure 16. (a) Overall rate of reaction versus external concentration with particle size as parameter (spheres). (b) Centre-line concentration versus external concentration. Curve A, $k_2r_o = 0.5$; curve B, $k_2r_o = 1.0$; curve C, $k_2r_o = 3.0$; curve D, $k_2r_o = 5.0$; curve E, $k_2r_o = 10.0$; curve F, $k_2r_o = 20.0$; curve G, $k_2r_o = 30.0$; curve H, $k_2r_o = 40.0$.

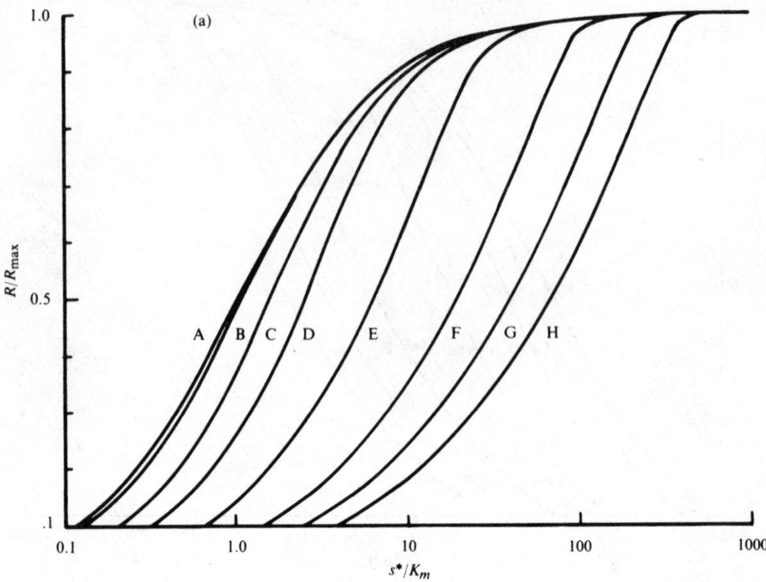

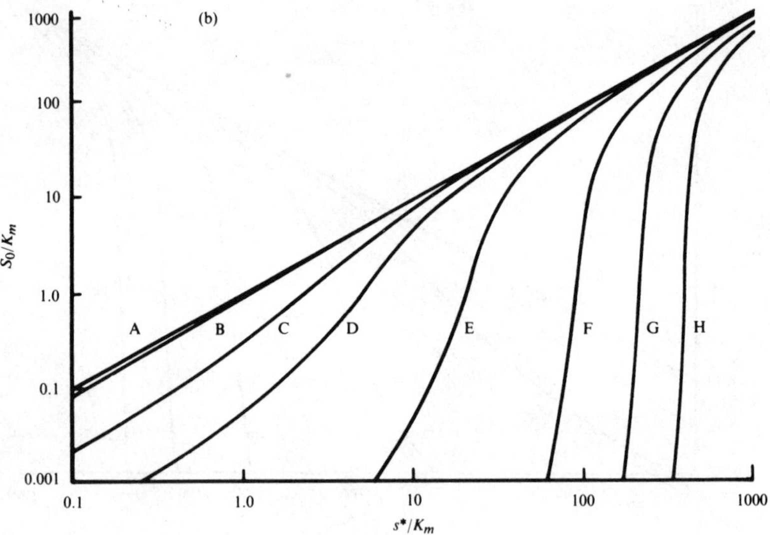

Figure 17. (a) Overall rate of reaction versus external concentration with particle size as parameter (cylinders). (b) Centre-line concentration versus external concentration. Curve A, $k_2r_o = 0.5$; curve B, $k_2r_o = 1.0$; curve C, $k_2r_o = 3.0$; curve D, $k_2r_o = 5.0$; curve E, $k_2r_o = 10.0$; curve F, $k_2r_o = 20.0$; curve G, $k_2r_o = 30.0$; curve H, $k_2r_o = 40.0$.

OPTIMUM PARTICLE SIZE FOR SUBSTRATE-INHIBITED SYSTEMS

Monod/Michaelis–Menten kinetics modified to account for substrate inhibition are described in Chapter 4 (*see* p.209) by the equation

$$\frac{R}{R_{\max}} = \frac{s^*/K_m}{1 + s^*/K_m + \alpha^2 (s^*/K_m)^2} \tag{25}$$

where

$$\alpha = \left(\frac{K_m}{K_i}\right)^{1/2}$$

Rahman (1982) showed that replacement of the Monod/Michaelis–Menten equation by Eq. (25) in problems associated with diffusion plus biological reaction results in Fig. 18 for spherical particles when $\alpha = 0.1$. Table 6 presents $k_2 r_0$ values and corresponding optimum substrate concentrations for the curves in Fig. 18.

Table 6. $k_2 r_0$ values and optimum substrate concentrations in Fig. 18.

Curve	$k_2 r_0$	Optimum substrate concentration $[(s^*/K_m)_{\text{opt}}]$
A	0.5	10.02
B	1.0	10.06
C	3.0	10.52
D	5.0	11.51
E	10.0	17.44
F	20.0	40.06
G	30.0	73.31
H	40.0	101.84

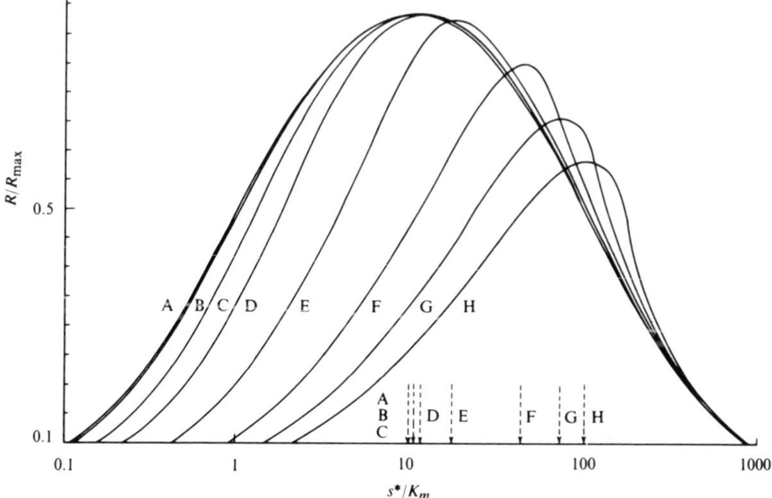

Figure 18. Effect of particle size on the overall rate of substrate uptake for diffusion-limited spherical particles ($\alpha = 0.1$).

Comparison of Fig. 18 with Fig. 5 (when $\alpha = 0.0$) shows the existence of rate maxima at the optimum substrate concentration when $\alpha > 0$ and that in the substrate-inhibited region the overall rate of substrate uptake can be *enhanced* by the diffusional limitation. Even more usefully (since the degree of enhancement is not great), it appears that large particles can be used without reduction in the substrate uptake rate at large s^*/K_m values.

Figs. 19 and 20 provide information on the maximum overall rate of substrate uptake that can be achieved when the optimum particle size is used at any substrate concentration. Fig. 19 gives the maximum overall rate of substrate uptake and Fig. 20 the optimum particle size. In practice, the optimum particle size mainly defines the upper limit of size before the diffusion effects reduce the rate of reaction below that given by Eq. (25). This factor is illustrated by curve 1A of Fig. 20, which corresponds to the largest particle that can be used for λ to remain greater than unity.

Reasons for the phenomena shown in Fig. 18 are illustrated in Figs. 21, 22 and 23, which are substrate and local rate of substrate uptake profiles within a particle consisting of a substrate-inhibited system ($\alpha = 0.1$). In all three figures, the substrate concentration falls with distance within the particle. The corresponding substrate uptake rates, calculated from Eq. (25) and evaluated with $s^* = s(r/r_o)$, show quite different behaviour. When s^*/K_m is low, $R_{local} < R^*$ i.e. $\lambda < 1$, but when s^*/K_m is high, $R_{local} > R^*$ i.e. $\lambda > 1$.

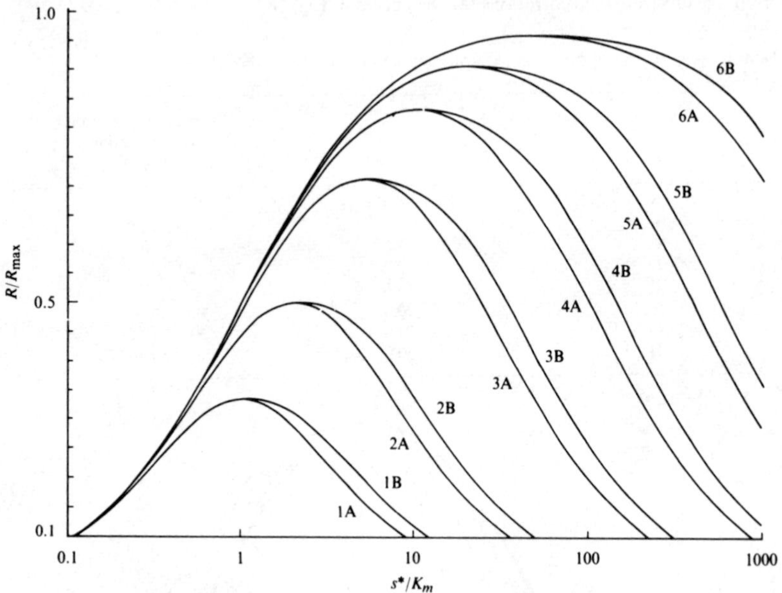

Figure 19. Overall rate of substrate uptake when using optimal particle size. α values used are presented in Table 7. Curves A, $k_2 r_o = 0$; curves B, $k_2 r_o =$ optimal values (given on Fig. 20).

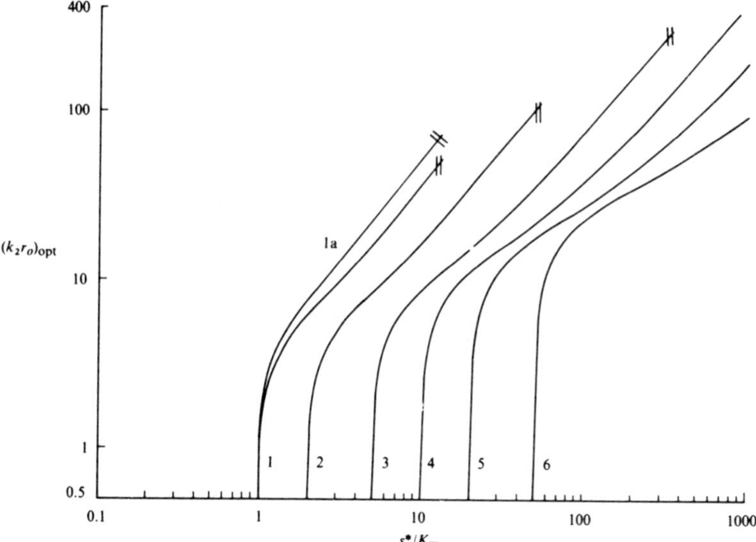

Figure 20. Optimal values of particle size corresponding to maximal overall rate of substrate uptake for a given substrate concentration (*see* Fig. 19). Curve 1A, maximum particle size for $\lambda \geqslant 1$.

Table 7. α values used in Figs. 19 and 20.

Curve	α
1	1.0
2	0.5
3	0.2
4	0.1
5	0.05
6	0.02

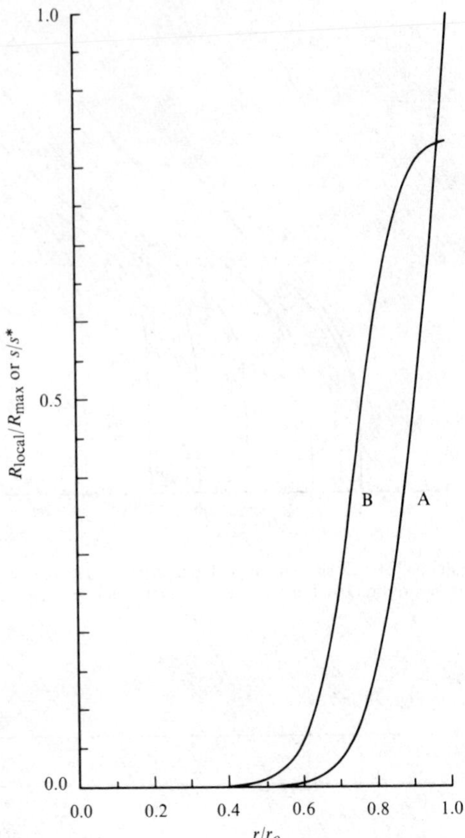

Figure 21. Substrate concentration (curve A) and local rate of reaction (curve B) profiles within a microbial particle subject to substrate inhibition at low external substrate concentration. $\alpha = 0.1$, $k_2 r_o = 20$, $s^*/K_m = 10$.

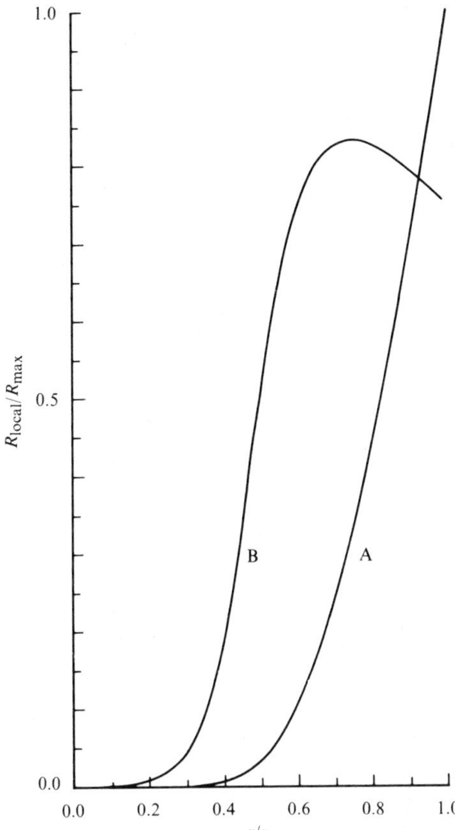

Figure 22. Substrate concentration (curve A) and local rate of reaction (curve B) profiles within a microbial particle subject to substrate inhibition at intermediate external substrate concentration. $\alpha = 0.1$, $k_2 r_o = 20$, $s^*/K_m = 30$.

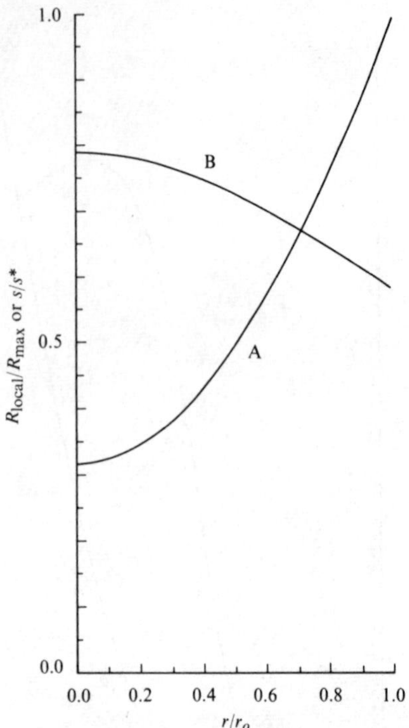

Figure 23. Substrate concentration (curve A) and local rate of reaction (curve B) profiles within a microbial particle subject to substrate inhibition at high external substrate concentration. $\alpha = 0.1$, $k_2r_o = 20$, $s^*/K_m = 70$.

PRODUCT EFFECTS

Figs. 24 and 25 show experimental data obtained by direct measurement of pH in the centre, i.e. pH_c, of a gel-immobilized urease particle, the effect of external concentration pH* being shown in Fig. 24 and of particle size being shown in Fig. 25. The differences between external and internal pH values (i.e. $pH_c - pH^*$) are considerable, and the effects of r_o and ρ_e suggest diffusional limitations on the egress of products. Since the activity of urease is highly pH-sensitive, it is more than likely that internal pH will significantly influence the overall rate of reaction, even when no diffusional limitations exist on the ingress of substrate.

Fig. 26 provides calculated data for gel-immobilized glucose oxidase (see Appendix A1, Chapter 6 for free solution kinetics) under zero-order conditions with respect to substrate. The data were obtained over a range of slab thicknesses and external pH values for a fixed enzyme concentration. The underlying 'bell-shaped' curve represents the free solution pH* characteristic. Clearly, $\lambda < 1$ at low pH* and $\lambda > 1$ at high pH*, with an optimum particle size (in this case 36 μm).

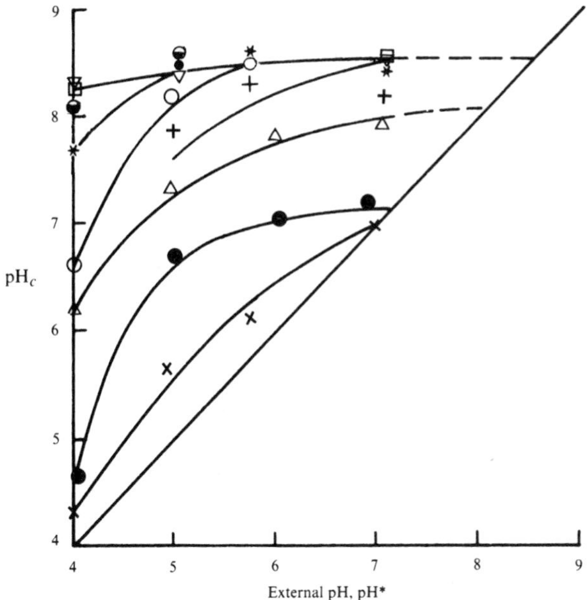

Figure 24. Influence of enzyme concentration on pH in the centre of a gel-immobilized urease particles pH$_c$. $r_o \sim 1$ mm, $s^* = 0.25$ M. Urease concentration before polymerization: x, 0.0 mg ml^{-1}; ●, 0.125 mg ml^{-1}; △, 0.25 mg ml^{-1}; +, 0.375 mg ml^{-1}; ○, 0.50 mg ml^{-1}; *, 0.625 mg ml^{-1}; □, 0.75 mg ml^{-1}; ▽, 0.875 mg ml^{-1}; ◒, 0.976 mg ml^{-1} [Atkinson *et al.* (1977)].

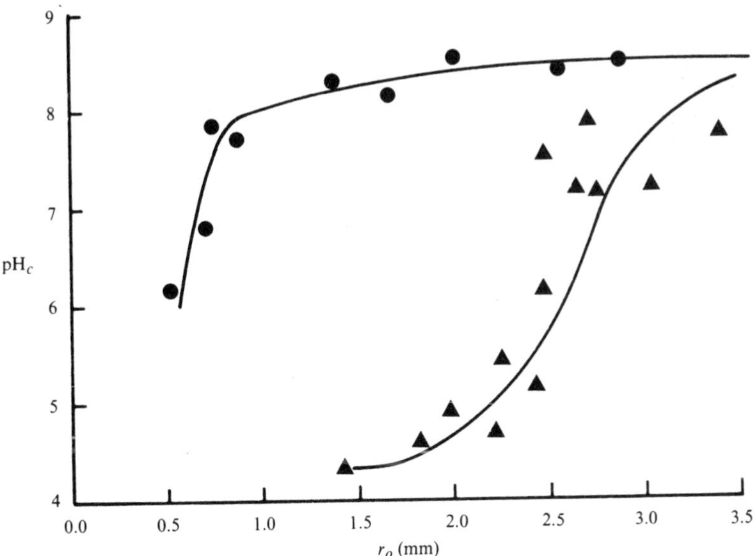

Figure 25. Effect of particle size on pH in the centre of gel-immobilized urease particles pH$_C$. pH* = 4.0, $s^* = 0.25$ M. Urease concentration before polymerization: ▲, 0.2 mg ml^{-1}; ●, 0.41 mg ml^{-1} [Atkinson *et al.* (1977)].

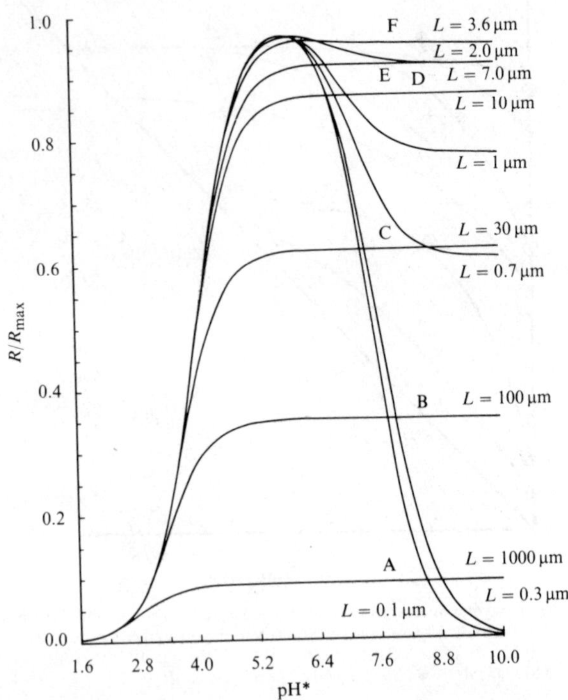

Figure 26. Overall rates of reaction versus pH* for gel-immobilized glucose oxidase particles (slab geometry) [Rahman (1982)].

Figs. 27, 28 and 29 show the internal conditions with various particles and conditions. These figures show qualitative similarity with Figs. 21, 22 and 23. Three conditions are evident.

1) When pH* > pH (free solution) > pH_c, as occurs in Fig. 27, a maximum occurs in the local rate of the reaction profile and λ may be greater or less than unity.
2) When pH_{opt} (free solution) > pH* > pH_c, as occurs in Fig. 28, the local rate of reaction is everywhere lower than that at the surface and $\lambda < 1$.
3) When pH* > pH_c > pH_{opt} (free solution), as occurs in Fig. 29, the local rate of reaction is everywhere greater than that at the surface and $\lambda > 1$.

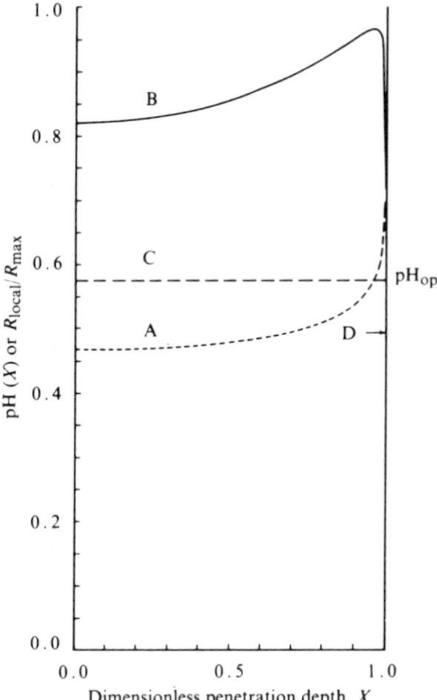

Figure 27. Internal condition within a gel-immobilized glucose oxidase particle (slab geometry). pH* = 7.5, L = 10 μm. Curve A, pH profile; curve B, local rate of reaction profile; line C, pH_{opt}; point D, rate of reaction at the external boundary.

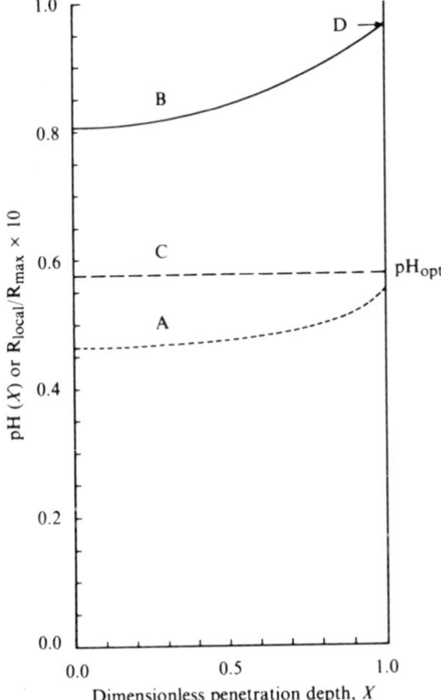

Figure 28. Internal condition within a gel-immobilized glucose oxidase particle (slab geometry). pH* = 5.5, L = 1 μm. Curve A, pH profile; curve B, local rate of reaction profile; line C, pH_{opt}; point D, rate of reaction at the external boundary.

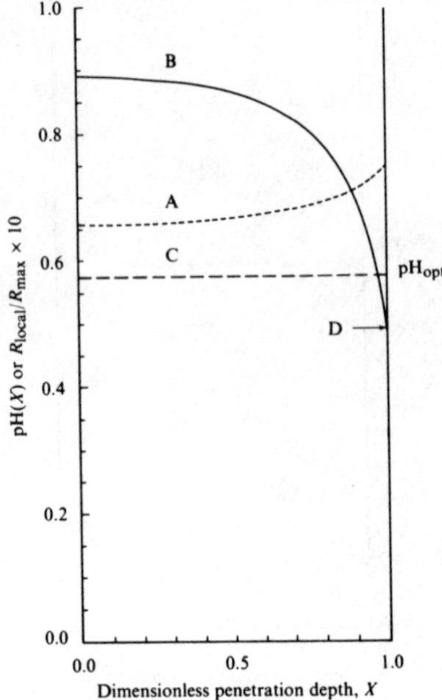

Figure 29. Internal condition within a gel-immobilized glucose oxidase particle (slab geometry). pH* = 7.5, L = 1 μm. Curve A, pH profile; curve B, local rate of reaction profile; line C, pH_{opt}; point D, rate of reaction at the external boundary.

NOMENCLATURE

A_p = external particle area
D_e = effective diffusion coefficient of substrate
d_p = particle diameter
h = liquid-phase mass transfer coefficient
h_{min} = minimum liquid-phase mass transfer coefficient
K_I = enzyme inhibition constant
K_m = Michaelis constant or Monod coefficient
k_2 = biological rate equation coefficient
L = microbial film thickness
L_{ideal} = ideal microbial film thickness
m = particle mass
N = flux of substrate
N_{max} = maximum flux of substrate
R = specific rate of substrate uptake
R_{max} = maximum specific rate of substrate uptake
r_o = microbial particle radius
Re = Reynolds number
Sc = Schmidt number
Sh = Sherwood number
s = substrate concentration
s^* = substrate concentration adjacent to particle
s^*_{CRIT} = critical substrate concentration adjacent to particle

s_0 = substrate concentration at the centre of particle
V_{max} = maximum rate of enzyme-catalyzed reaction
V_p = particle volume
X = dimensionless penetration depth
Y_s = biomass yield coefficient
α = particle shape coefficient
γ = ratio of diffusion coefficients (D_e/D)
λ = effectiveness factor
μ = fluid viscosity
μ_{max} = maximum specific growth rate
ρ = fluid density
ρ_e = enzyme density
ρ_o = biomass density
ρ_p = particle density
ϕ_P = dimensionless parameter

REFERENCES

Atkinson, B and Daoud, I.S. (1976) Microbial Flocs and Flocculation in Fermentation Process Engineering, in *Advances in Biochemical Engineering*, vol. 4, ed. by T.K. Ghose *et al.*, p.41 (Springer Verlag, Berlin).

Atkinson, B. and Daoud, I.S. (1968) The Analogy between Microbiological Reactions and Heterogeneous Catalysis. *Trans. Instn Chem. Engrs*, **46**, 19.

Atkinson, B. and Fowler, H.W. (1975) Significance of Microbial Film in Fermenters, in *Advances in Biochemical Engineering*, vol. 3, ed. by T.K. Ghose *et al.*, p.221 (Springer Verlag, Berlin).

Atkinson, B. and Knights, A.J. (1975) Microbial Film Fermenters — Their Present and Future Applications. *Biotechnol. Bioengng*, **17**, 1245.

Atkinson, B. and Lester, D.E. (1975) Enzyme Rate Equation for the Overall Rate of Reaction of Gel Immobilized Glucose Oxidase Particles under Buffered Conditions. I. Pseudo One Substrate Conditions. *Biotechnol. Bioengng*, **16**, 1299.

Atkinson, B., Rott, and Rousseau, (1977) Characteristics of Unbuffered Gel-Immobilized Urease Particle. I. Internal pH. *Biotechnol. Bioengng*, **19**, 1037.

Atkinson, B. and Rahman, F. (1979) Effects of Diffusion Limitations and Floc Size Distributions on Fermenter Performance and Interpretation of Experimental Data. *Biotechnol. Bioengng*, **21**, 221.

Frössling, N. (1938) Vaporization of the Falling Drop. *Gerland Beitr. Geophys.*, **52**, 170.

Kornegay, B.H. and Andrews, J.F. (1968) Kinetics of Fixed-Film Biological Reactors. *J. Water Poll. Contr. Fed.*, **40**, 460.

Parker, D.S., Kaufman, W.J. and Jenkins, D. (1971) Physical Conditioning of Activated Sludge Floc. *J. Water Poll. Contr. Fed.*, **44**, 414.

Rahman, F. (1982) Computational Studies in Diffusion-Limited Bioreactions, Ph.D. Thesis, University of Manchester.

Sanders, W.H. III (1966) Oxygen Utilization by Slime Organisms in Continuous Culture. *Air Water Poll.*, **10**, 253.

Tomlinson, T.G. and Snaddon, D.H.M. (1966) Biological Oxidation of Sewage by Films of Microorganisms. *Air Water Poll.*, **10**, 865.

CHAPTER 11

HEAT TRANSFER

IMPORTANCE OF HEAT EFFECTS IN BIOCHEMICAL PROCESSES

Within biological process engineering, heat transfer is involved in the following processes.

1. Heat sterilization of liquid media, air and the reaction vessels prior to use.
2. Removal of heat evolved during exothermic reaction processes.
3. Addition of heat when the optimal process temperature is high and the balance between heat losses and heat produced by the reaction fails to maintain this temperature (anaerobic digestion provides an example).
4. During the downstream processing of fermentation products, e.g., distillation, evaporation and drying (*see* Chapter 13).

The importance of these heat effects in the case of biological reactors is summarized in Fig. 1.

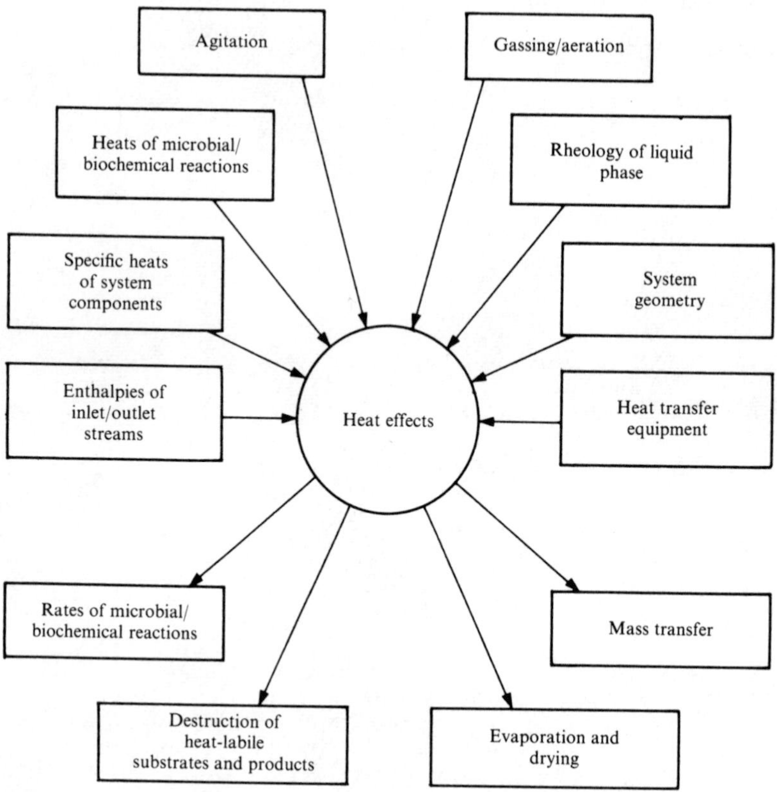

Figure 1. Factors contributing to and phenomena affected by heat effects in biological reactors.

GENERAL ENERGY BALANCE EQUATION

A general energy balance for a biological reactor can be expressed as

$$Q_{met} + Q_{ag} + Q_{gas} = Q_{acc} + Q_{exch} + Q_{evap} + Q_{sen} \tag{1}$$

where Q_{met} = rate of heat generation per unit volume due to microbial metabolism or enzyme activity
Q_{ag} = rate of heat generation per unit volume due to agitation (i.e. power input)
Q_{gas} = rate of heat generation per unit volume due to gassing/aeration
Q_{acc} = rate of heat accumulation per unit volume in the system
Q_{exch} = rate of heat transfer per unit volume to the surroundings or to heat-exchanger coolant
Q_{evap} = rate of heat loss per unit volume by evaporation
Q_{sen} = rate of sensible enthalpy gain per unit volume by the flow streams (exit − inlet)

In Eq. (1), Q_{exch} may also be considered as the heat exchange duty. Additional terms for consideration in conjunction with Eq. (1) are as follows.

1. The heats of solution of solids or concentrated liquids added during the process (*see* Tables 1 and 2).
2. The heats of absorption/desorption when gases, such as ammonia, are used.

Table 1. Heats of solution [Liley and Gambil (1973)].

Solute	Dilution[a]	Formula	Heat of solution [kg cal (g mol)$^{-1}$]
Calcium acetate	∞	$Ca(C_2H_3O_2)_2$	+ 7.6
Calcium chloride	∞	$CaCl_2$	+ 4.9
Cupric sulphate	800	$CuSO_4$	+ 15.9
Ferric sulphate	1000	$FeCl_3$	+ 31.7
Ferrous chloride	400	$FeCl_2$	+ 17.9
Magnesium chloride	∞	$MgCl_2$	+ 36.3
Magnesium sulphate	∞	$MgSO_4$	+ 21.1
Manganous chloride	400	$MnCl_2$	+ 16.0
Manganous sulphate	400	$MnSO_4$	+ 13.8
Phosphoric acid			
Ortho	400	H_3PO_4	+ 2.79
Pyro	aq	$H_4P_2O_7$	+ 25.9
Potassium acetate	∞	$KC_2H_3O_2$	+ 3.55
Potassium chloride	∞	KCl	− 4.4
Potassium hydroxide	∞	KOH	+ 12.91
Potassium iodide	∞	KI	− 5.23
Potassium nitrate	∞	KNO_3	− 8.63
Potassium sulphate	∞	K_2SO_4	− 6.32
Sodium acetate	∞	$NaC_2H_3O_2$	+ 4.09
Sodium chloride	∞	NaCl	− 1.16
Sodium hydroxide	∞	NaOH	+ 10.18
Sodium (di) phosphate	1600	Na_2HPO_4	+ 5.21
Sodium (tri) phosphate	1600	Na_3PO_4	+ 13.00
Sodium sulphate	∞	Na_2SO_4	+ 0.28

[a] Moles of water used to dissolve 1 g formula weight of substance; ∞, 'infinite dilution'; aq, aqueous solution of unspecified dilution.

Table 2. Heats of solution of organic compounds in water at infinite dilution and approximately room temperature [Liley and Gambil (1973)].

Solute	Formula	Heat of solution [kcal (g mol)$^{-1}$]
Acetic acid	$C_2H_4O_2$	− 2.25
Acetylacetone	$C_5H_8O_2$	− 0.64
Acetylurea	$C_3H_6N_2O_2$	− 6.81
Aconitic acid	$C_6H_6O_6$	− 4.21
Ammonium benzoate	$C_7H_9NO_2$	− 2.7
Ammonium succinate	$C_4H_{12}N_2O_4$	− 3.49
Aniline hydrochloride	C_6H_8ClN	− 2.73
Benzoic acid	$C_7H_6O_2$	− 6.50
Camphoric acid	$C_{10}H_{16}O_4$	+ 0.50
Citric acid	$C_6H_8O_7$	− 5.40
Dextrin	$C_{12}H_{20}O_{10}$	− 0.27
Fumaric acid	$C_4H_4O_4$	− 5.90
Hexamethylenetetramine	$C_6H_{12}N_4$	+ 4.78
m-Hydroxybenzamide	$C_7H_7NO_2$	− 4.16
m-Hydroxybenzamide hydrochloride	$C_7H_8ClNO_2$	− 7.00
o-Hydroxybenzamide	$C_7H_7NO_2$	− 4.34
p-Hydroxybenzamide	$C_7H_7NO_2$	− 5.39
o-Hydroxybenzoic acid	$C_7H_6O_3$	− 6.35
p-Hydroxybenzoic acid	$C_7H_6O_3$	− 5.78
o-Hydroxybenzyl alcohol	$C_7H_8O_2$	− 3.20
Inulin	$C_{36}H_{62}O_{31}$	− 0.096
i-Succinate	$C_4H_6O_4$	− 3.42
Itaconic acid	$C_5H_6O_4$	− 5.92
Lactose	$C_{12}H_{22}O_{11}.H_2O$	− 3.71
Maleic acid	$C_4H_4O_4$	− 4.44
Malic acid	$C_4H_6O_5$	− 3.15
Malonic acid	$C_3H_4O_4$	− 4.49
Mandetic acid	$C_8H_8O_3$	− 3.09
Mannitol	$C_6H_{14}O_6$	− 5.26
Menthol	$C_{10}H_{20}O$	0.00
Nicotine dihydrochloride	$C_{10}H_{16}Cl_2N_2$	+ 6.56
m-Nitrobenzoic acid	$C_7H_5NO_4$	− 5.59
o-Nitrobenzoic acid	$C_7H_5NO_4$	− 5.31
p-Nitrobenzoic acid	$C_7H_5NO_4$	− 8.89
m-Nitrophenol	$C_6H_5NO_3$	− 5.21
o-Nitrophenol	$C_6H_5NO_3$	− 6.31
p-Nitrophenol	$C_6H_5NO_3$	− 4.49
Oxalic acid	$C_2H_2O_4$	− 2.29
Oxalic acid. $2H_2O$	$C_2H_4O_4.2H_2O$	− 8.49
Phenol	C_6H_6O	− 2.61
Phthalic acid	$C_8H_6O_4$	− 4.87
Potassium benzoate		− 1.51
Potassium citrate		+ 2.82
Potassium tartrate		− 5.56
Pyrogallol	$C_6H_6O_3$	− 3.71
Pyrotartaric acid		− 5.02
Quinone	$C_6H_4O_2$	− 3.99
Raffinose	$C_{18}H_{32}O_{16}.\ 5H_2O$	− 9.70
Resorcinol	$C_6H_6O_2$	− 3.96
Silver malonate		− 9.80
Sodium citrate		+ 5.27
Sodium potassium tartrate		− 1.82
Sodium potassium tartrate. $4H_2O$		− 12.34
Sodium succinate		+ 3.39
Sodium succinate. $6H_2O$		− 10.99
Sodium tartrate		− 1.21
Sodium tartrate. $2H_2O$		− 5.88
Succinic acid	$C_4H_6O_4$	− 6.41

continued

(*Table 2 – continued*)

Solute	Formula	Heat of solution [kcal (g mol)$^{-1}$]
Succinimide	$C_4H_5NO_2$	− 4.30
Sucrose	$C_{12}H_{22}O_{11}$	− 1.32
D-Tartaric acid	$C_4H_6O_6$	− 3.45
Thiourea	CH_4N_2S	− 5.33
Urea	CH_4N_2O	− 3.61
Urea acetate		− 8.80
Urea formate		− 7.19
Urea nitrate		− 10.80
Urea oxalate		− 17.81
Vanillic acid	$C_8H_8O_4$	− 5.16
Vanillin	$C_8H_8O_3$	− 5.21

Table 3 illustrates the relative magnitudes of the terms in the general heat balance equation [(Eq. (1)].

In Table 4, it can be seen that the heat transfer costs can be comparable to the oxygen transfer costs.

Table 3. Relative magnitudes of energy balance terms during a *Bacillus subtilis* fermentation on molasses [Wang *et al.* (1979)].

Heat of metabolism	Heat of agitation	Heat accumulation (kcal ℓ$^{-1}$ h^{-1})	Heat lost to surroundings	Heat of evaporation	Sensible heat lost
1.12	3.32	3.81	0.61	0.023	0.005
8.65	3.31	11.3	0.65	0.045	0.010

Table 4. Effect of substrate and yield coefficients on operating costs of fermentation [Bailey and Ollis (1977)].

Substrate	Cost [$ (lb cells)$^{-1}$]			
	Substrate	Oxygen transfer	Heat removal	Total
Acetate	0.167	0.0062	0.01	0.184
Alkanes	0.040	0.0097	0.014	0.064
Ethanol	0.088	0.0075	0.013	0.11
Glucose equivalents (molasses)	0.039	0.0023	0.0054	0.047
Maleate (as waste)	0.000	0.0046	0.0075	0.012
Methane	0.016	0.033	0.037	0.086
Methanol	0.050	0.012	0.019	0.081
i-Propranol	0.116	0.027	0.031	0.174

Heat Evolution during Microbial Activity

During metabolism, heat evolution is a consequence of the thermodynamics of the overall microbial activity. Apart from anaerobic digestion and some other thermophilic microbial activity, the amount of heat produced is usually so high that if it is not removed it raises the temperature of the contents of the fermenter to a level beyond the optimum range for the system.

The evolution of heat during metabolic activity is related to the utilization of the carbon and the energy source. When the carbon source is being actively incorporated into biomass through anabolism during growth, about 40 to 50 per cent of the available enthalpy in the substrate is conserved in the biomass, the rest being given off as heat. When the carbon source is being catabolized to provide energy for cell maintenance, all the enthalpy associated with the oxidation of the substrate is released as heat. This is shown in the case of *Streptomyces niveus* in Fig. 2.

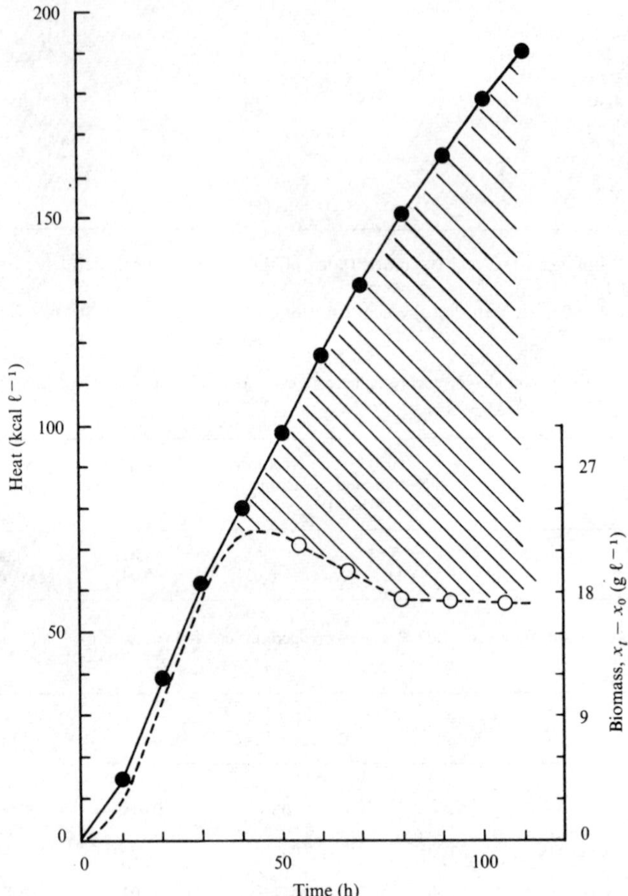

Figure 2. Heat evolution during *Streptomyces niveus* fermentation. Shaded area represents the heat evolved during culture maintenance. Clear area is the enthalpy conserved in the cell mass. ○, heat retained by growth and biomass; ●, total heat released by the substrate [Wang *et al.* (1979)].

If a biochemical product is formed, the heat evolved lies between the heat released during maintenance and that evolved during active growth. The amount of heat evolved is related to the stoichiometry for growth and product formation, while the rate of heat evolution is related to the rate of microbial activity.

If the stoichiometry of the microbial reactions is known (*see* Chapter 3), Q_{met} can be calculated from the standard enthalpies of substrates, biomass and products. For a chemical reaction

$$a_1 s_1 + a_2 s_2 + \ldots + a_n s_n \longrightarrow b_1 p_1 + b_2 p_2 + \ldots + b_m p_m \tag{2}$$

$$Q_{met} = \sum_{i=1}^{m} b_i H_{p,i} - \sum_{j=1}^{n} a_j H_{s,j} \tag{3}$$

where Q_{met} = total heat release by the microbial reaction
s = substrates
p = products (including biomass)
H_p, i = standard enthalpies of products
H_s, j = standard enthalpies of substrates

The standard enthalpies of a range of compounds are listed in Table 5.

The heat of reaction calculated from the enthalpy change depends on the stoichiometric numbers used. If, for example, each stoichiometric number is doubled, the heat of reaction is also doubled. Therefore, the stoichiometry is usually based on either moles or carbon-equivalents of substrate, biomass and products.

The standard enthalpies of various compounds are given in Table 5, but they can also be calculated from heats of formation H_f^0 or from heats of combustion H_{ox}^0, which are given for several compounds in Table 6 [*see* Chapters 2 and 3, and Smith and van Ness (1959)].

Table 5. Standard molar enthalpy and free energy of various compounds [Roels (1980)].

Component	Formula	Standard state	H^{0a} [kJ (C mol)$^{-1}$]	μ_f^{0b} [kJ (C mol)$^{-1}$]
Acetic acid	$C_2H_4O_2$	Aqueous	− 244.8	− 186.4
Ammonia	NH_3	Gas	− 46.3	− 16.7
		Aqueous	− 80.9	− 26.7
Ammonium ion	NH_4^+	Aqueous	− 133	− 79.6
Bicarbonate ion	HCO_3^-	Aqueous	− 692	− 588
Biomass	$CH_{1.8}O_{0.5}N_{0.2}$	Aqueous	− 91.4	− 67.1
Carbon dioxide	CO_2	Gas	− 394	− 395
Citric acid	$C_6H_8O_7$	Aqueous		− 195
Dodecane	$C_{12}H_{26}$	Liquid	− 29.3	2.5[c]
Ethane	C_2H_6	Gas	− 42.4	− 16.4
Ethanol	C_2H_6O	Aqueous	− 139	− 90.9
Formic acid	CH_2O_2	Aqueous	− 410.6	− 335.2
Fumaric acid	$C_4H_4O_4$	Aqueous		− 151.3
Glucose	$C_6H_{12}O_6$	Aqueous	− 211	− 153.1
Glycerol	$C_3H_8O_3$	Liquid	− 222.3	− 163.0
Lactic acid	$C_3H_6O_3$	Aqueous		− 173.0
Malic acid	$C_4H_6O_5$	Aqueous		− 141.0
Methane	CH_4	Gas	− 75.0	− 50.9
Methanol	CH_4O	Liquid	− 239	− 176.5
Nitric acid	HNO_3	Aqueous	− 173	− 110.7
Nitrogen	N_2	Gas	0	0
Oxalic acid	$C_2H_2O_4$	Aqueous	− 414	− 338
Oxygen	O_2	Gas	0	0
Pentane	C_5H_{12}	Aqueous	− 34.6	− 3.9
Propane	C_3H_8	Gas	− 34.7	− 7.8
Succinic acid	$C_4H_6O_4$	Aqueous		− 172.8
Water	H_2O	Liquid	− 286	− 238
	H^+	(pH = 7)		− 40.5

[a] H^0, the standard molar enthalpy, refers to the pure component in the standard state and is expressed per carbon mole of the component.
[b] μ_f^0 refers to 1 M aqueous solutions of the respective compounds.
[c] By extrapolation of data for lower hydrocarbons.

Table 6. Free energy and oxidation data for several compounds[a, b] [Servizi and Bogan (1964)].

Compound	Standard state	Heat of formation (kcal mol^{-1})	Heat of combustion (kcal mol^{-1})	Oxygen consumption (mol O_2 mol^{-1})
Acetaldehyde	Liquid	− 31.9	− 270	2.5
Acetic acid	Liquid	− 94.5	− 208	2.0
Alanine	Solid	− 88.9	− 161	1.5
Aniline	Liquid	+ 35.4	− 717	7.0
Arabinose	Solid	−180.5	− 573	5.0
Benzene	Liquid	+ 29.8	− 765	7.5
Benzoic acid	Liquid	− 59.2	− 773	7.5
n-Butanol	Liquid	− 40.4	− 621	6.0
Butyric acid	Liquid	− 91.5	− 514	5.0
Formaldehyde	Aqueous	− 31.0	− 120	1.0
Formic acid	Liquid	− 85.1	− 66	0.5
Fumaric acid	Solid	− 157	− 333	3.0
Glucose	Solid	−216	− 687	6.0
Glutamic acid	Solid	− 170.4	− 475	4.5
Glycerol	Liquid	− 113.6	− 407	35.0
Glycine	Solid	− 87.8	− 161	1.5
Heptanoic acid	Liquid	− 86.2	− 973	9.5
Lactic acid	Liquid	− 124	− 340	3.0
Malic acid	Solid	−211	− 336	3.0
Mannitol	Solid	−222.2	− 740	7.5
Methyl formate	Liquid	− 71.0	− 231	2.0
Naphthalene	Liquid	+ 45.9	−1216	12.0
Nitrobenzene	Liquid	+ 36.4	− 733	7.5
m-Nitrobenzoic acid	Solid	− 54.2	− 738	7.5
Octanoic acid	Liquid	− 84.6	−1124	11.0
Phenol	Solid	− 11.0	− 729	7.0
Propionic acid	Liquid	− 92.1	− 360	3.5
Pyrocatechol	Solid	− 51.4	− 684	6.5
Pyruvic acid	Solid	−114.1	− 282	2.5
Resorcinol	Solid	− 53.2	− 682	6.5
Succinic acid	Liquid	−178.8	− 369	3.5
Sucrose	Solid	−371.6	−1382	12.0
Toluene	Liquid	+ 26.5	− 914	9.0
m-Xylene	Liquid	+ 27.0	−1065	10.5

[a] Amino-nitrogen converted to ammonia. Nitro-nitrogen converted to nitrate.
[b] Terminal oxidant oxygen.

Table 7 gives the free energy and enthalpy changes of anaerobic glucose metabolism.

Table 7. Free energy and enthalpy change for various anaerobic product formation reactions starting with glucose [Roels (1980)].

Product	Free energy change [kJ (C mol glucose)$^{-1}$]	Enthalpy change [kJ (C mol glucose)$^{-1}$]
Acetic acid	−53.6	−34
Ethanol	−38	−13
Glycerol	− 9	+ 5
Methane	−67	−24
Methanol	−16	+16

In practice, stoichiometry of the form defined in Eq. (2) is rarely available, and recourse has to be made either to various assumptions regarding the degree of oxidation or to experimental data (*see* Chapter 2). Tables 8 and 9 provide some typical experimental data for the heat release per unit of cell growth.

Table 8. Influence of substrate and cell yield on oxygen requirement and heat production [Moss and Smith (1977)].

Microorganism	Substrate	Cell yield ($g\ \ell^{-1}$)	Oxygen required [g (100 g cells)$^{-1}$]	Heat released [k cal/ (100 g cells)$^{-1}$]	Heat released [kJ (100 g cells)$^{-1}$]
Bacteria	*n*-Alkanes	1.0	172	780	3266
Yeasts	Carbohydrates	0.5	67	380	1591
Yeasts	*n*-Alkanes	1.0	197	799	3345

Table 9. Heat released during the continuous culture of microorganisms using different substrates[a] [Moss and Smith (1977)].

Substrate	Cell yield (% of substrate used)	Heat released [kcal (100 g cell)$^{-1}$]	Heat released [kJ (100 g cell)$^{-1}$]	Rate of heat released (k cal h^{-1})	Rate of heat released (kJ h)$^{-1}$
n-Alkanes (C_{12}–C_{18})	100	780	3270	12	50
Methane	60	1800	7550	30	125
Sucrose	50	380	1590	6	25

[a] Dilution rate, 0.10 h^{-1}; cell density, 15 g ℓ^{-1}.

Figs. 3 and 4 suggest that the rate of heat evolution is proportional to the oxygen uptake rate, the latter providing a measure of the oxidation taking place.

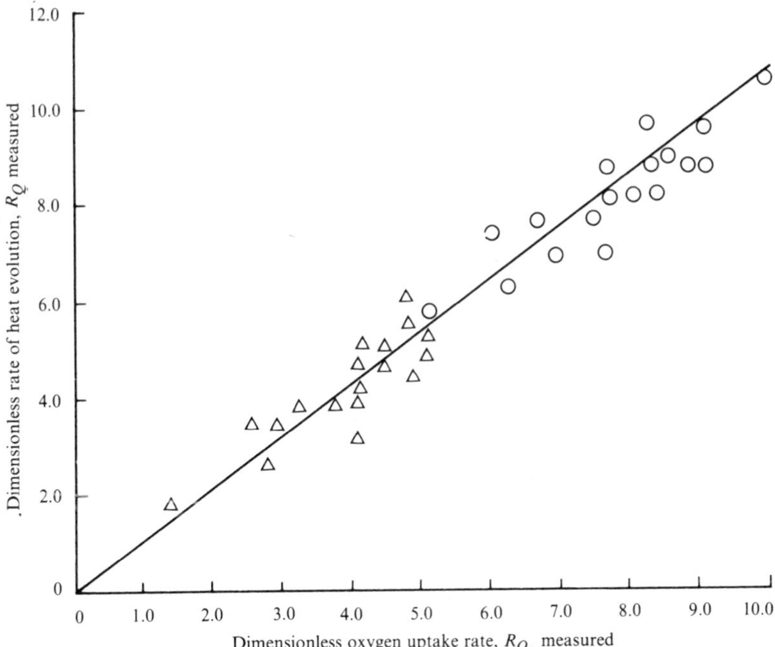

Figure 3. Relationship between heat evolution and oxygen uptake rates in fermentation processes. ○, yeast–hydrocarbon process; △, yeast–carbohydrate process [Ho (1979)].

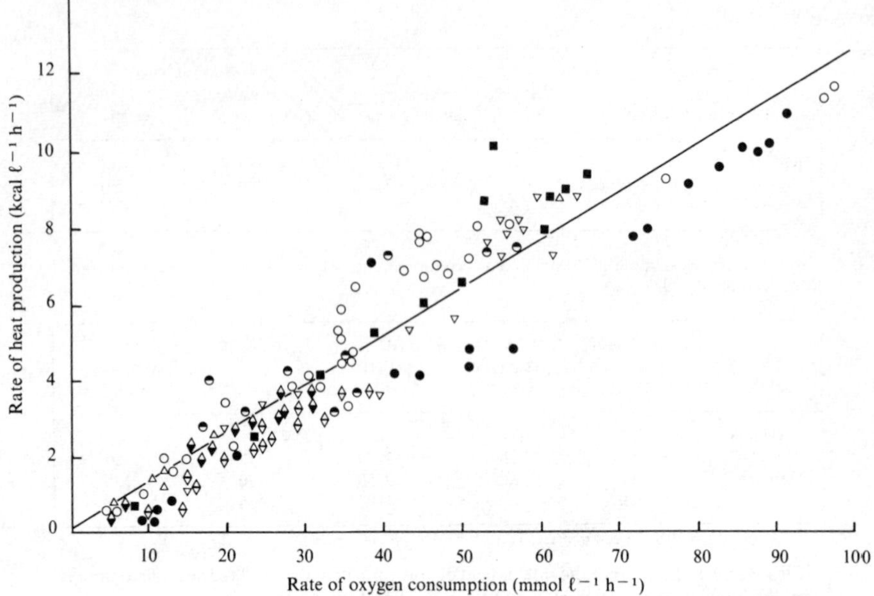

Figure 4. Correlation of heat evolution with oxygen consumption for a variety of microbial fermentations. ○, *Escherichia coli*, glucose; ◆ *Candida intermedia*, glucose; △, *C. intermedia*, molasses; ▽, *Bacillus subtilis*, glucose; □, *B. subtilis*, molasses; ◓, *B. subtilis*, sugar beet molasses; ⊖, *Aspergillus niger*, glucose; ●, *Asp. niger*, molasses [Wang *et al.* (1979)].

Estimated relationships between anticipated heat release and assumed yield coefficients for single-cell protein production from methane and methanol are illustrated in Figs. 5 and 6. The higher the biomass yield coefficient, i.e. the lower the level of substrate oxidation, the lower is the heat released. As in the case of Figs. 3 and 4, a linear relationship exists between heat production and cell formation (*see* Fig. 6).

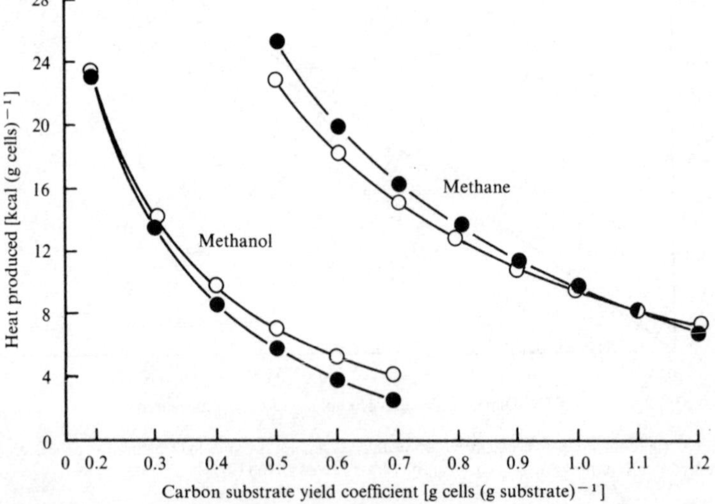

Figure 5. Estimated relationships between heat production (○) calculated from heats of combustion and yield coefficients (●) calculated from the correlation of Cooney in oxidation of methanol and methane by microorganisms [Hamer (1979)].

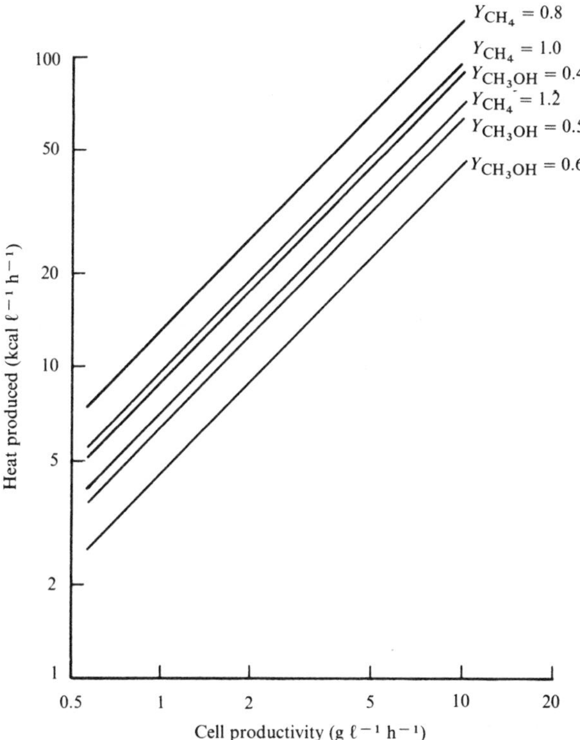

Figure 6. Estimated relationships between heat production and productivity at various yield coefficients [Hamer (1979)].

Heat produced by Power Input

The work performed on the contents of a fermenter by mechanical agitation and gassing/aeration appears as heat according to Joule's law and the mechanical equivalent of heat.

The power input to the liquid phase in a sparged system can be calculated from

$$\frac{P_g}{V} = \frac{F_g \rho_g}{V}\left(\frac{\alpha v_o^2}{2} + \frac{RT}{M}\ln\frac{p_o}{p}\right) \tag{4}$$

where P_g = power dissipated by sparged gas
V = clear liquid volume
F_g = gas flow rate
ρ_g = gas density
α = fraction of jet kinetic energy transmitted to bulk liquid, i.e. 0.06
v_o = gas velocity at the sparger hole
R = gas constant
T = absolute temperature
P_o = pressure at sparger
p = absolute pressure
M = molecular weight

This expression has been derived by Lehrer (1968). $v_o^2/2$ represents the jet kinetic energy developed at the sparger holes and transmitted to the bulk liquid. The second term within the brackets, i.e. RT/M In (p_o/p), is the energy required to move the gas through the static liquid head above the sparger. For well-designed spargers, the first term is often small and therefore can be neglected.

Mechanical power input in a sparged agitated system can be calculated using Eq. (5), which was recommended by Michel and Miller (1962).

$$\frac{P_m}{V} = \frac{0.706}{V}\left(\frac{P_{m,0}^2 N D^3}{F_g^{0.56}}\right)^{0.45} \tag{5}$$

where P_m and $P_{m,0}$ are aerated and unaerated power inputs, respectively, by mechanical agitation expressed in watts. The unaerated power input can be calculated from Eq. (6)

$$\frac{P_{m,0}}{V} = \frac{f N^3 D \rho_l}{V} \tag{6}$$

where ρ_l is the liquid density in kg m^{-3}, and N and D are impeller speed and diameter, respectively, expressed as s^{-1} and m, respectively. The power factor f can be obtained from the work of Rushton *et al.* (1950) and O'Connel and Mack (1950). Alternatively, graphs, such as shown in Fig. 7 and 8 can be used to calculate the unaerated mechanical power input.

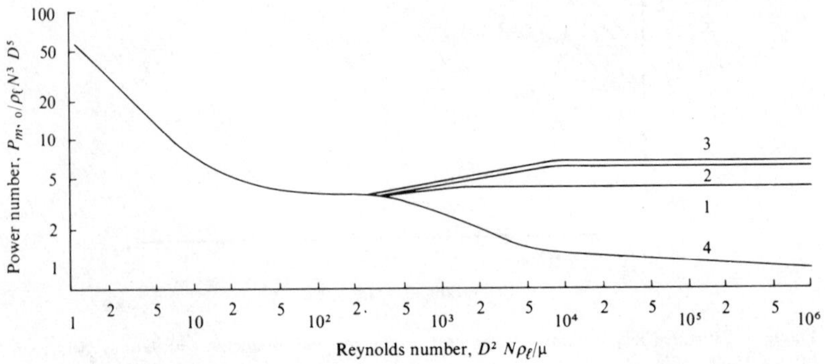

Figure 7. Power number as a function of Reynolds number of a 15.24 cm diameter 6-bladed turbine mixer. Tank diameter, 45.72 cm; turbine 15.24 cm above the bottom, liquid depth, 45.72 cm. Curve 1, baffles each 4 per cent tank diameter; curve 2, baffles each 10 per cent of tank diameter; curve 3, baffles each 17 per cent of tank diameter; curve 4, no baffles [Coulson and Richardson (1978)].

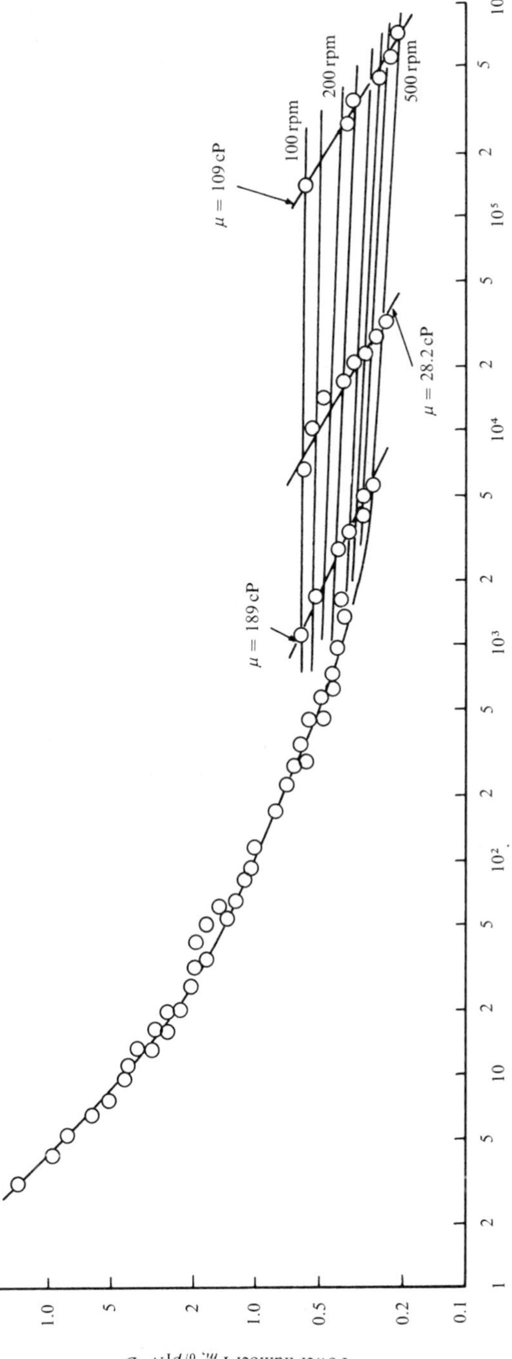

Figure 8. Power number as a function of Reynolds number for a 30.48 cm propeller mixer. Tank diameter, 137.16 cm; propeller 30.48 cm above tank bottom; liquid depth, 137.16 cm. [Coulson and Richardson (1978)].

Miller (1974) gives the following expression for the overall power input which is made up of the gassing power input [Eq. (4)] and mechanical power input [Eq. (5)], i.e.

$$\frac{P_i}{V} = \frac{P_m}{V} + c_1 \frac{P_g}{V} \tag{7}$$

Miller (1974) provides some values for c_1, which are presented in Fig. 9.

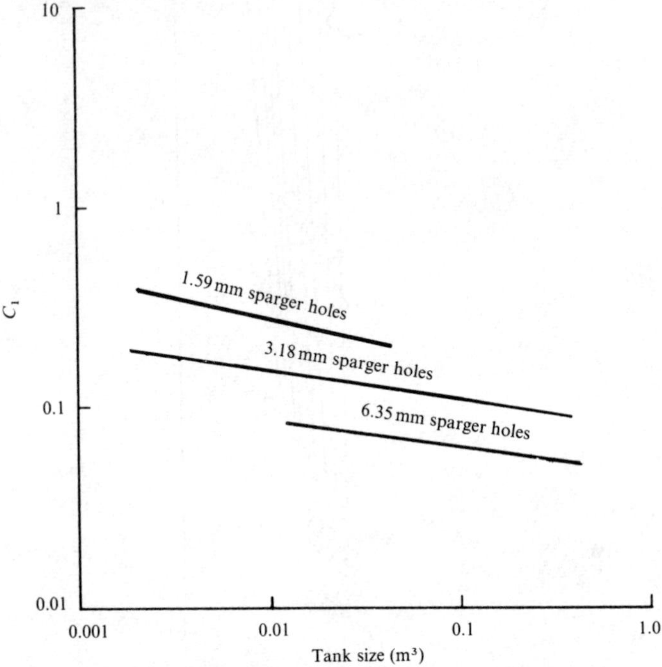

Figure 9. Power input correction factor [Miller (1974)].

If the power input to the contents of a fermenter comes from liquid jet energy, as in the case of loop reactors, the following expression proposed by Seipenbusch and Blenke (1980) may be used

$$P_l = F_l p_j = \frac{8}{\pi^2} \rho_l \frac{F_l^3}{D_j^4} \tag{8}$$

where P_l = liquid jet power
F_l = liquid flow rate
p_j = dynamic pressure (liquid nozzle)
D_j = diameter of the liquid jet nozzle.

Roels and van Suijdam (1980) have related the power input to the oxygen consumed in the fermenter as

$$P_i = k\phi_o \tag{9}$$

where P_i is the power input, ϕ_o the flow of oxygen through the system and k is a proportionality constant. For low-viscosity systems, k has a value of 88.4 kJ (mol O_2)$^{-1}$ and for high-viscosity systems a value of 523 kJ (mol O_2)$^{-1}$.

Sensible Enthalpy Gain by Flow Streams

When no phase change occurs, the sensible enthalpy gain Q_{sen} by a single inlet/outlet flow stream is given by

$$Q_{sen} = F\, c_p\, (T_e - T_i) \tag{10}$$

where F = flow rate
c_p = heat capacity of the stream
T_i = inlet stream temperature
T_e = exit stream temperature

Heat Loss due to Evaporation

Fig. 9 and Table 10 demonstrate the substantial quantities of water that can be lost due to evaporation during the course of a fermentation with concomitant heat losses.

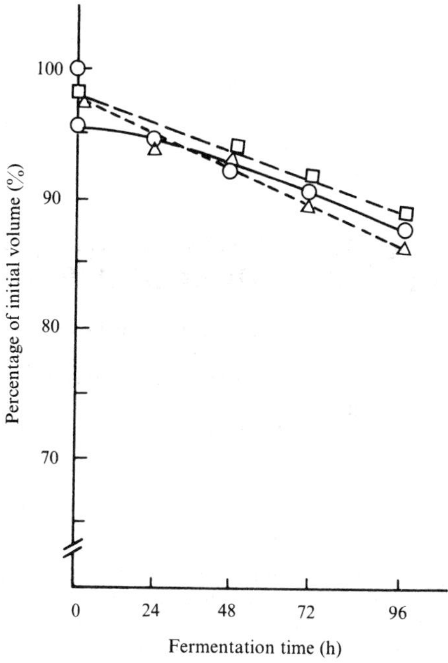

Figure 9. Evaporation loss in fermenters. Air flow rate, 0.6 vvm. □, 10 ℓ fermenter with condenser; △, 10 ℓ fermenter without condenser; ○, 200 ℓ fermenter without condenser [Yang (1979)].

Table 10. Evaporation losses from a 50 ml shake flask.[a]

Temperature (°C)	Losses (% of total volume)
30	11
32	13
34	15
37	19

[a] Duration of fermentation was 96 h.

The amount of heat lost due to evaporation is largely because of an increase in water vapour in the gas phase. If it is assumed that air enters and leaves in a saturated state but the air leaving the reactor is at a higher temperature, the heat loss due to evaporation can be expressed as follows

$$Q_{evap} \simeq G \lambda \Delta \mathscr{H} \tag{11}$$

where G is the dry air flow, λ is the latent heat of evaporation and $\Delta \mathscr{H}$ is the change in the water content (*see* Table 11).

Table 11. Water content of saturated air.

Temperature (°C)	Water content [mol water (mol dry air)$^{-1}$]
10	0.012
20	0.024
25	0.033
30	0.044
35	0.059
40	0.079

When the air streams are not saturated, recourse has to be made to heat and mass balances involving both sensible and latent heats as summarized on psychrometric charts [*see* Coulson and Richardson (1977)].

EXPERIMENTAL DETERMINATION OF HEAT EVOLUTION DURING MICROBIAL REACTION

Experimental determination has been conducted in small-scale systems. The dynamic calorimetric method involves the determination of the individual terms of the general energy balance [Eq. (1)]. In this method, the whole fermenter is an 'adiabatic' system [*see* Cooney *et al* (1968)]. The temperature control is turned off and the temperature of the system is allowed to rise by 0.5–1°C. The heat accumulation is then estimated from the slope of the temperature–time plot and the overall heat capacity of the system $c_{p,sys}$ according to

$$Q_{acc} = \frac{\Delta T}{\Delta t} c_{p,sys} \tag{12}$$

where ΔT is the change in the temperature of the system in the time interval Δt.

Q_{sen} is usually much less than Q_{met} and therefore can be neglected. Otherwise, in a batch, adiabatic, aerobic fermentation system, it is the sensible heat gain of the air stream. Q_{evap} can be eliminated by using air saturated with water at the temperature of the fermentation. If Q_{sen} and Q_{evap} cannot be neglected, they can be calculated from the wet- and dry-bulb temperatures of the entering and leaving gas.

$Q_{ag} + Q_{gas} + Q_{exch}$ can be measured before inoculation for the agitation and gassing rates that will be employed during the fermentation. Then the values of the terms Q_{acc}, Q_{sen}, Q_{ag}, Q_{gas} and Q_{exch} can be used in the general heat balance equation [Eq. (1)] to determine Q_{met}.

If the system is not treated adiabatically, but instead cooling water is used to remove the heat generated, the water inlet and outlet temperatures and the flow rate should be measured. If all the heat that would otherwise be accumulated in the system is removed by water, then

$$Q_{acc} = m c_p (T_e - T_i) \tag{13}$$

where m is the water flow rate, c_p the specific heat of water, and T_i and T_e the entering and leaving temperatures, respectively, of the water. The other heat terms can be calculated as previously described. This method is better applied to fermenters with low heat transfer surface to vessel volume ratios.

HEAT REMOVAL

A significant amount of heat has to be removed from fermentation equipment, i.e. Q_{exch}, into the coolant – usually water – through heat exchangers of the following types.

1. Jackets
2. Helical coils within the fermenter vessel
3. External helical coils
4. Vertical baffle-type coils within the fermenter
5. External heat transfer circuits, e.g., passage though a shell and tube exchanger
6. Combinations of these heat exchangers

Typical heat exchangers are schematically represented in Fig. 10.

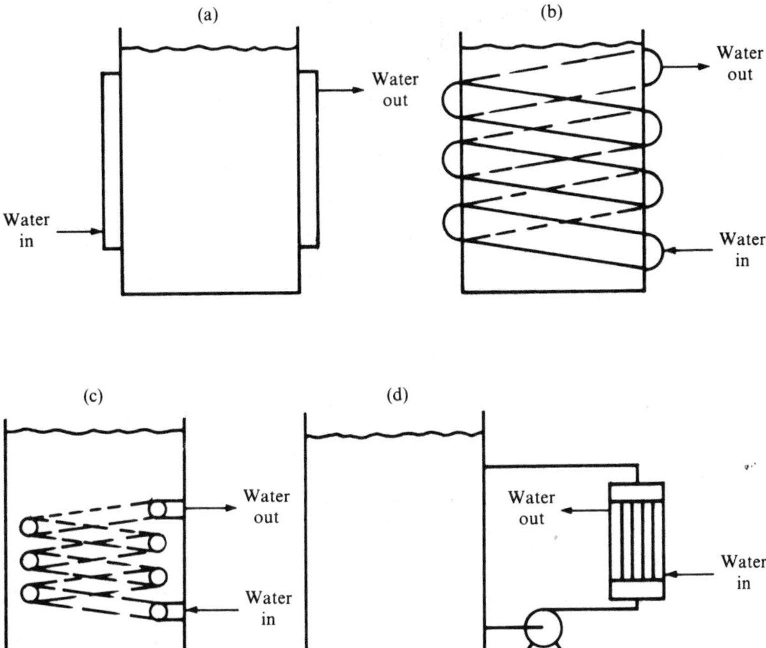

Figure 10. Methods of heat removal. (a) Jacket vessel; (b) external half coils (exaggerated dimensions); (c) internal coil, (d) external heat exchange.

The local heat flux q is related to the local temperature driving force ΔT by the equation

$$q = U \Delta T \tag{14}$$

where U is the local overall heat transfer coefficient. $1/U$ is the overall resistance to heat transfer and is composed of several resistances in series in the path of heat transfer. When the area normal to the direction of heat transfer is constant, as is the case for heat transfer across a plane wall (*see* Fig. 11)

$$U = \frac{1}{(1/h_i) + (1/h_{d,i}) + (l/k_w) + (1/h_{d,o}) + (1/h_o)} \quad (15)$$

where h_i and h_o are the individual film heat transfer coefficients of the inside and outside fluids, respectively, $h_{d,i}$ and $h_{d,o}$ are the fouling factors due to deposits (e.g., scale, dirt, microbial film, etc.) on the inside and outside surfaces, respectively, of the wall, k_w is the thermal conductivity of the wall and l is the thickness of the wall.

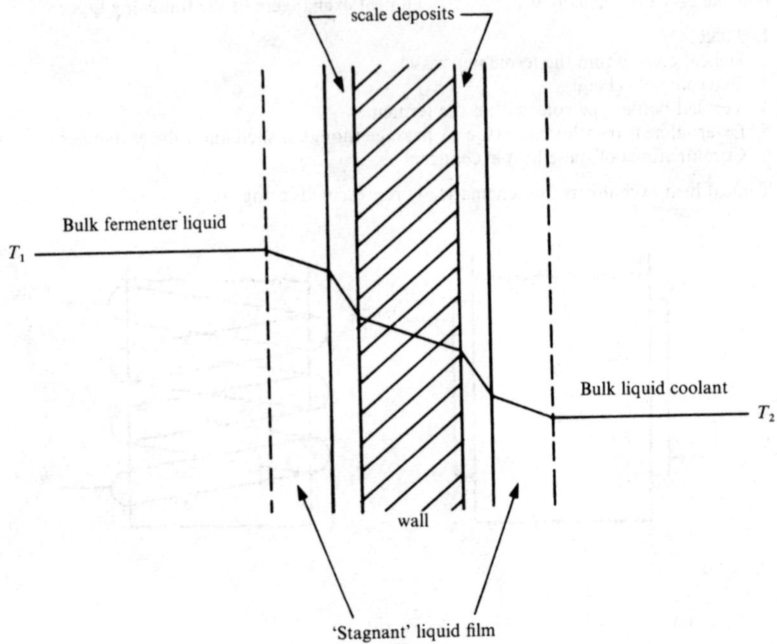

Figure 11. Schematic representation of the resistance to heat transfer.

When the transfer area normal to heat flow changes, as in the case of pipes or tubes, the overall heat transfer coefficient is referred to a specified area, e.g., that based upon the outside pipe diameter. Thus

$$U = \frac{1}{(D_o/D_i h_i) + (D_o/D_i h_{d,i}) + (D_o/k_w D_{lm}) + (1/h_{d,o}) + (1/h_o)} \quad (16)$$

where U is the overall heat transfer coefficient based on the outside surface area of the tube or pipe, D_o and D_i are the outside and inside tube diameters, respectively, and D_{lm} is the log-mean diameter of the tube (i.e. $D_{lm} = D_i/\ln(D_o/D_i)$. Thermal conductivities of various substances are given in Table 12.

In most practical applications, the temperature driving force $\triangle T$ varies from point to point over the heat transfer surface and an 'average' value has to be used.

Table 12. Thermal conductivities [Coulson and Richardson (1980)].

Substance	Temperature (K)	Thermal conductivity ($W\ m^{-1}\ K^{-1}$)
Solids–metals		
Admiralty metal	303	113
Aluminium	573	230
Bronze		189
Cadmium	291	94
Copper	373	377
Iron		
Wrought	291	61
Cast	326	48
Lead	373	33
Nickel	373	57
Silver	373	412
Steel 1% carbon	291	45
Tantulum	291	55
Solids–nonmetals		
Asbestos	273	0.16
Asbestos	373	0.19
Asbestos	473	0.21
Asbestos sheet	323	0.17
Bricks		
Alumina	703	3.10
Building	293	0.69
Cork	303	0.043
Cotton wool	303	0.029
Glass	303	1.09
Glass wool		0.041
Graphite	273	151.00
Magnesia 85%		0.07
Magnesite	473	3.80
Mica	323	0.43
Rubber (hard)	273	0.15
Sawdust	293	0.052
Liquids		
Acetic acid 50 per cent	293	0.35
Acetone	303	0.17
Aniline	273–293	0.17
Benzene	303	0.16
Calcium chloride 30 per cent	303	0.55
Ethanol 80 per cent	293	0.24
Glycerol 60 per cent	293	0.38
Glycerol 40 per cent	293	0.45
n-Heptane	303	0.14
Mercury	301	8.36
Sulphuric acid 90 per cent	303	0.36
60 per cent	303	0.43
Water	303	0.62
	303	0.66
Gases		
Air	273	0.024
Air	373	0.031
Carbon dioxide	273	0.015
Ethane	273	0.018
Ethylene	273	0.017
Hydrogen	273	0.17
Methane	273	0.029
Oxygen	273	0.024
Water vapour	373	0.025

1. If one side is at constant temperature, as is often the case in a stirred fermenter, and the coolant temperature rises in the direction of the coolant flow along a cooling coil, an arithmetic mean is appropriate

$$\Delta T_{am} = \frac{(T_f - T_e) + (T_f - T_i)}{2}$$

$$= \frac{T_f - (T_e + T_i)}{2} \tag{17}$$

where T_o is the bulk fluid temperature in the vessel and T_e and T_i are the temperatures of the coolant on entering and leaving, respectively.

2. If the fluids are in counter- or cocurrent flow and the temperature varies in both fluids, a log-mean temperature difference is appropriate

$$\Delta T_{lm} = \frac{\Delta T_e - \Delta T_i}{\ln \Delta T_e / \Delta T_i} \tag{18}$$

where ΔT_e and ΔT_i are the temperature driving forces of coolant entering and leaving, respectively.

3. If the flow pattern is more complex than either situation described above, the log-mean temperature difference defined in Eq. (18) is multiplied by a dimensionless factor. Such factors have been evaluated for a number of common heat exchange systems [McAdams (1954)].

When values for the overall heat transfer coefficient, area and temperature difference are known, the heat transfer rate can be calculated from Eq. (14).

BATCH HEATING AND COOLING

Fermenter contents are both heated and cooled under batch conditions. Examples of heating are associated with sterilization, while examples of cooling occur during post-fermentation stabilization.

Table 13 provides a summary of temperature–time course relationships resulting from the use of steam sparging, electrical heating and heat-exchanger systems. These relationships can be used to estimate the time required for heating/cooling operations in given systems under standard conditions.

Table 13. Temperature–time relationships in batch/cooling[a] [Deindorfer and Humphrey (1959)].

Type of heat transfer	Temperature–time relationship	Parameters
Coolant (heat exchanger)	$T = T_{co}(1 + be^{-at})$	$a = \frac{wc'}{M'c}(1 - e^{-UA/wc'})$
	(exponential)	$b = \frac{T_f - T_{co}}{T_{co}}$
Electrical heating	$T = T_f(1 + at)$	$a = \frac{q'}{M'cT_f}$
	(linear)	
Steam (heat exchanger)	$T = T_H(1 + be^{-at})$	$a = \frac{UA}{M'c}$
	(exponential)	$b = \frac{T_f - T_H}{T_H}$
Steam sparging	$T = T_0\left(1 + \frac{at}{1 + bt}\right)$	$a = \frac{HS}{M'cT_f}$
	(hyperbolic)	$b = \frac{S}{M'}$

[a] H = enthalpy differences between steam at sparger temperature and bulk liquid temperature, S = steam mass flow rate, M' = liquid mass, c = specific heat of bulk liquid, T_f = initial bulk liquid temperature, q' = rate of heat transfer [kcal (unit time)$^{-1}$], U = overall heat transfer coefficient (kcal m^{-2} h^{-1} °C^{-1}), A = heat transfer area (m^2), T_H = temperature of heat source, w = coolant mass flow rate, c' = coolant specific heat and T_{CO} = coolant inlet temperature.

HEAT TRANSFER COEFFICIENTS

Tables 14 and 15 provide an indication of the magnitude of heat transfer coefficients for various characteristic fluid properties and conditions. Table 16 gives some values of the thermal resistances due to formation of scale on heat transfer surfaces.

Table 14. General magnitude of heat transfer coefficient [Bailey and Ollis (1977)].

Fluid properties	Heat transfer coefficient (kcal m^{-2} h^{-1} °C^{-1})
Free convection	
Gases	3–20
Liquids	100–600
Boiling water	1000–20 000
Forced convection	
Gases	10–100
Viscous liquids	50–500
Water	500–10 000
Condensing vapours	1000–100 000

Table 15. Approximate film coefficients [Coulson and Richardson (1977)].

	Approximate film coefficients ($W\ m^{-2}\ K^{-1}$)
No change of state	
Gases	20–300
Oils	60–700
Organic solvents	350–3000
Water	1700–11 000
Condensing	
Ammonia	3000–6000
Heavy oils (vacuum)	120–300
Light oils	1200–2300
Organic solvents	900–2800
Steam	6000–17 000
Evaporation	
Ammonia	1100–2300
Heavy oils	60–300
Light oils	800–1700
Organic solvents	600–2000
Water	2000–12 000

Table 16. Thermal resistances of scale deposits [Coulson and Richardson (1977)].

Scale deposit source	Thermal resistance [$m^2\ K(kW)^{-1}$]
Water ($1\ ms^{-1}$ velocity, temperatures less than 320 K)	
Distilled	0.09
Sea	0.09
Clear river	0.21
Untreated cooling tower	0.58
Treated cooling tower	0.26
Treated boiler feed	0.26
Hard well	0.58
Steam	
Good quality–oil-free	0.052
Poor quality–oil-free	0.09
Exhaust from reciprocating engines	0.18
Liquids	
Treated brine	0.27
Organics	0.18
Fuel oils	1.0
Tars	2.0
Gases	
Air	0.25–0.50
Solvent vapours	0.14

Film heat transfer coefficients h in stirred tanks are dependent on the fluid properties both at the bulk liquid and the wall temperatures, the impeller speed and the geometry of the system. Thus

$$h = g(\rho, \mu_{\text{bulk}}, \mu_{\text{wall}}, c_p, k, N, D, D_T, \text{other geometric arrangements and dimensions}) \tag{19}$$

Dimensional analysis [Coulson and Richardson (1977)] allows Eq. (19) to be reduced to

$$\text{Nu} = g\left(\text{Re}, \text{Pr}, \frac{\mu}{\mu_m}, \text{geometric arrangements}\right) \tag{20}$$

where Nu is the Nusselt number ($\text{Nu} = hD_T/k$ or hd_c/k), Re is the Reynolds number ($\text{Re} = D^2N\rho/\mu$), Pr is the Prandtl number ($\text{Pr} = c_p\,\mu/k$) and μ/μ_m is the ratio of the viscosities of the bulk liquid and at the wall, d_c being the coil diameter.

Tables 17–24 list the ranges of Reynolds and Prandtl numbers for newtonian fluids in agitated vessels and Table 25 gives the corresponding correlations for a wide range of vessel geometries and internals using equations of the form

$$\text{Nu} = E\,\text{Re}^a\,\text{Pr}^b\,Vi^c\,X \tag{21}$$

where $Vi = \mu/\mu_m$ and X represents various geometric factors, etc. Similarly, Reynolds and Prandtl numbers for nonnewtonian fluids (*see* Chapter 8) in agitated vessels are presented in Tables 26–30 and the corresponding correlations are given in Table 31.

Table 17. Reynolds and Prandtl numbers for jacket heat transfer in newtonian fluids agitated using paddles [Edwards and Wilkinson (1972a)].

System number	Impeller	Vessel Inserts	Fluids	D_T (ft)	Re	Pr
1	Wide paddle	Coil	Water, aqueous glycerol solutions	1.0	15 000–70 000	~3–2000
2	Flat paddle	Coil	Water, oil, glycerol	1.0 5.0	284–240 000	~3–2000
3	Wide paddle	None	Oil, water, kerosene	2.0	600–600 000	~3–2000
4	Flat and angled paddles	No baffles	Mineral oils	1.0	100–100 000	300–8500
5	Wide paddle	Coil	Water, oils	3.3	800–800 000	2–2500
6	Flat and angled paddles	Coil, no baffles	Water, glycerine, castor oil	1.0	400–400 000	Not given
7	Narrow paddle	Baffles	Cylinder oil, linseed oil	2.0	20–4000	1400–550 000
8	Various paddles	None	Water, glycerol solutions	2.0	600–500 000	3–550 000

Table 18. Reynolds and Prandtl numbers for jacket heat transfer in newtonian fluids agitated using turbines [Edwards and Wilkinson (1972a)].

System number	Impeller	Vessel Inserts	Fluids	D_T (ft)	Re	Pr
9	8-bladed turbine	No baffles	Mineral oils	1.0	10–100 000	300–8500
10	Various turbines	Coil, no baffles	Water, glycerol, castor oil	1.0	400–400 000	Not given
11	3-bladed retreating	Baffles	Water, viscous organics	~1.2 ~3.5	40 000–2×10^6	Not given
12	6-bladed turbine	Baffles	Water	5.0	100 000–700 000	Not given
13	Flat and 45° turbines	Baffles	Water, lubricating oils	0.8	200–180 000	5–2000
14	Retreating, glass-coated	Baffles	Water, toluene, oil, water/toluene, water/lubricating oil	1.5	2000–700 000	Not given
15	Disc turbine flat blades	Baffled/no baffles	Water, oils, corn syrup	1.7	30–500 000	Not given
16	6-bladed turbine	Baffles	Lubricating oils, castor oil	1.0 2.3	30–40 000	Not given
17	6-bladed turbine	Coil, no baffles	Water, toluene, ethylene glycol, glycerol, oil	2.5	2 000–800 000	2–1200
18	4-bladed turbine	Coil, baffles	Water, aqueous suspensions	1.2	90 000–500 000	3.0–4.5
19	6-bladed turbine	Baffles	Oils, aqueous glycerol solutions	1.0	4 000–280 000	1.9–105
20	Angled and flat turbines	No baffles	Viscosity 5000–500 000 cP	1.2 2.5	Not given	Not given

Table 19. Reynolds and Prandtl numbers for jacket heat transfer in newtonian fluids agitated using propellers [Edwards and Wilkinson (1972a)].

System number	Impeller	Vessel Inserts	Fluids	D_T (ft)	Re	Pr
21	3-bladed propeller	No baffles	Mineral oils	1.0	100–40 000	300–8500
22	Marine propeller	Baffles	Water, lubricating oils	5.0	200–180 000	5–2000
23	Propeller	No baffles	Nitration liquors	5.0	5 500–37 000	218–1500
24	3-bladed propeller	Baffles	Various suspensions	2.0	270 000–2×10^6	1.9–6.2
25	Various propellers	Baffles	Water	1.0	200 000–10^6	2

Table 20. Reynolds and Prandtl number for jacket heat transfer in newtonian fluids agitated using anchors [Edwards and Wilkinson (1972a)].

System number	Impeller	Vessel Inserts	Fluids	D_T (ft)	Re	Pr
26	Anchor	No baffles	Mineral oils	1.0	200 000	300–8500
27	Various anchors	Coil, no baffles	Water, glycerol, castor oil	1.0	100–400 000	Not given
28	Anchor	No baffles	Sulphonation liquors	5.0	5400–37 000	218–1500
29	Anchors and ribbons	No baffles	Viscosity 500–500 000 cP	1.2 2.5	Not given	Not given
30	Various anchors	Coil, no baffles	Water, aqueous glycerol, aqueous molasses	1.0	130–4300	Not given
31	Anchors	Coil, no baffles	Water, aqueous glycerol	1.0 2.0	600–10^6	Not given
32	Anchors	None	Cylinder oils	3.8 0.8 2.0	30–5000	1400–70 000

Table 21. Reynolds and Prandtl numbers for coil heat transfer in newtonian fluids agitated using paddles [Edwards and Wilkinson (1972a)].

System number	Impeller	Vessel Inserts	Fluids	D_T (ft)	Re	Pr
33	Flat paddle	Helical coil, no baffles	Water, oil, glycerol	1.0 5.0	$283–3.2 \times 10^6$	3–2000
34	Wide paddle	Helical coil	Water, oils	3.3	30 000–100 000	2–2500
35	Various flat paddles	Helical coil, no baffles	Water, *i*-propanol	1.5 2.0	20 000–500 000	Not given
36	4 bladed flat paddle	Helical coil, no baffles	Water/toluene, water/kerosene	0.8	$37\,000–1.1 \times 10^6$	Not given
37	Paddle	Helical coil	Water, aqueous glycerol, castor oil	1.0	100–300 000	Not given

Table 22. Reynolds and Prandtl numbers for coil heat transfer in newtonian fluids agitated using turbines [Edwards and Wilkinson (1972a)].

System number	Impeller	Vessel Inserts	Fluids	D_T (ft)	Re	Pr
38	Flat, angled, retreating turbines	Helical coil, no baffles	Water, oil, toluene, glycol, glycerol, *i*-propanol	2.5	1500–770 000	2–1200
39	4-bladed flat turbine	Helical coil, baffles	Various suspensions	1.2	90 000–500 000	3–4.5
40	Turbine	Helical coil	Water, glycerol, castor oil	1.0	100–300 000	
41	6-bladed, 3-angled blade turbines	Helical coil, baffles, baffle coil	Water, polyalkylene glycol	1.5 2.0	300–400 000	4.4–520
42	4-flat blade turbine	Baffle coil	Water, oils	2.0 4.0	$1000–2 \times 10^6$	Not given
43	4-bladed turbine	Various coils, baffles	Water	1.2	Not given	Not given
44	6-bladed turbines	Single, multiple, helical coils, baffles	Water, lubricating oils	4.0	$5000–10^6$	Not given
45	Various turbines	Helical coil, baffles	Water, aqueous glycerol solution, kerosene	1.2	10 000–150 000	4–25
46	6 flat-bladed turbines	Helical coil, baffles	Water, turbine oils	4.0	$400–1.5 \times 10^6$	2.3–6300

Table 23. Reynolds and Prandtl numbers for coil heat transfer in newtonian fluids agitated using propellers [Edwards and Wilkinson (1972a)].

System number	Impeller	Vessel Inserts	Fluids	D_T (ft)	Re	Pr
47	Square pitch propeller	Coils and straight tubes	Water, oils, nitric acid	Not given	Not given	Not given
48	3-bladed 45° pitch propeller	Helical coil, baffles, baffle coil	Water, polyalkylene glycol solutions	1.5 2.0	300–400 000	4.4–520
49	3-bladed square pitch propeller	Helical coil, baffles	Not given	4.0	$1000–2 \times 10^6$	Not given
50	3-bladed square pitch propeller	Helical coil, baffles	Water	1.0	$25\,000–1.4 \times 10^6$	Not given
51	Various propellers	Helical coils, baffles	Water, oils	1.5	65 000–820 000	Not given

Table 24. Reynolds and Prandtl numbers for coil heat transfer in newtonian fluids agitated using anchors [Edwards and Wilkinson (1972a)].

System number	Impeller	Vessel Inserts	Fluids	D_T (ft)	Re	Pr
52	Anchor	Helical coil	Water, aqueous glycerol solutions	1.0	600×10^6	Not given
53	Anchor	Helical coil	Water, castor oil, aqueous glycerol solutions	1.0	100–300 000	Not given

Table 25. Correlations for newtonian fluids covering the systems listed in Tables 17–24 using Eq. (21) [Edwards and Wilkinson (1972a)].

System number	E	a	b	c	X	Comment
1	0.60	0.67	0.33	0.14		
2	0.36	0.67	0.33	0.14		
3	0.17	0.66	0.5			
4	0.74–1.6	0.5	0.24 or 0.33	0.14		E and b depend upon impeller geometry
5	0.36	0.67	0.33	0.14		
6	0.46	0.67	0.33	0.14	$\alpha^{0.15}(D_T/D)^{0.1}[1-0.211(0.63-\alpha)]$	$\alpha = n_b W \sin\theta/H'$, where n_b = number of baffles on impeller, W = blade width, θ = blade angle, H' = height of heat transfer surface
7	0.415	0.67	0.33	0.24		
8	0.112	0.75	0.44	0.25	$(D_T/D)^{0.4}(W/D)^{0.13}$	
9	1.0–1.2	0.5	0.33	0.14		E depends on impeller position
10	0.46	0.67	0.33	0.14	$\alpha^{0.15}(D_T/D)^{0.1}[1-0.211(0.63-\alpha)]$	*See* system number 6
11	0.33	0.67	0.33	0.14		For glassed stell impeller
	0.37	0.67	0.33	0.14		For alloy impeller
12	0.81	0.68	0.33	0.14	$[(H_A/D_T)^{0.67}+1-(H_A/D_T)]^{0.67}$	H_A = height of impeller from base
13	3.57	0.55	0.30			For flat turbines
	2.71	0.55	0.30			For 45° turbines
14	1.0–1.2	0.5	0.33	0.14		E depends on impeller position
15	0.54	0.67	0.33	0.14		Unbaffled
	0.74	0.67	0.33	0.14		Baffled
16	1.15	0.65	0.33	0.24	$(H_A/D_T)^{0.4}(H_L/D_T)^{-0.56}$	H_L = height of liquid
17	0.68	0.67	0.33	0.14		
18	0.66	0.65	0.33		$(H_A/D_T)^{0.25-0.00033N}$	Re and Pr are based on suspension properties
19	0.76	0.67	0.33	0.14		
20						No correlation
21	0.85	0.5	0.33	0.14		
	0.30	0.67	0.24	0.14		For systems with diffuser
22	0.64	0.67	0.30			
23	0.54	0.67	0.25	0.14		

24	0.575	0.6	0.26		$(D_T/D)^{0.33}\ (C_{p,s}/C_{p,f})^{0.13}\ \times$ $(\rho_f/\rho_s)^{0.16}\ (\varphi/1-\varphi)^{-0.04}$	C_p = specific heat of solid, φ = volume fraction of solids. Subscripts $_s$ and $_f$ refer to solid and fluid, respectively.
25	0.639	0.67	0.33	0.14		
26	1.38	0.5	0.28	0.14		
27	0.46	0.67	0.33	0.14	$\alpha^{0.15}(D_T/D)^{0.1}\ [1-0.211(0.63-\alpha)]$	*See* system number 6
28	0.55	0.67	0.25	0.14		
29						A model based on heat conduction through a stagnant film is used
30	0.35	0.67	0.33	0.14		Data correlated in terms of power requirement
31	1.00	0.50	0.33	0.18		For Re = 30–300
32	0.38	0.67	0.33	0.18		For Re = 300–5000
33	0.87	0.62	0.33	0.14		
34	0.87	0.62	0.33	0.14		
35	34	0.50	0.30		$(C_s/H_c)^{0.8}\ (W/D_c)^{0.25}\ (D^2 D_T/d_c^3)^{0.1}$	C_s = clearance between turns of coil, H_c = height of coil, D_c = coil helix diameter, d_c = diameter of coil tubing
36	4.04	0.42	0.33	0.14		
37	23	0.50	0.30	0.14	$\alpha^{0.15}\ (d_c/D_c)^{1.7}$	$\alpha = n_b\ W \sin\theta/H'$
38	1.4	0.62	0.33	0.14		
39	0.48	0.70	0.33		$(H_A/D_T)^{0.35-0.00049N}$	
40	23	0.50	0.30	0.14	$\alpha^{0.15}\ (d_c/D_c)^{1.7}$	$\alpha = n_b\ W \sin\theta/H'$
41						No correlation
42	0.09	0.65	0.30	0.40	$(d_c/D_T)^{-0.27}\ (2/n_B)$	n_B = number of baffles
43	0.18	0.67	0.33		$(d_c/D_T)^{-0.48}\ (D_c/D_r)^{-0.27}$ $(H_B/D_T)^{0.14}$	
44	0.026	0.65	0.28		$(D_T/D_C)^{0.98}$	For each coil in a triple coil system
45	1.04	0.67	0.33	0.14	$(D/D_T)^{0.18}\ (n_b/6)^{0.28}$	
46	0.17	0.67	0.37	Varies with viscosity (e.g., 0.97 at 0.3 cP, 0.18 at 1000 cP)	$(D/D_T)^{0.1}\ (d_c/D_T)^{0.5}$	

continued overleaf

(Table 25–continued)

System number	E	a	b	c	X	Comment
47	0.078	0.62	0.33	0.14		
48						No correlation
49	0.091	0.67	0.37	Varies with viscosity (e.g., 0.97 at 0.3 cP, 0.18 at 1000 cP)	$(D/D_T)^{0.1}\ (d_c/D_T)^{0.5}$	
50	0.0345	0.62			$(D_T/H_A)^{0.27}$	For 1 propeller, Re = 24 700 – 1.3 × 10⁶
	0.0886	0.55			$(H_A/D)^{0.0074}\ (H_A/D)^{0.0065}$	For 2 propellers, Re = 766 000 – 1.39 × 10⁶
51	0.0573	0.67	0.41	0.034	$(D_T/H_A)^{0.254}\ (S'/D)^{2.33}\ n_B^{-0.077}\ (D_T/J)^{0.058}\ (d_c/D_T)^{0.572}\ (C_s/d_c)^{-0.018}$	S' = pitch of propeller, J = width of baffle
52	0.77	0.67	0.33	0.14	$(D_T/d_c)^{0.48}\ (D_T/D_C)^{0.27}$	
53	23	0.50	0.30	0.14	$\alpha^{0.15}\ (d_c/D_c)^{1.7}$	$\alpha = n_b\ W \sin\theta/H'$

Table 26. Reynolds and Prandtl numbers for jacket heat transfer in nonnewtonian fluids agitated using paddles [Edwards and Wilkinson (1972 b)].

System number	Impeller	Vessel Inserts	Fluids	D_T (ft)	Re	Pr
1	Various paddles	Electric heater, no baffles	Water, spindle oil, compressor oil, aqueous sodium carboxymethyl cellulose and potassium chloride suspensions. $n = 0.6–1$	1.0	4–100 000	5–25 000
2	Paddle	Baffles	Water, glycerol, aqueous carbopol, $n = 0.36–1$	1.2	35–680 000	2–24 000

Table 27. Reynolds and Prandtl numbers for jacket heat transfer in nonnewtonian fluids agitated using turbines [Edwards and Wilkinson (1972 b)].

System number	Impeller	Vessel Inserts	Fluids	D_T (ft)	Re	Pr
3	6-bladed turbine	Electric heater, no baffles	Aqueous sodium carboxymethyl cellulose, potassium chloride suspensions $n = 0.6–1$	1.0	5–200 000	5–25 000
4	6-bladed turbine	Baffles	Aqueous carbopol $n = 0.36–1$	1.2	35–68 000	2–24 000
5	4-bladed 45° turbine	Baffles	Aqueous carboxymethyl cellulose, aqueous carbopol $n = 0.34–0.63$	2.5	100–5000	100–800
6	6-bladed turbine	Baffles	Chalk water slurries	0.6	200–80 000	2–700
7	6-bladed turbine	Baffles	Aqueous carbopol $n = 0.3–1$	0.6	80–93 000	2–650
8	Turbine	Coil	Cement slurries	1.0	10 000–200 000	Not given
9	Turbine	No baffles	Viscous liquids 5000–500 000 cP (at $5\,s^{-1}$)	1.2 2.5	Not given	Not given

Table 28. Reynolds and Prandtl numbers for jacket heat transfer in nonnewtonian fluids agitated using propellers [Edwards and Wilkinson (1972 b)].

System number	Impeller	Vessel Inserts	Fluids	D_T (ft)	Re	Pr
10	Propeller	Baffles	Water, glycerol, aqueous carbopol	1.2	35–680 000	2–24 000

Table 29. Reynolds and Prandtl numbers for jacket heat transfer in nonnewtonian fluids agitated using anchors [Edwards and Wilkinson (1972 b)].

System number	Impeller	Vessel Inserts	Fluids	D_T (ft)	Re	Pr
11	Anchor and ribbon	No baffles	Viscous liquids, 5000–500 000 cP (at 5 s^{-1})	1.2 2.5	Not given	Not given
12	Anchor	Coil	Chalk water slurries $n = 0.38$–1	1.0 2.0 3.8	200–10^6	Not given
13	Anchor	Electric heater	Water, oils, aqueous sodium carboxymethyl cellulose and potassium chloride suspension $n = 0.6$–1	1.0	8–300 000	5–25 000
14	Anchor	None	Water, glycerol, aqueous carbopol $n = 0.36$–1	1.0	35–680 000	2–24 000
15	Anchor	None	Chalk water slurries	0.6	336–95 000	2–621
16	Anchor	None	Aqueous carbopol $n = 0.3$–1	0.6	320–90 000	2–650
17	Anchor	Coil	Aqueous carboxymethyl cellulose, cement slurry	1.0	See Table 31	Not given
18	Anchor	Coil	Aqueous carboxymethyl cellulose, PVA, aqueous calcium carbonate suspension $n = 0.66$–1	1.8	100–550 000	2.5–5000

Table 30. Reynolds and Prandtl numbers for coil heat transfer in nonnewtonian fluids agitated using anchors, turbines or propeller [Edwards and Wilkinson (1972 b)].

System number	Impeller	Vessel Inserts	Fluids	D_T (ft)	Re	Pr
19	Anchors	Helical coil	Chalk water slurries $n = 0.3–1$	1.0 2.0 3.8	200–600 000	Not given
20	Various turbines, anchors	Helical coil	Aqueous carboxymethyl cellulose solution, cement slurries	1.0	*See* Table 31	Not given
21	3-bladed propeller	Helical coil, baffles	Aqueous carbopol $n = 0.53–1$	1.5	322–260 000	12–1100
22	Anchor	Helical coil	Aqueous carboxymethyl cellulose, aqueous calcium carbonate suspension	1.7	100–550 000	2.5–5000

Table 31. Correlations for nonnewtonian fluids covering the systems listed in Tables 26–30 using Eq. (21) [Edwards and Wilkinson (1972 a)].

System number	E	a	b	c	X	Comments
1	0.216	0.67	0.33		$(K_W/K)^{0.18}$	For double-bladed paddles where Re = 4–100 000, K_W = power-law parameter at wall temperature, K = power-law parameter
	0.477	0.45	0.33		$(K_W/K)^{0.18}$	For screw paddles where Re = 4–92
	0.207	0.67	0.33		$(K_W/K)^{0.18}$	For screw paddles where Re = 92–100 000
	0.912	0.33	0.33		$(K_W/K)^{0.55}$	For screw paddles in pipes where Re = 4–47
	0.41	0.50	0.33		$(K_W/K)^{0.25}$	For screw paddles in pipes where Re = 47–100 000
2	2.51	$1.28/(1+n)+0.15$	0.26		$(K_W/K)^{0.31}(D_T/D)^{-0.46}(W/D)^{0.46}n^{0.56}$	Where n = power-law index
3	0.215	0.67	0.33		$(K_W/K)^{0.18}$	
4	3.57	$1.25/(1+n)$	0.24		$(K/K_W)^{0.3}n^{0.78}$	
5	1.474	0.67	0.33	$\frac{0.24}{n}$		
6	0.534	0.67	0.33	0.14	$[\phi/(1-\phi)]^{0.65}$	ϕ = fraction volume of solids
7	0.482	0.67	0.33	0.12		
8			0.33	0.14	$0.4\,Re^{0.67}/(1+444\,He^{0.1}Pr^{0.33}Re^{-0.83})$	
9						No correlation. Conduction model proposed for heat transfer

10	0.55	$1.28/(1+n)$	0.30	0.32	$(D_T/D)^{-0.4}\, n^{1.32}$	
11						No correlation. Conduction model proposed for heat transfer
12	0.35	0.67	0.33	0.14		
13	0.636 0.374	0.50 0.67	0.33 0.33		$(K_W/K)^{0.18}$ $(K_W/K)^{0.18}$	For Re = 8–30 For Re = 30–300 000
14	0.56	$1.43/(1+n)$	0.30	0.34	$n^{0.54}$	
15	0.315	0.67	0.33	0.2	$[\Phi/(1-\Phi)]^{0.72}$	
16	0.315	0.67	0.33	0.12		
17	0.4	0.67	0.33	0.14		For carboxymethyl cellulose, Re = 300–800 000
			0.33	0.14	$0.4\,\mathrm{Re}^{0.67}/(1+444\,\mathrm{He}^{0.1}\,\mathrm{Pr}^{0.33}\,\mathrm{Re}^{-0.83})$	For slurry, Re = 10 000–200 000
18	0.63	0.67	0.30	0.18	$\delta^{0.67}$	$\delta = (3n+1)/4n$
19	0.077	0.67	0.33	0.14	$(D_T/d_c)^{0.48}\,(D_T/D_c)^{0.27}$	
20	$23n^{1.3}$	0.50	0.30		$\alpha^{0.15}\,(d_c/D_c)_1^{.7}\,(K/K_W)^{0.14}$	For pseudoplastic liquids, Re = 0.2–1000 $\alpha = n_b\, W \sin\theta/H'$
	23	0.50	0.30	0.14	$\alpha^{0.15}\,(d_c/D_c)^{1.7}$	For Bingham plastics, Re 10 000–80 000
21	0.258	0.62	0.32	0.20	$(D/D_T)^{0.1}\,(d_c/D_T)^{0.5}$	
22	0.835	0.67	0.30	0.18	$\delta^{9.5}$	$\delta = (3n+1)/4n$

LIMITING HEAT LOAD – SIGNIFICANCE OF FERMENTER CONFIGURATION

Increasing the scale of vessels, while preserving geometric similarity, increases the volume of fermentation broth in proportion to the cube of the linear dimensions, while the surface area of the vessel and fitments increases only as the square of the linear dimensions. This means that heat production, which is proportional to the volume of the fermentation fluid, increases more rapidly than the surface area available for jacketed heat transfer. The specific cooling area is then

$$A_c = \frac{A_s}{V_f} \simeq D_T^{-1} \tag{22}$$

where A_s is the cooling area surface and V_f is the working volume. Thus, a limiting fermenter size and production capacity exist at which the heat generated can just be removed.

The calculation of this limiting size and the influence of geometric and heat transfer parameters are given by Blenke (1979).

The propeller loop fermenter (*see* Fig. 12) is considered by way of an example. The areas

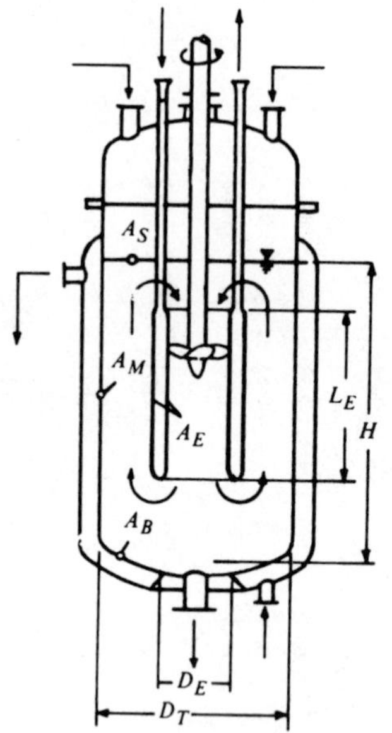

Figure 12. Propeller loop fermenter [Blenke (1979)].

available for cooling are (1) the external jacket A_M, (2) the bottom of the fermenter A_B and (3) internal cooling by the double-walled draught tube A_E. Then defining

$$s = \frac{H}{D_T} \quad d_E = \frac{D_E}{D_T} \quad l_E = \frac{L_E}{D_T}$$

the cooling areas and the fermenter volume are approximately

$$A_B \simeq \frac{\pi}{4} D_T^2 \qquad A_M \simeq \pi s D_T^2 \qquad A_E \simeq 2\pi d_E l_E D_T^2 \qquad V_f \simeq \frac{\pi}{4} s D_T^3$$

The specific cooling area of the reactor [Eq. (22)] is expressed as

$$A_c = \frac{\Sigma A_i}{V_f} = \frac{(\pi/4)\,D^2(1 + 4s + 8d_E l_E)}{(\pi/4)\,s D^3} = \frac{B}{D} \tag{23}$$

where B is defined as

$$\frac{1 + 4s + 8d_E l_E}{s}$$

The heat transfer required per unit volume of fermenter is

$$Q_{\text{exch}} = U\,A_c\,\Delta T_{lm} \tag{24}$$

where U is the overall heat transfer coefficient and ΔT_{lm} is the log-mean temperature difference between the fermenter contents (i.e. $T_f \simeq$ constant) and the mean value of the cooling medium, such as cooling water (i.e. $\overline{T}_c$).

Now

$$V_{f,\max} = \frac{\pi}{4} s\,D_{T,\max} = \frac{\pi}{4} s \frac{B^3}{A_c^3} \tag{25}$$

where A_c is given by Eq. (24) for the given heat transfer parameters U, ΔT_{lm} and Q_{exch}, i.e.

$$V_{f,\max} \propto \frac{\pi}{4} s\,B^3$$

$$\propto B' \tag{26}$$

For the propeller loop fermenter, the characteristic geometric number B' is defined as

$$B' = \frac{\pi}{4} s\,B^3 = \frac{\pi}{4}\frac{(1 + 4s + 8d_E l_E)^3}{s^2} \tag{27}$$

and depends only on dimensionless geometric parameters that are all referred to D_T.

Fig. 13 shows the magnitude of B', its dependence on the fermenter configuration and especially on the aspect ratio $s = H/D$. For a stirred tank fermenter with $s \simeq 1$ and external cooling, B' is approximately 100, while for a tubular loop fermenter with additional internal cooling on the draft tube with $s = 5$, $B' \simeq 2700$. Thus, according to Eq. (26), a 27-fold greater fermenter volume can be used with a high aspect ratio loop fermenter than with a conventional stirred tank fermenter for the same volumetric heat load (i.e. Q_{exch}) and cooling conditions (i.e. U and ΔT_{lm}).

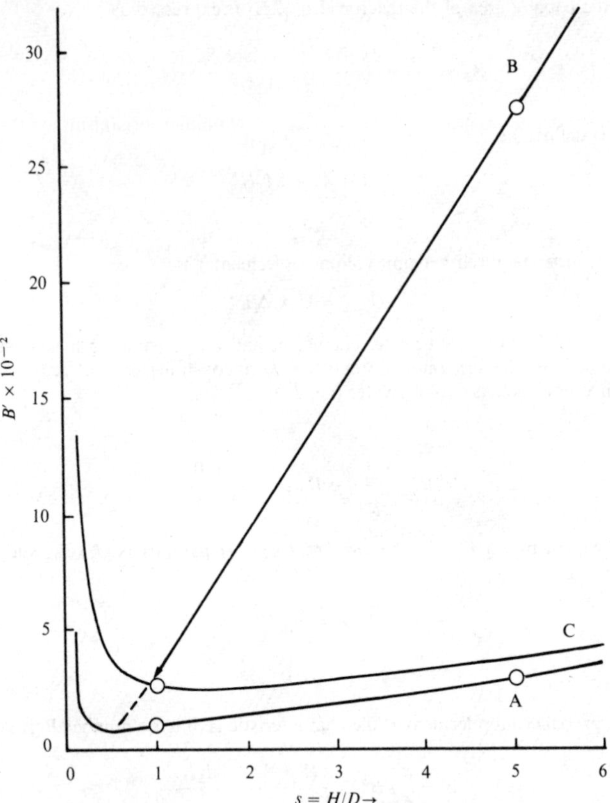

Figure 13. Characteristic geometric number B' for various reactor types and cooling systems dependent on the aspect ratio $s = H/D$. Curve A, jacketed stirred-tank fermenter; curve B, propeller loop fermenter; curve C, jacket stirred-tank fermenter with external heat exchange [Blenke (1979)].

Combining Eqs. (23) and (24) for appropriate values of the overall heat transfer coefficient and the logarithmic mean temperature difference, and given Q_{exch} [Eq. (1)], leads to the maximum possible reactor diameter D_{max}, i.e. the reactor diameter in which the heat generated can just be removed. Thus

$$D_{max} = B \frac{U \Delta T_{lm}}{Q_{exch}} \tag{28}$$

The corresponding maximum reactor volume is

$$V_{f,max} = \frac{\pi}{4} s D_{max}^3 = \frac{\pi}{4} s B^3 \left(\frac{U \Delta T_{lm}}{Q_{exch}} \right)^3$$

$$= B' \left(\frac{U \Delta T_{lm}}{Q_{exch}} \right)^3 \tag{29}$$

HEAT STERILIZATION

Rate of Death of Microorganisms

The two most common expressions for the death rate of microorganisms are the logarithmic and nonlogarithmic death rates. The logarithmic death rate can be expressed as

$$\frac{dn}{dt} = kn \tag{30}$$

where n is the concentration or number of viable microorganisms, k is the specific death rate and t is the time. If the initial number of viable microorganisms is n_0 at time zero, the integration of Eq. (30) gives

$$n = n_0 e^{-kt} \tag{31}$$

Eq. (31) is plotted for *Escherichia coli* at various temperatures in Fig. 14.

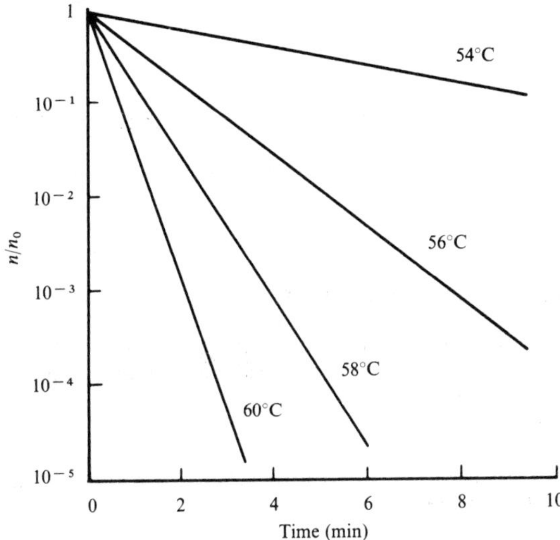

Figure 14. Typical death rate data for *Escherichia coli* in buffer [Aiba *et al.* (1965)].

The logarithmic death rate expression [Eq. (31)] does not always hold for bacterial spores, particularly during the short period immediately following exposure to heat. There are several nonlogarithmic death rate expressions to describe such phenomena.

Prokop and Humphrey (1970) suggested a sequential death model for bacterial spores

$$n_R \xrightarrow{k_R} n_S \xrightarrow{k_S} n_D$$

where k_R and k_S are the first-order rate constants for the transition from resistant to sensitive spores and from sensitive spores to death, respectively. This model yields

$$\frac{n}{n_0} = \frac{k_R}{k_R - k_S}\left[\exp\,(k_S t) - \frac{k_S}{k_R}\exp\,(-k_R t)\right] \tag{32}$$

Eq. (32) is expressed graphically in Fig. 15 with appropriate kinetic constants for spores of *Bacillus stearothermophilus.*

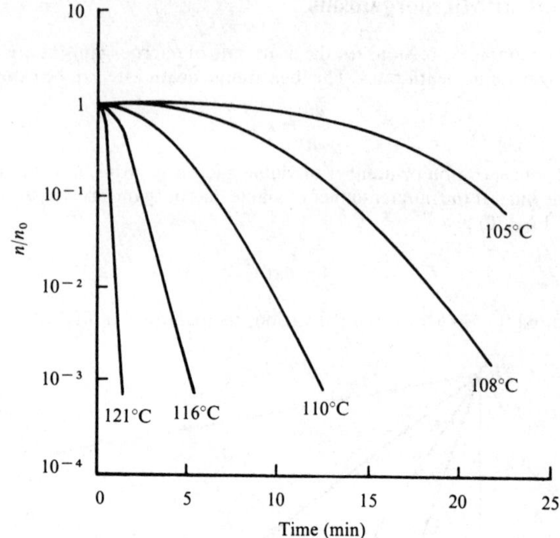

Figure 15. Typical death rate data for spores of *Bacillus stearothermophilus* FS 7954 in distilled water [Aiba *et al.* (1965)].

Effect of Temperature on Death Rate

Figs. 14 and 15 indicate that the death rates of microorganisms vary with temperature. The temperature-dependent effect on the specific death rate constant can be expressed as an Arrhenius equation as follows

$$k = ae^{-\Delta E/RT} \tag{33}$$

where a is a frequency factor and ΔE is the activation energy of death. The variation in the specific death rate constant is shown in Fig. 16.

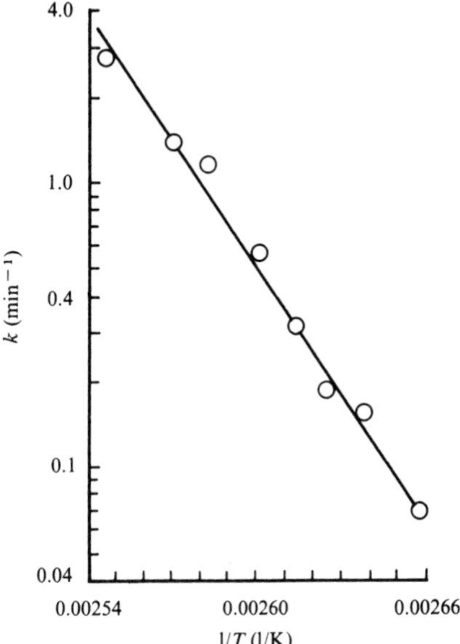

Figure 16. Correlation of isothermal death rate data for *Bacillus stearothermophilus* FS 7954 [Aiba *et al.* (1965)].

Definition of Various Terms used in Sterilization

Decimal reduction time

The decimal reduction time $\mathscr{D}$ is the time required to reduce the microbial population by one $\log_{10}$ cycle (*see* Fig. 17). In other words, it is the time required to destroy 90 per cent of the initial amount of microorganisms or spores, i.e.

$$\mathscr{D} = \frac{2.303}{k} \tag{34}$$

Typical values of $\mathscr{D}$ and k are given in Table 32.

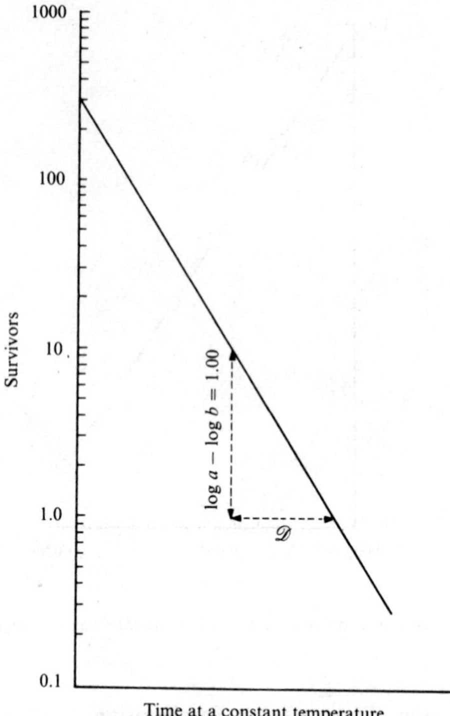

Figure 17. Logarithmic survivor curve expressed by ($t = \mathscr{D}$ (log a − log b) [Stumbo (1976)].

Table 32. Values of k and $\mathscr{D}$ for various bacterial spores at 121°C [Moss and Smith (1977)].

Culture	k (min^{-1})	$\mathscr{D}$ (min)
Bacillus subtilis FS 5230	2.6–3.8	0.6–0.9
B. stearothermophilus FS 1518	0.77	3
Clostridium sporogenes PA 3679	1.8	1.8

Z value

This value is numerically equal to the number of degrees fahrenheit required for the thermal destruction curve to move through one $\log_{10}$ cycle (*see* Fig. 18). Thus

$$\log \mathscr{D}_2 - \log \mathscr{D}_1 = \frac{1}{Z}(T_1 - T_2) \qquad (35)$$

where $\mathscr{D}_1$ and $\mathscr{D}_2$ are $\mathscr{D}$ values corresponding to temperatures T_1 and T_2 and the time to destroy 90 per cent of the cell population at a temperature of T_1 or T_2, respectively. The term Z shows the relative resistance of microorganisms to thermal death. To convert Z from degrees fahrenheit to degrees centigrade it should be multiplied by a factor of 0.555. Values of Z in degrees centigrade are provided for a variety of microorganisms in Table 33.

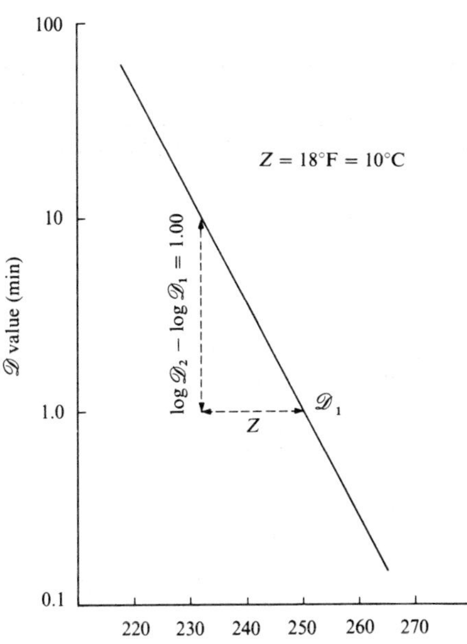

Figure 18. Thermal destruction curve passing through one minute at 250°F (moist heat) [Stumbo (1976)].

Table 33. Heat resistance of a variety of microorganisms [Moss and Smith (1977)].

Microorganism	Heating temperature (°C)	$\mathscr{D}$ (min)	Z (°C)	Suspension medium
Bacterial spores				
Clostridium thermosaccharolyticum	132	4.4	6.7	Water
	132	3.3	10	Molasses
Bacillus stearothermophilus	115	22.6	7.1	Water
B. megaterium	100	1	8.8	Water
Vegetative cells of bacteria				
Salmonella typhimurium	55.5	0.5	4.2	Ringer's solution
Actinomycete spores				
Thermoactinomyces vulgaris	100	11		0.001 M phosphate buffer (pH 7.0)
Yeast spores				
Saccharomyces cerevisiae	55	0.9		Phosphate buffer (pH 7.0)
Spores of filamentous fungi				
Aspergillus chevalieri	65	50	12.8	Sucrose/malt agar
Virus				
Adenovirus	55	17	2.5	Ice cream

Q value

The quotient describing how much more rapidly death proceeds at a centigrade temperature T_2 than it does at a lower centigrade temperature T_1 is defined as the Q value. For comparative purposes, it is common to give the coefficient for an increase of 10 degrees centigrade, i.e. $T_2 - T_1 = 10°C$ and this Q value is referred to as the Q_{10}, where

$$\log Q_{10} = \frac{18}{Z} \tag{36}$$

Batch Sterilization

In batch sterilization, thermal death of microorganisms occurs during heating, maintenance at a high temperature and cooling of the system (*see* Fig. 19). Therefore, since k varies with temperature

$$\nabla_{\text{total}} = \ln \frac{n_0}{n} = a \int_0^t e^{-\Delta E/RT}\, dt \tag{37}$$

where ∇_{total} is the required criterion of the sterilization and is determined by the acceptable level of survivors n at the end of the cycle. Thus

$$\nabla_{\text{total}} = \nabla_{\text{heating}} + \nabla_{\text{holding}} + \nabla_{\text{cooling}} \tag{38}$$

where

$$\nabla_{\text{heating}} = \ln \frac{n_0}{n_1} = a \int_0^{t_1} e^{-\Delta E/RT}\, dt \tag{39}$$

$$\nabla_{\text{holding}} = \ln \frac{n_1}{n_2} = a \int_{t_1}^{t_2} e^{-\Delta E/RT}\, dt \tag{40}$$

$$\nabla_{\text{cooling}} = \ln \frac{n_2}{n} = a \int_{t_2}^{t_3} e^{-\Delta E/RT}\, dt \tag{41}$$

the total cycle time being t_3. To solve Eqs. 39–41 one requires the temperature–time profiles. Such relationships are given on pp.881–882. Some typical values of the contribution of each portion of the sterilization cycle are given by Wang *et al.* (1979)].

as

$$\nabla_{\text{heating}}/\nabla_{\text{total}} = 0.2$$
$$\nabla_{\text{holding}}/\nabla_{\text{total}} = 0.75$$
$$\nabla_{\text{cooling}}/\nabla_{\text{total}} = 0.05$$

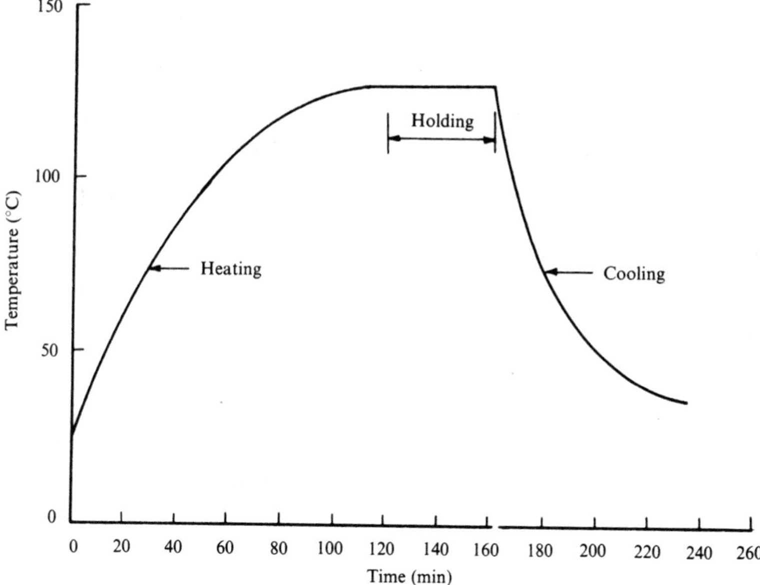

Figure 19. Typical heating, holding and cooling temperature–time profile for batch sterilization cycle [Wang *et al.* (1979)].

Continuous Sterilization

Continuous sterilization is achieved by either heating the medium indirectly using a tubular or plate and frame heat exchanger or directly by the injection of steam. Temperature–time profiles are shown in Fig. 20.

In the case of heat exchangers, the residence time and its distribution are important factors. The flow of the medium in the tubes of the heat exchanger may be either turbulent or laminar (*see* Fig. 21). The deviation from laminar plug flow can be expressed by a dimensionless Peclet number Pe. For ideal fluids in plug flow, Pe is infinite and for complete mixing – the result of turbulent flow – Pe is zero. Table 34 gives the effect of the residence–time distribution, in terms of Pe, on the performance of a continuous sterilizer.

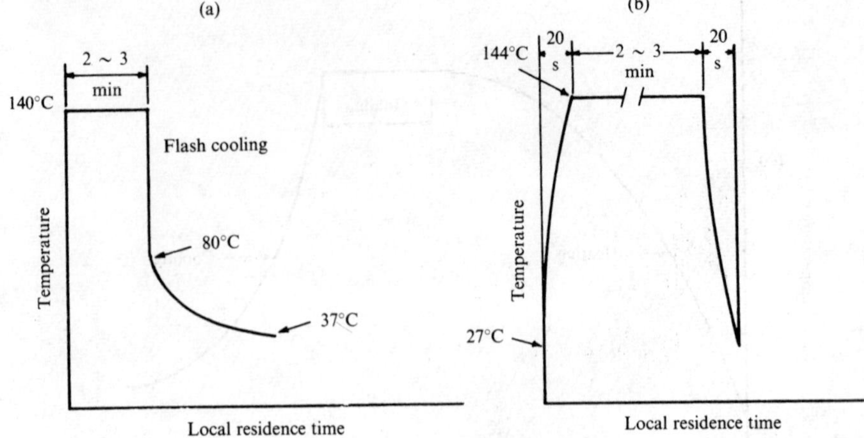

Figure 20. Temperature–time profiles in continuous sterilizers. (a) Steam injection sterilizer, (b) plate-exchanger sterilizer [Aiba *et al.* (1965)].

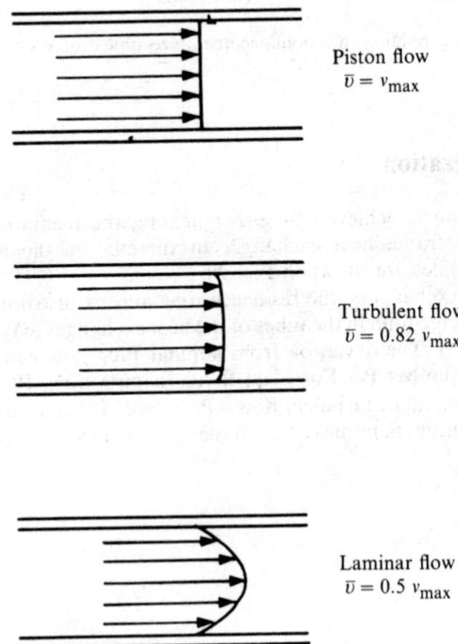

Figure 21. Distribution of velocities in fluids exhibiting different types of flow inside round pipes; $\bar{v}$ = mean velocity of the fluid.

Table 34. Effect of residence time distribution on continuous-culture sterilizer calculations [Wang *et al.* (1979)].

Specific death rate constant, k (min^{-1})	Average residence time (s)	Peclet number, Pe	Reduction of contaminant level, n/n_0	
			Plug flow assumption	Actual
10	5	200	1.8×10^{-22}	10^{-18}
10	5	100	1.8×10^{-22}	1.5×10^{-16}
10	5	70	1.8×10^{-22}	3×10^{-15}
10	5	50	1.8×10^{-22}	2×10^{-13}

Aiba *et al.* (1967) used a dispersion model to describe the performance of a continuous sterilizer in terms of Pe and the reaction number Ra as shown in Fig. 22. In this figure

$$\text{Ra} = \frac{kL}{\bar{v}}$$

$$\text{Pe} = \frac{\bar{v}L}{D_d}$$

where L = length of the sterilizer
$\bar{v}$ = average medium velocity
D_d = axial dispersion coefficient

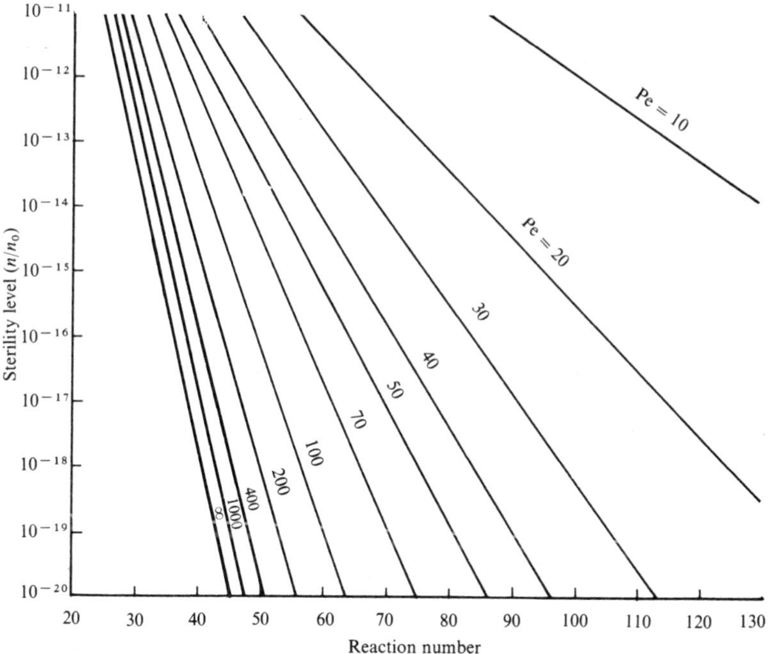

Figure 22. Prediction of the sterility level (n/n_0) at various reaction numbers (Ra = $kL/\bar{v}$ and flow conditions (Pe = $\bar{v}L/D_d$) for continuous sterilizers [Wang *et al.* (1979)].

NOMENCLATURES

A = heat transfer area
A_B = area of the bottom of a fermenter
A_c = specific cooling area
A_E = area of draught tube
A_M = area of jacket
A_s = actual cooling surface area
B' = characteristic geometric number of a system
C_s = clearance between coils
c_p = heat capacity of stream
$c_{p,\text{sys}}$ = total heat capacity of a system
D = impeller diameter
$\mathcal{D}$ = decimal reduction time
D_c = coil helix diameter
D_d = axial dispersion coefficient
D_i = inside tube diameter
D_j = diameter of liquid jet nozzle
D_{lm} = log-mean diameter of tube
D_{max} = maximum possible reactor diameter
D_o = outside tube diameter
D_T = tank diameter
d_c = coil tubing diameter
$\triangle E$ = activation energy of death
F = flow rate
F_g = gas flow rate
F_l = liquid flow rate
f = power factor
G = dry air flow
$\mathcal{H}$ = change in the water content
H^{O} = standard molar enthalpy
H_A = height of impeller from base
H_c = height of coil
$\triangle H_f^{\text{O}}$ = heat of formation
$\triangle H_{\text{ox}}^{\text{O}}$ = heat of combustion
$H_{p,i}$ = standard enthalpies of products
$H_{s,i}$ = standard enthalpies of substrates
h = film heat transfer coefficients
$h_{d,i}$ = heat transfer coefficient of deposit on the inside
$h_{d,o}$ = heat transfer coefficient of deposit on the outside
h_i = heat transfer coefficient of inside film
h_o = heat transfer coefficient of outside film
J = width of baffle
K = thermal conductivity
K_R = rate constant for the transition from resistant to sensitive microorganisms on heating
K_w = thermal conductivity of wall
k = specific death rate
k_S = rate constant from the transition from sensitive to dead spores on heating
L = length of sterilizer
L_E = height of cooling coil
l = wall thickness
M = molecular weight
M' = liquid mass
m = water flow rate
N = impeller speed
n = number of viable microorganisms
n_D = number of dead microorganisms
n_R = number of resistant microorganisms to heat
n_S = number of sensitive microorganisms to heat
Nu = Nusselt number
P_e = effective power input
P_g = power dissipated by sparged gas
P_i = power input
P_l = liquid jet power

P_m = aerated power input
$P_{m,0}$ = unaerated power input
p = absolute pressure
p_j = dynamic pressure
p_o = pressure at sparger
Pe = Peclet number
Pr = Prandtl number
Q_{acc} = rate of heat accumulation per unit volume in the system
Q_{ag} = rate of heat loss per unit volume by evaporation
Q_{exch} = rate of heat transfer per unit volume to the surroundings or to heat-exchanger coolant
Q_{gas} = rate of heat generation per unit volume due to gassing/aeration
Q_{met} = rate of heat generation per unit volume due to microbial metabolism or enzyme activity
Q_{sen} = rate of sensible enthalpy gain per unit volume by the flow
q = local heat flux
p' = rate of heat transfer
R = gas constant
Re = Reynolds number
S = steam flow rate
S' = pitch of baffle
s = aspect ratio
T = absolute temperature
T_e = outlet stream temperature
T_f = bulk fluid temperature of the fermenter
T_H = temperature of heat source
T_i = inlet stream temperature
t = time
U = local overall heat transfer
V = clear liquid volume
V_f = working volume
$V_{f,max}$ = maximum working volume
v = velocity
v_o = gas velocity at the sparger hole
W = impeller blade width
w = coolant mass flow rate

α = fraction of jet kinetic energy transmitted to bulk liquid
λ = heat of evaporation
μ = viscosity of bulk liquid
μ_m = viscosity at wall
ρ_g = gas density
ρ_l = liquid density
ϕ_o = oxygen flow
φ = volume fraction of solids

REFERENCES

Aiba, S., Humphrey, A.E. and Mills, N.F. (1965) *Biochemical Engineering* (Academic Press, New York).
Aiba, S., Nagatani, M. and Furuse, H. (1967) Some Analysis of Lag Phase in the Growth of Microbial Cells. *J. Ferment. Technol.*, **45,** 475.
Bailey, J.E. and Ollis, D.F. (1977) *Biochemical Engineering Fundamentals* (McGraw-Hill, New York).
Blenke, H. (1979) Loop Reactors, in *Advances in Biochemical Engineering*, vol.13, ed. by T.K. Ghose *et al.*, p.121 (Springer Verlag, Berlin).
Conney, C.L., Wang, D.I.C. and Mateles, R.I. (1969) Measurement of Heat Evolution and Correlation with Oxygen Consumption during Microbial Growth. *Biotechnol. Bioengng*, **11,** 269.
Coulson, J.M. and Richardson, J.F. (1977) *Chemical Engineering*, vol. 1 (Pergamon Press, Oxford).
Coulson, J.M. and Richardson, J.F. (1978) *Chemical Engineering*, vol. 2 (Pergamon Press, Oxford).
Deindorfer, F.H. and Humphrey, A.E. (1959) Analytical Method for Calculating Heat Sterilization Times. *Appl. Microbiol.*, **7,** 256.
Edwards, M.F. and Wilkinson, W.L. (1972a) Heat Transfer in Agitated Vessels. Part I. Newtonian Fluids. *Chem. Engr*, No. 264, 310.
Edwards, M.F. and Wilkinson, W.L. (1972b) Heat Transfer in Agitated Vessels. Part II. Non-Newtonian Fluids. *Chem. Engr*, No. 265, 328.

Hamer, G. (1979) Biomass from Natural Gas, in *Economic Microbiology*, vol. 4, Microbial Biomass, ed. by A.H. Rose, p.315 (Academic Press, London).

Ho, L. (1979). Process Analysis and Optimal Design of a Fermentation Process based upon Elemental Balanced Equations: Generalised Semi-Theoretical Equation for estimating Rate of Oxygen Demand and Heat Evolution. *Biotechnol. Bioengng*, **21**, 1289.

Lehrer, I.H. (1968) Gas Agitation of Liquids. *Ind. Engng. Chem., Proc. Des. Develop.*, **7**, 226.

Liley, P.E. and Gambil, W.R. (1973) Physical and Chemical Data, in *Perry's Chemical Engineering Handbook*, 5th edn. (McGraw-Hill, New York).

McAdams, W.H. (1954) Heat Transmission, 3rd edn (McGraw-Hill, New York).

Michel, B.J. and Miller, S.A. (1962) Power Requirements of Gas-Liquid Agitated Systems. *A.I.Ch.E.J.*, **8**, 262.

Miller, D.N. (1974) Scale-Up of Agitated Vessels. Gas-Liquid Mass Transfer, *A.I.Ch.E.J.*, **20**, 445.

Moss, M.O. and Smith, J.E. (1977) *Industrial Applications of Microbiology* (Surrey University Press, London).

O'Connel, F.P. and Mack, D.E. (1950) Simple Turbines in Fully Baffled Tanks. *Chem. Engng Prog.*, **46**, 358.

Prokop, A. and Humphrey, A.E. (1970) Kinetics of Disinfection, in *Disinfection*, ed. by M.A. Barnardo, p.61 (Marcel Dekker, New York).

Roels, J.A. (1980) Application of Macroscopic Principles to Microbial Metabolism. *Biotechnol. Bioengng*, **22**, 2457.

Roels, J.A. and van Suijdam, J.C. (1980) Energetic Efficiency of a Microbial Process with an External Power Input: Thermodynamic Approach. *Biotechnol. Bioengng*, **22**, 463.

Rushton, J.H., Costich, E.W. and Everett, H.J. (1950) Power Characteristics of Mixing Impellers. Part 2. *Chem. Engng Prog.*, **46**, 467.

Servizi, J.A. and Bogan, R.H. (1964) Thermodynamic Aspects of Biological Oxidation and Synthesis. *J. Water Poll. Contr. Fed.*, **36**, 607.

Seipenbusch, R. and Blenke, H. (1980) The Loop Reactor for Cultivating Yeast on n-Paraffin Substrate, in *Advances in Biochemical Engineering*, vol.15, ed. by T.K. Ghose *et al.*, p.1 (Springer-Verlag, Berlin).

Smith, J.M. and van Ness, H.C. (1959) *Introduction to Chemical Thermodynamics*, 2nd edn. (McGraw-Hill, New York).

Stumbo, C.R. (1976) Elements of Heat and Gaseous Sterilisation, in *Industrial Microbiology*, ed. by B.M. Miller and W. Litsky, p.412 (McGraw-Hill, New York).

Wang, D.I.C., Cooney, C.L., Demain, A.L. Dunnill, P., Humphrey, A.E. and Lilly, M.D. (1979) *Fermentation and Enzyme Technology* (John Wiley, New York).

Yang, H.Y. (1979). Volume Changes during Aerobic Fermentation. *Biotechnol. Bioengng.*, **21**, 525.

APPENDIX 1: CONSTANTS AND CONVERSION FACTORS

Gas law constant

$$
\begin{aligned}
R &= 1.987 \text{ cal g mol}^{-1}\text{K}^{-1} \\
&= 82.05\,\text{cm}^3 \text{ atm g mol}^{-1}\text{K}^{-1} \\
&= 8.314 \times 10^7 \text{ g cm}^2\,\text{s}^{-2} \text{ g mol}^{-1}\text{K}^{-1} \\
&= 8.314 \times 10^3 \text{ kg m}^2\,\text{s}^{-2} \text{ kg mol}^{-1}\text{K}^{-1} \\
&= 4.968 \times 10^4 \,\text{lb}_m \,\text{ft}^2\,\text{s}^{-2} \text{ lb mol}^{-1}{}^\circ\text{R}^{-1} \\
&= 1.544 \times 10^3 \text{ ft lb, lb mol}^{-1}{}^\circ\text{R}^{-1}
\end{aligned}
$$

Joule's constant (mechanical equivalent of heat)

$$
\begin{aligned}
J_c &= 4.1840 \times 10^7 \text{ erg cal}^{-1} \\
&= 778.16\,\text{ft lb, Btu}^{-1}
\end{aligned}
$$

Table A1. Conversion factors for heat transfer coefficients.

Given units	Conversion factor					
	$kg\ s^{-3}\ K^{-1}$ ($W\ m^{-2}\ K^{-1}$)	$lb_m\ s^{-3}\ {}^\circ F^{-1}$	$lb_f\ ft^{-1}\ s^{-1}\ {}^\circ F^{-1}$	$cal\ cm^{-2}\ s^{-1}\ K^{-1}$	$W\ cm^{-2}\ K^{-1}$	$Btu\ ft^{-2}\ h^{-1}\ {}^\circ F^{-1}$
$kg\ s^{-3}\ K^{-1}$	1	1.225	3.807×10^{-2}	2.390×10^{-5}	10^{-4}	1.761×10^{-1}
$lb_m\ s^{-3}\ {}^\circ F^{-1}$	8.165×10^{-1}	1	3.108×10^{-2}	1.951×10^{-5}	8.165×10^{-5}	1.438×10^{-1}
$lb_f\ ft^{-1}\ s^{-1}\ {}^\circ F^{-1}$	2.627×10^{1}	3.217×10^{1}	1	6.278×10^{-4}	2.627×10^{-3}	4.626
$cal\ cm^{-2}\ s^{-1}\ K^{-1}$	4.184×10^{4}	5.125×10^{4}	1.593×10^{3}	1	4.184	7.369×10^{3}
$Btu\ ft^{-2}\ h^{-1}\ {}^\circ F^{-1}$	5.678	6.955	2.162×10^{-1}	1.357×10^{-4}	5.678×10^{-4}	1

CHAPTER 12

DOWNSTREAM PROCESS ENGINEERING

INTRODUCTION

The functions involved in a microbiological process that are of concern to downstream processing are indicated in Fig. 1. Taking an alcohol fermentation as an example, the feed

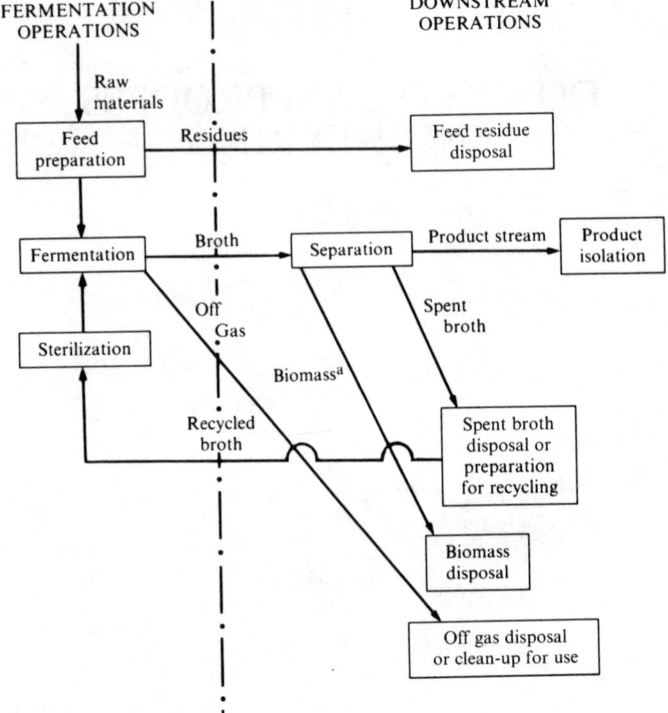

[a] Where biomass is not product stream.

Figure 1. Microbiological process functions.

preparation rejects the cellulosic, lignin, and much of the dextrin and protein fractions of the feed as the 'spent' grains of grain fermentations and the 'bagasse' of sugar fermentations. Of the fermentable carbon that goes to the fermenter 20–25 per cent is converted to carbon dioxide and 40–50 per cent appears as ethanol. Widely varying amounts (1–10 per cent) appear as higher alcohols and other volatile organic chemicals. Unfermentable soluble carbon compounds, which are rejected in the spent broth, may also make up a considerable part (1–10 per cent) of the soluble fraction of the feedstock. Similarly, a methanol fermentation tailored to produce biomass (i.e. single-cell protein) converts 45 per cent of the feed carbon to carbon dioxide and 55 per cent to biomass [*see* Dostalek and Molin (1975)]. While there are considerable variations between different fermentation processes, yields of this order are not unusual. In many processes, the conversion to the desired product is only a small part of the total fermentables consumed, e.g., Cooney (1979) showed that only about 10 per cent of the sugars is converted to penicillin G and Aunstrup (1979) demonstrated that in enzyme production 1–5 per cent appears as enzyme protein.

Current practice is to dispose of the very considerable quantities of byproducts as waste or to sell at low prices. Apart from yeast from alcohol production, biomass is generally incinerated or dumped, although it is a high-protein material of considerable potential value. Spent broth is usually treated as sewage and, as such, has a high oxygen demand, which necessitates expensive disposal treatments [Somejima *et al.* (1977)] and results in further high protein biomass which is also dumped.

Potentially, most of the soluble carbon in the feedstock, other than that converted to carbon dioxide, should be recoverable as useful products. Developments in this area of downstream processing are vital for microbial processes to compete with the chemical conversion of vegetable materials to commodity chemicals, particularly as microbiological processes start with the fundamental disadvantage of converting 20–50 per cent of the carbon compounds they consume to carbon dioxide.

Large-Scale Processes

Typical large-scale processes are listed in Table 1. The downstream processing of the products of large-scale industries is well understood. Gaseous products present no problems of purification for which established, cheap methods are not available. Recovery of alcohol and similar products requires distillation which can be made reasonably energy-effective — perhaps 600–900 kcal per litre of alcohol compared with 2000 kcal per litre for a conventional distillation train [*see* Remírez (1980a, 1980b) and Taffe (1980)]. Other processes, such as adsorption described by Remírez (1980a) or membrane separation described by Leiva *et al.* (1981), might offer advantages but, at an energy cost for distillation of under £0.02 per litre of alcohol, assuming energy at £0.60 per Therm, are unlikely to affect the success of the industry. Disposal of byproducts and wastes presents severe problems in processes other than that of alcohol production from grain. Surplus biomass (other than yeasts), spent broth and feed residues utilization or disposal represent problems for which economic solutions are required [Somejima *et al.* (1977)]. The waste problem is more severe if straw or woody material is used as feedstock, as is practical and necessary for large-scale alcohol production [Remírez (1980c)]. Day (1978) showed that the resulting lignin-rich residues represent a waste that is very difficult to process economically to acceptable standards.

Medium-Scale Processes

These processes are concerned with the production of commodity organic chemicals and substitution for natural products, generally in competition with other sources (*see* Table 1). The downstream-processing problems are essentially a matter of cost. The products themselves are chemically robust and conventional process operations can be used in their isolation and purification. Since microbiological processes form products in relatively dilute solution and in a slurry with difficult physical properties, the overall costs are sensitive to the cost per unit volume of the initial separation stages.

Surplus biomass is unlikely to be economically saleable except under the following circumstances.

1. The scale of operation must be large enough to justify the cost of obtaining regulatory approval for the use of the biomass as a feedstuff — reported figures for this cost vary from £12 million to £24 million — and of establishing the product in the market-place. The general concensus among producers of single-cell protein is that 50 000 ton yr^{-1} appears to be a worthwhile level for seeking approval [*see* Humphrey (1975)].
2. The biomass consists of yeast grown on a vegetable substrate. Such yeasts, provided that they are not contaminated, are traditional foodstuffs and approval for sale is readily achieved.

Table 1. Microbiological processes.

Process	Phase of product in broth	Typical broth volume	Typical product weight or quantity	Typical volume or weight reduction ratio from broth
Large-scale processes				
Ethanol and soluble products (conventional process)	Liquid	Widely varying scale	5–12% in broth	5–10
Polysaccharides	Liquid + solid	40–200 m^3 $batch^{-1}$	3% in broth	Unknown[a]
Single-cell protein from methanol	Solid	> 4000 m^3 day^{-1}	170 ton day^{-1}	
Single-cell protein + lipids from methanol	Solid	Unknown	3–5% in broth	5–7
Single-cell protein from gas oil	Solid	> 2000 m^3 day^{-1}		
Stable organic acids (e.g., citric acid)	Liquid	50–250 m^3 $batch^{-1}$	10% in broth	4
Medium-scale processes				
Cephalosporin	Liquid	50–200 m^3 $batch^{-1}$	3% in broth	5–10
Extracellular enzymes	Liquid	10–200 m^3 $batch^{-1}$	< 10.5% pure enzyme in broth	
Penicillin G (potassium salt)	Liquid	40–200 m^3 $batch^{-1}$	3% in broth	7–10
Streptomycin	Liquid	40–200 m^3 $batch^{-1}$	1.5% in broth	5–10
Small-scale processes				
Intracellular enzymes				
Conventional-route product	Solid	Not widely established		8
Glucose isomerase[b]	Solid	40–200 m^3 $batch^{-1}$	3% in broth	5–10
Two-phase aqueous process	Solid	Not widely established		8
Poliomyelitis vaccine	Liquid + solid	> 1 m^3 $batch^{-1}$	250 000 doses in 1 m^3	5–10
Vitamin B_{12}	Solid	40–100 m^3 batch	3% in broth	7

[a] Viscosity effects limit the concentration that can be achieved.
[b] Product is dried whole cells which exhibit enzymic activity.

For microbial single-cell protein, the cheaper petroleum substrates and other nutrients cost about £150 per ton of single-cell protein [costs from Dostalek and Molin (1975), Mateles (1975) and Moo-Young (1977) and updated]. A ton of single-cell protein involves processing about 33 tons of broth. Allowing for typical selling and distribution costs, the margins available for processing costs if the product is to compete with cheap natural products — at say £300 per ton — is about £4 per ton of broth. This figure is about the order of cost of a single centrifugal separation stage [Labuza (1975)]. These figures illustrate the extreme pressure on processing costs at the bulk end of the market.

Downstream costs currently represent 40 per cent of the 'added value', i.e. total costs less the cost of purchased materials [calculated from Matales (1975), Moo-Young (1977) and a report in *The Times* (1981) corrected for price changes]. Since fermentation efficiencies can be expected to continue their history of major improvement (*see* Chapter 3), and based on the suggestion that a five-fold increase in the output of a single-cell protein fermenter is conceivable [Sherwood (1981)], the proportion represented by downstream costs will increase. A five-fold increase in fermentation output would increase the downstream cost proportion from 40 to 75 per cent of the added value, assuming no corresponding improvements were made in downstream processing. These figures are based upon single-cell protein production from methanol, which has about the simplest conceivable downstream-processing sequence (*see* Fig. 15); for other products the proportion of downstream costs would be greater.

Spent broth must be considered for recycling to the fermenter. This is a 'proven' technique for single-cell protein processes, where the nature of the substrate used and the need to recover biomass without chemical contamination gives a clean spent broth stream suitable for immediate reuse. For other processes, however, whether the broth can be recycled will depend on what contaminants are introduced by the product-separation process, on the build-up of unconsumed fractions of the substrate and on the accumulation of inhibitory materials. For batch processes, recycling presents a storage problem, particularly if the spent broth is capable of supporting the growth of contaminating microorganisms.

Small-Scale Processes

This is concerned with production of pharmaceuticals, enzymes and other special products for which there are no alternative sources. A very wide range of materials and hence downstream processes are involved. Typical products of small-scale processes are given in Table 1.

World production of benzylpenicillin was 10 000 tonnes in 1979 [Dunnill (1981)]. The scale of the downstream units involved is consequently small. Penicillin and other major antibiotics are commonly fermented in 50–200 m³ batches, but the first extraction step gives a seven to ten times volume reduction. In many cases, the necessary production scale is closer to laboratory than plant equipment: a 650 ℓ fermentation batch produces 200 000 doses of poliomyelitis vaccine and four batches a week would satisfy the entire consumption in the United States of America, i.e. $30–40 \times 10^6$ dose yr^{-1} [van Wezel (1981)]. Consequently, recovery processes operate on a small scale and multistage purifications are economically acceptable (*see* p.924).

With the notable exception of glucose isomerase, most enzyme preparations include extracellular materials and are sold as crude concentrates with little purification. Recovery processes are thus relatively simple and involve removal of biomass solids, concentration and drying for powder formulation, with the addition of stabilizers for liquid products.

The small-scale recovery processes often necessitate treatment of spent aqueous solutions either for economic reasons or prior to disposal, e.g., steam stripping to recover small amounts of solvent carry-over from solvent extractions, pH adjustment and detoxification processes to remove traces of physiologically active materials or viable microorganisms.

DOWNSTREAM PROCESSING PROBLEMS

Feedstock produced by Fermentations

If the fermentation is carried out without any form of inert biomass support (*see* Chapter 1), the limit of fluidity for stirring or aeration is approached at 3–7 per cent (w/v) dry weight of biomass. No work has, as yet, suggested how this limit might be overcome. Physically, biomass is a compressible gelatinous solid with surface layers of polysaccharide that make it cohere and adhere, resulting in a viscous, highly nonnewtonian slurry as the feedstock for downstream processing.

In practice, a bacterial fermentation for single-cell protein will produce a broth of 3 per cent (w/v), at which concentration the slurry is about 60 per cent by volume wet biomass, 40 per cent being interparticle fluid. If biomass supports are used, somewhat higher operating biomass concentrations are possible and, in waste disposal fermentations, the concentrations have been raised from 2–5 g ℓ^{-1} to 10–40 g ℓ^{-1} [Cooper (1981) and Atkinson *et al.* (1980)].

By the standards of the feed streams to recovery processes aimed for in conventional chemical processing, fermenter broths are dilute aqueous systems (*see* Table 2). Thus it is particularly important to avoid thermal operations and to select processes that give large concentration increases in the first stage or stages. Additionally, the volume of the spent broth stream is large and spent broth treatment for disposal or recycling can be a major factor.

Volumes are reduced by at least one order of magnitude between the broth and final fluid stages of the recovery processes – in some cases by very much more, e.g., for vitamin B_{12} the ratio is over 1000:1. The style of plant and the range of economically acceptable unit operations consequently change markedly from the broth-handling stages to the final isolation.

Table 2. Typical product concentrations leaving fermenters.

Product	Concentration ($g\,\ell^{-1}$)
Acetone/butanol/ethanol mixture	18–20
Antibiotics by established processes (e.g., penicillin G)	10–30
Enzymes (e.g., serum protease)	2–5
Ethanol	70–120
Lipids	10–30
Organic acids (e.g., citric acid, lactic acid)	40–100
Riboflavin	10–15
Single-cell proteins (e.g., yeast where entire dry biomass is product)	30–50
Vitamin B_{12}	0.02

Many fermentation broths are unstable. Once they leave the controlled, aseptic environment of the fermenter, they are exposed to a severe change in the conditions. Thus, actively growing biomass from an aerated culture can be suddenly deprived of oxygen and experience a rapidly falling substrate concentration. This frequently produces rapid changes in physical properties and can lead to destruction of the desired products, e.g., the consumption of lipids as an alternative energy source for continuing metabolic activity and the destruction of enzymes by proteases released by the deteriorating cells. The broth also becomes open to contamination with foreign microorganisms which can have the same effects.

Similar problems can occur if the harvesting of a batch fermentation is delayed. The problems can be reduced by chilling to around 5°C. This is commonly used for enzymes and for other relatively small output processes, but is to be avoided, if possible, with larger outputs as chilling from a typical fermentation temperature of 35°C requires refrigeration energy of approximately 40 kWh per cubic metre of broth and therefore considerable capital expenditure. The time for appreciable deterioration to occur can be as little as 20 minutes at fermentation temperatures.

Properties of Products

Recovery operations and sequences which are practicable and necessary to use in downstream processing are governed largely by a number of simple factors.

If a product is intercellular, a cell-disruption step must precede its extraction. If the product is also water-soluble, it is then logical to perform this disruption while the biomass is still in a slurry form. Chemically stable, soluble materials permit a wide variety of recovery operations. A large proportion of microbial products are, however, chemically 'delicate', and this places severe limitations on the reactants, pH levels and temperatures that can be used. Solid-phase products present particular problems in that the solids must generally be recovered without contamination due to processing aids.

Many substances do not need to be isolated as pure materials, notably alcoholic beverages, enzymes and pharmaceuticals, such as some vaccines. An appropriate product in these cases is a complex mixture having the desired properties. Removal of specific products, e.g., pyrogens (in the case of pharmaceuticals given by injection) and nucleic acid (in the case of single-cell protein for human consumption) may, however, be necessary.

Major problems arise when separation of specific components from other chemically and physically similar materials is necessary, e.g., the isolation of enhanced-purity enzyme protein from other proteins, in which case highly specific physicochemical effects have to be used. Otherwise, it is unusual to find isolation procedures proscribed by a lack of availability of unit operations. Normally, there is a range of alternatives and a selection on the basis of cost and reliability criteria can be made. The one outstanding example of limitation on available methods proscribing the process choice is that of ethanol and other volatile soluble organic compounds where distillation is currently the only commercially practicable primary separation step.

Effluent Disposal and Spent-Broth Recycle

Although spent broth represents considerable value in treated water and unconsumed nutrients (particularly trace materials), it has not generally been found to be economic to treat it other than as sewage except in large-scale operations.

Volumes of spent broth are large (e.g., 4500 m^3 day^{-1} for a 50 000-ton yr^{-1} single-cell protein plant) and Somejima *et al.* (1977) showed that the BOD ranges from 10 000 to 30 000 ppm, which compares with 250 ppm for typical dry-weather flow to a municipal sewage plant. (The spent broth from a 50 000-ton yr^{-1} single-cell protein plant or a 150 000-ton yr^{-1} ethanol plant would be equivalent to the sewage from an industrial city of 300 000 inhabitants.) While they represent strong wastes in oxygen demand terms, spent broths are not difficult in other respects: pH is near neutrality and the liquor is generally free of specific highly toxic constituents and heavy metals. They may, however, depending on the recovery process used, be heavily contaminated with viable microorganisms from the fermentation which, even though they do not present any hazard, may interfere with effluent-treatment processes or produce undesirable effects if discharged through a river or sea outfall; in which case, specific pretreatment may be needed. A further factor, not yet encountered outside laboratory conditions, may be the presence of pathogenic organisms and traces of persistent physiologically active materials that will demand treatments and control procedures to ensure the hazard is eliminated prior to the normal effluent-treatment processes.

Recycling the spent broth to the fermenter is attractive in that it recovers unconsumed materials in the broth and eliminates most of the effluent. It appears to have been applied only to single-cell protein [*see* Labuza (1975)] and ethanol [*see* Wall *et al.* (1981)]. There are implicit restrictions on downstream processing which arise from recycling: the processing must not add materials that adversely affect the fermentation or the resterilizing step, either chemically or physically, e.g., surface-active agents may cause foaming and calcium ions may lead to the build up of scale on heat transfer surfaces. Also there have been fears expressed by Topiwala and Khosrovi (1978) that recycling may cause build up of inhibitory products. However, Labuza (1975), Topiwala and Khosrovi (1978), Gow *et al.* (1975), Murphy *et al.* (1981) and Khosrovi (1981) have not experienced such problems when recycling has been tried.

Common and Product-Specific Needs

The different technical targets of the broth-handling operations compared with the subsequent product recovery divide downstream processing into two areas.

Broth-handling stages

These are large-volume operations where cost per unit volume handled has a large effect on product cost because of the low concentrations of product in the flow stream. What can be done in terms of physical operations is dependent on the difficult physical properties of the broth. The fermentation process can be interrupted or delayed only at very considerable cost. A serious hold-up in the recovery stages will reduce production and generate a major problem in disposal of large volumes of highly abnormal fermenter burden. Consequentially, there is a very high incentive to make broth-handling stages simple and reliable.

A further problem with the broth-handling stages is the variable nature of the broth, which may be adventitious or deliberate. In batch fermentation processes particularly, mild infection is common and serious infection must occasionally be expected; even with good techniques and equipment designed for aseptic operation perhaps one batch in 20 will be slightly abnormal and one in 50 seriously contaminated. The recovery plant must be able to handle the slightly abnormal batches and, at least, make the seriously contaminated batches fit for disposal. Minor changes, such as might be expected from process modifications to improve fermenter performance, should also be within the flexibility of the broth-handling stages.

A consequence of these requirements has been that the primary broth-handling stages are generally assembled from a few well-established, versatile unit operations: essentially rotary or press filters or centrifuges to separate a clarified stream and a 'solids' stream, with distillation or adsorption of product on solid adsorbent from unfiltered broth. These processes are well developed and must be considered as being at a point on the 'learning curve' where major cost reductions are unlikely. A wider range of processes is needed.

Stages following a whole broth-handling stage

These stages are free of the constraints imposed by the physical properties of broth. If the initial stage is the separation of the broth into a liquid phase and a solids phase, and the product is in a liquid phase, there remain the problems of a dilute product stream that is liable to deteriorate if stored. Rapid processing to a stable concentrate or a concentrate that can be stabilized by cooling, for example, is consequently necessary. Again, to avoid any delay reflecting on the fermentation, high reliability is essential. The available choice of processes at this stage is, however, wide and dependent principally on the product properties.

In general terms, stable organic chemicals, e.g., organic acids, can be purified by conventional chemical processes, volatile materials by distillation and unstable compounds by diffusional, adsorption or other physical processes operating at ambient or low temperatures. Particularly intractable separation problems tend only to be encountered with high-value products and hence in the case of smaller-scale operations where almost any laboratory separation method can be applied to the production process.

BROTH PROCESSING

Criteria

The requirements for the primary downstream-processing stages are listed below.

1. High product recovery with desirably high rejection of unwanted components.
2. Reliable and robust processes and equipment that is insensitive to minor broth-quality variations.
3. Processes that give a large volume reduction in order to reduce capital and operating costs at subsequent stages.
4. Processes that leave a spent broth suitable for recycling or effluent treatment.
5. Low capital, low hold-up and low energy processes

As stated above, product concentrations are low and the cost per unit volume of broth handled has a large effect on the final costs. Typically, in processes for the replacement of natural products, such as protein feedstuffs or lipids, concentrations in the broth of 20–30 g ℓ^{-1} are foreseen. Broth-processing costs of £0.01 per gallon (i.e. £0.002 per litre) mean £70–100 per ton on the plant cost of the products – a very appreciable fraction of the competitive selling price of £300–900 per ton. At such processing-cost levels, evaporation or other processes with high-energy elements are unacceptable. The quantities of additives required for flocculation, pH control, as filter aids, etc. also become significant, e.g., filter-aid consumption represents £0.004 per gallon of broth at current bulk-purchase prices [*see* Dlouhy and Dahlstrom (1968)].

Table 3. Outline of downstream processes associated with typical products [Atkinson and Sainter (1981)].

Product	Typical process sequence	Known alternatives	Disposal of broth liquids	Disposal of biomass	Disposal of fermenter gas	Disposal of feed residues
Acetone/butanol/ethanol	1. Steam strip in presence of broth solid to separate volatiles 2. Distil volatiles to separate components	None demonstrated	Evaporated to dryness to give riboflavin-rich animal feed. Otherwise treat as waste		Hydrogen/carbon dioxide mixture saleable given local demand for hydrogen	Little residue, molasses as primary nutrient
Amino acids	1. Filter broth 2. Separate product by: a. Precipitation b. Direct crystallization c. Membrane or vacuum concentration followed by precipitation or crystallization 3. Purify by normal chemical processes	Wide variety of sequences depending on stability of product and concentration in broth	Treated as waste	Treated as waste or recovered as high-protein feed	To waste	None
Antibiotics Extraction processes (e.g., penicillin G)	1. Flocculate broth 2. Filter on rotary filters 3. Extract into 1/7 volume of butyl acetate at pH 2 with low residence time (e.g., centrifugal separators) 4. Purify and reduce volume by further solvent–aqueous solvent stages 5. Crystallize from aqueous solution by alcohol addition 6. Centrifuge out crystals 7. Dry — typically in stirred bed drier 8. Recover various solvents	Wide variety of extraction and/or adsorption stages possible	Treated as waste – may need steam stripping to recover entrained solvent. Effluent includes spent aqueous from extraction stages	Treated as waste (incinerated or dumped)	To waste	None

Adsorbtion processes (e.g. cephalosporin)	1. Flocculate broth 2. Filter on precoat rotary filters 3. Adsorb on to fixed bed of carbon or resin adsorbent 4. Desorb into 1/6 broth volume of acetone 5. Readsorb and extract into aqueous buffer solution using resin adsorbents, with further volume reductions 6. Precipitate by alcohol addition 7. Centrifuge 8. Dry 9. Recover alcohol	Wide variety of extraction and/or adsorbtion stages possible	Treated as waste. Effluents include adsorbent bed wash out and spent adsorbent regeneration solutions	Treated as waste	To waste	None
Whole broth adsorbtion (e.g., streptomycin)	1. Pass whole broth through resin adsorbent bed 2. Wash out retained broth and broth solids from bed 3. Elute product in aqueous buffer 4. Concentrate solution by vacuum evaporation 5. Crystallize by evaporation 6. Filter off and wash crystals, e.g., in agitated pressure filter 7. Dry 8. Regenerate adsorbent beds	Variety of multistage adsorbent procedures	Contains broth solids and diluted by adsorbent wash liquor. Treated as waste, may require filtration step to recover solids for separate disposal. Effluents also include resin-bed regeneration liquors	Treated as waste		

continued overleaf

(Table 2 – Continued)

Product	Typical process sequence	Known alternatives	Disposal of broth liquids	Disposal of biomass	Disposal of fermenter gas	Disposal of feed residues
Enzymes Extracellular	1. Flocculate broth and cool 2. Separate solids — rotary or press filter or centrifuge 3. Concentrate filtrate by ultrafiltration or vacuum evaporation 4. Formulate and standardize concentrate for sale or precipitate solid by alcohol or ammonium sulphate addition, filter off and dry. Formulate solid for sale in attractive physical form. Recover alcohol by distillation 5. For greater purity products, purify concentrate by adsorption/chromatographic processes (e.g., ion exchange resins, gel filtration)	1. Ultrafilter in presence of solids 2. Recover enzyme concentrate directly from broth by electrophoresis, selective adsorption, two aqueous phase partition, direct precipitation, etc. 3. Additional purification prior to concentration, e.g., heat treat to reduce nucleic acid content 4. Recover enzyme from concentrate as 'fixed' product on appropriate support material	Treated as waste	Treated as waste	To waste	None with normal primary nutrients
Intracellular	As above with cell disruption prior to solids removal. Mechanical disruption preferred due to loss of enzyme in chemical methods					
Whole cell preparations (e.g., glucose isomerase)	1. Flocculate broth 2. Separate solids by filtration or centrifuge 3. Dry and granulate solids (possibly with additives) to appropriate physical form		Treated as waste	As product	To waste	None with normal primary nutrients

Ethanol Organic solvents	1. Distil in presence of broth solids to alcohol concentrate 2. Redistil to separate fusel oils and further concentrate 3. Dehydrate by extractive distillation where required as anhydrous product	1. Wide range of distillation sequences 2. Dehydrate by adsorbtion (not commercially proven) 3. Dehydrate or separate at a dilute stage by membrane process (demonstrated in laboratory)	Recycle or treat as high-strength sewage. Some evaporated to dryness as feedstuff	Purge off part and dry as feedstuff. Recycle remainder	To waste	From sugar cane: burn as fuel From grain: dry and sell as feedstuff From starch/sugars: little residue
Alcoholic beverages	1. Separate fermentation solids by settling, flotation or filtering 2. Hold for varying periods for secondary reactions 3. Clarify and otherwise condition to give product stability	None which give traditionally accepted qualities	As product	Part recycled. Residue as protein-rich feed	To waste. Some recovered as liquid carbon dioxide	From grain: dry as a feedstuff From grapes: dump, referment to potable alcohol, or dry as feedstuff
Lipids (envisaged processes not yet commercially developed)	1. Flocculate broth 2. Separate solids as for single-cell protein 3. Recover oil from damp biomass by extrusion and/or solvent extraction 4. Recover solvent 5. Dry spent biomass 6. Deodorize and refine oil by normal edible oil processes	1. Direct separation of an oil and aqueous layer prior to filtration. (Cell lysis may be necessary as a prior step) 2. Solvent extraction of broth prior to filtration. (Cell lysis may be induced by solvent) 3. Deodorize and refine oil (may be avoidable by control of fermentation and rapid downstream processing)	Sterilize and recycle to fermenter or treat as waste	Sell as single-cell protein, (fermentation may be operated to produce both single-cell protein and lipids in economically optimal ratio)	To waste	None likely with envisaged primary nutrients

continued overleaf

(Table 2 – Continued)

Product	Typical process sequence	Known alternatives	Disposal of broth liquids	Disposal of biomass	Disposal of fermenter gas	Disposal of feed residues
Organic acids Chemically stable	1. Filter broth 2. Precipitate as insoluble salt, filter off 3. Purify by normal chemical processes	Variety of filtration/ centrifuge processes for the solid/liquid separation stages	Treated as waste (contaminated with processing chemicals)	Treated as waste	To waste	None
Acetic acid (potable)	1. Separate solids by sedimentation 2. Filter to clarify 3. For concentrated product, distil	None demonstrated. Membrane concentration process demonstrated in laboratory	As product	Small purge, dispose as waste	Waste	None–alcohol as primary nutrient
Polysaccharides	1. Condition broth by heat, chemical treatment, induce cell lysis (chemically) where necessary 2. Separate solids by precoat filtration or centrifuging where cell residues not wanted in product. Heating or dilution may be necessary to reduce viscosity 3. Precipitate product by alcohol addition, filter off and dry. Recover alcohol from drier exhaust 4. Distil spent broth to recover alcohol	1. Use of enzymes to render debris more easily removed 2. Isolation by separation of two aqueous phases by electrolyte addition 3. Isolation by acid or electrolyte addition to precipitate product 4. Separate cell debris after product isolation by redissolving and filtration. Sell clarified solution 5. Concentrate broth by membrane processes to fluidity limit, then dry concentrate	Treated as waste	Treated as waste	To waste	None

Single-cell protein Petroleum-based feedstock (methanol, *n* -paraffins or gas oil)	1. Flocculate solids 2. Separate solids as 15 per cent slurry by sedimentation flotation or centrifuging 3. Wash unconsumed nutrient from solids slurry in centrifuges (not necessary with volatile primary nutrient) 4. Dewater solids stream by further centrifugation step 5. Dry	None commercially proven. Range of possible dewatering methods (e.g., membrane, cross-flow filtration, electrokinetic)	Sterilize and recycle to fermenter. Purge, to control build-up of unwanted materials, to waste	As product	To waste after scrubbing to prevent odour emmission. Energy recovery on pressure let down from pressure fermentations possible	None
Paper mill waste liquor	1. Filter on rotary filter 2. Dewater filter cake by mechanical pressing 3. Dry		Sent to normal waste liquor treatment	As product	To waste	None
Vaccine Poliomyelitis (inactivated)	1. Wash growth tissue from support beads on which it has been grown, to give suspension in broth 2. Clarify by filtration 3. Gel filter to remove high-molecular weight materials 4. Concentrate by ultrafiltration 5. Purify in adsorbtion columns – selective removal of unwanted proteins 6. Deactivate by formalin treatment 7. Filter to clarify and sterilize, pack	A very wide variety of adsorbtion-type purification steps is possible	Treated as waste	Treated as waste	Waste	Waste

continued overleaf

(Table 2 – Continued)

Product	Typical process sequence	Known alternatives	Disposal of broth liquids	Disposal of biomass	Disposal of fermenter gas	Disposal of feed residues
Vitamin B_{12}	1. Flocculate broth 2. Filter or centrifuge to recover solids 3. Dry solids as vitamin-rich material for sale (or: extract solid with acid or cyanide solutions) 4. Adsorb product from extract or resin adsorbent 5. Elute into dilute aqueous butanol 6. Purify by partition between phenolic solvents and water 7. Crystallize from aqueous acetone solution	Biomass may be byproduct of other fermentation (e.g., the biomass from acetone/butanol process). Many different routes available based on physical separation methods after initial biomass extraction	Treated as waste	Product as a vitamin-rich material. Waste in manufacture of pure vitamins	Waste	None with normal feedstock

Broth Handling

The processes available for initial broth treatment are indicated in Fig. 2. In this figure, the term 'SELL' shows where the product might leave the sequence as finished material. Only the product paths are shown in Fig. 2, giving a spurious attractiveness to processes that treat whole, i.e. unfiltered, broth. This attractiveness only holds if the spent broth with its solids can be recycled or disposed of without separating the solids.

Most current processes follow the centre course shown in Fig. 2, with a solid-phase/liquid-phase separation stage, e.g., penicillin, traditional ethanol processes and processes where the solid phase is the required material (*see* Table 3). A few processes follow the left-hand course with product separation from the unfiltered broth, e.g., direct adsorption of streptomycin and power-alcohol processes. The right-hand course involves direct evaporation and is obsolete, although it has been used for semi-solid fermentations producing enriched animal feedstuffs.

Fig. 2 emphasizes the principal problem of evolving processes for handling fermentation broths, i.e. the dependence on the physical and physicochemical nature of the broths, either as they leave the fermenter or after whatever necessary operations, such as cell disruption, have been performed. High viscosity, the presence in the slurry of sticky, gelatinous material, 'solids' that are compressible, finely dispersed and of a density close to the suspending liquid are typical of fermented broths but are conditions that nonthermal separation processes do not handle well. These are conditions, in fact, that process engineers have been traditionally at considerable pains to avoid. Some attempts to alleviate these problems by treatment with flocculants, pH adjustment and other 'conditioning' processes are usually made but are of limited effectiveness.

Many of the physical properties of broths which control their behaviour in such operations as filtration, centrifugation and membrane separation are difficult to measure or express in terms that can be used in quantifying the mechanics of the process. Properties, such as the tendency to form nonmobile, relatively impervious deposits, of the solids to adhere to surfaces or structural degradation to adherent slimy deposits under mechanical stress, e.g., when removing filter cake from a filter cloth or discharging the solid phase from a horizontal bowl centrifuge, are not expressible in measurable numerical concepts. Development and scale-up are thus very dependent on experimentation. There is also a high risk that at full scale the fermentation may not reproduce the small-scale results, particularly the physical properties on which the recovery processes depend may vary. Added to this is the fact that the physical properties of a broth change markedly on storage so that experiments, if they are to be meaningful, must be performed using broth fresh from the fermenter. Facilities for both fermentation and recovery development must consequently be close together and on a scale sufficient to produce realistic results. Development thus becomes expensive in terms of operating costs and of equipment, and the range of skills that must be assembled on one site. It is also difficult under these circumstances for equipment manufacturers to make their full contribution; material dispatched to their laboratories for testing becomes seriously unrepresentative and the experimental results achieved with it are not to be trusted.

Fig. 2 also indicates the greater problems presented by solid-phase products compared with products that are in solution at the primary separation stage. With the former products, the solid must be separated from the fluid without contamination or denaturation. This places severe restrictions on broth pretreatments and the use of filter aids – factors that are essential features of most established solid/liquid separation processes used for fermentation broths. Removal of impurities from the solid phase also introduces difficulties in that any reslurrying operation to expose the solid to washing or chemical treatment involves diluting to relative low concentrations if the slurry is to be mobile. These slurries then involve a further high-volume processing step to recover the solid and leave a large volume of spent wash or reactant liquor as effluent, e.g., removal of trace unconsumed substrate from single-cell protein by washing with wetting agents at 15 per cent (w/v) solids concentration [*see* Humphrey (1975) and Labuza (1975)].

There is a comparatively limited choice of processes for the initial broth-handling stage.

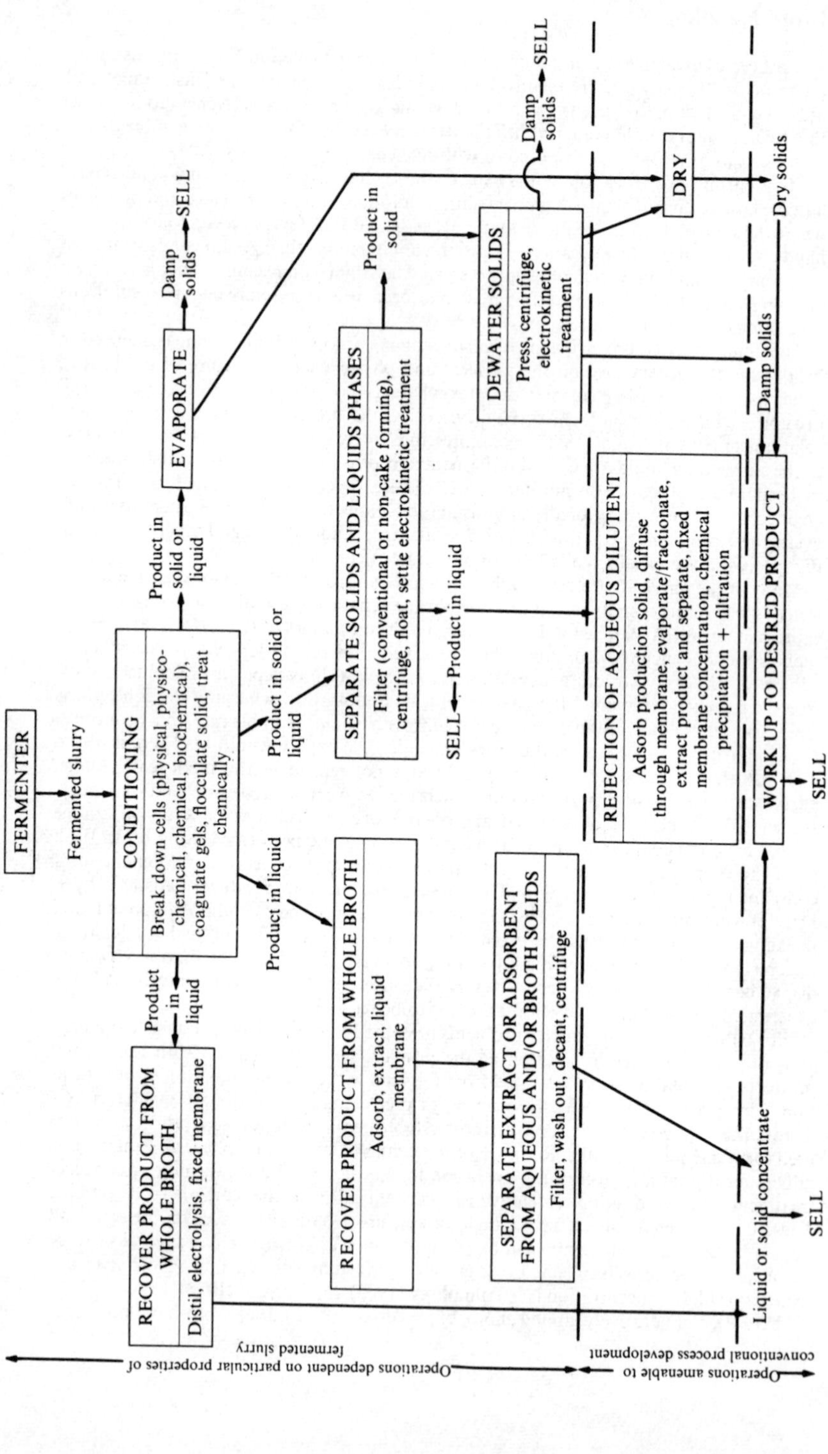

FERMENTER

Fermented slurry

CONDITIONING

Break down cells (physical, physico-chemical, chemical, biochemical), coagulate gels, flocculate solid, treat chemically

Product in liquid

RECOVER PRODUCT FROM WHOLE BROTH

Distil, electrolysis, fixed membrane

Product in liquid

RECOVER PRODUCT FROM WHOLE BROTH

Adsorb, extract, liquid membrane

SEPARATE EXTRACT OR ADSORBENT FROM AQUEOUS AND/OR BROTH SOLIDS

Filter, wash out, decant, centrifuge

Product in solid or liquid

SEPARATE SOLIDS AND LIQUIDS PHASES

Filter (conventional or non-cake forming), centrifuge, float, settle electrokinetic treatment

SELL ← Product in liquid

Product in solid

DEWATER SOLIDS

Press, centrifuge, electrokinetic treatment

Damp solids → SELL

Product in solid or liquid

EVAPORATE

Damp solids → SELL

DRY

Dry solids

Damp solids

REJECTION OF AQUEOUS DILUTENT

Adsorb production solid, diffuse through membrane, evaporate/fractionate, extract product and separate, fixed membrane concentration, chemical precipitation + filtration

WORK UP TO DESIRED PRODUCT

SELL

Liquid or solid concentrate

SELL

← Operations dependent on particular properties of fermented slurry →

← Operations amenable to conventional process development →

Figure 2. Product routes through primary recovery stages: available unit operations.

1. Where solid/liquid separation is required.
   a. Filtration — the solid is collected by a filter medium or concentrated as a thick slurry over a porous membrane.
   b. Settling — gravitational or centrifugal, producing a solids sludge or a concentrated slurry and a clarified liquid stream.
   c. Flotation — gas assisted, producing a high solids content froth that is skimmed off the clarified liquid.
   d. Electrokinetic — collection of the solids at an electrode as a high solids-content cake.

2. When the soluble product is separated from unfiltered broth.
   a. Extraction processes — product extracted in a second liquid phase under conditions such that the solids remain in the spent-broth phase.
   b. Membrane processes — product diffuses through a solid or liquid membrane, perhaps with assisting mechanisms, e.g., electrostatic or ion exchange.
   c. Distillation — for volatile products, using energy-effective sequences.

Broth Pretreatment to improve Processing

The minimum costs for the solid-separation stages will be achieved with broths that give high processing rates on simple equipment. Unless there is an unforeseen breakthrough in unit processes, this means broths that are free of gel material and in which solids are large and are rigid enough to separate and dewater freely without compressing to impervious films or smearing into the interstices of filter membranes. Such properties might be achieved in the fermenter or by conditioning the broth prior to processing. Current practice in broth-conditioning is based on the use of temperature and pH adjustments, the addition of surface-active agents, electrolytes and precipitants, and ageing periods. The development of such procedures is generally empirical.

The recognized need for better understanding of the physical chemistry of the cell, its membrane and surface films to improve fermentation technology is expected to lead to improved broth-conditioning techniques. Such developments could have a profound effect on the cost of downstream processing. They would also increase the degree of freedom open to fermentation development by relaxing from the fermentation the criteria on the broth properties required for economical downstream processing. For example, there has been a reluctance to use bacterial cultures because of the difficulty of separating small, dispersed cells.

Pretreatment, particularly the lysis of cells, can also have a considerable effect on downstream costs beyond the primary stages. The reduction in the bulk of cells which follows from the perforation of their external membranes reduces the drying costs for solids products or the liquor loss in wet solids for soluble products. It also increases the concentration of the slurry at the limit of fluidity and hence the concentration that can be reached without resort to expensive solids-handling procedures, e.g., addition of penicillin to the broth in glutamate production reduces the packed cell volume by 60–80 per cent [Kinoshita and Nakayama (1978)].

PRODUCT ISOLATION

The product concentrates that leave the broth-handling stages require a very wide range of processes to produce finished products. Examples of virtually every unit operation known to process engineering can be found in established processes for microbial products (*see* Table 3).

A considerable range of products requires only simple treatments, e.g., single-cell protein needs drying (under somewhat careful conditions to limit denaturation of the protein while, at the same time, reducing the number of viable microorganisms to an acceptable level), yeast requires pressing, bulk enzyme concentrates require formulation to adjust strength and prevent deterioration on storage, alcoholic beverages require clarification, etc. Other products (e.g., organic acids and glycols) are chemically robust and can be separated and purified by normal processing techniques. Frequently, the problems are such that normal process development techniques can be applied. Physical properties, reaction rates, stability limits and other variables are measurable in terms that can be applied to the design of full-scale plant from small-scale experiments using conventional process design procedures. Economical processing depends on identifying the full range of possible processes that can provide the required product quality and selecting the route that will be most appropriate to the full-scale operation. In addition to established microbiological product processes, a wide range of existing related technology can be drawn on, e.g., on the high-tonnage solid-phase area from starch products, food products and vegetable oil industries, in the small-tonnage area from recovering pharmaceuticals, enzymes, etc., from natural products and for unstable chemicals from chemical synthesis processes. Provided that processes that are scale-limited by physical factors, e.g., zone centrifugation and molecular distillation, are avoided, there are few laboratory separation processes for which an equivalent or closely related technology is not available on a commercial scale.

In contrast, the isolation of relatively pure, chemically delicate extra- and intracellular products involves bringing to a commercial scale separation processes that are currently only used in laboratories. It is also to be hoped that new separation processes will be developed which could replace those laboratory techniques that are unattractive or impracticable when scaled up. However, since the processes involved must usually be physical operations at, or near to, ambient temperatures that exploit molecular size or configuration, the range of phenomena available for use is limited.

A characteristic of the microbial product industry is that the required product is formed at the fermentation stage. Subsequent chemical changes are generally minor, e.g., conversion of an acid to a salt in order to achieve precipitation. Because of stability problems, physical separation processes, particularly nonthermal ones, are of much greater interest to the microbiological industry than to the chemical industry in general. Thus, while developments elsewhere can be expected to provide the basis, e.g., improved utilization of energy in alcohol distillation, there is an area of physical separation processes where the microbiological industry will have to provide the motivation for the development it needs.

Process choice is a matter of product quality, available development effort, available time and economics. The choice will depend considerably on the availability of appropriate equipment for and experience of particular methods.

A particular economic problem which is much affected by the availability of appropriate equipment on a range of scales is the time lost in achieving commercial return from new products that require regulatory approval. A change of method, such as may be forced on one as the scale increases, can lead to very long delays in obtaining approval for the product made by the new method.

Large- and Medium-Scale Production

The problem areas are almost all concerned with improving the economics of processing, particularly of solid-phase products.

1. Drying and reducing the viable microorganism count of the biomass.
2. Reduction of the nucleic acid content of single-cell proteins.
3. Removal of low levels of unwanted components of biomass such as traces of substrate and undesirable flavours.
4. Extraction of lipids from damp biomass.
5. Dehydration of ethanol by low-energy methods.
6. Purification of products that form very viscous solutions, e.g., polysaccharides.

The one identifiable area of medium- to large-scale processing for which there are currently no developed processes, but which are required, is the separation of the solid components of biomass, i.e. the separation of a wanted polymeric material from the remaining cell structure. Related technology is available in such processes as the recovery of cellulose from wood and in the corn-products industry.

Small-Scale Production

The problems lie almost entirely in the purification of physically and chemically delicate, large molecular weight materials to yield acceptable products.

Where compounds, which are finally isolated as a chemical entity, are produced, it is relatively simple to eliminate at the final purification stages traces of undesirable materials that may have originated in the fermenter or have been introduced during processing.

Protein products (e.g., enzymes and vaccines), steroids, etc., that are not separable as pure compounds offer greater problems. However, purification to a large extent, can be avoided, e.g., protease is sold as granulates with an enzyme protein content of 1–5 per cent or as a liquid concentrate with a content of 2 per cent.

PROCESS FLOWSHEETS

Outlines of process flowsheets giving indications of the quantities involved and the sizes of plant items are set out in Figs. 3–12. Most of the flowsheets are of established processes, while some (e.g., lipids and two aqueous-phase extraction of enzymes) have been demonstrated but have not been operated commercially at the scales indicated. The flowsheets only indicate the main process lines; subsidiary plant (e.g., plant concerned with wash-out and sterilizing-solution handling, solvent recovery, etc.) is omitted.

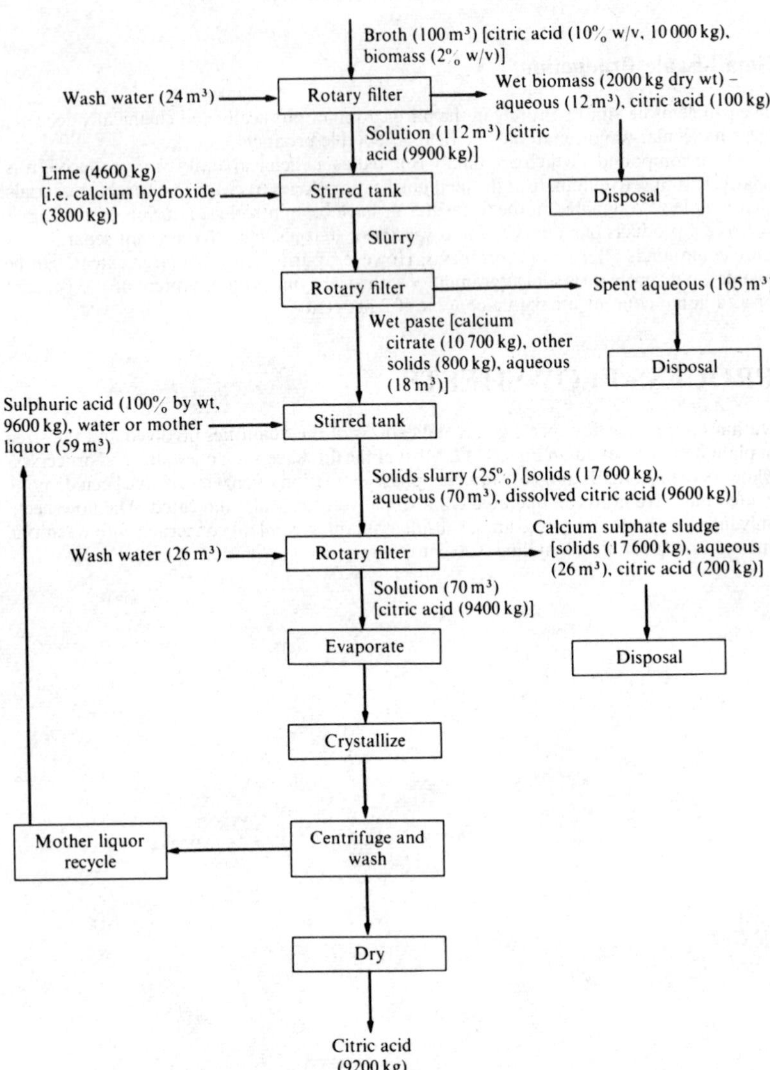

Figure 3. Citric acid flowsheet. Isolation of a chemically stable organic acid based on a 100 m³ batch producing 9200 kg acid (100 per cent by weight).

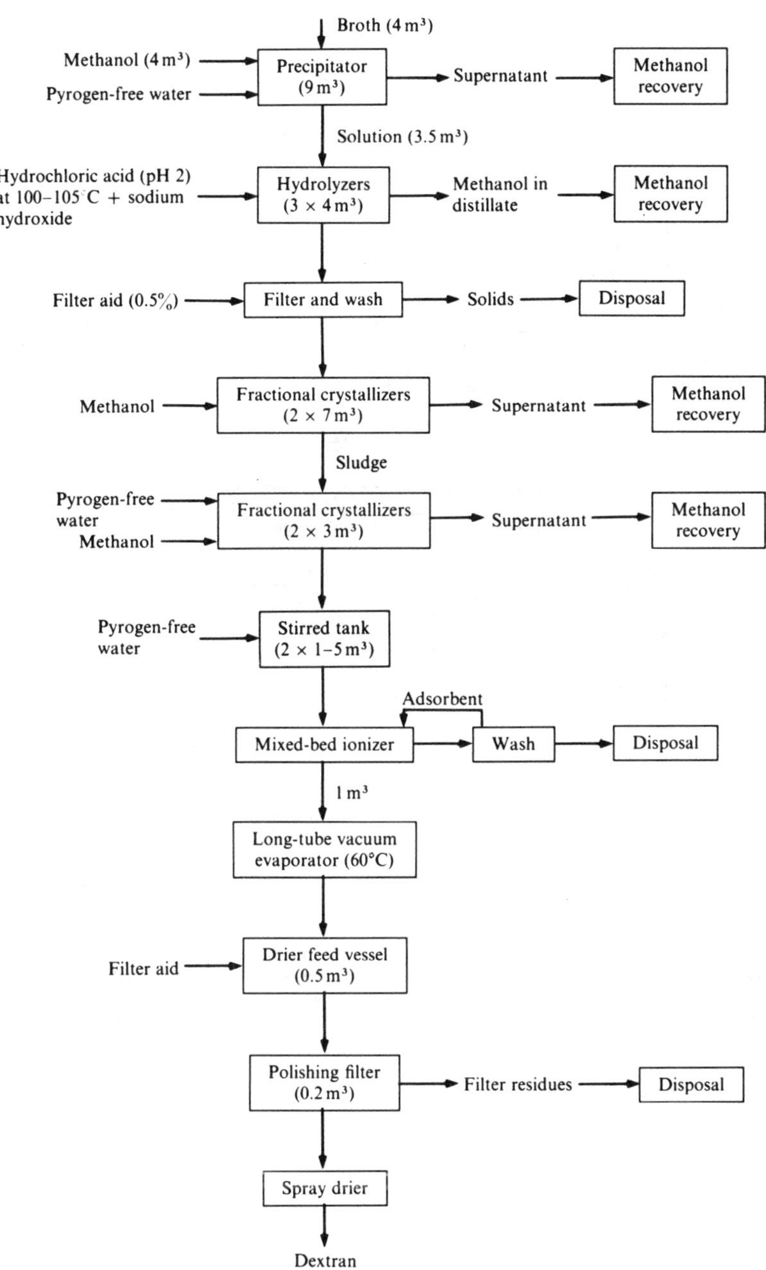

Figure 4. Dextran processing by fractional methanol precipitation. Dextran is formed in a broth with low-solids content, based on $4 \times 4\,m^3$ batches of broth every 24 hours. The process involves clean-down between batches and apparent occupation is thus low.

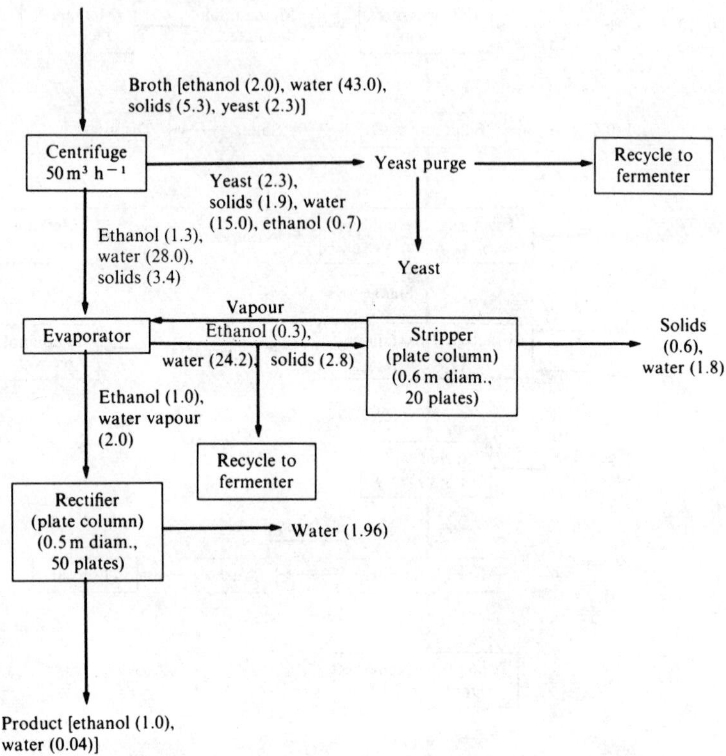

Figure 5. Ethanol processing, based on the Alfa Laval 'Biostill' process. Heat recovery and other heat exchangers are not shown. The quantities given are hourly flows (in tonnes) for 1 tonne h^{-1} 100 per cent (by wt) of ethanol output from 4.5 per cent ethanol broth. 'Solids' refer to dissolved nonvolatile material.

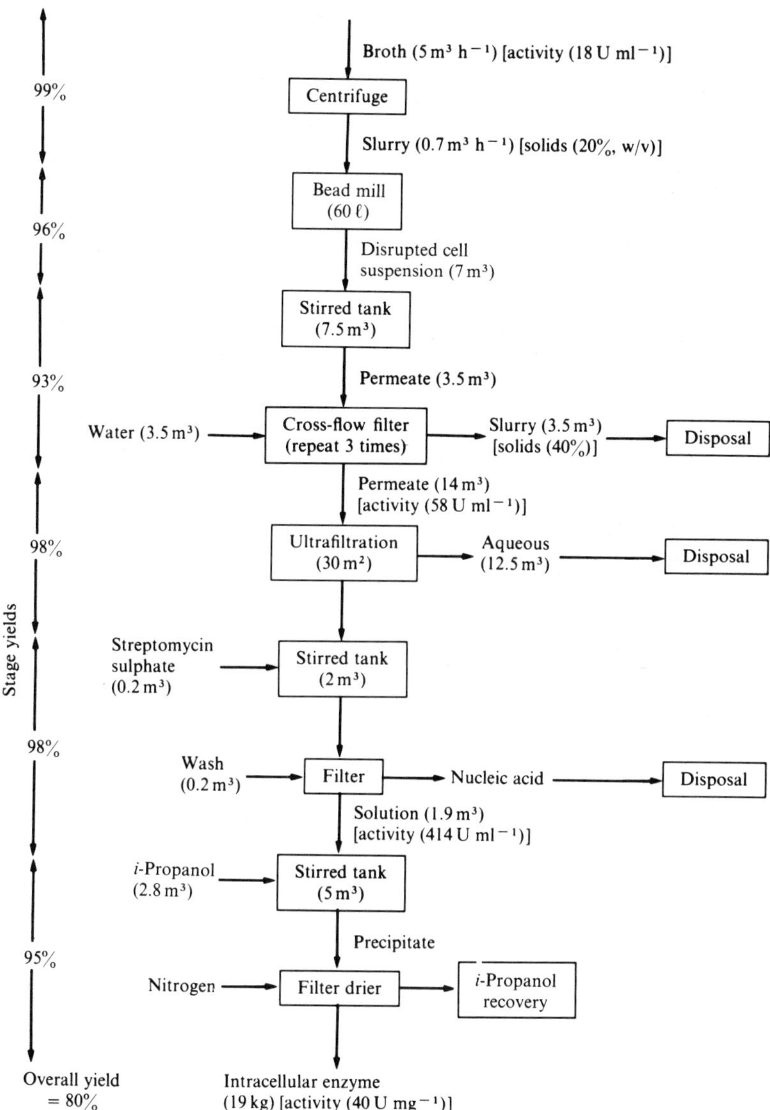

Figure 6. Indicative flowsheet for production of intracellular enzyme, based on a 50 m³ broth every 12 hours, comprising 10 hours' operating time and two hours' clean out. Activity figures are appropriate for amidase.

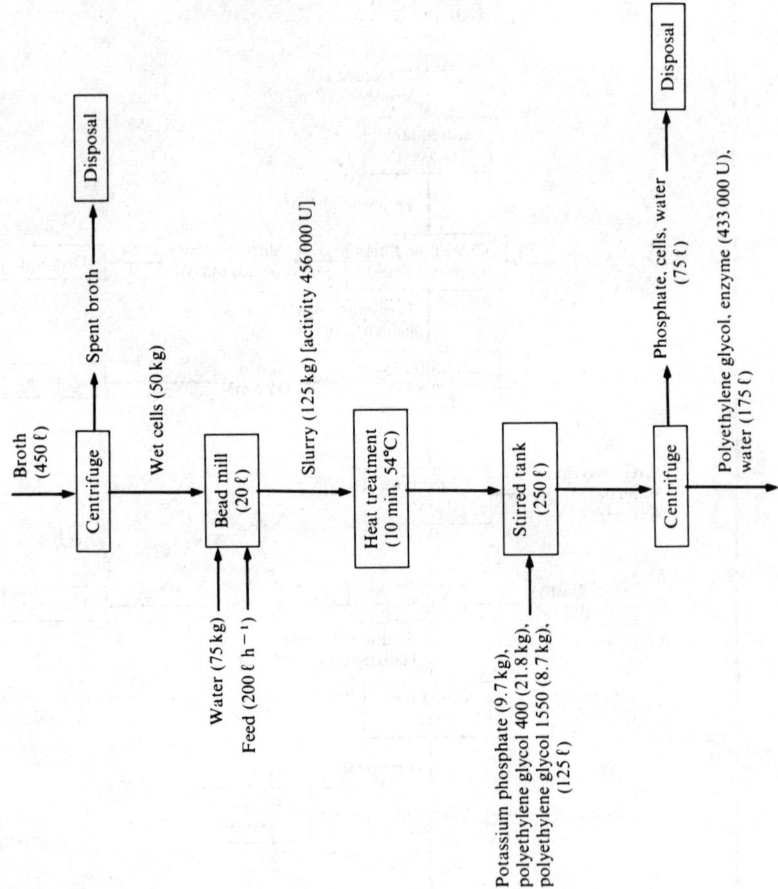
Broth (450 ℓ)
Centrifuge
Spent broth
Disposal
Wet cells (50 kg)
Bead mill (20 ℓ)
Water (75 kg)
Feed (200 ℓ h^{-1})
Slurry (125 kg) [activity 456 000 U]
Heat treatment (10 min, 54°C)
Stirred tank (250 ℓ)
Potassium phosphate (9.7 kg), polyethylene glycol 400 (21.8 kg), polyethylene glycol 1550 (8.7 kg), (125 ℓ)
Centrifuge
Phosphate, cells, water (75 ℓ)
Disposal
Polyethylene glycol, enzyme (433 000 U), water (175 ℓ)

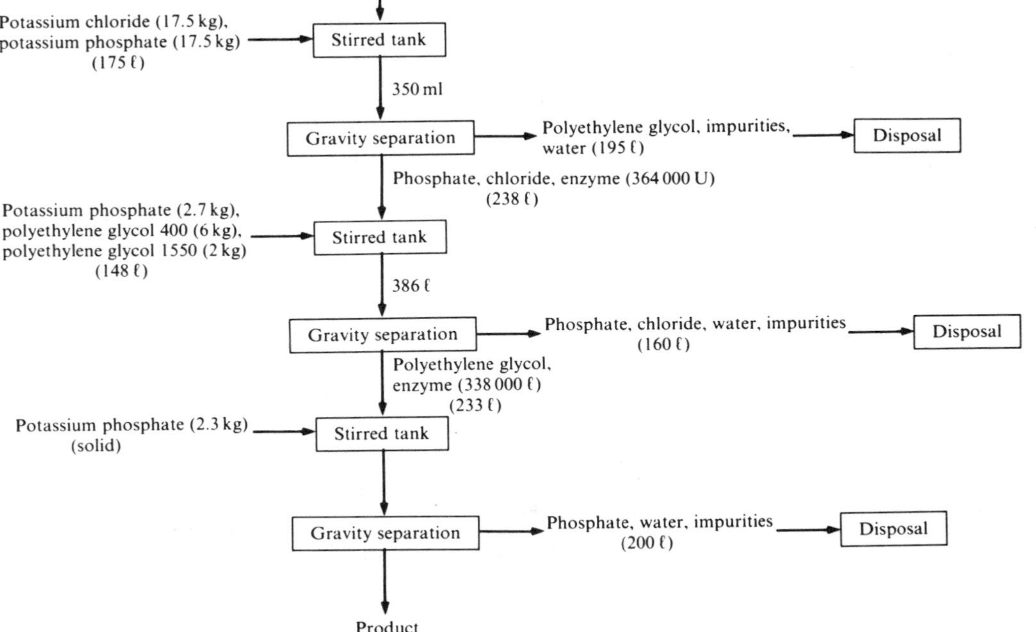

Figure 7. Semitechnical-scale example of two aqueous-phase extraction process for intracellular enzyme, based on a 0.5 m^3 fermenter producing 111 g enzyme protein. The enzyme is separated from polyethylene glycol in the product at the point of use by ultrafiltration.

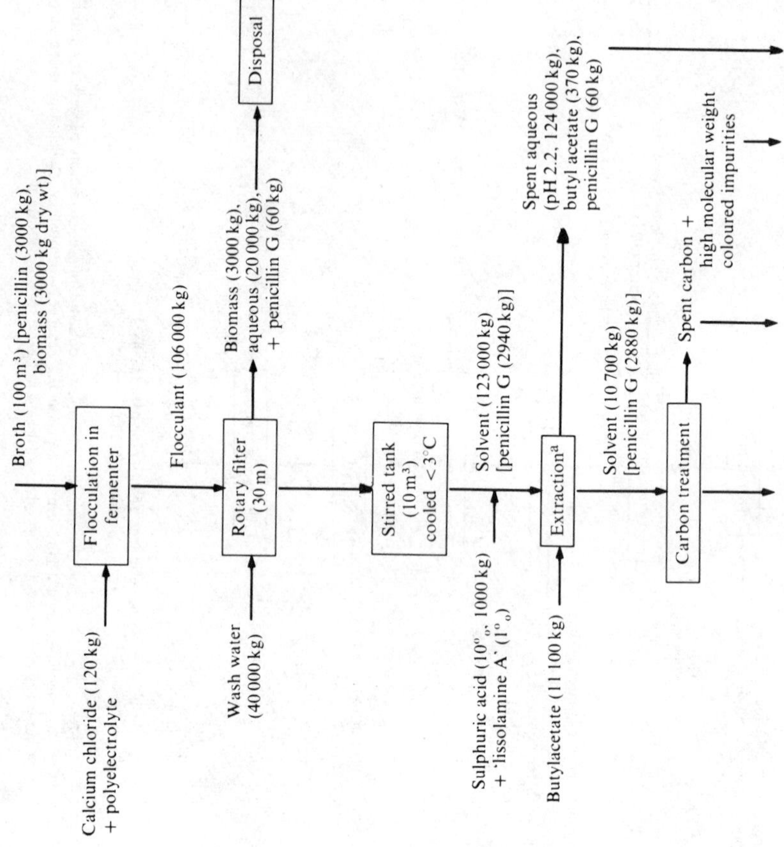
Broth (100 m³) [penicillin (3000 kg), biomass (3000 kg dry wt)]
Flocculation in fermenter
Calcium chloride (120 kg) + polyelectrolyte
Flocculant (106 000 kg)
Rotary filter (30 m)
Wash water (40 000 kg)
Biomass (3000 kg), aqueous (20 000 kg), + penicillin G (60 kg)
Disposal
Stirred tank (10 m³) cooled < 3°C
Sulphuric acid (10%, 1000 kg) + 'lissolamine A' (1%)
Solvent (123 000 kg) [penicillin G (2940 kg)]
Extraction[a]
Butylacetate (11 100 kg)
Spent aqueous (pH 2.2, 124 000 kg), butyl acetate (370 kg), penicillin G (60 kg)
Solvent (10 700 kg) [penicillin G (2880 kg)]
Carbon treatment
Spent carbon + high molecular weight coloured impurities

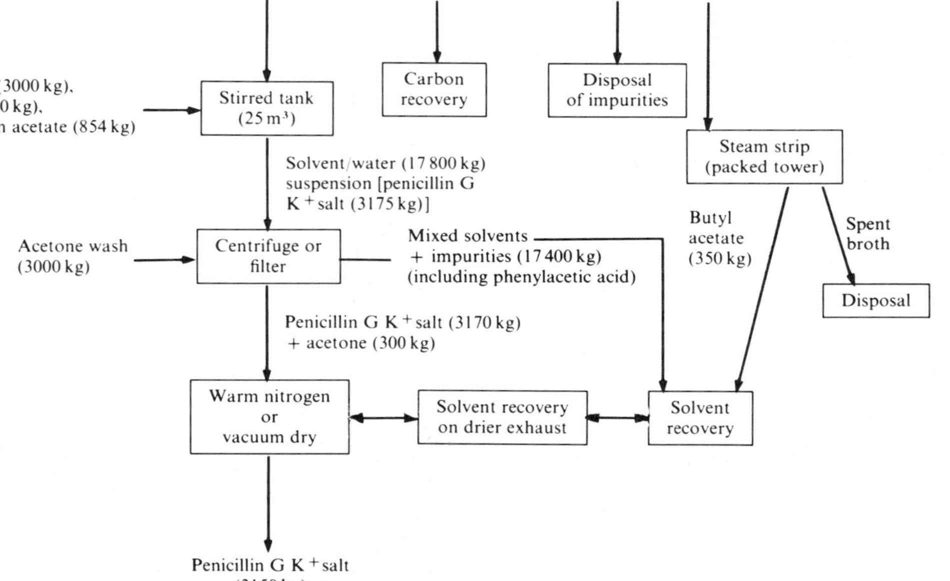

[a] Two or more stages involved. Two Podbielniak D36 machines would process 100 m^3 in 12 hours, including clean-down time. 'Lissolamine A', a quaternary ammonium, surface-active agent, acts as an emulsion breaker.

Figure 8. Extraction of penicillin G potassium salt, based on a 100 m^3 batch giving 3150 kg penicillin G potassium salt.

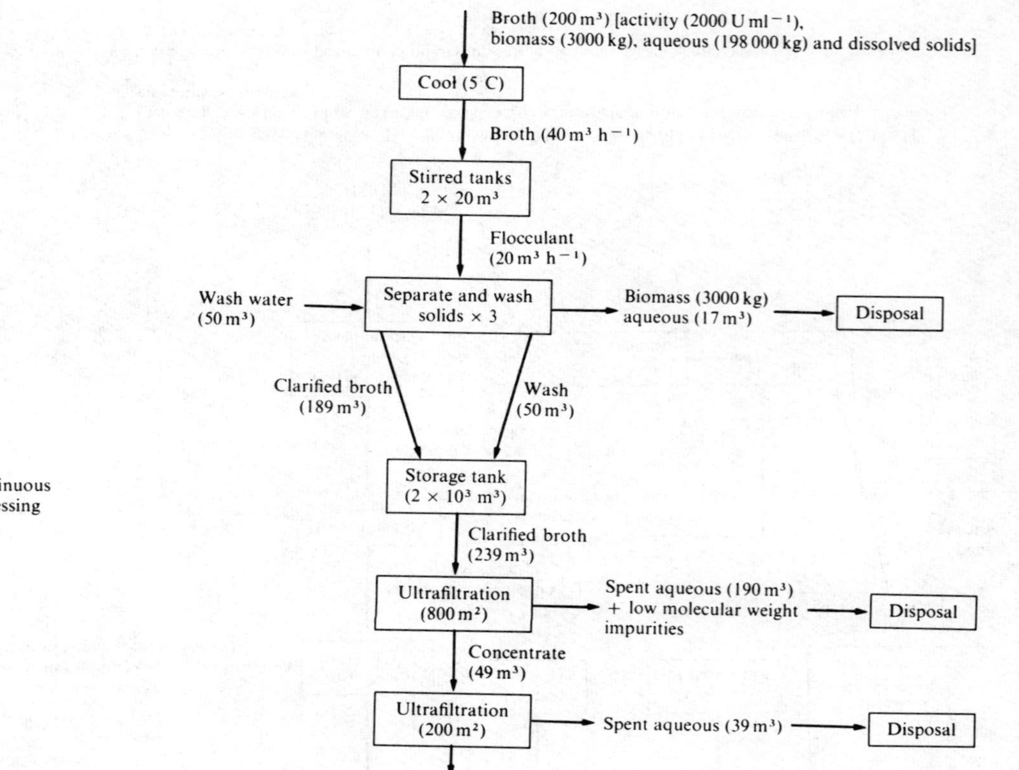

Broth (200 m³) [activity (2000 U ml⁻¹), biomass (3000 kg), aqueous (198 000 kg) and dissolved solids]
Cool (5°C)
Broth (40 m³ h⁻¹)
Stirred tanks 2 × 20 m³
Flocculant (20 m³ h⁻¹)
Wash water (50 m³)
Separate and wash solids × 3
Biomass (3000 kg) aqueous (17 m³)
Disposal
Clarified broth (189 m³)
Wash (50 m³)
Storage tank (2 × 10³ m³)
Clarified broth (239 m³)
Ultrafiltration (800 m²)
Spent aqueous (190 m³) + low molecular weight impurities
Disposal
Concentrate (49 m³)
Ultrafiltration (200 m²)
Spent aqueous (39 m³)
Disposal
Continuous processing

Concentrate
(10 m³)
Precoat slurry
Polishing filter
(150 m²)
Spent precoat
slurry
Disposal
Concentrate
(10 m³)
Storage tank
(10 m³)
or
Precipitator
(15 m³)
Anhydrous
sodium
sulphate
(4000 kg)
Disposal
Spent
aqueous
(9 m³)
Ultrafiltration
(40 m²)
Rotary filter
(7 m²)
Filtrate
Disposal
Concentrate
(1 m³)
Blend and
standardize
(1.5 m³)
Salt
(200 kg)
Bulk vacuum
drier
Polishing filter
Batch
processing
Dry solid product
(2000 kg, 165 000 U/g)
Sterilizing filter
Product solution
(1 m³, 330 000 U ml⁻¹)
Overall yield = 82%

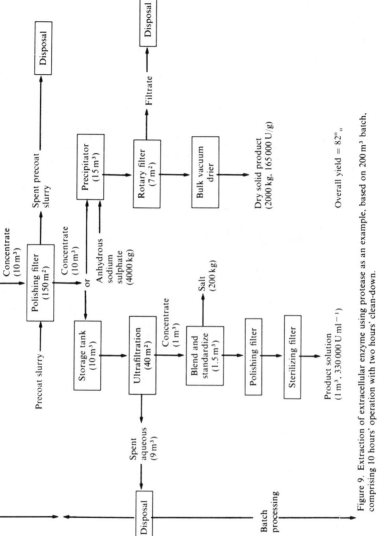

Figure 9. Extraction of extracellular enzyme using protease as an example, based on 200 m^3 batch, comprising 10 hours' operation with two hours' clean-down.

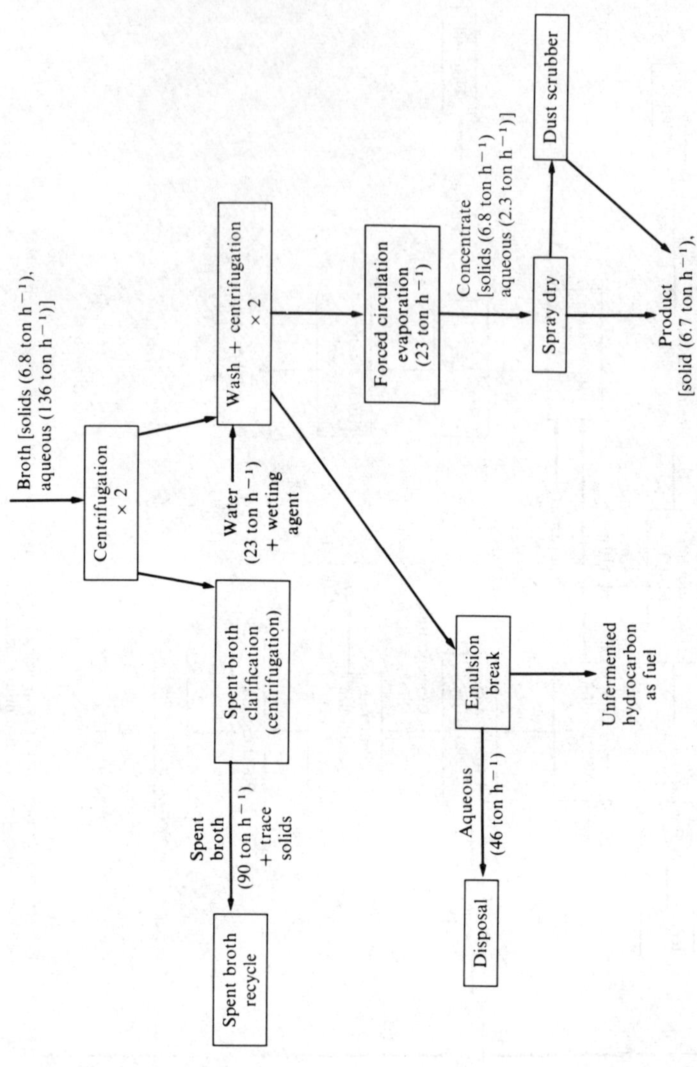

Figure 10. Extraction of single-cell protein produced from purified hydrocarbons, based on 7000-h yr^{-1} operation producing a 50 000-ton yr^{-1} product with 5 per cent moisture content.

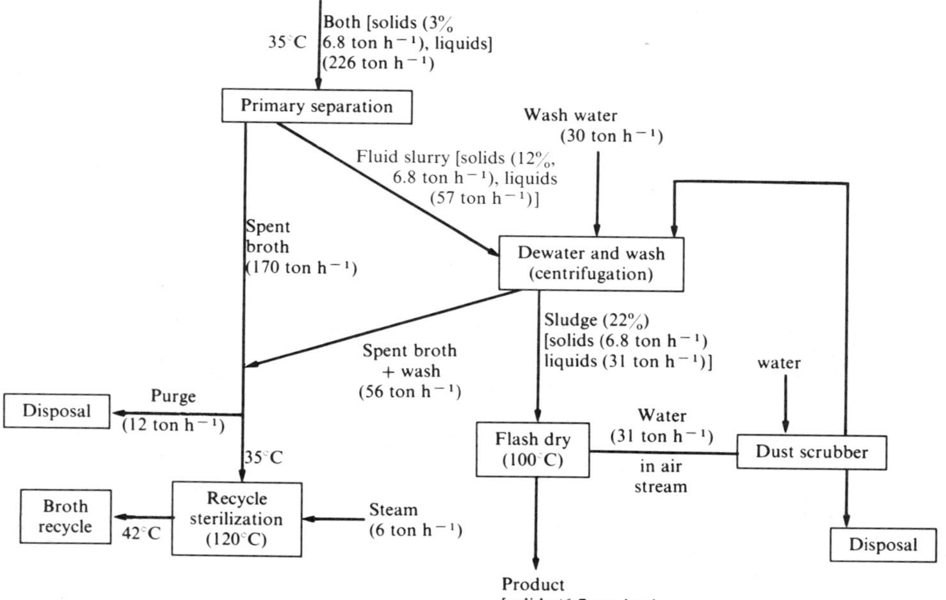

Figure 11. Flow quantities for the ICI single-cell protein production from methanol as implied in the published information, based on a 7000-h yr^{-1} operation producing a 50 000-ton yr^{-1} product with 5 per cent moisture content.

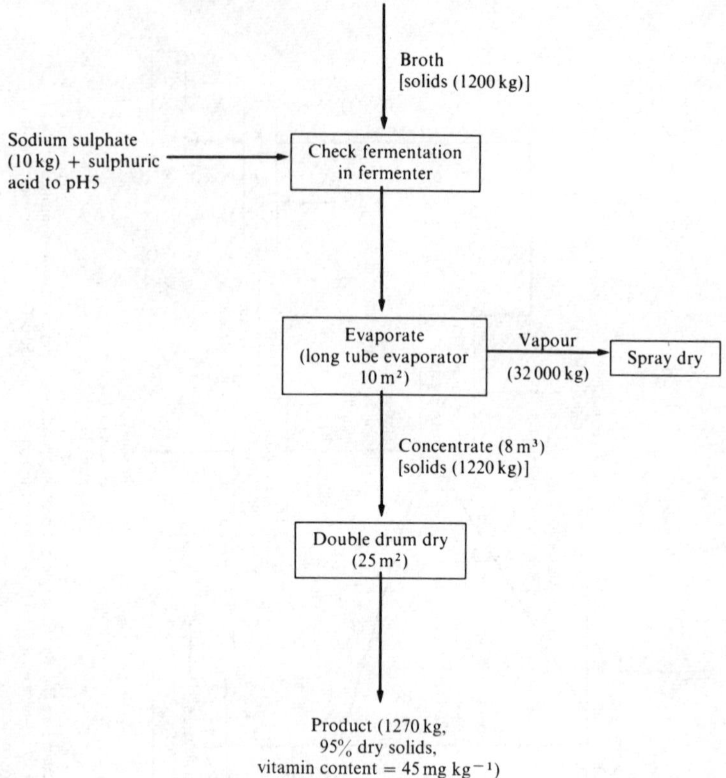

Figure 12. Vitamin B_{12} feed supplement extraction, based on a 40 m^3 batch with 20 hours' operating time. This process is outdated; modern equivalents would use nonthermal concentration and more efficient drying.

IMPACT OF SCALE AND CONTINUOUS FERMENTATION

Large-Scale Processes

Microbiological processes for protein feedstuffs, lipids, liquid fuels and chemical feedstocks representing any significant fraction of the total demand for these products results in very large volumes of downstream-processing feed streams. As little as 2 per cent of the world's forecasted protein demand of 2.8×10^{-6} ton yr^{-1} in 1990 [*see* Hoshiai (1981)] represents 250 000 ton day^{-1} of broth. For protein, lipids, liquid fuels and chemical feedstocks, total broth flows as big as the current world petroleum production (i.e. 6.5×10^6 ton day^{-1}) are conceivable. At these flow rates, very fine tuning of processes is an economic necessity.

Plants at individual sites are unlikely to be particularly large. Sugar cane, for example, is expensive to transport and the raw material deteriorates if stored, while the ethanol made from it is stable and readily transported. Ethanol plants are not technically difficult to operate and do not need a supporting infrastructure of special technical and scientific skills. Brazil has built 120 000 ℓ day^{-1} plants. The economic size of plants producing ethanol from more easily transported, storable substrate in the United States is very much larger, over 600 000 ℓ day^{-1} (i.e. about 12000 ton day^{-1} of broth).

The technically more difficult processes for products such as single-cell protein or lipids, which require aseptic conditions and the maintenance of the culture strain over long periods, are most economical at the technical limit of size assuming they operate on a readily transported, storable substrate. Currently, single-cell protein fermenters run up to 1500 m^3. Two per cent of world animal-feed protein needs between 10 – most optimistic figure based on five times the currently achieved output on methanol substrate predicted by (1981) – and 235 – design output figure for an *n*-alkane substrate plant proposed by Suzuki such fermenters and represents of the order of 250 000 ton day^{-1} of broth. Current individual single-cell protein plants have outputs of 50 000–60 000 ton yr^{-1}, i.e. about 5000 ton day^{-1} of broth.

Medium-Scale Processes

Outputs from plants in this range depend on sufficient demand for the products in order to achieve the economics of larger scale. Demand is, however, limited, for example, glutamic acid production of 400 000 ton yr^{-1} [*see* Dazai (1981)] is probably the largest, this being a very large figure for a commodity organic chemical.

Continuous Fermentations

Continuous fermentations are likely to be encountered more frequently in future (*see* Chapter 7) in response to the pressures discussed below. There are essentially three variants of continuous fermentations.

1. Biomass is the required product, either as such or as a source of intracellular materials. Such fermentations are operated to produce large weights of biomass, and the recovery process separates this and leaves spent aqueous phase for disposal or recycle.
2. The required material is formed extracellularly, e.g., glutamic acid. The fermentation is operated to maintain the required enzyme concentration outside the cells giving efficient substrate conversion, with a minimum of biomass formation. Recovery desirably extracts product and leaves the biomass and the dissolved enzymes to continue their activities. This has not been commercially achieved, however, the enzymes and products normally going to the recovery processes with the aqueous fraction.
3. The required material although extracellular is formed intracellularly and is then excreted by the cells (e.g., penicillin or ethanol).

In general, aseptic recycle of biomass is unattractive as the practical difficulties of maintaining asepsis through the solid/liquid separation processes external to the fermenter are great. Biomass recycle is consequently limited to those fermentations where strict asepsis is not necessary – notably ethanol and methane production. In the case of ethanol, the pH and presence of ethanol discourage competing microorganisms, while in the case of methane production, a mixed culture is used that is capable of maintaining itself against competition. Suspended, or otherwise retained, biomass avoids such problems (*see* Chapter 7). The consequences of such developments are as follows:

1. Processes are needed that will recover materials directly from the fermenter (e.g., membrane and solvent-extraction processes that can be introduced into the fermenter and operated without introducing infection). Possible methods have been demonstrated [*see* Lopez-Leiva *et al.* (1981)] but are not yet in commercial use.
2. Lower solids-contents broths will be produced. Biomass surplus to the process needs is rejected from the fermenter leading to biomass concentrations that are a fraction of those encountered in the broth from batch fermentations.
3. Most fermentations need reasonable concentrations of substrate if they are to proceed at an acceptable volumetric rate of reaction and many fermentations deteriorate due to cell lysis if substrate is not available. Broths from continuous fermentation carry useful and, occasionally, considerable amounts of unconsumed substrate.

COSTS

Cost Factors in Recovery Operations

Multistage operation

Downstream processing usually involves a multistage sequence. Multistage operations are expensive to an extent which is frequently not appreciated.

A material with a value of £200 per ton put through five stages, each costing £40 per ton of product processed and each giving a 95 per cent yield will have cost, at the end of the sequence, £494 per ton. If those five steps could be replaced by one with a 95 per cent yield, the same end cost would be realised at a processing cost for the stage of £270 per ton. Alternatively, if one step costing £40 per ton of product processed could be evolved, the yield of that step to achieve the same end cost could be as low as 50 per cent.

Additionally, capital and operating costs tend to be proportional to the number of stages.

There is consequently a prime economic target of using processes with as few stages as possible. Such processes are likely to be economic even if the individual stages are relatively expensive. A particular implication is that some operations that seem expensive (e.g., affinity chromatography or ion-pair extraction) can be economical if they show particular selectivity effects.

Capital costs

The type of processes most suited to the broth-handling stages are, because of the volumes involved, not energy-intensive. Capital cost-dependent components (i.e. depreciation, return on capital and maintenance) thus make up a large part of the overall cost (e.g., over 80 per cent for a rotary filtration or centrifugation step on a large scale). Even for processes usually considered energy-intensive on large-scale plant, the capital element is still very considerable, e.g., 30–55 per cent of the cost of a spray-drying operation. Capital factors become increasingly important as attempts are made to reduce energy and labour costs, the classical example being ethanol recovery where additional equipment is used to reduce the energy consumption of the traditional distillation train.

The above factors suggest that the most satisfactory primary downstream stages contain these features.

1. Process sequences that use the minimum number of stages to produce a small-volume product concentrate stream from broth.
2. The use of unit operations that give a high throughput per unit of capital cost by manipulation of the physical properties of the broth to enable conventional operations to operate at high rates.

Subsequent processing stages show a greater variation in characteristics, dependant on the particular products being made (*see* Table 3). Generally, however, the scale is smaller, and operating and reagent (or consumable materials, such as adsorbents) costs make up a greater part of the total cost.

Operating costs

Processes with low energy demands per unit volume handled are very desirable at the high-volume broth-handling stages. The restriction is severe on large-scale and commodity-chemical processes. The energy required to chill a broth from a fermenter temperature of 35°C to one that will prevent deterioration, say 5°C, will cost about £1 per cubic metre, i.e. perhaps £30 per ton of a product such as single-cell protein, which is 10 per cent of its value (assuming that there is no heat recovery).

Energy-intensive operations, such as mechanical cell disruption with an energy cost of approximately £1.5 per cubic metre [power figure from Rehacek and Schaefer (1977)] or drying with a fuel cost of about £13 per cubic metre of water evaporated in a spray drier, are restricted to the stages of the process where volumes have been reduced. It is important to distinguish between recoverable and nonrecoverable heat. Thus, the chilling of a broth represents a situation where heat recovery is unlikely to be economic unless local circumstances present a particular use for low-grade heat (e.g., a large space-heating load throughout the year). Ethanol distillation, on the other hand, presents wide opportunities for heat recovery as most of the operations are at, or above, temperatures at which heat can be readily used, either within the distillation plant itself or externally. Energy costs for downstream processing can then be reduced to around 4 per cent of the product cost compared with 12 per cent for a conventional distillation sequence, using figures calculated from Remírez (1980a, 1980b) assuming purchased fuel as the energy source, the product cost for fermentation ethanol being taken as £0.20 per litre.

The relatively dilute nature of broth means that reagents and consumable materials, even at comparatively low concentrations, can represent appreciable sums in terms of product cost, e.g. filter aid used by the economical method described by Dlouhy and Dahlstrom (1968) costs £9 per ton of a product present at 10 per cent concentration in the broth. Similarly, if adsorbent is used to extract a product from a broth, the quantity of adsorbent needed is large both to handle the volume in a reasonable time and because much of the adsorbent capacity is taken up by the other miscellaneous adsorbable materials present in broth and by fouling from colloidal materials. As a consequence, elution and regeneration liquor volumes and regeneration chemical usages can be high relative to the weight of product. An example is streptomycin adsorption, from whole broth, where the concentration of product on resin achieved is about 10 per cent of the capacity of the resin for pure streptomycin [Prasod *et al.* (1980)].

Reactants or solvents used at the broth stage can also lead to additional costs in handling the spent broth. Thus broth from a penicillin extraction requires stripping of trace solvent and pH adjustment.

Targets

While it is difficult to establish the economic circumstances of a particular plant (e.g., what internal transfer price to use for a petroleum-based substrate for a plant in a petrochemical complex) making some broad assumptions it is possible to establish the order of magnitude of the allowable downstream costs for a microbiological product to compete with other sources. The following assumptions are made.

1. Petroleum-based substrate materials are available at about 66 per cent of market price, carbohydrate in molasses at £50 per ton, and carbohydrate to factories adjacent to the fields at the values assumed in the Brazilian ethanol programme assessment (Goodrich (1982)] (i.e., total materials cost are £110 per ton of ethanol).
2. Fermentation rates are increased to three times the present rates.
3. Three broad types of process are considered, i.e. substitutes for the cheaper natural protein and lipids selling at £300 per ton, bulk organic chemicals at £500–900 per ton and volatile organics, such as ethanol, which cost about £220 per ton when made from petroleum.

For the three types of process the following margins are available for downstream processing.

1. For £300 per ton products made at 3 per cent (w/v) concentration in broth from petroleum-based substrate (e.g., single-cell protein or single-cell protein with lipids in co-production), the margin available for downstream processing is about £100 per ton of product.
2. For £700 per ton products made at 10 per cent concentration in broth from carbohydrate substrate (e.g., organic acids or glycol), the margin available for downstream processing is about £300 per ton of product.
3. For £220 per ton volatile organics made at 7 per cent (w/v) in broth from locally grown carbohydrate (e.g. ethanol production), the margin available for downstream processing is about £40 per ton of product.

These figures indicate, given the assumed increases in fermentation rates and hence costs, that:

1. For the cheapest materials, only the simplest downstream processing can be used, e.g. single-cell protein made by separating the broth aqueous phase and drying the solids is acceptable, but purification of the single-cell protein or the isolation of lipids requires major improvements to present techniques.
2. Mid-range products can use present recovery techniques.
3. Volatile organics are competitive with the best presently available recovery techniques compared with the synthetic routes to the same compounds (but not as liquid fuels).

Feedstock

While costs are highly dependent on such factors as operating know-how and practices, research and development attainments, site location and facilities, scale of operation, accounting practice, governmental policies on industrial assistance and on national and local taxation practices, it is desirable to have an assessment of the levels of cost involved in producing and processing broth. The following figures have been calculated on the basis set out in Appendix 1 and should at least be self-consistent.

Fermentation broth produced on a large scale (i.e. 1500-cubic metre fermenters), from materials purchased at open market prices to produce 3 per cent (w/v) of biomass, e.g. for single-cell protein production, costs in the region of £10 per cubic metre if made from methanol, which gives a very fast fermentation. Made from more slowly consumed materials, e.g., alkanes, the much greater fermenter volume required – approximately eight times greater – makes capital cost factors (i.e. depreciation and return on investment) dominant so that, although raw material costs are reduced, costs rise to around £20 per cubic metre. A similar fermentation from molasses carried out batch-wise on a small scale using a 25-cubic metre fermenter with a 72-hour cycle time costs £70–100 per cubic metre.

Fermentations producing large quantities of product (e.g., the typical 7–15 per cent (w/v) of soluble product in organic-acid production) cost rather more as they require more substrate. In this case, 10 per cent (w/v) of soluble product would typically cost a further £20 per cubic metre of broth using methanol and £30 per cubic metre using molasses at the yields generally encountered.

For slow, 72-hour cycle batch fermentations on a 25- or 50-cubic metre scale, depreciation, return on capital services and operating costs usually predominate, amounting to around £60 per cubic metre for broth from a 25-cubic metre fermenter. With continuous fermenters and fast fermentations, the proportion is much lower, possibly £5 per cubic metre for single-cell protein from methanol on a 1500-cubic metre scale.

The above figures assume fermentations using little in the way of additives besides the substrate, ammonia and phosphate. If precursors, such as the phenylacetic acid used in penicillin production, or special materials, such as meat extracts, are needed material costs can be much higher. Costs also escalate markedly as the scale decreases, so that broths for vaccine production can be of a different order of magnitude to the above figures. [Costs are calculated from information given by the Institution of Chemical Engineers (1977), Tannenbaum and Wang (1975), Moo-Young (1977), Suzuki (181) and numerous other sources descriptive of standard processes.]

Cost of Recovery Operations

The cost of recovery operations shows an enormous range from the relatively cheap bulk-processing sequences of single-cell protein or ethanol manufacture to the laboratory-scale separation processes involved in vaccine production. For those products where economic pressures are greatest (e.g., commodity chemicals, ethanol and single-cell protein) and where the scale is such that typical costs can be estimated, the costs of individual operations are highly product-dependent.

To illustrate this, the costs in Tables 4 and 5 have been calculated from plant manufacturers and literature-quoted throughputs for what are normally considered a low-cost step (i.e. the initial separation of broth to give a clarified liquid and a 'solids' stream, either sludge or fluid slurry) and an expensive process step (i.e. thermal drying of a somewhat temperature-sensitive product, e.g., yeast or single-cell protein).

Table 4. Broth separation on largest commercial centrifuges or large rotary filters.

Process step	Cost [£ (m^3 broth)$^{-1}$]	Cost (£ ton^{-1}) At 3 per cent in broth	At 10 per cent in broth
Rotary filter	0.2 –5.0	3–150	1–50
Centrifuge	0.15–4.5		

Table 5. Drying in gas-fired spray driers over an air temperature range of 370°C to 95°C to give a product of 95 per cent solid and 5 per cent moisture, without heat recovery.

Drier rate	Cost (£ ton^{-1}) At 15 per cent (w/v) solid in drier feed	At 30 per cent (w/v) solid in drier feed
At 7000 ton yr^{-1} water evaporated	260	130
At 40 000 ton yr^{-1} water evaporated	140	70

Table 6. Energy consumption of typical processes.

Operation	Energy source	Cost	Comment
Large-scale distillation, with varying degrees of energy recovery	Steam	£40–100 per ton of ethanol	Dependent on feed concentration
	Fuel oil	£10–20 per ton of ethanol	
Drying, flask or spray with 20 per cent heat recovery	Light oil	£10–20 per ton of water evaporated	Dependent on permissible air inlet temperature
Evaporation, with no temperature restriction, three-stage evaporator	Steam	£7 per ton of water evaporated	
Evaporation, 40°C temperature limit in liquors	Steam	£18 per ton of water evaporated	
Crossflow/ultrafiltration, at 17 ℓ m^{-2} h^{-1}	Electricity	£6 per cubic metre of permeate	Cost range, £0.03–18 per cubic metre, depending on permeate rate
Centrifuge, large vertical spindle at 80 m^3 h^{-1}	Electricity	£0.025 per cubic metre of broth	Cost range, £0.02–2 per cubic metre, depending on throughput
Cell disruption	Electricity	£1.5 per cubic metre of slurry	For 96 per cent disruption, cost very dependent on extent required and on type/history of cells
Electrodeposition	Electricity	£0.7–1.1 per ton of solid precipitated	Figure for clay in electrolyte-free liquor. Must be regarded as a minimum
Electrical/membrane process (bipolar exchange membrane cell)	Electricity	£0.6 per kilogram for simple monovalent electrolyte	Practical figures would range upwards from this, dependent on solution and membrane characteristics
Steam strip or sterilize broth, 80 per cent heat recovery	Steam	£3 per cubic metre	Assuming only trace of solvent to strip out
Distill solvent, 60 per cent heat recovery	Steam	£2 per cubic metre of solvent	Dependent on reflux ratio needed. Possible range, £1–4 per cubic metre
Stir slurry at 0.1 Hp per cubic metre	Electricity	£0.1 per cubic metre per day	Provides moderate agitation of thin slurry

ENERGY USE AND RECOVERY

Downstream processes generally operate at room temperature or below. The few that operate at elevated temperatures – drying, evaporation, distillation – have heat-rejection temperatures of 100°C or less. The plants in themselves have little use for low-grade heat. Thus there is no scope for the classical approach to energy use and recovery of establishing a thermal hierarchy of process steps. Generally, each process step must be regarded as an individual energy consumer, and the overall consumption is a summation of these.

Energy costs are most important at the broth-handling stages of the process, where the product concentration is low and the energy cost per unit weight of product is magnified. In the later stages, it is only with low-value products that energy is a major factor in the overall costs, although reductions even with high-value products are a desirable aim. Table 6 lists the order of cost figures of energy use for a series of operations on a plant scale sufficient to make heat recovery, where applicable, economic; on a small scale costs would be higher. Although on flow property-dependent operations the cost is highly throughput-sensitive, the table does indicate:

1. The importance of avoiding thermal processes on broth.
2. The potential of some novel processes to reduce energy consumption.
3. The importance of ancillary operations in energy consumption.

The figure for the power consumption of stirring is included to illustrate the point that process sequences involving many steps, even if they do not involve particularly energy-intensive operations, can add significantly to energy consumption.

REFERENCES

Atkinson, B., Black, G.M. and Pinches, A. (1980) Process Intensification using Cell Support Systems. *Process Biochem.*, **15**(5), 24.

Atkinson, B. and Sainter, P. (1981) Technological Forecasting for Downstream Processing in Biotechnology. Phase 1. Intermediate Forecast Report. *European Economic Community FAST Project Contract*, No. FJT/C/020/80/UK/H.

Aunstrup, K. (1979) Production, Isolation and Economics of Extracellular Enzymes, in *Applied Biochemistry and Bioengineering*, vol. 2, *Enzyme Technology*, ed. by L.B. Wingard, jun., *et al.*, p.27 (Academic Press, New York).

Cooney, C.L. (1979) Conversion Yields in Penicillin Production. *Process Biochem.*, **14**(5), 31.

Cooper, P.F. (1981) The Use of Biological Fluidised Beds for the Treatment of Domestic and Industrial Wastewaters. *Chem. Engr*, (371/2), 373.

Day, J.H. (1978) Impact of Environmental Constraints on the Pulp and Paper Industry, in *Proceedings of the International Symposium on Waste Treatment and Utilization*, ed. by M. Moo-Young, (Pergamon Press, Oxford).

Dazai, M. (1981) Transformation Recrystallization Method for Purification of L-Glutamic Acid. *1st Engineering Foundation Conference on Advances in Fermentation Recovery Process Technology*, Banff.

Dlouhy, P.E. and Dahlstrom, D.A. (1968) Continuous Filtration of Pharmaceutical Products. *Chem. Engng Progr.*, **64**(4), 116.

Dostalek, M. and Molin, N. (1975) Studies of Biomass Production by Methanol-Oxidizing Bacteria, in *Single Cell Protein II*, ed. by S.R. Tannenbaum and D.I.C. Wang, p.383 (MIT Press, Cambridge, Ms).

Dunnill, P. (1981) Biotechnology and Industry, *Chemy Ind.*, (7), 204.

Goodrich, R.S. (1982) Brazil's Alcohol Motor Fuel Programme. *Chem. Engng Prog.*, **78**(1), 29.

Gow, J.S., Littlehailes, J.D., Smith, S.R.L. and Water, R.B. (1975) SCP Production from Methanol Bacteria, in *Single Cell Protein II*, ed. by S.R. Tannenbaum and D.I.C. Wang, p.370 (MIT Press, Cambridge, Ms).

Hoshiai, K. (1981) Present and Future Protein Demands for Animal Feeding. International Symposium on Single-cell Proteins, APRIA, Paris.

Humphrey, A.E. (1975) Product Outlook and Technical Feasibility, in *Single Cell Protein II*, ed. by S.R. Tannenbaum and D.I.C. Wang, p.1 (MIT Press, Cambridge).

Institution of Chemical Engineers (1977) *A New Guide to Capital Cost Estimating* (I. Chem. E., Rugby).

Kinoshita, S. and Nakayama, K. (1978) Amino Acids, in *Economic Microbiology*, vol. 2, *Primary Products of Metabolism*, ed. by A.H. Rose, p.209 (Academic Press, New York).

Khosrovi, B. (1981) Water Recycle in SCP Production Processes. *1st Engineering Foundation Conference on Advances in Fermentation Recovery Process Technology*, Banff.

Labuza, T.P. (1975) Cell Collection, Recovery and Drying for SCP, in *Single Cell Protein II*, ed. by S.R. Tannenbaum and D.I.C. Wang, p.208 (MIT Press, Cambridge).

Leiva, M.L., Hagerdal, V. and Mattiasson, B. (1981) Membrane Technology in Bioconversion of Cellulose to Ethanol. *1st Engineering Foundation Conference on Advances in Fermentation Recovery Process Technology*, Banff.

Leiva, M.L., Hagerdal, B. and Mattiasson, B. (1981) Membrane Technology in Bioconversion of Cellulose to Ethanol. *European Congress of Biotechnology*, Eastbourne.

Mateles, R.I. (1977) Production of SCP in Israel, in *Single Cell Protein II*, ed. by S.R. Tannenbaum and D.I.C. Wang, p.208 (MIT Press, Cambridge, Ms).

Moo-Young, M. (1977) Economics of SCP Production. *Process Biochem.*, **12**(5), 6.

Murphy, T.K., Wilke, C.R. and Blanch, H.W. (1981) Water Recycle in Extractive Fermentation. *1st Engineering Foundation Conference on Advances in Fermentation Recovery Process Technology*, Banff.

Prasod, R., Gupta, A.K. and Bajpai, R.K. (1980) Adsorption of Streptomycin on Ion-Exchange Resins: Equilibrium Studies. *J. Chem. Technol. Biotechnol.*, **30**, 324.

Rehacek, J. and Schaefer, J. (1977) Disintegration of Microorganisms in an Industrial Horizontal Mill. *Biotechnol. Bioengng*, **19**, 1523.

Remirez, R. (1980a) Low Energy Processes vie for Ethanol Plant Market. *Chem. Engng*, **87**(6), 57.

Remirez, R. (1980b) New Ethanol Route wears a Low-Energy Label. *Chem. Engng*, **87**(23), 103.

Remirez, R. (1980c) Is the U.S. Ethanol Goal Feasible? Knock on Wood. *Chem. Engng*, **87**(21), 53.

Sherwood, M. (1981) A Century in the Chemicals' Industry, ICI Mag., Jan.–Feb., 4.

Somejima, H., Koike, T., Nazuchi, S. and Nagashime, M. (1977) Treatment of Fermentation Waste Waters, in *Recent Developments in Separation Science*, vol. IIIB, p.314 (CRC Press, Cleveland).

Suzuki, Y. (1981) Dainippon Ink and Chemicals Projects for producing SCP from n-Paraffin. *International Symposium on Single-Cell Protein*, APRIA, Paris.

Taffe, P. (1980) Biomass Route shows Lower Cost. *Chem. Age*, Nov., 11.

Tannenbaum, S.R. and Wang, D.I.C.(ed.) *Single Cell Protein*, vol.1 (MIT Press, Cambridge).

The Times (1981) 3 Jan., 2.

Topiwala, H.H. and Khosrovi, B. (1978) Water Recycle in Biomass Process. *Biotechnol. Bioengng*, **20**, 73.

van Wezel, A.L. (1981) Large-Scale Production of Inactivated Poliomyelitis Vaccine. *1st Engineering Foundation Conference on Advances in Fermentation Recovery Process Technology*, Banff.

Wall, J.S., Bothast, R.J., Lagoda, A.A. and Saxson, K.R. (1981) Effects of Recycling Distillers Solubles on Alcohol and Feed By-Product Production from Grain Fermentation. *1st Engineering Foundation Conference on Advances in Fermentation Recovery Process Technology*, Banff.

APPENDIX 1: BASIS OF COST CALCULATIONS

It has been considered desirable to introduce some cost calculations in Chapter 12. These are presented with reservations, since costs are very dependent on specific local knowledge and skills, and are much affected by site factors, scale, management and accounting practices, taxation, availability of government assistance and availability of services and materials.

The following assumptions are used in cost calculations.

1. Fixed capital cost (total) is four times the cost of the principal plant items except where a single major item that will need little piping, instrumentation, etc. is involved, where lower factors have been used.
2. Depreciation is 12 per cent of the capital per year.
3. Return on capital is 12.5 per cent per year.
4. No allowance is made for taxation.
5. No working capital is allowed for when calculating unit operation costs.
6. Labour and overheads are £12,000 per man year.
7. Service costs are as follows: steam £15 per tonne, cooling water £0.2 per cubic metre, town water £2.0 per cubic metre, power £28 per mWh, natural gas £0.3 per therm (i.e. £3.0 per gJ), fuel oil (light) £18 per barrel (i.e. £2.9 per gJ) and plant occupation 8000 hours per year.
8. Plant item costs are based on manufacturers' order of cost quotations for typical units and from data given by the Institution of Chemical Engineers (1977). Cost indices from Process Engineering (1981) were used to allow for inflation.

Material costs used are quoted open market prices for bulk delivery; the availability of local supplies, such as might be available to a plant on an integrated manufacturing site, could affect some figures markedly.

References

Institution of Chemical Engineers (1977) *A New Guide to Capital Cost Estimating* (I. Chem. E., Rugby).
Process Engineering (1981) 62(11).

CHAPTER 13

PRODUCT RECOVERY PROCESSES AND UNIT OPERATIONS

INTRODUCTION

The transition from laboratory to commercial scale is a matter of reducing the cost of achieving the desired effects. This is done by increasing the scale of operations, by modifying or changing the unit operations used, by changing the basic process or by permutations of these options. Increasing the frequency of operations on the same scale is hardly ever an attractive option in bulk processing – as opposed to mechanical processes – and is only used when no alternatives can be evolved. It is thus normally impossible to develop a commercial-scale method without altering variables that control the effects of the process. Understanding the nature and effects of these variables is thus essential for successful production-scale development.

For product recovery, there are available a wide range of laboratory-scale processing operations for which plant-scale equivalents or alternatives are required but are not currently available. In the financially important area of processing the fermentation broth, the controlling variables are not well understood or even quantifiable by present technology and current practice only uses a very limited part of the potential range of methods. There are also specific operations (e.g., biomass drying) which because of their high cost limit the development of the industry.

The processes concerned are essentially controlled by physical or physicochemical properties of the systems which have to be handled and the relation of these properties to the mechanisms of the unit operations used. Consequently, the physicochemical nature of broths and their constituents and of specifically microbiological products that cannot be isolated by conventional chemical processes are important, as are the physical mechanisms of unit operations in relation to these properties. Understanding of these factors is a necessary basis for the use of downstream processing technology.

Fig. 1 contains a listing of unit operations, specifies their range of application in terms of the molecular or particle size of the material to be recovered and identifies the physicochemical characteristics upon which their recovery mechanisms are based.

The technical development of a process to plant status, as opposed to the definition of a method in a laboratory, demands the evolution of know-how, experience and equipment for the unit operations involved to standards appropriate for the scale of the operation. While such evolution is possible on a one-off basis, the cost and time involved are seldom sustainable. Consequently, only operations applicable to a range of processes develop to a stage where the information on their characteristics and the equipment they require is widely available. Conversely, the applications must be sufficiently numerous to justify economically the process developers' and equipment manufacturers' efforts that are necessary for the establishment of a unit operation of wide application.

Downstream processing for microbiological products has available to it a relatively limited range of developed processes and unit operations.

At the broth-handling stages, where the scale is large and, consequently, a high standard of development is needed, the processes depend essentially on the application of techniques and of equipment developed elsewhere. For the large-scale development of the industry, the costs of these operations represent a major restraint. Specialist techniques more suited to the industry are needed. The scale of broth-handling operations makes equipment supply commercially attractive.

For the subsequent processing stages, there are a number of products where established commercial-scale unit operations are not applicable or not the most appropriate. The range of laboratory techniques that can be developed to plant scale, however, is extremely wide.

Appendix I to this chapter provides a listing of specific applications of particular unit operations. The corresponding references provide information and data on particular performance characteristics and operational features.

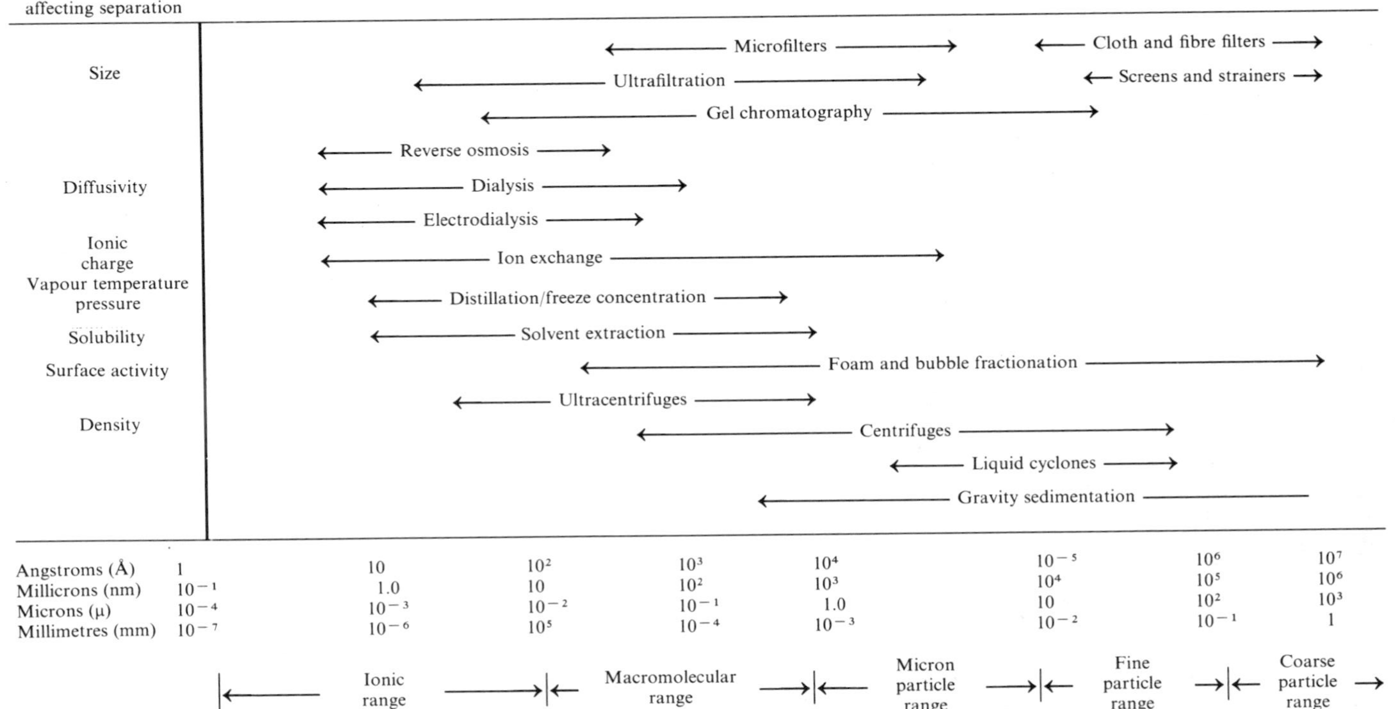

Figure 1. Ranges of applications of various unit operations.

BROTH-HANDLING STAGES

Broth Pretreatment

Broth pretreatment is used primarily to improve the broth-handling properties, typically by flocculating cells and using colloidal material. It is occasionally employed to modify the product properties, e.g., heat treatment of polysaccharide-containing broth.

Processes involved are:

1. Ageing – many bacteria tend to flocculate when active growth ceases [*see* Busch and Stumm (1968)].
2. Heat treatment.
3. pH changes.
4. Addition of flocculants, such as calcium chloride and polyelectrolytes.

As a process step, it is usually of very low cost: a typical dosage of calcium chloride would be 1 g ℓ^{-1} and of polyelectrolyte would be 0.1 mg ℓ^{-1} costing £0.05 per cubic metre of broth (*see* Table 1).

The effect of pretreatment can be spectacular, e.g., Gasner and Wang (1970) reported a 500–2000-fold improvement in the sedimentation rate of a range of microorganisms and Moss and Smith (1977) recorded an order of magnitude improvement in the filtration rate of a microbial culture (*see* Table 2).

Table 1. Flocculant dosages.

Agent	Flocculant dose [g (100 g dry cell wt)$^{-1}$] Glucose broth	Hydrocarbon broth	Resuspended cells in buffer	Penicillin broth	Dilute slurries	References
Polyelectrolytes					0.045–4.5	Spruce (1978)
Anionic polyelectrolytes						
Polystyrene sulphate	0.2	0.1	0.06			Gasner and Wang (1970)
Polyacrylamide	Ineffective	Ineffective	Ineffective			
Cationic polyelectrolytes						
Polyethylene imine	10		7.0			Gasner and Wang (1970)
Calcium chloride				200		Private communication
Colloidal clay, bentonite	2.0	20.0	0.6			Gasner and Wang (1970)
Inorganic coagulants					0.045–4.5	Spruce (1978)

Table 2. Effect of various pretreatments on the filtration rate of a culture of *Acinetoabacter calcoaceticus* [Moss and Smith (1977)].

Acidification	Heat	pH at time of filtration	Filtration temperature (°C)	Filtration rate (ml min^{-1})
−	−	7.0	25	0.53
+	−	3.5	25	2.22
+	+	3.5	85	10.00
−	+	7.0	85	0.16

The objectives of broth pretreatment are listed below.

1. Increase throughputs at a solid–liquid separation stage. Since the cost of solid–liquid separation stages is proportional to equipment throughput – within wide limits – effective flocculation can have an order of magnitude effect on these costs.
2. Reduce fouling in a whole-broth adsorption step.
3. Control solids behaviour in a whole-broth extraction step.
4. Combine with cell lysis, since cell lysis alters cell structure and is inducible by agents that may also be effective in flocculation, e.g., heat or quaternary ammonium wetting agents.

Current practice in broth pretreatment is highly empirical and lags behind the high level of similar technology developed in water and waste water treatment and with other difficult industrial solid–liquid separation steps where the basic science is well developed. When pretreatment can be made effective, at the low costs involved, it is preferable to the alternative of constraining the fermentation to less than the optimum conditions to meet the needs of the recovery processes.

Cell Separation

The separation of cells from the suspending medium (*see* Table 3) is carried out for a number of reasons.

1. To separate product(s), which may be either extracellular or intracellular, from unwanted material. For intracellular soluble products, cells may be lyzed prior to the separation.
2. To remove cell fractions after reslurrying during the extraction of intracellular products.
3. To remove traces of unconsumed substrate by successive cell-washing stages.
4. To remove cells from spent broth after the product has been extracted by adsorption/distillation from whole broth and prior to broth recycle or effluent disposal.

A very limited range of techniques is in common use for this operation.

1. Filter press and mechanized press devices for broths with difficult characteristics (*see* Fig. 2). This technique is only applicable on a relatively small scale.
2. Rotary filtration (*see* Figs. 3 and 4) with precoat for less easily filtered broths, producing solids as a sludge. Some typical operating results for the filtration of various fermentation broths using a rotary drum filter are given in Tables 4 and 5.
3. Centrifugation in vertical spindle separators (*see* Fig. 5) to produce the 'solids' phase as a cream or mobile sludge.
4. Centrifugation in horizontal bowl machines (*see* Fig. 6) to produce solids as a sludge (usually used as a second dewatering stage or with reslurried cells, as the separating force is not normally adequate for fermenter broth).
5. Settling and flotation – the traditional clarifying processes of the alcoholic beverages and in waste-water treatment.

	Cell lysis	Centrifugation/sedimentation	Direct adsorption	Distillation/evaporation	Electrokinetic separation	Filtration, conventional	Flotation	Liquid membrane extraction	Mechanical cell disruption	Solvent extraction	Ultrafiltration/cross-flow filtration
Properties of 'solids'											
Rigidity of particles						√			√		
3 potential of particles					√						
Cohesiveness of particles		√				√					
State of flocculation and particle size		√	√		√	√	√				
Density of flocs or particles		√					√				
Fluidity of slurry, limiting concentration of solids in fluid slurry		√	√	√				√	√	√	√
Emulsion formation, wetting properties							√	√		√	
Retention of solids by aqueous phase						√		√		√	
Fouling of surfaces, adhesiveness of solids			√	√		√					√
Cell membrane strength	√								√		
Cell membrane permeability/ chemistry	√										
Properties of product and/or 'liquids'											
Molecular weight (or particle size) of wanted fraction											√
Volatility of wanted fraction				√							
Distribution coefficient of wanted fraction								√		√	
Adsorption properties of wanted fraction			√								
Conductivity of aqueous phase					√						
Viscosity of aqueous phase		√	√		√	√	√	√		√	√

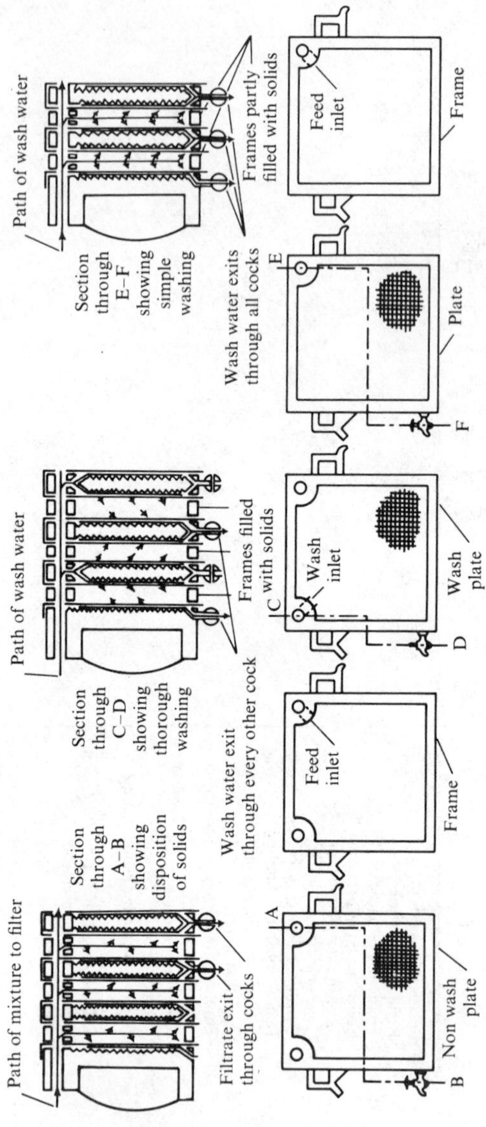

Figure 2. Filling and washing flow patterns in a filter press [Perry and Chilton (1973)]

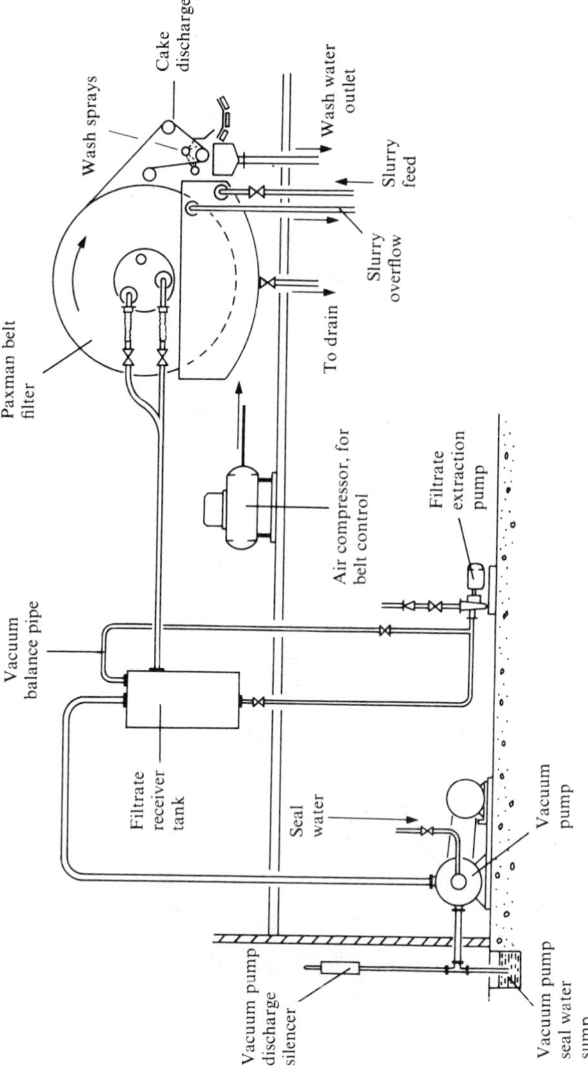

Figure 3. Paxman belt filter installation

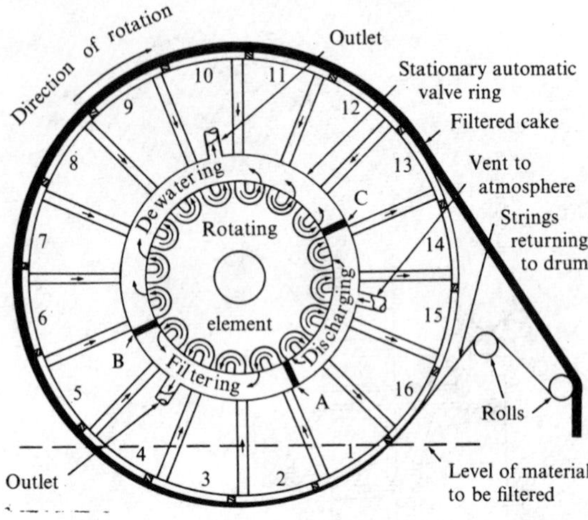

Figure 4. Diagram of continuous vacuum filter operation. Sections 1–4 are filtering, sections 5–12 are dewatering and section 13 is discharging the cake with the string discharge. Sections 14, 15 and 16 are ready to start a new cycle. The "hub" of this drawing illustrates the automatic valve, where *A*, *B* and *C* represent members in the valve [Perry and Chilton (1973)].

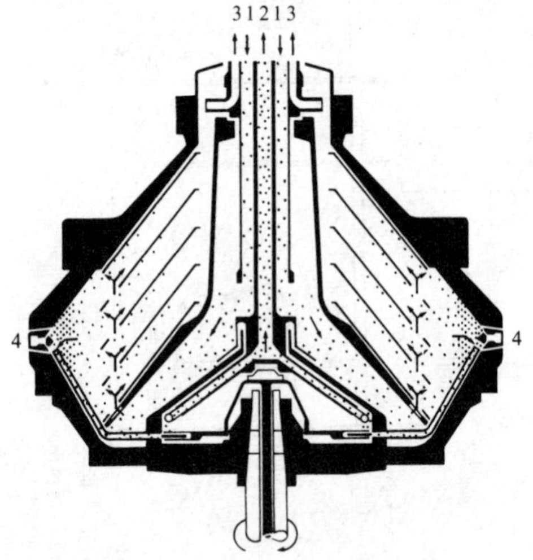

Figure 5. Vertical spindle separator. Longitudinal section through the bowl showing: inlet (1), outlet for concentrate (2), outlet for effluent (3), CIP nozzles (4).
The solids, being heavier, are forced outwards to the periphery of the bowl, through the concentrate tubes and the internal nozzles into the paring chamber where they are skimmed off by the paring tubes and discharged. The lighter effluent is displaced towards the centre, leaving the bowl through the paring disc pump.

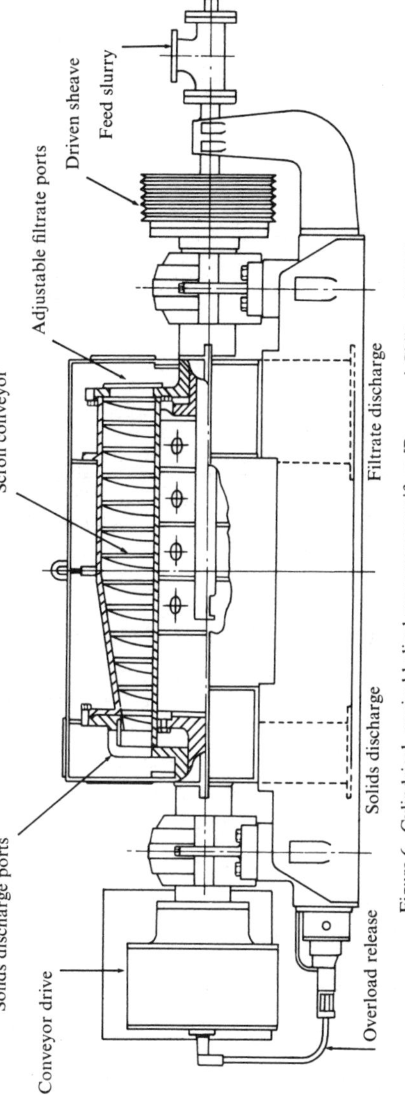

Figure 6. Cylindrical–conical helical conveyor centrifuge [Perry and Chilton (1973)].

Table 4. Operating results for filtration of various fermentation broths (vacuum 20–25 in. Hg) [Dlouhy and Dahlstrom (1968)].

	Bacillus licheniformis	*Streptomyces erythreus*	*Strep. kanamyceticus*	*Penicillium chrysogenum*
Filter type	Vacuum precoat	Vacuum drum	Vacuum precoat	Vacuum drum or precoat
Design filtration rate (gal ft^{-2} h^{-1})	4–8	10	2.1	35–45
Solids in slurry (% wt)	8	25	7	2–8
Slurry temperature (°C)		Ambient	40–50	
Submergence (%)	35	33	35	35–55
Cycle time (min)	0.5	3	3	3
Cake moisture (% wt)				60–70
Cake thickness (in.)		0.25–0.38		
Filter medium	Precoat	Nylon	Precoat	Polypropylene
Slurry admix required (% wt of slurry)	0.4–2.6	None	6	None
Precoat consumption [lb dry wt (1000 gal of filtrate)$^{-1}$]			72	
Cake discharge mechanism	Precoat	String	Precoat	String or precoat

Table 5. Throughputs of rotary filters.

Filter type	Liquor		Wet solid	
	Type	Throughput ($\ell\ m^{-2}\ h^{-1}$)	Throughput ($kg\ m^{-2}\ h^{-1}$)	Solid content [% (w/v)]
Drum, knife discharge	Yeast from beer	17–55	17–55	25
Belt	Flocculated sewage sludge	750–1200	20–25	25–30
Precoat	Waterworks lime sludge	1000	900	75
	Waterworks alum sludge	1000	25	30
	Various fermenter broths	120–400		
Strip discharge		400–1400		

The throughput of all the above processes is extremely dependent on the physical properties. The major costs are associated with the capital charges, power forming only a small proportion. Consequently, the range of cost per unit volume is very wide.

Centrifugal and settling processes depend on particle size and rigidity (i.e. the resistance of the particles to compaction with the formation of an impervious cake) which controls the resistance of the filter cake to flow. Fouling of filter membranes is controlled by cake adherence, adhesiveness, particle size and rigidity. Use of conventional equipment is primarily a matter of controlling the physical properties of the solids in the broth and the occurrence of colloids and gel material either in the body of the broth or on cell surfaces. These conditions have to be achieved either by fermentation or by pretreatment.

Centrifugation equipment is limited by the available mechanical strength of materials used in its construction. In conventional filtration, once an appreciable solids layer is deposited, the behaviour is controlled by this layer. The filter media or the replaceable media, e.g., paper or the precoat filter aid, have only limited effect.

Where difficult broths have to be handled different equipment is used. Different equipment types are considered in the following subsections.

Centrifugal separation in hydraulic devices

The hydrocyclone has replaced centrifuges in many starch separation processes.

Flotation processes

Traditionally, flotation is used in beer processes. It has also been demonstrated with some microbiological solids such as algae [*see* Dubinsky *et al.* (1979)] and is widely used in mineral separation [*see* Bietelshees *et al.* (1979)].

Depth filtration

Gelatinous precipitates and weak flocs are traditionally separated in the water-treatment industry by depth filters, with the collection of the solid material in a matrix rather than on a porous surface, such as a sand filter.

Noncake-forming filtration

This consists of cross-flow filtration over porous, coarse membranes. Cross-flow filtration, even with comparatively tight membranes, would be economic compared with a moderately slow rotary filtration. Beaton (1980) suggested that operation on a broth at a flow rate of 17 ℓ m^{-2} h^{-1} would be economic against rotary filtration with a flow rate of 400 ℓ m^{-2} h^{-1}. Much higher cross-flow rates are possible, e.g., Henry (1972) quotes a value of 67–118 ℓ m^{-2} h^{-1} for a bacterial suspension and flow rates as high as 900 ℓ m^{-2} h^{-1} for pigment slurries have been reported by Klein (1981).

Electrokinetic methods

Electrokinetic separation of solids can produce a cake of very high solids content. Tests on a bacterial suspension using electrokinetic deposition produced a cake comprising 40 per cent (w/v) solids, compared with a typical filter or centrifuge discharge with a solids content of 15–25 per cent (w/v) (private communication). Such a performance is extremely valuable where biomass has to be dried, saving possibly £100 of the drying cost per ton of product compared with a typical drier feed of 20 per cent (w/v), thus justifying an additional cost of £3 per cubic metre of broth above conventional filtration and dewatering processes for a typical 3 per cent solids broth.

An alternative application of electrokinetic methods is to keep the filtration surfaces clean by preventing solids deposition [*see* Wakeman (1981)]. This approach can have dramatic effects on filtration rates.

A problem with electrokinetic methods is that the voltages required i.e. 10–50 V cm^{-1} as proposed by Wakeman (1981) and Mantell (1960) cause electrolysis in the conducting solutions. The concentrations of the products of electrolysis need to be controlled in the case of most broths by circulating the liquors from the anode to the cathode. The current used in the electrolysis represents wasted power.

Direct Product Recovery from Broth

This process, which is summarized in Table 2, removes product constraints on the large-volume steps needed to recycle or dispose of broth. For instance, high temperatures can be used subsequent to product recovery to flocculate broth solids and facilitate solid–liquid separation.

Direct product recovery also introduces the possibility of simultaneous product recovery and fermentation if the recovery process can be operated in the fermenting liquor, i.e. aseptically, without affecting the fermentation. A large-volume process step, such as that associated with product recovery from the clear liquor from a solid–liquid separation stage, is eliminated by this process.

Many examples of direct product recovery from broth exist.

1. Direct adsorption of streptomycin from whole broth is achieved using an ion exchange resin following the batch fermentation.
2. Ethanol is recovered by vacuum distillation during continuous fermentation, with both the fermentation and the first-stage distillation carried out under vacuum conditions at temperatures and ethanol concentrations such that viable yeasts can be maintained while the ethanol is removed. This arrangement reduces the inhibition due to the alcohol (*see* Fig. 7).

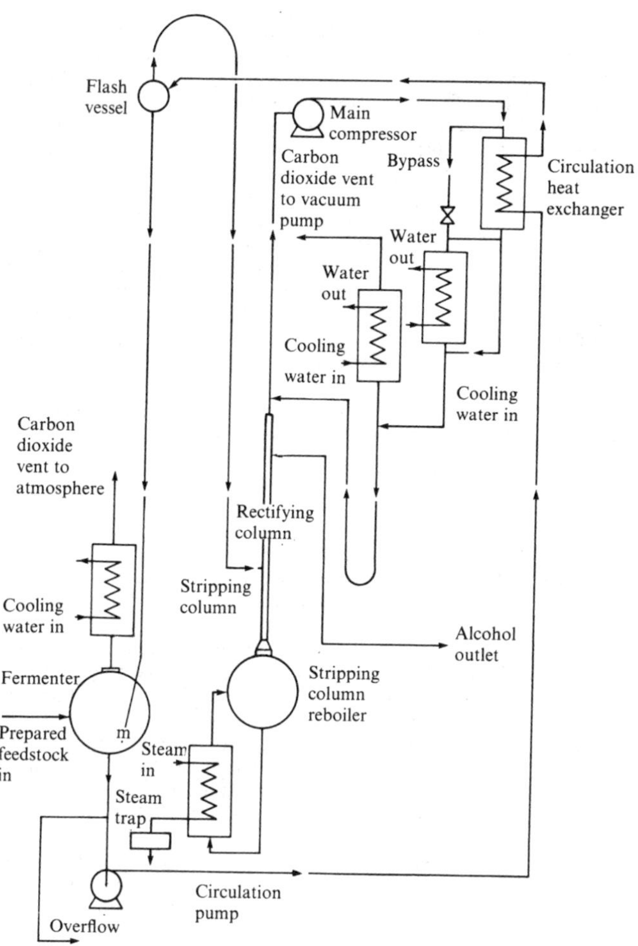

Figure 7. Atpal pilot plant (courtesy of W.S. Atkins group, Epsom, Surrey, UK).

3. Solvent recovery from whole broth is used for the recovery of penicillin and steroids. Difficulties, however, are encountered with emulsion formation as reported by West and Patterson (1981), solid deposition in the separators, solids in the solvent extract and excessive solvent in the spent broth. Two aqueous-phase extraction offers control of the solids behaviour and has been demonstrated on a reasonable scale for enzymes by Kroner *et al.* (1982). It is, however, an expensive process in terms of reagents (e.g., 30 kilograms of polyethylene glycol is used for 125 litres of broth) needed to create the phase split, whether they are used on a throw-away basis or recovered. Any extraction process can potentially be used for whole broth but control of the solids behaviour is critical to achieve a clean extract.

4. Some products may be directly adsorbed on adsorbents. The adsorbent particles must be such that the broth solids can be removed from the adsorbent mass, which means particles large enough and dense enough to form a bed from which the solids can be

washed. Particular problems are the presence of other adsorbable materials in the broth (e.g. unfermented sugars, proteins, etc.), which take up adsorbent capacity so that product loading on the adsorbent is restricted. Furthermore, the presence of colloidal and gel material fouls the adsorbent. Elution and clean-up of the adsorbent and regeneration thus require a succession of washes. (Typically, the sequence is wash out broth solids, elute less strongly adsorbed impurities, elute product, elute strongly adsorbed impurities, regenerate and wash out regenerant, with occasional treatments using more vigorous reagents to remove fouling.) Reagent consumption tends to be high, adsorbent life short and effluent generation considerable. The adsorbent must also be robust, which makes for difficulties in applying the more selective (e.g., surface-coated) adsorbents to whole broth.

5. Removal of products from the fermentation through semipermeable membranes is attractive, particularly in that it can be used during fermentation to remove products that are inhibitory to the fermentation, particularly ethanol. Fouling is a major problem where the membrane is within the aseptic volume of the plant and not accessible for cleaning. The control of surface film deposition achieved in cross-flow filtration and ultrafiltration can be combined with a wide range of membrane processes (ultrafiltration, reverse osmosis, facilitated transport membranes, electrokinetic processes using membranes, etc.) to achieve selective recovery of components from whole broth.

Cell Disruption and Lysis

Cell fracture by mechanical, microbiological or chemical means (*see* Table 2) releases intracellular material into the aqueous phase as solutes or colloids, with the reduction of the 'bulk' of the solids. Lyzed cell material has a packed volume that is much less than the whole-cell volume, e.g., the moisture content of separated lyzed cells from such equipment as horizontal-bowl centrifuges is much less than that for whole cells, thus reducing the drying costs when dried cell material is the required product. This bulk reduction can be very considerable, e.g., 60–80 per cent for a glutamic acid fermentation cell mass [*see* Kinoshita and Nakayama (1978)].

Also cell disruption allows further treatment of the cell solids material for purification or extraction of a product, such as lipids, and facilitates broth handling by changing the physical nature of the solids, e.g., by increasing the apparent density, by reducing the tendency to distort with the formation of impervious packed layers and by disrupting gelatinous surface films.

The susceptibility of cells to mechanical disruption or lysis is highly dependent on the cell type, their condition and their age [*see* Engler and Robinson (1981) and Asenjo (1981)]. The main problems occur with products that have stabilities that preclude the use of extreme chemical or thermal conditions.

Mechanical fracture

Mechanical fracture is a fine grinding operation in a fluid medium and is essentially a matter of fluid shear forces. (In this size range, mechanical impact has very little effect, even in gaseous media.) Unlike typical grinding operations, the individual particles need considerable distortion before they shear; distortions many hundred times greater than for a typical solid particle. To achieve this homogenizers or bead mills are normally used but the scale of operation is small, power consumptions are high and throughputs are relatively low. In a bead mill, which is apparently more efficient than an homogenizer, Rehacek and Schaefer (1977) quoted that the power consumption is 56 kWh per cubic metre of slurry and the throughput 13 litres of slurry per hour per litre of gross mill volume. Some very intensive laboratory shear devices, such as the 'X' press, are also used. Since cells in mechanically stirred fermenters withstand very high shear stresses (i.e. stresses which in liquid-dispersion terms would produce effects approaching that of the typical homogenizer) less intense mechanical devices are ineffective.

Chemical and microbiological fracture

On a large scale, chemical or microbiological techniques are necessary to achieve breakdown of the cells. Lysis may occur adventitiously once the broth is removed from the fermenter and is suffering from lack of nutrients and oxygen and is, perhaps, contaminated. (An ageing period is often included after fermentation for enzyme broths to improve their processing behaviour.) Lysis may be induced by a combination of the following factors.

1. Increases in temperature.
2. Extremes of pH.
3. Freezing the broth.
4. Exposure to osmotic shock.
5. Addition of cytotoxic chemicals (e.g., chlorine).
6. Addition of reagents that interfere with the functions that maintain the cell membrane or that directly affect the cell membrane (e.g., penicillin, quaternary ammonium wetting agents or proteases).
7. Autolysis due to release of enzymes within the cell (e.g., lysozyme) that attack the cell structure.

Ancillary Operations

The broth-handling stages of plants producing microbiological product, because of the large volumes handled, create ancillary operations on a considerable scale. These can be economically important in the choice of processes used and, occasionally, can create technical difficulties.

The primary and inevitable ancillary operations are the disposal of spent broth or biomass or their treatment to make them fit for recycle [*see* Atkinson and Sainter (1981)]. Depending on the broth-handling methods used, further problems may be introduced into these operations and additional operations created. The principal factors are the effects of the downstream processes on the spent broth, the nature of other effluents created and the need to recover and make fit for reuse the solvents and reagents used in the processes. Typical problems are listed below.

1. From broth pretreatment or cell lysis – the presence of surface-active agents, of metal salts used as flocculants and extremes of pH in the spent broth.
2. From whole-broth extraction processes – the presence of broth solids (and, consequently, very high viable microorganism levels) in the spent broth.
3. From solvent-extraction processes – the presence of trace solvent in the spent broth, with the necessity of the solvent having to be washed or redistilled prior to use or the necessity of the separation of mixed solvent (e.g., acetone/butyl acetate from the penicillin mixed solvent precipitation route).
4. From two aqueous-phase processes – high concentrations of polyethylene glycol, dextran, phosphates or similar materials in effluent streams, requiring treatment prior to disposal or recovery.
5. From liquid membrane processes – trace solvent and emulsions in the effluent, and surface-active agents, necessitating an emulsion-breaking, solvent-recovery step.
6. From adsorption operations – part-exhausted regeneration solutions have to be made fit for disposal with wash-out liquors as an additional effluent.
7. From cell-drying operations – dust and odours in spray or flash drier exhaust. Dust has a high protein content and may constitute a health hazard.

The cost of controlling the above effects is highly variable, but never negligible, e.g., if a spent broth has to be heated either to sterilize or to strip out a trace of solvent, steam consumption (assuming 80 per cent recovery) will represent a cost of about £3 per cubic metre of broth.

PRODUCT RECOVERY STAGES

The stages of downstream processing that produce a sales product from the concentrate leaving the broth-handling stages are highly product-dependent and, unlike the broth-handling stages, exhibit few general characteristics. Where the products are chemically stable and isolated as relatively pure chemical entities, processing parallels that of the normal chemical industry. Materials, such as organic acids and pharmaceuticals, including penicillin (which can be recovered and purified from the solvent extract of the broth by a wide range of typical chemical processes) offer no problems peculiar to the microbiological industry. Special problems, however, are encountered in the case of the microbiologically specialized products, many of such products being unstable and not amenable to the processing methods commonly encountered in the conventional chemical industry. The 'concentrate' leaving the broth-handling stages may also introduce special problems in that it may contain a miscellany of materials not commonly encountered in chemical feedstocks brought forward from the broth.

At this stage of the process, the choice of method is very product-dependent, e.g., it is no good considering a precipitation method unless the product is capable of forming an insoluble compound with a reagent of low cost relative to the product's end value and unless the precipitation is reversible with a low-cost reagent. There will, however, normally be some choice of methods available. The criteria of choice are then economic, being dependent on factors such as selectivity, yield, volume reduction achieved, need for subsidiary processes (e.g., solvent or reagent recovery, additional effluent treatment) and their capital- and operating-cost effect on the whole of the recovery sequence. Tables 6 and 7 set out some of these factors against the possible processes. The processes listed in Tables 6 and 7 may also be applicable to whole broth if the necessary control of solids behaviour can be achieved (e.g., in solvent extraction, retaining solids in spent aqueous phase and preventing interfacial build-up in separators).

Table 6. Technical aspects of unit operations.

Unit operation	Current state of development	Typical product range	Recovery	Selectivity	Volume reduction	Product output form
Distillation	Industrial	Volatile materials	High	All volatile materials recovered	High	Strong solution of product
Concentration						
Thermal	Industrial	Nonvolatile materials	Dependent on product stability	Nil	Typically 10:1	Concentrated broth
Ultrafiltration	Industrial	Molecular weights > 50 000	High	Rejects all low molecular weight compounds	Typically 10:1–100:1	Concentrated, partly purified broth
Extraction						
Solvent						
Conventional	Industrial	Solvent-soluble	Moderate	Rejects all non-solvent solubles	Typically 5:1–10:1	Solvent solution of product
Ion pairing	Semi-technical	Solvent-soluble, appropriate counterions must be available	High	Potentially very high	Potentially high	Solvent solution of product and counterions
Two aqueous phases	Small plant scale	High molecular weight, soluble materials, e.g., enzymes	Moderate	Moderate	Unknown	Solution of product in polyethylene glycol
Two aqueous phases + liquid membrane	Plant scale in other industries	Potentially versatile	Potentially high	Potentially very high	Potentially high	Emulsion of solvent and product solution
Precipitation and filtration/ centrifugation	Industrial	Products with convenient insoluble derivatives	High	Moderate to high > 20% product in filter cake	Very high	Sludge of insoluble derivative of product
Adsorption on solid adsorbent	Industrial	Large molecular weight and unstable materials	Moderate	High (special separation effects)	Typically 5:1–10:1	Solution in eluate
Product extraction through membrane (possibly with electrical assistance)	Laboratory and semi-technical scale	Low molecular weight materials	Probably moderate	Moderate	Unknown	Aqueous solution
Electrokinetic	Laboratory and semi-technical	High molecular weight	Probably moderate	Moderate	Unknown	Solution in water or broth

Table 7. Economic aspects of unit operations.

Unit operation	Operability	Applicability – product values	Effluents	Consequent operations
Distillation	Good	Volatile, low-value products	Spent broth recyclable	None
Concentration				
Thermal	Good; some fouling problems	Low-value 'mixture' products	None; clear condensate	None
Ultrafiltration	Membrane fouling and cleaning problems	Low-value products	Broth stripped of high molecular weight materials	Membrane cleaning
Extraction				
Solvent				
Conventional	Good, some fouling/ emulsification problems	Moderate-value products	Broth contaminated with solvent	Solvent recovery
Ion pairing	Probably good	Moderate-value products	Broth contaminated with solvent and counterions	Solvent recovery possible and stripping of counterions
Two aqueous phases	Probably good	High-value products	Broth contaminated with separating agent, e.g., phosphate and trace polyethylene glycol	Recovery of reagents might be economic
Two aqueous phases + liquid membrane	Probably good	High-value products	Some solvent and wetting agent contamination of spent aqueous phase	Emulsion and separation
Precipitation and filtration/ centrifugation	Probably good	Low-value products	Spent broth possibly contaminated with precipitant	Redissolving precipitate
Adsorption on solid adsorbent	Good; fouling may be a problem (absorbent life may be an important factor)	Moderate-value products	Spent broth uncontaminated	Regeneration of adsorbent
Product extraction through membrane (possibly with electrical assistance)	Unknown, fouling may be a serious problem (membrane life may be an important factor)	Low-value products	Spent broth uncontaminated	Cleaning membrane
Electrokinetic	Unknown	Moderate- to high-value products	Spent broth uncontaminated	None

Economic Characteristics

The nature of the economic constraints on process choice in the final stages of processing is different from those commonly encountered in the broth stages.

For a broth, with a value of perhaps £30 per cubic metre, a single filtration step costing £5 per cubic metre of the material processed represents a major increase in product value. For the concentrate which feeds the final stages, typically 10 per cent of the broth volume and worth probably (from a £30 per cubic metre broth) £500 per cubic metre, a £5 per cubic metre process step is of much less consequence. A yield loss is, however, more important as the product is of increased worth at this stage. For similar reasons, the cost penalty of multistep process sequences is particularly high at these later stages.

Economic processing consequently demands few, high selectivity stages, and high unit operation costs are likely to be tolerable if they give good yield and selectivity. Reliability of processing is important: the consequences of failure to maintain quality and contamination standards are severe in terms of product value, and the opportunities to correct quality and contamination are few.

Recovery from Clarified Aqueous Fraction

A wide range of unit operations has been established for this step (*see* Tables 6 and 7). The principal ones are considered below.

Distillation

This method of recovery is used for volatile organic compounds, in particular ethanol, butanol, acetone and acetic acid.

Precipitation with filtration or centrifugation

Organic acids are converted to the insoluble calcium salts which are then recovered using filtration or centrifugation.

Solvent extraction

This is the classical route for the recovery of antibiotics. The systems employed generally use plain solvents (i.e. without additives) in multistage equipment, usually with pH adjustment to give a favourable partition coefficient.

Thermal concentration or ultrafiltration

Typically, this approach is used to produce enzyme or polysaccharide concentrates. Thermal concentration is normally performed at about 40°C and gives no purification. Ultrafiltration rejects low molecular weight components with the permeate, usually retaining molecules larger than 10 000 daltons. With successive dilution, reasonably complete removal of low molecular weight fractions is possible.

Adsorption

Adsorption from clarified broth is used for a wide range of pharmaceuticals (e.g., streptomycin, cephalosporin) with the product being adsorbed onto fixed beds of materials such as carbon or ion exchange resins. It is also used as a stripping process, adsorbing unwanted materials prior to a concentrating step (e.g., in vaccine production) where adsorbents of the Sephadex-type are normal.

Ion pair solvent extraction

Solvent extraction enhanced by the addition of active ions to the system [*see* King (1981) and Verrall (1981)] offers the advantages of selectivity and application to a wide product range.

Two aqueous-phase extraction

This operation, which consists of either two aqueous phases achieved by relatively heavy dosing with for instance polyethylene glycol and phosphates [*see* Kroner *et al.* (1982)] or a liquid membrane where the second aqueous phase is encapsulated in a solvent film as an emulsion, offers advantages of selectivity. Furthermore, the extraction methods can be applied to materials that are not solvent-soluble and, with liquid membranes, offers the opportunity of incorporating partition effect with chemical changes, e.g., the dispersed aqueous phase can be at a different pH to that of the continuous phase.

Specific affinity adsorption methods

Chromatographic and other advanced methods can be applied if adsorbents with the right characteristics are available at an economic price. Adsorbents must be sufficiently robust to withstand the fouling (and hence severe cleaning procedures) likely to be encountered with the liquors, which contain colloidal and high molecular weight materials from the fermentation. Since volumes at this processing stage are large, and consequently adsorbent bed volumes are also large, the economic price needs to be of the order of that of bulk ion exchange resins rather than that of adsorbents used for the chromatographic separation of biological materials (*see* Appendix 2).

Large Molecule Purification and Recovery

A wide range of microbiological products are large, unstable molecules (e.g., the proteins of vaccines and enzymes). These are commonly sold as broth concentrates (or precipitates from broth concentrates) of mixed materials with the required activity. Pharmaceuticals of this type also frequently require selective removal of impurities (e.g., pyrogens, proteins with specific undesirable effects and other products from the broth, such as unfermented sugars, electrolytes and polysaccharides).

The chemical fragility of these materials limits the separation processes that can be used to physical operations dependent on molecular size and configuration or low-energy bonding to active materials (e.g., coordinated complexes with ligands on an adsorbent surface). Primary purification steps tend to depend on molecular size, while final purification depends on either molecular size fractionation or complex formation.

Typical processes are:

Fractional crystallization

This is achieved by addition of electrolytes – commonly, sodium sulphate and ammonium sulphate – or water-soluble solvents, e.g., ethanol, *i*-propanol, acetone, glycol, etc. This operation is often used for complete precipitation of enzyme solids or in the final crystallization of antibiotics. It can be made selective by successive additions of the salting-out material or operated continuously [*see* Lilly and Dunnill (1969)].

Extraction processes

Such procedures offer similar possibilities to those discussed under broth handling (*see* p.947). The scale of operation and concentrations of the product make it economically possible to use elaborate, expensive extractants (or additives to extractants, as in 'ion pair' systems) compared with what can be used at the broth stages.

Adsorptive processes

These are typically used in fixed-bed configuration. Adsorbents vary from simple, chemically inactive materials, such as silica gel and carbon, to highly specific resins with active surface coatings. The type of separation achieved can vary from simple molecular weight discrimination to the specific pick-up of a single antibody on a surface activated with a particular antigen. As a production operation, a number of problems exist.

1. It is difficult to make the operation continuous, which is important for large-scale products.
2. The life of the adsorbent is important. Generally, the more selective, elaborate adsorbents deteriorate more quickly due to fouling and deactivation of the surface than do the simple adsorbents.

Membrane processes

Membranes can be inactive, as in ultrafiltration. This essentially gives a molecular weight separation. When membranes are active, as in ion exchange membranes, some effect of chemical separation is possible. The driving force can be pressure, concentration or electrical potential (or a combination of these). For the typical large molecules that are of special interest in microbiological processes, inactive membranes offer the widest scope and are well developed [*see* Cooper (1972)]. Principal problems are the membrane life, controlled largely by fouling and, for active membranes, loss of activity. Resistance to cleaning techniques is important: for many microbiological products, cleaning includes sterilization by hypochlorite solutions or steam.

Electrokinetic transport processes

These may be purely electrokinetic, either in open solution or through some highly porous stream separation diaphragm (e.g., nylon cloth), or through semipermeable membranes [e.g., ion exchange *see* Nagasubramonian *et al.* (1980)]. A wide spectrum of processes of this type had semitechnical-scale equipment developed [*see* Bier and Egan (1981), Aitchison *et al.* (1981) and McRae and Leatz (1972)]. They are attractive because of their highly selective fractionating ability and their capacity for separating a very wide range of particle sizes, from small ions to macromolecules and even solid particles. A principal problem is that, at the voltages involved, electrolysis occurs if the solution is conducting, leading to the generation of high pH at the cathodes, low at the anodes as well as consuming power in unwanted effects.

High-gradient magnetic separation

High-intensity magnetic fields can transport molecules and particles at velocities that differ according to their inducible or inherent magnetic properties and thus offer some of the features of electrokinetic transport without the danger of electrolysis. This process has been described by Williams *et al.* (1981).

Other processes

While the above processes present separation opportunities suited to the separation of macromolecules, it must be recalled that the more normal purification steps of crystallization, chemical precipitation and redissolving, and precipitation of impurities may be applicable depending on the product properties. If so, they will probably represent the more economic alternatives (*see* Table 8).

Table 8. Protein precipitation.

Product	Precipitant	Amount	Comment	Reference
Amidase	Propylene glycol	6–20% (by vol.)	6% gives partial precipitation, 10% precipitates bulk, 20% used commercially	Lilly and Dunnill (1969)
	Ammonium sulphate	568 g ℓ^{-1} at 10°C		Lilly and Dunnill (1969)
	Sodium sulphate	110 g ℓ^{-1} at 15°C	Quantity of precipitant temperature-dependent	Lilly and Dunnill (1969)
Proteases	*i*-Propanol	200–400% (by vol.)	Many impurities also precipitated	Keay *et al.* (1972)

Biomass Processing

Biomass processing beyond the broth-handling stage is required only to a limited extent. On the large scale, washing stages are used to remove traces of insoluble substrate during the solid–liquid separation of single-cell protein processes. On a smaller scale, yeast is autolyzed and debittered, some intracellular materials are extracted at a comparatively small scale (e.g., enzymes and pharmaceuticals) and glucose isomerase is made by forming biomass with various additives into particles suitable for use as a catalyst. Extraction of materials from damp biomass, e.g., reduction of the nucleic acid content of single-cell protein recovery of lipids and recovery of a wider range of intracellular products lead to two conventional sequences.

1. Reslurry the biomass to a fluid slurry (20–30 per cent solids), lyze or otherwise disrupt the cells, chemically treat [e.g., alkaline treatment to extract nucleic acid from single-cell protein reported by Viikari and Linko (1977)], separate an aqueous stream (with the product or extracted impurity) from the solid, dry or dispose of solid and work up product from aqueous stream. Combinations of this sequence with a solvent treatment have been described by Wang *et al.* (1981) and Labuza (1975).
2. Dry the biomass, preform into robust, granular pellets, solvent extract, recover product and solvent from the extract, purify the product, recover solvent from biomass and dry the biomass (usually involving some steam stripping to clear the last traces of solvent from the biomass), redistil or otherwise clean up solvent for reuse.

These are long processing sequences, involving some high-energy steps, unattractive in operating costs for low-value products, such as single-cell protein and lipids [*see* Ratledge (1982)] and in operating costs and yield loss for high-value materials.

Cell Drying and Reduction of Viability

This is required for whole cell material to produce a physical form that gives an attractive product (e.g., yeast, single-cell protein) and to produce special physical forms (e.g., whole-cell glucose isomerase material) for use as a fixed-bed catalyst.

The dryness (probably under 5 per cent moisture) must check product deterioration in storage and the product must normally be free of viable organisms.

The feed to the drying step is usually in one of the following forms.

1. A cream [perhaps 15 per cent (w/v) solids].
2. A dewatered sludge [20–35 per cent (w/v) solids].
3. A damp solid [40 per cent (w/v) solids or more].

Dewatering is not necessarily pursued to the technical limits as it may deplete desirable solubles, e.g., vitamins in animal feedstuffs.

Biomass normally has a limit to the time and temperature to which it can be exposed without creating unacceptable decomposition. The choice is thus between long drying times at lower temperatures or brief exposure to more severe conditions. If the low-temperature route is chosen, as it may be if very temperature-sensitive materials are to be dried, reduction of the viable organism count must be performed by pasteurization prior to drying. This means a fluid slurry must be handled that is usually dried directly without an intermediate dewatering step, the considerable extra drying energy demand being accepted.

The normal choice is driers with a short exposure time: typically, surface film (drum or belt), spray or flash driers. Of these, the spray and flash tend to give a sufficient reduction in viability with product temperatures of about 100°C, whereas drum driers need somewhat higher temperatures and cause more product decomposition. To be efficient, spray and flash driers need to operate with a high temperature drop in the drying air feed: an air inlet temperature of over 300°C appears practicable, the limit being set by the risk of ignition of dry product deposits in the hotter zones [and the consequent risk of dust explosion discussed by Labuza (1975)].

While drying is an expensive step, within the limitations imposed by product decomposition and the need to reduce cell viability, it is difficult to envisage developments that will much reduce the cost unless heat recovery technology and economics alter. Since the waste heat must be rejected at around 100°C, the possibilities of heat recovery beyond the 20 per cent or so that can be used in air preheating must depend on having local use for low-grade heat (e.g., space heating).

EXAMPLES OF INNOVATION

That innovation in unit processes and operations is both possible and practicable is illustrated by a number of developments in recent years which have achieved both processing and commercial acceptance within downstream biological process engineering and by others for which acceptance appears inevitable.

Liquid Carbon Dioxide Extraction

This process (*see* Figs. 8 and 9) has been developed for the extraction of bittering substances from hops as an alternative to extraction based upon the use of organic solvents. The process has been described by Grimmett (1981).

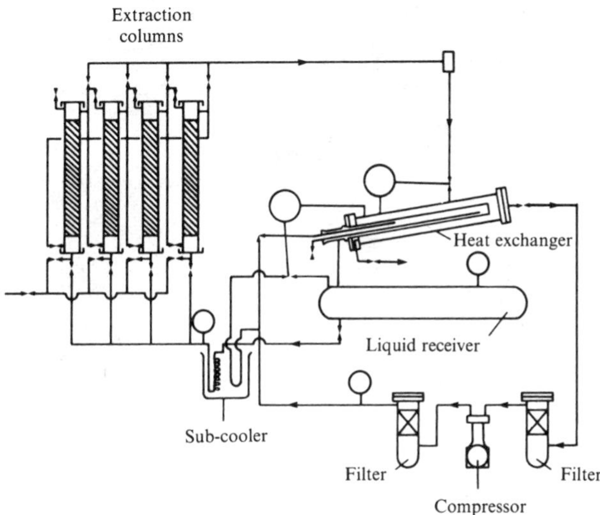

Figure 8. Liquid carbon dioxide hop extraction process (courtesy of Brewing Research Foundation, Nutfield, Surrey, UK).

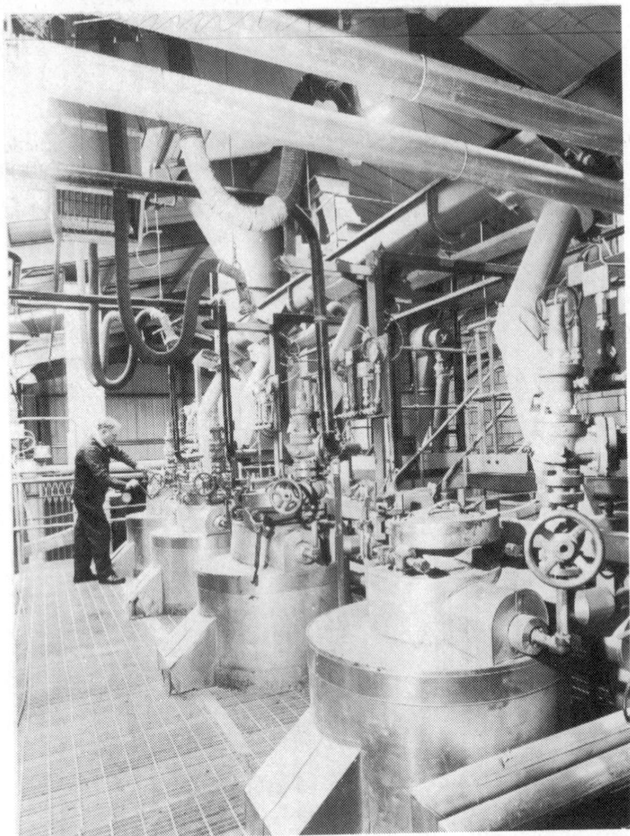

Figure 9. Liquid carbon dioxide hop extraction columns at Pauls and White International, Surrey, U.K.

Biomass Support Particles

Such particles, which are illustrated on p.32, allow high biomass concentrations in continuous fermenters and provide the basis for biomass recovery based on decanting and compression as an alternative to sedimentation, filtration, etc. (see Figs. 10–12). The use of these supports is described by Atkinson *et al.* (1980).

Whirlpool Separator

The whirlpool separator, as illustrated diagrammatically in Fig. 13, has been shown by Hudston (1969) to allow the batch processing of brewers' wort to remove suspended solids. The equipment produces a clarified liquor and concentrated slurry. The vortex principle established by Einstein (1926) describes the operation of the separator but the phenomena involved are those commonly experienced in the deposition of tea leaves following stirring.

Figure 10. Biomass support elements being continuously removed from a fermentation vessel.

Enhanced liquid/liquid extraction

Electrically enhanced extraction occurs when a dispersed phase is passed through a high-voltage nozzle. Thornton (1976) has shown that this technique results in small droplets moving at high velocities, with high mass transfer coefficients, i.e. conditions appropriate to the separation of materials that are sensitive to prolonged exposure to either heat or solvents.

Continuous Ion Exchange

The Vistec process for the continuous isolation of protein described by Jones (1974) and illustrated in Fig. 14 consists of three unit operations. Firstly, the protein is adsorbed onto the ion exchange medium, then the protein-loaded medium is washed to remove contaminants and finally the protein is desorbed either using salts or by pH change. The latter method of protein desorption is frequently preferable as it eliminates the problem of salt contamination. Each operation is performed in a stirred vessel followed by a settling cone, transfer of the medium being achieved by an air pump which exposes the ion exchange cellulose medium to minimal abrasion.

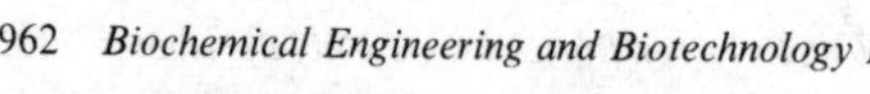

Figure 11. Excess biomass displaced from support elements at concentrations of up to 6 per cent.

Figure 12. Biomass being displaced from support elements which are subsequently returned to the fermenter.

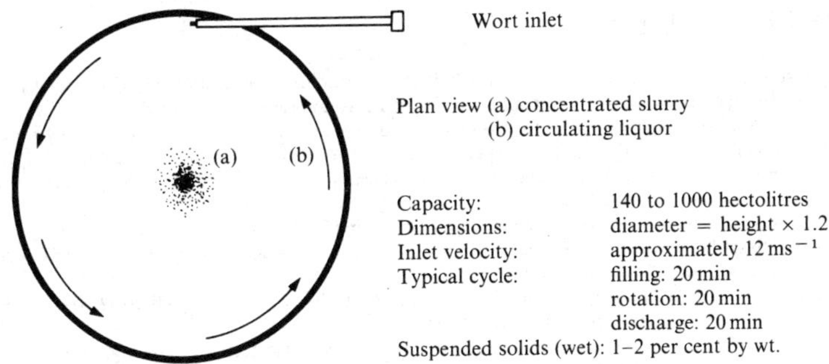

Figure 13. Whirlpool separator: plan view showing concentrated slurry (a) and circulating liquor (b).

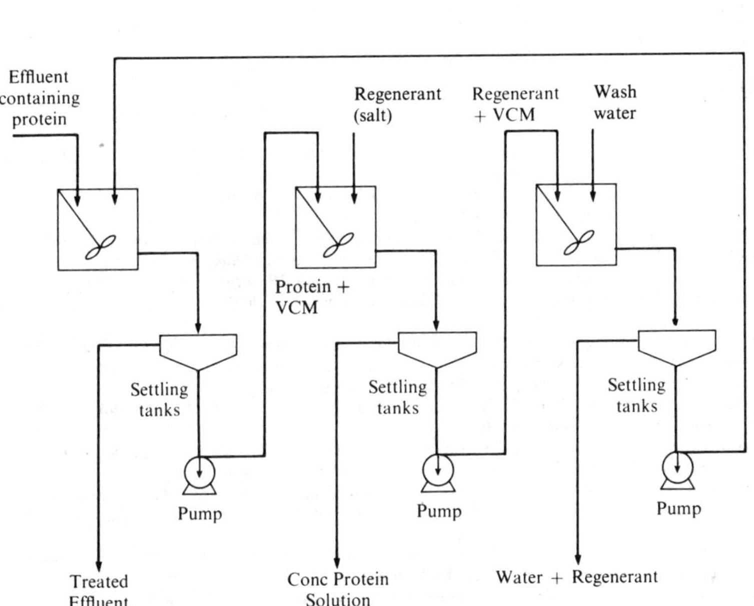

Figure 14. Flow diagram of Vistec continuous ion exchange system using viscose cellulose medium VCM (Viscose Developments Ltd, Swansea, UK).

REFERENCES

Aitchison, G.F., Thomson, A.R. and Maltuck, P. (1981) Fractionation of Biological Material by Large Scale Electrophoresis. 2nd European Congress of Biotechnology, Eastbourne.

Asenjo, J.A. (1981) Light Enzymes for the Disruption of Whole Yeast Cells. 1st Engineering Foundations Conference on Advances in Fermentation Recovery Process Technology, Banff.

Atkinson, B., Black, G.M. and Pinches, A. (1980) Process Intensification using Cell Support Systems. *Process Biochem.*, **15**(5), 24.

Atkinson, B. and Sainter, P. (1981) Technology Forecasting for Downstream Processing in Biotechnology, Phase I, Intermediate Forecast Report. European Economic Community FAST Project Contract No. FST/C/020/80/UK 11.

Beaton, N.C. (1980) The Application of Ultrafiltration to Fermentation Products. *Polymer Science Technology*, vol. 13 (Plenum Press, New York).

Bier, M. and Egan, J. (1981) Large Scale Protein Purification by Isoelectric Focussing. 1st Engineering Foundations Conference on Advances in Fermentation Recovery Process Technology, Banff.

Bietelshees, C.P., King, C.J. and Sephton, H.H. (1979) Precipitate Flotation for Removal of Copper, in *Recent Developments in Separation Science*, ed. by N.N. Li, p.43 (CRC Press, Cleveland).

Busch, P.L. and Stumm, W. (1968) Chemical Interactions in the Aggregation of Bacteria. *Environ. Sci. Technol.*, **2**, 49.

Cooper, A.R. (ed.) (1972) U.F. Membranes and Their Applications. *Polymer Science Technology*, vol. 13 (Plenum Press, New York).

Dlouhy, P.E. and Dalhstrom, D.A. (1968) Continuous Filtration of Pharmaceutical Products. *Chem. Engng Prog.*, **64**, 116.

Dubinsky, Z., Berner, T. and Aaronson, S. (1979) Potential of Large Scale Algal Culture for Biomass and Lipid Production. *Biotechnol. Bioengng Symp.*, **8**, 131.

Einstein, A. (1926) *Naturwissenschaften*, 223.

Engler, C.R. and Robinson, C.W. (1981) Effect of Organism Type and Growth Conditions on Cell Disruption by Impingement. *Biotechnol. Lett.*, **3**, 83.

Gasner, C.C. and Wang, D.I.C. (1970) Microbial Recovery Enhancement through Flocculation. *Biotechnol. Bioengng*, **12**, 873.

Grimett, C.M. (1981) The Use of Liquid Carbon Dioxide for Extracting Natural Products. *Chemy Ind.*, **10**, 359.

Henry, J.D. (1972) Cross Flow Filtration, in *Recent Developments in Separation Science*, vol. 2, ed. by N.N. Li, p.205 (CRC Press, Cleveland).

Hudston, (1969) The Story of the Whirlpool, *MBAA Tech. Quart.* **6**, 164.

Jones, D.T. (1974) Protein Recovery by Ion Exchange. *Process Biochem.*, **9** (12), 17.

Keay, L., Moseley, M.H., Anderson, R.G., O'Connor, R.J. and Wildi, B.S. (1972) Production and Isolation of Microbial Proteases. Enzyme Engineering. *Biotechnol. Bioengng Symp.*, **3**, 63.

King, C.J. (1981) Complexing Extractants for Recovery and Separation of Polar Organics. 1st Engineering Foundations Conference on Advances in Fermentation Recovery Process Technology, Banff.

Kinoshita, S. and Nakayama, K. (1978) Amino Acids, in *Economic Microbiology*, vol. 2, *Primary Products of Metabolism*, ed. by A.H. Rose, p.285 (Academic Press, London).

Klein, W. (1981) Cross Flow Filtration. Filtech Conference, London.

Kroner, K.H., Schutte, H., Stack, W. and Kula, M.R. (1982) Scale-Up of Formate Dehydrogenase by Partition. *J. Chem. Technol. Biotechnol.*, **32**, 130.

Labuza, T.P. (1975) Cell Collection, Recovery and Drying for SCP Manufacture, in *Single Cell Protein*, vol. 2, ed. by S.R. Tannenbaum and D.I.C. Wang (MIT Press, Cambridge).

Lilly, M.D. and Dunnill, P. (1969) Isolation of Intracellular Enzymes from Microorganisms, the Development of a Continuous Process, in *Fermentation Advances*, ed. by D. Perlman, p.225 (Academic Press, New York).

McRae, W.A. and Leatz, F.B. (1972) Electrophoresis and Other Ion Selective Processes, in *Recent Developments in Separation Science*, ed. by N.N. Li, p.157 (CRC Press, Cleveland).

Mantell, C.L. (1960) *Electro-Chemical Engineering*, 4th edn (McGraw-Hill, New York).

Moss, M.O. and Smith, J.E. (1977) *Industrial Applications of Microbiology* (Surrey University Press, London).

Nagasubramonian, K., Chlanda, F.P. and Liu, K.J. (1980) Bipolar Membrane Technology, an Engineering and Economic Analysis, in *Recent Advances in Separation Techniques*, vol. 2, p.97 (AIChE, New York).

Perry, R.H. and Chilton, C.H. (1973) *Chemical Engineers' Handbook*, 5th edn (McGraw-Hill Kogakusha, Tokyo).

Ratledge, C. (1982) Microbial Oils and Fats, an Assessment of Their Commercial Potential. *Prog. Ind. Microbiol.*, **16**, 119.

Rehacek, J. and Schaefer, J. (1977) Disintegration of Microorganisms in an Industrial Horizontal Mill. *Biotechnol. Bioengng.*, **19**, 1523.
Spruce, F. (1978) Private communication.
Thornton, (1976) Electrically Enhanced Liquid–Liquid Extraction. *Chem. Engr, Birmingham Univ.*, **27**, 6.
Verrall, M.S. (1981) Ion Pair Extraction of Microbial Metabolites. SCI Meeting on Solvent Extraction of Biological Molecules, London.
Viikari, L. and Linko, M. (1977) Reduction of Nucleic Acid Content of SCP. *Process Biochem.*, **12** (5), 17.
Wakeman, R.J. (1981) Effect of Solids Concentration and pH on Electrofiltration. Filtech Conference, London.
Wang, Y., Wang, D.C. and Li, K.-L., (1981) Further Studies on n-Paraffin Yeast as Fodder in China. International Symposium on Single-Cell Protein, Paris.
West, J.M. and Patterson, A. (1981) Whole Broth Extraction. 1st Engineering Foundations Conference on Advances in Fermentation Recovery Process Technology, Banff.
Williams, J.A., Baxter, W., Collins, J., Harding, K., Leslie, C.M. and Sills, R.J. (1981) The Application of High Gradient Magnetic Separation in Nuclear Fuel Reprocessing. *Nucl. Technol.*, **25**, 284.

APPENDIX 1: APPLICATION OF VARIOUS UNIT OPERATIONS

Unit operation	Relevant application	Reference
Adsorption	Streptomycin recovery	Prasad *et al.* (1980)
	Isolation of hydrophilic fermentation products	Voser (1982)
	Separation of glucose/fructose mixtures	Barker and Church (1981)
Chemical pretreatment	Filtration performance	Pearce and Allen (1981)
Complexing	Recovery and separation of polar organics	King (1981)
Continuous filtration	Pharmaceutical products	Dlouhy and Dahlstrom (1968)
Cross-flow filtration	Solid–liquid separation	Henry (1972)
	Solid–liquid separation	Klein (1981)
Coupled transport membranes	Organic acids	Smith (1981)
Drying	Biomass	Labuza (1975)
Electrofiltration	Effect of solids concentration and pH	Wakeman (1981)
Electrolysis	Ion selection	Li (1977)
Electrophoresis	Fractionation of biological materials	Aitchison *et al.* (1981)
	Protein precipitation	Bier and Egan (1981)
Extraction	Whole broth	West and Patterson (1981)
Ion pair	Microbial metabolites	Verrall (1981)
Near critical liquid		Helljarkle and deFilippi (1982)
Supercritical gas		Bolt (1980)
Two aqueous-phase	Formate dehydrogenase	Kroner *et al.* (1982)
Flocculation	Bacteria	Busch and Stumm (1968)
High-gradient magnetic separation		Williams (1981)
Liquid membrane processes	Recovery of soluble components	Li (1977)
	Whole broth	Drioli *et al.* (1981)
Membrane technology	Conversion of cellulose to ethanol	Lopez-Leiva *et al.* (1981)
	Whole broth	Drioli *et al.* (1981)
Recrystallization	Purification of glutamic acid	Dozoi (1981)

References

Aitchison, G.F., Thomson, A.R. and Maltuck, P. (1981) Fractionation of Biological Materials by Large Scale Electrophoresis. 2nd European Congress of Biotechnology, Eastbourne.

Barker, P.E. and Church, C.H. (1981) A Sequential Chromatographic Process for the Separation of Glucose/Fructose Mixtures. *Chem. Engr*, No. 371/2, 389.

Bier, M. and Egan, J. (1981) Large-Scale Protein Purification by Isoelectric Focusing. 1st Engineering Foundations Conference on Advances in Fermentation Recovery Process Technology, Banff.

Bolt, J.R. (1980) The Development of Separation Process involving Gas at near Critical Conditions, in *Research and Development Priorities in Separation Processes* (IChemE, Rugby).

Busch, P.L. and Stumm, W. (1968) Chemical Interactions in the Aggregation of Bacteria. *Environ. Sci. Technol.*, **2**, 49.

Dlouhy, P.E. and Dahlstrom, D.A. (1968) Continuous Filtration of Pharmaceutical Products. *Chem. Engng Prog.*, **64**, 116.

Dozoi, M. (1981) Transformation Recrystallization Method for Purification of L-Glutamic Acid. 1st Engineering Foundations Conference on Advances in Fermentation Recovery Process Technology, Banff.

Drioli, E., Serafin, G. and Rigoli, A. (1981) Pressure Driven Membrane Processes in the Separation of a Bioactive Compound from Fermentation Broths. 1st Engineering Foundations Conference on Advances in Fermentation Recovery Process Technology, Banff.

Helljarkle, G. and deFilippi, R.P. (1982) Extraction Processes using Solvents near Their Thermodynamic Critical Point. *Chem. Engr*, No. 379, 136.

Henry, J.D. (1972) Cross-Flow Microfiltration, in *Recent Developments in Separation Science*, vol. 2, ed. by N.N. Li, p.205 (CRC, Cleveland).

King, C.J. (1981) Complexing Extracts for Recovery and Separation of Polar Organics. 1st Engineering Foundations Conference on Advances in Fermentation Recovery Process Technology, Banff.

Klein, W. (1981) Cross-Flow Filtration. Filtech Conference, London.

Kroner, K.H., Schutte, H., Stack, W. and Kula, M.R. (1982) Scale-Up of Formate Dehydrogenase by Partition. *J. Chem. Technol. Biotechnol.*, **32**, 130.

Labuza, T.P. (1975) Cell Collection Recovery and Drying for Single-Cell Protein, in *Single-Cell Protein* , vol. 2, ed. by S.R. Tannenbaum and D.I.C. Wang (MIT Press, Cambridge).

Li, N.N. (ed) (1977) *Recent Developments in Separation Science*, vol. IIIA (CRC Press, Cleveland).

Lopez-Leiva, M., Hagerdal, B. and Mattiasson, B. (1981) Membrane Technology in Bioconversion of Cellulose to Ethanol. 2nd European Congress of Biotechnology, Eastbourne.

Pearce, M.J. and Allen, A.P. (1981) Chemical Treatments for Optimum Filtration Performance. Filtech Conference, London.

Prasad, R., Gupta, A.K. and Bajpai, R.K. (1980) Adsorption of Streptomycin on Ion Exchange Resins: Equilibrium Studies. *J. Chem. Technol. Biotechnol.*, **30**, 324.

Smith, B.R. (1981) Organic Acid Recovery with Coupled Transport Membranes. 1st Engineering Foundations Conference on Advances in Fermentation Recovery Process Technology, Banff.

Verrall, M.S. (1981) Ion Pair Extraction of Microbial Metabolites. SCI Meeting on Solvent Extraction of Biological Molecules, London.

Voser, W. (1982) Isolation of Hydrophilic Fermentation Products by Adsorption Chromatography. *J. Chem. Technol. Biotechnol.*, **32**, 109.

Wakeman, R.J. (1981) Effects of Solids Concentration and pH on Electrofiltration. Filtech Conference, London.

West, J.M. and Patterson, A. (1981) Whole Broth Extraction. 1st Engineering Foundations Conference on Advances in Fermentation Recovery Process Technology, Banff.

Williams, J.A., Baxter, W., Collins, G., Harding, K., Leslie, C.M. and Sillis, R.J. (1981) The Application of High Gradient Magnetic Separation in Nuclear Fuel Reprocessing. *Nucl. Technol.*, **25**, 284.

APPENDIX 2: AFFINITY CHROMATOGRAPHY MATERIALS

Adsorbant	Substance adsorbed	Adsorption capacity	Comment	Reference
5'-AMP-Sepharose	Lactate dehydrogenase	10 g (ℓ gel)$^{-1}$ [~50 g (kg dry gel)$^{-1}$]	Equilibrium affected by conditions, e.g. 0.05% Triton in 10 mm phosphate buffer increased adsorption 20-fold	Dunnill (1978)
Amicon matrix gel PBA	Protein	0.1 g (ℓ gel)$^{-1}$ [~0.05–50 g (kg dry gel)$^{-1}$]		*Amicon News* (1981)
	β-Galactosidase	19 g (ℓ gel)$^{-1}$ [~100 (kg dry gel)$^{-1}$]	18 ℓ column operable in 2 h cycle, producing 3.6 g pure enzyme per cycle	Dunnill and Lilly (1974)
Sephadex G-50 or G-75	Protein	0.33–0.83 g (ℓ gel)$^{-1}$ [~3.3–8.3 g (kg dry gel)$^{-1}$]	300 ℓ column is capable of fractionating 100–250 g protein per cycle	Porath (1972)
Sephadex G-25	Benzyl penicillin	40 g (ℓ gel)$^{-1}$ [~400 g (kg dry gel)$^{-1}$]	500 ℓ column, 50 ℓ feed solution, 500 ℓ saline eluant. Recovery > 98% for 375 ℓ liquor. Cycle time 110 min	Edwards (1969)
Sephadex	Chymotrypsin	35–90 g (kg dry gel)$^{-1}$		Baum and Wrobel (1975)
Polyacrylamide	Protein	67 g (kg dry gel)$^{-1}$		Baum and Wrobel (1975)
Agarose gel	Cyanogen bromide	20–400 g (kg dry gel)$^{-1}$	Activation of agarose gel	Baum and Wrobel (1975)

References

Amicon News (1981) November issue.

Baum, G. and Wrobel, S.J. (1975) Affinity Chromatography, in *Enzymology*, vol. 1, *Immobilized Enzymes, Antigens, Antibodies and Peptides*, ed. by H.H. Weetall, p.419 (Dekker, New York).

Dunnill, P. (1978) The State of Enzyme Isolation Technology, in *Enzyme Engineering*, vol. 3, ed. by E.K. Pye and H.H. Weetall, p.207 (Plenum Press, New York).

Dunnill, P. and Lilly, M.D. (1974) Recent Developments in Enzyme Isolation Processes, in *Enzyme Engineering*, vol. 2, ed. by E.K. Pye and H.H. Weetall, p.43 (Plenum Press, New York).

Edwards, V.H. (1969) Separation Techniques for the Recovery of Materials from Aqueous Solutions, in *Fermentation Advances*, ed. by D. Perlman, p.273 (Academic Press, New York).

Porath, J. (1972) Chromatographic Methods in Fractionation of Enzymes. *Biotechnol. Bioengng Symp.*, **3**, 145.

CHAPTER 14

PROCESSES

INTRODUCTION

Industrial biological processes encompass the following stages.

1. Raw material preparation.
2. Fermentation medium preparation.
3. Fermentation.
4. Product recovery and purification.
5. Preparation of a sales product.

Fig. 1 illustrates a typical process, indicating the various stages, the number of vessels and identifying the 'services' required in support of each stage. The material balances associated with the process contained in Fig. 1 are shown in Fig. 2. Figs. 1 and 2 together identify the range of factors that it is necessary to consider during the development or assessment of a biological process.

The sources of carbohydrate, nitrogen and growth factors used as industrial raw materials are listed in Tables 1, 2 and 3. An index of the major fermentation companies is given in Table 4.

Table 1. Sources of carbohydrate [Rhodes and Fletcher (1966)].

Carbohydrate	Sources
Glucose	Pure glucose monohydrate, hydrolyzed starch
Lactose	Pure lactose, whey powder
Starch	Barley, groundnut meal, oat flour, rye flour, soya bean meal
Sucrose	Beet molasses, cane molasses, crude brown sugar, pure white sugar

Table 2. Sources of nitrogen [Rhodes and Fletcher (1966)].

Nitrogen source	Nitrogen content [% nitrogen (by wt)]
Barley	1.5–2.0
Beet molasses	1.5–2.0
Cane molasses	1.5–2.0
Corn-steep liquor	4.5
Groundnut meal	8.0
Oat flour	1.5–2.0
Pharmamedia	8.0
Rye flour	1.5–2.0
Soya bean meal	8.0
Whey powder	4.5

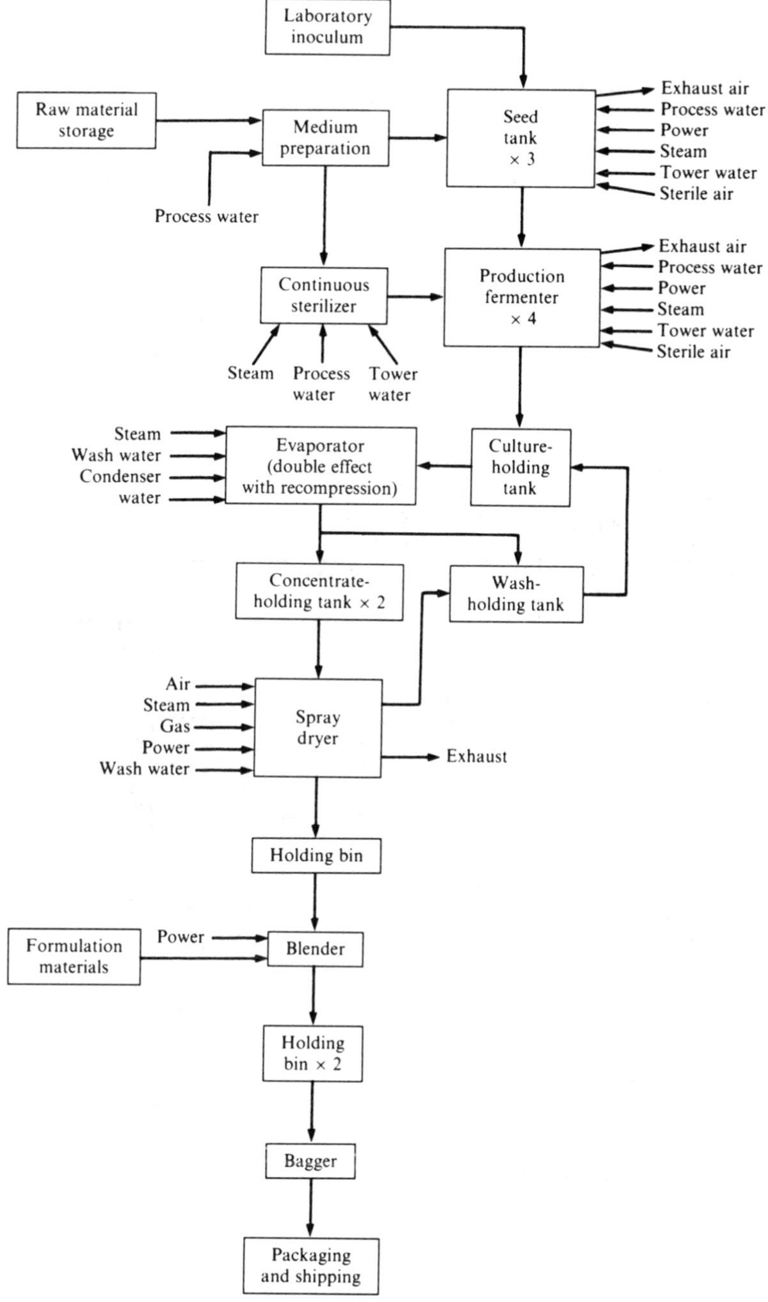

Figure 1. Process flow diagram [Bartholomew and Reisman (1979)].

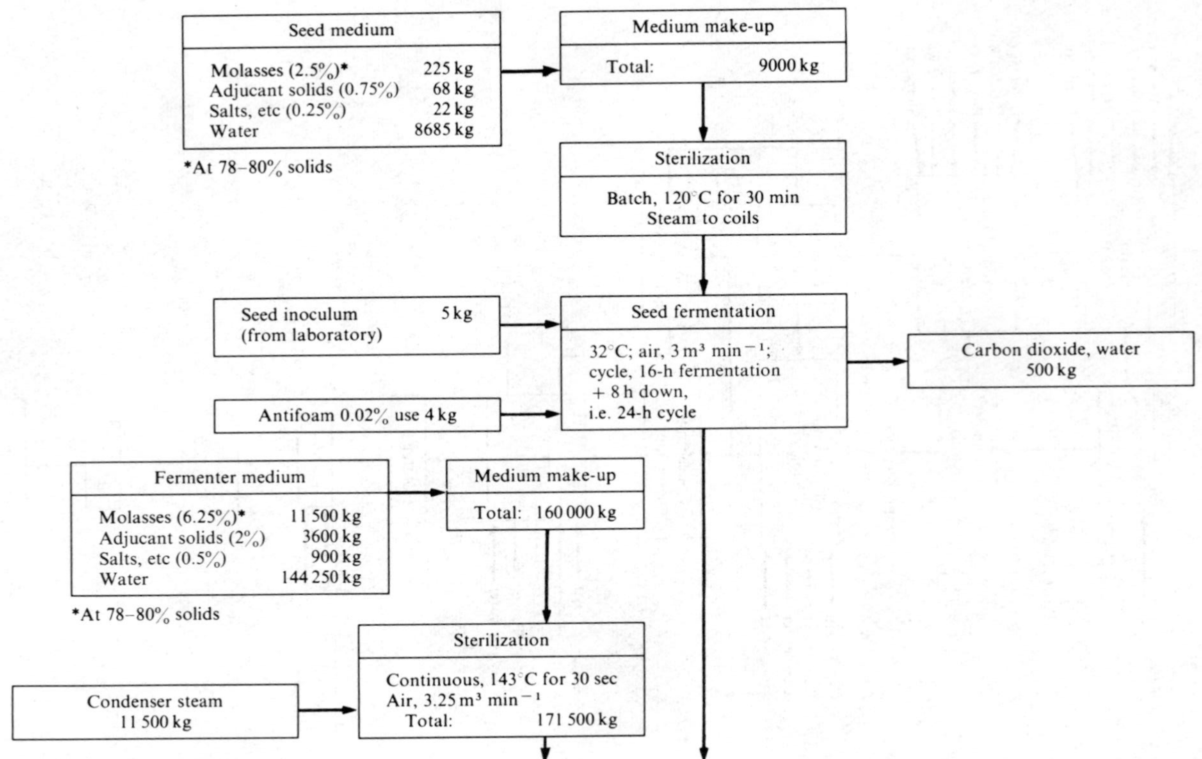

Seed medium
Molasses (2.5%)* 225 kg
Adjucant solids (0.75%) 68 kg
Salts, etc (0.25%) 22 kg
Water 8685 kg
*At 78–80% solids
Medium make-up
Total: 9000 kg
Sterilization
Batch, 120°C for 30 min
Steam to coils
Seed inoculum 5 kg
(from laboratory)
Antifoam 0.02% use 4 kg
Seed fermentation
32°C; air, 3 m³ min⁻¹;
cycle, 16-h fermentation
+ 8 h down,
i.e. 24-h cycle
Carbon dioxide, water
500 kg
Fermenter medium
Molasses (6.25%)* 11 500 kg
Adjucant solids (2%) 3600 kg
Salts, etc (0.5%) 900 kg
Water 144 250 kg
*At 78–80% solids
Medium make-up
Total: 160 000 kg
Sterilization
Continuous, 143°C for 30 sec
Air, 3.25 m³ min⁻¹
Total: 171 500 kg
Condenser steam
11 500 kg

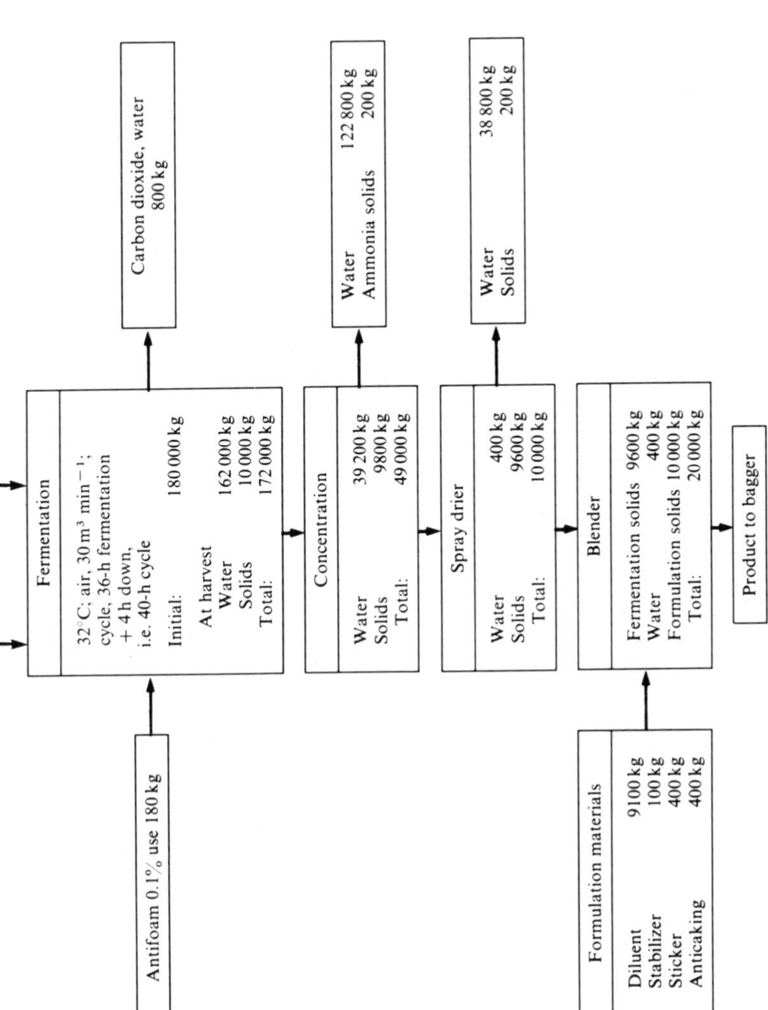

Figure 2. Material balance flow sheet (per batch) [Bartholomew and Reisman (1979)].

Table 3. Accessory food factors required by nutritionally exacting microorganisms.

Accessory growth factor	Active form	Chemical group transferred	Substance needed to fulfil metabolic requirement	Source of growth factor
Biotin	Biotin	Carbon dioxide fixation	Biotin	Cane molasses, corn-steep liquor, *Penicillium* spent mycelium
Choline	Phosphatides		Choline	Egg yolk, hops
Cyanocobalamin		Carboxyl displacement, methyl group synthesis	Cyanocobalamin, other cobalamins	Activated sewage sludge, cow faeces, liver, meat, silage, *Streptomyces griseus* mycelium
Folic acid	Tetrahydrofolic acid	Formyl group	Folic acid, *p*-aminobenzoic acid	Liver, *Penicillium* spent mycelium, spinach
Haemins	Cell haemins	Electrons	Haemins	Blood
Inositol	Phosphatides		Inositol	Corn-steep liquor
Lipoic acid	Lipoic acid	Hydrogen and acyl groups	α-Lipoic acid, thioctic acid	Liver
Nicotinic acid or nicotinamide	Nicotinamide adenine dinucleotide (NAD)	Hydrogen	Nicotinic acid or nicotinamide	Liver, *Penicillium* spent mycelium, wheat seeds
	Nicotinamide adenine dinucleotide phosphate (NADP)	Hydrogen	Nucleotides of nicotinamide	Liver, *Penicillium* spent mycelium, wheat seeds
	Nicotinamide mononucleotide	Hydrogen		Liver, *Penicillium* spent mycelium, wheat seed
Pantothenic acid	Coenzyme A (CoA)	Acyl group	Pantothenic acid	Beet molasses, corn-steep liquor, *Penicillium* spent mycelium
Purines	Purine nucleotides		Purines, nucleotides, derived from purines	Dried blood, meat
Pyridoxal	Pyridoxal phosphate	Amino group, carboxyl displacement	Pyridoxine, pyridoxamine, pyridoxal, pyridoxal phosphate	Corn-steep liquor, maize seeds, *Penicillium* spent mycelium, rice polishings, wheat seeds, yeast

Pyrimidines	Pyrimidine nucleotides		Pyrimidines, nucleotides derived from pyrimidines	Meat
Riboflavin	Flavin mononucleotide (FMN)	Hydrogen	Riboflavin	Cereals, corn-steep liquor
	Flavin adenine dinucleotide (FAD)	Hydrogen	Riboflavin	Cereals, corn-steep liquor
Thiamin	Thiamin pyrophosphate	Carboxy displacement, C_2-aldehyde groups	Pyrimidine and/or thiazole, thiamin	Rice polishings, wheat germ, yeast

Table 4. World index to fermentation companies [Perlman (1977)].

Abbott Laboratories, North Chicago, Illinois, USA
Ajinomoto Company, Tokyo, Japan
Aktiebolaget Astra, Södertalje, Sweden
Aktiebolaget KABI, Stockholm, Sweden
Aktiebolaget S.J.A., Stockholm, Sweden
Alembic Chemical Works Co., Ltd., Baroda, India
American Cyanamid, Wayne, New Jersey, USA
Anchor Yeast (Pty.), Ltd., Johannesburg, South Africa
Anheuser-Busch, Inc., St. Louis, Missouri, USA
Antibioticos, Madrid, Spain
Apothekernes Laboratori für Specialpraeparater A/S, Oslo, Norway
Asahi Chemical Industry, Osaka, Japan
Ayerst Laboratories, New York, USA
Banyu Pharmaceutical Company, Tokyo, Japan
Beecham, Inc., Clifton, New Jersey, USA
Beecham Research Laboratories, Brentford, UK
Biochemie GmbH, Kundl, Austria
Biogal, Debrecen, Hungary
Bowmans Chemicals, London, UK
Bristol Laboratories, Syracuse, New York, USA
Istituto Carlo Erba, Milan, Italy
Chinoin, Budapest, Hungary
Ciba-Geigy, Basle, Switzerland
Cipan, Lisbon, Portugal
Clinton Corn Processing Co., Clinton, Iowa, USA
Commercial Solvents Corporation, Terre Haute, Indiana
Compagnie Européenne de Fermentation, Villeneuve La Garconne, France
Compania Argentian de Levaduras S.A., Buenos Aires, Argentina
CPC International Inc., Argo, Illinois, USA
Dairyland Food Industries, Inc., Waukesha, Wisconsin, USA
Diaspa, Milan, Italy
Dista Products Ltd., Liverpool, UK
Distillers Company (Yeast) Ltd., Burgh Heath, Epsom, Surrey, UK
Dumex A/S, Copenhagen, Denmark
Etablissements Fould Springer, Maisons-Alfort (Seine), France
Farbenfabriken Bayer AG, Leverkusen, W. Germany
Farbwerke Hoechst AG, Frankfurt (Main), W. Germany
Farmitalia S.p.a., Milan, Italy
Fervet, S.p.a., Naples, Italy
Finnish State Alcohol Monopoly (Alko), Helsinki, Finland
Fujisawa Pharmaceutical Company, Osaka, Japan
H. Lundbeck & Company, Valby, Denmark
Mauri Brothers and Thomsen, Ltd., Sydney, Australia
Meiji Seika Kaisha Ltd., Tokyo, Japan
Merck & Company, Inc., Rahway, New Jersey, USA
Miles Laboratories, Inc., Elkhart, Indiana, USA
Nihon Kayaku Company, Tokyo, Japan
Nihon Nohyaku Company, Tokyo, Japan
Nikken Chemicals Company, Ltd., Tokyo, Japan
Norddeutsche Hefe Industrie GmbH., Hamburg-Wandsbeck, W. Germany
Novo Industri A/S, Copenhagen, Denmark
Oriental Yeast Company, Ltd., Tokyo, Japan
Orsan, S.A., Paris, France
Parke, Davis and Company, Detroit, Michigan, USA
S. B. Penick and Company, Orange, New Jersey, USA
Chas. Pfizer and Company, New York, New York, USA
Pfizer International, New York, New York, USA
Pharmacosmos, Valby, Denmark
Pierrel S.p.a., Milan, Italy
Premier Malt Products, Milwaukee, Wisconsin, USA
Proter S.p.a., Opera, Italy
Publicker Industries, Inc., Philadelphia, Pennsylvania, USA
Rachelle Laboratories, Inc., Long Beach, California, USA
Quimasa S.A. – Quimica Industrial Santo Amaro, Santo Amaro – São Paulo, Brazil
Recherche Industrie Therapeutique, Genval, Belgium
Rhône Poulenc, Paris, France
Rohm GmbH Chemische Fabrik, Darmstadt, W. Germany
Rohm & Haas, Philadelphia, Pennsylvania, USA
Roussel UCLAF, Romainville, France
Sandoz-Wander, Inc., Homestead, Florida, USA
Sankyo Company Ltd., Tokyo, Japan
Sanraku Ocean, Tokyo, Japan
Schering AG, Berlin, W. Germany
Schering Corporation, Inc., Bloomfield, New Jersey, USA
Searle Biochemics, Arlington Heights, Illinois, USA
Shionogi & Company, Ltd., Osaka, Japan
Société Industrielle le Saffre, Marcq-en-Baroeul, France
Societa Prodotti Antibiotici, Milan, Italy
Société Françise des Petroles BP, Levere, France

Table 4 – (continued)

Gist Brocades, N.V., Delft, Netherlands
Gist & Spiritusfabrieken, Ghent, Belgium
Glaxo Laboratories, Ltd., Greenford, UK
Grain Processing Corporation, Muscatine, Iowa, USA
Gruppo Lepetit, Milan, Italy
Hindustan Antibiotics, Ltd., Pimpri, India
Hoffmann-LaRoche, Inc., Nutley, New Jersey, USA
F. Hoffmann-LaRoche & Co. Ltd., Basle, Switzerland
Icar, S.p.a., Rome, Italy
Imperial Chemical Industries, Ltd., Manchester, UK
Kaken Chemical Company, Tokyo, Japan
Kanegafuchi Chemical Industry Company, Osaka, Japan
Kayaku Antibiotics Research Company, Tokyo, Japan
Kowa Company, Nagoya, Japan
Kyowa Hakko Kogyo Company, Tokyo, Japan
Laboratories Roger Bellon, Neuilly, France
Leo Pharmaceutical Products, Ballerup, Denmark
Eli Lilly and Company, Indianapolis, Indiana, USA
Lohmann & Company, AG, Cuxhaven, W. Germany
E. R. Squibb & Sons, Princeton, New Jersey, USA
Squibb International, New York, New York, USA
Standard Brands, Inc., Stamford, Connecticut, USA
Stauffer Chemical Company, Stamford, Connecticut, USA
Syndicat des Producteurs de Levure Aliment de France, Paris, France
Takeda Chemical Industries Ltd., Osaka, Japan
Tanabe Seiyaku Company, Ltd., Osaka, Japan
Toyo Jyozo, Shizuoka-ken, Japan
Union Yeast Products, Ltd., Johannesburg, South Africa
Universal Foods Corporation, Milwaukee, Wisconsin, USA
Upjohn Company, Kalamazoo, Michigan, USA
Vereinte Maunter Markhofsche Presshefe Fabriken, Vienna, Austria
Wallerstein Laboratories, Deerfield, Illinois, USA
Wyeth Laboratories, Philadelphia, Pennsylvania, USA
Yamanouchi Pharmaceutical Company, Tokyo, Japan
Zellstoffabrik Waldhof, Mannheim-Waldhof, W. Germany

ACETONE/BUTANOL

Raw Materials

Acetone and butanol are produced anaerobically by *Clostridium acetobutylicum* from starchy materials. In the fermentation, one, or more, of the three sources of carbohydrates, i.e. corn, blackstrap molasses or high-test molasses, is used. Other carbon sources include wheat, rice, horsechestnuts, Jerusalem artichokes, potatoes and beet molasses. Carbohydrates can also be derived from waste products, such as hydrolyzed wood or corn cobs, whey, sulphite liquor and hydrol, or as a byproduct in the manufacture of glucose from corn.

Sugar solutions prepared by the hydrolysis of maple, oak and fir are steam-stripped of furfural and other impurities, and the pH is adjusted to 6.5 using lime. Sulphite liquors are first treated with lime to remove sulphites, then with additional lime to precipitate lignin, followed by removal of excess calcium salts as the sulphate. The fermentation of corn does not require any additives. The valuable corn oil, which is not required for the fermentation, is recovered by steeping the corn, followed by grinding with roller mills which flatten the germ so that it can be removed by screens. Oil is then extracted from the germ and the germ-free meal is used for fermentation. The corn meal is added to 60 per cent water and 40 per cent stillage (i.e. fermented liquor from which solvent has been distilled) to a calculated concentration of 8.5 per cent of original dried corn.

Since a butanol concentration of 13 g ℓ^{-1} in the final fermentation causes inhibition of the metabolism, to obtain complete utilization of the starch the initial starch concentration used is approximately 6–8 per cent of the dry corn equivalent. Some microbial species require the addition of ammonia to the medium. When molasses is used, the sugar concentration in the starting mash is 5.5–7.5 per cent and the mash is supplemented with superphosphate and ammonia. The pH is maintained at 5.6–6.0. Hot-still residue from the solvent stripping (i.e. slopback) is added to the mash make-up in all butanol fermentations the amount being 25–50 per cent of the total mash. Tables 5 and 6 list the microorganisms and substrates used in acetone/butanol fermentations together with the yields and relative concentrations attainable.

Process outline

Fig. 3 illustrates the process flow sheet for acetone/butanol fermentations. Table 7 gives the time course of the fermentation and Table 8 a material balance.

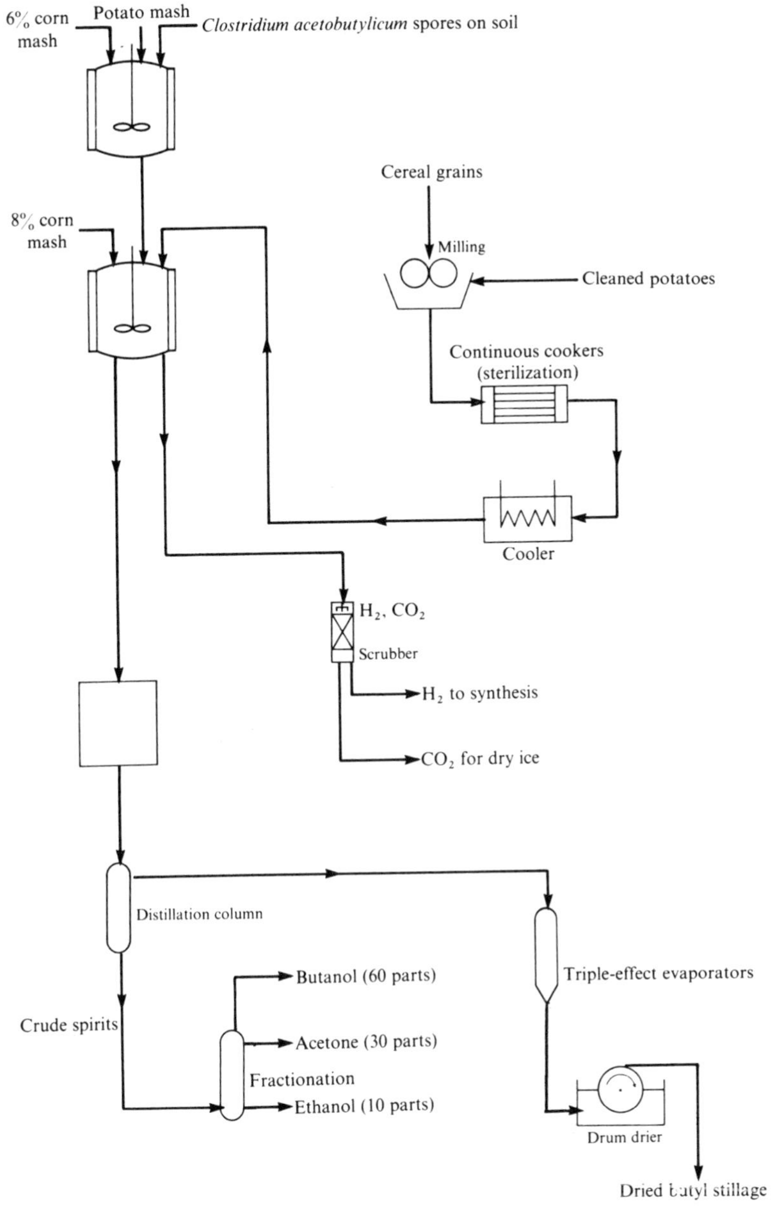

6% corn mash
Potato mash
Clostridium acetobutylicum spores on soil
Cereal grains
8% corn mash
Milling
Cleaned potatoes
Continuous cookers (sterilization)
Cooler
H_2, CO_2
Scrubber
H_2 to synthesis
CO_2 for dry ice
Distillation column
Triple-effect evaporators
Butanol (60 parts)
Crude spirits
Acetone (30 parts)
Fractionation
Ethanol (10 parts)
Drum drier
Dried butyl stillage

Figure 3. Flow diagram of a typical acetone/butanol fermentation using starch products [Beesch, (1953)].

Table 5. Acetone/butanol/ethanol fermentation of corn cobs, corn stalk and wood [Walton and Martin (1979)].

Culture	Substrate(s)	Time (h)	Yield (%)	Solvent composition (%)		
				Butanol	Acetone	Ethanol
Clostridium acetobutylicum S_{25}	Hydrolyzate of corn cobs (7%)		26.2			
	Hydrolyzate of sawdust (7%)		22.2			
C. acetobutylicum 314	Hydrolyzate of corn cobs (40–60%, 3.7% total), molasses (40–60% sugar)	48–72	40	67.5		
Butyl culture	1 part hydrolyzate of corn stalks, 3 parts molasses	50–55	31–37			
	Pentoses (13.5%)		25.4	67	33	
	Hydrolyzates of wood and plants (8%)		35.5	62	32	6

Table 6. Production of acetone/butanol in countries other than the United Kingdom and the United States [Walton and Martin (1979)].

Culture	Substrate	Time (h)	Yield (%)	Solvent composition (%)			
				Butanol	Acetone	Ethanol	*i*-Propanol
Clostridium acetobutylicum	Sugar solution (12%) in presence of active carbon (6%); carbon can be reused	36	66	33			
	Molasses plus rye or wheat meal	26.5	66	31	3		
	Addition of animal charcoal (1.0%) increases solvent yields						
C. acetobytylicum S_{25}	Millet (5%)		34.4				
	Corn (5%)		33.9				
	Sweet potatoes		33.7				
	Corn cobs		26.2				
	Molasses		24.9				
	Corn (6%), molasses (4%)		24.3				
	Sawdust (7%)		22.2				
C. acetobutylicum isolated from sugar-cane roots	Mash in 1 ℓ containing sugar cane juice (250 ml, 20° Brix), ground *Vicia sativa* (1 g), KH_2PO_4 (2.5 g), $CaCO_3$ (4 g)	32	73	19–23	3–4		
C. aurianticum	Glucose, fructose, sucrose and starch used with $(NH_4)_2SO_4$, $NaNO_3$, peptone or urea as nitrogen source		30–40	60.5	0.5	7.4	25.6
C. saccharoperbutylacetonicum	Blackstrap molasses (51.4 g sugar 55 g per ℓ of mash)		34.2	75.4	20.2	4.4	
C. saccharoperbutylacetonicum ATCC 13564	Sugar solution (4%), $(NH_4)_2SO_4$ (0.2%), calcium superphosphate (0.1%), calcium carbonate (0.3%)	60	28.5				

continued overleaf

Table 6 – (continued)

Culture	Substrate	Time (h)	Yield (%)	Solvent composition (%) Butanol	Acetone	Ethanol	*i*-Propanol
C. saccharoperbutylacetonicum N-1-41	Molasses with $(NH_4)_2\ SO_4$, $CaCO_3$ and Ca superphosphate	60		82.1	15	2.9	
C. saccharobutylacetonicum-liquefaciens	Molasses [sugar (5%)], protein nutrients; alkaline buffer added to mash	36–48	30	60	30	10	
Butyl culture	Calcium acetate and acetic acid increase fermentation yield	40					
	Rye flour with sterile air blown in for 2–3 min per hour, butanol output increased 3.4–9.1%						
Butyl culture	Starch (100%)		37.1				
	Starch (90%), beef molasses (10%)		36.7				
	Starch (50%), beet molasses (50%)		38.6				
	Starch (37%), beet molasses (63%)		37.9				
Culture 1 (100 000-ℓ tank)	Molasses [sugar (6.37%)] + rice bran (0.3%), $(NH_4)_2SO_4$(0.3%), $CaCO_3$ (0.4%). Temperature, 27–37.5°C	104	33.6				
SES-4	Molasses [sugar (5–6%)]		33	56.8	33.5	9.6	
SES-5	Temperature, 33°C		33	56.2	29.9	13.8	
SES-5 (100 000-ℓ tank)	Molasses [sugar (6.28%)] + rice bran (0.3%), $(NH_4)_2SO_4$ (0.3%), $CaCO_3$ (0.4%). Temperature 35–37°C	90	32.0				

Table 7. Acetone/butanol fermentation of blackstrap molasses[a] [Beesch (1952)].

Time (h)	pH	° Brix	Gas rate (Units)	Ammonium hydroxide, 28% added (ℓ)
1				265.0
2	5.55	8.7		
3				
4	5.45	8.6		
5				
6	5.45	8.6	2	
7			8	
8	5.40	8.5	10	
9			11	
10	5.25	8.0	12	113.6
11			13	
12	5.15	8.0	14	113.6
13			15	151.4
14	5.30	7.8	15	189.3
15			15	189.3
16	5.20	7.8	15	
17			16	
18	5.30	7.5	17	227.2
19			18	265.0
20	5.35	7.3	19	302.8
22	5.35	6.9	20	416.4
24	5.30	6.9	20	605.6
26	5.65	6.5	21	
28	5.65	6.5	21	
30	5.55	6.4	22	
32	5.55			
34	5.30	6.1		
36	5.35	5.6		
38	5.35	5.2		
40	5.35	4.6		
42	5.35	4.1		
44	5.40	3.4		
46	5.35	3.3		
48	5.40	3.0		

[a] Fermentation data: volume of mash, 757 m^3; sugar concentration, 6 per cent; culture (*Clostridium saccharoacetoperbutylicum*), 2 per cent; total ammonium hydroxide (28 per cent), 2763 ℓ; total mixed solids, 18.57 g $ℓ^{-1}$; acetone 20.93 per cent; *n*-butanol, 74.70 per cent; ethanol, 4.37 per cent; yield of total solids from sugar, 31.0 per cent; residual sugar, 4.5 g $ℓ^{-1}$; residual ammonia 11 ppm; fermentation temperature, 30.6°C.

Table 8. Material balance in acetone/butanol fermentation of blackstrap molasses[a] [Beesch (1952)].

Product	Yield (%)
Butanol	11.5
Acetone	4.9
Ethanol	0.5
Carbon dioxide	32.1
Hydrogen	0.8
Dry feed containing protein 2.72 kg, ash 2.72 kg	28.6

[a] Starting material: blackstrap molasses (45.36 kg) containing total solids (36.97 kg), sucrose and invert sugar (25.85 kg), protein (1.41 kg), ash (2.81 kg). Fermentation data: culture of saccharolytic bacteria used with ammonia (1 per cent) added in the form of ammonium hydroxide.

Product Recovery

The fermented liquor is sent to a beer well from which it is fed to a beer still at a constant flow. The still contains about 30 perforated plates that concentrate the solvents to 50 per cent solution in water. The mixed solvents are then separated by batch fractionation. The various crude fractions are further fractionated to remove traces of contaminating solvents. The butanol fraction is dried by taking off the distillate through a decanter.

The gas produced from a molasses culture yielding solvents with a butanol content of 70 per cent consists of 67 per cent carbon dioxide and 33 per cent hydrogen. About 4.7 cubic feet of gas is produced per pound of sucrose. Approximately 5.9 cubic feet of gas is produced per pound of dry corn, and the composition is about 40 per cent hydrogen and 60 per cent carbon dioxide, the carbon dioxide produced being solidified into 'dry ice'. The gases carry appreciable amounts of solvents that are removed by activated carbon or by a water-scrubbing system.

ANTIBIOTICS

General Aspects

Various antibiotics of clinical importance, the microorganisms that produce them and their antimicrobial spectrum are listed in Table 9. There are many similarities in the processes used for antibiotic production and the common features of research, development and manufacture are given in Table 10.

The media used favour both growth and the production of the antibiotic at high yields. They contain, in addition to the usual sources of carbon, nitrogen, minerals and buffers, precursors to increase the yield of the antibiotic, as in the case of penicillin.

The medium is sterilized either directly in the fermenter or by a batch or continuous method before its introduction into a previously sterilized fermenter. Sterilization is economically very important in the antibiotic industry; fewer than 2 per cent of the batches are lost as a result of contamination with antibiotic-resistant microorganisms or with phages.

Inoculum levels are large — as much as 10 per cent of the fermentation liquor. The fermenters which may be as large as 400 m^3, with a height-to-diameter ratio in the range of two to three, are constructed of stainless steel or nickel chrome alloys. The temperature and pH are controlled carefully and continuously. Adequate oxygen transfer is very important and foaming must be controlled using antifoams that are nontoxic to the microbes and that do not interfere with the recovery of the antibiotic. Certain ingredients may be added during the fermentation to increase the yields, e.g., sugar may be added continuously or in increments. Precursors, if used, are added in small amounts and at levels that are nontoxic to the microorganism.

The final concentration of an antibiotic is usually less than 1 g ℓ^{-1}, and a litre of medium may contain as much as 40 g of suspended solids. Fig. 4. contains a general flow sheet for the commercial recovery and purification of antibiotics.

An example of the evolution, over a 20-year period, of processes for the manufacture of antibiotics (in this case, bacitracin) is given in Table 11, which illustrates how technical changes have contributed to increased yields (i.e. a 40-fold increase). An industrial flow sheet for bacitracin manufacture is given in Fig. 5 (a).

The process for the production of penicillin is used in the following section (*see* pp.991–995) to indicate the factors of importance in the large-scale manufacture of antibiotics.

Table 9. Antibiotics of clinical importance [Porter (1976)].

Generic name	Type	Source	Antimicrobial spectrum; some important properties
Amphotericin B	Amphoteric heptaene	*Streptomyces nodosus*	Fungi
Bacitracin	Weakly basic polypeptide	*Bacillus subtilis*	Gram-positive bacteria
Cephalosporins	Derivatives of 7-aminocephalosporanic acid	*Cephalosporium* spp., *Emericellopsis* spp.	Generally penicillinase-resistant but inactivated by cephalosporinase
Cephalexin		Semisynthetic	Gram-positive bacteria, some gram-negative bacteria; orally absorbed
Cephaloglycin		Semisynthetic	Like cephalexin; orally absorbed
Cephaloridine		Semisynthetic	Like cephalexin; not orally absorbed
Cephalothin		Semisynthetic	Like cephalexin; not orally absorbed
Chloramphenicol	Aromatic structure with a nitro group, neutral	*Strept. venezuelae*, synthesis	Gram-positive bacteria, gram-negative bacteria, rickettsiae, large viruses, *Entamoeba*
Colistin	Basic polypeptide	*Aerobacillus colistinus*	Primarily gram-negative bacteria
Erythromycin	Basic macrolide	*Strept. erythreus*	Gram-positive bacteria, a few gram-negative bacteria, some protozoa
Gentamicin	Strong base; aminoglycoside	*Micromonospora* spp.	Gram-positive bacteria, gram-negative bacteria
Griseofulvin	A spirocyclohexenone benzofuranone, neutral	*Penicillium* spp., synthesis	Fungi
Kanamycin	Tetra-acidic base, aminoglycoside	*Strept. kanamyceticus*	Gram-positive bacteria, gram-negative bacteria, some protozoa
Lincomycin	Basic glycoside	*Strept. lincolnensis*	Gram-positive bacteria
Neomycin	Complex of B and C, which are isomeric, basic aminoglycosides	*Strept. fradiae*, other *Streptomyces* spp.	Gram-positive bacteria, gram-negative bacteria, mycobacteria
Novobiocin	Dibasic acid	*Streptomyces* spp.	Primarily gram-positive bacteria
Nystatin	Amphoteric tetraene	*Streptomyces* spp.	Fungi
Penicillins	Strong monobasic carboxylic acids, derivatives of 6-aminopenicillanic acid		
Ampicillin		Semisynthetic	Gram-positive bacteria, gram-negative bacteria; penicillinase-sensitive, acid-stable
Carbenicillin		Semisynthetic	Gram-positive bacteria, gram-negative bacteria including *Pseudomonas* spp.; penicillinase-sensitive, not orally absorbed

continued overleaf

Generic name	Type	Source	Antimicrobial spectrum; some important properties
Cloxacillin		Semisynthetic	Gram-positive bacteria; penicillinase-resistant, acid-stable
Dicloxacillin		Semisynthetic	Like cloxacillin
Methicillin		Semisynthetic	Gram-positive bacteria; penicillinase-resistant, acid-labile
Nafcillin		Semisynthetic	Like cloxacillin
Oxacillin		Semisynthetic	Like cloxacillin
Penicillin G		*Penicillium* spp., *Aspergillus* spp.	Gram-positive bacteria; penicillinase-sensitive, acid-labile
Penicillin V		Biosynthetic	Gram-positive bacteria; penicillinase-sensitive, acid-stable
Polymyxin	Basic polypeptide	*B. polymyxa*	Primarily gram-negative bacteria
Rifamycin	Complex of which derivatives of rifamycin B, containing an ansa-naphthohydroquinone group, are the most important	*Nocardia mediterranea*	Primarily gram-positive bacteria, mycobacteria
Rifampin		Semisynthetic	Gram-positive bacteria, gram-negative bacteria, mycobacteria, some viruses; orally absorbed
Spectinomycin	Basic aminocyclitol	*Streptomyces* spp.	Gram-positive bacteria, gram-negative bacteria, specifically *Neisseria gonorrhoeae*
Streptomycin	Strong base, aminoglycoside	*Streptomyces* spp.	Gram-positive bacteria, gram-negative bacteria, *Mycobacterium tuberculosis*
Tetracyclines	Amphoteric		Gram-positive bacteria, gram-negative bacteria, rickettsiae, large viruses, coccidia, amoebae and balanthidia, mycoplasms
Chlortetracycline		*Strept. aureofaciens*	
Demeclocycline		*Strept. aureofaciens* mutant	
Doxycycline		Semisynthetic	
Methacycline		Semisynthetic	
Minocycline		Semisynthetic	Also inhibits staphylococci resistant to other tetracyclines
Oxytetracycline		*Strept. rimosus*	
Tetracycline		Catalytic hydrogenation of chlortetracycline, *Strept. aureofaciens* mutant	
Tyrothricin	Mixture of polypeptides	*B. brevis*	Gram-positive bacteria
Vancomycin	Amphoteric glycopeptide	*Strept. orientalis*	Gram-positive bacteria

Table 10. Common features of antibiotic manufacture [Prescott and Dunn (1959)].

- Isolation or collection of cultures
- Screening of cultures to detect those with antimicrobial activity
- Development of methods for submerged-culture production
  - Development of media
  - pH adjustment and control
  - Optimum temperature
  - Optimum aeration conditions
  - Control of foam formation
  - Optimum time or length of fermentation
  - Other factors
- Development of methods for isolation and purification
  - Determination of properties of antibiotic
    - Physical
      - Adsorption and absorption
    - Chemical
      - Reactions
      - Solubility in solvents
      - Stability in acids, alkalis, heat, etc.
- Evaluation of antibiotic
  - Pharmacological tests:
    - Toxicity in animals (e.g., mice, rats, rabbits, guinea pigs, dogs, monkeys)
    - Therapeutic effect on experimental infections
  - Antimicrobial activity
    - Bacteriostatic action
    - Bactericidal action
    - Fungistatic action
    - Fungicidal action
    - Antiviral action
    - Other actions
  - Comparison with existing antibiotics
- Development of pilot-plant production methods
- Preparation of pure product for clinical trials
  - Determination of resistance
  - Determination of synergistic action
- Testing of purified antibiotic
  - Pharmacological
    - Potency
    - Sterility
    - Pyrogenicity
    - Toxicity and cross-reaction
    - Histamine response
  - Physical and chemical
    - Moisture
    - pH
    - Heat stability
    - Crystallinity
- Development of plant-scale production methods
- Other miscellaneous considerations
  - Development of methods to control production of antibiotic
  - Determination of chemical structure of antibiotic
  - Determination of mode of action of antibiotic
  - Development of new applications
  - Development of marketing and distribution system
  - Financing of business

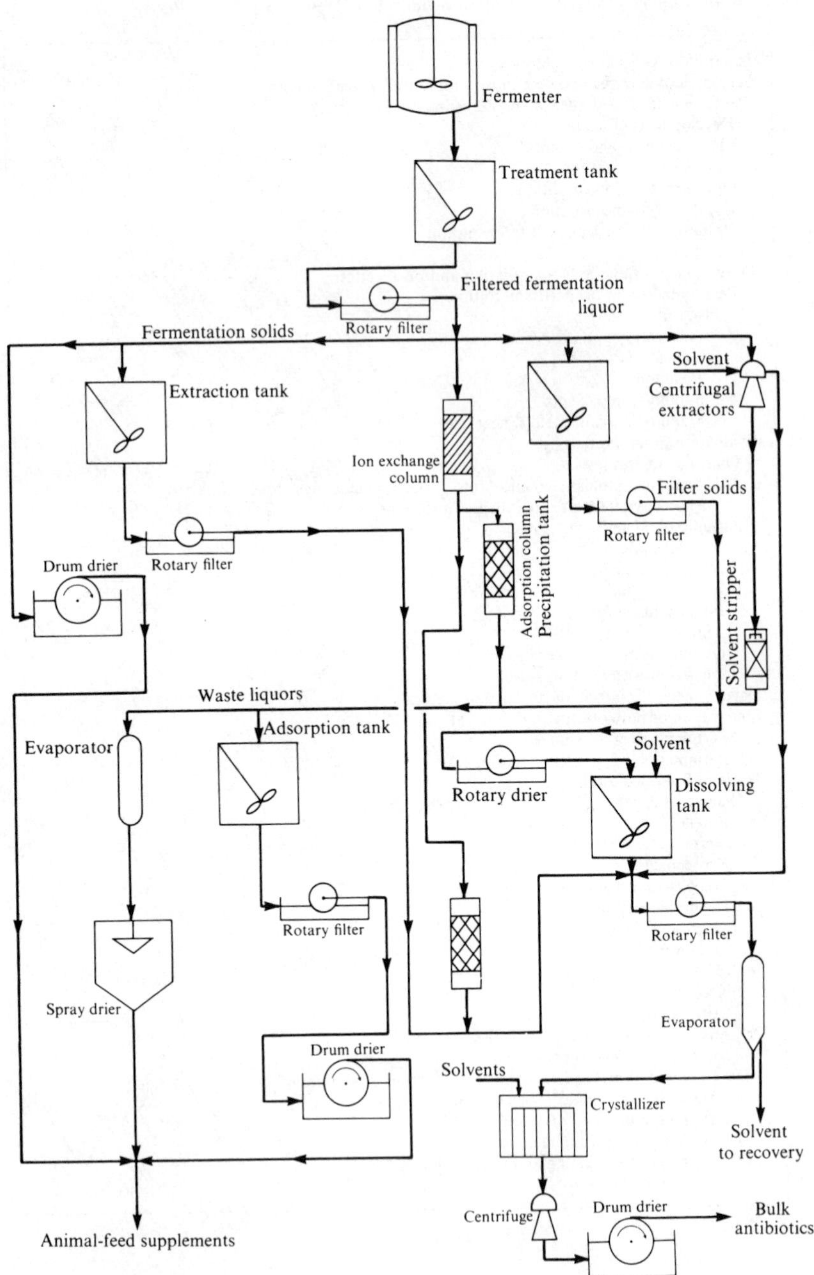

Figure 4. Generalized flow sheet for the commercial recovery and purification of antibiotics [Courtesy of Charles Pfizer and Co., Inc.] [Casida, (1968)].

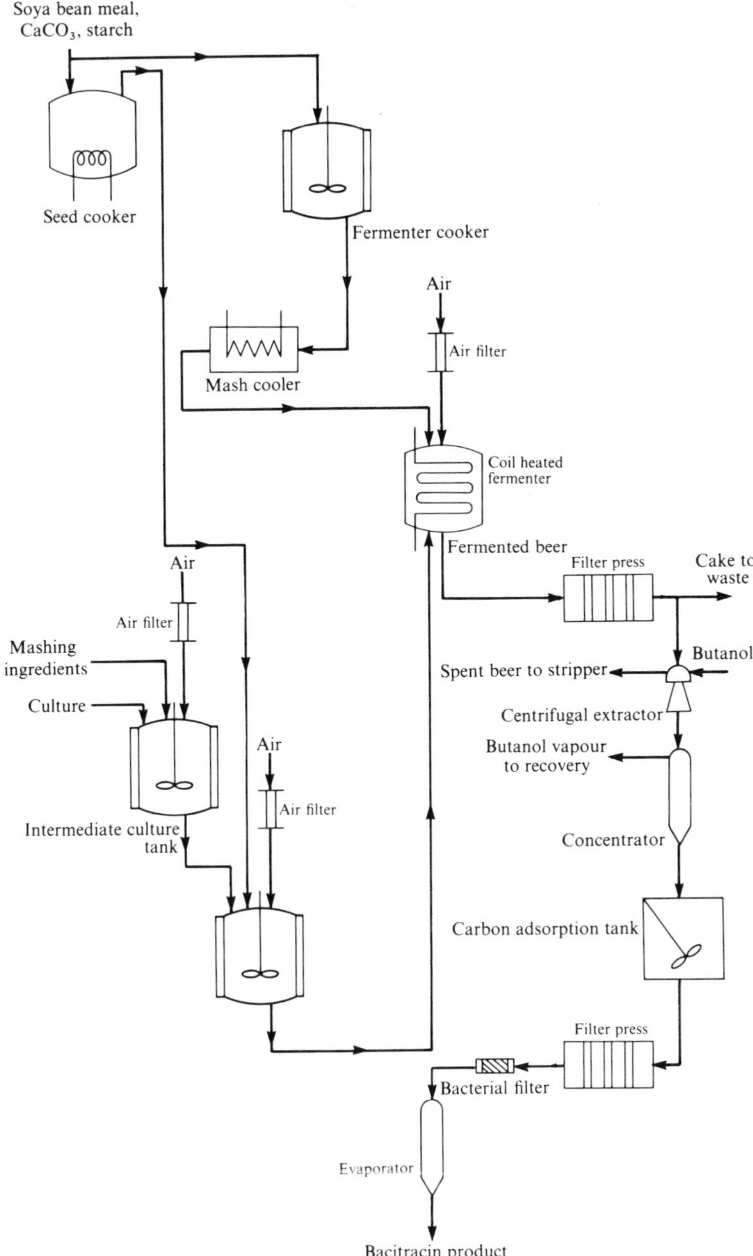

Figure 5(a). Bacitracin flow sheet.

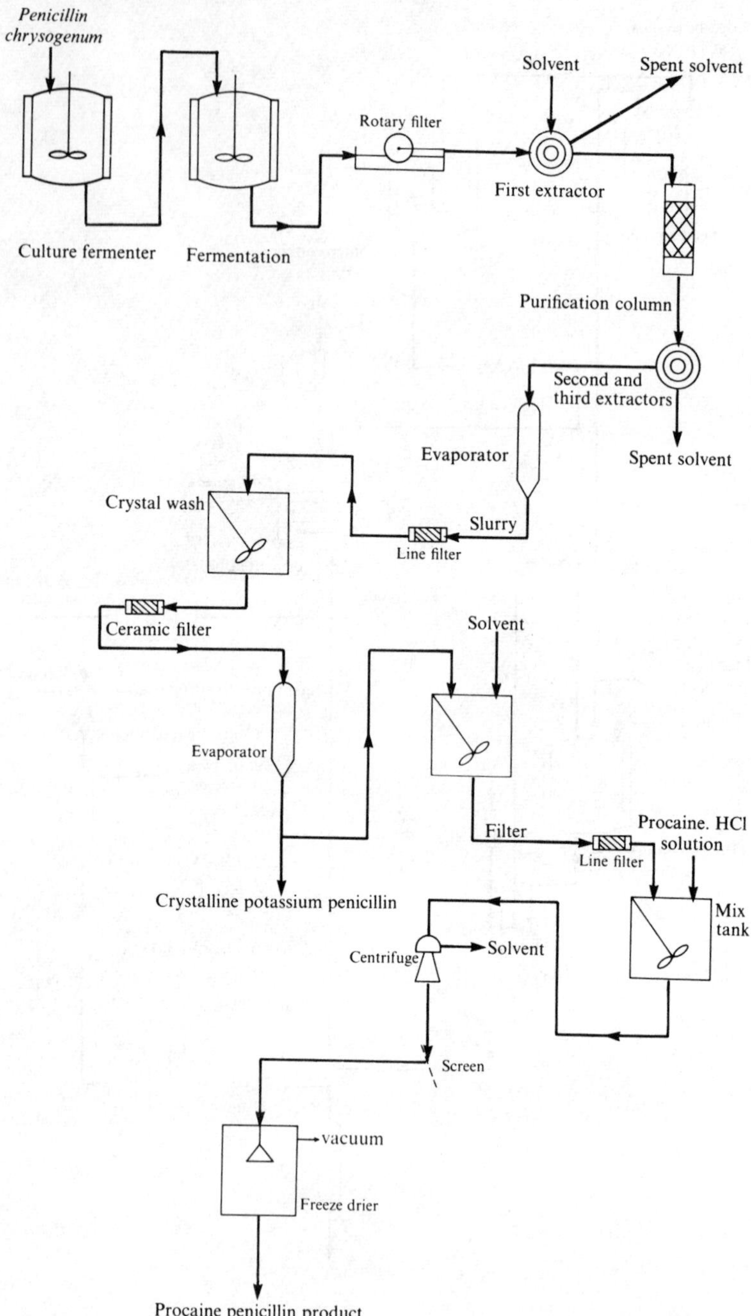

Figure 5(b). Flow diagram for penicillin production.

Table 11. Characteristics of processes used in the production of bacitracin [Perlman (1979)].

Medium composition	Aeration method	Incubation period (h)	Yield (U ml^{-1})
Tryptone, meat infusion, other broths	Surface culture	72–120	10–26
Soya bean meal, starch, calcium lactate	Surface culture	168	88
Soya bean meal, calcium lactate, calcium carbonate starch	Submerged culture	26	80–90
Cotton-seed meal, soya bean meal, calcium carbonate, dextrin	Submerged culture	24	125
Soya bean meal, starch, calcium carbonate	Submerged culture	24	325
Soya bean grits, sucrose	Submerged culture	30	400

Penicillin

Raw materials

Penicillium chrysogenum strains are used for the production of penicillins G and V. These highly developed mutant strains can also produce other penicillins provided that the appropriate carboxylic acid is added to the medium in sufficient concentration. The improvement and selection of high-yielding strains are crucial in the case of a highly competitive product. Table 12 contains a list of mutagens used for the improvement of the penicillin yields of *P. chrysogenum*.

A typical vegetative or inoculum medium contains an organic nitrogen source, such as corn-steep liquor, a sufficient concentration of a fermentable carbohydrate, such as 2 per cent (w/v) sucrose or glucose. Calcium carbonate at a concentration of 0.5–1.0 per cent (w/v) acts as a buffer, while other inorganic salts may be required for growth. If lactose is the carbon source in the fermentation stage, it is often included with glucose or sucrose in the final stage of inoculum development, and this stage is allowed to proceed until the readily used sugar is exhausted and formation of β-galactosidase is induced. A typical inoculum contains 20 g dry weight of cells per litre, at a level of 10 per cent (by vol.) of the fermentation liquor.

P. chrysogenum can utilize a variety of carbon and energy sources. Among those suitable for penicillin production are glucose, lactose, fructose, sucrose, maltose, raffinose, melibiose, inulin, invert sugar, dextrins and starches. Molasses, ethanol and fatty oils, such as soya bean oil and corn oil, can also be used. Typical concentrations are 3–6 per cent for lactose, 3–5 per cent for corn-steep liquor, 1 per cent for glucose and 1.5 per cent for starch. Corn-steep liquor, cotton-seed meal and wheat or rice bran are used as nitrogen sources. When ammonium sulphate is used as the nitrogen source, it is necessary to maintain a concentration of ammonia of 250–340 mg ℓ^{-1} for the continued synthesis of penicillin. Ammonium sulphate also provides sulphur for microbial growth. Phenethylamine, phenylacetic acid and phenoxyacetic acid are used as precursors. The theoretical requirement for sodium phenylacetate is 0.47 g per gram of penicillin G acid whereas for sodium phenoxyacetate it is 0.5 g per gram of penicillin V acid. These compounds, particularly sodium phenylacetate, are toxic to the microorganism. Phytic acid, which is contained in plant-derived nitrogen sources, is an important source of phosphorus for the fermentation. Phosphorus is also added in the form of inorganic salts. Potassium, magnesium, manganese, zinc, cobalt and copper are also found in media used for penicillin synthesis. Two examples of media used for penicillin are contained in Table 13.

Table 12. Mutagens used in the improvement of penicillin yields by *Penicillium chrysogenum* [Queener and Swartz (1979)].

Diepoxybutane
Ethyl methanesulphonate
Ethylenimine
Methyl-*bis*-(β-chloroethyl)amine
N-Methyl-*N'*-nitro-*N*-nitrosoguanidine
Nitrous acid
Ultraviolet light irradiation at 275 and 253 nm
X-irradiation
γ-irradiation

Table 13. Media used for penicillin production [Perlman (1967)].

Component	Corn-steep liquor medium (%)	Calam and Hockenhull's medium (%)
Main carbohydrate		
Lactose	3.0–4.0	3.0
Other carbohydrates		
Glucose	0.0–0.5	1.0
Starch		1.5
Nonreducing polysaccharides	Variable	
Specific precursors		
Phenethylamine[a] and other precursors	Variable	
Phenylacetic acid		0.05
Organic acids		
Acetic acid	~0.05	0.25
Citric acid		1.0
Lactic acid	~0.5	
Main nitrogen source		
Amino acids, peptides amines	Variable	
Ammonium sulphate		0.5
Other nitrogen sources		
Ammonia	Variable	
Ethylamine		0.3
Total nitrogen	0.15–0.2	0.2
Total solids	8.0–9.0	8.5

[a] Phenethylamine is often increased by adding pure phenethylamine or phenylacetic acid supplements.

Process outline

Fig. 5(b) shows the flow sheet of a typical process for penicillin production. Spores of a *P. chrysogenum* strain are used to inoculate 100 ml of the medium in a 500-ml flask. Incubation takes place on a rotary shaker at 250 rpm, 2-in. displacement, at 25°C. After four days' incubation, the contents are transferred to medium (2 ℓ). This second culture is incubated for two days and the contents are transferred to medium (500 ℓ) in an 800-ℓ stainless steel tank. After incubation for three days, this culture is used to inoculate a 180 000-ℓ fermenter. The final fermentation is completed in five to six days.

Generally, the pH of fermentation is around 6.5 and the temperature is in the range 23–28°C. The culture volume is typically 40 000–200 000 ℓ. The process is aerobic, having

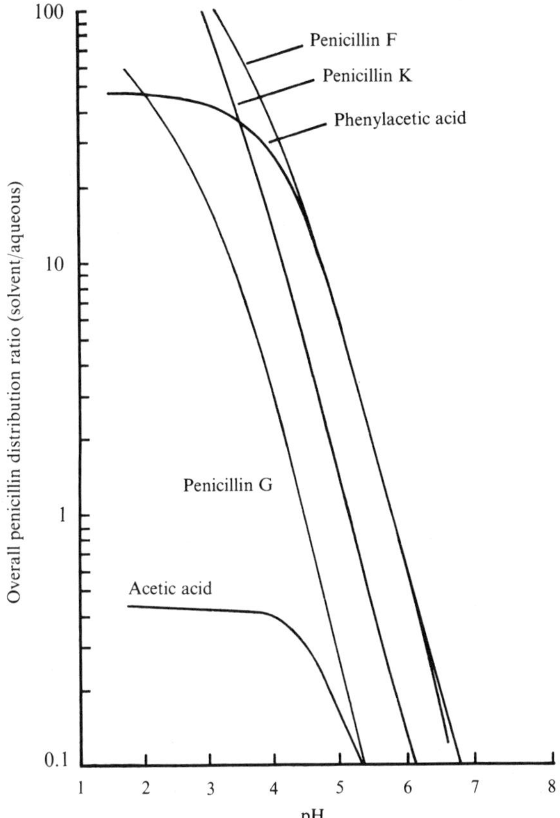

Figure 6. Partition of penicillins as a function of the pH of the extraction solvent [Queener and Swartz (1979)].

a volumetric oxygen uptake rate in the range 0.4–0.8 mmol ℓ^{-1} min^{-1} through the culture at a rate of 0.5–1.0 vvm. The broth is vigorously agitated using turbine agitators of various designs. Power input is generally of the order of 1–4 W ℓ^{-1}, including that introduced by the air stream.

A theoretical maximum yield for penicillin has been estimated to be 0.12 g penicillin per gram of carbohydrate (i.e. glucose). In practice, the final penicillin yield is 3–5 per cent based on the carbohydrate consumed.

Product recovery

The yield of penicillin is often expressed in terms of units, where a unit is equal to 0.5888 μg of sodium benzylpenicillin. A penicillin isolation and purification process is outlined in Table 14. The broth contains about 20–35 g of penicillin G or V per litre. The mycelia are separated from the liquid by filtration, occasionally using filter aids or precoats. Then, the penicillin-rich filtrate is cooled to 0–4°C to minimize chemical and enzyme degradation during solvent extraction. Sometimes, the filtrate is further clarified

Table 14. Outline of a penicillin purification process [Queener and Swartz (1979)].

Step	Purpose	Equipment	Basis
Filtration	Separation of mycelia from penicillin-containing broth	Continuous, rotary vacuum drum filter	Size
Extraction	Separation of penicillin from other soluble components	Continuous, multistage countercurrent extractors	Differential extraction: when $pH > pK_a$, penicillin more soluble in organic phase; when $pH < pK_a$, penicillin more soluble in aqueous phase
Crystallization	Further purification and stabilization	Tanks and gravitational separators	Via addition of Na^+ or K^+
Drying	Stabilization	Vacuum or warm-air driers	Anhydrous solvent

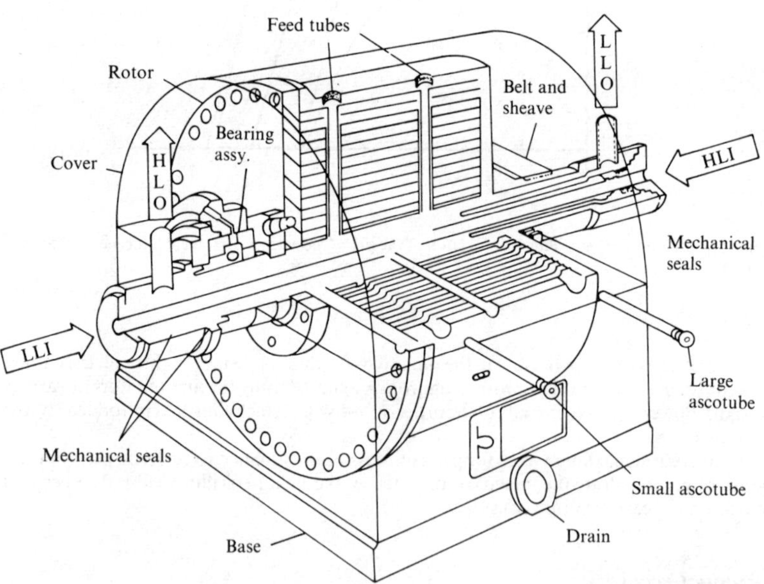

Figure 7. Diagram of the Podbielniak extractor. The cylindrical drum contains perforated, concentric shells and is rapidly rotated on the horizontal shaft (2000–5000 rpm). Liquids enter through the shaft; heavy liquid is led to the centre of the drum, light liquid to the periphery. The heavy liquid flows radially outward, displacing the light liquid inwardly and both are led out through the shaft. These extractors are especially useful for liquids of very small density difference where very short residence times are essential. HLI, LLI, HLO, LLO indicate heavy and light liquid in and out, respectively [Queener and Swartz (1979)].

by a second filtration with 1–1.5 per cent Hyflo, sometimes with the precipitation of proteinaceous material by addition of aluminium sulphate or tannic acid.

In the next stage, penicillin G or V is extracted into amyl acetate or butyl acetate by a continuous countercurrent process. Penicillin G and V are strong acids with pK_a values in the range 2.5–3.1 and molecular weights of 334 and 350 daltons, respectively. As the acid forms are soluble in several organic solvents, they are extracted with high efficiency into amyl acetate or butyl acetate at pH 2.5–3.0. Cyclic ketones can also be used as solvents at pH 4.0. Fig. 6 shows the basis for the extractive separation, using penicillins F, K and G as examples.

The extraction is carried out in continuous, countercurrent multistage centrifugal extractors (*see* Fig. 7). Efficient extraction uses a solvent-to-broth ratio of 0.1. The penicillin-containing solvent is then treated with 0.25–0.5 per cent carbon to remove pigments and other impurities, and then penicillin is crystallized by addition of potassium or sodium acetate. Penicillin can be back-extracted into water by addition of alkali (potassium or sodium hydroxide) or buffer at pH 5.0–7.5. The volume ratio of water-to-solvent is 0.1–0.2 in a continuous, multistage extractor. Penicillin can be precipitated from the resulting aqueous phase. A three-stage extraction process is illustrated in Fig. 8. Penicillin crystals are washed and predried with anhydrous *i*-propanol, *n*-butanol or another volatile solvent. Final drying is accomplished using vacuum, warm air or radiant heat on large horizontal belt filters.

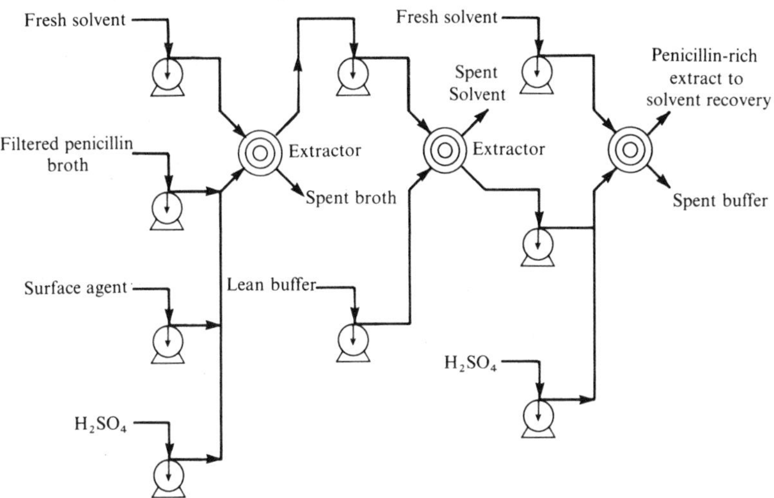

Figure 8. A three-stage extraction process used in penicillin production [Queener and Swartz (1979)].

CAROTENOIDS

Raw Materials

Carotenoids are one of the most important groups of natural pigments, being yellow to red in colour. Some carotenoids possess essential biological functions, in particular, as precursors of vitamin A. Although carotenoids are widely found in plants and animals, only microorganisms and plants have the necessary biochemical systems to synthesize them *de novo*. Different types of microorganisms produce different carotenoids, e.g., β-carotene and related carotenes are mainly found in fungi and algae, while xanthophylls are synthesized almost exclusively by bacteria and algae. The rates of carotenoid production vary considerably for different microorganisms.

β-Carotene is produced by species of the genera *Choanephora* and *Blakeslea* of the family Choanephoreaceae (class Phycomycetes) and other fungi, as well as by yeasts such as *Rhodotorula flava*, *R. gracilis* and *R. sannieli*. Industrially, *B. trispora* is the preferred species. The use of the two sexual forms [i.e. the (+) and (−) strains] of this species leads to an increase in the mycelial carotenoid content.

The culture medium used is viscous, being rich in vegetable oils, kerosene and surface-active agents, with addition of β-ionone during incubation and an antioxidant to prevent oxidation of the β-carotene. The medium normally contains 1–8 per cent vegetable or animal oils, which can be partially or even totally replaced by cheaper oils. Examples of alternative oils include linoleic and linolenic acid-rich oils (i.e. autoclave oils and soap-stock) which are byproducts of oil refining. As β-ionone is not incorporated into β-carotene, it can be partially or totally replaced by the bicyclic terpenes of turpentine and other natural terpenes. Other substitutes that are only partically efficient are isoprene dimers and trimers, cyclohexane, cyclohexanone and their trimethyl derivatives. Cheap

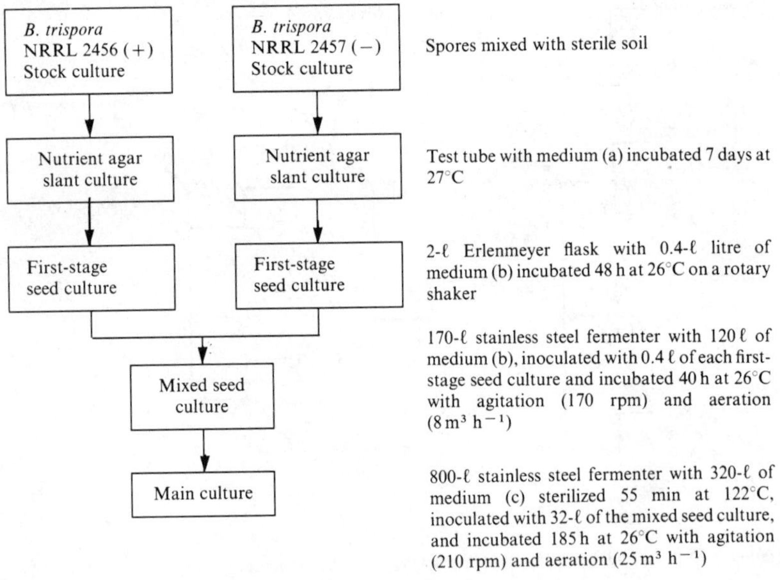

Medium (a): medium M_5 (from Hesseltine and Benjamin, 1959.)

Medium (b): Corn steep 7 g; corn starch, 50 g; potassium dihydrogen phosphate, 0.5 g; manganese sulphate, 0.1 g; thiamin hydrochloride, 0.01 g; with tap water to 1 ℓ.

Medium (c): Distillers' solubles, 70 g; corn starch, 60 g; soya bean oil, 30 ml; cotton-seed oil 30 ml; ethoxyquin, 0.35 g; manganese sulphate, 0.2 g; thiamin hydrochloride, 0.5 mg; isoniazid, 0.6 gm; kerosene 20 ml; with tap water to 1 ℓ and pH adjusted to 6.3. The last two substances are each sterilized separately by filtration. After 48 h growth 1 g ℓ^{-1} of β-ionone and 5 ml per litre of kerosene are added aseptically; then a total amount of 42 ml per litre of glucose, as an aqueous 53 per cent solution, is continuously injected up to the end of the culture.

Figure 9. A high-yielding β-carotene process using *Blakeslea trispora* [Ninet and Renaut (1979)].

citrus fruit peels can also be added to the medium instead of β-ionone, because of their citric acid content. Isoniazid, iproniazid and 4-formylpyridine at concentrations of 1, 1 and 2 g ℓ$^{-1}$, respectively, are added as activators and show an enhancing effect when used in combination with β-ionone or 2,6,6-trimethyl-1-acetylcyclohexene, giving a yield of β-carotene of 3 g ℓ$^{-1}$ or 30 mg per gram of dried mycelium, compared with the usual yield of about 1 g ℓ$^{-1}$. The exact role of the activators is not known.

The (−) sexual form of *B. trispora*, in the absence of the (+) form, is capable of synthesizing β-carotene in significant quantities if the hormonal substances – trisporic acids – produced in cultures of the mixed forms are added to the medium. However, trisporic acids are produced in a mixed culture of *B. trispora* when cultured in a medium containing distillers' solubles, starch, cotton-seed oil and kerosene with a yield of 3.5 g ℓ$^{-1}$ and can be extracted from the broth and added to a culture of the (−) strain in the same medium, supplemented with β-ionone (0.5 g ℓ$^{-1}$). This culture then yields β-carotene at a concentration of 1.2 g ℓ$^{-1}$.

Process Outline

The culture of *B. trispora* for the production of β-carotene is summarized in Fig. 9.

Product Recovery

β-Carotene is an endocellular product. For animal feedstuffs, the mycelia can be used

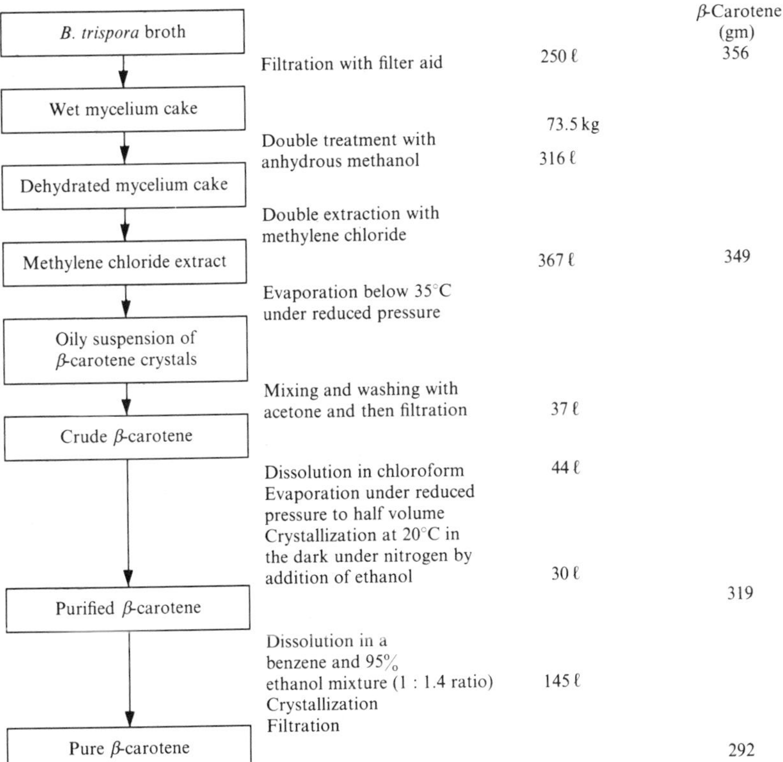

Figure 10. Recovery and purification of β-carotene from *Blakeslea trispora* [Ninet and Renaut (1979)].

directly after drying. However, the stability of β-carotene in biomass is poor and antioxidants, e.g., ethoxyquin at a concentration of $0.5\,g\,\ell^{-1}$, are added to the broth or to the recovered biomass. Fig. 10 shows the stages in the recovery and purification of β-carotene. The carotenoids can also be extracted from the mycelium using refined oils and are then steam stripped to remove odours and volatile impurities.

ENZYMES

General Aspects

Raw materials

Although enzymes are complex chemicals produced commercially in large quantities, the processes used for their production are relatively simple. A microorganism is cultured in a suitable medium with the concomitant production of enzyme and its subsequent recovery. The problems lie in the details and the product recovery stages of each process.

Table 15. Production of industrial enzymes [Aunstrup *et al.* (1979)].

Enzyme	Amounts produced per year (ton of pure enzyme protein)	Relative sales value (%)
Bacillus amylase	300	12
Fungal amylase	10	3
Glucoamylase	300	14
Glucose isomerase	50	12
Pectinases	10	10
Bacillus protease	500	40
Fungal protease	10	1
Microbial rennet	10	7
Other enzymes		1

Table 15 lists some of the industrial enzymes, mostly produced by microorganisms belonging to the two genera *Bacillus* and *Aspergillus*. *Bacillus* species are well suited for enzyme production. Most, except for *B. cereus*, are saprophytes, which do not produce toxins, and are grown relatively easily without the need for expensive growth factors. *Aspergillus* species are not pathogenic, apart from *Asp. fumigatus*, and are not producers of toxin, with the exception of *Asp. parasiticus*.

Enzyme-producing microorganisms can be grown on either semisolid or liquid media. Semisolid substrate, which is usually moist wheat bran with various additives, is used for the production of the enzymes listed in Table 16. The medium is prepared by mixing bran

Table 16. Enzymes produced using semisolid media.

Enzyme	Microorganism
α-Amylase	*Aspergillus oryzae*
Glucoamylase	*Rhizopus* sp.
Lactase	*Asp. oryzae*
Pectinase	*Asp. niger*
Protease	*Asp. niger, Asp. oryzae*
Rennet	*Mucor pusillus*

with water and additives, followed by steam sterilization of the mixture in an autoclave while stirring. The sterilized medium is aseptically transferred on to trays to a thickness of 1–10 cm or into rotating drums. Inoculation takes place either in the autoclave after cooling or in the cultivation equipment. Aeration is achieved by blowing humidified air over the culture.

The media used for submerged cultivation are inexpensive and support good growth. The raw materials are usually proteinaceous feed materials combined with starch sources, such as grain and corn or carbohydrates, such as sucrose, lactose and starch hydrolyzates. Phosphates, magnesium salts and ammonium salts (or nitrates) are supplemented. Some typical media used for the production of enzymes in submerged culture are given in Table 17.

Table 17. Typical fermentation media.

Enzyme	Media[a]
Submerged culture	
Bacillus amylase	Corn starch, 40; ground corn (hominy), 100; corn-steep liquor, 65
	Potato starch, 100; ground barley, 50; soya bean meal, 20; sodium caseinate, 10; $Na_2HPO_4.12H_2O$, 9
Fungal amylase	Corn starch, 24; corn-steep liquor, 36; Na_2HPO_4, 47; $CaCl_2$, 1; KCl, 0.2; $MgCl_2.6H_2O$, 0.2
Glucoamylase	Corn starch, 150; corn-steep liquor, 20; pH adjusted with gaseous ammonia
Bacillus protease	Starch hydrolyzate, 50; soya bean meal, 20; casein, 20; Na_2HPO_4, 3.3
	Starch hydrolyzate, 150; lactose, 4.3; cotton-seed meal, 30; brewers' yeast, 7.2; soya protein, 3.65; K_2HPO_4, 4.3; $MgSO_4.H_2O$, 1.25; trace metals
	Ground barley, 100; soya bean meal, 30; Na_2CO_3 (to adjust pH to 9–10)
Fungal protease	Corn starch, 30; corn-steep liquor, 5; soya bean meal, 10; casein, 12; gelatin, 5; distillers' dried solubles, 5; KH_2PO_4, 2.4; $NaNO_3$, 1; NH_4Cl, 1; $FeSO_4$, 0.01;
Semisolid culture	
Lactase	Wheat bran, 100; 0.2 N HCl (contains traces of Zn, Fe, Cu), 60
Lipase	Wheat bran, 3; soya bean meal, 1; water, 3

[a] Submerged-culture compositions are expressed as $g\,\ell^{-1}$, while semisolid culture compositions are expressed in parts.

Many enzyme fermentations are carried out near neutral pH, the pH being controlled by a buffer system, such as phosphates or calcium carbonate. Another method is to add substances the metabolism of which results in a change in pH in the desired direction, e.g., salts of organic acids and nitrates raise the pH while ammonium salts tend to lower the pH. All enzyme fermentations are aerobic. In some cases, e.g., amyloglucosidase production, oxygen-limiting conditions are used.

Glucose represses the formation of some enzymes, such as α-amylase, and to ensure very low glucose concentrations slowly decomposable carbohydrates, such as starch or lactose, are used. Alternatively, glucose is added gradually during the fermentation. Some enzymes, such as pectinase and lactase, require inducers for synthesis at high yields. Otherwise, mutants are developed.

In the formation of the medium, the enzyme recovery steps are taken into consideration and, at the end of fermentation, the aim is that the suspended solids and viscosity are low, as well as the free carbohydrate and amino acids concentrations. The concentrations at the end of fermentation depend on the initial medium composition. A typical concentration for enzyme protein is 1–5 per cent of the initial dry matter, for residual nutrients and metabolites 5–10 per cent and for cell mass 2–10 per cent.

Process outline

Important factors in processes for the production of enzymes are:

1. Microbial strain or plant or animal tissue source.
2. Growth rate and stage of the growth cycle at which the enzyme is produced.
3. Medium formulation.
4. Fermentation conditions, i.e. temperature, pH, aeration and foam control.
4. Regulatory mechanisms, i.e. induction, feedback repression, catabolic repression and gene dosage.

Fig. 11 provides a summary of the range of sources of commercial enzyme preparations, together with appropriate recovery and purification procedures for required sales products. Fig. 12 illustrates a process for the preparation of intracellular enzymes (*see* Table 18 and Chapter 6 for intracellular enzymes produced commercially).

Table 18. Some intracellular microbial enzymes of industrial importance.

Name	EC number	Source
L-Asparaginase	3.5.1.1	*Erwinia carotovora* *Escherichia coli*
Catalase	1.11.16	*Aspergillus niger*
Cholesterol oxidase	1.1.3.6	*Nocardia rhodochrous*
β-Galactosidase	3.2.1.23	*Kluyveromyces fragilis* *Saccharomyces lactis*
Glucose isomerase	5.3.1.5	*Bacillus coagulans* *Streptomyces* sp.
Glucose oxidase	1.1.3.4	*Asp. niger* *Penicillium notatum*
Glucose-6-phosphate dehydrogenase	1.1.1.49	Yeast
Invertase	3.2.1.26	*Sacch. cerevisiae*
Penicillin acylase	3.5.1.11	*E. coli*

General product recovery

Figs. 13 and 14 contain flow sheets of two processes for enzyme recovery and purification.

At the end of the fermentation, the broth is rapidly cooled to about 5°C to reduce deterioration. The refrigerated broth is filtered or centrifuged to remove the microorganisms after adjusting the pH and treatment with coagulating or flocculating agents to remove the colloidal particles. Inorganic salts, e.g., calcium phosphate precipitated from soluble calcium and phosphate salts in the broth, can be used to precipitate cells and colloids, or polyelectrolytes are used. Diatomaceous earth (2–4 per cent) may be added to the broth as a body feed before filtration. Suspended solids are removed by vacuum drum filtration or by a disc-type centrifuge with a self-cleaning bowl.

Fig. 11 shows the subsequent purification steps leading to commercial-grade enzymes. To obtain a higher degree of purity, the enzyme is precipitated with acetone, alcohols or inorganic salts, such as ammonium or sodium sulphate. Fractional precipitation gives higher purities than one-step precipitation. In large-scale operations, salts are preferred to solvents because of explosion hazards.

Intracellular enzymes are used whenever possible without extracting the enzyme. If the enzyme must be extracted, the cells are disrupted by a homogenizer or bead mill, with the same purification methods then being used as for extracellular enzymes.

When enzymes are sold as dusty powders with particle sizes of less than 10 μm, there is the danger of exposure to enzyme dust, and several granulation processes have been described. A common method is to embed the enzyme in waxy spheres, consisting of a

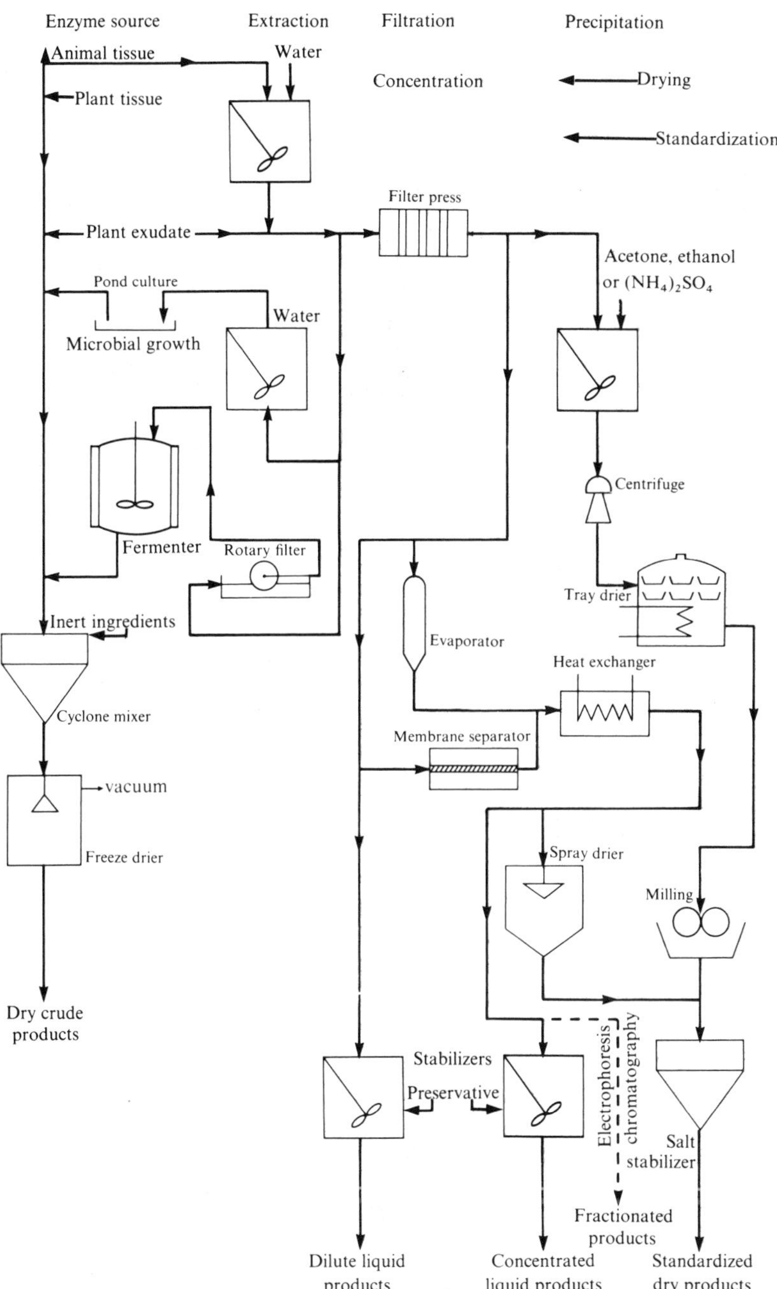

Figure 11. Summary of processing operations used in manufacture of commercial enzymes [Faith *et al.*, (1971)].

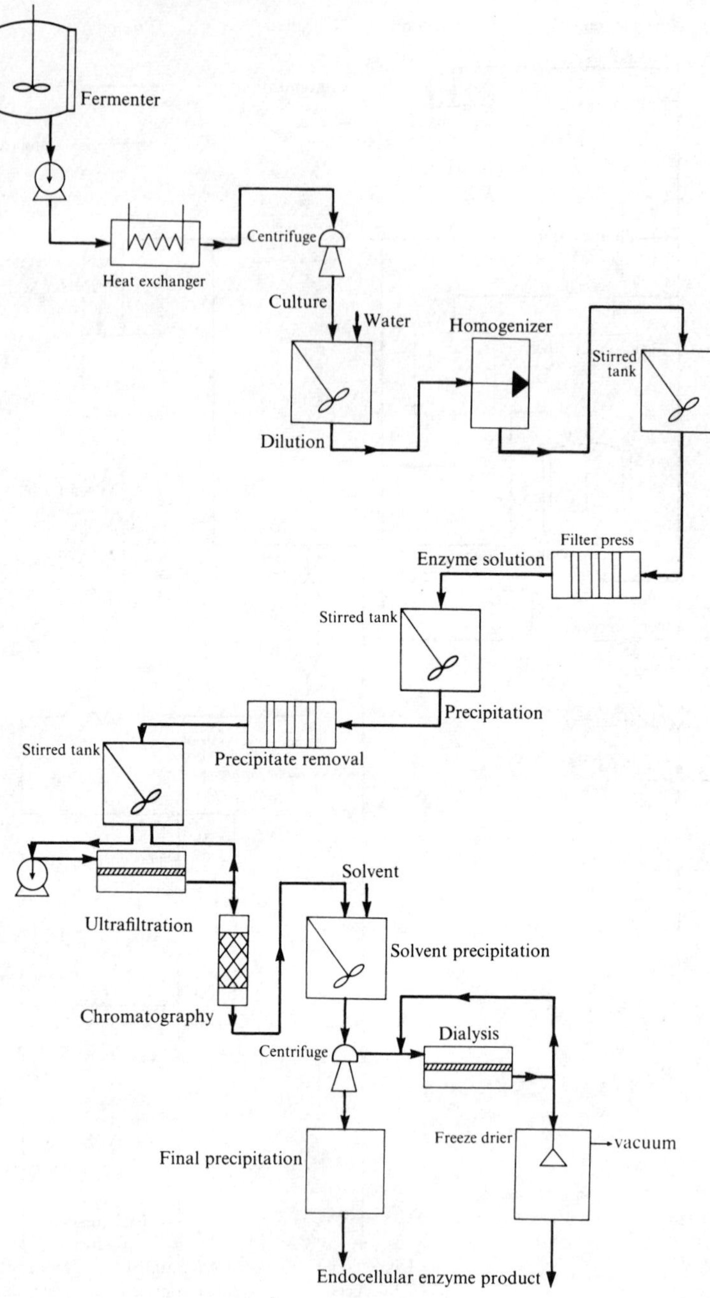

Figure 12. Production of purified endocellular microbial enzymes [Malby (1970)].

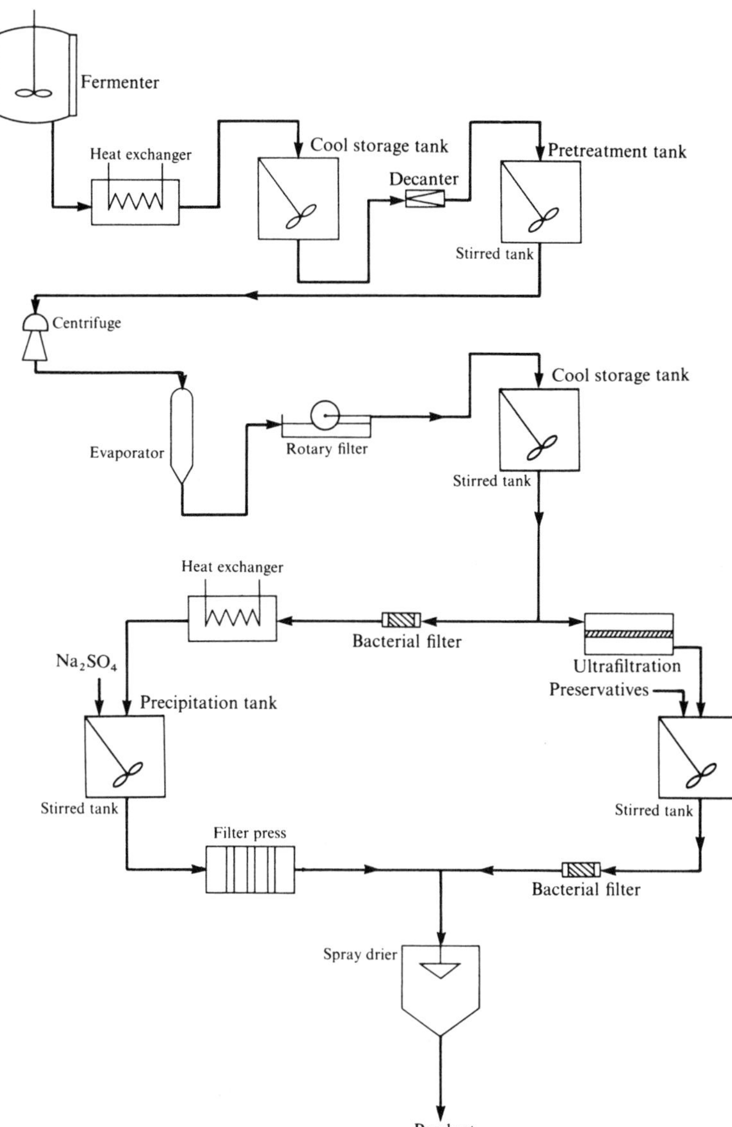

Figure 13. Enzyme recovery.

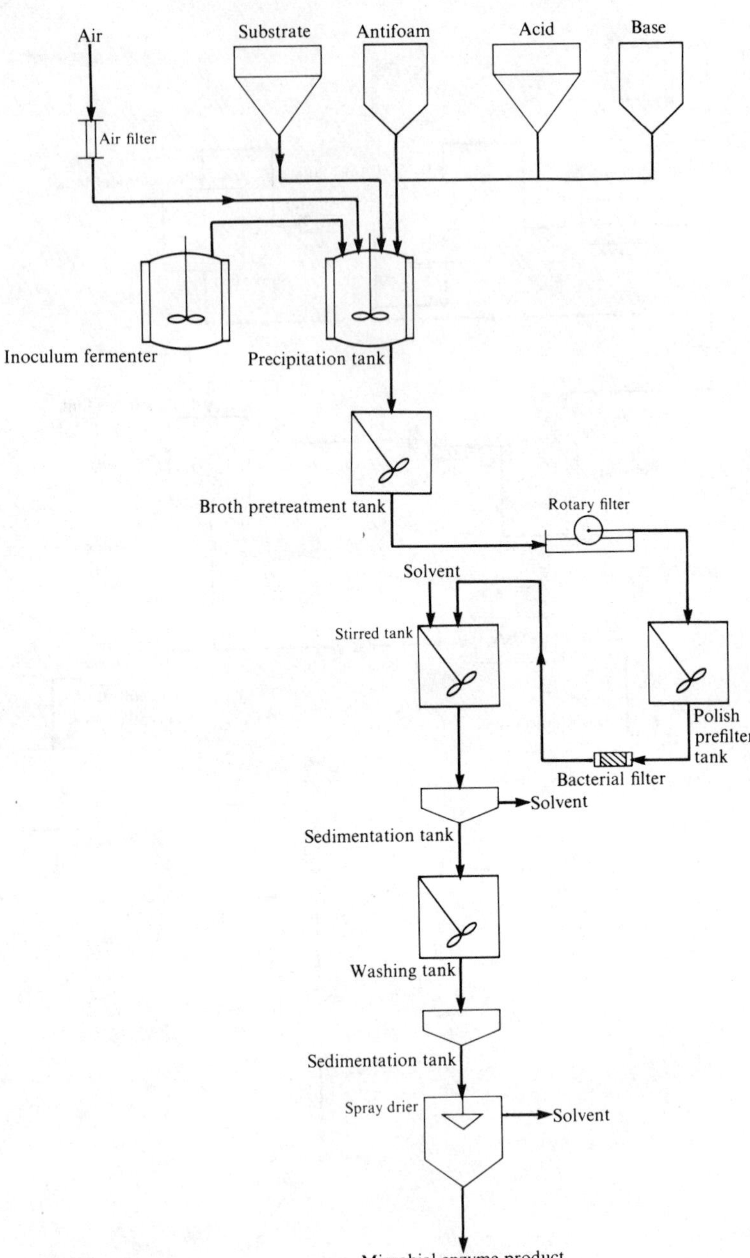

Figure 14. Solvent precipitation for production of microbial enzymes [Keay *et al.* (1972)].

nonionic surfactant, using spray cooling or prilling. In the marumerizer (*see* Fig. 15), the enzyme is mixed with a filler, a binder and water, extruded and then formed into spheres. The resulting rigid particles have a size range of 200–500 μm. The granules may be coated with an inert film.

All enzyme products must be checked for toxicity, allergenicity and microbial safety.

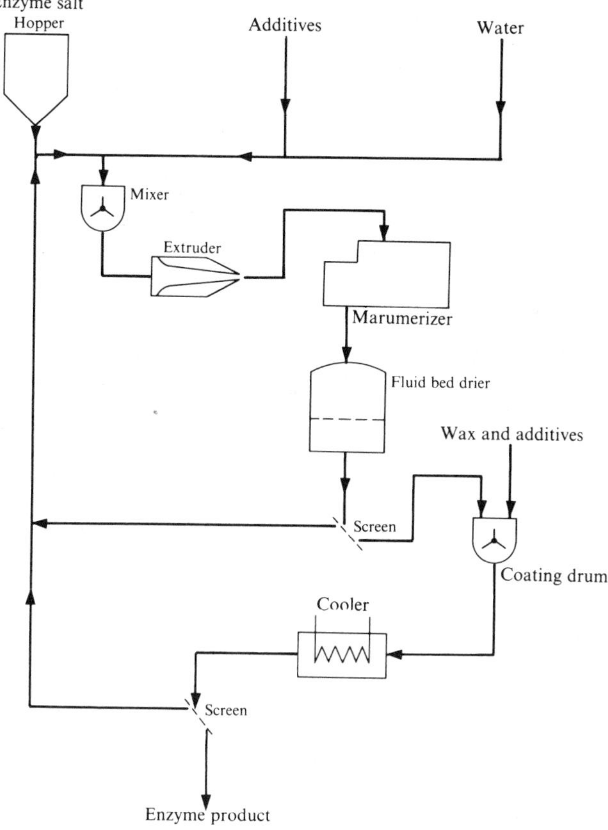

Figure 15. Preparation of a dust-free enzyme product [Aunstrup *et al.* (1979)].

Proteolytic Enzymes

Alkaline serine proteases

1. Subtilisin Carlsberg, the most widely used detergent protease, is produced by *Bacillus licheniformis*. The medium is given in Table 17. The temperature of fermentation is 30–40°C and the pH neutral. Enzyme production starts when maximum cell growth is reached after 10–20 hours and continues at an almost constant rate until the end of fermentation. The protease hydrolyzes all the proteins in the medium and, at the end of the fermentation, protease is practically the only protein dissolved in the broth. The yield may be 10 per cent of the initial protein content of the medium. The enzyme is sold

primarily in the form of dust-free granules containing 1–5 per cent enzyme protein. It is also stable in liquid form, the enzyme content of the liquid preparation being about 2 per cent.

The pH optimum for the hydrolysis of casein is about 10. Over 80 per cent of the activity is retained in the pH range 8–11. The enzyme is stable over the pH range 5–10 at 25°C and for 1 hour when exposed to temperatures of approximately 50°C at pH 8.5. Inactivation is rapid below pH 4 and above 11.5, at temperatures above 70°C. Only 30–35 per cent of the peptide bonds in casein remain unhydrolyzed by the enzyme. The enzyme concentration in most detergents is about 0.015–0.025 per cent active enzyme protein.

2. Proteases from alkalophilic *Bacillus* species are stable up to pH 12. The production methods are similar to those used for subtilisins, but the pH must be kept above 7.5. For this purpose, sodium carbonate or lactates are used, but starch hydrolyzates may be used for amylolytic strains. Proteases constitute over 90 per cent of the enzyme content of the broth. The enzymes are prepared in granulated form for detergents and as dust-free powder for dehairing of hides. The protease content of the preparations is generally 1–2 per cent active enzyme protein. They are stable in the pH range 6–12 and at temperatures up to 60°C.

Acid proteases

The majority of microbial acid proteases are produced by fungi.

1. *Mucor* proteases are used as milk-coagulating enzymes. The producing microorganisms are *Mucor pusillus* and *M. miehei*, which are thermophiles. The growth temperature range is 20–55°C for *M. pusillus* and 30–60°C for *M. miehei*.

The two species require different production methods. *M. pusillus* is grown on a semisolid medium consisting of 60 per cent wheat bran with water for 3 days at 30°C. The enzyme is extracted with water, the yield being 3200 Soxhlet units per gram of wheat bran. Ammonium salts are added to the bran to improve yields. *M. miehei* is cultivated in submerged culture using a medium containing 4 per cent starch, 3 per cent soya bean meal, 10 per cent ground barley and 0.5 per cent calcium carbonate for seven days at 30°C. The yield is about 3500 Soxhlet units per millilitre of broth.

During fermentation, other enzymes, such as lipase, esterases, amylase and cellulase, are also secreted. The mycelia are removed by filtration and a concentrated liquid product is obtained by vacuum filtration or reverse osmosis. To the final solution, sodium chloride at a concentration of 20 per cent is added as a preservative. Lipase may be removed by lowering the pH to 2.0–3.5 for a short time. Any nonspecific proteases are removed by adsorption on silica at pH 5. The commercial preparations are sold at concentrations of 10 000–150 000 Soxhlet units per ml. The amount of active enzyme in such preparations is 0.2–3.0 per cent.

2. *Aspergillus* proteases are produced by *Asp. oryzae* at high yields when grown on semisolid media. Wheat or rice bran is used in a medium that has a high ratio of inorganic nitrogen to carbon. During the fermentation, α-amylase, glucoamylase, cellulase and pectinase are formed. The protease is extracted with water, precipitated using a solvent and sold in 'solid' form.

Commercial protease preparations from *Asp. oryzae* contain acid, neutral and alkaline proteases and are active over the pH range 4–11. *Asp. phoenicis, Asp. niger* var. *Macropus* produce acid proteases with an activity range of pH 2–6.

Amylolytic Enzymes

Bacterial α -amylases

These amylases are produced by *Bacillus amyloliquefaciens* and *B. licheniformis*. Table 19 gives the composition of a medium used for bacterial amylase production. The fermenta-

Table 19. Medium for bacterial amylase production [Underkofler (1976)].

Component	Amount (%)
Ground soya bean meal	1.85
Amber BYF (autolyzed brewers' yeast fractions, Amber Laboratories)	1.50
Distillers' dried solubles	0.76
N-Z amine (enzymic casein hydrolyzate, Sheffield Chemical Co.)	0.65
Lactose	4.75
$MgSO_4.7H_2O$	0.04
Hodag KG-1 antifoam (Hodag Chemical Corp.)	0.05
Water	90.40

tion temperature is 30–40°C and the pH neutral. If pH falls below 6, the amylase is denatured and calcium carbonate is added as buffer. The production of α-amylase starts when the bacterial count reaches 10^9–10^{10} cells per millilitre after about 10–20 hours, and continues for another 100–150 hours.

Bacterial α-amylases are marketed usually as liquid preparations, which are preserved with 20 per cent sodium chloride. The most active preparations contain 2 per cent active amylase protein, and the most active solid preparation contains 5 per cent active amylase protein. *B. licheniformis* amylase has a wider pH range than the *B. amyloliquefaciens* amylase. Calcium ions affect the enzyme activity and the pH optimum occurs at 6.5–7.0.

Fungal α-amylases

These amylases are produced by *Asp. oryzae* and *Asp. niger*, which are grown on wheat bran. Some other enzymes are produced in addition to α-amylase. *Asp. oryzae* amylase can be produced in submerged culture employing media similar to those used for *Bacillus* amylases. Since the viscosity due to the mycelia is very high, aeration and agitation present problems. Glucose inhibits the production of amylases. *Aspergillus* amylase produces more sugars than *Bacillus* amylase, e.g., it is possible to obtain 50 per cent of maltose when starch is hydrolyzed by *Aspergillus* amylase.

Glucoamylase

Glucoamylase is produced industrially by *Aspergillus, Rhizopus* and *Endomyces* species. The optimum temperature for fermentation is 30–35°C. The *Aspergillus* species are cultivated in submerged culture using a medium with a high concentration of starch. A typical composition is 20 per cent corn and 2.5 per cent corn-steep liquor. The starch is liquefied with a heat-stable bacterial α-amylase prior to sterilization. Glucose does not repress the formation of the enzyme. Fermentation continues for four to five days, during which period, the pH drops to 3–4. The pH may be controlled either by sodium hydroxide or by introducing ammonia into the air stream. Glucose formed by the hydrolysis of starch is also used by *Endomyces*. The pH is not allowed to drop below 4.5, otherwise the α-amylase is denatured.

Glucoamylase is marketed in liquid form. The active enzyme protein in commercial preparations is about 5 per cent. The products contain small amounts of other enzymes, such as α-amylase, protease, cellulase, etc.

Other Enzymes

Other enzymes produced industrially include cellulases, dextranases, lactases, lipases, pullulanases, pectinases, β-glucanases, etc. As examples, the flow sheets of two processes for cellulase are given in Figs. 16 and 17. Fig. 18 depicts the important steps in the production of whole-cell glucose isomerase preparations.

Specifications

The significant operations associated with major process schemes in the starch industry are listed in Table 20. Fig. 19 contains a sequence of operations associated with starch liquefaction, saccharification and the subsequent enzymic isomerization of glucose. Fig. 20 provides additional process details on the isomerization and refining stages.

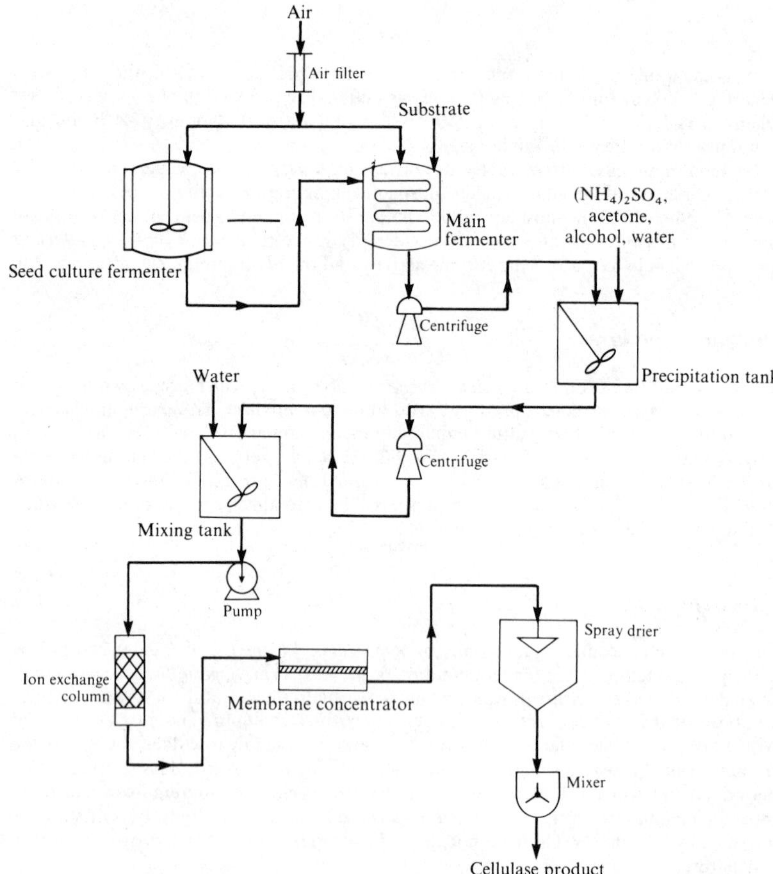

Figure 16. Submerged-culture process for cellulase production using *Trichoderma viride* [Ghose and Pathak (1973)].

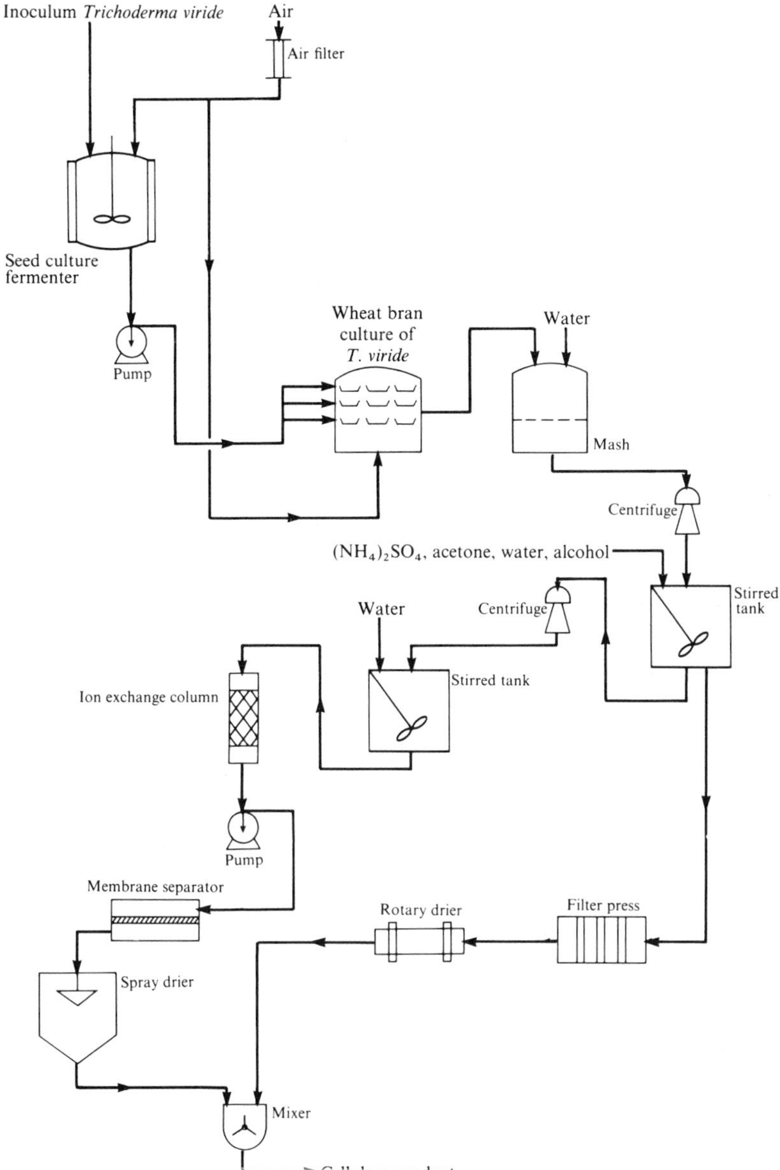

Figure 17. Cellulase production by *Trichoderma viride* using the Koji technique. [Ghose and Pathak (1973)].

Table 20. Examples of enzymatic processes in the starch industry [Reilly (1979)].

Process	Enzyme	Source	Reaction conditions pH	(°C)	Product
Liquefaction	α-Amylase	*Bacillus amyloliquefaciens*	5.5–7	90	Maltodextrins
		B. licheniformis	5.5–9	110	DE 10–20[a]
Debranching	Pullulanase	*Klebsiella pneumoniae*	6–7	50–60	Intermediate process in the manufacture of dextrose
Saccharification	α-Amylase	*Aspergillus oryzae*	5–7	50–55	High-maltose syrup
					High-DE syrup
Saccharification	Glucoamylase	*Asp. niger* (*Rhizopus* sp.)	4–5	55–60	High-DE syrup, crystalline dextrose
Isomerization	Glucose isomerase	(*Streptomyces* sp.)	6.5–8.5	60–65	Fructose syrup
		B. coagulans			
		Actinoplanes sp.			

[a] Dextrose equivalent.

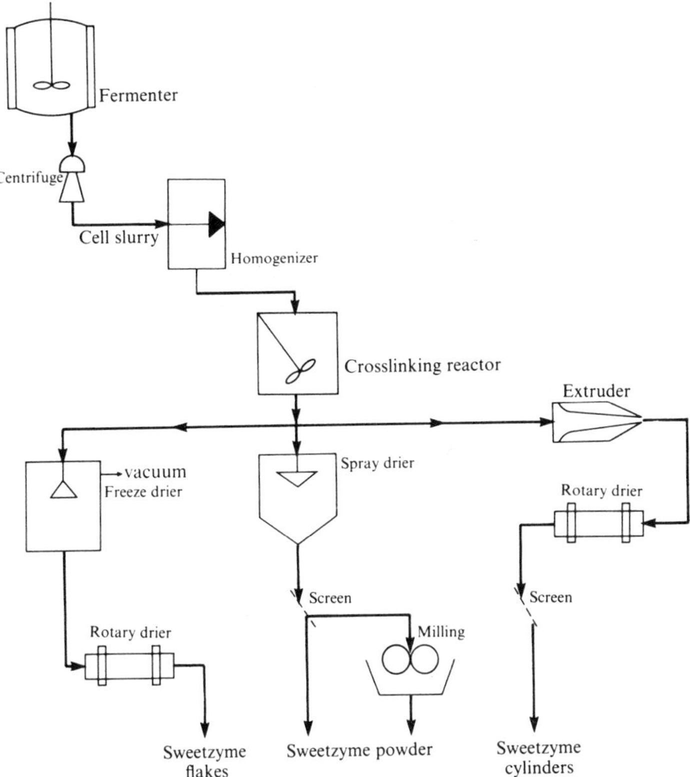

Figure 18. Important steps in the manufacture of Sweetzyme [Hemmingsen (1979)].

Tables 21–23 contain information on the activity of glucose isomerase preparations, operating ranges and stability. Figs. 21–23 shows the effect of pH, temperature, particle size and shape on the activity of glucose isomerase preparations.

Table 21. Typical operating ranges for immobilized glucose isomerase [Hemmingsen (1979)].

Substrate	40–50% dry solids [92–94% glucose (dry basis), 6–8% polysaccharides (dry basis)]
pH	7.0–8.5
Temperature	60–65°C
Metal	Magnesium ion

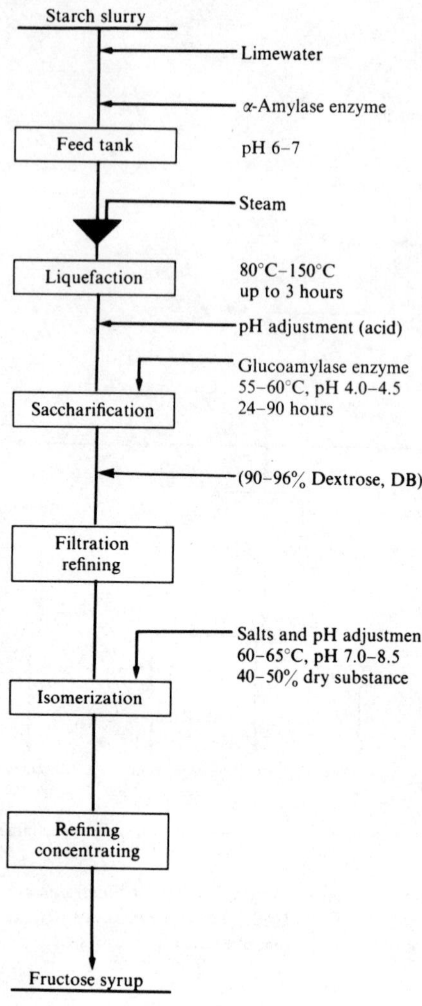

Figure 19. Starch liquefaction and saccharification processes [Antrim *et al.* (1979)].

Table 22. Productivity of some commercial glucose isomerases [Hemmingsen (1979)].

Manufacture of immobilized glucose isomerase	Productivity (g 42% HFCS[a] (g enzyme)$^{-1}$]
Clinton Corn Processing Company	7200–9000
Gist Brogades N.V.	1778
ICI Americas, Inc.	2000
Miles Kali-Chemie	1000
Novo Industri A/S	1000–1600 (plant scale), < 2300 (laboratory scale)

[a] HFCS = high fructose corn syrup.

Table 23. Half lives of Sweetzyme in laboratory experiments[a] [Hemmingsen (1979].

Temperature (°C)	$t_{\frac{1}{2}}$
60	1250
65	550
70	170

[a] 45 per cent (by wt) glucose; 60-ml columns; pH (inlet), 8.5; Mg^{2+}, 0.4 mM.

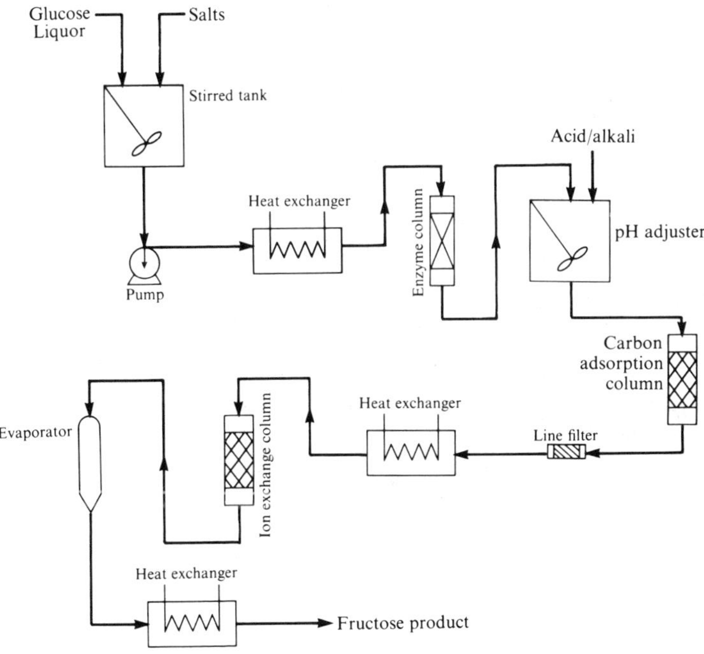

Figure 20. Enzyme isomerization of glucose to fructose [Schnyder (1974)].

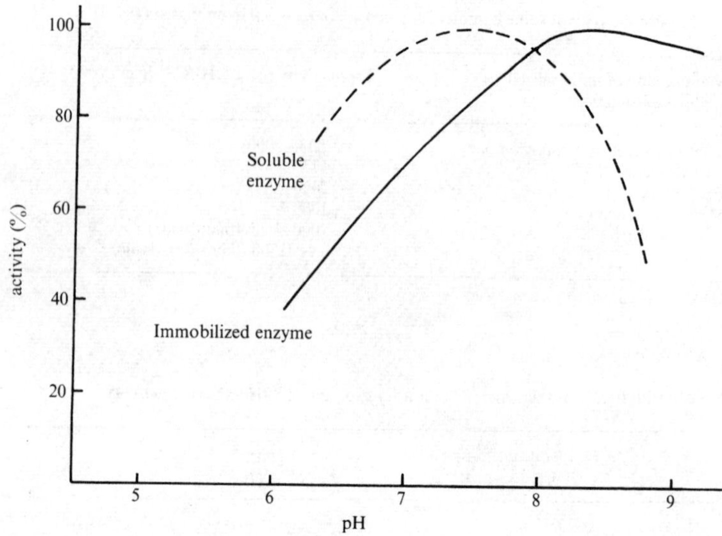

Figure 21. pH–activity profile of immobilized isomerase compared with soluble enzyme: batch reactor; temperature, 65°C; space time, 20 min; substrate, 5 per cent (by wt) glucose; Co^{2+}, 1 mM; Mg^{2+}, 0.1 M. Immobilized enzyme: column reactor; temperature, 65°C; space time, 5–10 min; substrate, 40 per cent (by wt) glucose; Mg^{2+}, 4 mM [Hemmingsen (1979)].

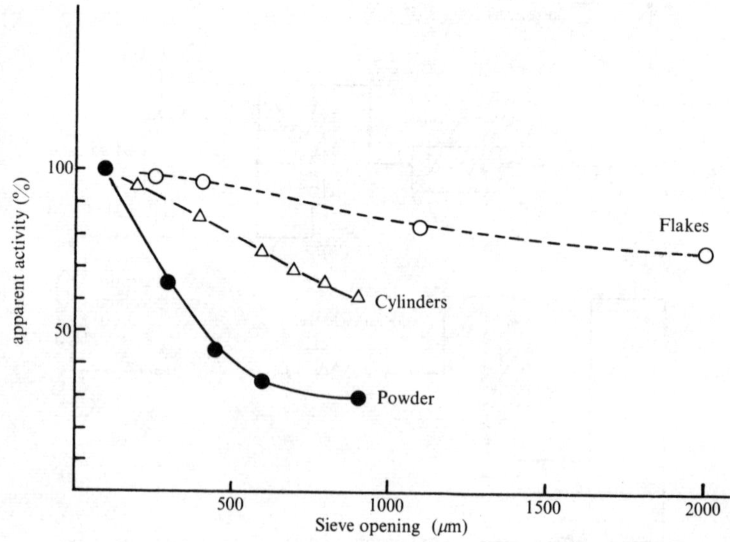

Figure 22. Apparent activity in relation to particle size [Hemmingsen (1979)].

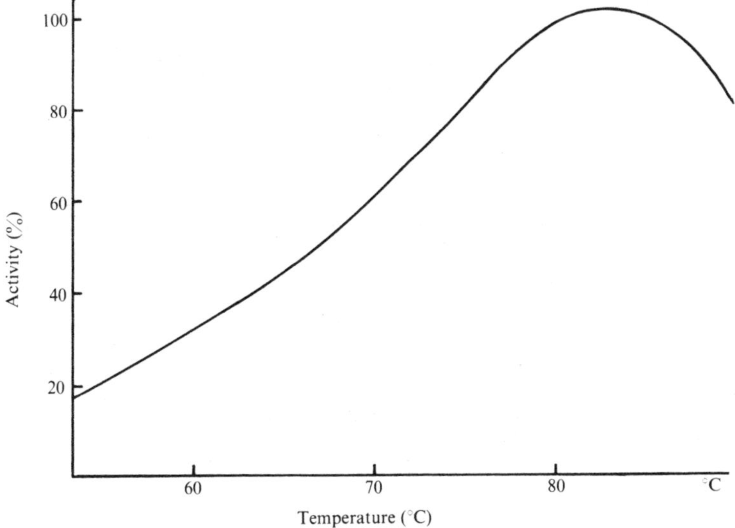

Figure 23. Temperature–activity profile of Sweetzyme. Conditions: substrate, 40 per cent (by wt) glucose; pH (inlet), 8.5; Mg^{2+}, 4 mM; 150-ml glass column; space time, 10–30 min [Hemmingsen (1979)].

ETHANOL

Raw Materials

Sugar, sucrose, glucose or fructose can be obtained for the production of alcohol from three major classes of raw materials, i.e. sugar-containing materials, starch materials and cellulose. Sugar cane juice, blackstrap molasses, sugar beet, sweet sorghum and fruit-cannery waste are all sugar-containing raw materials. The starchy raw materials are corn, wheat, oats, rice, etc., sweet potatoes and cassava. Cellulose is available as corn stalks, sugar cane bagasse, wheat and rice straw, saw mill waste, harvestable wood, waste newsprint, etc. Table 24 compares alternative sugar sources.

Table 24. Sugar sources and substrates in ethanol production [Maiorella *et al.* (1981)].

Source	Unit raw material cost [\$ (ℓ ethanol)$^{-1}$]	Unit byproduct credit [\$ (ℓ ethanol)$^{-1}$]	New raw material cost [\$ (ℓ ethanol)$^{-1}$]
Corn	0.25	0.095	0.155
Corn stover	0.23[a]		
Cull potatoes	0.34	0.008	0.172
Douglas fir	0.06[a]		
Grain sorghum	0.23	0.095	0.135
Molasses	0.26	0.032	0.228
Starch	0.32	0.00	0.320
Sugar beets	0.34	0.16	0.180
Wheat	0.34	0.12	0.220

[a] Maximum yield assuming 100 per cent conversion to glucose. Typical yields are one-third of this.

Sugar cane juice is prepared by grinding and extracting sugar cane and then precipitating the inorganics with dilute sulphuric acid followed by lime. During the production of table sugar, the sucrose from sugar cane is crystallized in multieffect evaporators. The noncrystallizable residue is blackstrap molasses and this is usually diluted to 10–20 per cent (w/v) before fermentation. The cereal grain is first milled and pressure cooled to solubilize the starch. After cooling, saccharifying enzymes are added. Potatoes and cassava require similar hydrolytic procedures. The conversion yields of cellulose hydrolysis are generally less than 60 per cent.

Fig. 24 contains a flow sheet for sugar cane milling and juice preparation for ethanol production. Table 25 gives some quantitative information about the acid pretreatment and enzymic hydrolysis of corn stover. Fig. 25 shows a schematic flow diagram of the malting process used in brewing, while Tables 26 and 27 indicate the additional nutrient requirements for ethanol production.

Glucose inhibits both aerobic and anaerobic yeast fermentations at concentrations above 150 g ℓ^{-1}. At glucose concentrations of 3–100 g ℓ^{-1}, catabolite repression of the oxidative pathways permits production of ethanol even in the presence of oxygen. At low sugar concentrations, oxygen concentrations of 0.05–0.1 mm Hg oxygen tension are adequate to ensure anaerobic fermentation. Ethanol at levels higher than 110 g ℓ^{-1} generally inhibit growth and ethanol production. Genetically modified *Zymomonas mobilis* tolerates ethanol concentrations without inhibition.

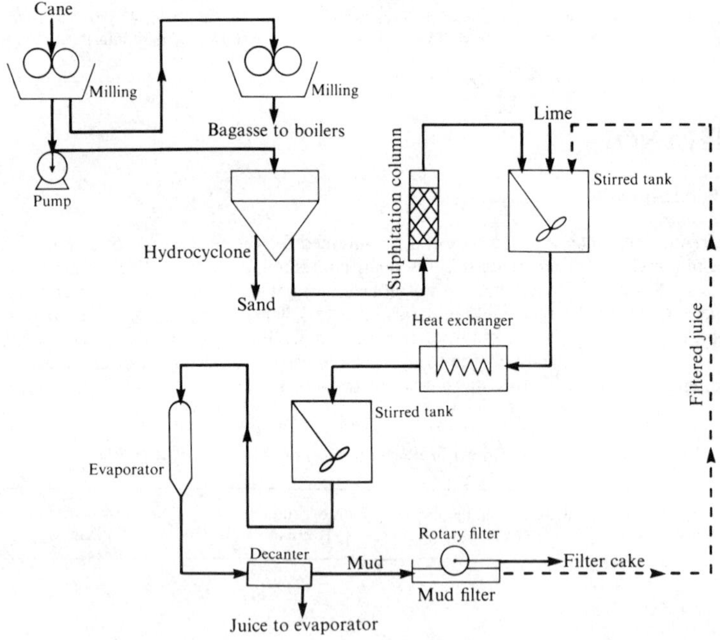

Figure 24. Sugar cane milling and juice preparation [Lindeman and Rocchiccioli (1979)].

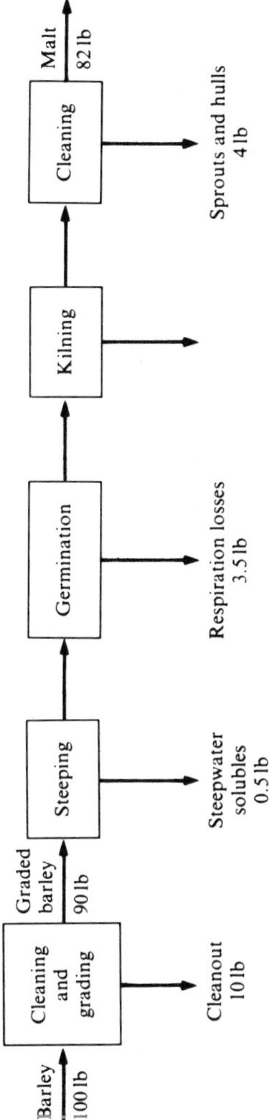

Figure 25. Schematic diagram of the malting process; material balance on a dry basis [Westermann and Huige [(1979)].

Table 25. Acid pretreatment and enzymic hydrolysis of corn stover [Wilke *et al.* (1981)].

Criterion	Quantity
Feed (2 mm corn stover)	1376 ton day^{-1}
Carbohydrate content	58% sugar equivalent
Acid pretreatment	0.09 M H_2SO_4; 7.5% suspension; 100°C; 5.5 h
Acid extracts (70% xylose)	181 ton day^{-1}; sugar/acid, ratio 2.4
Solids	885 ton day^{-1}
Enzymic hydrolysis	3.5 FPA (0.61 U ml^{-1}; 5% suspension; 45°C; 40 h)
Enzyme recovery	58%
Enzyme productivity	0.46 U ml^{-1} day^{-1}; Cell concentration, 7 g ℓ^{-1}
Cellulose conversion	40% to glucose
Hydrolyzate (90% glucose)	240 ton day^{-1}; glucose: enzyme ratio, 21
Sugar solution	2.6%

Table 26. General media trace element requirements for ethanol production.

Level of requirement	Trace elements
Most likely required	Mn, Co, Cu, Zn
Maybe required	B, Al, Si, Cl, V, Cr, Ni, As, So, Mo, Sn, I
Least likely	Be, F, Sc, Ti, Ga, Ge, Bi, Zr, W

Table 27. Composition of a typical synthetic medium for ethanol production.

Component	Amount ($g\,\ell^{-1}$)
Glucose	100.0
Ammonium sulphate	5.19
Potassium dihydrogen phosphate	1.53
Magnesium sulphate.$7H_2O$	0.55
Calcium chloride.$2H_2O$	0.13
Boric acid	0.01
Cobalt sulphate.$7H_2O$	0.001
Copper sulphate.$5H_2O$	0.004
Zinc sulphate.$7H_2O$	0.010
Manganous sulphate.H_2O	0.003
Potassium iodide	0.001
Ferrous sulphate.$7H_2O$	0.002
Aluminium sulphate	0.003
Biotin	0.000125
Pantothenate	0.00625
Inositol	0.125
Thiamin	0.005
Pyridoxine	0.00625
p-Aminobenzoic acid	0.001
Nicotinic acid	0.005

Process Outline

Table 28 summarizes a range of processes suggested and partially developed in attempts to develop an economic 'power alcohol' industry in the United States.

Table 28. Processes used for the production of ethanol.

Method	Outline	Comments
Berkeley Process	Derived from Natick process (*see* below); hemicellulose also used	Likely candidate for large-scale operations
Grain alcohol	Grain malted to hydrolyze the starch. Yeast produces ethanol. Stillage concentrated for cattle feed	Profitability dependent on low corn prices and a stable cattle feed market
Gulf Process	Enzymes added for simultaneous saccharification and fermentation	Hydrolysis yields not outstanding and good use of hemicellulose undeveloped
Iotech Process	Steam explosion fractures biomass for good hydrolysis	Lignin byproduct very valuable
M.I.T. Process	Mixed mould cultures hydrolyze biomass and produce ethanol	Simple, but effective, high potential
Natick Process	Cellulosic materials treated with *Trichoderma* sp. enzymes to yield degradable sugars	Pretreatment by grinding too expensive. Has not focused on using hemicellulose
Pennsylvania/ General Electric	Solvent extraction of lignin gives excellent hydrolysis	Recovery of organic solvents expensive
Purdue Process	Cellulose and hemicellulose removed, allowing excellent hydrolysis using acid or enzymes	Regeneration of solvent may be costly, but a very high-yield process
Sugar cane	Juices or molasses converted directly by yeast which is washed and recycled	Stillage too high in salt content for cattle feeding. Credits for cane fibre could be high

Figs. 26–32 cover a range of flow sheets that have been devised largely to meet the needs of particular raw materials. Tables 30–32 provide information on yields. The relevant tables and figures giving data for ethanol production from a particular raw material are summarized in Table 29.

Table 29.

Raw material	Figure	Table
Cassava	26	30
Corn	27	
Corn stover	28, 29	
Starch		31
Sugar cane		32
Sugar solution	30	
Wood	31	
Wood chips	32	

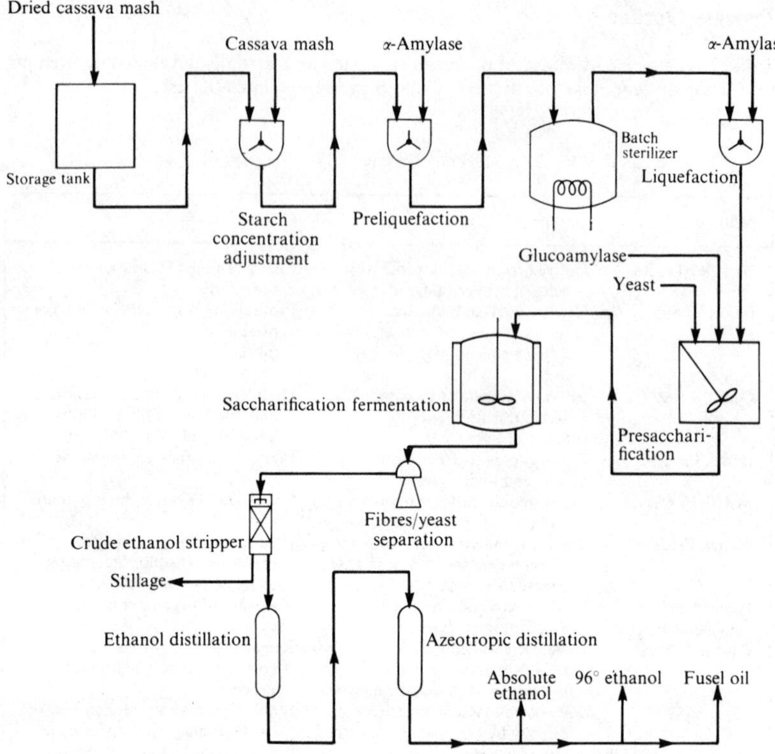

Figure 26. Production of ethanol from cassava root [Lindeman and Rocchiccioli (1979)].

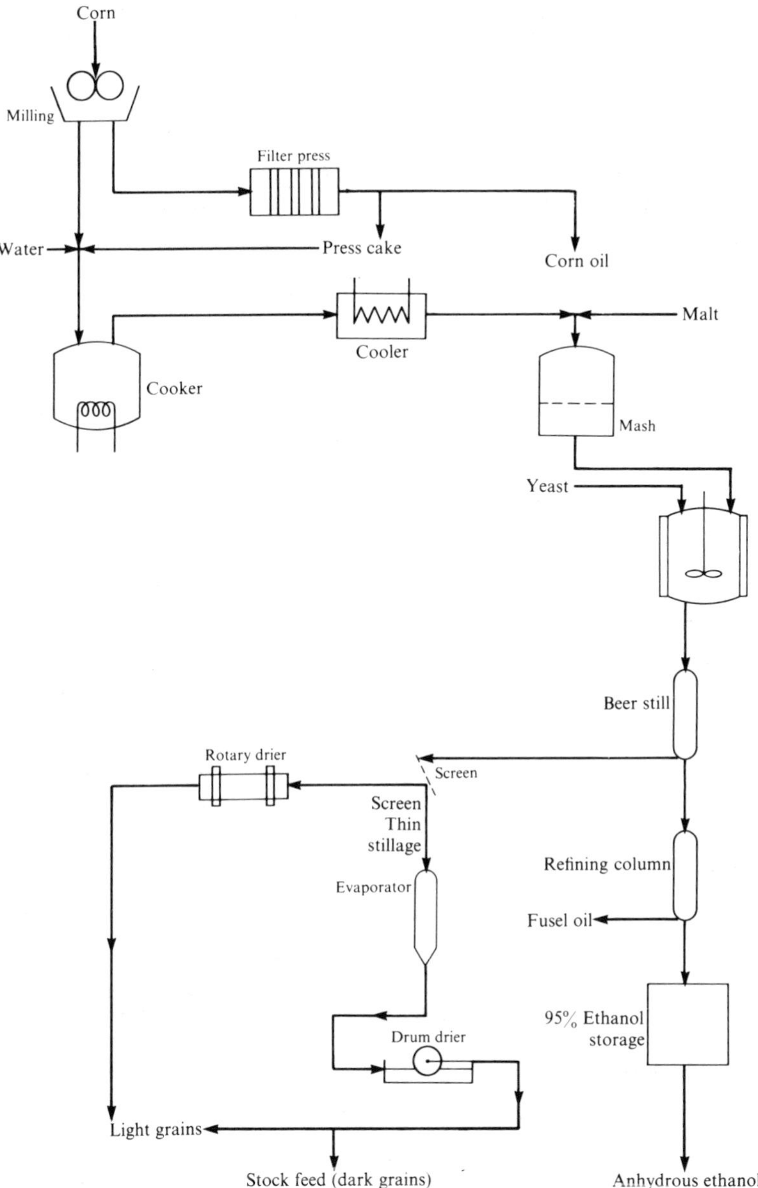

Figure 27. Process flow diagram for conversion of corn to alcohol using the Peoria pilot plant [Maiorella *et al.* (1981)].

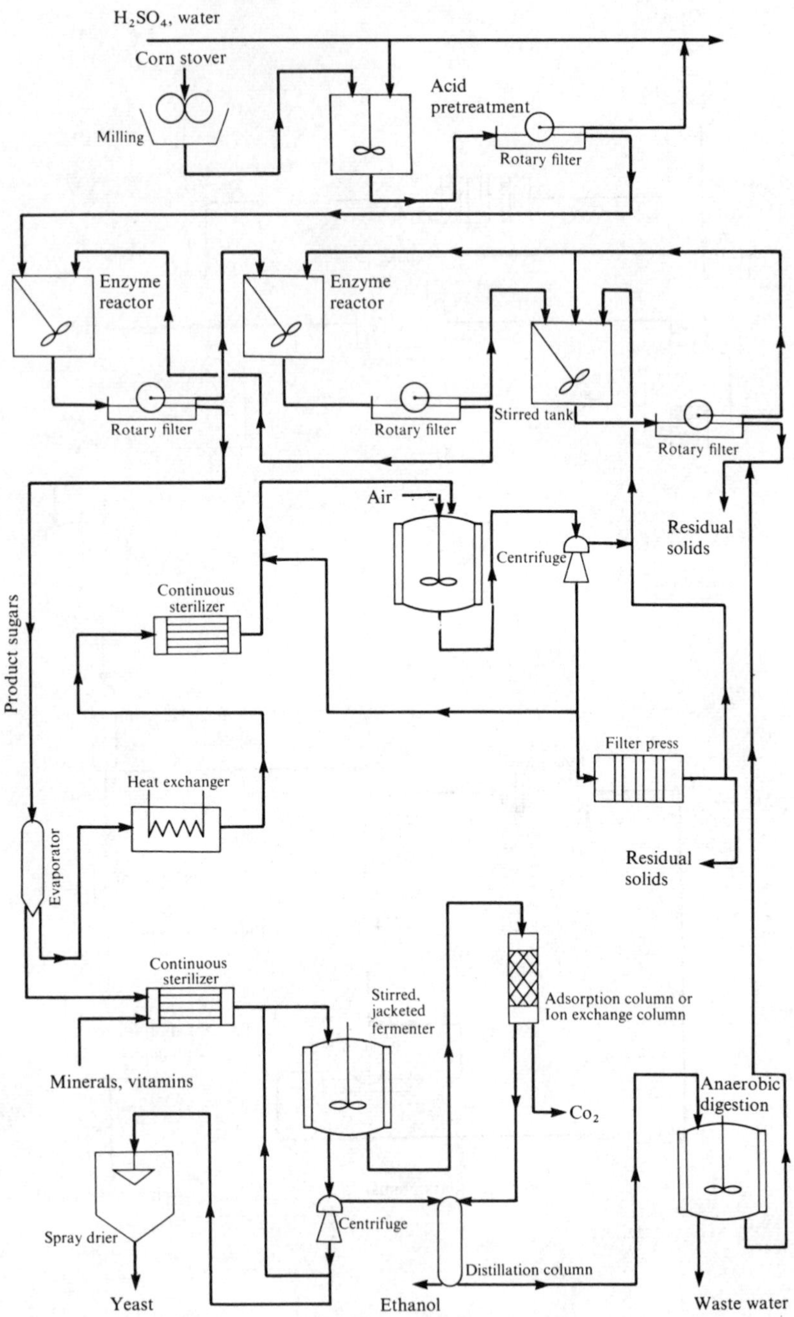

Figure 28. Material balance flow diagram for production of ethanol from corn stover [Wilke *et al.* (1981)].

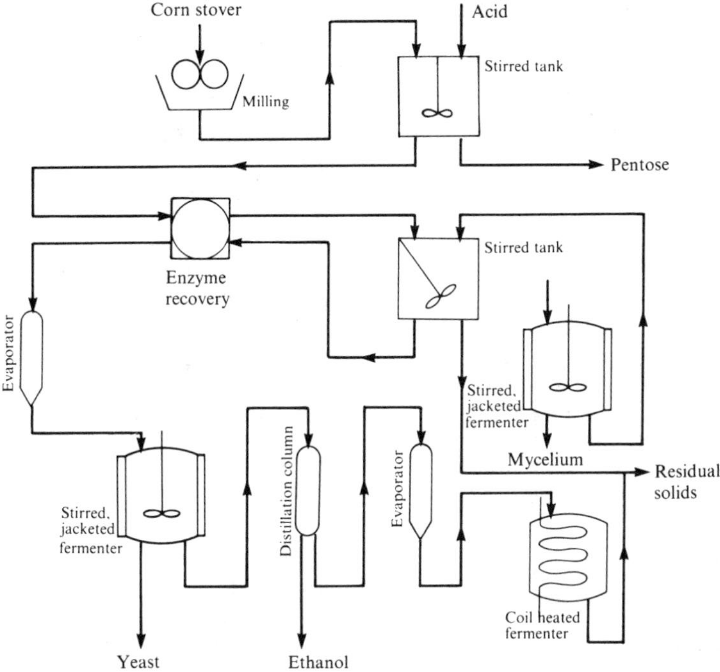

Figure 29. Simplified flow diagram for production of ethanol from corn stover.

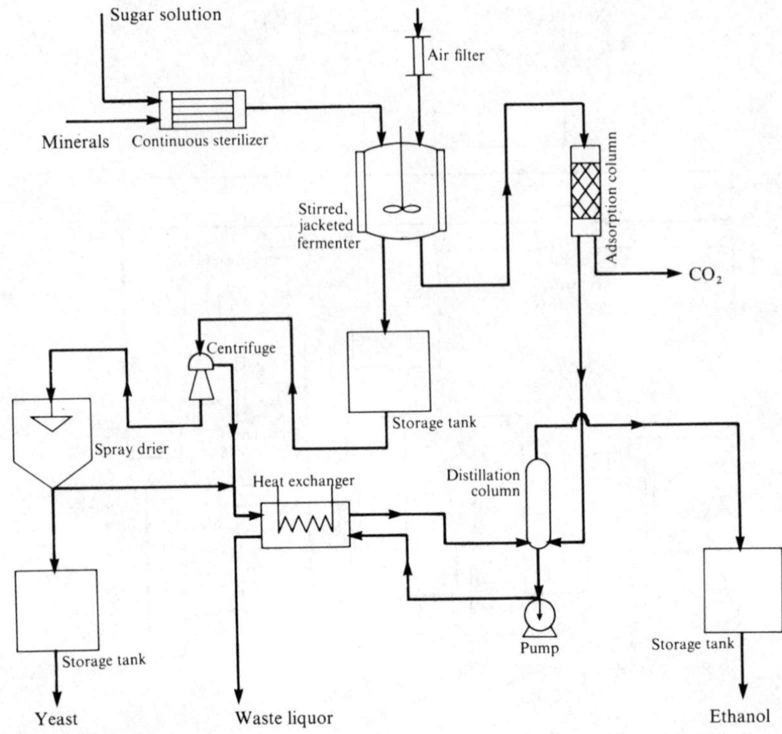

Figure 30. Production of ethanol from sugar solution [Cysewski and Wilke (1976)].

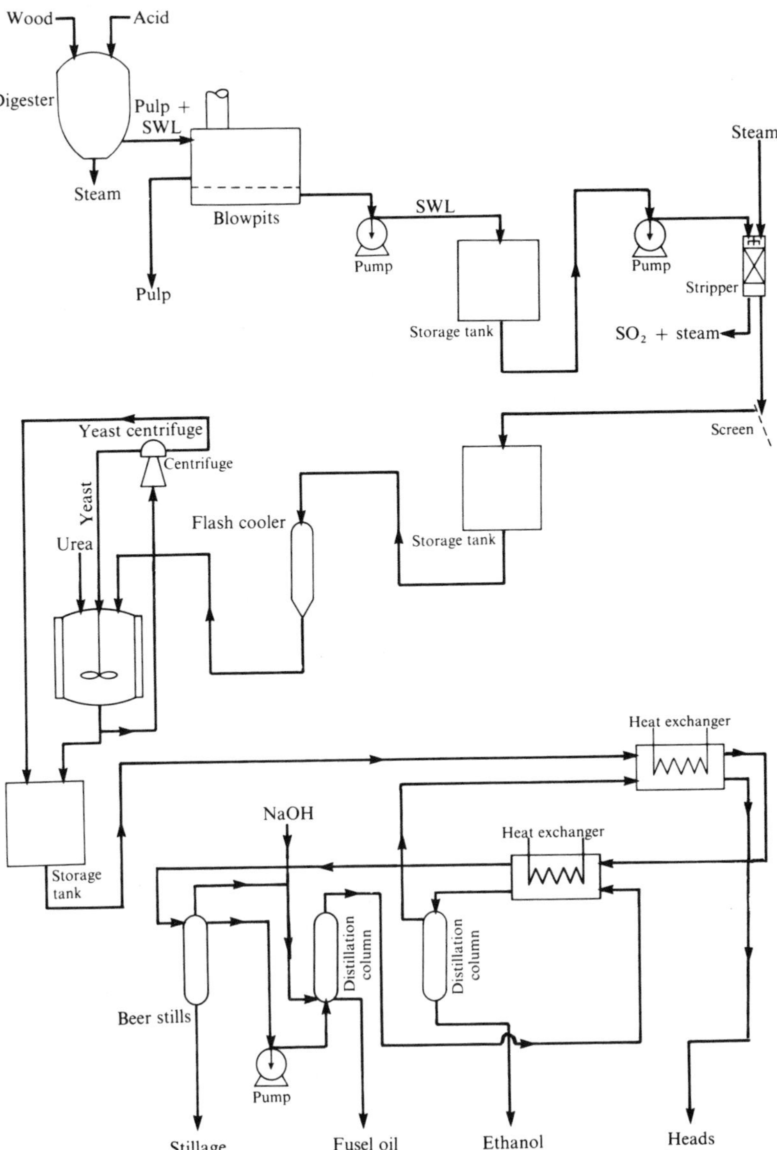

Figure 31. Production of ethanol from wood [Ericsson (1947)].

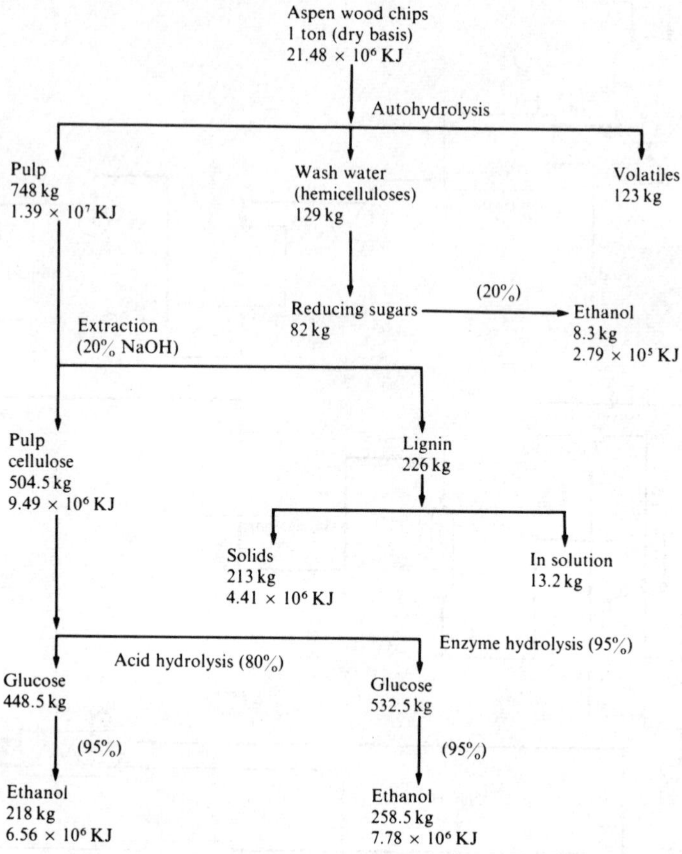

Figure 32. Material and energy balance for ethanol production from wood by acid and enzymic hydrolysis [Kosaric *et al.* (1981)].

Table 30. Productivity factors for ethanol production from cassava root [Lindeman and Rocchiccioli (1979)].

Factor	Production
Cassava root yield in 1.5-yr cycle	27 000 kg ha^{-1}
Starch content range	20–35% (by wt)
Average starch content	25% (by wt)
Ethanol yield for 100% efficient conversion and recovery	180 ℓ (1000 kg root)$^{-1}$
Reasonably attainable global efficiency	70–75%
Ethanol yield for 70% global efficiency	125 ℓ (1000 kg root)$^{-1}$, i.e. 3440 ℓ ha^{-1}

Table 31. Ethanol production from starch [Maisch *et al.* (1979)].

Remarks

Stoichiometry
Starch + water → glucose
$C_6H_{10}O_5$ (162) + H_2O (18) → $C_6H_{12}O_6$ (180)
Glucose → 2 ethanol + 2 carbon dioxide
$C_6H_{12}O_6$ (180) → 2 C_2H_5OH (92) + 2 CO_2 (88)

Theoretical yields — dry basis (70% starch for calculation)
100 kg dry corn → 70 kg dry starch + 30 kg dry feed
70 kg starch → 77.7777 kg glucose (7.777 kg hydrolyzed grain)
77.7777 kg glucose → 39.7531 kg ethanol + 38.0246 kg CO_2
Summary:
100 kg corn + 7.7777 kg hydrolyzed grain → 107.7777 kg
39.7531 kg ethanol + 38.0246 kg CO_2 + 30.0 kg feed

Typical analysis of distillery grain receipts

	Moisture (%)	Starch	Dry basis Protein	Fat
Barley malt	5.4– 6.0	57.0–61.5	12.2–13.8	1.6–2.1
Rye	11.6–12.9	64.2–67.7	13.3–14.6	1.3–2.0
Corn	13.5–16.1	71.2–73.0	9.4–10.3	3.9–4.9
Milo	13.1–14.6	72.4–75.9	9.7–11.9	2.2–3.4

Yield calculations
Moisture
Grain into fermenter × moisture in grain = moisture in mash
i.e. Corn 99 × 14.08 = 13.94%
Malt 1 × 6.00 = 0.06%
total = 14.00%

Proof gallon (PG)
Volume of mash × proof by sample distillation ÷ 100 = proof gallon
i.e. 98 115 gal × 16.43 = 16 120

Yield
Proof gallons/bushels[a] in fermenter = PG/bu
i.e. 16,120/3060 = 5.27

Dry basis
Proof gallons/dry matter
i.e. 5.27/0.86 (100 − 14) = 6.13 dry basis yield

Alcohol product per bushel (theoretical)
56 lb corn → 39.2 lb starch → 22.262 lb ethanol
i.e. 22.262/3.305 (conversion factor) = 6.76 PG

[a] A bushel is 56 pounds.

Table 32. Productivity factors for ethanol production from sugar cane [Lindeman and Rocchiccioli (1979)].

Factor	Value for ethanol obtained indirectly from final molasses	Value for ethanol obtained directly from sugar cane juice
Sugar cane yield in 1.5–2-yr cycle (kg ha^{-1})	63 000	63 000
Average sucrose [13.2% (by wt)] yield (kg ha^{-1})	8320	8320
Crystal sugar production (kg ha^{-1})	7000	
Final molasses or cane juice production (kg ha^{-1})	2210	6620
Fermentable sugar molasses or juice (kg ha^{-1})	1320	8730
Ethanol yield at 100% global efficiency	675	4460
Ethanol yield with reasonable 85% global efficiency	11.5 ℓ (1000 kg cane)$^{-1}$, i.e. 730 ℓ ha^{-1}	75 ℓ (1000 kg cane)$^{-1}$, i.e. 4800 ℓ ha^{-1}

Product Recovery

Ethanol production and purification is achieved through distillation (*see* Fig. 33). The separation of dilute ethanol/water solutions presents a problem, as such solutions form an azeotrope at 95.7 per cent (by wt) (i.e. 89 mol per cent) ethanol (*see* Fig. 34). Anhydrous ethanol is usually produced by azeotropic distillation, as shown in Fig. 35, i.e. by addition of a third component, such as benzene, to modify the azeotrope. Alternative processes for ethanol production are vacuum distillation, extractive distillation, vapour-phase water adsorption, solvent extraction, etc. Table 35 summarizes the energy requirements of various schemes for ethanol recovery. Generally, low energy demand is associated with increased capital cost.

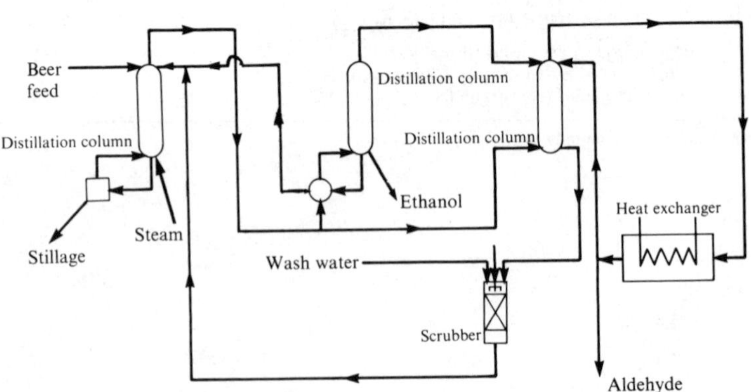

Figure 33. Three column ethanol/water distillation system [Maiorella *et al.* (1981)].

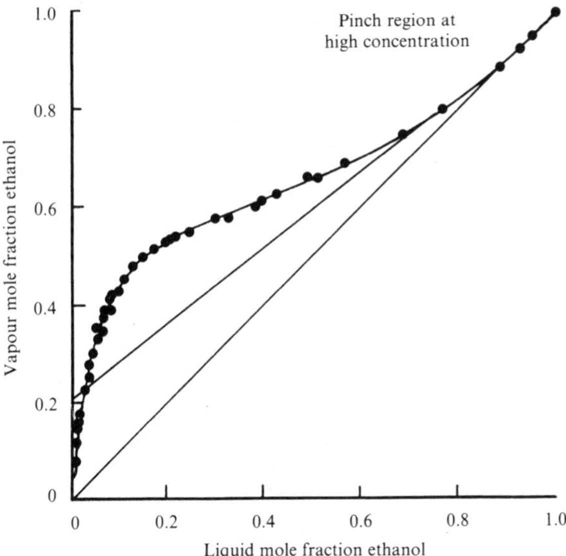

Figure 34. Atmospheric pressure — ethanol/water equilibrium [Maiorella *et al.* (1981)].

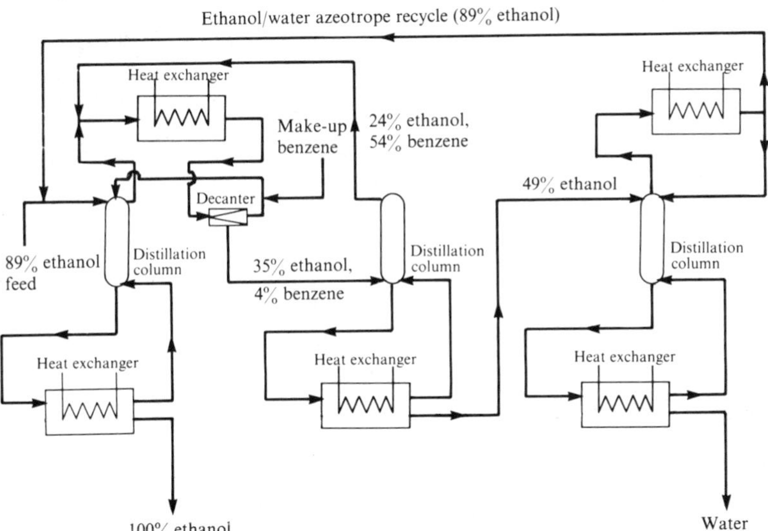

Figure 35. Azeotropic distillation with benzene for pure ethanol production [Maiorella *et al.* (1981)].

Table 33. Energy requirements of ethanol recovery [Maiorella *et al.* (1981)].

Process	Energy consumption (kg steam ℓ^{-1})	Comments
Simple 1- or 2-column distillation to 96 vol% ethanol from 6.25 vol%	2.4	20-tray beer still, 30-tray rectifying column
3-column vapour reuse process for high-quality 96 wt % industrial alcohol from 6.25 vol%	2.4	20-tray beer still, 45-tray aldehyde column, 54-tray rectifier
4-column Barbet unit for 96 vol% industrial alcohol from 6.25 vol%	4.1	20-tray beer still, 30-tray purifying column, 54-tray rectifying column
Benzene dehydration from 96 to 99.9 vol% (by vol.) ethanol	1.0	50-tray dehydrating column, 30-tray water–steam column, 45-tray supplementary rectifier
Combined 3-column vapour reuse and high-pressure ether distillation for 99.9% ethanol from 6.25%	2.5	20-tray beer still, 45-tray aldehyde column, 54-tray rectifier, 30-tray dehydrator, 20-tray water stripper
Pentane dehydration from 96 to 99.9 vol% ethanol	0.9	18-tray dehydration column, 14-tray water stripper column (supplementary rectifier also required)
Vacuum rectification to produce 95 wt % ethanol from 13 wt %	1.1	40-tray rectifier (assumes no aldehydes or fusel oil)
Extractive distillation using salts to produce 99.9 mole % ethanol from 5 wt % ethanol	1.1 (additional energy for salt recovery, depending on the process chosen)	Single 20-tray column (assumes no aldehydes or fusel oil) hinges on low-energy recovery of salts, not yet studied
Vapour-phase adsorption of water to produce 99.9% alcohol from 85% alcohol	Energy requirements for adsorbent regeneration have not been established	Only small laboratory-scale tests have been conducted
Extraction	Laboratory-scale process requires further study	
Membrane separation	Laboratory scale process	
Molecular sieve separation	Conceptual process requires further study	

ORGANIC ACIDS

Acetic Acid (Vinegar)

Raw materials

The main substrate for the production of acetic acid by *Acetobacter* species is ethanol. Therefore, wine, beer, cider and other alcoholic beverages, as well as distilled ethanol, can serve as raw materials. Alternatively, fermentation is conducted in two stages: firstly ethanol is produced from a suitable raw material and this is then fed to the acetator without intermediate distillation, as in the case of malt, fruit, sugar and grain vinegars.

In most of these raw materials, all the necessary nutrients exist for the acetic acid-

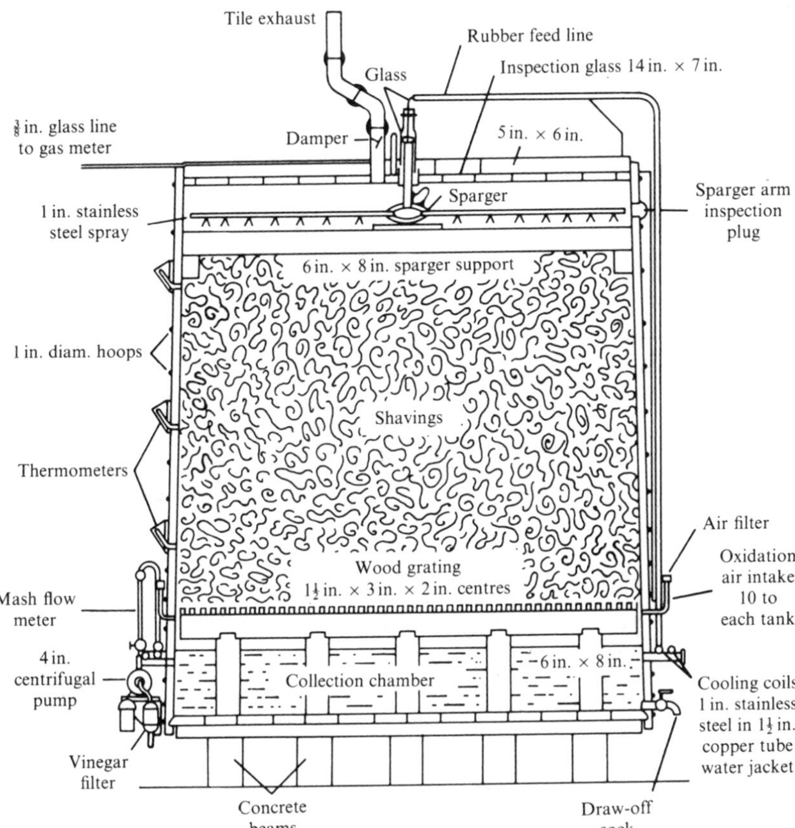

Figure 36. Trickling generator [Hansen (1935)].

producing bacteria. Supplementary nutrients are necessary when a pure ethanol/water/acetic acid mixture is employed. The ethanol concentration in the medium is less than 5 per cent (by vol.).

Process outline

Acetic acid can be produced either in trickling generators or in submerged-culture generators.

Fig. 36 represents the cross-section of a trickling generator. It consists of a tank made of cypress or redwood, which is packed with about 2000 cubic feet of curled beechwood shavings supported on a false bottom. The lower part of the generator with a volume of about 3500 gallons is used as storage space.

A pump circulates an ethanol/water/acetic acid mixture from the storage reservoir up through a cooler to a distributor at the top of the tank. The microorganisms – *Acetobacter* species – are retained on the surface of the wood shavings. A blower is used to force air through the packing, but care is taken to avoid excessive air rates which waste ethanol and acetic acid through stripping. A typical air flow rate is about $80\,ft^3\,ft^{-3}\,h^{-1}$. The air rate is regulated to maintain the outlet oxygen concentration slightly above 12 per cent.

When acidities exceed 12 per cent, rates of production tend to drop drastically. The

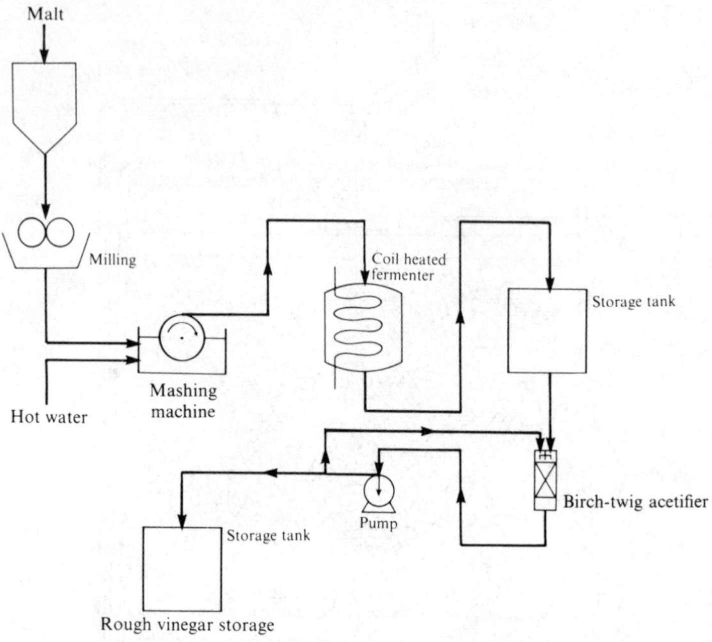

Figure 37. A process used for producing vinegar from barley malt

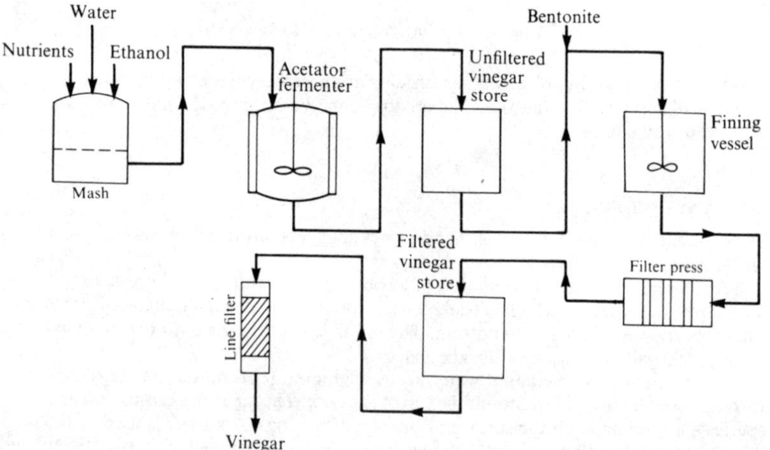

Figure 38. Flow sheet for production of vinegar using the Frings acetator [Greenshields (1978)].

amounts of finished vinegar withdrawn from the generator are such that, when replaced with ethanol-containing charges, the combined ethanol concentration is less than 5 per cent (by vol.). A properly operated generator has a packing life of 20 years and a good conversion efficiency is 88–90 per cent of the theoretical value.

A flow sheet for a process used for production of vinegar from barley malt employing a trickling generator is shown in Fig. 37.

The most common submerged-culture vinegar fermenter is the Frings acetator, which comprises a stainless steel fermenter with a high-speed agitator located at the bottom. The important feature of this fermenter is the high efficiency of oxygen transfer. The acetator is operated batch-wise, with a cycle time of 35 hours for 12 per cent acetic acid. A flow sheet for the production of vinegar using the Frings acetator is given in Fig. 38.

Other vinegar generators include the Yeomans cavitator, the Burgeois process, the Fardon process and the tower fermenter, all of which are described by Nickol (1979).

Product recovery and specification

Distilled vinegar from trickling generators is almost free of insoluble material. However, other vinegars from trickling generators and all vinegars from submerged-culture processes require filtering with the use of filter aids and fining agents to achieve the desired clarity. To prevent bacterial growth after the product has been bottled, vinegar is pasteurized. Vinegar may be concentrated, usually by a freezing process, as described by Nickol (1979). Table 34 gives the chemical composition of cider vinegars.

Table 34. Chemical composition of cider vinegars [Joslyn (1970)].

Constituent	Average	Concentration Maximum	Minimum
Total acid [% (w/v, as acetic acid)]	4.94	7.96	3.29
Total solids [% (w/v)]	2.54	4.52	1.37
Nonsugar solids [% (w/v)]	1.90	2.89	1.26
Reducing sugars in solids [% (w/v)]	19.6	45.0	5.6
Total ash [% (w/v)]	0.367	0.52	0.20
Alkalinity of water-soluble ash (ml 0.1 N acid[a])	35.7	56.0	21.5
Ash in nonsugar solids [% (w/v)]	18.8	26.5	11.2
Soluble phosphate (mg P_2O_5)	17.3	39.9	6.7
Insoluble phosphate (mg P_2O_5[a])	29.3	64.2	15.1
Ethanol [% (w/v)]	0.35	2.0	0.03
Glycerol [% (w/v)]	0.30	0.46	0.23
Polarization (direct; °V[b])	−1.46	−3.6	−0.2
Polarization (indirect; °V[b])	−1.69	−3.1	0.0

[a] Content from 100 ml vinegar.
[b] °V indicates one degree Vanzkescale and equals 0.34657° angular rotation, with sodium D light.

Citric Acid

Raw materials

For the production of citric acid, beet molasses, cane molasses, corn starch, corn syrup, glucose, sucrose, wheat bran or sweet potato wastes can be used as the major raw material. Metal ions, such as iron, copper, zinc, etc., have an important influence on the fermen-

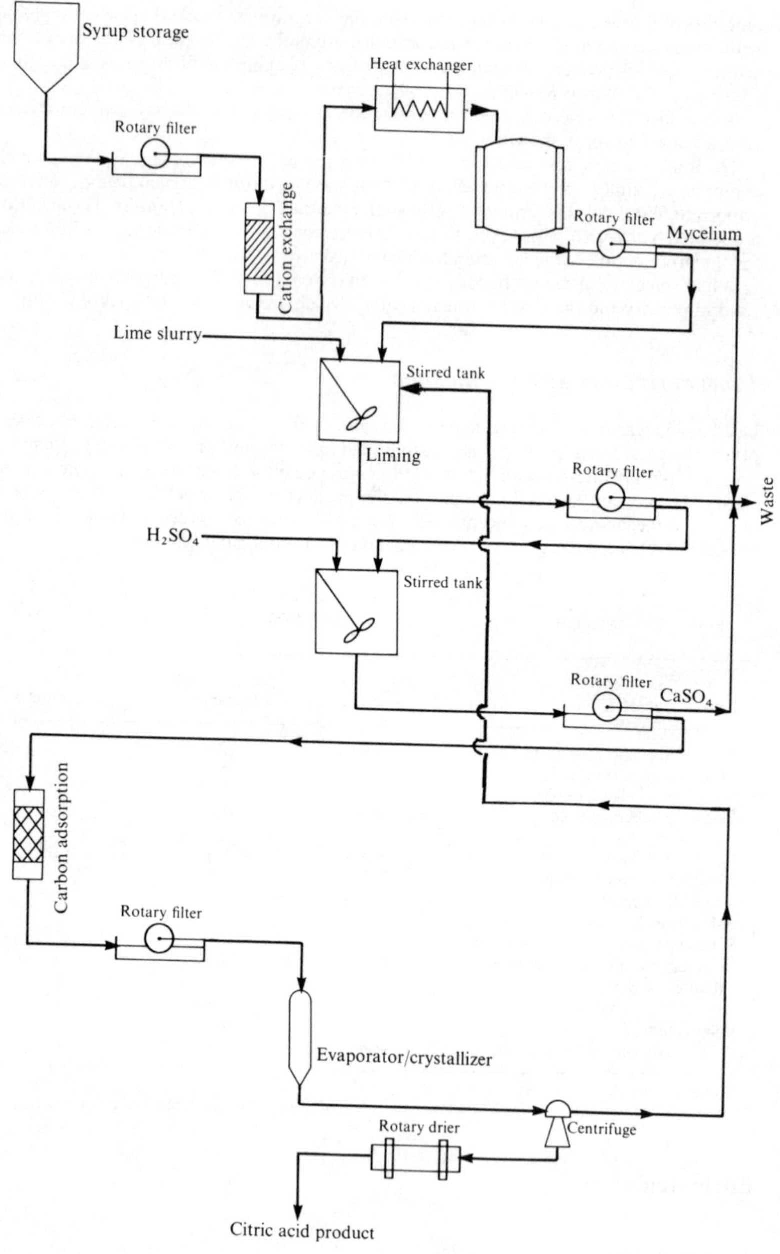

Figure 39. Citric acid flow sheet [Beaman (1967)].

Table 35. Production of citric acid by the submerged process [Prescott and Dunn (1959)].

Carbohydrate source	Reagent used in preparation of substrate	Yield, (%)	Duration of fermentation
Beet molasses	Potassium ferrocyanide	64 (max.)[a]	8–9 days
	Potassium ferrocyanide	72 (max.)[a]	70 h
	Potassium ferrocyanide	68.5[a]	116 h
	Potassium ferrocyanide	64 (average)[a]	80–116 h
Cane molasses	Methanol	63.3[b]	9 days
	Bone char + Suchar CSP + amberlites	61.9[b]	12 days
Corn starch	Methanol	71.5[b]	10 days
Corn syrup	Morpholine	79.9	8.67 days
Glucose	Methanol	71.2[b]	7 days
Sucrose		72 (average)[a]	9 days

[a] Based on available sugar.
[b] Based on sugar consumed.

tation. Since the concentration of trace elements affects citric acid production, these may have to be removed from the raw materials by chemical methods or using ion exchange resins [*see* Kapoor *et al.* (1982)]. Table 35 gives some data for the production of citric acid from industrial raw materials.

Process outline

Citric acid can be produced by either liquid surface-culture or submerged-culture processes.

In the surface-culture process, which is used by many manufacturers, the inoculated broth is placed in shallow pans of high-purity aluminium or stainless steel. Humidified air is blown over the surface of the broth for five to six days, after which time dry air is used. Spores germinate within 24 hours and mycelium covers the surface of the broth. The original sugar concentration of 20–25 per cent is reduced to 1–3 per cent 8–10 days after inoculation. Little citric acid is produced during growth. The initial pH is 5–6 and, upon spore germination, it rapidly approaches pH 1.5–2. The final yield of citric acid is in the range 80–85 per cent of the weight of the initial carbohydrate.

In the submerged-culture process, the inoculum usually consists of spores of a strain of *Aspergillus niger* grown on solid nutrient media. For decationized solutions, No. 316 stainless steel is used to avoid corrosion. Initially, the pH is about 4, and it falls to 1.5–2 by rapid uptake of ammonium ions. Little citric acid is formed in young cultures before pH 2 is reached. Air is continuously sparged at a rate of 0.5–1.5 vvm.

Fig. 39 represents the flow sheet of a process for citric acid production.

The amount of copper and iron added is very important. For this reason, steel fermenters cannot be used without glass or plastic coatings. However, the common blue glass coating is not effective since cobalt leaches from this glass.

Product recovery and specification

The culture solution is filtered to remove the mycelium. A polishing filtration may be necessary to remove residual antifoam, mycelium or oxalate. Calcium citrate is precipitated from the clear solution by addition of calcium hydroxide, which must have a very low magnesium content to avoid losses due to soluble magnesium citrate. Calcium citrate is then filtered off and the filter cake is transferred to a tank, where it is treated with sulphuric acid to precipitate calcium sulphate. The dilute filtrate containing citric acid is purified by passing over activated carbon, demineralized by ion exchangers and the purified solution is evaporated yielding crystals of citric acid. The crystals are removed by centrifugation.

Citric acid is marketed as an anhydrous crystalline chemical, as the crystalline monohydrate or as the crystalline sodium salt. Anhydrous and monohydrate crystals may be granular (15–50 mesh), fine or powdery (35–100 mesh for anhydrous acid and 60–200 mesh for the monohydrate). Crystals are supplied in moisture-proof bags or drums. Caking may occur if stored at high temperature and humidity; storage at 21°C and 50–70 per cent relative humidity is satisfactory.

Gluconic Acid

Raw materials

Gluconic acid is the free acid form of D-glucono-δ-lactone, which is produced by direct dehydrogenation of D-glucose by glucose oxidase present in a wide variety of bacteria and fungi. Commonly used microorganisms are *Aspergillus niger*, *Penicillium chrysogenum* and other *Penicillium* species, various *Acetobacter* species (especially *A. suboxydans*), various species of *Pseudomonas*, *Pullularia pullulans* and a species of *Moraxella*.

The medium consists of aqueous solutions of D-glucose, several inorganic salts and corn-steep liquor; the solution is not deionized. Table 36 gives the composition of a typical

Table 36. Media used for gluconic acid production by *Aspergillus niger* [Gastrock *et al.* (1938)].

Ingredient	Medium composition			
	Culture medium A	Sporulation medium	Germination medium	Fermentation medium
Refined corn sugar[a] ($g\,\ell^{-1}$)	30.0	50.0	100.0	Variable
$MgSO_4.7H_2O$ ($g\,\ell^{-1}$)	0.10	0.12	0.25	0.156
KH_2PO_4 ($g\,\ell^{-1}$)	0.12	0.144	0.30	0.188
$(NH_4)_2HPO_4$ ($g\,\ell^{-1}$)	0.00	0.56	0.80	0.388
NH_4NO_3 ($g\,\ell^{-1}$)	0.225	0.00	0.00	0.00
Peptone ($g\,\ell^{-1}$)	0.25	0.20	0.02	0.00
Potatoes ($g\,\ell^{-1}$)	200.0	0.00	0.00	0.00
Agar ($g\,\ell^{-1}$)	20.0	1.50	0.00	0.00
$CaCO_3$ ($g\,\ell^{-1}$)	4.0	0.00	37.50	26.00
Beer ($ml\,\ell^{-1}$)	0.00	45.0	40.0	0.00
Water source	Distilled	Distilled	Tap	Tap

[a] Contains 91.5 per cent glucose and corresponds closely to glucose monohydrate.

Table 37. Effect of air pressure on gluconic acid yields[a] [Wells *et al.* (1937)].

Gauge pressure ($kg\,cm^{-2}$)	Glucose consumed (g)	Gluconic acid produced (g)	Gluconic acid yield (%)	
			Based on glucose consumed	Based on glucose available
0.35	178	173	89.1	32.1
1.05	257	258	92.1	47.9
2.11	336	351	93.0	65.1
3.16	429	454	97.1	84.2

[a] Air flow, 1200 ml min^{-1}; speed, 13 rpm; medium volume, 3.2 ℓ; glucose available, 495 g; fermentation period, 18 h; temperature, 30°C.

medium. Most of the mineral requirements are supplied by corn-steep liquor, although addition of calcium carbonate to the medium may be used to increase the yield. Glucose concentrations are in the range 150–200 g ℓ^{-1}. *Asp. niger* can tolerate glucose levels of 200–350 g ℓ^{-1}, in the presence of excess calcium carbonate and in the presence of boron at 0.5–2.5 g ℓ^{-1}.

The equilibrium between the lactone and gluconic acid is controlled by the pH and the temperature of the medium. The pH of the medium should not be lower than 6.5 after inoculation. The fermentation is aerobic; increased air pressures (2–3 atm) result in more rapid fermentation. Table 37 shows the effect of air pressure on gluconic acid yields.

Process outline

Submerged fermentation is used for the production of gluconic acid. In Fig. 40, the flow sheet of a process for the production of sodium gluconate is represented.

Calcium gluconate or sodium gluconate is formed, depending on the mode of pH control, initial glucose concentration and final gluconate concentration. In the formation of the calcium salt, calcium carbonate slurry is added to control pH whereas, in the latter case, the pH is controlled by addition of sodium hydroxide. The solubility of calcium gluconate in water is about 4 per cent at 30°C. Supersaturation of calcium gluconate occurs in fermentation solutions containing about 15 per cent or more initial glucose concentration. Addition of the calcium carbonate slurry is delayed until enough glucono-δ-lactone has been formed. The sodium salt is much more soluble than calcium gluconate; a typical fermentation medium has an anhydrous glucose content of 250–350 g ℓ^{-1}. Air is introduced at the rate of about 1–1.5 vvm. Tables 38–40 give some information about the fermentation conditions and the yields attainable.

Table 38. Production of gluconic acid by different methods [Wells *et al.* (1937)].

Microorganism	Type of fermentation	Fermentation vessel	Theoretical yield[a] (%)	Fermentation period (days)
Penicillium chrysogenum	Submerged (pressure)	Glass bottle (sintered glass, false bottom)	80.4	8
	Submerged (pressure)	Rotary drum (aluminium)	80.0	2.2
P. luteum purpurogenum	Surface	Shallow pan (aluminium)	57.4	11

[a] From 20 per cent glucose solutions.

Table 39. Data on the production of gluconic acid and gluconates [Prescott and Dunn (1959)].

Microorganism	Type of Fermentation	Glucose concentration (%)	Temperature (%)	Duration	Principal products	Yield of principal products (%)
Acetobacter gluconicum	Aeration	5	30	4 days	Gluconic acid	71.8–77.2
Aspergillus niger NRRL 3	Submerged (pressure)	24–30	33.0–34.4	<40 h	Na gluconate	
Asp. niger NRRL 67	Submerged (pressure)	11.5	30	9 h	Gluconic acid + Ca gluconate	
	Submerged (pressure)	15.5	30	18 h	Gluconic acid + Ca gluconate	97.1[a]
Penicillium chrysogenum	Submerged (pressure)	20	30	2.2 days	Gluconic acid + Ca gluconate	80.0
	Submerged (pressure)	20	30	8 days	Gluconic acid + Ca gluconate	80–87
	Surface	20–25	30	8–10 days	Gluconic acid + Ca gluconate	60
P. luteum purpurogenum var. *rubrisclerotium*	Surface	20–25	25	9 days	Gluconic acid	55–65

[a] Based on glucose consumed.

Table 40. Production of 2-ketogluconic acid by *Pseudomonas* species[a] [Lockwood *et al.* (1941)].

Microorganism	Glucose consumed[b]	Reducing value of fermented liquor[c]	Optical rotation of fermented liquor[d]	2-Ketogluconic acid		Calcium		
				Found	Yield based on glucose consumed	Total in solution	Due to 2-ketogluconic acid	
	($g\,\ell^{-1}$)			($g\,\ell^{-1}$)	(%)	($g\,\ell^{-1}$)	($g\,\ell^{-1}$)	Percentage (%)
Ps. aeruginosa	81	114	−4.90	58	67	6.2	6.0	97
	78	124	−4.90	60	71	6.4	6.2	97
Ps. fluorescens	93	67	−2.29	30	30	9.7	3.1	32
	87	164	−6.05	77	82	8.4	7.9	94
Ps. fluorescens 142	58	166	−2.14	47	75	5.2	4.9	93
Ps. fluorescens 948	83	166	−6.32	79	88	8.4	8.2	97
Ps. fluorescens 949	83	161	−5.93	75	84	9.8	7.7	79
Ps. fragi 4973	97	165	−7.77	90	86	10.3	9.3	90
Ps. graveolens 4683	97	155	−6.82	80	77	9.3	8.3	89
Ps. graveolens 4684	75	166	−4.60	66	82	6.9	6.8	98
Ps. mildenbergii 795	97	186	−9.11	105	100	10.8	10.8	100
Ps. ovalis 950	76	117	−3.09	45	55	6.3	4.6	73
Ps. pavonacea 951	55	170	−1.77	46	77	5.8	4.8	82
Ps. putida 4359	87	162	−6.62	80	85	9.2	8.3	90
Ps. schuylkilliensis	86	166	−6.00	76	82	9.1	7.8	86
Ps. vendrelli 7700	71	168	−4.06	62	81	6.5	6.4	98

[a] Nutrient solution (200 ml) in gas washing bottles containing glucose (5 g), calcium carbonate (5 g), corn-steep liquor (1 g), urea (0.4 g), potassium dihydrogen phosphate (12 g), magnesium sulphate $.7H_2O$ (0.05 g) and oleic acid (1 drop), temperature, 30°C; air flow, 200 ml min^{-1}; duration, 8 days.

[b] Original concentration, 9.6–10 per cent.

[c] Expressed as milligrams of copper per millilitre of solution.

[d] Degree of rotation at 20–25°C in a 1-dm tube.

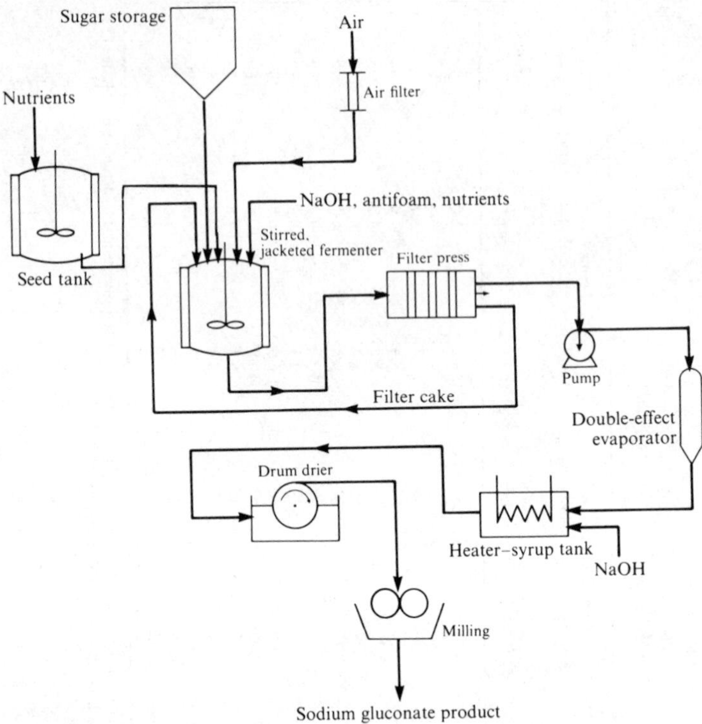

Figure 40. Flow sheet for the production of sodium gluconate [Blom *et al.* (1952)].

Product recovery and specifications

At the end of the fermentation, the microorganisms and other suspended solids are removed by conventional filtration methods. The filtrate is decolourized using activated carbon and the activated carbon is then removed by filtration. Calcium gluconate is concentrated to a 15–20 per cent solution by evaporation. Upon cooling to a temperature just above 0°C, most of the supersaturated calcium gluconate crystallizes, but seeding may be necessary. The spongy mass of needle crystals is passed through pressure filters or basket centrifuges to remove mother liquor. Sodium gluconate is removed by concentrating to 42–45 per cent solids, adjusting to pH 7.5 by sodium hydroxide and subsequent drum drying.

Aqueous solutions of gluconic acid are an equilibrium mixture of the free acid, glucono-δ-lactone and glucono-γ-lactone. Crystals separating out at 0–30°C are predominantly gluconic acid, at 30–70°C mainly glucono-δ-lactone and at temperatures greater than 70°C mainly glucono-γ-lactone. Standard 50 per cent gluconic acid solution may be obtained by treating the calcium gluconate solutions with sulphuric acid and heating to precipitate calcium sulphate, followed by filtration and concentration of the filtrate to 50 per cent solids. Table 41 gives some specifications for the products of gluconic acid fermentations.

Table 41. Gluconic acid specifications [Ward (1967)].

Gluconic acid, technical-grade 50 per cent	
Colour	Light yellow
Gluconic acid	49.0–51.0%
Specific gravity	1.23–1.24
Residue on ignition	≤ 0.45%
Sulphate	Trace
Chloride	Trace
Reducing substances	Not more than 3% on 100% acid basis
Toxicity	Nontoxic
Storage above 18°C is recommended to prevent crystallization of acid or lactone.	
Sodium gluconate, technical-grade	
Appearance	White or tan powder
Purity	98.8–99.8%
Moisture	0.1–0.7%
Reducing sugar	0.1–0.5%
Solubility, 25°C	
In water	590 g ℓ$^{-1}$
In alcohol	Sparingly soluble
In ether	Insoluble
pH in aqueous solution	5.3–6.3
Stability in aqueous solution, 0–100°C	Stable
Glucono-δ-lactone	
Specific rotation (freshly prepared)	$[\alpha]_D^{20}$ +64°C
Decomposition point	158–162°C
Solubility, 20°C	
In water	Very soluble
In alcohol	Slightly soluble
In ether	Insoluble
Appearance	White, crystalline powder
Taste	Sweet
pH (fresh aqueous solution)	3.6

Itaconic Acid

Raw materials

Itaconic acid (i.e. methylenesuccinic acid) accumulates in cultures of *Aspergillus terreus* and *Asp. itaconicus*. Current manufacturing processes are based on the use of molasses as the substrate. In early fermentation, sugar concentrations higher than 7 per cent were not fermented efficiently. It was later found that if beet molasses was used in the spore-germinating medium it was possible to ferment cane molasses at a sugar concentration of 15 per cent with a final itaconic acid concentration of 8.5 per cent, corresponding to a yield of 85 per cent of the theoretical maximum. The composition of a typical fermentation medium is given in Table 42.

Table 42. Medium composition for itaconic acid production.

Component	Concentration (g ℓ$^{-1}$)
Cane molasses (sugar content)	150
$CuSO_4.7H_2O$	3.0
$MgSO_4.7H_2O$	0.01
$ZnSO_4$	1.0

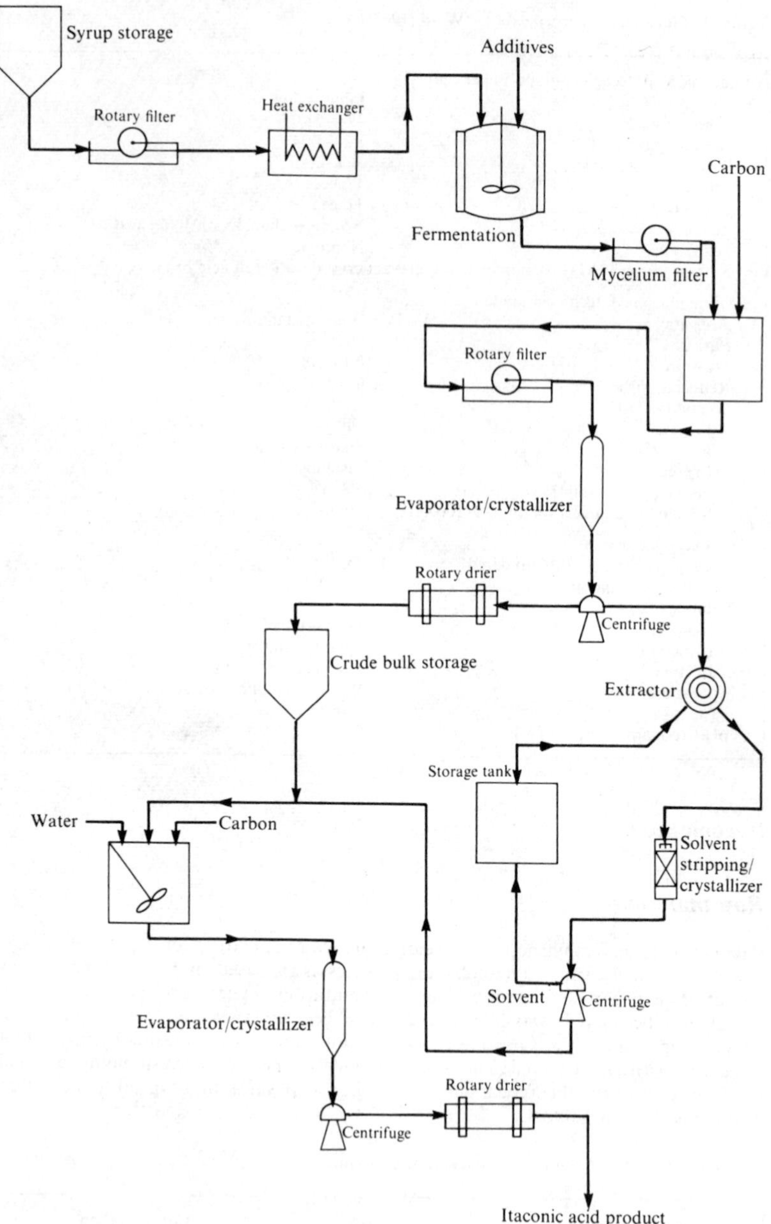

Figure 41. Flow sheet for itaconic acid manufacture [Lockwood and Schweiger (1967)].

Table 43. Data for itaconic acid fermentation by *Aspergillus terreus* strains [Prescott and Dunn (1959)].

Microorganism strain	Fermentation, scale, aeration, agitation	Sugar Source	Sugar Amount (g ℓ$^{-1}$)	Nitrogen Source	Nitrogen Amount (g ℓ$^{-1}$)	Corn-steep liquor (ml ℓ$^{-1}$)	Magnesium sulphate.$7H_2O$ (g ℓ$^{-1}$)	pH	Acid added to adjust pH	Temperature (°C)	Duration (days)	Yield (%)
ATCC 10020	Submerged, 20 ℓ glass fermenter, ~ lvvm	Sucrose	8	NH_4NO_3	2.5	1.5	1.5	1.8–1.9	H_2SO_4	30–31	7–8	38[a]
ATCC 10020 + NRRL 1960	Surface, 6 ℓ glass fermenter	Sucrose	14–17	NH_4NO_3	2.5	4	4.4	1.8–2.3	H_2SO_4	30–31	12	41[a]
ATCC 10029 + NRRL 1960	Surface, 230 ml glass fermenter	Sucrose	12–18	NH_4NO_3	5–8		6–9	3	H_2SO_4, HCl or HNO_3	30	21	~32[a]
NRRL 265	Surface, 200 ml glass fermenter	Glucose	25	NH_4NO_3	0.25	4	0.25	1.8–2.0	HNO_3	30	10–12	28–29[a]
NRRL 1960	Surface, 200 ml glass fermenter	Glucose	25	NH_4NO_3	2.5	4	4.5–5.0	1.9–2.3	HNO_3	30	12	~37[b]
	Surface, 12 ℓ aluminium fermenter	Glucose	16.5	NH_4NO_3	2.5	4	4.4	~2.0	HNO_3	30–32	12	30[a]
	Submerged, 300 ml glass fermenter, agitated	Glucose	6	NH_4NO_3	2.5	1.5	0.75	1.8–1.9	HNO_3	30	12	25.5[a]
	Submerged, 20 ℓ stainless steel No. 360 fermenter, 0.03 vvm, 15 psig	Glucose	6	$(NH_4)_2SO_4$	2.67	1.5	5	1.8–2	H_2SO_4	34	2–3	45–54[a]
	Submerged, 200- or 600-gal. stainless steel No. 302 or 347 fermenter; 0.125 vvm, 10–20 psig, 75–175 rpm	Glucose	6	$(NH_4)_2SO_4$	2.7	1.8	0.8	2.1–4.9	H_2SO_4 or itaconic acid	35	2–3	60[a]
Unknown	Submerged, 0.25 vvm, 15 psig, 100–125 rpm	Glucose	5–7	$(NH_4)_2SO_4$	3	1.5	0.8	2.5–5.0	H_2SO_4 or itaconic acid	34–35	~2.75–4	61–65[a]

[a] Based on sugar supplied.
[b] Based on sugar consumed.

Process outline

Table 43 gives some early data for itaconic acid fermentations and Fig. 41 contains the flow sheet of an itaconic acid process.

The fermentation medium is inoculated with one-fifth volume of a suspension of germinated spores. The temperature is maintained at 39–42°C and air is sparged to provide vigorous agitation. During the first 24 hours of incubation, the pH changes from 5.1 to 3.1 Then lime or ammonia is added to adjust the pH to about 3.8, and the fermentation is continued for two or more days. The final itaconic acid solution, starting with molasses with a sugar content of 150 g ℓ^{-1}, contains 85 g ℓ^{-1}.

Itaconic acid fermentations resemble those of citric acid, the copper concentrations being similar and excessive iron concentrations reduce the accumulation of itaconic acid.

Whenever the concentration of itaconic acid exceeds 7 per cent, the fermentation rate is reduced. However, if the titratable acidity is neutralized using ammonia or sodium hydroxide, the fermentation rate remains constant, and the final acid concentration may reach 15–18 per cent.

Product recovery

If some of the acid formed during fermentation has been neutralized, the resulting broth is acidified using an inorganic acid. Mycelia and other suspended solids are then removed by filtration, the filtrate then being treated with activated carbon, which is removed subsequently by filtration. The resulting filtrate is then evaporated and cooled to induce crystallization. When industrial-grade product is required, the treatment with activated carbon is omitted.

Lactic Acid

Raw materials

There are numerous species of bacteria and fungi that are capable of producing relatively large amounts of lactic acid from carbohydrates; the type of microorganisms selected depends on the carbohydrate fermented. *Lactobacillus bulgaricus*, *L. casei* or *Streptococcus lactis* may be used to ferment milk or whey, *L. bulgaricus* being favoured. *L. delbrueckii*, *L. pentosus*, *L. leichmannii* or *L. bulgaricus* may be used to ferment glucose or maltose. For hydrolyzed starches, *L. delbrueckii* is employed with another lactic acid producer, such as *L. bulgaricus* or *Strep. lactis*. Apart from these commercially preferred, homofermentative lactobacilli (i.e. the species that produce only cells and lactic acids in significant amounts), algae, yeasts and phycomycetous fungi are also capable of producing lactic acid. Yields comparable to those obtained using homofermentative lactic acid bacteria have been obtained only in glucose media using the fungus *Rhizopus oryzae*.

Lactic acid is generally produced from glucose, maltose, sucrose or lactose. Starches, especially those from corn and potatoes, are hydrolyzed by enzymes or by acid – preferably, sulphuric acid – to maltose and glucose before the lactic acid fermentation. Xylose is fermented by *L. pentoaceticus* to yield acetic acid as well as lactic acid. Molasses and whey can be used as sources of sugars. Sulphite waste liquor and Jerusalem artichokes are potential raw materials. For the best results using sulphite waste liquor, it must be steam stripped to remove sulphur dioxide and treated with alkali to remove lignin before fermentation.

Lactic acid bacteria have complex nutritional requirements, especially for B vitamins. These requirements are usually met by enrichment of the medium with crude vegetable sources, such as malt sprouts, i.e. the rootlets of germinating barley grains. However, care has to be taken not to overheat the enrichment materials during drying so that they do not loose their nutritional value.

The sugar concentration in the medium is initially adjusted to 5–20 per cent, but usually does not exceed 12 per cent. At higher sugar concentrations, calcium lactate may crystallize out in the fermenter with the formation of a solid mass that is difficult to handle.

The inoculum volume is usually about 5 per cent of the fermentation broth volume. The pH of the culture broth is kept in the range 5.5–6.5 by neutralization of the acid produced. Continuous control of pH is advantageous since the yields and rates can be increased thereby. If the lactic acid is not neutralized, high acidity will develop which the lactic acid bacteria cannot tolerate and the fermentation will not go to completion. Plant fermentation time is commonly 5–10 days, but with control of pH at 6.3–6.5 by continuous neutralization using a slurry of calcium hydroxide. This complete fermentation of 12–13 per cent glucose is achieved in 72 hours. Commercial fermentation yields are 93–95 per cent of the weight of glucose supplied.

Lactic acid fermentation is anaerobic, but the microorganisms used are facultative aerobes, thus making the strict exclusion of air unnecessary. The broth is usually gently stirred to keep calcium carbonate in suspension.

The fermentation is carried out at comparatively high temperatures, in fermentation using *L. delbrueckii* a temperature of 45°C, or higher, using *L. bulgaricus* 45–50°C and using *L. pentosus*, *L. casei* or *Strept. lactis* about 30°C.

Process outline

Fig. 42 is a flow sheet for the production of lactic acid using corn sugar as the raw material.

The pH of the medium used is maintained at 5.8–6.0. The culture tanks are charged with 375 gallons of mash. The 6600- and 30 000 gal. fermenters have working volumes of 4600 and 24 000 gal., respectively. The temperature is maintained at 49°C by circulating water through the stainless steel jackets of the culture tanks and the stainless steel coils of the fermenters.

The inoculum, *L. delbrueckii* is developed from a culture tube to a 500 ml flask, to a 6 ℓ flask containing 3 ℓ of medium and then to a culture tank. The contents of one culture tank are used to inoculate a 6600 gal. fermenter, those of three culture tanks a 30 000 gal. fermenter.

The fermentation is considered to be complete when the reducing sugar content drops to 0.1 per cent. The normal time of fermentation is four to six days. At the end of the fermentation, the broth is heated to 83°C to destroy the bacteria.

Table 44 gives lactic acid yields attainable using three different stains of bacteria, *Strept. lactis*, *L. casei* and *L. delbrueckii*, while Table 45 lists some data for lactic acid production from molasses by *L. delbrueckii*.

Table 44. *d*-Lactic acid production by various microorganisms [Tatum and Peterson (1935)].

Microorganism	Temperature (°C)	Lactic acid formed [g (540 g glucose)$^{-1}$]	Glucose converted (%)	Zinc lactate analysis: Water of crystallization (%)	Zinc lactate analysis: Specific rotation[a] ($[\alpha]_D^{20}$)
Lactobacillus casei	37	505	93	12.80	− 8.22
L. delbrueckii	37	517	95	12.92	− 8.33
Streptococcus lactis R	30	510	94	12.38	− 8.65

[a] 4 per cent concentration.

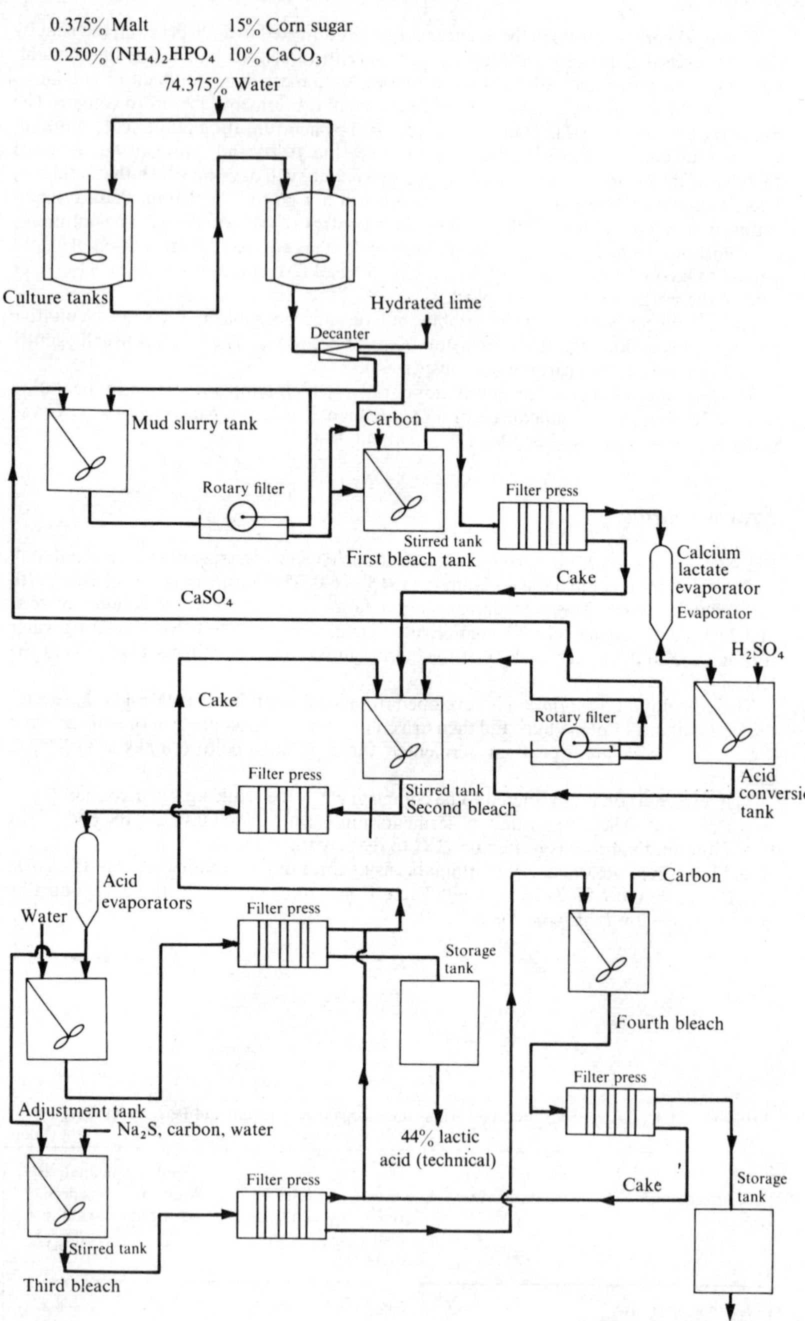

Figure 42. Flow sheet for the production of lactic acid at Hammond, Ind., plant of American Maize-Products Co. [Inskeep *et al.* (1952)].

Table 45. Large-scale *l*-lactic acid fermentation by *Lactobacillus delbrueckii*[a] [Pan *et al.* (1940)].

Factor	Run 1	Run 2
Puerto Rican blackstrap molasses (kg)	54.4	45.5
Sugar content of molasses (%)	55.9	59.5
Malt sprouts (kg)	6.7	6.6
Calcium carbonate (kg)	15.0	14.7
Total volume (ℓ)	241.5	238.5
Duration of fermentation (h)	21	16
Initial sugar concentration (g ℓ^{-1})	126	113.2
Final sugar concentration (g ℓ^{-1})	11.0	14.6
Fermentation (%)	91.3	87.0
Lactic acid (g ℓ^{-1})	110.0	93.5
Yield (% sugar fermented)	95.7	95.0
Yield (% sugar in molasses)	87.3	82.6
Specific rotation of calcium lactate[b] ($[\alpha]_D$)	+6.16	+6.06

[a] Temperature, 44–46°C.
[b] Actual value for calcium lactate is 6.13.

Product recovery and specification

The problems in the lactic acid process lie mostly in the recovery and not in the fermentation step. Both the *d*- and the *l*-forms of the acid can only be crystallized with great difficulty and in low yields; in their purest form the acids are usually colourless syrups that readily absorb water.

The suspended solids and most of the microorganisms are removed from the solution by conventional industrial precoat filters. For technical-grade products, calcium present is precipitated as calcium sulphate dihydrate, which can be filtered off, and then the filtrate is concentrated to 35–40 per cent lactic acid by evaporation. If more calcium sulphate precipitates, it is removed also by filtration.

Food-grade lactic acid is an aqueous solution of 50–65 per cent acidity. A medium with a higher-grade sugar source and a minimal protein content is used for this fermentation. Calcium is precipitated as calcium sulphate and washed. This wash is combined with the filtrate and is treated with activated carbon, evaporated to 25 per cent solids, retreated with activated carbon and finally evaporated to about 50–65 per cent total acidity. If the final product is discoloured, ferrocyanide can be used to precipitate iron or copper ions which are then filtered off. Alternatively, the broth is filtered to remove suspended solids, evaporated to recover calcium lactate, which is then washed with cold water, centrifuged and then decomposed with sulphuric acid. The rest of the recovery procedure is as described above.

Plastic-grade lactic acid can be obtained by esterification with methanol after concentrating. Another procedure is solvent extraction with *i*-propyl ether followed by extraction of the ether with water. The product is a colourless solution of 85 per cent acidity, with 76–78 per cent lactic acid, 2–3.5 per cent volatile acids and 0.5–1 per cent ash. Approximately 30 per cent of the acid is either lactide polymers or linear polymers or both.

A summary of commercially available derivatives of lactic acid and their uses is given in Table 46.

Table 46. Commercially available derivatives of lactic acid and uses [Prescott and Dunn (1959)].

Derivative	Use
Antimony lactate	In mordant dyeing
Calcium lactate	Calcium therapy
Copper lactate	Electroplating baths
Iron lactate	Treatment of anaemia
Sodium lactate	Plasticizer, humectant
Strontium lactate	Pharmaceutical
n-Butyl lactate	Solvent in lacquers
Ethyl lactate	Solvent, lubricant
Methyl lactate	Intermediate

POLYSACCHARIDES

General Aspects

Microbial polysaccharides can be classified as extracellular, structural or intracellular storage forms. Extracellular polysaccharides can be either exocellular capsules of the cell wall or loose slime components that accumulate outside the cell wall and then diffuse into the medium.

Although many polysaccharides are produced by bacteria, yeast and fungi, only a few have been utilized commercially (*see* Tables 47 and 48).

As can be seen from Table 48, the substrates for the microbial production of polysaccharides are mainly sucrose, glucose and other simple sugars. For some microorganisms, the carbon source determines both the quality and the quantity of the polysaccharide produced. For instance, *Bacillus polymyxa* when grown on a sucrose agar medium synthesizes a heteropolysaccharide consisting of glucose, mannose and fructose, whereas, when grown on a medium containing a monosaccharide, such as fructose, glucose, arabinose or galactose, the polysaccharide is made up of glucose, mannose and uronic acid. However, a strain of *B. polymyxa* produces a heteropolysaccharide when the

Table 47. Microbial polysaccharides of commercial importance [Lawson and Sutherland (1978)].

Polysaccharide	Current state of development	Future development	Trade name	Company involved
Bakers' yeast glucan	In development	Unknown	BYR	Anheuser-Busch Inc.
Curdlan	In development	Unknown		Takeda Chemical Ind.
Dextran	In production	Static	Various	Dextran Products, Polydex
Erwinia exopolysaccharide	In production in USA	Unknown	Zanflo	Kelco
Microbial alginate	In development	No decision yet made to commercialize		
Pullulan	In development	Commercialization announced	Pullulan	Hayashibara Corp.
Scleroglucan	In development	Unknown	Polytran F.S.	Pillsbury
Xanthan gum	In production	Expanding	Keltrol Rhodopol 23	Kelco Rhône Poulenc/ General Mills

Table 48. Processes used in the production of microbial polysaccharides of commercial importance [Lawson and Sutherland (1978)].

Polysaccharide	Microorganism	Substrate	Process	Sugar residues in polymer
Bakers' yeast glucan	*Saccharomyces cerevisiae*	Glucose	Batch	Glucose, mannose
Curdlan	*Agrobacterium* sp., *Alcaligenes faecalis*	Glucose	Batch	Glucose
Dextran	*Leuconostoc mesenteroides*	Sucrose	Cell-free enzyme	Glucose
Erwinia exopolysaccharide	*Erwinia tahitica*	Glucose, glucose syrup	Batch	Glucose, galactose, fucose, uronic acid (acetyl)
Microbial alginate	*Azotobacter vinelandii*	Sucrose	Batch and continuous	Mannuronic acid, guluronic acid (acetyl)
Pullulan	*Aureobasidium pullulans*	Glucose syrup	Batch	Glucose
Scleroglucan	*Sclerotium glucanicum*	Glucose		
Xanthan gum	*Xanthomonas campestris*	Glucose, glucose syrup	Batch	Glucose (acetate) glucuronic acid, mannose (pyruvate)

liquid medium contains glucose as the carbon source but a homopolysaccharide containing only fructose when sucrose is used as the carbon source. In contrast, the extracellular heteropolysaccharides synthesized by *Xanthomonas campestris* do not change significantly in composition or molecular weight as a result of changes in the nature of the substrate.

The concentration of the carbon source affects the efficiency with which it is converted into polysaccharide. For instance, an increase in glucose concentration can decrease the conversion efficiency of glucose to polysaccharide by *X. campestris*. A nitrogen source is essential for cell growth and the synthesis of the enzymes necessary for polysaccharide formation. However, an excess of nitrogen generally reduces conversion of the carbon source to extracellular polysaccharide.

All commercial-scale polysaccharide formation processes are aerobic. As the viscosity of the medium increases with polysaccharide formation, oxygen transfer to the cells becomes increasingly more difficult (*see* Chapters 8 and 9). Temperature is often a critical factor in polysaccharide synthesis, generally the optimum temperature for growth being that for product formation also. All commercial polysaccharide-producing microorganisms are mesophiles. In order to increase oxygen transfer, usually hot air is sparged through the medium with high degrees of aeration. The optimum pH for the synthesis of bacterial polysaccharides is 6–7.5. For fungi, the pH optimum is 4–5.5.

Many microorganisms have strict requirements for certain elements, such as potassium, phosphorus, magnesium and calcium. Other elements, such as molybdenum, iron, copper and zinc may also be necessary. However, depending on the microbial species, certain minerals can inhibit product formation.

Xanthan Gum

Raw materials

The efficient xanthan gum-producers are *Xanthomonas campestris, X. phaseoli, X. malvacearum, X. carotae, X. manihotis* and *X. juglandis*.

For commercial production, commercial-grade glucose or starch, thinned by a combination of acid and enzyme treatment, is used as the substrate. Sucrose and acid whey are also effective in polysaccharide production. The glucose in the whey hydrolyzate is used more rapidly than the galactose, but both sugars are almost completely utilized without diauxy. Glucose concentrations of 1–5 per cent have been found to give the best xanthan gum yields; at higher glucose concentrations, product yields decrease.

Nitrogen sources, such as corn-steep liquor, casein hydrolyzate and distillers' dried solubles, can be used but due to their relatively undefined and variable composition, they have been superseded by yeast and soya bean extracts. Ammonium chloride, dipotassium hydrogen phosphate, magnesium chloride or sulphate and some trace elements are also included in the media.

Process outline

Xanthan gum is produced commercially in a conventional batch process, which usually runs for approximately 80 hours. Continuous fermentation has only been investigated on a laboratory scale but can be operated at a dilution rate of $0.05\,h^{-1}$ to give a 20-hour fermentation. However the problem of maintaining sterility even in batch fermenters indicates that continuous processes would have a high risk of contamination. The process for xanthan gum production is summarized in Fig. 43.

The temperature of fermentation is about 28°C and the process is aerobic. During fermentation, the pH of the medium falls due to the formation of metabolic acids and xanthan gum, which contains acidic functional groups. If the pH reaches a critical value, i.e. 5.0, the gum production decreases greatly. A near neutral pH allows gum synthesis to continue until all the carbohydrate is exhausted. When fermentation is terminated, the broth usually has a pH of about 6.0 and a viscosity of greater than 30 000 cP [measured on the Contraves viscometer at 25°C] and a shear rate of $1\,s^{-1}$. High yields approaching 75–80 per cent could be expected but, due to difficulties encountered because of the high viscosity, the concentration of the gum is about 5 per cent.

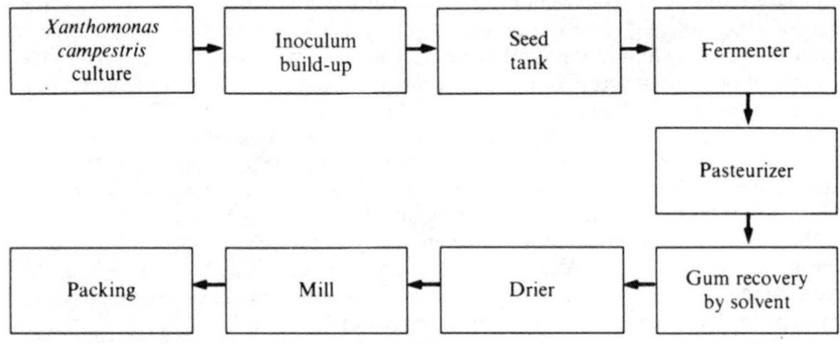

Figure 43. Flow sheet for xanthan gum process [Kang and Cottrell (1979)].

Product recovery and specification

Product recovery starts with the pasteurization of the broth to kill the bacteria. Cells may also be subjected to enzymic digestion.

Drum or spray drying of the fermentation liquor yields a crude polysaccharide product. Other grades of product are obtained by precipitating the gum from the fermentation liquor using methanol or *i*-propanol in the presence of potassium chloride. The added alcohol is then normally recovered from the spent liquor by distillation. Long-chain quaternary ammonium salts may also be used as precipitators but this method is not acceptable for a food-grade product. Alternatively, xanthan gum can be precipitated with calcium ions followed by washing the insoluble calcium complex with acid or salt.

The precipitated cake is recovered by filtration and centrifugation, followed by drying, milling, testing and packaging. Sometimes the product is shredded before drying on a moving-band drier.

A clear product can be obtained by diluting the fermentation liquor and then clarifying by filtration, by treating with hypochlorite or by heating at a high pH.

Xanthan gum has a high molecular weight (in the range 2×10^6–5×10^7 daltons) and contains D-glucose 2.8 mol), D-mannose (3.0 mol), D-glucuronic acid (2.0 mol), acetic acid (approximately 4.7 per cent) and pyruvic acid (approximately 3 per cent).

One of the most important properties of xanthan gum is its ability to control the rheological properties of fluids. It can be dissolved in either hot or cold water to produce a high-viscosity, pseudoplastic solution even at low concentrations. Aqueous solutions containing 0.75 per cent or more of the gum have a shear stress of 15 dynes cm^{-2} in the absence of salts and of 52 dynes cm^{-2} in the presence of 1 per cent potassium chloride at a shear rate of 0.01 s^{-1} (*see* Fig. 44).

Fig. 45 indicates that xanthan gum solutions have excellent thermal stability. Similarly, Fig. 46 implies that xanthan gum solutions, in the presence of a low level of salt, are unaffected by pH in the range 1.5–13.

Xanthan gum dissolves directly in acid solutions, such as 5 per cent sulphuric acid, 5 per cent nitric acid, 5 per cent acetic acid, 10 per cent hydrochloric acid and 25 per cent phosphoric acid, and the resulting solutions are reasonably stable at 25°C for several

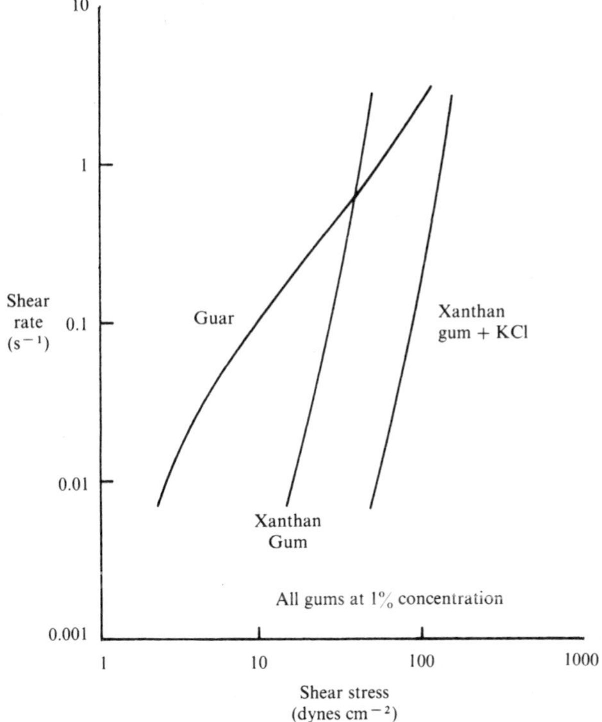

Figure 44. Rheological characteristics of guar gum, xanthan gum and xanthan gum with potassium chloride in aqueous solutions [Kang and Cottrell (1979)].

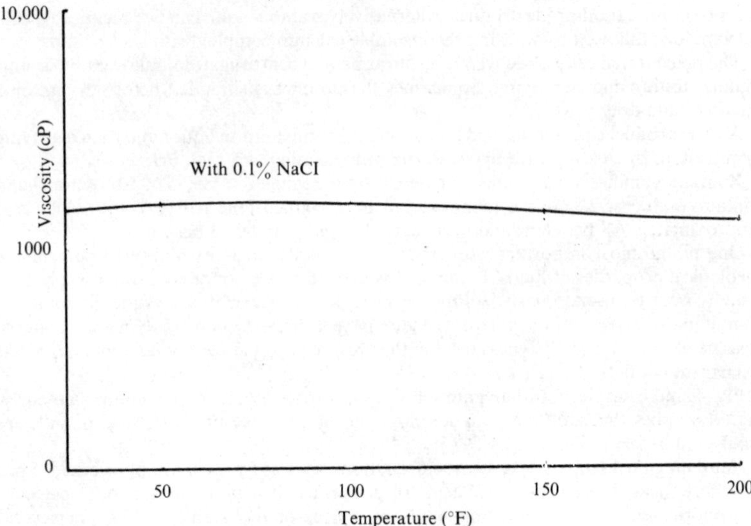

Figure 45. Effect of temperature on 1 per cent xanthan gum solution viscosity [Kang and Cottrell (1979)].

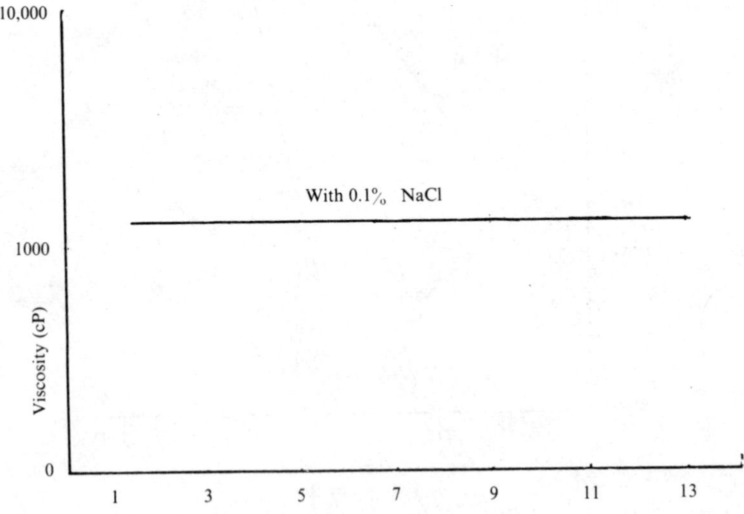

Figure 46. Effect of pH on viscosity of 1 per cent xanthan gum solution [Kang and Cottrell (1979)].

months. Highly alkaline solutions have excellent stability at 25°C. Most salts are compatible with xanthan gum solution. However, polyvalent metal ions cause gelation or precipitation at high pH values. Solutions of xanthan gum are compatible with ethanol, methanol, *i*-propanol and acetone at concentrations of up to 50–60 per cent. At higher concentrations, gelation or precipitation occurs. Xanthan gum is insoluble in most organic solvents, except for formamide at 25°C and ethylene glycol at 65°C.

Enzymes, such as cellulase, hemicellulase, pectinase and amylase, do not degrade the gum in solution. However, it is broken down by strong oxidizing agents, such as peroxides, persulphates and hypochlorites.

Dextran

Raw materials

Although a large number of bacteria are capable of producing dextran, commercially used species are limited to *Leuconostoc mesenteroides* and *Streptobacterium dextranicum*.

The medium contains 10 per cent sucrose, 0.5 per cent dipotassium hydrogen phosphate, 0.25 per cent Difco yeast extract and 0.02 per cent magnesium sulphate heptahydrate. The inoculum is increased over several stages in a similar medium at 25°C in an agitated fermenter to provide a 10 per cent fermentation solution. Molasses is usually used as the carbohydrate source.

Process outline

Fig. 47a contains a flow sheet for the production of clinical-grade dextran.

The raw materials are mixed and heated to 60°C in a slurry tank. The medium is sterilized continuously at 142°C and cooled to 25°C before being distributed between the inoculum tank (206 gal.) and the main 1300 gal. fermenter. The culture is built up over several stages at 25°C.

During the fermentation, the pH decreases and the viscosity of the medium increases, with a pH at the end of about 4.5 and a viscosity of 400–700 cP. pH adjustment is achieved by addition of alkali and further sucrose is also added at intervals. The production of dextran is rapid and basically involves the use of bacterial extracellular enzymes present in the culture broth. Contamination does not present a problem because the fermentation is relatively short. Cultures of *L. mesenteroides* do not require vigorous aeration.

Product recovery and specification

Dextran is recovered by precipitation using acetone or alcohol and then the product is purified to make it suitable for clinical use.

The molecular weight of the dextran ranges from 1×10^4 to 2×10^7 daltons or higher. A 2 per cent solution of dextran has a viscosity of about 150 cP. Dextran has good compatibility with salts, acids and bases, and is resistant to degradation at high temperatures. Table 49 provides an early example of the specification required for dextran.

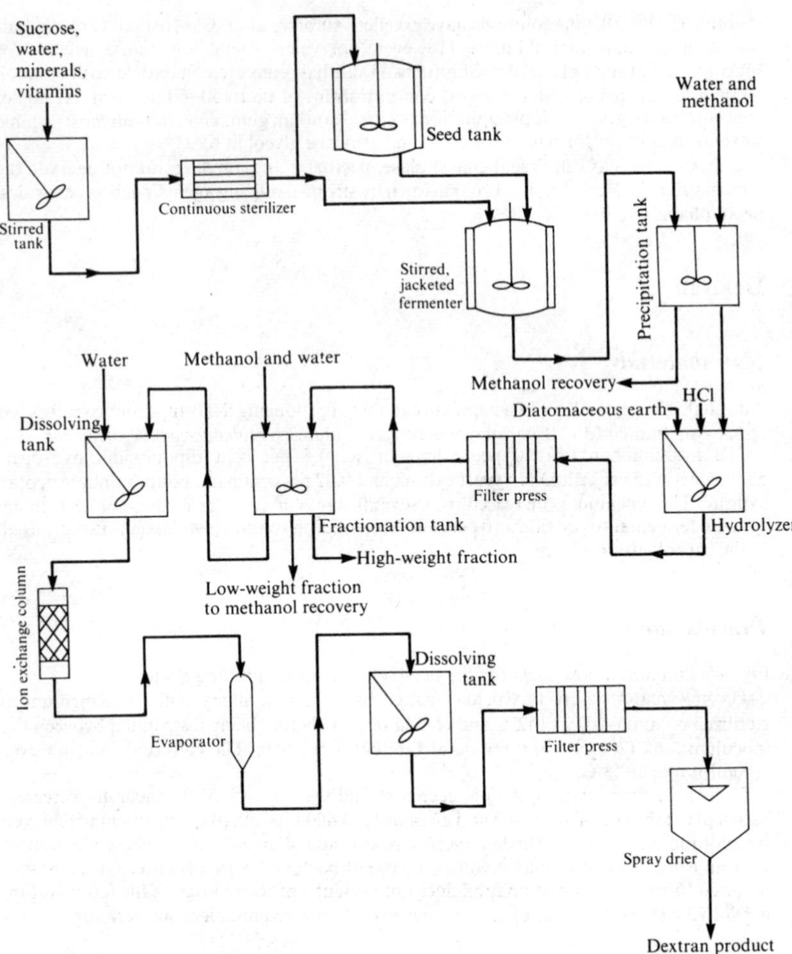

Figure 47a. Flow sheet for the production of clinical dextran: plant of the Commercial Solvents Corp. at the Terre Haute, Ind., USA [Bixler *et al.* (1953)].

Erwinia Exopolysaccharide (Zanflo)

Zanflo is the trade name of a heteropolysaccharide produced by an extensively mutated strain of the bacterium *Erwinia tahitica*, which is isolated from soil.

The medium contains lactose and hydrolyzed starch as carbon sources, which are preferable to glucose, sucrose and maltose. The other medium constituents are phosphate as a buffering agent, ammonium nitrate and soya protein as nitrogen sources, magnesium

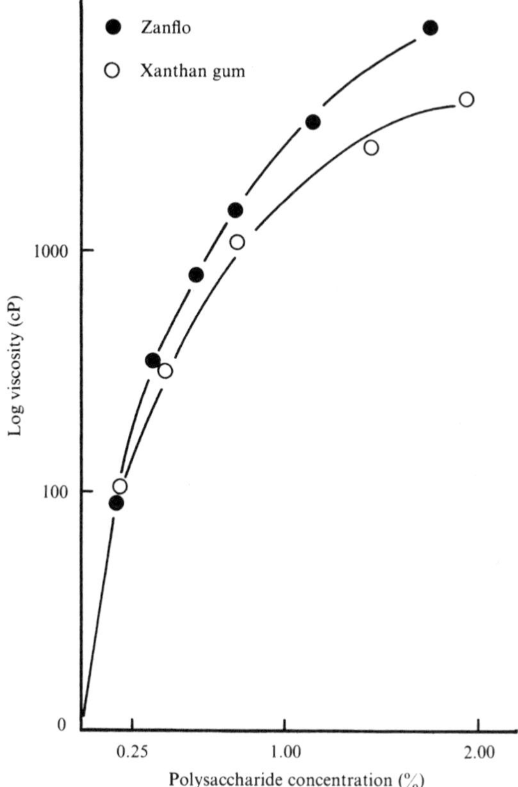

Figure 47b. Viscosity versus concentration for Zanflo [Kang and Cottrell (1979)].

Table 49. Specification for dextran[a] [Armed Forces Medical Procurement Agency (1952)].

Property	Value
Dextran content (g ℓ$^{-1}$)	57–63
Sodium chloride (g ℓ$^{-1}$)	8.5–9.5
pH	5.0–7.0
Viscosity at 25°C (cP)	2.5–3.5
Maximum nitrogen (mg ℓ$^{-1}$)	10
Maximum ash (residue on ignition) (mg ℓ$^{-1}$)	0.56
Average molecular weight[b, c]	75 000 ± 25 000
Intrinsic viscosity (ℓ g^{-1})	0.023 ± 0.005
Colour	Colourless
Maximum buffering capacity (ml 0.1 N NaOH per litre of dextran solution)	30
Pyrogenicity	Negative
Toxicity	Negative
Antigenicity	Negative
Sterility	Negative

[a] Values for 6 per cent solutions.
[b] Determined by light-scattering measurements.
[c] The high molecular weight fraction (5–10 per cent) should not exceed 200 000 and the low molecular weight fraction (5–10 per cent) should not exceed 25 000.

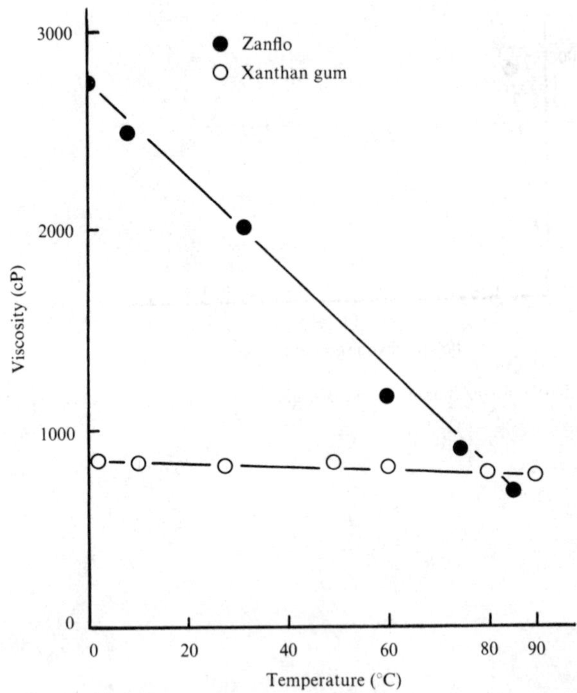

Figure 48. Effect of temperature on Zanflo (1 per cent) viscosity [Kang and Cottrell (1979)].

sulphate and trace minerals. The process is a submerged, aerobic fermentation carried out at about 30°C in the presence of added iron.

The product contains 97 per cent carbohydrate and 3 per cent protein. The carbohydrate portion contains glucose, galactose, glucuronic acid and fucose in the molar ratio of 3:2:1.5:1. Uronic acid is about 20 per cent (by wt) of the polysaccharide.

Fig. 47b shows the viscosity–concentration relationship for Zanflo. Fig. 48 demonstrates the effect of temperature and Fig. 49 the effect of pH on Zanflo viscosity.

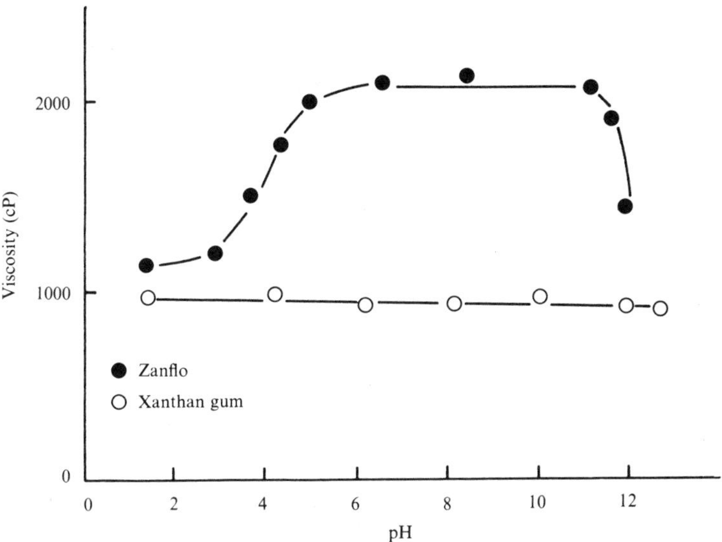

Figure 49. Effect of pH on Zanflo viscosity [Kang and Cottrell (1979)].

SINGLE-CELL PROTEIN

The aim of the numerous processes developed for the production of rapidly growing microorganisms, which are capable of utilizing readily available organic materials as their carbon and energy sources, is to convert cheap, inorganic nitrogen compounds into nutritionally valuable cellular proteins. Such proteins may be used as replacements for the more expensive, traditional plant and animal proteins used in human and animal diets.

Various groups of microorganisms, including algae, bacteria, yeasts, moulds and higher fungi, have been considered for use as sources of protein, the dried cells of these microorganisms being referred to as single-cell proteins.

Table 50 lists some chemical plants capable of producing single-cell proteins, together with the substrates and products. Figs. 50 and 51 depict the elements of typical single-cell protein processes.

Table 50. Feed-grade single-cell protein plants [Schwartz and Leathen (1976)].

Plant class	Company	Plant location	Substrate	Type of microorganism	Plant size (ton yr^{-1})
Demonstration	British Petroleum	United Kingdom	*n*-Alkane	Yeast	4000
	Chinese Petroleum	Taiwan	*n*-Alkane	Yeast	1000
	Dianippon	Japan	*n*-Alkane	Yeast	Unknown
	Imperial Chemical Industries	United Kingdom	Methanol	Bacteria	1000
	Kanegafuchi	Japan	*n*-Alkane	Yeast	5000
	Kohjin	Japan	Unknown	Yeast	2400
	Kyowa Hakko	Japan	*n*-Alkane	Yeast	1500
	Milbrew	United States	Whey	Yeast	5000
	Shell	Netherlands	Methane	Bacteria	1000
	Svenska–Socker	Sweden	Potato starch	Yeast	2000
Semicommercial	British Petroleum	France	Gas oil	Yeast	20 000
	Imperial Chemical Industries	United Kingdom	Methanol	Bacteria	50 000
	United Paper Mills	Finland	Sulphite waste	Yeast	10 000
	USSR State	USSR	Unknown	Yeast	20 000
Commercial	British Petroleum	Italy	*n*-Alkane	Yeast	100 000
	Liquichimica	Italy	*n*-Alkane	Yeast	100 000
Other systems	LSU–Bechtel		Cellulose waste	Bacteria	
	Tate and Lyle		Citric acid waste	Fungi	
	ICAITI (Guatemala)		Coffee waste	Fungi	
	IFP		Carbon dioxide, sunlight	Algae	
	General Electric		Feed lot waste	Bacteria	
	Mitsubishi		Methanol	Yeast	
	Finnish Pulp and Paper		Paper pulp waste	Fungi	

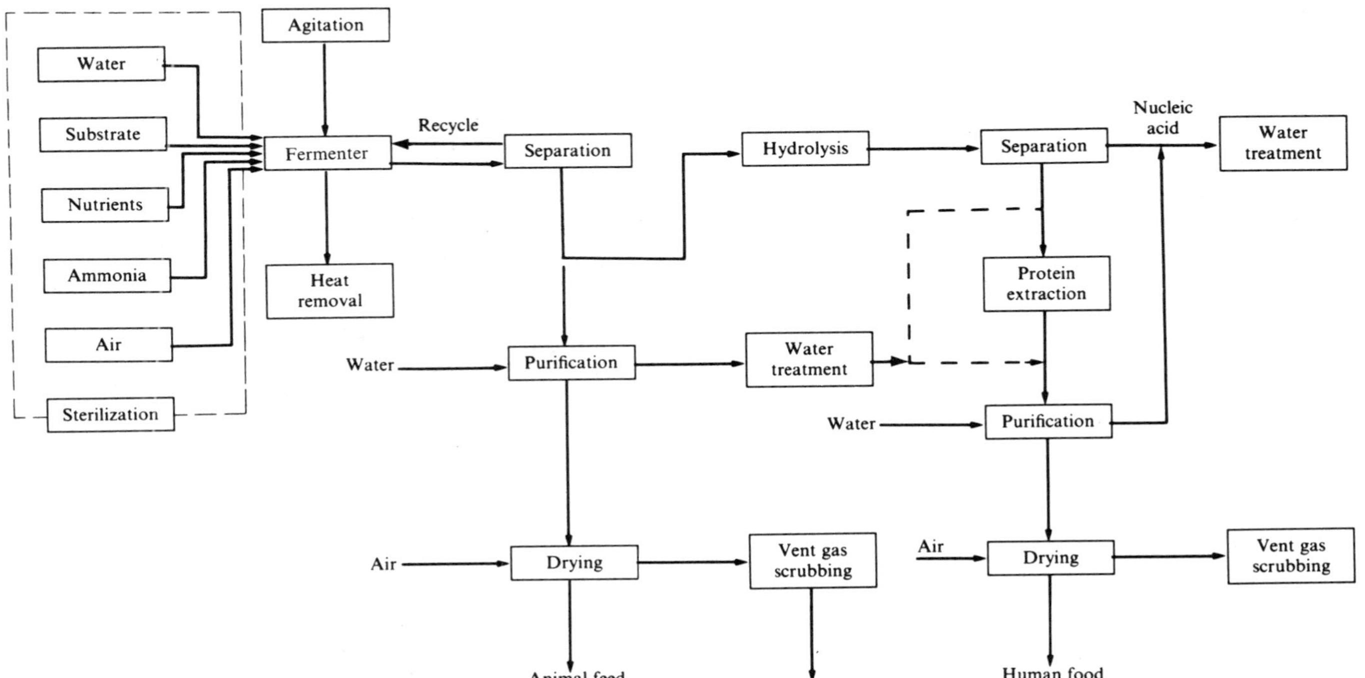

Figure 50. Elements of a typical single-cell protein process [Schwartz and Leathen (1976)].

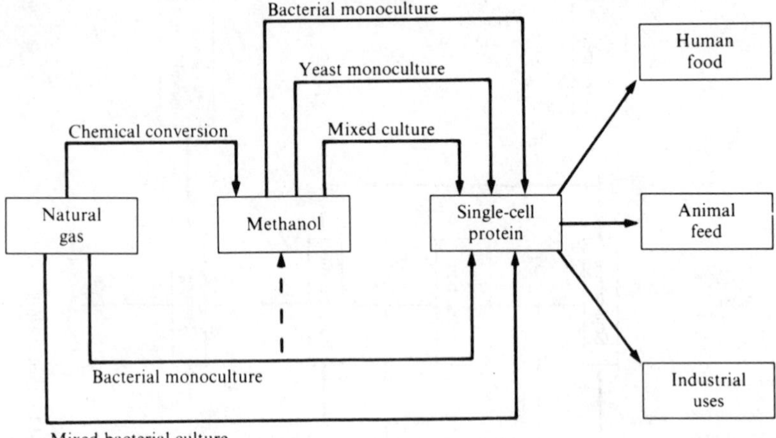

Figure 51. Alternative routes for single-cell protein production from natural gas [Topiwala (1974)].

Algae

Raw materials

Algae can be grown either photosynthetically or heterotrophically. Photosynthetic and autotrophic growth involves using either artificial or sunlight and carbon dioxide, while heterotrophic growth occurs in the dark with organic carbon and energy sources.

As illumination is the limiting factor in photosynthetic algal growth, outdoor cultivation is restricted to use of shallow ponds or 20–30 cm lagoons sited between latitudes 35°N and 35°S. The efficiency of light-energy conversion is low, resulting in a requirement of approximately 35 kWh for the production of 1 kilogram of algae. Algae use light with wavelengths in the region of 700 nm. *Chlorella sorokiniana* has been grown at dry weight concentrations of 25 g ℓ^{-1} at an illumination of 300 000 lm [*see* Litchfield (1979)], which is 30 times the intensity of sunlight as measured at the earth's surface.

The carbon dioxide content of air is low, being 0.03 per cent. Natural alkaline waters, which contain high concentrations of bicarbonates, may be used to enhance algal growth or additional carbon dioxide may be supplied using combustion gases, etc. For this purpose, the pH of the medium needs to be alkaline in order that bicarbonates, and then the dissolved carbon dioxide tension is independent of its partial pressure in the gas phase.

Table 51 gives some process conditions for the growth of algae at various levels of production.

Table 51. Growth of selected algae [Litchfield (1979)].

Microorganism	Carbon and energy source	Scale	Temperature (°C)	pH	Specific growth rate (h^{-1})	Culture density	Yield[a]
Chlorella pyrenoidosa 71105	CO_2(0.7% in air), urea (0.4 g ℓ^{-1}) fluorescent lamp (15 000 lm)	600-gal tank	39	4.5–5.8	0.0625	0.28% (by vol.)	1 g day^{-1}
C. pyrenoidosa	CO_2(10%), fluorescent lamp (52 000 lm)	2.7 ℓ	38	6.4–6.5		8.98 g ℓ^{-1}	36.5 g day^{-1}
Chlorella regularis	Autotrophic: CO_2 (5% in air), fluorescent lamp (40 klux)	1 ℓ	36	6.5	0.26	6 ml ℓ^{-1}	—
	Heterotrophic (dark): acetic acid (100 ml ℓ^{-1})	20-ℓ jar fermenter (13-ℓ medium)	36	6.8	0.28	15 ml ℓ^{-1}	0.48 g (acetate utilized)$^{-1}$
Chlorella sorokiniana	CO_2 (5% in air), 70 ml min^{-1} KNO_3 (20 mM) fluorescent lamps	405-ml annular chemostat	39				1.59 g day^{-1}
Oocystis polymorpha	CO_2 (5% in air), 70 ml min^{-1} fluorescent lamps	405-ml annular chemostat	39				1.25 g day^{-1}
Scenedesmus acutus 276-3A	CO_2, sunlight	Shallow tanks 55 m^2, 30-cm deep	Ambient	7–8			20 g m^{-2} day^{-1}
Sc. quadricauda	SO_2, sunlight	10-ℓ pond	25–35	8.5–9.5		1 g ℓ^{-1}	20 ton $acre^{-1}$ yr^{-1} i.e. (12.6 g/m^{-2} day^{-1})
Sc. quadricauda + *Sc. obliquus*	CO_2(0.5%), sunlight	54 000 ℓ (900 m^2)	34	6.5–7.0		1.5–2.0 g ℓ^{-1}	10 g m^{-2} day^{-1}

(continued)

Table 51 – (continued)

Microorganism	Carbon and energy source	Scale	Temperature (°C)	pH	Specific growth rate (h^{-1})	Culture density	Yield[a]
Spirulina maxima	CO_2(0.5%), combustion gases, bicarbonate, sunlight	700-m² pond	Ambient				15 g m^{-2} day^{-1}
	CO_2 (combustion gas), sunlight	20 × 5 × 0.1-m basin	Ambient				12 g m^{-2} day^{-1}
	CO_2-enriched air, 250-ml conical flasks (100 ml medium) 4000 lux, 2 vvm synthetic medium		30	9.5		2 g ℓ$^{-1}$ (12 days)	
	CO_2-enriched air, 250-ml conical flasks (100 ml medium) 4000 lux, 2 vvm sewage effluent		30	9.5		0.77 g ℓ$^{-1}$ (9 days)	
S. platensis	CO_2, 10-ℓ jar fermenter (20-cm diameter), 10 klux		35–37	8–10.5	0.019	4.2 g ℓ$^{-1}$	

[a] On dry weight basis.

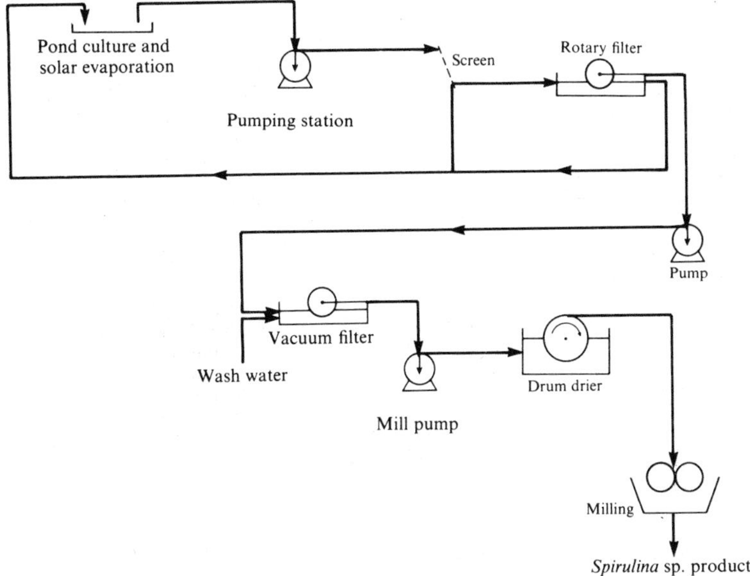

Figure 52. Flow sheet of the Sosa Texacoco pilot plant for producing *Spirulina* species [Benemann *et al.* (1979)].

Process outline

Fig. 52 shows the flow sheet for the Sosa Texacoco pilot plant for the cultivation, harvesting and processing of *Spirulina* species at the rate of 1 tonne per day. The cultivation unit is a 2–3 ft deep, 10 ha pond with a longitudinal baffle. The pond is fertilized using nitrate and iron.

Daily yields are quoted by Benemann *et al.* (1979), when averaged over the year, as 10 g m^{-2}. As can be seen from Table 51, daily yields on a dry weight basis are generally in the range 12–15 g m^{-2}.

Table 52 lists the values of some parameters and their control methods for the economically feasible operation of ponds involved in the production of algal biomass.

Product recovery

Separation of algal biomass from the suspending liquid presents a major problem due to the low dry weight biomass concentrations of 1–2 g ℓ^{-1} and the low settling rates of these cultures.

Cells are recovered by concentrating, dewatering and drying. Flocculants, such as aluminium sulphate, calcium hydroxide and cationic polymers, are effective but they cannot be subsequently separated from the biomass. Algae may flocculate in shallow ponds at pH 9.5 or above without the addition of flocculants [*see* Litchfield (1979)]. *Spirulina maxima* forms clumps that float to the surface when growth is maximal and these clumps can be harvested by skimming.

Table 52. Economically feasible pond operations in production of algal biomass [Benemann *et al.* (1979)].

Parameter	Normal limits	Control method(s)
Algal concentration	150–700 mg ℓ$^{-1}$	Harvesting, dilution, recycle, inoculation
Depth	20–50 cm	Dilution, harvesting
Hydraulic detention time	1.5–6.0 days	Dilution
Phytoplankton detention time	1.5–6.0 days	Biomass recycle, dilution
Zooplankton detention time	0.5–6.0 days	Harvesting-recycling with DSM screen
Hydraulic loading	2–20 cm^{-1} day	Dilution (applicable to oxidation ponds)
pH	6.0–10.15	Carbon dioxide, dilution
Nutrient additions	Should not be limiting	Add with dilution water or independently
Oxygen tension	0–25 mg ℓ$^{-1}$	Mixing, carbonation
Light absorption	Absorption of 99–99.9% of incident light	First four parameters, mixing

Bacteria and Actinomycetes

Raw materials

Various species of bacteria can utilize a wide range of carbon and energy sources, including sugars, starch, cellulose – either in pure form or as agricultural or forest product wastes – hydrocarbons and petrochemicals.

Carbon and energy substrate concentrations are in the range 1–5 per cent for batch cultures but can be considerably lower for continuous cultures. Carbon-to-nitrogen ratios in the growth medium are maintained at around 10:1. When the nitrogen content is lower, this can become the growth-limiting factor and lipids or, in some cases, poly-β-hydroxybutyrate are accumulated in the cells. Anhydrous ammonia or ammonium salts are suitable nitrogen sources. Phosphorus is supplied as feed-grade phosphoric acid to avoid contamination due to arsenic or fluoride found in crude industrial phosphoric acid. Mineral salts in water supplies are usually adequate for growth, although sometimes iron, magnesium and manganese as the sulphates or hydroxides, rather than as the chlorides to avoid corrosion, are added. The pH is controlled in the range 5–7 by addition of ammonia and phosphoric acid.

Oxygen transfer to growing cells is an important factor for single-cell protein production under aerobic conditions, especially from hydrocarbons, methanol, ethanol and hydrogen. The potential explosion hazards that exist when using gaseous carbon and energy sources set a limit on the concentration of oxygen in the gas phase. For example, when methane is used for growth, the oxygen concentration in the gas should be less than 12 per cent (by vol.).

Table 53 contains a list of various substrates and the corresponding bacteria and actinomycetes that can utilize them. Various growth conditions for bacteria and actinomycetes producing single-cell protein are presented in Table 54.

Process outline

Fig. 53 shows the flow sheet for a process developed by Imperial Chemical Industries for the continuous production of *Methylophilus methylotrophus* using sterile medium contain-

Table 53. Various substrates that can be used by bacteria and actinomycetes for the production of single-cell protein.

Substrates	Bacteria and actinomycetes
Liquid *n*-Alkanes	*Acinetobacter cerificans, Achromobacter delvacuate, Mycobacterium phlei, Nocardia* sp., *Pseudomonas* sp.
Gaseous *n*-alkanes (methane, ethane, propane, *n*-butane, *i*-butane, propylene, butylene, etc.)	*Arthrobacter simplex, Brevibacterium ketoglutamicum, Corynebacterium hydrocarbonoclastus, Nocardia paraffinica*
Cellulosic wastes	*Thermomonospora fusca*
Ethanol	*Acinetobacter calcoaceticus*
Hydrogen	*Alcaligenes eutrophus, A. paradoxus, Pseudomonas facalis, Ps. flava, Ps. palleronii, Ps. ruhlandii, Ps. saccharophila*
Methane	A mixed culture of *Acinetobacter* sp., *Flavobacterium* sp. and *Hyphomicrobium* sp., *Methylomonas methanica, Methylococcus capsulatus*
Methanol	*A mixed culture of Methylomonas methylovora* and *Flavobacterium* sp., *M. clara, M. methanolica, Methylophilus (Pseudomonas) methylotrophus, Pseudomonas* sp., *Ps. inaudita, Ps. utilis, Streptomyces* sp., *Xanthomonas* sp.
Sulphite waste liquor	*Pseudomonas denitrificans*

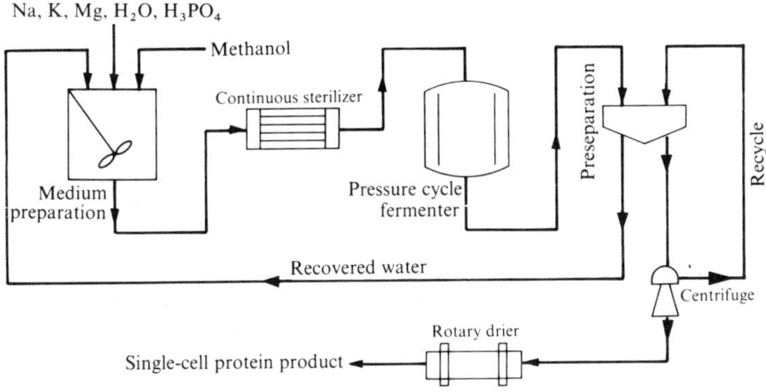

Figure 53. Production of bacterial protein from methanol [Moss and Smith (1977)].

ing methanol produced from natural gas, inorganic salts and ammonia. A suspension containing about 3 per cent dry weight bacterial cells is continuously removed. The cells are then concentrated by a flocculation and flotation process (the recovered liquid being recycled) and the product dried. The product, known as Pruteen, contains 72 per cent protein and 8.6 per cent total lipids with an amino acid profile high in lysine and methionine, being comparable to that of fish meal.

Another process that utilizes methanol to produce single-cell protein is outlined in Fig. 54, this time employing a loop-type fermenter. Fig. 55 shows the flow sheet of a process proposed by Louisiana State University for the production of single-cell protein from cellulosic waste materials.

Table 54. Growth of selected bacteria and actinomycetes on various substrates [Litchfield (1979)].

Microorganism	Carbon and energy source	Scale, aeration, agitation	Temperature (°C)	pH	Specific growth rate, μ or dilution rate, D (h^{-1})	Culture density[a] ($g\,\ell^{-1}$)	Yield[a] [g (gsubstrate)$^{-1}$]
Achromobacter delvacuvate	Diesel oil	6000-ℓ fermenter, 3000-ℓ medium, 1 vvm	35–36	7.0–7.2		10–15 (48 h)	
Acinetobacter (Micrococcus) cerificans	Gas oil	7.5-ℓ fermenter	30	7.0	$\mu = 0.4–1.0$	8–10	0.10–0.12
	n-Hexadecane	4.5-ℓ medium, 1 vvm 500 rpm			$\mu = 1.1–2.0$	8–10	0.80–0.90
	n-Hexadecane	7.5-ℓ fermenter, 3.5–7.0 mM $O_2\,\ell^{-1}\,min^{-1}$	30	6.8	$\mu = 1.33$		1.20
Bacillus megaterium	Collagen meat-packing waste	2.3-ℓ working volume fermenter, 3 vvm, 400 rpm (continuous)	34	7.0	$D = 0.25$		
Brevibacterium sp.	Mesquite wood	14-ℓ fermenter, 1.0–1.5 vvm, 1350–1500 rpm	30–37	6.45–7.2			0.444
Cellulomonas sp.	Bagasse	7-, 14-, 530-ℓ fermenters (batch or continuous)	34	6.6–6.8	$\mu = 0.20–0.29$ $D = 0.08–0.10$	16 (batch) 10 (continuous)	0.44–0.50
Methylococcus capsulatus	Methane (6.7%)	2.8-ℓ medium, O_2 [17.1–19.4% by vol)] 38.6–49.5 ml min^{-1}, 1450 rpm (continuous)	37	6.9	$\mu = 0.14$	0.4	1.00–1.03
Methylomonas clara	Methanol	1000-m^3 reactor volume (continuous)	39	6.8	$\mu = 0.5$ $D = 0.3–0.5$		0.50
Methylomonas methanolica	Methanol	4-ℓ working volume fermenter, 300–400 mM $O_2\,\ell^{-1}$, 1.5 vvm, 1200–1500 rpm (continuous)	30	6.0	$D = 0.24$	9.6	0.48
Methylophilus (Pseudomonas) methylotrophus methylotrophus	Methanol	Pressure cycle air-lift fermenter (continuous)	35–40		0.38–0.50	30	0.5

Nocardia sp. NBZ–23	*n*-Alkanes	7.5-ℓ fermenter, 6–8-ℓ medium, 1500 rpm	30	6.8	1.25	14.7	0.98
Pseudomonas sp. No. 5401	Fuel oil	6000-ℓ fermenter (batch)	36–38	7.0	$\mu = 0.16$	16 (24–26 h)	1.00
	Fuel oil	6000-ℓ fermenter (continuous)	36–38	7.0	$D = 0.12$	10	
	Fuel oil	6000-ℓ fermenter (continuous)	36–38	7.0	$D = 0.25$	8	
A mixed culture of *Hyphomicrobium* sp., *Acinetobacter* sp. and *Flavobacterium* sp.	Methane	10-ℓ fermenter, 0.9-ℓ medium, 1 atm, 470 rpm	32	5.7	$D = 0.06$	0.8	0.99
Rhodopseudomonas gelatinosa	Bicarbonate, wheat bran	4-ℓ working volume fermenter, 75-W incandescent lamp, 1100 rpm	40	0.2	$\mu = 0.31$ $D = 0.028$	4.33 (batch) 3.15 (continuous)	
Thermomonospora fusca	Cellulose-pulping fines	10-ℓ fermenter, 3 ℓ min^{-1}, 60 rpm	55	7.4			0.35–0.40

[a] On dry weight basis.

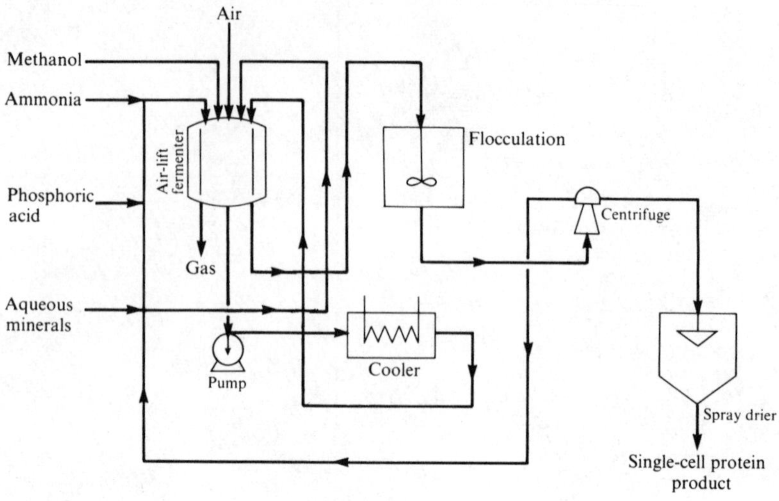

Figure 54. Schematic diagram of the Hoechst/Uhde single-cell protein process [Faust *et al.* (1977)].

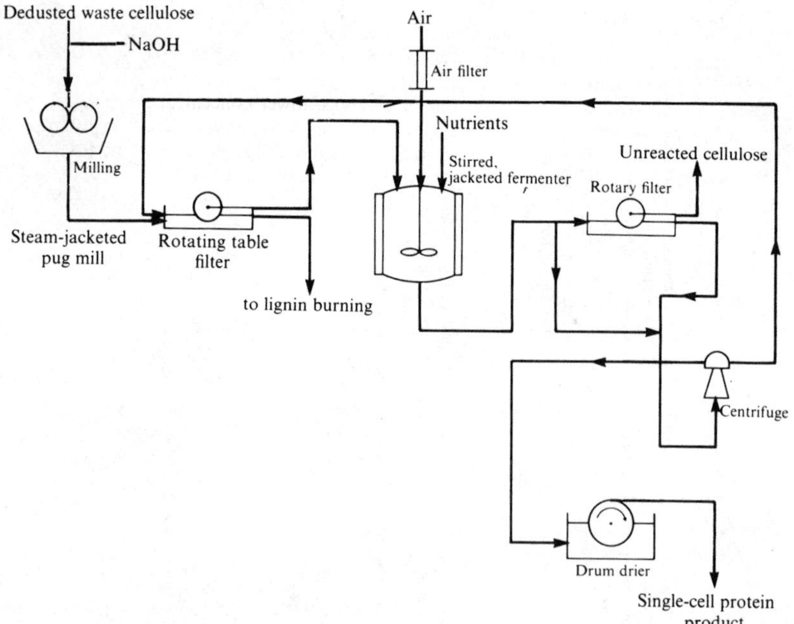

Figure 55. Flow sheet for producing single-cell protein (*Cellulomonas* sp.) in the Louisiana State University process [Callihan and Clemmer [1979)].

Product recovery

In most single-cell protein processes, the dry cell concentrations are in the range 10–20 g ℓ^{-1}. Therefore, large volumes of water must be handled. The bacterial size is about 1–2 μm and the bacterial cell density is about 1003 g m^{-3}. The cost of centrifugation may be four times as great as for yeast centrifugation.

Plate-and-frame filter presses are not amenable to continuous processing. In vacuum filters, the cell size and density cause compression on screens and filtration ceases as the void volume decreases to near zero. Filter aids and flocculants cannot be used as these would contaminate the product.

Acinetobacter cerificans has been concentrated by a two-zone froth flotation process described by Litchfield (1979). Imperial Chemical Industries has a proprietary process for the separation of *Methylophilus methylotrophus* without the use of flocculating agents, the biomass being subsequently concentrated in decanter-type centrifuges and then dried. The Hoechst/Uhde process separates *Methylomonas clara* by electrochemical coagulation and centrifugation followed by spray drying.

Yeasts

Raw materials

Table 55 lists, along with the growth conditions, some of the substrates that can be utilized by yeasts. These substrates include *n*-alkanes, methanol, ethanol, diesel oil, gas oil, brewery wastes, sulphite waste liquor, starch, anaerobic digester supernatant, molasses, cheese whey and domestic sewage. The species of yeasts most commonly used as single-cell protein include *Candida*, *Hansenula*, *Kluyveromyces*, *Rhodotorula* and *Torulopsis*. However, for food-related fermentations, strains of *Saccharomyces cerevisiae* are grown usually on sugar-containing media, such as molasses, wood sugars or spent sulphite liquor.

For single-cell protein production, the carbon-to-nitrogen ratio of the medium is in the range 7:1–10:1. Concentrations of carbohydrates in batch cultures are about 1–5 per cent. In continuous cultures with hydrocarbons or alcohols, lower concentrations are used. Carbohydrate substrates are readily soluble in aqueous media, while C_{10}–C_{18} hydrocarbons are only sparingly soluble in water. Anhydrous ammonia together with phosphoric acid is used to keep the pH in the range of 3.5–4.5. For aerobic growth on hydrocarbons, the oxygen requirement is 1 g per gram dry weight of biomass, and for growth on *n*-alkanes it is about 2 g per gram dry weight of cells.

Table 55. Growth of selected yeasts on various substrates [Litchfield (1979)].

Microorganism	Carbon and energy source	Scale, aeration, agitation	Temperature	pH	Specific growth rate, μ or dilution rate, D (h^{-1})	Culture density[a] ($g\ \ell^{-1}$)	Yield[a] [g (g substrate)$^{-1}$]
Candida enthanothermophilum ATCC 20380	Ethanol	30-ℓ fermenter, 17-ℓ medium, 17-ℓ min^{-1}, 500 rpm (continuous)	40	3.5	$D = 0.20$	8.0	0.95
C. lipolytica	*n*-Alkanes	1800-ℓ working volume fermenter, 1.5 vvm (continuous)	32	5.5	$D = 0.16$	23.6	0.88
	Gas oil	12-m^3 fermenter, 2400-ℓ medium, 1 vvm (continuous)	30	4.0		25	0.18
C. tropicalis	*n*-Alkanes	50 000-ℓ air-lift fermenter	30	3.0	$\mu = 0.15–0.24$	10–30	1.0–1.1
C. utilis	Sulphite waste liquor	Waldhof fermenter, 0.5 cfm, 100–140 mM $O_2\ \ell\ h^{-1}$ (continuous)	32	5.0	$\mu = 0.5$		0.39
	Ethanol	Waldhof fermenter, 0.5 cfm, 100–140 mM $O_2\ \ell^{-1}\ h^{-1}$ (continuous)	30	4.6	$\mu = 0.3$	6–7	
Hansenula polymorpha ATCC 26012	Methanol	1-ℓ fermenter, 375-ml medium	37–42	4.5–5.5	$\mu = 0.22$ $D = 0.13$	1.2	0.36
Kluyveromyces (Saccharomyces) fragilis	Cheese whey	1600-gal (6057-gal.) fermenter, 1.5 mM O_2 gal.$^{-1}$ min^{-1}	32	5.5	$\mu = 0.66$	14.6	0.55
Rhodotorula glutinis	Domestic sewage	14-ℓ fermenter, 0.7 vvm, 300 rpm (continuous)	28	7.4	$\mu = 0.16–0.21$	0.08	
Saccharomyces cerevisiae	Cane molasses	75–225-m^3 fermenters 2 mM $O_2\ \ell^{-1}\ min^{-1}$, 1 vvm	30	4.5–5.0	$\mu = 0.20–0.25$	40–45	0.50
	Cane molasses	5900-gal. (34068-ℓ) fermenter (continuous)	30	4.5–5.0	$\mu = 0.14$	70	

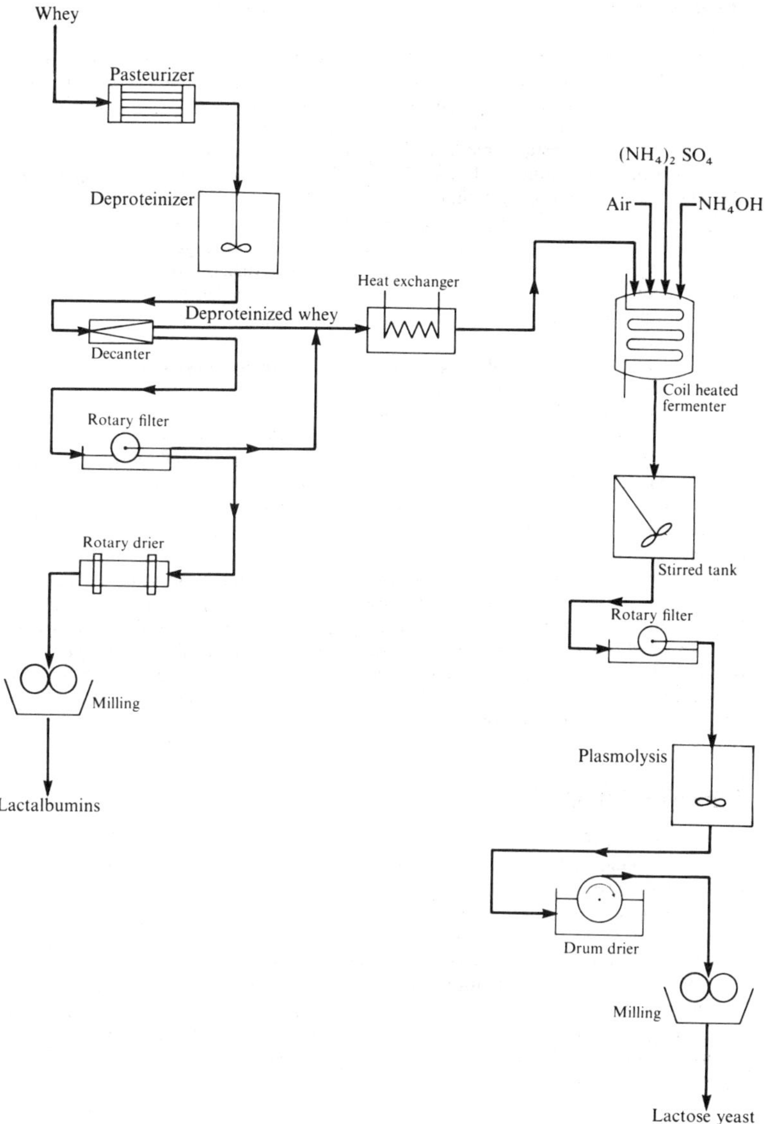

Figure 56. Flow diagram showing production of a whey protein and *Kluyveromyces lactis* biomass in the Bel Fromageries process [Meyrath and Bayer (1979)].

Process outline

The flow sheet for the Bel process developed to produce yeast utilizing whey is shown in Fig. 56. Whey is pasteurized as it emerges from the cheese plant and nonassimilable whey proteins are separated, 75 per cent of the whey proteins being recovered, which corresponds to a concentration of 5 g ℓ^{-1}. The lactose concentration is adjusted to 3.4 per cent and mineral salts are supplemented. In the fermenter, the steady-state culture of *Kluyveromyces lactis* is maintained at 38°C and pH 3.5. To a net fermentation volume of 22–23 m^3, air is supplied by compressors (1700–1800 $m^3\ h^{-1}$) with a substrate flow rate of 5600–6000 $\ell\ h^{-1}$. The residual sugar concentration is less than 1 g ℓ^{-1}.

The effluent from the fermenter is passed through centrifuges to recover the yeast, followed by resuspension and centrifugation. Further concentration is achieved by continuous rotary filters. The resulting yeast cake is plasmolyzed at 83–85°C, resulting in the liquefaction of the solid mass, which is roller dried to about 95 per cent solids, followed by packaging.

Fig. 57 contains the flow sheet of another process for the production of *Candida utilis* from paper pulp sulphite liquor waste.

Two British Petroleum processes for the production of yeast from oil are shown in Fig. 58. Ammonia is used in both these processes as the nitrogen source. In the Grangemouth process, *n*-alkanes (i.e. C_{10}–C_{18}) used as the carbon and energy substrate are separated from gas oil by molecular sieves, the residue being returned to the refinery. In the Lavera process, gas is injected directly into an air-lift fermenter which is open to the atmosphere. Contamination in this process is prevented by controlling the pH, temperature (30°C) and dilution rate.

The process steps in the production of bakers' yeast and food yeast are depicted in Fig. 59. Following the inoculation of pasteur flasks, the yeast is grown in progressively larger fermenters in which the biomass increases from 1 lb in the initial stage to 1500 lb in the reproduction-stage fermenter. The trade fermentation is the final stage and is carried out in large fermenters (200 m^3). The process streams for yeast separation in the process outlined in Fig. 59 result in three major end products: compressed yeast, active dry yeast and food yeast, a common form of inactive yeast.

Product recovery

Yeasts have a size in the range 5–8 μm and a density of 1.04–1.09 g cm^{-3}. The cells can be readily separated from the growth medium by continuous centrifugation. After subsequent centrifugations, the final washed cream contains 15–20 per cent solids.

If *n*-alkanes are used as the substrate, a wash with a surfactant may be required to remove traces of hydrocarbons after the initial centrifugation. If gas oil is used, decantation, phase separation with solvents, washing with surfactants and solvent extraction are employed alone or in combination. The solvents are then removed by steam.

For separation of *Kluyveromyces fragilis* after growth on cheese whey, the growth medium is passed through a three-stage evaporator to concentrate the solids from 8 to 27 per cent, resulting in a feed-grade product. Multistage centrifugation is also used. The separated cells are then drum or spray dried.

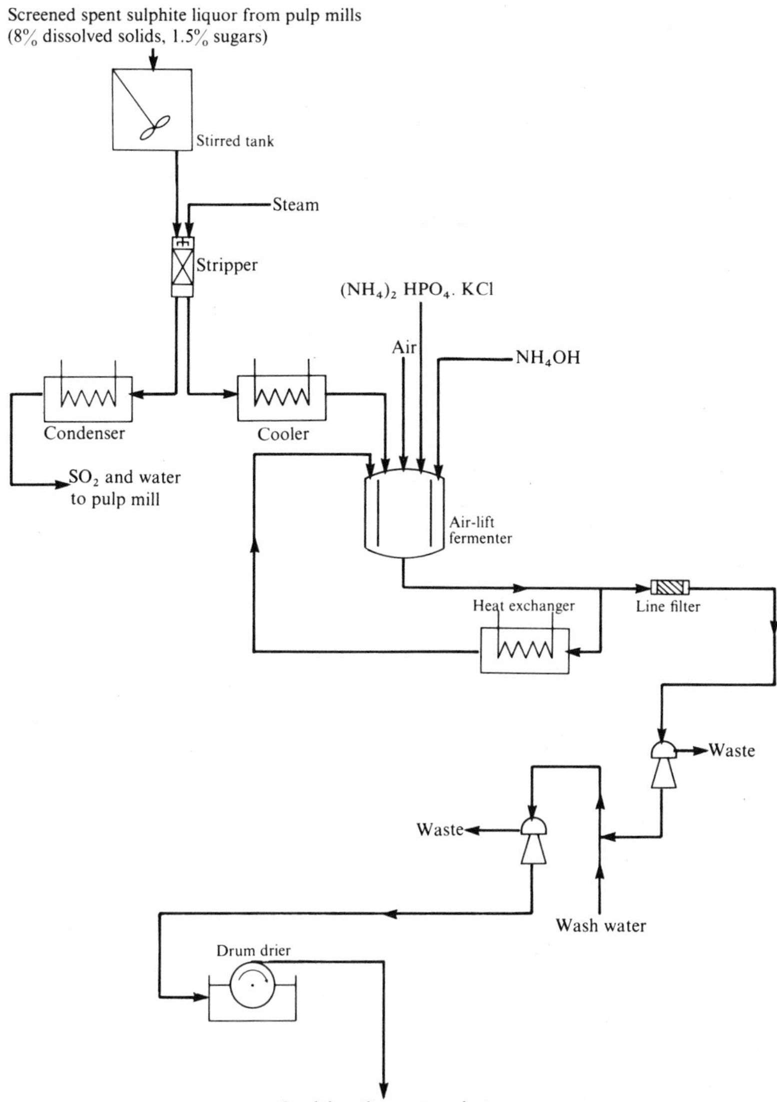

Figure 57. Flow sheet showing production of *Candida utilis* food yeast from paper pulp sulphite liquor waste [Inskeep *et al.* (1951)].

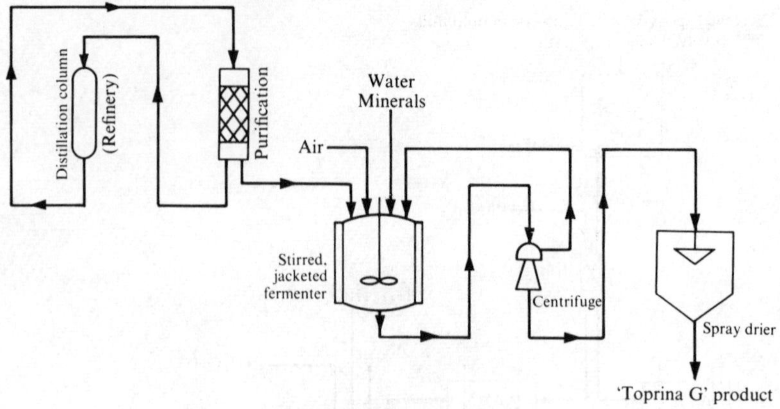

n-Alkane process, Grangemouth

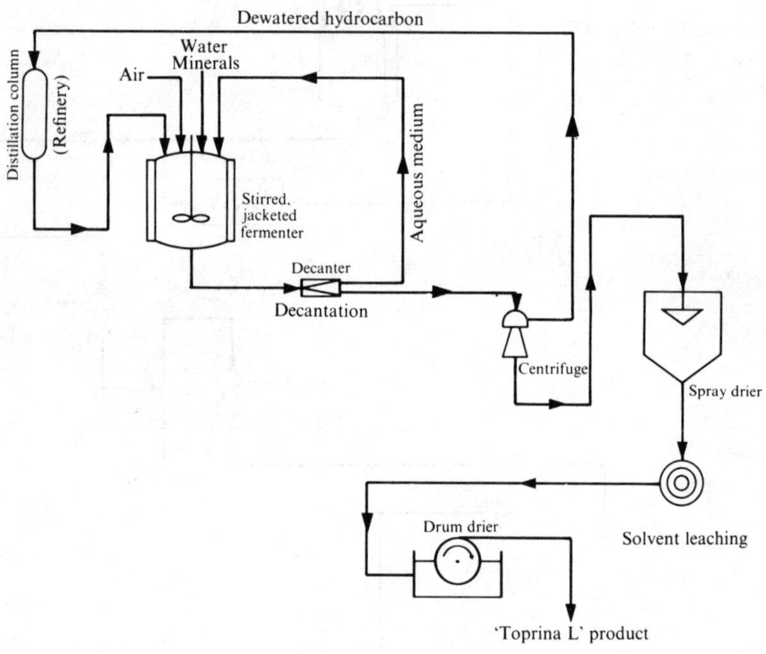

Gas oil process (Lavera)

Figure 58. Summaries of two BP processes for production of yeast protein feeds from oil [Moss and Smith (1977)].

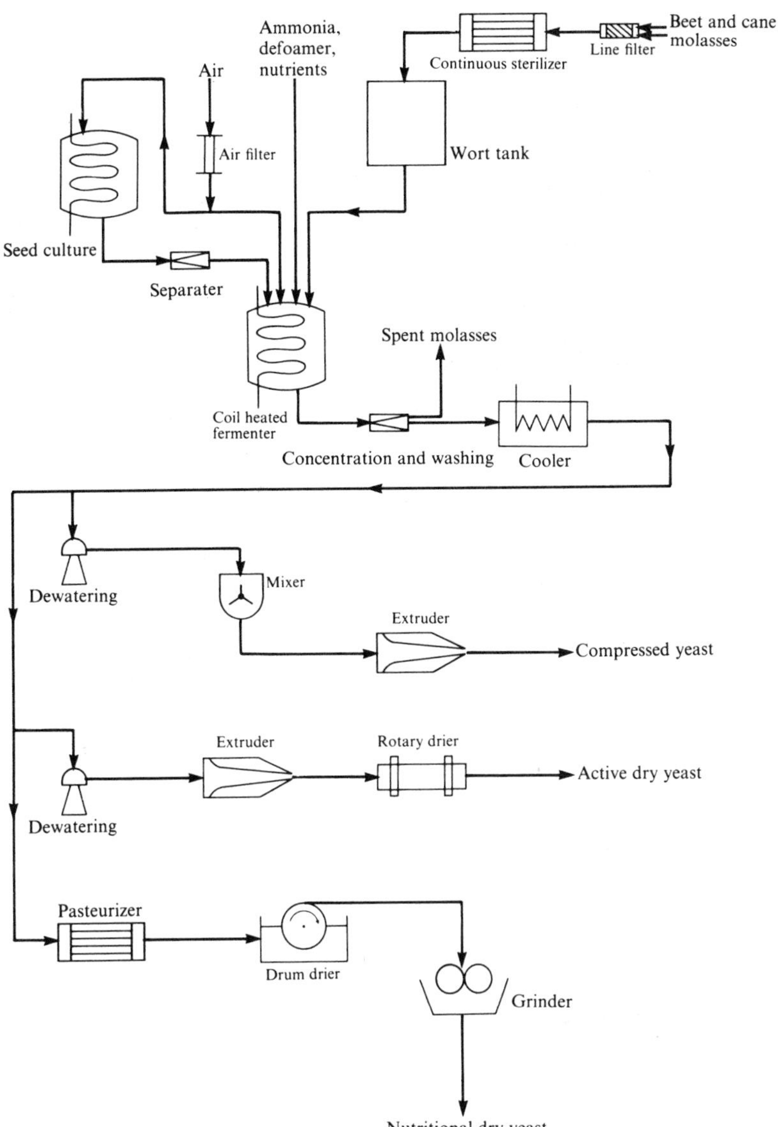

Figure 59. Process steps in the production of bakers' yeast and food yeast [Peppler (1979)].

Moulds and Higher Fungi

Raw materials

Certain higher fungi, e.g., mushrooms, are usually grown as the fruiting bodies on manure or synthetic compost beds. Table 56 lists some typical examples of organisms and substrates together with growth conditions for submerged-culture fermentations.

The concentration of the carbon substrate is the range 1–10 per cent and the carbon-to-nitrogen ratio may be as high as 20 : 1, but is usually 5:1–15:1. Anhydrous ammonia or ammonium salts are used as the nitrogen source, with phosphoric acid as the source of phosphorus. Mineral salts may have to be supplemented, depending on the nature of the main growth medium.

The pH range is 3.0–7.0 (e.g., in growing mycelia for morel mushrooms, a pH of 6.0–6.5, is used to maximize growth rate, yield and flavour). The oxygen requirements are usually complex.

Process outline

Fig. 60 contains a flow sheet for a mushroom farm, together with relevant material balances.

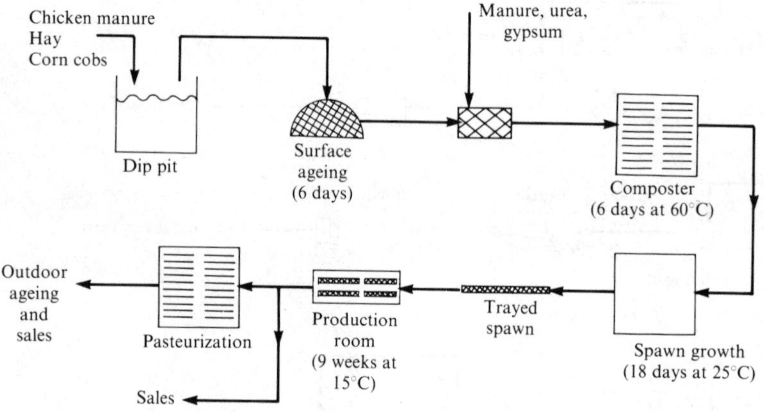

Figure 60. Mushroom process flow sheet, Butler County Mushroom Farm, Butler County, Pennsylvania, USA [Hatch and Finger (1979)].

Product recovery

Moulds and higher fungi are easily recovered from the growth media using basket centrifuges, rotary vacuum filters and inclined screens. The solids content of the mycelial product is in the range 22–45 per cent. Fungal mycelia are dried in steam-tube, rotary, tray or belt driers.

Table 56. Growth of selected moulds and higher fungi on various carbon and energy sources [Litchfield (1979)].

Microorganism	Carbon and energy source	Scale, aeration, agitation	Temperature	pH	Specific growth rate, μ or dilution rate, D (h^{-1})	Culture density[a] ($g\ \ell^{-1}$)	Mycelial yield[a] [g (g substrate)$^{-}$] Supplied	Used
Agaricus blazei	Glucose, citrus press water, orange juice	250-ml conical shake flask, 2-in. stroke, 80 cpm, 1-ℓ, 40-ℓ bottle	Ambient	3.5–7.5		15.2 18.8 20.6	30 31 34	42 46 41
A. campestris	Glucose	20-ℓ fermenter, 2-3 vvm, 400 rpm	25	4.5		20		
	Malt syrup-cane molasses	2-ℓ fermenter, 13 mM $O_2\ \ell^{-1}\ h^{-1}$, 400 rpm	27	5.0–5.5		7.2	15.8	44.6
Aspergillus niger	Carob bean extract	3000-ℓ fermenter (total capacity)	30–36	3.4	$\mu = 0.25$	31.5	45	
Asp. oryzae	Coffee wastewater	5000-gal. (18 927-ℓ) tank, 1 vvm	28	4.0–4.5	$D = 0.037$			45
Boletus edulis	Glucose	300-ml conical flask 100-ml medium, 150 rpm	25	4.5–5.5	$\mu = 0.0017$	2.5		25
Calvatica gigantea	Brewery waste, grain press	250-ml conical flask 100-ml medium, 200 rpm	25	6.0		6.25		24.8
	Brewery waste, trub press liquor	250-ml conical flask 100-ml medium, 200 rpm	25	6.0		27.72		74.9
Fusarium moniliforme	Carob bean extract	14-ℓ fermenter (8.5-ℓ working volume), 400–700 rpm, 0.25 vvm (semicontinuous)	30	5.5–6.5	$\mu = 0.18$	8.8		0.384

(continued)

Table 56 – (continued)

Microorganism	Carbon and energy source	Scale, aeration, agitation	Temperature	pH	Specific growth rate, μ or dilution rate, D (h^{-1})	Culture density[a] ($g\ \ell^{-1}$)	Mycelial yield[a] [g (g substrate)$^{-}$] Supplied	Used
Geotrichum sp.	Corn and pea waste	10 000-gal (37 854-ℓ) aeration pool (continuous)	Ambient	3.7		0.75–1.0		
Gliocladium deliquescens	Corn net milling waste	50 000-gal (189 270-ℓ) (continuous)	Ambient	4.6		1.2		
	Soya whey	18-ℓ fermenter	Ambient	4.6		3.2–3.5		
Lentinus edodes	Glucose	30-ℓ fermenter	25	5.5	$\mu = 0.12$	7.5		26
Morchella crassipes	Glucose, maltose, lactose, cheese whey, corn canning, waste, pumpkin canning, waste, sulphite liquor (ammonia)	10-ℓ carboy, 7-ℓ medium, 0.08 mM $O_2\ \ell^{-1}\ min^{-1}$	25	6.5		8.02 3.55 3.30 2.38 0.75 5.51	33.6 31.4 29.2 5.96 13.9 20.0	48.6 47.8 46.3 32.7 27.8 42.6
	Glucose, maltose, lactose, cheese whey, corn canning waste, pumpkin canning waste, sulphite liquor (ammonia)	5-gal. (18.93-ℓ) carboy, 16-ℓ medium, 0.25 vvm	Ambient	5.0–6.0		1.9		65
M. deliciosa	Sulphite waste liquor	5-gal. (18.93-ℓ) carboy, 16-ℓ medium	Ambient	6.0		10		32
M. esculenta	Glucose, maltose, lactase, cheese whey, corn canning waste, pumpkin canning waste, sulphite waste liquor (ammonia)	10-gal. (37.85-ℓ) carboy, 7-ℓ medium, 0.08 mM $O_2\ \ell^{-1}$ min^{-1}	25	6.5		7.85 3.40 1.65 1.28 0.85 7.82	32.8 30.1 14.6 3.21 15.3 28.8	48.1 47.5 43.4 32.1 33.3 50.4

	Glucose, maltose, lactose, cheese whey, corn canning waste, sulphite waste liquor (ammonia)	5-gal. carboy, 16-ℓ medium, 0.25 vvm	Ambient	5.0–6.0		1.9		65
M. hortensis	Glucose, maltose, lactose, cheese whey, corn canning waste, pumpkin canning waste	10-ℓ carboy, 7-ℓ medium, 0.98 mM O_2 ℓ$^{-1}$ min^{-1}	25	6.5		8.20	34.3	48.3
						3.75	33.2	49.0
						3.60	31.8	47.9
						8.65	23.5	43.6
						1.23	22.3	33.5
						8.25	27.5	48.1
Paecilomyces varioti (Pekilo)	Spent sulphite liquor	360-m³ fermenter			$D = 0.2$	17		55
Trichoderma harzianum	Coffee waste	5000-gal. (18 927-ℓ) 7500-gal. (28 391-ℓ) tank, 2 vvm (continuous)	Ambient	3.5	0.09–0.10			31–47
T. viride	Corn and pea waste	10 000-gal. (37 854-ℓ) aeration pool 11 000-gal (41 640-ℓ) ditch (continuous)	Ambient	4.6		1.2		
	Ball-milled cellulose	14-ℓ fermenter, 6.1-ℓ medium, 0.3 vvm, 350 rpm (continuous)	30	50	$D = 0.033$–0.08	4.0		70
Tricholoma nudum	Glucose	3-ℓ fermenter, 1-ℓ min^{-1}, 400 rpm	25	5.0		22.6		37.6
	Sulphite waste liquor	30-ℓ fermenter, 8.2-ℓ min^{-1}, 330 rpm	25	5.5		10.6		64.0

[a] On dry weight basis.

General Product Specifications for Single-Cell Protein

Important aspects of the product quality of single-cell protein are as follows.

1. Nutritional value of the product.
2. Safety of the product.
3. Production of functional protein concentrates and isolates free from nucleic acids and toxic factors.

The overall composition of a range of single-cell proteins from a variety of microbial sources is given in Table 57 and Table 58 lists the amino acid compositions of some proteins. Table 59 provides some data on the nutritional values of single-cell protein products defined by a number of factors – an explanation of these factors is given below.

In humans, a daily intake of algae of 30–40 g is possible, but the digestibility of these proteins is low and the taste is unpleasant. Also, algae are not good sources of B-group vitamins, but they do contain plentiful quantities of β-carotene and vitamin K, as well as some vitamin C.

The nucleic acid content on a dry weight basis is 4–6, 6–11, 2.5–6 and up to 16 per cent in algae, yeasts, moulds and bacteria, respectively [*see* Litchfield (1979)]. In man, a daily intake of 2 g of yeast nucleic acid is within the acceptable limits. However, intakes of greater than 3 g a day increase the risk of kidney stones or gout. Furthermore, gastro-intestinal disorders, including nausea and vomiting, have been reported in humans fed protein from algae, bacteria and yeast grown on ethanol.

Digestibility

The digestibility of biomass D is the percentage of the total nitrogen consumed that is absorbed from the alimentary tract. The total quantity of microbial protein ingested by animals is measured and the nitrogen content I is analyzed. Over the same period, faeces and urine are collected and their nitrogen contents F and U, respectively, are measured. Thus

$$D = \frac{I - F}{U} \times 100$$

Biological value

This is the percentage of the total nitrogen assimilated that is retained by the body, taking into account the simultaneous loss of endogenous nitrogen through urinary excretion. Thus

$$\text{Biological value, BV} = \frac{I - (F + U)}{I - F} \times 100$$

Protein efficiency ratio

This efficiency ratio PER is the proportion of nitrogen retained by animals fed the test protein compared with that retained when a reference protein, such as egg albumin, is fed.

Table 57. Composition of selected microorganisms of interest in single-cell protein production [Litchfield (1979)].

Microorganism	Substrate	Composition [g (100 g dry wt)$^{-1}$]						
		Nitrogen	Protein	Fat	Total carbohydrate	Crude fibre	Ash	Energy (kcal g^{-1})
Algae								
Chlorella sorokiniana	Carbon dioxide	9.6	60	8	22	3	9	5.2
C. regularis S-50	Carbon dioxide	9.3	58	16		4.4	6.7	
Spirulina maxima, synthetic medium	Carbon dioxide	10	62	3				
S. maxima, sewage	Carbon dioxide	8.5	53	4.8	28		15	
Bacteria and actinomycetes								
Acinetobacter (Micrococcus) cerificans	*n*-Hexadecane	11	72					
Cellulomonas sp.	Bagasse	14	87	8			7	
Methalomonas clara	Methanol	12–13	80–85	8–10			8–12	
Methylophilus (Pseudomonas) methylotrophus	Methanol	13	83	7		< 0.05	8.6	3.0
Thermomonospora fusca	Pulping fines	4.8–5.6	30–35					
Yeasts								
Candida lipolytica (Toprina)	*n*-Alkanes	10	65	8.1			6	
C. lipolytica	Gas oil	11	69	1.5			8	
C. utilis	Ethanol	8.3	52	7		5	8	
C. utilis	Sulphite	9	55	5			8	
Hansenula polymorpha	Methanol		50					
Kluyveromyces (Saccharomyces) fragilis	Cheese whey	9	54	1			9	
Saccharomyces cerevisiae	Molasses	8.4	53	6.3			7.3	
Trichosporon cutaneum	Oxanone wastes	8.6	54	8	31		7	11
Moulds and higher fungi								
Agaricus campestris white variety	Glucose		36	3	49	6.9	4.5	
A. campestris, brown variety	Glucose		45				5.2	

(continued)

Table 57 – (continued)

Microorganism	Substrate	Nitrogen	Protein	Fat	Total carbohydrate	Crude fibre	Ash	Energy (kcal g^{-1})
		Composition [g (100 g dry wt)$^{-1}$]						
Aspergillus niger	Molasses	7.7	50					
Fusarium graminearum	Starch	8.7	54					
Morchella crassipes	Glucose	5.0	31	3.1				
M. crassipes	Sulphite waste liquor	4.1	26	4.4	39		5.9	
M. esculenta	Glucose	5.0	31	1.9				
M. hortensis	Glucose	5.4	34	1.4				
Paecilomyces varioti (Pekilo)	Sulphite waste liquor	8.8	55	1.3	25	7	6	
Trichoderma viride	Starch	10.2	64					

Table 58. Amino acid content of selected organisms of interest for single-cell protein production (Litchfield (1979)].

Organism	Substrate	Ala	Arg	Asp	Cys	Glu	Gly	His	Ile	Leu	Lys	Met	Phe	Pro	Ser	Thr	Try	Tyr	Val
		Amino acid content [g (16 g N)$^{-1}$]																	
Algae																			
Chlorella sorokiniana	Carbon dioxide	5.9	5.6	5.9		9.3	4.8	1.4	3.4	4.0	7.8	1.8	2.7	4.0	2.2	3.2	1.4	2.7	5.1
C. regularis	Carbon dioxide	7.3	5.8	8.8	0.7	11.8	5.4	1.8	4.2	8.1	7.7	1.3	5.1	4.3	3.0	3.6	1.5	2.6	5.9
S-50 Autotrophic																			
S-50 heterotrophic	Glucose	7.4	10.2	7.7	0.9	9.9	4.9	3.0	3.4	7.0	9.4	1.8	3.2	2.2	3.4	4.0	1.4	3.0	5.1
Spirulina maxima	Carbon dioxide	6.5	6.6	8.6	0.4	13.6	4.6	1.7	5.8	7.8	4.8	1.5	4.6	3.8	4.3	4.6	1.3	3.9	6.3

Bacteria and Actinomycetes																			
Acinetobacter (Micrococcus) cerificans	*n*-Alkanes	6.7	4.7	8.0		10.7	5.0	1.7	4.3	6.5	5.2	1.8	3.6	3.1	2.8	4.1	2.8	1.3	5.4
Cellulomonas alcaligenes	Bagasse		6.5					7.8	5.4	7.4	7.6	2.0	4.7			5.5			7.1
Methylococcus capsulatus	Methane		6.2		0.6		5.1	2.2	4.3	8.1	5.7	2.7	4.6			4.6		3.8	6.5
Methylophilus (Pseudomonas) methylotrophus	Methanol	6.8	4.5	8.5	0.6	9.6	4.9	1.8	4.3	6.8	5.9	2.4	3.4	3.0	3.4	4.6	0.9	3.1	5.2
Thermomonospora fusca	Cellulose-pulping fines	13.9	5.6	6.7	0.4	18.0	4.4	2.0	3.2	6.1	3.6	2.0	2.6	6.1	2.6	4.0		1.9	13.0
Yeasts																			
Candida lipolytica (Toprina)	*n*-Alkanes	7.4	4.8	10.2	1.1	11.3	4.8	2.0	4.5	7.0	7.0	1.8	4.4	4.4	4.8	4.9	1.4	3.5	5.4
C. lipolytica	Gas oil	5.8	5.0	10.0	0.9	12.1	4.5	2.1	5.3	7.8	7.8	1.6	4.8	3.7	5.1	5.4	1.3	4.0	5.8
C. utilis	Sulphite waste	5.8	5.4	9.2		15.6	3.6	1.2	3.8	7.6	4.8	1.1	8.6	6.0	5.0	5.4	2.4	6.2	3.8
C. utilis	Ethanol	5.5	5.4	8.8	0.4	14.6	4.5	2.1	4.5	7.1	6.6	1.4	4.1	3.4	4.7	5.5	1.2	3.3	5.7
Hansenula polymorpha ATCC 26012	Methanol	6.1	5.6	10.5	0.7	14.4	5.2	2.4	5.1	8.3	8.1	1.5	5.0			5.2		4.8	6.2
Kluyveromyces (Saccharomyces) fragilis	Cheese whey							2.1	4.0	6.1	6.9	1.9	2.8			5.8	1.4	2.4	5.4
Rhodotorula glutinis	Domestic sewage	5.1	4.5	6.0		9.3	3.5	2.2	3.3	5.4	7.1		3.2	3.0	3.3	3.1		3.1	3.6
Saccharomyces cerevisiae	Molasses		5.0		1.6			4.0	5.5	7.9	8.2	2.5	4.5			4.8	1.2	5.0	5.5
Moulds and higher fungi																			
Aspergillus niger	Carob bean extract				1.1				4.2	5.7	5.9	2.6	3.8			5.0	2.1	3.2	5.2
Fusarium graminearum	Starch	6.6	6.8	10.4	1.2	11.0	6.5	3.3	4.0	6.0	6.9	1.8	4.0	4.8	5.3	4.6		3.8	6.6
Gliocladium deliquescens	Waste starch	6.1	4.8	7.9	1.0	8.6	4.7	2.3	3.8	6.2	5.3	1.3	3.0	4.2	3.9	4.3	1.4	3.6	4.7
Morchella crassipes	Glucose	6.2	3.0	4.8	0.4	14.1	3.1	2.0	2.9	5.6	3.5	1.0	1.9	5.1	3.1	3.0	1.5	1.7	3.0
	Sulphite waste liquor (ammonia)	5.3	3.9	5.9	0.6	8.3	3.8	2.0	2.9	6.1	5.6	1.2	5.8	4.0	4.0	4.1			4.1
M. esculenta	Glucose	4.8	8.0	5.0	0.3	14.8	2.9	2.1	2.7	5.1	3.8	0.9	2.5	4.2	3.1	3.0	0.9	1.7	3.4
M. hortensis	Glucose	4.5	4.0	4.6	0.4	15.4	3.0	1.9	2.4	5.0	3.0	0.7	2.3	4.5	2.8	2.7	1.0	1.9	2.9
Paecilomyces varioti	Spent sulphite liquor				1.1				4.3	6.9	6.4	1.5	3.7			4.6	1.2	3.4	5.1
Trichoderma harzianum	Coffee-processing wastes								4.0	5.8	4.7	1.1	3.5			3.8			4.9
T. viride	Barley straw	4.8	4.0	7.8	1.5	8.0	4.5	1.9	3.5	5.8	4.4	1.4	8.7	3.8	4.2	4.9		3.3	4.4
Trichosporon cutaneum	Oxanone wastewater				1.0				5.2	7.1	7.3	1.6	4.5			5.5	1.4	4.2	6.1

Table 59. Performance of selected single-cell proteins in animal feeding studies [Litchfield (1979)].

Single-cell protein	Substrate	Treatment	Amount in diet (%)	Animal	Protein digestibility, D (%)	Protein efficiency ratio, PER	Biological value, BV	Feed conversion ratio [kg (kg wt gain)$^{-1}$]
Algae								
Chlorella sorokiniana		Dried	10	Rat	86	2.19		
		Dried + 0.2 L-methionine	10	Rat	86	2.9		
Spirulina maxima		Dried	10	Rat	84	2.3–2.6	72	
Bacteria								
Acinetobacter (Micrococcus) cerificans	*n*-Hexadecane	Dried		Rat	83.4		67	
Bacillus megaterium	Collagen waste	Dried		Rat		1.88		
Methylophilus methylotrophus	Methanol	Dried	9.8	Broiler chicken				2.30
	Methanol	Dried	6.7	Pig				3.13
Pseudomonas sp.	Methanol	Dried	10	Broiler chicken	88			2.58
Yeast								
Candida lipolytica	*n*-Alkanes	Dried		Rat	96		61	
	n-Alkanes	Dried + 0.3 DL-methionine		Rat	96		91	
	n-Alkanes	Dried	10	Broiler chicken	88			2.58
	n-Alkanes	Dried	7.5	Pig	92			3.04
	Gas and oil	Dried		Rat	94		54	
	Gas and oil	Dried + 0.3 DL-methionine		Rat	95		96	
C. tropicalis	*n*-Alkanes	Dried	15	Broiler chicken				2.03
C. utilis	Sulphite waste liquor	Dried		Rat	85–88	0.9–1.4	32–48	
	Sulphite waste liquor	Dried + 0.5 DL-methionine		Rat	90	2.0–2.3	88	

Kluyveromyces fragilis	Cheese whey	Dried		Rat				1.5
Saccharomyces cerevisiae	Molasses	Dried		Rat	80–90	1.7	58–69	
Fungi								
Aspergillus niger	Carob bean extract	Dried	20	Broiler chicken		2.50		2.02
Fusarium graminearum		Dried	10.5	Rat		1.89		
		Dried	40	Pig				1.76
F. moniliforme	Carob bean extract	Dried		Rat		1.15		
	Carob bean extract	Dried + 6.3 DL-methionine		Rat		2.38		
Gliocladium deliquescens	Waste starch	Dried		Rat			49	
Trichoderma viride	Waste starch	Dried		Rat			48	

MICROBIAL TRANSFORMATION OF STEROIDS

General Aspects

The various types of microbial transformations are listed in Table 60. Fig. 61 shows some of the steroids that are produced from progesterone by transformations carried out using microorganisms. Fig. 62 is a similar diagram with Reichstein compounds as the starting material.

Table 60. Types of microbial transformations of steroids [Sebek and Perlman (1979)].

Transformation	Mechanism
Oxidation	Conversion of secondary alcohol to ketone Introduction of primary hydroxyl on steroid side chain Introduction of secondary hydroxyl on steroid nucleus Introduction of tertiary hydroxyl on steroid nucleus Dehydrogenations of ring A of steroid nucleus in positions 1,2 and 4,5 Aromatization of ring A of the steroid nucleus Oxidation of the methylene group to ketone group Cleavage of side-chain of pregnane at C-17 to form ketone Cleavage of side-chain of pregnane at C-17 and opening of D ring to form testololactone Cleavage of side chain of steroids to form carboxyl group Cleavage of side chain or pregnane steroids at C-17 to form secondary alcohol Formation of epoxides Decarboxylation of acids
Reduction	Reduction of ketone to secondary alcohol Reduction of aldehyde to primary alcohol Hydrogenation of double bond at position 1,2 of ring A Hydrogenation of double bond at positions 4,5 of ring A and at 5,6 of ring B Elimination of secondary alcohol Formation of homosteroids of the androstane series from pregnane derivatives
Hydrolysis	Saponification of steroid esters Acetylation
Esterification	

Process Outline

In order to carry out the desired transformation, the relevant microorganisms are grown on suitable media in fermenters with aeration and agitation. The composition of some media and other information about the fermentations and transformations are given in Table 61.

After the growth of the microorganism, a measured quantity of the steroid is dissolved in a suitable solvent, such as ethanol, acetone or propylene glycol, and is added to the medium. Enzymes secreted by the microorganisms act upon the steroid and perform the desired transformation under controlled conditions of temperature, agitation and pH. Since steroids are essentially insoluble or only sparingly soluble in water, they react as aqueous suspensions. The transformation period is usually 24–48 hours.

Table 62 lists some steroid transformations of commercial importance. The transformations can be performed by the following preparations.

1. Growing cells.
2. 'Resting' cells.
3. Cell-free extracts.
4. Immobilized whole cells.
5. Immobilized cell-free extracts.
6. Spore suspensions.

Table 61. Media and fermentation conditions used in microbial transformations of steroids [Sebek and Perlman (1979)].

Microorganism	Steroid substrate	Steroid product [approximate yields, % (by wt)	Composition of medium[a]	Length of incubation; temperature; aeration
Alcaligenes faecalis	Cholic acid	Ketocholic acids (90–100%)	A	2 days (monoketo acid), 4 days (diketo acid), 6 days (triketo acid); 37–39°C; surface culture
Corynebacterium mediolanum	21-Acetoxy-3β-hydroxy-5-pregnen-20-one	21-Hydroxy-4-pregnene-3,20-dione (30%)	B	6 days; 36–37°C; pure oxygen with agitation
Cunninghamella blakesleeana H334	Compound S	Cortisone (19%), cortisol (65%)	C	3 days, 28°C; rotary shaker (250 rpm)
Cylindrocarpon radicicola ATCC 11011	Progesterone	1-Dehydrotestolo-lactone (50%)	D	3 days, 25°C, reciprocating shaker (120 spm)
Fusarium solani	Progesterone	1,4-Androstadiene-3,17-dione (85%)	E	4 days; 25°C; rotary shaker (100 rpm)
Rhizopus arrhizus ATCC 11145	4-Androstene-3, 17-dione	11 α-Hydroxy-4-androstene-3,17-dione (25%)	F	4 days; 28°C; small aerated tank (6–7 mM $O_2\ \ell^{-1}\ min^{-1}$)
Streptomyces albus	Oestradiol	Oestrone (90–95%)	G	6 h of substrate oxidation with resting cells; 30°C
Strept. aureus	Progesterone	15α-Hydroxy-4-pregnene-3,20-dione (11%)	H	3 days; 25°C; rotary shaker (280 rpm)

[a] Composition of medium A: 4.7 g $(NH_4)_2SO_4$, 0.5 g asparagine, 5 g NaCl, 0.65 g NaOH, 2 g glycerol, inorganic salts, 5 g cholic acid, distilled water to ℓ medium; B: 60-ml yeast water, 10 ml of 0.2 M Na_2HPO_4, 1 g KH_2PO, 0.2 g 21-acetoxy-3-hydroxypregnan-20-one; medium C: 3 g $NaNO_3$, 0.5 g KCl, 30 g dextrin, 0.01 g $FeSO_4.7H_2O$, 10 mg Tween 80, 0.5 g $MgSO_4.7H_2O$, 1.3 g $K_2HPO_4.3H_2O$, distilled water to 1 ℓ, pH 7.2; medium D: 3 g corn-steep solids, 3 g $NH_4H_2PO_4$, 2.5 g $CaCO_3$, 2.2 g soya bean oil, 0.5 g progesterone, distilled water to 1 ℓ, pH 7.0; medium E: 15 g peptone, 6 ml corn-steep liquor, 50 g glucose, distilled water to 1 ℓ, pH 6.0, progesterone (0.25 g) added after 2 days of incubation; medium F: 20 g lactalbumin digest, 5 ml corn-steep liquor, 50 g glucose, tap water to 1 ℓ, pH 5.5–5.9; androstenedione (0.25 g) added after 27 hours of incubation; medium G: nutrient broth (to grow cells), phosphate buffer (pH 7.0) (to oxidize oestradiol); medium H: 2.2 g soya bean oil, 15 g soya bean meal, 10 g glucose, 2.5 g $CaCO_3$, 0.25 g progesterone, water to 1 ℓ.

Table 62. Some steroid transformations of commercial importance [Sebek and Perlman (1979)].

Reaction	Substrate	Product	Microorganism	Some industrial producers
11α-Hydroxylation	Progesterone	11α-hydroxyprogesterone	*Rhizopus nigricans*	Upjohn Company
11β-Hydroxylation	Compound S	Cortisol	*Curvularia lunata*	Pfizer, Inc.; Gist-Brocades
16α-Hydroxylation	9α-Fluorocortisol	9α-fluoro-16α-hydroxycortisol	*Streptomyces roseochromogenus*	E.R. Squibb and Sons; Lederle Laboratories
1-Dehydrogenation	Cortisol	Prednisolone	*Arthrobacter simplex*, *Corynebacterium simplex*	Schering Corporation
	Dienediol	Trienediol	*Septomyxa affinis*	Upjohn Company
1-Dehydrogenation, side-chain cleavage, and D ring expansion	Progesterone	1-dehydrotestololactone	*Cylindrocarpon radicicola*	E.R. Squibb and Sons
Side-chain cleavage	β-Sitosterol	Androstadienedione and/or androstenedione	*Mycobacterium* spp.	G.D. Searle and Company

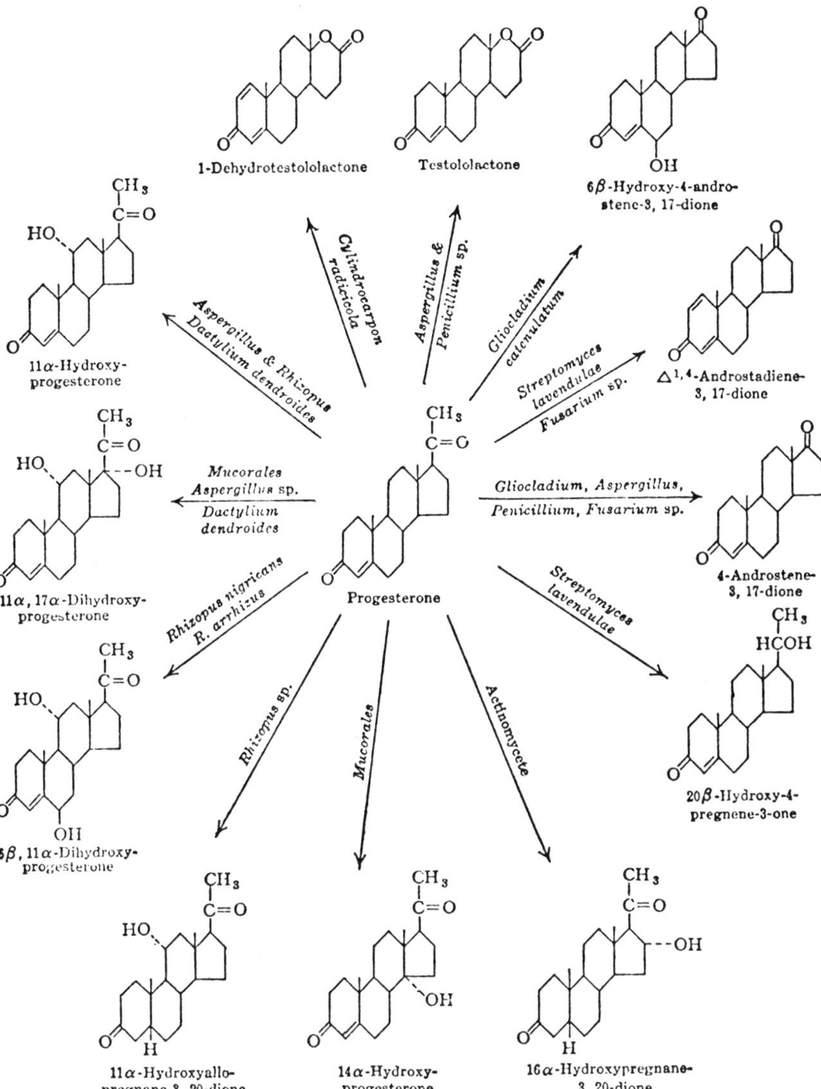

Figure 61. Some steroids produced from progesterone by microbiological transformations [Prescott and Dunn (1959)].

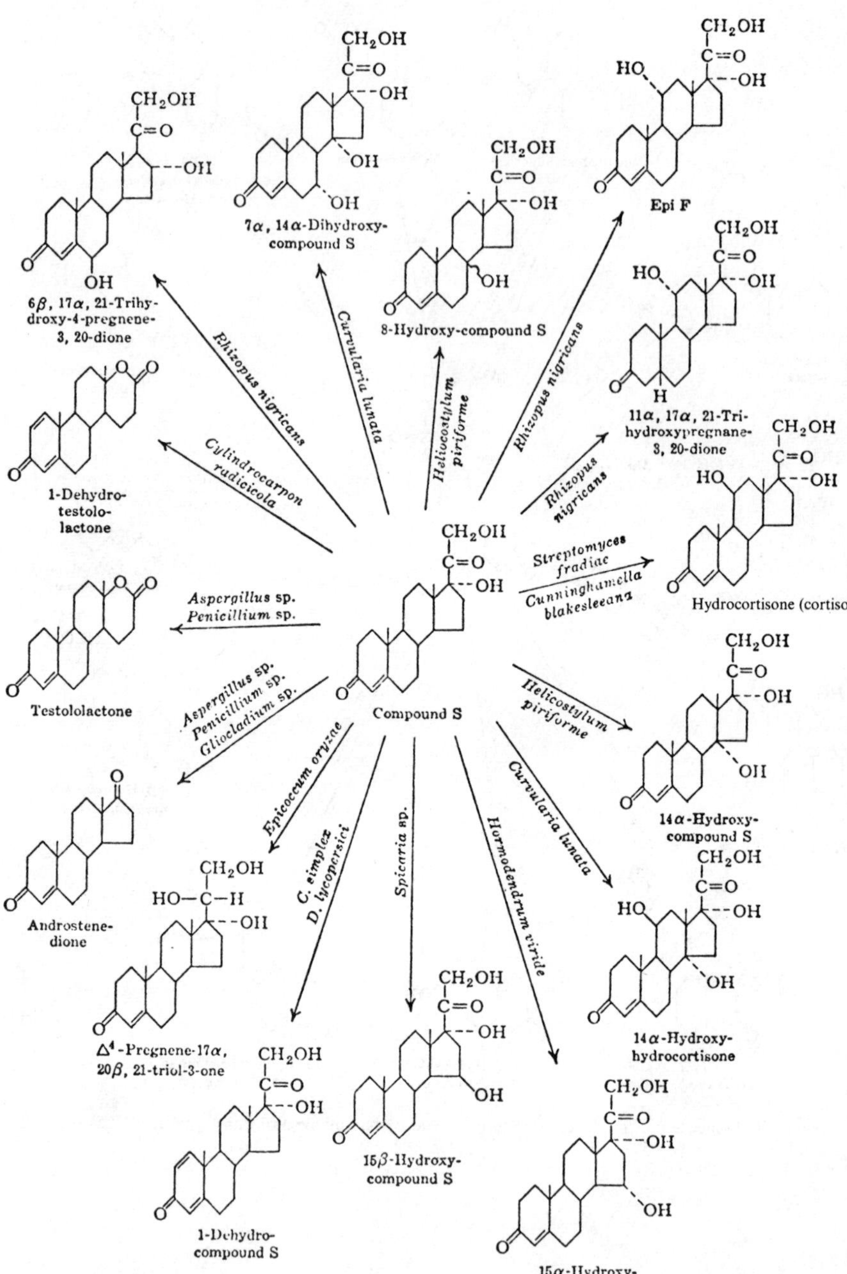

Figure 62. Some steroids formed from Reichstein compound S by microbiological transformations [Prescott and Dunn (1959)].

Product Recovery

After the transformation, the biomass is separated from the fermentation liquor and is extracted using a suitable solvent, such as methylene chloride, ethylene chloride or chloroform. The extracts are added to the liquor and the biomass is rejected. The liquor, plus the added extract, is subjected to further solvent extraction, purified and examined. The liquor may be extracted up to four times.

VITAMINS

Cyanocobalamin (Vitamin B_{12})

Raw materials

It seems probable that the only primary source of cyanocobalamin in nature is the metabolic activity of microorganisms. There is no convincing evidence for the vitamin's synthesis in tissues of higher plants or animals. Cyanocobalamin is produced chiefly by a wide range of bacteria and actinomycetes, although not to any significant extent by yeasts and moulds. Table 63 contains a list of microorganisms and carbon sources that can be used for the production of this vitamin. Because of their high growth rates and productivities, two species of *Propionibacterium* and one *Pseudomonas* species cultivated on carbohydrates are preferred industrially rather than *Streptomyces* and other genera.

Apart from the carbohydrates given in Table 63, other sources of carbohydrates, such as molasses, corn-steep liquor, distillers' solubles, alcohols, hydrocarbons and food industry wastes can also be used as raw materials. Table 64 gives some of these media.

The best microbial strains are spontaneous or induced mutants, screened for their resistance to agents, e.g., cobalt or manganese ions, antibiotics, etc. Induced mutants can be produced by treatment with UV or X-rays or chemicals, including *N*-methyl-*N'*-nitro-*N*-nitrosoguanidine, nitrosoethylurea, ethylenimine, diethylsulphate, etc.

Table 63. Cyanocobalamin production by different bacterial strains [Florent and Ninet (1979)].

Microorganism	Carbon source	Yield (mg ℓ^{-1})
Bacterium FM-02T	Methanol	2.6
Corynebacterium sp. + *Rhodopseudomonas* sp.	*n*-Alkanes	2.3
Methanobacillus omelianskii	Methanol	8.8
Mixed methanogenic bacteria	Methanol	35
Micromonospora sp.	Glucose	11.5
Nocardia gardneri	Hexadecane	4.5
N. rugosa	Glucose, cane molasses	14
Propionibacterium freudenreichii	Glucose	25
P. shermanii	Glucose	23–39
P. vannielli	Glucose	25
Protoaminobacter ruber	Methanol	2.5
Pseudomonas denitrificans	Beet molasses	59
Streptomyces olivaceus	Glucose, lactose	8.5

Table 64. Media used for synthesis of cyanocobalamin by various microorganisms.

Microorganism	Medium
Bacillus megaterium	Beet molasses, ammonium phosphate, cobalt salts, lactic acid
Flavobacterium devorans	Glucose, soya bean meal, corn-steep liquor, cobalt, other inorganic ions
F. solare	Yeast extract, malt extract, glucose, salts or maltose, penicillin mash residue
Propionibacterium freudenreichii	Glucose, casein hydrolyzate, yeast extract, cobalt salts, lactic acid
Pseudomonas sp, + *Proteus* sp.	Soya bean meal or other protein materials
Saccharomyces cerevisiae + *Aerobacter cloacae*	Grain ethanol stillage or penicillin fermentation residues, inorganic salts
Streptomyces fradiae	Glucose, brewers' yeast, soya bean meal, cobalt salts, inorganic salts
Strept. griseus	Beef extract, casein hydrolyzate, cobalt or glucose, soya bean meal, distillers' solubles, cobalt
Strept. olivaceus	Glucose, soya bean meal, distillers' solubles, cobalt salts, inorganic salts
Strept. vitaminicus	Glucose, animal-stick liquor or other protein materials, cobalt

Cyanocobalamin belongs to a large family of cobaltocorrinoids with the important radical being 5,6-dimethylbenzimidazole (DBI). It is necessary to add to the medium some essential elements, cobalt ions in all cases and often DBI. It can be beneficial if other potential precursors of corrinoids, such as glycine, threonine, δ-aminolaevulinic acid and aminopropanol are supplemented. It is important to select microbial species that synthesize the 5,6-dimethyl-α-benzimidazolylcobamide exclusively, because there are some microorganisms that give high yields of related cobamides including pseudovitamin B_{12} (adeninylcobamide) instead of true vitamin B_{12}.

Process outline

For a number of years, cyanocobalamin was recovered from the spent liquors of streptomycete antibiotic fermentations which produced streptomycin, chlortetracycline or neomycin. However, these dual-purpose fermentations could not meet the demand.

Table 65 gives some of the characteristics of various processes for the microbial production of cyanocobalamin. Several microaerophilic propionibacteria produce cobaltocorrinoids in carbohydrate media supplemented with cobalt and DBI without aeration. However, *Propionibacterium freudenreichii* ATCC 6207, *P. shermanii* ATCC 13673 and certain mutants of these strains synthesize their own DBI. Therefore, these microorganisms are preferred. Since aeration favours DBI formation, a two-stage culture is used to increase yields. The first stage is an anaerobic culture which promotes the growth of the bacteria and etiocobalamin biosynthesis, with almost complete depletion of sugar. The second, aerobic stage leads to DBI synthesis and conversion of etiocobalamin to deoxyadenosylcobalamin.

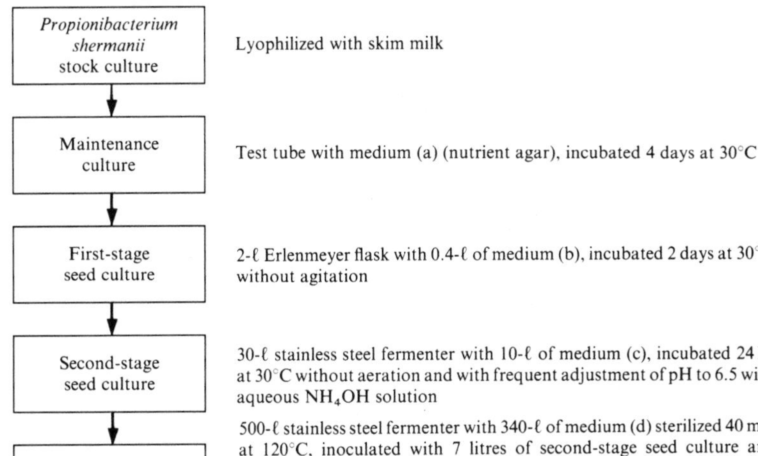

Medium (a): Tryptone, 10 g; yeast extract, 10 g; filtered tomato juice, 200 ml; agar, 15 gm; with tap water to 1 ℓ and pH adjusted to 7.2.
Medium (b): Medium (a) without agar.
Medium (c): Corn-steep liquor, 20 g; dextrose, 90 g; with tap water to 1 ℓ and pH adjusted to 6.5.
Medium (d): Corn-steep liquor, 40 g; dextrose (sterilized separately), 100 g; cobalt chloride, 20 mg; with tap water to 1 ℓ and pH adjusted to 7.0.

Figure 63. *Cyanocobalamin* from *Propionibacterium shermanii.* Semipilot-plant scale fermentation process [Florent and Ninet (1979)].

Fig. 63 shows an example of a process using *P. shermanii* where the main culture has two modes of aeration. The fermentation media consists of glucose or inverted molasses at concentrations of 10–100 g ℓ^{-1}, small amounts of ferrous, manganous and magnesium salts, in addition to cobalt salts at concentrations of 10–100 mg ℓ^{-1}, buffering or neutralizing agents and nitrogenous compounds. For the nitrogenous compounds, yeast preparation, casein hydrolyzates or corn-steep liquor (30–70 g ℓ^{-1}) is used. The corn-steep liquor contains lactic and pantothenic acids; the latter stimulates corrinoid synthesis. The culture is kept at 30°C and pH 6.5–7.0.

Table 65. Processes for microbial production of cyanocobalamin[a] [Perlman (1977)].

Microorganism	Ingredients of medium	Yield of vitamin B_{12} (mg ℓ^{-1})	Comments
Bacillus megaterium	Beet molasses, ammonium phosphate, cobalt salt, inorganic salts	0.45	Aerated fermentation (18 h)
Butyribacterium rettgeri	Corn-steep liquor; cobalt salt, glucose, maintained at pH 7 with ammonium hydroxide	5	4-day anaerobic fermentation
Micromonospora spp.	Soya bean meal, glucose, calcium carbonate, cobalt salt	11.5	7-day aerated fermentation
Propionibacterium freudenreichii	Corn-steep liquor, glucose, cobalt salt, maintained at pH 7 with ammonium hydroxide	19	3 days anaerobic and 3 days aerobic
P. freudenreichii	Corn-steep liquor (or autolyzed *Penicillium* mycelium), glucose, cobalt salt, maintained at pH 7 with ammonium hydroxide	8	Continuous, 2-stage fermentation; 33-h retention time
P. shermanii	Corn-steep liquor, glucose, cobalt salt, maintained at pH 7 with ammonium hydroxide	23	3 days anaerobic and 4 days aerobic
Streptomyces spp.	Soya bean meal, glucose, cobalt salt, K_2 potassium phosphate	5.7	6-day aerated fermentation
Streptomyces olivaceus	Soya bean meal, glucose, distillers' solubles, cobalt salt, inorganic salts	3.3	6-day aerated fermentation

[a] While the cultures are presumed to produce the coenzyme form of vitamin B_{12} (5,6-dimethyl-α-benzimidazolylcobamide-5′-deoxyadenosine), the vitamin is usually isolated in the cyanide form.

The final concentration of cyanocobalamin in *Propionibacterium* fermentations is about 25–40 mg ℓ^{-1}. However, concentrations as high as 216 mg ℓ^{-1} have been reported [Aries (1974)].

Fig. 64 identifies the main stages of a process using *Ps. denitrificans* for the production of cyanocobalamin. The synthesis of cobalamin occurs simultaneously with the growth of bacteria. The fermentation is carried out aerobically with agitation, batch-wise or continuously. The only requirements of this strain are a carbohydrate (e.g., sucrose) yeast extract and several metallic salts. At the beginning of the fermentation, DBI (10–25 mg ℓ^{-1}) and cobaltous nitrate (40–200 mg ℓ^{-1}) are added to the medium. In industrial fermentations, beet molasses at a concentration of 60–120 g ℓ^{-1}, along with ammonium phosphate (2–5 g ℓ^{-1}) and some oligoelements, are used. The optimum temperature is 28°C, with a pH of about 7.0. The final concentration of the vitamin is about 60 mg ℓ^{-1}.

Thermophilic methanogenic bacteria, such as *Methanobacillus* and *Methanobacterium* species, are capable of producing cyanocobalamin (2 mg ℓ^{-1}) from methanol (8 g ℓ^{-1}). *Nocardia gardneri* synthesizes vitamin B_{12} at a concentration of approximately 4.5 mg ℓ^{-1} from a hexadecane concentration of 20 g ℓ^{-1} and a mixed culture of *Corynebacterium* species.

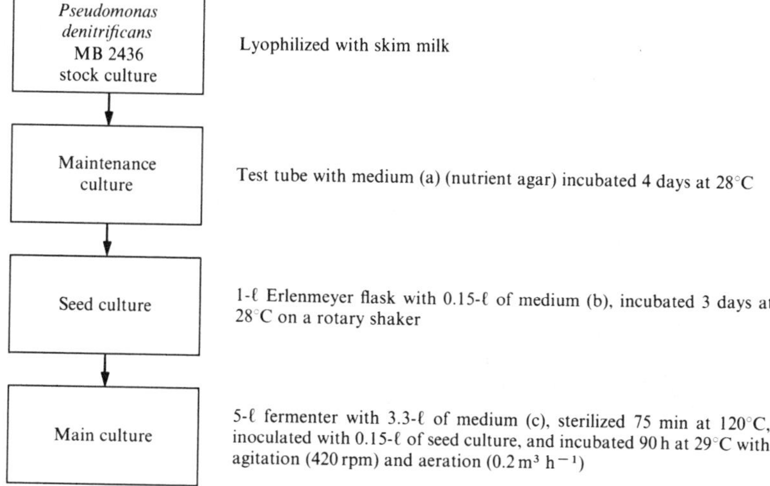

Medium (a): Beet molasses, 60 g; brewers' yeast, 1 g; casein hydrolyzate 1 g; diammonium phosphate, 2 g; magnesium sulphate, 1 g; manganese sulphate, 200 mg; zinc sulphate, 20 mg; molybate sulphate, 5 mg; agar, 25 g; with tap water to 1 ℓ and pH adjusted to 7.4.
Medium (b): Medium (a) without agar.
Medium (c): Beet molasses, 100 g; yeast, 2 g; diammonium phosphate, 5 g; magnesium sulphate, 3 g; manganese sulphate, 200 mg; cobalt nitrate, 188 mg; 5,6-dimethylbenzimidazole, 25 mg; zinc sulphate, 20 mg; sodium molybdate, 5 mg; with tap water to 1 ℓ and pH adjusted to 7.4.

Figure 64. Cyanocobalamin from *Pseudomonas denitrificans*. Laboratory-scale fermentation process [Florent and Ninet (1979)].

Product recovery

Almost all the cobamides formed during the fermentation are retained within the cells. The cells are separated from the fermentation liquor by high-speed centrifugation. The biomass is then drum dried for use as animal or poultry feed supplements. Otherwise, an aqueous solution of harvested cells or the whole broth is heated at 80–120°C at pH 6.5–8.5 for 10–30 minutes. In the production of cyanocobalamin, a mixture of cobalamin and some analogues is obtained. Since the isolation of each compound is impractical, all the compounds are converted to cyanocobalamin. This conversion is achieved by treating the heated broth or cell suspension with cyanide or thiocyanide, usually in the presence of sodium nitrite or chloramine B. If the cyanide treatment is to be deferred, all the compounds are converted to the more stable sulphato- or sulphito-cobalamin form. The cyanocobalamin is separated from the solution by adsorption on the ion exchange resin Amberlite IRC-50 or Amberlite XAD-2 and, subsequently, eluted. Alumina or activated carbon is used as an alternative adsorbent. Elution is achieved by hydroalcoholic or hydrophenolic mixtures. Another method used to recover the cyanocobalamin from aqueous solutions is to extract with phenol or cresol alone or mixed with benzene, butanol, carbon tetrachloride or chloroform. Finally, the product is recovered by precipitation or by crystallization following evaporation. An isolation process for cyanocobalamin is given in Fig. 65.

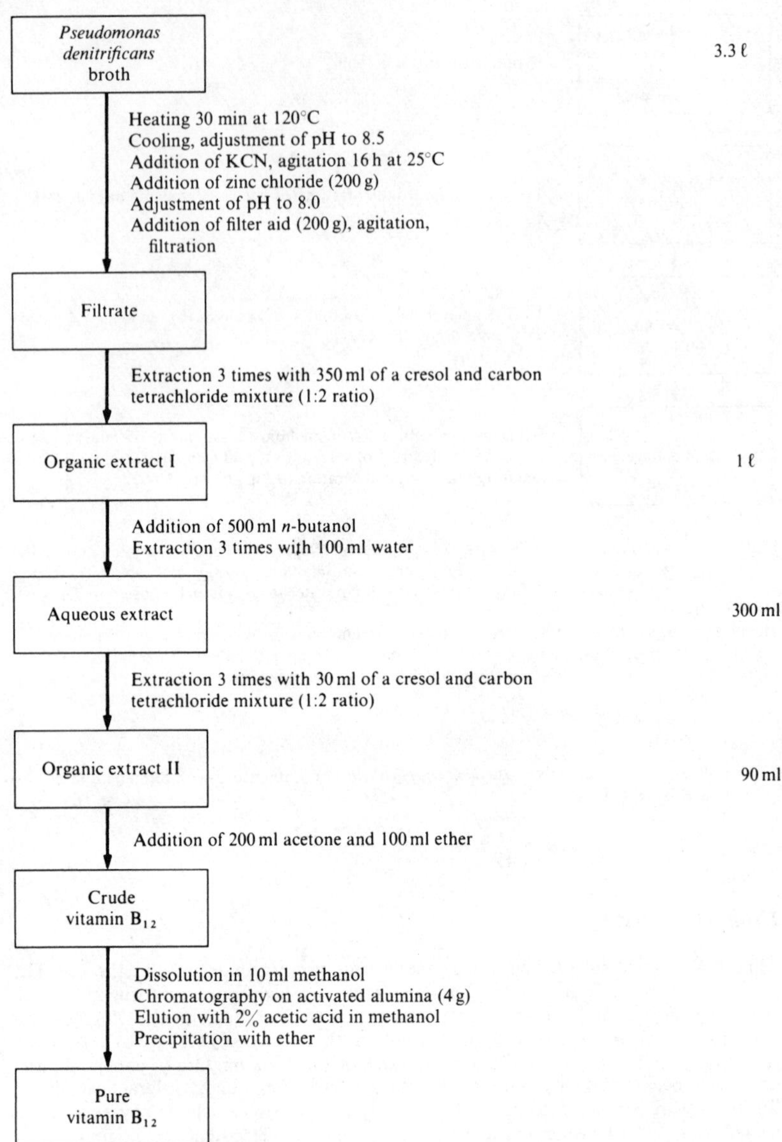

Figure 65. Crystalline vitamin B_{12} isolation process [Florent and Ninet (1979)].

Riboflavin (Vitamin B_2)

Riboflavin is produced by several microorganisms, especially yeasts, ascomycetes and bacteria. However, the industrially important microorganisms are ascomycetes, *Eremothecium ashbyii* and *Ashbya gossypii*, yeasts, in particular *Candida* species, and *Clostridium acetobutylicum*. Table 66 contains a list of microorganisms that can produce significant amounts of riboflavin.

Table 66. Microorganisms producing considerable amounts of riboflavin and the effects of iron on the biosynthesis of vitamin [Perlman (1979)].

Microorganism	Riboflavin yield (mg ℓ$^{-1}$)	Optimum iron concentration (mg ℓ$^{-1}$)
Ashbya gossypii	6420	Not critical
Candida flareri	567	0.04–0.06
Clostridium acetobutylicum	97	1–3
Eremothecium ashbyii	2480	Not critical
Mycobacterium smegmatis	58	Not critical
Mycocandida riboflavina	200	Not critical

A. gossypii requires certain nutrients for growth and maximum yields of the vitamin. Commercial glucose is the best carbon source, but it can be substituted with sucrose and maltose. Starch, molasses and the common pentoses, e.g., ribose and xylose, are not utilized by *A. gossypii* for riboflavin synthesis. For the nitrogen source, corn-steep liquor, animal-stick liquor, tankage or meat scraps or peptone is used. Enzymically degraded collagen and lipids, together with corn-steep liquor, distillers' solubles and brewers' yeast increase the yields. Table 67 shows the effect of supplementing the growth medium with peptones and glycine. Biotin, thiamin and *meso*-inositol are required for adequate growth of the micro-organisms. A sterilization time of less than 30 minutes, a dilute 'young' inoculum, e.g., 2 per cent (by vol.), and efficient aeration further contribute to increased yields.

Table 67. Effect of supplementing growth medium with peptones and glycine on riboflavin production by *Ashbya gossypii* [Perlman (1979)].

Nature of supplement	Riboflavin yield (mg ℓ$^{-1}$)
Peptic hydrolyzate of animal tissue	1520
Equal amounts of a peptic hydrolyzate of animal tissue and pancreatic digest of casein	1280
Pancreatic digest of lactalbumin	1000
Pancreatic digests of casein	340
Pancreatic digest of gelatin	3620
Papaic digest of soya bean meal	673
Glycine[a] (0 g ℓ$^{-1}$)	3280
Glycine[a] (1 g ℓ$^{-1}$)	3640
Glycine[a] (2 g ℓ$^{-1}$)	3980
Glycine[a] (3 g ℓ$^{-1}$)	4200

[a] Basal medium contained corn-steep liquor solid, 2.25 per cent (w/v); Wilson's peptone W-809, 3.5 per cent (w/v); soya bean oil, 4.5 per cent (w/v).

Riboflavin can be produced by *E. ashbyii* from substantially carbohydrate-free media when lipids are used as the energy source (for vitamin production by this strain grown on a variety of media, *see* Table 68). Based on the total weight of nutrients, a typical medium contains 10–90 per cent proteinaceous material, a metabolizable lipid and nutrients, i.e. peptone or a combination salts (0.05 per cent potassium dihydrogen phosphate, 0.07 per cent magnesium sulphate heptohydrate, 0.10 per cent sodium chloride and 0.004 per cent ferrous sulphate heptohydrate). *E. ashbyii* requires biotin, inositol, thiamin and some constituents of peptone for growth. Peptone also stimulates the production of riboflavin.

Candida species, such as *C. gluilliermondia*, can be grown in media containing glucose, mannose, levulose or sucrose. Asparagine and glycine are suitable nitrogen sources. Small amounts of potassium cyanide can increase the yields. The media also contain 'non-iron' inorganic salts; iron has a critical influence on the yields.

Cl. acetobutylicum can be grown in grain mashes or on low iron-containing whey.

Table 68. Riboflavin production by *Eremothecium ashbyii*.

Medium[a]	Initial pH	Incubation period (days)	Temperature (°C)	Riboflavin yield ($g\,\ell^{-1}$)
Glucose (0.5) + maltose (2.5) wheat flour stillage		7.0		1.35
Glucose (6.0), yeast extract (1.0)	4.6	5.0		0.412
Ground lentils (4.0), molasses (1.125), inverted molasses (0.25), ammonium succinate (0.5), sodium chloride (0.5)	6.8–7.5	5.7	28	2.48
Ground lentils (4.0), cane molasses (1.0), sodium chloride (0.5)		8	28	1.4
Ground lentils (4.0), cane molasses (1.0), sodium chloride (0.5)		7	28	1.1
Ground lentils (4.0), sucrose (0.75), inverted sucrose (0.3), ammonium succinate (0.3), sodium chloride (0.5)		8	28	1.7
Malt extract (1.75), glucose (0.5), casein (1.0), corn oil (1.0)		3.6		1.2
Molasses (1.0), malt extract (1.5), tankage (24)	5.5–6.5			0.50
High-malted wheat mash (12.0)				1.00
Brown sugar (5.0), neopeptone (3.0), wheat germ (1.0), beef extract (0.3)	6.0	7.0		1.80
Sucrose (5.0), ammonium hydrogen phosphate (3.0)	5.5–6.0	5.0		0.571
Sucrose (5.0) in whey + skimmed milk (1.1)	5.70–5.95	6.0		1.125–2.20
Thin-grain stillage, corn-steep liquor (0.3), glucose (0.2)	5.8	4.0		0.436

[a] Compositions are expressed as percentages (w/v).

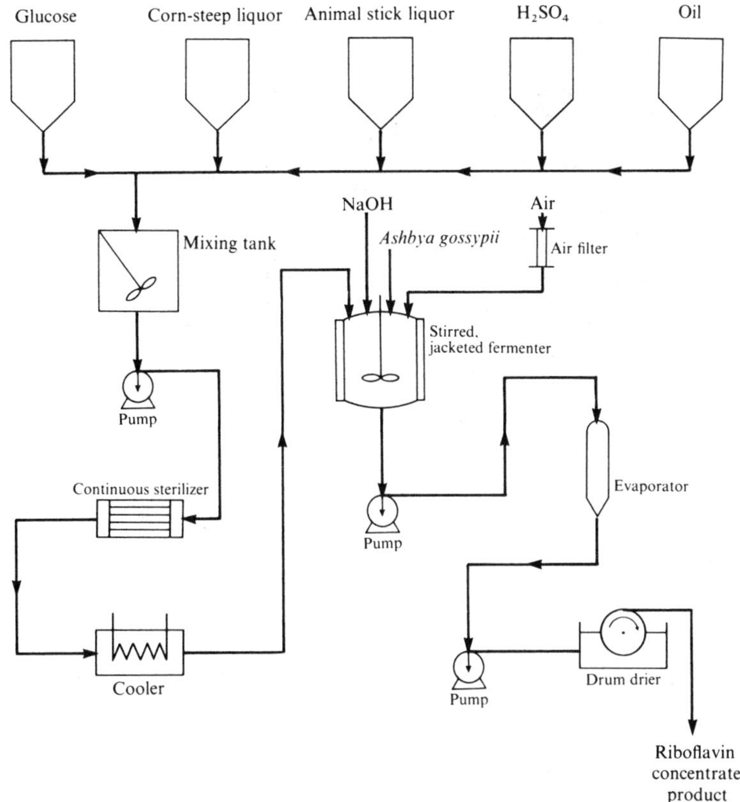

Figure 66. Flow sheet for riboflavin production by fermentation [Pfeifer *et al.* (1950)].

Process outline

A flow sheet of an early process for riboflavin production by *A. gossypii* is shown in Fig. 66. The medium contains 2 per cent glucose, 1.8–2.1 per cent corn-steep liquor, 1 per cent animal-stick liquor and a small amount of antifoam. The medium is sterilized continuously at pH 4.5 and 135°C for 5 minutes. The inoculum is 0.5–1 per cent of the main fermentation solution. Aeration is sufficient at an air flow rate of about 0.25 vvm with mild agitation. The fermentation is performed at about 29°C for 96–120 hours.

In a more recent process, the optimum requirements are an air flow rate of 0.33 vvm and agitation with three impellers at a power input of 1.0 HP per 1000 litres of medium. Excessive foaming is controlled by the initial addition of emulsified silicon antifoam followed by soya bean oil, which also acts as a nutrient. Sterilization of the medium is achieved by heating at 121°C for three hours. The optimum incubation temperature is 28°C over the seven-day incubation period. Using suitable mutants, the yields of riboflavin are in the region of $4\,g\,\ell^{-1}$.

Product recovery

When the riboflavin produced is to be used as an animal feed supplement, the pH is adjusted to 4.5 and the broth is concentrated to about 30 per cent solids and then dried in double drum driers. When a crystalline product is required, the riboflavin content of the cells is solubilized by heating the broth for one hour at 121°C, with the subsequent removal of the insoluble matter by centrifugation. The riboflavin in solution is then converted to a less soluble form by either chemical or microbiological methods. In the former case, a soluble reducing agent and a finely divided diatomaceous earth are used. Alternatively, the riboflavin is adsorbed on to Fuller's earth, silica gel or other adsorbents, and then eluted with an aldehyde, ketone or alcoholic solution of an organic base. Microbial methods involve the riboflavin being converted to a less soluble form by the action of reducing bacteria, e.g., *Streptococcus faecalis*. The precipitated vitamin treated by the chemical or microbiological methods is dissolved in water, polar solvents or an alkaline solution, oxidized by aeration and recovered by crystallization (the alkaline solutions are acidified to crystallize riboflavin).

REFERENCES

Antrim, R.L., Colilla, W. and Schnyder, B.J. (1979) Glucose Isomerase Production of High-Fructose Syrups, in *Applied Biochemistry and Bioengineering*, vol. 2, *Enzyme Technology*, ed. by L.B. Wingard jun., *et al.*, p.98 (Academic Press, New York).

Aries, R. (1974) *French Pat.*, 2 209 842.

Armed Forces Medical Procurement Agency (1952) Specification 1-161-890.

Aunstrup, K., Andersen, O., Falch, E.A. and Nielsen, T.K. (1979) Production of Microbial Enzymes, in *Microbial Technology*, vol. 1, *Microbial Processes*, ed. by H.J. Peppler and D. Perlman, p.282 (Academic Press, New York).

Bartholomew, W.H. and Reisman, H.B. (1979) Economics of Fermentation Process, in *Microbial Technology*, vol. 2, *Fermentation Technology*, ed. by H.J. Peppler and D. Perlman, p.463 (Academic Press, New York).

Beaman, R.G. (1967) Vinegar Fermentation, in *Microbial Technology*, ed. by H.J. Peppler, p.344 (Reinhold, New York).

Beesch, S.C. (1952) Acetone–Butanol Fermentation of Sugars. *Ind. Engng Chem.*, **44**, 1677.

Beesch, S.C. (1953) Microbial Process Report: Acetone–Butanol Fermentation of Starches. *Appl. Microbiol.*, **1**, 85.

Benemann, J.R., Weissman, J.C. and Oswald, W.J. (1979) Algal Biomass, in *Economic Microbiology*, vol. 4, *Microbial Biomass*, ed. by A.H. Rose, p.177 (Academic Press, London).

Bixler, G.H., Hines, G.E., McGhee, R.M., and Shurter, R.A. (1953) Dextran. *Ind. Engng Chem*, **45**, 692.

Blom, R.H., Pfeifer, V.F., Moyer, A.J., Traufler, D.H., Conway, H.F., Crocker, C.K., Farison, R.E. and Hannibal, D.V. (1952) Sodium Gluconate Production. Fermentation with *Aspergillus niger*. *Ind. Engng Chem.*, **44**, 435.

Callihan, C.D. and Clemmer, J.E. (1979) Biomass from Cellulosic Materials, in *Economic Microbiology*, vol. 4, *Microbial Biomass*, ed. by A.H. Rose, p.271 (Academic Press, London).

Casida, L.E., jun. (1968) *Industrial Microbiology* (John Wiley, New York).

Cysewski, G.R. and Wilke, C.R. (1976) Utilization of Cellulosic Materials through Enzymatic Hydrolysis to Ethanol and Single-Cell Protein. *Biotechnol. Bioengng*, **18**, 1297.

Ericsson, E.O. (1947) Alcohol from Sulphite Waste Liquor. *Chem. Engng Prog.*, **43**, 165.

Faith, W.T., Neubeck, C.E. and Reese, E.T. (1971) Production and Application of Enzymes, in *Advances in Biochemical Engineering*, vol. 1, ed. by T.K. Ghose and A. Fiechter, A., p.92 (Springer-Verlag, Berlin).

Faust, U., Prave, P. and Sukatsch, D.A. (1977) Continuous Biomass Production from Methanol by *Methylomonas clara*. *J. Ferment. Technol.*, **55**, 609.

Florent, J. and Ninet, L. (1979) Vitamin B_{12}, in *Microbial Technology*, vol. 1, *Microbial Processes*, ed. by H.J. Peppler and D. Perlman, p.497 (Academic Press, New York).

Gastrock, E.A., Porges, N., Wells, P.A. and Moyer, A.J. (1938) Gluconic Acid Production – Effect of Variables on Production of Submerged Mould Growths. *Ind. Engng Chem.*, **30**, 782.

Ghose, T.K. and Pathak, A.N. (1973) Cellulases. I. Sources, Technology. *Process Biochem.*, **8** (4), 35.

Greenshields, R.N. (1978) Acetic Acid: Vinegar, in *Economic Microbiology*, vol. 2, *Primary Products of Metabolism*, ed. by A.H. Rose, p.121 (Academic Press, London).

Hansen, A.E. (1935) Making Vinegar by the Frings Process. *Fd Ind.*, **7**, 277.

Hatch, R.T. and Finger, S.M. (1979) Mushroom Fermentation, in *Microbial Technology*, vol. 2, *Fermentation Technology*, ed. by H.J. Peppler and D. Perlman, p.179 (Academic Press, New York).

Hemmingsen, S.H. (1979) Development of an Immobilized Glucose Isomerase for Industrial Application, in *Applied Biochemistry and Bioengineering*, vol. 2, Enzyme Technology, ed. by L.B. Wingard, jun. *et al.*, (Academic Press, New York).

Hesseltine, C.W. and Benjamin, C.R. (1959) Microbiological Production of Carotenoids. Part VI. Factors affecting Sporulation and Cell Growth in Choanethecracea. *Mycologia*, **51**, 887.

Inskeep, G.C. and Wiley, A.J., Holderbury, J.N. and Hughes, L.P. (1951) Food Yeast from Sulfite Liquor. (1951) *Ind. Engng Chem.*, **43**, 1702.

Inskeep, G.C., Taylor, G.G. and Breitzke, W.C. (1952) Lactic Acid from Corn Sugar. *Ind Engng Chem.*, **44**, 1955.

Joslyn, M.A. (1970), in *Encyclopedia of Chemical Technology*, vol. 21, ed. by R.E. Kirk and D.F. Othmer, p.254 (John Wiley, New York).

Kang, K.S. and Cottrell, I.W. (1979) Polysaccharides, in *Microbial Technology*, vol. 1, *Microbial Process*, ed. H.J. Peppler and D. Perlman, p.418 (Academic Press, New York).

Kapoor, K.K., Chaudhary, K. and Tauro, P. (1982) Citric Acid, in Prescott and Dunn's Industrial Microbiology, 4th edn, ed. by G. Reede, p.709 (AVI Publishing, Westport).

Keay, L., Moseley, M.H., Anderson, R.G., O'Connor, R.J. and Wildi, B.S. (1972) Production and Isolation of Microbial Proteases. *Biotechnol. Bioengng Symp.*, **3**, 75.

Kosaric, N., Duvnjak, Z. and Stewart, G.G. (1981) Fuel Ethanol from Biomass: Production, Economics, and Energy, in *Advances in Biochemical Engineering*, vol. 20, ed. by A. Fiechter p.419 [Springer-Verlag, Berlin).

Lawson, G.T. and Sutherland, I.W. (1978) Polysaccharides, in *Economic Microbiology*, vol. 2, *Primary Products of Metabolism*, ed. by A.H. Rose, p.328 (Academic Press, New York).
Lindeman, L.R. and Rocchiccioli, C. (1979) Ethanol in Brazil: Brief Summary of the State of the Industry in 1977. *Biotechnol. Bioengng*, **21**, 1107.
Litchfield, J.H. (1967) Submerged Culture of Mushroom Mycelium, in *Microbial Technology*, ed. H.J. Peppler, p.107 (Reinhold, New York).
Litchfield, J.H. (1976) Food Microbiology, in *Industrial Microbiology*, ed. by B.M. Miller and W. Litsky, p.257 (McGraw-Hill, New York).
Litchfield, J.H. (1979) Production of Single-Cell Protein for Use in Food or Feed, in *Microbial Technology*, vol. 1, *Microbial Processes*, ed. by H.J. Peppler and D. Perlman, p.93 (Academic Press, New York).
Lockwood, L.B. and Schweiger, L.B. (1967) Citric and Itaconic Acid Fermentations, in *Microbial Technology*, ed. by H.J. Peppler, p.183 (Reinhold, New York).
Lockwood, L.B., Tabonkin, B.T. and Ward, G.E. (1941) Production of Gluconic Acid and 2-ketogluconic acid from Glucose by Species of *Pseudomonas* and *Phytomonas*. *J. Bacteriol.*, **42**, 51.
Maiorella, B., Wilke, C.R. and Blanch (1981) Alcohol Production and Recovery, in *Advances in Biochemical Engineering*, vol. 20, ed. by A. Fiechter, p.43 (Springer-Verlag, Berlin).
Maisch, W.F., Sobolov, M. and Petricola, A.J. (1979) Distilled Beverages, in *Microbial Technology*, vol. 2, *Fermentation Technology*, ed. by H.J. Peppler and D. Perlman, p.79 (Academic Press, New York).
Malby, P.G. (1970) Production of Endocellular Enzymes. *Process Biochem.*, **5** (8), 22.
Meyrath, J. and Bayer, K. (1979) Biomass from Whey, in *Economic Microbiology*, vol. 4, *Microbial Biomass*, ed. by A.H. Rose, p.208 (Academic Press, London).
Moss, M.O. and Smith, J.E. (1977) *Industrial Applications of Microbiology* (Surrey University Press, London).
Nickol, G.B. (1979) Vinegar, in *Microbial Technology*, vol. 2, *Fermentation Technology*, ed. by H.J. Peppler and D. Perlman, p.155 (Academic Press, New York).
Ninet, L. and Renaut, J. (1979) Carotenoids, in *Microbial Technology*, vol. 1, *Microbial Processes*, ed. by H.J. Peppler and D. Perlman, p.529 (Academic Press, New York).
Pan, S.C., Peterson, W.H. and Johnson, M.J. (1940) Acceleration of Lactic Acid Fermentation by Heat-Labile Substances. *Ind. Engng Chem.*, **32**, 709.
Peppler, H.J. (1979) Production of Yeasts and Yeast Products, in *Microbial Technology*, vol. 1 *Microbial Processes*, ed. by H.J. Peppler and D. Perlman, p. 157 (Academic Press, New York).
Perlman, D. (1967) Microbial Production of Therapeutic Compounds, in *Microbial Technology*, ed. by H.J. Peppler, p.251 (Reinhold, New York).
Perlman, D. (1977) Industrial Microbiology, in Benchmark Papers in Microbiology, ed. by R.W. Thomas, (Dowdenm Hutchinson and Ross, Stroudsburg).
Perlman, D. (1979) Microbial Production of Antibiotics, in *Microbial Technology*, vol. 2, *Microbial Processes*, ed. by H.J. Peppler and D. Perlman, p.241 (Academic Press, New York).
Perlman, D. (1979) Vitamins, in *Economic Microbiology*, vol. 2, *Primary Products of Metabolism*, ed. A.H. Rose, p.303 (Academic Press, London).
Pfeifer, F., Tanner, F.W. jun., Vojnovich, C. and Traufler, D.H. (1950) Riboflavin by Fermentation with *Ashbya gossypii*. *Ind. Engng Chem.*, **42**, 1776.
Porter, J.N. (1976) Antibiotics, in *Industrial Microbiology*, ed. by B.M. Miller and W. Litsky, p.60 (McGraw-Hill, New York).
Prescott, S.C. and Dunn, C.G. (1959) *Industrial Microbiology* (McGraw-Hill, New York).
Queener, S. and Swartz, R. (1979) Penicillins: Biosynthetic and Semisynthetic, in *Economic Microbiology*, vol. 3., *Secondary Products of Metabolism*, ed. by A.H. Rose, p.35 (Academic Press, London).
Reilly, P.J. (1979) Starch Hydrolysis with Soluble and Immobilized Glucoamylase, in *Applied Biochemistry and Bioengineering*, vol. 2, *Enzyme Technology*, ed. by L.B. Wingard, jun. *et al.*, p.185 (Academic Press, New York).
Rhodes, A. and Fletcher, D.L. (1966) *Principles of Industrial Microbiology* (Pergamon Press, Oxford).
Schwartz, R.D. and Leathen, W.W. (1976) Petroleum Microbiology, in *Industrial Microbiology*, ed. by B.M. Miller and W. Litsky, p.384 (McGraw-Hill, New York).
Sebek, O.K. and Perlman, D. (1979) Microbial Transformation of Steroids and Sterols, in *Microbial Technology*, vol. 1, *Microbial Processes*, ed. by H.J. Peppler and D. Perlman, p.483 (Academic Press, New York).
Tatum, E.L. and Peterson, W.H. (1935) Fermentation Method for Production of *d*-Lactic Acid *Ind. Engng Chem.*, **27**, 1493.
Topiwala, H.H. (1974) Application of Kinetics to Biological Reactor Design. *Biotechnol. Bioengng Symp.*, **4**, 681.
Underkofler, L.A. (1976) Microbial Enzymes, in *Industrial Microbiology*, ed. by B.M. Miller and W. Litsky, p.128 (McGraw-Hill, New York).

Walton, M.T. and Martin, J.L. (1979) Production of Butanol-Acetone by Fermentation, in *Microbial Technology*, vol. 1, *Microbial Processes*, ed. by H.J. Peppler and D. Perlman, p.188 (Academic. Press, New York).

Ward, G.E. (1967) Production of Gluconic Acid, Glucose Oxidase, Fructose, and Sorbose, in *Microbial Technology*, ed. by H.J. Peppler, p.200 (Reinhold, New York).

Wells, P.A., Moyer, A.J., Stubbs, J.J., Herrick, H.T. and May, O.E. (1937) Gluconic Acid Production – Effect of Pressure, Air Flow and Agitation on Glucomic Acid – Production by Submerged Mould Growths. *Ind. Engng Chem.*, **29**, 653.

Westermann, D.H. and Huige, N.J. (1979) Beer Brewing, in *Microbial Technology*, vol. 1, *Fermentation Technology*, ed. by H.J. Peppler and D. Perlman, p.2 (Academic Press, New York).

Wilke, C.R., Yang, R.D., Sciamanna, A.F. and Freitas, R.P. (1981) Raw Material Evaluation and Process Development for Conversion of Biomass to Sugars and Ethanol. *Biotechnol. Bioengng*, **23**, 163.

APPENDIX I: COMPOSITION OF SOURCES OF CARBOHYDRATES

Tables A1–A6 contain the specifications of various sources of carbohydrates used as the basis for fermentation media.

Table A1. Composition of beet molasses and cane molasses [Rhodes and Fletcher (1966)].

Component	Content (%) Beet molasses	Cane molasses
Sucrose	48.5	33.4
Raffinose	1.0	
Invert	1.0	21.2
Ash	10.8	9.8
Organic nonsugars	20.7	19.6
Nitrogen	1.5–2.0	
Water	18.0	16.0

Table A2. Composition of a beet molasses sample as percentage dry weight [Moss and Smith (1977)].

Component	Content
Water	16.5
Dry weight	81–83.5
Sucrose	48–50
Raffinose	1
Glucose + fructose	1
Nitrogen compounds	12–13
Glutamic acid	3.5
Other amino acids (asparagine, aspartic acid, alanine, glycine)	5.5
Betaine	3.25–4.25
Ash	11–12
K_2O	45[a]
Na_2O	15[a]
P_2O_5	0.7[a]
CaO	3[a]
MgO	1.8[a]
SO_4^{2-}	6.3[a]
SiO_2	1.1[a]
Cl^-	18[a]
Fe_2O_3	0.2[a]
Al_2O	0.8[a]

[a] Expressed as percentage of total ash content.

Table A3. Composition of grape musts [Moss and Smith (1977)].

Component	Content (%)
Glucose	8–13[a]
Fructose	7–12[a]
Sucrose	Traces
Tartaric acid	0.2–1.0
Malic acid	0.1–0.8[a]
Citric acid	0.01–0.05
Oxalic, gluconic, glucuronic, phosphoric acids	Traces
Tannins/tannic acid	0.05–0.4[b]
Catechols	Traces
Amino acids and other organic nitrogen	0.03–0.17
Ash	0.2–0.6
pH	3.1–3.9

[a] Depends on ripeness.
[b] More if the must contains skins, seeds or stalks.

Table A4. Composition of wood molasses [Moss and Smith (1977)].

Component	Content [% (by wt)]
Solids	60–52
Reducing sugars (as glucose)	48–50
Other carbohydrates	0.5–1.5
Noncarbohydrate organic compounds	6–8
Ash	2–3
Nitrogen	0.065
Volatile organic acids	1–2

Table A5. Composition of spent spruce sulphite liquor [Forss and Passinen (1976)].

Component	Content [% (w/v)]
Lignosulphuric acids	43
Hemilignin compounds	12
Incompletely hydrolyzed hemicellulose and uronic acids	7
Monosaccharides (total)	22
D-Glucose	2.6
D-Xylose	4.6
D-Mannose	11.0
D-Galactose	2.6
L-Arabinose	0.9
Acetic acid	6
Aldonic acids and other substrates	10

Table A6. Composition of agriculture residues [Wilke *et al.* (1981)].

Material[a]	Glucan	Hexosans [% (by wt)] Mannan	Galactan	Pentosans [% (by wt)] Xylan	Arabinan	Lignin [% (by wt)]	Ash [% (by wt)]	Azeotropic benzene/ ethanol extractives	Acid-insoluble material [% (by wt)	Other
Barley straw	37.5	1.26	1.71	15.0	3.96	13.8	10.8	9.7	2 ± 1	
Corn stover	35.1	0.25	0.75	13.0	2.8	15.1	4.3	5.5	1 ± 1	4 protein
Rice straw	36.9	1.6	0.4	13.0	4.0	9.9	12.4	4.4	2 ± 1	
Sorghum straw	32.5	0.8	0.2	15.0	3.0	14.5	10.1	6.2	1 ± 1	1 protein
Wheat straw	32.9	0.72	2.16	16.9	2.1	14.5	9.6	7.2	3 ± 1	3 protein

[a] Wiley milled 2 mm, 40–60 mesh fraction and 100 per cent dry.

References

Forss, K. and Passinen, K. (1976) Utilization of the Spent Liquor Components in the Pekilo Protein Process and the Influence of the Process upon the Environmental Problems of a Sulphite Mill. *Pap. Puu.*, **58**, 608.

Moss, M.O. and Smith, J.E. (1977) *Industrial Applications of Microbiology* (Surrey University Press, London).

Rhodes, A. and Fletcher, D.L. (1966) *Principles of Industrial Microbiology* (Pergamon Press, Oxford).

Wilke, C.R., Yang, R.D., Sciamanna, A.F. and Freitas, R.P. (1981) Raw Material Evaluation and Process Development for Conversion of Biomass to Sugars and Ethanol. *Biotechnol. Bioengng.*, **23**, 163.

APPENDIX II: COMPOSITION OF VARIOUS INDUSTRIAL MEDIA

Tables A7–A16 contain specifications of various industrial fermentation media.

Table A7. Composition of α-ketogluconic acid media [Prescott and Dunn (1959)].

Components	Amount ($g\,\ell^{-1}$)
Commercial glucose	118[a]
Corn-steep liquor	5
Octadecyl alcohol (antifoam)	0.3
Urea (sterilized separately)	2
$MgSO_4.7H_2O$	0.25
KH_2PO_4	0.60
$CaCO_3$ (sterilized separately)	27

[a] Amount necessary for a 10 per cent glucose concentration.

Table A8. Medium for β-carotene production [Hanson (1967)].

	Shake flasks	Pilot plant
Cotton-seed embryo meal (%)	5.0	4.0
Ground whole corn (%)	2.5	2.0
Vegetable oil (corn, soya bean, or cotton seed) (%)	5.0	3.0
Deodorized kerosene[a] (%)	5.0	3.0
Nonionic detergent (%)	0.12	
Thiamin hydrochloride ($mg\,\ell^{-1}$)	2.0	0.2
Citrus molasses (%)		5.0

[a] 'Deo-Base' (Sonneborn Chemical and Refining Corp., New York).

Table A9. Composition of citric acid media [Shu and Johnson (1948)].

Constituents	Medium A ($g\,\ell^{-1}$)	Medium B ($g\,\ell^{-1}$)
Domino sucrose	140	140
Bacto agar	20	
KH_2PO_4	1.0	2.5
$MgSO_4.7H_2O$	0.25	0.25
NH_4NO_3	2.5	2.5
HCl to pH 3.8		
Trace metal[a]		
Cu^{2+}	0.00048	0.00006
Zn^{2+}	0.0038	0.00025
Fe^{3+}	0.0022	0.0013
Mn^{2+}[b]	<0.0010	<0.0010

[a] Listed quantities of metals include amounts present as impurities in other constituents of the medium. Media were sterilized at 120°C for 15 minutes.
[b] The importance of low manganese concentrations in both sporulation and fermentation media for submerged citric acid production is demonstrated in work reported elsewhere.

Table A10. Composition of media used in the production of erythromycin [Moss and Smith (1977)].

Component	Sporulation medium	Medium for production of vegetative inoculum	Erythromycin production medium
Starch ($g\,\ell^{-1}$)	5.0	15.0	
Glucose ($g\,\ell^{-1}$)	5.0		40.0
Soya bean meal ($g\,\ell^{-1}$)		15.0	
Corn-steep liquor ($g\,\ell^{-1}$)		5.0	
Corn extract ($g\,\ell^{-1}$)			10.0
Tryptone ($g\,\ell^{-1}$)	5.0		
Betaine ($g\,\ell^{-1}$)	0.5		
Curloay B.G. ($g\,\ell^{-1}$)	2.0		
K_2HPO_4 ($g\,\ell^{-1}$)	0.2		
$(NH_4)_2SO_4$ ($g\,\ell^{-1}$)			6.0
NaCl ($g\,\ell^{-1}$)	10.0	5.0	2.5
$CaCl_2$ ($g\,\ell^{-1}$)	0.08		
$CaCO_3$ ($g\,\ell^{-1}$)		3.0	10.0
Mineral mixture (ml)	2.0		
Sperm oil ($g\,\ell^{-1}$)			6.0
Agar ($g\,\ell^{-1}$)	20.0		

Table A11. Media used for sodium gluconate production [Blom *et al.* (1952)].

Component	Stock culture medium	Sporulation medium	Germination medium	Fermentation medium
Glucose (commercial-grade) (g)	30	50	10 000	136 400
$MgSO_4.7H_2O$ (g)	0.1	0.12	31.0	94
KH_2PO_4 (g)	0.12	0.15	38.0	113
$(NH_4)_2HPO_4$ (g)		0.60	80.0	240
KCl (g)		0.20		
NH_4NO_3 (g)	0.25			
$CaCO_3$ (g)	4.0			
Iron tartrate (g)		0.01		
Potato extract (ml)[a]	500 +	30.0		
Beer (ml)		30.0		
Corn-steep liquor (ml)			750.0	2100
Urea (g)			20	60
Peptone (g)	0.25			
Agar (g)	25.0	1.5		
NaOH to pH 6.5 (g)			25.0	
H_2SO_4 to pH 4.5 (ml)				100
Tap water to make litres	1.0	1.0	170.3[b]	568[c]
Sterilization time at 121°C (min)	30.0	30.0	30.0	15–30

[a] Peeled and sliced potatoes (200 g) were autoclaved for 15 minutes in 500 ml of tap water and filtered through cheese cloth.
[b] After sterilization.
[c] After inoculation.

Table A12. Media used in lactic acid production [Ward *et al.* (1938)].

Component	Germination medium ($g\,\ell^{-1}$)	Fermentation medium ($g\,\ell^{-1}$)
Glucose (91.5%, commercial)	110.0	150.0
Urea	2.0	2.0
KH_2PO_4	0.60	0.60
$MgSO_4.7H_2O$	0.25	0.25
$ZnSO_4.7H_2O$	0.088	0.044
$CaCO_3$	10.0	
Octadecyl alcohol		0.03[a b]

[a] 200 g to each 3-litre portion – sterilized separately.
[b] Dissolved in 1.7 ml ethanol (added to prevent excessive foaming of the fermentation medium during the rotation of the drum).

Table A13. Substrates and concentrations for submerged-culture growth of mushroom mycelium [Litchfield (1967)].

Microorganism	Substrates	Concentration (%)
Agaricus campestris	L-Arabinose, D-fructose D-galactose, maltose	5
	D-Glucose	2.5–10
	D-Mannose, sucrose, D-xylose, dextrin, D-mannitol, soluble starch	5
	Asparagus butt juice, cane molasses, citrus molasses	5[a]
	Beet molasses	6[a]
	Corn syrup	3[a]
	Malt syrup – cane molasses	2.5, 4[a]
	Pear waste	5[a]
	Spent sulphite liquor	2.12[a]
	Vinasse	0.15–1.37[a]
A. blazei	D-Glucose	5
	Citrus press water, orange juice	6[a]
	Spent sulphite liquor	1.5[a]
A. placomyces *A. rodmanii* *Armillaria mellea*	D-Glucose	5
Boletus indecisus *Cantharellus cibarius* *Collybia velutipes* *Tricholoma nudum* *Xylaria polymorpha* *Morchella hybrida*	D-Glucose	4–5
	Beet molasses	6[a]
	Soya bean whey	0.47[a]
	Spent sulphite liquor	1.5–2.12[a]
Morchella crassipes *M. esculenta* *M. hortensis*	D-Glucose	2.5–5
	Lactose	2.5
	Maltose	2.5
	Corn-canning waste	0.54
	Pumpkin-canning waste	2.75
	Cheese whey	4[b]
	Sucrose, D-xylose	3
	Cane molasses	5

Table A13 – (continued)

Microorganism	Substrates	Concentration (%)
M. esculenta	Beet molasses	12
	Citrus molasses	5[a]
	Corn syrup	3[a]
Boletus indecisus	Vinasse	0.21–1.42[a]
Tricholoma nudum		1.41[a]
Collybia umbulata	D-Glucose	5
Coprinus comatus	D-Glucose	2–6
Hebeloma sinapizans	D-Glucose	5
Helvella gigas	D-Glucose	2.5
Lepiota naucina	D-Glucose	5
	Spent sulphite liquor	1.5[a]
L. procera	D-Glucose	5
L. rhacodes	D-Glucose	5
Lycoperdon umbrimum	D-Glucose	5
Morchella angusticeps	D-Glucose	2.5
M. conica	D-Glucose	2.5
M. rimosipes	D-Glucose	2.5
M. semilibera	D-Glucose	2.5
Pleurotus ostreatus	D-Glucose	5
Polyporus circinatus	D-Glucose	3
Polyporus sulphureus	D-Glucose	5
Psilocybe spp.	D-Glucose	5
Schizophyllium commune	D-Glucose	5

[a] Expressed as glucose.
[b] Expressed as lactose.

Table A14. Typical nitrogen sources and concentrations used for submerged – culture growth of mushroom mycelium [Litchfield (1967)].

Nitrogen Source	Mushroom	Concentrations (%) Compound basis	Nitrogen basis
Ammonium acetate	*Agaricus campestris*	0.55–1.9[a]	0.10–0.35[a]
	Boletus indecisus	0.31	0.56
	Collybia velutipes	0.31	0.56
	Tricholoma nudum	0.31	0.56
Ammonium chloride	*T. nudum*	0.38	0.10
Ammonium citrate (dibasic)	*A. campestris*	0.81–2.8	0.10–0.35
Ammonium nitrate	*A. campestris*	0.42	0.15
	T. nudum	0.29	0.1
Ammonium phosphate (monobasic)	*A. blazei*	0.1	0.012
	C. velutipes	0.1	0.012
	Lepiota naucina	0.1	0.012
	T. nudum	0.1	0.012
Ammonium phosphate (dibasic)	*A. campestris*	0.20–0.34	0.04–0.07
	Coprinus comatus	0.20–0.34	0.04–0.07
	Helvella gigas	0.20–0.34	0.04–0.07
	Morchella crassipes	0.20–0.34, 0.20–0.4	0.04–0.07, 0.04–0.08
	M. esculenta	0.3, 0.2–0.4	0.06, 0.04–0.08
	M. hortenis	0.2–0.4	0.04–0.08
Ammonium sulphate	*A. campestris*	0.5, 0.21–0.6	0.11, 0.04–0.12
	Boletus indecisus	0.5, 0.4	0.11, 0.08
	Cantharellus cibarius	0.5	0.11
	C. velutipes	0.5	0.11
	Morchella hybrida	0.5	0.11
	T. nudum	0.5, 0.4	0.11, 0.08
	Xylaria polymorpha	0.5	0.11
Ammonium tartrate	*A. campestris*	0.1–1.0	0.015–0.15
	B. indecisus	0.1–1.0	0.015–0.15
	C. velutipes	0.1	0.015
	M. hybrida	0.1–1.0	0.015–0.15
	T. nudum	0.1–0.8	0.015–0.12
Corn-steep liquor	*M. crassipes*		0.04–0.08
	M. esculenta		0.04–0.08
	M. hortensis		0.04–0.08
Malt syrup molasses	*A. campestris*		0.03–0.08
Malt sprout extract	*A. campestris*		0.07–0.23
Potassium nitrate	*A. blazei*	0.1	
	T. nudum	0.1	
Soya bean whey	*B. indecisus*		0.05–0.10
	C. velutipes		0.05
	Canth. cibarius		0.05
	M. hybrida		0.05
	T. nudum		0.05–0.10
	X. polymorpha		0.05
Urea	*A. campestris*	0.10–0.35[a]	0.05–0.16
	T. nudum	0.1	0.05
Yeast extract	*A. campestris*		0.03–0.07

[a] Per cent nitrogen per cent of carbohydrate.

Table A15. Media used in microbial transformation of steroids [Perlman (1967)].

Microorganism	Substrate	Product	Medium
Alcaligenes faecalis	Cholic acid	Keto-cholic acids	$(NH_4)_2SO_4$, 47 g; asparagine, 0.5 g; glycerol, 2 g; $HgCl_2$, 0.005 g; NaCl, 5 g; NaOH, 0.65 g; $CaCl_2$, 0.005 g, $FeCl_2$, 0.005 g, KH_2PO_4, 2.7 g; substrate, 5 g; distilled water to 1 ℓ
Corynebacterium mediolanum	21-Acetoxy-Δ^5-pregnene-3β-ol-20-one	Δ^4-Pregnene-21-ol-3, 20-dione	Yeast water, 60 ml; 0.2 M Na_2 HPO_4, 10 ml; 0.2 M KH_2PO_4, 10 ml; substrate, 0.2 g
Cunninghamsella blakesleeana H 334	Reichstein compounds	Cortisol and cortisone	$NaNO_3$, 3 g; KCl, 0.5 g; dextrin, 30 g; $FeSO_4$. $7H_2O$, 0.1 g, Tween 80, 10 ml; $MgSO_4$. $7H_2O$, 0.5 g, KH_2PO_4. $3H_2O$, 1.3 g; distilled water to 1 ℓ; pH adjusted to 7.2
Cylindrocarpon radicicola ATCC 11011	Progesterone	Δ^1-Testololactone	Corn-steep liquor solids, 3 g; $NH_4H_2PO_4$, 3 g; $CaCO_3$, 2.5 g; soya bean oil, 2.2 g; substrate, 0.5 g; distilled water to 1 ℓ; pH adjusted to 7.0 using NaOH
Fusarium solani	Progesterone	$\Delta^{1,4}$-Androstadiene-3,17-dione	Peptone, 15 g; corn-steep liquor, 6 ml; glucose, 50 g; distilled water to 1 ℓ; pH adjusted to 6.0; substrate (at 48 h) 0.25
Rhizopus arrhizus ATCC 11 145	Δ^1-Androstene-3,17-dione	Δ^4-Androstene-11α-ol-3,17-dione	Lactalbumin digest, 20 g; corn-steep liquor, 5 ml; glucose, 50 g; tap water to 1 ℓ; pH adjusted to 5.9; substrate (at 27 h), 0.25 g
Streptomyces albus	Oestradiol	Oestrone	Nutrient broth for growing cells; phosphate buffer (pH 7) for steroid oxidation
Strept. aureus	Progesterone	Δ^4-Pregnene-15α-ol-3,20-dione	Soya bean oil, 2.2 g; soya bean meal, 15 g; glucose, 10 g; $CaCO_3$, 2.5 g, water to 1 ℓ, substrate, 0.25 g

Table A16. Inoculation and fermentation media for vitamin B_{12} [Pfeifer *et al.* (1954)].

	Medium 1	Medium 2
Dextrose (%)	0.5	1.0
Corn-steep liquor solids (%)	0.5	
Distillers' solubles or soya bean meal[a]		4.0
$CoCl_2.6H_2O$ (ppm)	2.0	2–10
Calcium carbonate (%)		0.5
Soya bean oil (%)	0.1	0.1
Adjust pH to	7.0	7.0

[a] Extracted and unheated.

References

Blom, R.H., Pfeifer, V.F., Moyer, A.J., Traufler, D.H., Conway, H.F., Crocker, C.K., Farison, R.E. and Hannibal, D.V. (1952) Sodium Gluconate Production. Fermentation with *Aspergillus niger*. *Ind. Engng Chem.*, **44**, 435.

Hanson, A.M. (1967) Microbial Production of Pigments and Vitamins, in *Microbial Technology*, ed. by H.J. Peppler, p.222 (Reinhold, New York).

Litchfield, J.H. (1967) Submerged Culture of Mushroom Mycelium, in *Microbial Technology*, ed. by H.J. Peppler, p.107 (Reinhold, New York).

Moss, M.O. and Smith, J.E. (1977) *Industrial Applications of Microbiology* (Surrey University Press, London).

Perlman, D. (1967) Microbial Production of Therapeutic Compounds, in *Microbial Technology*, ed. by H.J. Peppler, p. 251 (Reinhold, New York).

Pfeifer, V.F., Vojnovich, C. and Heger, E.N. (1954) Vitamin B_{12} by Fermentation with *Streptomyces olivaceus*. *Ind. Engng Chem.*, **46**, 849.

Prescott, S.C. and Dunn, C.G. (1959) *Industrial Microbiology* (McGraw-Hill, New York).

Shu, P. and Johnson, M.J. (1948) Citric Acid Production by Submerged Fermentation with *Aspergillus niger*. *Ind. Engng Chem.*, **40**, 1202.

Ward, G.E., Lockwood, L.B., Tabonkin, B. and Wells, P.A. (1938) Fermentation Process for Dextrolactic acid. *Ind. Engng Chem.*, **30**, 1233.

Index of Topics